33ʳᵈ EUROPEAN SYMPOSIUM ON COMPUTER AIDED PROCESS ENGINEERING

VOLUME 3

COMPUTER-AIDED CHEMICAL ENGINEERING, 52

33rd EUROPEAN SYMPOSIUM ON COMPUTER AIDED PROCESS ENGINEERING

VOLUME 3

Edited by

Prof. Antonios C. Kokossis
National Technical University of Athens
School of Chemical Engineering
Athens, Greece

Prof. Michael C. Georgiadis
Aristotle University of Thessaloniki
School of Engineering
Department of Chemical Engineering
Greece

Prof. Efstratios Pistikopoulos
Chemical Engineering
Texas A&M Energy Institute
TX, USA

ELSEVIER

Amsterdam – Boston – Heidelberg – London – New York – Oxford
Paris – San Diego – San Francisco – Singapore – Sydney – Tokyo

Elsevier
Radarweg 29, PO Box 211, 1000 AE Amsterdam, Netherlands
The Boulevard, Langford Lane, Kidlington, Oxford OX5 1GB, UK
50 Hampshire Street, 5th Floor, Cambridge, MA 02139, USA

British Library Cataloguing in Publication Data
A catalogue record for this book is available from the British Library

Library of Congress Cataloging-in-Publication Data
A catalog record for this book is available from the Library of Congress

ISBN (Volume 3):   978-0-443-23555-9
ISBN (Set) :   978-0-443-15274-0
ISSN: 1570-7946

For information on all Elsevier publications visit our
website at https://www.elsevier.com/

Publisher: Candice Janco
Acquisition Editor: Anita Koch
Editorial Project Manager: Lena Sparks
Production Project Manager: Paul Prasad Chandramohan
Designer: Greg Harris

Typeset by STRAIVE

Contents

T3: Safe and sustainable products by design

T4: Green and sustainable processes for the circular economy

T5: Systems methods in industrial biotechnology and biomedical applications

Antonis Kokossis, Michael C. Georgiadis, Efstratios N. Pistikopoulos (Eds.)
PROCEEDINGS OF THE 33rd European Symposium on Computer Aided Process Engineering
(ESCAPE33), June 18-21, 2023, Athens, Greece

Residence Time Distribution characterization in a Continuous Manufacturing tableting line using PCA and PLS-DA modeling

Pau Lapiedra Carrasquer[a], Satyajeet S. Bhonsale[a], Liang Li[b], Jan F.M. Van Impe[a]

[a] BioTeC+, KU Leuven, Gebroeders De Smetstraat 1, Gent 9000, Belgium
jan.vanimpe@kuleuven.be
[b] Janssen Pharmaceutica, Turnhoutseweg 30, Beerse Antwerp 2340, Belgium
lli110@its.jnj.com

Abstract

In this work, a new approach to characterize the residence time distribution (RTD) using supervised and unsupervised learning techniques was investigated. The NIR spectral data from the feed frame was used to capture the dynamic response of a continuous tableting line.

The data-driven approaches investigated in this study using MPCA and PLS-DA were able to successfully capture the main variation of the step change performed during the RTD experiments. Its good performance compared with the traditional RTD characterization methodology offers the potential to be used to predict the RTD parameters. This could contribute to making the RTD determination procedure more efficient and reduce the need to make offline measurements.

Keywords: data-driven modeling, NIR, MPCA, continuous manufacturing, tableting

1. Introduction

The progressing adoption of continuous manufacturing in the pharmaceutical industry still faces a series of challenges to achieving its full potential. Tracing any disturbance along the line in real-time is essential to guarantee the quality of the final tablets, as it contributes to significantly reducing the amount of product to be discarded, making continuous operation a more sustainable option compared to the current batch manufacturing. The Residence Time Distribution (RTD) is a widely used method that defines the probability distribution of the time that a particle stays inside a system. The RTD can be determined experimentally by measuring the concentration of the product over time using process analytical tools (PAT) such as Near Infrared (NIR) (Bhalode et al.).

Most continuous tableting lines have NIR probes located in the feed frame of the tablet press. The spectral data obtained from these measurements are normally used to measure the blend uniformity of the powder material right before being compressed. This probe operates in-line, and it can measure every second. This high measurement frequency offers the possibility to extract more information about the dynamic response of the process compared with the information obtained from the API concentration measurements of the limited number of samples extracted at the end of the line. The use of in-line NIR measurements has the potential to provide a real-time prediction of the concentration of the drug substance in the tablet (Hetrick et al.). In this study, a novel approach was explored to exploit all the information from the spectral data to capture a more detailed dynamic response to a step change experiment. We propose to use the NIR spectral data to predict the dynamic response of a step change in the drug substance concentration to characterize the RTD. To achieve that, supervised and unsupervised data-driven approaches were used and compared. The unsupervised learning method, a variation of PCA called Multiway Principal Component Analysis (MPCA) and the supervised method, Partial Least Squares Discriminant Analysis (PLS-DA) were applied to the NIR spectral data. Both techniques were compared to see which one can better capture the dynamic response of the step change to be used for RTD modeling.

2. Theory and Methods

2.1. RTD model

Determining the RTD process model is a critical step in product and process development, and it is an essential part of the control strategy to detect non-conforming material at the outlet of the CM line. Developing a clear connection between the different inline measurements (i.e., process parameters and PAT data) and the product characteristics at specific locations is a key aspect. When a drop in the product quality is detected at a certain point, the affected fraction of the product needs to be discarded. This means that the travel time of the product through the line, and the amount of material needs to be known to accurately discard the affected material. This information is obtained from the RTD characteristics of the material in the specific production line. Traditional models that describe RTD are based on the exit-age distribution $E(t)$, described by Equation (1), where $C(t)$ is the time-dependent concentration of a tracer at the outlet of the unit operation.

$$E(t) = \frac{C(t)}{\int_0^\infty C(t)dt} \tag{1}$$

The RTD was generally modelled as an ideal PFR + 2 CSTRs in series. Its expression can be written as in Equation (2). Where θ is the time delay, and τ_1 and τ_2 represents time constants.

$$E(t) = \frac{e^{\frac{t-\theta}{\tau_1}} - e^{\frac{t-\theta}{\tau_2}}}{\tau_1 - \tau_2} \tag{2}$$

The calibrated RTD model is product specific and consists of two parts: determination of RTD model parameters and determination of RTD model uncertainty. Once calibrated, the RTD model parameters can be used to calculate the % potency at the tablet press outlet on the CM line. Additionally, the RTD model uncertainty is used to calculate the out-of-specification limits for tablet quality. A series of experiments are performed at different conditions to calibrate the RTD model for a given product.

2.2. Experimental data

The dataset used in this study was obtained from 13 RTD characterization runs for a Direct Compression setup. For each run, a step change was applied to the API mass flow and its response was measured in terms of API concentration (%). This concentration is what is usually used to fit the RTD model in Equation (2). To obtain these measurements, a number of tablets are sampled, and their concentration is measured offline. To develop and test the new approach, NIR spectral data of the same 13 runs obtained from the probe located in the feed frame of the tablet press was used. For every run, the measurement of absorbance over a certain wavelength range was measured during the duration of the step change study. This results in a three-dimensional data structure: runs (or batches), measurements over time, and variables (in this case the wavelengths), which can be written in the form of a three-way array.

Table 1. Step changes applied to the API mass flow for every run

Run name	Step change	Run name	Step change	Run name	Step change
Run 1	100%-->85%	Run 6	100%--130%	Run 11	100%-->70%
Run 2	85%-->100%	Run 7	100%--> 85%	Run 12	100%-->115%
Run 3	100%-->70%	Run 8	100%-->85%	Run 13	115%-->100%
Run 4	100%-->115%	Run 9	100%-->85%		
Run 5	115%-->100%	Run 10	85%-->100%		

2.3. MPCA and PLS-DA

Principal component analysis (PCA) is a commonly used unsupervised data-driven method for dimensionality reduction. PCA transforms the original data into a latent variable model and a residual error. In this model, the latent variables are oriented in the direction that gives the greatest variance of the scores. In other words, it offers a way to summarize the main trends of the data in a reduced number of variables. Due to the three-dimensional nature of the spectral data, a variation of the classic PCA called Multiway PCA (MPCA) was chosen. Commonly used for batch process monitoring, MPCA, unfolds the three-way array into a large 2-dimensional matrix. Due to different durations of the RTD runs, a variable-wise unfolding was chosen (Figure 1). In this way, PCA can be used for the resulting 2D unfolded matrix regardless of the run duration.

The MPCA method was compared to the PLS-DA. PLS is a supervised learning data-driven method that also achieves dimensionality reduction through the latent variable space. The main difference with PCA is that in PLS the latent variables are extracted by obtaining the directions of the largest covariance between input and output. The discriminant analysis variation of PLS is used for classification and the output vector is defined as the class labels to be classified. In this study, the class labels corresponded to the setpoint concentrations before and after applying the step change (Pedersen et al.).

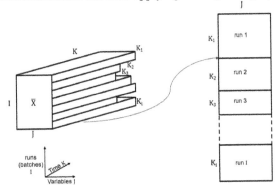

Figure 1. MPCA variable-wise unfolding of a dataset with unequal experiment duration

3. Results and discussion

3.1. MPCA and PLS-DA to NIR data

3.1.1. Pre-treatment

The first step in the data pre-treatment was to omit the measurements in which the line was stopped. To filter out the noise in the spectra data and find the most relevant information for the given product, extreme wavelength values were not considered.

A standard normal variate (SNV) was applied to the dataset. It is a common pre-processing step in spectral data analysis that neutralizes the offset of NIR signals and corrects scattering effects and makes all spectra from different runs comparable. In this method, a normalization of the data is performed by subtracting each spectrum by its own mean and dividing it by its own standard deviation.

Prior to applying MPCA and PLS-DA, and after applying the SNV for each run, mean centering and scaling were applied to normalize the data along the different runs.

3.1.2. MPCA vs PLS-DA

The MPCA model was applied to the pre-treated data. Following the Wold's R criteria, the optimal rank of the model was found at 3. However, with only 2 latent variables, the model was able to explain 94% of the system variation. Similarly, PLS-DA model was trained and cross-validated using only the data corresponding to the steady state, so the model could find the correlations between the data and the label vector. The optimal rank was found to be 2. Once the rank of the PLS model was fixed at 2, the model was applied to the whole spectral data. The results of both models can be visualized by plotting the 1st latent variable against the 2nd (Figure 2).

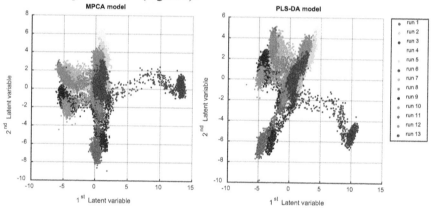

Figure 2. Scores of 1st LV against 2nd LV for both MPCA (left) and PLS-DA (right) models. The long shape of the clusters corresponds to the time evolution of the data during the step change, meaning that the edges of every cluster group the time points during steady state (before and after the step change) and the intermediate points correspond to the moment of the dynamic step change

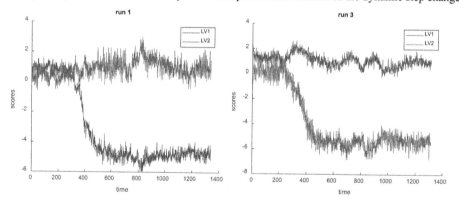

Figure 3. Scores over time for MPCA model of LV1 and LV2 in two cases: (left) when LV1 mostly summarizes the variation caused by the step change and (right) when it is LV2 that captures this variation

In both cases, the same phenomenon is observed: some of the run's variation is captured in a perpendicular direction from the other runs. However, MPCA makes a better job in capturing these variations aligned with each LV. For this reason, MPCA scores were chosen as the input signal for the RTD model fit. Figure 3 shows the concentration profiles of two of the runs, which exemplify the two behaviors identified in Figure 2. For runs 1, 2, 6, 7, 8, 9 and 10 it is LV1's scores that best capture the dynamic step change, for the other runs it is LV2 that best captures this variation.

3.2. RTD model fit

The scores of the different runs were used to fit the RTD model and it was compared with the model obtained using the traditional methodology using the experimental concentration data. The data was fit to the RTD model in terms of concentration C(t) by combining Equation (1) to Equation (2) using the fmincon MATLAB function. For every run and approach a set of three RTD parameters were obtained: θ, τ_1 and τ_2.

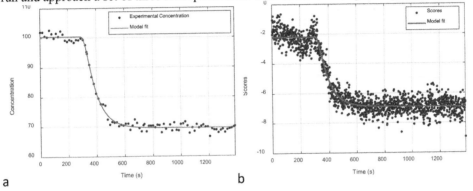

Figure 4. Example of input data and RTD model fit for experimental concentration data (a) and for scores data (b)

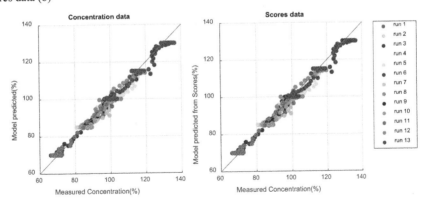

Figure 5. Parity plot of measured vs predicted tablet concentration using the RTD model for the traditional approach using measurements from tablet samples (left) and the new approach using the scores from the feed frame spectral data (right)

Using the set of RTD parameters of the run at target conditions (in this study, run 11) the predicted concentration was calculated with the RTD model. To assess the performance of the model prediction in this study, the root mean square error of prediction (RMSEP) was calculated for every run and for the two approaches using Equation (3), where $C_{RTD\ model}$ represents the tablet concentration calculated with the RTD model, $C_{measured}$ is the measured concentration, and N is the number of samples for every run (Hurley et al.). In the case of the experimental concentration data, the average RMSEP was 1.868

and for the scores data approach, the average RMSEP of all the runs was 1.873. A parity plot is shown in Figure 5 for both approaches, where the R^2 for the experimental concentration data and scores data is 0.9877 and 0.9875, respectively. This indicates that the new approach using the scores of the MPCA performs at the same level as using the measured tablet concentration data, and both approaches do a good job in predicting the tablet concentration.

$$RMSEP = \sqrt{\frac{C_{RTD\ model} - C_{measured}}{N}} \qquad (3)$$

4. Conclusion

The goal of this study was related to exploiting the spectral information from the PAT data. The raw spectral data from the NIR probe located in the feed frame was used to extract the dynamic response during the RTD studies in the tableting process. An unsupervised and supervised data-driven method was applied to the spectral data to summarize the dynamic information. MPCA and PLS-DA methods were able to successfully capture the dynamic response of the step change in a single latent variable for every run. This latent variable was used as the input signal for the RTD model fit, and its performance in determining the RTD parameters was compared to the current method using tablet concentration measurements. This new approach was able to approximate the RTD parameters for every run. Moreover, the model prediction given by the scores data performs equally well as when using the tablet concentration data. These results indicate the new approach may provide an alternative to approximate the quality of the tablets using the NIR measurements from the feed frame. This could contribute to making the RTD determination procedure more efficient and lowering the need to make offline measurements.

Acknowledgement

The authors want to acknowledge the financial support from the Flanders Innovation & Entrepreneurship (VLAIO) and Janssen Pharmaceutica NV in Belgium (Grant No. HBC.2022.0177 and HBC.2018.0355).

References

P.Bhalode et al., 2021, Using residence time distribution in pharmaceutical solid dose manufacturing – A critical review, International Journal of Pharmaceutics

E.Hetrick et al. 2017, Development of Near Infrared Spectroscopy-based Process Monitoring Methodology for Pharmaceutical Continuous Manufacturing Using an Offline Calibration Approach, Analytical Chemistry

T.Pedersen et al., 2021, Determination of Residence Time Distribution in a Continuous Powder Mixing Process With Supervised and Unsupervised Modeling of In-line Near Infrared (NIR) Spectroscopic Data, Journal of Pharmaceutical Sciences

N. Velez 2022, Challenges, opportunities and recent advances in near infrared spectroscopy applications for monitoring blend uniformity in the continuous manufacturing of solid oral dosage forms, International Journal of Pharmaceutics

S. Hurley et al., 2022, Development and Use of a Residence Time Distribution (RTD) Model Control Strategy for a Continuous Manufacturing Drug Product Pharmaceutical Process, Pharmaceutics

Antonis Kokossis, Michael C. Georgiadis, Efstratios N. Pistikopoulos (Eds.)
PROCEEDINGS OF THE 33rd European Symposium on Computer Aided Process Engineering
(ESCAPE33), June 18-21, 2023, Athens, Greece

Optimal power distribution in a P2A plant

Joachim Weel Rosbo[a], Tobias K. S. Ritschel[b], Steen Hørsholt[c], Anker D. Jensen[a] , John Bagterp Jørgensen[b], Jakob K. Huusom[a]

[a]*Dept. of Chemical and Biochemical Engineering, Technical University of Denmark,*
[b]*Dept. of Applied Mathematics and Computer Science, Technical University of Denmark,* [c]*2-control ApS*

Abstract

In Power-to-Ammonia (P2A) plants the ammonia reactor is required to operate over a wide operating window between 20% to 125% of the nominal load. We formulate a rigorous model for an ammonia synthesis loop with a three-bed adiabatic quench-cooled ammonia reactor. Given a steady-state solution to the ammonia synthesis loop, we evaluate the total power input for the P2A plant composed of the electrical utility for H_2 and N_2 production, and compression work. The total power input and the electrical efficiency for storing electrical energy in NH_3 are evaluated over a wide range of reactor H_2 flow, N_2/H_2-ratio and recycle-ratio of the reactants. In this way, we can identify the optimal reactor flow for a given power input to the P2A plant. The optimum is revealed to be at a significantly greater N_2/H_2- and recycle-ratio than used in conventional ammonia plants. This reflects the very expensive production of H_2 via electrolysis in P2A compared to conventional production via steam-methane reforming.

Keywords: Power-to-Ammonia, Optimization, Flexible operation.

1. Introduction

Ammonia is perceived as one of the most promising mediums for chemically based storage of renewable energy sources, e.g. wind and solar power. In P2A, electrical energy from renewable sources is used to generate the reactants, hydrogen via electrolysis and nitrogen via air separation. Due to the intermittent nature of renewable energy sources, P2A plants are required to operate over an operating window from 20% to 125% of the nominal load (Armijo & Philibert, 2020). This defines entirely new operation requirements for ammonia reactors, which are conventionally operated with a stable and reliable supply of reactants from steam-methane reforming. Due to the highly energy-intensive process of hydrogen production, ammonia plants are traditionally located in areas where natural gas is cheap and readily available. But in P2A the hydrogen is produced from electrical energy, which makes the hydrogen more costly to produce. In this study, we investigate the optimal operation of an ammonia synthesis loop in P2A with varying power inputs.

2. System description and specifications

Figure 1 shows a schematic illustration of the P2A plant considered in this work. The ammonia is produced by the Haber-Bosch process in an adiabatic quench cooled reactor (AQCR), which is a commonly used reactor in the industry (Inamuddin et al., 2020). Due to equilibrium limitations, the single-pass conversion of the reactor is around 25 %. Therefore, a relatively large recycling of reactants is required. Rosbo et al. (2022) describe mathematical models of the units in the synthesis loop (reactor, heat exchanger, purge, and separator). We use the case study defined by Rosbo et al. (2022) based on a

100MW plant with operating pressure, temperatures, and reactor flowrates scaled from a conventional ammonia reactor.

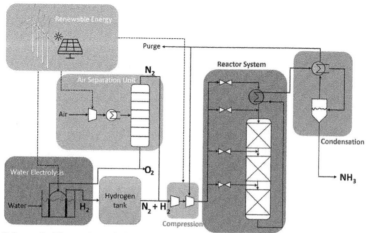

Figure 1: Schematic illustration of the main units in a P2A plant. Electricity is supplied from renewable sources and used in the electrolyser, air separation unit (ASU), and compressors.

2.1.1. Pressure drop

The reactor bed is assumed at constant pressure in the mathematical model as the pressure drop is not significant. However, the work of recycle compression relates to the pressure drop over the reactor. Thus, we assess the pressure drop over a bed via Ergun's equation,

$$\frac{\mathrm{d}P}{\mathrm{d}l} = \frac{180\mu^g}{d_p^2}\frac{(1-\epsilon)^2}{\epsilon^3}v_s + \frac{1.75\rho^g}{d_p}\frac{(1-\epsilon)}{\epsilon^3}v_s^2$$

in which P is the pressure, l is bed length, μ^g is the gas viscosity, ρ^g is the gas density, d_p is the catalyst particles hydraulic diameter and v_s is the superficial gas velocity.

3. Energy requirements for synthesis loop

3.1. Electrolyser

The production of hydrogen from water consumes most of the power input of the P2A plant. Hydrogen is produced in the electrolyser at a cost of 48 kWh/kg corresponding to an electrolyser efficiency of 67% compared with the low heating value of hydrogen (Armijo & Philibert, 2020).

3.2. Air separation unit

The nitrogen feed stream is produced in an air separation unit by cryogenic distillation. The power requirement of the air separation unit (ASU) is mainly from the air compression train. The electrical utility for air separation is set to 0.115 kWh/kg nitrogen (Morgan et al., 2014).

3.3. Compressors

The synthesis feed gas requires extensive compression from a few bars up to the operating pressure of the ammonia reactor at 200 bars. The work of an isentropic compression from pressure P_{in} to P_{out} is

$$W_{comp.}^{isen.} = H\left(P_{out}, T_{out}^{isen.}, N\right) - H(P_{in}, T_{in}, N) \qquad (1)$$

The outlet temperature for the isentropic compression, T_{out}^{isen}, can be found from solving

$$S\left(P_{out}, T_{out}^{isen.}, N\right) - S(P_{in}, T_{in}, N) = 0 \qquad (2)$$

which is done using Newton's method. We use the thermodynamic tool ThermoLib for evaluating the state functions H and S (Ritschel et al., 2016). The actual compressor work, $W_{comp.}$, can be found from

$$W_{comp.} = \frac{W_s}{\eta_{isen}} \tag{3}$$

where η_{isen} is the isentropic efficiency ($\eta_{isen} = 0.75$). The compression of the feed gas is carried out by multistage compression with a maximum compression ratio of 4.

3.4. Total power input and energy efficiency of the P2A plant

The total power is given as the sum of the individual contributions from H_2 production, W_{H_2}, N_2 production, W_{N_2}, and compressors

$$W_{tot} = W_{H_2} + W_{N_2} + W_{comp.} \tag{4}$$

We define the energy efficiency as

$$\eta_E = \frac{LHV_{NH_3} \dot{m}_{NH_3}}{W_{tot}} \tag{5}$$

Where LHV_{NH_3} is the lower heating value of NH_3. The energy efficiency, η_E, describes how well the electrical energy is stored as chemical energy in ammonia.

4. Results

4.1. Nominal case with conventional reactor feed stoichiometry

For the nominal case, based on a conventional reactor configuration, the total power requirement for the P2A plant is 104.2 MW and the electricity to ammonia efficiency is 50.35%. Figure 2 shows the division of the electrical utility between H_2 and N_2 production, and compressor work. The electrolyser is consuming the largest fraction of the total power input (91%), while the ASU and compressor consumes relatively equal amounts of electricity (4% and 5%).

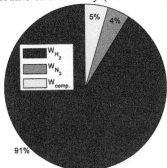

Case	W_{tot} [MW]	η_E
Nominal	104.2	50.35%
Optimal	118.1	50.52%
Opt. norm	104.2	51.29%

Left: Figure 2: Fractional energy requirements in nominal P2A plant.
Right: Table 1: Total work and energy efficiency for the nominal and optimised reactor configuration.

The reactor valve settings are optimised using the optimisation algorithm presented by Rosbo et al. (2022). Table 1 shows the total power input and energy efficiency with optimised reactor quench flows. The conversion is about 10% higher for the optimised reactor and, consequently, more reactants are consumed. Therefore, the optimised configuration uses more power at 118.1 MW. Compared to the nominal case, the energy efficiency is increased to 50.52 %. For comparison at an equal power input the optimised solution is scaled down to match the power utility of the nominal case. In this way, we obtain an energy efficiency of 51.29%.

4.2. Operating at larger N_2/H_2-ratio

By optimising the reactor system, we increase the energy efficiency by almost 1%. This was achieved by keeping N_2/H_2 at the stoichiometric ratio in the reactor feed, which reflects operation in a conventional ammonia plant. However, from the pie chart in Figure 2 it is evident that in P2A the hydrogen is very costly to produce. Thus, in P2A it may be more energy efficient to operate at a N_2/H_2-ratio larger than the stoichiometric ratio. By operating with N_2 in excess we facility greater conversion of H_2. We define the stoichiometry factor, S_{N_2/H_2}, based on the N_2/H_2-ratio in the reactor feed, N_{Rf}, as

$$S_{N_2/H_2} = \frac{3N_{Rf,N_2}}{N_{Rf,H_2}}$$

Figure 3a displays the NH_3 production rate as a function of the hydrogen reactor feed rate relative to the nominal case and stoichiometry factor, S_{H_2/N_2}. The total power input, W_{tot}, is overlayed by black contours.

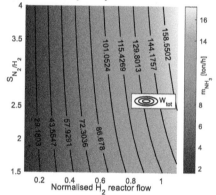

Figure 2a: NH_3 production rate overlayed with contours of total power input.

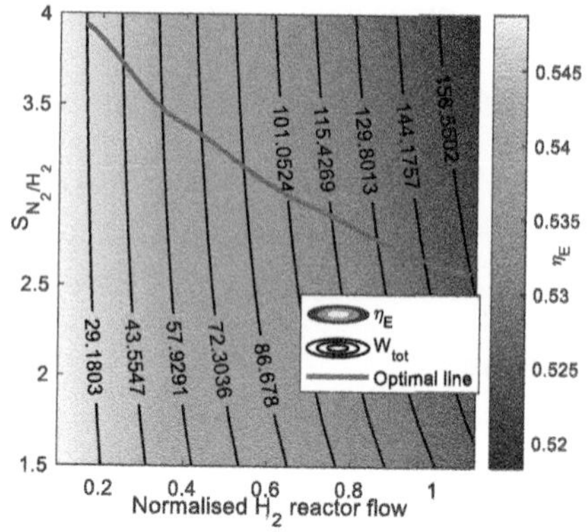

Figure 3b: Energy efficiency, η_E overlayed with contours of total power input.

Figure 3b shows the energy efficiency in the plane of relative H_2 reactor feed flow and S_{N_2/H_2}. Observe that the optimal stoichiometry factor is as high as 2.7 to 3.8 over the power range. Naturally, the energy efficiency is increasing for lower power input as the reactor is converting the reactants more efficiently at lower throughput. The red curve identifies the maximum energy efficiency for a given power. The optimal curve is moving towards higher stoichiometry factor for lower power input. At lower reactor throughput the reactor is slightly oversized, and it is advantages to operate with relatively more nitrogen. Operating with excessive flow of N_2 yields a significantly improved energy efficiency at approximately 53.5% for a power input around 104 MW. This clearly illustrates the difference in operation of an ammonia reactor for P2A compared with conventional operation. Figure 4a and 4b show accumulative graphs of the electrical power consumption and reactor flowrate, respectively. Although the reactor is operated at close to equal flowrate of N_2 and H_2, the power input for the ASU and compressors is still only a small fraction of the total power consumption.

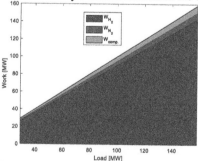

Figure 4a Accumulated energy.

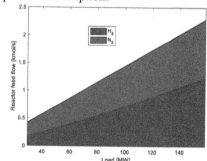

Figure 4b Accumulated flow rate.

4.3. Operating at a larger recycle-ratio

In the above analysis we kept the flow of argon in the reactor constant. From Figure 4a it is observed that the compressor work is relatively small compared to the amount of power used for hydrogen production. By reducing the purge fraction, we can facilitate more overall conversion of hydrogen. This is at the cost of a higher recycle stream and more compression work. In practise we vary the recycle by changing the reactor flowrate of argon. Figure 5 shows surfaces of constant power input as a function of normalised H_2 flowrate, S_{N_2/H_2} and Ar/N_2-ratio. The red curve illustrates the optimal reactor inlet flows for a given power consumption. The argon to nitration ratio in the reactor feed is ranging from 0.6 to 1.8, which is significantly higher than a conventional ratio of 0.13. Figure 5 shows, we can obtain an energy efficiency of around 56% by operating with a large recycle-ratio even for power inputs above 100 MW.

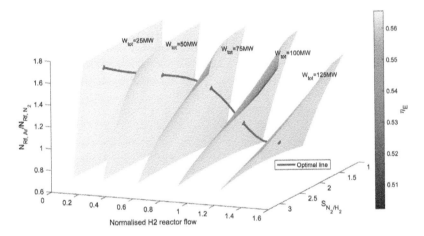

Figure 5 Surfaces of constant power input as a function of normalised H_2 flowrate, S_{N_2/H_2} and Ar/N_2-ratio. The surfaces color depicts the electrical efficiency.

Figure 6a displays the electrical utility of the main P2A units for achieving optimal energy efficiency. Note the compression work is increased and the ASU power input decreased compared with Figure 4a. The larger recycle-ratio causes excess N_2 to quickly build up, which lowers the requirement of fresh feed N_2 production. Figure 6b shows the reactor feed flow of H_2, N_2 and Ar achieving maximum energy efficiency. Clearly, a large amount of inert Argon is built up in the synthesis loop.

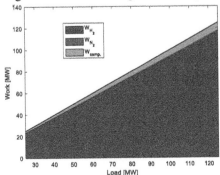 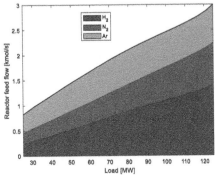

Figure 6a Accumulated energy.   Figure 6b Accumulated flowrate.

5. Discussion

This study suggests that in P2A the synthesis reactor should be operated at significantly higher N_2/H_2-ratio than stoichiometric to obtain maximum energy efficiency. Additionally, we found that larger recycle of reactants compared to conventional operation increased the energy efficiency. We assume that the effectiveness factor of the catalyst particles is independent of the gas concentrations. Most literature on modelling of ammonia reactors use either the Temkin-Pyzhev rate expression or Dyson's equation (Dyson & Simon, 1968). These rate expressions from 1940 and 1968 are naturally developed at conventional reactor conditions with a stoichiometric N_2/H_2-ratio. Therefore, the validity of these rate expressions is unclear when operating at large excess of N_2 and inert argon. Nielsen et al. (1964) investigated the reaction rate of ammonia synthesis on iron catalyst under various temperatures, pressures, flow rates and even five runs with changing N_2/H_2-ratio. These results suggested that the reaction rate may be negatively affected at lower flowrates, but the Temkin-Pyzhev equation yielded relatively accurate results at more industrially relevant flow velocities. Traditionally, argon is not mentioned as a poisonous species for ammonia catalysts and no literature seems to discuss the effect of argon on catalyst activity. Clearly, more experimental research needs to be conducted on how operating at large excess of N_2 and inert argon affects the ammonia catalyst. But this study indicates that better utilisation of the electrical energy in P2A can be achieved by operating the ammonia reactor at non-traditional conditions.

6. Conclusion

This paper has investigated the optimal operation of an ammonia synthesis loop for P2A using steady state optimisation. We identified the optimal distribution of electrical energy between H_2 and N_2 production, and compression work for power inputs ranging from around 20% to 125% of the nominal load. We found the optimum to be at significantly larger N_2/H_2- and recycle-ratios compared with conventional ammonia plants. Operating with large excess of N_2 and recycle-ratio ensures a high conversion of H_2, which is extremely costly to produce in P2A. However, the effect of operating at large excess of N_2 and inert argon on the ammonia catalyst needs to be further investigated experimentally.

References

Armijo, J., & Philibert, C. (2020). Flexible production of green hydrogen and ammonia from variable solar and wind energy: Case study of Chile and Argentina. *International Journal of Hydrogen Energy, 45*(3), 1541–1558.

Dyson, D. C., & Simon, J. M. (1968). A kinetic expression with diffusion correction for ammonia synthesis on industrial catalyst. *Industrial and Engineering Chemistry Fundamentals, 7*(4), '

Inamuddin, R. B., & Asiri, A. (2020). Sustainable Ammonia Production. In *Green Energy and Technology*. Springer Nature Switzerland.

Morgan, E., Manwell, J., & McGowan, J. (2014). Wind-powered ammonia fuel production for remote islands: A case study. *Renewable Energy, 72*, 51–61.

Nielsen, A., Kjaer, J., & Hansen, B. (1964). Rate equation and mechanism of ammonia synthesis at industrial conditions. *Journal of Catalysis, 3*(1), 68–79

Ritschel, T. K. S., Gaspar, J., Capolei, A., & Jørgensen, J. B. (2016). *An open-source thermodynamic software library. DTU Compute-Technical Report-2016 No. 12.*

Rosbo, J. W., Ritschel, T. K. S., Hørsholt, S., Huusom, J. K., & Jørgensen, J. B. (2022). Flexible operation, optimisation and stabilising control of a quench cooled ammonia reactor for Power-to-Ammonia. *Submitted to Computers & Chemical Engineering.*

Antonis Kokossis, Michael C. Georgiadis, Efstratios N. Pistikopoulos (Eds.)
PROCEEDINGS OF THE 33rd European Symposium on Computer Aided Process Engineering
(ESCAPE33), June 18-21, 2023, Athens, Greece

Blend-scheduling optimisation for continuous and batch mixtures: modelling and solving algorithms

Mahmoud A. Ahmednooh,[a,b] Brenno C. Menezes,[b] Mohammed Yaqot,[b] Jeffrey D. Kelly[c]

[a]*Division of Engineering Management and Decision Sciences, College of Science and Engineering, Hamad Bin Khalifa University, Doha, Qatar Foundation, Qatar*

[b]*Division of Production Planning and Scheduling, Um Said Refinery, Qatar Energy, Doha, Qatar*

[c]*Industrial Algorithms Ltd., 15 St. Andrews Road, Toronto M1P 4C3, Canada*
bmenezes@hbku.edu.qa

Abstract

Simultaneous blending and scheduling optimisation represents a mixed-integer nonlinear programming (MINLP) problem, whereby the binary variable relaxation that forms a nonlinear programming (NLP) in the first stage of a full-space algorithm (without dropping any decision variable) may lead to convergence issues. Furthermore, there is no guarantee to reach a global optimal solution since it solves an NLP problem to then address the mixed-integer linear programming (MILP) problem in outer-approximation algorithms. On the other hand, by neglecting quality variables and constraint of the blends or mixtures in the MILP problem, in a decomposed or non-full space problem (when dropping nonlinear constraints), infeasibilities or local optimal solutions may be found in the second stage NLP programs. In this case, there will be a change that from the MILP resulted assignments of the components to be blended, quality to specify the final blended material in the NLP problem may not be suffice. Or even if so, local optimal solution may occur. To skip these issues, an MILP-NLP decomposition is tailored to solve, in the first stage, the MILP logistics problem of the blend-scheduling optimisation whereby the blending relationship for each property quality is approached as amounts of quality balances. In such constraints, the blended quality variable is replaced by the bounds of the specifications plus the slack variable or less the surplus variable to close the linear quality balance. However, this MILP-NLP decomposition with a blending approximation in the MILP still cannot avoid convergence issues or local optimal inherent to the NLP stage and is dependent on the volume- or mass-based material flow and blended property governing rules. Then, we apply an optimisation-simulation algorithm to converge the MILP solution to a global optimal by considering a substitution of the current blending error successively in the next MILP solution. Nevertheless, this algorithm can be applied only in continuous mixtures, but we introduce the novel strategy to solve batch mixtures by using component quantities found in a continuous mixture topology as inputs.

Keywords: blending operations, scheduling optimisation, MILP, NLP, decomposition.

1. Introduction

Blending and scheduling processes are considered crucial steps in crude-oil and metallurgical refineries, food processing sites, pulp and paper productions, to name a few. The main goal in these processing industries is to transform different raw materials into useful products such as gasoline, diesel, and kerosene (from crude-oils) and copper alloys (from minerals). Stricter constraints from environmental regulations and lower profit margins from commodity falling prices have been considered for the blending of these streams to achieve better efficiency in the operations and to adhere to specifications (Lotero et al., 2016). The blending process is a significant stage that helps the processing sites to convert low-value raw materials to higher added-value products, with consideration to eliminate giveaway or excess of qualities and off-specification products.

Most (if not all) of the blend-scheduling solutions in the market are either simulation-based or simple linear programming (LP) approaches. These tools are limited in terms of capability since the simulation-based ones are entirely based on trial-and-error of the binary variables, which is time-consuming and difficult to change or readjust. Also, linear programming software is limited to a fixed or given operation, in general, addressing a single-period problem. A proposed solution that can provide multi-period solution with automated decision-making in a blend-scheduling perspective is a breakthrough in the quantity-logic-quality decisions in the aforementioned industries. This becomes explore in optimisation problems since the blending of streams presents a nonlinear (NLP) behavior, which brings difficulties on convergence and cannot avoid local optimal solutions. For an optimisation of a blendshop or blend-scheduling problem, the blending and assignment of equipment (and their modes) and of material streams represents a mixed-integer nonlinear programming (MINLP), whereby, to avoid MINLP convergence issues when scaling-up blend-scheduling problems, we proposed a successive substitution approach based on optimisation-simulation methods to iteratively reduce the error inherent by using factor-flows or amounts of quality balances from Kelly et al. (2018). In equality balances of input-output amounts of qualities, the outputted amounts of qualities are calculated by the quality specification multiplied by the total flow of material plus the slack variable or less the surplus variable to close the linear quality balance, whereby this additional variable, known as factor-flows, close the balance of each property.

2. Problem statement

In the continuous mixture of the blend-scheduling represented in Figure 1, it is produced diesel by blending light diesel gasoil in R1 and light cycle oil in R2. Both intermediate components have similar specific gravity (SG) and sulfur concentration (SC) to produce diesel, whereby specifications on the final diesel qualities should be maintained within bounds. Their qualities are: R1 with 0.80 of SG and 0.90 of SC (at \$ 85 cost); R2 with 0.95 of SG and 1.20 of SC (at \$ 70 cost); and D1 with 0.875 of SG and 1.05 of SC to specify (at \$ 100 as selling price per barrel). There are four cases to be addressed in this work to avoid an MINLP inherent to blend-scheduling problems. Cases 1 to 3 are applied to continuous blending (Figure 1) and case 4 to batch mixtures (Figure 1, small topology).

1) MILP-NLP decomposition: nonlinear quality balances are neglected in the MILP problem, then the assignments of the S1 to S4 tanks to the blender are fixed to 0 or 1 in the following NLP solution.

2) MILP-NLP decomposition: using factor-flow balances where the nonlinear quality constraints are approximated in the MILP problem as amounts of qualities (Kelly et al. 2018), whereby the blended quality variable is replaced by the bounds of the specifications plus the slack variable or less the surplus variable to close the linear quality balance. Then, the assignments of the S1 to S4 tanks to the blender are fixed to 0 or 1 in the following NLP solution.

3) SS-MILP iteration: successive substitution of an optimisation-simulation method considering the error between the factor-flow approximated property calculated by the quantities of the components found in the MILP and the volume- or mass-based blended property governing rules (from the simulation step) for the same amounts of the components. The quality error is used in the next MILP optimisation until convergence.

4) SS-MILP iteration and MILP-NLP batch: the amounts or quantity flows from the tanks S1 to S4 to the blenders converged in the SS-MILP algorithm are fixed in the blend-scheduling case without the blender (as seen in Figure 1, small topology)

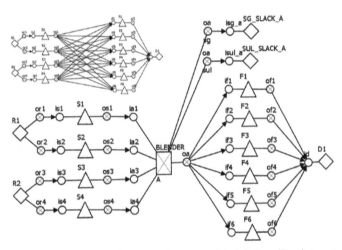

Figure 1: Blend-scheduling using a continuous and batch (small) mixture topology.

3. Mathematical modelling

The objective function in Eq.(1) maximises the diesel production, where $x_{j,i,t}$ represent the selling product and consumed components in the diesel production at time t. Following the unit-operation-port-state superstructure (UOPSS) formulation from Kelly (2005), in the MILP problem, all flows and holdups are governed by semi-continuous constraints (relating binary and continuous variables) of the shapes to themselves, such as $\bar{x}^L_{j,i,t} y_{j,i,t} \leq x_{j,i,t} \leq \bar{x}^U_{j,i,t} y_{j,i,t} \ \forall \ (j,i) \in JI, t$ considering outlet port set j to inlet set i. The sets I and J represent in- and out-ports, respectively, while the set JI defines connecting flows between out- and in-ports. The semi-continuous constraints for throughputs $x_{m,t}$ of the unit-operations m for the blender are controlled by $\bar{x}^L_{m,t} y_{m,t} \leq x_{m,t} \leq \bar{x}^U_{m,t} y_{m,t}$ and the holdup or inventory level $xh_{m,t}$ by $\overline{xh}^L_{m,t} y_{m,t} \leq xh_{m,t} \leq \overline{xh}^U_{m,t} y_{m,t}$. In the indices in the summations from Eq.(1) to (7), the subsets of the I, J, and JI follow the flowsheet in Figure 1, considering $i \in I_{do}$ as inlets of downstream unit-operations and $j \in J_{up}$ as outlets of upstream unit-operations. For $x_{j,i,t}, x_{m,t}, xh_{m,t} \geq 0$; $y_{j,i,t}, y_{m,t} = \{0,1\}$; $zsu_{m,t} = (0,1)$:

$$Max \ Z = \sum_t \left(\sum_{product} x_{j,i=D1,t} - \sum_{components} x_{j=R1 \ or \ R2,i,t} \right) \tag{1}$$

$$\frac{1}{\overline{x}_{m,t}^U} \sum_{j \in J_{up}} x_{j,i,t} \leq y_{m,t} \leq \frac{1}{\overline{x}_{m,t}^L} \sum_{j \in J_{up}} x_{j,i,t} \quad \forall \, (i,m) \in (M_{Blender}, M_{Product}) \, , t \tag{2}$$

$$\frac{1}{\overline{x}_{m,t}^U} \sum_{i \in I_{do}} x_{j,i,t} \leq y_{m,t} \leq \frac{1}{\overline{x}_{m,t}^L} \sum_{i \in I_{do}} x_{j,i,t} \quad \forall \, (m,j) \in (M_{Blender}, M_{Component}) \, , t \tag{3}$$

$$\frac{1}{\overline{x}_{m,t}^U} \sum_{j \in J_{up}} x_{j,i,t} \leq x_{m,t} \leq \frac{1}{\overline{x}_{m,t}^L} \sum_{j \in J_{up}} x_{j,i,t} \quad \forall \, (i,m) \in M_{Blender} \, , t \tag{4}$$

$$\frac{1}{\overline{x}_{m,t}^U} \sum_{i \in I_{do}} x_{j,i,t} \leq x_{m,t} \leq \frac{1}{\overline{x}_{m,t}^L} \sum_{i \in I_{do}} x_{j,i,t} \quad \forall \, (m,j) \in M_{Blender} \, , t \tag{5}$$

$$xh_{m,t} = xh_{m,t-1} + \sum_{j_{up} \in J} x_{j_{up},i,t} - \sum_{i_{do} \in I} x_{j,i_{do},t} \quad \forall \, (i,m,j) \in M_{TK}, t \tag{6}$$

$$\sum_{i} \sum_{j_{up}} x_{j_{up},i,t} = \sum_{i_{do}} x_{j,i_{do},t} \quad \forall \left((j_{up}, i), (j, i_{do}), t \right), m \in M_{Blender} \tag{7}$$

$$mu_{j,t}^L \, y_{m,t} \leq \sum_{i_{do}} y_{j,i_{do},t} \leq mu_{j,t}^U \, y_{m,t} \ \forall \, (m,j) \in M_{Blender} \, , t \tag{8}$$

$$mu_{i,t}^L \, y_{m,t} \leq \sum_{j_{up}} y_{j_{up},i,t} \leq mu_{i,t}^U \, y_{m,t} \ \forall \, (i,m) \in M_{Product}, t \tag{9}$$

$$y_{m_{up},t} + y_{m,t} \geq 2y_{j_{up},i,t} \ \forall \left(m_{up}, j_{up}, i, m \right), t \tag{10}$$

$$y_{m,t} - y_{m,t-1} - zsu_{m,t} + zsd_{m,t} = 0 \ \forall \, m \in M_{Blender}, t \tag{11}$$

$$y_{m,t} + y_{m,t-1} - zsu_{m,t} - zsd_{m,t} - 2zsw_{m,t} = 0 \ \forall \, m \in M_{Blender} \, , t \tag{12}$$

$$zsu_{m,t} + zsd_{m,t} + zsw_{m,t} \leq 1 \ \forall \, m \in M_{Blender}, t \tag{13}$$

Equations (2) and (3) represent the semi-continuous constraints relating neighboring flows of the JI connections to the throughputs of unit-operation setups they are connected to. Different from the semi-continuous constraints to itself (of the shapes to themselves), Eq.(2) and Eq.(3) allow the flows throughout the connections only if the nodes (the unit-operations) are true. On the other hand, Eq.(4) and Eq.(5) balance the in-out material flows of the connections to the throughputs of the unit-operations, whereby there is a process transformation such as in the blender. The unit-operations' inventory or holdup quantity balance of tanks are calculated in Eq.(6). Equation (7) is the material flow balance in the blender. Multi-use constraints control the use of the blender's outlet in Eq.(8) and the inlet of the product demand point $D1$ in Eq.(9).

Equation (10) represent the constraint for the structural transitions that allow the setup $y_{m,t}$ of connected out-port-states j and in-port-states i unit-operations. When the setup of unit-operations m and m' is equal to the unitary, by implication, the setup variable of the arrow stream $y_{j,i,t}$ between the neighbor unit-operations must be true, and vice-and-versa. These logic valid cuts reduce the tree search in branch-and-bound methods. The temporal transition in Equations (11) and (12) control the operations for semi-continuous blenders from Kelly and Zyngier (2007). The binary variable $y_{m,t}$ manages the start-up ($zsu_{m,t}$) switch-over ($zsw_{m,t}$) and shut-down variables ($zsd_{m,t}$), which are relaxed in the interval [0,1]. Equation (13) guarantees the integrality of the relaxed variables.

For the NLP problem, the binary variables found in the MILP are fixed. The quality balance constraints in the blender are defined in Eq.(14) for the volume-based property $v_{j,p,t}$ for $p \in P_v$ (as in SG) and in Eq. (15) for the mass-based property $w_{j,p,t}$ for $p \in P_w$ (as in SC). Equations (16) and (17) represent the quality balances for volume- and mass-based properties in tanks. The quality variable for the out-ports of a tank unit-operation ($m \in M_{TK}$) is the quality of the blend within the tank $v_{j,p,t} = v_{m,p,t}$ and $w_{j,p,t} = w_{m,p,t}$.

$$v_{j,p,t} \sum_{j_{up}} x_{j_{up},i,t} = \sum_{j_{up}} v_{j_{up},p,t} x_{j_{up},i,t} \quad \forall j \in M_{Blender}, p \in P_v, t \tag{14}$$

$$w_{j,p,t} \sum_{j_{up}} v_{j_{up}=sg,p,t} x_{j_{up},i,t} = \sum_{j_{up}} w_{j_{up},p,t} v_{j_{up}=sg,p,t} x_{j_{up},i,t} \quad \forall j \in M_{Blender}, p \in P_w, t \tag{15}$$

$$v_{m,p,t} xh_{m,t} = v_{m,p,t-1} xh_{m,t-1} + \sum_{j_{up}} v_{j_{up},p,t} x_{j_{up},i,t} - v_{m,p,t} \sum_{i_{do}} x_{j,i_{do},t} \tag{16}$$
$$\forall (i,m,j) \in M_{TK}, p \in P_v, t$$

$$w_{m,p,t} v_{m,p=sg,t} xh_{m,t} = w_{m,p,t-1} v_{m,p=sg,t-1} xh_{m,t-1} + \sum_{j_{up}} w_{j_{up},p,t} v_{j_{up},p=sg,t} x_{j_{up},i,t} - \tag{17}$$
$$w_{m,p,t} v_{m,p=sg,t} \sum_{i_{do}} x_{j,i_{do},t} \quad \forall (i,m,j) \in M_{TK}, p \in P_w, t$$

4. Results

The optimisation for the proposed problems (to produce 100 K barrels per day) for 10 days of time-horizon and 1 day as time-step uses an Intel Core i7 machine at 2.2 GHz (6 threads) with 32 GB of RAM. In the case 1, for the MILP, there are 552 constraints (122 equality) for 312 continuous variables and 210 binary variables with 400 degrees-of-freedom (variables minus equality constraints). For the NLP, there are 531 constraints (491 equality) for 621 continuous variables with 130 degrees-of-freedom. The results using GUROBI 10.0 and CPLEX 20.0 for the MILP blend-scheduling cases are solved within 1 second. And the NLP are solved by linking these solvers to sequential linear programming quadratic engine (SLPQPE) in the modelling platform Industrial Modeling and Programming Language (IMPL) used in this work (see Kelly and Menezes, 2019).

1) MILP-NLP decomposition: the MILP solution yields $ 440 and it gets infeasible in the NLP. Since the nonlinear quality balances are neglected in the MILP problem, the assignments to produce diesel selects the S3 or S4 tanks to the blender at a reduced cost. However, then the NLP problem is solved it is not possible to meet the specification since better SG and SC qualities from S1 or S2 tanks are needed.

2) MILP-NLP decomposition: the MILP solution yields $ 380 and the NLP $ 320. By using factor-flow balances with the nonlinear quality constraints approximated in the MILP as amounts of qualities, whereby the blended quality variable is replaced by the bounds of the specifications plus the slack variable or less the surplus, S1 or S2 tanks are assigned. Then, for the binary variables fixed in NLP problem, it is possible to meet the specifications of SG and SC qualities. However, as seen in the solution in Table 1, the flows are not optimal since in several days the blend-shop uses only S1 or S2 tanks.

3) SS-MILP iteration: successive substitution of an optimisation-simulation method considering the error between the factor-flow approximated property calculated by the quantities of the components found in the MILP and the volume- or mass-based oar ad hoc blended property governing rules (from the simulation step) for the same amounts of the components. The quality error is used in the next MILP optimisation until convergence. It reaches the global optimum solution within two SS iterations

4) SS-MILP iteration and MILP-NLP batch: the amounts or quantity flows from the tanks S1 to S4 to the blenders, converged in the SS-MILP algorithm, are fixed in the blend-scheduling case without blender. In the MILP-NLP batch case, with the SS-MILP quantities fixed in the outlets of the S1-S4 and inlets of the F1-F6 tanks (within an adjusted time-step that permits one batch flow at a time), the solution yields a feasible and optimal results with 0% gap between the MILP and NLP problems.

Table 1. Results of the cases in the blend-scheduling problems.

day	Case 2-MILP (380 K USD)				Case 2-NLP (320 K USD)				Case 3 (371.6 K USD)			
	S1	S2	S3	S4	S1	S2	S3	S4	S1	S2	S3	S4
1	50.0		50.0		58.7		41.3		54.3		45.7	
2	50.0		50.0		100.0				54.3		45.7	
3		50.0		50.0		100.0				54.3		45.7
4		50.0		50.0		100.0				54.3		45.7
5	50.0		50.0		100.0				54.3		45.7	
6		50.0		50.0		100.0				54.3		45.7
7		50.0	50.0			100.0				54.3	45.7	
8		50.0	50.0			91.3	8.7			54.3	45.7	
9		50.0		50.0		92.4		7.6		54.3		45.7
10	50.0			50.0	57.6			42.4	54.3			45.7

5. Conclusion

The importance of approximating an NLP blending as an LP factor-flow approach in the MILP stage is considered in this work, whereby factor-flow variables by equalizing in-out amounts of qualities in the blender allow global optimal solutions since it avoids NLP problems in a second stage of the MILP stage. For batch mixture topology, the drawback of the factor-flow approach by considering a convergent and continuous point of mixture can be circumvented when fixing the material flows of the components found in the SS-MILP (case 3) as inputs in the MILP-NLP batch case, adjusting of needed the time-step.

References

I. Lotero, F. Trespalacios, I.E. Grossmann, D.J. Papageorgiou, M.-S. Cheon, (2016). An MILP-MINLP decomposition method for the global optimization of a source based model of the multiperiod blending problem, Computers & Chemical Engineering, 87, 13–35.

J.D. Kelly, B.C. Menezes, I.E Grossmann, 2018, Successive LP approximation for nonconvex blending in MILP scheduling optimization using factors for qualities in the process industry, Industrial & Engineering Chemistry Research, 57(32), 11076-11093.

J.D. Kelly, 2005, The Unit-Operation-Stock Superstructure (UOSS) and the Quantity-Logic-Quality Paradigm (QLQP) for production scheduling in the process industries, In Multidisciplinary International Scheduling Conference Proceedings: New York, United States, 327-333.

J.D. Kelly, B.C. Menezes, 2019, Industrial Modeling and Programming Language (IMPL) for off- and on-line optimization and estimation applications. In: Fathi M., Khakifirooz M., Pardalos P. (eds) Optimization in Large Scale Problems. Springer Optimization and Its Applications, 152, 75-96.

Antonis Kokossis, Michael C. Georgiadis, Efstratios N. Pistikopoulos (Eds.)
PROCEEDINGS OF THE 33rd European Symposium on Computer Aided Process Engineering
(ESCAPE33), June 18-21, 2023, Athens, Greece

A machine learning dynamic modelling scheme for wastewater treatment plants using cooperative particle swarm optimization and neural networks

Teo Protoulis,[a] Ioannis Kalogeropoulos,[b,c] Ioannis Kordatos,[a] Haralambos Sarimveis,[b] Alex Alexandridis[a]

[a]Department of Electrical and Electronic Engineering, University of West Attica, Ancient Olive Grove Campus, Thivon 250 & P. Ralli, Aigaleo, 12244, Greece
[b]School of Chemical Engineering, National Technical University of Athens, Heroon Polytechneiou 9, Zografou, 15780, Greece
[c]Systemica - G.Vangelatos & Co L.P., Ethnikis Antistaseos 70, 16237, Athens, Greece

Abstract

This work introduces a framework for developing data-driven dynamic models for wastewater treatment plants (WWTPs) by using a limited amount of data. More specifically, a system identification problem is defined that estimates a number of critical parameters included in the COST/IWA benchmark simulation model No.1 (BSM1), which is adapted to a real WWTP. Identification of these parameters is performed by formulating an optimization problem, which is solved by utilizing a novel cooperative particle swarm optimization (CPSO) method. The identified model is then used to generate big volumes of dynamic data for training radial basis function networks (RBFNs), which are suitable for being integrated in process control configurations, like model predictive control schemes. The proposed optimization method is validated using statistical testing and is shown to compare favorably against alternative methods, while evaluation metrics indicate that the developed models predict the dynamic behavior of the plant with high accuracy.

Keywords: Cooperative particle swarm optimization; identification; modeling; radial basis function neural networks; wastewater treatment plants

1. Introduction

Wastewater treatment plants (WWTPs) are considered to be a major component in circular economy due to their ability of receiving large volumes of polluted water as influent and allowing the maximum percentage of it to be reused. Developing automatic control schemes for WWTPs is a challenging task, not only due to the multiple and highly complex biological and biochemical phenomena taking place in the various processes, but also because strict environmental regulations have to be met. The performance of WWTP control systems can be improved, by incorporating mathematical models capable of accurately capturing the system dynamics. On the other hand, developing such models that are also simple enough to be incorporated in control schemes, is a challenging task.

Data-driven technologies, such as artificial neural networks (ANNs) appear as a promising solution due to the ability to capture the nonlinear dynamics of WWTPs with satisfactory accuracy (Sadeghassadi et al., 2018), and the potential to be easily integrated within advanced model predictive control (MPC) methodologies (Papadimitrakis &

Alexandridis, 2022). On the other hand, training ANNs for WWTP modelling needs big volumes of data (Matheri et al., 2021). Unfortunately, due to the limited number of sensors available in WWTPs, the volumes of experimental data collected from the process are usually not sufficient; at the same time the collected data are of low quality due to several reasons (e.g. discontinuity, missing values, sensors failures, steady-state operation for long intervals), which lead to poor predictive performance of the developed models.

In this work, a novel approach is proposed for overcoming the problem of limited data availability in order to build highly accurate ANN models, suitable to be incorporated in MPC schemes. More specifically, a system identification problem is first defined, where the objective is to identify the critical parameters of the COST/IWA benchmark simulation model No. 1 (BSM1) (Alex et al., 2008), adapted to a real WWTP. The identification task takes the form of a complex nonlinear optimization problem, which is solved efficiently by introducing a novel cooperative particle swarm optimization (CPSO) (Kapnopoulos & Alexandridis, 2022) approach, specifically customized for WWTPs. The identified BSM1 model is used next to generate large volumes of dynamic data suitable for building high accuracy models that are based on radial basis function networks (RBFNs), trained with the fuzzy means (FM) algorithm (Alexandridis et al., 2013).

The rest of this work is structured as follows; in section 2, the formulation of the system identification problem is described and the proposed CPSO solver is presented. The next section presents the proposed method for developing dynamic WWTP models, including a brief description of the RBF architecture and the FM algorithm. The method is evaluated through a case study presented in section 4. Finally, the paper ends with concluding remarks and directions for future work.

2. System identification using cooperative particle swarm optimization

2.1. Parameter identification of the Benchmark Simulation Model No. 1 (BSM1)
BSM1 is a detailed WWTP model as described in (Alex et al., 2008). For the purposes of this work, the original BSM1 model has been adapted to a real WWTP, located in Greece. This procedure involved modifications to both structural and functional aspects of the original BSM1. The layout had to be adopted in terms of the number of process lines and the volume of the reactors and the secondary settlers. In addition, the capacity of aeration and pumping systems was modified. As far as the functional interventions are concerned, the average influent flow rate of the actual plant was used and correspondingly, the internal and external recirculation flow rates were also modified. Fig. 1 depicts a schematic representation of the particular WWTP.

The BSM1 model includes 14 kinetic and 5 stoichiometric parameters involved in the mass-balance equations of each reactor, the default values of which are fixed. Proper adaptation of this generic model to an actual WWTP requires an accurate estimation of these parameters. In this work, an identification scheme for estimating these parameters is proposed, using limited dynamic process data that can be directly collected from specific sensors of the WWTP, such as nitrate and nitrite concentration of the anoxic tanks, dissolved oxygen of the aerobic tanks and the ammonia and nitrite and nitrate concentration of the last aerobic tanks.

The identification process is carried out by formulating a nonlinear optimization problem, where the objective is to minimize the mean squared error (MSE) between the actual

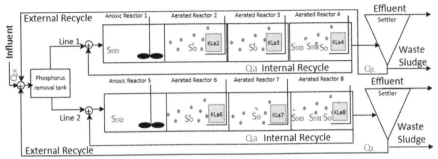

Fig. 1: Schematic Representation of the real WWTP under study

process data and the data generated at discrete time instances, in particular every 15 minutes, by imposing a limited number of steps on the manipulated variables of the identified BSM1 model. The resulting optimization problem exhibits several unwelcome properties, including high dimensionality, multimodality and noise contamination, which render the use of standard optimization methods inefficient.

2.2. Cooperative particle swarm optimization

Particle swarm optimization (Alexandridis et al., 2013), is a metaheuristic search method based on a population of potential solutions, called particles. Each particle explores the search space based on the personal and global best position achieved at each iteration of the algorithm. The standard PSO algorithm groups all the design variables into one swarm, optimizing them simultaneously. However, the high number of parameters involved in this particular problem makes it difficult for the solver to approach a satisfactory solution. To remedy this situation, we introduce a cooperative PSO method (CPSO), which takes into account correlations between the design variables and splits them accordingly into several groups, thus forming distinct swarms. Each swarm evolves independently in the search space while concurrently exchanging information with the rest of the swarms. In each iteration, the positions and velocities of the particles of each swarm are updated according to the following equations:

$$P_k v_{i,j}(t + 1) = w \cdot P_k v_{i,j}(t) + c_1 \cdot r_{1,i}(t) \cdot [P_k y_{i,j}(t) - P_k x_{(i,j)}(t)] + \cdots\cdots + c_2 \cdot r_{2,i}(t) \cdot [P_k \hat{y}_j(t) - P_k x_{i,j}(t)]$$

$$P_k x_{i,j}(t)(t + 1) = P_k x_{i,j}(t) + P_k v_{i,j}(t + 1) \tag{1}$$

where the position of the *i-th* particle in *j-th* dimension is denoted by $P_k x_{i,j}(t)$ for the *k-th* swarm. The velocity of each particle is denoted by $P_k v_{i,j}(t)$, $P_k y_{i,j}(t)$ represents the position of the best particle of each swarm and $P_k \hat{y}_j(t)$ denotes the position of the global best particle. Coefficients c_1, c_2 correspond to the local and global adjustment weights, w is the inertia weight and $r_{1,i}, r_{2,i}$ are uniformly distributed random numbers. As each swarm now contains only a part of the design vector, the value of the objective function is calculated by creating a context vector which combines the particles of each swarm with the global best particles of the remaining swarms. A more detailed description of CPSO can be found in (Kapnopoulos & Alexandridis, 2022). This customized CPSO methodology has a number of advantages over standard PSO, including a finer-grained assignment which avoids the two steps forward - one step back problem, and the fact that each swarm evolves independently in the search space, thus, improving the diversity of the solution. These advantages ultimately lead to a highly accurate estimation of the BSM1 parameters.

3. Dynamic modelling using radial basis function neural network models

RBF neural networks represent a special type of ANNs used for application to various problems of diverse nature, including MPC controller design (Papadimitrakis & Alexandridis, 2022), because of their simple architecture and their high approximation capabilities. For a typical structure of an RBF neural network with N input variables, L hidden nodes and one output variable, the output of the RBF network for an observation t is calculated as:

$$\hat{y}(t) = \sum_{l=1}^{L} w_l \varphi_l \left(\sum_{n=1}^{N} \left(x_n(t) - \hat{x}_{l,n} \right)^2 \right)^{1/2} \tag{2}$$

where φ represents the radial basis function used, x_n, is the input space vector of the network, $\hat{x}_l$, is the center of the lth hidden neuron, and $w_l{}^T = \left[w_1, w_2, \dots, w_l, \right]$ is the vector of synaptic weights that connect the hidden layer with the output layer. The radially symmetric function employed in this work was the thin plate spline (TPS) function:

$$\varphi(x) = x^2 \log(x) \tag{3}$$

In order to tune the structure of nodes of the hidden layer, the centers are selected employing the unsupervised FM algorithm, which has found numerous successful applications in different scientific fields (Papadimitrakis & Alexandridis, 2022). The FM algorithm uses fuzzy partitions of the input space to determine the number and locations of the hidden nodes. A detailed description of the algorithm can be found in (Alexandridis et al., 2013). After the estimation of the hidden node centers, weights are calculated by using linear least squares, implemented in matrix form:

$$w^{\mathrm{T}} = Y^{\mathrm{T}} \varPhi (\varPhi^{\mathrm{T}} \varPhi)^{-1} \tag{4}$$

where $\varPhi = [\varphi_1, \varphi_2, \dots, \varphi_L]$ and Y are the measured target values.

In order to produce dynamic RBF models with the aforementioned approach, past values of the process manipulated variables, the measured disturbances and the measured state variables - except from the output variable - are used as inputs to the RBF network.

4. Results and Discussion

4.1. System identification

The proposed identification scheme is employed by initially applying pseudorandom binary sequences on the BSM1 model input variables and disturbances. Data for the 12 measured state variables are collected for a 2-month period with a sampling period of $T_s = 15min$, which is a time interval suitable for controller design (Alex et al., 2008); the data are then corrupted with Gaussian noise $\sim N(0, \sigma)$, where σ is 2.5% of the mean measured value for each variable. This set of data is introduced to the CPSO identification scheme for estimating the values of the BSM1 kinetic and stoichiometric parameters.

The parameters to be estimated are separated into distinct swarms based on the different processes they affect during the operation of the plant. The first two swarms consist of the 8 and 5 parameters related to the heterotrophic and autotrophic phenomena, respectively. The 3 parameters affecting hydrolysis and the unique parameter influencing ammonification, form the 3rd and 4th swarm, respectively. The remaining parameters related to the COD in biomass and in products from the biomass, constitute the 5th swarm.

For comparison purposes, the employed CPSO solver was tested against standard PSO and a modified PSO methodology (Mezura-Montes et al., 2011). The algorithms are individually tuned using literature suggestions in conjunction with a trial-and-error procedure. Due to the inherent stochastic nature of the algorithms, 15 runs are executed for each algorithm and the superiority of CPSO is validated by running a two-sample *t*-test. Table 1 summarizes the results. As it can be seen, the performance of CPSO is superior in terms of the best or the average solution with a statistical significance higher than 90%, while it also manages to produce more consistent results, as indicated by the standard deviation values.

4.2. Dynamic modeling

The proposed method can produce dynamic models for different process variables; in this work, we present the results for the ammonia mass concentration (S_{NH}, g N.m^{-3}) in the last aerated tank (Reactor 4) in Line 1. The input features were: (A) lags for all the manipulated variables, i.e. oxygen coefficients ($K_L a$, d^{-1}) in all the aerated reactors and the internal recycling flows (Q_a, m^3 d^{-1}) for both lines, (B) lags for measured plant disturbances, i.e. the flow rate (Q_{input}, m^3 d^{-1}) and ammonia influent ($S_{NH\,input}$, g N.m^{-3}) and (C) lags for all the measured state variables, i.e. dissolved oxygen concentration (S_O, g (-COD).m^{-3}) in all aerated tanks, nitrite and nitrate nitrogen mass concentration (S_{NO}, g N.m^{-3}) for all the anoxic tanks and aerated reactors 4, 8, and ammonia mass concentration (S_{NH}, g N.m^{-3}) for reactor 8. The number of past values for each feature was decided empirically by performing step tests and observing the dynamic responses of the measured output variables. The total number of input features was equal to 46. Using the BSM1 model with parameters identified by the proposed CPSO scheme, a large dataset containing 10 years of simulated data with 15 minutes sampling was generated in order to train the RBFN with the procedure described in the previous section. To properly evaluate the RBFN model predictive abilities, a new testing dataset was produced, spanning 5 years of operation, using this time the BSM1 with the original parameters and contamination with noise; this dataset allows for an unbiased evaluation of the RBFN predictions, as it is built using the original model and thus is independent from the RBFN training data generation procedure. The mean absolute error (MAE) between RBFN predictions and the original BSM1 data and the R^2 coefficient were found equal to 0.88 g N.m^{-3} and 0.99, respectively. A visual comparison between the actual and predicted values is shown in Figure 2. It can be seen that the proposed framework manages to produce RBFN models that accurately track the data produced by the original BSM1 model, despite using a very limited amount of data from it (just two months in this case).

5. Conclusions

This work introduces a dynamic modelling scheme to derive highly accurate data-driven predictive models for WWTPs by using only a limited amount of data. The collected data are fed to a novel CPSO-based identification scheme, to accurately estimate the values of the kinetic and stoichiometric parameters of the BSM1 model. The performance of this method is verified through comparisons against two other PSO algorithms.

Table 1. Performance Metrics

Algorithm	Fitness Average	Fitness standard deviation	Best Fitness	Average Function Evaluations	p-value
CPSO	0.00068	0.00014	0.00041	2348	-
Standard PSO	0.00223	0.00244	0.00045	2726	0.0311
Modified PSO	0.00106	0.00076	0.00044	2535	0.0885

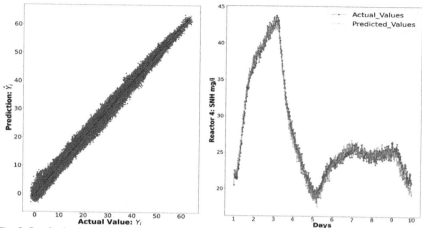

Fig. 2. Results for the RBF-based ammonia concentration model (a) actual vs predicted values for the whole testing dataset and (b) actual and predicted values vs time for an arbitrary selected window of the testing dataset

The identified model is used to produce large volumes of dynamic data for training discrete time RBF predictive models. The results indicate that the derived models are capable of predicting the dynamic behavior of the WWTP with high accuracy, while at the same time being much simpler than first-principle-based models; these properties make them ideal candidates for integration in MPC schemes, a step which will be pursued as feature research.

Acknowledgements: This research has been co-financed by the European Regional Development Fund of the European Union and Greek national funds through the Operational Program Competitiveness, Entrepreneurship and Innovation, under the call RESEARCH – CREATE – INNOVATE (project code: T2EΔK-02191).

References

J. Alex, L. Benedetti, J. B. Copp, et. al, 2008, Benchmark simulation model no.1 (BSM1), Report by the IWA Task group on benchmarking of control strategies for WWTPs, 1.
A. Alexandridis, E. Chondrodima, H. Sarimveis, 2013, Radial Basis Function network training using a non symmetric partition of the input space and particle swarm optimization, IEEE Transactions on Neural Networks and Learning Systems, 24(2), 219-230.
A. Kapnopoulos, A. Alexandridis, 2022, A cooperative particle swarm optimization approach for tuning an MPC-based quadrotor trajectory tracking scheme, Aerospace Science and Technology, 127, 107725.
A. N. Matheri, F. Ntuli, J. C. Ngila, T. Seodigeng, C. Zvinowanda, 2021, Performance prediction of trace metals and cod in wastewater treatment using artificial neural network, Computers & Chemical Engineering, 149, 107308.
E. Mezura - Montes, C. A. Coello Coello, 2011, Constraint – handling in nature-inspired numerical optimization: Past, present and future, Swarm and Evolutionary Computation, 1(4), 173-194.
M. Papadimitrakis, A. Alexandridis, 2022, Active vehicle suspension control using road preview model predictive control and radial basis function networks, Applied Soft Computing, 120.
M. Sadeghassadi, C. Macnab, B. Gopaluni, D. Westwick, 2018, Application of neural networks for optimal-setpoint design and MPC control in biological wastewater treatment, Computers & Chemical Engineering, 115, 150-60.

Antonis Kokossis, Michael C. Georgiadis, Efstratios N. Pistikopoulos (Eds.)
PROCEEDINGS OF THE 33rd European Symposium on Computer Aided Process Engineering
(ESCAPE33), June 18-21, 2023, Athens, Greece
© 2023 Elsevier B.V. All rights reserved. http://dx.doi.org/10.1016/B978-0-443-15274-0.50285-7

Can electrical utilities play a decisive role in generating knowledge for better use of energy in small and medium-sized industries?

José Luis Suarez Castaneda,[a] Mélissa Lemire,[a] Marc-André Richard[a]

[a]*Laboratoire des technologies de l'énergie, Institut de Recherche d'Hydro-Québec, 600 av de la Montagne, Shawinigan, Québec, G9N 7N5, Canada*

Abstract

This paper presents two visualization tools that use meter data to facilitate power management in the industrial sector. The first determines typical daily load profiles based on shape and magnitude with a clustering algorithm. A whole year's worth of electricity consumption can then be visualized on a calendar featuring typical daily profiles on a single screen. Different clustering algorithms and scaling techniques are compared for twelve plants from four different industrial sectors. Data scaled between 0 and 1 or with the normal distribution and the k-Means algorithm showed the best performance. The second tool specifically addresses the management of maximum power demand by suggesting power shedding levels and showing their impacts on billing.

Keywords: electricity consumption, power demand, cluster analysis, operational control, SME energy efficiency

1. Introduction

To achieve net-zero emissions by 2050, electrification is the preferred option for industrial decarbonization in Québec, Canada, since the vast majority of the power generated by its public electric utility, Hydro-Québec, comes from renewable resources and it is competitively priced (Hydro-Québec, 2022). Therefore some pathways to achieve climate targets predict an increase in electricity demand of the order of two-thirds more than current levels (Dunsky, 2021). The substantial increase in demand for both energy and capacity required for this transition could limit decarbonization potential or require major investments in infrastructure, which would in turn lead to rising energy costs and delays. It is therefore essential that every industrial sector optimize its energy use and maximize its flexibility. In this context, the current role of a utility is not only to supply energy, but also to facilitate such a transition in the most efficient way possible. With limited ressources and information, demand side management in small and medium-sized industries is particularly challenging (Thiede et al., 2013). The focus of programmes should be to facilitate knowledge creation among industries (Palm et al., 2020).

Previous work (Narciso and Martins, 2020; Richard et al., 2017; Zhou et al., 2013) confirmed the benefits of using machine learning strategies and visualization tools to facilitate the understanding of electricity consumption. However, most of these works focus on a single industrial activity do not confirm the method performance when analyzing different production processes. The proposed solution is a platform to help industries better understand their energy consumption and act accordingly. It comprises three modules: the first focuses on metering and public data; the second introduces production data and the third integrates an IoT solution. This paper focuses on the first module, more specifically the description of two visualization tools. The first one is a clustering algorithm that groups typical daily load profiles of a production plant based on shape and consumption level ("amplitude") from a year's worth of meter data. A whole year's electricity consumption can be visualized on a calendar featuring typical daily profiles on a single screen. Automatically generated typical profiles help industrial customers relate the shape of the profiles to their production schedule. This may stimulate a discussion about various demand side management measures, often related to operational control. Different clustering algorithms and scaling techniques are compared in Section 2 of this paper. The second visualization tool specifically addresses the management of maximum power demand by suggesting and evaluating power shedding levels and is discussed in Section 3. Finally, Section 4 presents the conclusion.

2. Typical daily load profiles (clustering)

2.1. Methodology

2.1.1. Data
Contrarily to Zhou et al. (2013) who use clustering algorithms for load classification purpose, *i.e.* segmentation of electricity consumers, this work aims to analyze electricity consumption of a single production plant by identifying typical daily load profiles, an therefore, energy consumption patterns. To verify the performances of the algorithms of the proposed solution, three plants from each of four different industrial processes were considered: pharmaceutical (plants A to C), chemical (plants D to F), food (plants G to I) and plastic and rubber (plants J to L). To avoid loss of information, no features were extracted and no outliers were removed so that both good and bad consumer practices could be identified if unusual load profiles recur. However, if missing data (or non-existent data) is greater than 10% of all data for a given day, this day is removed from analysis; otherwise, missing values are interpolated. Raw data corresponds to the plant's electrical energy consumption (kWh) for every 15 minutes for the whole year 2021. The data is organized such that each day, i, is represented by a vector with 96 dimensions ($L_i = [x_{i\,1}, x_{i\,2}, x_{i\,3}, ... x_{i\,N=96}]$). All the data is organized in a 365 × 96 matrix. However, the inclusion of a supplementary dimension representing this vector norm (as a representation of consumption level) is included as an option to evaluate the impact of this "load amplitude". The vector norm is calculated using Eq. (1):

$$\|L_i\| = \left(\sum_{j=1}^{N} |x_{ij}|^2\right)^{1/2}$$

Eq. (1)

As mentioned, this module aims to identify the typical consumption profile. Then, the average consumption in each time step for each cluster is considered as the expected one. Therefore, given a consumption profile and belonging to a cluster, it is possible to evaluate the consumption of a plant considering its deviation.

2.1.2. Clustering algorithm
Clustering analysis is a machine learning tool to identify patterns by forming groups of data that are similar to one another but different from other groups. This technique is an unsupervised learning method because target values are not known. Most of this work has been aimed at comparing the consumption of different plants, buildings and industries (Zhou et al., 2013). However, Richard et al. (2017) and Bourdeau et al. (2021) have shown the advantages of using clustering to find the typical profiles that represent the expected

behaviors of plant/building consumption. Before selecting the specific clustering method, the data to be analyzed had to be prepared. In addition to the treatment mentioned above, three different data scaling techniques were used: based on the vector norm (each day is scaled individually), min-max of the whole matrix (between 0 and 1) and normal (or Gaussian) distribution of the whole matrix. These scaling techniques are represented in the following equations:

$$L_i' = L_i \cdot [1/\|L_i\|] \qquad \text{Eq. (2)}$$

$$x_{ij}' = \frac{x_{ij} - min(X)}{max(X) - min(X)} \qquad \text{Eq. (3)}$$

$$x_i' = \frac{x_i - \mu}{\sigma} \qquad \text{Eq. (4)}$$

where L_i' is the scaling vector for day, i, x_{ij}' is the normalized value (j is the 15-minutes interval between readings) and μ and σ are respectively the mean and standard deviation of all data (whole year, X matrix). When the norm is included as the last dimension, it is scaled considering all the norms of all vectors.

This paper summarizes the performance evaluation of the most common clustering methods used for load classification: k-means (Bourdeau et al., 2021), k-shape (Yang et al., 2017), Gaussian mixture model (GMM) (Wang et al., 2022), and fuzzy c-means (FCM) (Zhou et al., 2013). Each clustering method is evaluated using the three scaling techniques mentioned above. The number of clusters is obtained by maximizing the silhouette score (Rousseeuw, 1987), which is also used to compare the performance of clustering strategies. The silhouette score for an object i belonging to cluster A is defined in Eq. (5).

$$S(i) = \frac{b(i) - a(i)}{max(a, b)} \qquad \text{Eq. (5)}$$

where a is the average intra-cluster distance and b is the average shortest distance to another cluster Euclidian distance is used for these calculations. The range of this score is [-1,1], therefore the value closest to 1 represents the best clustering strategy.

2.2. Results

In this section, we present the results of the clustering strategies. Figure 1 shows the average silhouette scores for all clustering methods and the three corresponding scaling techniques. In general, k-means, GMM and FCM algorithms performed the best. The number of clusters for GMM and k-means are higher than or equal to FCM (Table 1).

Table 1. Number of clusters for all clustering strategies (min-max scaling technique)

Plant/Algorithm	A	B	C	D	E	F	G	H	I	J	K	L
k-means	3	3	2	2	2	2	3	3	6	2	3	5
k-shape	4	2	2	3	2	2	2	4	2	3	2	2
GMM	3	3	2	2	3	2	2	3	6	3	3	6
FCM	3	2	2	2	2	2	3	3	5	2	3	2

A higher number of clusters allows for more information on plant performance and production behavior to be obtained. However, GMM is more sensitive to outliers, so unusual loads are identified as a cluster with one profile, which can affect the silhouette score and induce distortions. This is the case for the plant G, which explains why the number of clusters is higher for k-means than for GMM. On the other hand, it can be seen in Figure 1 that scaling based on the vector norm is less successful. Although adding the norm as an additional dimension increases the score, it is still worse than other techniques. The supplementary dimension for the other scaling techniques does not have a significant effect.

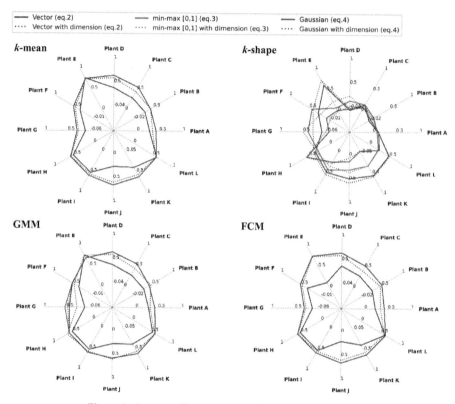

Figure 1. Average silhouette scores for all clustering strategies

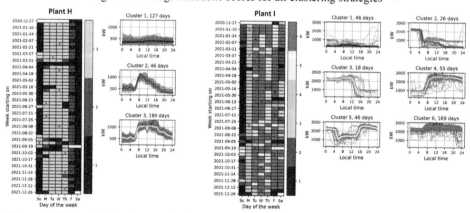

Figure 2. Cluster distribution on calendar and suggested clusters for plants H and I with *k*-means and min-max scaling technique.

Figure 2 presents the proposed display. The clustering strategy allows production days to be distinguished from non-production days and identifies the shape of some typical production profiles. For instance, plant I cluster 5 represents days when maintenance was being performed on major equipment. The module can incite customers to stop unnecessary loads during non-production hours. For example, plant H had a nighttime load that was half of its load during the day. The day and time of the peaks that induced power demand costs are indicated with a plus sign (+).

3. Peak power demand management tool

3.1. Methodology

The second visualization tool aims to raise interest in power demand management by comparing how three different levels of power shedding can impact electricity billing. The structure of the rate in Québec for business customers includes billing demand (CAD/kW) and energy charges (CAD/kWh) (Hydro-Québec, 2022).

The proposed indicator to demine load shedding levels is based on the marginal cost of electricity during peaks. The highest peaks typically account for a limited quantity of energy, but significantly affect billing demand. Hence, above a given level, if power demand costs are distributed over energy, the marginal energy cost increases to a few dollars per kWh. The power threshold, $P_{Threshold}$, (in kW), is calculated using the marginal energy, e_{marg} (in kWh) and the distributed power demand cost parameter, DPDC (in CAD/kWh), using:

$$e_{marg} = \int_{Billing\ period} (P(t) - P_{Threshold}) \cdot \left(H(P(t) - P_{Threshold})\right) dt \qquad \text{Eq. (6)}$$

$$P_{Threshold} = P_{max} - \frac{DPDC * e_{marg}}{PDC} \qquad \text{Eq. (7)}$$

where P_{max} is the maximum power demand of the billing period and PDC is the power demand cost, CAD15.154/kW (Hydro-Québec, 2022). Marginal energy is modeled by equation (6). Where H is the Heaviside step function. The proposed algorithm to resolve this problem is as follow. All power data is sorted in descending order based considering the accumulated value of time on the abscissa axis. This relationship between power and time allows us to obtain the time $t_{Threshold}$ according to a fixed $P_{Threshold}$. Then, equation (8) can be used to obtain the marginal energy. These equations are solved iteratively until reaching a convergence in the value of DPDC to CAD1.00/kWh (based on experience) to determine the minimum plausible load shedding power level. The two additional levels were evenly divided based on power demand.

$$e_{marg} = \int_{0}^{t_{Threshold}} (P - P_{Threshold}) dt \qquad \text{Eq. (8)}$$

Not only the DPDC parameter is significant to customers, but this indicator also indirectly allows for the shape of the profile and the cut-off time to be taken into account. It also facilitates price comparisons with alternative energy sources.

Results and discussion

Table 2 shows results of the peak power demand management tool. Figure 3 shows a month in winter for plants B and D.

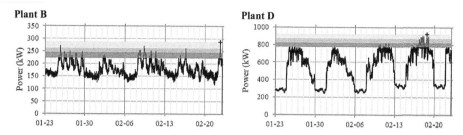

Figure 3. Peak power management tool example and impact on billing for plant B and plant D

Table 2. Summary of peak power management tool

Period		Winter month				Summer month			
Plant	Shedding	$P_{threshold}$ - P_{max} (kW)	Power demand savings (CAD)	% savings on total bill	Distributed power demand costs above threshold (CAD/kWh)	$P_{threshold}$ - P_{max} (kW)	Power demand savings ($ CAD)	% savings on total bill	Distributed power demand costs above threshold (CAD/kWh)
B	1	263-288	397	3.47%	38.60	301-324	359	3.01%	1.56
	2	244-288	685	5.99%	5.43	298-324	406	3.40%	1.40
	3	222-288	1028	8.98%	0.99	287-324	584	4.90%	0.93
D	1	869-921	806	2.46%	8.12	871-896	389	1.17%	6.36
	2	821-921	1562	4.77%	2.20	847-896	767	2.30%	2.08
	3	772-921	2334	7.13%	1.00	823-896	1139	3.42%	0.99

4. Conclusion

The combined use of clustering algorithm *k*-means and scaling data, either between 0 and 1 or using the normal distribution, for cluster analysis achieved the best performance in determining typical profiles in the plants analyzed, i.e., silhouette score, number of clusters, and number of profiles per cluster. For these scaling techniques, the vector norm (as a measure of consumption level) has no effect on the results. The silhouette score can clearly identify different typical profiles. However, more detailed divisions may be required to identify different groups of typical profiles with similar consumption levels but with different consumption shapes. An optional feature integrating a user choice is considered in the proposed solution. Future work could be aimed at improving and evaluating other metrics that would suggest efficient and meaningful clustering for the customer.

In this study, we have also confirmed the great variability in electricity consumption, even within the same industrial sector. This calls for further work on robust solutions to meet this challenge. In the same vein, although twelve different plants (representing four different industrial sectors) are evaluated in this study, a results validation is included in the following stages of the project to consider other types of industries, batch and continuous processes, as well as customer experience. The power demand cost savings calculated in the peak power management tool and the distributed power costs in kWh clearly illustrate the effect of shedding power and can encourage customers to manage peak power demand. They would be able to see the number of hours and quantity of the power to shed and the resulting savings. Future work will focus on the validation of the cut-off threshold and methodology as they relate to customer experience (technical and financial feasibility).

References

D.A.C. Narciso, F.G. Martins, 2020, Application of machine learning tools for energy efficiency in industry: A review, Energy Reports, 6, 1181–1199.

Dunsky, 2021, Dunsky Advises Government of Quebec on Pathways to Achieve Climate Targets,
https://www.dunsky.com/dunsky-advises-government-of-quebec-on-pathways-to-achieve-climate-targets

EU Commission, 2014, European Competitiveness Report, 190-195.

Hydro-Québec, 2022, Electricity Rates Effective April 1, 2022, 204 p.
https://www.hydroquebec.com/business/customer-space/rates/

H. Wang, Y. Tian, A. Li, J. Wu & G. Sun, 2022, Resident user load classification method based on improved Gaussian mixture model clustering, MATEC Web of Conferences 355, 2021 International Conference on Physics, Computing and Mathematical (ICPCM2021).

J. Palm, F. Backman, 2020, Energy efficiency in SMEs: overcoming the communication barrier, Energy Efficiency, 13:809–821.

J. Yang, C. Ning, C. Deb, F. Zhang, D. Cheong, S. E. Lee, C. Sekhar, K. W. Tham, 2017, k-shape clustering algorithm for building energy usage patterns analysis and forecasting model accuracy improvement, Energy and Buildings, 146:27-37.

K.-L. Zhou, S.-L.Yang, C. Shen, 2013, A review of electric load classification in smart grid environment. Renewable and Sustainable Energy Reviews, 24:103–110.

M.-A. Richard, H. Fortin, M.-A. Leduc, A. Poulin, M. Fournier, 2017, Daily load profiles clustering: a powerful tool for medium-sized industries demand side management, ACEEE Summer Study on Energy Efficiency in Industry 2017, Denver, Colorado.

M. Bourdeau, P. Basset, S. Beauchêne, D. Da Silva, T. Guiot, D. Werner, E. Nefzaoui. Classification of daily electric load profiles of non-residential buildings, Energy & Buildings, 233.

P.J. Rousseeuw, 1987, Silhouettes: a graphical aid to the interpretation and validation of cluster analysis, J Computational Appl Math, *20*:53-65.

S. Thiede, G. Posselt, C. Herrmann, 2013, SME appropriate concept for continuously improving the energy and resource efficiency in manufacturing companies, CIRP Journal of Manufacturing Science and Technology, 6:204–211

Antonis Kokossis, Michael C. Georgiadis, Efstratios N. Pistikopoulos (Eds.)
PROCEEDINGS OF THE 33rd European Symposium on Computer Aided Process Engineering
(ESCAPE33), June 18-21, 2023, Athens, Greece

Hybrid data-driven and first principles monitoring applied to the Tennessee Eastman process

Eduardo Iraola[a,b], José M. Nougués[a], Antonio Del Rio Chanona[c], Lluís Batet[b], Luis Sedano[b,d]

[a] Inprocess Technology and Consulting Group, Carrer de Pedro i Pons, 9, 08034 Barcelona, Spain
[b] ETSEIB, Universitat Politècnica de Catalunya, Avinguda Diagonal, 647, 08028 Barcelona, Spain
[c] Sargent Centre for Process Systems Engineering, Imperial College London, Roderic Hill Building, South Kensington Campus, London SW7 2AZ, United Kingdom
[d] FUS_ALIANZ Science, Engineering & Consulting. C/ Nord 19, Àtic, 43700. El Vendrell, Tarragona, Spain

eduardo.iraola@inprocessgroup.com

Abstract

In this work we present a hybrid monitoring approach for fault detection using the Tennessee Eastman (TE) process. We benchmark our proposed approach against previous methods in the available literature and analyze the benefits and shortcomings. The hybrid monitoring approach contains two steps. First, from a model-based perspective, a dynamic model of the TE plant is constructed using commercial dynamic simulation software and data is generated from the TE plant and its model. In the second step, a data-driven analysis is conducted. This involves training different fault detection models with the previously obtained datasets to detect plant faults as early as possible. The results show that combining datasets can improve the traditional pure data-driven monitoring performance with only plant data. This work highlights the usefulness of combining process modeling and machine learning in the monitoring and prognostics fields when data availability from the actual process is limited.

Keywords: Tennessee Eastman, monitoring, Aspen HYSYS, machine learning, fault detection.

1. Introduction

Process monitoring and fault detection is critical in the process industry in terms of improving process performance, reducing plant halts, and therefore improving return of investment. Generally, we can approach process monitoring from a model-based or a data-driven point of view. Model-based perspectives—for instance, observers and parity equations—are based on traditional control theory and are well-suited for small systems [1]. However, this approach has not transcended to industry-scale processes due to the difficulty in adapting the high complexity of these processes to this kind of models. Data-driven methods can overcome this issue by being trained using only output data of the system. Therefore, they can describe the complexity of a system without the need to know the intrinsic process details, which can be time-consuming and impractical to identify. However, data-driven approaches do not extrapolate well to regions of the state space not observed in training. In these cases, given judicious assumptions, model-based approaches can extrapolate and allow better interpretability. We can see that both approaches show advantages and disadvantages, and the choice will vary depending on the specific case.

In this work, we present a hybrid fault detection approach that merges data fed directly from the plant and model data derived from first principles. The goal is to evaluate if there is any benefit in using a model to boost the information we provide the fault detector. In this way, we take advantage of rigorous simulation to enhance the results of a pure data-driven approach. Early examples of this kind of hybrid approach can be found in Schwarte and Isermann (2002) [2], however, in our opinion, this research theme can be further improved.

The paper is organized as follows. Section 2 outlines a short review of data-driven fault detection methods and formalizes the hybrid approach. Section 3 describes the case study selected for this work, the Tennessee Eastman process. Then, Section 4 shows the results of the benchmark. Finally, Section 5 summarizes the conclusions of the work.

2. Fault detection

Fault detection is the branch of process monitoring that studies how to identify events that cause a variable or property of the system to deviate from an allowed range. Figure 1 summarizes a possible process monitoring and fault diagnostic framework based on Aldrich and Auret (2013) [3]. In this scenario, the monitoring system is trained with normal operation conditions (NOC) data only, therefore a semi-supervised approach is followed. A semi-supervised approach provides the advantage of not needing real fault data for training, which may be expensive or impractical to obtain—e.g., operating the plant to off-control states or introducing dangerous disturbances just to acquire data. Supervised classifiers may struggle with any kind of new fault that was not previously trained for; therefore, they can be easily outperformed by semi-supervised approaches.

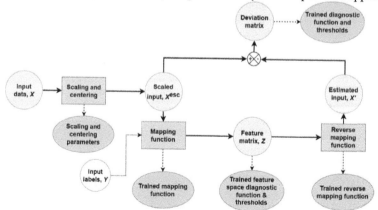

Figure 1. Semi-supervised fault detection framework, with data preprocessing, mapping, and reconstruction steps.

A monitoring system such as that in Figure 1 works as follows: it develops a series of features, $Z \in \mathbb{R}^{n \times n_z}$, based on a raw dataset, $X \in \mathbb{R}^{n \times m}$, which is usually preprocessed to improve performance, and tries to reconstruct the input into $X' \in \mathbb{R}^{n \times m}$. n is the number of instances of the dataset, and m and n_z are the size of the input and the latent space, respectively (typically $m > n_z$). Process monitoring compares the reconstructed input with the actual input and extracts the result of the in-control or out-of-control state of the plant.

2.1. Types of semi-supervised data-driven fault detection methods

One of the fundamental fault detection methods is *univariate statistical monitoring* [4], which obtains the operation limits using statistical theory and hypothesis testing. Given a level of significance, α, and assuming Gaussian distribution on the input data, our null hypothesis would assume normal operation, while the alternative hypothesis would imply the existence of faults. This way, if the mean and standard deviation of the tracked variable, x, is μ and σ respectively, we say that the process is in normal operation conditions (NOC) if x is inside the following range:

$$\mu - c_{\alpha/2}\sigma \leq x \leq \mu + c_{\alpha/2}\sigma, \tag{1}$$

where $c_{\alpha/2}$ is the standard normal deviate for the $1 - \alpha/2$ percentile. This approach considers serial relationships between a variable state in different time steps, but fails to include spatial relationships, i.e., the influence between different variables at the same instant in a complex process.

To consider this influence, we need to work with a *multivariate statistical monitoring* approach. In this case, instead, the Hotelling's T^2 test should be used [4]. To this end, the statistic is $T^2 = \mathbf{z}^T\mathbf{z}$, where the feature vector, $\mathbf{z} \in \mathbb{R}^{n_z}$, is $\mathbf{z} = \Lambda^{-1/2}V^T\mathbf{x}'$, with $\mathbf{x}' \in \mathbb{R}^m$. $\Lambda \in \mathbb{R}^{n_z \times n_z}$ is the diagonal matrix containing the n_z eigenvalues of the covariance matrix of the training dataset, X, and $V \in \mathbb{R}^{n_z \times m}$ is the matrix of the eigenvectors of the same eigenvalue decomposition. Equation (2) shows how to compute the T^2 detection threshold for an α level of significance when we use the sample covariance matrix,

$$T_\alpha^2 = \frac{m(n-1)(n+1)}{n(n-m)} F_\alpha(m, n-m) \tag{2}$$

The output of this fault detection approach is a confidence ellipse for which we accept a new measurement as NOC if it satisfies $T^2 < T_\alpha^2$. If we cannot assume a Gaussian distribution or the input dataset is too small, we would need to consider more robust hypothesis tests [5].

The multivariate detection approach can be further improved by dimensionality reduction techniques such as Principal Component Analysis (PCA), Fisher Discriminant analysis (FDA), Partial Least Squares (PLS), or Canonical Variate Analysis (CVA) [4]. These perform the same multivariate statistical testing but now apply it to a lower dimensional dataset that is more representative of the actual trends of the system, rejecting noise and non-useful information. For instance, PCA projects the raw data into the lower dimensional space created by the k eigenvectors corresponding to the highest eigenvalues of the covariance matrix, being $k < m$. Then, the detection threshold can be calculated using Equation (2), with k number of features instead of the original number of features m. If several timesteps are stacked in the same time window that is fed to the PCA algorithm to also account for serial relationships, we call this method Dynamic PCA (DPCA). The added time series analysis tends to outperform standard PCA.

More advanced techniques for fault detection include the use of machine learning models. Given that, we are focusing on semi-supervised approaches in which we have only NOC data from the plant, a vanilla neural network classifier architecture does not apply to a semi-supervised approach. It is here where auto-associative networks or autoencoders appear.

Autoencoders (see Figure 2) use an encoder set of layers to transform a single input instance $\mathbf{x}$ into a latent space representation $\mathbf{z}$, typically of lower dimensionality than the input data, and then use the decoder layers to return a reconstruction of the input $\mathbf{x}'$ from the latent representation obtained by the encoder [6]. As in Figure 1, the deviations between the actual input and the reconstruction are used to determine if the instance corresponds to normal operation or not. Unlike the previous approaches, autoencoders can capture nonlinear relationships between the different variables, therefore being useful for more complex processes. For this reason, these models are also sometimes called nonlinear PCA (NLPCA).

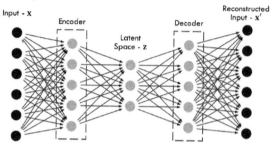

Figure 2. Autoencoder architecture.

A set of the different methods presented in this section will be tested with both pure data-driven and hybrid dataset approaches, the latter being explained next.

2.2. Hybrid fault detection approach

In this section, we present a hybrid fault detection approach that adds the advantages of model-based methods to data-driven approaches, while considering the tools already available in the process industry. This method leverages process simulation models to improve the dataset preprocessing step using software such as Aspen HYSYS.

Here, the plant and the model use the same high-level vector of control signals, u, as inputs; and their outputs, y and y^*, respectively, are compared. This results in the so-called residual generation, r. Residuals will make up the datasets for training and cross-validation, and therefore will be fed to the selected fault detection method: limit sensing, PCA, autoencoder, etc. Figure 3 summarizes the procedure.

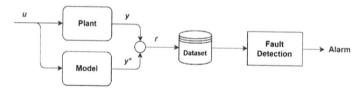

Figure 3. Hybrid fault detection framework.

The hybrid procedure is expected to show better invariant metrics with respect to faults and anomalies in the process: the plant output y can be affected by external disturbances, hence intrinsic changes and its comparison with the ideal behavior represented by the model y^* can be more efficient as a process monitoring indicator. Note that data harmonization between model and plant must be studied to minimize differences and account for model mismatches. This is assumed as a preliminary condition in this paper and will not be analyzed here.

3. Case study – The Tennessee Eastman process

The Tennessee Eastman (TE) process is a virtual plant that Downs and Vogel designed based on an Eastman Chemical Company process [7]. In this virtual plant, a series of reactants, A, C, D, and E are processed to obtain two products, G and H, as pure as possible, with an unavoidable inert component mixed in the feed, B, and a byproduct, F.

To obtain the desired products, the system is made up of five unit operations: a reactor, a condenser, a liquid-vapor phase separator, a compressor, and a stripper column. Figure 4 shows a process flow diagram of the complete system.

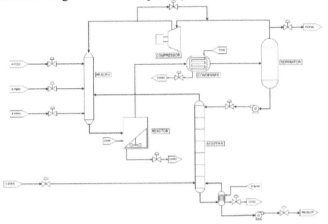

Figure 4. The Tennessee Eastman process.

In terms of the user interface, the TE process presents 41 measurable variables with noise and delay and 12 first-order dynamics actuators that allow for external interaction with the system. The process allows the activation of 20 types of disturbances at any point during the simulation. These can work as the faults to be detected by the monitoring system. The disturbances represent step changes, drifts, random noise, and component malfunctions in the supply conditions, cooling streams, and actuators of the plant.

For this work, we modeled the TE process in Aspen HYSYS, a dynamic simulator widely used in the process industry. We use the original TE script as the *plant* and the HYSYS simulation as the *model*, and collect data from the steady-state operation of both. The training dataset includes normal operation from both plant and model, and the validation dataset includes the activation of different disturbances for the plant and fault-free operation for the model. The new datasets are created according to the scheme in Figure 3, and we apply the monitoring methods summarized in Section 2 to compare them with pure data-driven approaches found in the literature.

4. Results

The results of the hybrid approach over the TE process data are shown in Figure 4 using previous literature results for comparison [4]. We selected the F_1 score,

$$F_1 = \frac{\text{TP}}{\text{TP} + 0.5(\text{FN} + \text{FP})} \qquad (3)$$

as the target performance metric to harmonize the different results. TP stands for the number of true positives, FN for the number of false negatives, and FP for the number of false positives predicted by the data-driven model. Therefore, F_1 represents the harmonic mean of precision (the percentage of instances predicted as positive that are correctly classified) and recall (the percentage of true positive instances that are correctly classified). F_1 is widely used because it works as an effective way to balance the effect of false negatives and false positives in classification models. Note that, in the framework of fault detection, positive instances correspond to faults, and negative instances correspond to normal operation.

Figure 5 shows the results. Note that all pure data-driven approaches have similar performance and the hybrid approaches tend to perform better. The DPCA pure data-driven score from the literature shows a small unexpected drop that can be explained by a high false positive ratio caused by the specific threshold calibration procedure used in [4]. The F_1 score heavily penalizes unbalances between false positives and false negatives, therefore the unexpected behavior. Results strongly suggest that accounting for first principles through rigorous simulation can improve data-driven approaches.

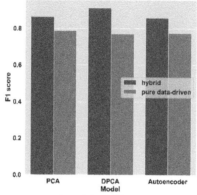

Figure 5. Comparison between pure data-driven (literature) versus hybrid dataset performance.

5. Conclusions

This work describes the hybrid fault detection approach and analyses its performance using the classical Tennessee Eastman process. To conduct this benchmark, the Tennessee Eastman original script is assumed to be the actual plant, and an Aspen HYSYS model is built to try to replicate it and generate a new residuals dataset. The results show that the hybrid approach is an interesting framework that tends to improve monitoring performance, even when applied to the most basic monitoring approaches. Future work will introduce more complex datasets, including dynamic plant operation, and perform fully supervised classification, taking advantage of the models to replicate faults and feed them into the fault detection methods and comparing them to the residual-based approach.

References

[1]   R. J. Patton, P. M. Frank, and R. N. Clark, *Issues of Fault Diagnosis for Dynamic Systems*. London: Springer-Verlag, 2000.

[2]   A. Schwarte and R. Isermann, *Neural network applications for model based fault detection with parity equations*, vol. 35, no. 1. IFAC, 2002. doi: 10.3182/20020721-6-es-1901.01502.

[3]   C. Aldrich and L. Auret, *Unsupervised Process Monitoring and Fault Diagnosis with Machine Learning Methods*. London: Springer London, 2013. doi: 10.1007/978-1-4471-5185-2.

[4]   L. H. Chiang, E. L. Russell, and R. D. Braatz, *Fault Detection and Diagnosis in Industrial Systems*. London: Springer London, 2001. doi: 10.1007/978-1-4471-0347-9.

[5]   R. Wilcox, *Introduction to Robust Estimation and Hypothesis Testing*, 5th ed. Elsevier, 2021. doi: 10.1016/C2010-0-67044-1.

[6]   M. A. Kramer, "Nonlinear principal component analysis using autoassociative neural networks," *AIChE J.*, vol. 37, no. 2, pp. 233–243, 1991, doi: 10.1002/aic.690370209.

[7]   J. J. Downs and E. F. Vogel, "A plant-wide industrial process control problem," *Comput. Chem. Eng.*, vol. 17, no. 3, pp. 245–255, Mar. 1993, doi: 10.1016/0098-1354(93)80018.

Antonis Kokossis, Michael C. Georgiadis, Efstratios N. Pistikopoulos (Eds.)
PROCEEDINGS OF THE 33rd European Symposium on Computer Aided Process Engineering
(ESCAPE33), June 18-21, 2023, Athens, Greece

Data-Driven Process Monitoring for Knowledge Discovery: Local and Global Structures

Estelle E. Seghers, José A. Romagnoli

Department of Chemical Engineering, Louisiana State University, Baton Rouge, Louisiana 70803, United States of America

Abstract

In industrial processes, the amount of raw data generated can add complexity in the analysis and understanding of the process dynamics. Being able to properly interpret this data can help improve plant operation. A platform is introduced for monitoring of industrial processes and optimization of the model-building process. FASTMAN-JMP (FAST MANual Data Manipulation implemented in JMP [1]) is a tool developed in Python to apply various data mining and machine learning techniques quickly and easily to better understand valuable patterns and hidden trends in process data. It is shown that local and global structures in the data set can be visualized and related to actual process operations though the identification of the variables responsible for the separation between selected given clusters. Furthermore, adequate comparisons of these algorithms can be difficult, having different loss functions with many parameters. We aim to decipher these algorithms, and how they work in the context of industrial data. Results are presented for an industrial case study of a pyrolysis reactor.

Keywords: process monitoring, data mining, knowledge discovery, machine learning, data-driven modeling.

1. Introduction

The digital data being created today has surpassed many projected values, reaching roughly 44 zettabytes (10^{21} bytes) generated every year as of 2020 [2]. This era of big data introduces an opportunity to improve industrial plant operation and control with increased productivity and safety. Data mining and machine learning algorithms present a way to achieve increased knowledge discovery in industrial processes, but it is also vital to understand how these methods function. Of particular interest is understanding the granularity of these analyses performed with industrial data, namely in how machine learning methods distinguish and illustrate the global and local structures present in data. In the context of industrial chemical engineering plant data, global structures generally suggest overall trends in the process (over a period of hours, or days, or longer) while local structures generally refer to short term fluctuations in certain process variables or controllers. Dimensionality reduction, also known as feature extraction, is a particularly important component within an unsupervised approach for analysing process data.

Most of the discussions in the computer science community deal with synthetic data and the main conclusion may or may not be relevant when dealing with industrial process data. In the author's opinion there is a need for a proper discussion on these issues in the context of industrial manufacturing process data. A Python-based data analysis graphical user interface, called FASTMAN-JMP (FAST MANual Data Manipulation implemented in JMP [1]), was developed using Python libraries (NumPy, pandas, sci-kit learn, etc.) and incorporated as an add-in into JMP, a statistical data analysis software, to facilitate the use and visualization of different methods for data mining and knowledge discovery. With the availability of such a powerful environment, the concepts of global and local data structures are first analyzed using data from an industrial pyrolysis reactor. It will be shown that local and global structures in the data set can be visualized within the Self Organizing Maps (SOM) [3] and related to actual process operations though a subspace greedy search (SGS) algorithm [4] allowing for the identification of the variables responsible for the separation between any two given clusters. Furthermore, when running alternative dimensionality reduction algorithms, it is not clear how these results yield trustworthy representations of the original data distribution and, more importantly, if they lead to proper characterization of the process states. Even comparisons of these algorithms can be difficult having different loss functions with many parameters. Consequently, another goal of this work is to decipher these algorithms, and how they work in the context of industrial chemical engineering process data.

2. Understanding the Data: Global and Local Structures

The goal of dimensionality reduction is to reduce the number of features in a dataset to obtain a compressed representation that is useful for analysis, by retaining as much "structure" in the data as possible. There are many algorithms for dimensionality reduction, and they can generally be discussed in the context of two principal archetypes: those that primarily aim to preserve global structure (overall trends in the process over a period of hours, or days, or longer) and those that primarily aim to preserve local structure (short-term fluctuations in certain variables or controllers). There is no strict definition of what constitutes global or local structure preservation, but global structure-preservation methods may be thought of as preserving overall placement of large clusters of observations, i.e., the data at the extremes in high-dimensional space are also at the extremes in low-dimensional space.

In contrast, local structure-preservation methods focus on preserving neighborhoods of observations, so that near neighbors in the high-dimensional space are still near neighbors in the low-dimensional space. If the data lie along manifolds or in clusters, that information will not necessarily be preserved when projecting to a very low number of dimensions. That is, a technique which preserves global structure does not necessarily preserve which points are close neighbors in the high-dimensional space.

3. Dimensionality Reduction: Global and Local Structure Preservation

In Principal Components Analysis (PCA) [5], a very popular very popular dimensionality reduction technique, the extracted principal component vectors graphically illustrate most of the variance in the data, and therefore primarily captures the global structure of the dataset while ignoring the detailed local structure information among the process data. Since the development of PCA, a class of dimensionality reduction techniques, known as manifold learning, has been developed in the pattern recognition area, including t-SNE [6], UMAP [7], and more recently TriMAP [8]. These methods are popular in biological applications and have been used in chemical manufacturing applications ([6], [9], [10]). However, one critical drawback of these methods is that they were either designed to preserve local structure or struggle with balancing the preservation of both global and local structures. Furthermore, these methods can be sensitive to initialization, and more importantly, they can be difficult to understand and tune.

Pairwise Controlled Manifold Approximation and Projection (PaCMAP) is a dimensionality reduction technique developed by Wang et al. [11] that uses a stochastic gradient descent algorithm, boasts effective and reliable performance with preserving local and global structure alike, and performs significantly faster than other contemporaneous dimensionality reduction methods. Of interest to note is that none of the previously described methods are able to smoothly transition from local structure preservation to global structure preservation or vice versa through manipulation of parameters, however, PaCMAP can and does so automatically by dynamically adjusting its own parameters over a series of iterations through three distinct phases using a unique loss function. PaCMAP performs this optimization of the low-dimensional embedding by using three kinds of pairs of datapoints: neighbor pairs, mid-near pairs, and further pairs. The PaCMAP algorithm applies the initialization procedure selected to set the initial values of the low-dimensional embedding (initialization can be done randomly or using PCA). Once initialized, the values are then optimized. The algorithm uses the Adam Optimizer to reduce the loss function. The overall loss function is defined as:

$$Loss^{PaCMAP} = \quad w_{NB} \cdot \sum_{i,j} \frac{\tilde{d}_{ij}}{10 + \tilde{d}_{ij}} \qquad (neighbors; strong\ attraction)$$

$$+ \quad w_{MN} \cdot \sum_{i,k} \frac{\tilde{d}_{ik}}{10,000 + \tilde{d}_{ik}} \qquad (mid\text{-}near\ pairs;\ weak\ attraction)$$

$$+ \quad w_{FP} \cdot \sum_{i,l} \frac{1}{1 + \tilde{d}_{il}} \qquad (further\ pairs; repulsion)$$

where d is the scaled distance between a pair of observations, i and j are neighbors, i and k are mid-near pairs, i and l are further pairs, and w refers to the weights of all the pairwise interactions that change over the course of iterations. In general, the goal is to strongly attract neighbor pairs, weakly attract mid-near pairs, and repulse further pairs to properly construct the global and local structures. In the first stage of the PaCMAP algorithm's optimization, the initial embedding of points is improved to preserve primarily the global structure. In the second phase, the focus eases to improving the local structure while maintaining the developed global structure by moving to a higher weighting of neighbor pairs. Finally, by more heavily weighting the further pairs and neighbor pairs, the third phase shifts to refinement of the local structure without adjusting the global structure.

4. FASTMAN-JMP Use-Case of an Industrial Pyrolysis Reactor

The implementation of this tool was tested using an industrial pyrolysis reactor case study. The use of an industrial dataset illustrates the suitability and effectiveness of using FASTMAN-JMP to apply the methodology and algorithms discussed to complex problems in the chemical engineering industry. The industrial pyrolysis reactor dataset analyzed covers a period of several months. The reactor, which is a very well-known chemical process unit, cracks heavy hydrocarbons into higher-value, lighter hydrocarbons. These datasets depict multiple modes of steady-state operation as well as process drift due to a buildup of equipment coke. The chemical reaction occurs in a fired furnace that is heated by burning fuel gas, as shown in *Figure 1*. This pyrolysis reactor takes a naphtha feed, injects it with steam, and cracks it into ethylene at extremely high temperatures. The entire process has 27 variables, including hydrocarbon flow, steam flow, cracking temperatures per coil, and coil outlet temperature. Also shown in *Figure 1* is the associated SOM plot, where the axes are the coordinates of the nodes present in the map. As illustrated by the distinct borders of contrasting colors, several main regions and subregions can be visualized in the SOM plot as an indication of the data structure i.e., global and local events. In this application the process is moving along different regions during the full cycle of operation.

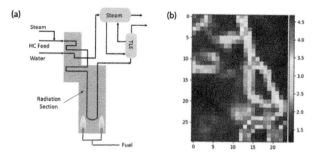

Figure 1: *(a) Pyrolysis reactor process scheme; (b) SOM plot*

Figure 2 illustrates the 2D projection results of initial stages of the PaCMAP algorithm using PCA initialization as the number of iterations increases. These results are at the very beginning of the first stage of the PaCMAP algorithm. There are very few clusters identified initially, categorizing most of the data into one large grouping. This is also revealed in the clusters projected on the SOM plots in **Figure 3** (corresponding to the clusters identified in **Figure 2**) as indicated by the dots overlaying the SOM. In **Figure 2.a**, "Cluster 1" includes datapoints that cover an area on the SOM with very distinct divisions (the bottom right corner in **Figure 3.a**) and "Cluster 2" encompasses the entire remainder of the area on the SOM (the projected cluster shown in **Figure 3.a**), with both clusters grouping together data that would otherwise be separated based on the more granular SOM coloring. Clearly, this dimensionality reduction method initially identifies the global structure of the data (i.e., the main process events) then gradually eases its global structure optimization and begins optimization of the local structure. Finally, in the later stages, the algorithm focuses on refining the local structure. In comparison to the early stages of the PaCMAP algorithm, **Figures 2.d** and **3.d** correspond to the results using PCA projection which focuses on global data structure. The PCA results are equivalent to those after just 60 iterations of the PaCMAP algorithm.

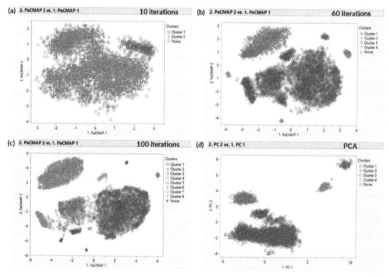

Figure 2: *(a) PaCMAP results after 10 iterations; (b) PaCMAP results after 60 iterations; (c) PaCMAP results after 100 iterations; and (d) PCA results*

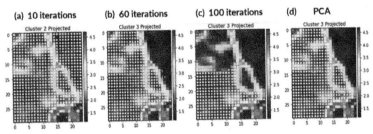

Figure 3: *PaCMAP cluster projection on the SOM: (a) after 10 iterations; (b) after 60 iterations; (c) after 100 iterations; and (d) PCA cluster projection on the SOM*

This is also corroborated in **Figure 4** where the time evolution of the hydrocarbon flow rate in coils 1 and 4 is shown (solid and dashed, respectively). The highlighted section of the data corresponds to the conditions identified as "Cluster 3" in **Figures 2.b** and **3.b**. The identification of key process variables responsible for the cluster formation is performed using the subspace greedy search (SGS) algorithm [4].

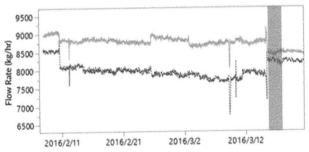

Figure 4: *Time evolution of hydrocarbon flows for coils 1 (solid; top) and 4 (dashed; bottom)*

Shown in **Figure 5** is a graph of the value of the PaCMAP loss function for the pyrolysis reactor dataset as a function of number of iterations. As discussed before, the loss function changes over the course of the PaCMAP algorithm stages based on the weights of neighbor pairs, mid-near points, and further points. As shown, the loss is slightly improved as the number of iterations increases in the second and third stages. However, the largest rate of the improvement (i.e., the most noticeable change in the embedding) occurs in the first stage, with the focus on the global structure development primarily focusing on mid-near pairs.

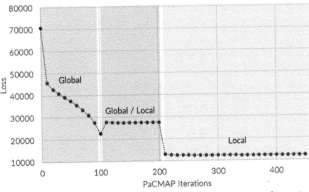

Figure 5: *PaCMAP loss function evolution as a function of iterations*

The sharp transitions in ***Figure 5*** are due to the distinct changes in the loss function weightings at the onset of the three optimization phases. Of course, since this is based on the relative amount of neighbor pairs, mid-near pairs, and further pairs present in a dataset and the distances between these points, the qualitative trend may differ between datasets.

5. Conclusions

The concepts of global and local data structures were analyzed using data from an industrial pyrolysis reactor. It was shown with FASTMAN-JMP that local and global structures in the data set can be visualized and related to actual process operations allowing for the identification of the variables responsible for the separation between any two given clusters. Furthermore, it was shown that PaCMAP is an effective and reliable dimensionality reduction technique that preserves both local and global structures and performs significantly faster than other methods. By building up an understanding of how these dimensionality reduction algorithms work and how to tune parameters – a process made easier with FASTMAN-JMP – we can more effectively use these powerful tools to visualize and extract knowledge from large, high-dimensional datasets in the chemical manufacturing industry.

References

[1]　*JMP®, Version 16.2,* Cary, NC: SAS Institute Inc., 1989–2021.

[2]　Y. Shi, Big Data and Big Data Analytics, Singapore: Springer, 2022.

[3]　T. Kohonen, "The self-organizing map," *Proceedings to the IEEE,* vol. 78, no. 9, pp. 1464-1480, 1990.

[4]　W. Zhu, W. Sun and J. Romagnoli, "Adaptive k-Nearest-Neighbor Method for Process Monitoring," *Industrial & Engineering Chemistry Research,* vol. 57, pp. 257-258, 2018.

[5]　H. Hotelling, "Analysis of a complex of statistical variables into principal components," *Journal of Educational Psychology,* vol. 24, pp. 417-441, 1933.

[6]　L. van der Maaten and G. Hinton, "Visualizing Data using t-SNE," *Journal of Machine Learning Research,* vol. 9, pp. 2579-2605, 2008.

[7]　L. McInnes, J. Healy and J. Melville, "UMAP: Uniform Manifold Approximation and Projection for Dimension Reduction," *arXiv:1802.03426v3 [stat.ML],* 2020.

[8]　E. Amid and M. Warmuth, "TriMap: Large-scale Dimensionality Reduction Using Triplets," *arXiv:1910.00204v2 [cs.LG],* 2019.

[9]　Z. Webb and J. Romagnoli, "Real-Time Chemical Process Monitoring with UMAP," *Computer Aided Chemical Engineering,* vol. 50, pp. 2077-2082, 2021.

[10] L. Briceno-Mena, M. Nnadili, M. Benton and J. Romagnoli, "Data mining and knowledge discovery in chemical processes: Effect of alternative processing techniques," *Data-Centric Engineering,* vol. 3, p. e18, 2022.

[11] Y. Wang, H. Huang, C. Rudin and Y. Shaposhnik, "Understanding How Dimension Reduction Tools Work: An Empirical Approach to Deciphering t-SNE, UMAP, TriMap, and PaCMAP for Data Visualization," *Journal of Machine Learning Research,* vol. 22, no. 201, pp. 1-73, 2021.

Antonis Kokossis, Michael C. Georgiadis, Efstratios N. Pistikopoulos (Eds.)
PROCEEDINGS OF THE 33rd European Symposium on Computer Aided Process Engineering
(ESCAPE33), June 18-21, 2023, Athens, Greece

Fault detection in high-pressure pumps in Low-Density Polyethylene Autoclave Reactors

Maria Giuliana F. Torraga,[a] Tahyna B. Fontoura,[b] Rodrigo M. Lima,[c] Luciana A. da Silva,[d] Tiago da Silva Osório[d]

[a]Braskem S.A., Rua da União 765, Jardim Sonia Maria, Mauá - SP 09380-900, Brazil
[b]Braskem S.A., Rua Marumbi, 1001, Campos Elíseos, Duque de Caxias - RJ, 25221-000, Brazil
[c]Braskem S.A., Av. Pres C. e Silva, 400, Capuava - Santo André - SP, 09270-000, Brazil
[d]Braskem S.A., BR 386, Rod. Tabaí, km 419, Via Oeste, Triunfo - RS, 95853-000, Brazil
maria.giuliana@braskem.com

Abstract

Low-density polyethylene is a polymer obtained via free-radical polymerization at high pressures in both tubular and autoclave reactors. At high pressures, ethylene can suffer an explosive decomposition ("decomp"), leading to runaway reactions. Several disturbances that can lead to "decomp", such as excess initiator in feed, feed impurity, controller failures, etc. The present work focus on the development of a fault detection algorithm to detect failures in high-pressure pumps in the autoclave process. These pumps inject the initiator and are prone to failures such as leaks, cylinder failures, reversal system failure and check valve failures. The fault detection algorithm was developed to early detect failures in the pumps. The methodology consisted in developing data-driven models to monitor the process. The models were developed using data from an industrial autoclave reactor unit and were tested against failures events that happened between 2019 and 2022, with at least 70% of the events successfully detected between 5 and 50 minutes prior to the failure.

Keywords: fault-detection, polyethylene, statistical process control, data-driven models.

1. Introduction

Low-density polyethylene (LDPE) is a polymer obtained via free-radical polymerization in bulk in both tubular and continuous autoclave reactors, at high pressure. Free-radical polymerization includes mainly four elementary reactions: initiation, propagation, bimolecular termination, and chain transfer reactions. In this type of polymerization, the polymer microstructure and properties are intrinsically related to the reaction. For instance, the long and short-chain branches observed in LDPE are due to chain transfer reactions to polymer species and backbiting reactions (Hamielec and Tobita, 2000). In order to generate the free radicals to initiate the polymerization, chemical initiators can be used, at low concentrations. These initiators decomposes thermally generating radicals for monomer addition. Commercial thermal initiators include azo and peroxides compounds, and in the case of LDPE, the initiation depends both on the temperature and the pressure of the system.

Free-radical polymerization is highly exothermic (ΔH_{poly} of 24 kcal mol^{-1} for LDPE), therefore, the heat removal and temperature control play an important role in both productivity, product quality and process safety. In industry, there are two main processes to produce LDPE, in tubular and in autoclave reactors. In both cases, the reaction is carried out at high pressures (1500-2000 atm) and high temperatures (150-230 °C). An important reaction that also occurs at high temperature and pressure is the explosive decomposition of ethylene (Zhang et al., 1996), forming hydrogen, carbon, methane, and other low-saturated and unsaturated hydrocarbons. This undesired reaction affects both tubular and autoclave reactors and can lead to a runaway ("decomp").

Several disturbances that can lead to "decomp", such as excess initiator in feed, feed impurities, feed inlet temperature, controller failures, mechanical frictions, inadequate mixing, etc (Zhang et al., 1996). Ultimately, all these disturbances can cause hot spots in the reactor, which can initiate the runway. As Zhang et al. (1996) have pointed out, the decomposition occurs in order of milliseconds, and it is irreversible once it has started. Therefore, several works have been developed to predict failures in LDPE reactors.

Sivalingam et al. (2015) developed a predictive model to detect decomposition for high-pressure ethylene/vinyl acetate copolymerization in an autoclave reactor. The energy balance error was used as a reference to normal operation, considering the heat released by reaction and the sensible heat absorbed by the feed and products. The heat of stirring and lost to the environment were not considered. A principal component analysis (PCA) on heat balance model components was performed, using data from normal operation, under steady state, with 1s time interval for calibration. Control charts based on heat balance allowed detecting impending decomposition of the reactor. Sharmin et al. (2008) also developed a data-driven model to detect failures in LDPE reactors. The methodology coupled the global energy balance terms (reaction, products and reactants enthalpy), calculated using the industrial data, and PCA method to detect decomposition in the reactor. The heat released by the reaction was obtained using the overall conversion in the reactor. The model could detect decomposition with a lead-time about 30 seconds.

The main objective of the present work was the development of a fault detection algorithm to detect failures in high-pressure pumps used in the autoclave reactors of LDPE polymerization. Particularly, the modeling approach presented here takes into account the initiator injection, which is known to be a potential cause of decomposition. Besides that, the heat released by the reaction was accounted using the definition from the free-radical polymerization. The fault detection tool was developed using plant data from an industrial autoclave reactor unit and was tested against pump leakages events that happened between 2019 and 2022.

2. Methodology

The methodology consisted in developing data driven models coupled with the energy balance to monitor the process. In order to do that, the following models were developed: (i) charts of the residuals for the energy balance model; (ii) control charts of a process KPI and; (iii) a PCA model using Hotelling's T-squared statistic using the energy balance as input.

2.1. Historical data for calibration and validation

The fault detection tool was developed using plant data from an industrial autoclave reactor unit and was tested against pump failure events that happened between 2019 and 2022. The reactor has three pumps that inject initiator at different reactor heights and has monomer feeds along the reactor. The process variables collected to monitor the energy balance were: monomer feed mass flow rates, monomer feed temperature, initiator mass flow rate and temperature along the reaction height, with 1 second sampling time. Another process variable collected specifically related to the pump operation is the "pump output" variable, that is an indirect measurement of the number of pulses of the pump piston, also with 1 second sampling time. The dataset was cleaned, and data recorded during start-up, shut downs, and containing corrupted values were removed, and the cleaned dataset was divided in two sets (calibration and validation). The calibration set contains the cleaned base without the 2h before and after the failure in each pump.

2.2. Charts of the residuals for the energy balance error model

The energy balance for each zone of the continuous autoclave reactor at steady-state operation and neglecting the heat transfer to the surroundings and the shaft work can be written as:

$$\dot{m}_{in}c_pT_{in} + \dot{m}_{feed}c_pT_{feed} - \dot{m}_{out}c_pT_{out} + R_P(-\Delta H_p)V = 0 \tag{1}$$

where, $\dot{m}_{in}$, $\dot{m}_{feed}$ and $\dot{m}_{out}$ are the mass flow rate entering the zone (in), of the fresh monomer entering the zone (feed) and the feed leaving the zone (out); c_p is the specific heat capacity of the mixture; T_{in}, T_{feed} and T_{out} are the temperature of the feed entering the zone (in), of the fresh monomer (feed) and the feed leaving the zone (out); R_P is the polymerization rate; $-\Delta H_p$ is the polymerization enthalpy; V is the zone volume.

The polymerization rate, R_P can be obtained from the free-radical polymerization kinetics, assuming pseudo steady-state for the radicals:

$$R_P = k_p[M]\sqrt{\frac{2fk_I[I]}{k_t}} \tag{2}$$

where, k_p, k_I and k_t are the propagation, initiation and termination rate constants; f the initiator efficiency; $[M]$ the monomer concentration; and $[I]$ the initiator concentration. Accounting the polymerization rate as a function of initiator concentration allows the model to be sensitive to changes in the initiator feed, which is associated with failures in the injection pumps. The kinetic rate constants used in the model are from the work of Zhang et al. (1996). During normal operation, the energy balance should be equal to zero at all times within reasonable limits (Sharmin et al., 2008), and when failures occurs in the pumps, the energy balance error will exceed the normal threshold limits, indicating that the system is moving away from the normal condition operation. The error balance in reactor zone i is:

$$Error_i = \dot{m}_{in}c_pT_{in} + \dot{m}_{feed}c_pT_{feed} - \dot{m}_{out}c_pT_{out} + R_P(-\Delta H_p)V \tag{3}$$

The error in the energy balance was calculated for the normal operation (calibration set) and can be applied to each reactor zone sequentially. The upper and lower control limits (UCL and LCL, respectively) for the error in each reactor zone are calculated from the average ($\overline{Error_i}$) and standard deviation (σ_i) of the in control dataset.

$$UCL_i/LCL_i = \overline{Error_i} \pm L\sigma_i \tag{4}$$

where L is the control limit width, typically, $L = 3$.

2.3. Monitor a process KPI

A KPI can be used to monitor the failures, using the process variables "pump output" and the temperature of the reactor zone in which the pump is injecting the initiator.

$$KPI_i = \frac{\left(T_{zone\ i}\Big/\overline{T_{zone\ \iota}}\right)}{\left(Output\ pump\ i\Big/\overline{Output\ pump\ \iota}\right)} \tag{5}$$

This KPI is normalizes and sensitive to the failures in the individual pumps, once when the pump is leaking, less peroxide is entering the reactor zone and, as consequence, the reactor zone temperature decreases and the "pump output" increases, because of the temperature reactor control policy. During normal operation, the KPI should vary within the normal limits, and when failures in the pumps occur, the KPI will exceed the normal threshold limits, indicating that the system is moving away from the normal condition. The UCL and LCL for the KPI were calculated in the same way as the energy balance error.

2.4. PCA model

PCA is a multivariate statistic technique commonly used to statistic quality control and fault detection. This method is ideal for handling a large number of highly correlated data and noisy process variables. PCA decomposes the observation matrix **Y** (process data) in a linear combination:

$$Y = TP^T = \sum_{i=1}^{q} t_i p_i^T \tag{6}$$

where, t_i are the principal component score, p_i are the loading vector (eigenvalues of the covariance matrix of Y) and q is the number of measured quality variables. Usually, most of the variability in the data is retained by the first principal components (MacGregor and Kourti, 1995). The data matrix used for the PCA model contains the scaled the elements of the energy balance (Equation 3). In order to build the quality control charts based on PCA, the model was developed using historical data from steady state, when only common cause variation was present. For new measurements, the observations can be projected using the loading vectors to obtain the scores, $t_{i,new}$. The Hotteling's T^2 is used for the multivariate control charts. The upper control limit (UCL) for T^2 is obtained according to MacGregor and Kourti (1995).

3. Results

3.1. Charts of the residuals for the energy balance error model

The energy balance error was evaluated for the calibration dataset, according to Equation 3, allowing the determination of the UCL and LCL of the control charts for each one of the three pumps present in the process. Table 1 shows the percentage of events detected, lead-time and percentage of false alarms for the model.

The lead-time is the warning time of the model before the event occur, although the average lead-time is presented in the results of Table 1, there is a distribution of warning times for the events, as depicted in the histogram of lead-time in Figure 1. Some events can be detected between 5-35 minutes before the failure, and others can be detected between 95-125 minutes before the event. The broad range of warning times is related to the different failures modes present in these pumps, as leaks, cylinder failures, reversal system failure and check valve failures. Depending on the failure mode, the fault can happen very fast, reducing the model warning time. Then, the broad warning time is associated to the inherent nature of the failures.

3.2. Monitor a process KPI

The percentage of events detected, lead-time and percentage of false alarms for the model using the process KPI (Equation 5) are presented in Table 1. Compared to the energy balance error model (Table 1), the KPI monitoring has smaller warning times, once the KPI model depends only on the actual value of the process variables and is not able to represent an energy accumulation in the process, as the energy balance is. Despite of that, the KPI model can detect more events than the energy balance error for the middle pump. Figure 2 presents the control chart for the KPI during a failure event; it is possible to observe the KPI value moving away from the control limits around 1 hour before the failure. After the pump maintenance, the KPI value returns to the normal operation limits.

Table 1: Percentage of events detected, lead-time (min) and percentage of false alarms for the models.

Method	Pump	Events detected (%)	Average lead-time (min)	False alarm (%)
Energy balance error model (L=3)	Top	42.86%	58.33	9.10%
	Middle	40.00%	13.75	5.56%
	Bottom	88.89%	6.88	6.79%
KPI model (L=3)	Top	42.9%	29.00	5.80%
	Middle	80.0%	8.38	7.70%
	Bottom	55.6%	13.40	4.50%
PCA model (alpha = 0.01)	Top	100.0%	51.43	17.35%
	Middle	70.0%	34.29	8.84%
	Bottom	88.9%	30.00	12.01%

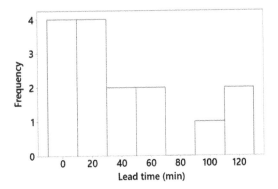

Figure 1: Histogram of lead-time (min) for the failures events for the energy balance error model.

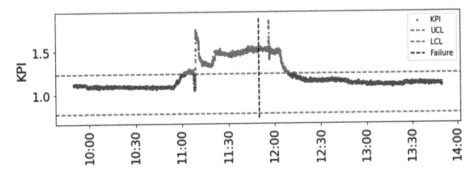

Figure 2: Control chart for the normalized KPI during a failure event. KPI (.) UCL/LCL (red --) and failure event (black --).

3.3. PCA model

Table 1 presents PCA model performance, using the energy balance terms as input. The results shows that combining the energy balance model with the multivariate statistic approach of PCA provides a fault detection algorithm that can detect more events with increased warning times for the failures in the peroxide pumps. The T^2 control chart in Figure 3 show the validation data for a failure event in one of the pumps.

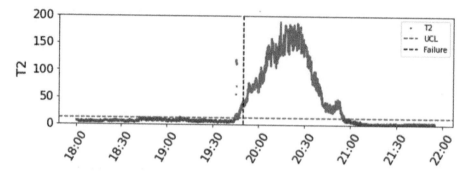

Figure 3: Control chart for the T^2 statistic during a failure event. T^2 (.) UCL (red --) and failure event (black --).

4. Conclusion

This work presented the study of fault detection models for LDPE autoclave reactors peroxide pumps. The failures in initiator pumps can lead to the explosive decomposition of ethylene and are critical for process control and safety. Data-driven models combined with phenomenological knowledge of the system play an important role to improve fault detection algorithms, as showed in this paper. The model developed was tested against failures events that happened between 2019 and 2022, with at least 70% of the events successfully detected between 5 and 50 minutes prior to the failure. The development of tools that can support the operation of this process is still a challenge, due to the non-linear and fast nature of the process.

References

Hamielec, A. E., & Tobita, H. (2000). Polymerization processes. Ullmann's Encyclopedia of Industrial Chemistry.

MacGregor, J. F., & Kourti, T. (1995). Statistical process control of multivariate processes. Control engineering practice, 3(3), 403-414.

Sharmin, R., Shah, S. L., & Sundararaj, U. (2008). A PCA based fault detection scheme for an industrial high pressure polyethylene reactor. Macromolecular Reaction Engineering, 2(1), 12-30.

Sivalingam, G., Soni, N. J., & Vakil, S. M. (2015). Detection of decomposition for high pressure ethylene/vinyl acetate copolymerization in autoclave reactor using principal component analysis on heat balance model. The Canadian Journal of Chemical Engineering, 93(6), 1063-1075.

Zhang, S. X., Read, N. K., & Ray, W. H. (1996). Runaway phenomena in low-density polyethylene autoclave reactors. AIChE journal, 42(10), 2911-2925.

Antonis Kokossis, Michael C. Georgiadis, Efstratios N. Pistikopoulos (Eds.)
PROCEEDINGS OF THE 33rd European Symposium on Computer Aided Process Engineering
(ESCAPE33), June 18-21, 2023, Athens, Greece

Using ultrasonic and process data to identify surging and gas lock phenomena during ESP operation

Tiago F. Souza[a], Cáio C. S. Araújo[a], Maurício M. F. Figueiredo[a], Ana M. F. Fileti[a]

[a] *School of Chemical Engineering, University of Campinas - UNICAMP, Albert Einstein Avenue, 500 – Campinas/SP, 13083-852, Brazil*
Corresponding author: t226161@dac.unicamp.br

Abstract

Electrical Submersible Pump (ESP) has deteriorated performance due to free gas in its intake, which may result in two undesirable phenomena: surging and gas lock. In this sense, this work investigates and determines features set — based on ultrasonic and process data — to identify these undesirable phenomena. The experimental data was obtained with a fixed ESP rotational speed (3500 rpm), varying the inlet water flowrate (1 to 60 kg/min) and air flowrate (0.02 to 0.1 kg/h). The ultrasonic measurements were carried out using a 2.25 MHz transducer. Using ultrasonic echo energy and time of flight signals, and the pump discharge pressure, it was possible partitioning the experimental data into two sets using *K*-means clustering tool. One of these groups is mainly composed of ESP operating points in surging or gas lock conditions, and another predominantly consists of ESP stable conditions.

Keywords: Electrical Submersible Pump; surging; gas lock; ultrasonic; clustering.

1.   Introduction

Electrical Submersible Pump (ESP) is widely applied to lift oil from offshore wells to production facilities (Takacs, 2017). The ESP performance is negatively affected due to the presence of free gas in pump intake. The free gas in ESP intake can result in two undesirable phenomena: surging and gas lock (Verde, 2016). Surging occurs when the pump head decreases as the liquid flowrate decreases. In turn, gas lock occurs when the liquid pumping stops due to blockage of free gas at ESP intake. Operating the ESP in surging or gas lock conditions can lead to severe instabilities, such as abrupt shutdowns due to fluctuations in the load of the motor coupled to the ESP.

Experimental investigations have been developed to predict the operating conditions that results in surging or gas lock during ESP operation under liquid-gas flow. Such investigations resulted in empirical or semi-empirical models to identify the ESP operating points in which surging or gas lock occurs (Turpin; Lean; Bearden, 1986; Zhu, 2017). However, these models require input variables not measured in real-time or continuously in individual oil wells equipped with ESP, such as liquid flowrate, gas flowrate, and free gas in ESP inlet. In order to develop an alternative technique, Oliveira (2014) carried out an experimental investigation whose results demonstrate the potential of ultrasonic measurement technique to identify the surging during ESP operation under liquid-gas flow.

In this context, this work aims to investigate and determine features set, based on ultrasonic and process data, capable of identifying the unstable operating conditions —

surging or gas lock — and stable operating conditions during ESP operation under liquid-gas flow.

Material and methods

1.1.   Test plant

The test plant was built to acquire ultrasonic and process data during ESP operation. It was composed of three flowlines: single-phase liquid flow line, single-phase gas flowline and two-phase liquid-gas flow line. An illustrative diagram of the setup is shown in Figure 1. The liquid and liquid-gas flow line are composed of a 23 mm inner diameter ASTM 240 – AISI 304 stainless steel. The gas flow line is a polyurethane tube of 8 mm inner diameter.

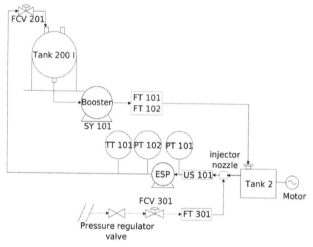

Figure 1 – Experimental setup diagram.

The liquid flow line begins at the 200 litters separator tank (Tank 200 l) which has the function of separating the gas and liquid phases from the two-phase mixture and feeds liquid to the booster pump. The booster, a centrifugal pump (KSB / Megablock 40-25-160 GG), was driven by 3 HP motor that was controlled by an inverter frequency drive (SY-101). The booster kept constant the pressure at the ESP intake (Schlumberger / D475N). After the booster, a Coriolis mass flowmeter (FT-101/FT-102) was installed to measure the liquid mass flow and liquid density. The ESP was driven by a 1 HP electric motor (WEG / W22 IR3 PREMIUM).

The gas flow line was fed by an external air supply system. The pressure regulator valve kept at the 5 bar g the air pressure upstream the control valve (FCV 301), which controlled the amount of air in ESP intake. To measure the gas mass flowrate, a Coriolis mass flowmeter (FT 301) was installed. The gas flow line ends at the injector nozzle, where the liquid-gas flow line begins.

The injector nozzle was installed at 35 cm upstream of the ESP intake. A porous medium was used as the injector nozzle to ensure uniform distribution of the gas in form of bubbles into the liquid flow. After the injector nozzle, an annular flow occurs with the outside inner diameter of 52 mm and the shaft diameter of 20 mm.

In the horizontal position, the inlet and outlet flow conditions of the ESP was monitored by one temperature transmitter (TT 101), one ultrasound structure (US 101) and two pressure sensors (PT 101 and PT 102). At the US 101, the ultrasound transducer was fixed in a non-invasive and non-intrusive way. Besides, the FCV 201 was place after the ESP to regulate the flowrate of liquid-gas flow. A sample rate of 4 Hz was used to acquire the process data.

1.2. Description of ultrasonic measurement technique

The ultrasonic measurement system consists of an ultrasound support (US 101), an ultrasonic transducer, and a device for generating and digitizing ultrasonic signals. The function of the US 101 support is to accommodate the ultrasonic transducers orthogonally to the flow in a non-invasive and non-intrusive way. This structure was made using polyacetal material, internal diameter of 52 mm, 10 mm delay line and a hole to accommodate transducers positioned perpendicular to the flow. Industrial coupling gel was used to perform the coupling of the ultrasonic transducer with the inner wall of the ultrasonic support. The ultrasonic measurements were performed using only one piezoelectric ultrasound transducer (Olympus - V106-SM) with a central frequency of 2.25 MHz.

The ultrasonic signal generation and digitization device (US Ultratek - DSPUT5000-2) was used in pulse-echo mode, that is, the system emit and capture waves through the same transducer. The device has 8-bit vertical resolution and a maximum sampling rate of 100 MHz. The parameters of the ultrasonic acquisition device were: pulse voltage of -300 V, a pulse width of 222 ns; sample rate of 25 MHz, a pulse repetition frequency of 1 kHz, a digital gain of 40 dB; damping of 620 Ω, an acquisition time of 15 s; and 720 points in depth per wave. Such values were defined according to the fundamentals of signal sampling (Holman, 2012), and the technical guidelines of the manufacturers of the ultrasonic acquisition board and ultrasonic transducers.

The Figure 2-a represents a schematic mode of the ultrasonic wave propagation through solid and liquid based medium. At the wall-flow interface, part of the wave intensity associated with the initial pulse (P_0) is reflected in the first wall pulse reflection (R_{WF}). The remainder is transmitted to the liquid based medium. After that, the transmitted pulse will propagate until reaches the liquid-gas (R_{LG}) or shaft-flow (R_{SF}) interface, in which the wave will be reflected to the ultrasound. A typical ultrasonic echo signal is illustrated in the Figure 2-b. The position of the liquid-gas (R_{LG}), shaft-flow (R_{SF}) interface and first wall pulse reflection (R_{WF}) is indicated.

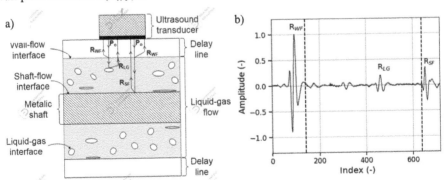

Figure 2 – Schematic ultrasonic mode of the wave propagation and reflection at the wall and two-phase flow (a). Echo signal containing two-phase flow reflection (b).

1.3. Experimental design and procedure

The test matrix has 9 experimental conditions for single-phase, and 26, 23 and 20 experimental conditions for two-phase with gas flowrate of 0.02 kg/h, 0.06 kg/h and 0.10

kg/h, respectively. Each experimental condition was experimentally measured in 4 replicates to evaluate the reproducibility, totalling 312 sample acquisitions. Each sample acquisition lasted 30 s for the process data under constant ESP rotation (3500 RPM) and intake pressure (100 kPa). The gas mass flowrate (m_G) were experimentally adopted according to the ESP capacity to deal with the total amount of free gas in its intake. The experiments were carried out with water as the continuous phase and air as the dispersed phase. During the experiments, the temperature remained in the range of 24 to 26 °C.

In summary, the procedure adopted for the execution of the matrix was: a) The first acquisition was obtained from the condition in which the intake pressure was 100 kPa and the pump head was equal to zero, which corresponds to the maximum liquid flowrate in that condition; ii) afterwards, the variables of interest corresponding to the next acquisition point were adjusted, waiting 2 minutes to stabilize the flux in the new condition; iii) ultrasonic measurements and process data were acquired; iv) the new experimental condition (acquisition) was adjusted, and ultrasonic and process data were collected until the experimental points of the test matrix finished.

2.   Results and discussion

2.1.   ESP performance curves

Figure 3 shows the ESP performance curves under single-phase and two-phase flow conditions investigated experimentally in this work. The ESP operation under two-phase flow conditions can exhibit two phenomena: surging and gas lock. The surging phenomenon implies an abnormal behavior, the pump head decreases as the liquid flowrate decreases. The beginning of surging (surging point) corresponds to the curve peak in the ESP performance curve under two-phase flow. It is noticeable that as the gas flowrate increases, the stable region — the right side of the curve peak — became narrower due to the less capacity of the ESP to deal with free gas in its intake. The gas lock occurs when the pump head is approximately null after beginning of surging. In face of that, we can divide the two-phase performance curves in two regions: stable and unstable.

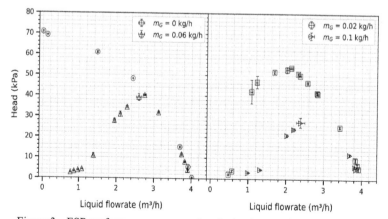

Figure 3 – ESP performance curves under single-phase and two-phase flow.

2.2.   Echoes signals and process data - Feature extraction

The features used to cluster the ESP operating points were obtained by processing the ultrasonic echoes signals and pump discharge pressure data. Four features were extracted: arithmetic mean of echoes energy signals from ultrasonic wave reflections in liquid-gas interface ($\bar{E}_{LG}$); arithmetic mean of echoes energy signals from ultrasonic wave reflections in liquid-shaft interface ($\bar{E}_{SF}$); time of flight standard deviation from ultrasonic wave reflections in liquid-gas interface (t_{LG}); and variation coefficient of ESP pressure discharge ($\bar{P}_D$). The features mathematical definitions are available in Table 1. These features were normalized between -1 and 1 before applying the *K*-means clustering.

Table 1 – Features definitions

$\bar{E}_{LG}$		$\bar{E}_{SF}$	
$mean\left[\left(\sum_{k=140}^{640} (A_k)^2\right)_i\right]$ $i = 1,2, \dots 15000$	(1)	$mean\left[\left(\sum_{k=640}^{720} (A_k)^2\right)_i\right]$ $i = 1,2, \dots 15000$	(2)
t_{LG}		$\bar{P}_D$	
$std[argmax(A_{120:640})_i]$ $i = 1,2, \dots 15000$	(3)	$std[(P_D)_i]/mean[(P_D)_i]$ $i = 1,2, \dots 120$	(4)

2.3. *K-means clustering*

To execute *K*-means clustering, the method *k*-means++ was adopted for handling centroids initialization (Guha *et al.*, 2003). In addition, two metrics were used to assess the clustering performance: silhouette coefficient and rand index (Rousseeuw, 1987; Hubert; Arabie, 1985). Figure 4-a shows these metrics score versus the number of clusters due to *K*-means application to ESP experimental data. Facing these metrics scores, it is reasonable to consider clustering the data in 2 generated clusters due to the highest values of the rand index and silhouette coefficient. Figure 4-b indicates the ESP experimental data distribution into 2 clusters. The amount of 279 samples data was partitioned into two clusters. Cluster 1 has grouped 138 samples, and Cluster 2 grouped has 141 samples. There are samples of data obtained experimentally at the same operating condition (acquisition) partitioned in different clusters. These samples are associated with acquisitions 14, 15, 16, and equal to 0.02 kg/h; acquisition 10 and equal to 0.06 kg/h; acquisition 6 and equal to 0.1 kg/h. These acquisitions in performance curves are identical or closer to surging points. Despite having samples associated with identical acquisitions in different clusters, it is notable that Cluster 1 is predominantly composed of ESP experimental data in stable conditions, and Cluster 2 is majorly formed by ESP experimental data in unstable conditions. These results indicate that features used in k-means clustering can be used to identify if the ESP is operating or not in surging or gas lock conditions (unstable conditions).

3. Conclusions

This work suggests that is possible identifying surging or gas lock phenomena (unstable conditions) in ESP operation under two-phase flow by using the following features extracted from ultrasonic echoes signals and process data: $\bar{E}_{LG}$; $\bar{E}_{SF}$; t_{LG}; and $\overline{P_D}$. *K*-means clustering was used to cluster the ESP experimental data by using the features previously mentioned. According to silhouette coefficient and rand index scores, it is reasonable to cluster the ESP experimental data in 2 clusters. The generated Cluster 1 is predominantly composed of ESP experimental data in stable conditions, and Cluster 2 is majorly formed by ESP experimental data in unstable conditions. The silhouette coefficient and rand index scores for this clustering were 0.5 and 0.93, respectively. Thus, these results indicate that features used in *K*-means clustering can be used to identify the two undesirable phenomena during ESP operation.

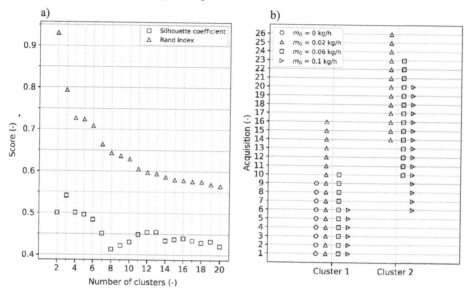

Figure 4 - Number of clusters analysis based on metrics score (a) and visual representation of the ESP experimental data grouped in only two clusters (b).

References

W. M. Verde et al., 2017. Experimental study of gas-liquid two-phase flow patterns within centrifugal pumps impellers, **Experimental Thermal and Fluid Science**, v. 85, p. 37-51.

G. Takacs, 2017. **Electrical submersible pumps manual: design, operations, and maintenance**, 2. ed. Gulf professional publishing.

J. P. Holman. **Experimental Methods for Engineers**. 8. ed. McGraw-Hill, 2012.

J. Zhu, X. Guo, F. Liang, H. Zhang. Experimental Study and Mechanistic Modeling of Pressure Surging in Electrical Submersible Pump. **Journal of Natural Gas Science and Engineering,** v. 45, p. 625-636, 2017.

J. L. Turpin, J.F. Lea, J.L. Bearden. Gas-Liquid Flow Through Centrifugal Pumps-Correlation of Data. In: **Proceedings of the 3rd International Pump Symposium**, p. 13–20, 1986.

B. D. Oliveira. **Monitoramento do Desempenho de uma Bomba Centrífuga Submersa com Auxílio da Técnica Ultrassônica**, 2014.

T. F. Souza. **Desenvolvimento de Uma Técnica Ultrassônica para a Estimativa da Fração de Vazio e Vazão da Fase Dispersa de Escoamentos Água-Ar**, 2021.

L. Hubert, P. Arabie. Comparing partitions. **Journal of classification**, v. 2, n. 1, p. 193-218, 1985.

J. P. Rousseeuw. Silhouettes: a graphical aid to the interpretation and validation of cluster analysis. **Journal of computational and applied mathematics**, v. 20, p. 53-65, 1987.

S. Guha, A. Meyerson, N. Mishra, R. Motwani, O'Callaghan, L. Clustering data streams: Theory and practice. **IEEE transactions on knowledge and data engineering**, v. 15, n. 3, p. 515-528, 2003.

Antonis Kokossis, Michael C. Georgiadis, Efstratios N. Pistikopoulos (Eds.)
PROCEEDINGS OF THE 33rd European Symposium on Computer Aided Process Engineering
(ESCAPE33), June 18-21, 2023, Athens, Greece
© 2023 Elsevier B.V. All rights reserved. http://dx.doi.org/10.1016/B978-0-443-15274-0.50290-0

A data-driven predictive model of gas flow rate from ultrasound and process variables using artificial neural networks

Cáio C. S. Araújo[a], Tiago F. Souza[a], Maurício M. F. Figueiredo[a], Ana M. F. Fileti[a]

a School of Chemical Engineering, University of Campinas - UNICAMP, Albert Einstein. Avenue, 500 – Campinas/SP, 13083-852, Brazil
Corresponding author: c228076@dac.unicamp.br

Abstract

The bubble flow is a flow regime present in various fields. A usual equipment that operates with the bubble flow is the bubble column. Many papers deal with determining flow regime, hold up, and mass transfer coefficient in a bubble column. However, little attention has been given to the measurement of the gas flowrate. In this context, this work presents a technique to predict the gas flowrate in a bubble column from ultrasound signals using a new set of ultrasonic variables combined with process variables using a feedforward artificial neural network (ANN). Experiments were carried out in a 2-meters-long-bubble column, with inner diameter of 52.5 mm using a single 2.25 MHz ultrasound transducer. The gas flow rate varied in the range of 0 to 4 liters per minute. The best architecture obtained for ANN has one hidden layer with 36 neurons. The test mean absolute error was 0,1 LPM.

Keywords: flow rate, artificial neural networks, monitoring, ultrasound, soft computing.

1.   Introduction

The bubble flow is a flow regime present in various fields, like the pharmaceutical, biochemical, and petrochemical industries (Shu et al., 2019). A usual equipment presents in these industries, that operates with the bubble flow, is the bubble column. A bubble column can be used as a reactor that operates with a stagnant liquid phase, in which the gas flows throughout. Many papers deal with determining flow regime, bubble size, liquid hold up, and mass transfer coefficient in a bubble column using machine learning (Biessey et al., 2021; Serraa et al., 2020; Supardan, 2003). However, little attention has been given to the measurement of the gas phase flowrate in a bubble column direct from the combination of ultrasonic and process signals without intermediate needed analysis, like calculation of velocity and liquid hold up. In addition, according to the findings of Mudde & Saito (2001), it is expected that the method developed here using bubble columns can be applied in two-phase liquid-gas bubble flow and get similar results.

In this context, this work proposed a new set of ultrasonic variables combined with pressure drop for gas flowrate prediction in a bubble flow using a 52.5 mm inner diameter vertical bubble column based on feedforward artificial neural network. The ultrasonic signals were acquired using just one ultrasound transducer.

2.   **Material and methods**

2.1.   Experimental setup

The experimental setup consists of a pipe with an internal diameter of 52.5 mm and a length of 2 m arranged vertically (Figure 1). The two-phase line was instrumented with two pressure sensors (PT 20 and PT 21) to measure the void fraction of the dispersed phase and a temperature sensor (TT 20) to obtain the physical properties of the two-phase mixture. Ultrasonic measurements were performed with a single inclined at 10° with the flow, located at support (UT 10).

The air flow line contains a pressure regulator, a rotameter (FI 10), a check valve, a pressure sensor (PT 10) and a temperature sensor (TT 10). The gas injection was performed through a porous structure positioned at the base of the bubble column (gas injection).

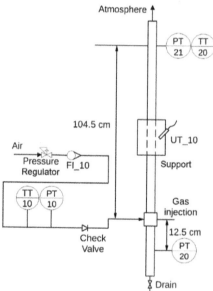

Figure 1 – P&ID diagram of the experimental setup.

2.2.   Ultrasonic instrumentation

The ultrasonic measurement system basically comprises an ultrasonic transducer and a ultrasound device (US Ultratek/DSPUT5000-2) for generating, receiving and digitalizing ultrasonic signals (Figure 2). In all experiments, the transducer remained confined between the support and a spring, which is pressed by a cable gland. Industrial Vaseline was used to perform the acoustic coupling between the transducer and the ultrasonic spool. The support, made with Delrin polymeric material, was installed 60 cm above the air injection point. During the experiments, the transducer was installed with inclination $\theta = 10°$. Besides, the transducer (Panametrics/ M106-SM) with a central frequency of 2.25 MHz was used.

The ultrasound device with 8 bits of vertical resolution was used for generation, reception and digitization of ultrasonic signals. The sampling frequency (50 MHz) was defined in order to respect Nyquist's theorem, which defines the minimum frequency as being at least twice the highest frequency to be observed. The damping (620 Ω), along with the gain (60 dB), will be experimentally defined in order to return the highest amplitude values without signal saturation. The pulse frequency (2 kHz) was defined based on the ability of the system to detect the same bubble at least twice throughout the ultrasonic measurement volume. The pulse width (222 ns) was defined based on the manual of manufacturer, based on the central frequency of the transducer. Furthermore, a transducer supply voltage of -300 V_{P-P} was used.

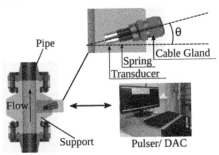

Figure 2 - Components of the ultrasonic measurement system.

2.3.   Test matrix and procedure

The test matrix was defined based on the flow pattern map of Shah et al. (1982), considering only the bubble flow pattern in all experiments. In this way, the minimum and maximum flow limit was defined from 0 to 4 liters per minute (LPM). Experiments were carried out varying the flowrate from 0 to 4 LPM with increments of 0.5 LPM, that is, 9 measurements. Each measurement was repeated 4 times, that is, 36 measurements. The experiments were carried out with water as the continuous phase and air as the dispersed phase. The mixture temperature of the experiments was maintained in the range of 20 to 22 °C.

In each experimental condition, 15 s of ultrasonic measurements and 120 s of process data at 1 Hz of acquisition frequency were acquired. All ultrasonic echoes had 1,000 points in depth, the necessary to capture half cross-section area of the pipe, resulting in one ultrasonic echo signal per experimental measurement. In summary, the experimental procedure adopted was: i) Fill the vertical pipe with tap water until it reaches 1.2 m above the gas injection point and keep it constant; ii) define the pre-defined gas flowrate randomly; iii) wait 60 s for flow stabilization; iv) acquire ultrasonic and process data; v) return to the ii step until the final of the text matrix.

2.4.   *Topology of artificial neural network*

The ANN developed and used to predict the gas flow rate was developed in Python programming using the scikit-learn library. The ANN was chosen because it has the ability map non-linear relationships between inputs and outputs. In the training step, the Adam optimization algorithm was used. The ANN input ultrasonic variables used were: mean energy signal ($\bar{E}$); standard deviation of the energy signal (σ_E); mean transit time ($\overline{ToF}$); standard deviation of transit time (σ_{ToF}). The ANN input process variable used was the mean pressure drop (ΔP). The ultrasonic echo signal (matrix of 30,000 rows by 1,000 columns) was processed to return energy (E) and transit time (ToF) signals in which were calculated the associated mean and standard deviation, according to definitions presented in Table 1. Indeed, the $\bar{E}, \sigma_E, \overline{ToF}, \sigma_{ToF}$ and ΔP was used to feed the ANN.

The output variable was the gas flow rate. The mathematical definition of E and ToF signal associated with echoes signals are presented in Table 1. The dataset was randomly divided in two groups: i) 75% for training; and ii) 25 % for test step. In training step, 10 % of the training dataset was used for validation. The best architecture was defined using GridSearchCV optimization method (Pedregosa et al., 2011). It was tested the number of neurons on hidden layer in the range of 1 to 50 and identity, relu, hyperbolic tangent and logistic as activation functions. Besides, the performance of the artificial neural network was evaluated using the mean square error (MSE), mean absolute error (MAE), and determination Coefficient (R^2) of the straight line that fits predicted versus actual values of gas flow metrics.

Table 1 - Definition of energy and transit time signals.

$E_i = \sum_{j=1}^{n} [e_{(i,j)}]^2$	$ToF_i = \text{argmax}_{e(i,j)flow} - argmax_{e(i,j)wall}$
$E = [E_1, E_1, ..., E_m]$	$ToF = [ToF_1, ToF_2, ..., ToF_m]$

Where: $e_{(i,j)}$ is the value of echo in row i and column j with m rows and n columns; E_i is the energy associated with row i; ToF_i is the transit time associated with row i; and $\text{argmax}_{e(i,j)flow}$ and $\text{argmax}_{e(i,j)wall}$ are the position of maximum echo amplitude value in the flow and wall region, respectively.

3.   Results and discussion

3.1.   Ultrasonic echoes

The Figure 3 illustrates the ultrasonic echoes referring to the 2,000 ultrasonic pulses (1 s) emitted for the single-phase (0 LPM) and two-phase simulated bubble flows. Dividing the wall signal from the signal in the position of the dashed line, where from the begging to the dashed line is the wall region, it is possible to note the sensitivity of the transducer with the passage of bubbles in the ultrasonic measurement region. In addition, it is observed that the reflected signals from the liquid-gas interfaces of the bubbles are not limited to a specific region of half cross-section area, showing the relative uniformity distribution of the bubbles across the half cross sectional area of the pipe.

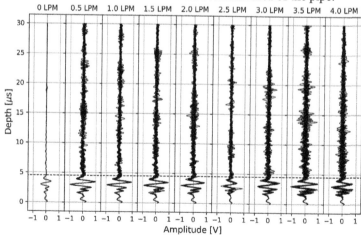

Figure 3 - Ultrasonic echoes for single-phase (0LPM) and two-phase flows in the vertical bubble flow.

The Table 2 shows the limits of values associated with ultrasonic ($\bar{E}$, σ_E, ToF, σ_{ToF}) and process (ΔP, Q) signals. These variables were used as input, except the variable Q that was uses as output.

Table 2 - Range of the ultrasonic and process variables.

Variable	$\bar{E}$	σ_E	$\overline{ToF}$	σ_{ToF}	ΔP	Q
Units	$[V^2]$	$[V]$	$[\mu s]$	$[\mu s]$	$[kPa]$	$[LPM]$
Minimum value	0.17	0.015	650.30	404.74	10.56	0
Maximum value	3.39	2.656	1018.06	495.55	12.64	4.0

3.2.   *ANN for prediction of gas flow rate*

The ANN which presented the best prediction of gas flow rate had an architecture with 36 neurons in only one hidden layer, and rectified linear unit (relu) as activation function. The ANN predictions and absolute error for training and test set are illustrated in Figure 4 and 5, respectively. The ANN score metrics calculated for the training set are the following: MSE (0.007); MAE (0.063); and R^2 (1.000). In turn, the score metrics calculated for the test set are: MSE (0.013); MAE (0.078); and R^2 (0.998).

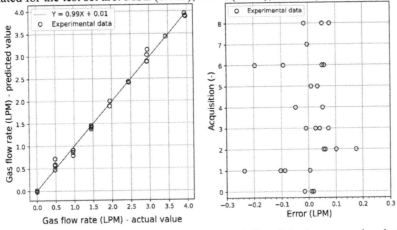

Figure 4 - ANN performance comparison for train step (left) and absolute error values between actual and prediction value of the gas flow rate for training (right).

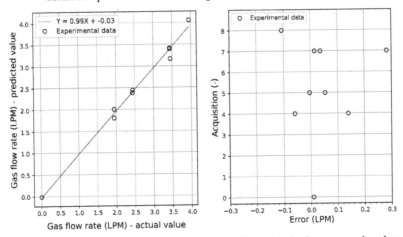

Figure 5 - ANN performance comparison for test step (left) and absolute error values between actual and prediction value of the gas flow rate for testing (right).

The results of test step, illustrated in Figure 5, showed that the distribution of the absolute error was close to the obtained in training step, which represents a good agreement between the prediction and actual gas flowrate for test and train step. As a consequence, the MAE, MSE and R^2 are also in good agreement in relation to the train step results. Therefore, the metrics score relative to test set suggest a satisfactory generalization capability of ANN developed to predict gas flowrate.

The results of gas flowrate prediction also indicates that the choose ultrasonic variables ($\bar{E}$, σ_E, ToF, σ_{ToF}) together with process variable (ΔP) have the ability to represent the physics of the flow. The energy (E) captures the frequency and intensity of the bubble reflection in the echoes, the transit time (ToF) captures the location of the bubbles reflection in the ultrasonic echoes and the pressure drop (ΔP) infer about the concentration of the bubble in the two-phase mixture. Indeed, the physical representativity of the choose variable contribute to the good agreement in the results. Further, the good results indicate that these ultrasonic and process variables could be used together to estimate the gas flow rate in bubble columns.

4.   Conclusions

The results showed the ability of the ANN to predict the gas flow rate. The mean absolute error from the reference was 0.1 LPM and the majority of the error in training and test was in the range of -0.1 to 0.1 LPM. From the agreement between the predicted and actual gas flow rate, we concluded that the ANN based model proposed is promising to estimate gas flow rate in bubble flows using ultrasonic combined with process data as input. Further, the selected ultrasonic variables ($\bar{E}$, σ_E, ToF, σ_{ToF}) and the process variable (ΔP) showed good ability to represent the bubble flow phenomena and predict the gas flowrate. To increase the robustness of the proposed ANN based technique we proposed tests with different fluid as continuous and dispersed phase and in other pipe inner diameters.

References

S. Shu, D. Vidal, F. Bertrand, J. Chaouki, 2019, Multiscale multiphase phenomena in bubble column reactors: A review, **Journal of Renewable Energy**, v. 141, p. 613-631.

P. Biessey, H. Bayer, C. Thebeling, E. Hilbrands, and M. Grunewald, 2021, Prediction of Bubble Sizes in Bubble Columns with Machine Learning Methods. **Journal of Chemie Ingenieur Technik**, v. 93, p. 1968-1975.

P. L. S. Serraa, P. H. F. Masottib, M. S. Rochab, D. A. de Andradeb, W. M. Torresb, R. N. de Mesquita, 2020, Two-phase flow void fraction estimation based on bubble image segmentation using Randomized Hough Transform with Neural Network (RHTN). **Journal of Progress in Nuclear Energy**, v. 118, p. 103133.

M. D. Supardan, A. Maezawa, and S. Uchida, 2003, **Determination of Local Gas Holdup and Volumetric Mass Transfer Coefficient in a Bubble Column by Means of an Ultrasonic Method and Neural Network**. Journal of Chemical Engineering Technology, v. 26, p. 1080-1083.

R. F. Mudde, T. Saito, 2001, Hydrodynamical similarities between bubble column and bubbly pipe flow, **Journal of Fluid Mechanics**, v. 437, p. 203-228.

Y. T. Shah, B. G. Kelkar, S. P. Godbole, and W. D. Deckwer, 1982, Design parameters estimations for bubble column reactors. **AIChE Journal**, v. 28, p.353–379.

F. Pedregosa et al, 2011, Scikit-learn: machine learning in Python. **Mach. Learn. Res Journal**.

Antonis Kokossis, Michael C. Georgiadis, Efstratios N. Pistikopoulos (Eds.)
PROCEEDINGS OF THE 33rd European Symposium on Computer Aided Process Engineering
(ESCAPE33), June 18-21, 2023, Athens, Greece

An NN-NMPC for Controlling a Crystallization Process in the Saturated and Undersaturated Zones

Fernando Arrais R. D. Lima[a], Marcellus G. F. de Moraes[b], Amaro G. Barreto[a],
Argimiro R. Secchi[a,b], Maurício B. de Souza Jr.[a,b]

[a] *School of Chemistry, EPQB, Universidade Federal do Rio de Janeiro, Av. Horácio Macedo, 2030, CT, Bloco E, 21941-914, Rio de Janeiro, RJ – Brazil*
[b] *PEQ/COPPE – Universidade Federal do Rio de Janeiro, Av. Horácio Macedo, 2030, CT, Bloco G, G115, 21941-914, Rio de Janeiro, RJ – Brazil*
farrais@eq.ufrj.br

Abstract

The crystallization process has been applied in pharmaceutical, fine chemicals and food industries and is the main process in the pharmaceutical industry for particle formation. This process is used for separation and purification, recovering as much solute as possible and producing crystals with the desired length. In this work, a nonlinear model predictive controller based on neural networks (NN-NMPC) was proposed to control the coefficient of variation and length of crystals for the potassium sulphate batch cooling crystallization. The neural networks were designed using an optimizer to choose the best hyperparameters with simulated and experimental data. They were trained to predict the first four moments and the concentration in order to calculate the coefficient of variation and the length of crystals. The potassium sulphate crystallization process was modelled using a population balance model solved by the method of moments, considering the nucleation, growth and dissolution mechanisms. This rigorous model was used to represent the real process for the controller's performance evaluation. Differently from previous studies, the neural networks were trained considering the saturated and the undersaturated thermodynamic conditions. The controller's performance was studied for servo cases, analyzing the efficiency of the NN-NMPC to reach different set-points using the process temperature as the manipulated variable. The operation in the undersaturated and saturated conditions guarantees more degrees of freedom to control both variables (coefficient of variation and length of crystals), manipulating only the temperature. The proposed approach efficiently achieved the desired set-points. The controller also presented good performance for a case considering plant-model mismatch. The NN-NMPC performance was compared to an NMPC based on the phenomenological model, showing similar behavior with reduced computational effort, evidencing that the NN-NMPC has great potential to be applied in real time for the crystallization process.

Keywords: Crystallization, neural networks, NMPC, servo control, dissolution.

1. Introduction

The crystallization process is used for product separation and purification, generating a solid crystalline product. The regulatory requirements must be respected for an efficient crystallization process, producing crystals with the desired size distribution. Therefore, the crystallization process needs effective control to achieve this goal[1]. Another challenge of crystallization processes consists of modeling this phenomenon, which is usually done using the population balance approach. However, this can take too much effort because of the necessity of using different models to describe the mechanisms of this process. Data-driven models, in particular, neural networks, have mainly been applied to model and control chemical processes because they can capture intrinsic nonlinearities[2]. Nevertheless, there is a lack of applications in the literature using neural networks to model the crystallization process under saturated and undersaturated conditions.

This work aims to model a crystallization process to predict the moments of the particle-size distribution and the solute concentration with neural networks. Unlike other works, this study considered data under saturated and undersaturated conditions, using only one model to predict each state variable under different conditions. Also, the current work proposes control strategies for the crystallization process different from the study of Moraes et al.[3], which focuses on modeling it. Five neural networks were trained using experimental and simulated data, and the hyperparameters were chosen considering an optimization approach. The capacity of the neural networks was tested to predict the state variables n sampling times in the future, making good predictions for large horizons. Moreover, the proposed networks were used as an internal model in a Nonlinear Model Predictive Controller (NMPC).

2. Neural Networks Models

Neural networks are mathematical models that try to represent the correlation between input and output patterns. The most applied neural network in the literature consists of the feedforward network, particularly the multilayer perceptron (MLP) network. An MLP network comprises an input layer, an output layer, and one or more hidden layers[4]. In the current work, five MLP networks were developed to predict the moments and the solute concentration of a potassium sulfate batch crystallization process.

The dataset consisted of ten experimental batches in the saturated condition and five in the subsaturated condition developed by Moraes et al.[3] in LABCADS/UFRJ. In these experiments, the moments and the solute concentration were determined for different temperatures, totaling 724 patterns. Also, simulated data were generated using a population balance model for the saturated condition and a population balance model for the undersaturated state, developed by Moraes et al.[3]. Therefore, 15 batches were simulated, considering different temperatures and initial and saturation conditions, producing a total of 6800 simulated patterns. The training dataset was composed of seven experimental batches in the saturated state, four experimental batches in the subsaturated condition, and eleven simulated batches, totaling 5726 patterns. The test dataset was composed of the 1798 patterns of the simulations and experiments that were not used for the training. A sampling time of 1.25 min was adopted in the experiments and simulations, considering the same value adopted in the experiments of Moraes et al.[3]. The batches chosen to compose the training and test datasets were randomly selected, using in the test dataset experimental and simulated data for the saturated and undersaturated states with conditions not used in training.

The inputs and outputs of each NN are shown in Table 1, in which T is the process temperature, C/C_{eq} is the ratio between the solute concentration and the concentration of the solute in equilibrium, μ_i is the ith-order moment, and k is the sampling time. Different combinations of inputs were investigated. The NNs were developed using the Scikit-learn library from Python. Furthermore, the hyperparameter of each neural network was chosen using an optimization approach from a Python library called Optuna. The optimization problem consisted of finding the activation function, number of hidden layers, number of neurons in the hidden layer, batch size, and learning rate that leads to R^2 values of the test samples close to one.

Table 1: Inputs and outputs of the developed neural networks.

Neural Network	Inputs	Outputs
1	$T(k-1), C/C_{eq}(k-1), \mu_0(k-1),$ $T(k), C/C_{eq}(k), \mu_0(k)$	$\mu_0(k+1), ..., \mu_0(k+n)$
2	$T(k-1), C/C_{eq}(k-1),$ $\mu_0(k-1), \mu_1(k-1),$ $T(k), C/C_{eq}(k), \mu_0(k), \mu_1(k)$	$\mu_1(k+1), ..., \mu_1(k+n)$
3	$T(k-1), C/C_{eq}(k-1),$ $\mu_1(k-1), \mu_2(k-1),$ $T(k), C/C_{eq}(k), \mu_1(k), \mu_2(k)$	$\mu_2(k+1), ..., \mu_2(k+n)$
4	$T(k-1), C/C_{eq}(k-1),$ $\mu_2(k-1), \mu_3(k-1),$ $T(k), C/C_{eq}(k), \mu_2(k), \mu_3(k)$	$\mu_3(k+1), ..., \mu_3(k+n)$
5	$T(k-1), C/C_{eq}(k-1), C(k-1),$ $T(k), C/C_{eq}(k), C(k)$	$C(k+1), ..., C(k+n)$

The hyperparameters used in the networks are presented in Table 2, and all neural networks were trained using the Adam algorithm.

Table 2: Hyperparameters of the neural networks.

Neural Network	Hidden Layers	Number of Neurons	Activation Function	Batch Size	Learning Rate
1	4	55; 55; 55; 55	Relu	15	Constant
2	3	45; 45; 45	Relu	5	Constant
3	4	45; 45; 45; 45	Relu	5	Constant
4	4	40; 40; 40; 40	Relu	5	Adaptive
5	1	15	Logistic	135	Constant

The R^2 values for predictions two, five, ten, fifteen and twenty steps forward for the test samples are presented in Table 3. It can be observed that the neural networks could efficiently predict the moments and the solute concentration of a batch crystallization process, even for large prediction horizons. Therefore, the neural networks to predict μ_0, μ_1, and μ_2 were used as the internal model of an NMPC.

Table 3: R^2 values, considering the test samples and different prediction horizons.

Neural Network	2 steps	5 steps	10 steps	15 steps	20 steps
1	0.9998	0.9996	0.9986	0.9987	0.9977
2	0.9997	0.9997	0.9989	0.9956	0.9941
3	0.9997	0.9995	0.9980	0.9963	0.9932
4	0.9998	0.9997	0.9990	0.9982	0.9970
5	0.9995	0.9987	0.9955	0.9947	0.9901

3. Nonlinear Model Predictive Control

The controlled variables chosen for the optimization problem were the coefficient of variation and the crystal length, which are calculated by Equations 1 and 2, respectively. The manipulated variable u is the process temperature. The MPC goal is to minimize the process performance index J, the objective function of the problem shown in Equation 3. The parameters δ and φ are the weights of the outputs, and γ is the weight of the increment input. The parameters δ, φ and γ were tuned as 1, 10 and 5, respectively. The superscript SP indicates the set-point value of the controlled variable. The prediction horizon P was defined as 10, and the control horizon was tuned as 1. The optimization constraint is given by Equation 4.

$$V = \sqrt{\frac{\mu_2 \mu_0}{\mu_1^2} - 1} \qquad (1)$$

$$L = \mu_1 / \mu_0 \qquad (2)$$

$$J = \sum_{j=1}^{P} \{\delta[V(k+j) - V^{SP}(k+j)]^2 + \varphi[L(k+j) - L^{SP}(k+j)]^2\} + \gamma[\Delta u(k)]^2 \qquad (3)$$

$$-1 \leq \Delta u(k) \leq 1 \qquad (4)$$

In the control loop, the process was simulated based on the model developed in the study of Moraes et al.[3], and a sampling time of 1.25 min was adopted. This optimization problem was solved using Successive Quadratic Programming (SQP) algorithm with absolute accuracy of 10^{-10} and relative accuracy of 10^{-8}. The performance of the controller based on the neural networks was compared to a controller based on the phenomenological model. For the NMPC based on the population balance model, the optimization problem was solved using the IPOPT solver, defining a tolerance of 10^{-8}.

The controller was tested for set-points equal to 16 and 14.5 μm for the crystal's length and 1.0 and 1.1 for the coefficient of variation. The results of the simulations for both controllers are shown in Figures 1 and 2, in which PB-NMPC is the NMPC based on the phenomenological model developed by Moraes et al.[3], and NN-NMPC is the NMPC based on the neural networks. In the first case, the PB-NMPC reached the set-point for the coefficient of variation but could not get the set-point for the length. However, the NN-NMPC could achieve both set-points during this simulation. In the second case, the

NN-NMPC could reach both set-points, but the PB-NMPC presented a small offset for the coefficient of variance. Even with the advantage of the NN-NMPC, the PB-NMPC presented a mean squared error (MSE) equal to 1.7437 for the crystal's length and 0.0787 for the coefficient of variation in the first scenario, while these values were 0.5731 and 0.0526 in the second case studied. In contrast, the NN-NMPC shown an MSE equal to 4.9396 for the crystal's length and 0.0970 for the coefficient of variation in the first case, while these values were 4.1703 and 0.0428 in the second scenario. In both cases, the ratio C/C_{eq} was higher than one almost all the time, indicating that the simulations happened basically in the saturated condition. This situation was expected because the process does not stay too long in the subsaturated condition and rapidly tries to reach equilibrium (C/C_{eq} equals to one).

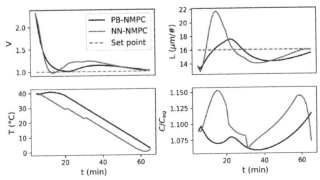

Figure 1: Control loop results for set-points equal to 16 μm and 1.0.

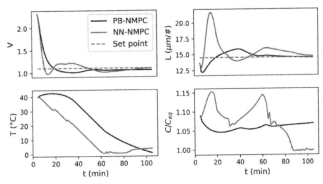

Figure 2: Control loop results for set-points equal to 14.5 μm and 1.1.

The performance of the controllers was tested to achieve the same set-points of 16 μm and 1.0 but considering mismatches in the growth coefficient and the dissolution coefficient, using a difference of 10% on their original values. The results for this case are presented in Figure 3, showing similar results to the previous case. The NN-NMPC could reach both set-points once again, while the PB-NMPC could not reach the set-point for the crystal's length. The MSE values were 4.9370 for the crystal's length and 0.0921 for the coefficient of variation considering the NN-NMPC, while these values were 1.5551 and 0.0792 for the PB-NMPC.

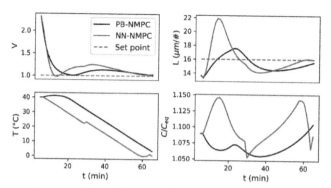

Figure 3: Control loop results considering mismatches.

4. Conclusion and Prospects

In this study, neural network models were developed to predict the first four moments and the solute concentration of a batch crystallization process. The hyperparameters of the neural networks were defined using an optimization approach. The capacity of the neural networks was tested to predict these values from two to twenty sampling times ahead, verifying R^2 greater than 99% for the test samples in all cases analyzed. Moreover, the neural networks have the advantage to use only one model to make predictions of a particular state variable under the saturated and undersaturated conditions. The neural network models were used as the internal model of an NMPC to control the crystal's length and the coefficient of variation manipulating the temperature. The proposed controller could efficiently maintain the controlled variables in the set-points, showing similar results to an NMPC based on the phenomenological model. The simulations remained mostly in the supersaturation condition, since the process remains quickly in the undersaturation region and seeks to reach equilibrium. Moreover, the controller could efficiently handle mismatches in the model. The proposed strategy reached the set-points efficiently and with low computational cost, showing a potential to be used in real crystallization processes. A similar application could be made to a continuous crystallization process for future work.

Acknowledgements

CAPES for providing scholarship (Finance Code 001), the financial support from CNPq (Grants No. 303587/2020-2 and 311153/2021-6), and FAPERJ (Grant No. E-26/201.148/2022).

References

1. Ahn B., Bosetti L., Mazzotti M., 2021. Secondary Nucleation by Interparticle Energies. II. Kinetics. Cryst. Growth Des., v. 22, p. 74–86.
2. Lima, F. A. R. D., Moraes, M. G. F., Secchi, A. R., Souza Jr., M. B., 2022. Development of a recurrent neural networks-based NMPC for controlling the concentration of a crystallization process. Digital Chemical Engineering, v. 5, 100052.
3. Moraes, M. G. F., Secchi, A. R., Vieira, G. M., Barreto Jr., A. G., Souza Jr., M. B., 2019. Measurement and Modeling of Crystal Size and Shape Distributions of Potassium Sulfate Through Dynamic Image Analysis in Batch Cooling Crystallization. I Brazilian Congress on Process Systems Engineering (PSEBR 2019), p. 170.
4. Himmelblau D. M., 2008. Accounts of Experiences in the Application of Artificial Neural Networks in Chemical Engineering. Ind. Eng. Chem. Res., v. 47, p. 5782–5796.

Antonis Kokossis, Michael C. Georgiadis, Efstratios N. Pistikopoulos (Eds.)
PROCEEDINGS OF THE 33rd European Symposium on Computer Aided Process Engineering
(ESCAPE33), June 18-21, 2023, Athens, Greece

A novel neural network bounds-tightening procedure for multiparametric programming and control

Dustin Kenefake,[a,b] Rahul Kakaodkar,[a,b] Moustafa Ali,[a,b] Efstratios N. Pistikopoulos,[a,b]

[a]*Texas A&M Energy Institute, Texas A&M University, College Station, United States*
[b]*Artie McFerrin Department of Chemical Engineering, Texas A&M University, College Station, United States*

Abstract
Recently, it has been shown that rectified linear unit (ReLU) based neural networks (NN) are mixed integer linear representable and therefore can be incorporated into mixed integer linear programming frameworks. In this work we propose a novel tightening procedure that is based on a multiparametric programming formulation of the corresponding ReLU-reformulated optimization problem. The bounding procedure features 1) the generation of valid tight bounds on the individual auxiliary variables introduced from the ReLU NN reformulation, and 2) bounds on groups of input and output variables to these ReLU NNs. The tightened bounds are valid for all parameter realizations, and thus can be reused between different problem realizations. This tightening only needs to be computed a single time and thus the overall procedure is once-and-offline. As this bounds-tightening procedure is applied offline, more computationally expensive methods for formulation tightening can then be applied to the online case. We demonstrate the effectiveness of this method in a case study of model predictive control of a nonlinear chemostat where the dynamics are approximated with a ReLU NN.

Keywords: Multiparametric Programming, Model Predictive Control, Neural Network

1. Introduction

Multiparametric programming has garnered much interest in the control literature since it was shown that Model Predictive Control (MPC) problems can be formulated as multiparametric programs (Bemporad et al., 2000; Pistikopoulos et al., 2021) where these restated MPCs are typically referred to as multiparametric MPCs (mpMPC). The feature of multiparametric programming often used in the control literature is that it allows for the generation of explicit algebraic solutions to the MPC problem ahead of time that are functions of the uncertain parameters, θ (Pappas et al., 2021; Pistikopoulos et al., 2021). Another feature of the multiparametric program formulation is that it represents the set of all possible optimization problems given a parametric uncertainty set Θ, and it is this aspect that will be explored in this work. A prototypical multiparametric program is shown in eqn. 1. With $x \in \mathbb{X}$ being the set of decision variables, and $\theta \in \Theta$ being the set of uncertain parameters. Here, $f(x, \theta)$ is the objective function, $g(x, \theta) = 0$ is the set of equality constraints, and $h(x, \theta) \leq 0$ are the inequality constraints.

$$\begin{aligned}
\min_{x} \quad & f(x, \theta) \\
\text{s.t.} \quad & g(x, \theta) = 0 \\
& h(x, \theta) \leq 0 \\
& x \in \mathbb{X}, \theta \in \Theta
\end{aligned} \qquad (1)$$

It has been shown that ReLU-based NNs are mixed integer linear representable using one of the various reformulation strategies that have been proposed in the literature, such as the partition-based reformulation approach by Tsay et al. (2021) and the Big-M reformulation as proposed by Grimstad and Andersson (2019) and Fischetti and Jo (2018). Various methods have been explored to accelerate the solution of these optimization problems, such as deriving tighter bounds and reformulations (Tjandraatmadja et al., 2020; Tsay et al., 2021; Grimstad and Andersson, 2019) and finding the explicit solution of the resulting multiparametric program (Katz et al., 2020). Here, bounds-tightening refers to finding tight ranges for the auxiliary variables introduced in the reformulation of NNs. In the case of bounds-tightening, the literature has focused on finding tighter bounds for individual optimization problems and has not thoroughly explored the case where one can tighten and reuse bounds over sets of optimization problems, such as those that can be represented by multiparametric programming formulations, e.g., eqn. 1.

In this work, we propose a multiparametric programming-based strategy for generating tight bounds for ReLU-based NN reformulations that can be reused for all realizations of optimization problems, $\theta \in \Theta$. These are tight bounds that are valid for all possible problem realizations and thus can be effectively reused between parameter realizations, $\theta \in \Theta$. This reusing of tight bounds between problems allows for the bounds-tightening to be calculated once-and-offline for any particular multiparametric program with ReLU-based NN constraints. The effectiveness of this approach is displayed on a ReLU-based NN MPC of an unstable nonlinear chemostat process.

2. Method

This section is separated into two subsections. The first subsection briefly describes the Big-M MILP reformulation of ReLU-based NN. The second subsection gives an overview of the proposed bounds-tightening procedure.

2.1. ReLU Neural Network Reformulation

A ReLU-based NN is in essence a continuous piecewise affine function and thus can be represented via mixed integer linear constraints, in the same way that any other continuous piecewise affine function can be. With ReLU-based NN, the only non-linearity that must be addressed is the ReLU activation function itself. This is a rectification function where if the input is less than 0, then it is rectified to 0; otherwise, the input is not altered by the activation function. This can be modeled with the max function, $max(0, x)$. This can be effectively linearized by introducing auxiliary variables.

The Big-M reformulation for layer k of a ReLU-based NN is shown in eqn. 2. Where the input to hidden layer k is z^{k+1}, the weight and bias values of the layer are W^k and b^k respectively, and the output of the hidden layer is z^{k+1}. Auxiliary slack variables s^k and selection variables y^k are introduced to linearize the max function exactly. In the case $y_i^k = 0$, then $(W^k z^k + b^k)_i > 0$ and thus does not to be rectified so $s_i^k = 0$; similarly, in the case of $y_i^k = 1$, then $(W^k z^k + b^k)_i < 0$ and thus z_i^{k+1} needs to be rectified to zero via s_i^k. Initial bounds for $M^{s_i^k}$ and $M^{z_i^k}$ are typically found via interval arithmetic (Grimstad and Andersson (2019)). As long as the Big-M values are significantly large, i.e. provide a true upper bound to the auxiliary variables, the resulting reformulation is valid. However, if these values are made overly large, the computational performance of the resulting optimization problem will be affected. More details on this reformulation can be found in the works of Grimstad and Andersson (2019), Fischetti and Jo (2018), and Tjandraatmadja et al. (2020).

$$
\begin{aligned}
z^{k+1} &= \max(0, W^k z^k + b^k) \\
z^{k+1} &\geq 0
\end{aligned}
\quad \rightarrow \quad
\begin{aligned}
z^{k+1} &= W^k z^k + b^k + s^k \\
z^{k+1} &\geq 0 \\
0 \leq s_i^k &\leq M^{s_i^k} y_i^k \\
0 \leq z_i^{k+1} &\leq M^{z_i^k}\left(1 - y_i^k\right) \\
y_i^k &\in \{0,1\}
\end{aligned}
\tag{2}
$$

2.2. Tightening Procedure

While the Big-M reformulation of the ReLU-based NN typically allows for the use of standard MIP solvers, such as Gurobi or CPLEX, the formulation does not typically lead to a tight mathematical program (Grimstad and Andersson, 2019; Tsay et al., 2021). Instead of proposing a bounds-tightening procedure for a particular problem realization, we propose tightening over all possible problem realizations, $\theta \in \Theta$.

By having an explicit parameterization of an optimization problem, as represented in eqn. 1, the parametric uncertainty space, Θ, can directly be used to find tight valid bounds on all optimization variables. By promoting the parameters, θ, into decision variables in the reformulated optimization problem, tight worst-case bounds can be found for each variable via MILP optimization, such as can be seen in eqn. 3. These bounds are valid for every possible parameter realization in the uncertainty space and allow for these bounds to be shared between optimization problems with different parameter realizations, $\theta \in \Theta$.

$$
\begin{aligned}
&\min_{x,z,s,y,\theta} \quad x_i \\
\text{s.t.} \quad &g_{\text{reformulated}}\left(x, z^1, \ldots, z^M, s^1, \ldots, s^M, y^1, \ldots, y^M, \theta\right) = 0 \\
&h_{\text{reformulated}}\left(x, z^1, \ldots, z^M, s^1, \ldots, s^M, y^1, \ldots, y^M, \theta\right) \leq 0 \\
&x \in \mathbb{X}, \quad \theta \in \Theta
\end{aligned}
\tag{3}
$$

A similar optimization problem, as shown in eqn. 3, is solved for every variable in the formulation, e.g., s_i^k, z_i^k, y_i^k and θ_i, for both the upper and lower bounds of each variable given the parameters, $\theta \in \Theta$.

3. Case Study: Unstable Nonlinear Chemostat

The proposed methodology is applied to a ReLU-based NN MPC for the unstable nonlinear chemostat process adopted by Eaton and Rawlings (1992). The model equations of the considered process are shown in eqn. 4, describing the dynamics of the biomass, x, and the substrate, s, with time along with the substrate inhibition model, $\mu(s)$. In this process the specific parameters are $s_f = 4$, $y = 0.4$, $\mu_{\max} = 0.53$, $k_m = 0.12$, $k_1 = 0.4545$. The dilution rate, D is the input variable that can be manipulated to change the trajectory of the states, x and s, of the process.

$$\frac{dx}{dt} = \left(\mu\big(s(t)\big) - D(t)\right)x(t)$$

$$\frac{ds}{dt} = \big(s_f - s(t)\big)D(t) - \frac{\mu\big(s(t)\big)x(t)}{y}$$

$$\mu(s) = \frac{\mu_{max}s(t)}{k_m + s(t) + k_1 s(t)^2}$$

(4)

The dynamics of the chemostat process are then approximated by a ReLU-based NN model with 3 hidden layers with 15 nodes in each hidden layer. The NN was generated via the following procedure. A data set was constructed by simulating 1,000,000 randomly generated initial states and input actions inside of the operating window, (x, s, D), for a single time step to build an input-output data set, $(x(t), s(t), D(t)) \rightarrow (x(t + \Delta t), s(t + \Delta t))$. The input action, $D(t)$, was held constant for the entire time interval, Δt, in a zero-order old fashion. This input-output data set was split into a train and test set with a 90/10 ratio. The NN model was then trained via stochastic gradient decent (sgd) to a fit performance of $9.16 \cdot 10^{-5}$ MSE, and $6.9 \cdot 10^{-3}$ MAE. This model was then validated on the test set with a performance of $9.3 \cdot 10^{-5}$ MSE and $6.3 \cdot 10^{-3}$ MAE, indicating that a predictive model was generated that has acceptable performance for this application.

The multiparametric MPC (mpMPC) for this process can be seen in eqn. 5. Here the objective is to maximize the biomass production at each time period, i. The predictive model of this MPC is based on the NN that was generated in the previous step. The states of the system are bounded above and below, as is the input action. For this case study, the control and output horizon are 7-time steps.

$$\max_{x,s,d} \sum_{i=0}^{N} x_i$$

$$
\begin{aligned}
\text{s.t.} \quad x_{i+1}, s_{t+1} &= NN(x_t, s_t, D_i), \quad \forall i \in \{0, \dots, N-1\} \\
0.1 &\leq x_i \leq 2.0, \quad \forall i \in \{0, \dots, N\} \\
0.1 &\leq s_i \leq 2.0, \quad \forall i \in \{0, \dots, N\} \\
0.05 &\leq D_i \leq 0.6, \quad \forall i \in \{0, \dots, N-1\} \\
x_0 &= \theta_0, s_0 = \theta_1, \theta \in \Theta
\end{aligned}
$$

(5)

Firstly, the proposed bounds-tightening procedure is applied to the NN separate from the overall optimization problem. Here tight upper and lower bounds on the auxiliary variables were found that are valid for all problem realizations, $\theta \in \Theta$. This tightening of the individual NN took only 11.5 seconds. These tightened bounds were then used when generating the Big-M reformulation of the mpMPC. It was found that 12 of the 45 auxiliary binary variables in each NN were only feasible at either 0 or 1 and thus could effectively be removed from every problem instance.

The MILP-reformulated mpMPC has 701 continuous variables, 315 binary variables, and 2,221 constraints. Tight upper and lower bounds were then found for x_i, s_i, and D_i, as well as sums of these variables, e.g., $\sum x_i$, which are valid for any problem realization, $\theta \in \Theta$. These bounds were added as constraints to the reformulated mpMPC. This once-and-offline bounds-tightening process took 2,941 seconds.

The effectiveness of the once-and-offline bounds-tightening procedure is shown in a control case study where the computational burden of the ReLU NN-based MPC is compared with and without the proposed bounds-tightening procedure. For these computational experiments, we used Gurobi 10.0 as the MILP solver and used a computer with an Intel i7-12700k and 32GB DDR4 RAM. Here, we take the system at the initial state realization of $\theta = (0.2, 0.5) = (x_0, s_0)$. Without the proposed bounds-tightening procedure applied, the MPC problem hit the time-out limit of 8,000 seconds, explored 21,976,708 nodes, and performed 334,045,961 simplex iterations with a gap of 9.9%. Utilizing the proposed bounds-tightening procedure where valid tight bounds are generated for all problem realizations ahead of time, the ReLU-based NN MPC problem was solved to proven optimality in only 5.36 seconds, explored 14,046 nodes, and performed 205,700 simplex iterations. The predictive performance of the NN model and the effect of applying the optimal inputs, D_t, that were generated by the MPC to the process can be seen in Figure 1.

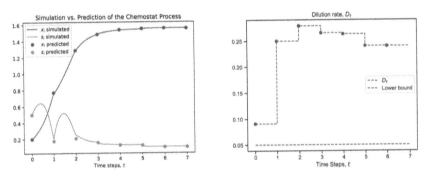

Figure 1. Evolution of the chemostat process with time given the dilution rates, D_t, from the MPC compared to the prediction of the NN inside the MPC.

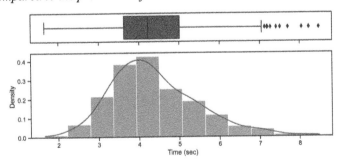

Figure 2. Histogram and box plot of time to solve each problem in the problem set.

Another computational study was performed, where 500 random initial state realizations of the chemostat system, $\theta = (x_0, s_0) \in \Theta$ were generated. The corresponding MPC problems for each initial state were solved utilizing the calculated tight bounds. Each problem was solved to optimality; if the resulting problem was found to be infeasible, the time to solve was reported as the time to prove infeasibility. The average time to solve the pre-tightened MPC problems was 4.37 seconds, with a maximum of 8.48 seconds.

The distribution of the solve times for these control problems is presented in Figure 2. As a comparison, 5 problems were attempted without the improved bounds; however, they all timed out at 8,000 seconds without a proven optimal solution. As shown, this proposed procedure leads to significant reductions in the time to solve, especially when compared to the case where these precomputed bounds are not provided.

4. Conclusions

In this work, a once-and-offline bounds-tightening procedure for ReLU-based NN multiparametric programming problems was developed. This method finds tight bounds for every auxiliary variable introduced via the Big-M reformulation that is valid for every realization of the parametric uncertainty set, $\theta \in \Theta$, allowing for the reuse of these bounds between problems. In addition, bounds on the input and output variables of the NN were computed, removing the need to generate new bounds for each optimization problem. The tightened bounds on the auxiliary, input and output variables furnish a much more computationally tractable optimization problem, and this was demonstrated in the case study where a ReLU-based NN MPC was implemented for a chemostat process. The proposed method made the controller computationally feasible, taking only 5.36 seconds compared to the untightened case which timed out at 8,000 seconds.

In the future, we plan to apply the proposed tightening procedure to the partition reformulation of ReLU-based NNs as proposed by Tsay et al. (2021) to analyze if similar computational acceleration can be realized in those ReLU-based NN reformulations. We plan to apply this to other realizations of multiparametric mixed integer programming formulations. Hierarchical MPC, which consists of layers of MPC and is typically stated as a multilevel optimization problem, is of keen interest to the authors. In the case of two controllers where the lower-level controller is a convex Quadratic Program, this can be reformulated as a multiparametric mixed integer program that could have the same tightening procedure applied to it.

References

A. Bemporad, M. Morari, V. Dua, E. N. Pistikopoulos, 2000. The explicit solution of model predictive control via multiparametric quadratic programming 2, 872–876 vol.2.

J. W. Eaton, J. B. Rawlings, 1992. Model-predictive control of chemical processes. Chemical Engineering Science 47 (4), 705–720.

M. Fischetti, J. Jo, 2018. Deep neural networks and mixed integer linear optimization. Constraints 23 (3), 296–309.

B. Grimstad, H. Andersson, 2019. Relu networks as surrogate models in mixed-integer linear programs. Computers & Chemical Engineering 131, 106580.

J. Katz, I. Pappas, S. Avraamidou, E. N. Pistikopoulos, 2020. Integrating deep learning models and multiparametric programming. Computers & Chemical Engineering 136, 106801.

I. Pappas, D. Kenefake, B. Burnak, S. Avraamidou, H. S. Ganesh, J. Katz, N. A. Diangelakis, E. N. Pistikopoulos, 2021. Multiparametric programming in process systems engineering: Recent developments and path forward. Frontiers in Chemical Engineering 2, 32.

E. N. Pistikopoulos, N. A. Diangelakis, R. Oberdieck, 2021. Multi-parametric optimization and Control. Wiley.

C. Tjandraatmadja, R. Anderson, J. Huchette, W. Ma, K. K. Patel, J. P. Vielma, 2020. The convex relaxation barrier, revisited: Tightened single-neuron relaxations for neural network verification. Advances in Neural Information Processing Systems 33, 21675–21686.

C. Tsay, J. Kronqvist, A. Thebelt, R. Misener, 2021. Partition-based formulations for mixed-integer optimization of trained relu neural networks. Advances in Neural Information Processing Systems 34, 3068–3080.

Antonis Kokossis, Michael C. Georgiadis, Efstratios N. Pistikopoulos (Eds.)
PROCEEDINGS OF THE 33rd European Symposium on Computer Aided Process Engineering
(ESCAPE33), June 18-21, 2023, Athens, Greece

Multistage Economic NMPC for Gas Pipeline Networks with Uncertainty

Sakshi Naik,[a,*] Robert Parker,[a] Lorenz T. Biegler,[a]

[a]*Carnegie Mellon University, Pittsburgh, PA 15289, USA*
*ssnaik@andrew.cmu.edu

Abstract

Gas pipelines form complex highly integrated networks to transport natural gas with dynamic operation due to time varying demands, composition, and ambient conditions. These can be modeled and optimized through non-linear optimal control problems with model equations, operation bounds and uncertainty descriptions. In this study, we use a multistage Economic Nonlinear Model Predictive Controller (eNMPC) to find optimal operational policies for networks with dynamic demands and uncertain parameters. This approach relies on constructing a scenario tree by generating extreme cases of the uncertain parameters with separate control sequences to address constraint violations in each case. For demonstration, we assume that the efficiency of the compressors is uncertain. Under this uncertainty we show that standard eNMPC, designed with a nominal value of compressor efficiency, violates constraints in the plant. We demonstrate that multistage eNMPC prevents constraint violations in all uncertain realizations and provides a promising robust control strategy for these networks.

1. Introduction

Natural gas is the cornerstone of the electric grid and significant increases in consumption are projected in the coming years. Gas is transported from suppliers to consumers using large pipeline networks, which operate under time varying conditions such as temperature, customer demands and gas composition. Pipeline networks typically consist of supply nodes, demand nodes, compressor stations and individual lines of various diameters and lengths. Compressor stations are used to supply boost pressure to the gas to compensate for friction losses, and this is a major operating cost (Liu et al. (2020)). Moreover, a major issue in pipeline operation is the presence of uncertainty. Liu et al. (2020) used a robust optimization approach to optimize gas networks with demand and composition uncertainty. Also, two-stage stochastic optimal control for pipeline networks has been performed by Zavala (2014) with a focus on uncertainty in gas demands. However, this work assumes that uncertain profiles will resolve in time after a fixed horizon, but it was not implemented in a moving horizon approach. In contrast, we consider a multistage MPC in which the value of the uncertain parameter is revealed at each sampling time. Lin et al. (2022) demonstrated the effectiveness of multistage NMPC to handle uncertainty in pumping treatment in hydraulic fracturing. The advantage of using an eNMPC over tracking NMPC for optimal pipeline operation was demonstrated in a deterministic setting by Gopalakrishnan and Biegler (2013). In this paper, we extend this approach through multistage eNMPC for optimal control of gas-pipeline networks under uncertainty.

There are multiple uncertain parameters that can be considered in the operation of gas pipelines such as uncertainty in demand, gas composition, compressor efficiency, etc.

Although the concept of multistage eNMPC can be generalized to any uncertain parameter, we choose the uncertainty in compressor efficiency to demonstrate the results of this work. The uncertainty in compressor efficiency is due to time-varying flow of gas through the compressors and ambient conditions of operation such as temperature and relative humidity. Moreover, compression unit failures are extremely challenging for gas transmission systems and a significant issue for gas operators (Tran et al. (2018)). Using multistage eNMPC to control gas-pipeline networks with uncertainty is therefore a promising direction for further research.

2. Model Formulation

We first define the pipeline momentum and mass balances, compressor equations, node balances and equations that describe the network interconnections (Zavala (2014)). We assume isothermal gas flow through a horizontal pipe. The momentum balance and mass balance equation for a given line $l \in L$ with a given value of parametric uncertainty $w \in \Omega:\{1,2,..,N_\Omega\}$ is given by:

$$\frac{1}{A_l}\frac{dF_l(t,x,w)}{dt} + \frac{dP_l(t,x,w)}{dx} + \frac{8\lambda_l c^2}{\pi^2 D_l^5}\frac{F_l(t,x,w)|F_l(t,x,w)|}{P_l(t,x,w)} = 0 \quad (1) \qquad \frac{dP_l(t,x,w)}{dt} + \frac{c^2}{A_l}\frac{dF_l(t,x,w)}{dx} = 0 \quad (2)$$

where $t \in T$ and $x \in X$ are points in time and space domains respectively. P_l and F_l refer to the pressures and flows in line l and c is the speed of sound of the gas. A_l is the cross-sectional area of the pipeline and D_l is the diameter of the pipeline l. Area and friction factor (λ_l) equations are:

$$A_l = \frac{\pi D_l^2}{4} \quad \forall l \in L \qquad\qquad (3) \qquad\qquad \lambda_l = \left[2\log_{10}\left(\frac{3.7\ D_l(mm)}{(\text{rugosity})}\right)\right]^{-2} \quad (4)$$

In the pipe network we consider the node set $\mathcal{N}$ and line set L. For each node $n \in \mathcal{N}$ we define a set of supplies S_n and demands D_n. L^{in} is the subset of lines bringing gas to node n and L^{out} is the subset of links which take out gas from node n. $s_i(t,w)$ represents the actual supply flowrate from supply i and $d_j(t,w)$ represents the actual demand flowrate from demand j. The flow balances at the nodes are given by:

$$\sum_{i\in S_n} s_i(t,w) + \sum_{l\in L_n^{in}} F_l^{in}(t,w) = \sum_{j\in D_n} d_j(t,w) + \sum_{l\in L_n^{out}} F_l^{out}(t,w) \qquad \forall\, n \in \mathcal{N} \quad (5)$$

F_l^{in} and P_l^{in} are dummy flows linked to the pipeline flows by the boundary equations:

$$F_l(t,\bar{L}_l,w) = F_l^{in}(t,w) \quad (6) \qquad F_l(t,0,w) = F_l^{out}(t,w) \qquad \forall l \in L, w \in \Omega, t \in T \quad (7)$$

where $\bar{L}_l$ is the length of line l. In addition to the pipelines and the nodes, we have compressor stations associated with some pipelines. We define a set of pipelines which have a compressor associated with them as L_a (active lines) and a set of pipelines which don't have a compressor as L_p (passive lines). The power consumed in the active lines L_a is given by:

$$\text{Power}_l(t,w) = T_{in}(t,w)c_p\frac{F_l(t,0,w)}{\eta_l(t,w)}\left[\left[\frac{P_{in,l}(t,w)+\Delta P_l(t,w)}{P_{in,l}(t,w)}\right]^\beta - 1\right] \quad \forall l \in L_a, w \in \Omega, t \in T \quad (8)$$

T_{in} is the temperature of the inlet stream. $\eta_l(t,\ w)$ is the efficiency of the compressor in line l which is an uncertain parameter. β is the compression coefficient which is calculated as $\beta = \frac{c_p-c_v}{c_p}$. The boundary conditions for the pipeline pressures are:

$$P_n^{rec}(t,w) = P_l(t,\bar{L}_l,w) \qquad\qquad\qquad \forall l \in L, t \in T, w \in \Omega \qquad (9)$$
$$P_l(t,0,w) = P_{in,l}(t,w) \qquad\qquad\qquad \forall l \in L_p, t \in T, w \in \Omega \qquad (10)$$
$$P_l(t,0,w) = P_{in,l}(t,w) + \Delta P_l(t,w) \qquad\quad \forall l \in L_a, t \in T, w \in \Omega \qquad (11)$$

P^{rec} are the pressures received at the nodes from the pipelines. The time domain t is discretized using an implicit Euler method with equally spaced points each having Δt length. We represent each point in the discretized time set by the index τ. The space domain along the length of each line is discretized using a forward discretization scheme with N_x equally spaced points with each segment having a length of Δx. We define $\bar{X} = \{1,2,..,N_x-1\}$. Spatial and temporal discretizations were chosen based on simulation results obtained from the model. The differential states from the resulting differential-algebraic equations (DAEs) correspond to spatially discretized flows and pressures. The algebraic state equations include the node balances, compressor power expression and the boundary conditions. A dynamic optimization problem is formulated to minimize the boost cost of the compressor while meeting the demands without depleting the linepack inventory (gas stored in the pipelines at the end of operation). Fixing the supply pressure and demand flows leaves us with the compressor boost pressures as the control variables.

2.1. eNMPC Formulation

eNMPC provides a sequence of controls for a given horizon. The first control input is implemented in the plant. The new state of the plant is then sampled and acts as a new starting point for the controller. The horizon of the controller is then advanced by one sample time and the process repeats again. We consider an eNMPC horizon of 24 hours and a total time of 48 hours. The points on the discretized horizon are defined by the set

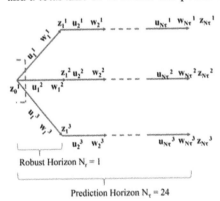

Robust Horizon $N_r = 1$

Prediction Horizon $N_\tau = 24$

Figure 1. Scenario tree with a robust horizon $N_r = 1$ for multistage eNMPC

$\hat{T} = \{1,2,..,N_\tau\}$. The goal of the optimization problem is to minimize the boost cost of the compressor while meeting the demands, without completely depleting the inventory. The boost cost is calculated based on the power of the compressor and the cost of electricity. Assuming the cost of electricity is C_{pr} \$/kWh, the boost cost can be calculated as:

$$J(w) = \sum_{l \in L_a} \sum_{\tau \in \hat{T}} C_{pr} \times Power_{l,\tau}(w)\Delta t \quad (12)$$

To meet the demands we have equality constraints between the target demands and the actual demands. To avoid depletion of the linepack inventory, we use a lower bound on linepack flow and pressure. The optimization problem for each control horizon is formulated as:

$$\min_{\Delta P_{l,\tau}(w)} \sum_{l \in L_a} \sum_{\tau \in \hat{T}} C_{pr} \times Power_{l,\tau}(w)\Delta t$$

$d_{j,\tau}^{target} - d_{j,\tau}(w) = 0$ $\forall \tau \in \hat{T}, \forall j \in D_n, \forall n \in \mathcal{N}, \forall w \in \Omega$] Demand constraints

$P_{l,N_\tau,k}(w) \geq P_{LB}$ bar $\forall l \in L, \forall k \in \{2,3,..,N_x\}, \forall w \in \Omega$] Terminal

$F_{l,N_\tau,k}(w) \geq F_{LB}$ kmol/hr $\forall l \in L, \forall k \in \bar{X}, \forall w \in \Omega$] bounds

$S_{LB} \leq s_{i,\tau}(w) \leq s_{UB}$ $\forall i \in S_n, \forall n \in \mathcal{N}, \forall \tau \in \hat{T}, \forall w \in \Omega$] Variable

$P_{LB}^{rec} \leq P_{j,\tau}^{rec}(w) \leq P_{UB}^{rec}$ $\forall j \in D_n, \forall n \in \mathcal{N}, \forall \tau \in \hat{T}, \forall w \in \Omega$ } bounds

$0 \leq Power_{l,\tau}(w) \leq Power_{UB}$ $\forall \tau \in \hat{T}, \forall w \in \Omega$

Constraints (1)-(12), discretized in $\hat{T}$ and $\hat{X}$

The above constraints are labelled as (13). This formulation holds for the dynamic optimization of a pipeline network where the value of the uncertain parameter is known. In case of uncertain efficiency it is useful to implement Multistage eNMPC as it considers explicit realization of uncertainty by considering a scenario tree. In accordance with the standard setup of multistage eNMPC (Lucia et al. (2013)) we choose three realizations of the uncertain parameter: minimum, nominal and maximum in set $\Omega = \{w^{min}, w^{nom}, w^{max}\}$. Non-anticipativity constraints enforce equal values of the control variables for branches b and $\bar{b}$ with the same parent node. Figure 1 shows a scenario tree with a robust horizon $N_r = 1$. The level of uncertainty stays the same after the robust horizon. This allows us to have only two non-anticipativity constraints and the problem size remains tractable. Since the efficiency at each time point is an uncertain parameter, we have N_τ uncertain parameters. The optimization problem for a multistage NMPC is then formulated as:

$$\min_{\Delta P(w_\tau^b)_{\{\tau \in \mathcal{T}\}}} \sum_{w_\tau^b \in \Omega} \frac{1}{|\Omega|} J(w_\tau^b)$$

$$\left.\begin{array}{ll} \Delta P_\tau^b = , \Delta P_\tau^{\bar{b}} & \text{if } , z_{\tau-1}^b = z_{\tau-1}^{\bar{b}} \quad \forall\, b, \bar{b} \in B, \forall\, \tau \in \{1, ..., N_r\} \\ w_\tau^b = w_{\tau+1}^b & \forall\, \tau \in \{N_{r+1}, ..., N_{\tau-1}\} \end{array}\right\} \quad \text{Nonanticipativity constraints}$$

Where B denotes the set of all branches in the scenario tree. $z_\tau^b, \Delta P_\tau^b, w_\tau^b$ are the states, controls, and uncertain parameters at stage τ in branch b. The objective function is the average of the objectives of the nominal NMPC. The system is modeled in IDAES (Lee et al. (2021)), with Pyomo DAE (Nicholson et al. (2018), Bynum et al. (2021)) and is solved using IPOPT version 3.13.2 (Wächter and Biegler (2006)).

3. Model Formulation

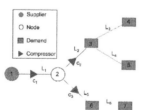

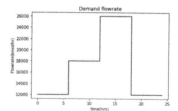

Figure 2(a) Gas pipeline network   Figure 2(b) Dynamic demand profile at
                                                      demand nodes

A pipeline network with five demands, one supply and three compressor stations is considered to demonstrate the proposed approach (Liu et al. (2020)). Figure 2(a) shows the network structure while the gas flow demand for each demand node is shown Figure 2(b). It is assumed that the profile repeats over a period of 24 hours. Node 1 is the supply node with a fixed gas supply pressure of 100 bar. Node 2 is a junction node with only inlet and outlet pipeline with no demand or supply. It is assumed that at the initial state, compressor c_1 provides a boost pressure of 20 bar. This is necessary to obtain a consistent initial condition for our dynamic system. We use an upper bound of 60 MW on compressor power. A standard NMPC is used on this network to find the optimal control

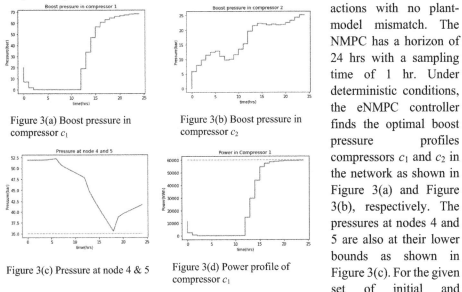

actions with no plant-model mismatch. The NMPC has a horizon of 24 hrs with a sampling time of 1 hr. Under deterministic conditions, the eNMPC controller finds the optimal boost pressure profiles compressors c_1 and c_2 in the network as shown in Figure 3(a) and Figure 3(b), respectively. The pressures at nodes 4 and 5 are also at their lower bounds as shown in Figure 3(c). For the given set of initial and

Figure 3(a) Boost pressure in compressor c_1

Figure 3(b) Boost pressure in compressor c_2

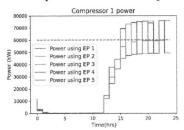

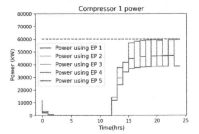

Figure 3(c) Pressure at node 4 & 5

Figure 3(d) Power profile of compressor c_1

boundary conditions, there is no boost pressure required in compressor c_3. Figure 3(d) shows that the power in compressor c_1 is at its bound and is likely to experience bound violations under uncertainty in compressor efficiency.

3.1. Introducing uncertainty in plant

Now we consider uncertainty in the efficiencies of all three compressors. Five randomly generated efficiency profiles are applied, with efficiency at each time point taking one of the three extreme values {0.55, 0.7, 0.85}. The standard eNMPC modeled with nominal compressor efficiency of 0.7 is unaware of this uncertainty and therefore is likely to violate the power bound in multiple efficiency realizations. It can be seen from Figure 4(a) that the power bound in compressor c_1 is violated beyond 15 hours with standard

Figure 4(a) Compressor c_1 power profiles using standard eNMPC with plant-model mismatch

Figure 4(b) Compressor c_1 power profiles using multistage eNMPC with plant-model mismatch

eNMPC. (The different colors indicate the power profiles obtained in each uncertain scenario). Multistage eNMPC as shown in Figure 4(b) doesn't violate the power bound in any case. The average cost obtained using standard eNMPC was US$37885 while a multistage eNMPC predicted a cost of US$42963. The optimal solution obtained using the multistage eNMPC is 13.4% more expensive compared to the standard eNMPC, this represents the cost of handling uncertainty. The multistage eNMPC finds alternate control

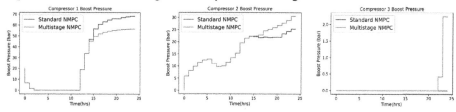

Figure 5. Boost pressure profiles obtained using standard eNMPC vs Multistage eNMPC

profiles which are robust to uncertainty and don't lead to bound violations in the plant. Figure(5) shows a comparison between the controls found by the multistage eNMPC and the standard eNMPC. We see that multistage eNMPC predicts the same control sequence as a standard eNMPC until 15 hrs. Beyond 15 hrs, multistage eNMPC suggests that a lower boost pressure should be used in compressor c_1 to avoid the power bound violation. But, to supply gas above the minimum demand pressure at nodes 4 and 5, a higher boost pressure is used in compressor c_2. Also, unlike the deterministic case, Compressor c_3 provides boost pressure, which is required to satisfy the lower bounds on terminal pressure.

4. Conclusions and Future Work

Standard eNMPC with fixed, nominal compressor efficiency finds the optimal control sequence without violating any bounds. On the other hand, when standard eNMPC encounters parametric uncertainty, we observe bound violations due to plant-model mismatch. The multistage eNMPC approach results in a control policy which does not violate any bounds in the uncertain plant. In the future, we plan to apply these eNMPC strategies to larger, more complex problems and investigate how they scale up to real world GASLIB networks (Schmidt et al. (2017)).

References

M.L.Bynum, 2021, Pyomo–optimization modeling in python, 3rd Edition. Vol. 67. Springer Science & Business
A. Gopalakrishnan, 2013. Economic Nonlinear Model Predictive Control for periodic optimal operation of gas pipeline networks. Computers & Chemical Engineering 52.
A. Lee, 2021, The IDAES process modeling framework and model library-Flexibility for process simulation and optimization. Journal of Advanced Manufacturing and Processing 3 (3), e10095
K.-H. Lin, 2022, Multistage nonlinear model predictive control for pumping treatment in hydraulic fracturing. AIChE Journal 68 (3), e17537,
K. Liu, 2020, Dynamic optimization of natural gas pipeline networks with demand and composition uncertainty. Chemical Engineering Science 215, 115449.
S. Lucia, 2013. Multi-stage Nonlinear Model Predictive Control Applied to a Semi-batch Polymerization Reactor under Uncertainty. Journal of Process Control 23, 1306–1319.
B. Nicholson, 2018, Pyomo.DAE: a modeling and automatic discretization framework for optimization with differential and algebraic equations. Mathematical Programming Computation 10 (2), 187–223.
M. Schmidt, 2017. GasLib—A Library of Gas Network Instances. Data 2 (4), 40, number: 4 Publisher: Multidisciplinary Digital Publishing Institute.
T. H. Tran, 2018. Linepack planning models for gas transmission network under uncertainty. European Journal of Operational Research 268 (2), 688–702.
A. Wachter, 2006. On the implementation of a primal-dual interior point filter line search algorithm for large-scale nonlinear programming. Mathematical Programming 106 (1), 25–57.
V. M. Zavala, 2014. Stochastic optimal control model for natural gas networks. Computers & Chemical Engineering 64, 103–113.

Antonis Kokossis, Michael C. Georgiadis, Efstratios N. Pistikopoulos (Eds.)
PROCEEDINGS OF THE 33rd European Symposium on Computer Aided Process Engineering
(ESCAPE33), June 18-21, 2023, Athens, Greece

A new time-bucket MILP formulation for optimal lot-sizing and scheduling of an industrial make-and-fill process

Roderich Wallrath [a,b], Florian Seanner [c], Matthias Lampe [a], Meik Franke [b]

[a]*Bayer AG, Kaiser-Wilhelm Allee 1, 51368 Leverkusen, Germany*
[b]*University of Twente, Faculty of Science and Technology, Sustainable Process Technology, Process Design and Optimization, Drienerlolaan 5,*
7522 NB Enschede, The Netherlands
[c]*SimPlan Systems GmbH, Sophie-Scholl-Platz 6, 63452 Hanau, Germany*

r.wallrath@utwente.nl

Abstract

We propose a new time-bucket MILP model for lot-sizing and scheduling problems arising in multistage production processes. The time-bucket model benefits from advantages of both continuous and discrete time representations while overcoming their shortcomings. In particular, we show how the time-bucket model allows to easily include a variety of typical, important real-world constraints, can be solved with moderate computer effort, and thus promotes MILP for large-scale, industrial problems. To illustrate that, we apply the time-bucket approach to the flow shop problem of a batch formulation and filling process from an industrial pesticide production. We reconcile the MILP solution with a validated discrete event simulation (DES) model of the process to obtain optimal and real-world feasible results. A comparison of the MILP-DES solution to a manually optimized solution for a one-month production data set shows that more than 17% of production capacity can be freed up and significant improvement in on-time delivery.

Keywords: Production planning and scheduling; mixed-integer linear programming; discrete event systems in manufacturing; modeling of manufacturing operations.

1. Introduction

Production scheduling of large-scale, industrial processes is widely recognized as a challenging optimization task both by academic and industrial researchers. With an increasing availability of real-time process data, ever-growing computational capabilities, and the advance of digital twins, optimal production scheduling has gained new relevance. However, there is still a large gap between industrial reality and academic research. As pointed out in Harjunkoski (2016), practitioners are confronted with many different challenges, when building their solutions in an industrial environment. While mixed-integer programming (MIP) is a well-established optimization technique in asset-heavy industries such as petrochemicals Castro et al. (2018), and discrete manufacturing industries such as semiconductors Qin et al. (2019) and automotive Gnoni et al. (2003), it is relatively new to industries such as specialty chemicals Borisovsky et al. (2019), pharma Sarin et al. (2014); Costa (2015), and consumer goods Sel et al. (2015); Clark et al. (2011). Alternatives such as discrete event simulation (DES) address natural drawbacks of MILP such as the curse of dimensionality but do not provide bounds on the solutions. Despite research efforts with MIP and DES Frazzon et al. (2016); Nikolopoulou and Ierapetritou (2012), real-world scheduling problems with dozens of products, multiple production lines, changeover costs, and lot-sizing still can only be solved to sub-optimality or require massive computational resources. Herein we propose a new time-bucket mixed-integer linear programming (MILP) model for lot-sizing and scheduling problems arising in multistage production processes. The main idea behind time-bucket models based on Fleischmann and Meyr (1997) is to partition time into fixed-length macro-periods and flexible length micro-periods that lie within the macro periods. This model formulation allows to easily include a variety of typical, important real-world constraints and can be solved with moderate computer resources. We apply our approach to a real-world case study of a large-scale industrial make-and-fill process of an agrochemical production plant, and show that our modelling approach is able to capture the different characteristics of the problem efficiently. We analyse the evolution of the bounds to ensure that the exact search procedure of a MIP-solver can be truncated after short time with near-optimal solutions. In this way, a 1-month production data set can be solved with a moderate computational resources and within 1 hour solution time.

2. Time-bucket MILP formulation

2.1 Time-bucket approach

In the time-bucket MILP model mass balances, resource allocations, shift schedules, and due dates are modeled in the fixed time grid, while production quantities and changeover times are modeled in the flexible time grid. The fixed time grid consists of macroperiods, while the flexible time grid consists of microperiods. Macroperiods are fixed time intervals with uniform length, that are defined globally. As shown in Figure 2, there is a configurable number of microperiods $s \in S_t$ in each macroperiod $t \in T$. Microperiods are flexible time intervals with variable length, that are defined per processing resource. Binary changeover variables $y_{pp'ls}$ indicate changeovers from product p to p' on processing resource l in microperiod s. Each microperiod consists of a fixed sequence of 4 phases, set-up start, production, idle, and set-up end as shown in Figure 1.

Figure 1: Microperiods consist of setup start, production, idle and setup end phase.

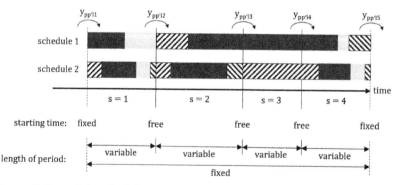

Figure 2: 2 possible schedules on a resource l in one macroperiod with 4 microperiods s.

The length of phase production is determined by the production quantity q_{pls} of a product p in microperiod s on a processing resource l. The lengths of phases set-up start and set-up end are determined by the sequence-dependent changeover times before and after the production. The phase idle provides an optional interval for idle time. *Example.* We consider a processing resource l in a macroperiod with 4 microperiods as shown in Figure 2. In *schedule1* two production phases take place. The same time structure also allows to model 3 production phases as shown in *schedule2*. The changeover times must be taken into account using the setup start and setup end phases. If no changeover time is required the production phase can cover full microperiods. Similarly, long changeover times and idle times can be modeled. Since the microperiods are defined per processing resource l and have variable lengths the time bucket formulation provides a highly flexible time representation for MILP models.

2.2 MILP formulation

A crucial variable of the MILP model are production orders q_{pls}. They express that a quantity of a product $p \in P$ is produced on processing resource $l \in L$ during the microperiod $s \in S_t$ of macroperiod $t \in T$. Production orders seize processing resources for a minimum duration which depends on (1) the product p because of product dependent processing parameters (for example viscosity), (2) the resource l because of resource dependent processing parameters (for example milling efficiency) and (3) the quantity of the production order, which must be greater than a single batch size $q_{min,l}$ at resource l. To describe (1) and (2) a production coefficient $a_{p,l}$ is defined, which corresponds to the production rate of product p on the resource l. A changeover time $ct_{pp'l}$ for subsequent products p and p' is defined for each resource l. The corresponding binary changeover variable $y_{pp'ls}$ indicates whether a changeover from product p to p' takes place during the microperiod s on resource l. Furthermore, we distinguish intermediate products $p \in IP \subset P$ and finished products $\hat{p} \in FP \subset P$. Since an intermediate product can be converted into different finished products, bill of material coefficients $b_{p\hat{p}}$ are defined as indicator parameters. Demands $D_{\hat{p}t}$ are defined with respect to finished product $\hat{p}$ in macroperiod t. Similarly, the backlog quantities $B_{\hat{p}t}$ are defined for finished product $\hat{p}$ in macroperiod t.

2.3 Objective function

We suggest a 2-step, monolithical solution approach with warm start. In the first optimization step, we define the minimization of all backlog quantities $B_{\hat{p}t}$ as the objective as shown in Equation 1.

$$\min \sum_{\hat{p} \in FP} \sum_{t \in T} B_{\hat{p}t} \tag{1}$$

In the second optimization step, the binary changeover variables $y_{pp'ls}$ are initialized with the result of the first step. The objective of the second step is the minimization of the total changeover time $y_{pp'ls} \, ct_{pp'l}$ as shown in Equation 2.

$$\min \sum_{\substack{s \in S_t \\ t \in T}} \sum_{\substack{p,p' \in P \\ l \in L}} y_{pp'ls} \, ct_{pp'l} \tag{2}$$

3. Case Study

We consider the scheduling problem of a flexible flow shop in an industrial manufacturing plant for crop protection chemicals as shown in Figure 3. The problem originates from the last two stages in the manufacturing process of pesticide chemicals, which are also known as product formulation and filling. The formulation and filling process takes place in a multi-product batch plant with 2 stages. In the first stage mixing, milling and reactions of raw materials take place, which results in bulk product. In the second stage, bulk product is filled into final containers. The first and second stage each consist of 7 individual processing lines. We consider 50 unpackaged products and 83 finished products. We assume product-sequence and line-specific changeover costs, and minimum lot-sizes. There are 7 optional buffer tanks to decouple the two stages and maximize their utilization. Different shift schedules for the two stages determine how many lines can be operated simultaneously. The bottleneck of the process varies with production demand and the shift schedules.

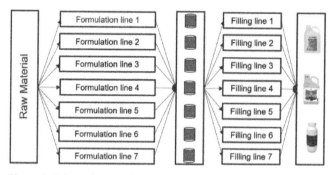

Figure 3: Schematic overview of the formulation and filling process.

4. Results

4.1 MILP Solution

We optimize the formulation and filling process for a 1-month production period. For the original demand data, we find a solution with no production backlog and 110 hours of changeover time with a remaining gap of 27.3% after a total solution time of 1.22 hours as further specified in Table 1 . Optimization step 1 shows that the 1-month production demand $D_{\beta t}$ can be satisfied without backlog in a makespan less than 31 days after 805 seconds solution time. Consequently a no-backlog constraint $\Sigma B_{\beta t} = 0$ is enforced in the second optimization step, which minimizes the total changeover time to a value of 110 hours after 3600 seconds solution time. For the alternative demand scenarios we observe similar results.

Table 1: MILP results for different 1-month demand scenarios.

Set	Changeover (h)	Gap (%)	1st (s)	2nd (s)
Org	110	27.3	805	3600
2	130	35.1	659	3600
3	134	38.8	1033	3600
4	120	26.0	739	3600
5	120	21.8	828	3600

Since the optimality gap of the second optimization step is greater than 0 for all instances, we investigate the evolution of bounds and gap in the second optimization step over the course of 4 hours. As shown in Figure 4, the remaining gap is decreasing further for all instances after 3600 seconds. However, that is mainly due to the improvement of the lower bounds. Near-optimal solutions are found relatively quickly as indicated by the upper bounds in Figure 5. Consequently, the MIP-search can be truncated early despite relatively large remaining gaps. We conclude that the proposed MILP formulation provides near optimal solutions within 30 minutes solution time, but a proof of optimality is costly due to weak lower bounds.

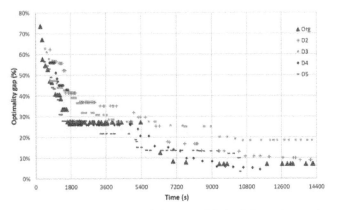

Figure 4: Evolution of the remaining gaps (%) for 5 demand scenarios.

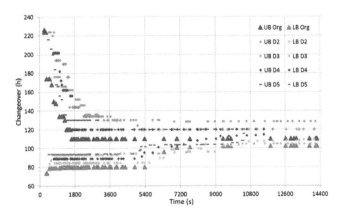

Figure 5: Evolution of the upper and lower bounds for 5 demand scenarios.

4.2 DES validation

For validation the MILP result is reconciled using a validated DES model of the process. We compare the resulting MILP-DES schedule with a manually optimized DES (manual-DES) schedule. To obtain the latter, production experts generate campaign lists and refine them iteratively using the DES model. From the makespans of the MILP-DES schedule and the manual-DES in Table 2, we conclude that the MILP approach is able to generate schedules that do not violate the delivery dates at the end of the month, while the manually optimized schedule contains backlog of approximately 2%. The adherence to delivery dates and demand quantities is ensured by minimizing the total backlog in the first optimization step (see Equation 1). When comparing the occupancy times of all lines in the formulation and filling stage, we observe that MILP-DES approach reduces the number of occupancy time by a total of 1218 machine hours. Approximately 10% formulation capacity and 27% filling capacity is freed as a result of minimizing the total changeover times in the second optimization step (see Equation 2).

Table 2: DES reconciled results.

Approach	Bcklg (%)	Mkspan (d)	Form (h)	Fill (h)
MILP-DES	0	31	3309	2251
manual-DES	1.988	33.75	3674	3104

A direct comparison of the schedules shows that the MILP-DES solution in Figure 7 contains a higher number changeovers on the filling lines to synchronize with the formulation stage. As a result of this synchronization less changeover costs are incurred on the formulation lines and no backlog is produced (see Table 2). A high number of changeovers on the filling lines also is justified by the fact that they are relatively inexpensive compared to formulation changeovers because filling lines consist of a smaller number of machines and less cleaning effort is needed.

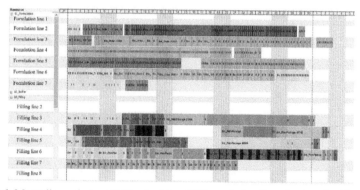

Figure 6: Manually-optimized DES schedule of 7 formulation lines (top) and 7 filling lines (bottom).

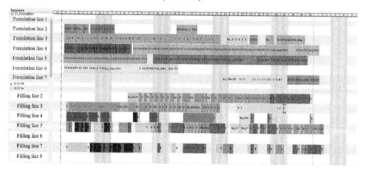

Figure 7: MILP-optimized DES schedule of 7 formulation lines (top) and 7 filling lines (bottom).

5. Conclusion

We showed that scheduling problems can be solved using a new time-bucket MILP formulation, which allows to include important real-world parameters such as lot-sizing, sequence dependent changeovers, buffer of intermediate bulk product, and operator schedules. We demonstrated the applicability of the time-bucket formulation with an industrial case study of a 2-stage agrochemical production process, for which we proposed a 2-step, monolithic solution procedure of the MILP model followed by a DES reconciliation step. We compared results of the MILP-DES approach with the results of a DES-aided, manual optimization approach, and observed that counterintuitive yet better schedules are found with respect to production backlog and resource utilization. We conclude that the time-bucket formulation represents viable industrial solution, that can be adopted for similar problems.

References

P. Borisovsky, A. Eremeev, J. Kallrath, 06 2019. Multi-product continuous plant scheduling: combination of decomposition, genetic algorithm, and constructive heuristic. International Journal of Production Research 58, 1–19.

P. M. Castro, I. E. Grossmann, Q. Zhang, 2018. Expanding scope and computational challenges in process scheduling. Computers & Chemical Engineering 114, 14–42, fOCAPO/CPC 2017.

A. Clark, R. Morabito, E. A. Toso, 02 2011. Production setup-sequencing and lot-sizing at an animal nutrition plant through atsp subtour elimination and patching. J. Scheduling 14, 119.

A. Costa, 2015. Hybrid genetic optimization for solving the batch-scheduling problem in a pharmaceutical industry. Computers & Industrial Engineering 79, 130–147. B. Fleischmann, H. Meyr, Mar. 1997. The general lotsizing and scheduling problem 19 (1), 11–21.

E. M. Frazzon, A. Albrecht, P. A. Hurtado, 2016. Simulation-based optimization for the integrated scheduling of production and logistic systems. IFAC-PapersOnLine 49 (12), 1050–1055, 8th IFAC Conference on Manufacturing Modelling, Management and Control MIM 2016.

M. Gnoni, R. Iavagnilio, G. Mossa, G. Mummolo, A. Di Leva, 2003. Production planning of a multi-site manufacturing system by hybrid modelling: A case study from the automotive industry. International Journal of Production Economics 85 (2), 251–262, supply Chain Management.

I. Harjunkoski, 2016. Deploying scheduling solutions in an industrial environment. Computers & Chemical Engineering 91, 127–135, 12th International Symposium on Process Systems Engineering & 25th European Symposium of Computer Aided Process Engineering (PSE-2015/ESCAPE-25), 31 May - 4 June 2015, Copenhagen, Denmark.

A. Nikolopoulou, M. Ierapetritou, 12 2012. Hybrid simulation based optimization approach for supply chain management. Computers & Chemical Engineering 47, 183–193.

W. Qin, Z. Zhuang, Y. Liu, O. Tang, 2019. A two-stage ant colony algorithm for hybrid flow shop scheduling with lot sizing and calendar constraints in printed circuit board assembly. Computers & Industrial Engineering 138, 106115.

S. C. Sarin, H. D. Sherali, L. Liao, 2014. Primary pharmaceutical manufacturing scheduling problem. IIE Transactions 46 (12), 1298–1314.

C. Sel, B. Bilgen, J. Bloemhof-Ruwaard, J. van der Vorst, 2015. Multi-bucket optimization for integrated planning and scheduling in the perishable dairy supply chain. Computers & Chemical Engineering 77, 59–73.

Antonis Kokossis, Michael C. Georgiadis, Efstratios N. Pistikopoulos (Eds.)
PROCEEDINGS OF THE 33rd European Symposium on Computer Aided Process Engineering
(ESCAPE 33), June 18-21, 2023, Athens, Greece

Study of Different Formulations for the Multiperiod Blending Problem Applied to Lithium Recovery from Produced Water

Daniel Ovalle[a], Joshua L. Pulsipher[a], Camilo Gomez[b], Jorge M. Gomez[c], Carl D. Laird[a], Markus G. Drouven[d], Ignacio E. Grossmann[a]

[a]*Department of Chemical Engineering, Carnegie Mellon University, Pittsburgh, PA 15213, USA*
[b]*Department of Industrial Engineering, Universidad de Los Andes, Bogota, 111711, Colombia*
[c]*Department of Chemical Engineering, Universidad de Los Andes, Bogota, 111711, Colombia*
[d]*National Energy Technology Lab, Pittsburgh, PA 15236, USA*
grossmann@cmu.edu

Abstract

Oil and gas production wastewater (i.e., produced water) may contain appreciable concentrations of rare-earth elements and critical minerals (REE/CMs), such as lithium, that can be recovered. However, each individual produced water source may have insufficient concentration or volume to meet the economic and operating requirements of the recovery facility. Therefore, there is the need to appropriately blend multiple sources to meet recovery and water reuse demands. Optimal stream mixing operation planning can be posed as a multiperiod blending problem (MPBP) to provide quantity and quality guarantees over time. We present several formulations to solve the MPBP for the recovery of REE/CMs from produced water and propose a decomposition approach that leverages strategies in general disjunctive programming to enhance its performance. We compare these proposed formulations/strategies via two illustrative case studies on recovering lithium from a network of produced water sources.

Keywords: Multiperiod Blending, Lithium Recovery, Mixed-Integer Nonlinear Programming, Decomposition

1. Introduction

Wastewater streams from oil and gas production (i.e., produced water) are expected to surpass 60 million barrels of water per day by 2030 in the U.S. alone (Wright, 2022). The treatment needed to mitigate the environmental impact of these wastewater streams is expensive due to high salinity and contaminant levels (Gaustad et al., 2021). Hence, a significant portion of produced water in the U.S. is simply disposed of via underground injection.

Notably, however, produced water sources may contain appreciable concentrations of critical minerals (CMs) such as lithium which are crucial for manufacturing electronics, pharmaceuticals, batteries, renewable energy generators, and more (Quillinan et al., 2018). Moreover, establishing a sustainable REE/CM supply chain is a critical concern of many industrial and governmental stakeholders. Thus, the recovery of REE/CMs from produced water has the potential to add sufficient economic value to incentivize the use of treatment technologies that mitigate the impact of produced water on the environment.

Operating a water network for the recovery of REE/CMs from multiple produced water sources is highly complex since REE/CM concentrations vary dynamically and geographically, there are limited storage locations, and treatment/recovery facilities have strict inlet-flow requirements (Gaustad et al., 2021). Such a system can be modeled as a classic multiperiod blending problem where source streams with time-varying compositions are blended in pooling tanks to satisfy quality and quantity demand requirements. Classic pooling formulations such as the one proposed in Haverly (1978) use bilinear terms to track compositions along the network. However, such formulations are often difficult to solve due to their inherent nonconvexity (Gounaris and Floudas, 2008). This difficulty is compounded by multiperiod models that incorporate binary variables.

In this work, we investigate several multiperiod blending problem (MPBP) formulations and solution strategies to optimally plan the delivery of lithium-rich streams for recovery and water reuse in development activities. In particular, we adapt the direct, generalized disjunctive programming, and decomposition MPBP approaches as described in Kolodziej et al. (2013) and Lotero et al. (2016) to a representative produced water network case study. It is shown that such strategies are key to guiding the treatment of a multi-enterprise produced water blending and determine whether such a system is feasible and economically viable.

2.   Problem Formulation and Solution Strategies

2.1. Basic Problem Setup and Nomenclature

The multiperiod blending problem (MPBP) is defined over a set of supply tanks $\mathcal{S}$, blending tanks $\mathcal{B}$, and demand tanks $\mathcal{D}$ interconnected by a set of edges $\mathcal{A}$. The problem seeks to determine the optimal mixing schedule over a discretized time horizon $\mathcal{T}$ that maximizes the operation profit, while meeting flow and concentration specifications for each component $q \in \mathcal{Q}$. The initial conditions for each tank $n \in \mathcal{N} := \mathcal{S} \cup \mathcal{B} \cup \mathcal{D}$ are specified via the initial level of every tank I_n^0 and its corresponding compositions C_{qn}^0. The incoming flow F_{st}^{IN} at each supply s is known with an incoming composition C_{qst}^{IN} that varies over t. The nodal inventories and flows are restricted by tank and pipeline capacities, which are denoted $[I_n^L, I_n^U]$ and $[F_{nn'}^L, F_{nn'}^U]$, respectively. Required demand varies with time and needs to satisfy specifications on the flow $[FD_{dt}^L, FD_{dt}^U]$ and concentration $[C_{qd}^L, C_{qd}^U]$. The MPBP formulations considered here assume that blending tanks do not operate at steady-state. Mixing requires that the tanks be charged at one period of time and discharged at another period (Lotero et al., 2016).

2.2. Direct MIQCP Formulation

We model the MPBP directly as a mixed-integer quadratically constrained program (MIQCP). The MIQCP formulation follows from extending steady-state pooling formulations to incorporate multiperiod scheduling. Here, binary variables $x_{nn't} \in \{0,1\}$ are introduced to indicate the existence of flow at each time period. This addition enables us to enforce a minimum flow in the pipeline, when the flow exists. We obtain the MIQCP formulation by adapting the formulation presented by Kolodziej et al. (2013) to handle time-varying concentration in the supply:

$$\max \quad \sum_{t \in \mathcal{T}} \left[\sum_{(n,d) \in \mathcal{A}} \beta_d^T F_{ndt} - \sum_{(s,n) \in \mathcal{A}} \beta_s^T F_{snt} - \sum_{(n,n') \in \mathcal{A}} \left(\alpha_{nn'}^N x_{nn't} + \beta_{nn'}^N F_{nn't} \right) \right] \tag{1a}$$

$$\text{s.t.} \quad F_{nn'}^L x_{nn't} \leq F_{nn't} \leq F_{nn'}^U x_{nn't}, \qquad\qquad (n,n') \in \mathcal{A}, t \in \mathcal{T} \tag{1b}$$

$$C_{qd}^L - M(1 - x_{bdt}) \leq C_{qbt-1} \leq C_{qd}^U + M(1 - x_{bdt}), \quad q \in \mathcal{Q}, (b,d) \in \mathcal{A}, t \in \mathcal{T} \tag{1c}$$

$$C_{qd}^L - M(1 - x_{sdt}) \leq C_{qst}^{IN} \leq C_{qd}^U + M(1 - x_{sdt}), \quad q \in \mathcal{Q}, (s,d) \in \mathcal{A}, t \in \mathcal{T} \tag{1d}$$

$$I_{st} = I_{st-1} + F_{st}^{IN} - \sum_{(s,n) \in \mathcal{A}} F_{snt}, \qquad\qquad s \in \mathcal{S}, t \in \mathcal{T} \tag{1e}$$

$$I_{bt} = I_{bt-1} + \sum_{(n,b) \in \mathcal{A}} F_{nbt} - \sum_{(b,n) \in \mathcal{A}} F_{bnt}, \qquad b \in \mathcal{B}, t \in \mathcal{T} \tag{1f}$$

$$I_{dt} = I_{dt-1} + \sum_{(n,d) \in \mathcal{A}} F_{ndt} - FD_{dt}, \qquad\qquad d \in \mathcal{D}, t \in \mathcal{T} \tag{1g}$$

$$x_{nbt} + x_{bn't} \leq 1, \qquad\qquad (n,b), (b,n') \in \mathcal{A}, t \in \mathcal{T} \tag{1h}$$

$$I_{bt}C_{qbt} = I_{bt-1}C_{qbt-1} + \sum_{(s,b)\in\mathcal{A}} F_{sbt}C_{qst}^{IN} +$$

$$\sum_{(b',b)\in\mathcal{A}} F_{b'bt}C_{qb't-1} - \sum_{(b,n)\in\mathcal{A}} F_{bnt}C_{qbt-1}, \qquad q\in\mathcal{Q}, b\in\mathcal{B}, t\in\mathcal{T} \qquad (1i)$$

$$I_n^L \le I_{nt} \le I_n^U, \qquad\qquad\qquad n\in\mathcal{N}, t\in\mathcal{T} \qquad (1j)$$

$$C_q^L \le C_{qbt} \le C_q^U, \qquad\qquad\qquad q\in\mathcal{Q}, b\in\mathcal{B}, t\in\mathcal{T} \qquad (1k)$$

$$FD_{dt}^L \le FD_{dt} \le FD_{dt}^U, \qquad\qquad d\in\mathcal{D}, t\in\mathcal{T} \qquad (1l)$$

$$F_{nn't} \ge 0, x_{nn't} \in \{0,1\}, \qquad\qquad (n,n')\in\mathcal{A}, t\in\mathcal{T} \qquad (1m)$$

Where β_d^T is the unit profit of demand d satisfied, β_s^T is the unit cost of supply stream s, $\beta_{nn'}^N$ is the unit transportation cost, $\alpha_{nn'}^N$ is the fixed transportation cost, F_{st}^{IN} is the incoming flow from supply tank s at time period t, $F_{nn't}$ is the flow between node n and node n' at time t, FD_{dt} is the flow directed to demand tank d at period t, and $M \in \mathbb{R}$ is a sufficiently large big-M constant. For simplicity in presentation, the edge set $\mathcal{A}$ can refer to a subset of edges depending on the indices being used where n refers to general nodes and s, b, and d refer to supply, blending, and demand nodes, respectively. The formulation allows flows to go directly from the supply to the demands without blending if and only if they meet the required specifications. Also note that Equation (1h) enforces that blending tanks cannot be charged, mixed, and discharged simultaneously. The key complicating factor of Problem (1) is that Equation (1i) involves bilinear terms.

2.3. Generalized Disjunctive Programming Formulation

Generalized disjunctive programming (GDP) is a framework for naturally posing mathematical programs that incorporate symbolic logic relationships between the variables (Grossmann and Trespalacios, 2013). In the MPBP, the bilinear mass balance shown in (1i) only needs to be considered when tanks are charging. Hence, this nonconvex equation can be omitted when a tank is discharging. Using GDP, we can define $Y_{bt} \in \{\text{True, False}\}$ to specify whether or not, at a given period, tank b is charging ($Y_{bt} = \text{True}$) or discharging ($Y_{bt} = \text{False}$). With this, we can replace Equation (1i) in Problem (1) with the following disjunction that accounts for the operational mode of the tank:

$$\begin{bmatrix} Y_{bt} \\ I_{bt} = I_{bt-1} + \sum_{(n,b)\in\mathcal{A}} F_{nbt} \\ I_{bt}C_{qbt} = I_{bt-1}C_{qbt-1} + \sum_{(s,b)\in\mathcal{A}} F_{sbt}C_{qs}^{IN} \\ + \sum_{(b',b)\in\mathcal{A}} F_{b'bt}C_{qb't-1}, \quad q\in\mathcal{Q} \end{bmatrix} \vee \begin{bmatrix} \neg Y_{bt} \\ I_{bt} = I_{bt-1} - \sum_{(b,n)\in\mathcal{A}} F_{bnt} \\ C_{qbt} = C_{qbt-1}, \quad q\in\mathcal{Q} \end{bmatrix} \qquad (2)$$

for each $b \in \mathcal{B}$ and $t \in \mathcal{T}$. Note that bilinear terms are only considered in the left disjunct. Also, the mass balances in both disjunctions have fewer terms since they leverage the tank operational mode to only consider active flows.

Formulation (1) has other constraints that enforce logical implications on variables, which are amenable for GDP reformulation. For instance, Equation (1b) activates the flow $F_{nn't}$ only if $x_{nn't} = 1$ during that period. Hence we can also redefine the flow existence as a Boolean variable $X_{nn't} \in \{\text{True, False}\}$ and rewrite Equation (1b) with the following disjunction:

$$\begin{bmatrix} X_{nbt} \\ F_{nb}^L \le F_{nbt} \le F_{nb}^U \end{bmatrix} \vee \begin{bmatrix} \neg X_{nbt} \\ F_{nbt} = 0 \end{bmatrix}, \quad (n,b)\in\mathcal{A}, t\in\mathcal{T}. \qquad (3)$$

Equations (1c) and (1d) can be reformulated via similar GDP disjunctions. Moreover, Equation (1h) can be reformulated using GDP logical propositions:

$$X_{nbt} \Rightarrow Y_{bt}, \qquad (n,b)\in\mathcal{A}, t\in\mathcal{T} \qquad (4a)$$

$$X_{bnt} \Rightarrow \neg Y_{bt}, \qquad (b,n)\in\mathcal{A}, t\in\mathcal{T} \qquad (4b)$$

All these GDP equations are substituted into Formulation (1) to yield the GDP-based MPBP, which generalizes the GDP formulation proposed in (Lotero et al., 2016) to account for time-varying concentrations (prevalent in produced water constituents). Note that this formulation extends the one presented in (1) to account for the tank operating mode to yield fewer bilinearities after operating modes are determined.

2.4. GDP with Redundant Constraints (RC)

Typically, mixed-integer solvers employ continuous relaxations of the problem (by relaxing variable integrality) to iteratively obtain an optimal solution. Tight relaxations often provide high-quality solutions relative to the mixed-integer formulation, which can accelerate solver convergence. One way to tighten relaxations is to include additional constraints therefore redundant that provide no additional modeling information but further restrict the feasible region.

We derive redundant constraints for our GDP formulation by tracking component flow origins:

$$F_{nn't} = \sum_{r \in \mathcal{R}} \tilde{F}_{rnn't}, \quad (n,n') \in \mathcal{A}, t \in \mathcal{T} \tag{5a}$$

$$I_{bt} = \sum_{r \in \mathcal{R}} \tilde{I}_{rbt}, \quad b \in \mathcal{B}, t \in \mathcal{T} \tag{5b}$$

where $\mathcal{R} := \mathcal{S} \cup \hat{\mathcal{B}}$ is the set of possible initial origins, $\hat{\mathcal{B}}$ is the set of blending tanks with nonzero initial inventory, and $\tilde{F}, \tilde{I}$, and $\widetilde{C^0}$ are the flows, inventories, and initial composition identified by origin, respectively. Following Lotero et al. (2016), we employ other redundant constraints to strengthen the GDP formulation, and we refer the reader to that work for more details and analysis on the tightness of the relaxation.

2.5. Two-Stage MILP-MIQCP Decomposition

We can pose a two-stage decomposition to the above GDP formulations for complex MPBPs that incur high computational cost. We define an upper-level problem that omits the bilinear species balances to obtain a linear formulation that provides an upper bound UB on the optimal solution and candidate operating states of Y_{bt}. Then we define a lower-level problem that fixes Y_{bt} in the full formulation to the values provided by the upper-level problem and removes the bilinear terms corresponding to idle/discharging tanks. When feasible, the lower-level problem provides a lower bound on the optimal solution. Integer cuts are added to the upper-level problem after each run of the lower-level problem as shown in (Lotero et al., 2016):

$$Z \leq -\left(UB - Z^i\right)\left(\sum_{b \in \mathcal{B}, t \in \mathcal{T} | \hat{y}_{bt}^i = 1} y_{bt} - \sum_{b \in \mathcal{B}, t \in \mathcal{T} | \hat{y}_{bt}^i = 0} y_{bt}\right) + \tag{6a}$$

$$\left(UB - Z^i\right)\left(\sum_{b \in \mathcal{B}, t \in \mathcal{T}} \left(\hat{y}_{bt}^i\right) - 1\right) + UB, \quad i \in \mathcal{I}_O$$

$$\sum_{b \in \mathcal{B}, t \in \mathcal{T} | \hat{y}_{bt}^i = 1} (1 - y_{bt}) + \sum_{b \in \mathcal{B}, t \in \mathcal{T} | \hat{y}_{bt}^i = 0} y_{bt} \geq 1, \quad i \in \mathcal{I}_F \tag{6b}$$

where Z is the objective of the upper-level problem, y_{bt} are the binary variables associated with Y_{bt}, $\hat{y}_{bt}$ are the fixed values of the tank modes at the ith iteration, and $\mathcal{I}_O$ and $\mathcal{I}_F$ contain the feasible and infeasible iteration indices, respectively. Here, Equation (6a) is added based on a feasible lower-level solution, and Equation (6b) is added based on infeasible solutions. Iterating between upper- and lower-level problems leads to convergence to a globally optimal solution of the MPBP. The gap of the decomposition is calculated between the upper and lower bounds (LB and UB) as shown in Equation (7) where ϵ is a small tolerance. A formal analysis is provided in (Lotero et al., 2016).

$$\text{Gap} = \frac{UB - LB}{LB + \varepsilon} \tag{7}$$

3. Case Study

We compare the formulations in Section 2 via two case studies focused on delivering a lithium rich stream from a network of produced water sources. Case 1 is a smaller system composed by a subset of the tanks considered in Case 2. Produced water with varied lithium concentrations is produced at different wells and a source of fresh water is available. Several demand nodes are considered with their respective schedules for feed quality and quantity. We choose parameters based on those reported by Dworzanowski (2019) and Figure 1 details the topology of Case 2.

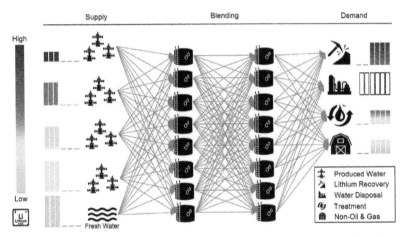

Figure 1: Topology of Case Study 2. Edges that connect sources with demands directly are considered in the case study but are not shown in the figure for simplicity.

GDP problems are often solved via big-M reformulation (BM) or Hull reformulation (HR) (Grossmann and Trespalacios, 2013; Lee and Grossmann, 2000). Previous work conducted by Lotero et al. (2016) and Ovalle Varela et al. (2021) only use BM reformulations. We explore solving these formulations using both BM and HR to study their impact on solution performance. We implement all approaches in Pyomo.GDP on a Linux machine with 8 Intel® Xeon® Gold 6234 CPUs running at 3.30 GHz with 128 hardware threads and 1 TB of RAM with Ubuntu. We use Gurobi v9.5.1 and BARON v22.7.23 as appropriate to solve all formulations. We impose a target optimality gap of 1% and 3% for case studies 1 and 2, respectively; moreover, we set a wall-time of 3600s.

Table 1: Approach comparison for the lithium recovery case studies

		Case 1		Case 2	
Upper-level	Lower-level	Time [s]	Gap[%]	Time [s]	Gap[%]
MIQCP	-	843.87	0.77	3,600	20.02
GDP (BM)	-	3,600	61.10	3,600	47.91
GDP (HR)	-	3,600	-	3,600	-
RC (BM)	-	69.05	1.00	3,600	26.33
RC (HR)	-	3,600	1.30	3,600	-
RC (BM)	RC (BM)	160.57	**0.03**	3,600	-
RC (HR)	RC (BM)	**12.54**	0.08	**54.1**	**1.08**

Table 1 shows the solution time and final optimality gap of each solution method and case study. The columns corresponding to the levels indicate the selected formulation and GDP reformulation (BM or CH) when applicable. Rows without an entry in the lower-level column are solved directly without the two-stage decomposition. Note that the two-stage decomposition using RC (BM) and RC (BM) was originally proposed by Lotero et al. (2016), and decomposition using RC (HR) and RC (BM) is proposed in this work. The results from Case 2 suggest that the monolithic formulations readily become intractable, which justifies the use of decomposition strategies. In fact, only our proposed decomposition is able to achieve an optimal solution for Case 2 within the wall-time. Hence, we observe that the choice of GDP reformulation strategy can significantly affect solution performance.

4. Conclusions

The multiperiod blending problem is a nonconvex mixed-integer quadratically constrained program that is challenging to solve for real life applications such as lithium recovery from produced water. We observe that the previous solution approaches considered in this work are extremely expensive computationally for this problem class. Moreover, our adaptation of the two-stage decomposition approach to use hull reformulations significantly enhances scalability. These results motivate further investigation into how combinations of possible formulations, GDP solution techniques, fixing strategies, and solver tunings can accelerate the convergence of the two-stage decomposition strategy considered in this work. Future work also includes integrating these formulations within the PARETO framework (Drouven et al., 2022) for REE-CM recovery extensions.

Acknowledgements: We gratefully acknowledge support from the U.S. Department of Energy, Office of Fossil Energy and Carbon Management, through the Environmental Prudent Stewardship Program.

Disclaimers: This project was funded by the United States Department of Energy, National Energy Technology Laboratory, in part, through a support contract. Neither the United States Government nor any agency thereof, nor any of its employees, makes any warranty, expressor implied, or assumes any legal liability or responsibility for the accuracy, completeness, or usefulness of any information, apparatus, product, or process disclosed, or represents that its use would not infringe privately owned rights. Reference herein to any specific commercial product, process, or service by trade name, trademark, manufacturer, or otherwise does not necessarily constitute or imply its endorsement recommendation, or favoring by the United States Government or any agency thereof. The views and opinions of authors expressed herein do not necessarily state or reflect those of the United States Government or any agency thereof.

References

M. G. Drouven, A. J. Caldéron, M. A. Zamarripa, K. Beattie, 2022. Pareto: An open-source produced water optimization framework. Optimization and Engineering, 1–21.

M. Dworzanowski, 2019. Preliminary economic assessment of LANXESS smackover project. URL https://www.sec.gov/Archives/edgar/data/1537137/000119312521204052/d194326dex991.htm

G. Gaustad, E. Williams, A. Leader, 2021. Rare earth metals from secondary sources: Review of potential supply from waste and byproducts. Resources, Conservation and Recycling 167, 105213.

C. Gounaris, C. Floudas, 2008. Convexity of products of univariate functions and convexification transformations for geometric programming. Journal of Optimization Theory and Applications 138 (3), 407–427.

I. E. Grossmann, F. Trespalacios, 2013. Systematic modeling of discrete-continuous optimization models through generalized disjunctive programming. AIChE Journal 59 (9), 3276–3295.

C. A. Haverly, 1978. Studies of the behavior of recursion for the pooling problem. ACM SIGMAP bulletin (25), 19–28.

S. P. Kolodziej, I. E. Grossmann, K. C. Furman, N. W. Sawaya, 2013. A discretization-based approach for the optimization of the multiperiod blend scheduling problem. Computers & Chemical Engineering 53, 122–142.

S. Lee, I. E. Grossmann, 2000. New algorithms for nonlinear generalized disjunctive programming. Computers & Chemical Engineering 24 (9-10), 2125–2141.

I. Lotero, F. Trespalacios, I. E. Grossmann, D. J. Papageorgiou, M.-S. Cheon, 2016. An MILP-MINLP decomposition method for the global optimization of a source based model of the multiperiod blending problem. Computers & Chemical Engineering 87, 13–35.

D. Ovalle Varela, et al., 2021. Systematic solving study for the optimization of the multiperiod blending problem: a multiple mathematical approach solution guide. Universidad de los Andes.

S. Quillinan, C. Nye, M. Engle, T. T. Bartos, G. Neupane, J. Brant, D. Bagdonas, T. McLing, J. F. McLaughlin, E. Phillips, et al., 2018. Assessing rare earth element concentrations in geothermal and oil and gas produced waters: A potential domestic source of strategic mineral commodities. Univ. of Wyoming, Laramie, WY (United States).

B. Wright, 2022. The produced water conundrum grows across unconventionals. Journal of Petroleum Technology 74 (01), 38–45.

Antonis Kokossis, Michael C. Georgiadis, Efstratios N. Pistikopoulos (Eds.)
PROCEEDINGS OF THE 33rd European Symposium on Computer Aided Process Engineering
(ESCAPE33), June 18-21, 2023, Athens, Greece

State and Parameter Estimation in Dynamic Real-time Optimization with Closed-Loop Prediction

Jose Matias,[a] Christopher L.E. Swartz,[a]

[a]*Department of Chemical Engineering, McMaster University,1280 Main St W, Hamilton, Canada, {assumpcj, swartz}@mcmaster.ca*

Abstract

In this paper, we propose the integration of a state and parameter estimation step with closed-loop dynamic real-time optimization (CL-DRTO). Different from other dynamic real-time optimization strategies, CL-DRTO uses a dynamic model that considers the closed-loop response of the plant under the action of a controller. So far, plant feedback has been incorporated into CL-DRTO using an additive noise paradigm by a bias updating strategy. In this work, we perform the feedback step in CL-DRTO updating model states and parameters via Moving Horizon Estimation. As an additional contribution, we compare CL-DRTO with two other production optimization schemes, Steady-state Real-time Optimization (SRTO) and Real-time Optimization with Persistent Parameter Adaptation (ROPA). In the latter, a dynamic model is used for only the state and parameter estimation step in SRTO, eliminating the requirement for the plant to be at steady state prior to the economic optimization execution. We show that CL-DRTO outperforms the other two schemes in terms of economic and control performance when applied to systems under control performance limitations.

Keywords: real-time optimization; dynamic real-time optimization; closed-loop; state estimation; parameter estimation.

1. Main Text

Dynamic real-time Optimization (DRTO) is a useful tool for improving plant economic performance in an increasingly dynamic market environment. DRTO computes optimal trajectories for the plant decision variables by maximizing an economic performance index subject to the system dynamics and operational/quality constraints. The optimal decisions are then sent to the plant controllers, which are responsible for implementing the computed values. In contrast to the traditional Steady-state RTO (SRTO) (Chen and Joseph, 1987), DRTO utilizes a dynamic plant model, making it better suited for systems that exhibit slow dynamics and/or frequent transitions.

The dynamic models employed at the DRTO level are often derived under a "perfect control" assumption. That is, the control response time for disturbance rejection and setpoint tracking is assumed negligible when compared to the DRTO sampling time. However, depending on the controller tuning and the system characteristics (e.g., long dead time, inverse response, etc.), this assumption may not hold. To address this issue, Jamaludin and Swartz (2017) proposed a closed-loop DRTO (CL-DRTO) scheme in which future control actions of an underlying constrained MPC system are considered. By rigorously predicting the interaction between controllers and system response, CL-DRTO improves the system performance compared to the open-loop counterpart.

One important challenge when applying CL-DRTO is how to use plant information to update the model. In the previous CL-DRTO implementations, the plant condition is represented in the model by an additive model uncertainty. Here, the difference between the measurements at the current time and the predicted output by the model is computed as a ``bias", which is then added to the model outputs and kept constant over the CL-DRTO prediction horizon (similar to the typical DMC/QDMC updating scheme). Despite presenting good results for a variety of case studies, this feedback strategy is not flexible in terms of different noise structures, which is critical for the model updating step (Kolås et al. 2008). For example, if a certain parameter drifts away from its nominal value, simply adding a bias to the model outputs may not suffice to rigorously represent the plant condition in the model.

Therefore, we propose the use of a Moving Horizon Estimator (MHE) for updating the model in CL-DRTO. With this model updating formulation, we can represent the model uncertainty using time-varying parameters by augmenting the states with a parameter vector. Therefore, we are not limited to an additive noise representation, which may lead to an improper model updating depending on the nature of the model uncertainty (Kolås et al. 2008). To benchmark this new model updating strategy, we compare CL-DRTO + MHE with two other production optimization schemes that also deal with model uncertainty via online parameter estimation, namely Steady-state Real-time Optimization (SRTO) and Real-time Optimization with Persistent Parameter Adaptation (ROPA) (Matias and Le Roux, 2018). In ROPA, a dynamic model is used for only the state and parameter estimation step in a SRTO, which eliminates the requirement for the plant to be at steady state prior to an RTO execution. We show that the combination of CL-DRTO and MHE yields a superior performance over the other two schemes in terms of economic and control performance when the controllers are detuned or tuned too aggressively.

2. Case Study: Distillation Column

The distillation column described in Skogestad (1997) is used as the case study. The column has 40 theoretical stages and the goal is to separate an ideal binary mixture and obtain a top product with 99% purity. A diagram of the distillation column and of the control structure is shown Figure 1. The model follows these assumptions: binary mixture; constant relative volatility; equilibrium on all stages; total condenser; constant molar flows; no vapor holdup; and linearized liquid dynamics.

For the control structure, we assume perfect pressure control, i.e. constant pressure, and we use the LV-configuration, where the reboiler and condenser levels are controlled using the bottoms (B) and distillate (D) flows, respectively. For composition control, we use the one-point control strategy, where the reflux (L) is used to control the top composition x_D and

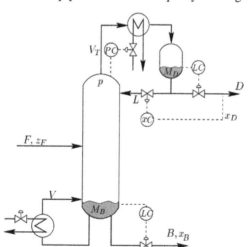

Figure 1: Distillation column diagram adapted from Skogestad (1997)

the bottom composition is left uncontrolled. Hence, the boilup (V) is the only degree of freedom (independent variable) of the system.

Ideally, these controllers should be tuned such that they yield a fast control loop response. However, in practice, this fast-responding tuning may lead to aggressive behavior to disturbances and setpoint changes. Consequently, the controlled variable may oscillate and overshoot before settling around the setpoint, which is undesired from a practical point of view. One alternative is to detune the controllers and change the gains to obtain a slower and smooth response, i.e., less oscillatory, and more tolerant to changes in process conditions. The problem is that this strategy goes against the "perfect control" assumption at the production optimization layer.

3. Simulation Setup

We tuned the sump and condenser level controllers for tight control. On the other hand, the parameters of the top composition control were chosen to achieve a smooth and slow response subject to acceptable disturbance rejection. The controller sampling time is set to $\Delta t_{cont} = 5\ s$.

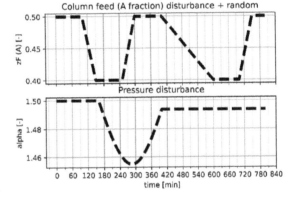

The plant is affected by disturbances (Figure 2) in the feed fraction (z_F), which we assume as unknown, and in the relative volatility (α) of the mixture. The rationale behind the disturbances is the following: the feed

Figure 2: Disturbance scenarios

fraction may vary depending on the unit load and its actual value may be hard to determine exactly due to an upstream tank layering, or the variations of the raw material. The oscillation in α may be connected to the vapor-liquid equilibrium and caused by changes in the column pressure setpoint.

For the economic optimization problem, we want to maximize the difference between the product revenues, considering their purity, and the utility costs (steam for the reboiler, and cooling water for the condenser). The heating and cooling costs are assumed to be proportional to the flows. The objective function can then be written as:

$$\emptyset = c_1 D x_D + c_2 B x_B - (c_3(D + L) + c_4 V) \qquad (1)$$

in which, c_i are the respective prices and costs in [\$/mol], and $c = [1.5, 0.2, 0.03, 0.02]\frac{\$}{kmol}$. Since the top stream contains the product of interest, we specify $c_1 \gg c_2$. For the economic optimization problem, we also want to consider a constraint on the top purity $(x_D \geq 0.99)$.

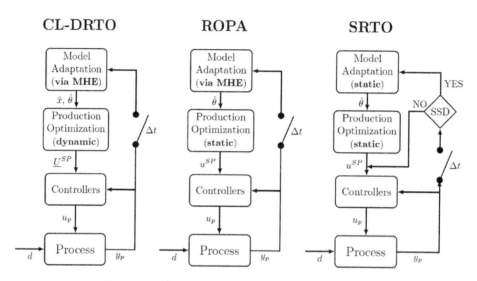

Figure 3: Block diagram of the production optimization methods

4. Production Optimization Methods

The production optimization methods studied in this paper are implemented as shown in Figure 3. All methods are executed at the same rate ($\Delta t = 15$ min), which is approximately the open-loop composition response time to changes in internal flows. Then, SRTO and ROPA optimize the future steady-state behavior of the plant and send this information to the controllers as the optimal setpoints (u^{SP}), which are kept constant until the next cycle execution. On the other hand, CL-DRTO optimizes the behavior of the system (plant + controllers) for the next 75 minutes, discretized in intervals of 3 minutes. Next, a sequence of 5 setpoints ($\underline{U}^{SP}$) is sent to the controllers. Similar to the other two methods, $\underline{U}^{SP}$ is updated every 15 minutes.

In the SRTO case, the steady state detection (SSD) was performed based on the comparison of total variance of a signal in a fixed length window of measurements with the total variance of subsequent measurements within the same window length (see Tool A in Camara et al. (2016)). If this ratio of variances (R-statistics) is below a chosen threshold, we assume that the two variances come from the same distribution and, consequently, the process is at steady-state. We use a composition in the rectifying section as a steady-state representative measurement. The other system measurements y_p are 4 compositions along the distillation column and the sump and reboiler levels. For the formulation of the Production Optimization (dynamic and static) and Model Adaptation (static) block, please refer to Matias et al. (2022). The model adaptation via MHE is shown next.

4.1. Dynamic State and Parameter Estimation via Moving Horizon Estimator (MHE)
We assume that the uncertainty structure is known, and the plant-model mismatch can be compensated by estimating the model parameters θ (here, z_F and α), which represent the disturbances d from Figure 2. Hence, the MHE formulation can be written as:

$$\min_{\theta_k, x_k} \left(\begin{bmatrix} x_{t_0-N_e} \\ \theta_{t_0-N_e} \end{bmatrix} - \begin{bmatrix} \hat{x}_{t_0-N_e} \\ \hat{\theta}_{t_0-N_e} \end{bmatrix} \right)^T \Pi_{x,\theta} \left(\begin{bmatrix} x_{t_0-N_e} \\ \theta_{t_0-N_e} \end{bmatrix} - \begin{bmatrix} \hat{x}_{t_0-N_e} \\ \hat{\theta}_{t_0-N_e} \end{bmatrix} \right) +$$

$$\sum_{k=t_0-N_e}^{t_0} (y_{p,k} - y_k)^T Q_y (y_{p,k} - y_k) + \sum_{k=t_0-N_e+1}^{t_0} (\theta_k - \theta_{k-1})^T Q_\theta (\theta_k - \theta_{k-1})$$

subjected to:

$$x_{k+1} = F(x_k, u_{p,k}, \theta_k) \qquad\qquad k = \{t_0 - N_e, \cdots, t_0 - 1\} \qquad (2)$$
$$y_k = H x_k \qquad\qquad\qquad\qquad\quad k = \{t_0 - N_e, \cdots, t_0\}$$
$$x_k \in \mathcal{X}, \quad \theta_k \in \Theta \qquad\qquad\quad k = \{t_0 - N_e, \cdots, t_0\}$$

where, the subscript k represents a given sampling time, t_0 is the current time, and N_e is the estimation window length. x and y are the model states and outputs. u_p and y_p are the inputs implemented in the plant and plant measurements. $F(\cdot,\cdot,\cdot)$ is the discretized dynamic model, which considers the underlying PI controllers. H is a selector matrix. Q_y and Q_θ are symmetric positive definite tuning matrices. $\mathcal{X}$ and Θ are the feasible state and parameter set. $\hat{x}_{t_0-N_e}$ and $\hat{\theta}_{t_0-N_e}$ are the most likely prior value of states and parameters at the beginning of the estimation window, and $\Pi_{x,\theta}$ is the arrival cost matrix. The latter is used for summarizing the information before the estimation window and we use an extended Kalman Filter to compute it. Note that in this MHE formulation we assume that the parameters change within a sampling time following a random walk model.

5. Results and Discussion

The results are shown in Figure 4. Regarding the system disturbances, Figure 4a shows that MHE is able to estimate the uncertain parameters/disturbances accurately, even though they drift away from their nominal value with time. Moreover, we note that SRTO is seldom executed (executions are indicated by red circles) because the plant is at steady-state only for short periods due to the nature of the disturbances. This influences how often the boilup rate V (the free manipulated variable) is updated in the SRTO simulation. In Figure 4b, we note that V is updated only twice, which decreases SRTO economic benefits. On the other hand, the values of V computed by ROPA are a good approximation of the CL-DRTO, where the transients are rigorously optimized.

The main benefits of CL-DRTO, i.e., predicting the underlying behavior of the controllers, can be seen in Figures 4c and Figure 4d. In Figure 4c, we show the profile of the distillate rate determined by the condenser level controller, which was tuned to obtain fast level control. The consequence is the highly oscillatory behavior of the control variable D in ROPA simulations. On the other hand, CL-DRTO takes the aggressive control actions into account and the resulting behavior of D presents only minor oscillations, which are necessary to maintain the level constant due to the relatively small holdup of the condenser. For the top constraint controller (Figure 4d), we see the detrimental effect of the slow control tuning in ROPA. Even though the manipulated variable profile V is relatively well approximated as shown in Figure 4b, the slow composition control is not able to maintain x_D above 0.99. Moreover, the implementation of the setpoints computed by ROPA and SRTO overpurify the top stream, which causes a negative effect on the economic performance. The cumulative profit achieved by the methods in [\$] are $\emptyset_{cumm.SRTO} = 451$, $\emptyset_{cumm.ROPA} = 446$, and $\emptyset_{cumm.CL-DRTO} = 455$. Hence, ROPA and SRTO lead to smaller cumulative profits than CL-DRTO.

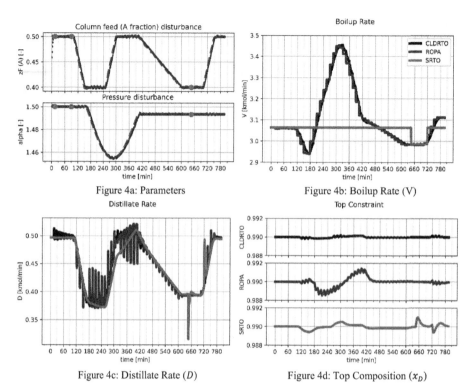

Figure 4a: Parameters

Figure 4b: Boilup Rate (V)

Figure 4c: Distillate Rate (D)

Figure 4d: Top Composition (x_D)

Figure 4: Simulation Results. CL-DRTO, ROPA and SRTO results are represented by black, blue, and red lines, respectively. In Figure 4a, the dashed line represents the true parameter values.

6. Conclusion

We explored the use of state and parameter estimation in Closed-loop Dynamic Real-time Optimization and evaluated the benefits of using this plant feedback strategy when compared to two other production optimization methods that incorporate plant information via online parameter estimation: Steady-state Real-time Optimization and Real-time Optimization with Persistent Parameter Adaptation. The simulations were conducted using a distillation system case study. The results show that the performance limitations caused by detuning or tightly tuning the controllers can be overcome by CL-DRTO, while they have a detrimental effect on the performance of the other two methods.

References

M. Câmara, et al., 2016. Performance evaluation of real industrial rto systems. Processes 4 (4), 44.

C. Y. Chen, B. Joseph, 1987. On-line optimization using a two-phase approach: An application study. Industrial & engineering chemistry research 26 (9), 1924–1930.

M. Z. Jamaludin, C. L. Swartz, 2017. Dynamic real-time optimization with closed-loop prediction. AIChE Journal 63 (9), 3896–3911.

S. Kolås et al., 2008. State estimation is the real challenge in nmpc. In: International workshop on assessment and future directions of nonlinear model predictive control.

J. Matias, G. A. Le Roux, 2018. Real-time optimization with persistent parameter adaptation using online parameter estimation. Journal of Process Control 68, 195–204.

J. Matias, et al., 2022. Steady-state real-time optimization using transient measurements on an experimental rig. Journal of Process Control 115, 181–196.

S. Skogestad, 1997. Dynamics and control of distillation columns: A tutorial introduction. Chemical Engineering Research and Design 75 (6), 539–562.

Antonis Kokossis, Michael C. Georgiadis, Efstratios N. Pistikopoulos (Eds.)
PROCEEDINGS OF THE 33rd European Symposium on Computer Aided Process Engineering
(ESCAPE33), June 18-21, 2023, Athens, Greece

Unsupervised anomaly detection model for diesel off-spec color change triggered by flooding

A. Eren Vedin[a], Sadık Odemis[b], Aysegul Sener[b], Gizem Kayar[b], Mammad Aliyev[b]

[a]SOCAR Turkey, Vadistanbul, Istanbul 34485, Turkey
[b]SOCAR STAR Oil Refinery, Aliaga, Izmir 35800, Turkey

Abstract

Crude oil feed instability can cause significant problems in refineries, where various types of crude oils are processed and frequent feed changes are taken place. In particular, if crude oil changes are performed in a short time, it can create various operational disruptions in the Crude Distillation Unit (CDU). In case of a sudden increase or decrease of light ends in the CDU feed, pressure fluctuation or flooding problems can occur in the atmospheric distillation column. In order to prevent potential upset cases caused by crude changes, an unsupervised machine-learning model was developed. The model which is presented in this article analyzes the high-correlated relationship between pressure of column flash zone, upper and lower diesel trays in steady-state operation. Principal Component Analysis (PCA) is performed with these three inputs. The PCA reconstruction error is calculated with new process values in every minute and the pressure profile in the column is instantly detected by the error. A soft sensor was developed by applying the unsupervised anomaly detection method to detect off-spec color.

Keywords: Crude oil distillation, diesel off-spec color, unsupervised anomaly detection, Principal component analysis

1. Introduction

The crude distillation unit separates the components in crude oil by utilizing the difference in boiling points (McCaffrey, 2021). Product yields differ depending on the choice and origin of crude oil. Due to market conditions, refineries today prefer to process crude oil from different sources. However, operational conditions like the blending of different crude oils, rapid crude-oil tank changes, or operation capacity maximization can lead to challenges for column operation. These challenges make the operation more susceptible to potential operational faults or other process disruptions. In such cases, flooding and weeping problems can be seen in the atmospheric column. In this article, diesel off-spec color change triggered by flooding prediction is submitted. Although there are other reasons causing off-spec color like pH change or preheat-train heat exchanger leaks, these conditions are not taken into account for prediction (Kister, 2006).

A flooding prediction model was developed because there is no online ASTM color analyzer system in the process that automates the color measurements (ASTM D1500-12(2017)). When a color change of the product is detected by routine laboratory samples or by visual controls, the product has already become off-spec. Therefore, an analytical model development was considered necessary. Since flooding is a rare anomaly in normal operation, historical data obtained from the process becomes also imbalanced. Overfitting and inaccurate results can be observed in traditional supervised approaches which apply train-validation separation in the dataset, and these models can be inadequate against the problem. For this reason, an unsupervised anomaly detection model was adopted to overcome this weakness.

It is known in the literature that the temperature diagram, fraction yields, or column pump around flows can play a role in flooding identification. Since these parameters are also intervened in steady state operating conditions, flash zone pressure, diesel lower and upper stage pressures were used instead of these process parameters. An unsupervised anomaly detection model based on Principle Component Analysis (PCA) and PCA Reconstruction Error was deployed on Distributed Control System (DCS) for flooding. The model was developed by Python programming language and Python Anomaly Detection Tool Kit (ADTK) library was used for PCA method. After this study, the model detected a real diesel off-spec color case triggered by flooding 100 minutes before happening. With the early warning system, operators took action to adjust the operation to prohibit color changes in products

2. Methodology

2.1. Flooding and Off-Spec Mechanism

The objective of the crude unit (CDU) is to process a crude oil blend into its fractions by means of boiling point differences. In distillation process, flooding and weeping are common problems that can occur and cause loss of separation. The flooding problem is a complicated abnormality whose mechanism is based on liquid or excessive vapor phase accumulation and liquid-vapor separation inability. Although flooding reduces separation efficiency in the column dramatically, there can be additional and more crucial effects in the process like feed rate reduction, operational interruption, or off-spec product (Kister and Olsson, 2019). As can be seen both in the refinery incidents and in the literature, flooding causes a sudden increase in delta pressure in the column and can cause off-spec color diesel which is also called as black diesel. (Bird et al., 2018).

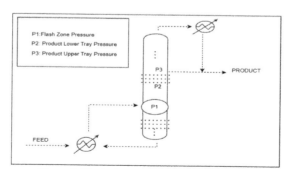

Figure 1: Typical Distillation Column Schema

A typical distillation column scheme was given in Figure 1, where P1 refers to flash zone pressure, P2 and P3 refer to diesel lower tray pressure and diesel upper tray pressure. In case rapid pressure difference changes between P1 and P2 or P3, product draw-off amounts from column trays are also changed. This causes heavy components to come along with the bottom stream and turns the diesel color black (Bird et al., 2018). Fortunately, flooding and off-spec color prediction are possible by examining the pressure at these stages. This study gives a chance to the operator sufficient time to eliminate the risk of flooding.

2.2. PCA Analysis and Reconstruction Error

Principle Component Analysis (PCA) is a classical statistical method that creates principle components that refer to the data set. Firstly, data is converted to a matrix, and vectors are generated by the decomposition of this matrix. As a second step, the original data is reconstructed by the first principle components, and reconstruction error is calculated with the difference between the new reconstructed data and the original data. The difference between the new matrix and the original data matrix is compared at each model run and gives an error. Calculated reconstruction error is examined by anomaly detection method, to expose at which level of error there is abnormal or not. With the increase in error, anomalies that occur in the process are detected (McCaffrey, 2021). Although there is no supervised learning included in this method, both high correlation in the dataset and strong coherence in the time series, generate a natural knowledge for the model. Therefore, the help of reconstruction error rise in value can detect anomalies in the data effectively. Unsupervised Anomaly Detection Machine Learning models based on PCA analysis are widely applied to industrial problems (Takeishi, 2019).

Another benefit that enables PCA analysis valuable in the industry is dimension reduction. Today, the management and control of any distillation unit are operated more efficiently and stable by engineers and operators with the help of transmitters and process control systems. The presence of many transmitters such as temperature, pressure, and flow, allows the system to be controlled with more assurance. However, having many measurement points, which are highly correlated to each other, makes data analysis more complicated and model data larger. The PCA algorithm eliminates this situation, which is known as multicollinearity. In PCA method, new independent orthogonal variables are created. At this point, PCA analysis is used as a rapid and efficient multivariate statistical method with ease of PCA reconstruction error that can be successfully applied in industry to detect abnormal situations or deviations in the process (Thomas et al., 1996).

2.3. Data Flow from Business Network to DCS and Application

Model outputs were integrated into the Industrial control system (ICS) so that control operators can use the results. Process Data History (PHD) in PHD Historian is operated to collect and pull the live process data and the results are calculated with Python code. After calculation, results are sent to Distributed Control System (DCS) for operator control. Purdue Enterprise Reference Architecture (PERA) model by ISA-99 was used as a concept model for ICS network segmentation (ANSI/ISA-TR99.00.01-2007) Deployed anomaly model was observed to predict flooding 100 minutes in advance. By adding an alarm to the tags created in the DCS system, audible and written warnings are created for the operator to take action. When the operator receives an alarm from the system, they try to reduce the risk of flooding by interfering with the crude oil column product drafts, pump around flows, or return temperatures.

3. Results and Discussion

Pressure differentiation is not the only reason for product color change. Incorrect process control tunings, abnormal column temperatures, abnormal operation of condenser or furnaces, physically damaged inner-column parts, leaks in exchangers or high nitrogen levels in the crude mix may also cause darker color. Each of these parameters were examined one by one before the study. Since sudden pressure changes and sudden increase in column level was detected during off-spec product, it was estimated that the problem was caused by flooding and a pressure-induced model can help to detect off-spec color. An unsupervised anomaly detection model based on PCA analysis was studied over the pressure parameters in the CDU column to predict the flooding abnormality and off-spec diesel color. Pressure transmitters at the column flash zone, diesel stage lower tray, and upper tray were used with the help of their highly correlated interactions.

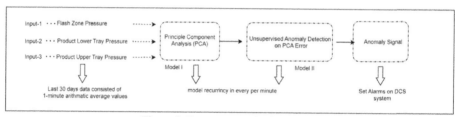

Figure 2: Data structure in the model

Not only early flooding detection but also taking fast action is significant for unit operation. Therefore, the model was scheduled to run once per minute. By processing data from the last 30 days to the model at each run, the normal fluctuation and noise level in the steady state process are understood by the model despite unsupervised learning. The data structure was created with 1-minute average values to increase sensitivity against small new changes in operation. The data architecture drawing was given in Figure 2.

A real diesel off-spec color change case, which was detected by the analytical model, is seen in Figure 3. Pressure parameters used in the model, atmospheric distillation column level, PCA reconstruction error, and delta pressure are the process parameters shown in the figure after rescaling between 0 and 1. Hairline I indicates that the model alarmed at 5:45 (see figure 3c) when incipient flooding occurred, and hairline II indicates runaway flooding occurred at 7:30 with column level increase (see figure 3 b).

Before hairline I, there was a high correlate movement of flash zone pressure, diesel lower and upper tray pressures, but the movement seems to have ceased just before the onset of the flooding initiation (see figure 3a). PCA analysis used this behavior between model parameters. The deterioration of the relationship between the parameters caused the increase in PCA error. However, it is seen that there was no increase in column level until hairline II, where flooding increased its severity. With this time difference, It was aimed to give the operator the necessary time to act.

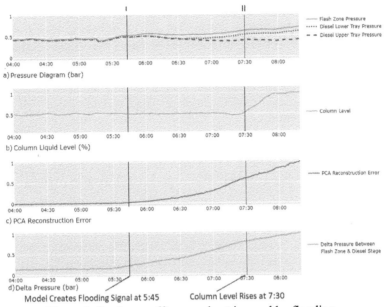

a) Pressure Diagram (bar)

b) Column Liquid Level (%)

c) PCA Reconstruction Error

d) Delta Pressure (bar)

Model Creates Flooding Signal at 5:45   Column Level Rises at 7:30

Figure 3: Real case off-spec color triggered by flooding

In literature knowledge, delta pressure between product tray and flash zone is used as another method. In this method, delta pressure is used as a stand-alone indicator to detect flooding. (Deviation, 2009). During flooding, responses of delta pressure and PCA error can be seen as similar to each other (see figure 3d). However, this method was not preferred in the study. There are always continuous and small deviations in delta pressure (DP). Although a rapid increase of these deviations can be used to create an alarm condition, DP is usually manually controlled and changed by operators during ordinary operations too. This means, deviations coming from operational set point changes or process deviations can be confused by the flooding model and this can create false alarms.

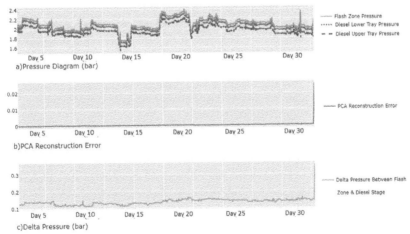

a) Pressure Diagram (bar)

b) PCA Reconstruction Error

c) Delta Pressure (bar)

Figure 4: Historical data prior to flooding. a) Pressure diagram, b) PCA reconstruction error, c) Delta pressure between flash zone and diesel stage

One-month pressure changes of diesel flash zone and diesel trays are given in figure 4a. The operational changes and process noises were seen to create bigger noises during both flooding case and steady-state operation period (see figure 4c). On the other hand, PCA reconstruction error seems more stable during steady-state period and started a rapid increase during only flooding (see figure 4b). For this reason, PCA analysis was preferred as an alternative model in this application. Instead of using the delta pressure alone, a PCA error-based unsupervised anomaly detection model was developed with flash zone, upper and lower diesel pressures.

Nevertheless, delta pressure calculation was also used for understanding if there was a positive or negative differentiation in the column. Inferential was developed based on PCA analysis, and DP was added as a supportive decision mechanism for the operator to see to filter flooding conditions among all abnormal conditions like weeping.

4. Conclusion

Crude oil feed instabilities can create instability in atmospheric distillation column and can cause upsets like flooding and weeping. In the article unsupervised anomaly detection model based on PCA analysis was used for early flooding detection which can cause off-spec color changes. It has been seen that the developed model predicted a real flooding and color change case 100 minutes in advance. The off-spec color formation was seen by a diesel laboratory sample. The study was proven and confirmed by users. Any off-spec problem has not been observed in the refinery for one year.

References

ANSI/ISA-TR99.00.01-2007 Security Technologies for Industrial Automation and Control Systems.

ASTM D1500-12(2017), DOI: https://dx.doi.org/10.1520/D1500-12R17.

Bird, J., Brown Burns, J., Racette, Y.,J, Beaulieu, 2018, Use advanced predictive analytics for early detection and warning of column flooding events, *Hydrocarbon Processing, June 2018.*

Deviation, D. S. (2009). Distillation Column Flooding Diagnostics with Intelligent Differential Pressure Transmitter.

Henry Z. Kister,2006, Distillation Troubleshooting,John Wiley & Sons, Inc.,576.

Kister, Z., Olsson, M. (2019). An Investigation of Premature Flooding in a Distillation Column,Chemical Engineering Journal,https://www.chemengonline.com/investigation-premature-flooding-distillation-column

McCaffrey, J. (2021). Anomaly Detection Using Principal Component Analysis (PCA). Visual Studio Magazine, https://visualstudiomagazine.com/articles/2021/10/20/anomaly-detection-pca.aspx.

Takeishi, N.,2019, Shapley values of reconstruction errors of pca for explaining anomaly detection. In 2019 international conference on data mining workshops (icdmw) (pp. 793-798). IEEE.

Thomas, C., Wada, T., & Seborg, D. E. (1996). Principal component analysis applied to Process monitoring of an industrial distillation column. IFAC Proceedings Volumes, 29(1), 5859-5864.

Antonis Kokossis, Michael C. Georgiadis, Efstratios N. Pistikopoulos (Eds.)
PROCEEDINGS OF THE 33rd European Symposium on Computer Aided Process Engineering
(ESCAPE33), June 18-21, 2023, Athens, Greece

Artificial neural networks-based identification of the WWTP DO sensor types of faults

Norbert-Botond Mihály[a], Alexandra-Veronica Luca[a], Vasile Mircea Cristea[a]*

[a]*Babes-Bolyai University of Cluj-Napoca, 1 Mihail Kogalniceanu Street, 400028 Cluj-Napoca, Romania*

Abstract

Efficient and sustainable operation of the wastewater treatment plants using the activated sludge technology strongly depends on the aeration control. The role of the Dissolved Oxygen (DO) sensor for achieving the desired performance of the control loop is critical and the faults of the sensor have to be identified with promptitude. The paper presents an effective artificial neural networks (ANN)-based diagnosis and identification methodology for seven types of DO sensor faults, revealing the advantages of the proposed identification approach. The classification ANNs trained with the Bayesian regularization method showed high accuracy in identifying the fault classes, with an overall accuracy of 99.5% for the testing dataset. The developed ANNs allow the efficient and early detection of the sensor fault presence, while performing the correct identification of the sensor fault type.

Keywords: Artificial Neural Networks, fault type diagnosis, Dissolved Oxygen sensor.

1. Introduction

Increased awareness of the eutrophication negative impact on water quality and advances in environmental protection technologies have resulted in more stringent wastewater treatment requirements. Regulations developed during recent decades have followed this trend. Tightening treatment regulations are driving the addition of new processing units and renewal of existing ones in the wastewater treatment plants. The subsequent increase in operational and management investments, primarily related to energy consumption and sustainable development, encouraged modern WWTPs to meet the challenges of preserving and enhancing effluent quality while ensuring efficient and safe operation, associated to cost optimization (Haimi et al. 2013).

For comprehensive process monitoring and efficient control, accurate measurement of the composition, flow rate and temperature of the influent, effluent and process streams are obviously required. Furthermore, satisfying the needs for safety and effective process control asks for the ability to detect sensor faults as early as possible. Monitoring and control at the WWTP are focused on specific equipment, such as biological reactors, settlers, pumps, air blowers, control valves, diffusers, and the pipeline system (Morera et al. 2017). Composition measurements make use of sensors that monitor Dissolved Oxygen (DO), nitrates and nitrites, ammonia, phosphorous, suspended solids, and organic matter (Mamandipoor et al. 2020). The precise information provided by these sensors is essential for the efficient and secure operation of the entire plant.

Development of numerous methods have been aimed for fault detection. Due to their inherent detection capacity, multivariate statistical methods have the ability to reveal faults in time-varying, ill-defined, and non-linear systems (Li and Yan 2019). The role of the Dissolved Oxygen sensor for achieving the desired performance of the nitrification control loop is critical and the faults of the sensor have to be identified with promptitude. Few studies have been conducted to investigate the issue of DO sensor failures. PCA methodology was used to detect deviations due to sensor clogging (Samuelsson et al. 2018), while another study implemented PCA for data driven detection of oxygen sensor faults (Luca et al. 2021).

Artificial intelligence (AI) techniques have been aimed to perform sensor fault detection and error type diagnosis in a variety of fields. They span over different applications, such as error diagnostic strategies for vapor compression refrigeration system (Kocyigit 2015), on-line fault detection of a sensor in a nuclear reactor core (Messai et al. 2015), fault detection and isolation in sensors of an internal combustion engine (Cervantes-Bobadilla et al. 2023), and fault diagnosis of vibration signals (Zhang et al. 2022).

AI was proven to be an effective tool, intended to target various implementations in WWTPs for: prediction of process parameters (Zaghloul and Achari 2022), prediction of plant performance indices and operation optimization by finding the control loop setpoints (Mihály et al. 2022a), optimization of aeration in the aerobic reactor (Mihály and Cristea 2022b), and optimization of recycle flowrates for better effluent quality and energy savings (Mihaly et al. 2022c). One fault diagnosis using ANNs was successfully applied to sludge bulking measurements with the intention of ensuring process safety and WWTP effluent quality. The developed model could differentiate five types of sludge bulking faults (Han et al. 2019). However, artificial neural networks (ANNs) were scarcely applied to the identification of WWTP sensor faults.

The scope of the current paper is to design, train and test the performance of an ANN based tool that is capable of detection and identification of seven types of Dissolved Oxygen sensor faults. The novelty consists in the application of proposed ANN diagnosis tool for the municipal A²O WWTP, having implemented the two main control loops, one for nitrification and another for denitrification.

2. Methodology

2.1. Data generation

The data implied by the development of the ANN fault identification tool were obtained using a previously calibrated first-principle Activated Sludge Model no.1 for the Anaerobic-Anoxic-Oxic reactors configuration (Várhelyi et al. 2019) of the WWTP considered as case study. Seven types of sensor faults were simulated for the Dissolved Oxygen sensor: bias, also known as shift or off-set of the signal; complete failure, which can be maximum or minimum and occurs when the measured value is either the highest or the lowest value of the sensor calibration range; drift, which is an irregularly varying deviation in time of the measured value, compared to the true DO one; fixed value, when the DO signal value is constant; loss of accuracy, meaning that the value provided by the sensor is affected by imprecision around the true value; wrong gain, known also as a typical calibration error (Luca et al. 2021). They were denoted with class numbers ranging from two to eight, while normal fault-free case was considered as class number one.

The first 100 days of the simulations were performed in normal (nominal) operation and were considered for obtaining the regular (quasi-stationary) state of the WWTP. The simulation continued in normal operation mode until day 140, where the different types of DO sensor faults were implemented until the 160th day of operation. Data obtained during these 20 days of either normal or faulty operation periods, with a sampling time of 15 minutes, was used in the development of the classification ANNs. The same data generation approach, for a different fault identification methodology, was implemented in a previous study and where additional information can be found (Luca et al. 2021).

2.2. ANN development

A set of process parameters were utilized in the development of the ANNs. They were the clean effluent concentration of the variables: total nitrogen, total Kjeldahl nitrogen, chemical oxygen demand, nitrate and nitrite nitrogen, free and saline ammonia, and total suspended solids. The following concentration variables of the bottom effluent from the secondary clarifier were also taken into consideration: nitrate and nitrite nitrogen, free and saline ammonia, total suspended solids, slowly biodegradable substrate, heterotrophic biomass, autotrophic biomass, inert particulate products, particulate biodegradable organic nitrogen, soluble biodegradable organic nitrogen, Dissolved Oxygen concentration, readily biodegradable substrate, alkalinity, associated to waste flow rate, and the temperature variables. For the identification of the different DO sensor faults an ANN with two layers was considered, the first hidden layer with tangent-sigmoid transfer function, while the second (classification) layer with SoftMax as transfer function. The implementation of the first-principle model and development of ANNs were carried out in MATLAB & Simulink software environment. The hidden layer size (number of hidden neurons) was optimized by a trial-and-error process, searching around the number of ANN inputs, while the training efficiency was studied with both Levenberg-Marquardt and Bayesian regularization backpropagation training algorithms with variable learning rates.

3. Results and discussion

The dataset was divided in the following parts when the Levenberg-Marquardt training algorithm was employed: 70% for training, 15% for validation, and 15% for testing.

The training results showed good results, with the exception of the highest obtained confusion value of 58.7%, due to the confusion between the complete failure maximum (class 3) and fixed value (class 6) fault types, characterized by comparable effects. This can be observed for the confusion matrix of the testing dataset presented in Figure 1.

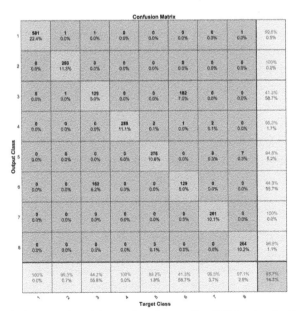

Figure 1. Confusion matrix of the testing dataset for ANN trained with Levenberg-Marquardt algorithm

However, the trained classification ANN properly detects the different fault classes, only in 0.5% of the cases classifying the faulty operation as normal (class 1). At the same time, the identification is fairly accurate except the previously mentioned confusion, while the next highest confusion percentage being of 5.2%, when loss of accuracy (class 7) and wrong gain (class 8) fault types are wrongly classified as drift faults (class 5).

In comparison to the previous training approach, the classification performed by the ANN trained with the Bayesian regularization algorithm showed better results. Results obtained for the testing data set are presented in Figure 2. This method of training did not require a validation dataset. Consequently, 85% of the data were used for training, while the remaining 15% constituted the testing dataset. In terms of training, the highest confusion value was of 1.1% in case of the wrong gain fault, where out of 1920 data points only 21 were classified as other fault classes. It is worth mentioning that neither at training, nor at testing were any of the faulty data points classified as normal (faulty-free) operation, and the previously confused complete maximum and fixed value sensor faults were correctly classified by this latter ANN.

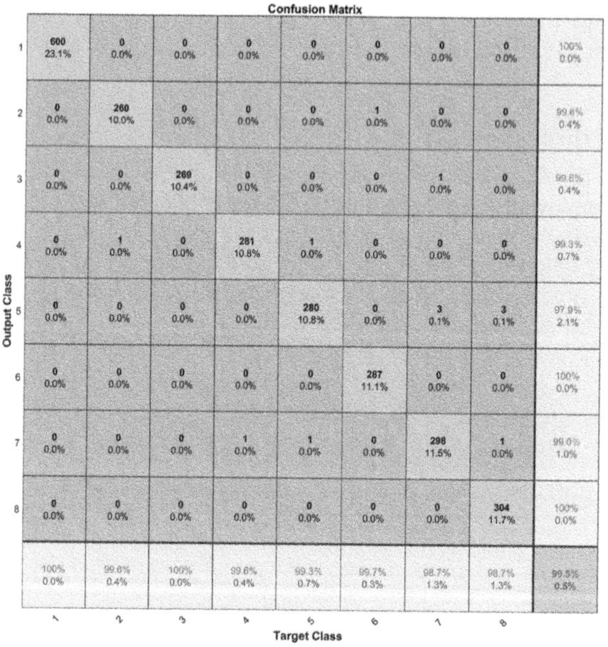

Confusion Matrix

Output Class \ Target Class	1	2	3	4	5	6	7	8	
1	600 / 23.1%	0 / 0.0%	0 / 0.0%	0 / 0.0%	0 / 0.0%	0 / 0.0%	0 / 0.0%	0 / 0.0%	100% / 0.0%
2	0 / 0.0%	260 / 10.0%	0 / 0.0%	0 / 0.0%	0 / 0.0%	1 / 0.0%	0 / 0.0%	0 / 0.0%	99.6% / 0.4%
3	0 / 0.0%	0 / 0.0%	269 / 10.4%	0 / 0.0%	0 / 0.0%	0 / 0.0%	1 / 0.0%	0 / 0.0%	99.6% / 0.4%
4	0 / 0.0%	1 / 0.0%	0 / 0.0%	281 / 10.8%	1 / 0.0%	0 / 0.0%	0 / 0.0%	0 / 0.0%	99.3% / 0.7%
5	0 / 0.0%	0 / 0.0%	0 / 0.0%	0 / 0.0%	280 / 10.8%	0 / 0.0%	3 / 0.1%	3 / 0.1%	97.9% / 2.1%
6	0 / 0.0%	0 / 0.0%	0 / 0.0%	0 / 0.0%	0 / 0.0%	287 / 11.1%	0 / 0.0%	0 / 0.0%	100% / 0.0%
7	0 / 0.0%	0 / 0.0%	0 / 0.0%	1 / 0.0%	1 / 0.0%	0 / 0.0%	298 / 11.5%	1 / 0.0%	99.0% / 1.0%
8	0 / 0.0%	0 / 0.0%	0 / 0.0%	0 / 0.0%	0 / 0.0%	0 / 0.0%	0 / 0.0%	304 / 11.7%	100% / 0.0%
	100% / 0.0%	99.6% / 0.4%	100% / 0.0%	99.6% / 0.4%	99.3% / 0.7%	99.7% / 0.3%	98.7% / 1.3%	98.7% / 1.3%	99.5% / 0.5%

Figure 2. Confusion matrix of testing dataset for ANN trained with Bayesian regularization algorithm

Similar to the previous ANN, at the testing dataset we can observe that loss of accuracy and wrong gain fault types (classes 7 and 8) are the toughest to identify correctly, exhibiting the highest confusion percentage revealed by the reduced value of 2.1%. Regarding the training time of the two ANNs, the Levenberg-Marquardt algorithm was less time-consuming, as it required 1.3 minutes to run, while the training with Bayesian regularization lasted for 15.8 minutes. Even though there is a significant difference in computational resources needed between the two training algorithms, the superior classification accuracy offered by the second ANN leads to its ranking as the best and recommended classification model.

The classification with the second ANN for the testing data set and case of fault class 7 (loss of accuracy), which was collectively the most difficult to identify, are presented in Figure 3. The faulty operation of the WWTP is detected without time lag, from the very first failure point of the testing data (following day 140, when the sensor fault begins). The classification network quickly and correctly identified the different fault types in: 3 hours for bias, 2.5 hours in case of complete fail maximum, 2.5 hours at complete fail minimum, 3 hours for drift, 3.5 hours for fixed, 4 hours at loss of accuracy, and the longest identification of 11 hours in case of wrong gain also shown in Figure 3.

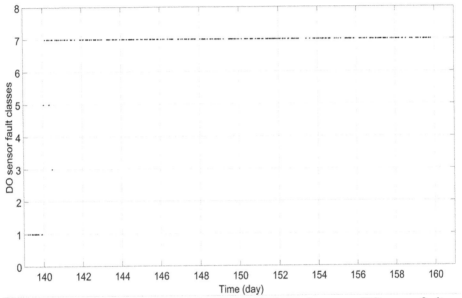

Figure 3. Classification of the testing dataset for loss of accuracy DO sensor fault

The practical implementation of the ANN based identification tools relies on the dedicated measuring instrumentation required to obtain in real time the variables used as ANN inputs. The classification offered by this ANN is of great importance for the efficient management of the WWTP and mitigation of both the environmental and economic negative impact that a faulty sensor causes by the late detection and identification of the fault. Depending on the type of the sensor fault, the CO_2 and N_2O green House Gases emissions released during the faulty sensor operation may show increased values, characterized by factors reaching one order of magnitude higher than the values of the normal WWTP operation. They are also associated to the significant degradation of the discharged effluent water quality and harmfully affect the downstream river basin. The proposed ANN diagnosis for DO sensor fault types is useful for metrological maintenance of measuring equipment, producers of intelligent sensors and their integration in digital twins.

4. Conclusions

In this study, the ANN-based sensor fault type identification method was proposed for diagnosing seven types of DO sensor faults, typical to the operation of a municipal case study WWTP. The DO measured values are of vary large importance for the aeration control loop, and they significantly influence the effluent quality, energy consumption and sustainable operation of the plant. The investigated fault types (bias, complete failure maximum, complete failure minimum, drift, fixed value, loss of accuracy and wrong gain) were successfully identified by the classification ANN. Assessed for the testing datasets, a confidence level of 85.7% was achieved in case of the ANN trained with Levenberg-Marquardt algorithm, while this value was of 99.5% for the ANN trained with Bayesian regularization algorithm. The potential of the methodologies had been demonstrated by the efficiency of the ANN-based fault identification approach. Major practical importance can be attributed to the developed classification ANNs for the accurate monitoring and high-performing control of the WWTP complex and non-linear intrinsic processes, in support of their safe and efficient operation.

References

M. Cervantes-Bobadilla, J. García-Morales, Y.I. Saavedra-Benítez, J.A. Hernández-Pérez, M. Adam-Medina, G.V. Guerrero-Ramírez, R.F. Escobar-Jímenez, 2023, Multiple fault detection and isolation using artificial neural networks in sensors of an internal combustion engine, Eng Appl Artif Intell, 117:105524.

H. Haimi, M. Mulas, F. Corona, R. Vahala, 2013, Data-derived soft-sensors for biological wastewater treatment plants: An overview, Environ Model Softw, 47:88–107.

H-G. Han, H-X. Liu, Z. Liu, J-F. Qiao, 2019, Fault detection of sludge bulking using a self-organizing type-2 fuzzy-neural-network, Control Eng Pract, 90:27–37.

N. Kocyigit, 2015, Fault and sensor error diagnostic strategies for a vapor compression refrigeration system by using fuzzy inference systems and artificial neural network, Int J Refrig, 50:69–79.

Z. Li, X. Yan, 2019, Ensemble model of wastewater treatment plant based on rich diversity of principal component determining by genetic algorithm for status monitoring, Control Eng Pract, 88(130):38–51.

A.V. Luca, M. Simon-Várhelyi, N.B. Mihály, V.M. Cristea, 2021, Data driven detection of different dissolved oxygen sensor faults for improving operation of the wwtp control system, Processes, 9(9):1633.

B. Mamandipoor, M. Majd, S. Sheikhalishahi, C. Modena, V. Osmani, 2020, Monitoring and detecting faults in wastewater treatment plants using deep learning, Environ Monit Assess, 192(2):148.

A. Messai, A. Mellit, I. Abdellani, A. Massi Pavan, 2015, On-line fault detection of a fuel rod temperature measurement sensor in a nuclear reactor core using ANNs, Prog Nucl Energy, 79:8–21.

N.B. Mihály, M. Simon-Várhelyi, V.M. Cristea, 2022a, Data-driven modelling based on artificial neural networks for predicting energy and effluent quality indices and wastewater treatment plant optimization, Optim Eng, 23:2235–2259.

N-B. Mihály, V.M. Cristea, 2022b, Optimization of the Wastewater Treatment Plant Aeration Using Artificial Neural Networks Models, Computer Aided Chemical Engineering, 51:1375–1380.

N-B. Mihaly, M. Simon-Varhelyi, A.V. Luca, V.M. Cristea, 2022c, Optimization of the Wastewater Treatment Plant Recycle Flowrates Using Artificial Neural Networks, 2022, IEEE International Conference on Automation, Quality and Testing, Robotics (AQTR), 1–6.

S. Morera, L. Corominas, M. Rigola, M. Poch, J. Comas, 2017, Using a detailed inventory of a large wastewater treatment plant to estimate the relative importance of construction to the overall environmental impacts, Water Res, 122:614–623.

O. Samuelsson, A. Björk, J. Zambrano, B. Carlsson, 2018, Fault signatures and bias progression in dissolved oxygen sensors, Water Sci Technol, 78(5):1034–1044.

M. Várhelyi, V.M. Cristea, M. Brehar, E.D. Nemeș, A. Nair, 2019, WWTP model calibration based on different optimization approaches, Environ Eng Manag J, 18(8):1657–1670.

M.S. Zaghloul, G. Achari, 2022, Application of machine learning techniques to model a full-scale wastewater treatment plant with biological nutrient removal, J Environ Chem Eng, 10(3):107430.

Y. Zhang, L. He, G. Cheng, 2022, MLPC-CNN: A multi-sensor vibration signal fault diagnosis method under less computing resources, Measurement, 188:110407.

Antonis Kokossis, Michael C. Georgiadis, Efstratios N. Pistikopoulos (Eds.)
PROCEEDINGS OF THE 33rd European Symposium on Computer Aided Process Engineering
(ESCAPE33), June 18-21, 2023, Athens, Greece

Energy Out-of-distribution Based Fault Detection of Multivariate Time-series Data

Umang Goswami,[a] Jyoti Rani,[a] Deepak Kumar,[a] Hariprasad Kodamana[a,b]
Manojkumar Ramteke[a,b]

[a]Department of Chemical Engineering, Indian Institute of Technology, New Delhi-110016, India
[b]Yardi School of Artificial Intelligence, Indian Institute of Technology, New Delhi-110016, India

Abstract

A major challenge faced by the chemical process industry is carrying out operations safely and safely. The proposed work entails a fault detection approach for a multivariate time series dataset by utilizing the energy scores instead of the traditional approach. This work proposes a loss function which utilizes the concept of in-distribution and out of the distribution of data. Energy scores are more theoretically aligned with the probability density of the inputs and can be used as a scoring function. For a pre-trained neural network, energy can be utilized as a scoring function and can also be used as a trainable cost function. The concept of out-of-distribution is similar to that of any outlier identification method. Similarly, for energy out of distribution, an energy value which falls below a certain threshold can be considered an outlier and is addressed as out-of-distribution. The values within the range are in-distribution. Higher energy values imply a lower likelihood of occurrence and vice versa. The proposed approach is compared with different deep learning approaches like Auto-encoders (AEs), LSTMs and LSTM-AEs that are traditionally used for anomaly detection and utilize the softmax scores. The Proposed methodology is also compared with some state-of-the-art fault detection methods, such as the PCA and DPCA and returns encouraging results. Energy based out of distribution is coupled with various deep learning methods to identify faulty and normal points. When teamed with the Auto-encoder network, energy-based scoring proved to be of significant dominance compared to other methods. The study was validated for the benchmark Tennessee Eastman data for fault detection.

Keywords: Fault detection, softmax score, energy score, outliers, Out-of distribution

1. Introduction

Anomalies or Outliers constitute a significant hindrance in the smooth functioning of the process industry. Fault detection or anomaly identification is one of the exercises that need to be carried out not only for safety purposes but also from a financial standpoint. Abnormal behaviour of the systems is a significant challenge not only for the chemical process plant but also for manufacturing, production and pharmaceutical industries. (Kodamana; 2017). Process models can be used to identify such faults, which are obtained either by first-principles-based models or data-based models. In systems with inherent non-linearity and non-Gaussianity, first-principle models are tougher to build. On the other hand, data-based models can account for all the difficulties faced by the first-principles models and can be applied due to their flexibility.

Over the years, several data-based methods have been developed for fault detection. Various machine learning approaches, such as Principal Component Analysis and Dynamic Principal Component Analysis (Rani, 2023), have been thoroughly used by the Process industries. Similarly, deep learning frameworks have also found their feet in the fault detection domain. Neural network architectures such as Auto-encoders (Sakurada, 2014), long short-term memory (LSTMs) (Nguyen, 2021; Yang, 2019) and an amalgamation of LSTM-AEs have been widely used. Deep learning architectures utilize neural networks to understand and learn a pattern for identifying faults. This work proposes an approach where the traditional neural networks are combined with the energy scores for fault classification. The loss function of the energy-based out-of-distribution is utilized for fault identification such that the problem of vanishing gradients does not occur in the system.

Various researchers have proposed out of distribution-based methods (Du, 2019) with different scoring functions. Scoring methods such as softmax score, Mahalanobis distance score, OE (Hendrycks, 2019) and ODIN (Liang, 2017) score are some of the distribution methods that the researchers have proposed. The energy-based scoring method proposed by (Liu, 2020) uses an energy-based score to identify the out-of-distribution data. In contrast, little to no work is done concerning the multi-variate time-series data for identifying faults using energy out-of-distribution methods. Compared to the softmax scores, energy scores are more theoretically aligned with the probability density of the inputs. The higher dimensional data can be mapped to a single dimension data with the help of energy scores. Based on the application, energy can be used as a trainable cost function for fine-tuning the classification model. A high energy value corresponds to a lower likelihood of occurrence and vice versa. Values exceeding a particular threshold can be considered under the umbrella of out-of-distribution and can be regarded as a fault. This serves as a classification method for faulty and normal points. The present work is carried out for Multivariate Time-series data and validated for the Tennessee Eastman Dataset benchmark. Energy-based scoring functions, when coupled with Deep learning frameworks, produced promising results.

The remainder of the article is structured as follows: Section 2 contains the proposed methodology, Section 3 deals with the results and discussions, and Section 4 holds the conclusion and the significance of the proposed work.

2. Proposed Methodology

2.1. Deep-Learning Frameworks for Fault Detection

As discussed in the previous section, deep-learning-based fault detection methods have traditionally been used for fault detection. These architectures differ from each other in the way the neural network is designed for each of them. Different layers and number of neurons for each of the framework is used. Auto-encoder is a reconstruction-based operator. On the other hand, LSTM is not a simple feed-forward network but also has feedback connections for the neurons, and LSTM-AE utilizes the architecture of Auto-encoders in an LSTM model. The proposed methodology is presented in figure 1, where an auto-encoder network is considered, wherein an energy-based scoring method is used to detect the faults. Similarly, this methodology also considers different neural networks, as presented in the sub-sections below.

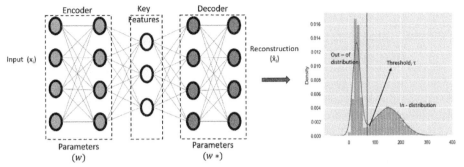

Figure 1: A schematic of the proposed Methodology.

2.1.2. Auto-Encoder (AE)

Auto-encoder network is a feed-forward neural network which has an encoder and decoder entailed it. The encoder tries to learn the data and converts it to a hidden space of lower dimension, and the decoder then captures the hidden space and reconstructs the original data. In a traditional auto-encoder, the mean squared error is used as a loss function, and the loss function is computed between the network output and the original input.

2.1.3. Long-Short Terme Memory (LSTM) Networks

LSTM networks are an extension of the Recurrent Neural networks (RNN), which can handle long-term dependencies. LSTM is not a simple feed-forward network; instead, it is a particular type of neural network that considers the previous neurons and takes inputs from them. LSTM considers sequential data and returns output for a specific window of time steps. It has three significant parts; in the first part, the model is trained to forget the irrelevant information, the second part learns the new information, and the third part returns the updated information from the current time step to the next time step.

2.1.4. LSTM-AE

LSTM-AE considers the encoder-decoder structure for an input sequence. The salient feature of the auto-encoder network is to reconstruct the input, and the LSTM-AE uses this to tackle the sequential data.

2.1.5. Energy Function

The main objective of Energy-based models is to map higher dimensional data to a single non-probabilistic scalar quantity called energy. The function E(x): $R^D \rightarrow R$ maps each input x to a single value, and collection of these can be turned into a probability distribution function through Gibbs distribution. The probability density P(x) can be represented as:

$$P(y \mid x) = \frac{e^{-E(x,y)/T}}{e^{-E(x)/T}} \tag{1}$$

The parameter T is a temperature function, and its value can be considered according to the data. The energy-based model can be linked to machine learning, especially in any discriminative model where the high dimensional data, converted to a single dimensional value, called logits. Softmax function can be used to derive the categorical distribution of logits, which can be seen in equation (2).

$$P(y \mid x) = \frac{e^{f_y(x)/T}}{\sum_i^K e^{f_i(x)/T}} \tag{2}$$

Combining equations (1) and (2), it can be inferred that the free energy function E(x;f) can be written as:

$$E(x:f) = -T.\log \sum_i^K e^{f_i(x)/T}$$

(3)

In this work, energy could not serve as a sole function that could help identifying the distribution. Energy scores can then be used as trainable cost function and can be treated as a regularization term along with the cross-entropy loss function. It helps to learn better the in-distribution and out-distribution data, and an energy gap can be created where the out-of ditribution data is segregated from the in-distribution data. In equation (4), the energy regularization term is added to the existing cross-entropy loss. $F_y(x)$ is the traditional softmax scoring function and α is the regularization parameter. Equation (5) which represents the cost function with two squared hinge loss terms which penalize any value exceeding the in-distribution and any value below the out-distribution margin., where are marginal values for the in-distribution and out-distribution, respectively.

$$\min_w E_{(x,y)\sim D_{in}^{train}} \left[-\log F_y(x)\right] + \alpha. L_{energy}$$

(4)

$$L_{energy} = E_{(x_{in},y)\sim D_{in}^{train}} (\max(0.01 * (E(x_{in}) - n_{in}), E(x_{in}) - n_{in}))^2 +$$
$$E_{x_{out}\sim D_{in}^{train}} (\max(0.01 * (n_{out} - E(x_{out})), n_{out} - E(x_{out})))^2$$

(5)

In this work, we have updated the regularization term in the loss function (Liu, 2020) to avoid the vanishing gradient problem. We have used a $0.01 * (E(x\_in) - n\_in)$ term in both the hinge losses on the lines of leaky ReLU. The loss function now penalizes the samples in energy values in the range of n_{in} and n_{out}.

3. Numerical Simulation

3.1. Datasets

The benchmark Tennessee Eastman dataset is used for testing the proposed algorithm for fault detection. There are 41 measured and 12 controlled variables in the benchmark Tennessee Eastman dataset and twenty-one faults are available for fault detection. In particular, we have considered Faults 6, 9, 15 and 19 for this study.

3.2. Simulation

For training the neural networks, we have employed the novel energy loss coupled with the cross-entropy loss. For the encoder, 64 neurons were considered in the first layer, followed by 32,16 and 8 neurons in the subsequent layers. Similarly, 16,32 and 64 neurons were considered for the decoder. Adam optimizer was used for the auto-encoder network, and the newly framed loss function was used. For the auto-encoder network, the energy threshold was found to be 0.030, and any value above the threshold value was considered a fault. Similarly, for the LSTM network, which was used for prediction in a sliding window approach, 40 neurons were used in the first layer, 20 in the second layer and the Adam optimizer was used. A dropout of 0.2 was utilized. LSTM-AE had one hundred neurons, followed by 50 neurons for the encoder section and 50, and 100 for the decoder section with a dropout of 0.5 and Equation (5) was used as the loss function. Similarly, the proposed methodology was also compared to the existing machine learning models such as the PCA and DPCA for the faults considered below.

3.3. Results and Discussion

From the parameters taken into consideration, it can be seen from Table 1 and Table 2 that the Auto-encoder network with energy out of distribution (EOD) performed better as compared to other fault detection methods. Table 1 represents the data for Faults 6 and 9 (average of both) for the Tennessee Eastman dataset, and Table 2 is the average of the faults 15 and 19 considered for the Tennessee Eastman dataset. Faults 6, 9, 15 and 19 were considered for the study, as they capture different types of faults in the Tennessee Eastman data. Faults 1-7 represent the step type, 8-12 are random variations, 13 is slow drift, 14 and 15 are sticking types, whereas 16-20 are unknown. Fault 21 is a constant position fault. In our work, we have tried to incorporate the step, random variations, sticking and unknown fault types.

Table 1: Performance Metrics for Faults 6 and 9 using the proposed loss function.

Methodology	Recall	Precision	F1-Score	AUC	FP	FN
Auto-Encoder EOD	0.9405	0.98	0.9598	0.944	5	8
LSTM EOD	0.8695	0.93	0.8988	0.862	12	18
LSTM-AE EOD	0.9008	0.921	0.9107	0.918	8	13
PCA	0.961	0.71	0.8166	0.831	27	3
DPCA	0.985	0.69	0.816	0.845	31	2

Table 2: Performance Metrics for Fault 15 and 19 of Tennessee Eastman dataset using the proposed loss function.

Methodology	Recall	Precision	F1-Score	AUC	FP	FN
Auto-Encoder EOD	0.918	0.96	0.9385	0.941	7	10
LSTM EOD	0.81	0.88	0.843	0.828	14	22
LSTM-AE EOD	0.87	0.92	0.8943	0.871	10	14
PCA	0.912	0.49	0.6374	0.656	51	7
DPCA	0.978	0.53	0.6928	0.765	47	3

Auto-encoder EOD, the network coupled with the proposed loss function had a higher Recall and Precision value, thus accounting to lower False Positives and False Negatives, which is a desirable result. Auto-encoder networks in the traditional environment perform better when compared to the other networks. LSTM-AE, a combination of two frameworks, turns out to be the second best as indicated in Table 1 and 2.

4. Conclusion

The proposed work utilizes the concept of energy out-of-distribution for fault detection. To this extent, higher dimensional data is converted to a vector of a single dimension. This data mapping helps to formulate the energy values, which are then identified as in-distribution and out-distribution based on a suitable threshold. The energy values exceeding the threshold are classified as faults. Traditional methods such as Auto-encoders, LSTMs and LSTM-AEs were coupled with the proposed loss function for fault identification. The methodology was also validated on the benchmark Industrial Tennessee Eastman Dataset. The results indicate that the proposed approach, Auto-encoder EOD, when compared to the other deep learning and machine learning techniques yielded significant results.

Acknowledgements

The authors would like to gratefully thank the Indian Institute of Technology, Delhi, for providing the necessary research facilities to carry out the proposed work.

References

Kodamana, H., Raveendran, R., & Huang, B. (2017). Mixtures of probabilistic PCA with common structure latent bases for process monitoring. *IEEE Transactions on Control Systems Technology*, *27*(2), 838-846.

Liu, W., Wang, X., Owens, J., & Li, Y. (2020). Energy-based out-of-distribution detection. *Advances in Neural Information Processing Systems*, *33*, 21464-21475.

Du, Y., & Mordatch, I. (2019). Implicit generation and generalization in energy-based models. *arXiv preprint arXiv:1903.08689*.

Sakurada, M. and Yairi, T., 2014, December. Anomaly detection using autoencoders with nonlinear dimensionality reduction. In *Proceedings of the MLSDA 2014 2nd workshop on machine learning for sensory data analysis* (pp. 4-11).

Nguyen, H.D., Tran, K.P., Thomassey, S. and Hamad, M., 2021. Forecasting and Anomaly Detection approaches using LSTM and LSTM Autoencoder techniques with the applications in supply chain management. *International Journal of Information Management*, *57*, p.102282.

Yang, J., Guo, Y. and Zhao, W., 2019. Long short-term memory neural network based fault detection and isolation for electro-mechanical actuators. *Neurocomputing*, *360*, pp.85-96.

Liang, S., Li, Y., & Srikant, R. (2017). Enhancing the reliability of out-of-distribution image detection in neural networks. *arXiv preprint arXiv:1706.02690*.

Hendrycks, D., Mazeika, M., & Dietterich, T. (2018). Deep anomaly detection with outlier exposure. *arXiv preprint arXiv:1812.04606*.

Rani, J., Roy, A. A., Kodamana, H., & Tamboli, P. K. (2023). Fault detection of pressurized heavy water nuclear reactors with steady state and dynamic characteristics using data-driven techniques. *Progress in Nuclear Energy*, *156*, 104516.

Antonis Kokossis, Michael C. Georgiadis, Efstratios N. Pistikopoulos (Eds.)
PROCEEDINGS OF THE 33rd European Symposium on Computer Aided Process Engineering
(ESCAPE33), June 18-21, 2023, Athens, Greece

Development of a Centralized Classifier for Decentralized Decision Making

Marco S. Reis,[a,*] Eugeniu Strelet,[a] Joel Sansana,[a] Margarida J. Quina,[a] Licínio M. Gando-Ferreira,[a] Tiago J. Rato,[a]

*a*Univ Coimbra, CIEPQPF, Department of Chemical Engineering, Rua Sílvio Lima, Pólo II – Pinhal de Marrocos, 3030-790 Coimbra, Portugal
*marco@eq.uc.pt

Abstract

With the current development pace of IoT technology and analytical instrumentation, the acquisition of information from devices or operations that are geographically distributed is becoming increasingly common. In some applications, these devices/operations are essentially alike but executed with slightly distinct machines/measurement systems. Such differences have been handled by constructing local models, in a fully decentralized and independent way. However, it is now possible to conceive improved centralized schemes to derive a unified model that takes advantage of the entire data lake. In this work, we demonstrate how such an endeavor can be achieved in the scope of a real case study of predicting an important property (coagulation) of waste lubricant oil (WLO) using FTIR spectra collected at different locations (laboratories). The unified classifier uses a compound mapping of Partial Least Squares for Discriminant Analysis (PLS-DA) and the Bayesian linear classifier. The best prediction models were selected by screening over 36 potential models. The best candidate models to be implemented in practice showed an accuracy greater than 98 %, while retaining less than four latent variables.

Keywords: Decentralized classification; Waste lubricating oil; Standardization methods; Partial Least Squares for Discriminant Analysis.

1. Introduction

The development of analytical solutions for decision-making and operational management using data collected from geographically distributed devices is becoming increasingly common (Marr, 2015, Qin, 2014, Reis et al., 2021). The conventional approach adopted consists of deriving local models to handle the specificities of the devices or operations. However, this poses several problems of scalability, maintenance, and comparability. Therefore, new centralized methodologies have been developed. The overarching goal is to integrate the information collected from multiple sources and derive a unified model that takes advantage of the entire data lake. A variety of model transfer approaches were developed for handling spectroscopic data (Mishra et al., 2021, Nikzad-Langerodi et al., 2021). Techniques such as Partial Least Squares (PLS) model inversion (Jaeckle et al., 2000), Joint-Y PLS (García Muñoz et al., 2005), calibration transfer (Diaz et al., 2022), and more recently transfer learning (Mishra et al., 2021, Nikzad-Langerodi et al., 2021) and domain adaptation (Pan et al., 2011), offer different paths to achieve the aforementioned goal. In general, they assume the existence of a master or reference device, whose model needs to be transferred to the remaining devices. However, we will consider here the case where there is no such master device or operation, all of them having symmetrical roles and importance, but where a common model still needs to be found for the final classification task. We demonstrate how such an endeavor can be achieved in the scope of a real case study, briefly described below.

Sogilub is the national organization that manages the entire supply chain for the regeneration of waste lubricant oil (WLO) in Portugal. WLOs are collected from different locations but should comply with certain specifications to proceed for regeneration. Currently, WLO characterization is made at four different accredited laboratories, using equipment that, even though similar (Fourier-Transform Infrared spectrometers), show some differences, that originate slightly different spectra even for the same samples. The goal of the present work is to develop a centralized classifier that all laboratories can use, despite the local differences on the devices. In other words, instead of developing a classifier for each laboratory, with much less data and representativeness of the entire population of WLO, we aim at constructing a unified, centralized classifier, that benefits from the information collected in all laboratories. The proposed centralized approach is based on a combination of data pre-processing, a projection operation to a common space (enabling data integration) and finally modeling of the projected data.

The rest of this article is organized as follows. In Sections 2 the standardization techniques, pre-processing techniques and modeling methodologies considered in this work are briefly introduced. Afterwards, in Section 3, the results are presented. Finally, a summary of the conclusions is provided in Section 4.

2. Methodology

The proposed methodology aims to build a centralized classifier based on FTIR spectra of WLO to predict a key property in multiple locations (laboratories). The methodology is composed by three main stages addressing (Stage 1) spectral standardization, (Stage 2) spectral pre-processing and (Stage 3) training of the unified classification model as described in the following subsections. A schematic representation of the processed procedure is presented in Figure 1.

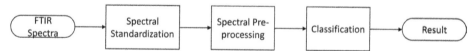

Figure 1 Diagram of the main stages of the centralized classifier.

2.1. Stage 1: Spectral Standardization

In this framework, samples are collected from the same population, but present distinct correlation structures due to the use of different measurement equipment. Therefore, they must first be brought to a common space so that the spectra become source-independent. To achieve this, standardization techniques are used to relate the spectra measured in a reference equipment ($S_1 \in \mathbb{R}^{N \times M_1}$, where N is the number of samples and M_i is the number of wavelengths measured in equipment i) to the spectra of the same samples obtained on a secondary equipment ($S_2 \in \mathbb{R}^{N \times M_2}$) by a transformation matrix ($F \in \mathbb{R}^{M_2 \times M_1}$) composed of linear relationships, such that $S_1 = S_2 F$. In this work two standardization techniques were considered to determine the transformation matrix:

- Direct Standardization (DS) (Wang et al., 1991);
- Piecewise Direct Standardization (PDS) (Feudale et al., 2002).

2.2. Stage 2: Spectral Pre-Processing

After projecting the spectra to a common space, it is necessary to correct for undesirable artifacts and enhance the relevant features in the spectra. To this end, several pre-processing techniques are currently available. In this work, four pre-processing techniques were considered:

- Mean Centering (MC);
- Multiplicative Scatter Correction (MSC) (Martens et al., 1983);
- Standard Normal Variate (SNV) (Barnes et al., 1989);
- Savitzky-Golay Differentiation (SGD) (Savitzky et al., 1964).

Different combinations of these pre-processing techniques, as well as different parameterizations of SGD (derivative order, "D#"; window size, "W#"; and polynomial order, "O#") were examined (see Table 2), leading to 12 distinct pre-processing variations.

2.3. Stage 3: Classification in the Common Space

The final stage of the proposed procedure comprises the training of a classification model. In this framework, classification was made through a two stages procedure by application of Partial Least Squares - Discriminant Analysis (PLS-DA) (Barker et al., 2003), followed by Fisher Discriminant Analysis (FDA) (Hastie et al., 2009). In this case, the response variable of PLS-DA is an indicator variable that codifies the class labels. Afterwards, PLS is applied to extract the latent variables with greater discriminative power. Next, the extracted latent variables are fed to FDA to obtain the final classifier. As the performance of the classifier may be affected by the wavelengths included in the model, three PLS-based approaches were considered to select the relevant wavelength intervals:

- PLS (Wold et al., 2001);
- Forward interval PLS (FiPLS) (Xiaobo et al., 2007);
- Backward interval PLS (BiPLS) (Xiaobo et al., 2007).

3. Results

3.1. Case Study

The case study concerns the regeneration process of WLO conducted in four facilities with different geographical localizations. Before processing, the coagulation potential of the collected WLO is determined through a coagulation test using an alkaline treatment with KOH. If a certain WLO sample gives a positive result in the coagulation test, the associated lot must follow an alternative treatment other than regeneration. Otherwise, the WLO may clog the equipment, causing a premature stop of the process for cleaning and subsequent disposal of the entire production. Therefore, it is highly desirable to have a fast and reliable prediction of the coagulation potential to direct the WLO to a suitable treatment. To achieve this, it is proposed to predict the coagulation potential using the FTIR spectra of WLO samples.

This study uses two datasets: (i) a dataset composed by WLO samples (spectra) analyzed in four laboratories, which is used for selecting the standardization technique and (ii) another dataset composed by WLO samples (spectra and the respective coagulation label) used to train and test the calibration models. In all laboratories, the spectra were collected in the transmittance mode for 700 wavenumbers in the range of 4000 to 500 cm^{-1} at a resolution of 5 cm^{-1}.

3.2. Setting up the Spectral Standardization to the Common Space

The first dataset is composed of six samples of WLO whose FTIR spectra were measured in all four laboratories with three replicates each. By applying Principal Component Analysis (PCA) (Jolliffe et al., 2016) to the spectra, the differences between laboratories can be inspected thought analysis of the scatterplot of the first two principal components (explaining 92 % of the variability in the spectra) represented in Figure 2. In Figure 2 (a) it is verified that even though the spectra are from the same set of samples, there are systematic deviations in the spectra due to the use of different equipment. This motivates the use of a standardization technique to bring the spectra to a common space. In this regard, Figure 2 (b) shows a significant improvement in homogeneity between the different laboratories after standardization.

In this work, DS and PDS were adopted using Laboratory II as the reference laboratory. To avoid overfitting, the transformation matrices of DS and PDS were obtained by PLS. To determine the best standardization technique (DS or PDS), their quality was assessed using the root mean squared error (RMSE) between the standardized and reference spectrum of each sample. These results are shown Table 1 along with a two-sided paired t-test. The results show that DS tends to produce standardized spectrum with lower RMSE. Furthermore, visual inspection of the standardized spectra shows that PDS introduced some artifacts in the transformed spectra. Thus, DS was selected as the best standardization technique for the current case study.

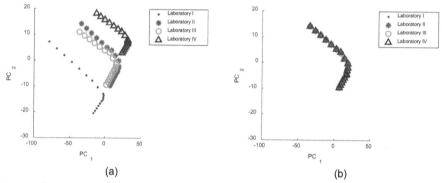

Figure 2 Scatterplot of the first and second principal components of samples measured in four different laboratories: (a) original spectra; (b) spectra after DS standardization.

Table 1 Comparison of the RMSE of the standardization techniques in each laboratory using a two-sided paired t-test.

Laboratory	RMSE		RMSE Comparison (DS – PDS)	
	DS	PDS	Test statistic (t)	p-value
I	0.2324	0.3814	-4.1525	<0.001
III	0.3559	0.4317	-0.5264	0.6054
IV	0.4727	0.4934	-1.8912	0.0758

3.3. Setting up and Selection of Calibration Model

The second dataset is used to train and test the centralized classification model. This dataset is composed of 120 samples (16 samples from Laboratory I; 58 samples from Laboratory II; 30 samples from Laboratory III; and 16 samples from Laboratory IV). Among these samples, 60 of them are from WLO that coagulate, and the other 60 samples are from WLO that do not coagulate.

Following the previous results, the samples were standardized by DS into a common space defined by Laboratory II. The samples were then used to train and select the best classification model from a pool of 36 distinct models (resulting from the combination of 12 pre-processing techniques and three modeling approaches). For FiPLS and BiPLS the spectra were divided into 15 intervals with approximately 47 wavenumbers each.

The raw dataset was randomly split into a training dataset with 80 % of the samples and a test dataset with the remaining 20 % of the samples. The models were then trained on the training dataset using Monte Carlo Cross-Validation (MCCV) for tuning the model's hyperparameters. Afterwards, their performance was assessed on the test dataset. In both cases, the key performance indicator (KPI) to assess the model's performance was the prediction accuracy given by,

$$H = \frac{TP + TN}{n} \tag{1}$$

where *TP* in the number of true positives, *TN* is the number of true negatives and *n* is the number of samples in the training/test dataset. The accuracy varies from 0 to 1, where values closer to 1 relate to better classification capabilities.

The accuracy and number of retained latent variables of each model under consideration is presented in Table 2. The results show that all models attain high accuracy. However, several models also retain a considerably large number of latent variables, indicating that some overfitting may be taking place. Nevertheless, a few models present a good balance between the number of retained latent variables and accuracy. For instance, SGD-D2-W9-O2 in combination with FiPLS has an accuracy of 100 % with four latent variables, while SNV-SGD-D1-W7-O2 in combination with BiPLS has an accuracy of 98 % with only two latent variables.

Table 2 Accuracy and number of retained latent variables of each model.

Pre-Processing Technique	Accuracy			Number of Retained Latent Varaibles		
	FiPLS	BiPLS	PLS	FiPLS	BiPLS	PLS
MC	0.9960	1.0000	0.9710	5	9	4
SNV	0.9360	0.9757	0.9757	4	5	5
MSC	0.9998	0.9707	0.9640	5	4	4
SGD-D1-W7-O2	1.0000	1.0000	1.0000	7	8	8
SGD-D1-W15-O2	0.9452	0.9525	0.9527	7	8	8
SGD-D2-W9-O2	1.0000	0.9993	0.9400	4	7	3
SNV-SGD-D1-W7-O2	0.9843	0.9836	0.9990	6	2	2
SNV-SGD-D1-W15-O2	1.0000	0.9998	0.9827	6	5	4
SNV-SGD-D2-W9-O2	0.9683	0.9571	0.9788	6	4	4
MSC-SGD-D1-W7-O2	0.9843	1.0000	0.9610	4	7	4
MSC-SGD-D1-W15-O2	0.8866	0.8736	0.8818	5	2	2
MSC-SGD-D2-W9-O2	0.9752	0.9995	0.9948	6	5	4

4. Conclusions

A centralized classification model was developed to predict the coagulation potential of WLO during the recovery process. Predictions are made from FTIR spectra collected in four laboratories, which introduces laboratory-specific variability in the data. The proposed methodology was able to handle this heterogeneity, by projecting all sources to a common space. In this regard, DS provided the best results. The best prediction model was then selected by screening over 36 potential models. The results showed that FiPLS and BiPLS in combination with Savitzky-Golay Differentiation led to prediction accuracies greater than 98 %, while retaining a low number of latent variables. Thus, these models are the best candidates to be implemented in practice. In future work, we will evaluate the performance of the centralized classifier over time and assess the impact of instrument and/or sample degradation on model quality.

Acknowledgments

The authors gratefully acknowledge the financial support of SOGILUB – Sociedade de Gestão Integrada de Óleos Lubrificantes Usados, Lda. The authors also acknowledge the support from the Chemical Process Engineering and Forest Products Research Centre (CIEPQPF), which is financed by national funds from FCT/MCTES (reference UIDB/EQU/00102/2020).

References

M. Barker, W. Rayens, 2003, Partial least squares for discrimination, Journal of Chemometrics, 17, 3, 166-173.

R. J. Barnes, M. S. Dhanoa, S. J. Lister, 1989, Standard Normal Variate Transformation and Detrending of Near-Infrared Diffuse Reflectance Spectra, Applied Spectroscopy, 43, 5, 772-777.

V. F. Diaz, P. Mishra, J.-M. Roger, W. Saeys, 2022, Domain invariant covariate selection (DiCovSel) for selecting generalized features across domains, Chemometrics and Intelligent Laboratory Systems, 222, 104499.

R. N. Feudale, N. A. Woody, H. Tan, A. J. Myles, S. D. Brown, J. Ferré, 2002, Transfer of multivariate calibration models: a review, Chemometrics and Intelligent Laboratory Systems, 64, 2, 181-192.

S. García Muñoz, J. F. MacGregor, T. Kourti, 2005, Product transfer between sites using Joint-Y PLS, Chemometrics and Intelligent Laboratory Systems, 79, 1, 101-114.

T. Hastie, R. Tibshirani, J. Friedman, 2009, The Elements of Statistical Learning: Data Mining, Inference, and Prediction, New York, NY, Springer.

C. Jaeckle, J. F. MacGregor, 2000, Product Transfer Between Plants Using Historical Process Data, AIChE Journal, 46, 10, 1989-1997.

I. T. Jolliffe, J. Cadima, 2016, Principal component analysis: a review and recent developments, Philosophical Transactions of the Royal Society A: Mathematical, Physical and Engineering Sciences, 374, 2065, 20150202.

B. Marr, 2015, Big Data - Using Smart Big Data Analytics to Make Better Decisions and Improve Performance, Chichester, Wiley.

H. Martens, S. A. Jensen, P. Geladi, 1983, In Proc. Nordic Symp. Applied Statistics(Ed, Christie, O. H. J.) Stokkland Forlag, Stavanger, Norway, pp. 205-234.

P. Mishra, D. Passos, 2021, Deep calibration transfer: Transferring deep learning models between infrared spectroscopy instruments, Infrared Physics & Technology, 117, 103863.

R. Nikzad-Langerodi, E. Andries, 2021, A chemometrician's guide to transfer learning, Journal of Chemometrics, 35, 11, e3373.

S. J. Pan, I. W. Tsang, J. T. Kwok, Q. Yang, 2011, Domain Adaptation via Transfer Component Analysis, IEEE Transactions on Neural Networks, 22, 2, 199-210.

S. J. Qin, 2014, Process data Analytics in the Era of Big Data, AIChE Journal, 60, 9, 3092-3100.

M. S. Reis, P. M. Saraiva, 2021, Data-Centric Process Systems Engineering: a Push Towards PSE 4.0, Computers & Chemical Engineering, 155, 107529.

A. Savitzky, M. J. E. Golay, 1964, Smoothing and Differentiation of Data by Simplified Least Squares Procedures, Analytical Chemistry, 36, 8, 1627-1639.

Y. Wang, D. J. Veltkamp, B. R. Kowalski, 1991, Multivariate Instrument Standardization, Analytical Chemistry, 63, 23, 2750-2756.

S. Wold, M. Sjöström, L. Eriksson, 2001, PLS-Regression: A Basic Tool of Chemometrics, Chemometrics and Intelligent Laboratory Systems, 58, 109-130.

Z. Xiaobo, Z. Jiewen, L. Yanxiao, 2007, Selection of the efficient wavelength regions in FT-NIR spectroscopy for determination of SSC of 'Fuji' apple based on BiPLS and FiPLS models, Vibrational Spectroscopy, 44, 2, 220-227.

Antonis Kokossis, Michael C. Georgiadis, Efstratios N. Pistikopoulos (Eds.)
PROCEEDINGS OF THE 33rd European Symposium on Computer Aided Process Engineering
(ESCAPE33), June 18-21, 2023, Athens, Greece

Fault detection using Fourier neural operator

Jyoti Rani[a], Tapas Tripura[c], Umang Goswami[a], Hariprasad Kodamana[a,b], Souvik Chakraborty[b,c]

[a] *Department of Chemical Engineering, Indian Institute of Technology Delhi, New Delhi-110016, India*
[b] *Yardi School of Artificial Intelligence, Indian Institute of Technology Delhi, New Delhi-110016, India*
[c] *Department of Applied Mechanics, Indian Institute of Technology Delhi, New Delhi-110016, India*

Abstract

In order to generate higher-quality products and increase process efficiency, there has been a strong push in the processing and manufacturing sectors. This has called for the creation of methods to identify and fix faults to ensure optimal performance. As a result, it is essential to develop monitoring systems that can effectively detect and identify these faults so that operators can quickly resolve them. This article proposes a novel fault detection method that adopts a deep learning approach using a Fourier neural operator (FNO) in a probabilistic way, an operator learning model that aims to learn the distribution of multivariate process data and apply them for fault detection. Herein, the historical data under normal process conditions were first utilized to construct a multivariate statistical model; after that, the model was used to monitor the process and detect faults online. The proposed FNO combines the integral kernel with Fourier transformation in a probabilistic way. As the Fourier transform helps in the time-frequency localization of time series, FNO takes advantage of them to discover the complex time-frequency characteristics underlying multivariate datasets. On the benchmark Tennessee Eastman process (TEP), a real-world chemical manufacturing dataset, the performance of the proposed method was demonstrated and compared to that of the widely used fault detection methods.

Keywords: Fault detection; Fourier Transform; Neural operator; Probability distribution

1. Introduction

A fault in industrial manufacturing processes is any anomalous deviation from normal operating conditions (NOC). Faults are a reason for concern since even minor flaws in a industrial system can set off a cascade that reduces efficiency and dependability. On the other hand, numerous cutting-edge fault detection techniques are accessible to ascertain whether the issue has occurred in the system (also known as anomaly detection in other applications).In recent decades, multivariate statistical techniques for process monitoring, especially fault detection, have developed rapidly (Kodamana et al., 2019). The principal component analysis (PCA) is mainly used in linear data processing algorithms and has proven to be highly effective in fault detection. Dynamic principal component analysis (DPCA) has been suggested in the literature to cope with serially correlated multivariate observations. Both PCA and DPCA have a linear structural model and correlation constrain while dealing with in-process dynamics (Rani et al., 2023). However, extremely non-linear and highly correlated process data pose challenges. Simultaneously, Neural network (NN) based techniques have gained considerable attention due to their flexibility and capability in modelling complex structures and temporal dynamics. Researchers have applied deep learning to the field of process monitoring, using techniques such as long short-term memory (LSTM), autoencoder (AE), Recurrent neural network (RNN), and LSTM-based AE (Lee et al., 2019). It is challenging for data-driven neural networks (NNs) to generalize beyond the training data. As an efficient and successful strategy, the trained networks should generalize to unseen inputs and predict the output precisely to address these challenges; neural operators (NOs) have been developed recently. Since NOs only need to be trained once, they can learn the mapping among two infinite-dimensional function spaces (Kovachki et al., 2021). After being trained, they can be applied to predict the outcome of any particular input function.

The global integral operators are passed through the non-linear local activation functions via NOs, similar to NNs, to learn the complex non-linear dynamics of process data. Among the recently developed NOs, the DeepONet architecture is the first to learn such infinite-dimensional function spaces (Lu et al., 2022). Similarly, a graph neural operator (GNO) is proposed that gave new ideas for using graphs but became unstable with the increased number of hidden layers. In order to learn the network parameters in Fourier space, the Fourier neural operator (FNO) is invented, simultaneously (Li et al., 2020). The fast Fourier transform (FFT) spectral decomposition of the input and computing the integral convolution kernel in the Fourier space constitute the basis of the FNO (Li et al., 2020). The present work extends the FNO in a probabilistic sense and proposes the Fourier neural operator (FNO) for learning the distribution of finite-dimensional multivariate input. The proposed FNO is then applied for fault detection.

2. Methodology

2.1. Proposed Framework

This paper presents the FNO algorithm for learning the distribution of provided time series. To learn the distribution of the framework, the output space data was created by splitting the input data set. The input-output pairs were denoted as $\mathbf{Y} \in \mathcal{U}$ and $\bar{\mathbf{Y}} \in \mathcal{U}$, where $\mathcal{U} \in \mathbb{R}^N$ is a completely normed vector space. The distribution of the input was characterized by using the FNO, and a new channel was created by augmenting an m-dimensional random variable $\mathbf{Z} = \{\xi_1, \dots, \xi_k, \dots, \xi_T\}$ with $\xi_k \in \mathbb{R}^N$. The purpose of this framework is to develop a network that can utilise the information to learn about the distribution employing the operator.

To do this, first, lift the inputs $a(y_t) \in \mathcal{A}$ to a high-dimensional space via a local transformation $M(a(y_t)): \mathbb{R}^{d_a} \mapsto \mathbb{R}^{d_v}$ and denote the high-dimensional space as $v_0(y_t) = M(a(y_t))$, where $v_0(y_t)$ has values in $\mathbb{R}^{d_v}$. In a neural network framework, the local transformation $M(a(y_t))$ can be accomplished by building a shallow, fully connected neural network (FNN). In addition, it was assumed that a total of n steps are necessary to achieve convergence. After that, n updates $v_{j+1} = F(v_j); \forall j = 1, \dots, n$ was applied to $v_0(y_t)$, where the function F returns a value in $\mathbb{R}^{d_v}$. Once all of the n-iterations were performed, then another local transformation $G: \{v_n(y_t) \in \mathbb{R}^{d_v}\} \to \mathbb{R}^{d_u}$ was applied using an FNN to transform the lifted space into the output space $\check{\mathbf{Y}} \in \mathcal{U} = G(v_n(y_t))$. The step-wise update is given in Equation 1.

$$v_{j+1}(y_t) = Q\left(\left(\mathcal{K}(a;\phi) * v_j\right)(y_t) + \check{\partial}v_j(y_t)\right); \ t \in [0,T] \tag{1}$$

where $Q(\cdot)$ denotes a non-linear activation function, $\check{\partial}: R^{d_v} \to R^{d_v}$ is a linear transformation, $*$ is the convolution operator, and $\mathcal{K}$ denotes the wavelet integral operator parameterized by $\phi \in \Theta_{NN}$. The convolution in the integral kernel operator is represented as,

$$\left(\mathcal{K}(\phi \in \Theta_{NN}) * v_j\right)(y_t) := \int_T \kappa(y_t - \tau; \phi)v_j(\tau)\mathrm{d}\tau; \ t \in [0,T] \tag{2}$$

where $\kappa(\cdot, \cdot)$ is the kernel of the neural network parameterized by $\phi \in \Theta_{NN}$ in the Fourier domain. The overarching framework of the FNO consisting of the Fourier integral layer (FIL) is illustrated in Figure 1, which is further discussed in sub-section 2.1.

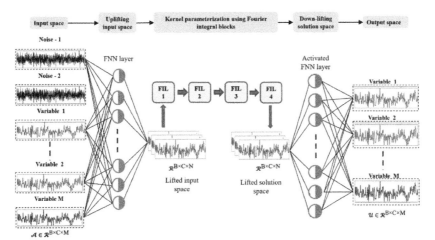

Figure 1: Architecture of Probabilistic Fourier Neural Operator. It has M+2 input consisting of M variables and 2-time series noise inputs. First, a higher dimensional projection of the inputs was made by utilizing an FNN space. After this, 4 Fourier internal blocks were applied to the projected data. The results from the other FNN were then used to transform the fourth Fourier integral block back to the original domain. This way, it will learn the distribution, and at the output, it will reconstruct only the M original variable without the noise.

2.2. Fourier Integral Block

In this section, the kernel was discovered in the frequency domain by constructing a finite-dimensional space with a Fourier transform. After that, the convolution operator was substituted in Equation 2, defined in Fourier space for the kernel integral operator. Assume $\mathcal{W}$ represent the Fourier transform (Equation 3) of a function $f \in \mathbb{R}^{d_v}$ and $\mathcal{W}^{-1}$ denotes the inverse, then

$$(\mathcal{W}f)_l(\omega) = \int_D f_l(y)e^{-2i\pi\langle \tau, \omega \rangle}d\tau, \quad (\mathcal{W}^{-1}f)_l(\tau) = \int_D f_l(\omega)e^{2i\pi\langle \tau, \omega \rangle}d\omega \tag{3}$$

for $l = 1, \dots, d_v$ where $i = \sqrt{-1}$ is the imaginary unit. By letting $\kappa_\phi(y, p, a(y), a(p)) = \kappa_\phi(y - p)$ in Equation 2 and applying the convolution theorem, the convolution was defined in the Fourier domain as Equation 4.

$$(\mathcal{K}(\phi)v_j)(\tau) = \mathcal{W}^{-1}\left(\mathcal{W}(\kappa_\phi) \cdot (\mathcal{W}v_j)\right)(\tau) \tag{4}$$

As parameterizing of κ_ϕ occurs directly in Fourier space, so introducing a weight tensor within the Fourier domain makes better sense as $R_\phi = \mathcal{W}(\kappa_\phi)$. With the new information, Fourier integral layer is defined in Equation 5.

$$(\mathcal{K}(\phi)v_j)(\tau) = \mathcal{W}^{-1}\left(R_\phi \cdot (\mathcal{W}v_j)\right)(\tau) \tag{5}$$

In the frequency mode, ω have $(\mathcal{W}v_j)(\omega) \in \mathbb{C}^{d_v}$ and $R_\phi(\omega) \in \mathbb{C}^{d_v \times d_v}$, and discrete modes $\mathbb{Z}^d$ can be used since κ is periodic, as it admits. Thus, a Fourier series expansion and continuous modes become more redundant and computationally costly. Therefore, a finite-dimensional parameterization was selected by truncating the Fourier series at the maximum number of modes, $\omega_{\max} = |Z_{\omega_{\max}}| = |\{\omega \in \mathbb{Z}^d : |\omega_l| \leq \omega_{\max,l}, \text{ for } l = 1, \dots, d\}|$. After this, the parameterization of R_ϕ occurs directly as a set of complex-valued ($\omega_{\max} \times d_v \times d_v$)-tensors made up of truncated Fourier modes as a result. The ℓ_1-norm of $\omega \in \mathbb{Z}^d$ is typically given an upper bound in order to characterize the low-frequency modes. Thus, computational and memory bottleneck was overcome by deciding $Z_{\omega_{max}}$ since it enables a practical implementation. In this work, the Fourier integral layers were constructed by using $Z_{\omega_{max}} = 16$.

2.3. FNO-based Process Monitoring
Once the FNO model is constructed, an FNO provides a reconstructed version of the input together with the distribution mean and standard deviation. Firstly, define the likelihood distribution $p_{X|\bar{X}}(Y \mid \bar{Y})$ as Gaussian distribution with sample mean μ_s and sample standard deviation σ_s. From the Gaussian representation of the likelihood function, a closed form for the reconstruction loss (Equation 6) was derived in terms of the log-likelihood as follows,

$$\log(p_{\mathbf{Y}|\bar{\mathbf{Y}}}(Y \mid \bar{Y})) = -\left(\log(\sigma_s) + \frac{1}{2}\left(\frac{y-\mu_s}{\sigma_s}\right)^2\right) \tag{6}$$

where μ_s and σ_s are the mean and sample standard deviation of $\bar{\mathbf{Y}}$. The details steps leading to Eq. 6 are omitted for brevity. Further, a Monte Carlo framework was adopted to learn the parameters of the input distribution. For $\bar{\mathbf{Y}}^{(l)}$ where $l = 1, ..., L$ were computed from the samples of $P_\theta(\bar{y} \mid y)$. After this, the standard deviation and mean were estimated for all variables. In the end, the presence of a fault in the data was tested using the condition standard deviation greater than the suitable threshold.

2.4. Datasets used in Validation Studies
Tennessee Eastman Process (TEP) dataset is a chemical simulation model based on a real-world chemical production process, and it has been widely implemented in the fields of process control, process optimization, and fault detection. There are five significant unit operations: a reactor, a product condenser, a recycle compressor, a vapor-liquid separator, and a product stripper. The TE simulator can simulate 21 different fault operation conditions in addition to the typical operating environment. Twenty-two continuous process variables, 19 composition variables, and 12 controlled variables make up the data set. In this investigation, 52 variables were monitored with fault IDs named ID1, ID6, ID7, ID8, ID16, ID18, and ID19, as all show different behaviour.

3. Simulation Studies

3.1. FNO Architecture
The four FNO layers that make up the FNO architecture were all triggered using the GeLU activation function. In order to improve the network settings, the ADAM optimizer was used. For this optimizer, the weight decay factor was set at 10^{-6}, and the initial learning rate was set at 0.001. Depending on the data loader's underlying technology, the batch size ranges from 10 to 25. The numerical experiments were carried out on a Quadro RTX5000 32 GB GPU. Each dataset was normalized using Z-score and Min-Max scaling before training the models to prevent training bias toward metrics with higher values.

3.2. Fault Detection Results
The FNO model established in the preceding section was used in the suggested process monitoring technique. The process data under normal operating conditions were gathered and used as a training dataset after normalizing to zero mean and unit variance. Variability in data reconstruction was monitored to determine the probability that the online data have a distribution similar to that of normal data, and it was calculated by employing standard deviation. The monitoring index was computed for each process data in online monitoring, and the control limit was compared using normal data distribution. The process was declared out of control when a high abnormality occurred in the data distribution, and because the standard deviation exceeded the limit, the process was considered faulty. Firstly, compare the multivariate data samples produced by FNO to the

real samples supplied for the stable data to evaluate the model's effectiveness. As shown in Figure 2, the FNO could learn the distribution and accurately reconstruct TEP data after a sufficient number of epochs.

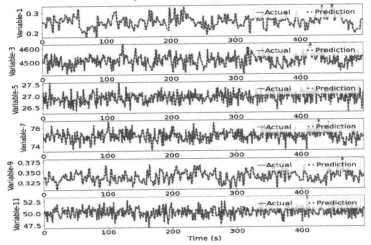

Figure 2: Reconstruction plot of TEP data for normal conditions using FNO

The standard deviation plot showed different behaviour when the fault was introduced within the dataset. The contribution plots of the factors contributing to the standard deviation at samples were computed to identify the variable responsible for any anomalous event, as shown in Figure 3. Therefore, this method aids in locating the variable also responsible for the unusual behaviour. In accordance with that, precision, recall, accuracy, and F1-score were calculated.

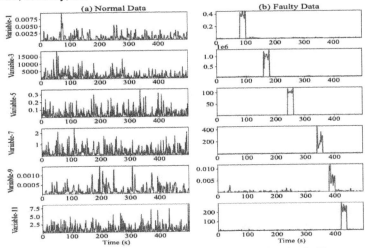

Figure 3: Standard deviation plot of normal data vs. faulty data for the Tennessee Eastman Process dataset

3.3. Performance Comparison of FNO with Baseline Methods
The fault detection performance of FNO was compared with other popular models such as DPCA, LSTM, and AE. When comparing the same number of noise levels among the other algorithms for each metric in Table 1, we see that the FNO has the best performance. Overall, FNO outperformed AE, LSTM, and other fault detection algorithms for the TEP dataset, as depicted in Table 1.

Table 1: Average performance metrics for Tennessee Eastman data set in the presence of the fault

Methods	Recall	Precision	F1-Score	AUC	FP	FN
AE	0.93	1.0	0.96	0.94	2	11
LSTM	0.80	0.87	0.83	0.81	11	21
LSTM-AE	0.86	0.91	0.88	0.87	7	16
DPCA	0.96	0.52	0.68	0.74	4	8
FNO	0.98	1.0	0.99	0.98	0	6

4. Conclusion

In this work, a novel fault detection method using FNO architecture was developed and evaluated on an enlarged TEP dataset. A broad range of recurrent and convolutional architectures such as LSTM, AE, LSTM-based AE, and DPCA was compared to establish the baseline using F1-score and Accuracy values. The novel Fourier neural operator (FNO) technique showed its effectiveness in distribution learning and anomaly detection in multivariable time series data by showing F1 score value as 0.99, AUC value of 0.98 with no FP values which show the superiority of suggested method. The proposed model brings in threefold advantages. (i) It accurately learns the operator dynamics, (ii) the introduction of Fourier transform brings in time to frequency localization, and (iii) the probabilistic framework helps to work in a noisy environment.

5. Acknowledgement

The authors acknowledge the Indian Institute of Technology, Delhi(IIT Delhi) for providing computational resources and a place to carry out this work.

References

Kovachki, N., Li, Z., Liu, B., Azizzadenesheli, K., Bhattacharya, K., Stuart, A., & Anandkumar, A. (2021). Neural operator: Learning maps between function spaces. *ArXiv Preprint ArXiv:2108.08481.*

Lee, S., Kwak, M., Tsui, K.-L., & Kim, S. B. (2019). Process monitoring using variational autoencoder for high-dimensional nonlinear processes. *Engineering Applications of Artificial Intelligence, 83*, 13–27. https://doi.org/https://doi.org/10.1016/j.engappai.2019.04.013

Li, Z., Kovachki, N., Azizzadenesheli, K., Liu, B., Bhattacharya, K., Stuart, A., & Anandkumar, A. (2020). Fourier neural operator for parametric partial differential equations. *ArXiv Preprint ArXiv:2010.08895.*

Lu, L., Meng, X., Cai, S., Mao, Z., Goswami, S., Zhang, Z., & Karniadakis, G. E. (2022). A comprehensive and fair comparison of two neural operators (with practical extensions) based on FAIR data. *Computer Methods in Applied Mechanics and Engineering, 393,* 114778. https://doi.org/10.1016/j.cma.2022.114778

Kodamana, H., Raveendran, R., & Huang, B. (2019). Mixtures of Probabilistic PCA With Common Structure Latent Bases for Process Monitoring. *IEEE Transactions on Control Systems Technology, 27*(2), 838–846. https://doi.org/10.1109/TCST.2017.2778691

Rani, J., Roy, A. A., Kodamana, H., & Tamboli, P. K. (2023). Fault detection of pressurized heavy water nuclear reactors with steady state and dynamic characteristics using data-driven techniques. *Progress in Nuclear Energy, 156,* 104516. https://doi.org/https://doi.org/10.1016/j.pnucene.2022.104516

Antonis Kokossis, Michael C. Georgiadis, Efstratios N. Pistikopoulos (Eds.)
PROCEEDINGS OF THE 33rd European Symposium on Computer Aided Process Engineering
(ESCAPE33), June 18-21, 2023, Athens, Greece

Proximal policy optimization for the control of mAB production

Nikita Gupta,[a] Shikhar Anand,[a] Deepak Kumar,[a] Manojkumar Ramteke,
[a,b]Hariprasad Kodamana[a,b*]

[a]*Department of Chemical Engineering, Indian Institute of Technology, New Delhi-110016, India*
[b]*Yardi School of Artificial Intelligence, Indian Institute of Technology, New Delhi-110016, India*
*kodamana@iitd.ac.in

Abstract

The increasing demand of monoclonal antibodies (mAB) has been the motivation to develop methods to improve its yield of its production. The product yield for any chemical process highly depends on the fidelity of the process automation used. To this extent, deployment of model-based controllers results in heavy computational load, and its performance highly dependent on the model's accuracy. To address some of these issues, the role of model-free controllers like reinforcement learning (RL) based controllers seems to be promising. Proximal policy optimization (PPO) is one of the stochastic deep RL methods that can be used to achieve this task. In this work, we have studied the PPO algorithm for the control of mAB concentration in the bioreactor. In this process, mAB concentration is the control variable and feed flowrate is the manipulated variable. Our simulation study shows that by application of the PPO algorithm for bioreactor, we are able to track the end-point concentration of mAB, by providing the optimal flowrate strategy.

Keywords: Reinforcement Learning (RL), Proximal Policy Optimization (PPO), monoclonal antibodies (mAB)

1. Introduction

For many years monoclonal antibodies (mAB) have been used to treat several complex human diseases like arthritis, cancers, Alzheimer's, COVID-19, and HIV as they have less antagonistic effects. Due to these benefits, research has been going on to improve the yield of the mAB. Therefore, a reasonable control strategy needs to be developed for this purpose. The controller performance is highly affected even by small changes in process parameters. As stated in the literature, many model-based control strategies like iterative learning control (ILC) and model predictive control (MPC) proved to overcome these issues (Gupta et al.,2022, Kern et al.,2015, Lee et al.,2007). Model-based controllers are highly dependent on the model and moreover, they require more computational time as the optimization problem has to be performed at each time step. However, many data-driven modeling strategies for chemical processes to overcome the limitations of developing a first-principle-based model (Jiang et al.,2019). Data-driven techniques, such as model-free reinforcement learning (RL), technique tackle the major drawbacks of model-based controllers (Joshi et al.,2021, Singh et al.,2020, Nikita et al., 2021). In contrast to conventional controllers, a controller based on RL allows the agent, which is equivalent to the controller, to learn the best course of action to take in response to input from the environment (Yoo et al., 2021, Bao et al., 2021, Sutton et al., 2018). One of the promisng algorithms that belong to the category of deep reinforcement learning for is proximal policy optimization (PPO). PPO greatly reduces the complexity by adopting the clipping mechanism. The objective function uses specific clipping which prevents new policy to diverge away from the old policy (Schulman et al., 2017).

84% of all mAbs (monoclonal antibodies) are generated from hamster ovary tissue. These cells have been altered to grow in a media that has been chemically produced in order to produce the target protein of interest. The principal source of carbon and energy for cultured cells is glucose, which they primarily utilize to maintain the cell's general homeostasis (Kumar et al., 2021). That's why varying glucose concentration in media affects the growth of mammalian cell culture performance. Inlet glucose concentration can be varied by changing the inlet flow rate. Temperature also affects overall mammalian cell culture. Therefore, temperature or inlet glucose concentration or flowrate can be considered as a manipulated variable to enhance the protein (mAb) concentration, which is the control variable. But in the present work, only feed flowrate has been used as a manipulated variable for the same. In the present work, the PPO algorithm has been applied to identify the optimal operating points that enhances the antibody production in the bioreactor.

2. Methodology

2.1 Process description

The fermentation process is carried out in a bioreactor in which the inlet stream consists of feed and media having glucose as raw material. Glucose, dead cells, lactate, viable cells, and protein (monoclonal antibodies) are the constituents of the outlet stream (Kumar et al., 2021). The six states are viable cells (X_v), monoclonal antibodies (mAB), glucose, lactate, dead cell density (X_d) and viability. The detailed schematics of the process can be shown below:

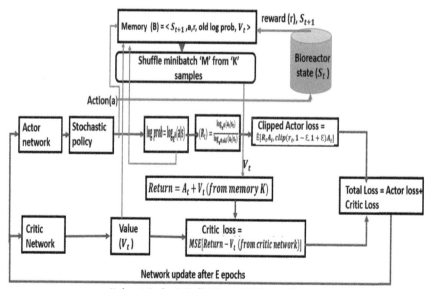

Figure 1: Schematic of the process

In this process, our main objective is to maximize the concentration of monoclonal antibodies (mAB) by controlling the glucose concentration by varying the inlet flowrate.

Therefore, for this process mAB concentration is our state (S) and flowrate is our action (a). Here we have used proximal policy optimization (PPO), one of the RL techniques to obtain the optimal control policy for obtaining the desired concentration. PPO is an actor-critic method, which acts as an agent and provide optimal action (a_t) to the bioreactor i.e., our environment at current state (S_t) and provide us with next state (S_{t+1}) and reward (r). Based on the reward earned, the agent learns on continuous interaction with the environment to provide optimal control strategy as indicated in Figure 1. In this schematic, we have two networks, actor and critic network. Actor networks take state (s) as input and provide distribution of actions and one action is sampled from it. Critic network also take state (s) as input to give value function. This value function evaluates, weather the action is good or bad. R_t is evaluated as ratio of old log of probability of actions and new log probability of actions. If the value of ratio is 1, then current action policy is better but if it is between 0 and 1, then current action policy is not good. Advantage function is also a good measure of if action is good or bad. If value of advantage is positive, then probability of selecting those set of actions is good and if value of advantage is negative, then those actions are discarded. Product of ratio and advantage is very good optimization function. This ratio is clipped between a minimum and maximum value, so that new policy will not diverge much away from the old policy. This clipped function is termed as actor loss. Critic loss is evaluated using mean square error of Return and value (i.e., output from the critic network). Total loss is the sum of critic loss and actor loss, which is used to update the networks based on stochastic gradient descent method. We shuffle the minibatch size of 'M' from the 'K' samples selected from the memory 'B'. The networks are updated after every 'E' epochs, then memory 'K' is discarded. This process is repeated till the old policy weights converges with new policy weights.

2.2 Proximal policy optimization (PPO) algorithm

Reinforcement learning (RL) is one of the branches of machine learning (ML) in which the agent (analogous to the controller) learns to give optimal actions by maximizing the reward through interaction with the environment (bioreactor). RL-based algorithms can be divided into categories, value-based method in which agent learn action based on learned value function and policy gradient method in which policy is optimized using gradient descent method. PPO is an on-policy method that can be implemented in an actor-critic style. Here critic can be implemented as the value function approximator, and the actor can be implemented as a policy approximator. Therefore, performance of this method proved better than the plain policy gradient method (Byun et al.,2020, Wang et al.,2020).

PPO performs multiple gradient descent to update the neural network for each iteration. The stability of the system is ensured by calculating the ratio between old policy and the new policy. Here the main objective is to reduce the divergence between old and the new policy (Hämäläinen et al.,2020). The basic algorithm steps are stated below:

1. There are two networks i.e., actor network ($\pi^\Theta(s_t)$) and critic network ($Q^\phi(s_t)$) with Θ and ϕ as respective weights for current state 's_t'.
2. Initialize the memory B.
3. Collect set of tuples of <state (s_t), action (a_t), reward (r_t), log probs, values (V_t)> at time instant 't' from memory B. Here 'log probs' is the logarithmic values of the probability of actions and values are the output from the critic network.

$$\log prob = \log_{\pi\Theta}(a|s) \tag{12}$$

4. Sample the set of 'K' samples from the memory 'B'.
5. Shuffle the minibatch size of 'M' from the 'K' samples collected form 'B' and perform 'E' epochs to update the network for each batch.
6. Calculate ratio $(R_t) = \frac{\log_{\pi\Theta}(a_t|s_t)}{\log_{\pi\Theta,old}(a_t|s_t)}$ i.e., ratio of log of probability of actions from memory and old log probabilities from the memory.

7   Update the actor network based on following clipped loss function:

$$L_{\pi\theta} = \check{E}[R_t A_t, clip(r_t, 1 - \varepsilon, 1 + \varepsilon) A_t] \tag{13}$$

$$A_t = r + \text{ß}V(s_{t+1}) - V(S_t) \tag{14}$$

Here A_t is the advantage, which considers for the goodness of state, ε (epsilon) is the clipping parameter and ß is the smoothing parameter.

8   Critic network is updated based on mean square error (MSE) value:

$$Return = A_t + V_t \ (from \ memory \ K) \tag{15}$$

$$L_{Q\phi} = MSE[Return - V_t \ (from \ critic \ network)] \tag{16}$$

9   Total loss = Actor loss + Critic loss

10 Update weights of the actor and critic networks (Θ and ϕ respectively) via stochastic gradient descent (SGD) method.

11 Perform iteration till the Θ_{old} converges to Θ.

3. Results and Discussions

Proximal policy optimization (PPO) based control have been applied to the bioreactor and their simulation results has been shown in this section. The efficacy of the algorithm can be shown by the root mean square error (RMSE), reward, and tracking plots. The feed flowrate mainly affects the desired protein concentration (mAB). Therefore, protein concentration is our control variable, and flowrate (ml/min) is our manipulated variable which is analogous to action. Python 3.7 version with Pytorch package has been used to train the deep neural network (DNN) for the application of the PPO algorithm. The actor and critic networks are DNN with 3 hidden layers and 256 hidden nodes each. A ReLU activation function has been used as an activation function for both actor and critic networks. Here the reference mAB concentration (mAB$_{ref}$) i.e., end-point concentration is 590 mg/l. Our main objective is to reduce the error (e = mAB - mAB$_{ref}$) with the help of the reward function as given below:

$$Reward = -|e| \tag{15}$$

The tracking performance of PPO can be shown in Figure 1(a) and the corresponding control policy (flowrate) can be seen in Figure 1(b). The PPO-based controller was trained for 200 episodes in presence of noise of 20 % in the input concentration of viable density. The controller performance can be evaluated with the help of average rewards (average of last 10 episodic rewards) as shown in Figure 2(a) and root mean square error (RMSE) values plot as shown in Figure 2(b). The results indicate that the proposed PPO based RL agent can effectively track the desired operating point.

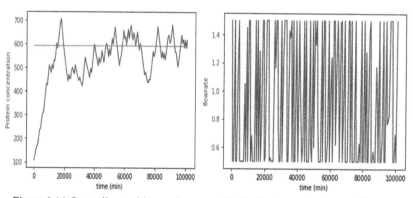

Figure 1 (a) Controller tracking performance (b) Optimal control strategy (flowrate)

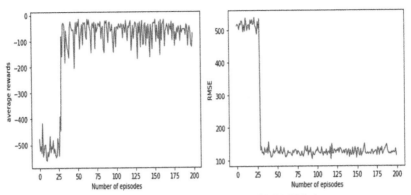

Figure 2 (a) Average reward plot (b) RMSE plot

4. Conclusions

Proximal policy optimization (PPO), an actor-critic RL-based algorithm has been implemented for the bioreactor system (environment) to control the protein concentration (mAB) to obtain the optimal flowrate (action) to achieve the desired results. PPO can track the desired mAB concentration which is 590 mg/l. The average rewards and RMSE plot show that the controller has performed well. Therefore, PPO can be a good candidate for control the complex dynamics of the bio reactor process.

5. Acknowledgement

The authors gratefully acknowledge Indian Institute of Technology, Delhi and DST-SERB India with file number CRG/2022/003722, SERB, INDIA for providing us resources to carry out experiments and simulations.

6. References

[1] Kern, Richard, and Yogendra Shastri. "Advanced control with parameter estimation of batch transesterification reactor." Journal of Process Control 33 (2015): 127-139. https://doi.org/10.1016/j.jprocont.2015.06.006

[2] Lee, Jay H., and Kwang S. Lee. "Iterative learning control applied to batch processes: An overview." IFAC Proceedings Volumes 39, no. 2 (2006): 1037-1046.https://doi.org/10.1016/j.conengprac.2006.11.013

[3] Jiang, Qingchao, Xuefeng Yan, Hui Yi, and Furong Gao. "Data-driven batch-end quality modeling and monitoring based on optimized sparse partial least squares." IEEE Transactions on Industrial Electronics 67, no. 5 (2019): 4098-4107. 10.1109/TIE.2019.2922941

[4] Yoo, Haeun, Boeun Kim, Jong Woo Kim, and Jay H. Lee. "Reinforcement learning based optimal control of batch processes using Monte-Carlo deep deterministic policy gradient with phase segmentation." Computers & Chemical Engineering 144 (2021): 107133.https://doi.org/10.1016/j.compchemeng.2020.107133

[5] Bao, Yaoyao, Yuanming Zhu, and Feng Qian. "A deep reinforcement learning approach to improve the learning performance in process control." Industrial & Engineering Chemistry Research 60, no. 15 (2021): 5504-5515. https://doi.org/10.1021/acs.iecr.0c05678

[6] Sutton, Richard S., and Andrew G. Barto. Reinforcement learning: An introduction. MIT press, 2018.

[7] Schulman, John, Filip Wolski, Prafulla Dhariwal, Alec Radford, and Oleg Klimov. "Proximal policy optimization algorithms." arXiv preprint arXiv:1707.06347 (2017). https://doi.org/10.48550/arXiv.1707.06347

[8] Wang, Yuhui, Hao He, and Xiaoyang Tan. "Truly proximal policy optimization." In Uncertainty in Artificial Intelligence, pp. 113-122. PMLR, 2020.

[10] Kumar, Deepak, Neelesh Gangwar, Anurag S. Rathore, and Manojkumar Ramteke. "Multi-objective optimization of monoclonal antibody production in bioreactor." Chemical Engineering and Processing-Process Intensification (2021): 108720.https://doi.org/10.1016/j.cep.2021.108720

[11] Byun, Ha-Eun, Boeun Kim, and Jay H. Lee. "Robust dual control of batch processes with parametric uncertainty using proximal policy optimization." In 2020 59th IEEE Conference on Decision and Control (CDC), pp. 3016-3021. IEEE, 2020. 10.1109/CDC42340.2020.9304019

[12] Hämäläinen, Perttu, Amin Babadi, Xiaoxiao Ma, and Jaakko Lehtinen. "PPO-CMA: Proximal policy optimization with covariance matrix adaptation." In 2020 IEEE 30th International Workshop on Machine Learning for Signal Processing (MLSP), pp. 1-6. IEEE, 2020. 10.1109/MLSP49062.2020.9231618

[13] Gupta, Nikita, Riju De, Hariprasad Kodamana, and Sharad Bhartiya. "Batch-to-Batch Adaptive Iterative Learning Control— Explicit Model Predictive Control Two-Tier Framework for the Control of Batch Transesterification Process." ACS omega 7, no. 45 (2022): 41001-41012. https://doi.org/10.1021/acsomega.2c04255

[14] Joshi, Tanuja, Shikhar Makker, Hariprasad Kodamana, and Harikumar Kandath. "Twin actor twin delayed deep deterministic policy gradient (TATD3) learning for batch process control." Computers & Chemical Engineering 155 (2021): 107527. https://doi.org/10.1016/j.compchemeng.2021.107527

[15] Singh, Vikas, and Hariprasad Kodamana. "Reinforcement learning based control of batch polymerisation processes." IFAC-PapersOnLine 53, no. 1 (2020): 667-672. https://doi.org/10.1016/j.ifacol.2020.06.111

[16] Nikita, Saxena, Anamika Tiwari, Deepak Sonawat, Hariprasad Kodamana, and Anurag S. Rathore. "Reinforcement learning based optimization of process chromatography for continuous processing of biopharmaceuticals." Chemical Engineering Science 230 (2021): 116171. https://doi.org/10.1016/j.ces.2020.116171

Antonis Kokossis, Michael C. Georgiadis, Efstratios N. Pistikopoulos (Eds.)
PROCEEDINGS OF THE 33rd European Symposium on Computer Aided Process Engineering
(ESCAPE33), June 18-21, 2023, Athens, Greece
© 2023 Elsevier B.V. All rights reserved. http://dx.doi.org/10.1016/B978-0-443-15274-0.50303-6

Process design and economic evaluation of biomass-based negative emission technologies

Wei Wu[a*], Rasa Supankanok[a,b], Walairat Chandra-Ambhorn[b**], Nattapat Pongboriboon[a,b]

[a]Department of Chemical Engineering, National Cheng Kung University, Tainan 70101, Taiwan, ROC
[b]Department of Chemical Engineering, School of Engineering, King Mongkut's Institute of Technology Ladkrabang, Bangkok, 10520, Thailand
Corresponding weiwu@gs.ncku.edu.tw

Abstract

A palm oil-based polygeneration system (POPS) is simulated to produce the main product of high-purity green diesel as well as the liquefied petroleum gas (LPG) as a by-product. The CO_2-negative design includes approaches of (i) a series of cryogenic separators for the recovery of approximately 65.5% of hydrogen feedstock, (ii) the evacuated tube solar collector (ETSC) for reducing 35% flue gas from the furnace, (iii) the amine-based CO_2 capture process for pursuing the high-purity CO_2 product. The economic analysis of POPS shows that the process becomes economically attractive if the diesel price and crude palm oil should be around 1.98 and 0.47 kg^{-1}, respectively.

Keywords: Palm oil; Carbon capture; Polygeneration; Negative CO_2 emission; Economic analysis

1. Introduction

Oil palm is the major source of vegetable oil in the world, mainly produced in Southeast Asia. Palm oil not only contains triglycerides and many free fatty acids, but also it is a proper feedstock for green diesel production in Thailand due to Thailand being currently the third-largest palm oil producer in the world. A palm fatty acid distillate (PFAD) is a by-product generated from refining crude palm oil, so it can be treated as the feedstock of the green diesel production process. A study showed that the green diesel produced from PFAD achieved 84% GHG emissions reduction as compared to the GHG emissions of petroleum diesel (Alhajji and Demirel, 2016). In the process simulation of the production of green diesel from the hydrotreatment of vegetable oil, a model of the hydrotreatment of soybean oil for producing green diesel was validated according to experimental data, and the green diesel produced from Brazilian soybean oil could meet quality standards (Cavalcanti et al., 2022). The negative emissions technologies will become more important in the next decades due to pricing the carbon emissions. The bio-energy with carbon capture and storage technology was one of the primary negative carbon emission methods. A study showed that the process design of a biomass-based polygeneration system for power generation and ammonia synthesis could achieve negative CO_2 emissions (Xu et al., 2021).

2. Study Design

2.1 Palm oil-based polygeneration system (POPS)

A palm oil-based polygeneration system (POPS) named Design 1 is depicted in **Fig. 1** which mainly consists of a furnace, an FBHTR, a three-phase separator, and a series of cryogenic separators (CSep1, CSep2) to produce green diesel and green LPG (propane), notably the CO_2-rich gases are emitted from the furnace and CSep1, respectively. Two feedstocks of palm oil and hydrogen are heated up to 324°C by a furnace and a combination of a heater and a compressor, respectively, and they are fed into the fixed bed hydrotreating reactor (FBHTR) at the prescribed operating pressure. Notably, 65.5% of the feedstock of hydrogen is recovered through a series of cryogenic separators in Design 1, and the FBHTR is covered with a cooling jacket where the recycled water is produced from the three-phase separator. The outlet stream of FBHTR is cooled down and fed into a specific three-phase separator. The vapor-phase stream is fed into a sequence of cryogenic separators (CSep1 and CSep2) at the prescribed operating temperature and pressure to separate green LPG, CO_2, and hydrogen, where the hydrogen flow is fully recycled and mixed with fresh hydrogen flow. The simulation results show that the purity of green diesel is up to 98.3wt% and the yield of green diesel is around 84.4%, but the purities of CO_2 and green LPG from CSep1 and CSep2 are close to 80.6% and 84.6%, respectively. Besides water and hydrogen being fully recycled, green LPG is used to meet the LPG fuel demand of the furnace.

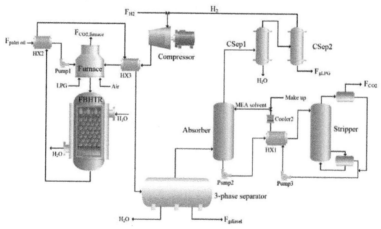

Figure 1 Palm oil-based polygeneration system

2.2 Performance evaluation

To address the performances of the POPS design, the indicators like net energy efficiency and net CO_2 emissions are addressed as follows.

(i)   Net energy efficiency

$$\eta_{net} = \frac{E_{net}}{F_{palm\ oil} \times HHV_{palm\ oil} + F_{H2} \times HHV_{H2}} \tag{1}$$

and E_{net} is the net energy shown by

$$E_{net} = F_{gdiesel} \times x_{gdiesel} \times LHV_{gdiesel}$$
$$+ (F_{LPG} \times x_{LPG} \times LHV_{LPG} - W_{tot}) \tag{2}$$

where $F_{palm\ oil}$ and F_{H2} are the inlet mass flowrates of palm oil and hydrogen, respectively. $F_{gdiesel}$ and F_{LPG} are the outlet mass flowrates of green diesel and green

LPG, respectively. $x_{gdiesel}$ and x_{LPG} represent mass fractions of green diesel and green LPG in the outlet streams of $F_{gdiesel}$ and F_{LPG}, respectively.

(ii)   Net CO_2 emissions

The net CO_2 (equivalent) emissions ($CO_{2,net}$) of each design including the total CO_2 emissions ($CO_{2,tot}$) of all utilities are used in each design and net CO_2 sequestration ($CO_{2,seque}$) of producing palm oil from oil palm plantations. It is noted that CO_2 (equivalent) emissions of equipment are evaluated by using the carbon tracking tool in Aspen Plus®. The detailed calculations have been shown by the previous work (Wu et al, 2023). **Table 1** shows that η_{net} is up to 68.14%, and the corresponding $CO_{2,net}$ is negative.

Table 1 Energy efficiency and CO_2 emissions

E_{net}	kW	8696.47
η_{net}	%	68.14
$CO_{2,net}$	kg/h	-113.39

Fig. 2 shows the combination of the duty of each unit in the POPS, the corresponding CO_2 (equivalent) emissions of each equipment is shown by **Fig. 3**.

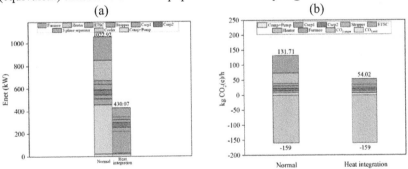

Figure 2 POPS in terms of (a) E_{tot} and (b) CO_2 (e) emissions

2.3 Economic evaluation

The economic evaluation is based on a combination of capital expenditure, operating expenditure, and carbon revenue. Through the Aspen Process Economic Analyzer (APEA), the capital expenditure (CAPEX) and operating expenditures (OPEX) of POPS are 8.5 M$ and 11.7M$/year, respectively.

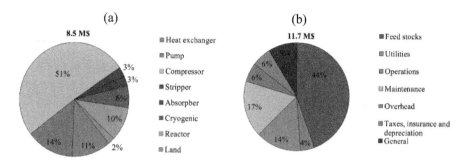

Figure 3 Pie chart of POPS in terms of (a) CAPEX and (b)OPEX

Fig. 3(a) shows that the equipment cost of the compressor dominates the total investment cost due to high hydrogen partial pressure, and **Fig. 3(b)** shows that the palm oil price dominates the total operating cost since it takes up to 44%.

Carbon Revenue

Referring to the emission trading scheme (ETS) for estimating CO_2 price, the carbon revenues (ΔP_i) is based on the following calculation (Wu et al., 2023)

$$\Delta P_i = D_i \times P_{t,i} + B_i \tag{3}$$

where D_i represents the amount of CO_2 reduction through the CO_2 capture and utilization (CCU) approaches. B_i represents the government incentives for the CCU, which here equals to 15000 USD. Moreover, the profile of annual carbon revenues of the POPS are shown in **Fig. 4** according to the carbon price from 2021 to 2050.

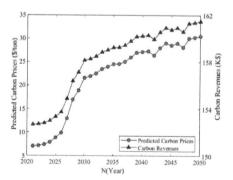

Figure 4 Carbon revenue of the POPS

Net present value

To evaluate the cumulative discounted cash flow of this project, it is noted (i) an interest rate of 15%, (ii) depreciation in five years, (iii) the plant construction for the first three years, and (iv) the working capital included in the third year. The cumulative discounted cash flow is shown in **Fig. 5**, notably, the net present value (NPV) remains negative until 2037. The payback period is more than 12 years from the start of plant operation.

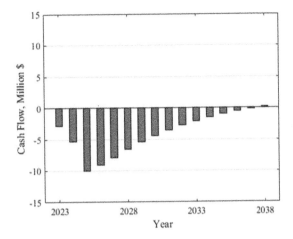

Figure 5 Cumulative discounted cash flow

Assuming an internal rate of return (IRR) of 15% and. The actual diesel and crude palm oil (CPO) prices are approximately 1.13 and 0.88 $kg^{-1}. In September 2019, the purchase price of CPO was set at 16.2 THB per kilogram (0.47 $kg^{-1}). In order to make this process economically attractive, the diesel price and CPO should become 1.98 and 0.47 $kg^{-1}, respectively.

3. Conclusions

The process design of the green diesel produced from the hydrotreating reaction of palm oil is addressed by using Aspen Plus®. The outlet compositions of green diesel, water, propane, carbon dioxide, and hydrogen from the FBHTR are effectively separated and purified through a three-phase separator and a series of cryogenic separators. The amine-based CO_2 capture process integrated with two cryogenic separators in the post-separation framework not only increases two high-purity (over 99.9%) products of carbon dioxide and propane but also it ensures the maximum negative net CO_2 emissions. The economic scenarios show that the prices of crude palm oil and carbon revenue are key factors to promote the commercial potential of the POPS.

References

Alhajji, M., Y.Demirel, Energy intensity and environmental impact metrics of the back-end separation of ethylene plant by thermodynamic analysis, Int J Energy Environ Eng. 7 (2016) 45–59. https://doi.org/10.1007/s40095-015-0194-9.

Cavalcanti, CJS, Ravagnani, MASS, Stragevitch, L, Carvalho, FR, PimentelMF. Simulation of the soybean oil hydrotreating process for green diesel production. Clean Chem Eng 2022;1:100004.

Xu, C, Y.Liu, Q.Zhang, T.Xin, R.Zhao, M.Wang, W.Saw, S.Lu, Thermodynamic analysis of a novel biomass polygeneration system for ammonia synthesis and power generation using Allam power cycle, Energy Convers Manag. 2021;247:114746.

Wu, W, Supankanok, R, Chandra-Ambhorn, W, Taipabu, MI, Novel CO_2-negative design of palm oil-based polygeneration systems. Renewable Energy 2023;203:622-633.

Wu, W, Xu, H, Shi, B, Kuo, PC, Techno-economic analysis of plastic wastes-based polygeneration processes. Chemical Engineering & Processing: Process Intensification 2023;184:109297.

Antonis Kokossis, Michael C. Georgiadis, Efstratios N. Pistikopoulos (Eds.)
PROCEEDINGS OF THE 33rd European Symposium on Computer Aided Process Engineering
(ESCAPE33), June 18-21, 2023, Athens, Greece

Design and intensification of the production process of propylene glycol as a high value-added glycerol derivative

Jahaziel Alberto Sánchez-Gómez [a], Fernando Israel Gómez-Castro [a], Salvador Hernández [a]

[a] Departmento de Ingeniería Química, División de Ciencias Naturales y Exactas, Campus Guanajuato, Universidad de Guanajuato, Noria Alta S/N, Guanajuato, Gto. 36050, Mexico. fgomez@ugto.mx .

Abstract

Glycerol is a by-product obtained in the biodiesel production. Such by-product can be used as raw material to obtain high added value derivatives. However, the reduction in energy requirements and capital costs for the conversion routes is necessary to enhance the economic indicators. This can be achieved through the intensification of the conventional glycerol processing routes. This work proposes the design and intensification of processes to produce propylene glycol (PG) from crude glycerol. Four different PG production schemes are analyzed. The conversion processes are intensified by means of thermal coupling and reactive distillation systems. Each system is designed by rigorous simulation in Aspen Plus™. As results, it has been determined that implementing a thermally coupled system in the PG purification section, a clear decrease in the required reboiler duty (37%) is obtained.

Keywords: process intensification, crude glycerol valorization, process simulation, propylene glycol production.

1. Introduction

In the biodiesel production, the transesterification reaction generates large amounts of glycerol; for instance, in 2020, 46 billion liters of biodiesel were produced worldwide, with an estimated generation of more than 4.6 billion liters of crude glycerol (REN 21, 2021). The increasing glycerol excess has motivated the development of chemical processes which employ it as the main raw material. Among the different processes for the valorization of glycerol, oxidation, hydrogenolysis, dehydration, esterification, polymerization, carboxylation and pyrolysis can be mentioned (Pagliaro and Rossi, 2010; Monteiro et al., 2018). In the oxidation reaction, the main chemical derivatives to be obtained from glycerol are glyceric acid, dihydroxyacetone, lactic acid, mesoxalic acid and oxalic acid. Some intermediate products are also obtained, such as glyceraldehyde and glycolic acid (Monteiro et al., 2018). On the other hand, the reduction of glycerol generates mainly 1,2-propylene glycol, 1,3-propylene glycol, ethylene glycol and other by-products such as propanol, methanol and carbon dioxide (Muraza, 2019).
Propylene glycol (PG) is an important chemical product that can be used as surfactant, humectant, antifreeze, solvent, and preservative. Conventionally, PG is obtained by hydrolysis of propylene oxide, which is generated from petrochemicals. Alternatively, PG can be produced by a more environmentally friendly route by using glycerol as a

carbon and hydrogen source, requiring fewer fossil sources. Although cost competitiveness remains a major constraint, the conversion of glycerol to PG is a promising route. Evidence of this is the recent announcement by ORLEN Południe about the opening by the end of 2021 of a commercial green PG plant with a production capacity of 30,000 tons per year (ORLEN Południe, n.d.).

A strategy to enhance the economic indicators on the PG production process with glycerol as raw material implies applying process intensification strategies. Process intensification (PI) is a branch of chemical engineering by means of which it is possible to obtain more efficient processes. The main objective of PI is to achieve a technological improvement in a process, so that it occupies less space, it is more efficient with energy management and is cleaner and safer (Kiss, 2019). To achieve this objective, the aims are to reduce the size of the equipment and/or the number of basic operations of the process using multifunctional equipment. In this way, PI emerges as a strategy for generating highly efficient processes.

The present work has as main objective, the simulation, intensification and assessment of the hydrodeoxygenation process to produce high purity PG (99% mol) from glycerol using thermally coupled distillation columns and reactive distillation. The performance of the conventional and intensified processes is evaluated in terms of the energy consumption and the size of the equipment, measured through the required number of stages.

2. Methodology

2.1. Thermodynamic and kinetic modelling

Considering the polar nature of glycerol, the presence of hydrogen in the reaction systems, and the presence of organic compounds in a large amount of water, the NRTL model turns out to be the most suitable approach to describe the liquid-vapor equilibrium. Also, this model has been used in the literature to describe similar mixtures (Sun et al., 2022). The binary interaction parameters for the NRTL model were obtained from the Aspen Plus database.

The production of propylene glycol ($C_3H_8O_2$) from glycerol ($C_3H_8O_3$) has as by-products ethylene glycol ($C_2H_6O_2$), methanol (CH_3OH) and water (H_2O). The reactions involved in the generation of PG are shown in Eqs. 1 and 2.

$$C_3H_8O_3 + H_2 \underset{k_1}{\rightarrow} C_3H_8O_2 + H_2O \tag{1}$$

$$C_3H_8O_3 + H_2 \underset{k_2}{\rightarrow} C_2H_6O_2 + CH_3OH \tag{2}$$

The hydrodeoxygenation of glycerol to produce PG occurs in the liquid phase, using Cu/ZrO_2 as catalyst. For this catalytic reaction, the kinetic parameters for a first-order power-law kinetic model obtained by Gabrysch et al. (2019) have been used; such kinetic model is given by the following equations:

$$r_1 = k_1 c_{H_2} \tag{3}$$

$$r_2 = k_2 c_{H_2} \tag{4}$$

Variation of the kinetic constant (k_l) with temperature is given by the Arrhenius equation:

$$k_I = k_{0,I} e^{-E_{A,I}/RT}$$
(3)

where $k_{0,I}$ is the pre-exponential factor, and $E_{A,I}$ is the activation energy. The parameters required for the calculation of reaction rates are summarized in Table 1.

Table 1.- Pre-exponential factor and activation energy for hydrodeoxygenation reactions.

I	$k_{0,I}$	$E_{A,I}$ (kJ/mol)
1	2.3×10^{10} L mol^{-1} s^{-1}	106
2	9×10^{7} L mol^{-1} s^{-1}	97

2.2. Process design and simulation

The analyzed processing schemes for PG production from glycerol are the conventional scheme (CS), a thermally coupled scheme (TCS), a reactive distillation scheme (RDS) and a thermally coupled reactive distillation scheme (TCRDS). The studied configurations are shown in Fig 1. The conventional process scheme (CS) is considered as the base case for comparison with the proposed intensified processes. The schemes have been rigorously simulated in Aspen Plus™, and the PG production process alternatives were jointly evaluated using the energy consumption (associated with operation costs) and the number of stages (associated with capital costs) as decision criteria.

Most of the components have a very low boiling point compared to the product of interest and the rate reactions are fast, then it may be appropriate to use heuristic strategies for the design of the intensified process schemes (Keller, 2014). The feed stream has a flow rate of 100 kmol/h of glycerol and water (1:1 wt.). The distillation columns operate with a pressure drop of 10 psia (0.68 atm) and the reactions are defined in Aspen Plus™ as REAC-DIS type reactions, with residence times for the gas phase of 0.8 min for the hydrodeoxygenation reaction. The number of stages is determined through a sensitivity analysis, aiming to obtain the lowest number of stages necessary to obtain PG with a minimum purity of 99% in mol.

In the design of the thermally coupled alternatives, a sensitivity analysis in the interconnection streams was carried out. This procedure consists in the manipulation of the interconnection mole flowrate (vapor or liquid) to reduce the energy requirements of such schemes while the purity of PG is maintained at 99 % mol. Additionally, in the simulation of the proposed processing schemes, it has been necessary to fix the molar flowrate at the bottoms, since, based on a previous study, it was observed that these operating variables have a greater influence on the purity of PG in the output stream. In addition, by manipulating the reflux ratio it is possible to control the energy load in the reboiler.

3. Results and discussion

Figure 2 shows the required reboiler duty for the conventional and intensified processes. It is worth to mention that all the designs are still non optimal. However, all the designs satisfy the requirements in purity and recovery of PG. The results indicate that the conventional scheme (CS) have energy consumption of around 2169 kW. Additionally, the direct thermal coupling of CS to TCS represents a more significant decrease in reboiler duty than the intensification of the process using the reactive distillation system RDS. In the case of the TCS, this process shows the lowest energy consumption (1362 kW). However, the reactive distillation thermally coupled (TCRDS) has a very similar energy consumption to the TCS. The difference is only 13 kW.

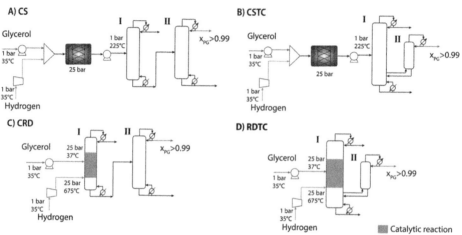

Fig 1.- Schemes for the process of propylene glycol production from glycerol, a) conventional scheme (CS), b) thermally coupled scheme (TCS), c) reactive distillation scheme (RDS), d) thermally coupled reactive distillation scheme (TCRDS).

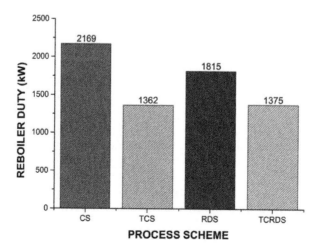

Fig 2.- Reboiler duty required in the processes for the production and purification of PG.

To complement the analysis, Table 2 presents all the design specifications, including the number of stages, the reboiler duty and the obtained savings. It has been found that for a similar value of energy requirements (1375 kW), the TCRDS requires a lower number of total stages (15) than the TCS scheme. Moreover, the TCS scheme requires a separate reactor, which would increase the equipment cost. In this sense, the TCRDS scheme would be presented as the most promissory processing scheme for PG production and purification, since it would represent the scheme associated with the lowest operating and capital costs.

Table 2.- Design specification for the conventional and intensified processes schemes.

Scheme	CS	TCS	RDS	TCRDS
Number of Stages				
Column I	20	40	20	35
Column II	40	20	35	10
Total Stages	**60**	**60**	**55**	**45**
Feed Stage				
Column I	5	5,20	11,19	9,21,22
Column II	20	20	5	10
Reflux Ratio				
Column I	0.5	0.21	0.1	0.1
Column II	2.6	0.35	2.6	0.15
Distillate Flow, kmol/h				
Column I	99.56	99.75	100	99.73
Column II	15	15.15	15	15.27
Bottoms Flow, kmol/h				
Column I	16.44	1.1	16	1
Column II	1.44	4.85	1	2.13
Reboiler Duty, kW	2169	1362	1815	1375
Savings, %	-	**37.21**	**16.32**	**36.61**

4. Conclusions

This work presents different alternatives to produce and purify PG using glycerol as raw material. Based on the topology of the conventional process, different intensified alternatives are proposed in the reaction and purification zones. The intensified processes have been contrasted with the conventional scheme. The results indicate that a maximum reduction of 37% in energy requirement can be achieved in comparation with the conventional scheme. The reduction in energy requirements lead to a more sustainable process and minor production costs of PG from glycerol. Additionally, the TCRDS shows important reductions in the number and size of processing equipment.

References

A.A. Kiss, 2019, Novel catalytic reactive distillation processes for a sustainable chemical industry, Top. Catal., 62, 1132–1148.

M.R. Monteiro, C.L. Kugelmeier, R.S. Pinheiro, M.O. Batalha, 2018, Glycerol from biodiesel production: Technological paths for sustainability, Renew. Sustain. Energy Rev., 88, 109–122.

M. Pagliaro, M. Rossi, 2010, The Future of Glycerol, RSC Publishing, 2nd Edition, Cambridge.

O. Muraza, 2019. Peculiarities of glycerol conversion to chemicals over zeolite-based catalysts, Front. Chem., 7, 1–11.

ORLEN Południe, n.d., in https://www.orlenpoludnie.pl/en/media/news/Pages/ORLEN-Południe-finalises-its-investment-in-green-glycol.aspx.

P Sun, W Zhang, X Yu, J Zhang, N Xu, Z Zhang, M Liu, D Zhang, G Zhan, Z Liu, C Yang, W Yan and X Jin, 2022, Hydrogenolysis of Glycerol to Propylene Glycol: Energy, Tech Economic, and Environmental Studies; Front. Chem, 9:778579.

REN 21, 2021, Renewables 2021 Global Status Report - REN21 in https://www.ren21.net/wp-content/uploads/2019/05/GSR2021_Full_Report.pdf.

T. Gabrysch, M. Muhler, B. Peng, 2019, The kinetics of glycerol hydrodeoxygenation to 1,2-propanediol over Cu/ZrO2 in the aqueous phase, Appl. Catal. A Gen., 576, 47–53.

T. Keller, 2014. Reactive Distillation, in A. Górak, Z. Olujić (Eds.), Distillation: Equipment and Processes, Academic Press, Boston, 261–294.

Antonis Kokossis, Michael C. Georgiadis, Efstratios N. Pistikopoulos (Eds.)
PROCEEDINGS OF THE 33rd European Symposium on Computer Aided Process Engineering
(ESCAPE33), June 18-21, 2023, Athens, Greece

Optimal Design of Heat Integrated Hybrid Dividing Wall Columns by Vapor Recompression

Dian Ning Chia, Fanyi Duanmu, Eva Sorensen[*]

Department of Chemical Engineering, University College London (UCL),
Torrington Place, London WC1E 7JE, United Kingdom
[*]e.sorensen@ucl.ac.uk

Abstract

A hybrid dividing wall column (HDWC) is a highly integrated design that combines a dividing wall column with a hybrid distillation-membrane process into a single process. It has previously been shown that HDWC can reduce the total annualized cost compared to its conventional DWC counterpart for the separation of an azeotropic mixture. The energy efficiency of HDWC can be further enhanced by vapor recompression, which is studied in this work. Since the HDWC has two reboilers, three different variations of heat integration are investigated, with the heat exchanger in the compression cycle replacing the reboiler in either the prefractionator, the main column, and for both columns, respectively. For the case study considered, it was found that the total annualized cost for the heat integrated HDWC can be up to 35% lower than the HDWC design, however, although heat integration always reduces operating costs, it may not always reduce total annualized costs.

Keywords: Process Intensification, Distillation, Hybrid Dividing Wall Column, Vapor Recompression, Optimization

1. Introduction

Process intensification (PI), the concept of integrating two or more processing units or methods within a single unit to enhance operation, is an important approach when designing more energy-intensified processes. Two of the most prominent examples of PI separations are the dividing wall column (DWC), where a ternary separation takes place within a single column shell, and the hybrid distillation-membrane process, where a membrane stage is incorporated into the design to help separate a close-boiling or azeotropic mixture. The DWC achieves energy savings through the reduction of mixing/remixing effects due to the central wall, while the incorporation of the membrane in a hybrid process provides savings by separating the mixture via a mechanism that is not based on VLE (e.g., solution-diffusion). Recently, Chia et al. (2022) introduced a hybrid dividing wall column (HDWC) that integrates both of these PI alternatives. In their work, the HDWC was compared with various standard hybrid distillation-pervaporation processes, and it was shown that the HDWC could separate an azeotropic mixture with a lower total annualized cost (TAC). The HDWC could be further improved by applying suitable heat integration methods to reduce energy consumption and TAC even further. It was also shown that heating and cooling requirements in the column part were about four times higher than that of the total energy required in the membrane part of the process. Therefore, further enhancement of energy efficiency is mainly dependent on the distillation column. There is no guarantee that heat integration methods will benefit the design (i.e., reduce TAC), as this depends on the process and the mixture, however,

Duanmu and Sorensen (2022) showed that vapor recompression can have great success in reducing both energy consumption and TAC for DWCs, thus vapor recompression will be considered and studied in this work. The HDWC process is here used to separate a binary minimum-boiling azeotropic mixture rather than a ternary mixture as for "regular" DWC, and the DWC has the central wall extending all the way to the bottom. The DWC is considered in a Petlyuk arrangement, i.e. a prefractionator and a main column.

2. Methodology

To apply vapor recompression heat integration, the condenser of the column is removed and the vapor distillate is pressurized by a compressor. A superheater before the compressor is required to maintain the vapor phase even after increasing the stream pressure. The pressurized high-temperature vapor is then utilized to vaporize the boil-up flow by replacing the reboiler with a heat exchanger. The standard hybrid dividing wall column (HDWC) is shown in Figure 1a. The design incorporates one condenser and two reboilers, which means there are a few variations that can be considered for vapor recompression-assisted HDWC. The high-temperature vapor can be used to vaporize the boil-up flow from the prefractionator (VR-HDWC-Pre; Figure 1b); from the main column (VR-HDWC-Main; Figure 1c); or from both columns (VR-HDWC-Both; Figure 1d). In the following, these three variations will be developed, optimized, and compared with the base case without heat integration (HDWC). All four structures shown in Figure 1 contain two separation units, a DWC (wall extended from the middle to the bottom) and a membrane network. For the DWC, the feed, with composition assumed to be below the azeotropic point, enters the prefractionator. A stream with composition close to the azeotropic point will leave from the top of the prefractionator (i.e., top of the wall) and enters the main column as the thermal coupling stream, while the remaining component (almost pure heavy component) will leave the system from the bottom of the prefractionator as product. The azeotropic stream will leave the DWC and pass to the membrane network. Unlike in the standard hybrid distillation-pervaporation process, where the purpose of the membrane is to purify the azeotropic mixture to achieve the product specification, here the purpose of the membrane is to assist in crossing the azeotropic point. Next, the retentate (with composition shifted to the other side of the azeotropic point compared to the feed composition) will be further separated in the main column section and almost pure light component will leave from the bottom of the main column as product.

The membrane network model proposed by Marriott and Sorensen (2003) is applied, where the membrane network contains n membrane stages connected in series, and in each membrane stage there are i membrane modules connected in parallel. It should be noted that the number of membrane modules in parallel in each membrane stage can be different and is determined through optimization. The procedure for constructing this membrane network and the overall hybrid distillation-pervaporation process can be found in Chia and Sorensen (2022). Although not shown in this work, before each membrane stage, a heater may be installed to improve the membrane separation performance and the requirement of each membrane stage heater can be optimized.

For the vapor recompression assisted HDWCs shown in Figures 1b to 1d, the condenser is replaced with a series of superheater, condenser, and trim cooler. The superheater and condenser create a pressurized vapor with high temperature, which allows heat exchange between the vapor and the liquid bottom flow from the distillation column. Three VR-

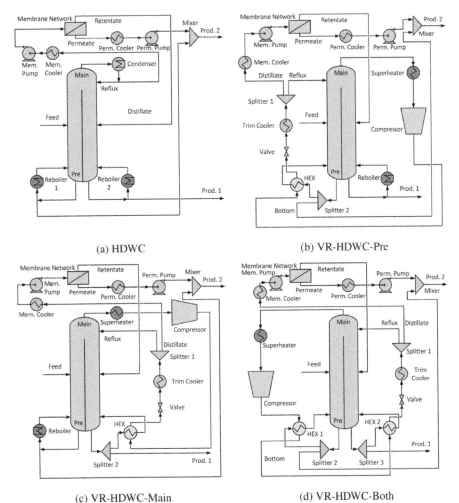

(a) HDWC

(b) VR-HDWC-Pre

(c) VR-HDWC-Main

(d) VR-HDWC-Both

Figure 1. Flowsheets of (a) the base hybrid dividing wall column (HDWC) and vapor recompression assisted HDWCs with the heat exchanger replacing the reboiler in (b) prefractionator, (c) main column, and (d) both columns.

HDWC structures are named based on the location of the heat exchanger, i.e., VR-HDWC-Pre is the system with the reboiler in the prefractionator replaced with a heat exchanger, and VD-HDWC-Main where the heat exchanger is replacing the reboiler on the main column side. In VR-HDWC-Both, the pressurized vapor heats up the liquid bottom stream from the prefractionator first (as it has a higher boiling point compared to that of the main column), and then heats up the liquid bottom stream from the main column. For all, the hot outlet from the last heat exchanger passes through a valve to reduce its pressure back to the column operating pressure. Then, the stream is cooled down to the saturated liquid as both reflux and the feed to the pervaporation membrane need to be in the liquid phase.

3. Case Studies

The case study considered is the separation of $0.2/0.8 \ mol \ mol^{-1}$ of ethyl acetate and ethanol. The feed is assumed to be saturated liquid at $1\ atm$, and is supplied at $200\ kmol\ h^{-1}$. The product purity specifications are 0.99 mol%. UNIQUAC is used to describe the liquid phase thermodynamics, while ideal gas is assumed for the vapor phase. The minimum temperature approach is set as 10 K in the heat exchangers.

All models (rigorous equilibrium-based) are built in gPROMS Process (Process Systems Enterprise, 2021). As there are no available libraries for dividing wall columns, the corresponding Petlyuk structure (not shown) is constructed using the built-in distillation column library, while the pervaporation membrane is modeled using a user-defined model. The cost equations for the distillation column can be found in Duanmu et al. (2022b), while the membrane costs can be found in González and Ortiz (2002). The optimizer chosen is the Particle Swarm Optimization (PSO), and the details of the settings can be found in Duanmu et al. (2022a). The constraints for the optimization include the product purities, using the same number of stages on both sides of the wall, a minimum approach temperature in the heat exchanger, and no phase change in the compressor (i.e., the vapor fraction in the stream after the compressor equal to one).

The key results for each optimal design are shown in Table 1. For the column design, all designs have a similar total number of stages for both the prefractionator (25-28) and the main column (27-29). The feed locations in HDWC are slightly lower compared with the VR-HDWC designs. There are no clear reasons for this, as there are no clear mixing effects observed from the liquid composition profiles in each design (not shown), indicating that the feed locations are reasonable. The changes in the feed locations may be dominated by other factors, which will be discussed later. Moving to the specifications in the main column, it is clear that the side liquid flow rate in the main column (i.e., the liquid thermal coupling stream at the top of the wall) for both VR-HDWC-Pre and VR-HDWC-Both is higher compared with the other two designs, and this is a result of the reboiler in the prefractionator being replaced with a heat exchanger. It can be seen that the trim cooler duty in both VR-HDWC-Pre and VR-HDWC-Main ($-1164\ kW$ and $-971\ kW$, respectively) are large, indicating that the energy capacity of the pressurized vapor is much higher than the energy required in the heat exchanger. Therefore, increasing the side liquid flowrate will increase the flow rate of the cold liquid in the heat exchanger, which will, in turn, maximize the energy transfer between the pressurized vapor and cold liquid, and at the same time develops a better composition profile in the prefractionator (i.e., the top vapor compositions from the prefractionator are closer to the azeotropic point than the other two designs).

Next, the smaller number of membrane modules required in VR-HDWC-Main and VR-HDWC-Both can easily be explained. In order to minimize the energy wasted (i.e., to minimize the trim cooler duty), VR-HDWC-Main and VR-HDWC-Both tend to maximize the energy exchange in the heat exchanger by prioritizing the separation of the pure component from the azeotrope in the main section of the distillation column rather than in the membrane, thus the number of membrane modules required are fewer. In other words, the sloppier membrane separation will cause the heat duty required in the heat exchanger to increase, thus utilizing more of the energy from the pressurized vapor (i.e., decreasing the trim cooler duty) and, at the same time, reducing the membrane demand, which is a win-win situation for VR-HDWC-Main and VR-HDWC-Both. This is reflected in the higher boil-up flow rate and lower pure retentate composition in VR-HDWC- Main and VR-HDWC-Both when compared to the other two designs (not shown). Considering the energy consumption, it is clear that VR-HDWC-Both has the

Table 1. Optimal results for the base hybrid dividing wall column (HDWC) and the vapor recompression assisted HDWC with the heat exchanger replacing the reboiler in the prefractionator (VR-HDWC-Pre), main column (VR-HDWC-Main) and both columns (VR-HDWC-Both).

Items	HDWC	VR-HDWC Pre	VR-HDWC Main	VR-HDWC Both	Units
Prefractionator					
Total stages	26	26	28	25	-
Feed stage	11	14	14	15	-
Bottom product *	118	128	125	128	$kmol\,h^{-1}$
Boil-up flow	201	231	214	237	$kmol\,h^{-1}$
Main Column					
Total stages	28	29	29	27	-
Feed stage (from pre.)	2	2	2	2	-
Feed stage (from mem.)	12	17	18	18	-
Side liq. stage	2	2	2	2	-
Side liq. flow rate	142	186	164	194	$kmol\,h^{-1}$
Reflux *	230	268	286	322	$kmol\,h^{-1}$
Bottom product *	39	39	39	39	$kmol\,h^{-1}$
Boil-up flow	63	54	98	105	$kmol\,h^{-1}$
Vapor Recompression System					
Superheater temp.	-	345.25	345.20	345.16	K
Compressor pressure	-	1.74	1.68	1.74	bar
Total Duty/Power (Excluding Membrane Network)					
Reboiler (pre)	2113	-	2252	-	kW
Reboiler (main)	567	488	-	-	kW
Condenser / Trim cooler	-3456	-1164	-3209	-971	kW
Compressor	-	221	234	271	kW
Heat exchangers (total)	-	2429	875	3442	kW
Membrane Network					
No. membrane stages [†]	4	4	5	4	-
Total no. modules	87	77	60	58	-
Total membrane area	522	462	360	348	m^2
Total heater duty	359	292	281	232	kW
Permeate cooling duty	-602	-463	-505	-462	kW
Capital Cost					
Column	1.21	1.35	1.38	1.31	$M\$$
Reboiler / Superheater	0.66	0.34	0.58	0.13	$M\$$
Condenser / Trim cooler	0.52	0.24	0.49	0.22	$M\$$
Heat exchanger	-	0.36	0.20	0.57	$M\$$
Compressor	-	3.56	3.62	3.78	$M\$$
Membrane	2.86	2.53	1.97	1.91	$M\$$
Others	0.81	0.79	0.92	0.79	$M\$$
Operating Cost					
Heating	1.29	0.33	1.08	0.10	$M\$y^{-1}$
Cooling	0.28	0.20	0.23	0.19	$M\$y^{-1}$
Electricity	0.00	0.11	0.12	0.14	$M\$y^{-1}$
Mem. replacement	0.08	0.07	0.06	0.05	$M\$y^{-1}$
Overall					
Capital cost	6.06	9.17	9.16	8.70	$M\$$
Annualized capital cost	0.76	1.15	1.14	1.09	$M\$y^{-1}$
Operating cost	1.65	0.71	1.49	0.48	$M\$y^{-1}$
Total annualized cost	2.40	1.86	2.63	1.57	$M\$y^{-1}$

*   For VR-HDWCs, the bottom products are controlled by Splitters 2 and 3, and the reflux is controlled by Splitter 1.

†   Upper bound for the number of membrane stages is 8.

highest duty in the heat exchanger and lowest duty in the trim cooler, indicating a higher energy efficiency of the heat integration structure.

Comparing the economic performances of each design, the capital cost (CAPEX) is much lower for HDWC (about 34 % reduction compared with VR-HDWC-Pre), and the main reason is the high cost of installing the compressor in VR-HDWCs (contributing to about 40% of the CAPEX). It should be noted that the membrane cost in VR-HDWC-Main and VR-HDWC-Both on the other hand drops significantly (up to 33%) compared with HDWC due to the smaller membrane network required, as explained earlier. Considering the operating cost (OPEX), it is not surprising that all three VR-HDWCs have lower OPEX due to the energy saving in the reboiler heating cost, especially for VR-HDWC-Both. In terms of the TAC, the objective function in the optimisation, VR-HDWC-Both is the cheapest design with about 35% TAC reduction compared to HDWC. VR-HDWC-Main, however, is about 10 % more expensive than HDWC. This indicates that the VR-HDWC designs may not always yield a lower TAC, as the VR-HDWC designs are closely related to the trade-off between the capital cost of the compressor and the operating cost. The TAC is also strongly influenced by individual costs (not shown).

4. Conclusions

In conclusion, this work proposed three vapor recompression assisted hybrid dividing wall column designs (VR-HDWC) to further improve energy efficiency compared to just a hybrid dividing wall column (HDWC). Compared with the base HDWC design, all VR-HDWC designs save operating costs, although not always total annualized costs (TAC) as the results of the vapor recompression assisted design relies heavily on the trade-off between the more expensive compressor capital cost vs the cheaper electricity cost.

References

D. N. Chia, F. Duanmu, E. Sorensen, 2022. Optimal Design of Hybrid Dividing Wall Columns for Azeotropic Separations. In: The 12th International Conference Distillation & Absorption 2022.

D. N. Chia, E. Sorensen, 2022. Optimal Design of Hybrid Distillation/ Pervaporation Processes. In: Y. Ya-mashita, M. Kano (Eds.), Proceedings of the 14th International Symposium on Process Systems Engineering - PSE2021+. Elsevier B.V., pp. 313–318.

F. Duanmu, D. N. Chia, E. Sorensen, 2022a. A Combined Particle Swarm Optimization and Outer Approximation Optimization Strategy for the Optimal Design of Distillation Systems. In: Y. Yamashita, M. Kano (Eds.), Proceedings of the 14th International Symposium on Process Systems Engineering - PSE2021+. Elsevier B.V., pp. 1315–1320.

F. Duanmu, D. N. Chia, E. Sorensen, 2022b. A shortcut design method for complex distillation structures. Chemical Engineering Research and Design 180, 346–368.

F. Duanmu, E. Sorensen, 2022. Optimal Design of Heat Integrated Reduced Vapor Transfer Dividing Wall Columns. In: Y. Yamashita, M. Kano (Eds.), Proceedings of the 14th International Symposium on Process Systems Engineering - PSE2021+. Elsevier B.V., pp. 175–180.

B. González, I. Ortiz, 2002. Modelling and simulation of a hybrid process (pervaporation-distillation) for the separation of azeotropic mixtures of alcohol-ether. Journal of Chemical Technology & Biotechnology 77 (1), 29–42.

J. Marriott, E. Sorensen, 2003. A general approach to modelling membrane modules. Chemical Engineering Science 58 (22), 4975–4990.

Process Systems Enterprise, 2021. gPROMS Process version 2.2.

Antonis Kokossis, Michael C. Georgiadis, Efstratios N. Pistikopoulos (Eds.)
PROCEEDINGS OF THE 33rd European Symposium on Computer Aided Process Engineering
(ESCAPE33), June 18-21, 2023, Athens, Greece
© 2023 Elsevier B.V. All rights reserved. http://dx.doi.org/10.1016/B978-0-443-15274-0.50306-1

Efficient use of energy in distillation sequences.

José A. Caballero,[a] Juan A. Labarta,[a] Laureano Jiménez,[b]

[a]Institute of Chemical Process Engineering, University of Alicante, Carretera de S. Vicente s.n. 03690, Alicante, Spain.
[b]Department of Chemical Engineering, Universitat Rovira I Virgili, Av. Països Catalans 26, 43007, Tarragona, Spain.

Abstract

In this work, we show that the systematic and simultaneous implementation of some heat integration alternatives like direct reboiler condenser heat exchange, implementation of vapor (re)compression cycles, leverage of heat in high-pressure steam utilities at lower pressure/temperatures, multiple effect distillation, etc. can eventually produce considerably reductions in energy and total annualized cost. While none of these alternatives is new, they are rarely simultaneously implemented in a distillation sequence, however, we show that there is great potential in the synergic effects of implementing simultaneously some of them. We illustrate the procedure with the separation of a mixture of acetone and chloroform using pressure swing distillation. Results show that we can get impressive reductions in energy and total annualized costs.

Keywords: add three to five keywords here.

1. Introduction

Even though in the chemical industry there are a good number of alternatives to separate components, distillation accounts for around 90-95% of all the separations and purification (Humphrey and Keller, 1997), and it seems that this situation will not change shortly. Notwithstanding, it is not uncommon the claim that distillation must be substituted by other, more efficient separation technologies (National Academies of Sciences, 2019). The main reason is based on the fact that distillation vaporizes a mixture in the reboiler and then liquifies a vapor stream in the condenser. While this is true in a single distillation column, Agrawal and Tumbalam Gooty, (2020) have shown that energy required in distillation is not necessarily proportional to the reboiler heat duty and that distillation can be much more efficient than what intuition seems to suggest. Even more, they presented an industrial separation example -separation of propylene from propane using distillation and membranes- that contradicts this believe. They showed that, in this case, distillation needs only a fraction of the energy requirement for membrane separation.

Different alternatives have been proposed to increase the energy efficiency in distillation. If we are dealing with a single separation (i.e. a single distillation column) we can consider alternatives like multi-effect distillation, and intermediate heat exchangers either in the rectifying or in the stripping section, -the optimal location, which can be done using column grand composite curves-, (Bandyopadhyay et al., 1998), pre-fractionations, internally heat integrated distillation columns (HIDiC) and heat pump assisted distillation (Vapor recompression, Mechanical or Thermal Vapor recompression -VRC- or absorption heat pumps) (Kiss et al., 2012) and combinations of some of the previous alternatives. Kiss et al., (2012) also gave an algorithm for selecting a technology for heat integration in binary distillation.

In multicomponent distillation, the number of alternatives increases because we can integrate the alternatives for heat integration between different columns with those of a single column and the alternatives for column sequencing inherent in multi-component distillation. Therefore, if the number of possible sequences is large, the first stage must be selecting a sequence of distillation columns with inherent good energy performance and then systematically checking for the different alternatives to increase energy efficiency.

In the case of zeotropic distillation, the structural characteristics of the search space were defined by different authors around 10-15 years ago (Caballero and Grossmann, 2006; Giridhar and Agrawal, 2010) Those authors, established the basis, and the structural considerations to take into account for generating valid sequences of columns that range from sequences based on conventional distillation columns (a column with one condenser, one reboiler, and two products) to fully thermally coupled distillation sequences (a single reboiler and a single condenser for all the separation sequence). From these sequences, it is possible to systematically generate intensified alternatives like Divided Wall Columns (DWC), and Kaibel Columns – which separate 4 components in a single column-, among others. A recent review of the advances in intensification in distillation can be found in the work by Jiang and Agrawal, (2019).

In the case of separations involving azeotropes, the number of alternatives is considerably lower than that of the zeotropic counterpart, and there are other considerations to take into account. For example, in extractive distillation, the selection of the extractant is likely the decision with the highest influence on the energy consumption of the system.

In multi-component distillation, the direct heat integration between the condenser of a column and the reboiler of another, modifying the pressure(s) in the column(s) if necessary is the most obvious alternative for saving energy, and different optimization strategies have been proposed since the 70s of the last century.

However, while there are a good number of alternatives for improving the energy efficiency in distillation it is uncommon to implement two or more alternatives simultaneously. In this work, we show that the simultaneous combination of different alternatives can produce important savings in energy (and in total annual cost). Besides we present a framework to combine all those alternatives, and of course, the new ones that could eventually appear.

2. An algorithm for efficient energy integration in distillation.

The implementation of an optimization model (i.e. through a superstructure) to simultaneously design/optimize the distillation sequence and each column in it together with all the considered alternatives for heat integration, is likely to produce very large and complex models, that could be, in the best case very hard to solve. Instead, we propose an approach that sequentially tries to improve efficiency in the selected configuration:

1. Select the best sequence of distillation columns. If we are considering a zeotropic mixture we can use any of the approaches based on mathematical programming to get the optimal configuration. For example, (Nallasivam et al., 2016) used an algorithm that ensures the global optimal solution. In those cases, it is also common that there are some alternatives with similar performance, so a priori we should select a sub-set of alternatives. In azeotropic (extractive, reactive…) distillation, the starting point is an optimized feasible separation sequence.

2. In the first place, we must consider the possibility of heat integration with the rest of the process. In some cases, there is a waste heat stream, that can be used in distillation and no further action is needed. (We do not consider such a case in this paper)

3. For each column in the sequence, we determine the lower and upper pressure limits. This can be based on thermodynamics, product stability, security or operational issues, etc. The pressure determines which are the highest and lowest temperatures in each column in the condenser and reboiler.

     3.a.   The change of temperature (pressure) allows the development of strategies for direct heat integration between condensers and reboilers of different columns by manipulating the pressure of the columns if necessary. In the literature, there are different models to do this step rigorously (Caballero and Grossmann, 2006; Floudas and Paules, 1988)

     3.b.   The limits in pressure and temperature allow for determining the possibility of multi-effect distillation which has a special interest in difficult separations.

4. When we stack two or more columns -this is the case of divided wall columns and some thermally coupled sections- it is usually necessary to increase the internal flows in some column sections with the corresponding extra energy consumption. In those cases, an intermediate heat exchanger can help to recover the system's optimal operating conditions, and in some cases, it is possible to implement a VRC that reduces, even more, the column duty (Navarro-Amorós et al., 2013)

5. If the sequence requires heat at very different temperatures, in some situations it is possible to recover energy from the exhausted high-temperature utility to use them at lower temperatures. For example, from high-pressure steam (say at 250 °C, 40 bar) that produces 100 kW when condensates in a reboiler, it is possible to recover 18.9 kW at 180 °C by expanding the exhaust water up to 10 bar and 34 kW if we expand to 1.2 bar (~120 °C).

6. The use of intermediate heat exchangers (reboilers in the stripping section and condensers in the rectifying section) can, in some cases, reduce the condenser/reboiler heat duties, which can be substituted with cheaper utilities. Besides, these new heat exchangers can eventually be used in the heat integration strategy. Column Grand Composite curves (CGCC) can be used to determine the optimal location of those heat exchangers (Bandyopadhyay et al., 1998)

7. Heat pump-assisted configurations (Kiss et al., 2012) (vapor compression, mechanical or thermal vapor recompression, absorption heat pumps, bottom flashing, etc.) can be implemented not only between the condenser and reboiler of a given column but between different heat exchangers (source and sinks of heat) whose temperature difference is not too large (typically no more than 30°C)

The systematic application of all those alternatives can yield a rich space of alternatives with an important potential for energy reduction.

Finally, considering the intensification of the sequence of columns could allow extra savings in investment. This last point is out of the scope of this work, but a review of different alternatives can be found in Jiang and Agrawal, (2019).

3. Discussion on Results

To illustrate some of the characteristics of the proposed algorithm we consider the separation of a mixture of Acetone and Chloroform using pressure swing distillation (PSD). The objective is to obtain each product with a purity of at least 0.995 mol fraction. Table 1 shows the data for the example.

Table 1. Data for the example.

Components	Composition (mol fraction)			
Acetone	0.75	Feed Flow	80 kmol/h (5872 kg/h)	
Chloroform	0.25	Pressure	101.3 kPa	

Cold Utilities	Cost ($/kW·y)	Hot Utilities	Cost ($/kW·y)
water (20-15 °C)	11.4	LP Steam (~2 bar 120 °C)	277.5
		HP Steam (~10 bar 180 °C)	292.18

Thermodynamics NRTL (default Aspen-HYSYS parameters)
Electricity 0.067 $/kWh
interest = 10% in 10 years
Cost estimation based on correlations by (Turton et al., 2013)

The base case distillation sequence is shown in Figure 1. The simulation of the PSD separation was done in Aspen-Hysys V.11, including the sizing of the columns. The base case has a total module cost of k$1,710. The total cost of utilities is 2,901.7 k$/year with a cost of manufacturing (COM) of 3,900.3 k$/year, which yields a TAC (total annualized cost) of 4,198.7 k$/y. The feeds to all the columns are forced to be saturated liquids, so it is necessary to include some heat exchangers in all the streams entering the distillation columns.

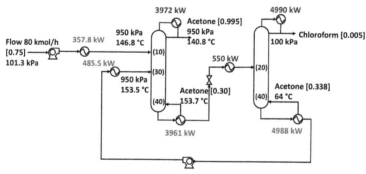

Figure 1. Separation of Acetone and Chloroform by PSD. Base case for comparison. Numbers in the square brackets refer to acetone mole fractions, and numbers in brackets refer to tray number.

Even though there is some degree of freedom in the selection of the pressures of both columns, in this case, the pressures are mainly determined by the variation of the azeotrope composition, so column pressures will be fixed in all the alternatives.

It is possible to leverage part of the energy of the exhausted MPS (water around 10 bar and 180 °C) by expanding it to 1.4 bar (~110°C). If we separate the vapor fraction it is possible to recover 593.8 kW by condensing this new low-pressure steam than can be used in the low-pressure column. Figure 2a shows the resulting process. The utility cost is reduced by 5.8 %.

Figure 2a. suggests that it is possible to increase energy efficiency and the manufacturing cost by the heat integration of the different hot and cold streams (heaters, coolers, condensers, and reboilers.) The most obvious is the integration of the condenser of the high-pressure column with the reboiler of the low-pressure column -Figure 2b-. A pinch analysis shows that it is possible to reduce the cost of utilities to 1,433 k$/year with a cost of manufacturing of 2,094.7 k$/year and a TAC of 2,392.8 k$/y. This is a reduction in the cost of utilities of 50.1% and a reduction in TAC of around 43%.

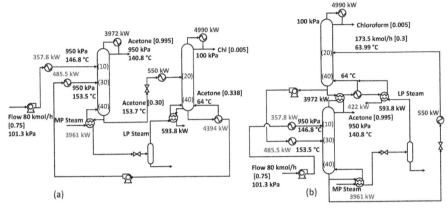

Figure 2. (a) PSD acetone and Chloroform in which the exhausted MP Steam is expanded to 1.4 bar to recover a fraction of energy in form of LP-Steam that is used in the reboiler of the LP-column. (b) include the integration between the condenser of the HP column and the reboiler of the LP column.

The temperature difference between the reboiler and condenser in the HP column is lower than 15 °C in both columns. In these conditions, a VRC usually allows for reducing energy costs, even though the capital cost can increase considerably. In this case, it is possible to implement a VRC in each column and at the same time maintain the possibility of heat integration between the condenser or HP column and the reboiler of the LP column (Figure 3).

Capital cost increases to 4,040 k$, around 120%, but energy cost decreases also considerably. A pinch analysis shows that total energy cost is reduced to 533.0 k$/y, which yields a COM of 1,382.8 k$/y and a TAC of 2,037 k$/y. Savings in energy costs and TAC are around 81% and 51.5 % respectively compared to the base case and around 63% and 14.9 % compared to the best heat-integrated sequence without VRCs.

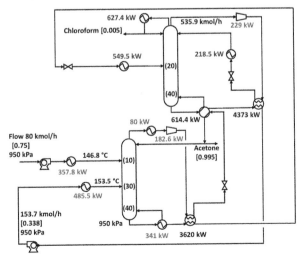

Figure 3. Integration of a Vapor Recompression Cycle in each column. Note that heat integration between the condenser in the HP column and the reboiler in the LP column is still possible.

4. Conclusions

In this paper, we have shown that it is possible to increase the efficiency of distillation columns by sequentially and iteratively adding different alternatives of heat integration. While no one of those alternatives is new, the simultaneous consideration of all/some of them can produce an impressive increase in energy efficiency that is reflected in a considerable reduction in total costs.

We have illustrated the procedure with the separation of a mixture of acetone and chloroform using pressure swing distillation. The systematic consideration of different available alternatives allows for reducing the energy cost by an impressive 81% when compared with a non-heat integrated based case or 51.5% compared with a base case using only direct heat integration. In terms of TAC, reductions are around 63% and 14.9% respectively.

5. Acknowledgments

The authors acknowledge financial support to the Spanish «Ministerio de Ciencia e Innovación» under project PID2021-124139NB-C21

References

Agrawal, R., Tumbalam Gooty, R., 2020. Misconceptions about efficiency and maturity of distillation. AIChE J. 66, 2–4. https://doi.org/10.1002/aic.16294

Bandyopadhyay, S., Malik, R.K., Shenoy, U. V., 1998. Temperature-enthalpy curve for energy targeting of distillation columns. Comput. Chem. Eng. 22, 1733–1744. https://doi.org/10.1016/S0098-1354(98)00250-6

Caballero, J.A., Grossmann, I.E., 2006. Structural Considerations and Modeling in the Synthesis of Heat-Integrated–Thermally Coupled Distillation Sequences. Ind. Eng. Chem. Res. 45, 8454–8474. https://doi.org/10.1021/ie060030w

Floudas, C.A., Paules, G.E., 1988. A mixed-integer nonlinear programming formulation for the synthesis of heat-integrated distillation sequences. Comput. Chem. Eng. 12, 531–546. https://doi.org/10.1016/0098-1354(88)87003-0

Giridhar, A., Agrawal, R., 2010. Synthesis of distillation configurations: I. Characteristics of a good search space. Comput. Chem. Eng. 34, 73.

Humphrey, J., Keller, G., 1997. Separation Process Technology. McGraw-Hill Education.

Jiang, Z., Agrawal, R., 2019. Process intensification in multicomponent distillation: A review of recent advancements. Chem. Eng. Res. Des. 147, 122–145. https://doi.org/10.1016/J.CHERD.2019.04.023

Kiss, A.A., Flores Landaeta, S.J., Infante Ferreira, C.A., 2012. Towards energy efficient distillation technologies – Making the right choice. Energy 47, 531–542. https://doi.org/http://dx.doi.org/10.1016/j.energy.2012.09.038

Nallasivam, U., Shah, V.H., Shenvi, A.A., Huff, J., Tawarmalani, M., Agrawal, R., 2016. Global optimization of multicomponent distillation configurations: 2. Enumeration based global minimization algorithm. AIChE J. 62, 2071–2086. https://doi.org/10.1002/aic.15204

National Academies of Sciences, 2019. A Research Agenda for Transforming Separation Science, A Research Agenda for Transforming Separation Science. https://doi.org/10.17226/25421

Navarro-Amorós, M.A., Ruiz-Femenia, R., Caballero, J.A., 2013. A new technique for recovering energy in thermally coupled distillation using vapor recompression cycles. Aiche J. 59, 3767–3781.

Turton, R., Bailei, R.C., Whiting, W.B., Shaeiwitz, J. A., & Bhattacharyya, D., 2013. Analysis, Synthesis and Design of Chemical Processes., 4th editio. ed. Pearson Education, Inc., Upper Saddle River, NJ. USA.

Antonis Kokossis, Michael C. Georgiadis, Efstratios N. Pistikopoulos (Eds.)
PROCEEDINGS OF THE 33rd European Symposium on Computer Aided Process Engineering
(ESCAPE33), June 18-21, 2023, Athens, Greece

The use of Design Thinking in the innovation of artificial connected reefs

Justine Michel[a], Manon Arnaud[a], Adrien Roudaut[a], Albane Tenneson[a], Lucie Mace[a], April Schnaffner[a], Mame Cheikh Sow[a], Nicolas Chaton[a], Nader Wehbe[b], Paul-Emeric Roger[b], Anne Asensio[b], Arnaud Coutu[a]

[a]Institut Polytechnique UniLaSalle, Rue Pierre Waguet, 60000 Beauvais, France
[b]Dassault Systèmes, 10 Rue Marcel Dassault, 78140 Vélizy-Villacoublay, France
arnaud.coutu@unilasalle.fr

Abstract

The climate changes that have occurred in recent years are modifying the characteristics of the environment. Seas and oceans swallow, each year, a part of the CO_2 we emit, acidifying it in consequence (*Francour and al., 2006*). To respond to this environmental alert the Dassault Systèmes Foundation launched a call for projects to UniLaSalle students to help protecting the oceans and reduce the anthropic impact. Using the Design Thinking method, the students developed solutions that links the working methods of designers and engineers. This method is a perfectly adapted approach to answer this type of problem because designers work for human beings, while engineers work for the nature. It was use in order to well understand the perimeter of the project. Its exercises allow you to ask questions about the general appearance of a project while keeping in mind how humans and/or the environment can impact it or be impacted.

Keywords: Artificial Reef, Design Thinking, Fluid simulation, Innovation, Modelling.

1. Introduction

The Dassault Systèmes Foundation had invited UniLaSalle students to participate to a call for projects. Their goal was to bring together Design and Engineering in order to mix, within the framework of science, different ways of thinking. To do that, the Foundation asked students to create a scientific project with the aim of protecting the oceans and offered them a training in Design Thinking to organize their ideas in a new way.

Climate change, which currently affects 45% of threatened marine species, is not the only stressor threatening the marine environment. It is the fifth most important factor after overfishing, transport, urban development and pollution (Wait, 2022). Therefore, it is imperative to start to focus on the preservation of the ecosystem.

During this project, Design Thinking allowed the team to highlight the importance of public awareness and the progress of research. In order to deal with this problem, the student chose to develop a project based on the creation of artificial reefs connected to sensors in order to follow the progress of the environment in which they are located. The goal was to create a database that could be used to create animations in museums or to support researchers working on specific marine environments. Step by step, the main idea switch from the development of artificial coral reefs to intelligent 'reefed' mooring posts. This change expanded the study area to regions of high anthropogenic stress and promoted the presence of fauna and flora in ports (Luff et al., 2019). On this file, it will be explained the processes of design thinking, modulization and simulation that permitted this reconversion.

2. Methodology

2.1. The Design Thinking process

> « *Most people mistakenly think that design is about appearance (...)*
> *That's not how we define design. (...) Design is how it works*»
> - Steve Jobs in 2003

This approach, based on people, was oriented towards innovation in order to create innovative services or products. The particularity of this approach is the modification of the working method (Usabilis, 2018) by asking the right questions at the beginning of the project in order to guide its development into the right direction.

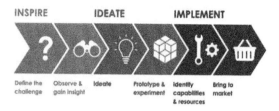

During the creation phase of the project, three workshops had been set up by the Design Studio in order to guide UniLaSalle students through the learning and use of Design Thinking. This approach made it possible to co-create projects and to determine future problems and needs by following a precise

*Figure 1: Three steps to simplify the process (**Usabilis, 2018**)*

thought process (Figure 1). In this context, the supervisor prepared exercises on a remote collaboration space called Miro. This tool consists of a whiteboard that allows you to work and share your ideas with images, post-its, mind maps, tables, text boxes, and other tools.

Design Thinking exercises allow to ask questions about the general aspect of a project while keeping in mind how humans and/or the environment can impact it or be impacted. These are lighthearted questions that specify each parameter of the project, thus allowing the team to grasp most of the difficulties it will encounter in the future (Figure 2). They follow a pattern that always involves: actors, priorities, problems and proposed solutions.

Figure 2: Exercise questions of the fortress realized by the Design Studio

To complete these exercises, the engineering students answered the questions using post-its. This method allowed everyone to express themselves and to understand the opinions of others in order to discuss the problems of the project.

Following this step, each group made a pitch explaining their project. Three meetings were held in order to answer new questions:

- <u>What is the problem?</u> Context, understanding of human problems, environmental impact, opportunities and trade-offs?
- <u>What solution is proposed?</u> To do what? At what cost? What is needed to implement it? What will be tested to verify its relevance?

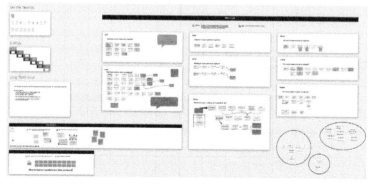

Figure 3: Example of Design Thinking work done on Miro

These pitches were used to distinguish which project would be about to be developed for an internship. The chosen subject had been the one that is presented on this paper: the "intelligent artificial reef". Therefore, the 4 members that imagined this solution constituted a team of students coming from different specialities (biology, environment, mechanics and geology) in order to develop it.

In order to create a group spirit and to communicate everyone's ideas, the new team had a formation of Design Thinking with the Design Studio supervisor. This period allowed students to bond during the important choices of the project by taking votes. Each student had post-its that they could place on whatever their choice was. The 3 choices with the most votes were then put aside to make a second round. The main objective was to make everyone agree with the development of the project by letting them discuss and debate (Figure 3).

This period also gave the students personal goals and a problem to solve. From this stage on, the students worked semi-autonomously as a small research office would. The Design Studio accompanied the group conducting weekly meetings on Friday mornings, for 30 minutes to 1 hour, to monitor the project and advise its members. These meetings began early and ended on the day of the Fondation Dassault Systèmes visit, at the end of the internship. The framing of these meetings allowed the team to refocus in this research in order to avoid misdirection and not waste time. Moreover, these project reviews allowed students to use Design Thinking by reflex, they no longer need to consult each other before a meeting because everyone follows the role of the others and knows where the project is at.

2.2. Digital designs and simulations

The Ocean ReGen team worked on the creation of an artificial reef and "reefed" mooring post, both composed of sensors to monitor the physico-chemical and biological criteria of the marine environment (sound, image, turbidity, temperature, pollution and population tracking). The sensors that make up the electrical system of both solutions could be modified according to the user's needs by slightly modifying the proposed system.

To carry out this project, the students used Dassault Systèmes solutions such as the 3DEXPERIENCE platform. This platform is a collaborative environment that allows the creation of innovative products using virtual experiences. It also provides a real-time view of project activity through its 3D Dashboards (Figure 4), which connect people, ideas and data.

This collaborative environment also allows the use of some Dassault Systèmes software, including Catia, Creative Experience and Drafting. The most important is the Catia software. It allowed students to model their reefs and moorings spots in 3D. These models were then used in the Drafting application to create constructions plans for each shape and then in Creative Experience to make virtual enhancement videos. In addition, these models were imported into XFlow, a fluid simulation software, also produced by Dassault Systèmes, but not included into the 3DEXPERIENCE platform. In order to optimize to the maximum their creation, numerous 3D models had been produced.

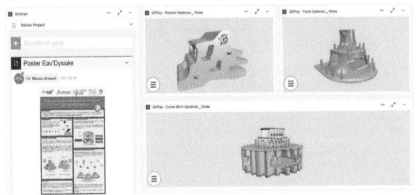

Figure 4: Example of a 3D Dashboard presenting the 3D models of the reefs made

XFlow permitted to choose the most appropriate shape for the team's needs by following some criteria... In order to attract as many species and entities as possible, reefs and moorings must had a complex shape (Charbonnel et al., 2001) and had to own shadow zones, turbidity, an upward current and a dead zone (where the current is weak or non-existent) within or behind the structure (Figure 5).

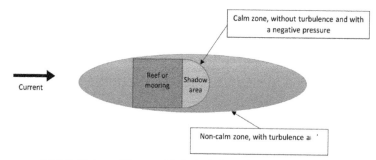

Figure 5 : Diagram of the current behavior in front of an immersed structure

Ocean ReGen students realized meetings with 6 scientists working in the domain of artificial reefs in order to have opinions on their achievements and to have new proposals. Those meetings helped the students to choose the ideal location, the shape and the materials of the reef.

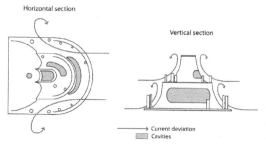

Figure 6 : Deflecting the current with an artificial reef (realised on Adobe Illustrator)

This particularly form (Fig. 6) had been developed in order to correspond to the criteria previously described in figure 5.

This form can further disperse the sand like you can see with the red arrows. This allows lateral movement of the sediment which helps slow down the waves at the back of the reef. However, this also creates an acceleration of the current on each side of the shape and creates erosion which can cause the structure to bury.

This burial is mainly due to their weight and shape. Indeed, most of the structures are made of concrete and the classic concrete weighs 250 kg/m² for 10 cm thickness (Béton, 2019). It happens that some structures are buried by 60% in only 5 years. Unfortunately, it remains one of the biggest challenges that encounter compagnies or researcher that try to create new reefs and the students weren't an exception. Indeed, because of this burial rate, the structures must be removed after few years.

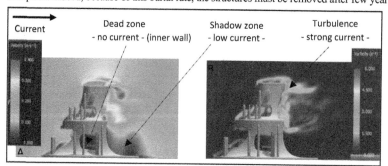

Figure 8 : Simulation of current speed (A) and turbidity (B) on XFlow® (Yacht)

Thanks to a fluid simulation software, it was possible to see that this shape respects the presence of a dead zone (without current), a shadow zone (weak current) and a zone with strong turbulences. These selection criteria were chosen because the turbulence makes noise and attracts crustaceans, the dead zones are useful for species that like areas without current and the shadow zone at the back of the structure allows a mix between the two. The objective in gathering all these criteria is to attract as many different ~~fish~~ species as possible. These three 3D models are therefore the latest versions proposed by the students (Figure 7).

3. Analysis

3.1. Project development through collaboration between Design & Engineering

The confrontation between Design & Engineering allowed the emergence of many ideas. The initial concept developed by the selected team proposed the implementation of an intelligent artificial coral reef. However, following the collective reflexion activities orchestrated by the Design Studio, the training of the different Dassault Systèmes tools and the exchanges with professionals such as Ifremer, ALR, the MIO... the project underwent many changes.

A final visit to the Fondation Dassault Systèmes permitted the Ocean ReGen project team to propose two different scenarios for the implementation of reefs. First of all, the aspect of artificial coral reefs had disappeared. Indeed, the project had split into two parts, one of which allows the implementation of an intelligent artificial reef and the other one of a reefed mooring post, also intelligent. An artificial reef would be a structure that would be voluntarily immersed in water to help the biomass to develop. A traditional mooring post is a massive concrete platform that is immersed. This structure, connected to a chain and a buoy, allows the anchoring of boats in order to avoid the ravages caused by anchors on the sea floor. The difference with the reefed mooring post would come from the shape of the structure. Unlike a massive block, the goal is to complexify its shape to create a mini reef while keeping the basic utility of a traditional mooring post. This idea came from exchanges with researchers who worked on the Port-Cros reefs immersed in 1985. Their feedback and advices guided the choices made by the group of students regarding the shape of their structures and their location.

On the other hand, the weekly design meetings also helped to guide the choice of sensors for the intelligent system while linking them to the human. To do this, the student had to answer why and to whom? In others terms: Why use these sensors and for whom would they be useful? The chosen ones are hydrophones, cameras, a tracking system, turbidity, temperature and pollution sensors. (Table 1)

Sensors		for whom?
Hydrophones	• Lack of knowledge about marine sounds. • Identify day/night organisms. • Locating disturbing anthropogenic sounds. • Monitoring against illegal fishing ...	**Museums**: diffusion of videos and aquatic sounds to show marine life to the general public
Video cameras	• Ecological monitoring of species. • Surveillance against illegal fishing. • Information on the transparency of the water and the possibility of diving.	**Researchers**: ecological monitoring of the environment and scientific discoveries
Turbidity sensors	• Detection of turbidity anomalies and thus detect certain pollutions. • Information on the transparency of the water and the possibility of diving.	**Authorities**: coastal surveillance for the protection of marine organisms by fighting against illegal fishing
Temperature sensors	• Information on ocean warming. • Understand the impact of temperatures on marine organisms.	**Swimming monitoring**: turbidity and pollution levels can be used to prohibit swimming and protect the public.
Capteurs de pollution	• Monitoring of pollution levels in ports and bays. • Monitoring the impact of pollution on marine populations	
Tracing system	• Ecological monitoring of fish species (interaction and movements).	

Table 1: Table of sensors use

In order to illustrate the process a little better... After a meeting with Atlantique Landes Récifs, the association expressed a need for cameras in order to obtain information on the environment around the area where their reefs are located. The usefulness would allow to know the level of transparency of the water thanks to the recordings video in order to be able to organize their diving missions. Public awareness was one of the main missions of the Ocean ReGen team. For this purpose, the data from the tracking systems, the sounds recorded by the hydrophones as well as the images from the cameras could be transmitted and used by museums or schools. The same is true for the data collected by the temperature and pollution sensors. Moreover, those sensors would also help the scientific progress and the development of nature. Human being knows little about marine sounds. It would have the possibility to increase the amount of data concerning the degradation of the biomass and the impact of the implantation of the structures that the ReGen Project team proposed.

The Design Studio's methods were reused many times by the team members. When it came to gathering everyone's ideas, Miro was the main tool used to create, deliberate and choose the best idea. This was helpful when it came to choosing the shape of the logo, the slogan phrase and the project name. The same was true for the choice of the reef shapes and mooring post. Dozens of 3D models were made and only 3 were chosen.

3.2. Students' feedback on the collaboration between Design & Engineering

This experience was filled with questioning, misdirection and success. The Design Thinking process allowed the group of students to see and revisit their ideas, project progress, and goals many times. Originally, the students aimed to incorporate salinity and oxygen sensors into the system in addition to temperature and turbidity... However, after a meeting with the deputy director of the MIO (Mediterranean Institute of Oceanology), instead he advised them to implement hydrophones. This technology is not very developed nowadays and marine sounds are still not well known in the scientific world. The same is true for cameras. After discussions with ALR, the usefulness of incorporating them into the system would be beneficial to them. Consequently, after having worked for several weeks on the installation of the wrong sensor device, the mechanical pole had to radically reorient its research.

The team of 9 students managed to be quick in their decision making allowing the project to develop quickly. This was possible thanks to their motivation, the advice of external professionals and the Design Thinking methods taught by the Design Studio at the beginning of their internship. The team became more cohesive and was able to focus on the task at hand.

Design Thinking allows different domains to mix and move forward together. Indeed, this method allows to build a project link by link, by considering future problems. This way of thinking is the very source of research. Generally, the students particularly appreciated the use of Design Thinking and planed of using it for their future research projects.

4. Conclusion and perspectives

Finally, the use of Design Thinking helped to direct and organize the project. The collaboration with the designers allowed the engineering students to learn a new approach to project management and the objectives of innovation. Thanks to it, the students were able to propose two scenarios that meet the needs of the environment and humans. Their work on the reefs already allows the 3D printing of the chosen shape thanks to the different construction plans made. As for the electrical system, it still needs improvements to be immersed and functional.

In order to finalize the project, UniLaSalle was considering organizing a new summer internship. This internship would focus on the development of the electronic part and would aim to realize a prototype of the reef and its intelligent system. The current team continues the exchanges with the various contacted organizations... The exchanges with ALR aimed to confirm their collaboration with the GeoLab. If this were to happen, ALR would help ReGen members with the prototyping of the electrical part and would implement it on it reefs in the Atlantic. The students were also keeping in touch with Xtree, in hopes of getting them to print the modeled reefs.

Acknowledgments

The authors gratefully thank the Design Studio, the educational research platform Le GéoLab and La Fondation Dassault Systèmes for the support provided for this work. The authors want to thank also Maxime Denis and Julien Duquennoy for their precious help.

References

Béton Direct., 2017. Quel est le poids du béton ? Tout sur le béton Date de consultation : 04/02/2023. Disponible sur : <https://www.toutsurlebeton.fr/le-ba-ba-du-beton/quel-est-le-poids-du-beton/>.

E. Charbonnel, D. ODY, L. LE DIREAC'H, 2001, Effet de la complexification de l'architecture des récifs artificiels du Parc National de Port-Cros (Méditerranée, France) sur les peuplements ichtylogiques | Parc national de Port-Cros et Porquerolles, date of consultation : 14/11/2022. http://www.portcros-parcnational.fr/fr/rapports-scientifiques/effet-de-la-complexification-de-larchitecture-des-recifs-artificiels-du-parc.

P. Francour, J.-F. Magréau, P.-A. Mannoni, J.-M. Cottalorda, J. Gratiot, 2006, Ancrages écologiques permanents.Guide d'Aide à la Gestion des Aires Marines Protégées. Université de Nice-Sophia Antipolis & Parc Nationalde Port-Cros, Nice : 68

A.L. Luff, E.V. Seehan, M. Parry, N.D. Higgs, 2019, A simple mooring modification reduces impacts on seagrass meadows, date of consultation: 30/11/2022. https://www.nature.com/articles/s41598-019-55425-y

Usabilis., 2018, Design Thinking Definition. Date of consultation : 10/11/2022. https://www.usabilis.com/quest-ce-que-le-design-thinking.

L. Wait, 2022, Unchecked global emissions on track to initiate mass extinction of marine life. In : Princeton University, date of consultation : 30/11/2022. https://www.princeton.edu/news/2022/04/28/unchecked-global-emissions-track-initiate-mass-extinction-marine-life.

Antonis Kokossis, Michael C. Georgiadis, Efstratios N. Pistikopoulos (Eds.)
PROCEEDINGS OF THE 33rd European Symposium on Computer Aided Process Engineering
(ESCAPE33), June 18-21, 2023, Athens, Greece

Substitution of hexane in vegetable oil extraction using Computer Aided Molecular Design

Mohamad Nehmeh,[a,b] Ivonne Rodriguez-Donis,[a] Vincent Gerbaud,[b] and Sophie Thiebaud-Roux.[a]

[a] *Laboratoire de Chimie Agro-Industrielle, Université de Toulouse, INRAE, INP, Toulouse (Fr)*
[b] *Laboratoire de Génie Chimique, Université de Toulouse, CNRS, INP, UPS, Toulouse (Fr)*

Abstract

Hexane is currently the preferred solvent for the extraction of oilseed oils. Despite a good selectivity towards lipids and an easy separation from the oil by distillation, it raises toxicity and non-renewable origin issues. Hence its replacement is investigated using Reverse Engineering methodology based on Computer Aided Molecular Design (CAMD). It starts by defining target values for selected physicochemical properties (boiling point, flash point, etc.) of the existing solvents used for oil extraction (Nehmeh et al., 2022). Then molecular structures which properties match these targets are sought: thousands of solvents were evaluated by means of our CAMD tool InBioSynSolv (IBSS) (Heintz et al., 2014). Results show that hexane is not the most performant solvent, impaired by a low flash point. Further refining of the candidates generated with IBSS is done by comparing their sigma-potential using COSMO-RS. In fine, alternative and greener solvents are identified for the substitution of hexane.

Keywords: Computer Aided Molecular Design, Reverse Engineering, Rapeseed oil, Oil extraction.

1. Introduction

The use of hexane in oil extraction is a widely spread practice. It is in fact the solvent of choice for industrial processes due to its selectivity towards lipids and ease of separation (Nehmeh et al., 2022). Nevertheless, this highly flammable solvent with an unpleasant gasoline-like odour, forms an explosive mixture with air (Pohanish, 2017). Besides, hexane is toxic to the environment (hazard statement H411) and is suspected of being reprotoxic (H361f) and fatal (H304) (Regulation (EC) No 1272/2008). Hitherto, its use in the production of foodstuff is still allowed under strict regulations, i.e. a maximum residue of 1 ppm in the oil (Directive 2009/32/EC).

The time and resource consuming basic "trial and error" methodology for finding solvent substitutes is nowadays replaced by a Reverse Engineering approach based on Computer Aided Molecular Design (CAMD). The methodology starts by defining a set of physicochemical properties related to the main solvent functionalities, along with the corresponding target values for each property. It is then followed by a search or a design of solvents meeting these requirements.

In this work, CAMD is used for the substitution of hexane in the extraction process of rapeseed (canola) oil. The solubility of the oil in the alternative solvents is first assessed using Hansen Solubility Parameters. Chem21 solvent selection guide (Prat et al., 2016) is then used to evaluate the safety, health and environmental impacts of the solvents, based on the hazard statements (GHS) available to date. A second validation is done by comparing the sigma-potential profile of the solvents and the oil, using COSMO-RS.

2. Methodology

Reverse Engineering is a 3-phase process. The first phase, the intelligence phase, consists of identifying the problem and defining a comprehensive list of solvent functionalities, described by target physico-chemical properties (e.g. solubility power described by Hansen Solubility Parameters). Afterwards, an individual performance for each property as well as a global performance function are defined. In the second phase, the design phase, two paths can be considered: the evaluation of solvents or the design of molecular structures using a CAMD tool. The last phase is about the choice, and deals with the selection of the best candidates, either pure components or mixtures.

2.1. Hansen Solubility Parameters

Grounded on the "like dissolves like" rule, the interactions between a solute and a solvent can be described using the Hansen Solubility Parameters (HSP), δD, δP and δH (Hansen, 2007). The cohesive energy density δ_T^2 (Hildebrand solubility parameter) is defined as the amount of energy required to break a solvent into gas molecules, divided by its molar volume. It is further approximated as the sum of the energy densities: δD^2, δP^2 and δH^2, for London atomic dispersion, Keesom polar forces and hydrogen bonds. This can be visualized by plotting the parameters as coordinates into a 3D-space. The HSP values for a given solute, as well as a sphere radius encompassing solvents that solubilize the solute, can be determined experimentally by a series of solubility tests in solvents with known parameters. The sphere is constructed as to include a maximum of the solvents miscible with the solute. According to the theory, similar HSP values for a solvent and a solute indicate a higher affinity between them. Furthermore, the Relative Energy Difference RED for a solvent can be calculated as the ratio of the relative distance D between the solute and the given solvent, and the sphere radius R. A solvent "inside" the sphere, thus having a RED<1, should dissolve the solute.

2.2. CAMD method by using IBSS

IBSS (InBioSynSolv) is a CAMD tool developed in the Laboratoire de Génie Chimique (LGC), based on a predictive and screening process. The prediction of physicochemical properties takes place through various implemented group contribution methods. As defined in the intelligence phase step, several physicochemical properties must be matched, each having a performance function *Perf* (eq.1) based on the target value for the property. For a given property *p*, the function compares the target value *P* with the predicted value *x*. The performance is set to 1 if the predicted value meets the target value, or is computed by a Gaussian function having a tolerance *tol* for a deviation *val*. A global performance function *GloPerf* (eq. 2) then transforms the multi-objective problem into a single-objective problem. During the molecular design, an optimization function *OF* (eq. 3) subject to *i* inequality and *k* equality is used in the aim of maximizing the *GloPerf*. The optimization variables are the molecular graph structure MG_i of the component *i* and the conditions $cond_j$ (conditions under which the properties calculated e.g. temperature, pressure) (Heintz et al., 2014).

$$Perf_p = [\ln(val)] \times \left(\frac{P-x}{tol}\right)^2 \ \textit{(eq.1)} \qquad GloPerf = \frac{\sum_{p=1}^{np} w_p \times Perf_p}{\sum_{p=1}^{np} w_p} \ \textit{(eq.2)}$$

$$OF = \max\left(GloPerf(MG_i, cond_j)\right) \ \textit{(eq.3)} \qquad \text{Such that } P_k\left(MG_i, cond_j\right) = P_{k,fixed}$$

$$P_{l,lowerbound} \leq P_l\left(MG_i, cond_j\right) \leq P_{l,upperbound}$$

Table 1: Solvent properties, target values and used models

Functionality	Property	Target value	Parameters	Models and individual weights
Solvency power	δD	$12.6 < \delta D < 20$ Based on the values of the test solvents that solubilized the oil	$\delta D < 12.6$ Val=0.8, Tol=0.5 $\delta D > 20$ Val'=0.5, Tol'=2.0	MB2010: 1.5 HSKASG2012:1.5
	δP	$\delta P < 12$ Based on the values of the test solvents that solubilized the oil	$\delta P > 12$ Val=0.7, Tol=2	MB2010: 1.5 HSKASG2012: 1.5
	δH	$\delta H < 19$ Based on the values of the test solvents that solubilized the oil	$\delta H > 19$ Val=0.5, Tol=2.0	MB2010: 1.5 HSKASG2012:1.5
	RED	$RED < 1$ A solvent with a RED < 1 should theoretically solubilize the oil	$RED > 1$ Val=0.4, Tol=0.5	MB2010: 4
Liquid state	Melting point	Melting point = 0 °C to keep solvent in the liquid form in normal temperature conditions	$MP > 0$ Val=0.8, Tol=2.0	HSKASG2012: 1 JR1987: 1
	Vapor pressure	Vapor pressure < 1×10^{-4} bar This value corresponds to the limit of what is a VOC (Directive 1999/13/EC)	$VP > 1 \times 10^{-4}$ Val=0.7, Tol=0.005	Riedel1954: 0.5
	Acentric factor	Acentric Factor < 1 Based on the values of the test solvents that solubilized the oil	$AF > 1$ Val= 0.8, Tol= 0.2	CDJ1993: 0.5 HSKASG2012: 0.5
Low flammability	Flash point	$FP > 23$ °C Solvents with Flash Point <23°C are said highly inflammable	$FP < 23$ Val=0.6, Tol=2.0	CPN2006: 1.5 HSKASG2012: 1.5
Energy related property	Boiling point	60 °C $< BP < 150$ °C These values are in the set of specifications of the project	$BP < 60$ Val=0.5, Tol=10 $BP > 150$ Val'=0.5, Tol'=20	HSKASG2012: 1 MG2001: 1 JR1987: 1
	Vaporization Enthalpy	Vaporization Enthalpy < 50 kJ/mol The value is based on the enthalpy of vaporization of water	$VE > 50$ Val=0.8, Tol=2.0	Riedel1954: 1 CG1994: 1
Transport property	Density	0.65 g/cm³ $< \rho < 1.3$ g/cm³ Based on the values of the test solvents that solubilized the oil	$\rho < 0.65$ Val=0.1, Tol=0.1 $\rho > 1.3$ Val'=0.1, Tol'=0.075	MB2010: 1
	Viscosity	Viscosity < 5 mPa.s Based real data of the test solvents that solubilized the oil	$\mu > 5$ Val=0.5, Tol=0.5	JR1987: 0.5 CMMG2008: 0.5
	Surface tension	15 mN/m < ST < 40 mN/m Based on the values of the test solvents that solubilized the oil	$ST < 15$ Val=0.8, Tol=1 $ST > 40$ Val'=0.8, Tol'=1	Pitzer1995: 0.5 ZuoStenby1997: 0.5
Water miscibility	K_{OW}	Log $(K_{OW}) < 4.5$ Bioconcentration in living organism is a concern for high values of K_{OW}	Log $(K_{OW}) > 4.5$ Val=0.4, Tol=0.6	MG2002: 0.5 HSKASG2012: 0.5
	W_S	Log $(W_S) < 4$ log(mg/L) Low aqueous solubility is sought as the seeds contain some moisture	Log $(W_S) > 4$ Val=0.4, Tol=0.5	MG2002: 1

2.3. Alternative solvents design methodology

The aforementioned three phases of Reverse Engineering methodology based on CAMD can be further refined in practice into five steps. The first step is the definition of the design problem by determining all the functionalities of the alternative solvents e.g. transport properties. The second step consists of converting the functionalities into physio-chemical properties that can be predicted e.g. volatility is converted to boiling point. For each property, a target value is determined based on the value of the existing solvents, or the process requirements (e.g. high flash point). The tolerance for each property performance function *Perf* is then defined, along with a weight for each property to compute the global performance *GloPerf*. The third step is the identification and choice of the required property models (group contribution methods selected as to provide the smallest relative error with experimental data collected). Table 1 summarizes these three steps. The fourth step can be divided into 2 parts: (1) evaluation of solvents and (2) molecular design and structural optimization of compounds based on pre-defined chemical moieties implemented in the tool. This step provides a list of the best candidates along with their corresponding *GloPerf* values. The last step is the choice step. During this step, the best candidates are screened based on various criteria. Herein, the Chem 21 Solvent selection guide was employed for the screening of candidates based on safety, health and environment criteria. Each criterion is given a score between 1 and 10, 1 being the least impacting in the corresponding category. The safety criterion is mainly based on the flash point of the solvent, along with other properties that were ignored herein due to the lack of data (electrical conductivity, etc.). The health criterion takes into consideration the hazard statement given by REACH as well as the boiling point (as an assessment of volatility). The last criterion is also based on the environment hazard statement (H4xx) as well as the boiling point, as an estimation of VOC emissions but also the ease of recycling and separation from the product. A solvent is then classified as "recommended", "problematic" or "hazardous" based on the three scores.

2.4. COSMO-RS computational method

COSMO-RS is a powerful theory based on a quantum-chemical approach, combined with statistical thermodynamics, for molecular description and thermodynamic properties prediction based on the chemical potential μ of the molecule (Klamt, 1995; Eckert and Klamt, 2002). For molecules not found in the COSMObase, COSMO files are generated using a BP-TZVPD-FINE basis set, after geometry optimization using Turbomole V7.4 (GmbH). Statistical thermodynamics assure the transition from the microscopic molecular interactions, described by sigma (σ)-profile, to macroscopic thermodynamic properties, by means of the (σ)-potential. The latter is a measure of the system's affinity to a surface of polarity σ. The (σ)-potential similarity (SPS) is calculated as the sum of the differences between the 2 molecules using COSMOtherm (BIOVIA). It corresponds to the similarities of the compounds with respect to their properties as solvent. The smaller the compound's chemical potential profiles overlapping, the smaller the SPS. Rapeseed's bio-chemical composition varies widely depending on the seed's variety and other factors such as environmental conditions (Nehmeh et al., 2022). Triolein (OOO), is one of the main triglycerides present in the rapeseed oil with concentrations up to over 70% (Guan et al., 2016) and will be used as the base for the SPS calculations.

3. Results and Discussion

3.1. Oil Sphere

After a series of solubility tests (25 wt% of oil) using a list of solvents carefully selected to fill and represent the Hansen space, the HSP of rapeseed oil were determined using HSPiP software. As the sphere did not have a perfect fit (around 0.9), the calculations were repeated 40 times in order to get a representative mean and standard deviation of the sphere centre coordinates as well as the radius (δD: 15.7 ± 0.24, δP: 2.8 ± 0.44, δH: 1.6 ± 0.62, R: 10.5 ± 0.64), as illustrated in figure 1. As can be predicted, the oil shows very low polarity and hydrogen bonding ability, mainly due to the presence of the triple ester groups ($-CH_2OOC-$) on the glycerol backbone.

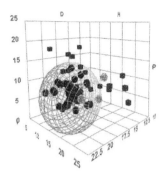

Figure 1: Hansen sphere for rapeseed oil.

3.2. IBSS Evaluation

A list of over 3900 molecules (real and theoretical) were evaluated using IBSS. No molecular structure optimization was carried out. Over 2000 molecules showed a global performance higher than 0.8. This result is somewhat expected, as the oil sphere radius, 10.5, is quite large, and many molecules are within the solubility sphere (RED<1: the property with the most weight). Therefore, the focus was put on the molecules that are closer to the sphere centre (RED<0.5), with acceptable boiling point (<160 °C), decreasing the selection down to 738 molecules. Over half of the remaining molecules were low polar hydrocarbons, like hexane, and are known to be usually toxic; they were then discarded. Special attention was given to oxygenated hydrocarbons, as their biosynthesis is generally possible, along with molecules containing heteroatoms e.g. nitrogen and fluorine. The Chem21 solvent selection guide was then applied to the 299 molecules left using all the hazard statement available to date, resulting in a list of 72 so-called "recommended" molecules, as per the same guide.

Name	Similarity	GloPerf	RED	BP (°C)	FP (°C)	Safety criteria	Health Criteria	Environment criteria	Final Criteria
Hexane	0.33	0.88	0.22	69	-24	7	7	7	Hazardous
Mol. 1	0.76	0.98	0.24	142	42	3	2	5	Recommended
Mol. 2	0.76	0.94	0.21	118	14	4	2	3	Recommended
Mol. 3	0.75	0.92	0.15	166	48	3	2	5	Recommended
Mol. 4	0.7	0.88	0.30	123		5	1	3	Recommended
Mol. 5	0.64	0.94	0.31	103	5	4	2	3	Recommended
Mol. 6	0.53	0.99	0.12	142	25	3	2	5	Recommended
Mol. 7	0.53	0.94	0.47	116	20	4	2	3	Recommended
Mol. 8	0.52	0.99	0.37	139	32	3	4	3	Recommended
Mol. 9	0.51	0.94	0.44		32	3	1	5	Recommended
Mol. 10	0.51	0.97	0.45	110	24	3	4	3	Recommended

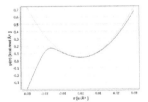

Figure 2: Molecules with the highest SPS compared to hexane (BP & FP are experimental data).

Figure 3: Triolein (red) vs hexane (green) sigma-potential.

3.3. Sigma Profile and solubility calculation

The sigma potential similarity of triolein with the 72 recommended molecules (GloPerf>0.8, RED<0.5) from IBSS evaluation was then calculated. Figure 2 represents the most interesting molecules along with hexane. Based on the RED value, hexane seems like a good solvent for the solubilization of oil. Nevertheless, its SPS is very low, as its σ-potential profile shows a quite different shape than that of the oil (figure 3). Furthermore, hexane's global performance is low (0.88), mainly due to its very low flash point, but also its relatively high vapour pressure (>0.1 bar), and it is classified as hazardous. Meanwhile, 10 of the studied molecules had a SPS higher than 0.5, and 40 other got a SPS higher than that of hexane, meaning that they potentially have better interactions with the oil.

4. Conclusion

Alternative solvents for the extraction of rapeseed oil have been evaluated and compared with the conventional hexane, using Reverse Engineering approach. The Hansen Solubility Parameters of the oil were determined experimentally. Then, an evaluation and screening of thousands of molecules was achieved using IBSS, resulting in 299 molecules with a global performance higher than 0.8, and a Relative Energy Difference lower than 0.5. Among those, 72 are thought to have no or very little impact on safety, health, and environment. The σ–potential of these solvents as well as that of hexane were then compared to that of triolein. It can be clearly and safely said that hexane is theoretically not the most suitable solvent for oil extraction. Further experimental validation of the performance of the evaluated molecules is essential. Moreover, further research on the availability, price, and route of production of each alternative solvent is needed. Coupling Reverse Engineering with CAMD has allowed the efficient selection of new potential candidates for the substitution of hexane.

Acknowledgments

This research was funded by the 3BCAR Carnot Institute.

This work has benefited from state aid managed by the Agence Nationale de la Recherche under the Investissements d'Avenir program bearing the reference ANR-18-EURE-0021.

References

BIOVIA COSMOtherm, Release 2022; Dassault Systèmes. http://www.3ds.com.

F. Eckert, A. Klamt, 2002. Fast solvent screening via quantum chemistry: COSMO-RS approach. AIChE Journal 48, 369–385.

European Parliament, Council of 11 March 1999. Directive 1999/13/EC on the limitation of emissions of volatile organic compounds due to the use of organic solvents in certain activities and installations. URL http://data.europa.eu/eli/dir/1999/13/oj/eng (accessed 11.28.22).

European Parliament, Council of 23 April 2009. Directive 2009/32/EC on the approximation of the laws of the Member States on extraction solvents used in the production of foodstuffs and food ingredients. URL https://eur-lex.europa.eu/eli/dir/2009/32/oj (accessed 28.11.22).

M. Guan, H. Chen, X. Xiong, X. Lu, X. Li, F. Huang, C. Guan, 2016. A Study on Triacylglycerol Composition and the Structure of High-Oleic Rapeseed Oil. Engineering 2, 258–262.

C. Hansen, 2007. Hansen Solubility Parameters: A User's Handbook, 2nd ed. CRC Press, Boca Raton.

J. Heintz, J.-P. Belaud, N. Pandya, M. Teles Dos Santos, V. Gerbaud, 2014. Computer aided product design tool for sustainable product development. Computers & Chemical Engineering 71, 362–376.

A. Klamt, 1995. Conductor-like screening model for real solvents: a new approach to the quantitative calculation of solvation phenomena. The Journal of Physical Chemistry 99, 2224–2235.

M. Nehmeh, I. Rodriguez-Donis, A. Cavaco-Soares, P. Evon, V. Gerbaud, S. Thiebaud-Roux, 2022. Bio-Refinery of Oilseeds: Oil Extraction, Secondary Metabolites Separation towards Protein Meal Valorisation—A Review. Processes 10, 841.

R.P. Pohanish, 2017. Sittig's Handbook of Toxic and Hazardous Chemicals and Carcinogens. William Andrew.

D. Prat, A. Wells, J. Hayler, H. Sneddon, C. Robert McElroy, S. Abou-Shehada, P. Dunn, 2016. CHEM21 selection guide of classical-and less classical-solvents. Green Chemistry 18, 288–296.

Regulation (EC) No 1272/2008 of the European parliament and of the council of 16 December 2008, 2008. Reach Classification, Labelling and Packaging of Substances and Mixtures.

TURBOMOLE V7.4, TURBOMOLE GmbH, Karlsruhe, 2019, http://www.turbomole.com.

Antonis Kokossis, Michael C. Georgiadis, Efstratios N. Pistikopoulos (Eds.)
PROCEEDINGS OF THE 33rd European Symposium on Computer Aided Process Engineering
(ESCAPE33), June 18-21, 2023, Athens, Greece

Innovative process for manufacturing pharmaceutical mini-tablets using 3D printing

Varun Sundarkumar,[a] Wanning Wang,[a] Zoltan K. Nagy,[a] Gintaras Reklaitis[a]

[a]*Davidson School of Chemical Engineering, Purdue University, West Lafayette Indiana 47906 USA*

Abstract

Providing drug products for pediatric patients is a challenging problem for the pharmaceutical industry. Children often require flexible low-dose medication with features like taste-masking and ease of swallowing. In recent years, mini-tablets have emerged as an attractive dosing solution that can meet these requirements. They are small form oral dosages around 2-4 mm in diameter that can be dispensed individually or in combination. Conventionally, they are made using methods like direct compression and hot melt extrusion. This study introduces a new technique to make mini-tablets: drop-on-demand 3D printing. Here the active ingredient is suspended in a liquid excipient, the formulation is printed as droplets and each drop is solidified to yield a mini-tablet. An optimal solvent bath that can uniformly capture mini-tablets is designed and dosages of Atorvastatin (active ingredient) are produced as a test case. Quality of these dosages is determined by measuring their content uniformity.

Keywords: pharmaceuticals, mini-tablets, 3D printing, drop-on-demand, low dose high precision drug products

1. Introduction

Pediatric medicines make up less than 10% of the overall drug market. Serving this patient group however poses a major challenge for the pharmaceutical industry. Factors like limited excipient compatibility, prevalence of rare diseases and smaller pediatric patient populations contribute to this (Milne & Bruss, 2008). Drug products made for children must also cater to their specific needs like having low dose strengths, taste masking and ease of swallowing etc., which in-turn make their production difficult.

Currently, two categories of drug products are popular for medicating children: liquid oral dosages and compounded dosages. Liquid oral drug products are attractive as they satisfy many of these requirements: they are flexible to dose, easy to consume and can be dispensed in low dose strengths. Compounded dosages are made in special pharmacies (called compounding pharmacies) where drug products are made at non-commercial specifications, by either altering the dose or changing the ingredients, to meet the needs of the patient. Both these products, however, face some limitations: 1) many active pharmaceutical ingredients (APIs) are unstable in aqueous media and prone to degradation 2) liquids are also susceptible to inaccurate dosing leading to adverse drug reactions. 3) In compounded dosages, achieving drug content uniformity and release profile matching commercial products is difficult (Zuccari et al., 2022).

In recent years, mini-tablets have emerged as an attractive alternative for pediatric medication. These are small-size solid oral drug products (2-4 mm in diameter) that are easy to swallow and can be dispensed individually or in combination in flexible doses. Two broad categories of techniques have been used to make them: powder based and 3D printing based. Powder based techniques include conventional methods like granulation and direct compression (Stoltenberg & Breitkreutz, 2011). Here the API is blended with excipient powders to yield a powder formulation. In granulation this formulation is processed into coarse aggregates. Each granule aggregate is then dispensed as a mini-tablet. In direct compression the powder formulation is compacted using small dies into mini-tablets. These methods rely on existing process technology and thus are easy to implement. They also can achieve high production rates. However, manufacturing low dose drug products is challenging for poorly flowing API powders. The second category of techniques is 3D printing, this is a manufacturing process in which the product is made in a layer-wise fashion. Hot melt extrusion is the most used 3D printing method for making mini-tablets (Krause et al., 2021). Here the API-excipient blend is extruded into a thin filament which is then melted and formed into the desired shape. Drug loading in the dosage can be easily customized by varying the number of printed layers. A key limitation of this method is the need for high operating temperatures which can cause degradation of the API or change its polymorphic form.

This study introduces a new method for making mini-tablets: using a drop-on-demand (DoD) inkjet printer. DoD falls under the umbrella of 3D printing methods, it processes a liquid formulation where the API is either dissolved or suspended in an excipient carrier fluid (Içten et al., 2017). The formulation is printed as droplets onto substrates like placebo tablets or capsules to make the drug product. This study aims to manufacture mini-tablets by solidifying individual drops generated by the DoD system. Printed droplets are discharged into an inert and viscous solvent bath where they settle slowly and solidify. Solid droplets are then washed and air dried to yield ready-to-use mini-tablets. The main advantage of this technique is producing dosages having low drug loadings with high precision. To develop this manufacturing route, first an optimal solvent bath is designed to uniformly capture and solidify the printed drops and next, mini-tablets for an API called atorvastatin are printed and tested for content uniformity.

2. Drop-on-demand printing of mini-tablets

As discussed before, DoD is a novel pharmaceutical 3D printing technique that builds dosages by printing multiple drops of an API containing formulation onto inert substrates. It has a versatile operation and can print a variety of formulation (melts, suspension etc.) and produce dosages with broad range of drug loadings. It can also incorporate emerging developments in pharmaceutical manufacturing like continuous processing and end-to-end operation (Sundarkumar et al., 2022).

Its working principle is as follows: the formulation ink for printing is held in the reservoir with constant agitation to ensure concentration homogeneity. Then a high-precision positive displacement pump dispenses a packet of fluid with constant volume through a nozzle in the form of a drop. This drop is then deposited on the desired substrate. The central idea in this study is to capture printed droplets and process them into individual mini-tablets. To achieve this a melt formulation is used, this formulation is then printed into an inert solvent bath (silicon oil) where the droplets self-solidify at room temperature (Figure 1). These solidified drops are then washed dried and collected as ready to use mini-tablets. Details of how the bath is designed is discussed in the results section.

Drug products have many critical quality attributes, such as, content uniformity, residual solvent content, dissolution behavior etc. For 3D printed drug products, content uniformity is an important metric as precise dosing is one of its key features. Thus, to test the quality of printed mini-tablets, uniformity in shape, weight and drug loading content across dosages are measured.

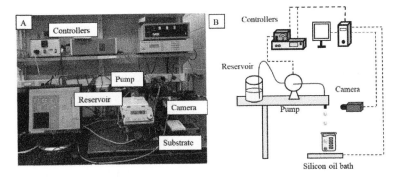

Figure 1. A] Apparatus and B] Schematic of the DoD printing system.

3. Materials

3.1. Chemicals
Atorvastatin (Dr. Reddy's Laboratories), Polyethylene glycol 2000 (PEG) (Fisher chemicals), Silicon oil (polysiloxane, PMX 200 1000 cSt, Dow Chemical Company), Hexamethyldisiloxane (HMDSO) (Fisher chemicals). Before designing the process, the excipient material needs to be selected. PEG is an excipient widely used in drug products and acts as a dissolution enhancer. It melts at ~ 65 °C and is a solid at room temperature. To solidify the printed droplets, silicon oil is used as a bath solvent as it is inert and does not interact with either the API or PEG. HMDSO is used as a washing fluid for the mini-tablets as it is a lighter inert silicon oil that can remove heavier oil adhering to the tablets.

3.2. Instruments
Nikon Eclipse E600 microscope is used to image and size mini-tablets, Waters ultra-pressure liquid chromatography (UPLC) is used to measure drug loading in mini-tablets.

4. Results

4.1. Designing the silicon oil bath
The first step in process development for this route is to design the inert solvent bath. Heat transfer (convective) due to the motion of the drop in the bath needs to be sufficient to solidify the mini-tablet. The critical process parameters here are bath viscosity and final drop temperature. Having low viscosity or high final drop temperature can lead to incomplete solidification or pooling of the droplets in the settling chamber respectively. On the other hand, high viscosities can lead to droplet aggregation (Figure 2). Optimal bath design is the one with the smallest settling chamber height (to reduce solvent amount required) and with the fastest rate of heat transfer.

To achieve this, a multi objective optimization problem can be posed with the goal of minimizing chamber height and solidification time (higher the heat transfer rate, shorter the solidification time). Decision variables here are bath viscosity η and final drop temperature T_f. Both variables are continuous: continuum in viscosity can be achieved by blending together base silicon oils (commercially available products) in required proportions. The bath is assumed to be at room temperature and thus temperature dependence of oil viscosity is ignored. To prevent pooling, an upper bound constraint is introduced on the final drop temperature; and to prevent drop aggregation, a minimum value is set for the distance between successive drops. Problem formulation is as follows:

$$min \ l_{chamber}$$

$$min \ t_{solidification}$$

$$st: \quad T_f < 35^o\ C, \quad \eta < \frac{5}{18} g d_{MT}(\rho_{MT} - \rho_{Si})$$

$$\eta \in [0.1, 10]\ Pas, \quad T_f \in [23, 35]\ ^\circ C$$

$l_{chamber}$ and $t_{solidification}$ are calculated as follows:

1. Physical properties of the system: $d_{MT} = 4\ mm$, $\rho_{MT} = 1124\ kgm^{-3}$, $C_{p_{MT}} = 2135\ Jkg^{-1}K^{-1}$, $k_{MT} = 0.31\ Wm^{-1}K^{-1}$, $T_i = 70^oC$, $\rho_{Si} = 950\ kgm^{-3}$, $k_{Si} = 0.15\ Wm^{-1}K^{-1}$, $C_{p_{Si}} = 1250\ Jmol^{-1}K^{-1}$, $T_{bath} = 23\ C$, $\Delta H_{melt} = 100\ kJkg^{-1}$, $T_{melting} = 45^oC$

2. First, calculate terminal settling velocity of the tablets. $v_t = \frac{g d_{MT}^2}{18\eta}(\rho_{MT} - \rho_f)$.

3. Next, compute heat transfer rate (assume convection dominated). $h = Nu\frac{k}{d}$, $Nu = 2 + \left(0.4Re^{0.5} + 0.06Re^{\frac{2}{3}}\right)Pr^{0.4}$, $Re = \frac{\rho_f v_t d}{\eta}$.

4. Finally, calculate solidification time $t_s = \frac{\rho_{MT} V C_P}{hA}\ln\left(\frac{T_i - T_b}{T_f - T_b}\right) + \frac{\Delta H \rho V}{hA(T_m - T_b)}$ and $l_{chamber} = v_t * t_s$.

5. Verify if convection driven heat transfer assumption is valid, i.e., if $\frac{hd}{k} \gg 1$.

The first constraint acts as a variable bound, 35 °C is chosen because it is half the initial temperature and the formulation is fully solid at this temperature. For the second constraint, we assume that for aggregation to be arrested a gap of at least one drop diameter between successive drops is necessary $l_{inter\ drop} > d_{MT}$. As drop generation rate is constant (one every 5 sec), the constraint can be framed in terms of settling velocity: $v_t > \frac{d_{MT}}{5}$ which is equivalent to $\eta < \frac{5}{18} g d_{MT}(\rho_{MT} - \rho_{Si})$.

The problem is solved by building a Pareto front (figure 2) using the Non-dominated Sorting Genetic Algorithm II (NSGA2) available in the Python library 'pymoo'. Although this algorithm does not guarantee global optimality, it provides a quick and efficient way to generate diverse non-dominated solutions that can represent the best trade-offs between objectives. From this Pareto set, the point closest to having a bath viscosity of 0.95 Pas is chosen for operation as it is closest to the material at hand. The operating point chosen is as follows:

$$\eta = 0.95\ Pas, T_f = 34.9, l_{chamber} = 6.5\ cm\ and\ t_{solidification} = 40.8\ sec.$$

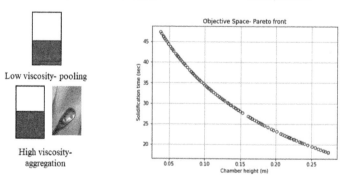

Figure 2. Design of solvent bath and settling chamber- optimization solution.

4.2. Consistency measurements for the printed mini-tablets

After determining bath viscosity, mini-tablet manufacturing is carried out. Drug products with two dose levels of atorvastatin are targeted: 0.1 and 1 mg. In each case formulation drops are printed into the bath, solidified, washed with HMDSO and air dried (figure 3).

To test whether this operation yielded consistent mini-tablets, uniformity of three parameters is measured across dosages: shape, weight and drug loading. To measure variation in shape, the diameter of each mini-tablet is determined under a microscope. Drug loading in each mini-tablet is measured by dissolved it in methanol and analyzing on an offline UPLC. 25 dosages of each dose strength are used for this analysis and the results are summarized in tables 1 and 2.

The results show that for both dose strengths highly precise mini-tablets are obtained. The variation in all three parameters is very low and is well under the regulatory threshold (5% relative standard deviation for drug loading as specified by the United States Food and Drug Administration). Drop size can be altered by changing the DoD printing conditions like ejection rate, volume per drop etc. Based on concentration of the active ingredient in the formulation, drop size can be adjusted to deliver the desired dose strengths in mini-tablets. High consistency seen in the process can be explained by two reasons: 1) it is easier to disperse API in liquid excipients compared to powder excipients leading to more uniform drug loadings; and 2) each mini-tablet is made of exactly one droplet; thus, uniformity of drop formation translates into uniformity in shape and weight of mini tablets. In this study, polymorphic transformation of the API crystals is not considered. Dissolving the API in a molten polymer like PEG and solidifying it yields amorphous solid dispersions. If polymorph control is critical then the API crystals can be suspended in an inert excipient that does not dissolve it (like triglyceride oils). This method also faces some limitations: the mini-tablets produced lack hardness which would make their transportation difficult. Manufacturing high drug loading mini-tablets is also challenging, as increasing particle loading beyond a certain concentration yields a paste-like formulation that does not generate uniform droplets.

To realize the full potential of this DoD approach (flexible dosage production rate, quick product changeovers etc.), the mini-tablet manufacturing process needs to be made entirely continuous. Currently, a part of the process (up to drop solidification) is continuous while the subsequent washing and drying steps are carried out batch-wise. Continuous operation can be achieved by integrating the DoD system with a continuous filter dryer unit. To support emerging initiatives in pharmaceutical manufacturing like digitalization and quality-by-design, sensors can be installed into the platform to track drug loading in the manufactured mini-tablets in real time. This will allow for building quality into the product during the manufacturing stage itself, and thus will mitigate the need for post manufacturing quality checks.

Table 1. Consistency results for 0.1 mg mini-tablets.

Parameter	Mean	Relative stdev
Diameter	3.14 mm	1.04 %
Weight	21.27 mg	1.54 %
Drug loading	0.107 mg	2.31 %

Table 2. Consistency results for 1 mg mini-tablets.

Parameter	Mean	Relative stdev
Diameter	2.84 mm	1.22 %
Weight	16.26 mg	2.18 %

Figure 3. Printed atorvastatin mini-tablets.

Drug loading	1.01 mg	2.44 %

5. Conclusions

This study introduces a novel technique to manufacture pharmaceutical mini-tablets. API is dissolved in a molten PEG formulation and its droplets are printed and solidified in an inert silicon oil bath. These mini-tablets are seen to show a high degree of consistency in shape, weight and drug loading across dosages. To assess if these dosages are equivalent to commercial products, their dissolution behavior and long-term stability need to be determined. This technique fills an important gap in the pediatric drug market- producing high-precision low dose mini-tablets. In conjunction with the other manufacturing techniques, this method can accelerate the adoption of pharmaceutical mini-tablets.

6. Acknowledgements

The authors would like to thank the Food & Drug Administration for partially funding this work, in part, through grant (U01FD006738). Views expressed in written materials or publications and by speakers and moderators do not necessarily reflect the official policies of the Department of Health and Human Services; nor does any mention of trade names, commercial practices, or organization imply endorsement by the United States Government. This material is based upon work partially supported by the National Science Foundation under Grant No. 2132142.

References

Içten, E., Purohit, H. S., Wallace, C., Giridhar, A., Taylor, L. S., Nagy, Z. K., & Reklaitis, G. V. (2017). Dropwise additive manufacturing of pharmaceutical products for amorphous and self emulsifying drug delivery systems. *International Journal of Pharmaceutics*, *524*(1–2). https://doi.org/10.1016/j.ijpharm.2017.04.003

Krause, J., Müller, L., Sarwinska, D., Seidlitz, A., Sznitowska, M., & Weitschies, W. (2021). 3D printing of mini tablets for pediatric use. *Pharmaceuticals*, *14*(2). https://doi.org/10.3390/ph14020143

Milne, C. P., & Bruss, J. B. (2008). The economics of pediatric formulation development for off-patent drugs. *Clinical Therapeutics*, *30*(11). https://doi.org/10.1016/j.clinthera.2008.11.019

Stoltenberg, I., & Breitkreutz, J. (2011). Orally disintegrating mini-tablets (ODMTs) - A novel solid oral dosage form for paediatric use. *European Journal of Pharmaceutics and Biopharmaceutics*, *78*(3). https://doi.org/10.1016/j.ejpb.2011.02.005

Sundarkumar, V., Nagy, Z. K., & Reklaitis, G. V. (2022). Small-Scale Continuous Drug Product Manufacturing using Dropwise Additive Manufacturing and Three Phase Settling for Integration with Upstream Drug Substance Production. *Journal of Pharmaceutical Sciences*. https://doi.org/10.1016/j.xphs.2022.03.009

Zuccari, G., Alfei, S., Marimpietri, D., Iurilli, V., Barabino, P., & Marchitto, L. (2022). Mini-Tablets: A Valid Strategy to Combine Efficacy and Safety in Pediatrics. In *Pharmaceuticals* (Vol. 15, Issue 1). https://doi.org/10.3390/ph15010108

Antonis Kokossis, Michael C. Georgiadis, Efstratios N. Pistikopoulos (Eds.)
PROCEEDINGS OF THE 33rd European Symposium on Computer Aided Process Engineering
(ESCAPE33), June 18-21, 2023, Athens, Greece
© 2023 Elsevier B.V. All rights reserved. http://dx.doi.org/10.1016/B978-0-443-15274-0.50310-3

Process simulation coupled with life cycle assessment of an alternative polyamide precursor

Andrea Mio,[a*] Maurizio Fermeglia,[a]

[a] *MolNBL@UniTS, Dipartimento di Ingegneria e Architettura, Università degli studi di Trieste, Piazzale Europa 1, 34127 Trieste, Italy*

Abstract

A detailed process simulation for the production process of a valuable alternative reactant to produce a polyamide, *i.e.*, m-xylylenediamine (MXDA) using Aspen Plus has been performed. A comparative life cycle assessment of MXDA with an equivalent polymer precursor, *i.e.*, hexamethylenediamine (HMDA), has been carried out using the material and energy balances obtained through process simulation. Since neither of the two compounds outperformed the other in the impact assessment's environmental analysis, a general trade-off must be determined, even though slightly better performances are gained by m-xylylenediamine. The uncertainty related to the energy demand, which is one of the most influencing parameters, has been assessed to identify the implications of energy losses. The outcomes are actually reversed when the energy requirement for MXDA is increased by 25%, suggesting a general preference for HMDA. Therefore, particular care should be taken on energy optimization when implementing MXDA manufacturing process industrially.

Keywords: LCA, Process Simulation, Multiscale Modeling, polyamide, xylylenediamine, hexamethylenediamine

1. Introduction

After the pandemic, the manufacturing of polymers is facing significant challenges as a result of the enormous rise in demand, which is hardly being met by the supply. In this panorama, a major role is played by polyamide production, whose global market size for Nylon 6 and Nylon66 was estimated at 32.6 B$ in 2022 and is expected to reach 53.04 B$ in 2030 (Grand View Research, 2020). The tremendous growth in Nylon global market can mainly be attributed to an increasing demand of China market, along with a consistent growth of the automotive industry. The principal application of Nylon6 is in the textile industry (Deopura and Padaki, 2015), *e.g.,* carpets, fabrics and clothes, while Nylon 66, which is characterized by greater wear, thermal and structural resistances, is commonly employed in the transport sector (Ishikawa et al., 2018) and for metal replacements in advanced applications, *e.g.,* marine engine covers (Mio et al., 2021).

Due to the disruption of the supply chains, the sharp rise in the price of raw materials and the growing market demand for sustainable products, producers are being compelled to find substitute precursors. To prevent wasting time, effort, and money on a promising new synthetic route that might become more burdensome when applied at an industrial scale, the introduction of novel synthetic pathways for polyamides needs to be evaluated from an environmental standpoint already at an early-stage design. The environmental implications of a product or a manufacturing process are accurately evaluated by LCA methodology (International Standards Organisation, 2006), which allows to address the environmental impacts of the entire life cycle of goods, from raw materials extraction (cradle) to disposal at their end of life (grave). However, it is frequently challenging to gather data on the materials and energy consumptions, the gaseous and liquid emissions, the wastes, and the by-products connected to the whole lifecycle of the substance under examination during product development in the chemical sector (Mio et al., 2018). Besides the usage of well-established databases, it is common practice to resort to process simulation for filling the gaps of life cycle inventories. Focusing on the polymer production sector, Gracida-Alvarez et al. (2019), coupled a carbon footprint analysis with a process simulation of fast pyrolysis for converting polyethylene plastic waste to monomers and aromatics, while Bello et al. (2020) conducted a life cycle assessment of a precursor of poly(ethylene furanoate), *i.e.,* 2,5-furandicarboxylic acid, using process simulation to perform sensitivity analyses and identify the major environmental hotspots. In this paper, a detailed process simulation for the production process of a valuable alternative reactant to produce a polyamide, *i.e.,* m-xylylenediamine (MXDA), has been assessed from an environmental standpoint.

2. Materials and Methods

Hexamethylenediamine (HMDA) and adipic acid are combined to form Nylon 66, following a polycondensation mechanism. While adipic acid is created via a multitude of syntheses that all start with benzene, industrial production of HMDA uses either butadiene or propylene via acrylonitrile. By comparing them from an economic and environmental perspective, Dros et al. (2015) have outlined and evaluated either the conventional production pathway and novel bio-based ones.

With the aim of providing a new reactant for producing a polyamide that can replace traditional Nylon 66, the production pathway for m-xylylenediamine (MXDA) has been modelled. In fact, being an aliphatic diamine with two reactive amine groups at the opposite side of the molecule, MXDA can easily replace HMDA in the polycondensation mechanism (Hugo et al., 2008a), or serve as an epoxy resin curing agent (Kutsuna and Kihara, 2008).

The production of MXDA has been described by several patented processes owned by different companies, *e.g.,* BASF (Ernst et al., 2008b, 2008a; Hugo et al., 2008a, 2008b, 2004) and Mitsubishi Gas Chemical Company (Kumano et al., 2013, 2011; Nakamura et al., 2003). As a result of a thorough investigation of the patents declared performance, the most recent process design suggested by Kumano et al. (2013) has been selected as the best available technique (BAT), due to a greater yield and lower energy consumption, as a result of the limited use of distillation columns. Moreover, this patent provides detailed information regarding the reaction duration, conditions, and, sometimes, waste by-products, which can be assumed as emissions. The published production process of MXDA follows five main steps, which have been modelled in the Aspen Plus flowsheet shown in Figure 1.

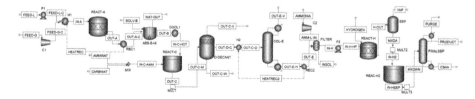

Figure 1: Flowsheet of the production of m-xylylenediamine

(i) The chain of reactions starts with an ammoxidation stage in REACT-A, where m-xylene interacts with ammonia and oxygen to produce isophthalonitrile (IPN) at 420 °C and 2 bar. In the first separation step, m-tolunitrile (TN) is used to fill unit ABS-B to conduct a liquid-vapor absorption process that directly injects the liquid-vapor stream into the solvent at 140 °C and 0.57 bar, aiming to separate IPN from by-products and contaminants. (ii) The subsequent addition of an aqueous solution of ammonia and ammonium carbonate in REACT-C at 110 °C and 2 bar neutralizes the resultant solution. (iii) In decanter D-DECANT, the mixed solution of organic and aqueous phases is separated at atmospheric pressure. (iv) The organic phase is then further separated under vacuum (0.35 bar) in COL-E to isolate the IPN from the solvent TN and the major part of volatile components. (v) After an ammonia addition that precipitates the insoluble components, IPN is hydrogenated using a nickel catalyst under medium temperature (80 °C) and high pressure (80 bar) in REAC-H2 to produce MXDA, which is then purified in the FINAL-SEP rectifier under vacuum. Due to the formation of undesired chemicals from a partial hydrogenation, a polymerization, or cleavage of the product, ideal reaction conditions (Nakamura et al., 2003) and proper solvent (Row et al., 2007) are essential in this hydrogenation reaction to yield a high grade of MXDA.

Due to the nature of the chemicals involved, the NRTL thermodynamic model has been set, using UNIFAC group contribution method to generate missing binary interaction parameters. High-pressure hydrogenation required an Equation of State (EOS)-based model, *i.e.,* Peng-Robinson, instead. The experimental process follows a semi-batch regime and has been scaled-up considering an initial charge of 350 kg of xylene to meet the needs of an industrial scale production. Heat integration has been implemented in order to transfer the heat from outlet hot streams to inlet cold streams. Since there were no known kinetic models for the reactions involved, conversion reactors were employed in an effort to achieve the yield specified in the patent, while reactions enthalpies were calculated by means of the enthalpies of formation of the components. When the yields were not specified, a conversion of 95% of the reactants were assumed, as reported within ecoinvent guidelines for conservative scenarios (Weidema et al., 2013).

Aiming at determining the most environmentally friendly polymer precursor, openLCA was used to compare the life cycle assessments of the two alternatives, after obtaining the material and energy balances through process simulation. The functional unit was defined as *1 kg of polymer precursor in the European market*, using a cradle-to-gate approach, *i.e.,* including raw materials extraction and refining, transportations and production of the substances. For the manufacture of reactants (such as xylene, ammonia, oxygen, etc.) secondary data have been retrieved using ecoinvent v3.8, with allocation cut-off method. Moreover, the unidentified byproducts and the unreacted chemicals have been modelled as aqueous emissions of analogous substances with a characterization factor specified within ReCiPe 2016 Midpoint (H), which has also been used during the life cycle impact assessment phase.

3. Results and Discussion

The outcomes of the comparative LCA between the alternative precursors are shown in Figure 2 and Table 1, where the contributions from several sources (raw materials, utilities and production process) have been highlighted.

According to the results of the impact category analysis, MXDA is less burdensome than HMDA in 10 of the 18 impact categories that were analyzed, with MXDA scores ranging from 50 to 80% of HMDA's ratings. The environmental performances of MXDA are worse when considering freshwater, marine and terrestrial ecotoxicity, marine and freshwater eutrophication, water consumption, land use, ionizing radiation and human toxicity (non-carcinogenic).

Table 1: *Life Cycle Impact Assessment results for HMDA and MXDA*

Impact category	Unit	HMDA				MXDA			
		Raw Materials	Utilities	Production	Total	Raw Materials	Utilities	Production	Total
Fine particulate matter formation	kg PM2.5 eq	3.27E-03	2.55E-03		5.82E-03	3.34E-03	2.25E-03		4.17E-03
Fossil resource scarcity	kg oil eq	1.49E+00	1.72E+00		3.20E+00	2.21E+00	7.34E-01		2.69E+00
Freshwater ecotoxicity	kg 1,4-DCB	4.34E-02	3.50E-02	1.31E-02	9.16E-02	6.61E-02	7.29E-02	7.23E-02	1.71E-01
Freshwater eutrophication	kg P eq	2.01E-04	4.54E-04		6.55E-04	1.03E-03	7.85E-04		8.87E-04
Global warming	kg CO2 eq	2.16E+00	3.25E+00	6.16E-01	6.02E+00	3.86E+00	2.32E+00	5.52E-01	5.77E+00
Human carcinogenic toxicity	kg 1,4-DCB	6.79E-02	7.92E-02		1.47E-01	1.32E-01	7.85E-02		1.40E-01
Human non-carcinogenic toxicity	kg 1,4-DCB	8.12E-01	7.72E-01		1.58E+00	1.64E+00	1.21E+00		1.65E+00
Ionizing radiation	kBq Co-60 eq	5.45E-02	1.61E-01		2.15E-01	5.15E-01	3.57E-01		3.90E-01
Land use	m2a crop eq	1.04E-02	2.41E-02		3.45E-02	2.71E-02	2.64E-02		6.13E-02
Marine ecotoxicity	kg 1,4-DCB	5.78E-02	5.30E-02	1.86E-04	1.11E-01	8.97E-02	9.54E-02	6.31E-03	1.38E-01
Marine eutrophication	kg N eq	3.20E-05	3.27E-05		6.47E-05	2.99E-04	5.60E-05		2.87E-04
Mineral resource scarcity	kg Cu eq	1.20E-02	1.77E-03		1.38E-02	7.26E-03	2.13E-03		8.00E-03
Ozone formation, Human health	kg NOx eq	3.32E-03	4.22E-03		7.54E-03	5.84E-03	3.20E-03		7.28E-03
Ozone formation, Terrestrial ecosystems	kg NOx eq	3.52E-03	4.37E-03		7.89E-03	6.13E-03	3.26E-03		7.62E-03
Stratospheric ozone depletion	kg CFC11 eq	5.97E-07	1.18E-06		1.77E-06	1.05E-06	7.06E-07		1.31E-06
Terrestrial acidification	kg SO2 eq	9.91E-03	7.42E-03		1.73E-02	9.36E-03	6.18E-03		1.20E-02
Terrestrial ecotoxicity	kg 1,4-DCB	1.88E+00	3.98E+00	2.31E-05	5.86E+00	4.86E+00	3.09E+00	4.20E-01	7.59E+00
Water consumption	m3	3.33E-02	6.49E-03		3.98E-02	9.92E-02	1.25E-02		6.19E-02

The ionizing radiation, human non-carcinogenic toxicity, marine ecotoxicity, and freshwater eutrophication are primarily influenced by the electricity consumption, which is higher for MXDA than HMDA. This is counterbalanced by a greater heat demand for HMDA, resulting in lower impacts for MXDA in terms of global warming potential, fossil resource scarcity, fine particulate matter formation, human carcinogenic toxicity and ozone-related impact categories. Freshwater ecotoxicity is affected by the m-xylylenediamine production due to the presence of a significant quantity of contaminants, primarily byproducts and impurities in wastewater, that must be further purified before being discharged into the environment. Raw materials for MXDA mainly affects marine eutrophication, due to the usage of ammonia and ammonium carbonate, as well as terrestrial ecotoxicity due to the use of m-xylene, while water consumption has been negatively impacted by both xylene and ammonia production. Raw materials for HMDA are more burdensome in terms mineral resource scarcity and terrestrial acidification, due to sulfur-based reactants. While the impacts of raw materials among the different impact categories are equally distributed between the two precursors, the impacts related to the energy consumptions are generally higher for MXDA.

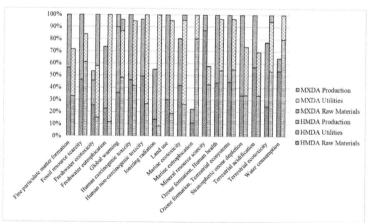

Figure 2: *Comparison between the Life Cycle Impact Assessment results for hexamethylenediamine (HDMA) and m-xylylenediamine (MXDA)*

Even though there is a limited use of the distillation column in the MXDA process, resulting in lower heat required, the electricity consumptions of compressors and pumps is still demanding from an environmental viewpoint. Since process simulation provided limited information due to a lack of available data for MXDA, the assumptions on energy consumptions have been subjected to a sensitivity analysis. Therefore, in order to emphasize the effects of a different energy demand from simulated data, both the heat and electricity demands have been adjusted by a factor of ±25%. The results of the sensitivity analysis are reported in Figure 3.

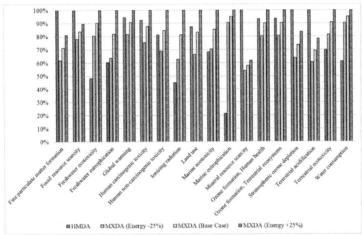

Figure 3: Sensitivity analysis results for different MXDA energy demand scenarios (±25%)

As expected, an energy demand increment of 25% worsens the environmental performance of MXDA for a multitude of impact categories, being the most burdensome precursor for 13 out of 18 impact categories. In particular, Global warming, Human carcinogenic toxicity, Land use, and Ozone formations gained worse scores than HMDA, degrading the MXDA base case that achieved superior performances in comparison with the other monomer. The precursors' relative positions do not change when the MXDA energy demand is reduced by 25% (apart from their non-carcinogenic toxicity to humans), but their impacts do get closer in situations where the MXDA results were higher than HMDA in the base case scenario, while their difference increases in impact categories where the MXDA results were lower than HMDA for base case scenario.

4. Conclusions

This paper investigated the environmental performances of a polyamide precursor, *i.e.*, m-xylylenediamine, generating its life cycle inventory using a publish patent coupled with process simulation. The outcomes of the study have been compared with an analogous assessment performed on a traditional polyamide precursor, *i.e.*, hexamethylenediamine. The outcomes showed how the environmental performances of the two precursors are mostly equivalent, with slightly better performances for m-xylylenediamine. However, the production energy demand (both heat and electricity) played an essential role to determine the most environmentally friendly precursor. Moving to an industrial scale to validate the process simulations outcomes is therefore necessary in order to determine the correct energy usage and modify the life cycle inventory accordingly.

This paper used process simulation along with life cycle assessment to fill in the existing gaps in life cycle inventories of novel materials and substances, which is a well-known example of the application of multiscale modeling for the early-design assessment of products and processes.

References

Bello, S., Méndez-Trelles, P., Rodil, E., Feijoo, G., Moreira, M.T., 2020. Towards improving the sustainability of bioplastics: Process modelling and life cycle assessment of two separation routes for 2,5-furandicarboxylic acid. Sep. Purif. Technol. 233, 116056.

Deopura, B.L., Padaki, N. V., 2015. Synthetic Textile Fibres: Polyamide, Polyester and Aramid Fibres. Text. Fash. Mater. Des. Technol. 97–114.

Dros, A.B., Larue, O., Reimond, A., De Campo, F., Pera-Titus, M., 2015. Hexamethylenediamine (HMDA) from fossil- vs. bio-based routes: an economic and life cycle assessment comparative study. Green Chem. 17, 4760–4772. https://doi.org/10.1039/c5gc01549a

Ernst, M., Hahn, T., Wenz, K., Hugo, R., Melder, J., 2008a. Method for producing a Xylylenediamine. US 2008/0214871 A1.

Ernst, M., Wenz, K., Hugo, R., Melder, J., Hahn, T., 2008b. Method for producing a Xylylenediamine. US 2008/0154061 A1.

Gracida-Alvarez, U.R., Winjobi, O., Sacramento-Rivero, J.C., Shonnard, D.R., 2019. System Analyses of High-Value Chemicals and Fuels from a Waste High-Density Polyethylene Refinery. Part 2: Carbon Footprint Analysis and Regional Electricity Effects. ACS Sustain. Chem. Eng. 7, 18267–18278. https://doi.org/https://doi.org/10.1021/acssuschemeng.9b04763

Grand View Research, 2020. Nylon Market Size, Share & Trends Analysis Report.

Hugo, R., Jourdan, S., Wenz, K., Preis, T., Weck, A., 2008a. Preparation of Xylylenedlamine (XDA). US 7,323,598 B2.

Hugo, R., Jourdan, S., Wenz, K., Preiss, T., Weck, A., 2008b. Preparation of Xylylenediamine (XDA). US7368610.

Hugo, R., Wenz, K., Wambsganss, R., Jourdan, S., Preiss, T., 2004. Method for the Production of Diaminoxylene By Continuous Hydrogenation of Liquid Phthalonitrile. WO2005026099A1.

International Standards Organisation, 2006. ISO 14044 Environmental management - Life cycle assessment - Requirements and guidelines.

Ishikawa, T., Amaoka, K., Masubuchi, Y., Yamamoto, T., Yamanaka, A., Arai, M., Takahashi, J., 2018. Overview of automotive structural composites technology developments in Japan. Compos. Sci. Technol. 155, 221–246.

Kumano, T., Ibi, Y., Arasuna, K., Nagao, S., 2013. Method for Producing Xylylenediamine. EP2671864 A1.

Kumano, T., Nakaya, K., Kato, K., Shigematsu, R., 2011. Production of Xylenediamines. EP 1873137 B1.

Kutsuna, T., Kihara, S., 2008. Adhesive of epoxy resin and curing agent with xylylenediamine structure. US 7425598 B2.

Mio, A., Bertagna, S., Cozzarini, L., Laurini, E., Bucci, V., Marinò, A., Fermeglia, M., 2021. Multiscale modelling techniques in life cycle assessment: Application to nanostructured polymer systems in the maritime industry. Sustain. Mater. Technol. 29, e00327. https://doi.org/10.1016/J.SUSMAT.2021.E00327

Mio, A., Limleamthong, P., Guillén-Gosaíbez, G., Fermeglia, M., 2018. Sustainability Evaluation of Alternative Routes for Fine Chemicals Production in an Early Stage of Process Design Adopting Process Simulation along with Data Envelopment Analysis. Ind. Eng. Chem. Res. 7946–7960. https://doi.org/10.1021/acs.iecr.7b05126

Nakamura, K., Amakawa, K., Shitara, T., 2003. Method for producing high purity Xylylendiamine. US 2003/0013917 A1.

Row, S.W., Chae, T.Y., Yoo, K.S., Lee, S.D., Lee, D.W., Shul, Y., 2007. Effect of reaction solvent on the hydrogenation of isophthalonitrile for meta-xylylendiamine preparation. Can. J. Chem. Eng. 85, 925–928. https://doi.org/10.1002/cjce.5450850614

Weidema, B.P., Bauer, C., Hischier, R., Mutel, C., Nemecek, T., Reinhard, J., Vadenbo, C.O., Wenet, G., 2013. Data quality guideline for the ecoinvent database version 3. Ecoinvent Report 1 (v3). Swiss Cent. Life Cycle Invent. 3, 169.

Antonis Kokossis, Michael C. Georgiadis, Efstratios N. Pistikopoulos (Eds.)
PROCEEDINGS OF THE 33rd European Symposium on Computer Aided Process Engineering
(ESCAPE33), June 18-21, 2023, Athens, Greece
© 2023 Elsevier B.V. All rights reserved. http://dx.doi.org/10.1016/B978-0-443-15274-0.50311-5

Connectivity Matrix-based Descriptors with Deep Learning for Estimation of Pure Component Properties

Qiong Pan,[a] Xiaolei Fan,[b] Jie Li,[a]

[a]*Centre for Process Integration, Department of Chemical Engineering, School of Engineering, The University of Manchester, Manchester M13 9PL, UK*
[b]*Department of Chemical Engineering, School of Engineering, The University of Manchester, Manchester M13 9PL, UK*

Abstract

Physicochemical properties are fundamental for the design of chemical products and processes. Various approaches including cheminformatic and graph-based methods have been applied for the estimation of pure component properties. In this work, we coupled our connectivity matrix-based molecular structure representation method with a deep learning neural network to develop molecular structure-property relationship models. Molecular structure information is represented by the connectivity matrix, while substructures information is represented by the extracted submatrix. This extraction does not cause any loss of structural information, substructural information can be recovered using the stored matrix. Matrices are transferred into features and characterised based on their eigenvalues. Statistical and mathematical approaches, followed by deep learning neural network models, are developed to correlate the structure-property relationship. Results on the case studies of normal boiling point show that the accuracy of the deep neural network model improved by 9.48% over the previous neural network model.

Keywords: molecular descriptor, deep neural network, property estimation, connectivity matrix

1. Introduction

Molecular properties such as the physical, chemical, and structural properties are fundamental characteristics of molecules. Subtle structural differences may lead to significant property differences. Therefore, accurate property determination is essential for effective chemical compound design and utilisation (Katritzky et al., 1997). However, measuring molecular properties by experimental approaches is laborious and expensive. As a result, estimating molecular properties by developing mathematical models based on existing experimental data is a feasible approach (Wen et al., 2021). Generally, current approaches for property prediction include correlation-based approaches, the group contribution (GC) method, quantitative structure-activity/property relationship (QSAR/QSPR), and graph convolution methods. Correlation-based approaches are the easiest because of the concise format of correlation equations(Korsten, 1997). One of the shortcomings is that prediction accuracy is usually not guaranteed, especially for compounds with heteroatoms. The GC method is under the assumption that a molecule's properties are determined by solving a set of equations that maps the number of each group to experimental property value, a property of a molecule is then calculated through the summation of all group contributions (Hukkerikar et al., 2012). One significant challenge of the GC method is that no method reported in the literature can be used for the systematic generation of a molecule's structural groups (or molecular fragments). The definition of groups usually has a significant effect on the performance of the prediction accuracy. It can lead to an inaccurate estimation if any of the predefined groups cannot describe one part of its molecular structure (Gani et al., 2005). Although reported GC methods predefined hundreds of groups that can represent most of the common structural characteristics of a compound, an automatic group generation approach tailored to the unique molecular pool can also improve the accuracy of structure-property correlation. Let alone the population of chemical compounds increases substantially because of the rapid progress of chemical synthesis techniques (Yoshida et al., 2011). Chemical graph theory is a branch of mathematical chemistry that models chemical topology phenomena mathematically. The basic idea of chemical graph theory is that atoms and bonds of a molecule can be thought of as nodes and edges in a graph (Randić, 2008). Therefore, topological indices representing molecular structure properties can be generated from the graph. Different to GC groups, which assume that each group contributes independently of other groups in the structure, many topological indices are a function of the entire graph and thus reflect the fundamental nature of the molecular structure. One important class of topological indices called connectivity indices considers how a graph is connected (Randic, 1975). The QSPR method correlates properties with molecular descriptors derived from molecular characteristics. Technically, molecular descriptors for QSPR models can be derived from GC groups, topological indices, and other defined descriptors (Austin et al., 2016). The major problem is that interpretation of the developed descriptors could be weak, provisional, or utterly lacking because some of the descriptors are chemically meaningless (Todeschini and Consonni, 2000). Recently, machine learning methods, especially graph neural networks (GNN), have been used to predict the structure-property relationship of molecules (Coley et al., 2019). GNN utilises a graph representation of molecules where atoms correspond to nodes and bonds to edges containing information about the molecular structure. However, the convolution process of GNN may result in the loss of information, making it difficult to trace the substructure's contribution to the property.

In this work, we applied the novel connectivity matrix-based molecular substructural descriptor generation method to normal boiling point property prediction. Different to connectivity indices which are usually derived from the adjacency matrix of a molecule, the connectivity matrix is a variant of the adjacency matrix. It includes the atomic number and actual bond type information other than the adjacency information. An extraction strategy is proposed to extract a plethora of submatrices (also called molecular structural fragments), each representing the environment of an atom/bond automatically and systematically from this connectivity matrix, rather than using predefined groups like GC methods. This extraction does not cause a loss of molecular information compared to methods with convolution steps. In addition, we increased the number of hidden layers of the neural network to increase model accuracy (Pan et al., 2023).

2. Methodology

2.1. Datasets

The normal boiling point (NBP) dataset used in this work is from the literature (Alshehri et al., 2021). As far as we know, it is the largest NBP database with molecular structural information represented in SMILES format and property values.

2.2. Generation of features

Our recent article describes the detailed method (Pan et al., 2023). The general idea is that the environment of an atom/bond can be represented by extracting entries in the matrix considering the connectivity information. All molecules stored in SMILES format are canonicalised before structural information extraction. Rdkit package and in-house Python scripts were implemented for the feature generation. Figure 1 shows the structure, and reference number of each atom in acetic acid, each atom in the molecule is labelled a number, which is used as row/column indexes when generating the connectivity matrix. In Table 1, a square matrix in which the number of rows equal to the number of atoms is generated. Diagonal entries store the atomic number of each corresponding atom, while off-diagonal entries store the connectivity information.

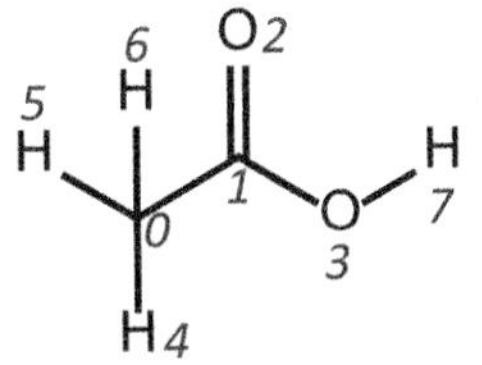

Figure 1 Structure of acetic acid and reference number of each atom

The environment of each substructure is represented by the extracted submatrix. For example, atom 1 can be represented by extracting intersections of rows 0, 1, 2, 3 and columns 0, 1, 2, 3 from the connectivity matrix. Eigenvalues and determinants of such extracted submatrices are calculated, which were used as feature names to classify the structural fragments and count the number of each feature (number of same fragments). Features were then characterised using statistical methods such as Pearson's correlation method and principal component analysis, as shown in Figure 2d and 2e. Although the determinants and eigenvalues of submatrices are chemically meaningless, substructural information can be recovered using the stored matrix, because atoms and connecting information were fully stored.

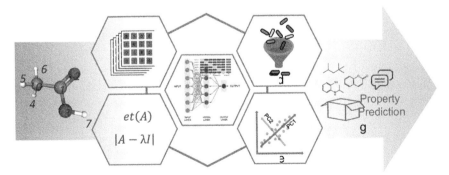

Figure 2 Flow chart of deep learning structure-property model development based on connectivity matrix

2.3. Development of structure-property models using DNN

Based on the previous work, we upgraded our neural network to a deep neural network (DNN) with multiple hidden layers and multiple nodes in each hidden layer. The dropout rate in each hidden layer is set as 25%. The number of neurons in the first hidden layer is optimised, while the number of neurons in the second layer is fixed to 80. Other features of the model remain the same in as our previous work, such as the early stopping and checkpoints step of callback methods are adopted to avoid overfitting, cross-validation to evaluate model performance over different subsets of the training dataset.

Table 1 Connectivity matrix representation of acetic acid

Index	0	1	2	3	4	5	6	7
0	6	1	0	0	1	1	1	0
1	1	6	2	1	0	0	0	0
2	0	2	8	0	0	0	0	0
3	0	1	0	8	0	0	0	1
4	1	0	0	0	1	0	0	0
5	1	0	0	0	0	1	0	0
6	1	0	0	0	0	0	1	0
7	0	0	0	1	0	0	0	1

3. Results and discussion

Root mean squared error (RMSE) and mean absolute percentage error (MAPE) are used to evaluate the performance of the constructed model. The investigated number of the first hidden layer units varies from 10% to 130% of the number of input features with a 20 % step. The effects of hidden layer units of the DNN model on RMSE and MAPE are presented in Table 2. The DNN model with 209 neurons in the first hidden layer performs best with a minimum RMSE of 31.36 and MAPE of 4.20%. However, there is no noticeable trend regarding the number of neurons with the model performance.

Table 2 Effects of hidden layer units of the DNN model on RMSE and MAPE

Neurons	RMSE			MAPE (%)		
	training set	validation set	test set	training set	validation set	test set
41	29.66	32.56	33.73	3.63	4.45	4.83
125	24.47	27.53	31.83	2.99	3.94	4.43
209	22.69	26.29	31.36	2.76	3.80	4.20
293	23.07	26.71	32.55	2.86	3.82	4.29
377	24.12	27.49	31.88	2.94	3.92	4.25
460	23.48	26.99	31.49	2.88	3.86	4.26
544	24.98	27.83	31.93	3.12	4.00	4.43

The results of the DNN model over the previous artificial neural network (ANN) model are presented in Table 3. The average RMSE of the 6-fold cross-validation DNN model on the test set is 31.36, which is reduced by 6.50% compared to 33.54 of the ANN model. The MAPE (%) of the DNN model is reduced by 9.48%, from 4.64% to 4.20%.

Table 3 Comparison of the DNN model with the ANN model

Model	RMSE			MAPE (%)		
	training set	validation set	test set	training set	validation set	test set
ANN	30.02	39.76	33.54	3.28	4.23	4.64
DNN	22.69	36.33	31.36	2.76	3.80	4.20

The detailed RMSE and MAPE values for the DNN model generated using the 6-fold cross-validation method are presented in Table 4. The RMSEs on the test set stayed within a relatively narrow band, ranging from 28.52 to 36.71.

The parity plot (i.e., predicted values vs. experimental values) is provided in Figure 3a, where red dots represent the predicted boiling point over the practical property value in the test set (at iteration 6). It can be observed that most of the 792 data points in the test set distribute along the diagonal line, indicating that the predicted values are close to the experimental values. Figure 3b shows the distribution of relative errors (REs) between the predicted and the experimental property values at iteration 6. The number of data points with a predicted RE between (–5 %, 0) and (0, 5 %) are 330 and 325, which are higher than the ANN model (265, 217). Our previous ANN model outperformed the popular GC method, and the results suggest that the DNN model has a better prediction accuracy than the ANN model.

Table 4 RMSE and MAPE values of the DNN model using the k -fold method

	RMSE			MAPE (%)		
Iteration	training set	validation set	test set	training set	validation set	test set
1	16.63	24.28	28.56	2.17	3.29	3.75
2	18.42	70.34	31.67	2.63	3.35	4.01
3	41.07	35.00	36.71	4.39	5.17	5.35
4	14.90	28.19	30.33	2.37	3.67	4.14
5	31.29	32.68	32.35	2.98	3.72	4.31
6	13.80	27.50	28.52	2.03	3.62	3.68

Figure 3 Parity plot from the test set (iteration 5)

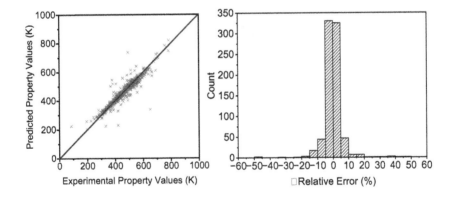

4. Conclusion

Structure information representation and mathematical modelling of structure-property relationships are critical factors in pure component properties prediction. In this work, based on the molecular structural features of molecules created using the novel connectivity matrix-based framework, we developed deep neural network (DNN) property prediction models. The effect of the number of hidden layer neurons was investigated. Compared to artificial neural networks, the increase in the number of hidden layers improved the accuracy of results. However, there was no apparent trend between the number of neurons and the model performance. The average RMSE of the DNN model is reduced by 6.50% compared to the ANN model (31.36 vs 33.54), while the MAPE (%) of the DNN model is reduced by 9.48% (4.20% vs 4.64%).

Acknowledgements

The authors gratefully acknowledge the financial support from China Scholarship Council (CSC) (No. 202006440006).

References

A.S. Alshehri, A.K. Tula, F. You, R. Gani, 2021. Next generation pure component property estimation models: With and without machine learning techniques. AIChE Journal, e17469.

N.D. Austin, N.V. Sahinidis, D.W. Trahan, 2016. Computer-aided molecular design: An introduction and review of tools, applications, and solution techniques. Chemical Engineering Research and Design 116, 2-26.

Connor W. Coley, W. Jin, L. Rogers, T.F. Jamison, T.S. Jaakkola, W.H. Green, R. Barzilay, K.F. Jensen, 2019. A graph-convolutional neural network model for the prediction of chemical reactivity. Chemical Science 10, 370-377.

R. Gani, P.M. Harper, M. Hostrup, 2005. Automatic creation of missing groups through connectivity index for pure-component property prediction. Industrial & Engineering Chemistry Research 44, 7262-7269.

A.S. Hukkerikar, B. Sarup, A. Ten Kate, J. Abildskov, G. Sin, R. Gani, 2012. Group-contribution+ (GC+) based estimation of properties of pure components: Improved property estimation and uncertainty analysis. Fluid Phase Equilibria 321, 25-43.

A.R. Katritzky, M. Karelson, V.S. Lobanov, 1997. QSPR as a means of predicting and understanding chemical and physical properties in terms of structure. Pure and Applied Chemistry 69, 245-248.

H. Korsten, 1997. Characterization of hydrocarbon systems by DBE concept. AIChE Journal 43, 1559-1568.

Q. Pan, X. Fan, J. Li, 2023. Automatic creation of molecular substructures for accurate estimation of pure component properties using connectivity matrices. Chemical Engineering Science 265, 118214.

M. Randic, 1975. Characterization of molecular branching. Journal of the American Chemical Society 97, 6609-6615.

M. Randić, 2008. On history of the Randić index and emerging hostility toward chemical graph theory. MATCH Commun. Math. Comput. Chem 59.

R. Todeschini, V. Consonni, 2000. Handbook of molecular descriptors. Wiley-VCH, Weinheim ;.

H. Wen, Y. Su, Z. Wang, S. Jin, J. Ren, W. Shen, M. Eden, 2021. A systematic modeling methodology of deep neural network - based structure - property relationship for rapid and reliable prediction on flashpoints. AIChE Journal, e17402.

J. Yoshida, H. Kim, A. Nagaki, 2011. Green and sustainable chemical synthesis using flow microreactors. Chemsuschem 4, 331-340.

Antonis Kokossis, Michael C. Georgiadis, Efstratios N. Pistikopoulos (Eds.)
PROCEEDINGS OF THE 33rd European Symposium on Computer Aided Process Engineering
(ESCAPE33), June 18-21, 2023, Athens, Greece

Novel biorefinery ideas for conversion of biomass to biofuel

Niloufar Ghavami,[a] Karhan Özdenkçi,[a] Simeone Chianese,[b] Dino Musmarra, [b]
Cataldo De Blasio [a]

[a] *Faculty of Science and Engineering, Åbo Akademi University, Rantakatu 2, 65100 Vaasa, Finland*
[b] *Department of Engineering, University of Campania "Luigi Vanvitelli", Via Roma 29, 81031 Aversa, Italy*

Abstract

The current economic model produces large amounts of waste causing environmental problems. In contrast, circular economy focuses on the efficient utilization of resources, increasing materials' lifecycle, recycling, and decreasing environmental damage. Biorefineries are the foundation for accomplishing sustainable development goals and transitioning to the circular economy. Biomass wastes are sustainable feedstock for biorefineries and are increasing due to population growth and urbanization.

Thermochemical conversion technologies specifically supercritical water gasification, hydrothermal liquefaction, and hydrothermal carbonization are viable technologies for wet biomass feedstock. Therefore, the objective of this article is to investigate a possible biorefinery integration scenario with the mentioned processes and the feasibility of the integrated processes. A study on biorefinery processes could facilitate the transition to a circular economy and sustainability and provide a holistic view for decision-makers.

Keywords: Circular economy, Biomass waste, Biorefinery, Thermochemical conversion methods.

1. Introduction

Growing waste production due to urbanization and population growth results in several environmental risks. For instance, sewage sludge (SS) is one of the major waste streams. The generation of SS in the 27 European Union member countries, (EU-27), varies between 0.1 to 30.8 kg per population equivalent and year (Havukainen et al., 2022). SS contains organics content, inorganic compounds, various microorganisms and pathogens, nitrogen and phosphorous and toxic pollutants (Tiwari et al., 2022). Therefore, energy recovery and the disposal process of SS are quite challenging.

Different methods are applied for processing SS including supercritical water gasification (SCWG) at temperatures higher than 400 °C, hydrothermal liquefaction (HTL) at 250–400 °C, and hydrothermal carbonization (HTC) at 180–280 °C (Park et al., 2019). The superiority of these processes is the removal of costly drying pre-treatment (Özdenkçi et al., 2019). Water's physical characteristics have an important role in hydrothermal processes. The physical properties change significantly in sub-critical and critical regions: water behaves like a non-polar solvent and a catalyst in hydrothermal processes (Ghavami et al., 2021). In HTL, the organic matter is liquefied to biocrude oil and other by-products as aqueous phase (AP), bio-char, and gaseous products (Nallasivam et al., 2022). HTL process consists of three main steps: depolymerization, decomposition and recombination (Gollakota et al., 2018). HTC is a proper approach to converting wet biomass to a carbonized solid fuel (hydrochar) with a moderately high heating value and nutrient-rich aqueous phase (Son Le et al., 2022; Xu et al., 2022). The HTC reactions include hydrolysis, dehydration, decarboxylation, aromatization, condensation, and polymerization (Ghavami et al., 2022). It is worth mentioning that the aqueous phase from HTL and HTC contains a huge amount of organic and inorganic compounds that need further processing (Merzari et al., 2019; Watson et al., 2020). SCWG is one of the effective technologies for converting biomass to hydrogen-rich gas products. Modell used it for the first time in the 1970s (Reddy et al., 2014; Wei et al., 2021). SCWG resulted in char formation being inevitable, especially with feedstocks involving unhydrolyzed large molecules, causing reactor plugging (Ghavami et al., 2021). Therefore, in order to reduce the char formation, the aqueous phase from HTL and HTC is processes in SCWG because of involving lighter and dissolved organics (Ong et al., 2020).

Integration of mentioned processes as a biorefinery concept is considered a sustainable approach to move towards a circular bioeconomy, carbon neutrality, waste management, and biofuel production (Culaba et al., 2022). Biorefineries are integrations of various processes producing multiple value-added products, in contrast with the traditional approach to a single product. Moreover, multiple products increase the rate of return on investment and protect against market uncertainties and fluctuations (Giwa et al., 2023). Designing a biorefinery requires a comprehensive understanding of integrated processes. In our previous article, we studied the integration of the HTC process into a biogas plant (Ghavami et al., 2022). Therefore, the objective of this study is to investigate the integration of hydrothermal processes with a biogas plant.

2. Process description and feedstock selection

Figure 1 depicts a complete view of the proposed biorefinery. SS feedstock is divided between anaerobic digestion (AD) and HTL unit operations, and the digestate from the AD process is directed to the HTC process. The aqueous phase of HTC and HTL processes enters the SCWG process for syngas production. The HTC process of SS digestate is at 200 °C and autogenous pressure with 30% solid load. The HTL process is simulated under the condition of 350 °C and autogenous pressure with 19% solid load. The aqueous phase (AP) from HTL and HTC gasified in 600 °C and 250 bar conditions.

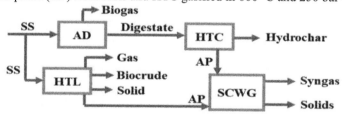

Figure 1. Flow diagram of the proposed biorefinery

Table 1 reports the dry basis compositions and the heating values (HHV) of the feedstocks and products based on the experiments (Parmar and Ross, 2019; Rahman et al., 2021).

Table 1. Analysis of feedstocks and products (Parmar and Ross, 2019; Rahman et al., 2021)

Sample	HHV (MJ/kg)	C	H	N	S	O	Ash	Yield
SS	14.1	33.1	5.5	5	0.7	25.9	29.8	-
SS digestate	14.9	28.7	3.1	3.4	1.5	16.4	46.9	-
Biocrude	28.3	53.3	6.01	3.9	1.1	29.9	4.7	37.1
Hydrochar	15.4	34.5	4.3	2.9	1.2	12.8	44.3	78

HTL, HTC, and SCWG processes are simulated in Aspen Plus V11.1 according to the following assumptions. The physical and chemical properties of non-conventional solids (NC solids) are estimated by 'HCOALGEN' and 'DCOALIGT'. Heating values for non-conventional components are calculated with a unified correlation valid for a wide range of substances (Channiwala and Parikh, 2002). Gas phase in HTL contains 85.3 mol% CO_2, 0.9 mol% H_2, 6.4 mol% CH_4, 4.4 mol% CO, and 2.9 mol% N_2. The gas phase in HTC contains 90 mol% CO_2, 1 mol% H_2, 3 mol% CH_4, and 6 mol% CO.

Prior to the simulation, it is needed to define components, thermodynamic method, and capacity. The component list is a combination of conventional and non-conventional substances including water, hydrogen, oxygen, carbon dioxide, carbon monoxide, methane, and nitrogen as conventional besides SS, digestate, hydrochar, biocrude oil, dissolved organics, and ash as non-conventional. The thermodynamic method is selected as PSRK for HTC and HTL sections because of accuracy at close-to-critical temperatures. Meanwhile, Peng-Robinson is used for the reactor and separation units in the SCWG section when converting non-conventional components to conventional ones and simulating the phase separation. In addition, mass balance is required to define yields on a non-inert basis for the RYIELD reactors in HTC and HTL processes. Yield reactors are used due to the non-conventional substances in the feedstocks. The HTC yields are calculated as 0.15, 0.81, 0.04, 0.005, 2.64e-6, 4.22e-5, and 0.00015 for hydrochar, water, organics, CO_2, H_2, CH_4, and CO, respectively. Similarly, the HTL yields are 0.05, 0.02, 0.83, 0.02, 0.06, 3.14e-5, 0.0016, 0.0019, and 0.0013 for biocrude oil, solid, water, organics, CO_2, H_2, CH_4, CO, and N_2, respectively. The SCWG process is simulated as a yield reactor decomposing the non-conventional components into stable elemental components, separating ash, and a RGIBBS as an equilibrium reactor minimizing Gibbs free energy. Figure 2 depicts the process simulation of the proposed biorefinery.

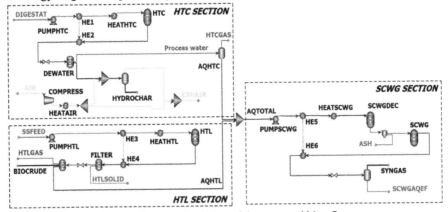

Figure 2. Process simulation of the proposed biorefinery

In the HTC section, the digestate feedstock (DIGESTAT) is pumped to 45 bars and is heated to 200 °C to reach the conditions in the HTC reactor. The stream exchanges heat with the reactor downstream by units 'HE1' and 'HE2' up to 180 °C and then reach the temperature of 196 °C by the external source in the HEATHTC heater. The product mixture goes through expansion and separation after heat exchange with the feedstock. Hydrochar is separated from the process water in the dewatering section and is dried down to 20% moisture by compressed air at 110 °C and 1.2 bars as described more in detail by Ghavami et al. (Ghavami et al., 2022). The HTC process water is conducted to a flash separation for removing gas products and then mixed with the HTL process water. In the HTL section, SS feedstock (SSFEED) is pumped to 170 bars and is heated to 350 °C to be prepared for the reaction condition, and same as the HTC part the reactor product exchange heat with the feedstock by units 'HE3' and 'HE4', then it reaches reaction condition by HEATHTL. The product mixture is passed through a filter to remove the

HTLSOLID after heat exchange with the SS feedstock. HTL products (gas, biocrude oil, and aqueous phase) are separated via a drum after expansion. The mixture of aqueous effluents (AQTOTAL) is pumped and heated with the same procedure as HTC and HTL to 250 bars and 720 °C. After the heat exchange with the aqueous mixture (units 'HE5' and 'HE6') and expansion, the syngas is separated in a flash unit.

3. Result and discussion

3.1. Mass balance

Different methodologies are applied for mass balance calculations due to using various feedstock and data availability for each process. For the HTC and HTL sections, mass balance calculations were conducted by hydrochar, biocrude oil, liquid, and gas yields as well as proximate and ultimate analysis. Moreover, oxygen and hydrogen in organics are calculated by experimental chemical oxygen demand (COD) and biochemical methane potential (BMP). Ash content in biocrude oil is assumed as 4.72% based on error bars of the dry-ash-free elemental composition. In the SCWG section, mass balance calculations in the decomposition reactor were based on elemental compositions. Figure 3 depicts mass balances for HTC, HTL, and SCWG reactors.

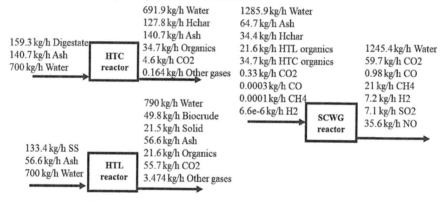

Figure 3. Mass balance in HTC, HTL, and SCWG reactors

3.2. Energy balance

Table 2 reports the power and heat requirements for the proposed biorefinery as well as energy obtained from HTL solid and syngas.

Table 2. Power and heat requirement for the biorefinery

Electricity requirement (kW)		Heat requirement (kW)	
PUMPHTC	1.7	HEATHTC	16.1
PUMPHTL	7.8	HEATHTL	50.2
PUMPSCWG	15.1	HEATAIR	162.1
COMPRESS	56.4	DEWATER	5.9
0.25 x HEATSCWG	41.9	0.75 x HEATSCWG	125.6
Total	122.9	Total	359.9
Electricity outcome (kW)		Heat outcome (kW)	
Syngas	152.8	Syngas	458.3
HTL solid	38.5	HTL solid	115.5
Total	191.3	Total	573.8

The electricity requirement results from the pumps and the compressor while heat requirement results from the external heat supplied to air and HTC and HTL reactor inlets. Since the SCWG inlet is at a high temperature, it is compromising the combined heat and power (CHP) production. Therefore, the duty of the "HEATSCWG" unit is considered as

25 % electricity and 75 % heat requirement. Similarly, the CHP obtained from the combustion of syngas and HTL solid is also assumed as 25 % electricity and 75 % heat. Biocrude oil can also be used as a fuel in the process or be upgraded to a transportation fuel. Meanwhile, the hydrochar can be used as a fertilizer, benefiting from nutrients and minerals. In addition, the remaining fertilizer elements can be recovered from the SCWG process as the solid outlet.

3.3. Feasibility and future aspects

The proposed biorefinery process is applicable to biogas plants with various feedstocks. Hydrothermal processes are suitable for converting wet biomass. In addition, the nutrients and minerals can be recovered on the solid downstream regarding the usage as a fertilizer. The proposed process leads to the generation of 68.4 kW of electricity and 213.9 kW of heat as the net amounts.

The proposed process has high potential in terms of feasibility since the combination improves economic performance and overcomes the operational issues of each conversion method. HTC is industrially applied for sewage sludge (Buttmann, 2017). Meanwhile, HTC alone requires energy, e.g. 58 kW electricity and 61 kW heat consumption to process 1000 kg/h of agricultural residue digestate with 30 % moisture (Ghavami et al., 2022). Similarly, HTL alone also requires energy, despite being stated as a feasible process for large-scale (Seiple et al., 2020). Moreover, SCWG was stated to be feasible for a more challenging feedstock (black liquor) provided that the operational issues are addressed well (Özdenkçi et al., 2019). The proposed process provides net CHP production via SCWG of aqueous effluents while producing hydrochar and bio-oil as well as addressing the aqueous discharge issues and char formation in SCWG.

4. Conclusion

The scheme of a biorefinery is proposed in this article and integrated processes including HTC, HTL, and SCWG are simulated in Aspen Plus. Moreover, the process is able to produce CHP, especially via syngas after covering the energy requirements as well as the biocrude oil product and solids to be used as fertilizers.

5. Acknowledgement

We gratefully acknowledge the financial support of the Kaute Foundation, the Finnish Foundation for Technology Promotion (Tekniikan edistämissäätiö), The University Foundation in Ostrobothnia (Högskolestiftelsen i Österbotten), and the Regional Council of Ostrobothnia for their support on the EU-ERUF project named "Implementation of a hydrothermal carbonization process for the disposal sludge and wet organic streams".

6. References

Buttmann, M., 2017. Industrial scale plat for sewage sludge treatment by hydrothermal carbonization in Jinin/China and phosphate recovery by Terranova ultra HTC process. Presented at the European Biosolids and Organic Resources Conference.

Channiwala, S.A., Parikh, P.P., 2002. A unified correlation for estimating HHV of solid, liquid and gaseous fuels. Fuel 81, 1051–1063. https://doi.org/10.1016/S0016-2361(01)00131-4

Culaba, A.B., Mayol, A.P., San Juan, J.L.G., Ubando, A.T., Bandala, A.A., Concepcion II, R.S., Alipio, M., Chen, W.-H., Show, P.L., Chang, J.-S., 2022. Design of biorefineries towards carbon neutrality: A critical review. Bioresource Technology 128256. https://doi.org/10.1016/j.biortech.2022.128256

Ghavami, N., Özdenkçi, K., Chianese, S., Musmarra, D., De Blasio, C., 2022. Process simulation of hydrothermal carbonization of digestate from energetic perspectives in Aspen Plus. Energy Conversion and Management 270, 116215. https://doi.org/10.1016/j.enconman.2022.116215

Ghavami, N., Özdenkçi, K., Salierno, G., Björklund-Sänkiaho, M., De Blasio, C., 2021. Analysis of operational issues in hydrothermal liquefaction and supercritical water gasification processes: a review. Biomass Conv. Bioref. https://doi.org/10.1007/s13399-021-02176-4

Giwa, T., Akbari, M., Kumar, A., 2023. Techno-economic assessment of an integrated biorefinery producing bio-oil, ethanol, and hydrogen. Fuel 332, 126022. https://doi.org/10.1016/j.fuel.2022.126022

Gollakota, A.R.K., Kishore, N., Gu, S., 2018. A review on hydrothermal liquefaction of biomass. Renewable and Sustainable Energy Reviews 81, 1378–1392. https://doi.org/10.1016/j.rser.2017.05.178

Havukainen, J., Saud, A., Astrup, T.F., Peltola, P., Horttanainen, M., 2022. Environmental performance of dewatered sewage sludge digestate utilization based on life cycle assessment. Waste Management 137, 210–221. https://doi.org/10.1016/j.wasman.2021.11.005

Merzari, F., Langone, M., Andreottola, G., Fiori, L., 2019. Methane production from process water of sewage sludge hydrothermal carbonization. A review. Valorising sludge through hydrothermal carbonization. Critical Reviews in Environmental Science and Technology 49, 947–988. https://doi.org/10.1080/10643389.2018.1561104

Nallasivam, J., Prashanth, P.F., Vinu, R., 2022. Chapter 4 - Hydrothermal liquefaction of biomass for the generation of value-added products, in: Varjani, S., Pandey, A., Taherzadeh, M.J., Ngo, H.H., Tyagi, R.D. (Eds.), Biomass, Biofuels, Biochemicals. Elsevier, pp. 65–107. https://doi.org/10.1016/B978-0-323-88511-9.00018-5

Ong, B.H.Y., Walmsley, T.G., Atkins, M.J., Walmsley, M.R.W., 2020. A Kraft Mill-Integrated Hydrothermal Liquefaction Process for Liquid Fuel Co-Production. Processes 8, 1216. https://doi.org/10.3390/pr8101216

Özdenkçi, K., De Blasio, C., Sarwar, G., Melin, K., Koskinen, J., Alopaeus, V., 2019. Techno-economic feasibility of supercritical water gasification of black liquor. Energy 189, 116284. https://doi.org/10.1016/j.energy.2019.116284

Park, M., Kim, N., Lee, S., Yeon, S., Seo, J.H., Park, D., 2019. A study of solubilization of sewage sludge by hydrothermal treatment. Journal of Environmental Management 250, 109490. https://doi.org/10.1016/j.jenvman.2019.109490

Parmar, K.R., Ross, A.B., 2019. Integration of Hydrothermal Carbonisation with Anaerobic Digestion; Opportunities for Valorisation of Digestate. Energies 12, 1586. https://doi.org/10.3390/en12091586

Rahman, T., Jahromi, H., Roy, P., Adhikari, S., Hassani, E., Oh, T.-S., 2021. Hydrothermal liquefaction of municipal sewage sludge: Effect of red mud catalyst in ethylene and inert ambiences. Energy Conversion and Management 245, 114615. https://doi.org/10.1016/j.enconman.2021.114615

Reddy, S.N., Nanda, S., Dalai, A.K., Kozinski, J.A., 2014. Supercritical water gasification of biomass for hydrogen production. International Journal of Hydrogen Energy 39, 6912–6926. https://doi.org/10.1016/j.ijhydene.2014.02.125

Seiple, T.E., Skaggs, R.L., Fillmore, L., Coleman, A.M., 2020. Municipal wastewater sludge as a renewable, cost-effective feedstock for transportation biofuels using hydrothermal liquefaction. Journal of Environmental Management 270, 110852. https://doi.org/10.1016/j.jenvman.2020.110852

Son Le, H., Chen, W.-H., Forruque Ahmed, S., Said, Z., Rafa, N., Tuan Le, A., Ağbulut, Ü., Veza, I., Phuong Nguyen, X., Quang Duong, X., Huang, Z., Hoang, A.T., 2022. Hydrothermal carbonization of food waste as sustainable energy conversion path. Bioresource Technology 363, 127958. https://doi.org/10.1016/j.biortech.2022.127958

Tiwari, N., Garua, B., Bansal, M., Sharma, J.G., 2022. 20 - Biorefinery and waste management by co-digestion of sewage sludge with organic wastes, in: Shah, M.P., Rodriguez-Couto, S., Shah, N., Banerjee, R. (Eds.), Development in Waste Water Treatment Research and Processes. Elsevier, pp. 365–386. https://doi.org/10.1016/B978-0-323-85584-6.00021-2

Watson, J., Wang, T., Si, B., Chen, W.-T., Aierzhati, A., Zhang, Y., 2020. Valorization of hydrothermal liquefaction aqueous phase: pathways towards commercial viability. Progress in Energy and Combustion Science 77, 100819. https://doi.org/10.1016/j.pecs.2019.100819

Wei, N., Xu, D., Hao, B., Guo, S., Guo, Y., Wang, S., 2021. Chemical reactions of organic compounds in supercritical water gasification and oxidation. Water Research 190, 116634. https://doi.org/10.1016/j.watres.2020.116634

Xu, Z.-X., Ma, X.-Q., Zhou, J., Duan, P.-G., Zhou, W.-Y., Ahmad, A., Luque, R., 2022. The influence of key reactions during hydrothermal carbonization of sewage sludge on aqueous phase properties: A review. Journal of Analytical and Applied Pyrolysis 167, 105678. https://doi.org/10.1016/j.jaap.2022.105678

Antonis Kokossis, Michael C. Georgiadis, Efstratios N. Pistikopoulos (Eds.)
PROCEEDINGS OF THE 33rd European Symposium on Computer Aided Process Engineering
(ESCAPE33), June 18-21, 2023, Athens, Greece

Waste integration for sustainable operation of WWTP's

Elena C. Blanco,[a] Mariano Martín,[a] Pastora Vega,[b]

[a] *Department of Chemical Engineering, University of Salamanca, 37008 Salamanca, Spain*
[b] *Department of Computer Science, University of Salamanca, 37008 Salamanca, Spain*
mariano.m3@usal.es

Abstract

The water-energy nexus is an important topic nowadays. Within it, the wastewater treatment plants (WWTPs) energy consumption is becoming an issue. WWTPs typically consume more energy than the one that can be obtained from the biogas produced from sludge anaerobic digestion. In this work, a process-level analysis is presented to study the feasibility of integrating both wastewater and municipal solid waste (MSW) treatment to achieve an energetically self-sustainable operation of a WWTP. The influence of the climate of different regions across the Iberian Peninsula on the energy requirements has also been evaluated. Mesophilic and thermophilic digestion is compared in Salamanca as a case base. Next, the optimal digestion temperature of 30 °C is found. Moreover, in all cities it is necessary for MSW to provide between 37-40% of the facility energy consumption. From an economic point of view, an investment between 0.09-0.16 €/kg of sewage sludge is obtained for the integrated process. Therefore, this techno-economic assessment demonstrates the feasibility of integrating these two treatments for a fully self-sufficient and sustainable process.

Keywords: energy, wastewater, municipal solid waste.

1. Introduction

Wastewater treatment plants are crucial facilities for reducing the potential environmental impact of wastewater discharges into receiving water bodies. However, they consume large amounts of energy. Until recently, energy was relatively affordable internationally, and many wastewater treatment facilities were not designed and operated with the goal of minimizing energy consumption, only to satisfy certain effluent quality requirements. Nowadays, rapid population growth and urbanization have led to an increase in wastewater production, so the number of WWTPs has also increased and effluent quality requirements have also become more demanding. This fact has resulted in a large energy consumption devoted to the operation of WWTPs, and the water-energy nexus has started to attract increasing attention from both environmental and economic points of view. On the one hand, WWTPs can represent 15–20% of the total energy consumption of municipal public structures and facilities, with energy also constituting anywhere 25-40% of the total operating costs in a conventional WWTP (Yang, et al., 2020). On the other hand, the operation of WWTPs involves greenhouse emissions due to the need for non-renewable energy sources; moreover, they have grown in recent years due to the increase in the volume of waste treated and the implementation of new processes aimed at achieving higher effluent quality. This water-energy nexus is promoting a series of studies on the relationship between energy and water for sustainable development, in particular, the potential for energy self-sufficient WWTPs has become an area of increasing research and innovation.

Anaerobic digestion (AD) is a common technology for sludge treatment at WWTPs and allows the production of biogas, which is the main energy source in a WWTP. The use of biogas for digester heating and electricity generation is a sustainable way to recover energy from WWTPs, as it could replace fossil fuels, with consequent sludge reduction. However, while the use of AD for sludge processing and biogas production is widely reported in the literature (Bin Khawer, et al., 2022), in most cases, the biogas cannot cover the energy requirements, thermal and electrical, of the plant. Even in developed countries, there is still a large gap for the energy self-sufficient WWTPs, so it is necessary to investigate and apply new trends in energy production (renewables) and new technologies to achieve it. An alternative to generate clean energy that has gained importance in recent years is its production from waste, known as waste-to-energy (WTE). MSW can be used to provide such energy helping manage this residue at the same time. Several waste-to-energy technologies, both conventional and non-conventional, have been described in recent literature (Munir, et al., 2021). Conventional waste treatment or disposal techniques include composting, anaerobic digestion, and landfilling, while incineration, pyrolysis, gasification, and hydrothermal processing are considered non-conventional waste-to-energy technologies. Among all of them, incineration is regarded as the most mature.

This work proposes the integration of sludge and MSW treatment for self-sufficient operation. The incineration of MSW produces steam to provide the thermal energy requirements in the AD process of the sewage sludge that cannot be covered by the biogas generated, and, also, to produce the total electricity requirements of the WWTP.

2. Process description and modeling

The integration of wastewater and MSW treatment combines two sections: sewage sludge treatment, that is the main thermal energy consumer within wastewater treatment, and the incineration of MSW, that is the provider of additional thermal energy and supplies power to the complete wastewater treatment, aiming at an energetically self-sufficient operation. The sewage sludge treatment is divided into four sections: biogas production, biogas purification, digestate conditioning and energy production (biogas combustion) (Figure 1). First, the sewage sludge is heated up and fed into a bioreactor where it is anaerobically digested to produce biogas, which contains methane, carbon dioxide, nitrogen, hydrogen sulfide, ammonia and moisture, and a decomposed substrate (digestate). The two main anaerobic digestion conditions are considered, mesophilic and thermophilic, which operate between 30-37°C and 50-55°C, respectively. Then, the digestate produced is conditioned in a reboiler, removing the excess of water and ammonia. This step is included so that the digestate can later be used as a fertilizer. The third step consists of the biogas purification by removing the CO_2 (and traces of NH_3) and the H_2S in a Pressure Swing Adsorption (PSA) system and in a fixed-bed reactor, respectively. Once the biogas is mainly methane, it is burned with air in a furnace to provide the thermal energy requirements of the process. They include heating up the combustion air and the sludge feed, and to produce steam to heat up the bioreactor and the heat exchanger of the sludge conditioning step.

The incineration of MSW is used to produce the additional thermal energy needed in the anaerobic digestion and the power requirements of the WWTP. A fraction of the MSW is burned in a furnace to produce the steam to heat the bioreactor and the heat exchanger of sludge conditioning, and the rest is used to produce steam to feed a Rankine cycle to generate power. In the Rankine cycle, three different sections of the steam turbine are considered (high, medium and low-pressure turbines). The treatment of the combustion

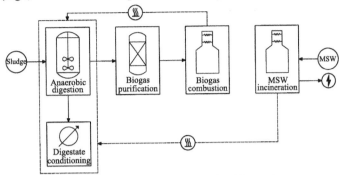

gases generated in the incineration is not considered in this work.

Figure 1. Process diagram of the integrated process of sewage sludge and MSW treatment.

Both processes, stand-alone digestion in Salamanca and the integrated process in all cities, have been modelled using and equation-based approach. These models are formulated using mass and energy balances, thermodynamics, chemical and vapor-liquid equilibria, etc., to evaluate their performance and are based on the work of León and Martín (2016). The entire superstructure is formulated as a non-linear programming problem and is implemented in GAMS using CONOPT 3.0 as preferred solver in a multistart optimization approach. The objective function is the minimization of the external energy requirements, given by (Eq.(1)):

$$Z = \left(E_{dig} - E_{biogas}\right) - E_{MSW} \qquad \text{Eq.1}$$

where E_{dig} is the total thermal energy required in the digestion process, E_{biogas} is the thermal energy supply by the generated biogas and E_{MSW} is the external thermal energy that needs to be provided by MSW.

This study is divided into two stages:

1) The evaluation of the operating conditions and the optimization of the operating temperature of the AD, minimizing in both cases the external supply of thermal energy. Firstly, mesophilic (35°C) and thermophilic (55°C) operation are compared. The variation of the water and air temperature that exists along the year in the city of Salamanca is considered, as a case base, because the thermal energy requirements depend on these temperatures. And, secondly, the operating temperature of the sludge digestion over a year has also been optimized, due to its effect on the amount of biogas produced and the energy consumption to heat up the sludge and dewatering step.

2) The study of the influence of the location on the thermal energy requirements of the optimized process. Different regions of the Iberian Peninsula have been evaluated considering the different climates that influence the air and water temperature each month, particularly: Oceanic in La Coruña (population: 244850, hab.eq.: 600000), Coastal Mediterranean in Barcelona (population: 1637000, hab.eq.: 1706250), Continental Mediterranean in Salamanca (population: 144000, hab.eq.: 260042) and Madrid (population: 3266000, hab.eq.: 1335000), and Oceanic Mediterranean in Sevilla (population: 689000, hab.eq.: 950000). The size of each WWTP is related to the amount of organic matter that is treated, considering that a person produces 26.9 kg of dry matter per year (Bianchini, et al., 2016). Also, it conditions its electricity consumption, considering that the average consumption of a WWTP in Spain is 5.6 kW/hab.eq. (IDEA, 2010).

3. Results

For the base case, the city of Salamanca, first, the operating conditions of the two different main types of AD have been studied. For the mesophilic and thermophilic digestion modes, the most energy intensive processes are the digestion itself and the digestate dewatering. The digester consumes 914.5 kW in both cases because it is considered as an isothermal process, so it only depends on the enthalpy of reaction. Nevertheless, the digestate dewatering consumes more energy in the mesophilic case, 467 kW vs. 358 kW, because the outlet temperature of the heat exchanger has always the maximum value (150°C), so the gradient temperature of the thermophilic case is lower. The other two duties are the sludge heating before digestion and the air heating before combustion and these values change over the months. The sum of these energy requirements goes from 103.5 kW in warm months to 175.4 kW in the coldest ones, in the case of mesophilic digestion, and, from 215.9 kW to 288.0 kW for thermophilic digestion. Obviously, they are higher as the digestion temperature increases. The amount of additional thermal energy is higher in the case of mesophilic digestion, although the digestion temperature is lower, an average of 39% vs. 37%. This fact is due to the significant contribution of digestate dewatering to the total thermal energy requirements of the process, that is higher in the mesophilic digestion (30% vs. 23%). In fact, the extra thermal energy without considering digestate conditioning is higher in the thermophilic case.

Then, digestion temperature has been optimized within the range of 30-55°C. The optimal operating temperature is 30°C, although it is shown that the additional thermal energy requirements is reduced with the increasing of digestion temperature, the problem tends to choose the lowest digestion temperature of the range to reduce the thermal energy involved in the process since the additional yield of biogas is not enough to mitigate the difference. This optimal temperature is the value used to study the influence of the climate, because it determines the value of the energy that heats up the sludge before digestion and the air before combustion. The change over the months and the different contribution on the total energy are shown in next section, where they are compared to other climates.

Once optimized the digestion temperature, the integrated process of wastewater a MSW treatment is modeled to make the process self-sufficient in terms of energy. The influence of the climate on the energy requirements has been studied for different cities (Salamanca, Madrid, Barcelona, La Coruña and Sevilla). The value of energy required in the digester and the one required to dehydrate the digestate per kilogram of sludge is the same in all cities, 680 kJ/kg, and 364 kJ/kg, respectively. In Figure 2, the energy requirements to heat sludge and air for each city over the months are shown. For all cases, June, July, and August present the lowest values because are the warmest months. Salamanca presents the maximum values, around 110 kJ/kg, because the winter in the Continental Mediterranean climate is the coldest one; nevertheless, in the warmest months, La Coruña exceeds the energy requirements of Salamanca, because the Oceanic climate tends to present mild summers, so water and air temperatures are lower, around 61-70 kJ/kg vs. 56-66 kJ/kg. The minimum values are obtained in Sevilla, from 37 kJ/kg to 86 kJ/kg, because the city is located in the most southern part of the Iberian Peninsula and the Continental Mediterranean climate presents the warmest temperatures, both air and water. The largest difference between the maximum and minimum value is shown in Madrid, 58%, because it presents the highest gradient temperature over the months and the minimum in La Coruña, 32%, where the thermal amplitude is lower due to ambient humidity; while, in Salamanca, this difference is 49%, in Sevilla, 56%, and, in Barcelona, 50%. This difference can also be seen in the shape of the curve presenting the data for the extra energy (Figure 2), which is more pronounced in the case of Madrid. The amount of additional energy to be provided by the MSW slightly varies between the different cities and is between 37% and 40% of the total energy required for the digestion process for all of them. This percentage is similar because the main contributor to the total energy required in the process is the energy for the digestion and for the digestate conditioning, which represent around 91% of the total thermal energy. Its value increases as thermal energy requirements increase, so the same trend is observed as in the case of energy needs throughout the year between cities. Therefore, the extremes are Seville, with the lowest percentage of additional energy, and Salamanca, with the highest, except in summer, when it is surpassed by La Coruña. From January to April, between Madrid, Barcelona and La Coruña, the first one is the major consumer, and the same occurs from October to December; however, this trend changes between April and October, where the energy requirements of La Coruña increase, becoming even higher than those of Salamanca.

The average amount of MSW that is fed to the integrated process is higher as the size of the WWTP increases because the energy and ranges from 0.26 kg/s in Salamanca to 1.67 kg/s in Barcelona. However, the amount of MSW per kilogram of sludge is almost the same for all cities, around 0.19 kg/kg, with a very slight difference between cities, being lower as the average of extra thermal energy decreases. Around 12% of the total MSW fed is devoted to the production of thermal energy, and the rest goes to the production of electricity. The databases show that that there is enough MSW in each city to cover the energy requirements of the WWTP, considering that, a person produces 455 kg of MSW per year (Eurostat, 2022).

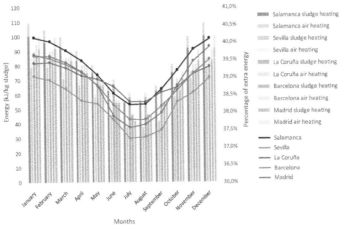

Figure 2. Energy contribution of sludge and air heating to digestion and percentage of extra energy needed in Salamanca, La Coruña, Sevilla, Madrid, and Barcelona per month.

An economic analysis of the integrated process has also been carried out for the different cities. Capital and production cost are obtained. Figure 3 shows the breakdown of the capital and the production cost for the different cities. The digester is the main item in the distribution of the capital cost in all cities, with around 51% of the total investment, followed by the steam turbines, with 25%. Regarding production costs, the main item is capital charges, with about 60% in all cities, followed by maintenance, with 20%. There is a slight difference in these percentages between cities, so they are higher as the size of the WWTP increases. Although, both capital and operating cost are higher in the larger WWTP, they benefit from economies of scale. The investment for the integrated process goes from 0.16 €/kg of sewage sludge in Salamanca, the smallest WWTP, to 0.09 €/kg in

Barcelona, which is the largest. The same is true for the investment per kilogram of MSW fed, which ranges from 0.86 €/kg in Salamanca to 0.46 €/kg in Barcelona.

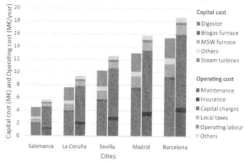

Figure 3. Capital and production cost for the different cities

4. Conclusions

In this work, the energy requirements of the WTTP have been studied. The thermal energy requirements in sewage sludge treatment are a challenging problem because the anaerobic digestion of sludge does not produce enough biogas to cover them. Therefore, an integrated wastewater and municipal solid waste treatment process has been evaluated to make the WWT self-sufficient in terms of energy. Moreover, the influence of the climate in different region of the Iberian Peninsula has been studied. The results show that it is feasible to integrate the treatment processes because there is enough MSW to provide energy to the WWTP in all cities. An amount of MSW per kilogram of sludge is around 0.19 kg/kg, with a very slight difference between cities, being lower as the average of extra thermal energy decreases. The investment cost ranges from 0.09 €/kg to 0.16 €/kg of sewage sludge, depending on the size of the WWTP. This analysis allows the successful integration of these treatment processes with the aim of a fully self-sufficient and sustainable system.

Acknowledgments

The authors acknowledge MICINN Spain grant PID2019-105434RB-C31 and Programa Investigo JCYL to Elena C. Blanco.

References

Bianchini, A., Bonfiglioli, L., Pellegrini, M. & Saccani, C., 2016. Sewage sludge management in Europe: a critical analysis of data quality. Int. J. of Environment and Waste Management, 18(226).

Bin Khawer, M. U., Raza Naqvi, S., Ali, I., Arshad, M., Juchelková, D., Waqas Anjum, M., Naqvi, M., 2022. Anaerobic digestion of sewage sludge for biogas & biohydrogen production: State-of-the-art trends and prospects. Fuel, 329(125416).

Eurostat, 2022. Municipal solid waste generated 2005 and 2020.

IDAE, 2010. Estudio de prospectiva. Consumo energético en el sector del agua.

León, E. & Martín, M., 2016. Optimal production of power in a combined cycle from manure based biogas. Energy Conversion and Management, Volume 114, pp. 89-99.

Munir, M., Mohaddespour, A., Nasr, A. & Carter, S., 2021. Municipal solid waste-to-energy processing for a circular economy in New Zealand. Renewable and Sustainable Energy Reviews, 145(111080).

Yang, X., Wei, J., Ye, G., Zhao, Y., Li, Z., Qiu, G., Li, F., WeI, C., 2020. The correlations among wastewater internal energy, energy consumption and energy recovery/production potentials in wastewater treatment plant: An assessment of the energy balance. Science of The Total Environment, 714(136655).

Antonis Kokossis, Michael C. Georgiadis, Efstratios N. Pistikopoulos (Eds.)
PROCEEDINGS OF THE 33rd European Symposium on Computer Aided Process Engineering
(ESCAPE33), June 18-21, 2023, Athens, Greece
© 2023 Elsevier B.V. All rights reserved. http://dx.doi.org/10.1016/B978-0-443-15274-0.50314-0

Multiscale design methodology for flow reactors supporting rapid prototyping

Tomoyui Taguchi,[a] Shigeru Kado,[b] Toshiyuki Watanabe,[b] Yoshiyuki Yamashita,[c]

[a]Chiyoda Corporation, Pharmaceutical Engineering Department, 4-6-2, Minatomirai, Nishi-ku, Yokohama-shi, Kanagawa 220-8765, Japan
[b]Chiyoda Corporation, Research and Development Center, 3-13, Moriya-cho, Kanagawa-ku, Yokohama-shi, Kanagawa 221-0022, Japan
[c]Department of Chemical Engineering, Tokyo University of Agriculture and Technology, Tokyo 184-8588, JAPAN
CorrespondingAuthor:taguchi.tomoyuki@chiyodacorp.com

Abstract

Continuous manufacturing has been incorporated in the pharmaceutical industry to achieve a beneficial production system for scale-up or scale-out, flexibility of demand response, and sustainability of material utilization (Yu, 2008). This paper proposes a rapid prototyping approach based on a mathematical model that integrates experiment and simulation. It is important to segregate the parameter identified for new manufacturing items from the parameter applied to common items based on the accumulation of experimental data. As an example case of rapid prototyping, a trickle bed reactor (TBR) as a multiphase heterogeneous catalytic reactor with complex interdependencies is presented. This design approach enables a rapid prototyping of production system supporting multiphase catalytic reaction. Both the hardware design, such as catalysts and filling systems in the equipment in multi-stage reactors, and the design of operating conditions can be decided considering the constraints of the equipment, materials and reagents. The cycle of data acquisition and model evaluation in laboratory scale can accelerate the actual manufacturing system establishment.

Keywords: Continuous pharmaceutical manufacturing, Modeling and simulation, Optimization, Process design and Trickle bed reactor

1. Introduction

TBR is used for applications ranging from large-scale manufacturing centered on petrochemicals to small-scale manufacturing, including pharmaceutical manufacturing. When compared to other types of reactors such as bubble column, fluidized bed, slurry type reactor, and so on, TBR is preferable as a multiphase catalytic reactor because of the simple separation process and an easy replacement of catalyst (Duduković et. al., 1999). In conventional large-scale production, a robust design of the wetting efficiency of the catalyst is essential for the prevention of catalyst and equipment burnout, owing to localized severe temperature fluctuations. Process modeling is often applied to TBR design in various industrial fields, as shown in Fig.1 (Al-Dahhan et. al., 1997).

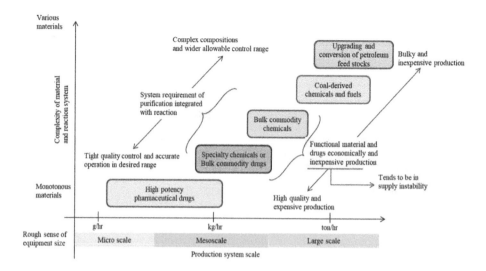

Fig.1   Conceptual map of a TBR application in various industrial fields

In the mesoscale region, engineering technology is required to establish a rapid and high-quality manufacturing system. It is necessary to determine the characteristics of the target product and establish a new database about the thermodynamic and transport properties (Masson et. al., 2022). The mesoscale production system tends to be in supply instability of the functional material and drugs because of the inexpensive production considering the complicated manufacturing process. Frequently consumed mesoscale products need to be manufactured economically and inexpensively in robust facilities designed with process integration. In a mesoscale TBR design, it is necessary to select a good catalyst size to compensate for its deterioration during transport ability in the laminar flow area, to suppress the pressure drop, and to enhance the reaction rate at a high solid-liquid ratio during transport of the reactants. Furthermore, the target material must be recovered at high purity when compared to the conventional production of petroleum products, where the amount of impurities should be small. The other motivation is to increase the synthetic route of organic synthesis through applying continuous manufacturing. The reductive amination and removal of benzyl-type protecting groups are two major examples of reduction reactions that frequently use catalytic hydrogenation over precious metal catalysts (Carey et al., 2006). The basic concept of the proposed method is to utilize the pre-determined information about the process and equipment and to incorporate the reaction kinetics. It can reduce the experimental work for the preparation of the TBR system, concerning the characteristics of the reaction and catalysts.

2. Concept of rapid prototyping

The rapid prototyping approach consists of three steps: 1) parameter adjustment of the reaction kinetics through laboratory experiments; 2) verification of the integrated model parameters through bench-scale experiments under suppressed material consumption; and 3) determination of production scale and system specifications.

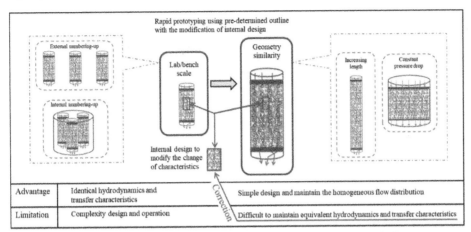

Fig.2　　Concept of rapid prototyping of fixed bed multiphase catalytic reactor

With the pre-determined models in Sections 2 and 3, we can focus on the exploration of reactions and material properties characteristic, and quickly determine the required design prototypes for different manufactured products. Figure 2 shows the morphology of the scale-up policy (Dong, et. al., 2021). Numbering-up has the advantage of retaining the production conditions established from the lab/bench scale. However, there are disadvantages that it is difficult to establish design, operation and control philosophy for multiple reactors. On contrary, scaling up has the advantage concerning simple configuration and more flexible design space of internal packed features and operating condition, although the homogeneity of flow and transport phenomena are required to modify. However, it can adjust the reaction performance, heat and mass transport characteristics to the desired condition that is similar to the condition of the lab/bench scale. The navigation system consisted of segregating data base both material features and transport phenomenon can help to determine the appropriate process design. Fig.3 shows the concept of navigation system. New material features need to be created for the new target production repeatedly, although a data base of transport phenomenon is usually to provide the best route to achieve a high-quality production. The experience of development cycle will enhance this navigation system according to the accumulated properties.

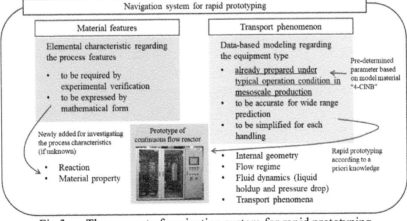

Fig.3　　The concept of navigation system for rapid prototyping

3. Integrated modeling of trickle bed reactor

The modeling of TBR can be separated into three sections: 1) reaction kinetics; 2) the size of the equipment and catalyst; and 3) transport phenomena. If the model parameters in sections 2 and 3 are pre-determined, the reaction model with different fixed bed catalysts can be applied in a simple and rapid way. The parameters of the process model are adjusted considering the reaction kinetics, catalyst properties (size and metal loading), flow regime, axial dispersion, pressure drop, transport phenomena, and liquid holdup. Establishment of the TBR system requires the comprehension of complex independent phenomena by conducting careful experimentation and deconvolution of various parameters (Azarpour et. al., 2021). The hydrogenation reaction of 4-chloronitrobenzene using a Pd/TiO2 catalyst is adopted as a mockup substance and reaction system. The intermediate 4-chloronitroaniline is set as the product for the evaluation of reaction pathways by overreaction, by-products (nitrobenzene), and impurities (nitrosamines). To build this integrated model, 300 experimental data points were used under the following conditions:

Table 1   The reference data for providing mockup integrated model of TBR

Process conditions	Filling conditions	Equipment size
Substrate concentration: 50 mol/m3, 100 mol/m3 Liquid flow rate: 0.2~80 mL/min. G/L ratio: 16.7~100 (volume ratio) Temperature: 35~60 ºC Pressure: 0~0.5 MPaG	Catalyst amount: 1~50 g (based on Pd/TiO2) Metal content: 0.2~5 wt%	Catalyst filling length: 10~320 mm Diameter (I.D.): 12.2mm, 30mm

Fig. 4 shows an overview of the integrated simulation model.

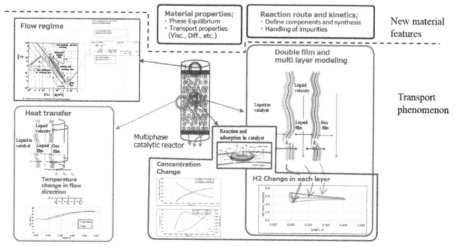

Fig.4   Integrated modeling of interdependency phenomenon

Based on the formula determined for the flow regime in a TBR (Gianetto at. al., 1978), the formula is modified to adapt the process conditions to the packing size applied in the mesoscale region. As a result, the liquid holdup and pressure drop change, which affects the heat and mass transport phenomena and reaction conditions in the integrated simulation model.

4. Model-based design approach

To provide an accurate integrated model, the acquisition of high-quality experimental data is important to provide a map constituting the accuracy of process characteristics. On the other hand, scale design experiences will update a compass to predict the right reactor scale with the interior design. The integrated model can aim to design the following aspects;

4.1. The internal design

The size of the catalyst is set at a few hundred μm, which does not change significantly from the viewpoint of reactor pressure loss and liquid holdup capacity. However, the catalyst amount and metal loading are deemed the design parameters that affect the balance between the reaction heat generation and heat removal by the jacket coolant. The different combinations of catalyst amount and metal loading are evaluated for reaction performance. Fig. 5 shows the conversion of starting material (SM) and Fig.6 shows the selectivity of target material (TM). It can be confirmed that both Fig.5 and Fig.6 give a good estimation. In other words, it is possible to predict the desired conditions using the integrated model and to determine design conditions rapidly.

4.2. The reactor scale design

The reactor scale is an important design parameter. Higher velocity makes the transport ability better, but it makes the pressure drop worse. The liquid flow and Gas/Liquid ratio also have a significant impact on liquid holdup. Therefore, different sets of catalyst packing heights are verified. Fig.7 shows the conversion of SM, and Fig.8 shows the selectivity of TM, respectively. It can be confirmed that both Fig.7 and Fig.8 give good estimations. The integrated model can predict the TBR performance for different diameters and catalyst packing heights; therefore, the desired operating condition can be determined by the selection of the appropriate size of reactor. By combining equipment size selection and internal design, it is possible to rapidly prepare a multiphase catalytic reactor system that satisfies the required manufacturing scale and specifications.

5. Conclusion

This paper introduces a rapid prototyping method, which is a conceptual model based on a comparison between mesoscale experimental data and simulation. In the design of multiphase catalytic reactors in the mesoscale region, the basic data on transport phenomena is still insufficient. Therefore, the fluid parameter and transport phenomena are incorporated into the integrated TBR mathematical model. Improvement of the integrated TBR database can be achieved through repeated experiments and simulation.

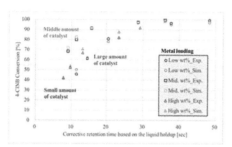

Fig.5   Prediction of conversion change based on the internal design

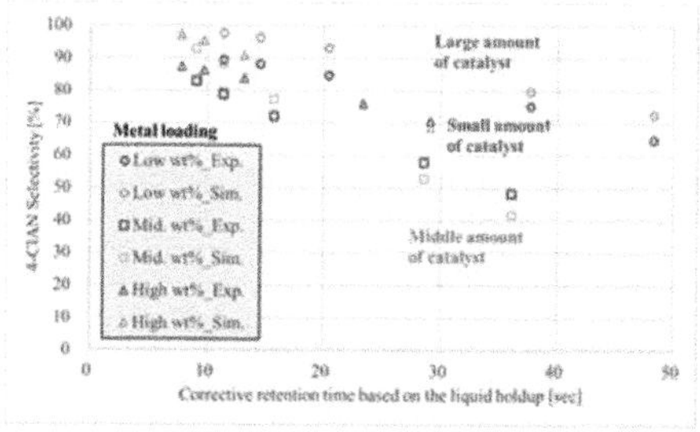

Fig.6   Prediction of yield change based on the internal design

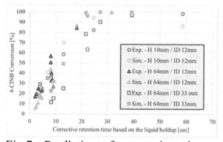
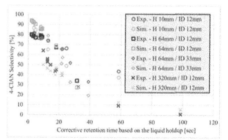

Fig.7   Prediction of conversion change based on reactor scale design

Fig.8   Prediction of yield change based on reactor scale design

This study has made it possible to evaluate the catalytic reaction performance and to quickly identify the parameters for the reaction characteristics. According to the reaction parameter incorporation into the integrated model, which can help the determination of reactor internal and scale design. This method can be used as a common platform to combine chemical and catalytic reactions, which can help researchers with the chemist and process engineer. This design approach can also make the frequent change of production items possible. Following that, while establishing the mesoscale continuous pharmaceutical manufacturing system, this approach can be used to study other processes such as synthesis and separation.

References

A. Azarpour et. al., 2021, Performance analysis and modeling of catalytic trickle-bed reactors: a comprehensive review, Journal of Industrial and Engineering Chemistry, 103, 1-41.

J. Carey et. al., 2006, Analysis of the reactions used for the preparation of drug candidate molecules, Organic & Biomolecular Chemistry, 4, 2337-2347.

M. H. Al-Dahhan, et. al., 1997, High-pressure trickle-bed reactors: A review, Ind. Eng. Chem. Res., 36, 3292-3314.

Z. Dong, Z. Wen, F. Zhao, S. Kuhn, and T. Noël, 2021, Scale-up of micro-and milli-reactors: An overview of strategies, design principles and applications, Chemical Engineering Science: X, 10, 100097.

M. P. Duduković, F. Larachi, P. L. Mills, 1999, Multiphase reactors–revisited, Chemical Engineering Science, 5413-14, 1975-1995.

A. Gianetto, G. Baldi, V. Specchia, S. Sicardi, 1978, Hydrodynamics and solid ‑ liquid contacting effectiveness in trickle ‑ bed reactors, AIChE Journal, 24, 6, 1087-1104.

E. Masson et. al., 2022, Fixed bed continuous hydrogenations in trickle flow mode: A pharmaceutical industry perspective, Org. Process Res. Dev. 2022, 26, 2190–2223

L. X. Yu, 2008, Pharmaceutical quality by design: product and process development, understanding, and control, Pharmaceutical research, 25(4), 781-791.

Antonis Kokossis, Michael C. Georgiadis, Efstratios N. Pistikopoulos (Eds.)
PROCEEDINGS OF THE 33rd European Symposium on Computer Aided Process Engineering
(ESCAPE33), June 18-21, 2023, Athens, Greece

Computer-aided multi-scale simulation of MOF-based membrane separation for CO_2 removal via integration of molecular and process simulation through machine learning

Xi Cheng,[a] Yangyanbing Liao,[b] Xiaolei Fan,[b] Jie Li,[a]

[a]Centre for Process Integration, Department of Chemical Engineering, School of Engineering, the University of Manchester M13 9PL, UK
[b]Department of Chemical Engineering, School of Engineering, the University of Manchester M13 9PL, UK
jie.li-2@manchester.ac.uk

Abstract

MOF membranes have shown high productivity in CO_2 removal from gas mixtures. In this work, we present a new integrated framework for reliable prediction of the process performance of MOF-based membranes (e.g. IRMOF-1). Firstly, molecular simulations are conducted to investigate the adsorption isotherms, self-diffusivity, activation energy, permeability and selectivity of IRMOF-1 for CO_2/CH_4 separation under different operating conditions. The simulated isotherms at 298 K are in good agreement with the experimental results. As the operating conditions vary, the permeability of CO_2 can scale from 40898 to 381776 barrer. A similar observation is made regarding selectivity, highlighting the dramatic effect of operating conditions on membrane performance. Further, predictive models are developed with machine learning, allowing the calculation of permeability and selectivity. As the mean squared error is 0.0086 and the R-Square is 0.9822, the predictive models are shown high reliability. The predictive models integrated with the tanks-in-series model of a hollow fiber membrane separation process are simulated by the finite-volume method. The feasibility and competence of the proposed framework are illustrated in three case studies.

Keywords: Metal-organic framework (MOF), CO_2/CH_4 separation, Molecular simulation, Machine learning, Membrane separation.

1. Introduction

As energy consumption increases due to large-scale industrial production, more and more greenhouse gases (mainly CO_2 and CH_4) are released, leading to serious greenhouse effects such as global warming. In the meantime, the extraction of raw natural gas fields is increasing where the separation of CO_2 and CH_4 is becoming particularly crucial. For industrial applications of CO_2/CH_4 gas mixture separation, the feed composition of CO_2 is widely distributed, covering a range of 30 to 90%. This makes it urgent to develop technologies for the efficient CO_2 removal. In order to reduce energy costs, membrane separation techniques are employed for the enrichment of lean feed gas. Membrane separation takes advantage of the selective separation function of materials. Therefore, operating conditions are typically moderate, leading to much lower energy consumption.

Furthermore, the membrane separation is generally easy to operate, control and scale up with high separation efficiency.

There are two parameters that characterize the efficiency of membrane separations, including permeability and selectivity. Conventional polymer membranes typically have a trade-off relationship between permeability and selectivity, failing to achieve high permeability and high selectivity simultaneously. Porous membranes are considered to be the alternative to polymer membranes due to their adsorption behavior and pore size sieving effect, which allow them to break through the limitation of Robeson trade-off upper bound of the polymer membrane (Robeson 2008). MOF stands for a new type of porous materials which have shown high productivity in CO_2 removal from gas mixtures. Various MOF materials have been synthesized experimentally, however, unfortunately their properties are generally evaluated at ambient temperature and 1 bar pressure (Erucar and Keskin 2013). Due to the fact that these properties can vary significantly depending on the operating conditions, this evaluation is no longer sufficient for industrial applications under varying operating conditions. On one hand, experimental studies are usually complex and expensive because of the uncertainty of uncontrollable errors and the variability of samples with impurities and defects. On the other hand, some MOF membrane performance have been evaluated by molecular simulations, most of these are carried out at ambient temperature and a fixed pressure (Li, Kresse et al. 2001). Therefore, existing work is insufficient to support a process optimization approach to identify the optimal design for CO_2/CH_4 membrane separation.

Although molecular simulation is extensively applied to study gas separation in MOFs, it is computationally expensive in terms of resources and time costs (Pronk, Pall et al. 2013). For this reason, it is a challenge to integrate molecular simulations with process simulations. Through molecular simulation it is possible to generate the big dataset on the performance of MOFs under different operating conditions. A prediction model can be developed based on this big dataset by machine learning. This incorporation shows great potential as it can efficiently realize integration design. To the best of our knowledge, there is currently no framework developed for the integration of molecular simulation into membrane process simulation via machine learning. Consequently, the main objective of this study is to develop a novel multi-scale design framework via the integration of molecular and process simulation through machine learning, which is evaluated by exploring the potential of a MOF-based membrane for the separation of CO_2/CH_4 mixture.

2. Methodology

2.1. Molecular Simulation

In this work, we firstly apply grand canonical Monte Carlo (GCMC) and molecular dynamics (MD) simulation to investigate the adsorption, diffusion, and permeation of a MOF-based membrane (e.g. IRMOF-1) for CO_2/CH_4 mixtures separation under different operating conditions. We choose IRMOF-1 as the MOF-based membrane material because it is one of the earliest MOFs that can be stably synthesized and industrialized and has been successfully fabricated as film. It is also the parent of the classical IRMOF series with a large number of available literature and experimental data. The CO_2/CH_4 binary mixtures with varied feed CO_2 compositions from natural gas is used in this work. The molecular simulation is carried out at temperatures of 273, 298, 323, 348, 373 and 423 K under pressure range of 0–20 bar. The adsorption isotherms of pure CO_2, CH_4 and their mixtures in IRMOF-1 are evaluated by GCMC simulation. MD simulation is then performed in NVT ensemble to investigate the self-diffusivity and diffusion activation energy. The permeability and membrane selectivity of IRMOF-1 are also calculated based on the solution-diffusion model. For detailed parameters and validation refer to the work (Cheng, Liao et al. 2023).

2.2. Machine Learning (e.g. Artificial Neural Network)

Based on the GCMC & MD simulation for the CO_2/CH_4 mixture separated in IRMOF-1 membrane under different operating conditions, two ANN models are developed and trained to describe the adsorption capacity and self-diffusivity, respectively. There are approximately 70% of the data selected randomly from dataset (1340 points) for model training, 15% of the data for validation and 15% of the data for testing. The ANN models use the feed CO_2 composition, temperature, and partial pressure as the input layer to create the input vector. The output layer of the network is a four-dimensional output vector containing adsorption capacity and self-diffusivity of each component. For detailed configuration of ANN refer to the work (Cheng, Liao et al. 2023).

2.3. Mathematical Modeling of Hollow Fiber Membrane

This work adopts a tanks-in-series model (Katoh, Tokumura et al. 2011) of the hollow fiber membrane as shown in Fig. 1, feeding from the shell side, capturing permeate CO_2 from the inside of fibers and yielding a product from the residue side. The tanks-in-series model consists of a series of equal sized and well-mixed tanks which are numbered in both sides from the feed end to the residue end. The number of tanks in the permeate side is set to double of that in the residue side. The mathematical model includes the relationship governing transport across the membrane, mass balance equations, pressure drop equation for permeate/feed sides and boundary and initial conditions that reflect the configuration and operation of the hollow fiber membrane module. For detailed expressions and models refer to the work (Cheng, Liao et al. 2023).

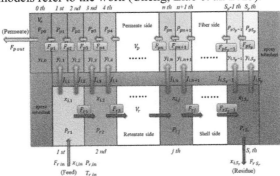

Figure 1 Schematic of tanks-in-series model for a counter-current hollow-fiber

2.4. Implementation

The GCMC and MD simulations are conducted in Materials Studio software. The ANN training, validation and testing are conducted in MATLAB providing early stopping applied in this work. The mathematical model of hollow fiber membrane separation process is simulated using the finite volume method implemented in MATLAB.

3. Results

3.1. GCMC & MD Simulation

Fig. 2a-b show the adsorption isotherms of pure CO_2 and CH_4 while Fig. 2c–d show adsorption isothermal surfaces of mixed CO_2 and mixed CH_4 in varied composition binary mixtures in IRMOF-1 from the GCMC simulation. On one hand, the pure gas GCMC results show that the maximum adsorption capacity of CO_2 is as high as 27.91

mmol g^{-1}, while the maximum adsorption capacity of CH$_4$ reaches 15.86 mmol g^{-1}. On the other hand, the mixture GCMC results demonstrate that CO$_2$ plays a dominant role in the mixed gas adsorption behavior, with a much larger adsorption capacity (24.80 mmol g^{-1}) than that of CH$_4$ (6.60 mmol g^{-1}).

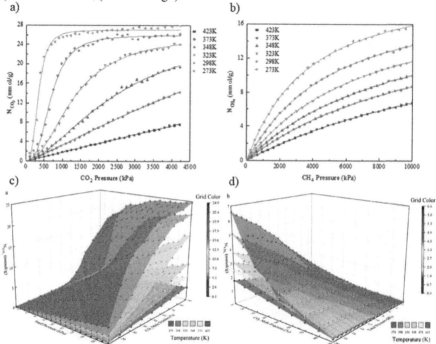

Figure 2 Adsorption isotherms of pure CO$_2$, CH$_4$ and CO$_2$/CH$_4$ binary mixture at different temperatures: a) pure CO$_2$ b) pure CH$_4$ c) mixed CO$_2$ d) mixed CH$_4$

Table 1 Diffusion activation energy of pure CO$_2$ and CH$_4$ at 1, 5, 10 bar

p (bar)	Pure CO$_2$ E$_a$ (kJ mol^{-1})	Pure CH$_4$ E$_a$ (kJ mol^{-1})
1	9.2	6.6
5	11.6	5.6
10	15.0	5.4

As observed from Table 1, the MD results for pure gases suggests that the activation energy of CO$_2$ diffusion in IRMOF-1 is greater than that of CH$_4$ under the same operating conditions, indicating that the diffusion of CO$_2$ is more difficult. This conclusion is consistent with the fact that the self-diffusivity of CO$_2$ is smaller than that of CH$_4$. On the other hand, the mixture MD results are completely opposite to the pure gas results, which can be seem from Fig. 3. The activation energy of CO$_2$ diffusion in IRMOF-1 is much lower than that of CH$_4$ in the mixture, which is also confirmed by experimental phenomena of CO$_2$/CH$_4$ binary mixture diffusion in other porous materials (Zhao, Feng et al. 2016). The combined effect of competitive adsorption and synergistic diffusion leads to a significant reduction in activation energy of mixed CO$_2$ diffusion. CH$_4$ carries CO$_2$ along with it through the membrane, while large amounts of CO$_2$ get stuck in the pores of membrane, blocking the pathway and making the diffusion of CH$_4$ difficult.

Based on the solution-diffusion model, permeability and selectivity could be calculated. The mixture results show that CO$_2$ permeability in IRMOF-1 is ranging from

40898 to 381776 barrer with the variation in operating conditions, the same phenomenon is also reflected in the selectivity. These indicate the great effect of operating conditions on the performance of MOF membrane. The rest of available data and analysis of GCMC and MD results are presented in detail in the work (Cheng, Liao et al. 2023).

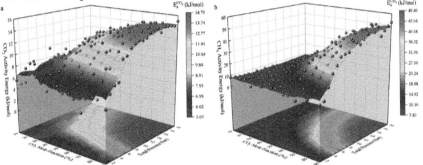

Figure 3 Diffusion activation energy of CO_2/CH_4 mixtures:
a) E_a of mixed CO_2, b) E_a of mixed CH_4

3.2. ANN Prediction

Prediction models for adsorption and diffusion are developed through ANN. The MSE of prediction test is 0.0086, and R^2 is 0.9822, which indicate the high prediction accuracy of the developed ANN models. Fig. 4 reveals the parity plot for adsorption capacity and self-diffusivity of CO_2 and CH_4 in mixtures from ANN prediction and GCMC & MD results under various operating conditions such as temperature (273–423 K), pressure (0–20 bar), and feed CO_2 composition (0.1–0.9).

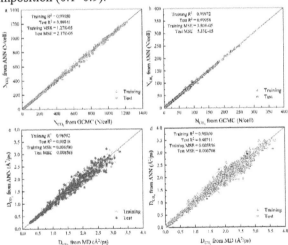

Figure 4 Parity plot for prediction from ANN vs. GCMC & MD

3.3. Integrated Process Simulation

Three case studies by process integration and simulation of membrane separation of CO_2/CH_4 mixture from raw natural gas, landfill gas and shale gas indicate that IRMOF-1 membrane can produce greater recovery with significantly smaller membrane area than the polymer PTMSP membrane from Ref. (Robeson 2008) to achieve the desired target purity. The computational results using IRMOF-1 membrane with the developed ANN model and using PTMSP membrane are provided in Table 2. All these demonstrate the proposed multi-scale design framework is feasible, practical, and superior. It integrates the variation of permeability with operating conditions into the dynamic simulation of membrane separation process. The rest of available data and analysis of integration results are presented in detail in the work (Cheng, Liao et al. 2023).

Table 2 Computational results of case study

Criterion	Case 1		Case 2		Case 3	
	This work	Reference work [9]	This work	Reference work [9]	This work	Reference work [9]
$F_{permeate}$ (Nm^3/h)	652	716	686	800	1186	1482
$F_{residue-product}$ (Nm^3/h)	1348	1284	1314	1200	814	518
Fiber number	6062	45600	7408	40500	17964	50480
Membrane area A_m (m^2)	14.47	108.9	17.69	96.70	42.89	120.5
$Purity_{CH_4}$ $(mole\ \%)$	90	90	80	80	65	65
$Recovery_{CH_4}$ $(\%)$	76	72	81	74	88	56

4. Conclusion

In this paper, the integrated design framework we initially proposed is well demonstrated in the gas separation hollow fiber membrane process with the example of IRMOF-1, which provides an integration idea for future optimization of process conditions and innovative design of optimal MOF membrane material.

References

Cheng, X., Y. Liao, Z. Lei, J. Li, X. Fan and X. Xiao (2023). "Multi-scale design of MOF-based membrane separation for CO2/CH4 mixture via integration of molecular simulation, machine learning and process modeling and simulation." Journal of Membrane Science: 121430.

Erucar, I. and S. Keskin (2013). "High CO2 Selectivity of an Amine-Functionalized Metal Organic Framework in Adsorption-Based and Membrane-Based Gas Separations." Industrial & Engineering Chemistry Research 52(9): 3462-3472.

Katoh, T., M. Tokumura, H. Yoshikawa and Y. Kawase (2011). "Dynamic simulation of multicomponent gas separation by hollow-fiber membrane module: Nonideal mixing flows in permeate and residue sides using the tanks-in-series model." Separation and Purification Technology 76(3): 362-372.

Li, X.-G., I. Kresse, Z.-K. Xu and J. Springer (2001). "Effect of temperature and pressure on gas transport in ethyl cellulose membrane." Polymer 42(16): 6801-6810.

Pronk, S., S. Pall, R. Schulz, P. Larsson, P. Bjelkmar, R. Apostolov, M. R. Shirts, J. C. Smith, P. M. Kasson, D. van der Spoel, B. Hess and E. Lindahl (2013). "GROMACS 4.5: a high-throughput and highly parallel open source molecular simulation toolkit." Bioinformatics 29(7): 845-854.

Robeson, L. M. (2008). "The upper bound revisited." Journal of Membrane Science 320(1-2): 390-400.

Zhao, Y., Y. Feng and X. Zhang (2016). "Selective Adsorption and Selective Transport Diffusion of CO2–CH4 Binary Mixture in Coal Ultramicropores." Environmental Science & Technology 50(17): 9380-9389.

Antonis Kokossis, Michael C. Georgiadis, Efstratios N. Pistikopoulos (Eds.)
PROCEEDINGS OF THE 33rd European Symposium on Computer Aided Process Engineering
(ESCAPE33), June 18-21, 2023, Athens, Greece
© 2023 Elsevier B.V. All rights reserved.   http://dx.doi.org/10.1016/B978-0-443-15274-0.50316-4

Decision making software for cosmetic product design based on an ontology

Alex Gabriel[a], Juliana Serna[a,b]*, Valentin Plantard–Wahl[c], Antoine Le Jemtel[c], Vincent Boly[a], Véronique Falk[b]

[a]Université de Lorraine, Equipe de Recherche des Processus Innovatif, ERPI-ENSGSI, 8 Rue Bastien-Lepage, 54000 Nancy, France
[b]Université de Lorraine, Laboratoire Réactions et Génie des Procédés, CNRS-LRGP-ENSIC, 1 Rue Grandville, 54000 Nancy, France
[c]Université de Lorraine, TELECOM Nancy, 193 Avenue Paul Mulle, 54600 Villers-lès-Nancy, France

*jsernaro@unal.edu.co

Abstract

The design of formulated chemical products such as cosmetic emulsions is normally carried out by interdisciplinary teams that need versatile and reliable tools to support them in the decision-making process. In view of this, this article presents a tool to support product designers during the conceptual design of emulsified cosmetic products. The tool is based on a previously developed ontology for cosmetic product design. It integrates concepts from emulsion science, cosmetic formulation, expert knowledge, and design heuristics in a systematic and flexible way. Additionally, it has a modern architecture based on mobile applications which makes it easy to use. The tool support three types of decisions: screening of ingredients, selection of ingredients considering multiple criteria (ingredient performance, origin, price, etc.), and evaluation of a formulation according to design heuristic. As a result, the tool allows the visualization of ingredient properties and a better selection of ingredients, as well as an early evaluation of formulations prior to experimentation.

Keywords: Product design, Cosmetic emulsions, Software application, Ontology

1. Introduction

The design of cosmetic products is a complex task that involves the management of multiple and heterogeneous information: raw material data, mathematical models, design heuristics, emulsions science principles, experimental data, among others (Zhang et al., 2020) (Kontogeorgis et al., 2022). Thereby, tools capable of handling varied information and comprehensively representing the solution space of design options are needed. They would have the potential to assist product designers in making decisions to seek original and context-specific solutions. Those tools should be flexible to be adapted to different design problems, easy to use and understand to help rather than hinder the design process, and reliable to be effective in supporting decision making.

There have been extensive contributions from the academy to produce tools for chemical product design. Some of the most notable are ProCAPD with an emphasis on the prediction of physicochemical properties for product optimization (Kalakul et al., 2018) and the virtual process-product design laboratory (VPPD-Lab) which contains a wide database of property and molecular design models to support design decisions (Kontogeorgis et al., 2022). Previously mentioned tools are generic and not specific for cosmetic design. They are mostly applicable for the definition and resolution of optimization design problems, i.e., they can be used when the design problem is or can be well and explicitly defined. Additional examples of practical tools to support formulators in the design of cosmetics are the app-based resources from Abbott (2022), which shows how scientific principles of colloidal science can be used to explore design solutions for different formulated products, mainly paints, and the commercial software Ansys Granta MI for material design and selection (Ansys, 2022).

However, the drawbacks of the contributions are their complexity, their lack of specificity for cosmetics and their low accessibility and ease of use. In view of the above, this article presents a cross platform decision-making tool for the design of emulsion-based cosmetics at early stages. It is called Formultools. Its architecture was inspired by the OntoCosmetic ontology (Serna et al., 2022), whose scope is the formulation of emulsion-based cosmetic products.

2. Design methodology

Considering the premise that the tool must be accessible, a great importance was given to the interaction with the user/formulator. The design process was inspired by the 5-plans user-centered design approach (Garrett, 2011). It involved cosmetics experts and computer scientists in co-design The development was done iteratively with intermediate usability tests with non-expert users. Each experience was assessed through a questionnaire using the method AttrakDiff (Lallemand et al., 2015) to follow usability improvement at each iteration. The problems encountered were progressively corrected until the current version was released.

2.1. Tool scope and functionality

The decision-making tool targets cosmetic product designers (the user). It is to be implemented at the early design stages. Three purposes were defined: the exploration of ingredients according to their properties; the selection of ingredients considering specific criteria, the evaluation of a possible formulation according to design heuristics.

3. Formultools: an application for the design of cosmetic products

3.1. Ontology as a foundation

The ontology OntoCosmetic is a knowledge representation containing all the concepts used for the development of the application. The ontology scope is the formulation of emulsion based cometic products. It is based on four main concepts: ingredients, formulations (list of ingredients with composition), properties, design heuristics, and their interrelations. Figure 1 presents an overview of the ontology:

- Ingredients and formulation (dosage): These concepts correspond to the variables that designers can select to achieve expected product properties, i.e., they are the independent variables because designers can select ingredients and their composition (dosage). Ingredients can be emollients, surfactants, thickeners, actives, other ingredients (stabilizers, preservatives, etc.).
- Properties: There are product properties and ingredients properties, and they characterize the behavior of the product and of the different ingredients respectively. Some properties are specific to a category of ingredient such as the HLB for surfactants. Some properties are quantitative (as the HLB) or qualitative (as the origin of ingredients, which can be natural or synthetic). As an example, Table 1 presents a list of the properties of emollients and their characteristics.

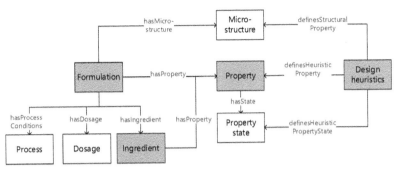

Figure 1. Overview of OntoCosmetic

Table 1. Main emollient properties in OntoCosmetic

CAS	Meaning	Range
INCI name	Abbreviation for International Nomenclature of Cosmetic Ingredients	String
Chemical class	chemical nature of an emollients: ester, fatty alcohol, hydrocarbons, silicones, triglyceride, fatty acid	String
Polarity (qualitative)	It determines the types of inter-molecular forces of the substance.	
Viscosity (qualitative)	Resistance of a fluid to the deformation	
Spreading (qualitative)	It identifies the easy of a fluid to spread on a surface	
Emollience (qualitative)	It indicates if an emollient produces a light skin feel or a rich skin feel	
After feeling (qualitative)	It indicates if an emollient leaves an oily residue after application	
State	indicates if the substance is liquid or solid.	
Biodegradability	It is the capacity for biological degradation of organic materials by living organisms	
Origin	It makes refer to the source of the substance: if it is of petrochemical origin or natural origin.	
Price	Ingredient cost	Float
HLBr	It represents the HLB needed by an oil or a mixture of oils to be more likely to be incorporated in a stable O/W emulsion.	Float
MW	Mass of a compound divided by the number of moles	Float

- Design heuristics: They are design rules based on expert knowledge and they relate ingredients, ingredients properties and the formulation with product properties and their values. They enable to predict the effects of design decisions in product properties. As an example, Table 2 presents a list containing some of the design heuristics.

Table 2. Example of design heuristics in OntoCosmetic

Product property	Heuristic
Stability	An emulsion tends to be more stable using a suitable combination of surfactants with a HLB equivalent to the HLB required by the emollients
Stability	It is recommended to use more than one surfactant
Stability	The quantity of surfactant should be between 14 to 20% of the total oily phase
Viscosity	Viscosity is low (lotion texture) if a low proportion of thickener is used
Viscosity	Viscosity is high (cream to gel texture) if a high proportion of thickener is used
Viscosity	Viscosity increases if the content of one of the following fatty alcohol increases: cetearyl alcohol, stearyl alcohol or cetyl alcohol
Viscosity	Viscosity is decreasing if a branched-chain higher alcohol is added

3.2. Formultools: a cross platform tool based on OntoCosmetic

Ontologies are great for formalizing knowledge but they are not intended for end users. Formultools aims to provide user-friendly interface to product designers who wants to explore the design space at early design stages looking for product concepts to be tested. It was developed to be simple, accessible, useful, and versatile. Its data architecture is directly inspired by the OntoCosmetic ontology as the knowledge graph used is an instantiated version of OntoCosmetic with ingredients, heuristics, and properties. The structure of the application is split into two main sections as represented in **Error! R eference source not found.**. An exploration section, which is focused on ingredients searching within a collection, visualization, and ranking. And a formulation section calculating product properties based on heuristics for their verification. Its interface is inspired on mobile applications.

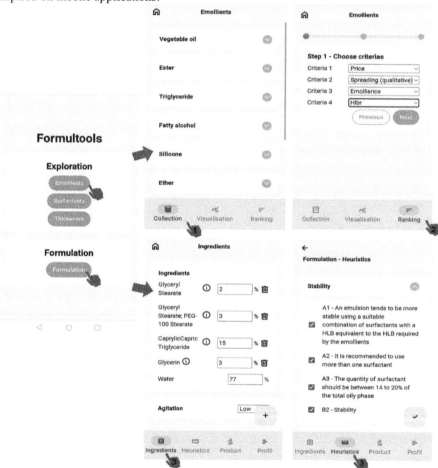

Figure 2. General interface of Formultools

The main functions of Formultools are explained below:

• Ingredient searching and visualization: The application has three categories of ingredients according to their functions in cosmetic products: emollients, surfactants, and thickeners, which can be accessed in the home page. Subsequently, the collection of ingredients of each category are listed and presented according to their chemical class. Ingredients can be explored in the following two ways: they can be viewed with their individual properties on the collection tap, they can be explored based on the values of their properties and compared graphically on the visualization tap. The latter also presents the filter option, where ingredients can be screened according to their properties.

- Figure **3** presents the exploration options.

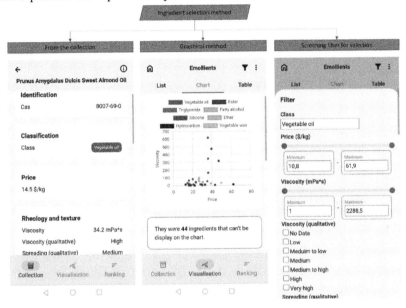

Figure 3. Exploring emollients using Formultools (collection, visualization, filter)

- Ingredient selection: The application enables the ranking of ingredients considering the properties that the designer wants to optimize. For this, the method Analytic Hierarchy Process (AHP) (Saaty, 1987) is used. With the aid of the tool, the formulator first defines the proprieties considered as criteria of selection in the first tap. Then, he defines which proprieties must be minimized and maximized in the second tap. Thirdly, he makes a pairwise comparison of proprieties to calculate their weights. The AHP method implies the calculation of a consistency ratio to evaluate the pairwise comparison, which can be also visualized. The last step is the visualization of the ranking. The four steps of ingredient selection are illustrated in Figure 4.

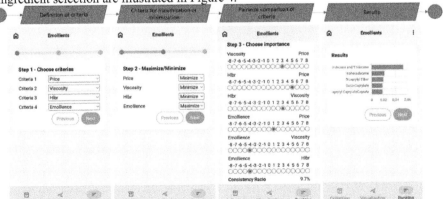

Figure 4. Ranking process of cosmetic ingredients using Formultools

- Formulation evaluation: FormulTools enables the evaluation of formulations suggested by designers according to design heuristic and recommendations contained in the data base. He defines a formulation (ingredients and composition), select design heuristics from the data base and based on the heuristics, the tool calculate product properties. In this way, the designer can propose solutions to design specifications and check if those solutions would correspond to specifications according to heuristics before testing them experimentally.

4. Conclusions and perspectives

The article presents Formultools, a tool to support decision-making process for the design of cosmetic emulsions at early stages. Currently, the tool can perform three functions: exploration and selection of ingredients and verification of cosmetic formulations. It was designed to make it as practical as possible which is why it has the form of a mobile application. In a near future, it will be tested with experts and improved for its subsequently application in real design cases.

Formultools was developed in co-design between experts on cosmetic formulation and computer science. Additionally, it was continuously tested with users and iteratively improved. As a result, the tool was developed in a very agile process.

In the long term, the tool will be enhanced to assist the design process in a more holistic way. This will require the implementation of new features to enable information management, learning from each successful design, use by various design agents, among others, while maintaining its simplicity and usability.

References

Abbott, S., 2022. Science for the Real World [WWW Document]. URL https://www.stevenabbott.co.uk/index.php

Ansys, 2022. Materials: Leading Material Intelligence [WWW Document]. URL https://www.ansys.com/products/materials#tab1-2

Garrett, J. J. (2011). The elements of user experience : User-centered design for the Web. In Interactions. New Riders. https://doi.org/10.1145/889692.889709

Kalakul, S., Zhang, L., Choudhury, H.A., Elbashir, N.O., Eden, M.R., Gani, R., 2018. ProCAPD – A Computer-Aided Model-Based Tool for Chemical Product Design and Analysis, in: Eden, M.R., Ierapetritou, M.G., Towler, G.P.B.T.-C.A.C.E. (Eds.), 13 International Symposium on Process Systems Engineering (PSE 2018). Elsevier, pp. 469–474. https://doi.org/https://doi.org/10.1016/B978-0-444-64241-7.50073-2

Kontogeorgis, G.M., Jhamb, S., Liang, X., Dam-Johansen, K., 2022. Computer-aided design of formulated products. Curr. Opin. Colloid Interface Sci. 57, 101536. https://doi.org/10.1016/j.cocis.2021.101536

Lallemand, C., Koenig, V., Gronier, G., & Martin, R. (2015). Création et validation d'une version française du questionnaire AttrakDiff pour l'évaluation de l'expérience utilisateur des systèmes interactifs. European Review of Applied Psychology, 65(5), 239-252. https://doi.org/10.1016/j.erap.2015.08.002

Saaty, R.W., 1987. The analytic hierarchy process—what it is and how it is used. Math. Model. 9, 161–176. https://doi.org/10.1016/0270-0255(87)90473-8

Serna, J., Rivera-gil, J.L., Gabriel, A., Arrieta-Escobar, J.A., Boly, V., Falk, V., Narvaez Rincon, P.C., 2022. Application of an ontology-based decision support system for the design of emulsion-based cosmetic products, in: ESCAPE 32. Toulouse.

Zhang, L., Mao, H., Liu, Q., Gani, R., 2020. Chemical product design – recent advances and perspectives. Curr. Opin. Chem. Eng. 27, 22–34. https://doi.org/10.1016/j.coche.2019.10.005

Antonis Kokossis, Michael C. Georgiadis, Efstratios N. Pistikopoulos (Eds.)
PROCEEDINGS OF THE 33rd European Symposium on Computer Aided Process Engineering
(ESCAPE33), June 18-21, 2023, Athens, Greece

Evaluation of amines for the removal of acid gases from the liquid hydrocarbons

Siddig S. Khalafalla [a], Umer Zahid [a, b]

[a] *Chemical Engineering Department, King Fahd University of Petroleum and Minerals, Dhahran 31261, Saudi Arabia.*
[b]*Interdisciplinary Research Center for Membranes & Water Security, King Fahd University of Petroleum and Minerals, Dhahran 31261, Saudi Arabia.*
uzahid@kfupm.edu.sa

Abstract

Liquefied Petroleum Gas (LPG) is a product made from crude oil and has several domestic and industrial applications. Continuous increases in energy demand around the world and strict environmental regulations require clean fuel for various applications. This requires the cleaning of liquid hydrocarbons from acid gases which are toxic to human health and disastrous to industrial equipment. Currently, most of the liquid hydrocarbon treatment is done using caustic solutions to remove acid gases. However, depending on the concentration of impurities in the feed and the amount of feed to be processed, a large amount of caustic solution is required which poses another disposal issue. This study proposes to replace the conventional caustic cleaning process with the amine process for liquid hydrocarbon treatment. The process consists of a liquid-liquid contactor where impurities are removed using the amine and then rich amine is regenerated in the stripper column. This study's goal is to assess various amine types for removing acid gases from liquid hydrocarbon feeds. The study is conducted using commercial simulation software ASPEN HYSYS® for the design and analysis of process performance. The studied variables include the concentration of the amine, the amine circulation rate, the effect of LPG feed temperature, and the lean amine temperature. The outcomes demonstrate that the removal of acid contaminants from the hydrocarbons can be a competitive process to achieve the required product purities.

Keywords: LPG sweetening; acid gases; process design; amines.

1. Introduction

LPG is typically a mixture of propane and butane in various proportions derived from the processing of crude oil or natural gas. LPG has a wide range of domestic and industrial usage. The global demand for LPG has been raised over the past decades due to the need for clean energy. Additionally, a compound annual growth rate (CAGR) of 2.2% is anticipated from 2021 to 2026 (*Global Five-Year LPG Demand Fuelled by New Petchems Plants, Res-Com; Supply to Overwhelm*, n.d.). Moreover, at the same time, strict environmental policies are in place to curtail environmental and health issues, therefore the demand for liquid hydrocarbon sweetening facilities has increased. The most common impurities present in LPG are carbon dioxide (CO_2), hydrogen sulfide (H_2S), carbon sulfides (COS), and mercaptans (RSH). These impurities, if not removed, can cause corrosion, operational inefficiencies, and other personnel health issues. Currently, many industries are using caustic solutions to treat the contaminants. However, usually, a large amount of caustic solution is required for the hydrocarbon sweetening which poses another environmental issue. Another method to remove impurities from the liquid hydrocarbon is to treat it with the amine solution. The typical amine process consists of contactor-stripper columns in series where the acid impurities are removed with amine solvent in a liquid-liquid extractor (LLE) (Jayakumar et al., 2017).

Amines have been used for many years in the natural gas industry for the sweetening of natural gas (Zahid et al., 2020). However, the application of amines for the treatment of liquid hydrocarbon is rarely studied. Most of the existing studies have focused on hydrocarbon treatment using caustic solutions and various ways how the process efficiency can be improved. However, in the literature, there are only a few studies that have investigated the use of amines for liquid hydrocarbon treatment. Qeshta et al. (Qeshta et al., 2015) studied the LPG sweetening process using MDEA for the Takreer refinery in Abu Dhabi, UAE where the current LPG sweetening process is the MEROX process. They concluded that amine can be a good alternative process for the treatment of LPG which can reduce significant losses in the caustic process. Zoghi et al. (Zoghi et al., 2022) investigated the effect of 30 wt.% DEA solution for the treatment of liquid propane containing both H_2S and CO_2.

The primary purpose of this research work is to focus on the design and simulation of the liquid hydrocarbon sweetening process using different amines in a liquid-liquid extractor (LLE). Three amine solvents namely methyl diethanolamine (MDEA), diglycolamine (DGA), and diethanolamine (DEA) are evaluated to analyze the sweetening performance. The goal is to keep H_2S and CO_2 in the sweet hydrocarbon stream leaving at the top of the contactor below 5 ppm. Finally, a parametric analysis has been performed to study the effect of various operational parameters.

2. Process Details

2.1. Thermodynamic

Aspen HYSYS was used to build the steady-state simulation model. For the calculation of the properties of the gas phase, the Peng-Robinson (PR) equation of state has been employed while the E-NRTL model for electrolytic thermodynamics with binary and electrolyte pair parameters optimized for liquid-liquid applications has been selected. Aspen HYSYS has a built-in library for all the equilibrium and kinetic reactions taking place in the acid gas removal (AGR) system which has been utilized from the acid gas-liquid treatment package. The available acid gas-liquid treatment thermodynamic package has been validated over a wide range of operating conditions with good accuracy, it can be used with confidence (Zahid et al., 2020).

2.2. Process Description

Figure 1 shows a typical LPG sweetening unit. Initially, LPG is fed at the bottom of the liquid-liquid extractor. As the amine solution flows countercurrent to the sour LPG, the components of the acid gas react with the amine to produce regenerable salts. The rich amine leaves at the bottom of the LLE, while the sweet LPG leaves at the top. Following this, any absorbed hydrocarbons are removed in a flash drum. The rich amine is then fed to the top of the stripper column after being heated in the rich/lean amine heat exchanger. H_2S and CO_2 are stripped from the rich amine as it flows down the column, leaving the lean amine at the bottom. It is then recycled to the top of the absorber where it passes through a lean/rich exchanger and repeats the cycle.

2.3. Process Simulation

This study explored three simulation cases with different impurities in the sour LPG feed as shown in Table 1. The first one has both H_2S and CO_2, while the second and the third have only either CO_2 or H_2S respectively. For each sour feed, three different amine solutions have been tested MDEA (35 wt.%), DGA (50 wt.%), and DEA (30 wt.%). The input variables to the simulation models are listed in Table 2. Trayed columns have been used to model the absorber and stripper due to high solvent circulation flow. For towers with a low-pressure drop requirement and a diameter of fewer than 2.5 feet, packed columns are typically advised. For single amines, the circulation rate is adjusted according to the required product purity, which is not more than 5 ppm of H_2S in the sweet gas. The stripper column's vent rate and reflux ratio have been specified to achieve column convergence, whereas the absorber column has a degree of freedom equal to zero.

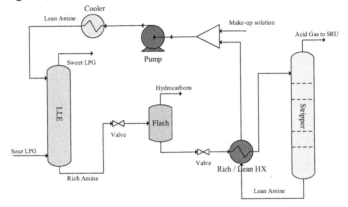

Figure 1: Process flow diagram of LPG sweetening unit.

Table 1: Sour LPG stream composition.

	Mole fraction		
Component	Case 1	Case 2	Case 3
Methane	0.009	0.006	0.011
Ethane	0.028	0.443	0.132
Propane	0.839	0.218	0.267
i-Butane	0.009	0.000	0.414
n-Butane	0.001	0.129	0.168
n-Pentane	0.001	0.068	0.005
n-hexane	0.000	0.060	0.000
CO_2	0.019	0.077	0.000
H_2S	0.093	0.000	0.002

Table 2: Input specifications used in the simulation for each LPG feed.

Parameter		Case 1	Case 2	Case 3
LPG Temperature	° C	38	15.6	51
LPG Pressure	bar	34.5	55.2	40.2
LPG flow rate	m³/h	20.9	159	24.4
LLE Number of stages		20	20	10
Amine Temperature	° C	54.44	45	51
Amine Pressure	bar	28.3	55.2	40.2
Rich amine Reg. Feed Temp.	° C	110	100	100
Regeneration column number of stages		10	10	20

3. Results

The simulation results such as rich loading, lean loading, and circulation rate have been summarized in table 3. Since each scenario has different lean loading and rich loading that reflect to have different circulation rates for the amine. For case 1 and case 3, DGA shows promising results while for case 2 MDEA was the best as illustrated in the next section.

Table 3: Main simulation results for the three cases.

	Case 1	Case 2	Case 3
MDEA 35 wt.%	Rich Loading: 0.45 Lean loading: 0.01 Circulation rate: 825 kmol/h	Rich Loading: 0.249 Lean loading: 0.01 Circulation rate: 543 kmol/h	Rich Loading: 0.273 Lean loading: 0.002 Circulation rate: 25 kmol/h
DGA 50 wt.%	Rich Loading: 0.35 Lean loading: 0.035 Circulation rate: 600 kmol/h	Rich Loading: 0.35, Lean loading: 0.01 Circulation rate:263 kmol/h	Rich Loading: 0.35 Lean loading: 0.02 Circulation rate: 12.3 kmol/h
DEA 30 wt. %	Rich Loading: 0.35 Lean loading: 0.015 Circulation rate: 1210 kmol/h	Rich Loading: 0.43 Lean loading: 0.03 Circulation rate: 355 kmol/h	Rich Loading: 0.35 Lean loading: 0.015 Circulation rate: 26 kmol/h

3.1. Regeneration energy requirements

Energy analysis for the regeneration process has been studied, the results show that the developed simulation models are in good agreement with the available literature data. The result shows that the DGA is preferred in the presence of H_2S and CO_2, with 62% and 38% less energy compared to MDEA and DEA, respectively. However, for case 2 in the presence of CO_2 only, MDEA has lower energy requirements with a slight improvement of approximately 3% and 11% compared to DEA and DGA. While in the case of H_2S only (Case 3), DGA has a lower energy requirement similar to case 1. Figure 2 illustrates the energy requirement of three cases for amine regeneration.

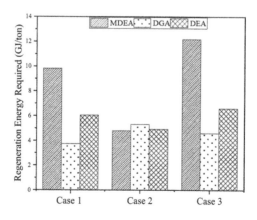

Figure 2: Energy required for the regeneration process.

3.2. Effect amine concentration

Acid gas removal in the LLE depends directly on amine concentrations. Figure 3 shows the effect of different amine concentrations on the acid gas in the sweet LPG and the energy required for the regeneration. It was observed that as the amine concentration increased, the acid gas in the sweet LPG decreased while there is an increase in the energy required for the regeneration. Similar trends were observed for Cases 2 and 3.

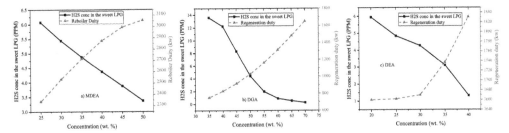

Figure 3: Effect of amines concentration

3.3. Effect of lean amine temperature

As shown in figure 4, the lean amine temperature influences both acid gas concentration in the sweet LPG and regeneration duty. As the lean amine temperature increased, the acid gas in the sweet LPG gradually increased due to the exothermic nature of the extraction of the acid gas with amines. However, the regeneration duty slightly decreases with the increase in the lean amine temperature. Similar trends were observed in other cases.

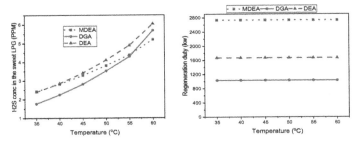

Figure 4: Effect of lean amine temperature

3.4. Effect of Lean amine circulation rate

One of the major parameters that directly affect the extraction of the acid gas from the LPG is the lean amine circulation rate. Figure 5 shows the effect of circulation rate on acid gas concentration and regeneration energy requirement. As the circulation rate increased, the acid gas concentration increased because of the lower residence time. On the other hand, the duty of the regeneration is also increased due to the high flow rate.

Moreover, the effect of the LPG feed temperature has been studied for Case 3. LPG feed temperature has been changed in temperature between 10 °C to 50 °C while keeping the difference between the lean amine temperature and the sour LPG feed fixed at 5 °C. The results show that the H_2S concentration in the sweet LPG increases as the temperature of the feed increases while the regeneration duty slightly decreases.

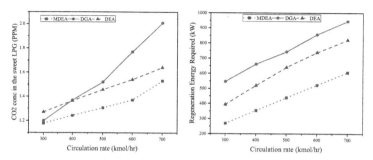

Figure 5: Effect of lean amine circulation rate

4. Conclusion

In conclusion, the design specification of the sweetening of the sour LPG should be maintained in such a way that the concentration of the acid gas in the sweet LPG achieved the product specification. Sensitivity analyses have been conducted for a few of the significant process variables that could have an impact on the process performance. The results show that the acid contaminants removal from the hydrocarbons can be a competitive process to achieve the required product purities

Acknowledgment: The author(s) would like to acknowledge the support provided by the Deanship of Research Oversight and Coordination (DROC) at King Fahd University of Petroleum & Minerals (KFUPM).

References

Global five-year LPG demand fuelled by new petchems plants, res-com; supply to overwhelm. (n.d.). Retrieved November 29, 2022, from https://www.energyaspects.com/lpg-demand-insight

Jayakumar, K., Panda, R. C., & Panday, A. (2017). *A Review: State-of-the-Art LPG Sweetening Process. 9*(2), 175–206.

Qeshta, H. J., Abuyahya, S., Pal, P., & Banat, F. (2015). Sweetening liquefied petroleum gas (LPG): Parametric sensitivity analysis using Aspen HYSYS. *Journal of Natural Gas Science and Engineering, 26*, 1011–1017. https://doi.org/10.1016/j.jngse.2015.08.004

Zahid, U., Sakheta, A., & Lee, C. J. (2020). Techno-economic analysis of acid gas removal from associated and non-associated sour gas using amine blend. *International Journal of Greenhouse Gas Control, 98*(November 2019), 103078. https://doi.org/10.1016/j.ijggc.2020.103078

Zoghi, A. T., Shokouhi, M., Naderi, F., Abbasghorbani, M., Fatehi, A., Pouladi, B., & Adhami, M. A. (2022). Investigation of Aqueous Diethanolamine Performance in Prediction of Hydrogen Sulfide and Carbonyl Sulfide Removal from Liquefied Propane. *Journal of Solution Chemistry, 51*(1), 84–96. https://doi.org/10.1007/s10953-021-01131-1

Antonis Kokossis, Michael C. Georgiadis, Efstratios N. Pistikopoulos (Eds.)
PROCEEDINGS OF THE 33rd European Symposium on Computer Aided Process Engineering
(ESCAPE33), June 18-21, 2023, Athens, Greece

Embedding Operating Flexibility in Process Design

Steven Sachio, Cleo Kontoravdi, Maria M. Papathanasiou

Sargent Centre for Process Systems Engineering, Department of Chemical Engineering, Imperial College London, London, United Kingdom SW72AZ

Abstract

Inherently flexible processes are able to handle disturbances better than point-specific designed processes. However, there is a lack of systematic framework to embed operating flexibility in process design. In this work, we propose a model-based framework for integrated analysis of process flexibility and performance based on the identification and assessment of process design spaces. The proposed framework enables the identification of operating regions, quantification, within which the product and process meet the specifications. The operational flexibility and sensitivity of the process to design and operating parameters is also quantified. We demonstrate the capabilities of the framework on a Protein A chromatographic separation, used in biopharmaceutical manufacturing. We identify the feed flow rate as the most influential process parameter, while we quantify an acceptable range for the feed stream variability (concentration: $0.37 - 0.43$ mg ml^{-1} and flowrate: $0.72 - 0.88$ ml min^{-1}) for a fixed process design.

Keywords: computing and systems engineering, process design, chromatography

1. Introduction

Embedding flexibility into process design can enable more agile and sustainable processes. This is particularly true in the case of (bio-) pharmaceutical production processes, where disturbances resulting from variability in processing streams are dominant. For example, in the case of monoclonal antibodies (mAbs), an increase in the impurity content (e.g., concentration of host cell proteins, antibody aggregates or fragments) may lead to additional separation columns and/or purification cycles required to reach the desired product purity. Subsequently, this can increase the overall downstream cost and processing times, leading to less economically and environmentally efficient processes (Papathanasiou et al, 2020; Narayanan et al, 2020). In addition to this, manufacturers need to demonstrate that the processes can meet consistently product quality criteria with respect to functionality and purity for regulatory authorities to grant market authorization. For this, there is an imminent need to design processes that are adaptive to mixture variabilities and can guarantee eco-efficient, sustainable production within specifications.

Traditionally, product quality is confirmed through a Quality by Testing (QbT) approach before batch release, which incurs high experimentation costs and waiting times, delaying market release. Furthermore, manufacturers are not able to take in-process corrective actions with this approach leading to 'off-spec' products and hence lost batches. To this end, the Quality by Design (QbD) initiative aims to shift the focus in process development to revolve around product quality (ICH, 2011) moving away from QbT. A key step in QbD-supported initiatives is the identification of a design space defined by good candidate condition sets where the process is guaranteed to satisfy the target specifications. This approach offers greater operational flexibility, in contrast to point-specific operation (Ding et al, 2021; Laky et al, 2022). Nonetheless, identification of a design space can lead to time- and cost- intensive exhaustive wet-lab experimentation. For this reason, computer-aided modelling tools are exploited for the pre-identification of good candidate operating condition sets (Kotidis et al, 2019; Diab et al, 2020; Kusumo et al, 2020). Here, we present a machine learning-aided design space identification framework that harnesses the potential of Sobol sequences for single-pass parallelizable sampling, allowing for rapid determination of acceptable operating conditions and space-specific sensitivity analysis.

2. Framework

The framework comprises of three main steps: 1) model development and problem formulation, 2) design space identification and 3) design space analysis.

2.1. Model Development and Problem Formulation

The process of interest is characterized using mathematical models, which can be mechanistic, hybrid, or fully data driven and are validated through experimental data. Following that, the manipulated variables (MVs), the KPIs and their associated bounds are identified. Finally, the KPIs of interest are selected and constraints with respect to both MVs and KPIs are defined.

2.2. Design Space Identification

First, virtual experimentation using Sobol sequence based on the formulated problem is performed to collect the dataset. Then, the constraints defined are applied to screen the dataset separating the satisfied and violated points. A collection of satisfied points, also known as a point cloud, is then set as the basis for the design space identification. The identification of the design space relies on geometrical hulls, known as alpha shapes which are a generalization of convex hulls with a parameter (alpha radius, α_r), determining the non-convexity of the resulting hull. As $\alpha_r \to \infty$, the formed hull is the convex hull, while as $\alpha_r \to 0$, the hull becomes increasingly non-convex until it reduces to the set of points at alpha radius equal to zero. The value of the alpha radius is critical for the identification of a continuous design space and is proportional to the number of points available in the space. In the case of non-linear problems, the design space shape is often non-convex and therefore at large alpha values, violated points may be found lying inside of the design space. However, at small alpha radius values, disjointed spaces start to be formed. To mitigate this, we propose two methods: namely the tolerance and the resolution support method.

2.2.1. Tolerance Method

Here, we allow for a percentage of violated points to be included in the design space to allow for the identification of a design space at low resolution with less disjointed spaces. To do this, we set a tolerance value for the allowed number of violations inside the design space is set. The alpha radius should not be too small so that the hull does not reduce to thin disjointed spaces, but not too large either as this would result in many points inside the design space violating the constraints. Here, a bisection search is used to find the alpha radius. In each iteration, a check for how many violations lie inside of the design space is performed and the iteration is terminated once the number of violations tolerance is met. This method allows for the rapid identification of the design space which can be useful provided that the resulting violations in the space are analyzed further to ensure that the violation of the KPIs meets the required accuracy.

2.2.2. Resolution Support Method

An alternative to the tolerance method is the resolution support method. For this, we train artificial neural networks (ANNs) using the manipulated variables as inputs and the KPIs as outputs. The trained ANN is used as an interpolator to increase the resolution of the dataset. Provided that it is accurate enough, this allows for small alpha radius values to be used, resulting in very fine design spaces which capture non-convexity well.

2.3. Design Space Analysis

The space defined in 2.2 is used for process flexibility analysis typically based on a nominal operating point (NOP) of interest. Starting from the NOP, a uniform cube with respect to the parameter axes is formed and expanded until one of its vertices intersects the design space boundary. The resulting region is defined as the uniform acceptable operating region (AOR). This allows for the extraction of multivariate proven acceptable ranges (MPARs) of the operating parameters with respect to the NOP used, which is an underpinning element for analyzing process flexibility and controllability. This analysis is highly dependent on the number of data points used to construct the design space. Therefore, design spaces formed using low-resolution datasets are coarser. Comparing the tolerance based and resolution support design spaces, the latter results in a larger AOR, owing to the richer dataset.

2.4. Sobol Sensitivity Indices

In addition to the systematic flexibility analysis, the framework also offers information on the sensitivity of the monitored KPIs with respect to the manipulated variables. For this, the obtained dataset within the framework can be directly used to calculate Sobol sensitivity indices (Sobol, 2001). With the resolution support method, where ANNs are trained for interpolation, the sensitivity analysis can also be performed in any acceptable operating region without further simulation of the original model. This allows for the identification of the most impactful manipulated variables and their second order interactions.

3. Case Study: Protein A Chromatography

3.1. The System

The framework presented in section 2 is applied to the semi-continuous multicolumn capture process presented by Steinebach et al (2016). The process considers two chromatography columns operated in a semi-continuous cyclic manner. Figure 1 shows the process schematic. In step A, both columns are interconnected, and fresh feed is introduced into column I. While the breakthrough from column I is fed into column 2. Next (step B), column I is washed while column

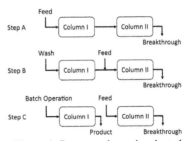

Figure 1. Process schematic adapted from Steinebach et al (2016).

II is fed the breakthrough from column I and the fresh feed simultaneously. In step C, the columns operate in batch mode with column I eluting the products and regenerating while column II is fed with the fresh feed. The process is repeated with the columns position switched. The process is modelled with Partial Differential and Algebraic Equations (PDAEs) with 50 discretization points over the column length. This results in a total of 808 ordinary differential equations (ODEs) and 812 algebraic equations. The process was validated by Steinebach et al (2016) for feed mAb concentration of $0.2 - 0.77$ mg/ml and feed flowrate of 0.5 – 1.5 ml/min. Both the process modelling and post-modelling analysis are carried out in Python.

3.2. Problem Formulation & Data Sampling

In this study, the objective is to investigate the process performance and flexibility of the process under the influence of feed concentration variations. The variables considered in this study are feed mAb concentration (c_{feed}) [disturbance], feed volumetric flowrate (Q_{feed}) [control variable], and the column switching time (T_{switch}) [design variable]. The mAb concentration is considered as a disturbance due to variability in the performance of the upstream bioreactor. On the other hand, the volumetric flowrate is a control variable while the column switching time is a design variable. The manipulated variable bounds for the study are based on the model validation range (Table 1). The KPIs considered are yield and productivity. Following standard practice, minimum constraints of 99% and 4 mg ml^{-1} h^{-1} are assumed for yield and productivity, respectively. The formulated problem is used to collect a total of 4096 points via quasi-random Sobol sampling.

Table 1. Manipulated variable bounds used in the study based on the validated model range (Steinebach et al., 2019).

Manipulated Variable	Lower Bound	Upper Bound
c_{feed} (mg ml^{-1})	0.21	0.63
Q_{feed} (ml min^{-1})	0.5	1.5
T_{switch} (min)	40	120

4. Results and Discussion

4.1. Design Space Identification

The collected dataset is screened using the constraints ($> 99\%$ yield and > 4 mg ml^{-1} h^{-1} productivity) specified to classify satisfied and violated points. Figures 2 (A), (B) and (C) illustrate the knowledge space (unconstrained), and design spaces (constrained) using the tolerance and the resolution support method, respectively. The proposed framework enables quantification of the volume of the design space including good candidate operating points. As discussed previously, in

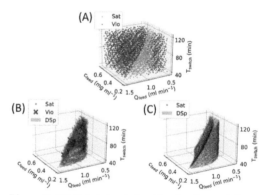

Figure 2. (A) Point cloud classification by applying constraints, (B) tolerance-based method, (C) resolution support method.

nonlinear problems as the one presented here, the design space may include points that violate one or more of the predefined constraints. A tolerance of 1% resulted in 12 out of 1281 points (0.94%) (large blue crosses) which violated the constraints inside of the design space. On the other hand, the resolution support method arrives at a zero-violation design space with a smoother surface. The framework also allows for the quantification of the size of the design space for exact comparison. In this case, the size is the volume of the space defined by the manipulated variables, hence the units are in mg $\left(\frac{mg}{ml} \times \frac{ml}{min} \times min = mg\right)$. The design space size of the tolerance-based method is 5.98 mg, while that of the resolution support method is 8.43 mg (41% larger). This is due to the higher resolution allowing for the use of a finer alpha radius which can capture the non-convexity of the problem better.

4.2. Design Space Analysis

The operational flexibility is then assessed by analyzing the identified design space. For this, we use the resolution support method (Figure 2 A(C)) as it provides the design space without any violated points. For the analysis, we choose two arbitrary nominal points of interest, as shown in Figure 3. The first nominal point lies in the center of the design space, while the second point is in the higher productivity region of the design space. The comparison between the two designs is detailed in Table 2.

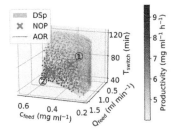

Figure 3. Good candidate operating point analysis with resolution support.

Table 2. Detailed comparison of NOP1 & NOP2

	NOP1	NOP2
AOR size (mg)	0.134	0.116
MPAR c_{feed} (mg ml^{-1})	0.40 ± 0.033	0.55 ± 0.032
MPAR Q_{feed} (ml min^{-1})	0.80 ± 0.080	0.75 ± 0.076
MPAR T_{switch} (min)	70.0 ± 6.34	50.0 ± 6.05
Average Productivity (mg ml^{-1} h^{-1})	5.91	7.54

The size of NOP1 is 15% larger than that of NOP2 indicating greater flexibility. This is further expressed in the respective MPARs of the manipulated variables which reduce by 5% going from NOP1 to NOP2. The average expected productivity when operating at NOP2 is 27% larger than that of NOP1. Using this framework, such trade-offs between flexibility and performance gain can be quantified systematically. Quantification of such metrics is essential for the design of robust and controllable processes that are crucial in biopharmaceutical manufacturing, bounded by stringent regulations.

4.3. Sobol Sensitivity Indices

We further investigate and quantify the impact of the studied variables and disturbances on the chosen KPIs. This is achieved through calculation of the Sobol sensitivity indices. For this, we use the dataset as generated in section 2. The Sobol sensitivity indices highly depends on the chosen bounds of the manipulated variables and the non-linearity of input output relationship within the chosen bounds. Hence, it is vital to be able to identify the sensitivity indices at different spaces especially for integration process units and process control. Figure 4 illustrates the Sobol sensitivity indices calculated for the acceptable

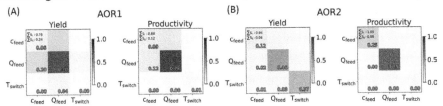

Figure 4. Sobol indices in (A): AOR1 & (B): AOR2.

operating region of NOP1 (AOR1) and NOP2 (AOR2). Whereby first order interactions are observed by looking at the main diagonals and second order interactions on the off-diagonal elements. First order indices refer to how the factional contribution of a variable on the variance of the KPI while the second order indices show the contribution of variable-variable interactions to the variance of the KPI. We assume that any interaction with an index below 10% is non-significant. There is a significant difference in the indices within AOR1 and AOR2 (Figure 4 (A) & (B)). The results indicate that for AOR1, the feed flowrate has the largest impact on both KPIs with indices larger than 0.70. AOR2 illustrates a different trend, where the switching time plays a bigger role in determining the yield of the product (from $S_i = 0.00$ in AOR1 to $S_i = 0.37$ in AOR2). On the other hand, the second order interactions observed between the feed concentration and flowrate in productivity of AOR1 ($S_{ij} = 0.20$) are not present in AOR2 ($S_{ij} = 0.02$). This is due to the nonlinear relationship between the KPIs and the manipulated variables which is a challenge for identifying the design space with zero violations. This framework allows for the unravelling of such complex relationships in a computationally efficient manner.

5. Conclusion

We presented a novel framework for the identification and analysis of process design spaces. The capabilities of the presented framework were illustrated through its application on a downstream purification unit used in biopharmaceutical production. We considered design (switching time) and operating decisions (flow rate), as well as disturbances (feed composition) simultaneously. Based on the process feasibility constraints and the target KPIs, we identified and quantified a design space that includes good candidate condition sets, predicted to meet the KPIs of interest. We further demonstrated how this space can be used for the evaluation and quantitative comparison of different operating conditions and their flexibility. This can assist manufacturers by offering a measurable estimate of the variability that a process at hand can handle without any further mediation and remaining within spec. In addition to this, we demonstrated how the framework enables space-specific sensitivity analyses without additional high-fidelity model simulation which is crucial for process control and integration. Importantly, the presented framework is generic, owing itself to the assistance of simultaneous design and flexibility assessment of different processes. In the future, the advantages of the framework will be further enhanced to incorporate formal optimization and control methodologies that can benefit from the process knowledge and good candidate initial guesses generated by the presented methodology.

References

Papathanasiou & Kontoravdi (2020): "Engineering challenges in therapeutic protein product and process design". Current Opinion in Chemical Engineering (27).

ICH (2011): "ICH Harmonised Tripartite Guideline: Pharmaceutical Development Q8, Step 4".

Ding and Ierapetritou (2021): "A novel framework of surrogate-based feasibility analysis for establishing design space of twin-column continuous chromatography". Int. J. Pharm. (609).

Laky et al (2022): "Determination of probabilistic design spaces in the hybrid manufacture of an active pharmaceutical ingredient using PharmaPy". Computer Aided Chem. Eng. (49).

Kotidis et al. (2019): "Constrained global sensitivity analysis for bioprocess design space identification". Computers and Chem. Eng. (125).

Diab and Gerogiorgis (2020): "Design Space Identification and Visualization for Continuous Pharmaceutical Manufacturing". Journal of Pharmaceutics (235).

Kusumo et al. (2020): "Bayesian approach to probabilistic design space characterization: A nested sampling strategy". Ind. & Eng. Chem. Research (59).

Narayanan et al. (2020): "Bioprocessing in the Digital Age: The Role of Process models". Biotechnol J (15).

Steinebach et al. (2016) "Model based adaptive control of a continuous capture process for monoclonal antibodies production". J Chromatogr A. (1444).

Antonis Kokossis, Michael C. Georgiadis, Efstratios N. Pistikopoulos (Eds.)
PROCEEDINGS OF THE 33rd European Symposium on Computer Aided Process Engineering
(ESCAPE33), June 18-21, 2023, Athens, Greece

Transfer learning for process design with reinforcement learning

Qinghe Gao[a], Haoyu Yang[a], Shachi M. Shanbhag[a], Artur M. Schweidtmann[a,*]

[a]*Process Intelligence Research Team, Department of Chemical Engineering, Delft University of Technology, Van der Maasweg 9, Delft 2629 HZ, The NETHERLANDS*
Corresponding author. Email: a.schweidtmann@tudelft.nl

Abstract

Process design is a creative task that is currently performed manually by engineers. Artificial intelligence provides new potential to facilitate process design. Specifically, reinforcement learning (RL) has shown some success in automating process design by integrating data-driven models that learn to build process flowsheets with process simulation in an iterative design process. However, one major challenge in the learning process is that the RL agent demands numerous process simulations in rigorous process simulators, thereby requiring long simulation times and expensive computational power. Therefore, typically short-cut simulation methods are employed to accelerate the learning process. Short-cut methods can, however, lead to inaccurate results. We thus propose to utilize transfer learning for process design with RL in combination with rigorous simulation methods. Transfer learning is an established approach from machine learning that stores knowledge gained while solving one problem and reuses this information on a different target domain. We integrate transfer learning in our RL framework for process design and apply it to an illustrative case study comprising equilibrium reactions, azeotropic separation, and recycles, our method can design economically feasible flowsheets with stable interaction with DWSIM. Our results show that transfer learning enables RL to economically design feasible flowsheets with DWSIM, resulting in a flowsheet with an 8% higher revenue. And the learning time can be reduced by a factor of 2.

Keywords: Reinforcement learning, process design, transfer learning

1. Introduction

The transition of chemical engineering to a sustainable and circular future requires new methods of process design (Fantke et al., 2021). Currently, methods for process design are mainly manual work with long development times and superstructure methods are also limited to predefined process configurations (Chen et al. 2017, Mitsos et al. 2018). Recently, reinforcement learning (RL), a branch of machine learning (ML), has shown promising results in process design (Midgley et al., 2020, Khan et al., 2020, Göttl et al., 2021, Stops et al., 2022, Kalmthout et al., 2022). One major challenge in RL for process design is the training process as it is trial-and-error based. Thereby, the learning process typically requires a large number of process simulations, which demands expensive computational power. Previous work (Khan et al., 2020, Göttl et al., 2021, Stops et al., 2022) mostly leverages short-cut process simulation methods to simulate the processes efficiently, which can lead to inaccurate results. Recent works employ rigorous process simulators such as COCO and Aspen Plus with rigorous models (Midgley et al., 2020, Kalmthout et al., 2022). Nevertheless, the problem of long simulation times hinders further developments (Kalmthout et al., 2022).

We propose to utilize transfer learning in process design with RL to facilitate the effectiveness and efficiency of the learning process. Transfer learning is a technique to improve learning performance by transferring knowledge from different but relevant domains to the target domain (Zhu et al., 2020). Adapting the concept of transfer learning for process design with RL, we first pre-train our recently proposed RL agent on a short-cut process simulation from our previous work (Stops et al., 2022). Then, we transfer the pre-trained agent to a rigorous process simulator DWSIM for further training. Finally, we illustrate the impact of transfer learning through one case study.

2. Methods

The RL problem can be formulated as a Markov decision process (MDP): $M = \{S, A, T, R\}$, which includes states $s \in S$, actions $a \in A$, transitions $T: S \times A$, and reward functions R. The agent aims to maximize the reward in the environment by literately taking action, evaluating the current reward, and updating the states. Specifically, in process synthesis tasks, states correspond to flowsheets. Actions are composed of two parts: Discrete and continuous actions. The discrete actions include the selections of a unit operation and its location in the flowsheet. The continuous actions define the design and operation variables of the corresponding unit operation. After the agent has performed actions, the states are updated. Based on the current state, a reward is calculated by the environment, e.g., the process simulation software, and returned to the agent as feedback on its actions. For process design, this reward is typically a design goal such as the process revenue. By repeating the process of performing actions and receiving rewards multiple times, the agent is trained to perform actions that result in a higher reward, corresponding to flowsheets with higher revenue.

2.1. Agent and environment

We adapt the RL framework from our previous work (Stops et al., 2022), where the states are presented as directed flowsheet graphs. Within the directed flowsheet graphs, nodes represent the unit operations and edges correspond to the process streams. Each node and edge is assigned a feature vector, respectively, storing information about the unit operation, e.g., type or size, and stream, e.g., thermodynamic data or flow rate of the stream. Furthermore, the agent architecture consists of three major parts: Graph encoder, actor networks, and critic networks. The graph encoder takes current flowsheet graphs as input and utilizes graph convolution in GNNs to learn information about the flowsheet graphs, in form of a vector representation, also referred to as flowsheet fingerprint. Actor networks are responsible for taking actions during the training process. There are three action levels: Selecting an open stream, selecting a unit operation, and selecting a corresponding design variable. Moreover, taking the flowsheet fingerprint as input, critic networks are used to estimate the reward of actions, and then actor networks will learn to take actions with the highest estimated reward. Specifically, the reward is calculated by using the DWSIM process simulator (Medeiros et al., 2018). The RL framework is implemented in Python including an interface for the agent to actively interact with DWSIM.

2.2. Transfer learning

We extend our RL framework by transfer learning. Specifically, we add a pre-training phase to the training of the agent. In the pre-training phase, we use a short-cut process simulation environment (Stops et al., 2022) and train the agent over 10000 episodes. Then, we transfer the pre-trained agent to a fine-tuning phase in which the agent is trained against a rigorous process simulator DWSIM for further 15000 training episodes. For the comparison, we directly train another agent with DWSIM environment over 15000 episodes. Note that in each episode, the agent generates a complete flowsheet. Both pre-training and fine-tuning processes are adapted from Proximal Policy Optimization (PPO) by OpenAI (Schulman et al., 2017). Then, the agent architecture is updated by gradient descent for the total loss function derived from summing up losses of actor networks, loss of critic networks, and corresponding entropies.

The agents are trained on a Windows server with a 3.5 GHz 24 cores Intel(R) Xeon(R) W-2265 CPU, NVIDIA GeForce RTX 3090 GPU and 64 GB memory.

3. Illustrative case study

The production of methyl acetate (MeOAC) is chosen as an illustrative case study.

3.1. Process simulation

The short-cut process methods for pre-training are illustrated in our previous work (Stops et al., 2022). Here, the process simulation with DWSIM is introduced. In this case study, the agent can choose reactors, distillation columns, and heat exchangers as unit operations. Besides, the agent can also decide to add recycles or claim open streams as products. The types of unit operations and corresponding design variables are defined as follows.

Reactor is deployed to convert reactants to the desired product (MeOAc). The reactor is modeled as a plug flow reactor (PFR) where the following reversible reaction takes place:

$$HOAc + MeOH \rightleftharpoons MeOAc + H_2O \tag{1}$$

For operational simplicity, the reactor is simulated isothermally, in which the temperature is kept constant regarding the inlet stream temperature. Besides, catalyst loading is not considered. The reaction kinetics are obtained from (Xu et al, 1996), with the equilibrium being related to temperature. The reactor cross-sectional area is determined by the relation: N/10, where N is the inlet molar flow (Stops et al., 2022). The design variable is the reactor length, which will be determined by the agent in the third-level continuous decision process. The range is from 3 to 10 m.

Distillation column is applied to separate MeOAc from the quaternary system. Rigorous distillation columns are used instead of shortcut columns in the previous work (Stops et al., 2022) to account for more realistic factors such as intermolecular interactions. The rigorous column models provide multiple possible choices of design parameters, from which the distillate to feed ratio (D/F) is selected as the third-level decision. Other adjustable parameters such as the number of stages and reflux ratio are set as fixed values (35 and 1.5, respectively). The D/F ratio can range from 0.4 to 0.6.

Heat exchanger is a DWSIM heater model. In the proposed framework, the heat exchanger is simulated based on the outlet temperature which is determined by the third-level decision. A temperature range from 278.15 K to 330.05 K is applied, where the upper limit refers to the lowest boiling point of the components which is MeOAc.

Recycle action consists of additional units including a splitter and a mixer. Firstly, the process stream is split into a recycle stream and a purge. Secondly, the recycled stream is merged with the selected feed with a mixer. Thereby, the split ratio is the third-level decision of the agent, which lies in the range of 0.1 to 0.9 for pragmatic consideration.

3.2. Reward

The reward determines the economic viability of generated flowsheets and teaches the agent to take feasible actions. First, a reward of 0 € is given when the incomplete flowsheets can converge after every single action, because the economic value is difficult to assess for an incomplete flowsheet. Second, whenever the agent fails the simulation by taking infeasible actions, the episode will be terminated immediately and a negative reward -10M€ is given. Finally, when a flowsheet is completed, we calculate the reward according to Equation 2:

$$r = \sum P_{product} - \sum C_{feed} - \sum \left(C_{operation} + 0.15 \cdot C_{invest} \right)_{units} \qquad (2)$$

where $P_{product}$ is the revenue of the sold product (Seider et al., 2008), C_{feed} is the costs of feeds, $C_{opeartion}$ is the operation costs (Smith, 2016) and C_{invest} is the total capital investment which is multiplied by factor 0.15 (Seider et al., 2008). In the case of negative rewards, a reduction factor 10 is applied to encourage the exploration of the agent.

4. Results and discussion

Figure 1 shows the learning curves for the agents with and without transfer learning. The scores represent the moving average rewards, i.e., the economic viability of the flowsheets, over 100 episodes. The training for the agent without transfer learning took 72 hours over 10000 episodes. For the agent with transfer learning, the pre-training took 2 hours over 10000 episodes and the further training took 72 hours over 10000 episodes. During the first 3500 episodes, the agent without transfer learning generates predominantly infeasible lengthy flowsheets, resulting in the learning curves rising slowly. In fact, due to the complexity of the design space, the agent has difficulty in learning from the previous failed flowsheets, leading to negative scores. After 3500 episodes, the agent mainly produces flowsheets with positive scores, which indicates that the designed processes are economically viable. Besides, within the training episodes, the learning curve slowly converges and reaches maximally to about 42. In comparison, the agent with transfer learning shows a quicker learning process. At the beginning of the learning process, the agent is able to mostly produces positive scores, and then the learning curve rises steeply. This demonstrates that the agents successfully leverage the pre-trained information from a short-cut process environment to make favorable decisions even in the early training stages. After about 4500 episodes, the score begins to fluctuate between 30 and 40 and maximumly reaches 46 which is higher than the agent without transfer learning.

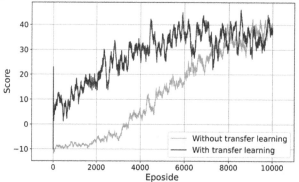

Figure 1: Learning curves of the agent with and without transfer learning. The blue line depicts the learning curve with transfer learning and the orange line indicates the learning curves without transfer learning. The scores are moving average rewards over 100 episodes and each learning curve runs over 10000 episodes.

Figure 2 displays the best flowsheet generated by the transfer learning agent and the continuous design variables are shown in Table 1. First, the feed (F1) is fed directly into three consecutive reactors (R1, R2, and R3) where the MeOAC and H_2O are produced by the esterification of HOAC and MeOH. Then, the resulting quaternary mixture is split up in a column (C1). The mixture is distilled from the top part of the column and sent to one heat exchanger (Hex1) to get the product (P1), containing enriched MeOAC and residues of HOAC and H_2O. The bottom product of the first column is further split up into the second column (C2) to produce pure H_2O in the distillate (P2) and a mixture of MeOH and H_2O in the third product stream (P3). And 90 % of the bottom product is recycled and mixed back into the feed.

While the resulting flowsheet has a positive reward, it is still far from a realistic engineering solution and future research is required. In particular, there are three major issues observed in the optimal flowsheet solution: (1) In industrial applications, MeOAC is primarily produced in reactive distillation (Huss et al., 2003). As our agent does not include reactive distillation as a unit operation, this cannot be identified. In future work, intensified unit operations can be added to our framework or the agent could operate on a phenomena level rather than a unit operation level. (2) The best flowsheet generated by the agent in this work contains three consecutive PFRs. The reason is that the length of a single PFR is limited to 10 m, which is not sufficient to finish the reaction. Therefore, the agent learns to choose multiple PFRs to fulfill the reaction and maximize the product. (3) One unnecessary heat exchanger is added after the distillation column C1 and before the product P1. We believe this is due to the small operations cost of the heat exchangers and the minor impact on the overall reward. Future research should further investigate possible mitigation strategies such as longer training or further hyperparameter tuning.

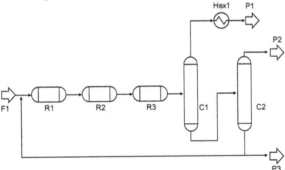

Figure 2: Best flowsheet generated by the transfer learning agent. First, MeOAC and H2O are produced from the feed (F1) in three consecutive reactors (R1, R2, and R3). Then the mixture is separated in the first column (C1). The first product (P1) is enriched with MeOAc but also contains residues of HOAC and H2O after one heat exchanger (Hex1). Then the bottom mixture of MeOH and H2O is further separated in the second column (C2). Pure H2O is in the distillate (P2) and the third product (P3) is the mixture of MeOH and H2O. And 90% of P3 is recycled and mixed with the feed stream.

Table 1. Design variables selected by the agent with transfer learning in the best flowsheet.

Unit operation	Design variable	Unit	Best run
Reactor 1 (R1)	Length	m	9.42
Reactor 2 (R2)	Length	m	9.25
Reactor 3 (R3)	Length	m	8.38
Column 1 (C1)	Distillate to feed ratio	-	0.58
Column 2 (C2)	Distillate to feed ratio	-	0.4
Heat exchanger 1 (Hex1)	Outlet temperature	K	315
Recycle	Recycled ratio	-	0.9

5. Conclusion

We propose to deploy the transfer learning for process design in RL to accelerate the learning process of the agent. The GNNs-based agent is first pre-trained with a short-cut simulation environment and then transferred to the rigorous process simulator environment for further training. In the illustrative case study, the agent is able to design economically feasible flowsheets in the process simulator DWSIM environment. Furthermore, the learning curves demonstrate that transfer learning indeed improves the efficiency of the learning process significantly and thus can be used to reduce the overall training time significantly. This work thus demonstrates that transfer learning can accelerate the learning process of graph-based RL with rigorous process simulator environments.

References

Chen, Q., & Grossmann, I. E. (2017). Recent developments and challenges in optimization-based process synthesis. Annu. Rev. Chem. Biomol. Eng, 8(1), 249-283.

Fantke, P., Cinquemani, C., Yaseneva, P., De Mello, J., Schwabe, H., Ebeling, B., & Lapkin, A. A. (2021). Transition to sustainable chemistry through digitalization. Chem, 7(11), 2866-2882.

Göttl, Q., Grimm, D. G., & Burger, J. (2022). Automated synthesis of steady-state continuous processes using reinforcement learning. Frontiers of Chemical Science and Engineering, 16(2), 288-302.

Huss, R. S., Chen, F., Malone, M. F., & Doherty, M. F. (2003). Reactive distillation for methyl acetate production. Computers & chemical engineering, 27(12), 1855-1866

Khan, A., & Lapkin, A. (2020). Searching for optimal process routes: a reinforcement learning approach. Computers & Chemical Engineering, 141, 107027.

Midgley, L. I. (2020). Deep Reinforcement Learning for Process Synthesis. arXiv preprint arXiv:2009.13265.

Mitsos, A., Asprion, N., Floudas, C. A., Bortz, M., Baldea, M., Bonvin, D., ... & Schäfer, P. (2018). Challenges in process optimization for new feedstocks and energy sources. Computers & Chemical Engineering, 113, 209-221.

Schulman, J., Wolski, F., Dhariwal, P., Radford, A., & Klimov, O. (2017). Proximal policy optimization algorithms. arXiv preprint arXiv:1707.06347.

Seider WD, Lewin DR, Seader JD, Widagdo S, Gani R, Ng KM. Product and process design16 principles: Synthesis, analysis, and evaluation (4th edition). Hoboken, NJ: Wiley . 2017

Smith R. Chemical process: Design and integration (2nd edition). Chichester, West Sussex, United Kingdom: Wiley . 2016.

Stops, L., Leenhouts, R., Gao, Q., & Schweidtmann, A. M. (2022). Flowsheet generation through hierarchical reinforcement learning and graph neural networks. AIChE Journal, e17938.

van Kalmthout, S. C., Midgley, L. I., & Franke, M. B. (2022). Synthesis of separation processes with reinforcement learning. arXiv preprint arXiv:2211.04327.

Xu, Z. P., & Chuang, K. T. (1996). Kinetics of acetic acid esterification over ion exchange catalysts. The Canadian Journal of chemical engineering, 74(4), 493-500.

Zhu, Z., Lin, K., & Zhou, J. (2020). Transfer learning in deep reinforcement learning: A survey. arXiv preprint arXiv:2009.07888.

Antonis Kokossis, Michael C. Georgiadis, Efstratios N. Pistikopoulos (Eds.)
PROCEEDINGS OF THE 33rd European Symposium on Computer Aided Process Engineering
(ESCAPE33), June 18-21, 2023, Athens, Greece

Data augmentation for machine learning of chemical process flowsheets

Lukas Schulze Balhorn[a], Edwin Hirtreiter[a], Lynn Luderer[a], Artur M. Schweidtmann[a,*]

[a] Process Intelligence Research, Department of Chemical Engineering, Delft University of Technology, Van der Maasweg 9, Delft 2629 HZ, The Netherlands
*Corresponding author. Email: a.schweidtmann@tudelft.nl

Abstract

Artificial intelligence has great potential for accelerating the design and engineering of chemical processes. Recently, we have shown that transformer-based language models can learn to auto-complete chemical process flowsheets using the SFILES 2.0 string notation. Also, we showed that language translation models can be used to translate Process Flow Diagrams (PFDs) into Process and Instrumentation Diagrams (P&IDs). However, artificial intelligence methods require big data and flowsheet data is currently limited. To mitigate this challenge of limited data, we propose a new data augmentation methodology for flowsheet data that is represented in the SFILES 2.0 notation. We show that the proposed data augmentation improves the performance of artificial intelligence-based process design models. In our case study flowsheet data augmentation improved the prediction uncertainty of the flowsheet autocompletion model by 14.7%. In the future, our flowsheet data augmentation can be used for other machine learning algorithms on chemical process flowsheets that are based on SFILES notation.

Keywords: Data Augmentation, Flowsheet Autocompletion, SFILES, Transformers

1. Introduction

The design of a flowsheet topology is an important step in early process synthesis. This step consists of selecting and arranging unit operations for a chemical process. Artificial intelligence (AI) methods have the potential to learn from previous flowsheets and support engineers in process development (Hirtreiter et al., 2022; Oeing et al., 2022; Schweidtmann, 2022; Vogel et al., 2023). For instance, Vogel et al. (2023) proposed an algorithm for the autocompletion of flowsheets. This autocompletion algorithm is inspired by text-autocompletion from natural language processing (NLP) that is based on generative transformer models (Radford et al., 2019). In addition, Hirtreiter et al. (2022) showed that the prediction of control structure elements from Process Flow Diagrams (PFDs) can be interpreted as a translation task between PFDs and Process and Instrumentation Diagrams (P&IDs). Hence, they deployed a sequence-to-sequence transformer architecture which is commonly used for translation of text between different languages. These flowsheet transformers rely on machine-readable flowsheet representations.

To represent flowsheets in a machine-readable format, we depict them as graphs or as text using unique, i.e., canonical, SFILES 2.0 strings (Vogel et al., 2022b). In general, flowsheets are drawings of chemical processes. Chemical engineers use flowsheets for the communication, planning, operation, simulation, and construction of these processes. An example flowsheet is given in Figure 1.

An intuitive way to represent flowsheets is via graphs with unit operations as nodes and stream connections as directed edges. Besides the graph representation, flowsheets can also be represented as strings. D'Anterroches (2005) introduced the Simplified Flowsheet Input-Line Entry-System (SFILES) notation, which we recently extended to include control structures and other features in (Vogel et al., 2022b). When creating the SFILES, we traverse the graph by starting at an input node and following the stream direction until we reach a product node or a recycle. In case the stream branches at a node, i.e., a splitter, we need to decide which stream to follow first. To determine the order of the branches in the linear string, the SFILES algorithm assigns each node a unique rank. The SFILES string for the flowsheet from Figure 1 is given by:

(raw)(hex){1}(r)<&|(raw)(pp)&|(mix)<1(v)(dist)[{tout}(prod)]{bout}(splt)1(prod)n|(raw)(hex){1}(prod).

While AI models require big training data, machine-readable flowsheet data is typically limited. The reason for limited data is that flowsheets are mainly depicted as images and therefore not machine-readable (Schweidtmann, 2022). Recently, we propose to automatically find flowsheet images in literature (Schulze Balhorn et al., 2022) and to make them machine-readable via computer vision (Theisen et al., 2023). However, this methods need to be implemented at a large scale. In addition, the majority of flowsheets are not publicly available due to company's intellectual property protection. Chemical process datasets with machine-readable flowsheets are rare and often contain only dozens of flowsheets (Hirtreiter et al., 2022; Oeing et al., 2022; Vogel et al., 2023; Zhang et al., 2018; Zheng et al., 2022). However, AI methods like transformers usually require big training data.

One promising approach to overcome limited training data for artificial intelligence is data augmentation. Data augmentation builds on the idea to generate additional artificial training data by masking, modifying, perturbating the available original training data at hand. For example, in computer vision, data augmentation is an established method to improve the model performance by adding modified copies of already existing images (Shorten & Khoshgoftaar, 2019). However, no data augmentation method for flowsheets exists yet. Thus, a augmentation method is needed which adds modified copies of already existing flowsheets to mitigate the issue of limited available flowsheet data.

We propose a novel data augmentation method for process flowsheets. Specifically, we use a text-based augmentation method for SFILES which is inspired by the augmentation of SMILES strings for molecules (Bjerrum, 2017). Our approach is to randomize the branching decisions in the SFILES generation. We demonstrate the proposed flowsheet augmentation method in the context of a flowsheet autocompletion model from (Vogel et al., 2023).

2. Data augmentation methodology

To augment the flowsheet data sets, we modify the branching decision in the SFILES generation algorithm to create multiple SFILES strings representing the same flowsheet. In the case of determining canonical SFILES the branching decisions are made by assigning every node of the graph to a unique rank. Hence, canonical SFILES are a unique mapping of a flowsheet graph to a string. When generating augmented SFILES, the branching decisions are made randomly, resulting in a non-canonical form. The difference lies in the order of branches in the linear string. The resulting augmented SFILES contain the identical information as the canonical SFILES, thus describing the same process flowsheet. Hence, all augmentations can be translated back to the original canonical SFILES. In Figure 1 we show the augmentation of the example flowsheet.

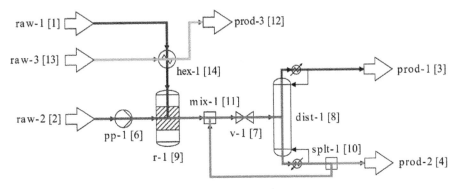

canonical SFILES

(raw)(hex){1}(r)<&|(raw)(pp)&|(mix)<1(v)(dist)[{tout}(prod)]{bout}(splt)1(prod)n|(raw)(hex){1}(prod)

augmentation SFILES

(raw)(hex){1}(r)<&|(raw)(pp)&|(mix)<1(v)(dist)[{bout}(splt)1(prod)]{tout}(prod)n|(raw)(hex){1}(prod)

(raw)(hex){1}(r)<&|(raw)(pp)&|(mix)<1(v)(dist)[{bout}(splt)1[(prod)]]{tout}(prod)n|(raw)(hex){1}(prod)

(raw)(hex){1}(r)<&|(raw)(pp)&|(mix)<1(v)(dist)[{tout}(prod)]{bout}(splt)1[(prod)]n|(raw)(hex){1}(prod)

Figure 1: Augmentation example for the flowsheet from Figure 1. The node rank is given in square brackets after the node name.

During augmentation, only the uniqueness of the SFILES representation is lost while the full flowsheet topology information is preserved. Notably, we do not change the order of the input branches during data augmentation. Otherwise, flowsheets with disconnected sub-graphs cannot be translated back to the original canonical form. This is for example important for independent processes with heat integration, such as *(raw)(hex){1}(prod)* in the example flowsheet from Figure 1.

The number of augmented SFILES that can be derived from a single flowsheet graph is limited. Specifically, the number of potential augmented SFILES depends on the number of available branching points for a given flowsheet, with more branching points leading to an exponential increase of augmentation possibilities. For example, the flowsheet in Figure 1 contains two branching points. This results in three augmented SFILES representations and four SFILES representations in total.

The two branching points in this case are the distillation column *dist-1* and the splitter after the distillation column *splt-1*. At the first branching point, we switch the order of the top and bottom outlet streams. The second augmentation affects only the bottom product stream *prod-2*. Recycles without unit operations always appear directly after the splitting unit, here that leads to *(splt-1)1* in all cases. In case the product branch is visited before the recycle branch, additional squared brackets around *prod-2* are used. To make the methodology for flowsheet augmentation openly accessible, we include it in our public SFILES 2.0 Github repository (Vogel et al., 2022a).

3. Case study and Results

3.1. Data and Data augmentation

We use two SFILES datasets which were created by (Vogel et al., 2023). Firstly, we use their proposed flowsheet generator to generate a large-scale dataset of about 8,000 artificial flowsheets. Here, we can flexibly scale the dataset size. Secondly, we use a dataset made from 223 Aspen and DWSIM chemical process simulations. We call this the real dataset. Before starting the training runs, we split each dataset into a training dataset, a validation dataset, and a test dataset. For the two training datasets we created a maximum of five augmentations for each SFILES, which roughly increases the dataset five-fold. By limiting the number of augmentations to five, we ensure that larger flowsheets are not over-represented.

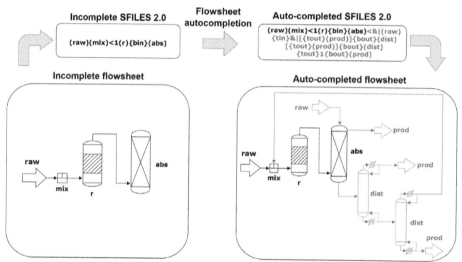

Figure 2: Example prediction of the flowsheet autocompletion model. Figure adapted from (Vogel et al., 2023).

3.2. Flowsheet autocompletion

We consider a process design case study and use the flowsheet autocompletion model from (Vogel et al., 2023). The objective of the model is to support chemical engineers in the design of a new process topology. It can suggest the following unit operation for an incomplete flowsheet, similar to sentence completion in messenger apps. Specifically, the start of a SFILES is given and the autocompletion model predicts how the sequence ends by iteratively predicting the following building block, token, of the SFILES (Figure 2). It should be highlighted that the prediction is only based on the process topology and does not consider operating points, components, and material flows. The autocompletion model is built on the transformer architecture in a decoder-only approach (Radford et al., 2019). Because transformer models are very data-intensive, the generated dataset is used to pretrain the flowsheet autocompletion. The real dataset is then used to fine tune the flowsheet autocompletion. For a more detailed description of the model we refer to (Vogel et al., 2023).

To study the effectiveness of data augmentation we retrain the model for flowsheet autocompletion from (Vogel et al., 2023). Specifically, we train the model both with augmented and non-augmented flowsheet data. Overall, we consider three different training scenarios. First, we train the model with the non-augmented generated dataset and fine-tune this model with the real dataset (i). This model is used to reproduce the results from (Vogel et al., 2023). Secondly, we augment the real data and use them to fine-tune the pretraining model, resulting in model (ii). We thus only alter the fine-tuning. Finally, we train the flowsheet autocompletion with augmented data for both pretraining and fine-tuning, yielding model (iii). For a fair comparison of training runs with and without data augmentation, we use the same hyperparameters.

3.3. Results

We use perplexity to measure the model performance. Perplexity describes the uncertainty of a model in its predictions. Therefore, a low perplexity is desirable. Here, a lower perplexity means that the model is more confident that the suggested unit operation is reasonable. Perplexity is the exponential of the negative average log-likelihood of the next token prediction. It is also equivalent to the exponential of the cross-entropy loss obtained during model training. The perplexity is computed as

$$PP(T) = \exp\left(-\frac{1}{n} \sum_{i}^{n} \log P(t_i | t_{1:i-1}) \right),$$
(1)

where $T = (t_1, \ldots, t_n)$ is a sequence of n tokens and $P(t_i | t_{1:i-1})$ describes the predicted probability for the next token t_i given the sequence $t_{1:i-1}$.

Table 1: Perplexity PP results of data augmentation. Best test result in bold font. The column "Augm." shows whether the training data were augmented or not. Pretraining and fine-tuning test perplexity are both evaluated after the model training is completed with the fine-tuning dataset.

Model	PP pretraining		PP fine-tuning				Training Time
	Augm.	Test	Augm.	Train	Val	Test	
(i)	No	5.38	No	3.13	4.23	5.02	51 min
(ii)	No	6.33	Yes	3.32	4.12	4.69	57 min
(iii)	Yes	**5.16**	Yes	3.07	3.80	**4.28**	1 h 31 min

The results of the different training runs are shown in Table 1. For the fine-tuning, model (ii) performs slightly better than model (i) on the test set. We explain this improvement by the fact that the fine-tuning data are limited in size (i) and with the data augmentation we can make the model more robust (ii). Even though the test perplexity is lower with data augmentation, the training perplexity is higher. This shows that the model trained with data augmentation is less prone to overfitting and generalizes better to unseen data. With data augmentation also applied to the pretraining (model (iii)), we see the best performance in all categories, improving the fine-tuning perplexity by 14.7% compared to model (i). We conclude that data augmentation can also improve the pretraining with a relatively large, generated dataset, resulting in a better fine-tuning performance.

In general, we see that the augmented SFILES are valid flowsheet representations and that they help the flowsheet autocompletion to learn the SFILES grammar and chemical process structure. For future work it would be interesting to investigate if it is more favorable for pretraining to generate more artificial data, to increase the dataset with data augmentation, or to combine both methods. It is worth noticing, that the training and test perplexity are in every case more similar for the models trained with augmented data. Due to the higher variance in the training data, these models are less prone to overfitting.

4. Conclusions

We propose a data augmentation of chemical flowsheet data by randomizing the branching decisions in the graph traversal for producing SFILES. We thereby demonstrate a way to increase the available flowsheet data for subsequent training of AI models. We apply augmented SFILES to the problem of flowsheet autocompletion and show that the augmentation improves the model performance in low data regimes. In future research, we aim to apply the data augmentation methodology to further NLP models to improve their performance. The long-term goal should be to increase the flowsheet dataset size by mining additional flowsheets from literature (Schweidtmann, 2022; Schulze Balhorn et al., 2022) and companies and, if necessary, digitizing them (Theisen et al., 2023). A combination of both, data augmentation and increased dataset, size will be necessary and most beneficial.

References

Bjerrum, E. J. (2017). Smiles enumeration as data augmentation for neural network modeling of molecules. arXiv preprint arXiv:1703.07076.

d'Anterroches, L. (2005). Process flowsheet generation & design through a group contribution approach. [CAPEC], Department of Chemical Engineering, Technical University of Denmark.

Hirtreiter, E., Schulze Balhorn, L., & Schweidtmann, A. M. (2022). Towards automatic generation of Piping and Instrumentation Diagrams (P&IDs) with Artificial Intelligence. arXiv preprint arXiv:2211.05583.

Oeing, J., Welscher, W., Krink, N., Jansen, L., Henke, F., & Kockmann, N. (2022). Using artificial intelligence to support the drawing of piping and instrumentation diagrams using dexpi standard. Digital Chemical Engineering, 4, 100038.

Radford, A., Wu, J., Child, R., Luan, D., Amodei, D., & Sutskever, I. (2019). Language models are unsupervised multitask learners. OpenAI blog, 1(8), 9.

Schulze Balhorn, L., Gao, Q., Goldstein, D., & Schweidtmanna, A. M. (2022). Flowsheet Recognition using Deep Convolutional Neural Networks. In Computer Aided Chemical Engineering (Vol. 49, pp. 1567-1572). Elsevier.

Schweidtmann, A. M. (2022). Flowsheet mining. Manuscript, In preparation. TU Delft.

Shorten, C., & Khoshgoftaar, T. M. (2019). A survey on image data augmentation for deep learning. Journal of big data, 6(1), 1-48.

Theisen, M. F., Flores, K. N., Schulze Balhorn, L., & Schweidtmann, A. M. (2023). Digitization of chemical process flow diagrams using deep convolutional neural networks. Digital Chemical Engineering, 6, 100072.

Vogel, G., Schulze Balhorn, L., Hirtreiter, E., & Schweidtmann, A. M. (2022a). Process-intelligence-research/sfiles2: V1.0.0 (Version Release). Github. https://github.com/process-intelligence-research/SFILES2

Vogel, G., Schulze Balhorn, L., Hirtreiter, E., & Schweidtmann, A. M. (2022b). SFILES 2.0: An extended text-based flowsheet representation. arXiv preprint arXiv:2208.00778.

Vogel, G., Schulze Balhorn, L., & Schweidtmann, A. M. (2023). Learning from flowsheets: A generative transformer model for autocompletion of flowsheets. Computers & Chemical Engineering, 171, 108162.

Zhang, T., Sahinidis, N. V., & Siirola, J. J. (2019). Pattern recognition in chemical process flowsheets. AIChE Journal, 65(2), 592-603.

Zheng, C., Chen, X., Zhang, T., Sahinidis, N. V., & Siirola, J. J. (2022). Learning process patterns via multiple sequence alignment. Computers & Chemical Engineering, 159, 107676.

Antonis Kokossis, Michael C. Georgiadis, Efstratios N. Pistikopoulos (Eds.)
PROCEEDINGS OF THE 33rd European Symposium on Computer Aided Process Engineering
(ESCAPE33), June 18-21, 2023, Athens, Greece

Systematic synthesis of pathways for hydrogenation and hydrogenolysis of acetylene over catalytic surfaces

Manjusha C. Padole, Swayam P. Misra, V. Sai Phani Kumar, Parag A. Deshpande

Quantum and Molecular Engineering Laboratory, Department of Chemical Engineering, Indian Institute of Technology Kharagpur, Kharagpur 721302, India

Abstract

Search for a good catalyst to catalyse surface reactions with high product selectivity has always been an important aspect of heterogeneous catalyst design. To this end, detailed insights into the mechanism of a reaction need to be developed, which is at times difficult for heterogeneous reactions. Efficient techniques are required which accurately provide all possible pathways over a heterogeneous catalytic surface given a set of chemically plausible species present. Linear algebra along with graph theoretical techniques can be used for identifying different possible reaction pathways from a set of elementary reactions. This study demonstrates the use of one such technique resulting in a reaction route diagram for acetylene hydrogenation-hydrogenolysis reaction system. A systematic protocol for the identification of all possible pathways is presented. Multitude of reaction pathways influencing the product selectivity of the reaction was observed with the possibility of existence of pathways with the desired selectivity.

Keywords: elementary reactions; surface species; reaction graph; incidence matrix; reaction route diagram.

1. Introduction

Rational design of a heterogeneous catalytic reactor requires a detailed understanding of the surface chemistry of the associated reaction. An important issue in the engineering of heterogeneous catalytic reactions is the identification of all possible reaction pathways leading to a specified product given an initial set of reactants. This problem can be quite challenging even for macroscopically simple reactions like acetylene hydrogenation, as has been highlighted in this study. Due to the possibility of existence of a large number of surface species on catalytic surfaces and their simultaneous interactions, on-surface synthesis has been reported by several investigators to yield a range of products based on the reaction and surface conditions. All these observations motivate the necessity of knowledge of all possible reaction pathways which can be observed over a heterogeneous surface. Recent developments in theoretical and computational frameworks for the analysis of reactive systems have proved to be helpful in gaining insights into on-surface systems. Such methods for the systematic reaction pathway search for the automated generation of reaction mechanisms work efficiently for small molecules only, and are time intensive for large molecules. Theoretical techniques reported till date require an efficient technique for the synthesis of reaction pathways. This work provides an efficient approach towards the systematic search of reaction pathways considering as the test case acetylene hydrogenation and hydrogenolysis. Since the reactions are carried out heterogeneously, as mentioned earlier, the possibility of existence of a large number of surface intermediates exists leading to varied macroscopically observed products. Apart from the desired reaction, namely the selective hydrogenation, complete hydrogenation to C_2H_6 is a possibility. Further, conversion of all three forms of C_2 gases to CH_4 by hydrogenolysis cannot be ruled out. Therefore, it is desirable to obtain all possible pathways during the aforementioned conversions. This remains the central issue addressed in this study.

With an aim of addressing the aforementioned issues, we propose a combination of mathematical techniques involving linear algebra and graph theory to get the correct picture. For acetylene-H_2 surface reactions, attempts have been made in the past to implement graph theory to complex reaction systems. Christiansen [1] and Temkin et al. [2] introduced the use of reaction graph for catalytic and noncatalytic reactions. Fishtik et al. [3] confirmed the use of this graph theory approach to be suitable for linear as well nonlinear kinetic mechanisms for single and multiple overall reactions. The success of such methods, however, depends on the identification of reaction pathways. In this study, we demonstrate a novel and generic approach for the analysis of surface catalytic systems which is capable of handling systems with a large number of surface species and macroscopic parallel reactions. Our approach considers the presence and influence of every possible surface species contributing towards the overall reaction. We demonstrate its applicability using hydrogenation and hydrogenolysis of C_2H_2 as the test case.

2. Problem formulation and methodology

Macroscopically observable chemical species involved in the system under consideration are C_2H_2, C_2H_4, C_2H_6, CH_4 and H_2. These five chemical species in different adsorbed states over a catalytic surface can give rise to three categories of macroscopically observable chemical reactions given below in the scheme of Figure 1. Application of linear algebra tells us that in the above system of six series-parallel reactions with five chemical species, only three reactions are independent. This can be arrived at by finding the rank of the stoichiometric matrix with reactions as rows and chemical species as columns. However, chemical insights invoke some fundamental questions to consider all six reactions for further analysis. Analysis of the sub-set consisting of reactions (1-3) will show one of equations (2) and (3) to be redundant. This means that C_2H_6 formation can take place by sequential hydrogenation of C_2H_2 via C_2H_4. However, reaction (2) chemically means that C_2H_6 formation takes place from C_2H_2 and we ask this question: is it possible to obtain C_2H_6 from C_2H_2 without observing C_2H_4 in any of the intermediate steps. For this, it is required to identify a pathway from C_2H_2 to C_2H_6 such that no intermediate species is C_2H_4. Similar analysis is required for hydrogenolysis reactions also.

The proposed algorithm for synthesis of pathways given an initial reactant and a final product is shown in Fig. 1.

Selective hydrogenation:

$$C_2H_2 + H_2 \rightleftharpoons C_2H_4 \ldots\ldots\ldots \text{ (1)}$$

Complete hydrogenation:

$$C_2H_2 + 2H_2 \rightleftharpoons C_2H_4 \ldots\ldots\ldots\text{(2)}$$

$$C_2H_4 + H_2 \rightleftharpoons C_2H_6 \ldots\ldots\ldots \text{ (3)}$$

Hydrogenolysis:

$$C_2H_2 + 3H_2 \rightleftharpoons 2CH_4 \ldots\ldots\text{(4)}$$

$$C_2H_4 + 2H_2 \rightleftharpoons 2CH_4 \ldots\ldots\text{(5)}$$

$$C_2H_6 + H_2 \rightleftharpoons 2CH_4 \ldots\ldots\ldots\text{(6)}$$

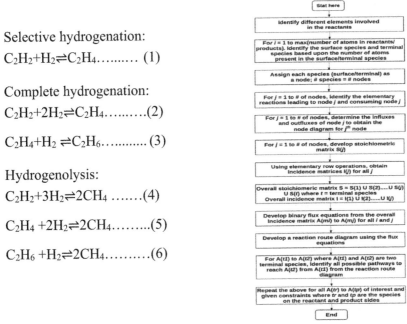

Fig. 1: Reactions involving the hydrogenation and hydrogenolysis are shown in left and the flowchart depicting the algorithm developed for systematic synthesis of pathways for a heterogeneously catalysed surface reactions is shown in right.

3. Results and discussion

The incidence matrix is a vital tool for the development of reaction route diagram as it gives a precise way of converting a set of reactants into a set of products. To obtain the stoichiometric sub-matrices, S(i) (i= 1 to 33), the elementary reactions are written in the mathematical form as

$$\sum_{n=1}^{33} [a_j n_j] = 0 \forall r; r = 1 \text{ to } 119 \ldots\ldots\ldots\ldots\ldots\ldots\text{.(7)}$$

where n_j in Eq. 7 represents the chemical species. We identified a total of thirty-three surface species and five terminal species for the system under consideration. Correspondingly, thirty-three node diagrams were developed. We have prepared predefined pathways before analysing them and this process of synthesis of pathways can be automated. A sample node diagram for one of the surface species is shown in Figure 2a. Based on the surface species and node diagrams, a hundred and nineteen elementary reactions were identified which would explain hydrogenation and hydrogenolysis of acetylene over a catalytic surface. We have verified that no other possible surface species and elementary reaction exists.

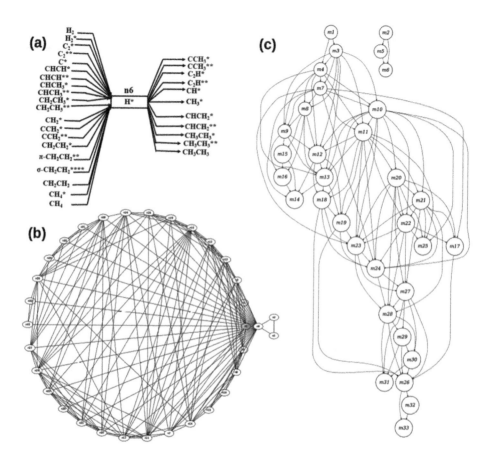

Fig. 2: Catalytic hydrogenation and hydrogenolysis of C_2H_2. Fig.2a: Node diagrams for the different chemical species identified. Fig. 2b: Reaction network diagram. Nodes n_i represent species connected by an undirected arrow, Fig. 2c: Reaction route graph.

A node appearing in the reaction scheme as a result of a surface reaction can be characterised distinctly by the appearance of both influxes and outfluxes in its node diagram. This can be seen in the node diagram of (H^*), for example. The conversion of the given species to other species and its formation from other species can be seen from the influxes and outfluxes of this node. It is to be noted that it is possible to obtain a given species from a combination of other species resulting in treatment of such interactions as influxes while the same species can as well be treated to be resulting into a combination of other species resulting in the appearance of such species as outfluxes. This happens because of the inherent reversibility of surface reactions. This, however, does not affect our analysis in any manner as we define the interconnections in the reaction network diagram as direction-less and the appearance of a given species as influx or outflux in a given node results in the same network. The set of all node diagrams indicating individual influxes and outfluxes can be collated as a reaction network diagram which has no directionality (Figure 2b) and further as a reaction route graph which has directionality (Figure 2c). The reaction route diagram can be used to obtain the reaction pathway diagram, as shown in Fig. 3. The two pathways for C_2H_2 hydrogenation to C_2H_4, for example, can be observed to involve two different modes of adsorption of C_2H_2, one being the single site adsorption ($C_2H_2^*$) while the other being the two site adsorption mode ($C_2H_2^{**}$). The intermediates following the adsorption steps can also be seen to be distinct. Both of these pathways have been reported previously for C_2H_2 hydrogenation to C_2H_4 [4, 5]. This shows the success of our method in capturing different pathways. Pathways for further hydrogenation of C_2H_4 to C_2H_6 were also identified, the conversion of C_2H_2 to C_2H_6 via C_2H_4 is shown in Fig. 3.

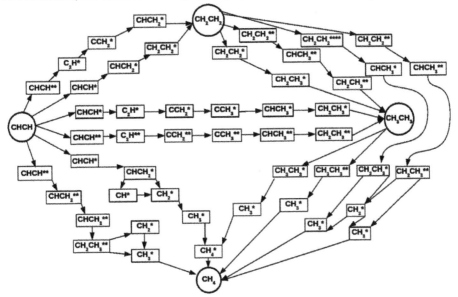

Fig. 3: Two sample pathways between any two terminal specified

Another important observation from the figure is the existence of distinct pathways for C_2H_2 hydrogenation to C_2H_6 without observing C_2H_4. These pathways have never been discussed in the literature and the existence of such pathways shown by our study signifies the importance of our systemtic synthesis protocol in identification of pathways. Identification of hydrogenolysis of different C_2 gases with the aforementioned constraints of observing or not observing a particular species can similarly be done, three instances for example are given in Fig. 3.

Conclusions

A novel and efficient technique utilising linear algebra and graph theory was developed for the identification of all possible pathways for heterogeneously catalysed surface reactions. The applicability of the method was demonstrated successfully for C_2H_2 hydrogenation-hydrogenolysis reaction. The macroscopically simple reactions involving C_2H_2, C_2H_4, C_2H_6, CH_4 and H_2 were found to have complex surface interactions resulting in thirty-three chemical species and a hundred and nineteen elementary surface reactions. A reaction route graph, developed from elementary reactions via the stoichiometric matrix and the incidence matrix, demonstrated complex surface interactions among the surface species and a multitude of reaction pathways for partial hydrogenation, complete hydrogenation and hydrogenolysis with existence of direct reaction routes from C_2H_2 to rest all terminal species. This method is capable of determining all possible pathways in short time frame compared to other expensive methods, for example techniques like quantum chemical calculations. Although the presented model has demonstrated for hydrogenation of ethylene, the same can be explored for other surface reactions as well.

References

[1] J. A. Christiansen, 1953, Elucidation of reaction mechanisms by the method of intermediates in quasi-stationary concentrations, AdV. Catal. 5, 311-353.
[2] O. N. Temkin, A. V. Zeigarnik, D. G. Bonchev, 1996, Chemical reaction networks: A graph theoretical approach, CRC Press: New York.
[3] I. Fishtik, C. A. Callaghan, R. Datta, Reaction route graphs. II. Examples of enzyme- and surface-catalyzed single overall reactions, J. Phys. Chem. B. 108 (2004) 5683-5697.
[4] D. Mei, M. Neurock, C. M. Smith, 2009, Hydrogenation of acetylene-ethylene mixtures over Pd and Pd-Ag alloys: First-principles-based kinetic monte carlo simulations, J. Catal. 268, 181-195.
[5] D. Mei, P. A. Sheth, M. Neurock, C. M. Smith, 2006, First-principles-based kinetic monte carlo simulation of the selective hydrogenation of acetylene over Pd(111), J. Catal. 242, 1-15.

Antonis Kokossis, Michael C. Georgiadis, Efstratios N. Pistikopoulos (Eds.)
PROCEEDINGS OF THE 33rd European Symposium on Computer Aided Process Engineering (ESCAPE33), June 18-21, 2023, Athens, Greece

New Methodology Integrating Tray Efficiency Predictions with Internals Design for Valve-Tray Distillation Columns

Chenguang Zhu,[a] Megan Jobson,[a] Nan Zhang[a]

[a]*Centre for Process Integration, Department of Chemical Engineering, The University of Manchester, Manchester, M13 9PL, United Kindom*

Abstract

Column internals design is a complex problem with many degrees of freedom and hydraulic constraints. This work proposes a new automatic design methodology for the internal design of valve-tray columns, iteratively applying results from rigorous distillation simulations, hydraulic correlations and tray efficiency predictions for valve trays. The feasibility of the proposed internals is checked against hydraulic constraints, including jet flooding, downcomer flooding and weir loading. The methodology is applied to an existing industrial valve-tray column separating C4 hydrocarbons; the results show the impact of the internals design approach on column hydraulics and tray efficiency. The proposed design can reduce the column diameter while avoiding hydraulic bottlenecks. Future work aims to implement the methodology for column design optimisation.

Keywords: Hydraulic analysis; Hydraulic Design; Tray efficiency;

1. Introduction

Distillation is a widely used, effective fluid separation process in industrial practice but requires substantial energy input and capital investment. Several published papers have addressed the optimal design of distillation columns, i.e. the column diameter, and number of trays in each section, to meet given separation specifications with minimum total annualised cost (TAC). However, these optimal column designs are still conceptual designs; details of column internals are not accounted for. The traditional approach to dimensioning column internals is to use trial and error: Kister et al., 1992; Chuang and Nandakumar,2000; Towler and Sinnott, 2021. No definitive procedure to guide the design of column internals is available in the open literature. Typically, assigning hardware to a conceptual design depends on the designer's experience using the trial-and-error method, without a comprehensive search for internals that meets hydraulic limits and operating constraints and targets with the lowest capital cost. As a result, the design is time-consuming and a new column is excessively conservative and the effect of the column's internal design on the column efficiency and sizing has not been considered.

Valve tray is used in this study as valve tray columns are the most popular choice for trayed distillation column internals in the industry since lower pressure drop than bubble cap trays and have a wide range of applicability (Kister et al., 1992). The capital cost of a valve-tray column depends on column or section diameter, the number of trays, and the column internals – namely, type of tray; type, dimensions and layout of valves; tray spacing, pass configuration, downcomer dimensions, etc. At the same time, the column internal dimensions affect the stage efficiency, and hence the theoretical number of stages, and thus the reflux ratio. The reflux ratio will in term affect the required dimensions of the internals, which implies that design optimisation should consider column internal dimensions and stage efficiency iteratively (or simultaneously).

Due to the development of computer-aided process engineering, limited research in the open literature has developed optimisation approaches for internals design for sieve-tray columns (Ogboja and Kuye (1990); Lahiri, 2014; Souza et al., 2022). However, these papers optimised the sieve tray column internals dimension by giving fluid information. Interaction between tray efficiency and column internals dimensions are ignored.

Lahiri (2014) used particle swarm optimisation and the Aspen Plus simulator to optimise the design of sieve-tray distillation columns, including tray specifications. The continuous design variables considered are: downcomer dimensions, tray dimensions and tray spacing, while the discrete variables are the number of stages, feed location and pass configuration. The objective function is the total annualised cost; the constraints are hydraulic limits and pressure drop. Lahiri (2014) only considers the continuous variables during the optimisation, while the discrete variables are specified manually. Significantly, tray efficiency is not evaluated within the optimisation. In practice, neglecting the tray efficiency could lead to different internal flows, and violation of hydraulic constraints.

Souza et al. (2022) formulated a mixed-integer nonlinear problem for optimising the internals dimensions (diameter, tray spacing, hole diameter, weir length, etc.) of sieve-trayed columns using column capital cost or the mass of shell and trays as an objective function, applying hydraulic (including jet flooding, downcomer flooding) and pressure drop constraints. Souza et al. (2022) first obtain the flow rate and physical properties from the Aspen Plus simulator with operating conditions and overall efficiency specified manually without any updating.

Ogboja and Kuye (1990) developed a formulation considering tray efficiency for sieve tray optimisation. Chan and Fair's (1984) correlation was used to calculate tray efficiency; the objective is then to maximise the tray efficiency with rigorous consideration of hydraulic limits and pressure drop. However, Ogboja and Kuye (1990) used a set of flow rates and fluid properties presented by literature to conduct the design, which means flow rate, fluid properties and the total number of stages are fixed in the optimisation. Although this work uses an exhaustive method to analyse all sets of variables (internal dimensions) to determine tray efficiencies, the effect of tray efficiency on the flow rate and fluid properties was not considered during the iteration of the tray efficiency evaluation.

Meanwhile, published research has not explored details of the hydraulic evaluation of valve-tray columns with cost as a design criterion, and the impact of the tray efficiency of the key tray in each section on the flow rate, and fluid properties should be assessed.

This work develops a systematic and automatic column design methodology for valve-tray columns considering column internals dimensions design, hydraulic performance of the trays, pressure drop and tray efficiency, with a focus on minimising column internals dimensions with the consideration of tray efficiency, where the operating limits (jet flooding, downcomer flooding and downcomer backup) in the column should be avoided.

2. Valve-trayed column design methodology

Figure 1 represents the proposed methodology which employed converged simulation of a simple distillation column (using Aspen Plus V11) and Python code which can exchange inputs and outputs with the process simulator automatically.

This work proposes a design methodology including an equilibrium-based 'rigorous' simulation of a column considering section efficiency (Murphree efficiency) to achieve a specified separation of a given feed; calculation of column internal dimensions (diameter, deck thickness, downcomer sizing, tray spacing, valve dimensions, tray spacing) under specified flooding limits and fractional hole area; and prediction of section efficiency of mean stage efficiencies within the section. The column is simulated using an established process simulation software assuming the stage efficiencies (as input variables), tray spacing, and the number of passes. The simulation results (stage-flow rates and fluid properties) are used as inputs for column hydraulic analysis and tray and downcomer calculations for every stage. The feasibility of the design is checked in terms of hydraulic limits and pressure drop, and the internal dimensions are then modified to avoid hydraulic limits. Next, the corresponding internal dimensions (active area, deck thickness, fractional hole area) and fluid properties are input to the stage efficiency prediction model of Chen and Chuang (1993). The calculated section efficiencies are re-input to the process simulator to recalculate the material and the energy flows and thus update the internal dimensions and section efficiencies, which means the column simulation and internal dimensions calculation are repeated until converged. The three steps are repeated until the calculated and assumed efficiencies are equal.

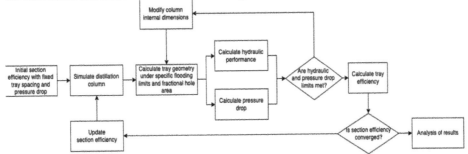

Figure 1. Proposed design methodology for valve-tray distillation column

2.1. Hydraulic analysis: methodology

Distillation columns can only operate efficiently with containing trays within certain hydraulic limits. A smaller diameter means a smaller active area or downcomer area for vapour-liquid pass-through, which will cause jet flooding or downcomer flooding. Thus, the selection of jet flooding and downcomer flooding upper limits have a significant effect on the diameter.

A maximum value of 80% is assumed since this figure is widely used in design textbooks. Jet flooding is predicted using the Glitsch correlation, also known as the 'Equation 13' (Branan, 2005) for conventional floating valve trays. Downcomer choke flooding is evaluated in terms of the volumetric flow rate at the downcomer entrance following the Glitsch correlation (Kister et al., 1992). Another typical constraint weir loading, the ratio of liquid volume flow rate to weir length, is assumed to limit 107 m3m-1h-1, as the value is typically applied in the practice (Kister et al., 1992). The clear liquid height, downcomer backup and pressure drop analysis are evaluated following correlations of Koch-Glitsch (2013); the vapour capacity factor and approximate tower diameter are calculated using the correlations of Branan (2005).

Downcomer backup is evaluated knowing the aerated liquid height in the downcomer, which is calculated by an aerated factor, the ratio of the clear liquid height to the aerated liquid height. A correlation is proposed, regressed from the aerated factor chart for valve trays of Klein (cited by Kister et al., 1992). The regressed model is shown in Eq. (1), where F_{va} is the superficial F-factor based on the active area.

$$\beta = -1.31F_{va} + 2.07F_{va}^2 - 1.71F_{va}^3 + 0.72F_{va}^4 - 0.15F_{va}^5 + 0.01F_{va}^6 + 0.97 \ (1)$$

Eq. (1) is only valid between the F_{va} from 0 to 3.05 $kg^{0.5}s^{-1}m^{-0.5}$.

2.2. Tray efficiency evaluation: methodology

The correlation of Chen and Chuang (1993) was developed based on a sieve tray in the hydrocarbon systems. However, Vennavelli et al. (2014) validated this correlation using FRI valve tray efficiency data (hydrocarbon test data) and showed that Chen and Chuang's (1993) model predicted the FRI valve tray efficiency better than the four existing valve tray efficiency models presented in published literature. Vennavelli et al. (2014) suggested that the sieve tray models present reasonable alternatives if they are developed from a wide range of data and operating conditions. Thus, there is good evidence that the Chen and Chuang (1993) model is valid for valve-tray columns in hydrocarbon systems.

3. Case study: separation of i-butane/n-butane mixture

The distillation system to demonstrate the proposed methodology is an industrial-scale valve-tray column separating the i-butane and n-butane mixture. However, the composition of distillate and bottom stream, and the section efficiency presented by Ilme et al. (2001) are not self-consistent. Thus, a self-consistent distillate and bottom stream composition, and section efficiency simulated and calculated following the methodology above are used instead of the results presented by Ilme et al. (2001), while the rest of the operating conditions and column internals dimensions are the same as that presented by Ilme et al. (2001). Thus, the i-butane/n-butane column based on the modified distillate and bottom stream composition, and section efficiency is used as a base case. The detailed operating conditions, stage distribution and product specifications are shown in Figure 2. Standard valve trays (Koch-Glitsch Ballast V-1 trays) are assumed for both sections. Section efficiency is applied instead of the tray efficiency of each tray, which is the average value of the tray efficiency of each tray in rectifying or stripping section predicted using Chen and Chuang (1993) model. The design methodology was applied automatically via the Python code to meet different flooding limits (jet flooding and downcomer flooding). The fractional hole area, which is the ratio between the hole area and the active area, was also varied. The impact of different flooding conditions and fractional hole area on the column diameter, section efficiency, pressure drop, and heat duty of the condenser and reboiler were analysed. The feed conditions, light key component purity in the distillate stream and distillate rate were maintained throughout the study. Table 1 presents the details of column internal dimensions and operating conditions under different flooding conditions (from 70% to 80%) and fractional hole area (from 16.6% to the lower limit).

The percentage of jet and downcomer flooding calculated based on the base case internal dimensions given by Ilme et al. (2001) are 69% and 52%. It is found that the diameter can reduce by 10% when the column operates under 80% of both jet and downcomer flooding and the fractional hole area is 16.6%. However, higher flooding condition means higher froth liquid height on the tray, leading to a slightly higher pressure drop than the base case, while section efficiencies are almost the same as the base case.

It is also found that the section efficiency is mainly dominated by the fractional hole area. The section efficiency can increase by around 7 to 10% in both rectifying and stripping sections under the same flooding condition. However, it shows that the lower limits of fractional hole areas are different under different flooding conditions due to downcomer backup constraints. Thanks to the increasing of section efficiency, the reflux ratio is smaller than the base case, leading to a low heat duty demand in both the condenser and reboiler.

New Design Methodoloy Integrating Tray Efficiency Predictions with Internals Design for Valve-Tray Distillation Columns

2027

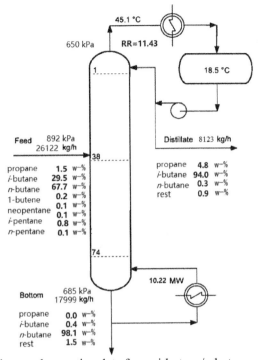

Figure 2. Flowsheet and operating data for an i-butane/n-butane column

Table 1. Column internal dimensions, hydraulic performance and pressure drop performance under different flooding limits and fractional hole area.

Flooding condition	base case	70%			80%		
% hole area	16.6%	16.6%	13.0%	8.7%	16.6%	13.0%	10.1%
Diameter	2,900 mm	2,796 mm	2,765 mm	2,721 mm	2,593 mm	2,564 mm	2,539 mm
Pressure drop	0.48 kPa	0.51 kPa	0.55 kPa	0.73 kPa	0.54 kPa	0.64 kPa	0.89 kPa
Heat duty	10.60MW; 10.22 MW	10.59MW; 10.21 MW	10.38MW; 10.00 MW	10.09MW; 9.71 MW	10.52MW; 10.15 MW	10.31MW; 9.94 MW	10.13MW; 9.76 MW
Section efficiency	110%, 115%	110%, 115%	114%, 119%	121%, 124%	112%, 116%	115%, 120%	119%, 123%

Since the section efficiency is integrated into the column internals design, the column diameter can reduce an extra 2% than only changing the flooding condition. However, the reduction of fractional hole area allows the vapour velocity through the valve tray to be increased, which results in a significant increase in pressure drop. As shown in the table, the pressure drop can be increased by up to 60%, from 0.54 kPa to 0.89 kPa, under the same flooding conditions.

4. Discussion

In this work, the complex tray internals design and tray efficiency prediction can be integrated automatically by computer-aided design. Thus, the size of the distillation column no longer needs to be too conservative based on the designers' experience, and the separation efficiency is also improved, resulting in substantial savings in capital investment. At the same time, it helps engineers clearly understand the current production capacity and the potential to increase production, which is also important for safe production.

5. Conclusion

The case study illustrates the benefit of the proposed automatic design methodology to design and evaluate design results to reduce the column diameter while satisfying the hydraulic limits (jet flooding, downcomer flooding, downcomer backup, and weir loading).

The methodology can be applied to trayed distillation columns and accounts for the impact of the change of hydraulic limits and fractional hole area on the pressure drop, downcomer backup and section efficiencies. In this case study, the distillation column diameter could be reduced by up to 12%, compared to the base case, by increasing the flooding conditions, without violating hydraulic constraints. Since the tray efficiency is integrated into the internal design, additional benefits can be obtained while the fractional hole area is changed. For the tray column, the fractional hole area is contributed by many variables, including valve leg length or hole diameter, deck thickness, and pitch of the hole, which will also affect the tray efficiency according to Chen and Chuang (1993) correlation. Thus, there is still a wide range to explore an optimal design.

However, the feed location and the number of actual stages are fixed in this work; these may not be optimal for the total annual cost. Cost-benefit analysis would be needed to further evaluate the design solutions.

References

Branan, C. R. (2005). Rules of Thumb for Chemical Engineers: A manual of Quick, Accurate Solutions to Everyday Process Engineering Problems. Elsevier

Chen, G.X. & Chuang, K.T. (1993), Prediction of point efficiency for sieve trays in distillation. Industrial & Engineering Chemistry Research, 32(4), 701–708.

Chuang, K. T., & Nandakumar, K. (2000). Tray Columns: Design. In I. D. Wilson (Ed.), Encyclopedia of Separation Science, pp. 1135–1140. Academic Press.

Ilme, J., Klemola, K., Aittamaa, J., & Nystrom, L. (2001). Calculating distillation efficiencies of multicomponent i-butane/n-butane column. Chemical Engineering Communications, 184(1), 1–21.

Kister, H. Z., Haas, J. R. ., Hart, D. R. ., & Gill, D. R. . (1992). Distillation Design (Vol. 1). McGraw-Hill New York.

Koch-Glitsch. (2013). Ballast tray design manual bulletin 4900. [Online] https://www.koch-glitsch.com/getattachment/ebcafced-ed20-419a-994b-b05ef37cf682/attachment.aspx [Accessed October 2022]

Lahiri, S. K. (2014). Particle swarm optimization technique for the optimal design of plate-type distillation column. In J. Valadi & P. Siarry (Eds.), Applications of Metaheuristics in Process Engineering (pp. 153–182). Springer International Publishing.

Ogboja, O., & Kuye, A. (1990). A procedure for the design and optimisation of sieve trays., Trans. I. Chem. E., 68(5), 445–452.

Souza, A. R. C., Bagajewicz, M. J., & Costa, A. L. H. (2022). Globally optimal distillation tray design using a mathematical programming approach. Chemical Engineering Research and Design, 180, 1–12.

Stichlmair, J. G., Klein, H., & Rehfeldt, S. (2021). Distillation: Principles and Practice. John Wiley & Sons. (pp. 487-533)

Towler, G., & Sinnott, R. (2021). Chemical Engineering Design: Principles, Practice and Economics of Plant and Process Design, 3rd ed. Butterworth-Heinemann.

Vennavelli, A. N., Whiteley, J. R., & Resetarits, M. R. (2014). Predicting valve tray efficiency. Chemical Engineering Research and Design, 92(11), 2148–2152.

Antonis Kokossis, Michael C. Georgiadis, Efstratios N. Pistikopoulos (Eds.)
PROCEEDINGS OF THE 33rd European Symposium on Computer Aided Process Engineering
(ESCAPE33), June 18-21, 2023, Athens, Greece
© 2023 Elsevier B.V. All rights reserved. http://dx.doi.org/10.1016/B978-0-443-15274-0.50323-1

Process Design and Intensification of Circulating Catalytic Fluidized Bed Membrane Reactor for Oxidative Coupling of Methane

Moustafa Ali[a], Yuhe Tian[b], Dustin Kenefake[a,c], Efstratios N. Pistikopoulos[a,c]*

[a]Texas A&M Energy Institute, Texas A&M University, College Station, Texas, USA, 77845.
[b]Department of Chemical and Biomedical Engineering, West Virginia University, Morgantown, West Virginia, United States, 26505.
[c]Artie McFerrin Department of Chemical Engineering, Texas A&M University, College Station, TX, USA, 77840
*Corresponding Author: Prof. Efstratios N. Pistikopoulos - stratos@tamu.edu

Abstract

The catalytic oxidative coupling of methane (OCM) process has received intense interest in the literature due to the potential to directly convert natural gas or methane to value-added chemicals at a reduced cost, energy consumption and carbon emissions when compared to conventional processes. However, major challenges such as low yield, catalyst deactivation, and reactor scale-up still challenge the commercialization of this process. A promising solution to address these challenges is to develop innovative OCM reactor designs leveraging the recent advances in modular process intensification, e.g. membrane reactors. The aim of this work is to design an optimal OCM process at commercial scale leveraging the concept of modular process intensification. We investigate an intensified fluidized bed membrane reactor (FBMR) catalyzed by La_2O_3/CaO and compare its performance to a conventional fluidized bed reactor (FBR). The use of membrane for oxygen feed distribution has been reported to result in better C_2+ yield and selectivity by selectively enhancing the desired reactions.

Keywords: OCM, Process Intensification, Fluidized Bed Membrane Reactors

1. Introduction

The oxidative coupling of methane has been the target of intense scientific and commercial interest for more than forty years due to the tremendous potential of the technology to reduce costs, energy, and environmental emissions in the production of ethylene. In OCM, methane (CH_4) and oxygen (O_2) react over a catalyst exothermically to form ethylene (C_2H_4), ethane (C_2H_6) water (H_2O) and heat, according to the following reactions (1 to 4) (Penteado et al., 2016). While the benefits of OCM have been known since the early 1980s, past efforts did not result in a viable process design with acceptable performance needed for commercialization. Despite the pioneering work in the catalysis of OCM that has managed to bring the commercialization realization of OCM (Galadima & Muraza, 2016; Onoja et al., 2019), there is need for a novel reactor design to achieve highly economically and environmentally relevant optimal OCM process design with optimal C_2+ yield, selectivity and CH_4 conversion (Barteau, 2022). These catalysts, while at times achieving promising yield and selectivity, were hampered by very high operating temperatures, low activities, and short lifetimes on the order of hours to days.

$$CH_4 + \frac{1}{2}O_2 \rightarrow C_2H_6 + H_2O \qquad \text{Main Reaction - (Reaction 1)}$$

$$C_2H_6 + \frac{1}{2}O_2 \rightarrow C_2H_4 + H_2O \qquad \text{Main Reaction - (Reaction 2)}$$

$$CH_4 + O_2 \rightarrow CO_2 + 2H_2O \qquad \text{Unwanted Side Reaction - (Reaction 3)}$$

$$2C_2H_4 + 3O_2 \rightarrow CO_2 + 2H_2O \qquad \text{Unwanted Side Reaction - (Reaction 4)}$$

The paper by (Cruellas et al., 2017) outlined that there are more than 20 types of reactors that were designed for experiments or only modeled for the optimized OCM production. In the aforementioned paper, it outlined that the most popular and widely applied reactors that, produce the highest C_2+ selectivity% CH_4 conversion%, and the highest C_2+ yield%. The most common types of conventional reactors used were the packed bed reactors and fluidized bed reactors. The advantages of fluidized bed reactors (FBR) are: (i) Perform almost isothermally and (ii) Stable and efficient performance especially concerning the reactor operability and the safety aspects (Chen et al., 2020; Schwarz et al., 2014; Yaghobi & Ghoreishy, 2008). Concerning the packed bed reactors (PBR), issues associated with the performance of the reactor were identified. In the application of OCM, the reactor suffers from the following: (i) Hot spot formations on the catalyst causing the reduction of C_2+ formation drastically due to high operation temperatures. (ii) Fast degradation of catalyst. This in fact leads to low yield of C_2+. While fluidized bed reactors overcame the catalyst deactivation and high heat management issues, still FBR showed that only low yield of 19% could be achieved (Jašo et al., 2011). Several works have tried to overcome some of the bottlenecks in both conventional reactors such as reducing the operating temperatures while improving the yield and for better conversion, however, the highest yield% reached was approximately 20% (Chen et al., 2020; Yaghobi & Ghoreishy, 2008). Membrane reactors offer a great advantage to overcome those issues, however, the overall conversion, yield and selectivity were relatively low. The maximum methane conversion C_2+ yield and selectivity achieved from membrane reactors were 24%, 20% and 47% respectively (Onoja et al., 2019). Although membrane reactors demonstrated an improved selectivity of C_2+, those systems suffered from low methane conversion due to the low flux of oxygen throughout the membrane which hindered the overall performance. While this issue could be avoided by raising the operating temperature, it negatively affected the selectivity of C_2+(Farrell & Linic, 2016). Cruellas et al., 2020 introduced the concept of bubbling fluidized bed reactor with a membrane installed and loaded with catalyst of interest. The advantages behind this system are the following (Cruellas et al., 2017): (i) Distributed oxygen feeding results in a local lower O_2 concentration, which allows to obtain higher yields, (ii) Stability of the catalyst particles. Typically, the advantages of traditional fluidized beds and use of membranes are combined. Thus, the objective of this paper is to design a novel intensified circulating Fluidized Bed Membrane Reactor (FMBR) for the application of OCM to overcome all the above-mentioned issues leveraging the use of membrane for better oxygen distribution and the internal catalyst circulation to achieve higher C_2+ yield. This paper summarizes the main findings behind the oxidative coupling of methane (OCM) system, main key elements to model an intensified OCM Circulating Fluidized Bed Membrane Reactor system. This rest of the paper is organized as follows: section 2 covers the kinetics of oxidative coupling of methane. Section 3 presents the design and validation of the conventional Fluidized Bed Reactor (FBR). Section 4 showcases the design and intensification of circulating Fluidized Bed Membrane Reactor followed by results and technical insights.

2. Kinetics of Oxidative Coupling of Methane

The reported kinetic model is considered the most used kinetic OCM model for the development of an OCM reactor (Ching et al., 2002; Cruellas et al., 2020). A comprehensive 10-step kinetic model of the oxidative coupling of methane to C_2+ hydrocarbons over a 10 % La_2O_3/CaO catalyst was developed on the basis of kinetic measurements in a micro catalytic fixed bed reactor covering a wide range of reaction conditions by Stansch et al., 1997. The reaction scheme shown in Figure 1 contains three primary and seven consecutive steps that involves many species. The conversion of hydrocarbons and of carbon monoxide with oxygen were described by applying Hougen-Watson type rate equations. For the remaining reaction, power law equations were applied. The kinetics model's parameters were reported by (Ching et al., 2002; Stansch et al., 1997)

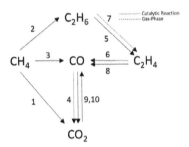

Figure 1: Reaction scheme of the kinetic model proposed by (Stansch et al., 1997)

3. Modeling of Conventional Fluidized Bed Reactor (FBR)

To systematically identify the optimal design solution, 2D high fidelity conventional Fluidized Bed Reactor (FBR) was adapted from (Cruellas et al., 2020a; Cruellas et al., 2020b) and modeled utilizing gPROMS ModelBuilder (PSE, 2020), which comprise partial differential algebraic equations accounting for mass balances, hydrodynamics, catalyst solid distribution and radial velocities across the reactor. The model is a core-annulus system which describes the two reactor regions that are characterized by different fluid dynamic conditions and catalyst concentrations. Figure 2 illustrates the hydrodynamic mechanistic of the 7 meters length and 0.6 meters diameter adapted reactor design. The model assumes that gas is transported upward in the core section, and it is stagnant in the outer region and accounts for radial gas mass transfer between the two sections. It was assumed that undiluted catalyst particles are fed to the reactor. The reactor is operated at 800 °C and 2 bar with methane to oxygen feed of 4:1. The reactor was simulated with a terminal velocity of 2 m/s and the value of mass transfer coefficient was fixed at 0.05 m/s. Results shows that C_2+ yield was found to be 15% with a selectivity of 50% and a methane conversion of 26% which matches the adapted model from (Cruellas et al., 2020a). The validation of FBR hydrodynamics is shown in Figure 3.

4. Design and Intensification of Circulating Fluidized Bed Membrane Reactor (FBMR)

Oxygen permeation only membranes are integrated in the circulating FBR design to investigate the performance the proposed intensified FBMR design. A configuration had been selected where the membrane tube is placed in middle of the reactor along the axial direction and the catalyst bed is positioned around the membrane tubes. Due to the high

M. Ali et al.

operating temperature of the process, $Ba_{1-x}Sr_xCo_{1-y}Fe_yO_{3-\delta}$ (BSCF) perovskite membrane is selected to be integrated with the conventional fluidized bed reactor with a diameter of 0.01 meters, whereas the membrane length is equal to the reactor length to maximize the oxygen distribution (Spallina et al., 2015).

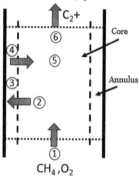

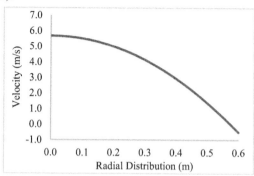

Figure 2: FBR Hydrodynamics Mechanism   Figure 3: Hydrodynamics Validation - FBR Design

This membrane can withstand high temperature ranges up to 1000 °C. In addition, the selected membrane is an O_2 permeation only meaning that air could be used as inlet along with methane for better economics (CAPEX and OPEX). The utilization of membrane along the axial distribution in the reaction offers a uniform distribution of O_2 thus, it helps to better control O_2 concentration along the reactor, and eventually improves $C_{2}+$ selectivity%, yield% and an increase in the CH_4 conversion%. The equations were modified to accommodate the placement of the membranes in the FBR design. The model includes the membrane characteristics, governing equations of both fuel and air sides, hydrodynamics, catalyst solid distribution and radial velocities across the reactor. The membrane characteristics including permeabilities, mass and heat coefficients were adapted from (Spallina et al., 2015; Xu & Thomson, 1999). The Fluidized Bed Membrane Reactor (FBMR) governing equations are presented in table 1.

Table 1: Fluidized Bed Membrane Reactor (FBMR) Governing Equations	
Mass Balance in the core	$\dfrac{dc_i^c v^c}{dx} = r_i(c_j^c, \gamma^c) - K_i^{ca}(c_i^c - c_i^a) + h_i^f a^f(c_i^c - c_i^f)$
Mass Balance in the anulus	$r_i(c_j^a, \gamma^a) + \dfrac{V^c}{V^a} K_i^{ca}(c_i^c - c_i^a) = 0$
Air Side Equations	$\dfrac{dc_i^{air} v^{air}}{dx} = -h_i^{air} a^{air}(c_i^{air} - c_i^{m,air})$
Total Mass Balance	$c^c = \displaystyle\sum_{i=1}^{N} c_i^c$
Average Solid Fraction	$\gamma^c = 1 - \varepsilon^c = \dfrac{1}{r_c^2} \displaystyle\int_0^{r_c} 2r\varepsilon(r)dr$
	$\gamma^a = 1 - \varepsilon^a = \dfrac{1}{r_a^2 - r_c^2} \displaystyle\int_{r_c}^{r_a} 2r\varepsilon(r)dr$
	$\varepsilon(r) = 1 - \dfrac{G(r)}{\rho_s V_p(r)}$
Radial Velocity Profile – Solid reflux	$\dfrac{G(r)}{G_s} = a\left(1 - \left(\dfrac{r}{r_a}\right)^5\right) + 1 - \dfrac{5a}{7}$
Radial Velocity Profile – Particle velocity	$V_p(r) = \left(1.5\dfrac{U_{g,c}}{\varepsilon^c} - V_t\right)\left(1 - \left(\dfrac{r}{r_c}\right)^2\right)$

Process Design and Intensification of Circulating Catalytic Fluidized Bed
Membrane Rector for Oxidative Coupling of Methane

2035

5. FBMR Results and Technical Insights

The fluidized bed membrane reactor was simulated with the same conditions selected for the conventional fluidized bed reactor with catalyst particle size of 70 μm and net solids flux of 100 kg m^{-2} s^{-1}. The amount of oxygen has been varied from 10% to 30% whereas the catalyst concentration was fixed at 10%. The results showed that the optimal feed ratio CH_4/O_2 of 4 with a bed temperature of 800 and pressure of 2 bar. The computed axial mole fraction profiles of methane and C_2+ are given in Figure 4. The overall process performance resulted in a C_2+ yield of 35%, C_2+ selectivity of 67% and methane conversion of 67%. The axial C_2+ selectivity, C_2+ yield and conversion profiles are presented in Figure 5. The integration of membrane in the FBR design along with the catalyst distribution offered feeding the oxygen along the axial direction. While this integration helps in keeping the oxygen partial pressure low, it does consequently help in enhancing the reactor's performance towards the yield of more C_2+. At the entrance of the reactor, the kinetics

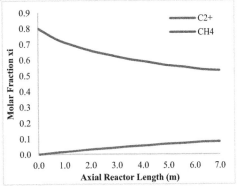

Figure 4: FBMR molar fraction profile of CH_4 and C_2+

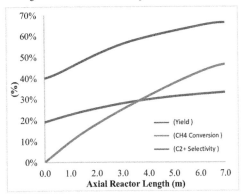

Figure 5: C_2+ Selectivity, Yield and CH_4 Conversion of FBMR

dominates over mass transfer allowing all the reaction to take place in the core only. It is important to note that both the kinetics resistance and mass transfer resistance are comparable as the reactions progress along the axial direction in both the core and annulus. This in fact could help protecting the C_2+ formed inside the intensified FMBR. Finally, one of the drawbacks of the OCM kinetics is the enormous production of carbon dioxide. It is one of most critical issues faced as it has a significant impact on the energy consumption, the environment and the product's purity (Galadima & Muraza, 2016). As the implementation of the membrane in the intensified FMBR design shifted the kinetics towards the production of more C_2+, a CO_2 reduction was achieved from 6.8% in the FBR configuration to 1.3% in the intensified FBMR design.

Conclusion

In this paper, we present the new intensified reactor design system for the simulation of oxidative coupling of methane. The conventional fluidized bed reactor was simulated and validated with a C2+ yield, selectivity, and methane conversion of 15%, 54% and 26% respectively. A membrane was integrated with the conventional reactor coupling a perovskite oxygen selective membrane along the axial direction to improve on the three parameters mentioned above. The reactor's design shows a noticeable improvement with a C_2+ yield of 35%, C_2+ selectivity of 67% and improved conversion of methane up to 50% with a reduced CO_2 production by almost 81%. Finally, the intensified fluidized bed membrane reactors could be a solution to improve the C_2+ yield%, C_2+ selectivity% and

CH₄ conversion%. Through this paper, it has been demonstrated that the concept of the intensified FMBR design can outperform the conventional FBR design.

6. Future Work

Although the new proposed intensified FBMR design solves some of the limitation for the OCM reaction to take place, it opened few research questions to uncover the role of membrane in achieving high C_2+ selectivity%, C_2+ yield% and CH₄ conversion%. The effect of these parameters (e.g. oxygen vacancy dilution coefficient (D_V), species diffusion coefficient (D_i) and perovskite membrane mass transfer coefficients (k_f & k_a)) will be investigated to enhance the economics of the OCM reactions and performance of FBMR. While the investigation of OCM reaction systems has mostly been focused on steady-state conceptual designs, ignoring the process dynamics limits the applicability of evaluating the implementation for commercial use.

7. Acknowledgement

We acknowledge the financial support from the Texas A&M Energy Institute, Artie McFerrin Department of Chemical Engineering at Texas A&M University, Shell USA and RAPID SYNOPSIS Project.

References

Barteau, M. A. (2022). Is it time to stop searching for better catalysts for oxidative coupling of methane? *Journal of Catalysis, 408*, 173–178. https://doi.org/10.1016/j.jcat.2022.03.006

Chen, L., Pannala, S., Broekhuis, R., Gautam, P., Gu, T., West, D., & Balakotaiah, V. (2020). Three-dimensional CFD simulation of pattern formation in a shallow packed-bed reactor for oxidative coupling of methane. *Chemical Engineering Journal, 400*(March), 125979. https://doi.org/10.1016/j.cej.2020.125979

Ching, T. T., Mohamed, A. R., & Bhatia, S. (2002). Modeling of catalytic reactor for oxidative coupling of methane using La2O3/CaO catalyst. *Chemical Engineering Journal, 87*(1), 49–59. https://doi.org/10.1016/S1385-8947(01)00191-7

Cruellas, A., Melchiori, T., Gallucci, F., & van Sint Annaland, M. (2017). Advanced reactor concepts for oxidative coupling of methane. *Catalysis Reviews - Science and Engineering, 59*(3), 234–294. https://doi.org/10.1080/01614940.2017.1348085

Cruellas, A., Melchiori, T., Gallucci, F., & van Sint Annaland, M. (2020). Oxidative Coupling of Methane: A Comparison of Different Reactor Configurations. *Energy Technology, 8*(8), 1–15. https://doi.org/10.1002/ente.201900148

Cruellas, Aitor, Heezius, J., Spallina, V., van Sint Annaland, M., Medrano, J. A., & Gallucci, F. (2020). Oxidative coupling of methane in membrane reactors; A techno-economic assessment. *Processes, 8*(3), 1–25. https://doi.org/10.3390/pr8030274

Farrell, B. L., & Linic, S. (2016). Oxidative coupling of methane over mixed oxide catalysts designed for solid oxide membrane reactors. *Catalysis Science and Technology, 6*(12), 4370–4376. https://doi.org/10.1039/c5cy01622c

Galadima, A., & Muraza, O. (2016). Revisiting the oxidative coupling of methane to ethylene in the golden period of shale gas: A review. *Journal of Industrial and Engineering Chemistry, 37*, 1–13. https://doi.org/10.1016/j.jiec.2016.03.027

Jašo, S., Arellano-Garcia, H., & Wozny, G. (2011). Oxidative coupling of methane in a fluidized bed reactor: Influence of feeding policy, hydrodynamics, and reactor geometry. *Chemical Engineering Journal, 171*(1), 255–271. https://doi.org/10.1016/j.cej.2011.03.077

Onoja, O. P., Wang, X., & Kechagiopoulos, P. N. (2019). Influencing selectivity in the oxidative coupling of methane by modulating oxygen permeation in a variable thickness membrane reactor. *Chemical Engineering and Processing - Process Intensification, 135*(August 2018), 156–167. https://doi.org/10.1016/j.cep.2018.11.016

PSE. (2020). *Process System Enterprise,gPROMS*. www.psenterprise.com/products/gproms

Schwarz, H., Geske, M., Franklin Goldsmith, C., Schlögl, R., & Horn, R. (2014). Fuel-rich methane oxidation in a high-pressure flow reactor studied by optical-fiber laser-induced fluorescence, multi-species sampling profile measurements and detailed kinetic simulations. *Combustion and Flame, 161*(7), 1688–1700. https://doi.org/10.1016/j.combustflame.2014.01.007

Spallina, V., Melchiori, T., Gallucci, F., & Van Sint Annaland, M. (2015). Auto-thermal reforming using mixed ion-electronic conducting ceramic membranes for a small-scale H2 production plant. *Molecules, 20*(3), 4998–5023. https://doi.org/10.3390/molecules20034998

Stansch, Z., Mleczko M, L., & Baerns. (1997). Comprehensive Kinetics of Oxidative Coupling of Methane over the La2O3/CaO Catalyst. *Industrial and Engineering Chemistry Research, 36*(7), 2568–2579. https://doi.org/10.1021/ie960562k

Xu, S. J., & Thomson, W. J. (1999). Oxygen permeation rates through ion-conducting perovskite membranes. *Chemical Engineering Science, 54*(17), 3839–3850. https://doi.org/10.1016/S0009-2509(99)00015-9

Yaghobi, N., & Ghoreishy, M. H. R. (2008). Oxidative coupling of methane in a fixed bed reactor over perovskite catalyst: A simulation study using experimental kinetic model. *Journal of Natural Gas Chemistry, 17*(1), 8–16. https://doi.org/10.1016/S1003-9953(08)60019-5

Antonis Kokossis, Michael C. Georgiadis, Efstratios N. Pistikopoulos (Eds.)
PROCEEDINGS OF THE 33rd European Symposium on Computer Aided Process Engineering
(ESCAPE33), June 18-21, 2023, Athens, Greece

Solvent pre-selection for extractive distillation using Gibbs-Helmholtz Graph Neural Networks

Edgar Ivan Sanchez Medina,[a*] Kai Sundmacher,[a,b]

[a] *Chair for Process Systems Engineering, Otto-von-Guericke University, Universitätsplatz. 2, Magdeburg, 39106, Germany*
[b] *Process Systems Engineering, Max Planck Institute for Dynamics of Complex Technical Systems, Sandtorstraße 1, Magdeburg, 39106, Germany*
[*] *sanchez@mpi-magdeburg.mpg.de*

Abstract

When selecting candidate solvents for extractive distillation, activity coefficients at infinite dilution are commonly employed. With them, the selectivity of different solvents for a given mixture can be estimated, and the solvents can be ranked accordingly. However, given the large chemical space of potential solvents and the limited experimental data for activity coefficients at infinite dilution, predictive methods become necessary to perform a solvent pre-selection process across a broad solvent space. In this work, a method for selecting solvents using the Gibbs-Helmholtz Graph Neural Network as predictive model for infinite dilution activity coefficients is presented. Different case-studies are given to illustrate the efficacy of this methodology ranging from aliphatic-aromatic separations to the separation of olefin-paraffin mixtures. The extended Margules equation is employed to estimate the vapor-liquid equilibria of the system of interest allowing the minimum solvent-to-feed ratio to be estimated. The results show that industrially relevant solvents are selected for different type of mixtures while exploring a much larger solvent space compared to the one delimited by the available experimental data. Also, it is shown that the pre-selection of solvents based on infinite dilution conditions differs from the one performed using the minimum solvent-to-feed ratio criterion. The latter is recommended because it approximates the conditions usually encountered in a real extractive distillation column.

Keywords: Graph Neural Network, Activity coefficient, Extractive distillation, Solvent selection.

1. Introduction

Separation processes in general, and distillation in particular, account for a significant part of the whole world's energy consumption (Sholl et al., 2016). Despite the efforts in alternative separation technologies, distillation is envisioned to remain as the predominant separation technology in practice during the coming decades. When dealing with close-boiling or azeotropic mixtures, extractive distillation is of particular interest. The reason is that by using an external agent, the entrainer, the thermodynamic equilibrium of the original mixture is conveniently modified, and the separation becomes possible or much more energy-efficient. However, the process of selecting the most appropriate solvent to enhance the separation is far from being straightforward. The solvent not only has to be thermodynamically powerful, but also economically viable and environmentally responsible.

The process of selecting the best solvent for a given mixture is particularly difficult because, typically, a large number of solvents need to be screened or novel solvents need to be designed from the vast chemical space. Therefore, an efficient pre-selection methodology that reduces the number of candidate solvents which need to be considered for more detailed evaluation is valuable. Brouwer et al. (2021) recently proposed a simple methodology for solvent pre-selection that does not rely on specialized software (e.g., COSMO-RS). Instead, it only uses experimental infinite dilution activity coefficients and the extended Margules equation. This makes it accessible to a broader range of engineers and researchers around the globe that might not have access to specialized software. However, its application is limited by the availability of experimental values for activity coefficients which in turn limits the number of solvents that are considered during the screening. Here, we extend this methodology by using our recently proposed Gibbs-Helmholtz Graph Neural Network approach (GH-GNN) (Sanchez Medina et al., 2022b) as a predictive method for enlarging the solvent search space. This removes the limitation of experimental activity coefficients availability and increases the number of solvents that can be considered during early-stage process design. Moreover, the original method's assumption of partly symmetric Margules parameters is here removed given that all binary mixtures can be potentially calculated by the GH-GNN.

2. Background

The relative volatility of the key components i, j in a mixture indicates its ease of separation. At low or moderate pressures, it is defined as the ratio of the pure component vapor pressures $P°$ multiplied by their activity coefficients γ. Since the ratio of the vapor pressures $(P_i°/P_j°)$ within a small temperature range is constant, the modification of the relative volatility, at the same operating pressure, can only be accomplished by the introduction of a solvent that effectively modifies the ratio of the activity coefficients (γ_i/γ_j). This ratio in the presence of a solvent is called selectivity. The maximum impact that a solvent has on the mixture's relative volatility occurs at infinite dilution conditions where the original mixture is infinitely diluted in the solvent. As a result, it is a common practice to rank the efficiency of different solvents according to the value of the selectivity at infinite dilution achieved by such solvents.

Recently, graph neural networks (GNNs) have shown promising performance on predicting activity coefficients at infinite dilution (Sanchez Medina et al. 2022a, Sanchez Medina et al., 2022b, Rittig et al. 2022). In this context, molecular graphs are constructed that contain atomic information (such as the type and hybridization of the atom) in their nodes and chemical bond information (such as the type of bond) in their edges. This information is then processed via a message-passing scheme. Moreover, in the GH-GNN model, global-level features (molecular descriptors describing the polarity and polarizability) are used as part of the learning process of activity coefficients at infinite dilution (Sanchez Medina et al. 2022b). First, the GH-GNN model uses two molecular graphs (corresponding to the solute and solvent in the mixture) that are passed through a molecular-level GNN. Then, a two-node mixture graph is constructed in which each node is defined by the solute and solvent embeddings learned by the first GNN. Intermolecular interaction information is gathered in the mixture graph using hydrogen-bonding information and the global-level embeddings of the solute and solvent graphs. This mixture graph is passed through a second GNN which learns the molecular interactions. Finally, an integrated Gibbs-Helmholtz expression is used to model the temperature-dependent activity coefficients at infinite dilution. This GH-GNN model is here used to estimate the necessary activity coefficients to perform the solvent pre-selection.

3. Methodology

3.1. Pseudo-binary relative volatility at infinite dilution (inf-RV)
The relative volatility between the key components i and j in the presence of the solvent is called pseudo-binary relative volatility and, at infinite dilution conditions, it is defined as

$$\alpha_{ij}^\infty = \frac{P_i^\circ}{P_j^\circ}\left(\frac{\gamma_i^\infty}{\gamma_j^\infty}\right)_S \tag{2}$$

where the activity coefficients at infinite dilution γ^∞ refer to the solute (i.e., key component i or j) infinitely diluted in the solvent. These coefficients were calculated using the GH-GNN model. The pure component vapor pressures were calculated using the Antoine equation with parameters available from the NIST Chemistry WebBook (2022). The screening was performed over 911 molecular solvents obtained from the original GH-GNN work (Sanchez Medina et al. 2022b) which contrasts with the screening performed by Brouwer et al. (2021) where only 69 molecular solvents were screened due to the restrictions of limited experimental data availability. The pseudo-binary relative volatility at infinite dilution (*inf-RV*) with respect to all solvents was calculated at different temperatures of industrial relevance, and the solvents were ranked according to the mean value across all utilized temperatures. The applicability domain prediction of the GH-GNN model based on chemical classes and Tanimoto similarity was used to only consider predictions within the applicability domain of the model.

3.2. Minimum solvent to feed ratio (min-SF)
As pointed out by Brouwer et al. (2021), the selection of solvents based on the pseudo-binary relative volatility at infinite dilution corresponds to situations where the solvent-to-feed ratio is very large. However, often much smaller solvent-to-feed ratios are used in practice. Therefore, the pre-selection of solvents was also carried out according to the minimum solvent-to-feed ratio (*min-SF*) that is necessary to have a pseudo-binary relative volatility equal to 3. This relative volatility value has been recommended as a threshold for which the energy savings due to employing extractive distillation are still considered attractive (Blahušiak, et al. 2018). Typically, if the relative volatility is increased beyond 3 the energy-savings are not as significant and extractive distillation becomes less attractive from an energetic perspective.

The extended Margules model (Eqs. 3-7) (Mukhopadhyay et al., 1993) is here used to estimate the activity coefficients at finite dilution. If one assumes that the ternary interaction parameter A_{123} is equal to zero, this equation has as its only parameters the activity coefficients at infinite dilution of the species involved which are here predicted by the GH-GNN model.

$$\ln(\gamma_1) = 2(x_1x_2A_{21} + x_1x_3A_{31}) + x_2^2A_{12} + x_3^2A_{13} + x_2x_3B_{123} - 2G^E \tag{3}$$
$$\ln(\gamma_2) = 2(x_2x_3A_{32} + x_2x_1A_{12}) + x_3^2A_{23} + x_1^2A_{21} + x_3x_1B_{123} - 2G^E \tag{4}$$
$$\ln(\gamma_3) = 2(x_3x_1A_{13} + x_3x_2A_{23}) + x_1^2A_{31} + x_2^2A_{32} + x_1x_2B_{123} - 2G^E \tag{5}$$
$$G^E = x_1x_2(x_2A_{12} + x_1A_{21}) + x_1x_3(x_3A_{13} + x_1A_{31}) + x_2x_3(x_3A_{23} + x_2A_{32}) + x_1x_2x_3B_{123} \tag{6}$$
$$B_{123} = 0.5(A_{12} + A_{21} + A_{13} + A_{31} + A_{23} + A_{32}) - A_{123} \text{ with}$$
$$A_{ik} = \ln(\gamma_{ik}^\infty) \tag{7}$$

Therefore, by employing only the infinite dilution activity coefficients together with the extended Margules model, the vapor-liquid equilibrium behavior of a given mixture, in the presence of a solvent, can be estimated. And the minimum solvent-to-feed ratio (*min-SF*) to reach a relative volatility of 3 can be used as the ranking criteria for pre-selecting solvents.

4. Results

4.1. Aliphatic-aromatic separations

We studied a collection of binary systems composed of benzene and one of the following alkanes: n-hexane, n-heptane, n-octane, n-nonane and n-decane. The above mentioned two metrics (i.e., *inf-RV* and *min-SF*) were used to rank the set of solvents considered during the screening. It is possible that, these two metrics lead to different solvents during the selection process depending on how the solvent's impact varies at the finite dilution. For instance, Figure 1 shows the top 10 solvents selected by each metric for the n-heptane-benzene system. While some solvents were selected by both metrics (indicated with "(Both)"), most solvents differ depending on the metric used for the selection.

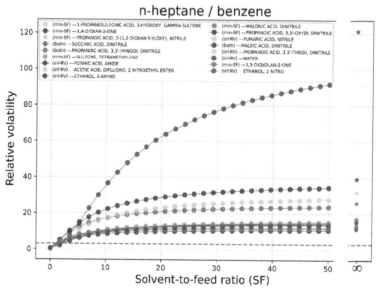

Figure 1. Pseudo-binary relative volatility of the top 10 solvents selected for the mixture n-heptane-benzene using min-SF and inf-RV. The horizontal red dotted line indicates a pseudo-relative volatility of 3. The star markers denote the corresponding pseudo-binary relative volatilities at infinite dilution.

Table 1. Top 5 solvents selected for the binary mixture benzene and the indicated alkane based on the minimum solvent-to-feed ratio (*min-SF*) to reach a pseudo-binary relative volatility of 3. The range of *min-SF* for the selected solvents is also indicated for each mixture.

Rank	n-hexane *min-SF*: 0.3	n-heptane *min-SF*: 1.0-1.2	n-octane *min-SF*: 2.3-3.7	n-nonane *min-SF*:3.3-8.4
1	1,2,3-Tris(2-cyanoethoxy)propane	(2Z)-2-Butenedinitrile	(2Z)-2-Butenedinitrile	Water
2	1,3-Propane sultone	Succinonitrile	Succinonitrile	(2Z)-2-Butenedinitrile
3	(Methylsulfonyl)ethane	Ethylene carbonate	Water	Succinonitrile

| 4 | 3,3'-(oxybis(ethyleneoxy))dipropionitrile | Sulfolane | 3,3'-Iminodipropanenitrile | Carbamoyl |
| 5 | Sulfolane | 1,3-Propane sultone | Carbamoyl | (2E)-2-Butenedinitrile |

The above occurs whenever the intersection of the pseudo-binary relative volatility trajectory of one solvent intersects the one of another solvent above the threshold of relative volatility equal to 3. For this reason, even though water was selected as the best solvent for this system based on *inf-RV*, it is actually the worst among the 17 solvents depicted on Figure 1 based on *min-SF* (i.e., its impact decays more rapidly as the solvent-to-feed ratio decreases compared to the other solvents). Given that in practice very large solvent-to-feed ratios are not plausible, a better pre-selection criterion should be *min-SF*. Table 1 shows the top 5 solvents selected for each of the considered aromatic-aliphatic binary systems.

As can be seen in Table 1, industrially relevant solvents are often selected which strongly indicates that the proposed methodology is able to pre-select promising candidates. For example, sulfolane (Tilstam, 2012), ethylene carbonate (Mohsen-Nia et al., 2010) and succinonitrile (Tan et al., 2022) are often reported as promising solvents for aliphatic-aromatic separations. For the mixture n-decane/benzene the top 5 selected solvents are similar to the ones for the system with n-nonane with a *min-SF* ranging from 5.6 for the top solvent to 30.5 for 1,4-Butynediol.

4.2. Olefin-paraffin separations

To exemplify olefin-paraffin separations the separation of the following binary mixtures was studied: n-hexane/1-hexene, n-butane/2-butene, n-heptane/1-heptene and n-propane/propene. Table 2 contains the top 5 performing solvents based on the *min-SF* metric along with the ranges needed to achieve a pseudo-binary relative volatility of 3. In the case of the n-heptane/1-heptene mixture only three solvents were found to increase the relative volatility up to 3 (corresponding to the top 3 solvents for the mixture n-hexane/1-hexene). This shows the difficulty of separating olefin-paraffin mixtures. Dimethyl sulfoxide stands out from the selected solvents for the n-propane/propene mixture agreeing with an extractive process already patented (Carter, 1981).

Table 2. Top 5 solvents selected for the shown olefin/paraffin mixtures based on the minimum solvent-to-feed ratio (*min-SF*) to reach a pseudo-binary relative volatility of 3. The range of *min-SF* for the selected solvents is also indicated for each mixture.

Rank	n-hexane/1-hexene *min-SF*: 5.2-439	n-butane/2-butene *min-SF*: 1.0-1.3	n-propane/propene *min-SF*: 1.0-2.5
1	Fumaronitrile	Carbamoyl	Glycerin
2	(2Z)-2-Butenedinitrile	Cyanogen	Dimethyl sulfoxide
3	Succinonitrile	Fumaronitrile	Carbamoyl
4	m-Phenylenediamine	2,4-Dinitrobenzaldehyde	Cyanogen
5	Morpholine	Succinonitrile	1,3-Benzenediamine

5. Conclusions and future work

The pre-selection of good candidate solvents for extractive distillation is an important step during chemical process development. Obtaining experimental data for all potential solvents in a large chemical space is practically impossible given the limitations of time and resources to which the exploration is bounded. Therefore, accurate and efficient predictive methods that assist experts in selecting suitable solvents are necessary. Here, we extend the methodology proposed by Brouwer et al. (2021) by using the Gibbs-

Helmholtz Graph Neural Network (GH-GNN) (Sanchez Medina et al., 2022b) together with the extended Margules equation to estimate the activity coefficients of all chemical species in the mixture of interest. With this, the minimum solvent-to-feed ratio needed to increase the relative volatility of the mixture to a certain point can be estimated. The impact of assuming an infinite solvent-to-feed ratio is also discussed by comparing the solvent selection obtained by the relative volatility at infinite dilution to the one obtained by using the minimum solvent-to-feed ratio metric. The latter is recommended because it approximates the real conditions of an extractive distillation column better. Industrially relevant solvents were selected using the proposed methodology for both aliphatic/aromatic and olefin/paraffin separations showing the validity of the proposed framework. The proposed methodology has been made open-source at https://github.com/edgarsmdn/SolvSelect_GHGNN.

Acknowledgments: Edgar Ivan Sanchez Medina is also affiliated with the International Max Planck Research School for Advanced Methods in Process and Systems Engineering - IMPRS ProEng at the Max Planck Institute for Dynamics of Complex Technical Systems Magdeburg.

References

Blahušiak, M., Kiss, A. A., Babic, K., Kersten, S. R., Bargeman, G., Schuur, B., 2018. Insights into the selection and design of fluid separation processes. Separation and purification technology, 194, pp. 301-318.

Brouwer, T., Kersten, S.R., Bargeman, G. and Schuur, B., 2021. Solvent pre-selection for extractive distillation using infinite dilution activity coefficients and the three-component Margules equation. Separation and purification technology, 276, pp. 119230.

Carter, C. O., 1981. U.S. Patent No. 4,267,034. US4267034A. Washington, DC: U.S. Patent and Trademark Office.

Mohsen-Nia, M., and Doulabi, F. M., 2010. Separation of aromatic hydrocarbons (toluene or benzene) from aliphatic hydrocarbon (n-heptane) by extraction with ethylene carbonate. The Journal of Chemical Thermodynamics, 42(10), pp. 1281-1285.

Mukhopadhyay, B., Basu, S., Holdaway, M. J., 1993. A discussion of Margules-type formulations for multicomponent solutions with a generalized approach. Geochimica et Cosmochimica Acta, 57(2), pp. 277-283.

NIST Chemistry WebBook, 2022. NIST Standard Reference Database Number 69. Accessed on 02 October 2022 at https://doi.org/10.18434/T4D303.

Rittig, J.G., Hicham, K.B., Schweidtmann, A.M., Dahmen, M. and Mitsos, A., 2022. Graph neural networks for temperature-dependent activity coefficient prediction of solutes in ionic liquids. arXiv preprint arXiv:2206.11776.

Sanchez Medina, E. I., Linke, S., Stoll, M., Sundmacher, K., 2022a. Graph neural networks for the prediction of infinite dilution activity coefficients. Digital Discovery, 1(3), pp. 216-225.

Sanchez Medina, E. I., Linke, S., Stoll, M., Sundmacher, K., 2022b. Gibbs-Helmholtz Graph Neural Network: capturing the temperature dependency of infinite dilution activity coefficients. arXiv preprint arXiv:2212.01199.

Sholl, D.S. and Lively, R.P., 2016. Seven chemical separations to change the world. Nature, 532(7600), pp. 435-437.

Tan, T., Cheng, H., Chen, G., Song, Z. and Qi, Z., 2022. Prediction of infinite-dilution activity coefficients with neural collaborative filtering. AIChE Journal, 68(9), pp.e17789.

Tilstam, U., 2012. Sulfolane: A versatile dipolar aprotic solvent. Organic Process Research & Development, 16(7), pp. 1273-1278.

Yang, D., Zhang, S., & Jiang, D. E., 2019. Efficient absorption of SO_2 by deep eutectic solvents formed by biobased aprotic organic compound succinonitrile and 1-ethyl-3-methylimidazolium chloride. ACS Sustainable Chemistry & Engineering, 7(10), pp. 9086-9091.

Antonis Kokossis, Michael C. Georgiadis, Efstratios N. Pistikopoulos (Eds.)
PROCEEDINGS OF THE 33rd European Symposium on Computer Aided Process Engineering
(ESCAPE33), June 18-21, 2023, Athens, Greece

Dhrushti-AI: A multi-screen multi-user eye-tracking system to understand the cognitive behavior of humans in process industries

Thasnimol Valuthottiyil Shajahan,[a] Rahul Madbhavi,[a] Mohammed Aatif
Shahab,[a] Babji Srinivasan,[a,c] Rajagopalan Srinivasan[b,c]

[a]Department of Applied Mechanics, Indian Institute of Technology Madras, Chennai,
600036, India
[b]Department of Chemical Engineering, Indian Institute of Technology Madras,
Chennai, 600036, India
[c]American Express Lab for Data Analytics, Risk and Technology, Indian Institute of
Technology Madras, Chennai, 600036, India
babji.srinivasan@iitm.ac.in, raj@iitm.ac.in

Abstract

Operator performance is key to ensuring safety in process industries. Therefore, a comprehensive assessment of their performance is critical for smooth and efficient plant operation. Traditional performance assessments are not comprehensive as these ignore cognitive aspects of performance. On the other hand, while eye-tracking-based approaches do provide a cognitive assessment during operating training, their applicability in a real-time real setting is limited. Existing eye-tracking systems come with many constraints affecting users' mobility (restricted movement in all directions) in their environment. In addition, it is beyond the scope of current eye trackers to track multiple users working in an environment, such as in control rooms. Satisfying the above requirements makes these eye-trackers expensive. In this work, we demonstrate the capabilities of an in-house developed cost-effective eye-tracking system to track users' eye movement in an unconstrained environment while giving freedom of head movement. Human subject studies are conducted to compare operators' gaze patterns and the quality of data with that obtained using commercial eye trackers. The results generally agree with data quality obtained using commercial eye trackers. Hence, the performance of the developed eye-tracking system is comparable to existing commercial eye-trackers while overcoming their limitations, such as restricted user movement, single-user tracking, and high cost.

Keywords: Operator performance, Eye-tracking, Cognitive behavior, Human error, Safety

1. Introduction

The use of innovative automation methods and safety management systems has helped the process industries become safer. Nonetheless, accidents of varying severity continue to occur. Studies indicate that human error is the primary cause of 75% of industrial accidents (Jung et al., 2020). The safe running of the plant is guaranteed by cooperation between the human and automation system. Operators must constantly keep track of the functioning of several control mechanisms and step in when the automation fails to provide as expected. However, the rise of digitization and the complexity of automation have changed the operator's role and added new difficulties (Nazir et al., 2014). To ensure optimum operating conditions, operators must collect, prioritize, and utilize data from several information sources. This adds to their cognitive workload and worsens if the operators are not well trained. Therefore, keeping track of the operator's performance is crucial to ensure safe plant operation.

Addressing operators' performance issues involves improving their competency and continuously observing their behavior in the field (Dai et al., 2016). There are hardly any studies focusing on monitoring operator performance in real time. For operators to become more competent, research has largely focused on providing them with the proper feedback based on their performance. Most performance assessment methods depend on subjective expert opinion that disregards the operators' cognitive behavior (Nazir et al., 2015). There is widespread agreement that cognitive behavior plays a significant role in unusual circumstances and accident causation.

In recent years, researchers have used eye-tracking to get insights into human cognitive behavior. It has been extensively used in understanding human cognitive processes such as situation awareness, expert level, learning ability, and mental state of the operator in several safety-critical domains such as healthcare, aviation, automobile driving, and nuclear power plant (Srinivasan et al., 2019). This work shows the potential of an in-house developed eye-tracker (Dhrushti-AI) capable of monitoring human performance in an unconstrained environment. Human subject studies were conducted to compare the quality of data obtained from Dhrushti-AI and commercial eye trackers.

Dhrushti-AI can enhance safety by addressing the challenges associated with operator performance during training and in real-time plant operation. Firstly, it can be used along with operator training simulators to identify flaws in the operator's understanding of the process during training. Secondly, it can be deployed in real control room settings to monitor the operator's performance in real-time. In addition, we can enhance operators' situational awareness based on the insights from eye-tracking data. For instance, if operators miss any critical information from HMI (Human Machine Interface) appropriate feedback can be provided. Adequate situational awareness and skills are crucial, especially when dealing with abnormal situations (Bhavsar et al., 2017).

2. Literature Review

Eye tracking is a technique to record a person's eye movement. It provides eye gaze coordinates (where the person is looking) and pupil size. There are two primary categories of eye trackers in the market: remote and wearable. The latter must be worn, whilst the former may be positioned close to the test screen. Wearable eye trackers are appropriate when the scenario is not constrained to a narrow space and the test participants are allowed to walk about the test setting. However, it may create discomfort for the user to wear a device for a long time, such as in a control room setting where operators work for long hours. On the other hand, a remote eye-tracker is set up in the user environment and it doesn't hinder the subject. Current key market players are Tobii, Smart Eye, SMI (SensoMotoric Instruments), and SR Research (Hermens et al., 2013).

Eye-tracking measures have been found to correlate with various aspects of human performance in safety-critical domains. An eye-tracking study using SMI Eye-tracking Glasses was conducted with air traffic controllers to capture their mental workload (Rodríguez et al., 2015). The study reveals that increased pupil diameter is associated with an increased mental workload. Another study found that changes in mental workload can be measured using frequency domain analysis of pupil size measurements. An increase in workload is connected with an increase in power level at low frequencies (Peysakhovich et al., 2015). The experiments are conducted by using a single-screen setup. During the whole experiment, participants' gaze position and pupil diameter were recorded with a remote SMI RED eye-tracker (SensoMotoric Instruments GmbH, Germany) at a sampling rate of 120 Hz. Wu et al. (2020) used eye-tracking to analyze the difference in cognitive behavior of 32 novices and 7 expert operators when doing activities involving standard operating procedures of accident scenarios in nuclear power plants (NPPs). The study uses iView X head-mounted eye tracking device (SMI, Germany) at a sampling rate of 50 Hz to capture the eye movements of operators. It was found that novice operators had a greater fixation rate, illustrating their intense search in various places and inexperience in tackling process disturbances.

Our previous studies with control room operators reveal distinct mental models of expert and novice operators (Shahab et al., 2021). During an abnormal situation, expert operators have higher dwell time on important regions of HMI than novices (Sharma et al., 2016). Further, operators who failed to complete a disturbance task have higher gaze entropy indicating that their eye gaze is distributed in several regions on the HMI unrelated to the disturbance (Bhavsar et al., 2017). These approaches provide interesting insights into operators' cognition and help identify the reason behind human failure. These eye-tracking studies were conducted using Tobii TX300 eye tracker to record participant eye movements at sampling rates of 120 Hz.

In summary, existing commercial eye-tracking systems support single-user and single-screen HMI with limited user movement. However, this is not the case in real control rooms where the operator needs to gather information displayed on multiple screens and involves a lot of head movements. In addition, satisfying the above requirements makes these eye-trackers expensive. In this work, we demonstrate the potential of an in-house developed cost-effective eye-tracker system (Dhrushti-AI) to record multiple users' eye movements from an unconstrained environment.

3. Methodology

In this work, we demonstrate the capabilities of an eye-tracker (Dhrushti-AI) developed to track users' gazes in the presence of head movements in unconstrained environments. The eye-tracker comprises an array of cameras fitted with lenses and filters, illuminators, an optics controller, power sources, and a computing unit. The camera array observes a region of the environment to detect users' faces and track their gazes, referred to as the headbox. The eye-tracking system illuminates the headbox with light in the near-infrared region using the illuminators. The optics controller generates the trigger signals to control the camera exposure and the illuminators. The compute unit detects the faces in the images obtained from the camera array using Histogram of Oriented Gradients (HOGs) based face detectors. Face recognition algorithms are utilized to identify faces and enable

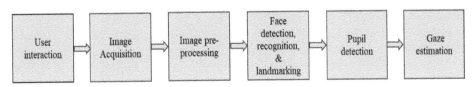

Fig. 1. Steps involved in the proposed eye-tracking system

the application of user-specific calibrated gaze models. Key landmarks in the faces (such as corners of the eyes, nose, etc.) are extracted and are used to estimate the users' head orientation. The pupils from these images are detected using several image processing techniques, such as thresholding, closing, and contour extraction. The eye-tracker utilizes the head orientation, pupil center locations, and the user-specific calibrated gaze model to estimate the gaze direction. The output of the developed eye-tracker is the user's gaze location (coordinates on screen) and pupil size variation. The overall methodology to get the output is shown in Fig. 1. The raw gaze data from Dhrushti-AI is processed further to obtain fixation and saccades using the Velocity-Threshold Identification (I-VT) filter.

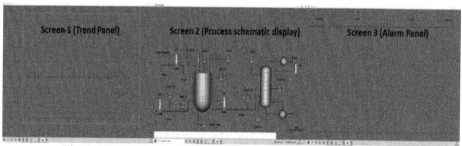

Fig. 2. HMI with the schematic and display units for three screens.

To demonstrate the applicability of the developed multi-screen eye-tracker (Dhrushti-AI), we used a simulated chemical process plant with which operators interacted via HMI. The HMI is presented to the operator on a display comprising three screens. Eight participants volunteered to play the role of the control room operator. They were asked to monitor and intervene during abnormalities. The participants first interacted with the process via a single-screen HMI of the chemical process displayed on the center screen. During this period, the other two screens were blank. In the next study, we displayed the HMI on all three screens. A typical task involves bringing a process variable(s) to within the acceptable limits as specified by the region between the upper limit and lower limit in Fig. 2. Eye-tracker calibration for each participant was performed at the beginning of the experiment. Next, we compare the results obtained from Dhrushti-AI and a commercial eye-tracker.

4. Result & Discussion

In this section, we illustrate the potential of Dhrusht-AI in capturing user eye movements in an unconstrained environment (typical control room setup) without restricting their head movements. The eye-tracker was set up to monitor user head movements at a distance of 45-60 cm from the center camera while allowing 25 cm horizontal movement and 35 cm vertical movement. Initial results have been obtained with image capture rates of 30 frames per second (FPS), while the hardware capabilities allow image capture at rates higher than 120 FPS. Gaze angle errors of less than 1.5° were obtained.

As discussed earlier, we conducted experiments with human subjects using Dhrushti- AI and commercial eye-tracker Smart-eye Aurora. We compared the quality of data obtained from the eye-tracker using data loss (proportion of missing gaze samples including blinks;(Holmqvist, 2017) and proportion of all fixation durations (dwell time) outside of various information sources, i.e. AOI's (Areas of interest) on the HMI. When dealing with abnormal situations operators primarily focus on various information sources which are relevant to the situation at hand (Sharma et al.,2016). Therefore, it is expected that the proportion of fixation on non-information sources obtained from Dhrushti-AI should be minimum and comparable to commercial eye trackers.

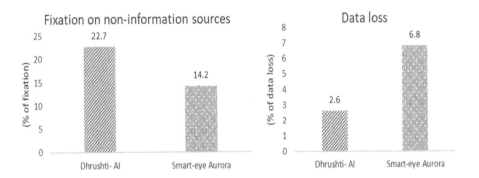

Fig. 3. Comparison of results obtained from Eye-trackers (Dhrushti-AI and Smart Eye Aurora): based on (a) Fixation on non-information sources, (b) Data loss

The results for both Dhrushti-AI and Smart Eye Aurora obtained from eight participants are shown in Fig. 3a and 3b. As observed from Fig. 3, the data loss for Dhrushti-AI is lesser compared to the smart-eye aurora (Fig.3b). This signifies that even with a significant head movement of participants, Dhrushti-AI is able to capture the operator's gaze position from all three screens. Similarly, the proportion of time on non-information sources obtained from Dhrushti-AI is also similar to that of Smart Eye Aurora (Fig.3b). It is important to note that few of the AOIs on the HMI were smaller and even with such lower size, Dhrushti-AI indicates that operators were looking at the AOIs most of the time.

In addition to the data quality, we also compare the distribution of dwell time on AOIs for operators who successfully completed the task and those who failed. The results indicate that experts have higher dwell time on key variables and exhibit lower gaze entropy than novices. These results are in general agreement with earlier works (Sharma et al., 2016; Bhavsar et al., 2017; Shahab et al., 2021) conducted using commercial eye trackers. In light of these results, the performance of the developed eye-tracking system (Dhrushti-AI) is comparable to existing commercial eye-trackers while overcoming their limitations such as restricted head movement, single screen setup, and high cost. Thus, the developed eye tracker has the potential to be deployed in control rooms for assessing operator performance.

5. Conclusion

In this work, we demonstrated the potential of a multi-screen- multi-user eye tracking system (Dhrushti-AI) in capturing operator cognitive behavior in a process plant setting.

There is widespread agreement that cognitive behavior plays a substantial role in the causes of abnormal situations and accidents. The capabilities of Dhrushti-AI are demonstrated with a 3- screen workstation as the HMI. We conducted experimental studies to validate the quality of the developed eye tracker with that of the commercially available eye-trackers. Results indicate that the performance of the Dhrushti-AI is lesser than the existing commercial eye trackers, in terms of data loss. The operator's cognitive behavior obtained from Dhrushti-AI is also in general agreement with our earlier works (Sharma et al., 2016) with research-grade eye trackers (smart eye Aurora). Thus, the developed multi-screen eye-tracker can be deployed in actual control rooms to infer the cognitive behavior of operators previously confined to controlled environments. Additionally, it can identify flaws in the thought process of novice operators during training and help them in acquiring knowledge about the process dynamics. In the future, we can increase the system's potential to meet the requirements of large-scale industrial studies.

References

Bhavsar, P., Srinivasan, B., & Srinivasan, R. (2017). Quantifying situation awareness of control room operators using eye-gaze behavior. Computers & chemical engineering, 106, 191-201.

Dai, Y., Wang, H., Khan, F., & Zhao, J. (2016). Abnormal situation management for smart chemical process operation. Current opinion in chemical engineering, 14, 49-55.

Holmqvist, K. (2017). Common predictors of accuracy, precision, and data loss in 12 eye-trackers. In The 7th Scandinavian Workshop on Eye Tracking.

Jung, S., Woo, J., & Kang, C. (2020). Analysis of severe industrial accidents caused by hazardous chemicals in South Korea from January 2008 to June 2018. Safety Science, 124, 104580.

Nazir, S., Kluge, A., & Manca, D. (2014). Automation in the process industry: Cure or curse? How can training improve operator's performance. In Computer Aided Chemical Engineering (Vol. 33, pp. 889-894). Elsevier. doi: 10.1016/B978-0-444-63456-6.50149-6

Nazir, S., Øvergård, K. I., & Yang, Z. (2015). Towards effective training for process and maritime industries. Procedia Manufacturing, 3, 1519-1526. doi: 10.1016/j.promfg.2015.07.409

Peysakhovich, V., Causse, M., Scannella, S. and Dehais, F., 2015. Frequency analysis of a task evoked pupillary response: Luminance-independent measure of mental effort. International Journal of Psychophysiology, 97(1), pp.30-37.

Rodríguez, S., Sánchez, L., López, P., & Cañas, J. J. (2015, September). Pupillometry to assess Air Traffic Controller workload through the Mental Workload Model. In Proceedings of the 5th International Conference on Application and Theory of Automation in Command and Control Systems (pp. 95-104). ACM.

Shahab, M. A., Iqbal, M. U., Srinivasan, B., & Srinivasan, R. (2022). HMM-based models of control room operator's cognition during process abnormalities. 1. Formalism and model identification. Journal of Loss Prevention in the Process Industries, 76, 104748.

Shahab, M. A., Srinivasan, B., & Srinivasan, R. (2021, December). Analysis of Control Room Operators' Competence using Cognitive Engineering Approaches to Improve Process Safety. In 2021 International Conference on Maintenance and Intelligent Asset Management (ICMIAM) (pp. 1-6). IEEE.

Sharma, C., Bhavsar, P., Srinivasan, B., & Srinivasan, R. (2016). Eye gaze movement studies of control room operators: A novel approach to improve process safety. Computers & Chemical Engineering, 85, 43-57.

Srinivasan, R., Srinivasan, B., Iqbal, M. U., Nemet, A., & Kravanja, Z. (2019). Recent developments towards enhancing process safety: Inherent safety and cognitive engineering. Computers & Chemical Engineering, 128, 364-383.

Wu, Y., Liu, Z., Jia, M., Tran, C. C., & Yan, S. (2020). Using artificial neural networks for predicting mental workload in nuclear power plants based on eye tracking. Nuclear Technology, 206(1), 94-106.

Antonis Kokossis, Michael C. Georgiadis, Efstratios N. Pistikopoulos (Eds.)
PROCEEDINGS OF THE 33rd European Symposium on Computer Aided Process Engineering
(ESCAPE33), June 18-21, 2023, Athens, Greece

Digital Twin- A System for Testing and Training

Michael Schueler[a], Tanja Mehling[b]

[a] *Siemens AG, DI PA SW ID MBS OA , Industriepark Hoechst, B598, 65926 Frankfurt*
michael.schueler@siemens.com
[b] *Siemens AG, DI PA SW ID MBS DS, Industriepark Hoechst, B598, 65926 Frankfurt*
tanja.mehling@siemens.com

Abstract

An Operator Training System (OTS) is a digital twin of a process plant, that enables to study the plant before it really exists. That does not just give an opportunity to train the operators of the plant, before plant startup, but it also helps to get a better understanding of the process.

In this work, an OTS of a novel process to produce biochemicals from wood was developed. Aiming to train the staff while the plant is still under construction and to gain a deep knowledge of this plant, a digital twin of the plant was built. According to the requirements for the training, a mid-fidelity process model was combined with a high-fidelity model.

To finalize the OTS, the process model was connected to the original Distributed Control System (DCS) environment.

It can be shown that the development of a process plant digital twin is highly effective in accelerating the commissioning of a new plant. On the one hand, the operators are enabled to gain a deep knowledge of the process and the operation of the plant even before the plant is finalized. On the other hand, the detailed simulation gives valuable insights to the process, which help to understand the process, the conditions, and the regulation concept.

Keywords: Digital Twin, Operator Training System, dynamic Simulation.

1. Introduction

In the recent years, Digital Twins became omnipresent in a huge variety of applications. They are of major contribution to the digitalization of the industries, still one of the actual megatrends. Of course, also the process industry applies Digital Twins over the life cycle of a production plant like described by Labisch (2019). An Operator Training System (OTS) is a digital twin of a process plant, that enables to study the plant even before it really exists. That does not just give an opportunity to train the designated operators of the plant, before plant startup, but it also helps to get a better understanding of the process and the plants functionalities as investigated by NAMUR (2006).

The aim of an OTS as a Digital Twin is, to reproduce the physical plant as well as the real drive and control mechanisms, the automation and operation. Both, the real plant and the virtual plant, i.e. the Digital Twin are compared in Figure 1.

The overall plant including the automation can be separated in five layers. For the real plant the first layer is the real process plant. The process plant contains all units like columns, reactors, vessels pumps, valves, pipes and so on.

The second layer of the real system contains all actuators and sensors. Actuators are typically pump motors and valve drives. The sensors measure all relevant process data like temperature, pressure, flow, and composition.

The third layer with the remote IOs (Input/Output) describes the operations or assets, transferring the data from the second layer (sensors and actuators) to the automation system (AS), the fourth layer.

Finally, the automation system is connected to the human machine interface (HMI). The fifth layer. By means of the HMI the state of the process and the plant is visualized, and the operator can control the plant.

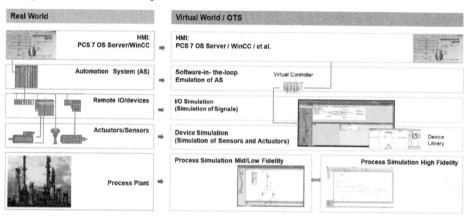

Figure 1: structure of a real plant and an OTS

Analogue to the overall real plant, an operator training system as a digital twin of a real plant has the same five layers. But in contrast to the real plant, most of the five layers are virtualized in an operator training system.

The basis for an operator training system is the process simulation. This virtual plant represents the real process plant (layer one). By means of the process model the behavior of the plant is simulated. For an operator training system mainly dynamic process models are used. With a dynamic process model the transient behavior of the process can be described. Dynamic simulations can be quite computationally intensive. Thus, it is aimed, to model as accurate as possible but only as accurate as needed, to save time and resources. Based on the needs for the training and considering the computational effort the fidelity of the process model can be selected. Even the combination of low or mid fidelity models with high fidelity models is possible. Within the process model all necessary process measurements are calculated and transferred to the next layer.

In this layer, the device simulation, the behavior of the measurement devices and actuators is simulated. The time dependent behavior of valve drives and motors as well as all necessary feedback signals for the automation system are calculated. Based on a library containing the necessary device modules, the relevant device simulation model can be composed.

The device Simulation is connected to the IO simulation (third level). In this level the signals from the device simulation are transferred to the fourth level, the emulation of the automation system and vice versa. The IO-level can also be used without the first two layers (process simulation and device simulation) to generate signals for testing the automation system.

The automation system itself is emulated in the fourth layer. With virtual controllers running on standard personal computers (PCs) the behavior of the automation system is simulated. The configuration of the automation system of the real plant can be loaded to the virtual controller without any change. The automation is executed and behaves like the real automation system.

Finally, as in the real plant control, the automation system of the Digital Twin is connected to the original human machine interface (HMI) of the distributed control system (DCS). In this fifth layer the process can be controlled with the very same operating screens like in the control room. Thus, with the OTS, the operator will not realize the differences between the Digital Twin and the real plant. Neither the operating system, i.e. the HMI, nor the effect of operating actions differ from the real plant. The only difference might be the actual surrounding of the operator, which might be anywhere but is not limited to the real control room.

In the current work, an example of an OTS is given as develop and implemented for a real process. The process itself is described, the proceeding of the development and the benefits are shown.

We think the correct split of model fidelity is an essential decision in the design of an OTS. Furthermore the use of tools that can direct reuse the original data of the automation system reduces the effort of design.

2. Development of the Operator Training System.

In this work, an OTS was realized, representing a novel process to produce biochemicals from wood. Aiming to train the staff while the plant is still under construction and to gain a deep knowledge of this plant and the corresponding process, a digital twin of the plant was built. According to the requirements for the training, to reproduce the real process and to consider the complexity of the process, a mid-fidelity process model was combined with a high-fidelity model. To finalize the OTS, the process model was connected to the original DCS environment. Thus, including the sensor and actor level, virtual controllers are applied to train the effect of interferences while the process is running, but also to train the startup, shutdown, disturbances, and an emergency case.

In the next section the process itself is briefly presented. Furthermore, the modeling of the process and it challenges are described.

2.1. Biochemical Process

The biochemical process is divided in several main steps (see Figure 2). In a first step the tree trunks are grinded into chips. These chips are washed and prepared for the next process steps. The wood is then enzymatically converted into intermediate products. These intermediate products are separated from the slurry and purified. Further, these intermediate products are chemically converted to the final product. In a last step the bio-chemicals are purified in a distillation train to achieve the required purities. The final purified products are stored in several tanks.

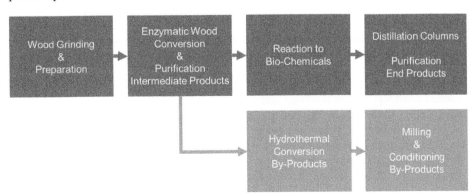

Figure 2: Block flow diagram of the process

The residual components separated from the slurry after the enzymatic conversion are further processed. This complex mixture is converted hydrothermally. These side products are further conditioned by drying and milling and the final byproducts are stored in tanks for further distribution.

2.2. Process modelling

To create a detailed but economic OTS, the process was divided in many individual process steps. Some of the process steps require an intensive, high-fidelity simulation, while others can be represented by less intensive mid-fidelity models. Some process steps were not considered at all, since their complexity is very low, the operation is well known and the benefit of a Digital Twin is comparatively low.

Therefore, the wood unloading, storage and coarse chopping are not represented in the Digital Twin. Furthermore, since the process model is focused on the main operation of the plant, cleaning facilities (cleaning in place (CIP) systems) and safety relief valves, sampling points and drain valves were neglected.

Simple models, thus, a mid-fidelity simulation was chosen for plant components such as tank farms, truck unloading stations and storage. Furthermore, utility systems, if relevant for the process simulation, are also simulated with simple models. Complex chemical processes such as reactions or phase changes typically do not take place in these parts of the plant, so that it is reasonable, to apply the simpler models in these cases.

In contrast, plant sections with reactions and apparatuses in which the phase equilibrium plays a decisive role (e.g. distillation columns) are simulated with high fidelity models. But not only the necessary complexity was decisive for the division into mid fidelity or high-fidelity models, but also that there are not too many interfaces between the two simulators.

2.2.1. High fidelity process model

Decisive for each process model is the application of reliable and evaluated physical properties of the participating components. Within this work, property packages adjusted to the requirements of the single process steps were defined for the high-fidelity models. On the one hand, all the necessary components for the various process steps had to be available. On the other hand, the number of components should be kept low, to keep the simulation simple and therefore not too slow. Some of the physical property packages contain up to about 50 different components. This is necessary to consider all the relevant educts and products of a reaction as well as solvents and components effecting the equilibrium or the reaction conditions like pH value and so on. Especially the many different materials, which are part of the natural raw material wood behave differently throughout the single process steps and must be accounted for individually. Nevertheless, the component list was simplified, wherever possible, to keep the simulation as simple as possible. In any case, the property packages were optimized to reduce the simulation effort, but precisely describe the phase behavior and the heat and mass balance at the same time. This was achieved by using the non-random-two-liquid model (NRTL model). Another challenge beside the large number of components was to get all the necessary parameters in the NRTL model. For well-known components like water these parameters are known. For other components like the components of wood these parameter, not only the pure component parameters but also the binary interaction parameters, must be adapted to the existing measurements.

An accurate simulation also requires a detailed description of the reaction kinetics in dependence of time and conditions (e.g. temperature, pressure or pH value). In the described process, several conversions (enzymatical, chemical, and hydrothermal) need to be considered. All of the reactions are complex and depend on the detailed composition of the raw material, which cannot be assumed to be constant. Accordingly, the detailed mechanism for these very complex reactions is not yet investigated in such a detail, as the input for the simulation requires. Thus, based on lab trials and data from pilot plans, the key influencing parameters for each conversion step was identified. In simplified descriptions, the effect of those parameters on the conversion rate was defined and the respective kinetics was implemented into the simulation.

Beside a reliable physical property package and adequate reaction mechanisms, models for the simulation of the single unit operations are required. For many parts of the plant, the model could be created with the standard elements from the simulation library. The library contains, for example, devices such as pumps, valves, tanks, heat exchangers and distillation columns.

However, there are also devices/plant parts that could not be created directly with devices from the library. For example, a model for a filter press had to be developed in-house. The filter press runs continuously, operating in different steps. Thus, the filling with suspension, dehumidifying by pressing, washing with fresh water, repeated pressing, and eventually further drying with compressed air need to be covered by the developed model. All these steps were implemented with a simplified substitute model, using the available library models like component splitters, pumps, and tanks, linked, controlled, and coordinated with an adjusted logic.

A tailor-made model also had to be implemented in case of the reactors. Although different reactors are available in the library, the kinetics derived for the reactions could not simply be integrated into them. In most cases the reactor was represented by a simple tank. The reaction is then calculated in a spreadsheet, taking the current state of the influencing reaction conditions into account. The actual material conversion and heat of reaction is then implemented by a virtual recirculated material stream.

In total, the entire high fidelity dynamic process model includes about 200 process and instrumentation diagrams (PIDs) with the corresponding devices. The total process model has been divided into around 15 individual models, so that the complete model can also be calculated in real time during the transitions from one state to the other with sufficient accuracy and with sufficiently small simulation steps. These individual models are simulated in parallel. The individual models are connected to each other so that the streams are transmitted from one model to the other at the interfaces.

2.2.2. Mid fidelity process model
Compared to the simulation with the high-fidelity model, the mid-fidelity model uses simplified approaches. A maximum of 20 different components are taken into account in the mid fidelity model, only. In addition, ideal gases or ideal liquids are assumed for the physical properties. Phase changes are not taken into account. These simplified assumptions are sufficient for the plant parts to be modeled like storage tanks, truck unloading und loading stations. All of the used models are part of the available libraries.

2.2.3. Model integration
The individual sub models of the high-fidelity process model are connected to the mid fidelity simulation. The corresponding information for the streams is exchanged at the transitions. The entire plant is thus simulated with the two types of process models. In addition, the process models are linked to the device level (layer two). In the device level the manipulated variable for the actuators in the process model is calculated, such as the valve position or the engine speed. In the same way, the corresponding signals for the control technology are calculated from the measured values of the process model.

3. Benefits of the OTS

The aim of developing the OTS for the described process, was to be able to train the future operators already while the plant is still under construction. Nevertheless, additional benefits were generated also on the way to develop the OTS. Important insights into the process could already be gained in the first phases of the project. On the one hand, the process design was improved for the distillation columns. The column design was based on the heat and mass balance generated form the steady state simulation, fitted to lab experiments and pilot plant data. Since manual adjustments, applied in the steady state simulation, cannot be used for dynamic simulation, a more general and robust approach was chosen. With improved physical properties, the simulation could be better adapted to the existing measurement results. However, it turns out that the required purities could not be achieved in one separation step with the chosen design of the column. Based on this knowledge, the design of the column could be adjusted just before ordering and a potentially expensive redesign during commissioning could be avoided. Normally such a redesign is not done, and the equipment is specified based on the heat and mass balance. In addition, the control concept could be adapted and improved based on the dynamic simulation, especially in the area of the distillation columns. Based on the dynamic behavior of the system, the control concepts could be optimized. Above all, the interactions between several columns in a row could be considered. Furthermore, it was possible to define the necessary number and position of complex online measurement methods to determine the purity.

4. Conclusion

In this work, an example for a successful development of an OTS is given. By the combination of mid- and high-fidelity models, an accurate, fast simulation of a novel, high complex and comprehensive process was realized. The OTS was successfully applied, to train the operators prior to the commissioning.

Furthermore, it was shown, that a digital twin in the form of an operator training system (OTS) can not only be used to train the plant operator. The normal operation of the plant as well as the start-up and shutdown of the plant can be practiced without risk. In addition, a wide variety of malfunctions and emergency situations can be simulated and the optimal behavior in these cases can be practiced again and again. All of this can also be done before the system is actually put into operation.

In addition, however, the dynamic simulation for an operator training system also provides valuable information in relation to system design and control concepts. In this way, knowledge about the later operation of the system can be gained in the early phases and can be incorporated into the construction of the system.

References

D. Labisch, Ch. Leingang, O. Lorenz, M. Oppelt, B-M. Pfeiffer, F. Pohmer, 2019, Evolution eines Digital Twin am Beispiel einer Ethylen-Anlage, atp magazin, 06/07 2019

Schulze, K., 2014, Trainingssimulation in der Prozessindustrie. Atp magazin, 56(01-02),

Cox, R. K., Smith, J. F., und Dimitratos, Y., 2006, Can simulationtechnology enable a paradigm shift in process control?: Modeling for the rest of us. Computers & chemical engineering, 30(10-12), (pp.1542-1552).

NAMUR, 2006, NA 60: Management von Trainingssimulatorprojekten.NAMUR: www.namur.net

Siemens AG, Operator Training System (OTS) – Für effizientere Analgennutzung, https://assets.new.siemens.com/siemens/assets/api/uuid:4a5f99a1-8f40-4614-b468-0ca9d9a2f645/285-operator-training-systems-de.pdf

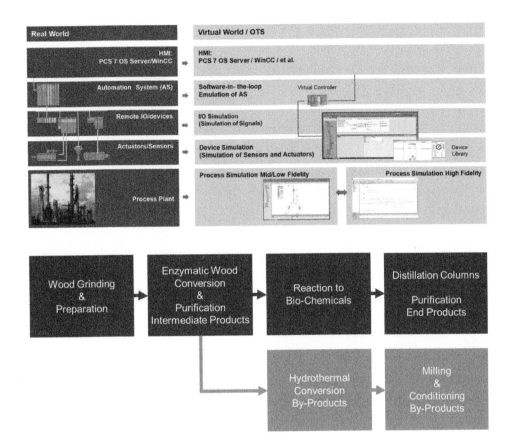

Antonis Kokossis, Michael C. Georgiadis, Efstratios N. Pistikopoulos (Eds.)
PROCEEDINGS OF THE 33rd European Symposium on Computer Aided Process Engineering
(ESCAPE33), June 18-21, 2023, Athens, Greece
© 2023 Elsevier B.V. All rights reserved. http://dx.doi.org/10.1016/B978-0-443-15274-0.50327-9

Process Validation in Modular Plants considering the Scale-Up and Prior Knowledge

J. Mädler[a]*, A. Koch[a], A. Bamberg[b], L. Urbas[a]

[a]*Chair of Process Control Systems & Process Systems Engineering Group,
TU Dresden, Dresden 01062, Germany*
[b]*Merck KGaA, Frankfurter Str. 250, Darmstadt 64293, Germany*
jonathan.maedler@tu-dresden.de

Abstract

Modular plants built from process equipment assemblies (PEAs) provide a promising solution to challenges like increasing requirements of flexibility and faster time to market. To leverage the potential in processes under good manufacturing practice (GMP) conditions new methods are required to support the process validation in modular plants. In this paper, we present an integrated approach to Quality by Design and Process Performance Qualification between lab and production scale in modular plants applying the Product-Process-Resource model by Bamberg et al. (2021). Furthermore, we reformulate the Extended Flexibility Analysis by Ochoa et al. (2021) adapting the terminology to the Quality by Design context and adding the control inputs in the form of the CPPs to the optimization problem. The problem is solved by means of a grid search algorithm. The approach is studied and illustrated in a simulation-based case study with a simple reaction system.

Keywords: Modular Plant, Quality by Design, Design Space, Process Performance Qualification, Scale-Up

1. Introduction

Modular Plants, in accordance with VDI 2776 (VDI, 2020) and VDI/VDE/NAMUR 2658 (VDI/VDE/NAMUR, 2019) consist of Process Equipment Assemblies (PEAs) enabling increased flexibility and faster time to market. These properties make this concept attractive to industry sectors which often operate under the good manufacturing practice (GMP) conditions. Although aspects like PEA selection (Schindel et al., 2021) und orchestration (Klose et al., 2022) have been considered in literature before, process validation and quality by design in modular plants remain rather underdeveloped. However, Mädler et al. (2022) did present a digital twin-concept for smart process equipment assemblies supporting process validation in modular plants. This concept aims to make prior knowledge Findable, Accessible, Interoperable, and Reusable (FAIR) and relates different kinds of information to a set of methods for process validation in modular plants. In this paper we build upon these results and propose a method for Quality by Design (QbD) and Process Performance Qualification (PPQ) in the production scale based on knowledge gained in process development with lab scale PEAs. The remainder of this paper is structured as follows: section 2 provides background information about Process Validation, Quality by Design, and the Product-Process-Resource model. In section 3, a method for the Design Space Identification and Design of Experiments in the

PPQ is introduced. This method is researched in a simulation-based case study with a reactor PEA in section 4. Section 5 provides a conclusion and an outlook.

2. Background

2.1. Process Validation

According to the FDA Guidance (FDA, 2011), Process Validation consists of the (1) Process Design, the (2) Process Qualification (PQ), and the (3) Continued Process Verification stage. In the Process Design stage, the process development takes place which is frequently carried out according to the Quality by Design (QbD) paradigm. Furthermore, the scale-up of the process from lab scale to production scale must be achieved. The Process Qualification stage consists of two parts: the "Design of Facilities & Qualification of Equipment and Utilities" and the Process Performance Qualification (PPQ) (Katz and Campbell, 2012). The PPQ aims to confirm the capability of the process and plant for reproducible production of sufficient quality (Katz and Campbell, 2012). The Continued Process Verification aims to assure that the process is kept under control and to facilitate continued process optimization (Katz and Campbell, 2012).

2.2. Quality by Design and Process Performance Qualification

The Quality by Design paradigm proclaims the application of scientific methodologies and risk management in order to assure product quality. An important sub-concept is the Design Space (DS), which can be interpreted as a mathematical model relating the Critical Process Parameters (CPPs), the Critical Material Attributes (CMAs) of the feed streams, and sufficient values for the Critical Quality Attributes (CQAs) of the product. The Design Space is enclosed by the Knowledge Space (KS), which "contains all the information about all regions of the process that have been investigated" (Boukouvala et al., 2017) in the process development. To assure product quality under disturbances, usually a Normal Operating Range (NOR) is chosen within - but smaller than - the Design Space (cf. Boukouvala et al., 2017). The identification of the Design Space frequently requires extensive experimentation which is often very expensive at the production scale. In such cases, the identification of the Design Space is done using lab scale experiments. Afterwards, the results are scaled to production scale and tested in the PPQ.

2.3. Product-Process-Resource Model

The Product-Process-Resource model of outlines the distribution of knowledge (Bamberg et al., 2021). The Capability Model (CM) is a product-independent description of the PEA capabilities like e.g., cooling capacity. The Transformation Model (TM) is a plant-independent description of the process. Combining these two models yields the Operation Model which can be used to characterize, identify, or predict the Design Space.

3. Methodology

The method presented in this paper is based on a scale-down concept employing behavioral models of PEAs in lab scale and production scale. Experiments for the identification of the Design Space are carried out utilizing a lab scale PEA which mimics the capabilities of the production scale PEA (e.g., Services with related Process Parameters). The gained knowledge is then used to derive an Operation Model for the production scale which is applied to estimate the Design Space in the production scale PEA and plan experiments for the PPQ.

3.1. A behavioral model for the Knowledge Space

In this paper, we consider a system of nonlinear algebraic equations (NAEs) for behavioral models of Knowledge Space

$$\mathbf{0} = \mathbf{h}\left(\mathbf{x}, \mathbf{u}_{\text{CMA}}, \mathbf{u}_{\text{CPP}}, \boldsymbol{\theta}_{\text{TM}}, \boldsymbol{\theta}_{\text{CM}}\right) \tag{1}$$

where $\mathbf{x}$ denotes the states of the system, $\mathbf{u}_{\text{CMA}}$ the independent inputs related to the CMAs, $\mathbf{u}_{\text{CPP}}$ the independent inputs related to the CPPs, $\boldsymbol{\theta}_{\text{TM}}$ the parameter values related to the Transformation Model of the material system, and $\boldsymbol{\theta}_{\text{CM}}$ the parameters related to the Capability Model of a PEA in lab scale or in production scale.

3.2. Identification of a Knowledge Space in Lab Scale

In the lab scale, a Knowledge Space can be identified applying prior knowledge from Product Development and the Capability Model of the lab scale PEA. This Capability Model provides one or multiple potential behavioral models according to equation (1). Furthermore, it does provide the parameters $\boldsymbol{\theta}_{\text{CM}}$ including their values which have been captured during a separate characterization of the PEA and/or through other previous experimental data. In addition, it provides information about $\mathbf{u}_{\text{CPP}}$ of the PEA, which relate to the services of the PEA and their service parameters. Furthermore, the characteristics of the educts ($\mathbf{u}_{\text{CMA}}$) are assumed to be controllable during lab experiments. The Knowledge Space is identified by applying statistical DoE to obtain initial parameter estimates and afterwards model-based design of experiment (MBDoE) for parameter estimation and for model discrimination. The results of these experiments are used to identify the parameter values $\boldsymbol{\theta}_{\text{TM}}$ including uncertainty estimates, applying e.g., the maximum likelihood method and confidence intervals.

3.3. Identification of the Design Space and Process Performance Qualification in the Production Scale

The Transformation Model and its parameters $\boldsymbol{\theta}_{\text{TM}}$ are extracted from the Operation Model of the lab scale PEA and combined with the Capability Model of the characterized production scale PEA. The resulting Operation Model is used to identify the Design Space in the production scale and optimally design experiments for the PPQ.

For the identification of the Design Space from a Knowledge Space described based on a behavioral model according to equation (1), the Extended Flexibility Analysis approach by Ochoa et al. (2021) is adopted. In contrast to this publication, we consider the control inputs in our approach, which relate to the Critical Process Parameters $\mathbf{u}_{\text{CPP}}$ (or z in Ochoa et al. (2021)). In the Flexibility Analysis, the permitted ranges of the CQAs are described by the feasibility functions

$$g_j\left(\mathbf{x}, \mathbf{u}_{\text{CMA}}, \mathbf{u}_{\text{CPP}}, \boldsymbol{\theta}_{\text{TM}}, \boldsymbol{\theta}_{\text{CM}}\right) \leq 0 \tag{2}$$

These functions are combined with the behavioral model from equation (1) and constraints to the independent inputs and parameters. The following formulation the flexibility index can be calculated

$$F = \max_{\delta_{\text{CMA}} \in \mathbb{R}^+} \delta_{\text{CMA}}$$

$$\text{s.t.} \quad \chi(\boldsymbol{\theta}_{\text{CM}}) = \max_{\mathbf{u}_{\text{CMA}} \in T_{\text{CMA}}} \min_{\mathbf{u}_{\text{CPP}} \in T_{\text{CPP}}} \max_{\boldsymbol{\theta}_{\text{TM}} \in T_{\text{TM}}} \max_{j \in J} g_j\left(\mathbf{x}, \mathbf{u}_{\text{CMA}}, \mathbf{u}_{\text{CPP}}, \boldsymbol{\theta}_{\text{TM}}, \boldsymbol{\theta}_{\text{CM}}\right) \leq 0 \tag{3}$$

$$\mathbf{h}\left(\mathbf{x}, \mathbf{u}_{\text{CMA}}, \mathbf{u}_{\text{CPP}}, \boldsymbol{\theta}_{\text{TM}}, \boldsymbol{\theta}_{\text{CM}}\right) = \mathbf{0}$$

$$T_{\text{CMA}}\left(\mathbf{u}_{\text{CMA}}\right) = \left\{\mathbf{u}_{\text{CMA}} : \mathbf{u}_{\text{CMA}}^{\text{N}} - \delta_{\text{CMA}} \Delta \mathbf{u}_{\text{CMA}}^- \leq \mathbf{u}_{\text{CMA}} \leq \mathbf{u}_{\text{CMA}}^{\text{N}} + \delta_{\text{CMA}} \Delta \mathbf{u}_{\text{CMA}}^+\right\}$$

$$T_{\text{CPP}}\left(\mathbf{u}_{\text{CPP}}\right) = \left\{\mathbf{u}_{\text{CPP}} : \mathbf{u}_{\text{CPP}}^{\text{LB}} \leq \mathbf{u}_{\text{CPP}} \leq \mathbf{u}_{\text{CPP}}^{\text{UB}}\right\}$$

$$T_{\text{TM}}\left(\boldsymbol{\theta}_{\text{TM}}\right) = \left\{\boldsymbol{\theta}_{\text{TM}} : \boldsymbol{\theta}_{\text{TM}}^{\text{LB}} \leq \boldsymbol{\theta}_{\text{TM}} \leq \boldsymbol{\theta}_{\text{TM}}^{\text{UB}}\right\}$$

where δ_{CMA} denotes the scalar flexibility index of a CMA and the superscripts LB and UB denote lower and upper bounds respectively. The optimization problem reads as follows: the maximum among the feasibility functions g_j is maximized based on the range of the uncertain transformation variables $\boldsymbol{\theta}_{\text{TM}}$, which have been identified within the lab scale. Hence, the multistage optimization problem generates the worst-case scenario in terms of the CQA values. The CPPs are then used to derive minimized values for the feasibility functions in the worst-case. Afterwards, the feasibility function values are again maximized based on the allowed range of CMA values. If the feasibility function values remain below 0, the range of CMAs is feasible, and the flexibility test is passed. Finally, the range of CMAs is maximized by maximizing the flexibility index δ_{CMA}. This multistage optimization problem is known to be very computationally expensive (cf. Ochoa and Grossmann, 2020). Thus, Ochoa and Grossmann (2020) list two reformulations for special cases like e.g., the vertex enumeration approach for cases with 1-D quasi-convex feasibility functions, and two general reformulations.

For this publication, we use a similar algorithm to the vertex enumeration approach according to Ochoa and Grossmann (2020), but instead of only considering the vertices, a grid of points is used which is spanned by the CMAs and Transformation Model parameters. This is done to get an approximation to the global solution in cases of non-convex feasibility functions. This approach is computationally very expensive, but easy to implement and therefore suits the demonstration of the approach.

The result of this Extended Flexibility Analysis is a hyperrectangle formed by the CMAs. In PPQ, the vertexes of this hyperrectangle with some safety margin should be tested in combination with the corresponding CPPs to span a Normal Operating Range. Furthermore, the Operational Model can be used to search for an optimal operation point.

4. Case Study

4.1. Example case

Let us assume that an owner/operator aims to scale-up a synthesis of a product B in an exothermic, homogeneous second order reaction from educt A, but B degrades in another reaction to C:

$$2\,A \rightarrow B \rightarrow C$$

Two stirred tank reactor (STR) PEAs (one in lab scale and one in production scale) are available to the owner/operator. Both PEAs can be operated in a continuous STR (CSTR) operation mode, but their minimal residence times and cooling capacities may differ. The behavioral models under sufficient homogenization can be described as follows

$$0 = \frac{1}{\tau}\left(c_{i,in} - c_i\right) + \sum_{j=1}^{NR}\left[\nu_{i,j}\,k_{0,j}\exp\left(-\frac{E_{A,j}}{RT}\right)\prod_{i=1}^{NC}c_i^{\gamma_{i,j}}\right] \tag{4}$$

where the subscripts i and j denote the species and the reactions with the number of components NC and the number of reactions NR respectively, c are the molar concentrations, ν the stoichiometric coefficients, $k_{0,j}$ the Arrhenius constants, $E_{A,j}$ the activation energies, R the universal gas constant, , and γ the reaction orders regarding a species. Assuming a dosing and tempering service of ideal behavior, the residence time τ and the temperature T can usually be classified as CPPs. The concentration of species A in the feed $c_{A,in}$ is a CMA. The owner operator aims to achieve the following CQAs:

$$\text{Profit} = \frac{100\,c_B - 25\,c_{A,in}}{\tau} \geq 2 \tag{5}$$

$$\text{Purity} = \frac{c_B}{c_A + c_B + c_C} \geq 0.8 \tag{6}$$

4.2. Identification of a Knowledge Space in Lab Scale

For the sake of simplicity, we assume that the Transformation Model parameter values including uncertainties in Table 1 have been identified in lab scale applying MBDoE.

Table 1: Assumed parameter values including uncertainty measures identified in lab scale

Parameter	Unit	Nominal value	Minimum	Maximum
$k_{0,1}$	L/(mol×min)	3.2e+06	3.19e+06	3.21e+06
$k_{0,2}$	1/min	5.2e+06	5.19e+06	5.21e+06
$E_{A,1}$	J/mol	4.0e+04	3.99e+04	4.01e+04
$E_{A,2}$	J/mol	5.5e+04	5.49e+04	5.51e+04

4.3. Identification of the Design Space in Production Scale

Figure 1 visualizes the CQAs (Profit, Purity) in dependence of the CMA ($c_{A,in}$) and the CPPs (τ,T) for the nominal parameter values θ_{TM} according to Table 1. Applying the Flexibility Analysis formulation in equation (3), the relation for the flexibility index according to Figure 2 is calculated. A conservative estimate of the flexibility index can be derived to be 0.175 assuming a Δu_{CMA} value of 1 mol/L for $c_{A,in}$. This can also be seen in the Figure 2 on the right side, where it can be recognized that the Profit and Purity values barely reach the sufficient level for combination of CPPs ($\tau = 44$ min, $T = 300$ K).

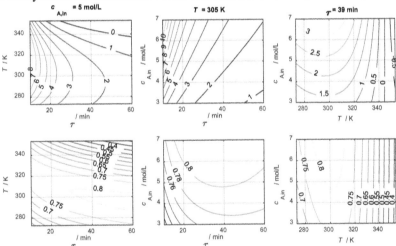

Figure 1: Profit (top row) and Purity (bottom row) in dependence of the CMA (molar concentration of species A in the feed $c_{A,in}$) and the CPPs (residence time τ, temperature T)

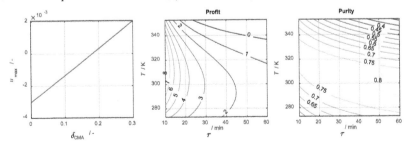

Figure 2: Maximum value of the feasibility functions u_{max} in dependence of the flexibility index δ_{CMA} (left); and the corresponding the Profit (central) and Purity (right) at $c_{A,in} = 4.825$ mol/L

5. Conclusions

In this paper, we introduced a methodology for Quality by Design and Process Performance Qualification in modular plants. This approach is based on similar PEAs in lab scale and production scale, which are integrated with Capability Models including behavioral models and Capability Model parameters. To avoid expensive experimentation in the production scale, experiments are carried out in the lab scale PEA initially based on statistical design of experiments and afterwards based on MBDoE. The identified Transformation Model including the Transformation Model parameter values with uncertainty measures are then integrated with the Capability Model of the production scale PEA. The resulting Operation Model is used to identify the Design Space and plan experiments for the PPQ. Furthermore, we suggested a reformulation of the Extended Flexibility Analysis problem proposed by Ochoa et al. (2021) by adding the CPPs (or control inputs) to the optimization problem. This allows the consideration of control actions based on measured deviation of the CMAs in the operation phase.

In future work, the approach should be applied to more complex and real-world problems. In addition, the design of experiments, experimentation and identification in lab scale should be further elaborated. Furthermore, the solution algorithm for the Extended Flexibility Analysis should be reworked adopting the general formulations by Ochoa and Grossmann (2020), or surrogate-based formulations.

References

Bamberg, A., Urbas, L., Bröcker, S., Bortz, M., Kockmann, N., 2021. The Digital Twin – Your Ingenious Companion for Process Engineering and Smart Production. Chem Eng Technol 44, 954–961. https://doi.org/10.1002/ceat.202000562

Boukouvala, F., Muzzio, F.J., Ierapetritou, M.G., 2017. Methods and Tools for Design Space Identification in Pharmaceutical Development, in: Comprehensive Quality by Design for Pharmaceutical Product Development and Manufacture. pp. 95–123. https://doi.org/10.1002/9781119356189.ch6

Katz, P., Campbell, C., 2012. FDA 2011 process validation guidance: Process validation revisited. Journal of GXP Compliance 16, 18.

Klose, A., Lorenz, J., Bittorf, L., Stark, K., Hoernicke, M., Stutz, A., Weinhold, H., Krink, N., Welscher, W., Eckert, M., Unland, S., Menschner, A., Da Silva Santos, P., Kockmann, N., Urbas, L., 2022. Orchestration of modular plants: Procedure and application for orchestration engineering. atp 63, 68–77. https://doi.org/10.17560/atp.v63i9.2599

Mädler, J., Rahm, J., Viedt, I., Urbas, L., 2022. A digital twin-concept for smart process equipment assemblies supporting process validation in modular plants, in: Montastruc, L., Negny, S. (Eds.), Computer Aided Chemical Engineering. Elsevier, pp. 1435–1440. https://doi.org/10.1016/B978-0-323-95879-0.50240-X

Ochoa, M.P., García-Muñoz, S., Stamatis, S., Grossmann, I.E., 2021. Novel flexibility index formulations for the selection of the operating range within a design space. Computers & Chemical Engineering 149, 107284. https://doi.org/10.1016/j.compchemeng.2021.107284

Ochoa, M.P., Grossmann, I.E., 2020. Novel MINLP formulations for flexibility analysis for measured and unmeasured uncertain parameters. Computers & Chemical Engineering 135, 106727. https://doi.org/10.1016/j.compchemeng.2020.106727

Schindel, Polyakova, Harding, Weinhold, Stenger, Grünewald, Bramsiepe, 2021. General approach for technology and Process Equipment Assembly (PEA) selection in process design. CEP:PI 159, 108223. https://doi.org/10.1016/j.cep.2020.108223

VDI, 2020. Modulare Anlagen - Grundlagen und Planung modularer Anlagen - Blatt 1 (VDI 2776:2020-11).

VDI/VDE/NAMUR, 2019. Automatisierungstechnisches Engineering modularer Anlagen in der Prozessindustrie - Allgemeines Konzept und Schnittstellen - Blatt 1 (VDI/VDE/NAMUR 2658-1:2019-10).

Antonis Kokossis, Michael C. Georgiadis, Efstratios N. Pistikopoulos (Eds.)
PROCEEDINGS OF THE 33rd European Symposium on Computer Aided Process Engineering
(ESCAPE33), June 18-21, 2023, Athens, Greece

Digital Twins for Scale-Up in Modular Plants: Requirements, Concept, and Roadmap

A. Koch[a*], J. Mädler[a], A. Bamberg[b], L. Urbas[a]

[a]*Chair of Process Control Systems & Process Systems Engineering Group, TU Dresden, Dresden 01062, Germany*;
[b]*Merck KGaA, Frankfurter Str. 250, Darmstadt 64293, Germany,*

Abstract

The application of Digital Twins for scale-up in Modular Plants is envisioned to significantly reduce the necessary time-to-process for new-to-market products and address scale-up challenges. To identify application specific requirements of a Digital Twin for scale-up in Modular Plants, expert interviews were conducted. The spectrum of identified requirements as well as plant life cycle aspects are incorporated into a workflow concept which illustrates the partial models of Digital Twins, as well as the interfaces for the exchange of information. Also presented is a roadmap illustrating the cluster-specific development pathways for Digital Twins for scale-up in Modular Plants.

Keywords: Digital Twin, Modular Plants, process development, scale-up

1. Introduction

In order to maintain a competitive edge within dynamic markets, such as pharmaceuticals and specialty chemicals, Modular Plants (MP) must address the following requirements: (1) reduce the time-to-market, (2) provide increased flexibility, and (3) have increased efficiency (Bernshausen et al., 2019). An additional hurdle is the scale-up from laboratory to production. The integration of Digital Twins (DT) into process development workflows could potentially significantly speed up the equipment selection of a manufacturing process. For this purpose, the combination of standardized, predesigned equipment such as Process Equipment Assemblies (PEA) with DTs is poised to be an integral tool to facilitate rapid time-to-process (Mädler et al., 2022a; Mädler et al., 2022b).

2. Background

2.1. Digital Twins

According to Rosen et al. (2019), a Digital Twin is a semantically linked collection of all digital artefacts, including design and engineering data as well as operational data and behavior descriptions. Moreover, a Digital Twin is intended for a specific purpose (Rosen et al, 2019). Bamberg et al. (2021) view the Digital Twin as consisting of different partial models which can be structured in the **Product-Process-Resource-Model** (Bamberg et al., 2021). The model of Bamberg et al. (2021) was applied to Modular Plants by Mädler et al. (2022b) to illustrate how the knowledge distribution in a Digital Twin is divided between the PEA manufacturers and the owners/operators (O/O). The **capability model** is a product independent description of a PEA containing structural, operational, and behavioral information about the PEA, independent of a concrete process. The

transformation model is a plant independent and product specific description of the process used to derive information for a concrete PEA such as residence times. Combining the capability model and the transformation model yields the **operation model,** which is used for recipes and optimization of process parameters (Mädler et al., 2022b).

2.2. Modular Plants

The structure of Modular Plants is outlined in the VDI 2776 Part 1. Modular Plants (MPs) consist of functional equipment assemblies (FEAs), which are combined into process equipment assemblies (PEAs) (VDI, 2020). PEAs can be further characterized as intelligent modules which are computerized and (static) self-describing (Mädler et al., 2022a). Automation in Modular Plants is described in VDI/VDE/NAMUR 2658 (VDI/VDE/NAMUR, 2019). The operation of multiple PEAs in a modular plant is carried out via the process orchestration layer (POL). Here, the Module Type Package (MTP) serves to integrate PEA into the POL, yielding a standardized and non-proprietary means to describe the automation in a PEA (Klose et al., 2019).

2.3. Process Design in Modular Plants

Applying modular-based plant design methodology results in decreased planning and construction times by incorporating predefined modules and devoting time to the selection process instead of to individual design of each piece of equipment (Bramsiepe et al., 2012). Methodologies for matching PEAs in process design can be found in literature. One example is Schindel et al. (2021), who utilize Matching Matrices to match the technical attributes of a PEA with the process requirements (Schindel et al., 2021).

3. Methodology: Expert Interviews

In order to delineate the task-specific requirements of a Digital Twin for Modular Plants for process development and scale-up, semi structured interview questionnaires based on a set of working hypotheses were conducted with experts from industry and academia with experience ranging from 3 to 45 years (cf. Koch, 2022). To elicit requirements, the interview transcripts were then analyzed using the coding agenda methodology of Mayring (Mayring, 2021). The coded requirements were grouped into five clusters, which were adapted from Oppelt (2016). These clusters, shown in Figure 1, are defined as follows:

(1) **Acceptance**: demonstrate DT and gain acceptance by Owners/Operators (O/O)
(2) **Efficiency:** address bottlenecks in process development workflows with DTs
(3) **Integration:** exchange of information using interfaces
(4) **Reusability:** reusability of models and information for new products
(5) **Usability:** accessibility and ease of use of DT by O/O

To illustrate how these requirements are applied to the DT workflow concept and roadmap, four selected requirements from the integration cluster are discussed: interfaces, semantically linked models, CAE System, and DEXPI. These requirements are well suited to describe the DT workflow in Section 5. A detailed explanation of all requirements can be found in Koch (2022). The requirement of a CAE System serves as a tool to store the digital artefacts including design and engineering data (cf. Rosen et al. (2019)), which are part of the essential information contained in a DT. The requirement of interfaces refers to the interfaces, which are needed as part of a DT in order to transfer information between engineering tools, such as a simulation tool and a CAE system. Interface also refers to interfaces between a simulation tool and operating data contained in a POL. One specific identified bottleneck in the process development workflow is the difficulty in transferring information between engineering tools.

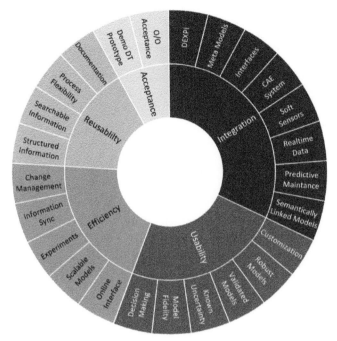

Figure 1: Clustered, expert identified requirements of Digital Twin for scale-up (Koch, 2022).

The integration of interfaces to support the obligatory semantic linkage between CAE tools could address this specific bottleneck, resulting in a DT concept that is in line with Rosen et al. (2019). The last requirement explained here, is DEXPI (Wiedau et al., 2019), which refers to a specific, openly available information model which could be used to transfer data between engineering tools and should be integrated into a DT concept.

4. RoadMap: Vision for Digital Twins in Scale-Up

4.1. Background and Prerequisites

In order to create a DT concept, the requirement catalog (see Figure 1) was workshopped and refined with software experts from industry. The final requirements catalog was used to create a Roadmap, shown in Figure 2, which outlines the necessary steps to enable this Digital Twin Vision for scale-up in Modular Plants. As a starting point for this specific Digital Twin, a minimum baseline was defined with two prerequisites: (1) MTP is integrated into both the lab and production PEAs and (2) the lab and production PEA setup makes both scale-up and scale-down possible.

4.2. Explanation

The roadmap depicts five different pathway lines with a series of different symbols. Each pathway line aligns with a requirement cluster explained in Section 3, and each symbol refers to a different requirement identified from the expert interviews shown in Figure 1. Additionally, each requirement was allocated to a specific tool or interface, which are categorized as follows: simulation tool (circle), plant engineering tool (triangle), interface (star), both simulation tool and plant engineering tool (diamond), and undefined/other (square). Moreover, each requirement is allocated to a specific time frame represented by the background color: short term (dark grey), medium term (medium grey), and long term (light grey).

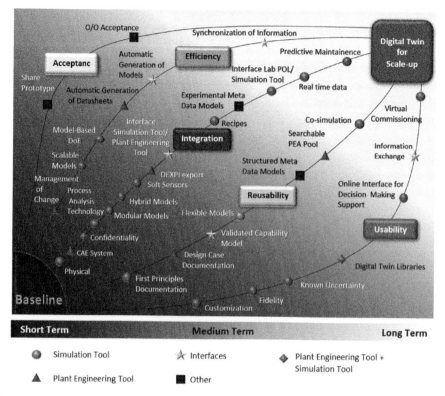

Figure 2: Roadmap for a Digital Twin for Scale-up using Modular Plants. Adapted from Koch (2022).

The requirements are ordered along each cluster pathway and allocated to a time frame based on both the importance to experts as well as the feasibility of the required technology needed to implement a requirement. Addressing each of these requirements yields the vision of a Digital Twin for scale-up potentially facilitating faster time-to-process.

5. Digital Twin Concept

5.1. Workflow Concept

Short- and medium-term requirements illustrated in the roadmap in Figure 2 are addressed with a concept workflow for a Digital Twin for scale-up, which is shown in Figure 3. This presented DT concept workflow assumes that information such as design and engineering data is contained in a Plant Engineering Tool, behavior descriptions in a simulation tool, operational data from the production PEA in the Production POL, and experimental data in the Lab POL. The partial models within the simulation tool are aligned with Bamberg et al. (2021). Here, the operation model for the Lab PEA consists of a capability model, representing the characterized capability of the Lab PEA resource and a transformation model for a specific product. This transformation model is shared with the production PEA and also contains relevant information from product development. This product specific transformation model is combined with the capability model from the production PEA into an operation model. Operational data from the Production PEA is used to improve the Operation Model for the Production PEA, illustrating the semantic linkage between models. The result of this workflow is an in-spec product for the supply chain.

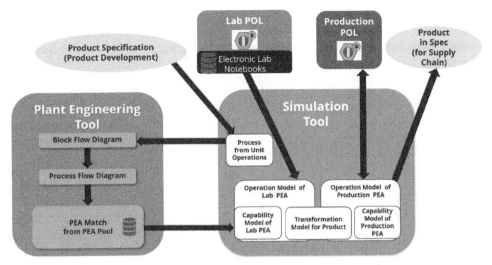

Figure 3: Workflow for a Digital Twin for Scale-up. Adapted from Koch (2022).

5.2. Requirements of an Interface to Support PEA Matching

To demonstrate how the expert requirements were integrated into the workflow shown in Figure 2, the interface between the simulation tool and plant engineering tool is explained. This Digital Twin concept centers around having a repository of resource models, or PEA Pool, stored in a searchable databank of a plant engineering tool containing design and engineering data. This addresses the requirement of structured and searchable information as well as CAE system. Here, the PEA matching methodology of Schindel et al. (2021) is referenced to identify the requirements of an interface that supports matching process models (from a simulation with unit operations) to resource models in a PEA Pool. This workflow begins with a resource independent simulation of a process for a new product based on a simulation created from unit operations (reaction, distillation, etc.). Schindel et al. (2021) assume that the information about a process provided by a simulation for matching PEAs to processes is low. Moreover, technical criteria (e.g., temperature, pressure, and viscosity) are identified as the most import information for successful matching. These technical criteria are found to be unit operation and material flow specific; thus, important technical criteria for the selection of a heat exchanger are not the same as the criteria for the selection of a distillation column. Hence, for a match to be successful it is essential that the simulation tool propagates these necessary technical criteria through an interface to a plant engineering tool. The exchange of this information across this interface could be supported by an information model such as DEXPI (Wiedau et al., 2019). A detailed explanation of this workflow concept, interfaces, and partial models as well as how the requirements were used to create this concept can be found in Koch (2022).

6. Discussion and Conclusion

In this paper, expert interview derived requirements of Digital Twins for scale-up in Modular Plants are presented and discussed. The expert interviews yielded purpose specific requirements that must be integrated into a Digital Twin for scale-up, thereby supporting process development and potentially facilitating faster time-to-process. Considering these requirements and challenges along with plant life cylce aspects, a concept DT workflow was presented. This DT concept was used to illustrate the linkage of the capability models and transformation models of a Digital Twin as well as the integration of information containted in the plant engineering tool, Lab POL, and Production POLs. This semantic linkage enables efficient engineering workflows, resulting in faster time-to- process and ultimately model reuse when PEAs are integrated into new processes. These interfaces support PEA matching, which is critical in order to support faster time-to-process. Moreover, the presented roadmap illustrates clear pathways to enable this vision of a Digital Twin and the necessary steps which must be taken to achieve this goal. Using this DT concept and roadmap, the next step is to develop prototypical implementations demonstrating the added benefit of a Digital Twin, resulting in owner/operator acceptance of a Digital Twin and integration into existing engineering workflows within process development. Moreover, prototypes should address the uncertainty and challenges in scale-up, specifically focusing on the linkage between scalable simulation models and information models. Here, alliance with existing standards is necessary to support the exchange of essential and potentially proprietary information between the partial models of the Digital Twin.

References

A. Bamberg, L. Urbas, S. Bröcker, M. Bortz, and N. Kockmann, 2021, The Digital Twin–Your Ingenious Companion for Process Engineering and Smart Production, Chem. Eng. Technol., 44, 6, 954–961.

J. Bernshausen, A. Haller, H. Bloch, M. Hoernicke, S. Hensel, A. Menschner, M. Stutz, M. Maurmaier, T. Holm, C. Schäfer, L. Urbas, U. Christmann, C. Fleischer-Trebes, & F. Stenger, 2019. Plug & Produce auf dem Sprung in den Markt: Neuerungen in Spezifikation und Implementierung des MTP. Atp Magazin, 61(1–2), 56–69.

B. Bramsiepe, S. Sievers, T. Seifert, G. D. Stefanidis, D.G. Vlachos, H. Schnitzer, B. Muster, C. Brunner, J.P.M. Sanders, M.E. Bruins, G. Schembecker, G., 2012. Low-cost small scale processing technologies for production applications in various environments—Mass produced factories. Chem. Eng. Process, 51, 32-52.

A. Klose, S. Merkelbach, A. Menschner, S. Hensel, S. Heinze, L. Urbas, C. Schäfer, S. Szmais, M. Eckert, T. Ruede, T. Scherwietes, P. Santos, F. Stenger, T. Holm, W. Welscher, N. Krink, T. Schenk, A. Stutz, M. Maurmaier, & F. Apitz, 2019. Anforderungen an die Orchestrierung für modulare Prozesseinheiten, Chem. Eng. Technol., 42, 11, 2282-2291.

A. Koch, 2022. Digital Twins for Lab2Plant in Modular Process Plants. Diploma Thesis. Unpublished

J. Mädler, J. Rahm, I. Viedt, and L. Urbas, 2022a, A digital twin-concept for smart process equipment assemblies supporting process validation in Modular Plants, In Computer Aided Chemical Engineering, 51, 1435-1440, Elsevier.

J. Mädler, I. Viedt, J. Lorenz, and L. Urbas, 2022b, Requirements to a digital twin-centered concept for smart manufacturing in Modular Plants considering distributed knowledge. In Y. Yamashita & M. Kano (Hrsg.), Comput. Aided Chem. Eng. 49, 1507–1512.

P. Mayring, P. (2021). Qualitative content analysis: A step-by-step guide. SAGE Publications.

M. Oppelt, 2016. Towards an integrated use of simulation within the life-cycle of a process plant.

R. Rosen, J. Fischer, & S. Boschert, 2019, Next generation digital twin: An ecosystem for mechatronic systems?, IFAC-Paper, 52, 15, 265-270.

A.-L. Schindel, M. Polyakova, D. Harding, H. Weinhold, F. Stenger, M. Grünewald, and C. Bramsiepe, 2021, General approach for technology and Process Equipment Assembly (PEA) selection in process design, CEP:PI, 159, 108223.

VDI/VDE/NAMUR, 2019, VDI/VDE/NAMUR 2658-Automation engineering of modular systems in the process industry-General concept and interfaces – Part 1, Beuth Verlag GmbH.

VDI 2776-1:2020-11, 2020. Modular Plants - Fundamentals and planning Modular Plants.

M. Wiedau, L. von Wedel, H. Temmen, R. Welke and N. Papkonstantinous, 2019, ENPRO Data Integration: Extending DEXPI Towards the Asset Lifecycle, Chem. Eng. Tech., 91, 249-255.

Antonis Kokossis, Michael C. Georgiadis, Efstratios N. Pistikopoulos (Eds.)
PROCEEDINGS OF THE 33rd European Symposium on Computer Aided Process Engineering
(ESCAPE33), June 18-21, 2023, Athens, Greece

Modeling and Simulation of a PSA Process for CO_2/Syngas Separation

Magno Fonseca Santos,[a] Stefano Ferrari Interlenghi,[b] Antonio Esio Bresciani,[a]

Newton Libanio Ferreira,[a] Gabriel Soares Bassani,[c] Rita Maria de Brito Alves[a]

[a]*Universidade de São Paulo, Av. Prof. Luciano Gualberto, n. 380, trav. 3, São Paulo 05508-010, Brazil*
[b]*Unidade SENAI CETIQT Parque, Rua Fernando de Souza, n. 120, Rio de Janeiro 21941-857, Brazil*
[c]*Repsol Sinopec Brazil, Praia de Botafogo, 300, Rio de Janeiro 22250-040, Brazil*

Abstract

This work aims to model and simulate a pressure swing adsorption (PSA) unit for CO_2/syngas separation using rigorous multi-bed and single-bed approaches. The PSA unit was designed to purify syngas from reverse water-gas shift (RWGS) reaction to application in the Fischer-Tropsch synthesis. Process simulations were performed in Aspen Adsorption® v10 software, using the Peng-Robinson equation of state. The association of 7 adsorption beds was considered, allowing the continuous production of purified streams, in a concept of cyclic steady-state operation. The single-bed approach reduces the system complexity and improves the processing power/time without losing precision. A sensitivity analysis showed that integrated RWGS and PSA processes can produce syngas with purity above 94.65% with an H_2/CO molar ratio of 1.50. The total recovery was around 99.98% and 99.57% for H_2 and CO, respectively, for an adsorption pressure of 20.00 bar, and a desorption pressure of 1.50 bar.

Keywords: Pressure swing adsorption, Syngas purification, Adsorption of CO_2, Process simulation.

1. Introduction

Syngas is a mixture composed mainly of carbon monoxide (CO) and hydrogen (H_2), whose importance as a building block for oxygenates production and Fischer-Tropsch (FT) synthesis to hydrocarbons (Theampetch et al., 2021) has increased. Syngas composition depends on its production technology, which includes methane reforming processes and reverse water-gas shift (RWGS) reaction (Dasireddy et al., 2021), which converts CO_2 and H_2 into CO. For the syngas to achieve the specification of the desired derivative, for instance, FT synthesis, major impurities such as CO_2 must be removed from the gas stream. Methods such as membranes, absorption, and adsorption have been used to separate CO_2 from syngas. Among these, adsorption is a well-established purification technology due to its efficiency, operational flexibility, low costs, and absence of by-products when combined with an effective regeneration process (Wibowo et al., 2021). In parallel adsorbent beds, alternated adsorption/desorption steps are used to maintain continuous feeding and high productivity. Adsorption performances depend on the operation parameters, such as gas type, mixture properties, temperature, and pressure (Kacem et al., 2015). The CO_2/syngas separation can potentially be carried out, at moderate temperatures and pressures, using adsorbent materials, such as zeolites and activated carbon (Wilson et al., 2020), which affects the unit process design.

These processes may be categorized according to the desorption technique employed, such as temperature swing adsorption (TSA), vacuum swing adsorption (VSA), pressure swing adsorption (PSA), and combined options. Only a few applications of TSA to CO_2 capture have been found and large-scale VSA is highly electricity-demanding, which seriously impacts the electrical system. PSA is widely used in gas separation applications (Golmakani et al., 2017). This work aims to model and simulate a PSA unit for CO_2/syngas separation using rigorous multi-bed and single-bed approaches. The PSA unit was designed to purify syngas from the RWGS reaction to application in the FT synthesis. The novelty of the proposal lies in the application of PSA technology with a new arrangement of adsorption beds and operational stages to separate large amounts of impurities (~64% wt of CO_2).

2. Methods and Modeling

PSA is a separation process whereby the gaseous components with weak physical interaction with the considered fixed phase, a solid material of high porosity, are separated from other substances. Highly interactive substances are adsorbed at higher pressures, and, after an adequate interval, the adsorption bed is intermittently depressurized to recover the product and regenerate the fixed phase at lower pressures (Durán et al., 2022). In this proposal, CO_2 must be adsorbed, allowing syngas to pass through the adsorbent bed. The desorbed gas mixture, rich in CO_2, is recycled to the RWGS reactor previously evaluated by Santos *et al.* (2022), considering a purge of 1% (Figure 1).

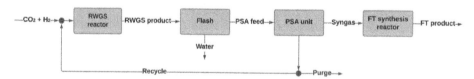

Figure 1. Process Flow Diagram of the general topology of the process.

2.1. Model implementation and parameters

Aspen Adsorption® v10 software was used to simulate the PSA unit. The Peng-Robinson equation of state was used for simulation. The extended Langmuir adsorption isotherms (Equation 1) and adsorbent parameters by Liu et al. (2020), and linear driving force adsorption kinetics (Equation 2) by Xiao et al. (2016) were used to predict the adsorption equilibrium behavior on a packed bed of activated carbon as the fixed phase. The same H_2 adsorption isotherms were used for H_2O. This assumption is justified by the fact that H_2O does not significantly affect the CO_2/syngas separation, similar to H_2, due to the hydrophobic character of activated carbon (Kumita et al., 1994).

$$n_i^* = \frac{IP_{1,i}e^{\frac{IP_{2,i}}{T}}p_i}{1 + (\sum_{i=1}^{N} IP_{3,i}e^{\frac{IP_{4,i}}{T}}p_i)}, i = 1, ..., N \tag{1}$$

$$\frac{\partial n_i}{\partial t} = k_i(n_i^* - n_i), i = 1, ..., N \tag{2}$$

where n_i^* is the equilibrium adsorption amount of species i (mol/kg); $IP_{k,i}$ are parameter k (k=1,4) of the Extended Langmuir model of species i, (mol/kg bar), (K), (1/bar), (K); n_i is the dynamic adsorption amount coefficient of species i (mol/kg), k_i is the mass transfer coefficient of species i (1/s).

2.2. PSA unit simulation

The association of 7 adsorption beds was considered for the PSA unit, allowing the continuous production of purified streams, in a concept of cyclic steady-state operation. The operational phases considered were: 1- Adsorption (A): Retention of higher adsorbent affinity species; 2- Equalizations (depressurizations) (EnD): Sequential bed depressurization to recovery product; 3- Proving purge (PP): Product utilization to clean adsorption bed interstitial spaces; 4- Blowdown (B): Depressurization for regeneration; 5- Purge (P): Bed interstitial clean-up for regeneration; 6- Equalizations (pressurizations) (EnP): Product recovery through sequential bed pressurization; and 7- Final pressurization (FP) with the clean product. Figure 2 brings the time-dependent diagram, showing the integration of each adsorption bed during a complete operating cycle.

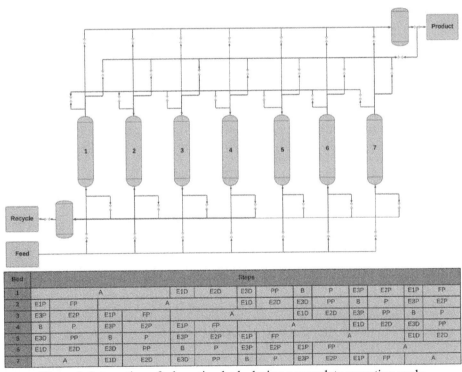

Bed	Steps															
1	A	A	A	A	A	A	E1D	E2D	E3D	PP	B	P	E3P	E2P	E1P	FP
2	E1P	FP	A	A	A	A	A	A	E1D	E2D	E3D	PP	B	P	E3P	E2P
3	E3P	E2P	E1P	FP	A	A	A	A	A	A	E1D	E2D	E3P	PP	B	P
4	B	P	E3P	E2P	E1P	FP	A	A	A	A	A	A	E1D	E2D	E3D	PP
5	E3D	PP	B	P	E3P	E2P	E1P	FP	A	A	A	A	A	A	E1D	E2D
6	E1D	E2D	E3D	PP	B	P	E3P	E2P	E1P	FP	A	A	A	A	A	A
7	A	A	E1D	E2D	E3D	PP	B	P	E3P	E2P	E1P	FP	A	A	A	A

Figure 2. Integration of adsorption beds during a complete operating cycle.

To simulate the multi-bed column adsorption, rigorous multi-bed and single-bed approaches were used. In the first approach, all existing beds were added to the simulation and connected via material streams by interaction valves and holdup tanks. In the single-bed approach, only one bed was modeled, and an interaction block was used. This block records the material profile from the rigorously modeled column and reruns it at user-defined points to simulate returning material. To make this approach possible, each adsorbent bed must be identical and undergoes the same steps in a given cycle.

Since the adsorption process is exothermic, the enthalpy variation is small, and the simulations were considered isothermal. The temperature of the adsorbent bed was kept constant at 298.15 K during the complete operating cycle. The pressure of the adsorption step was set at 20.00 bar for integration with the FT synthesis. The pressure of the purge step was set at 1.50 bar to avoid vacuum operation. Identical dead volumes before and after the adsorbent bed were considered. Table 1 summarizes the pressure and duration of the steps of an operating cycle.

Table 1. Conditions of the PSA process operation cycle.

Step	Pressure (bar)	Time (s)
A	20.00	360
E1D	20.00 → 15.42	60
E2D	15.42 → 8.57	120
E3D	8.57 → 5.58	60
PP	5.58 → 3.82	120
B	3.82 → 1.50	60
P	1.50	120
E3P	1.50 → 3.60	60
E2P	3.60 → 7.71	120
E1P	7.71 → 10.26	60
FP	10.26 → 20.00	120

A sensitivity analysis was performed to find the appropriate residence time for the syngas production respecting the length/diameter ratio of 4 for the adsorbent bed. To assess the performance of the operating cycle when the system reaches the steady state, the process performance indicators were estimated: syngas ($CO + H_2$) purity (Equation 3), H_2/CO molar ratio in the product, and CO, H_2, CO_2, CH_4 and H_2O recovery per pass (Equation 4) and global recovery (Equation 5).

$$Pur_s = \frac{M_{p,s}}{M_p} \tag{3}$$

$$Rec_{i,pp} = \frac{M_{p,i,pp}}{M_{p,pp}} \tag{4}$$

$$Rec_{i,o} = \frac{M_{p,i}}{M_p} \tag{5}$$

where Pur_s is the syngas purity (%), $M_{p,s}$ is the syngas in the product (kmol), M_p is the total product (kmol), $Rec_{i,pp}$ is the recovery per pass of species i (%), $M_{p,i,pp}$ is species i in the product per pass (kmol), $M_{p,pp}$ is the total product per pass (kmol), Rec_o is the overall recovery (%), $M_{p,i}$ is species i in the product (kmol).

3. Results and Discussion

3.1. Rigorous multi-bed approach

The rigorous multi-bed approach allows the accurate calculation of column interaction and the more precise inclusion of advanced control systems. However, systems with over two beds are extremely complex. The system initialization is extremely detailed, and convergence is slow and hard. Each element of each bed must be independently included in the cycle manager, generating huge data tables with exact specifications. In this approach, when two different pressure streams are separated by a block, the simulator automatically and instantly equalizes the pressures, preventing the step-by-step pressure reduction. For instance, if the top of two columns, at 1.50 bar and 20.00 bar, respectively, are separated by a closed valve, the calculation proceeds as normal. The moment the valve is slightly opened, both columns tops instantly go to around 15.00 bar. This creates unrealistic pressure profiles with huge increases in pressure. Thus, the modeling of each interaction is not recommended for multi-bed/column adsorption, since it is very time-consuming, a hard-to-converge system, and a power demanding processing.

3.2. Single-bed approach

The single-bed approach reduces the system complexity and improves the processing power/time without losing precision. Table 2 shows the effect of the residence time on the outcome of the PSA cycles with the single-bed approach. As expected, an opposite trend is observed in the pattern of syngas purity and CO and H_2 recovery per pass. Therefore, it is usually necessary to reach a compromise when selecting the optimum cyclic configuration. CO_2/syngas separation also depends on the requirements of the intended application.

Table 2. Simulation results in terms of process performance indicators.

Residence time (s)	27	37	47	57
Adsorbent bed diameter (m)	1.83	2.03	2.20	2.34
Adsorbent bed length (m)	7.31	8.12	8.79	9.38
Syngas purity (%)	74.33	84.66	95.50	97.86
H_2/CO molar ratio	1.51	1.67	2.14	8.91
CO recovery (%)	81.45	73.12	56.74	13.48
H_2 recovery (%)	99.02	98.61	97.97	96.65
CO_2 recovery (%)	40.14	20.13	5.96	1.66
CH_4 recovery (%)	60.10	43.93	18.47	2.10
H_2O recovery (%)	99.02	98.61	97.97	96.65

The results of the sensitivity analysis showed that in the residence time of 47 s the purified syngas presented the molar composition of 0.03% CH_4, 30.04% CO_2, 5.53% CO, 64.39% H_2, and 0. 32% H_2O. This composition is similar to that presented by Graciano et al. (2018) for FT synthesis in a slurry bubble column reactor. After the recycle convergence of the CO_2-rich stream to the RWGS reactor, the proposed system to treat 50 tons/h of CO_2 was able to produce syngas with 94.65% purity and an H_2/CO molar ratio of 1.50. The overall recovery was 99.98% and 99.57% for H_2 and CO, respectively. Table 3 shows the main streams of the proposed system shown in Figure 1. Each adsorbent bed in the PSA unit is 13.08 and 3.27 m in length and diameter, respectively.

Table 3. Component mass flow rate (ton/h).

Parameter	$CO_2 + H_2$	RWGS product	PSA feed	Syngas	Recycle
Total	54.99	163.05	145.20	36.04	108.06
CO_2	50.00	93.04	93.03	5.45	86.70
H_2	4.99	3.00	3.00	2.95	0.05
CO	-	48.46	48.46	27.39	20.86
H_2O	-	18.00	0.15	0.15	0.00
CH_4	-	0.55	0.55	0.10	0.45

4. Conclusion

A PSA unit for CO_2/syngas separation integrated into the RWGS process was studied. The single-bed approach was the best option for a multi-bed/column system, capable of reducing the system complexity and improving processing power/time without losing precision. Considering syngas purity and CO recovery, a residence time of 47s was in good agreement with the literature data. After integrating RWGS and PSA units, the proposed system can properly treat and separate the RWGS reaction products for a CO_2 load of 50 tons/h to produce 94.65% pure syngas. The total recovery was 99.98% and 99.57% for H_2 and CO, respectively, at adsorption and desorption pressures of 20 bar and 1.5 bar. Thus, PSA units using activated carbon as adsorbent material for CO_2/syngas separation is a technically feasible process.

5. Acknowledgment

The authors gratefully acknowledge Repsol Sinopec Brazil for its financial and technical support and ANP (Brazilian National Oil, Natural Gas, and Biofuels Agency) for the strategic importance of its support through the R&D levy regulation. The authors also thank the National Council for Scientific and Technological Development - CNPq (#314598/2021-9). This study was financed in part by the Coordination for the Improvement of Higher Level Personnel - CAPES - Finance Code 001.

References

A. Golmakani, S. Fatemi, and J. Tamnanloo, 2017, Investigating PSA, VSA, and TSA methods in SMR unit of refineries for hydrogen production with fuel cell specification, Separation and Purification Technology, 176, 73–91.

A. Theampetch, C. Prapainainar, S. Tungkamani, P. Narataruksa, T. Sornchamni, L. Árnadóttir, G. N. Jovanovic, 2021, Detailed microkinetic modelling of syngas to hydrocarbons via Fischer Tropsch synthesis over cobalt catalyst, International Journal of Hydrogen Energy, 46, 48, 24721-24741.

B. Liu, X. Yu, W. Shi, Y. Shen, D. Zhang, and Z. Tang, 2020, Two-stage VSA/PSA for capturing carbon dioxide (CO2) and producing hydrogen (H2) from steam-methane reforming gas, International Journal of Hydrogen Energy, 45, 46, 24870–24882.

H. Wibowo, H. Susanto, N. Grisdanurak, D. Hantoko, K. Yoshikawa, H. Qun, M. Yan, 2021, Recent developments of deep eutectic solvent as absorbent for CO2 removal from syngas produced from gasification: Current status, challenges, and further research, Journal of Environmental Chemical Engineering, 9, 4, 105439.

I. Durán, F. Rubiera, and C. Pevida, 2022, Modeling a biogas upgrading PSA unit with a sustainable activated carbon derived from pine sawdust. Sensitivity analysis on the adsorption of CO2 and CH4 mixtures, Chemical Engineering Journal, 428, 132564.

J. E. A. Graciano, B. Chachuat, and R. M. B. Alves, 2018, Conversion of CO2-Rich Natural Gas to Liquid Transportation Fuels via Trireforming and Fischer–Tropsch Synthesis: Model-Based Assessment, Industrial & Engineering Chemistry Research, 57, 30, 9964–9976.

J. Xiao, Y. Peng, P. Bénard, and R. Chahine, 2016, Thermal effects on breakthrough curves of pressure swing adsorption for hydrogen purification, International Journal of Hydrogen Energy, 41, 19, 8236–8245.

M. Kacem, M. Pellerano, and A. Delebarre, 2015, Pressure swing adsorption for CO2/N2 and CO2/CH4 separation: Comparison between activated carbons and zeolites performances, Fuel Processing Technology, 138, 271–283.

M. Kumita, F. Watanabe, and M. Hasatani, 1994, Effect of water vapor on CO2-PSA separation performance, in Ecomaterials, Elsevier, 329–332.

S. M. W. Wilson, D. A. Kennedy, and F. H. Tezel, 2020, Adsorbent Screening for CO2/CO Separation for Applications in Syngas Production, Separation and Purification Technology, 236, 116268.

V. D. B. C. Dasireddy, D. Vengust, B. Likozar, J. Kovač, and A. Mrzel, 2021, Production of syngas by CO2 reduction through Reverse Water–Gas Shift (RWGS) over catalytically-active molybdenum-based carbide, nitride and composite nanowires, Renewable Energy, 176, 251–261.

Antonis Kokossis, Michael C. Georgiadis, Efstratios N. Pistikopoulos (Eds.)
PROCEEDINGS OF THE 33rd European Symposium on Computer Aided Process Engineering
(ESCAPE33), June 18-21, 2023, Athens, Greece

Simultaneous design of integrated cyanobacteria-based biorefinery and its heat exchanger network

Matías Ramos[a,b], Romina Lasry Testa[a,b], Fernando Ramos[a,b], Vanina Estrada[a,b], Maria Soledad Diaz[a,b]

aPlanta Piloto de Ingeniería Química (PLAPIQUI CONICET-UNS), Camino La Carrindanga km. 7, Bahía Blanca, Argentina
bDepartamento de Ingeniería Química, Universidad Nacional del Sur (UNS), Bahía Blanca, Argentina

Abstract

In this work we propose a mixed integer nonlinear programming (MINLP) model for the optimal design of an integrated cyanobacteria-based biorefinery and its heat exchanger network (HEN), considering alternative *in-vivo* and *in-silico* *Synechocystis* sp. PCC 6803 strains, designed in our previous work. The objective is to potentially produce pigments, biopolymers, bioethanol, biogas and biofertilizers. The production process includes different alternatives embedded within a superstructure. To assess the feasibility of the biorefinery, we considered the Sustainability Profit (SP), a monetary-based sustainability metric as the objective function. The Net Present Value (NPV) was also calculated. Both indexes give positive values and numerical results provides useful insights on advanced biofuels production with metabolic engineered cyanobacteria. Comparison with suboptimal solutions using *in-vivo* strains is also provided.

Keywords: MINLP, *Synechocystis* sp. PCC 6803, Cyanobacteria-based biorefinery, *In-silico* cyanobacteria

1. Introduction

Cyanobacteria are promising candidates to become cell factories, as these photoautotrophic microorganisms are capable of growing on atmospheric and/or industrial carbon dioxide, inorganic phosphorus (P) and nitrogen (N) as macronutrients, and solar and/or artificial light as energy source. One interesting high added-value product for food and nutraceutical industries produced by cyanobacteria is the photosynthetic pigment phycocyanin, which possesses an intense blue color. Also, some cyanobacteria are able to store carbon as poly-β-hydroxybutyrate (PHB), a biopolymer similar to polypropylene with a wide range of applications. On the other hand, photoautotrophic fourth generation bioethanol production by cyanobacteria has been widely studied *in-vivo* (Dienst *et al.*, 2014) and *in-silico* (Lasry Testa *et al.* 2019, 2022) with mutant strains of *Synechocystis* sp. PCC6803 (*Synechocystis*).

In this work, we aim to assess the efficiency of *in-silico* strains of *Synechocystis* sp. PCC6803, developed in previous work (Lasry Testa *et al.* 2019, 2022), in a large scale biorefinery. We propose a mixed-integer nonlinear programming (MINLP) model for the simultaneous optimal plant design and heat exchanger network synthesis (HEN) of a cyanobacteria-based integrated biorefinery for pigments (phycocyanin and zeaxanthin), PHB, biofuels (bioethanol and biogas) and biofertilizers. To enhance cyanobacterial biomass valorization and biofuels production, and to close the nutrient (N and P) cycle in the integrated biorefinery, an anaerobic digestor (AD) is included in the superstructure. Four *Synechocystis* strains are considered as production options: three *in-silico* strains previously developed by our group (Lasry Testa *et al.* 2019, 2022) and one *in-vivo* strain (Vidal Vidal, 2009; Delpino *et al.* 2014). The idea behind evaluating the performance of the developed *in-silico* strains is to determine if it is worth the effort to design them, based on metabolic engineering in the laboratory. The objective function is the sustainability profit (Zore *et al.*, 2017) for the assessment of economic, environmental and social benefits

2. Process Description

2.1 *Synechocystis* strains

In this work, four different *Synechocystis* strains constitute the main "cell factories" for this biorefinery superstructure, as we aim at assessing *in-silico* developed strains efficiency within the frame of an integrated biorefinery. S1 (Vidal Vidal, 2009; Delpino *et al.*, 2014) is an *in-vivo* genetically engineered strain that harvests the insertion of *pdc* and *adh* genes from *Zymomonas mobilis*, to create an ethanol production pathway (as *Synechocystis* do not naturally produce ethanol). The remaining strains are *in-silico* ones, we have designed in previous work. S2 is a wild type strain (no ethanol production) represented by its genome scale model. It comprises 784 reactions and 535 metabolites, with 80 exchange reactions that include cytoplasm, carboxisome, tillacoidal lumen, tillacoidal membrane, cytoplasmatic membrane, periplasm, extracellular space (Lasry Testa et al., 2019). S3 includes the reactions codified by the genes *pdc* and *adh* and produces ethanol coupled to growth (Lasry Testa *et al.*, 2019), and S4 produces PHB coupled to growth (Lasry Testa *et al.*, 2022). The coupled strains were designed by formulating a bilevel optimization problem that identifies gene deletions to achieve the desired coupling. Coupling production to growth has the objective of turning the desired product production into a subproduct of growth, so that it becomes necessary to the microorganism's metabolic function. In the bilevel optimization problem, the outer objective function is to maximize product production rate, while setting an upper bound to the number of gene deletions (represented through binary variables), and the inner optimization problem minimizes product production rate subject to the metabolic network model (LP). The problem has been reformulated as a single level optimization problem by applying dual theory; i.e., replacing the inner LP by a set of equations comprising the primal LP constraints, its dual problem constraints and the strong duality condition (primal LP objective function equal to dual problem objective function). In this case, the dual problem objective function has bilinear continuous-discrete terms, which have been replaced by exact linearizations (Glover, 1975). The resulting Mixed Integer Linear Programming (MILP) model renders a genetically engineered strain that couples ethanol production to cell growth through fourteen genetic intervention in the case of S3 (Lasry Testa *et al.*, 2019) and with sixteen genetic interventions for PHB production in the case of S4.

2.2 Process Superstructure

The proposed superstructure of a cyanobacteria-based biorefinery for the potential production of pigments (phycocyanin and zeaxanthin), PHB, bioethanol, biogas and biofertilizers is shown in Fig. 1. It includes three main processing stages: production, separation and purification.

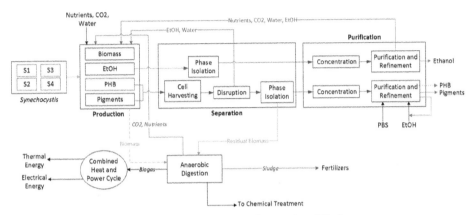

Figure 1. Cyanobacteria-based biorefinery simplified superstructure

The process begins with the production stage, where the four *Synechocystis* strains are considered for cultivation in open ponds at 30 °C, using CO_2 as carbon source, water, and N and P as nutrients. If S1 or S3 are chosen for cultivation, the harvesting process is carried out by microfiltration membranes to separate ethanol and culture media from the biomass. On the other hand, if S2 or S4 are selected, a centrifugation step is required to remove and recycle the culture media from the biomass.

A fraction of the harvested ethanol and water mixture (containing CO_2 and nutrients) from the microfiltration membranes (S1 or S3), is recycled to the open ponds in order to reach a minimal ethanol concentration threshold of 5 g/L, to attain an efficient performance of pervaporation membrane units in the subsequent ethanol separation and purification stages. The stream fed to the ethanol separation section is pre-heated to 50 °C in a heat exchanger before entering the polydimethylsiloxane membrane (PDMS) for pervaporation (where a concentration of 55 g ethanol/L is achieved). Next, ethanol is fed to a vapor compression distillation system where it reaches a concentration of 76%. The compressed vapor (ethanol/water mixture) is sent to a ceramic membrane followed by a cooling exchanger (25 °C) in order to obtain bioethanol fuel-grade concentration of 99.5% (Lopes *et al.*, 2019). Removed culture media streams from PDMS, ceramic membranes and vapor compression distillation system are also recycled to the cultivation system in order to enhance the sustainability feature of the proposed cyanobacterial biorefinery. For the four strains, the main stream of harvested biomass can be directly fed to an anaerobic digestor for the production of fertilizers and biogas, and to recycle nutrient streams to the cultivation stage (García Prieto *et al.*, 2017). Also, it can be led to the separation section for pigments or PHB (S4 strain only) extraction. At the pigment separation stage, the wet biomass is dried in a spray drier with hot air (110 °C) in a 5.03 air-to-water ratio. In this step, most of the water and ethanol remnants are removed and recycled to the production stage after a cooling (40 °C) and flash operation. The dried biomass is disrupted in a bed mill, and then, a phosphate buffer solution (PBS) in a 17.25 PBS-to-phycocyanin ratio is used in a diafiltration system to separate phycocyanin from the residual biomass and zeaxanthin. In order to obtain commercial phycocyanin, the purification of this pigment is conducted by freeze drying, removing mostly all of the water containing phosphates. For the case of the hydrophobic pigment, zeaxanthin, it is extracted in a solid mixer-settler extractor, using ethanol as solvent in a 7.43 ethanol-to-zeaxanthin ratio. Finally, the residual biomass is sent to the anaerobic digestor and the zeaxanthin is purified by chromatographic columns where the removed ethanol is reused in the solvent-extractor. The separation and purification stage for PHB, when strain S4 is cultivated, is described in detail in Ramos *et al.* (2019).

3. Mathematical Model

The proposed superstructure is formulated as a mixed-integer nonlinear programming (MINLP) problem and implemented in GAMS 35.2.0 (McCarl *et al.*, 2021) in order to determine the optimal design of an integrated cyanobacteria-based biorefinery and its HEN. The objective function is to maximize the monetary-based sustainability metric, Sustainability Profit (SP) (Zore *et al.*, 2017). The proposed superstructure includes mass and energy balances for the integrated biorefinery process, as well as its heat exchanger network (HEN) design and connection equations to link process design variables with HEN variables. Binary variables are associated to potential units and to heat exchanger matches.

4. Numerical Results

The MINLP model formulated for sustainability optimization and simultaneous process and HEN design includes 4,522 discrete variables, 24,840 continuous variables and 34,905 constraints. It was solved to an objective function value of SP=50.77 MMUS$/y using DICOPT, with CONOPT and CPLEX as nonlinear and linear subsolvers, respectively (Grossmann *et al.*, 2003). Table 1 shows the alternative strain yields, used as input for the model.

Table 1. *In-vivo* (S1) and *in-silico* (S2, S3, S4) strain yields

Strain	Biomass yield (g/(L.d))	Ethanol yield (g/(L.d))	PHB yield (g/(L.d))
S1	0,339	0,288	0
S2	1,375	0	0
S3	0,669	0,875	0
S4	0,229	0,00	0,119

A required production of 180 t/y of phycocyanin/h (industrial level) and 1700 t/y of ethanol is fixed. The optimal scheme corresponding to SP maximization includes the use of *Synechocystis* strains S2 and S3 for the production of 20.55 kg phycocyanin/h, 0.53 kg zeaxanthin/h, 200 kg /h ethanol and 10.86 kg biofertilizer/h. The flowsheet, with the optimal mass flowrates of the main biorefinery streams is presented in Fig. 2. It must be noted that the production of PHB is not selected in the optimal biorefinery scheme.

Regarding heat integration, additional runs were carried out without heat integration and numerical results show that the optimal integrated scheme increased the number of heat exchangers required, as compared to the optimal base case without HEN (10 vs 7) and it also reduced the cost of external utilities by 60 %. Moreover, the reduction of utilities matches from 7 to 5 accounts for significant utility savings that improve the sustainability objective function. As it occurs with other sustainability analysis papers from the literature, the most important contribution to the objective function is given by the economic pillar (Pedrozo *et al.*, 2022). According to the results shown in Table 2, negligible difference of 0.53 % is found in this pillar between the two proposed cases[a].

This is due to the increase in the number of heat exchangers for the heat integration case (10 HXs) with respect to the one without heat integration (7 HXs). However, there is a meaningful improvement of nearly 22 % in the environmental pillar, while only an enhancement of 2.70 % can be reached in the social pillar. Regarding the economic pillar, a further analysis demonstrates that utilities reduction does not impact this pillar as other variables, such as maintenance and repairs cost. The social pillar difference is associated with the increase in the number of heat exchangers after heat integration, representing more workers' contracts. A decrease of fossil fuels requirements (external utilities), positively benefits the environmental pillar in the objective function.

In order to gain more insight into the study of the genetically engineered *in-silico* strains, we run an additional optimization case, imposing that S3 (*in-silico* ethanol producer) cannot be selected. *In-vivo* (S1) and *in-silico* (S2) strains are selected for ethanol and pigment production. A comparison between the latter case with the optimal one, can be found in Table 2 (fourth column). The sustainability objective function has a 60% decrease (from 50.78 to 20.55 MMUS$/y) by the use of the *in-vivo* strain. The main difference is due to higher capital and operating costs associated to the suboptimal solution (S1 and S2) that negatively impact the economic pillar. Moreover, a decrease in the environmental pillar was also found. The obtained results are associated with the *in-vivo* lower biomass and bioethanol yields, which result in higher process flowrates that demand larger equipment size and energy requirements.

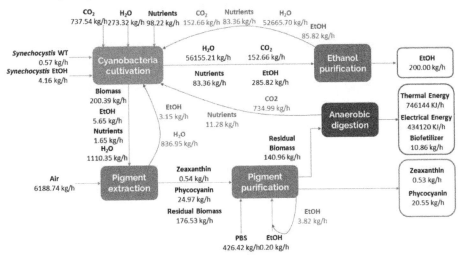

Figure 2. Integrated biorefinery optimal scheme for SP maximization

Table 2. Sustainability Profit pillars comparison

	Without HEN	With HEN	
	S2 and S3 (optimal)	S2 and S3 (optimal)	S1 and S2
SP (MMUS$/y)	50.44	50.78	20.55
P^{economic} (MMUS$/y)	47.43	47.18	19.14
P^{environmental} (MMUS$/y)	2.64	3.22	1.02
P^{social} (MMUS$/y)	0.37	0.38	0.39

5. Conclusions

In this work, we have addressed the production of pigments, biopolymers, bioethanol, biogas and biofertilizers with alternative strains of *Synechocystis* sp. PCC 6803 by a mixed integer nonlinear programming (MINLP) model for the optimal design of an integrated biorefinery and its heat exchanger network (HEN). A superstructure with different process alternatives and different *Synechocystis in-vivo* and *in-silico* strains was formulated. *In-silico* strains S2 and S3 are the ones selected in the optimal solution. Both the SP and the NPV for the biorefinery are positive. A production cost of 0.38 USD/kg for fourth generation bioethanol is achieved due to incomes from pigments (phycocyanin and zeaxanthin) sales. Overall, the numerical results obtained are encouraging into further improving and testing the superstructure model, as well as they show that the use of *in-silico* S2 and S3 strains increases up to 60% the SP, as compared to the *in-vivo* ethanol producer strain S1. Strains S2 and S3 would be possible candidates to be designed by metabolic engineering through gene editing tools such as CRISPR-Cas systems.

References

I. E. Grossmann, J. Viswanathan, A. Vecchietti, R. Raman, E. Kalvelagen, 2003. GAMS/DICOPT: A discrete continuous optimization package. Washington, DC, USA.

C. Delpino, V. Estrada, J. Laglecia, R. Vidal Vidal, F. Florencio, M. García Guerrero, M. S. Diaz, 2014. Dynamic flux balance analysis in cyanobacteria for ethanol production with simultaneous optimization approaches. Comp. Aided Chem. Eng. 33, 1165-1170.

D. Dienst, J. Georg, T. Abts, L. Jakorew, E. Kuchmina, T. Börner, A. Wilde, U. Dühring, H. Enke, W.R. Hess, 2014. Transcriptomic response to prolonged ethanol production in the cyanobacterium Synechocystis sp. PCC6803. Biotechnol. Biof. 7, 21.

C. V. García Prieto, F. D. Ramos, V. Estrada, M. A. Villar, M. S. Diaz, 2017. Optimization of an integrated algae-based biorefinery for the production of biodiesel astaxanthin and PHB. Energy. 139, 1159-1172.

F. Glover, 1975. Improved linear integer programming formulations of nonlinear integer problems. Manag. Sci. 22, 455–460.

R. Lasry Testa, C. Delpino, V. Estrada, M. S. Diaz, 2019. Bioethanol Production with Cyanobacteria by a Two-Stage Fermentation Strategy. Comp. Aided Chem. Eng. 46, 499-504.

R. Lasry Testa, C. Delpino, V. Estrada, M. S. Diaz, 2022. Development of *in silico* strategies to photoautotrophically produce poly-β-hydroxybutyrate (PHB) by cyanobacteria. Algal Research. 62, 102621.

T. F. Lopes, C. Cabanas, A. Silva, D. Fonseca, E. Santos, T. L. Guerra, C. Sheahan, A. Reis, F. Girio, 2019. Process simulation and techno-economic assessment for direct production of advanced bioethanol using a genetically modified *Synechocystis* sp. Bioresource Technology Reports. 6, 113-122.

B. A. McCarl, A. Meeraus, P. van der Eijk, M. Bussieck, S. Dirkse, P. Steacy, F. Nelissen, 2017. McCarl GAMS user guide, Washington, DC, USA.

H. A. Pedrozo, A. I. Casoni, F. D. Ramos, V. Estrada, M. S. Diaz, 2022. Simultaneous design of macroalgae-based integrated biorefineries and their heat exchanger network. Comp. Chem. Eng. 164, 107885.

H. Shokravi, M. Heidarrezaeri, Z. Shokravi, H. C. Ong, W. J. Lau, M. F. Md Din, A. F. Ismail, 2022. Fourth generation biofuel from genetically modified algal biomass for bioeconomic development. J. Biotechnol. 360, 23-36.

R. Vidal Vidal, 2009. Producción fotosintética de etanol por la cianobacteria *Synechocystis* sp. PCC6803. PhD Dissertation, University of Sevilla, Spain.

Z. Zore, L. Čuček, Z. Kravanja, 2017. Syntheses of sustainable supply networks with a new composite criterion – Sustainability profit. Comp. Chem. Eng. 102, 139-155.

Antonis Kokossis, Michael C. Georgiadis, Efstratios N. Pistikopoulos (Eds.)
PROCEEDINGS OF THE 33rd European Symposium on Computer Aided Process Engineering
(ESCAPE33), June 18-21, 2023, Athens, Greece
© 2023 Elsevier B.V. All rights reserved. http://dx.doi.org/10.1016/B978-0-443-15274-0.50331-0

Sustainable Conceptual Design of a Hydrogen Production via Natural Gas pyrolysis

Patience B. Shamaki,[a] Galo A.C Le Roux[a]

[a]*Department of Chemical Engineering, Politechnic school, universidade de Sao Paulo,Av. Prof. Luciano Gualberto, trav.3,n380, Sao Paulo,Brazil*
galoroux@usp.br

Abstract

Methane pyrolysis involves a simple 1-step endothermic reaction with solid carbon (C) and hydrogen (H_2) as the main products. The aim of this study is to perform a sustainable conceptual process design including heat integration and sustainability analysis of hydrogen production through natural gas (NG) thermal decomposition. Natural gas was used as process feed instead of pure methane and five component products: H_2, C, C_2H_4, C_2H_2, and C_6H_6. The process design, thermodynamic analysis, heat integration and sustainability analysis considering the economics and emissions using different energy sources was performed. As expected, results show conversion at lower temperatures for NG as compared to pure methane feed. It also showed the potential of utilizing heat from the process to improve energy efficiency in an integrated process. The cost of NG, C sales and C credit incentive significantly influence the process profitability. The economics also showed operating cost reduction by 21.6% after heat integration and about $78 kgCO_2/kgH_2$ was avoided using this process to produce H_2.

Keywords: methane pyrolysis, process design, hydrogen.

1. Introduction

The global pressure for sustainable processes and fuels to mitigate carbon footprint has garnered considerable attention to hydrogen. As a versatile and valuable raw material for industrial processes, hydrogen can also serve as clean fuel, being an energy carrier. Moreover, it can easily be mixed with one of its main raw materials CH_4 for cleaner and better combustion (Ishaq & Dincer, 2020). Even though hydrogen itself is a clean and abundant element present in most compounds, its production is usually not as clean. Hydrogen is primarily produced through the reforming of hydrocarbons, which produce high CO_2 content as a process by-product besides the emission from the process energy source, about 8-12 $kgCO_2/kgH_2$ is produced from the steam methane reforming process (SMR). A cleaner alternative is the use of water splitting, but the process is thermodynamically limited given that only the hydrogen in the water compound contains energy, the oxygen byproduct is essentially not so useful. This is why thermal decomposition of methane (TDM) also known as methane pyrolysis (MP) has received considerable attention in recent years as an economical means of producing low to zero emission hydrogen, in comparison to other hydrogen alternatives. It involves a simple 1-step endothermic reaction primary process to produce hydrogen and solid carbon as the main products. Due to the stability of methane, the decomposition process involves some secondary and auxiliary reactions producing other higher hydrocarbons at increasing temperature and consequent considerable heat consumption before it fully decomposes to the primary products (Sánchez-Bastardo et al., 2021). However, most literature only report the main products H_2, C without considering the impact of the intermediate products. Furthermore, the two main challenges limiting its commercialization involves the complexity of dealing with high temperature production, as well as the production of excess carbon beyond market demand. This is why several researches are ongoing to exploit this potential low to zero emission hydrogen production route in the current clime faced with increasing energy demand and declining climatic conditions due to increased emissions. (Naikoo et al., 2021) provided a detailed analysis on the different types of MP such as thermo-catalytic decomposition to reduce reaction temperature however, limited by catalyst degeneration due to high temperature, the use of molten metals or salt to promote heat integration(Leal Pérez et al., 2021) or the use of plasma torch and regenerative heat exchanger reactor (RHER) (Keipi Tiina, 2017; Kerscher et al., 2021) at very high temperature. Additionally, more studies of heat integration to reduce energy waste from the process are necessary. In this study, the thermodynamical feasibility of a sustainable process design is investigated and analyzed, even though the kinetics could be also limiting it is important to investigate the boundaries of the process to enable further research and innovative process design. Therefore, they are not going to be considered in this contribution.

2. Methodology and Process Description

2.1. Methodology

The process of hydrogen production using natural gas thermal decomposition was simulated using the Peng-Robinson- Boston-Mathias (PR-BM) as the property model method. The Gibb's energy minimization as the model for the reactor in ASPEN Plus. The base process as presented by Cheon, et al. (2021) was simulated, although in this work a RHER was not used. One of our objectives is to study the thermodynamically feasible extent of process heat integration, so as to investigate the important modalities required for a sustainable process design. The process data from ASPEN Plus simulation was used to perform heat integration via the PINCH analysis in the ASPEN Energy Analyzer, given that the reaction is slightly endothermic, but is favored by high temperatures. The analysis enabled minimum utility requirement/maximum exchanger requirement and the cheapest design was selected as seen in figure 1. The thermodynamic and sustainability analysis of the heat integrated process design considering the economics and the carbon footprint from energy sources was performed since the process itself does not produce CO_2 directly. The main process components include a pyrolysis reactor, a cyclone for solid separation and a PSA separation unit as shown in the process flow diagram from fig 1. The analysis based on the process thermodynamic properties performed enabled the choice of initial process flow design operating conditions.

2.2. Process simulation description and assumptions

As shown in the base process flowsheet diagram in figure 1, the natural gas feed is pre-heated to reactor temperature and fed into the reactor at 1195°C and 1 bar, this provides a global conversion rate of 95% and the products consisting of H_2, solid carbon, ethylene, acetylene and benzene, are sent to a cyclone (dust collector) with high efficiency to remove the solid carbon before it is cooled for hydrogen purification in the PSA unit (compressor, cooler and SEP component). For the separation 90% of the product gas is considered hydrogen, 8% unconverted methane and other hydrocarbons are recycled and 2% purged as flue gas. 5% of the recovered hydrogen is fed back to the process to improve the feed quality and the 95wt % pure hydrogen contains some gases from the product mixture. The base heat supply for the process is natural gas, other sources of heat supply (grid electricity (coal), renewable sourced energy) was also investigated for economic and emission footprint, given that process itself does not produce CO_2 directly but through the heating sources. The simulation was performed for a H_2 capacity of 414 kg/hr this equals a medium scale hydrogen production process.

Given the peculiarity of the high temperature favored pyrolysis process, the economic feasibility analysis was performed using the order of magnitude costs curve methods of capital and operating cost estimation techniques proposed in Towler and Sinnott (2012). Here the total invested cost (TIC) is divided by the total hydrogen produced to obtain the unit cost of hydrogen, and an optimization was performed to minimize the unit cost of hydrogen produced subject to the plant capacity. The itemized cost data was obtained from literature (Chen et al., 2021; Cheon et al., 2021; Towler & Sinnott, 2013) and ASPEN Plus software, and scaled to the predefined plant capacity; with preliminary capital cost estimation accuracy of +/- 30%. Table 1 shows the simulated process flow information.

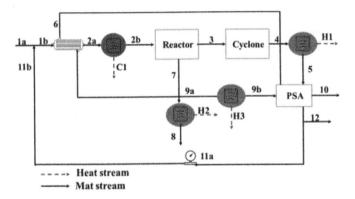

Figure 1: Heat Integrated methane pyrolysis process

3. Results and Discussions

3.1. Thermodynamic Analysis

The influence of the presence of other hydrocarbons in natural gas can be seen in the decomposition at lower temperatures when compared to utilization of pure methane feed from figure 2. The secondary products ethylene, benzene and acetylene were produced in very little quantities, with acetylene produced at higher quantity than the ethylene and benzene at very high temperature. Given that the quantity produced is very small, the secondary products considered were separated from the H_2 in the PSA unit and sent together with the unconverted methane to recycle, this significantly improved conversion rate to primary products. Figure 2 also showed that conversion rate is favored by low pressure. The consequence of utilizing methane as opposed to natural gas for process feed is evident in the emission rate as shown in table 1. This is due to the heat duty required to raise the temperature high enough to decompose pure methane.

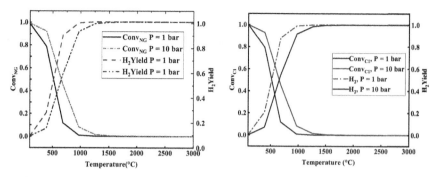

Figure 2: NG decomposition vs Methane conversion rate and product yield

Table 1: Material balance of thermal decomposition process

Param/Stream No	1a	1b	2a	2b	3	4	5	6
Temperature (°C)	25	23	440	1195	1195	1195	50	50
Pressure (bar)	1	1	1	1	1.01	1.01	1	11
Mol flow (kmol/h)								
NG	99	118	118	118	342	230	230	205
C1	112	132	132	132	357	244	244	219

Param/Stream No	7	8	9a	9b	10	11a	11b	12
Temperature (°C)	1195	30	487	139	50	50	26.3	50
Pressure (bar)	1.01	1	11	11	11	11	1.11	11
Mol flow(kmol/h)								
NG	112	112	230	230	230	19	19	6
C1	112	112	244	244	244	20	20	5

3.2. Sustainability analysis

3.2.1. Economics

For the economic analysis, the heat integration performed reduced the utility cost by 52% and consequently the operating cost by 21.6%. The heat integration for the single process of thermal decomposition was constrained by the process thermodynamics, more heat could be recovered when this process is integrated with other processes depending on the end use of the produced hydrogen and carbon. Furthermore, there is added technological complexity due to the high temperature required for the process as well as solid carbon recovery from the reactor, one challenge is the ability to recover heat from the solid carbon produced. This process requires a special heat exchange design as it cannot be simulated with the common shell and tube heat exchanger. (Péreza, et al. 2021) utilized a plate heat exchanger designed for solid-fluid (liquid or gas) heat exchange. These challenges reflect on the total cost of production and consequently on the price of hydrogen from MP process. Sensitivity analysis was performed on the designed process plant to ascertain parameters that can facilitate the point of 'break even', that is the point where the net profit value (NPV) equals zero. The selling price of hydrogen, carbon, discounted cash flow rate of return (DCFROR) from 0-35%, as well as carbon credit for

CO_2 emission avoidance was evaluated varying hydrogen price from 1 – 10$/kg, NG feed cost 0.1-1$/kg, carbon price from 0.00 – 2 $/kg, and carbon credit from 0-0.01$/kgCO_2 avoidance. The analysis showed that hydrogen selling price, considering NG price at 0.4651$/kgNG (being the highest spot price with NG cost hike), without any carbon sales and carbon credit must be 2.5$/kg to break even. This hydrogen selling price is a slightly above the price of hydrogen from SMR but less than and competitive with hydrogen produced via SMR with carbon capture sequestration (CCS). Carbon sales at 0.1618$/kg could enable hydrogen from MP to be competitive or at par with SMR with selling price of 2$/kg and feed cost of 0.4651$/kg. Moreover, carbon credit at 0.0051$/kgCO_{2avoided} could also lead to break even resulting in the 2$/kgH_2 and feed cost of 0.4651$/kgNG. However, for the price 0.3$/kgNG (the average spot price) used for this work, H_2 selling price of 1.8$/kg breaks even, the DCFROR analysis also showed that at 29.48% interest rate, the NPV equals zero as shown in figure 4.

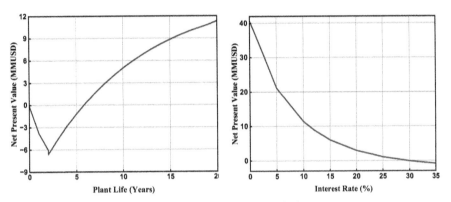

Figure 3: DCFROR and Hydrogen selling price sensitivity analysis

3.2.2. Carbon footprint

For the carbon footprint, we analyzed of the CO_2 emitted and CO_2 emission avoided when using NG as process feed against methane as process feed with three different energy sources (Natural gas, Coal electricity from the grid (US) and renewable energy sources). As expected, the process with methane as a feed emits slightly more than natural gas feed given that the main source of CO_2 emission is the energy source, and the heat required for the decomposition is higher with methane feed. On the average the process produces 3.3kg CO_2/kgH_2 and 0.04 kgCO_2/kgH_2 when renewable energy source is used, noting that the process with NG gas presents some CO_2 in the waste stream, accounting for the CO_2 limit in pipeline gas. Notwithstanding, this stream waste is insignificant as the process with pure methane still emitted more than that with NG feed. When comparing to other hydrogen production processes such as steam reforming with an average of 10kg CO_2/kgH_2 without carbon capture, the thermal decomposition of NG/methane significantly reduced the carbon footprint.

The total heating duty required for the base case was 5,474kW and 5,874kW, for NG feed and CH_4 feed respectively, and the cooling duty was 4,167kW and 4,353kW respectively for each feed. And after heat integration, the heating duty was reduced to 4,699kW and 5,054kW for each feed, and cooling duty was also reduced to 3387kW and 3715kW for each.

Table 2: CO$_2$ emission and CO$_2$ avoidance per heat source

CO$_{2e}$ (kg CO$_2$/kg H$_2$)	NG source	Grid elect (coal)	Renewable source
NG feed	3.42	9.99	0.046
C1 feed	3.53	10,282	0.015
Tot.CO$_{2av}$ (kg CO$_2$/hr)	NG source	Grid elect source	Renewable source
NG feed	-33885	-31160	-35284
C1 feed	-43437	-40448	-44993

4. Conclusion

A sustainable design of the thermal decomposition of natural gas was considered in this work, an optimal process heat integration design was made and the process analyzed considering thermodynamics, economics and carbon footprint mitigation. The results showed that selling carbon and carbon tax/credit is essential for plant profitability as enable an economical low emitting hydrogen production process. Furthermore, results show that the methane pyrolysis process enables better energy efficiency when integrated with other processes than as a stand-alone based on process complexity from thermodynamics and technological limitations. Given the economic potential, it will be beneficial to develop novel designs to overcome these challenges. Further works will focus on integrating the process with the end use process, conceptual heat exchanger network design that could enable recovery of heat from the hot carbon

Acknowledgement

This study was financed in part by the Coordenação de Aperfeiçoamento de Pessoal de Nível Superior – Brasil (CAPES) – Finance Code 001 and Petroleum Technology development fund – Nigeria (PTDF).

Reference

Chen, Z., Zhang, R., Xia, G., Wu, Y., Li, H., Sun, Z., & Sun, Z. (2021). Vacuum promoted methane decomposition for hydrogen production with carbon separation: Parameter optimization and economic assessment. Energy, 222.

Cheon, S., Byun, M., Lim, D., Lee, H., & Lim, H. (2021). Parametric study for thermal and catalytic methane pyrolysis for hydrogen production: Techno-economic and scenario analysis. Energies, 14(19). https://doi.org/10.3390/en14196102

Ishaq, H., & Dincer, I. (2020). Performance investigation of adding clean hydrogen to natural gas for better sustainability. Journal of Natural Gas Science and Engineering, 78.

Keipi Tiina. (2017). Technology Development and Techno-Economic Analysis of Hydrogen Production by Thermal Decomposition of Methane.

Kerscher, F., Stary, A., Gleis, S., Ulrich, A., Klein, H., & Spliethoff, H. (2021). Low-carbon hydrogen production via electron beam plasma methane pyrolysis: Techno-economic analysis and carbon footprint assessment. International Journal of Hydrogen Energy, 46(38), 19897–19912. https://doi.org/10.1016/J.IJHYDENE.2021.03.114

Leal Pérez, B. J., Medrano Jiménez, J. A., Bhardwaj, R., Goetheer, E., van Sint Annaland, M., & Gallucci, F. (2021). Methane pyrolysis in a molten gallium bubble column reactor for sustainable hydrogen production: Proof of concept & techno-economic assessment. International Journal of Hydrogen Energy, 46(7), 4917–4935.

Naikoo, G. A., Arshad, F., Hassan, I. U., Tabook, M. A., Pedram, M. Z., Mustaqeem, M., Tabassum, H., Ahmed, W., & Rezakazemi, M. (2021). Thermocatalytic Hydrogen Production Through Decomposition of Methane-A Review. Frontiers in Chemistry, 9(October), 1–24.

Sánchez-Bastardo, N., Schlögl, R., & Ruland, H. (2021). Methane Pyrolysis for Zero-Emission Hydrogen Production: A Potential Bridge Technology from Fossil Fuels to a Renewable and Sustainable Hydrogen Economy. Industrial and Engineering Chemistry Research, 60(32), 11855–11881. https://doi.org/10.1021/acs.iecr.1c01679

Towler, G. P., & Sinnott, R. K. (2013). Chemical engineering design : principles, practice, and economics of plant and process design. Butterworth-Heinemann.

Antonis Kokossis, Michael C. Georgiadis, Efstratios N. Pistikopoulos (Eds.)
PROCEEDINGS OF THE 33rd European Symposium on Computer Aided Process Engineering
(ESCAPE33), June 18-21, 2023, Athens, Greece

The influence of biomass characteristics and their uncertainties on the production of sustainable aviation fuel

Moaaz Shehab,[a,b,c] Diego Freire Ordóñez,[c] Mai Bui,[c] Kai Moshammer,[a] Edwin Zondervan[b]

[a]Physikalisch Technische Bundesanstalt (PTB), Bundesallee 100, Braunschweig 38116, Germany
[b]Twente University, Drienerlolaan 5, Enschede 7522 NB, Netherlands
[c]Imperial College London, Exhibition Rd, South Kensington, London SW7 2BX, UK.

Abstract

Sustainable aviation fuel (SAF) plays an important role in decarbonizing the aviation sector. The ASTM D7566 dictates several pathways to produce sustainable fuels that share the same characteristics as the conventional jet fuel. One of the common pathways to produce SAF is via the gasification of biomass to generate syngas. In this work a model was developed for the process to evaluate the use of different biomass sources on the SAF yield. Moreover, the influence of the measurement uncertainty in the biomass characteristics on the process performance was investigated. The study has shown that the hydrogen content is the most crucial element in the biomass to obtain a higher SAF yield. On the other hand, the uncertainty of the biomass characteristics such as the moisture content causes a 1.5% variation in the final SAF yield. Such analysis shows the urge for an accurate and reliable measurement of the biomass characteristics, this allows for a structural embedding of the uncertainty while making decisions during the process design and evaluation stage.

Keywords: Biomass gasification, sustainable aviation fuel, biomass characteristics.

1. Introduction

Aviation fuel accounts for 3% of the EU's total greenhouse gas emissions. To tackle this problem, the EU proposed a mandate to include a sustainable aviation fuel (SAF) uptake as a drop-in fuel, with 2% SAF starting from 2025 and increase up to 63% by 2050. SAF can be produced via different pathways which are regulated by the ASTM standard D7566. These pathways vary for different feedstocks. The most common feedstocks are waste oil and fats, cover crops, agriculture and forest residuals and the organic fraction in municipal solid waste (MSW). The Hydro processed Esters Fatty Acids (HEFA) process uses waste oil and fats to produce SAF, while for the agriculture and forest residual, gasification coupled with fischer tropsch (GFT) or alcohol to jet (ATJ) could be used. As gasification technology is widely and commercially established on coal gasification, the focus of this study is on utilizing different types of biomass in the gasification process to produce syngas. This syngas will be upgraded by the Fischer-Tropsch process (FT) to produce liquid hydrocarbons. To describe and understand which potential yields of SAF can be obtained from biomass, data on the biomass characteristics is required. However, such attributes are subject to enormous variations from source to source and from season to season. Therefore, the aim of this study is to analyze the effect of the different biomass types, their characteristics such as moisture content, elemental analysis and the energy content as well as their respective uncertainties on the production process of SAF. The biomass data used in the analysis was provided by the BIOFMET project where different biomass types were characterized and their respective uncertainties were determined.

2. Methodology

A block diagram was developed for the process of biomass conversion into sustainable aviation fuel based on the available literature data for biomass gasification and the Fischer-Tropsch process as shown in figure 1. (Almena et al. 2022; König et al. 2015; Larson et al. 2020).

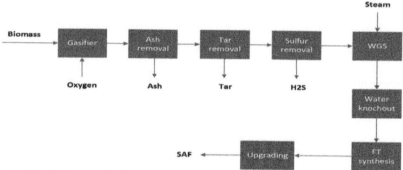

Figure 1. Block diagram for biomass to SAF conversion process.

The biomass is fed to a gasifier to produce syngas using oxygen. The syngas produced from the gasifier contains ash, tar and H_2S which need to be removed before being supplied to the water gas shift reactor (WGS) as those impurities can deactivate the catalyst in both the WGS and FT. In the WGS, the clean gas reacts with a super-heated steam to increase the H_2-CO ratio to 2:1 according to equation 1.

$$\text{WGS reaction: } CO + H_2O \rightarrow CO_2 + H_2 \tag{1}$$

The steam will react with the carbon monoxide to produce more hydrogen and carbon dioxide. The syngas will be directed to the FT to produce long chain hydrocarbons according to equation 2.

$$\text{Alkane production: } (2n + 1)\, H_2 + nCO \rightarrow C_nH_{2n+2} + nH_2O \tag{2}$$

After the formation of hydrocarbons, the syngas enters a series of flash drums where the temperature and pressure are adjusted accordingly to distil the SAF fraction. ASPEN Plus V10 was used to simulate the process and calculate the mass- and energy balances of the process streams. As a property method, the Peng-Robinson equation of state in combination with the Boston-Mathias equation was used. Oxygen gasification was selected as the air separation unit produces a nitrogen-free syngas which results in a syngas that is better suitable for the Fischer-Tropsch process. FT can be performed at lower as well as at higher temperatures. For the synthesis of long chain hydrocarbons low temperature operation of FT is recommended. In the FT reactor a closed loop design was opted to maximize the CO conversion which resulting in a higher SAF yield. The scale of the model was set around 1 GW which is equivalent to a feed rate of 300 ton/hour biomass. Three different types of biomass were analsyed, wood chips of high quality (WC-HQ), wood chips of industrial quality (WC-IQ) and wood pellets (WP). Within the framework of the BIOFMET project, the characteristics of those samples were measured, and their respective uncertainties were determined and minimized.

Table 1. Biomass properties.

Property	Wood chips - IQ	Wood chips – HQ	Wood pellet
C	47.13	47.08	50.22
H	5.69	5.61	5.84
N	0.198	0.074	0.093
S	0.01245	0.0058	0.0069
O	44.99	46.99	43.56
Ash	1.97	0.24	0.28
Moisture	10.76	11.24	5.93
Volatile Matter (VM)*	17–25	17–25	17–25
Fixed Carbon (FC)*	70–90	70–90	70–90

* The values of the volatile matter and fixed carbon are assumed based on the literature data for woody biomass. An average value of for VM and FC were used (Dai et al. 2015).

The model was used to test how the different biomass sources and their characteristics could influence the production of SAF. Operating conditions such as the temperature, pressure and the flow rate need to be optimized and fixed to have a fair comparison.

2.1 Model optimization

The gasification process decomposes the biomass and leads to a syngas that contains CO, CO_2, H_2, CH_4 along with undesired impurities such ash, tar and H_2S. Therefore, optimizing the gasifier is crucial to maximize the CO and H_2 yields, while lowering the concentration of the impurities. An effective gasification process operates at an equivalence ratio (ER) of 0.2 to 0.4. A value less than 0.2 lead to a dominant pyrolysis reaction, while for an ER of more than 0.4, the combustion reaction will dominate and more CO_2 and H_2O produced. In figure 2, a temperature of 840 °C was selected for the gasifier and different flowrates of oxygen were tested to determine the optimum ER ratio.

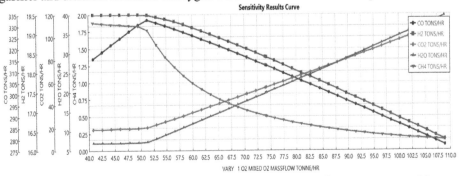

Figure 2. Sensitivity analysis for the oxygen flowrate against the syngas composition.

An oxygen flow rate of 78 ton/hr was opted which corresponds to an ER ratio of 0.2. This flow rate guarantees that more CO and H_2 are produced. For FT, different catalysts such as Ru, Fe, Ni and Co can be used to produce hydrocarbons. Ni has a high selectivity towards methane, while Ru is too expensive for use at a commercial scale. An iron catalyst is relatively tolerant to sulfur as compared to cobalt and can be used to adjust the H_2/CO ratio thanks to it is water gas shift reaction capability. However, it has a low selectivity towards long chain hydrocarbons. Therefore, iron is not suitable for the production of sustainable aviation fuel. Cobalt based catalysts produce less unsaturated hydrocarbons and alcohols, they have a long lifetime and are suitable to produce long chain alkanes. The Anderson–Schulz–Flory distribution, with a propagation probability α (growth factor = 0.95) as the only parameter was used to predict the fractional distribution of the different alkanes. The temperature and pressure of the FT reactor was adapted from the experimental results performed by König et al. (König et al. 2015). The reactor operates at 225 °C and a pressure of 2.5 MPa.

3. Results and discussion

The Aspen model was used to analyse the differences between the wood chips and wood pellet in terms of their influence on SAF production as shown in figure 3.

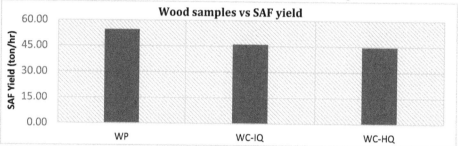

Figure 3. SAF yield of different biomass types.

WP provides the higher SAF yield compared to WC because of having a higher hydrogen and carbon content, lower moisture and ash content.

3.1. The influence of moisture uncertainty on the SAF yield

After drying the biomass and before feeding it to the gasifier, a sample is analysed to determine it is final moisture content. Determining the moisture content and it is respective uncertainty is scientifically challenging as a lot of factors influence the final results, starting from the experimental procedure, the surrounding conditions and the biomass composition. Typically, the uncertainty of the moisture content of the biomass can vary from ± 1-5%. In the model, the operating parameters were maintained fixed to analyse the influence of the moisture change on the final SAF yield as shown in figure 4.

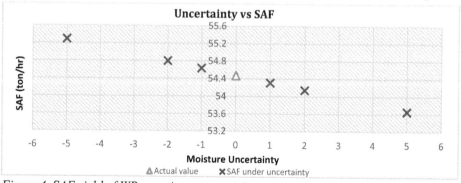

Figure 4. SAF yield of WP vs moisture uncertainty.

Wood pellet was used to perform the above-mentioned analysis. The actual value represents the obtained yield by using the measured moisture of 5.93% as given in table 1. It can be concluded that the final yield varies depending on the uncertainty. At an uncertainty of ± 5%, the SAF production will increase or decrease by 1.5 % (around 0.83 ton/hr). Theoretically, 1.5 % could be accepted, in reality a decrease of 0.83 ton/hr is significant as this corresponds to thousands of tons per year. Moreover, unforeseen higher/lower level of the moisture content due to the uncertainty leads to problems in the gasifier. In case of an increase in the moisture content, the temperature of the gasifier will decrease which influences the syngas quality and lowers it is heating value. Furthermore, the rate of tar cracking would decrease due to lower temperatures. In that case, the operator has to adjust the gasifier temperature. Therefore, accurately measured data with specified uncertainty is important to help the operator in deciding how to minimize the impact of any unpredicted variation in the process.

3.2. The influence of C,H,N,S,O uncertainty on the SAF yield

As syngas contains primary CO and H_2, the percentage of carbon, hydrogen and oxygen in the biomass would largely determine the SAF yield. Generally, biomass has an average ratio of 45–55% carbon, hydrogen of 5-7% and oxygen of 35 – 50% (Dai et al. 2015). During the process design phase it is important to consider the uncertainty of the different parameters to correctly size the instruments and have an accurate average for the final yield. For example, hydrogen has a very low density which leads to a large specific volume. For the given values above, the average uncertainties as provided from the BIOFMET data is for H around ± 0.15 %, while for both C and O is ± 1%. Small changes in composition can influence the yield as shown in figure 5.

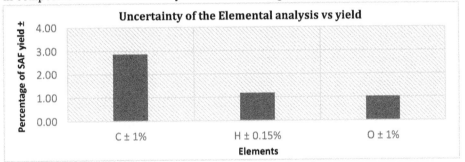

Figure 5. SAF yield vs the uncertainty of the elemental analysis.

A change of ±1% in C and O would influence the SAF yield by up to ± 3 and ± 1% respectively. In case of hydrogen a change of ± 0.15 % influences the yield by ± 1.3%. Hydrogen is the most influential factor in the process. Therefore, the choice for a biomass source should be based on it is hydrogen content and not the carbon content. The average range of carbon in the different biomass types is enough to provide the process with a sufficient CO.

3.3 The influence of the calorific value on the gasification performance

The energy content of the biomass could be determined experimentally by a bomb calorimeter or by using the elemental analysis. According to a survey performed within the scope of the BIOFMET project, the uncertainty of the energy content supplied to industry reaches up to 5 % (Shehab et al. 2022). The energy content influences the calculations of the cold gas efficiency (CGS) of the gasifier and the overall process performance. Therefore, a small error in the measurement raises the overall uncertainty of the process, as the uncertainty distribution might not be concentrated around the average. To calculate the cold gas efficiency, the lower heating value (LHV) of biomass should be calculated as (LHV) = HHV (Higher heating value) – 10.55(W + 9H). *W* is the moisture weight fraction and *H* is the hydrogen percentage in the biomass. An example is given below for the wood pellet. Shehab *et al*. 2022, provided an improved technical protocol to determine the energy content of the biomass with low uncertainty of ± 1% Based on those experimental data, the LHV of wood pellet is 19.92 MJ/kg, while the resulting LHV of the syngas provided by the ASPEN model is 13.56 MJ/kg. Equations 3 was used to calculate the CGS of the gasifier in case of 5% uncertainty and 1% uncertainty.

Cold gas efficiency: $\dfrac{LHV\ of\ syngas}{LHV\ of\ Biomass} * 100 = 68.08\ \%$ (3)

An uncertainty of ± 5 % causes inaccuracy in the efficiency of the gasifier by over ± 3.2%. While an uncertainty of ±1% in the energy content will lead to a substantially lower uncertainty of ± 0.6 %. Such an approach shows the importance of having a reliable testing mechanism for the biomass as it directly influences the process performance and efficiency. This information helps the process designer to provide a realistic range of uncertainty to the unit operation and the process efficiency.

4. Conclusions

An Aspen model was developed to simulate the conversion of biomass into sustainable aviation fuel. Optimizing the model was crucial to provide an accurate prediction of the syngas composition when using oxygen as a gasifying agent. Using different biomass types on the model proved that a slight variation in the biomass characteristics could substantially influence the SAF yield. Hydrogen is the most sensitive element in the biomass, therefore, selecting the biomass should be based on hydrogen content. Knowledge about the uncertainty of the biomass characteristics is crucial, so the process designer could provide a better estimation of the unit operation uncertainty and the final SAF yield

Acknowledgments

This BIOFMET research (19ENG09) was funded from the EMPIR programme (European Metrology Programme for Innovation and Research) co-financed by the Participating States and from the European Union's Horizon 2020 research and innovation programme.

5. References:

Almena, Alberto; Thornley, Patricia; Chong, Katie; Röder, Mirjam (2022): Carbon dioxide removal potential from decentralised bioenergy with carbon capture and storage (BECCS) and the relevance of operational choices. In *Biomass and Bioenergy* 159, p. 106406. DOI: 10.1016/j.biombioe.2022.106406.

Dai, Jianjun; Saayman, Jean; Grace, John R.; Ellis, Naoko (2015): Gasification of Woody Biomass. In *Annual review of chemical and biomolecular engineering* 6, pp. 77–99. DOI: 10.1146/annurev-chembioeng-061114-123310.

Kluh Daniel; Gaderer Matthias (2022): Integrating a Fischer-Tropsch Fuel Production Process into Pulp Mills. In *Chemical Engineering Transactions* 94, pp. 13–18. DOI: 10.3303/CET2294002.

König, Daniel H.; Baucks, Nadine; Dietrich, Ralph-Uwe; Wörner, Antje (2015): Simulation and evaluation of a process concept for the generation of synthetic fuel from CO2 and H2. In *Energy* 91, pp. 833–841. DOI: 10.1016/j.energy.2015.08.099.

Larson, Eric D.; Kreutz, Thomas G.; Greig, Chris; Williams, Robert H.; Rooney, Tim; Gray, Edward et al. (2020): Design and analysis of a low-carbon lignite/biomass-to-jet fuel demonstration project. In *Applied Energy* 260, p. 114209. DOI: 10.1016/j.apenergy.2019.114209.

Shehab, Moaaz; Stratulat, Camelia; Ozcan, Kemal; Boztepe, Aylin; Isleyen, Alper; Zondervan, Edwin; Moshammer, Kai (2022): A Comprehensive Analysis of the Risks Associated with the Determination of Biofuels' Calorific Value by Bomb Calorimetry. In *Energies* 15 (8), p. 2771. DOI: 10.3390/en15082771.

Antonis Kokossis, Michael C. Georgiadis, Efstratios N. Pistikopoulos (Eds.)
PROCEEDINGS OF THE 33rd European Symposium on Computer Aided Process Engineering
(ESCAPE33), June 18-21, 2023, Athens, Greece

Translation Ontology of optimal Decision Making along a Distributed Production Chain by Example

Peter Klein[a], Heinz A. Preisig[b], Natalia Konchakova[c]

[a]*Fraunhofer Institute for Industrial Mathematics, Kaiserslautern, Germany*
[b]*Norwegian University of Science and Technology, Trondheim, Norway*
[c]*Helmholtz-Zentrum Hereon, Geesthacht, Germany*

Abstract

Traditionally, product design is done from within a company. From a purely economic point of view, it will focus on optimising product performance while minimising costs and waste. The optimisation uses key performance indicators, which often result from modelling the production processes and the product's properties. Increasingly, though, product design and product manufacturing involve groups of companies, each of which has its own objectives and strategies. The central idea is to couple the single company-based silo optimisation to a collaborative optimisation and decision-making to unlock hereto hidden product innovation potentials on a digital platform. This platform supports the modelling process from describing the distributed manufacturing and the involved materials to product design, which we call *Translation*. In this contribution, we discuss a joint translation ontology, capturing the translation process and the recursive business-to-business process implemented in an H2020-funded innovation platform VIPCOAT.

Keywords: translation ontology, optimally distributed production, innovation environment, product design.

1. Introduction

The European Materials Modelling Council (EMMC) has developed a Translation process performed by so-called Translators to support the industry in utilising materials modelling and facilitating innovation. The delivered Translation Guide (Hristova-Bogaerds et al. 2019) describes in detail a six-step sequential translation process: business case identification → industrial case understanding → analysis of available data space → identification (translation) to modelling workflow → model execution and validation → back-translation of modelling results to business case. An extension of this guide, in particular of its last step, establish Business Decision Support Systems (BDSS), which introduce business-related key performance indicators (KPI) into the picture (Dykeman et al. 2020) and systematise the original back-translation process. Recent developments in this area are summarised in a report on the progress of implementing the translation and BDSS concept in various running European projects, see (Klein, Konchakova, et al. 2021).

In parallel to the standardisation of translation processes, first steps in materials modelling standardisation and ontologisation have been made by the EMMC community. These attempts are documented in a CEN workshop agreement on Modelling Data, MODA CWA (CEN workshop agreement, CWA 17284:2018 (E) 2018), and Elemental

Multiperspective Material Ontology (EMMO) (Ghedini et al. 2021). An ontologisation of the MODA CWA, compatible with the EMMO, has been described for one particular industrial process in (Klein, Preisig, et al. 2021), where the development of corrosion-inhibiting coatings served as an example of a translation process initially documented in a MODA and a set of roles to automatically map this MODA description to an EMMO compatible format of materials modelling workflows. The construction principle used is now systematised in a further CEN workshop agreement, see (CEN workshop agreement, CWA 17960:2022 (E) 2022).

In this paper, we describe the basic steps for the ontologisation of the corresponding translation process suitable for coating development and take into account that the development and production processes of coatings are distributed (chained) among different actors in the industry.

2. Industrial challenges and requirements

In general, new innovative product designs within a company aim to optimise product performance and customer acceptance while minimising costs and waste. On top, most companies implement long-term strategies like maintaining good supplier and customer relations, technological leadership, and, nowadays, sustainable, resilient, and green production and supply. The optimisation uses key performance indicators (KPIs), often resulting from modelling the production processes and the product's properties.

Increasingly, though, product design and product manufacturing involve groups of companies, each of which has its own objectives and strategies. The central idea of this work is to couple the single company-based silo optimisation to a collaborative optimisation and decision-making process to unlock hereto hidden product innovation potentials on a digital platform. This approach necessarily calls for interoperability along the value-added chain, which is greatly supported by ontologies. In this contribution, we discuss a joint translation ontology, capturing the translation process and the recursive business-to-business (B2B) process implemented in an H2020-funded innovation platform VIPCOAT ("Virtual Open Innovation Platform for Active Protective Coatings Guided by Modelling and Optimization") for the development of new coating systems.

The best-known inhibiting pigments in metal coatings contain hexavalent chromium-based compounds, or Cr(VI) compounds for short, as pigments. Acknowledging human health and environmental issues has led these Cr(VI) compounds to be heavily regulated, so a replacement in corrosion-inhibiting coatings, to be developed by coating suppliers, is necessary.

The project involves three major production companies: a nanoscale corrosion inhibitor producer, a coating formulation producer and a large consumer with very tight and demanding specifications related to the performance of the applied coating formulations on metal surfaces. In this B2B2B innovation environment, collaborative decision support is deployed on a cloud platform with endpoints in the companies, representing our ontology's first use case. An ontologisation of the materials modelling support to be used within each company is provided in ref. (Klein, Preisig, et al. 2021).

3. Ontology construction principles

The first step towards ontologisation of the VIPCOAT platform was done for the modelling layer in ref (Klein, Preisig, et al. 2021), compatible with the EMMO (Ghedini et al. 2021). As an intermediate step in this works, we first construct an ontologisation of the traditional translation process in a Translator to Business (T2B) relation, reflecting the EMMC Translation Guide (Hristova-Bogaerds et al. 2019). This ontology is consistent with a developed crosswalk from an initial VIPCOAT MODA workflow of the three industrial applications to the EMMO.

Translation as an interactive process is conceptually described in the ontology as subclasses of the translation class: partners and performance. While the Partners' subclass model basically the interaction between humans involved in a translation process, the Performance subclass models the realisation of business goals.

4. T2B and T2B2B Translation ontology

This initial Translation ontology models a one-to-one relationship between the two triples (Translator, Modelling, Simulation) and (Client, Business Case, Industrial Case) - T2B relationship. In the next step, this ontology will be extended to handle a one-to-many relationship (T2B2B) between the triple (Translator, Modelling, Simulation) and a chain of (Client, Business Case, Industrial Case) triples, thus ontologically modelling a B2B2B environment.

The approach used in this works implements a somewhat different perspective on ontologies. Usually, ontologies are constructed to systematise existing knowledge. Consecutively, we use the ontology-defined syntax and semantics to construct correspondingly structured knowledge, like ontology-driven tools and knowledge graphs. This approach is our paradigm for implementing ontology-based interoperable processes and tools.

4.1. Ontology for T2B Translation Processes

Translation as an interactive process using the original T2B pattern, see (Klein, Konchakova, et al. 2021), consist of two important concepts which are described as subclasses of an ontological Translation class: Partners and Performance. While the Partners model the interaction basically between humans involved in a Translation process, the Performance class models the realisation of business goals in an industrial Case or process. RDF-type data file for further processing of this ontology can be made available upon request.

One Translator- one client relationship structure has been added to the ontology and mapped into an RDF-type data file (see Figure 1).

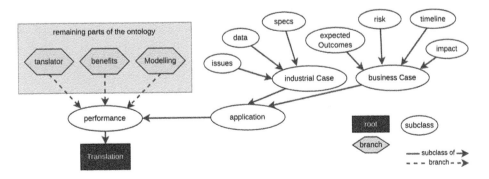

Figure 1: Section of the T2B ontology

The BDSS system deployed at the industrial partner sites may be used to visualise the results of the tasks and/or of experimental data individually at each site.

Decision Support is implemented on the VIPCOAT platform by multi-criteria optimisation and interactive explorers of the solutions of corresponding optimisation models. While multi-criteria optimisation is a well established mathematical technology used to find best balanced compromises between conflicting objectives (in the technological area usually called Key Performance Indicators), decision makers may interactively explore what-if-scenarios and add context information not modelled in the multi-criteria optimisation in order to take informed final decisions on products and processes.

4.2. Ontology for one-to-many Translation Processes

Applied to the example, VIPCOAT project picks this challenge up as depicted in Figure 2.

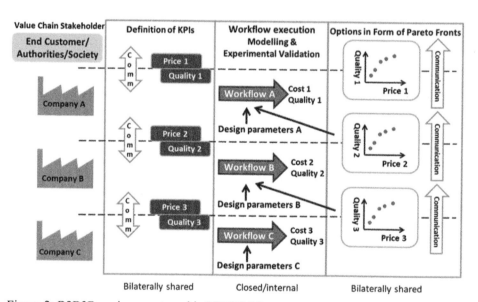

Figure 2: B2B2B environment used in VIPCOAT.

Partner C develops new pigments containing regulation-compliant inhibiting molecules. Partner B adds these pigments to formulate coatings, and Partner C applies this product on surfaces and tests their inhibiting performance to meet the End Customer's expectations. This VIPCOAT platform demonstration use case is thus structured along a value/production chain in a B2B2B environment. Typical for such environments is that the requirements flow is upstream of the production chain, starting at the End Customer. In VIPCOAT, the requirements specification on coatings is formulated to be applicable for aircraft manufacturers. The KPIs are very detailed and contain many more technical details to be fulfilled by a paint producer's coating formulation besides the corrosion-inhibiting properties. These requirements generally persist while the Cr(VI) compounds are to be replaced, a typical B2B supplier - OEM (original equipment manufacture) relation: The paint producer buys the pigments from his supplier. However, many of the requirements are not that hard specified. Also, the interaction between the different involved materials is not fully predictable. Together this provides a certain level of freedom of choice, allowing the complete chain to choose an optimal combination satisfying the coating-applying partner's requirements.

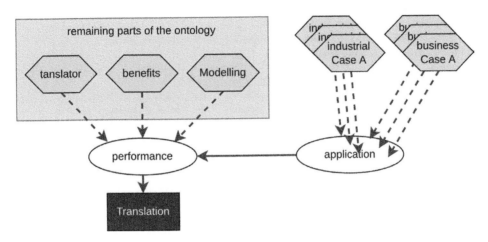

Figure 3: Sketch of the T2B2B ontology: relationship of one Translator and three industrial and business Cases (A, B, C)

VIPCOAT addresses these development challenges directly through its modelling tasks serving the most important steps in coating development by utilisation of integrated simulation and optimisation tools.

Further developments towards B2B2B collaborative decision-making, one of the VIPCOAT platform goals, call for a good understanding of B2B2B relations to implement such a new product development style. An ontology segment for the One-Translator and a B2B2B environment applied to active protective coating creation example is presented in Figure 3.

5. Conclusion

This contribution presents an ontological approach along a distributed production chain designed for protective coating developments. In this example, the materials modelling approach has already been ontologised, so a complete set of ontologies, including business-related ontological classes, are created. In its current implementation stage, the VIPCOAT platform takes advantage of these ontologies to support inter-company collaboration for fast innovations of corrosion-inhibiting coatings used in the aeronautic industry. The next step will be the introduction of domain ontologies to capture more general coating system formulation. These ontologies will get constructed in parallel to the VIPCOAT platform developments to support cross-fertilisation at least in the coating industry.

Acknowledgements: VIPCOAT H2020-NMBP-TO-IND-2020, Grant Agreement No. 952903; MARKETPLACE H2020-NMBP-25-2017, Grant Agreement No. 760173.

References

CEN workshop agreement, CWA 17284:2018 (E). 2018. "Materials Modelling: Terminology, Classification and Metadata."

CEN workshop agreement, CWA 17960:2022 (E). 2022. "ModGra - Graphical representation of physical process models."

D. Dykeman, A. Hashibon, P. Klein, and S. Belouettar, 2020, Guideline Business Decision Support Systems (BDSS) for Materials Modelling, Technical report. EMMC. https://doi.org/10.5281/zenodo.4054009.

E. Ghedini, G. Goldbeck, J. Friis, et al., 2021, Elementary Multiperspective Material Ontology (EMMO), GitHub. https://github.com/emmorepo/410EMMO.

D. Hristova-Bogaerds, P. Asinari, N. A. Konchakova, L. Bergamasco, A. M. Ramos, G. Goldbeck, D. Höche, O. Swang, and G. J. Schmitz, 2019, EMMC Translators' Guide, Technical report. EMMC. https://doi.org/10.5281/zenodo.3552260.

P. Klein, N. Konchakova, D. G. Hristova-Bogaerds, M. Noeske, A. Simperler, G. Goldbeck, and D. Höche, 2021, Translation in Materials Modelling: Process and Progress, Technical report. OntoTRANS – FORCE. https://doi.org/10.5281/zenodo.4729917.

P. Klein, H.A. Preisig, M. T. Horsch, and N. Konchakova, 2021, Application of an Ontology Based Process Model Construction Tool for Active Protective Coatings: Corrosion Inhibitor Release, In Proceedings of JOWO 2021, 26. CEUR-WS.

Virtual Open Innovation Platform for Active Protective Coatings Guided by Modelling and Optimization (VIPCOAT, Grant Agreement No 952903). 2021. https://cordis.europa.eu/project/id/952903/de, https://ms.hereon.de/vipcoat/.

Antonis Kokossis, Michael C. Georgiadis, Efstratios N. Pistikopoulos (Eds.)
PROCEEDINGS OF THE 33rd European Symposium on Computer Aided Process Engineering
(ESCAPE33), June 18-21, 2023, Athens, Greece

The Integration of Drug Substance and Drug Product Manufacturing Models: The Missing Link for Model-based End-to-End Process Development

Charalampos Christodoulou[a,*], Samir Diab[b], Gabriele Bano[c], Magdalini Aroniada[b], Neil Hodnett[a], Simeone Zomer[b]

[a] GlaxoSmithKline, Gunnels Wood Road, Stevenage SG1 2NY, United Kingdom
[b] GlaxoSmithKline, Park Road, Ware SG12 0DP, United Kingdom
[c] GlaxoSmithKline, 1250 S Collegeville Rd, Collegeville (PA) 19426, United States
*Corresponding author email: charalampos.x.christodoulou@gsk.com

Abstract

Demonstrations of pharmaceutical process modelling have been made in the literature in the pursuit of accelerated development. Modelling activities for drug substance (DS) and product (DP) manufacturing are often conducted separately, as project time constraints often require that the DS and DP processes be developed simultaneously, each with fixed decision criteria and requirements. We illustrate how connecting DS and DP process models is essential for a full process understanding and how key attributes of an isolated DS can propagate through a DP tableting line to impact the DP dissolution. We do so by performing an uncertainty analysis on a flowsheet model of a crystallization and direct compression line. The dissolution sensitivity highlights the need for end-to-end flowsheet modelling to realize the full benefits of modelling. We also discuss the key technical and regulatory obstacles that must be overcome to achieve this.

Keywords: Flowsheet model; Pharmaceutical manufacturing; Drug Substance; Drug Product

1. Introduction

The development of robust pharmaceutical manufacturing processes that ensure product quality, manufacturability and material efficiency requires a sound understanding of the impact of the process' inputs (e.g., raw material properties, process operating conditions) on product quality (e.g., purity, content uniformity, dissolution) across the operating space. Consideration of the end-to-end process in a rigorous manner can be challenging due to the large number of inputs to account for in a given process, as well as their multivariate interactions that potentially impact product quality and manufacturability.

Mathematical modelling can assist this by describing the relevant physical and chemical phenomena occurring in each unit operation and connecting these sub-models into an integrated flowsheet model to assess the criticality of multiple input factors *in-silico* (Diab et al., 2022a; 2022b). In industrial practice, modelling activities for drug substance (DS) and drug product (DP) manufacturing can occur independently, as the short lifespan of a patent means that DS and DP process developments usually must occur simultaneously and risk not being perfectly harmonized, e.g., the selected DS solid form may change under certain DP process conditions. Additionally, most DP models, besides those employing physics-based methods such as Discrete Element Method (DEM) are semi-empirical and thus the benefit of connecting them to other first-principles models more commonly available in DS processes are then not as significant. This does not allow for a full consideration of the impact of DS input variability on the DP process and quality. This "disjointed" consideration of DS and DP models misses the full potential of end-to-end modelling to accelerate development.

2. Pharmaceutical Manufacturing Process Modelling

Various efforts have been made towards the modelling of DS system models. Examples include the use of models to predict input factor impacts on related impurity levels (Diab et al., 2022a; 2022b), avoidance of undesirable solid state forms and control of particle size distribution, PSD (Montes et al., 2018). Examples of models for DP manufacturing include wet granulation (Boukouvala et al., 2013), dry granulation (Bano et al., 2022a) and direct compression (García-Muñoz et al., 2018).

There are few literature cases of end-to-end process modelling that span the interfaces between DS, DP manufacture and *in-vitro* product performance. Benyahia et al. (2012) were the first to model an entire continuous pilot plant but did not consider product performance. Nagy et al. (2021) modelled a DS synthesis and crystallization process with *in vitro* dissolution, but without explicit modelling of the DS isolation and DP manufacturing unit operations. Most examples omit isolation, even though this can drastically impact the DS physical properties (Guo, 2009). Fig. 1 shows an example end-to-end process and how each unit operation potentially impacts a quality attribute. While many DS process models predict purity and PSD, they often do not output responses that are key inputs to DP models. For example, changes in crystallizer operation could impact the DS PSD and shape, which may then impact the powder (blend) flowability, tablet content uniformity and weight variability, etc. We now illustrate how variation in the DS properties that are DP model outputs can propagate through to the final DP quality for a tablet manufactured in a direct compression (DC) line, using a flowsheet model.

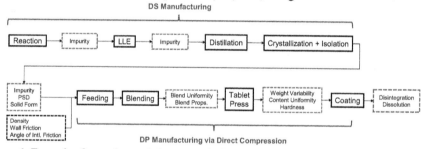

Figure 1. Example of an end-to-end pharmaceutical process. Black solid = Unit operation. Red = quality attribute. Black dashed = Material property not predicted from DS process model.

3. Case Study

The case study in this paper considers a seeded antisolvent crystallization from the final stage of a DS manufacturing process, a DC process model for DP manufacturing and subsequent *in vitro* product performance. The considered sub-models are all parameterized vs. industrial process data.

3.1 Drug Substance Crystallization

We consider a seeded, antisolvent DS crystallization. The supersaturated solution of DS in organic solvent is seeded at a fixed temperature and the resulting mixture is then aged. Following isothermal de-supersaturation, antisolvent is added and the suspension is further aged. The API mass balance is given by Eq. 1, where C_b = bulk solution concentration, ρ_c = crystal density, k_v = crystal volume-shape factor, G = overall growth rate, L = DS particle size, n = number size distribution. Growth contributions come from both pure API (G_{API}) and impurity (G_{Imp}) in Eqs. 2–4, where C_{sat} equilibrium solubility (a function of temperature, T, and antisolvent content, x_{AS}), ρ = solvent density, C_{imp} = impurity concentration in the feed to the crystallization, and model parameters, k_g, g, s, k_i and l are fit to de-supersaturation and impurity content (Taylor et al., 2021).

$$dC_b^{API}/dt = 3k_v\rho_c \int GL^2 n(L)dL \quad (1)$$

$$dP_d/dt = \Lambda d_h^2\, \varepsilon\, p_c\, /\, S_P\, \mathcal{T}_{or}^2 \mu P_d \quad (11)$$

$$G_{API} = k_g\left(\frac{C_b^{API} - C_{sat}}{\rho}\right)^g (x_{AS} + 1)^s \quad (2)$$

$$\frac{dM_t}{dt} = [2\rho_s(1-\varepsilon)\Omega(t)]\frac{dP_d}{dt} \quad (12)$$

$$G_{imp} = G_{API}k_i(x+1)^l C_{imp} \quad (3)$$

$$p_c = (\sigma\cos\theta_e)/d_h \; ; \; d_h = [\varepsilon/(1-\varepsilon)](d_p/6) \quad (13)$$

$$\frac{dL}{dt} = G = G_{API} + G_{imp} \quad (4)$$

$$\frac{\partial N_i}{\partial t} = B_i\delta(l - l_{0,i}) + R_{i,l}\frac{\partial N_i}{\partial l} \quad (14)$$

$$K = 1.5 V_b^{1/3} F_h \omega_b t_{blend} \quad (5)$$

$$B_i = \frac{1}{\rho_p}\left(\frac{x_i}{S_p l_{0,i}^3}\right)\frac{dM_t}{dt} \quad (15)$$

$$sf = a_{sf}(1 + b_{sf}P)/(1 + a_{sf}b_{sf}P) \quad (6)$$

$$R_{i,l} = \dot{M}_l/\rho_p \quad (16)$$

$$TS = TS_0\big((1-\beta) + \beta\exp(-\gamma K)\big) \quad (7)$$

$$\dot{M}_l = K_D(C_{sat} - C_b)^{n_{si}} \quad (17)$$

$$TS_0 = a_1\exp(b_1(1-sf)) \quad (8)$$

$$K_D \approx 1/K_r + 1/K_d = (K_d K_r)/(K_d + K_r) \quad (18)$$

$$\beta = a_2(1 - sf) + b_2 \quad (9)$$

$$C_b^{API} = x_{API}\rho_t\left(\frac{V_{t,0} - V_t}{V_t}\right) - \left(\frac{m_{API,3}\rho_p S_p}{V_m}\right) \quad (19)$$

$$V_c = (H_c - \dot{\varepsilon}t)A_t \quad (10)$$

$$\%LC = 100 C_b^{API} V_m / x_{API}\rho_t V_{t,0} \quad (20)$$

3.2 Blending and Direct Compression of Drug Product

The DC model is comprised of a blender and a tablet press. The Extended Kushner model (Nassar et al., 2021) is used to describe the tablet tensile strength (TS) as a function of the extent of blend lubrication, K, as per Eqs. 5–9, where V_b = blender volume, F_h = fractional headspace in blender, ω_b = blend speed, t_{blend} = blending time, sf = attained tablet solid fraction, P = pressure applied in tablet press, TS_0 = tablet tensile strength at zero-porosity. Parameters a_{sf}, b_{sf}, a_1, b_1, a_2, b_2 and γ are estimated from experimental data.

3.3 In-Vitro Product Performance: Disintegration and Dissolution

Both erosion and swelling are considered in the tablet disintegration (Markl et al., 2017). Erosion is described by Eq. 10, where V_c and H_c = coating layer volume and thickness, respectively, A_t = tablet surface area, $\dot{\varepsilon}$ = erosion rate. The penetration depth due to swelling is given by Eq. 11, where P_d = penetration depth, d_h = pore hydraulic diameter, $\mathcal{T}_{or}$ = tablet tortuosity, μ = liquid viscosity, p_c = capillary pressure. Model parameters S_p and Λ are to be estimated from experimental data (Peppas and Colombo, 1989). The change in tablet mass, M_t, is given by Eqs. 12 and 13, where ε = average porosity of the swollen product, ρ_s = solid true density, σ = liquid surface tension, θ_e = solid-liquid contact angle, d_p = tablet equivalent spherical volume diameter (Kuentz and Leuenberger, 1998) from disintegration time (DT). The DT is defined when $dP_d/dt = 0$.

Dissolution is described by Eqs. 14–20, where N_i = number size distribution, B_i = rate of particle release, l and $l_{0,i}$ = particle size at t and $t = 0$, respectively, ρ_p = particle density, x_i = mass fraction, M_t = tablet mass, $R_{i,l}$ = particle dissolution rate, $\dot{M}_l$ = overall mass flux, K_D = overall dissolution coefficient, which is calculated from the surface integration constant, K_r, and the diffusion mass transfer coefficient, k_d (estimated from diffusion coefficient, D, and unstirred layer thickness, h_l), n_{si} = order of dissolution. Model parameters K_r and n_{si} are to be estimated from dynamic dissolution data for tablets with different solid fractions. The dissolution extent is defined as the percentage label content, $\%LC$, where V_m = liquid volume in the USP <711> (2011) test vessel.

4. Uncertainty Analysis

We quantify the impact of DS property variability propagation through to DP dissolution in two steps: (1) uncertainty analysis on the crystallization process model to understand the expected variance in DS properties, (2) uncertainty analysis to quantify the impacts of DS property variability propagating through the DC process to the DP dissolution using Sobol sampling in gPROMS FormulatedProducts. The input factors for the crystallization uncertainty analysis were the seeding and age temperature and age duration, seed loading, antisolvent addition duration and quantity. The responses were DS d_{90} and purity.

Fig. 2 shows the variation in DS d_{90} and purity as a function of age temperature and age duration, both of which impact the selected responses when varied over the ranges investigated. The total range of the simulated DS d_{90} was approximately ±20% vs. the nominal value. The range of attained DS purity = 99.38–99.87 mol%. Even low levels of key DS impurities can be severe with respect to product quality, e.g., if they are genotoxic or mutagenic, and can also impact the crystallization kinetics and the resulting PSD, which is a direct input to the DP process. Modelling the impacts of impurities on crystallization kinetics is certainly not trivial (Capellades et al., 2022); further research in this area will be an important addition to the modelling literature.

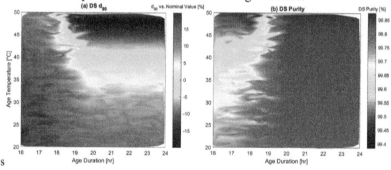

Figure 2. Age temperature and duration vs. DS properties: (a) d_{90}, (b) purity.

For the DC analysis (step 2), the input was the DS d_{90} (whose range is defined from the observed responses in step 1). The considered response was the DP dissolution, whose variation is shown in Fig. 3. The extent of dissolution at 20 min varies significantly (~90–95 %LC), while at 30 min the variation is slightly less (~94.4–96.7 %LC) vs. DS d_{90} variation. Of course, the selection of the timepoint(s) at which dissolution extent is examined impacts how sensitive the dissolution behavior is.

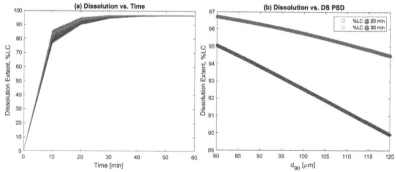

Figure 3. (a) Variation in dissolution extent vs. time. (b) Dissolution extent at 20 and 30 min vs. DS d_{90}.

5. Conclusions and Discussion – Industrial and Regulatory Considerations

The presented case study exemplifies how variation in DS properties can propagate to the final DP product performance and why modelling an end-to-end process is valuable. The modelling approach we have shown could also be used to directly inform the questions posed in the International Council for Harmonization (ICH) quality guideline Q6A (2000) Decision Tree #3, to assess whether an acceptance criterion is required for DS PSD. However, one must acknowledge that there remain certain obstacles to realizing end-to-end process modelling in general.

Material Science: Advancements in proven methods for DS bulk property prediction from the DS particle properties need to be made. A first principles description for this needs computational approaches to interpret and predict physical properties of crystalline solid forms incorporated into an integrated process model, including DS isolation steps – such studies are few in the literature (Destro et al., 2021).

Blend Property Prediction: Predicting blend properties from its composition and individual component properties is essential to describe the formulation-dependence of a process. Modelling the behavior of granular solid blends without DEM is not arbitrary. Many efforts have been made in the literature, only some of which are industrially pragmatic. Statistical approaches are limited by the range of data collated and cannot be reliably extrapolated outside of the considered calibration data space – such models that are currently available consider only a small range of flow characteristics for OSD forms. Approaches using the granular Bond number, although based more on fundamental particle behavior and interactions, require measurements that are not routinely collected in an industrial environment (e.g., particle surface energy). Bano et al. (2022b) proposed mixing rules for blend flow properties based on the van der Waals rules for fluid mixtures, while Jolliffe et al. (2022) proposed a protocol for modelling powder blend compaction properties. More efforts such as these for a wider set of blend properties are needed.

Standardized, Systematic Model Evaluation: Process modelling is still in its relative infancy in pharmaceutical manufacturing, and thus there can be a lack of widespread confidence and uptake of models by the pharmaceutical industry. Specifically, there may be a lack of scientists with the appropriate expertise for model development in a company, as well as resource constraints and a lack of management and stakeholder support. One component of the latter point is that there is no standardized workflow for systematic model development and evaluation that also considers uncertainty propagation. A recent publication by Braakman et al. (2021) describes a workflow for the systematic evaluation of quantitative systems pharmacology models – a similar workflow for integrated process models that is supported by both industry and regulators alike would be invaluable.

Regulatory Guidance: While there have been an increasing number of literature contributions from industrial research groups on modelling applications, there is not a clear consensus or guidance to what extent these approaches will be accepted by global regulatory agencies. While the ICH quality guidelines provide some guidance on model development, the level of detail that should be included in a marketing application on model usage to define a control strategy is less clear. Consequently, this leaves industrial practitioners with uncertainty on the likelihood of success; and therefore, resorting to an empirical approach based on Design of Experiments (DoE) and statistical analysis, which is a more widely applied aspect of the Quality by Design (QbD) framework, is thus the safer strategy. This highlights the need for better engagement between regulators and industry on model development and usage.

The increasing number of modelling applications by pharmaceutical companies is certainly promising. We believe that overcoming these barriers are critical to realizing the full benefits of end-to-end modelling for accelerated pharmaceutical development. Continued demonstration of successful case studies of model applications on active projects are key to move towards the systematic application of modelling by first intent for pharmaceutical process development.

Acknowledgments

This study was funded by a Digital Design capability project at GlaxoSmithKine (GSK).

References

G. Bano, R.M. Dhenge, S. Diab, D.J. Goodwin, L. Gorringe, M. Ahmed, R. Elkes, S. Zomer, 2022a. Streamlining the development of an industrial dry granulation process for an immediate release tablet with systems modelling. Chem. Eng. Res. Des., 178, 421–437.

G. Bano, M. Aroniada, Y. Vueva, 2022b. A model-based approach to predict the flowability of directly compressed pharmaceutical blends. Comput.-Aided Chem. Eng., 51, 31–6.

F. Boukouvala, A. Chaudhury, S. Sen, R. Zhou, L. Mioduszewski, M.G. Ierapetritou, R. Ramachandran, 2013. Computer-aided flowsheet simulation of a pharmaceutical tablet manufacturing process incorporating wet granulation. J. Pharm. Innov., 8, 11–27.

S. Braakman, P. Pathamanathan, H. Moore, 2022. Evaluation framework for systems models. CPT Pharmacometrics Syst. Pharmacol., 11, 264–289.

G. Capellades, J.O. Bonsu, A.S. Myerson, 2022. Impurity incorporation in solution crystallization: diagnosis, prevention, and control. CrystEngComm, 24, 1989–2001.

F. Destro, I. Hur, V. Wang, M. Abdi, X. Feng, E. Wood, S. Coleman, P. Firth, A. Barton, M. Barolo, Z.K. Nagy, 2021. Mathematical modeling and digital design of an intensified filtration-washing-drying unit for pharmaceutical continuous manufacturing. Chem. Eng. Sci., 244, 116803.

S. Diab, C. Christodoulou, G. Taylor, P. Rushworth, 2022a. Mathematical modeling and optimization to inform impurity control in an industrial active pharmaceutical ingredient manufacturing process. Org. Process Res. Dev., 26, 10, 2864–2881.

S. Diab, G. Bano, C. Christodoulou, N. Hodnett, A. Benedetti, M. Andersson, S. Zomer, 2022b. Application of a system model for continuous manufacturing of an active pharmaceutical ingredient in an industrial environment. J. Pharm. Innov. DOI: 10.1007/s12247-021-09609-7

S. García-Muñoz, A. Butterbaugh, I. Leavesley, L.F. Manley, D. Slade, S. Bermingham, 2018. A flowsheet model for development of a continuous process for tablets. AIChE J., 64, 511–25.

Y. Guo, 2009. Impact of solid-state characteristics to the physical stability of drug substance and drug product. Handbook of Stability Testing in Pharmaceutical Development, 241–246.

International Council on Harmonization, 2000. ICH Topic Q6A Specifications: Test procedures and acceptance criteria for new drug substances and new drug products: chemical substances.

H.G. Jolliffe, E. Ojo, C. Mendez, I. Houson, R. Elkes, G. Reynolds, A. Kong, E. Meehan, F.A. Becker, P.M. Piccione, S. Verma, A. Singarju, G. Halbert, J. Robertson, 2022. Linked experimental and modelling approaches for tablet property predictions. Int. J. Pharm., 626, 122116.

M. Kuentz, H. Leuenberger, 1998. Modified Young's modulus of microcrystalline cellulose tablets and the directed continuum percolation model. Pharm. Dev. Technol., 3, 13–19.

D. Markl, S. Yassin, D.I. Wilson, D.J. Goodwin, A. Anderson, J.A., Zeitler, 2017. Mathematical modelling of liquid transport in swelling immediate release tablets. Int. J. Pharm., 526, 1–10.

F.C.C. Montes, K. Gernaey, G. Sin, 2018. Dynamic plantwide modeling, uncertainty, and sensitivity analysis of a pharmaceutical synthesis: ibuprofen case study. Ind. Eng. Chem. Res. 57, 30, 10026–37.

B. Nagy, B. Szilágyi, A. Domokos, B. Vészi, K. Tacsi, Z. Rapi, H. Pataki, G. Marosi, Z.K. Nagy, Z.K. Nagy, 2021. Dynamic flowsheet model development and digital design of continuous pharmaceutical manufacturing with dissolution modeling. Chem. Eng. J., 419, 129947.

J. Nassar, B. Williams, C. Davies, K. Lief, R. Elkes, 2021. Lubrication empirical model to predict tensile strength of directly compressed powder blends. Int. J. Pharm., 592, 119980.

N.A. Peppas, P. Colombo, 1989. Development of disintegration forces during water penetration in porous pharmaceutical systems, J. Control. Release, 10, 245–250.

G. Taylor, H. Yao, B. Williams, R.E. Yule, N. Mitchell, 2021. Kinetic modelling and design space exploration of impurity inclusion in an API crystallization. International Symposium on Industrial Crystallisation.

USP <711> Dissolution. The United States Pharmacopeial Convention. 2011.

Antonis Kokossis, Michael C. Georgiadis, Efstratios N. Pistikopoulos (Eds.)
PROCEEDINGS OF THE 33rd European Symposium on Computer Aided Process Engineering
(ESCAPE33), June 18-21, 2023, Athens, Greece

Direct Compression Flowsheet Modelling to Assess Tablet Manufacturability and *In Vitro* Dissolution

Magdalini Aroniada[a,*], Charalampos Christodoulou[b], Houda Khaled[a]

[a]*GlaxoSmithKline, Park Road, Ware SG12 0DP, United Kingdom*
[b]*GlaxoSmithKline, Gunnels Wood Road, Stevenage SG1 2NY, United Kingdom*
**Corresponding author email: magdalini.x.aroniada@gsk.com*

Abstract

Direct Compression (DC) is a desirable manufacturing route for pharmaceutical tablets, having fewer unit operations than granulation processes. Assessment of the impact of process operating set points and lot-to-lot raw material property variation on manufacturability and product quality attributes is essential for the development of a robust process. In this study, we describe the development of a DC process model to assess the influence of process operating parameters and material properties on *in vitro* dissolution behavior. We summarize the model equations, followed by its calibration and validation vs. industrial data. We then use Global Sensitivity Analysis to investigate the effects of process operating parameters and material properties on *in vitro* dissolution.

Keywords: Direct compression; Flowsheet model; Pharmaceutical manufacturing, Global Sensitivity Analysis.

1.   Introduction

Detailed understanding of the impact of critical material attributes (CMAs) and critical process parameters (CPPs) that belong to different unit operations of the manufacturing process on tablet intermediate and/or critical quality attributes (iCQAs and CQAs, respectively) requires quantifying how their combinations and interactions of CMAs and CPPs affect product performance. This detailed understanding can be obtained by developing an integrated (flowsheet) model that links together the different unit operations of the process, often termed a "system" model in the pharmaceutical industry. Such a model is based on mechanistic and/or empirical sub-models, each simulating different unit operations of the process and can be used to enhance process understanding.

In this paper, we have developed a system model for an industrial direct compression (DC) line. We analyze the system by performing a variance-based global sensitivity analysis to identify and rank the critical process parameters, gain process understanding and support process development. The paper is structured as follows: (a) materials and methods used, (b) models used to develop the system model and validation, (c) Global Sensitivity Analysis (GSA) and input to the control strategy definition.

2.   Materials and methods

All data presented and used in this study were performed with formulations containing a GlaxoSmithKline (GSK) proprietary active pharmaceutical ingredient (API), microcrystalline cellulose (MCC), croscarmellose sodium and lubricant. For the formulations used, another solid form, "*API Form X*" can be present. The levels of this form depend on the incoming drug substance material and the tablet press operation. The software used for all modelling activities was gPROMS®.

3.   Models

This section describes the approach used to develop the system model for the DC process, how unit operations and product performance were modelled individually and how the sub-models were connected in the validated system model.

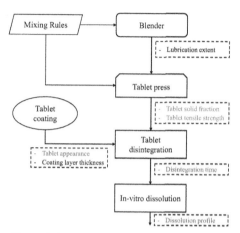

3.1.   Blender and Tablet press

The purpose of the semi-empirical blender model is to calculate the extent of lubrication in the blender (K_b). The extent of lubrication, introduced by Kushner et al. (2012), is a metric to establish the direct correlation between mixing history and the amount of lubrication observed in the system. We considered that the extent of lubrication in the feed frame (K_f) is also significant (Ketterhagen, 2017).

Figure 1. *System model flow diagram (highlighted with orange: iCQAs, and red: CQAs).*

The tablet press model is an empirical model which combines the Kawakita and Lüdde (1971) and the Extended-Kushner (Nassar et al., 2021) models for the calculation of the tablet porosity (i.e., solid fraction) and tensile strength, respectively, as a function of extent of lubrication. These models require calibration with data for the relevant process equipment. The tablet tensile strength can be calculated from Eq. 1:

$$TS/TS_0 \equiv (1 - \beta) + \beta \exp(-\gamma K) \tag{1}$$

where K = total extent of lubrication, γ = tensile strength decay rate, β = material lubrication extent coefficient, TS_0 = tensile strength of the unlubricated blend. The total extent of lubrication, $K = K_b + K_f$ and $K_f = v_{tip}\tau$ where v_{tip} is the tip speed of the feed frame paddle wheel, and τ is the mean particle residence time in the feed frame. The tablet tensile strength is then related to the hardness (the practically measured quantity) as per Pitt and Heasley (2013). The generation of API Form X in the tablet press is captured by a data-driven model whose inputs are API d_{90}, level of API form X and tablet hardness.

3.2.   Tablet disintegration

The tablet disintegration model includes two mechanisms: (1) liquid penetration kinetics and (2) swelling of the tablet (Markl et al., 2017). According to Desai et al. (2016), the break-up of inter-particle bonds can also be caused by strain recovery. Strain recovery is the forcing/transitioning of macromolecules (polymers/disintegrants) into a metastable configuration, e.g., due to spontaneous crystallisation during compaction. In contrast to the omni-directional enlargement of the swollen particles, strain recovery causes a unidirectional increase in the size of the particles. All the mentioned mechanisms of disintegration are interlinked and influenced strongly by the microstructure of the tablet. In this study, since water absorption leads to the particle expansion (swelling) which in turn causes the breakup of the tablet structure, we assume that the penetration rate is the disintegration rate-limiting step (Markl et al., 2017). The fluid flow in a tablet is described by the volume averaged velocity vector $\boldsymbol{u_p}$. This can be viewed as the volumetric flow rate per unit area. Following Bear (1988), to calculate $\boldsymbol{u_p}$, one can solve the continuity and momentum (Darcy's equation) balance equations:

$$\partial_x \boldsymbol{u_p} = 0 \; ; \; \boldsymbol{u_p} = -(\mathcal{K}/\mu_l)\,\partial_x p_p - \rho_l \boldsymbol{g} \tag{2}$$

where p_p = pressure inside the pores, $\mathcal{K}$ = porous medium permeability, $\boldsymbol{g}$ = gravitational acceleration, ρ_l and μ_l = density and viscosity of the penetrating liquid, respectively. In Eq. 2, $\mathcal{K}$ can be found from the Carman-Kozeny permeability equation, $\mathcal{K} = d_h^2 \epsilon / (S_p \mathcal{T}_{or}^2)$ where d_h = pore hydraulic diameter. To account for the fact that $\mathcal{K}$ varies within the same tablet, we introduced an empirical correction factor, Λ, which is fitted to tablet disintegration data. The corrected permeability is given by $\mathcal{K}^* \equiv \Lambda\mathcal{K}$.

To calculate average volume velocity vector, one needs to estimate the average porosity of the tablet. According to Markl et al. (2017), the latter is given by $\epsilon = \epsilon_0 (R_c/R_{co})^2$ where R_c and R_{co} = wet and dry tablet region pore radius. As the tablet swells, it increases in size and this is causing the capillary radius to decrease, approximated as follows:

$$R_c = R_{c0} - \left(\frac{d_p}{2}\right)\left[\left(\frac{\delta_0 + \lambda t}{\delta_0}\right)^{\frac{1}{3}} - 1\right] \tag{3}$$

where δ_0 = initial dry pore length, and λ = swelling rate which can be calculated following Peppas and Colombo (1989). To calculate the wetting front inside the pores, P_d, the wetting front profile can be defined as $F_p(r, z, t) \equiv P_d(r, t) - z$ (Leal, 2007). The substantial derivative of F_p is zero and this allows us to write:

$$\frac{dF_p}{dt} = 0 \Leftrightarrow \partial_t F_p + \boldsymbol{u_p}\,\partial_x F_p = 0 \Rightarrow \partial_t P_d + u_r|_{h_p}\,\partial_r P_d - u_z|_{h_p} = 0 \tag{4}$$

As stated above, the penetration rate is the disintegration rate-limiting step, thus, we calculate the rate of disintegration using the following expression:

$$\frac{dM_t}{dt} = [2\rho_s(1 - \epsilon)\Omega(t)]\frac{dP_d}{dt} \tag{5}$$

where M_t = mass of the tablet at time t, Ω = tablet cross-sectional surface area, and ρ_s = solid true density.

3.3.   *In-vitro dissolution*

For the *in-vitro* dissolution model, we have used the population balance model by Wilson et al. (2012) to simulate the change in the particle size distribution during dissolution. We calculate the rate of particle release from the mass concentration and the released particles' number size distribution. The extent of dissolution is defined as the percentage Label Content (%LC) in solution, calculated as:

$$\%LC = 100 C_b^{API} V_m / x_a \rho_t V_{t0} \tag{6}$$

where C_b = mass concentration of the API in the bulk liquid, V_m = USP <711> (2011) dissolution test vessel liquid volume, V_t = volume of the tablet at $t = 0$ and ρ_t = tablet density which is estimated from $\rho_t \approx \rho_s(1 - \epsilon)$. When $\%LC = 100\%$ the dissolution process is complete.

3.4.   *Coating*

To capture the influence of the coating layer, we used the "erosion rate" approach of Wilson et al. (2012) and Bano et al. (2022a). The main assumption of the model is that the liquid does not penetrate the coating layer, but the latter erodes with a constant erosion rate $\dot{\varepsilon}$. The coating volume V_c on the surface of the tablet with surface area $= A_t$ at any time is given by $V_c(t) = [H_c(t) - \dot{\varepsilon}t]A_t$ where H_c = coating thickness.

3.5.   *Mixing rules*

We have also used a model-based approach to predict the fill bulk density of the blends from individual components. The model is that by Bano et al. (2022b) and used in the blender and tablet press models.

3.6.   *Calibration and validation of the models*

All the above models were calibrated using experimental data from industrial DC campaigns. The system model was then validated vs. experimental data are shown in Fig. 2. The predictions and experimental data are in good agreement.

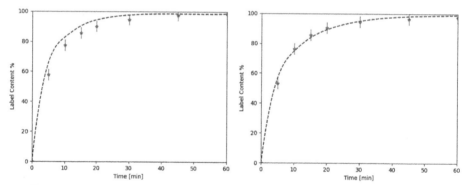

Figure 2. *Model validation (left = nominal blend; right = different blender fill level and compaction force). Red points + error bars = experimental data; Blue curve = model predictions.*

4.   **Results**

Using the validated model, we performed a Global Sensitivity Analysis (GSA) to assess the impacts of different process parameters and material properties on tablet *in-vitro* dissolution. The input factors are: API particle size, form X level in API, moisture content, compaction force, MCC bulk density, lubrication time, feed frame rotational speed, tablet production rate, dissolution medium volume, blender rotational speed, paddle speed, MCC particle size, batch size and medium temperature. The output response is *%LC* at 30 min based on the USP <711> (2011) dissolution test.

4.1.   *Global Sensitivity Analysis*

The first step on GSA is to perform the Elementary Effects (EE) method (Morris, 1991). This identified that the factors that have the biggest effect on the response of interest are the API particle size, level of API form X, the moisture content, the compaction force and the MCC bulk density. We then used the Sobol (1993) variance-based method.

There are two types of Sobol sensitivity indices: the first-order effect index (S_i), that represents the main effect contribution of each input to the variance of the output, and the total effect index (S_{Ti}), which measures the effect of both first- and higher-order effects (interactions). For this study, the total effect indices for the response that the most important factors affecting the Label Content (LC)% are the API Form X and the API particle size (d_{90}). Moreover, the interactions between MCC density, compaction force and water content (or moisture content) are expected to be important (Table 1).

4.2. Assessment of Process Operating Space

In this study we have used the validated model to run simulated experiments to assess the factor criticality (influential factors found in section 4.1). With this analysis we can capture scenarios in the considered process operating space where the tablet dissolution approaches or passes a threshold limit. Using the model, we can observe overall trends across the considered operating space by visualising simulated experiments as a function of

Table 1. *Variance-based sensitivity analysis. Factor ranking; response: LC% @30min.*

Ranking	Input Factors	S_i	S_{Ti}
1	Form X level [in API]	0.510	0.560
2	API particles X90	0.265	0.414
3	Moisture content	0.034	0.061
4	Compaction force	0.000	0.051
5	MCC density	0.000	0.026

different input factors. For example, in Fig. 3 we show a scatter plot that allows us to see the trend when increasing the API particle size. We also observe that some simulations are very close to the threshold limit ($\%LC < 85\%$ at 30 min). For these cases close to the threshold, the level of form X in API and water content are high. This highlights how the model could help to highlight points in the operating space which may pose a product quality issue, which may have been challenging from experiments alone.

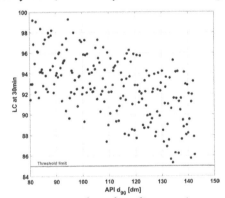

Figure 3. *Scatter plots of in-silico experiments API d_{90} effect on LC at 30min*

Figure 4. *Factor-factor interaction plots, %LC at 30min*

In addition, we performed simulations to visualize factor-factors interactions. In Fig. 4, we performed simulations with high and low values of API particle size vs. high and low MCC densities. In this case, the lines are almost parallel (highlighting that factor-factor interactions are almost negligible) and the limit is not crossed. This factor-factor interaction (API particle size and MCC density) is negligible.

5. Conclusions

In this study, we developed a system model for a DC tabletting line. The system model was calibrated and validated against data collected from industrial campaigns. At this stage, the validation showed that the model is predictive vs. *in vitro* dissolution data and has a very good prediction fidelity. Later, the model was used to perform *in-silico* calculations of dissolution profiles under USP <711> conditions. Based on this, we identified the critical process parameters which affect tablet dissolution (main CQA). In this work, it is highlighted that the modelling approach can be used to provide actionable insight into control strategy definition discussions and to help industrial practitioners develop robust pharmaceutical manufacturing processes.

Acknowledgments

This study was funded by a Digital Design capability project at GlaxoSmithKline (GSK).

References

G. Bano, R.M. Dhenge, S. Diab, D.J. Goodwin, L. Gorringe, M. Ahmed, R. Elkes, S. Zomer, 2022a. Streamlining the development of an industrial dry granulation process for an immediate release tablet with systems modelling. Chem. Eng. Res. Des., 178, 421–437.

G. Bano, M. Aroniada, Y. Vueva, 2022b. A model-based approach to predict the flowability of directly compressed pharmaceutical blends. Comput.-Aided Chem. Eng., 51, 31–6.

J. Bear, 1988. Dynamics of fluids in porous media. Courier Corporation.

P.M. Desai, C.V. Liew, P.W.S. Heng, 2016. Review of disintegrants and the disintegration phenomena. J. Pharm. Sci., 105, 9, 2545–2555.

K. Kawakita, K. Lüdde, 1971, Some considerations on powder compression equations, Powder Technology, 4(2), pp.61-68

W. Ketterhagen, A. Aliseda, M. am Ende, A. Berchielli, P. Doshi, B. Freireich, A. Prpich, 2017. Modeling tablet filmcoating processes. In Predictive Modeling of Pharmaceutical Unit Operations (pp. 273-316). Woodhead Publishing.

J. Kushner, 2012. Incorporating Turbula mixers into a blending scale-up model for evaluating the effect of magnesium stereate on tablet tensile strength and bulk specific volume. Int. J. Pharm., 399, 19–30.

L.G. Leal, 2007. *Advanced transport phenomena: fluid mechanics and convective transport processes*. Cambridge University Press.

D. Markl, S. Yassin, D.I. Wilson, D.J. Goodwin, A. Anderson, J.A., Zeitler, 2017. Mathematical modelling of liquid transport in swelling immediate release tablets. Int. J. Pharm., 526, 1–10.

M.D. Morris, 1991. Factorial sampling plans for preliminary computational experiments. Technometrics, 33, 2, 161–174.

J. Nassar, B. Williams, C. Davies, K. Lief, R. Elkes, 2021. Lubrication empirical model to predict tensile strength of directly compressed powder blends. Int. J. Pharm., 592, 119980.

N.A. Peppas, P. Colombo, 1989. Development of disintegration forces during water penetration in porous pharmaceutical systems, J. Control. Release, 10, 245–250.

K.G. Pitt, M.G. Heasley, 2013. Determination of the tensile strength of elongated tablets. Powder Technology, 238, pp.169-175.

I. Sobol', 1993, Sensitivity analysis for non-linear mathematical models. Mathematical Modeling & Computational Experiment (Engl. Transl.), 1, pp. 407–414.

USP <711> Dissolution. The United States Pharmacopeial Convention. 2011.

D. Wilson, S. Wren, G. Reynolds, 2012. Linking dissolution to disintegration in immediate release tablets using image analysis and a population balance modelling approach. Pharm. Res., 29, 198–208.

Antonis Kokossis, Michael C. Georgiadis, Efstratios N. Pistikopoulos (Eds.)
PROCEEDINGS OF THE 33rd European Symposium on Computer Aided Process Engineering
(ESCAPE33), June 18-21, 2023, Athens, Greece

Enhanced hot-liquid water pretreatment of biomass with recovery and valorization of side products

Tamara Janković,[a] Adrie J. J. Straathof,[a] Anton A. Kiss [a,b]

[a] *Department of Biotechnology, Delft University of Technology, van der Maasweg 9, 2629 HZ Delft, The Netherlands*
[b] *Department of Chemical Engineering, Delft University of Technology, van der Maasweg 9, 2629 HZ Delft, The Netherlands*
A.A.Kiss@tudelft.nl

Abstract

Lignocellulosic biomass potentially represents a great feedstock for biofuel production, but its' pretreatment needs to be enhanced in order to make biorefineries competitive with fossil fuel based alternatives. One way to make biorefineries more economically viable is to recover and valorize all generated by-products during the biomass pretreatment step. The main goal of this original research is to design an optimal process for recovering valuable by-products after hot-liquid water pretreatment of poplar biomass. Rigorous models for all operations included in the recovery process are developed using Aspen Plus as a CAPE tool. An optimal downstream processing sequence, consisting of multiple distillation steps, is designed to recover several valuable components, such as acetic acid, formic acid, furfural and 5-hydroxymethyl-furfural (HMF).

Keywords: biorefineries, lignocellulosic biomass, valorization, distillation

1. Introduction

The increasing concerns on environmental pollution and reduction of available fossil carbon sources lead to a need for a transition towards biofuels and bio-based chemicals from renewable sources. Technologies for the production of the first generation biofuels are already well established. On the other hand, their implementation directly competes with global food production as the first generation biomass can also be used for food supply. The potential solution is usage of lignocellulosic biomass as a source for biorefineries. As the second generation feedstock, it is a non-food crop that can be easily grown on land not suitable for food production. Therefore, the production of biofuel and biochemicals from lignocellulosic biomass does not threaten the food production process (Chiaramonti et al., 2013).

However, technologies and markets for conversion of the second generation feedstocks are still not well established. In order to become competitive with fossil-based production, biorefineries need to be more economically viable (Kiss & Infante Ferreira, 2016). Due to the complexity of the lignocellulosic biomass structure, the pretreatment step is needed to facilitate further conversion of lignocellulosic components to fermentable sugars. Since this process step significantly contributes to the total costs (up to 30-40%) (Ramaswamy et al., 2013), its' enhancement can drastically increase the viability of biorefineries. One way to improve the competitiveness of biorefineries is to recover and valorize valuable by-products generated during the biomass pretreatment step, which has not been the subject of many studies so far. This research gap is covered in this study.

2. Problem statement

2.1. Enhanced biomass pretreatment step

A common problem with different pretreatment methods is usage of additional chemicals. These chemicals might facilitate the pretreatment step but complicate downstream processing, as well as valorization of generated by-products. In that context, hot-liquid water pretreatment potentially presents an eco-friendly solution due to relatively mild process conditions, lack of additional chemicals, low process costs and a possibility to recover and valorize formed by-products (Chen et al., 2022).

2.2. Recovery of valuable by-products from the pretreatment step

After the biomass pretreatment step by hot-liquid water, the solid fraction is sent to enzymatic hydrolysis while the potential value of the liquid fraction is usually neglected. This research focuses on downstream processing of the liquid fraction after hot-liquid water pretreatment of poplar biomass. The main goal is to fully recover valuable by-products (e.g. acetic acid, formic acid, furfural, HMF). These components are not only economically valuable, but they can also inhibit fermentation reactions. Therefore, their removal will result in higher bioethanol yields. A crucial problem regarding by-product recovery is a very dilute solution after the pretreatment step (Kiss et al., 2016). Consequently, improved fluid separation processes are needed in order to recover by-products in a high-purity form that can be valorized on the market.

3. Methodology/Approach

The main objective of this research is to develop an optimal way to recover by-products from the liquid obtained after hot-liquid water pretreatment. The composition of the liquid, which is the feed stream for the designed downstream processing sequence, is obtained from literature (Jiminez-Gutierrez, 2023) and is presented in Table 1. This stream is diluted solution of different products generated by pretreatment of poplar biomass. Acetic acid is formed by hydrolysis of acetyl groups in the hemicellulose chain. C-5 sugars (mainly xylose and arabinose) from hemicellulose can degrade to furfural, while C-6 sugars (mainly glucose, mannose and galactose) from both cellulose and hemicellulose can degrade to HMF. Formic acid can be generated from both furfural and HMF, while levulinic acid can be formed by degradation of HMF. Rigorous Aspen Plus simulations are developed for every process operation. Components are defined in Aspen Plus according to the literature (Aspen Technology, 2021). Due to presence of different carboxylic acids in the initial mixture, NRTL property method coupled with Hayden-O'Connell model with correction for vapor phase association of carboxylic acids is used (Aspen Technology, 2020). Properties of the components and possible formation of azeotropes (Gmehling et al., 2004) are given in Table 2.

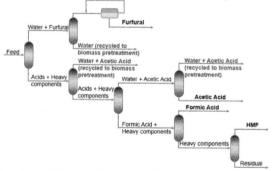

Figure 1. Simplified process flow scheme of the by-products recovery process

Table 1. Feed stream

Component	%wt
Water	96.568
Acetic acid	1.747
Formic acid	0.427
Furfural	0.297
HMF	0.128
Levulinic acid	0.005
Glucose and xylose	0.097
Other heavy components	0.730

Table 2. Components and azeotropes (1 atm)

Pure components		Azeotropes		
Component	T_b (°C)	Component	Molar fraction	T (°C) / type
Water	100.02	Water	0.9066	97.79
Formic acid	100.55	Furfural	0.0934	/heterogeneous
Acetic acid	118.01	Water	0.3607	106.50
Furfural	161.35	Formic acid	0.4998	/homogenous
		Acetic acid	0.1395	
Levulinic acid	256.98			
HMF	281.85	Water	0.4566	106.81
Glucose	343.85	Formic acid	0.5434	/homogenous

Different methods for recovery of present components from diluted solutions are proposed in the literature (Li et al., 2016), but recovery yields are usually insufficient and the usage of additional chemicals complicates downstream processing. Research is often done on lab-scale and scale-up is not considered. To avoid adding extra chemicals, distillation is chosen and component separation is proposed in a defined order (Figure 1). Since the feed is very diluted, the first step is to concentrate it by separating as much water as possible. Due to the formation of light heterogeneous azeotrope (Figure 2), furfural is separated together with water. It can further be recovered by simple phase separation in a decanter. Recovery of acetic and formic acids from an aqueous solution is complex due to existence of four distillation areas and two azeotropes (Figure 3). As acetic acid concentration is the highest of all by-products in the feed, its' recovery is crucial. Therefore, most of acetic acid and water are separated from formic acid and heavier components (HMF, levulinic acid, dissolved sugars and solubilized lignin components) at temperature lower than needed for ternary azeotrope formation. Further recovery of acetic acid is complex due to the tangent pinch at high water concentrations. Formic acid, plus the remaining water and acetic acid, can be separated from HMF and other high boiler components by distillation. The obtained mixture can be further treated in order to obtain a high-purity formic acid product. HMF can be recovered in a high-purity form, while other components can be sent to the fermentation step.

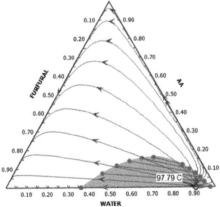

Figure 2. Ternary diagram for system water (W) - acetic acid (AA) - furfural (FURFURAL), mole basis

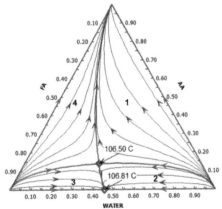

Figure 3. Residue Curve Map for system water (W) - formic acid (FA) - acetic acid (AA), mole basis

4. Results and discussion

4.1. By products recovery process design

The proposed optimal process design for recovery of valuable by-products after biomass pretreatment by hot-liquid water recovers acetic acid, furfural, HMF and formic acid. This process (Figure 4) consists of seven distillation columns and one decanter. Light azeotrope water-furfural is separated in column C-1, after which furfural is purified by additional distillation in column C-2 and phase separation in decanter DEC-1. This results in 100% furfural recovery, while purity of product stream is 97.9%wt. After separation of the water – furfural azeotrope, the solution is still very dilute and is further concentrated in column C-3. Some acetic acid is separated together with water, resulting in top product from this column with 0.5%wt of acetic acid (13.8% from acetic acid present in the feed stream). Since addition of acetic acid in pretreatment step improves digestibility of biomass by decreasing pH to mild-low values (Jiminez-Gutierrez, 2023), this stream can be recycled back to the biomass pretreatment step. Bottom product from column C-3 is further distilled in column C-4. Acetic acid, together with most of the water, is separated as a top product, while formic acid, levulinic acid, HMF and other heavy components are obtained in the bottom stream. Acetic acid is purified in column C-5 and obtained in a high purity form (99.8%wt), with total recovery of 78.8 %. HMF and heavier components are separated from formic acid, levulinic acid and remains of acetic acid and water in column C-6. HMF is further purified in column C-7 and obtained as high-purity (100%wt) top product with total recovery of 99.6 %, while heavy components that are obtained at the bottom can be recycled back to the fermentation step. Top product from column C-6 is stream with 73.6%wt of formic acid. Since composition of this stream is in distillation region 3 in Figure 3, high recovery of pure formic acid cannot be obtained by simple distillation due to the thermodynamic limitations of the water – formic acid azeotrope formation. Therefore, reactive distillation with methanol could be applied to completely recover formic acid (Painer et al., 2015). Moreover, additional purification of formic acid might not be necessary as there are applications of 70% formic acid solution (Singhal et al., 2015). Condition and composition of feed stream and product streams are given in Table 3.

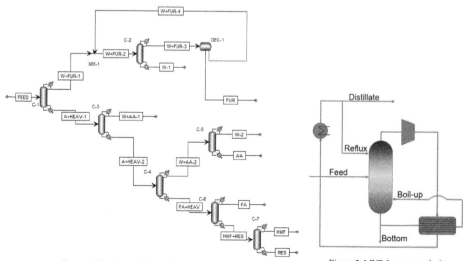

Figure 4. Flowsheet of by-products recovery process   Figure 5. MVR heat pump design

Table 3. Condition and composition of feed stream and product streams for by-products recovery process

Stream	FEED	FUR	AA	FA	HMF
Temperature [°C]	99.70	27.36	26.00	34.82	26.00
Pressure [bar]	1	1	1	0.05	0.01
Mass Flow [kg/h]	110000	333.89	1517.00	635.11	140.70
Mass Fraction					
Water	0.9657	0.0206	0.0003	0.2187	0
Acetic Acid	0.0175	0	0.9982	0.0360	0
Formic Acid	0.0043	0	0.0015	0.7358	0
Furfural	0.0030	0.9794	0	0	0
HMF	0.0013	0	0	0.0003	1
Levulinic acid	0.0001	0	0	0.0091	0
Glucose, xylose and other heavy components	0.0083	0	0	0	0

Table 4. Design parameters of distillation columns in by-products recovery process

Column	C-1	C-2	C-3	C-4	C-5	C-6	C-7
No. Stages	52	30	60	48	66	20	20
Feed Stage	35	12	40	36	33	10	13
Pressure [bar]	1	1	1	1	1	0.05	0.01

4.2. Distillation columns design

Design parameters (number of stages, feed stage and column pressure) for each of the columns are given in Table 4. As the feed stream is very dilute, large amounts of water need to be evaporated in C-1, C-3, C-4 and C-5. Hence these columns have larger number of stages and higher reboiler duties than the other columns. However, the top vapor from these columns can be used to evaporate bottom liquid by implementing mechanical vapor recompression as heat pump. Columns C-6 and C-7 operate at lower pressure to reduce temperatures and avoid thermal degradation.

4.3. Energy analysis

Mechanical vapor recompression (MVR) and heat integration (HI) are implemented to enhance base process design. MVR is a heat pump design that can be applied to decrease energy use when separating close boiling components (Kiss & Infante Ferreira, 2016). This implies compressing top vapor and using it to evaporate bottom liquid (Figure 5). In the designed optimal recovery process, MVR is applied to the columns C-1, C-3, C-4 and C-5. Comparison of key performance indicators for processes before and after enhancing process design is given in Table 5. The inefficiencies in power generation are taken into account for calculation of the total energy requirements through an electric to thermal conversion factor of 2.5. Total investment cost (TIC), total operating cost (TOC) and total annual cost (TAC) are calculated according to a published procedure (Kiss, 2013), with a payback time of 10 years. A Marshall and Swift cost index of 1638.2 (end of 2018) is used (Mantingh & Kiss, 2021). Emissions of CO_2 are calculated using the equations from the literature (Mantingh & Kiss, 2021).

Table 5. Key performance indicators for processes with and without MVR

Performance indicator	Base case	Enhanced case	Savings (%)
Thermal energy requirements (kW$_{th}$ h/kg$_{product}$)	75.13	0.18	99.8
Electrical energy requirements (kW$_e$ h/kg$_{product}$)	n/a	3.53	n/a
Primary energy requirements (kW$_{th}$ h/kg$_{product}$)	75.13	8.99	88.0
Total Investment Cost (k$)	24923	48284	n/a
Total Operating Cost ($/kg$_{product}$)	2.29	0.33	85.5
Total Annual Cost ($/kg$_{product}$)	2.41	0.56	76.6
CO_2 emissions (kg$_{CO2}$/kg$_{product}$)	10.84	1.88/ 0.03*	82.7/ 99.7*

*electricity from non-renewable sources / green electricity

5. Conclusions

This original work proposes an optimal process design for recovering by-products after biomass pretreatment by hot-liquid water. Improved fluid separation processes are used to obtain valuable products from very dilute feed stream (>96%wt water). Several biomass pretreatment products are effectively recovered, including: acetic acid, formic acid, furfural and HMF, with recoveries of around 78.8%, 99.5%, 100% and 99.6% respectively. Acetic acid, furfural and HMF are obtained in high-purity forms (99.8%wt, 97.9%wt and 100%wt respectively), while formic acid is recovered as a 73.6wt% solution. MVR and HI are efficiently implemented to further decrease the energy usage. Due to the use of additional compressors, TIC after implementation of MVR and HI is about 93.7% higher, but TOC and TAC are decreased for 85.5% and 76.6% (from 2.29 to 0.33 \$/kg$_{product}$ and from 2.41 to 0.56 \$/kg$_{product}$ respectively). Moreover, usage of MVR resulted in 88.0% energy savings and significantly lower CO_2 emission leading to a more environmentally friendly process.

References

Aspen Technology, *Aspen Physical Property System* (2020)

Aspen Technology, *Characterization of Lignocellulosic Biomass with Specification in Aspen Properties.* (2021). https://knowledgecenter.aspentech.com

Chen, W. H., Nižetić, S., Sirohi, R., Huang, Z., Luque, R., M.Papadopoulos, A., Sakthivel, R., Phuong Nguyen, X., & Tuan Hoang, A. (2022). Liquid hot water as sustainable biomass pretreatment technique for bioenergy production: A review. *Bioresource Technology, 344*(10). https://doi.org/10.1016/j.biortech.2021.126207

Chiaramonti, D., Giovannini, A., Janssen, R., & Mergner, R. (2013). *Lignocellulosic ethanol production plant by Biochemtex in Italy Lignocellulosic Ethanol Process and Demonstration A Handbook Part I.* WIP Renewable Energies. www.biolyfe.eu

Gmehling, J., Menke, J., Krafczyk, J., & Fischer, K. (2004). *Azeotropic Data* (Second Add). Wiley-VCH.

Jiminez-Gutierrez, J. M. (2023). *Pretreatment of lignocellulosic biomass for acetic acid co-valorization.* Delft University of Technology.

Kiss, A. A. (2013). *Advanced Distillation Technologies: Design, Control and Applications.* Wiley. https://doi.org/10.1002/9781118543702

Kiss, A. A., & Infante Ferreira, C. A. (2016). Heat Pumps in Chemical Process Industry. In *Heat Pumps in Chemical Process Industry.* CRS Press. https://doi.org/10.1201/9781315371030

Kiss, A. A., Lange, J. P., Schuur, B., Brilman, D. W. F., van der Ham, A. G. J., & Kersten, S. R. A. (2016). Separation technology–Making a difference in biorefineries. *Biomass and Bioenergy, 95*(5), 296–309. https://doi.org/10.1016/j.biombioe.2016.05.021

Li, Q.-Z., Feng, X.-J., Zhang, H.-B., Liu, H.-Z., Xian, M., Sun, C., Wang, J.-M., & Jiang, X.-L. (2016). Recovery Processes of Organic Acids from Fermentation Broths in the Biomass-Based Industry. *Journal of Microbiology and Biotechnology, 26*(1), 1–8. https://doi.org/10.4014/jmb.1505.05049

Mantingh, J., & Kiss, A. A. (2021). Enhanced process for energy efficient extraction of 1,3-butadiene from a crude C4 cut. *Separation and Purification Technology, 267*(3). https://doi.org/10.1016/j.seppur.2021.118656

Painer, D., Lux, S., & Siebenhofer, M. (2015). Recovery of Formic Acid and Acetic Acid from Waste Water Using Reactive Distillation. *Separation Science and Technology, 50*(18), 2930–2936. https://doi.org/10.1080/01496395.2015.1085407

Ramaswamy, S., Huang, H.-J., & Ramao, B. V. (2013). *Separation and Purification Technologies in Biorefineries.* Wiley. https://doi.org/10.1002/9781118493441

Singhal, N., Kumar, M., Kanaujia, P. K., & Virdi, J. S. (2015). MALDI-TOF mass spectrometry: An emerging technology for microbial identification and diagnosis. *Frontiers in Microbiology, 6*(8), 1–16. https://doi.org/10.3389/fmicb.2015.00791

Antonis Kokossis, Michael C. Georgiadis, Efstratios N. Pistikopoulos (Eds.)
PROCEEDINGS OF THE 33rd European Symposium on Computer Aided Process Engineering
(ESCAPE33), June 18-21, 2023, Athens, Greece

Development of deep learning framework to predict physicochemical properties for Ionic liquids

Sadah Mohammed[a], Fadwa Eljack[a], Saad Al-Sobhi[a], Monzure-Khoda Kazi[b]

[a]*Qatar University, Department of Chemical Engineering, College of Engineering, P.O. Box 2713, Doha, Qatar*
[b]*Texas A&M Energy Institute, Texas A&M University, College Station, TX 77843-3372, United States*
Fadwa.Eljack@qu.edu.qa

Abstract

In this paper, a deep learning-based group contribution approach has been developed to identify the optimum structure for ionic liquids (ILs) and to maximize the CO_2 absorption capacity. The suggested methodology demonstrates the steps required to build two deep-learning-based group contribution models, DNN-GC and ANN-GC, separately for IL viscosity and CO_2 solubility in the programming language 'Python' by using the two widely used modules, 'Scikit-Learn' and 'TensorFlow'. The CO_2 solubility of IL candidates may not result in the best CO_2 absorption process performance because many other properties of IL, such as viscosity, may affect process performance. To resolve this issue, the two developed models were merged to increase the model accuracy and to predict CO_2 solubility and IL viscosity at the same operating pressure and temperature. Based on the latter steps, an illustrative post-combustion CO_2 capture case study was performed to assess the applicability of the model to select the best IL candidate. With this method, a trade-off between CO_2 solubility and viscosity of ILs can be explicitly studied.

Keywords: Deep learning, Group contribution, Ionic liquids, Physiochemical properties

1. Introduction

Recently, ionic liquids (ILs) have been the subject of intensive research due to their tunability to fit the needs of a certain application. Also, these solvents possess several advantages including low energy required for energy regeneration, low vapor pressure, and non-corrosive (Nath & Henni, 2020). Despite these merits, their tunability also makes it difficult to identify the optimum candidate for CO_2 capture from the thousands of potential candidates. The number of novel ILs is on the rise and yet the experimental data on these novel ILs are scarce or not readily available in the open literature. Therefore, a robust and less time-consuming computational method such as predictive models, especially deep learning (DL) algorithms is needed for the optimal design of ILs and to accurately understand the dynamics of their physiochemical properties (Jian et al., 2022). Deep neural networks (DNNs) have a greater depth in terms of network layers and a larger number of network parameters compared to conventional shallow neural networks (Deng et al., 2020). Deep learning (DL) is the primary machine-learning model for predicting structure properties and is composed of deep neural networks (DNN) and artificial neural networks (ANN) (Schmidhuber, 2015). DL which is a subset of Machine learning (ML) are empirical models that apply multiple structures of a neural network to significantly enhance the computing efficiency leading to an increase in neuronal complexity (Pezhman et al., 2022). Artificial neural networks (ANNs), employ most DL algorithms, consisting of an input layer, an output layer, and one hidden layer (Pezhman et al., 2022). Both the DNN and ANN model receives inputs as artificial features from raw data, where they get correlated between layers until they reach the property prediction at the final layer. However, in DNN the hidden layer comprises a multi-layer to allow the neural network to reach deeper levels and to have more network parameters. Moreover, they are computationally advanced in terms of finding non-linear correlations (Deng et al., 2020). The chief objective of this work is to develop two deep learning-based group contribution (GC) models in python to predict ILs viscosity and their CO_2 absorption capacity at different pressure and temperature based on their cation-anion skeletons and functional groups. Furthermore, the two models are merged to optimize new IL molecule structure based on viscosity, cation-anion skeleton, and cation-substituents to achieve maximum CO_2 solubility in a post-combustion carbon capture case study.

2. Methodology

2.1 Data collection

In this study, a total of 3466 data points on viscosity and CO_2 solubility for 48 ILs, 1733 data points each, as well as their critical properties (T_C, P_C, and ω) were collected from Ionic liquids Database-ILThermo (v2.0) (*Ionic Liquids Database - ILThermo*, n.d.) at different temperature and pressure. The classifications of the cations in the selected ILs are Imidazolium (41%), Ammonium (15%), Phosphonium (15%), and Pryridinium (29%).

2.2 DL-GC model development

GC model is widely used in IL solvent design and predicting their thermophysical properties, and for many various decomposition approaches of IL molecules (Chen et al., 2022). A total of 28 different functional groups were obtained from IL molecular structure consisting of 4 cations, 13 anions, and 11 cation substituents (e.g. organic and inorganic groups) as shown in Table 1. So far, many DL-based non-linear GC models have been suggested for property estimation, and the results show that these non-linear models have higher prediction accuracy compared to linear models. Therefore, here the aim is to incorporate DNN and ANN with the GC model to construct a predictive model for the IL viscosity and CO_2 solubility. DNN was selected to predict IL viscosity due to better performance over ANN during model development. The optimum DL-GC architecture and hyperparameters will be identified through the grid search method in TensorFlow for DNN and ANN to determine the optimum number of neurons in each hidden layer.

Table 1: Functional group considered for ionic liquid design

IL functional group	GC parameters
Cation core	$[Im_{13}]^+$, $[Im_{123}]^+$, $[Pyr_{11}]^+$, $[Pyr_{13}]^+$, $[N]^+$, $[P]^+$
Anion core	$[BF4]^-$, $[Cl]^-$, $[DCA]^-$, $[PF_6]^-$, $[SCN]^-$, $[C(CN)_3]^-$, $[Tf_2N]^-$, $[MeSO_4]^-$, $[CH_2SO_4]^-$, $[TfO]^-$, $[CF_3COO]^-$, $[HCOO]^-$, $[NO_3]^-$, $[DCN]^-$, $[AC]^-$, $[(CH_2)_2PO_2]^-$
Cation substituents	$[CH_3]$, $[CH_2]$, $[OH]$, $[aC-CH_3]$, $[aN-CH_3]$, $[aN-CH_2]$, $[cN-CH_3]$, $[cN-CH_2]$, $[N-CH_2]$, $[N-CH_3]$, $[P-CH_2]$

2.3 Data pre-processing

As a first step, the collected data need to go through the pre-processing step to remove anomalies within the data such as outliers or missing data values. After that, feature selection was applied to select the independent variables for model inputs. In the DNN-GC model, certain features were dropped to improve the performance of the model. In total, 39 and 46 independent variables including the group contribution parameters (see Table 1) for DNN-GC and ANN-GC models, respectively, were identified to predict both dependent variables, the viscosity and the CO_2 solubility at different temperatures and pressure, resulting in 39 and 46 input nodes in the input layer and one output node in the output layers in both models. The following step is splitting the acquired dataset into two sets, the training set (90%) and the test set (10%), by using an in-built function in the Scikit-learn python library called train_test_split function.

2.4 Building and training the DL model

To continue with DL model development, both DNN and ANN were constructed and trained separately. A general structure for DNN-GC consists of multiple hidden layers and each layer comprises several neural elements. ANN-GC model consists of only one hidden layer that contains many neural elements. Both models were built using the Keras-Dense layer available in the TensorFlow package. A commonly used activation function named Rectified linear unit (ReLU) was selected for this study because this function acts as an excellent predictor that activates all neurons at the same time (Zarra et al., 2019). To speed up the optimization process during the training stage, the adaptive moment estimation (Adam) optimizer of TensorFlow was used. The loss function 'mean_absolute_error' (MAE) was used during training to identify the best regression based on the value of MAE. To assess the predictive performance of both models, three standard metrics were chosen namely Mean Absolute Error (MAE), Root Mean Square Error (RMSE), and coefficient of correlation (R^2).

3. Results and discussion

3.1 Identifying the optimum structure for DNN and ANN

To increase the accuracy of the two models, the grid search hyperparameters tuning method was applied to search for the optimum number of hidden layers, neurons, and epochs. When the grid search was used for the DNN-GC model, it was observed that 2 hidden layers with 30 and 256 neurons each, with a number of epochs 10,000, and activation function ReLU and sigmoid for the hidden layer and the output layer, respectively, were required to increase the model accuracy Also, MAE is used for training and validation loss function as stated earlier to validate the accuracy of the model. Furthermore, regularization and dropout methods were also used to avoid overfitting or underfitting during the training. A similar approach was followed for the ANN-GC model, only one hidden layer with 40 neurons was enough to increase the accuracy of the model and neither regularization nor dropout was required for the ANN-GC model. After training both models for nearly an hour, and based on the optimum structures, both models were validated graphically and statistically. Figure 1 depicts the correlated viscosity and CO_2 solubility values for both models versus the corresponding experimental data. Both predicted values for both models are accumulated around the 45° line which indicates a good fit. As for the statistical evaluation, the results have shown high R^2 of 95% and 96%, RMSE of 0.052 and 0.065, and MAE of 0.042 and 0.021 for DNN-GC and ANN-GC, respectively.

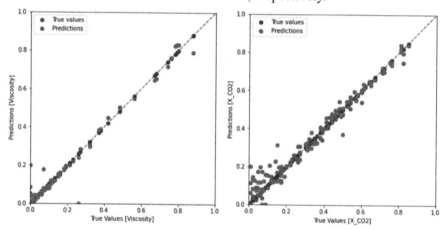

Figure 1: Cross-plot of true values versus predictions for Viscosity using DNN (Left) and CO_2 solubility using ANN (Right)

3.2 Illustrative case study

To explicitly study the trade-off between the CO_2 solubility and viscosity of ILs, both trained models were merged by making the output of the DNN-GC model (viscosity) as an input feature to the ANN-GC model to simultaneously understand the behavior of the viscosity of the ILs and their CO_2 solubility capacity by varying temperature and pressure. The 'Keras.Model' option available in TensorFlow was used to merge both models. To find the optimum parameters (activation function and epochs), a grid search was applied.

Optimizers ReLU and sigmoid for hidden and output layers, respectively, gave better performance, and an epoch of 8000 improved model accuracy (training time = 25 minutes). The previously used metrics were also applied to the merged model. As for graphical validation, a good fit is depicted and almost all predicted data are accumulated around the 45° line as shown in Figure 2. For statistical validation, results showed that the merged model has better accuracy with an R^2 of 97%. RMSE and MAE are 0.041 and 0.022, respectively. This indicates that by merging models, the performance of the model is enhanced.

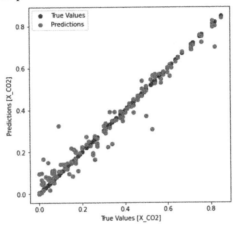

Figure 2: Cross-plot of true values versus predictions for CO_2 solubility using the merged model – DNN-ANN

In this illustrative study, the objective is to demonstrate the capability of the developed models to find the optimum structure of IL with maximal CO_2 absorption capacity using the given functional group in Table 1. Two constraints were set for the feasibility of ILs: (1) the maximum number of cation substituents [CH2] was set to be 8 and (2) viscosity less than 0.1 Pa.s. Also, only cation core $[Im_{13}]^+$ and $[Im_{123}]^+$ were considered in this example. To illustrate a post-combustion CO_2 capture, absorption conditions were pre-specified as 293.15 K and 1.5 MPa. A python code was developed to predict the optimum combination of a functional group to obtain an IL with high CO_2 capacity and low viscosity. Table 2 presents the best IL candidates, with the following IL structure cation $[Im_{13}]^+$ anion $[Tf_2N]^-$, 1OH, 8CH2, 1 aN-CH3, 1aN-CH2 having the highest CO_2 removal and low value of viscosity (< 0.1 Pa.s). However, IL with anion $[PF_6]^-$ might not be a viable option for the CO_2 capture process due to its high viscosity.

Table 2: Merged model results of the optimal ILs

Optimal IL group combination	IL Viscosity (Pa.s)	CO₂ Solubility
1OH, 8CH2, 1[Im13]⁺, 1 aN-CH3, 1aN-CH2, [Tf2N]⁻	0.0111	0.642
1OH, 8CH2, 1[Im13]⁺, 1 aN-CH3, 1aN-CH2, [PF6]⁻	0.553	0.543
1CH3, 7CH2, 1[Im123]⁺, 1 aN-CH3, 1aN-CH2, [AC]⁻	0.080	0.647

4. Conclusion

This paper presents a new modeling framework for the design of the optimal molecular structure of IL for CO_2 solubility, from a given set of functional groups. One important target property considered in this study is viscosity since it affects the pumping power required and ultimately the operating cost. To achieve this, two predictive deep learning models, DNN and ANN, were built and trained separately for IL viscosity and CO_2 solubility, respectively. The grid search method for hyperparameters tuning was employed to find the optimum DL-GC structure with regularization and dropout applied to avoid overfitting and underfitting. Results of validation metrics R^2 were found to be 95% and 96% for DNN and ANN, respectively. Then, both models were merged, with the output of the DNN model (viscosity) used as an input feature to the ANN model (CO_2 solubility). This improved the performance of the model ($R^2 = 97\%$). Furthermore, to investigate the applicability of the merged model, a case study was carried out to design optimum ILs based on a set of structural and viscosity constraints. Results showed that IL with $[Im_{13}]^+$, $[Tf2N]^-$, and cation substituents (1OH) is a good candidate in terms of viscosity and CO_2 solubility.

5. Acknowledgment

The authors acknowledge the paper was made possible by grant QUHI-CENG-22-23-465 from Qatar University. The statements made herein are solely the responsibility of the author[s].

6. Reference

Chen, Y., Peng, B., Kontogeorgis, G. M., & Liang, X. (2022). Machine learning for the prediction of viscosity of ionic liquid–water mixtures. *Journal of Molecular Liquids, 350*, 118546. https://doi.org/10.1016/j.molliq.2022.118546

Deng, T., Liu, F. hai, & Jia, G. zhu. (2020). Prediction carbon dioxide solubility in ionic liquids based on deep learning. *Molecular Physics, 118*(6), 1–8. https://doi.org/10.1080/00268976.2019.1652367

Ionic Liquids Database - ILThermo. (n.d.). Retrieved September 11, 2022, from https://ilthermo.boulder.nist.gov/

Nath, D., & Henni, A. (2020). Fluid Phase Equilibria Solubility of carbon dioxide (CO 2) in four bis (tri fl uoromethyl-sulfonyl) imide ([Tf 2 N]) based ionic liquids. *Fluid Phase Equilibria, 524*, 112757. https://doi.org/10.1016/j.fluid.2020.112757

Pezhman, S., Atashrouz, S., Nakhaei-kohani, R., & Hadavimoghaddam, F. (2022). Modeling of H2S solubility in ionic liquids using deep learning : A chemical structure-based approach. *Journal of Molecular Liquids, 351*, 118418. https://doi.org/10.1016/j.molliq.2021.118418

Schmidhuber, J. (2015). Deep Learning in neural networks: An overview. *Neural Networks, 61*, 85–117. https://doi.org/10.1016/j.neunet.2014.09.003

Zarra, T., Galang, M. G., Jr, F. B., & Belgiorno, V. (2019). Environmental odour management by arti fi cial neural network – A review. *Environment International, 133*(May), 105189. https://doi.org/10.1016/j.envint.2019.105189

Antonis Kokossis, Michael C. Georgiadis, Efstratios N. Pistikopoulos (Eds.)
PROCEEDINGS OF THE 33rd European Symposium on Computer Aided Process Engineering
(ESCAPE33), June 18-21, 2023, Athens, Greece

Economic optimization of the Northern Italian supply chain for residual plastic packaging waste treatment

Fabio Cieno[a], Daniel Crîstiu[a], Federico d'Amore[a,b], Fabrizio Bezzo[a]

[a]CAPE-Lab – Computer-Aided Process Engineering Laboratory, Department of Industrial Engineering, University of Padova, via Marzolo 9, Padova IT-35131, Italy
[b]Politecnico di Milano, Department of Energy, via Lambruschini 4, IT-20156 Milano (Italy)
fabrizio.bezzo@unipd.it

Abstract

Plastics are present in many products and nearly every commercial sector. However, the management of plastic waste is still an open challenge. This work proposes a mixed integer linear programming optimization model of the Northern Italian supply chain for mixed plastic waste treatment. The model aims at maximizing the economic performance of the overall supply chain, taking into account both the selection of the treatment technologies (i.e. incineration, gasification and pyrolysis) and finding the optimal distribution of the material flows. The best economic scenario is achieved by a supply chain where only incineration plants are selected, yielding a gross profit of 31.8 M€/y for the overall supply chain. However, by varying oil price and specific treatment costs both pyrolysis and incineration plants are selected, showing that pyrolysis could be an economically competitive technology for chemical recycle of the mixed plastic waste, leading to a gross profit of 21.6 M€/y (- 32 % w.r.t. base case scenario).

Keywords: mixed plastic waste, plastic recycling, optimization, supply chain.

1. Introduction

Plastics are key and ubiquitous materials whose increasing consumption poses a challenge with regard to the management of plastic waste, as significant fractions of plastics still end up in landfills or leak into the ocean (Li et al., 2022). The largest fraction of plastic waste is made of plastic packaging; unfortunately, only a limited fraction can be mechanically recycled. The residual fraction, called mixed plastic waste (MPW), is either used as an ingredient for secondary solid fuel production, or directly incinerated or landfilled. The challenges in managing MPW are due to the variability in its composition, and incineration, which is an industrially well-established pathway, can help reducing the amount of waste that needs to be landfilled, and recover electric energy as well. Pyrolysis and gasification represent potentially attractive emerging technologies for MPW management; however, pyrolysis only preserves some molecular complexity and can be considered the first step of a chemical recycle path. Pyrolysis and gasification were indeed identified as the most mature and commercially viable chemical recycling technologies suitable for MPW treatment, according to the Threshold Readiness Level (TRL) assessment of Solis and Silveira (2020).

Supply Chain (SC) optimization raises as an important research topic in achieving an efficient plastic circular economy, and it allows assessing both the optimal selection of

MPW treatment technologies and an optimal distribution of the material flows among the different stages of the SC. The combinatorial complexity of the MPW SC is given by its multi-stage, spatially-explicit nature, and mixed integer linear programming (MILP) models are regarded as powerful mathematical optimization tools that can deal with such systems (Kallrath, 2000). This study aims at proposing a spatially explicit MILP framework for the economic optimization (in terms of maximizing the annual gross profit [€/y]) of MPW SCs, focusing on the Northern Italian region and considering multiple technological options.

2. Modelling framework

This section will describe the stages of the optimized MPW SC in terms of key model input parameters. The problem is formulated as an annualized spatially-explicit MILP modelling framework based on the location (in terms of latitude and longitude) of Italian provinces, sorting centres, incineration plants and oil refineries. The model optimization outcomes provide quantitative information on the SC in terms of economic results and technology selection under different scenarios. Both waste-to-energy (i.e. incineration) and chemical recycling technologies (i.e. pyrolysis and gasification) are considered as treatment options. The block diagram of the MPW SC is represented in Figure 1.

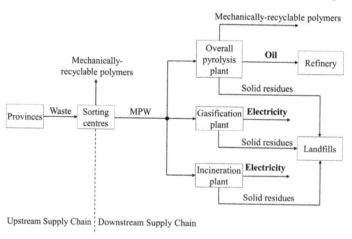

Figure 1. Block diagram of the MPW Supply Chain.

The stages of the MPW SC are divided into upstream SC and downstream SC. The upstream SC deals with the transport of plastic packaging waste (PPW) from provinces to sorting centres, separating the PPW share that can be mechanically recycled from the remaining part, i.e. the MPW. The downstream SC contains the MPW treatment options (incineration, gasification, pyrolysis), whose outputs are: electric power in the case of incineration and gasification, and oil in the case of pyrolysis plants, to be transported to a refinery for further processing (the system boundary is at the gate of the refinery in the case of pyrolysis oil). MPW is a heterogenous mixture, and, to comply with the quality requirements for the pyrolysis oil produced, an additional sorting step is coupled with the pyrolysis plant; its costs and sorting performance are taken into account together with those of pyrolysis. Two options of pyrolysis technology are implemented: *Po. Pyrolysis* – the reactor feed consists of the polyolefin fraction, producing an oil similar to Brent oil. *PoPs. Pyrolysis* – using polyolefin and polystyrene fractions, producing an oil similar to that of Western Canadian Select (WCS) (Mastellone, 2020). Road transport is considered as the only transport option since it reflects current practice for plastic waste.

2.1. Modelling assumptions and inputs

The spatially-explicit nodes are described in the model through a set $n = \{n_{1-86}\}$ comprising 47 provinces, 12 sorting centres, 24 incinerators and 3 refineries. Each node is characterized by its exact geographic coordinates. The incineration plants are the plants currently available in Northern Italy (Utilitalia, 2019). It is assumed that gasification plants can be installed only in proximity of incineration plants; similarly, pyrolysis plants can be sited either by incineration plants or by refineries. The yearly PPW flowrate of each province are taken from COREPLA (2020) and are referred to 2019. Sorting centres are characterized in terms of plant capacity and sorting performances (their PPW-to-MPW conversion factor). All treatment plants generate solid residues that are sent to nearby landfills. For each technology (except incineration) three plant sizes have been considered (*Small, Medium* and *Large*). Each technology has been characterized in terms of relevant mass balances, energy balances, and each plant size and each incinerator has been characterized by its Specific Treatment Costs (*STC* [€/kt]) defined as:

$$STC = \frac{CAPEX + OPEX}{C} \tag{1}$$

where C is the annual plant capacity [kt/y], *OPEX* is the Operational Expenditure [M€/y] and *CAPEX* is the annualized Capital Expenditure [M€/y] calculated as:

$$CAPEX = TCI \cdot \frac{i \cdot (1 + i)^N}{(1 + i)^N - 1} \tag{2}$$

TCI [M€] is the Total Capital Investment (updated using CEPCI, 2019), i is the interest rate, and N is the plant life. For all technologies and plant sizes, it is assumed that $i = 15\%$ and $N = 15$ years.

The decision variables in the optimization problem are:

- the PPW flowrates distribution from the provinces to the sorting centres;
- the MPW flowrates distribution from the sorting centres to the treatment plants;
- the selection of technology, plant size and location of the treatment plants.

2.2. Mathematical Formulation

The objective function of the MILP aims at maximizing the SC annual gross profit:

$$Objective = \max\{GP\} \tag{3}$$

Gross profit *GP* [€/y] is defined as the difference between the total annual revenues (*TR* [€/y]) and total annual costs (*TC* [€/y]):

$$GP = TR - TC \tag{4}$$

Total revenues *TR* [€/y] of Eq.(4) sums up revenues from gate tariffs cashed in $Rev^{T,tot}$ [€/y], and from selling mechanically-recyclable polymers sorted out in the SC Rev^{Mech} [€/y], pyrolysis oil Rev^{Oil} [€/y], and electric power produced Rev^{El} [€/y].

$$TR = Rev^{T,tot} + Rev^{Mech} + Rev^{Oil} + Rev^{El} \tag{5}$$

A gate tariff value of 210 €/t of treated MPW is considered. Revenues from selling the mechanically-recyclable polymers from pyrolysis plants are computed based on 2019 market price 378 €/t. To calculate the revenues from the pyrolysis oil, the oil market price is referring to 2019 average prices, and based on the choice of pyrolysis, either only from polyolefin fraction producing an oil similar to Brent oil (64.3 US$/bbl) or from polyolefin and polystyrene fractions producing an oil similar to WCS (44.3 US$/bbl). Revenues from electric power consider a market price of 52346 €/GWh, which is the average 2019 selling price in the Italian Power Exchange trading market.

The total cost TC [€/y] of Eq. (4) is defined as the sum of total MPW treatment cost ($TPTC$ [€/y]), total landfilling cost (TLC [€/y]) and total transport cost (TTC [€/y]):

$$TC = TPTC + TLC + TTC \tag{6}$$

Total MPW treatment cost ($TPTC$ [€/t]) is calculated as the sum between the overall plant specific treatment cost (STC [€/kt]) and incineration costs for remaining MPW PTC [€/y]:

$$TPTC = \sum m_{k,s}^{MPW} \cdot STC_{k,s} \cdot c_{k,s} + PTC \tag{7}$$

where $m_{k,s}^{MPW}$ [kt/y] is the MPW flowrate treated in a plant k of size s, while $c_{k,s}$ is a factor accounting for cost saving in the case a pyrolysis plant co-exists with a refinery.

The total landfilling cost (TLC [€/y]) is computed from the total residues produced in the SC (m^R [kt/y]) and landfill gate tariff T^l that equals 210 €/t of residues:

$$TLC = m^R \cdot T^l \tag{8}$$

Total transport cost (TTC [€/y]) is the sum of transporting waste, MPW, remaining MPW, oil and residues (TC^W, TC^{MPW}, TC^{RMPW}, TC^{Oil}, TC^R, respectively, all in [€/y]).

$$TTC = TC^W + TC^{MPW} + TC^{RMPW} + TC^{Oil} + TC^R \tag{9}$$

Each term is computed based on Unitary Transportation Cost (UTC [€/kt/km]) specific to the transported material (for solids: UTC^{MPW}, and for oil: UTC^{Oil}), and the matrix of linear distances between nodes calculated with the spherical law of cosines (d'Amore and Bezzo, 2017).

3. Results

The model has been implemented in GAMS software and optimized by means of CPLEX solver. Two different scenarios are considered: a base case scenario containing all treatment options, and an alternative scenario excluding incineration.

The economic performance of the MPW SC for the different scenarios, and a cost breakdown showing are presented in Table 1.

Table 1. Total supply chain cost TC [M€/y] and the contribution of each component into the total cost: total MPW treatment cost $TPTC$ [M€/y], total landfilling cost TLC [M€/y], total transport cost TTC [M€/y]. Annual gross profit GR [M€/y] are also presented.

Scenario	TC	TPTC		TLC		TTC		GP
	[M€/y]	[M€/y]	[%]	[M€/y]	[%]	[M€/y]	[%]	[M€/y]
Base Case	59.1	38.6	65.3	5.4	9.1	15.1	25.5	31.8
1-Pyro	68.2	45.5	66.7	7.1	10.4	15.6	22.9	
Alternative	99.4	66.0	66.4	11.7	11.8	21.7	21.8	21.6

3.1. Base Case Scenario Results

The base case scenario considers all MPW treatment technologies discussed for the economic optimization of the supply chain. It is characterized by a gross profit for the overall SC of 31.8 M€/y. Electric power contributes to about 21 % of total revenues. However, electricity revenues do not compensate for the total treatment costs. The largest contribution to total revenues (79 %) is given by the gate tariff received by treatment plants for treating MPW. The supply chain configurations are shown in Figure 2, for both the upstream and downstream sections.

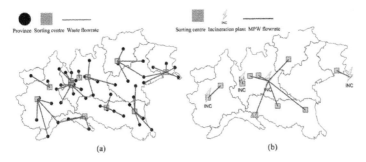

Figure 2. Base case supply chain configuration: (a) upstream SC, (b) downstream SC.

From Figure 2(a) it can be observed that the waste flowrate of each province is sent to the nearest sorting centre. This result suggests that whatever the advantages of sending the waste flowrates to sorting centres that are close to the final treatment plants, they are outmatched by the waste flowrates transport costs. From Figure 2(b), only the Northern Italian incineration plants are selected while no pyrolysis nor gasification plants are selected.

We verified that for the solver to select a different technology the oil price should be increased to a value of 72.7$/bbl for Brent oil and to 50.1$/bbl for WCS oil. The resulting configuration is shown in Figure 3, and it can be observed that both a pyrolysis plant and incineration plants are selected. The selected pyrolysis plant is of the largest plant size (benefiting from economies of scale) and it is located where a refinery exists, suggesting that the cost savings due to the coexistence of an oil refinery play a significant role.

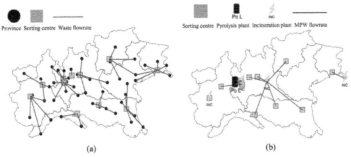

Figure 3. 1-Pyro Scenario MPW SC configuration: (a) upstream SC, (b) downstream SC.

3.2. Alternative Scenario Results

The alternative scenario excludes incineration (except for treating the remaining MPW from pyrolysis). The MPW SC configuration is shown in Figure 4, and it can be observed that only pyrolysis plants are selected, while gasification appears to not be economically competitive w.r.t to the other technologies. From Table 1, the gross profit related to the alternative scenario is 21.6 M€/y, i.e. about 32 % lower w.r.t. the base case scenario.

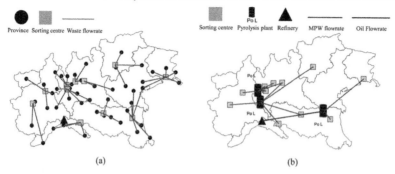

Figure 4. Alternative scenario MPW SC configuration: (a) upstream SC, (b) downstream SC

From Table 1 it can be observed that the total MPW treatment cost (*TPTC*) has the largest contribution representing more than 65 % of the to the total cost (*TC*) in all analysed scenarios. It can be observed that all cost components increase when pyrolysis plants are selected. In particular, the total landfilling cost (*TLC*) increases due to increasing flowrate of total solid residue generated by the pyrolysis in the SC.

4. Conclusions

This study presented a MILP framework for the optimal design of an MPW supply chain in Northern Italy. Both waste-to-energy (i.e. direct incineration) and chemical recycling technologies (i.e. pyrolysis and gasification) were considered as possible MPW treatment options. The economic optimization of the base case scenario gives a configuration in which only incineration is selected, exhibiting an annual gross profit of 31.8 M€/y. However, it was proven for a higher oil price (comparable to 2022 oil prices), pyrolysis becomes economically competitive. Gasification plants are never a viable alternative. If incineration is excluded, only pyrolysis is selected, giving a gross profit of 21.6 M€/y (- 32 % w.r.t. base case scenario). Future work will address a multi-objective optimization accounting for both economic and environmental performances.

References

F. d'Amore, F. Bezzo, 2017, Economic optimization of European supply chains for CO_2 capture, transport and sequestration. International Journal of Greenhouse Gas Control, 65, 99-116.

CEPCI, 2019, Chemical engineering plant cost index [WWW Document, Accessed 21/11/2022], https://www.chemengonline.com/pci-home.

COREPLA, 2020, Management Report 2019, [WWW Document, Accessed 21/11/2022] https://www.corepla.it/sites/default/files/documenti/relazione_sulla_gestione_2019.

J. Kallrath, 2000, Mixed Integer Optimization in the Chemical Process Industry: Experience, Potential and Future Perspectives, Chemical Engineering Research and Design, 78, 6.

H. Li, et al., 2022, Expanding Plastics Recycling Technologies: Chemical Aspects, Technology Status and Challenges, Green Chemistry Journal, 1463-9262.

M.L. Mastellone, 2020, Technical description and performance evaluation of different packaging plastic waste management's systems in a circular economy perspective, The Science of the Total Environment, 718, 137233.

M. Solis, S. Silveira, 2020, Technologies for chemical recycling of household plastics – A technical review and TRL assessment, Waste Management, 105.

Utilitalia, 2019, [WWW Document] https://www.utilitalia.it/pdf/9ea3b0ab-6079-4975-88b0-814663c3fb15 [Accessed 21/11/2022].

Antonis Kokossis, Michael C. Georgiadis, Efstratios N. Pistikopoulos (Eds.)
PROCEEDINGS OF THE 33rd European Symposium on Computer Aided Process Engineering
(ESCAPE33), June 18-21, 2023, Athens, Greece

Optimisation modeling for decision support in the industrial circular economy activities

K. G. Stylianopoulou[a], E. M. Kondili[a], C. M. Papapostolou[a], J. K. Kaldellis[b]

[a]*Optimisation of Production Systems Lab., Mechanical Engineering Department, University of West Attica, Greece*
[b]*Soft Energy Applications and Environmental Protection Lab., Mechanical Engineering Department, University of West Attica, Greece*

Abstract

Recently circular economy has attracted a lot of attention in its ability to handle global challenges related to resource scarcity, sustainability and supply chain unpredictability. Designing circular products is probably one of the most critical factors that manufacturers need to consider when it comes to implementing a closed-loop production strategy. In continuation to our in-progress research related to circular economy applications in the manufacturing and process industry, the present work focuses in the development of optimisation models for the circular economy activities in the industry that will assist substantially in the decision making for implementing the most efficient and sustainable activities. The model development for such complex processes has very significant added value for the evaluation of the economic, environmental and technical circular economy options available. In this context, optimisation models are presented as well as the main problem variables and parameters are identified and the problem constraints are developed. Furthermore, the optimisation criteria for different industrial systems are also described. The results and conclusions of the work highlight the characteristics of the complex circular economy alternatives and develop the basis for the continuation and the expansion of the work in different industrial units.

Keywords: Circular economy modelling, Optimisation, industrial activities.

1. Introduction

The notion of circular economy (CE) the last few years has been widely accepted from manufacturing industries in order to transform their products processing and consumption to more sustainable throughout their supply chain. A transition from linear to a closed loop or circular system requires a shift away from the traditional model of producing products. Manufacturing industries have a significant impact on engineering process that must be adaptable, sustainable and efficient while maintaining high quality and low cost. Nonetheless, the big market demand for a long period of time is resulting in depleting natural resources and making products less sustainable. Over the past few decades efforts have been initiated to reduce our environmental impact through clean manufacturing and life cycle management initiatives (Stylianopoulou et al., 2021). Thus, this necessitates an innovative approach in researching the challenges and limitations of traditional linear economy to circular economy by making products and the supply chain of products more sustainable. One of the activities and focus is extending to a second life products by remanufacturing/recycling used products which are discarded by end-users when they are no longer needed. Advances towards CE mean efforts to improve the circulation of resources, materials and energies by closing loops. Thus, the objective of the present work is to highlight the relevance of CE with optimisation and make a concise review of the current status in optimisation CE models, also, indicating the problem parameters and variables.

2. Tools for Circular Economy

Industries play an important role in implementing the CE notion. Product innovation, redesign, remanufacture, or value chain reconfiguration, can have a significant impact on how resources are managed and consumed. A remanufacturing industry must increase its quantity to effectively control variation in its supply chain and reduce the processing times required to intelligently incorporate remanufactured materials into new ones, such as the use of Industry 4.0 (I4.0). As a result, remanufacturing should be planned in such a way that allows high quantity content in the shortest amount of time when the products are produced (Baratsas et al., 2021, Nascimento et al., 2019). The use of harmonized CE standards and tools can be critical in the implementation of CE. This section presents selected CE tools which promote circularity in industries. Following, selective tools are presented which main goal is to promote circularity in industrial business.

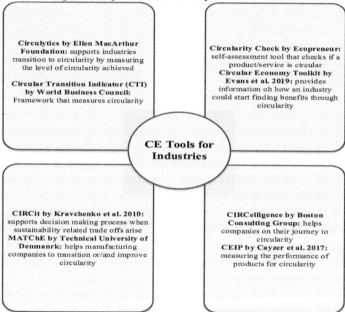

Figure 1: Existing CE Tools for Industries

However, these tools fail to give a quantitative solution in the CE approach by making them more of a framework than quantitative tool. Their approach is on a small amount of product life cycle without giving solution to the entire product lifecycle. Thus, there is a need for more accurate methodologies such as mathematical optimisation to begin implementing CE in industries.

3. Optimisation Models in CE

Circular economy in industries has been gaining interest form researchers and the industries themselves in view of transitioning from linear to circular economy. However, how an industry can approach circularity depends on the framework available to each country and how they can implement circularity.

Reference	Optimisation Model	Description	Field of Research
Zhao et al. 2012	Empirical study	Propose CE model for coal mine enterprise	Mining Industry
Li 2012	Quantitative evaluation model	Waste input-output analysis table	Circular Economy Enterprise
Wu et al. 2014	DEA window analysis	CE efficiency ranking	Country level
Ma et al. 2014	Evaluation system	Estimate circular Economy performance	Iron&steel industry
George et al. 2015	Theoretical model	CE model with two types of resources, polluting input & a recyclable input	CE activities
Igarashi et al. 2016	Multi-criteria optimization	Lower disassembly cost, higher recycling, & CO_2 saving rate	Recycling factories
Rada & Cioca 2017	MSW modelling	Planning waste management for CE	Municipal Solid Waste
Lin 2018	Empirical study	Define product design for smart manufacturing	Glass Industry
Jabbour et al. 2018	ReSOLVE model	Use of I4.0 to enhance CE principal application	Manufacturing Industries
Angouria-Tsorochidou 2018	Techno-economic assessment	Optimised collection of EoL electronic products	Electronic Industry
Rajput & Singh 2019	PCA-DEMATEL	Link between I4.0 & CE	Supply Chain networks
Nascimento et al. 2019	Qualitative research	Integrating I.4 & CE	Waste recovery to the supply chain
Velazquez Mar et al. 2019	Statistical entropy analysis	Optimizing a lithium-ion battery waste system	Battery waste sieving system
Garcia-Barragan et al. 2019	Mathematical model	Measuring the Circular Economy	Virgin & recycled materials
Bal & Badurdeen 2020	IoT & PSS model	Achieve CE through closed-loop material flow	EoL product recovery facilities
Atabaki et al. 2020	Multi-objective MILP model	Cost, CO_2 emissions & energy consumption for durable Products	Closed-loop supply chain
Yadav et al. 2020	Hybrid Best-Worst Method Elimination& Choice Expr. Reality	Framework for sustainable supply chain	Automotive sector
Baratsas et al. 2021	Framework &decision-making tool	CE PSE framework & decision-making tool for the modeling & optimisation of food SC	Food Supply Chains
Karayilan et al. 2021	Linear single-objective optimisation models	Environmental benefits & circularity in plastics wastes supply chains	Plastic Packaging

Table 1: Selected literature review of CE optimisation

For instance the EU supports recycling and recovery targets where in the US there is no relevant framework (Saidani et al., 2019, Bjørnbet et al., 2021). Performance indicators are also required in order to monitor targets in CE models in the manufacturing sector. Our research was conducted in over 100 articles relevant to CE implementation in manufacturing, food and electronic industries. However, only a few of them are related to the optimisation models development in CE. In the context of our focus in developing optimisation models expressing CE strategies for the industry, we have completed a detailed review in the suggested CE relevant optimisation models. The analysis presented in Table 1 is mainly focused in the structure of the optimisation models, its criterion and constraints.   . Through this research it was identified that Industry 4.0 (I4.0) with CE enables a sustainable innovation model that improves end-of-life activities, reduces waste and tracks production and consumption operations. On the other hand, publications that are not using I4.0 are either generally referring to optimisation models or are not giving solutions across the entire product lifecycle from manufacturing to use/reuse to end of life. What has been discovered is that product design optimisation is a rather unexplored area. Some publications proposed frameworks so as to analyse trade-offs between profit, environmental impact and circular performances but without optimizing the physical parameters of a product.

4. Mathematical Model to be developed

The problem considered here is to approach   CE and cleaner production with a mathematical model. In the publications reviewed, there are plenty of quantitative models which are focusing in I 4.0, smart manufacturing, Industrial Internet of Things (IIoT), but its integration with the optimisation models focusing on CE and cleaner production is attempted by only a few researchers.

Parameters	*Variables*
Demand for new product	Qty of product
Demand of new & refurbished product	Qty of product shipped from plant to warehouse to market
Demand for used products	Qty of returned product of least reusable
Demand for remanufacturing products	Qty of new products to be produced
Capacity for processing/production	Qty of remanufactured products to be produced
Energy consumption	Qty of products to be recycled
Production time	Qty of products to be refurbished
Number of returned products	Qty of products to be disposed
Cost of remanufactured products	*Constraints*
Cost of used products	Warehouse for returned product
Cost for new products	Qty of used products allocated from least reusable level
Setup Cost for production	Warehouse for products
Waste disposal cost	*Sets*
Warehouse cost	Set of products
Returned products to warehouse cost	Set of plant
Transportation Cost	Set of warehouses
Cost of disassembling & inspecting returned products	Set of market
Cost of products to be manufactured, cost of refurbished products	Set for time period
Collection and logistics costs	

Table 2: Description of proposed mathematical model parameters, variables and sets

Thus, there is a need of more focused and advanced optimisation models modelling circularity to fulfill demand, clean production and zero waste. Table 2 presents model assumptions, sets, parameters, and variables to formulate a mathematical optimisation model for industries to implement CE.

Currently thee are three dimensions that the circularity in industry is approached in our work.

1. Feasibility models are developed to investigate in detail all the investment and operational costs of potential remanufacturing/ reuse investments. The feasibility model is generic; however specific case study is implemented for the chemical recycle of plastics.
2. Environmental performance software is being implemented to provide various performance indices that are exploited then as inputs in the optimisation models.
3. Optimisation models are currently developed for various CE alternatives in specific industrial cases.

5. Conclusion

The industrial / manufacturing sector plays an important role in transitioning from linear to circular economy. Industries can influence the way resources are consumed and managed through implementing circular economy in their supply chains. Most researchers use the 4Rs strategies, reduce reuse recycle and recover, to approach this topic, however the 10Rs (refuse, rethink, reduce, reuse, repair, refurbish, remanufacture, repurpose, recycle and recover) of CE is more preferable solution in waste scale so as to prevent waste to landfill and achieve zero waste. In order to achieve this there is a must of implementing strategies for product and supply chain design and material selection. However, there are several barriers that backtrack such a transition to a more circular model. Those barriers are the lacking of a unifying definition of CE and a common understanding of the concept, lack of technological innovations in the supply chains of industries that are focusing in CE. In addition, optimisation models presented in the publications are more general and they are not giving solutions across the entire product lifecycle and supply chain of a product. Thus, more research is needed for better understanding of how CE industries can contribute to sustainable production and consumption. Furthermore, the optimisation models for CE product design needs to aim to more pertinent analysis for the whole product life cycle of an industry and comprehend the CE strategies. For this the use of harmonized CE standards and tools can play an important role. Further research needs to be done for optimisation models in material section from larger material databases which are going to benefit the CE product life cycle.

References

E. Angouria-Tsorochidou, C. Cimpan, K. Parajuly, 2018, Optimised collection of EoL electronic products for Circular economy: A techno-economic assessment, Procedia CIRP, 69, 986 – 991

M.S. Atabaki, M. Mohammadi, B. Naderi, 2020, New robust optimisation models for closed-loop supply chain of durable products: Towards a circular economy, Computers & Industrial Engineering, 146

A. Bal, F. Badurdeen, 2020, A Multi-Objective Facility Location Model to Implement Circular Economy, Procedia Manufacturing, 51, 1592-1599

S.G. Baratsas, E.N. Pistikopoulos, S. Avraamidou, 2021, Circular Economy Systems Engineering: A case study on the coffee supply chain, Computer Aided Chemical Engineering, 50, 1541-1546

M.M. Bjørnbet, C. Skaar, A.Ma. Fet, K. Ø. Schulte, 2021, Circular economy in manufacturing companies: A review of case study literature, Journal of Cleaner Production, 294

Ecopreneur, 2022, Circularity Check. https://ecopreneur.eu/circularity-check-landing-page/.

Ellen MacArthur Foundation, 2019, Circulytics: measuring circularity, https://www.ellenmacarthurfoundation.org/resources/apply/circulytics-measuring-circularity

J. Evans, N. Bocken, Univerisity of Cambirdge, 2022, Circilar Economy Toolkit, http://cicrulareconomytoolkit.org/Toolkit.html

J.F. García-Barragán, J. Eyckmans, S. Rousseau, 2019, Defining and Measuring the Circular Economy: A Mathematical Approach, Ecological Economics, 157,369–372

D.A.R. George, B.C. Lin, Y. Chen, 2015, A circular economy model of economic growth, Environmental Modelling & Software, 73, 60-63

K. Igarashi, T.Yamada, S.M. Gupta, M. Inoue, N. Itsubo, 2016, Disassembly system modeling and design with parts selection for cost, recycling and CO2 saving rates using multi criteria optimisation, J. Manuf. Syst, 38, 151–164

A.B.L. Jabbour, C.J.C. Jabbour, , M.G. Filho, D. Roubaud, 2018, Industry 4.0 and the circular economy: a proposed research agenda and original roadmap for sustainable operations, Ann. Oper. Res, 270, 273-286

S. Karayılan, O. Yılmaz, C. Uysal, S. Naneci, 2021, Prospective evaluation of circular economy practices within plastic packaging value chain through optimisation of life cycle impacts and circularity, Resources, Conservation & Recycling, 173

M. Kravchenko, D.C.A. Pigosso, T.C. McAloone, 2020, A Guidance for Navigating Trade-Offs to Support Sustainability-Related Decision-Making, https://doi.org/10.11583/DTU.12933431

S. Li, 2012, The Research on Quantitative Evaluation of Circular Economy Based on Waste Input-Output Analysis, Procedia Environmental Sciences, 12, 65 – 71

K.Y. Lin, 2018, User experience-based product design for smart production to empower industry 4.0 in the glass recycling circular economy, Comput. Ind. Eng., 125, 729-738

S. Ma, Z. Wen, J. Chen, Z. Wen, 2014, Mode of circular economy in China's iron and steel industry: a case study in Wu'an city, Journal of Cleaner Production, 64, 505-512

D.L.M. Nascimento, V. Alencastro, O.L.G. Quelhas, G.G.C. Rodrigo, A.G.R. Jose, L.R. Lona, G. Tortorella, 2019, Exploring Industry 4.0 technologies to enable circular economy practices in a manufacturing context: a business model proposal, J. Manuf. Technol. Manag. 30, 3, 607-627

S. Rajput, S.P. Singh, 2019, Identifying Industry 4.0 IoT enablers by integrated PCA-ISM-DEMATEL approach, Manag. Decis, 57, 8, 1784-1817

E.C. Rada, L., Cioca, 2017, Optimizing the Methodology of Characterization of Municipal Solid Waste in EU Under a Circular Economy Perspective, Energy Procedia, 119, 72-85

H. Rubel, A. Meyer zum Felde, J. Oltmanns, C. Lanfer, L. Bayer, 2020, CIRCelligence by BCG it's Time to close our future resource loops, https://www.bcg.com/de-de/circelligence-by-bcg-close-future-loops

M. Saidani, B. Yannou, Y. Leroy, F. Cluzel, A. Kendall, 2019, A taxonomy of circular economy indicators, Journal of Cleaner Production, 207, 542-559

K.G. Stylianopoulou, E.M. Kondili, J.K. Kaldellis, 2022, Process Systems Engineering prospects in Circular Economy implementation in industry, Computer Aided Chemical Engineering, 51, 1309-1314

Technical University of Denmark, 2019, MATChE - making the transition to circular economy, https://ready2loop.org/en/about

O.V. Velazquez Martínez, K.G.Van Den Boogaart, M. Lundstrom, A. Santasalo-Aarnio, M. Reuter, R. Serna-Guerrero, 2019, Statistical entropy analysis as tool for circular economy: Proof of concept by optimizing a lithium-ion battery waste sieving system, Journal of Cleaner Production, 212, 1568-1579

WBCSD, 2022, CTI tool, https://ctitool.com/

H. Wu, Y. Shi, Q. Xia, W. Zhu, 2014, Effectiveness of the policy of circular economy in China: A DEA-based analysis for the period of 11th five-year-plan, Resouces, Conservation and Recycling, 83, 163-175

G. Yadav, S. Luthra, S.K. Jakhar, S.K. Mangla, D.P. Rai, 2020, A framework to overcome sustainable supply chain challenges through solution measures of industry 4.0 and circular economy: an automotive case, J. Clean. Prod, 254

Y. Zhao, L. Zang, Z. Li, J. Qin, 2012, Discussion on the Model of Mining Circular Economy, Energy Procedia, 16, 438 – 443

Antonis Kokossis, Michael C. Georgiadis, Efstratios N. Pistikopoulos (Eds.)
PROCEEDINGS OF THE 33rd European Symposium on Computer Aided Process Engineering
(ESCAPE33), June 18-21, 2023, Athens, Greece

Model-Based Design of Experiments for the identification of microalgae growth models with limiting nutrients

Alberto Saccardo, Beatriz Felices-Rando, Eleonora Sforza, Fabrizio Bezzo

Department of Industrial Engineering, University of Padova, Via Marzolo 9, Padova 35131, Italy
fabrizio.bezzo@unipd.it

Abstract

Nutrients play a major role in several microalgal industrial processes and their efficient exploitation has an impact on environmental sustainability. The Droop model is suitable for the description of nutrients effect on microalgal growth but presents challenges in parameter identification for the high experimental effort required. In this work, a Model-Based Design of Experiments (MBDoE) was employed to plan information-rich experiments for precise parameter estimation. An initial dynamic sensitivity analysis showed that parameters of the Droop model can be practically identified with measurements of biomass and nitrogen content of the medium only. Finally, MBDoE was employed to plan *in silico* experiments, proving that two optimal experiments are sufficient to achieve accurate parameters estimation.

Keywords: Model Based Design of experiments; Droop model; microalgae; continuous photobioreactors; parameter estimation.

Introduction

Microalgae are a wide variety of photosynthetic microorganisms, both prokaryotic and eukaryotic, that can be found in all ecosystems. Thanks to their ability to exploit photosynthesis to fix CO_2 and consume nutrients as inorganic nitrogen and phosphorous from wastewater, they represent a sustainable alternative for multiple industrial applications (Kumar et al., 2020; Olaizola, 2003). Modelling the effect of nutrients on microalgal growth is of particular interest due to their strong relevance in different microalgal industrial processes and their impact on environmental sustainability. Even though Monod equations are conventionally used to describe microalgal growth, the Droop model better describes the growth rate as a function of the internal quota of the limiting nutrient, since microalgae still exhibit growth when dissolved nutrients are exhausted (Bernard, 2011). The main hurdle when working with the Droop model is related to the high number of measurements and experiments needed to reliably determine the kinetic parameters, resulting in long and time-consuming experimental campaigns (Turetta et al., 2022). In addition, working in steady state continuous cultures, which are more suitable for microalgal production, is even more time intensive, up to 11 weeks for each nutrient.

Based on those reasons, in this work, we propose a Model-Based Design of Experiments (MBDoE) approach to optimize the quality of experimental data while minimizing time and resources consumption (Franceschini and Macchietto, 2008). Accordingly, the objective is to exploit MBDoE to design experiments for maximising the information required for accurate parameter estimation of the Droop model for continuous microalgal cultivation from measurements of biomass and nitrogen concentrations only. After the introduction of the Droop model and the definition of the mathematical foundation of MBDoE, a dynamic sensitivity analysis will be presented to highlight possible issues in parameter estimation. To show the capability of MBDoE of designing experiments for accurate parameter estimation, iterative *in silico* experiments were performed, according to the MBDoE procedure.

1. Model and methodology

1.1. The Droop model

In the Droop model, the specific biomass growth rate μ (d^{-1}) is a function of both light intensity I (μmol m^{-2} s^{-1}) and the internal quota q (g$_N$ g$_x$$^{-1}$) of the limiting nutrient: in this work, nitrogen is considered as the limiting nutrient. Thus, the specific biomass growth can be expressed as (Bernard 2011):

$$\mu = \mu_{max} \frac{I}{I + K_I\left(\frac{I}{I_{opt}} - 1\right)^2}\left(1 - \frac{q_{min}}{q}\right) - k_d \quad , \tag{1}$$

where μ_{max} (d^{-1}) is the maximum specific growth rate of the microorganism, K_I (μmol m^{-2} s^{-1}) is the half saturation constant of light response curve, I_{opt} (μmol m^{-2} s^{-1}) is the optimal light intensity for growth, k_d (d^{-1}) is the specific decay rate and q_{min} (g$_N$ g$_x$$^{-1}$) is the minimum internal cell quota. Light intensity varies along the depth of the reactor and for a flat plate reactor it can be expressed through the Lambert-Beer law:

$$I = I_0 \exp(-k_a c_x z) \quad , \tag{2}$$

where I_0 (μmol m^{-2} s^{-1}) is the incident light intensity, k_a (m^{-2} g$_x$$^{-1}$) is the biomass light absorption coefficient, c_x (g$_x$ m^{-3}) is biomass concentration, and z (m) is the spatial coordinate along the thickness of the reactor. The biomass growth rate r_x (g$_x$ m^{-3} day^{-1}) can be expressed as

$$r_x = \mu c_x \quad , \tag{3}$$

and the average biomass growth rate r_x^{avg} (g$_x$ m^{-3} day^{-1}) along the culture depth W (m) is calculated as:

$$r_x^{avg} = \frac{1}{W}\int_0^W r_x \, dz \quad . \tag{4}$$

The nutrient uptake rate ρ (g$_N$ g$_x$$^{-1}$ d^{-1}) depends both on the external dissolved nutrient concentration c_N (g$_N$ m^{-3}) and on its internal quota (Turetta et al., 2022):

$$\rho = \rho_N \frac{c_N}{K_N + c_N}\left(1 - \frac{q}{q_{max}}\right) \quad , \tag{5}$$

with ρ_N (g$_N$ g$_x$$^{-1}$ d^{-1}) as the maximum nutrient uptake, K_N (g$_N$ m^{-3}) as the nutrient half saturation constant and q_{max} (g$_N$ g$_x$$^{-1}$) as the maximum nutrient quota in the biomass. For a continuous photobioreactor material balances for biomass are:

$$\frac{dc_x}{dt} = \mu(q)c_x - Dc_x \quad , \tag{6}$$

$$\frac{dc_N}{dt} = -\rho(c_N)c_x - Dc_N + Dc_{N,in} \quad , \tag{7}$$

where D (d^{-1}) is the dilution rate and $c_{N,in}$ (g$_N$ m^{-3}) the inlet nitrogen concentration. The internal quota is maintained by nutrient uptake from external medium and consumed by cells duplication and thus its material balance can be expressed as:

$$\frac{dq}{dt} = \rho(c_N) - \mu(q)q \quad . \tag{8}$$

In this work a MBDoE was implemented for planning of experiments to estimate Droop parameters K_N, q_{max}, q_{min}, ρ_N, while values of light parameters μ_{max}, K_I, I_{opt}, k_d, k_a were retrieved from Saccardo et al. (2022).

Model-based design of experiments for the identification of microalgae
growth models with limiting nutrients
2139

1.2. MBDoE

MBDoE is a mathematical technique that, building on *a priori* knowledge embedded in a mechanistic mathematical model, aims at obtaining the maximum information from an experimental apparatus that will yield the most informative data for modelling activity, in a statistical sense, avoiding waste of time and resources (Franceschini and Macchietto, 2008). This approach consists of exploiting the model equations and current parameter values to estimate the information content of the experiment by evaluating a certain objective function. An optimisation framework is then applied to the solution of the resulting numerical problem. The methodology was successfully applied in several fields, including model identification in microalgae systems (Bernardi et al., 2016).

For MBDoE application, Equations 1-8 can be rearranged in a compact form:

$$\begin{cases} \mathbf{f}(\dot{\mathbf{x}}(t), \mathbf{x}(t), \mathbf{u}(t), \boldsymbol{\theta}, \mathbf{k}, t) = 0 \\ \qquad \hat{\mathbf{y}}(t) = \mathbf{h}(\mathbf{x}(t)) \end{cases} \tag{9}$$

where $\mathbf{x}$ is the N_x-dimensional vector of all the model state variables and, $\hat{\mathbf{y}} = [c_x, c_N]^T$ is the N_y-dimensional vector of the measured response variables, $\mathbf{u} = [D, c_{N,in}]^T$ is the N_u-dimensional set of time-varying inputs to the process, $\boldsymbol{\theta} = [K_N, q_{max}, q_{min}, \rho_N]^T$ is the N_θ-dimensional vector of parameters to be determined, determined, $\mathbf{k} = [\mu_{max}, K_I, k_d, k_a, I, W]^T$ the N_k-dimensional vector of constants of the model and $0 < t < \tau$ is time, with τ as the duration of the experiment. The symbol $\wedge$ is used to define the estimate of a variable (e.g. $\hat{\boldsymbol{\theta}}$ contains the best available parameters estimation). Variable measurements are collected at time instant defined in the N_{sp}-dimensional vector $\mathbf{t}_{sp}$. Each experiment is defined by the experiment design vector $\boldsymbol{\phi} = [\mathbf{u}(t), \mathbf{t}_{sp}, \tau]^T$.

A key point of MBDoE is the formulation of a metric for the measurement of experimental information content. In mathematical terms, decreasing the size of the inference regions of the model parameters allows to improve parameter precision. This is equal to decrease the value of the elements of the parameter variance-covariance matrix $\mathbf{V}$ defined as (Franceschini and Macchietto, 2008):

$$\mathbf{V}(\hat{\boldsymbol{\theta}}, \boldsymbol{\phi}) = \left[\sum_r^{N_y} \sum_s^{N_y} \tilde{\sigma}_{rs} \mathbf{Q}_r^T \mathbf{Q}_s \right]^{-1}, \tag{10}$$

with

$$\mathbf{Q}_r = \frac{\partial \hat{y}_r}{\partial \theta_m}\bigg|_{t_l} \quad l = 1, \dots, n_{sp} \quad m = 1, \dots, N_\theta \quad . \tag{11}$$

$\mathbf{Q}_r$ is the (n_{sp} x N_θ) dynamic sensitivity matrix of the rth response. The term $\tilde{\sigma}_{rs}$ in Equation 10 is the (r, s) element of the inverse of variance-covariance matrix of the experimental error. The final designed experiment $\boldsymbol{\phi}^{opt}$ is the solution the following minimisation/maximisation problem:

$$\boldsymbol{\phi}^{opt} = \text{argmin}\{\det[\mathbf{V}(\hat{\boldsymbol{\theta}}, \boldsymbol{\phi})]\} = \text{argmin}\{\det[\mathbf{V}^{-1}(\hat{\boldsymbol{\theta}}, \boldsymbol{\phi})]\} \tag{12}$$

defined with the so-called D-criterion (Franceschini and Macchietto, 2008). In other words, the optimum experiment is the one that minimises the parametric uncertainty or maximises the experimental information content.

2. Results

2.1. Sensitivity analysis

A local dynamic sensitivity analysis was performed to highlight possible parameter identifiability issues. The dynamic trajectories of measured variables were perturbed by one-at-a-time variations of the parameters to be estimated. Nominal values of the parameters to be estimated are shown in Table 1. Additional parameters and constants (fixed light parameters from Saccardo et al. (2022), light intensity and reactor thickness)

A. Saccorda et al.

are shown in Table 2. The dynamic sensitivity $s_{i,p}(t)$ for the ith variable perturbed by the pth parameter at time t was computed as:

$$s_{i,p}(t) = \frac{\hat{y}_i(t) - \hat{y}_{i,p}^{pert}(t)}{\Delta\% \ \hat{y}_{i,REF}} \tag{13}$$

where $\hat{y}_i(t)$ is the value calculated with the nominal set of parameters, $\hat{y}_{i,p}^{pert}(t)$ the perturbed value calculated with the variation in the pth parameter, $\Delta\%$ is the percentage perturbation of the parameter (equal to 1% for all parameters) and $\hat{y}_{i,REF}$ is a reference value for the normalisation. Results of the sensitivity analysis are shown in Figure 1, where the effects of step wise changes in the inputs for all the perturbed parameters on the biomass and nitrogen concentrations are reported. Figure 1(a) highlights that q_{min} is the parameter with the highest impact on the biomass concentration and thus its estimation is expected to be linked to this measurement. Even though K_N, q_{max}, ρ_N have a minor impact on biomass concentration, Figure 1(b), shows that their main effect is on the nitrogen concentration and thus this variable is expected to be mainly responsible for their estimation. Figure 1(b) also evidences that some degree of correlation between q_{max} and ρ_N exists as well as anticorrelation between K_N and q_{max}, thus suggesting possible issues in the estimation of those parameters. However, Figure 1 also suggests that measurements of biomass and nitrogen concentration alone are sufficient for precise estimation of all Droop parameters if experiments are planned adequately.

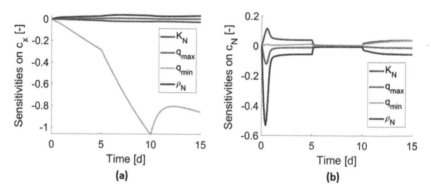

Figure 1: Dynamic sensitivity analysis: sensitivities of model parameters on biomass (a) and nitrogen concentration (b).

Table 1: Reference values of parameters of Droop model.

Parameter	Value
K_N	14.23 g_N m^{-3}
q_{max}	0.90 g_N g_x^{-1}
q_{min}	0.005 g_N g_x^{-1}
ρ_N	0.62 g_N g_x^{-1} d^{-1}

2.2. In silico experiments

In silico iterative experiments have been performed to inspect the capability of MBDoE to plan experiments for precise parameter estimation.

Table 2: Fixed light parameters (Saccardo et al., 2022). and constants for simulations

Name	Value
μ_{max}	2 d^{-1}
K_I	110 μmol m^{-2} s^{-1}
I_{opt}	405 μmol m^{-2} s^{-1}
k_d	0.45 d^{-1}
k_a	0.09 m^{-2} g_x^{-1}
I	222 μmol m^{-2} s^{-1}
W	0.035 m

Reference parameter values of Table 1 have been perturbed by 50% and used to plan a first MBDoE. Then, a simulation using the "true" parameters of Table 1 has been performed with the input trajectories planned by the MBDoE. White noise has been then applied on these simulation results to obtain *in silico* data, employed for a subsequent parameter estimation. If the estimation was not precise the last parametric set was then used for a new MBDoE and the entire procedure was repeated until a precise estimation was achieved. It was found that two planned experiments of 5 and 9 days respectively are sufficient for a precise parameter estimation. Dynamic profiles of the first and second experiment are shown in Figure 2. Notice that in the first experiment input variables were allowed with impulse profiles. Moreover, all the input variables were allowed to vary from 9 AM to 5 PM to reflect real working conditions. Table 3 shows the initial perturbed parameters, parameters fitted after the first and second *in silico* experiment and the corresponding 95% *t*-values. All parameters from the second fit are close to those in Table 1 and have *t*-values above the reference one all above the reference *t*-value, showing high parametric accuracy. Results of the second fitting are shown in Figure 3 for both experiment I and II.

Table 3: Model parameters for *in silico* experiments. Reference *t*-value is equal to 1.67.

Parameter	Perturbed value	First fit	First fit 95% *t*-value	Second fit	Second fit 95% *t*-value
K_N [g_N m^{-3}]	7.115	14.74	4.9	14.73	5.01
q_{max} [g_N g_x^{-1}]	0.45	0.8662	2.5	0.83	2.95
q_{min} [g_N g_x^{-1}]	0.0075	0.00285	0.15	0.0055	1.72
ρ_N [g_N g_x^{-1} d^{-1}]	0.93	0.64	3.29	0.67	3.73

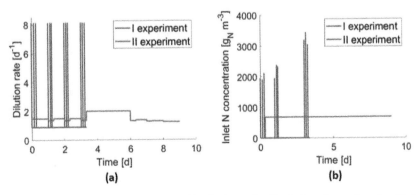

Figure 2: Dynamic profiles of input variables in the optimally designed experiments: dilution rate (a), inlet nitrogen concentration (b).

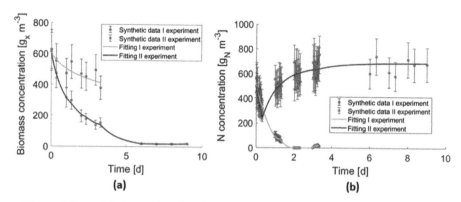

Figure 3: Second fitting results of synthetic data from the two optimal *in silico* experiment: biomass (a) and nitrogen concentration (b).

Conclusions

In this work, the effectiveness of MBDoE techniques for precise parameter estimation of the Droop model was assessed. An initial local sensitivity analysis showed that the biomass concentration is mainly affected by q_{min} and the nitrogen concentration by K_N, q_{max}, ρ_N. Although some degree of correlation (between q_{max} and ρ_N) and anticorrelation (between K_N and q_{max}) exists, parameters are expected to be practically identifiable from measurements of biomass and nitrogen concentrations only. MBDoE was iteratively employed to plan *in silico* experiments, showing that two optimal experiments (of 5 and 9 days) can be sufficient to estimate all the parameters with accuracy. Moreover, MBDoE allowed to reduce the experimental time by over 80%, from about 77 days to 14 days for both experiments. Future work will aim at assessing the effectiveness of the proposed methodology experimentally.

Acknowledgements

This project has received funding from the European Union's Horizon 2020 programme under the Marie Skłodowska-Curie grant agreement No. 955520 "DigitAlgaesation".

References

Bernard, O. (2011). Hurdles and challenges for modelling and control of microalgae for CO2 mitigation and biofuel production. Journal of Process Control, 21(10), 1378–1389.

Bernardi, A., Nikolaou, A., Meneghesso, A., Morosinotto, T., Chachuat, B., Bezzo, F. (2016). High-fidelity modelling methodology of light-limited photosynthetic production in microalgae. PLoS One 11, 1–20.

Franceschini, G., Macchietto, S. (2008). Model-based design of experiments for parameter precision: State of the art. Chemical Engineering Science, 63(19), 4846–4872.

Kumar, M., Sun, Y., Rathour, R., Pandey, A., Thakur, I. S., Tsang, D. C. W. (2020). Algae as potential feedstock for the production of biofuels and value-added products: Opportunities and challenges. In Science of the Total Environment (Vol. 716). Elsevier B.V.

Olaizola, M. (2003). Commercial development of microalgal biotechnology: from the test tube to the marketplace. Biomolecular Engineering, 20(4–6), 459–466.

Saccardo, A., Bezzo, F., Sforza, E. (2022). Microalgae growth in ultra-thin steady-state continuous photobioreactors: assessing self-shading effects. Front. Bioeng. Biotechnol. 10, 1–9.

Turetta, M., Barbera, E., Trentin, G., Bertucco, A., Sforza, E. (2022). Modeling the production of cyanophycin in Synechocystis sp. PCC 6803 cultivated in chemostat reactors. Bioresource Technology Reports, 19, 101132.

Antonis Kokossis, Michael C. Georgiadis, Efstratios N. Pistikopoulos (Eds.)
PROCEEDINGS OF THE 33rd European Symposium on Computer Aided Process Engineering
(ESCAPE33), June 18-21, 2023, Athens, Greece

Enabling circularity: Analysis of factors influencing MSW sorting behaviour in Central Norway

Tuva Grytli,[a] Cansu Birgen,[b]

[a] *SINTEF Industry, P.O. Box 4760 Torgarden, 7465 Trondheim, Norway*
[b] *SINTEF Energy Research, P.O. Box 4760 Torgarden, 7465 Trondheim, Norway*

Abstract

A comprehensive waste composition analysis was undertaken in 2021, covering ten municipal waste companies in Central Norway. The results are used in a multiple linear regression model to explore factors influencing the municipal solid waste sorting behaviour. The aim of the study is to better understand what affects waste sorting, to aid in the progress of achieving circular economy targets. The share of food waste and mixed waste in the mixed waste bin is analysed considering three variables: Food waste collection system, glass and metal packaging waste collection system and type of settlement. The waste collection systems are found to have a significant effect on the sorting behaviour, where the *possibility* for sorting and the *user-friendliness* and *transparency* of the sorting system are important factors. Type of settlement is not found to have a significant effect on the sorting behaviour. We recommend further in-depth studies to confirm and expand on the findings in this study, and to better utilise the potential of waste composition analyses to enhance understanding of waste sorting behaviour. This understanding is crucial to reach a more circular waste system.

Keywords: Municipal solid waste, sorting behaviour, waste management, regression analysis, circular economy.

1. Introduction

To achieve a more circular future, municipal solid waste (MSW) management plays a key role, as large quantities of materials with great potential for reuse and recycling are handled. The 2021 Norwegian material recovery rate for MSW was 43 % (Statistics Norway 2022), far from the 2025 target in line with the EU of 55 % (Directive (EU) 2018/851 2018). Effective sorting at source is vital to increase material recovery, and there is room for improvement as more than 60 % of the waste in the mixed waste bin is recyclable materials (Fagerheim et al 2021). Effective waste strategy planning relies on good data (Thomas 2004), and this paper aims to contribute by studying the effects of various factors on the sorting behaviour.

This work was initiated to test an assumption that there is a difference in MSW sorting behaviour between different size settlements, i.e., cities versus more sparsely populated areas. To test this assumption, consistent and comparable data was needed. In 2021 the largest waste composition analysis in Norway to date was carried out, for the circular waste cluster CIVAC (Fagerheim et al 2021). The analysis was carried out for eight MSW companies in Central Norway, further including results from waste composition analyses performed in the same time period by two other MSW companies in the region. The resulting analysis covers an area of more than 750 000 inhabitants, and more than ten tonnes of waste was analysed. This data set made it possible to compare variations between cities and villages/towns, as well as variations between different sorting systems, on a sample area basis.

2. Methodology

Using the results from the waste composition analysis, a multiple linear regression analysis was carried out, to study what significantly affects waste sorting.

The waste composition analysis investigated the contents of the mixed waste (MW) bin, identifying the share (weight percentage) of the 11 fractions shown in Table 1. The share of the fraction was used as dependent variable, and the regression was run for each fraction. Using a 5 % significance level, significant results were found for five fractions (in italic in Table 1). The focus of this study is the two largest fractions: Food waste including paper towel (FW) and MW which comprises the correctly sorted MW.

Table 1 Overview of dependent variables: waste fractions in the mixed waste bin.

Waste fractions
Paper and cardboard
Food waste incl. paper towel
Garden waste
Bags for waste
Plastic packaging
Glass packaging
Metal packaging
Other metal
Recyclable textiles
Hazardous waste and WEEE
Mixed waste

Twenty distinct sample areas were analysed, nine in cities and 11 in smaller settlements. 15 of the areas have separate collection of FW, and nine have kerbside collection of glass and metal packaging waste (G&M). One of the two additional MSW companies included reported only the totals, which cover eight sample areas, but in our analysis only counts as one sample area: a city without separate FW and kerbside G&M collection. See Table 2 for an overview of the MSW companies' solutions for sorting at source. This

Table 3 Overview of MSW companies' source sorting solutions: basis for independent variables.

	MSW 1	MSW 2	MSW 3	MSW 4	MSW 5	MSW 6	MSW 7	MSW 8	MSW 9	MSW 10
Mixed waste										
Food waste		×							×	×
Paper/cardboard										
Plastic										
Glass/metal										

Kerbside collection	Central coll./recycling point
Optibag	No separate collection: ×

methodology made it possible to isolate three independent variables for the regression analysis, shown in Table 3. As the independent variables are categorical, they were all initially coded as dummy variables (Wooldridge 2009).

Table 2 Overview of independent variables and their coding in the model.

Independent variables	Short name	Values		
		-1	0	1
Separate FW collection	sepFW	No separate collection	Optibag	Separate FW bin
Kerbside G&M collection	kerbG&M		Central recycling point or civic amenity site	Kerbside G&M collection
Settlement			Town/village	City

For separate FW collection, two systems were represented in the sample areas: A separate bin, and the optibag system. In the optibag system, several waste fractions are sorted in colour-coded bags and collected in the same bin, to be machine sorted centrally. Using a scatter plot, in Figure 1, we found that the results from the optibag system lie between no separate collection and a separate FW bin. To corroborate this finding, the share of FW in the MW bin for a city which also uses the optibag system, is included in the figure as *Alt. optibag*. The coding of the dummy variable was changed to linear, increasing the explanatory power of the model (R^2) by 10 percentage points.

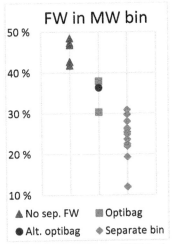

Figure 1 Actual shares of FW in the MW bin, by source sorting solution.

This study has a small number of observations for a multiple regression analysis, which can increase uncertainty. This is because it was carried out at a sample area level rather than a household level. A sample area is however aggregated waste from typically 30-40 households, and we expect that the analysis represents around 600-800 households. Each bin collected is a snapshot in time, which may not be representative of the households' average waste composition. Aggregating into sample areas may thus make a more representative waste composition, but for an area rather than a household. As we do not have any information about each household, the additional resolution would give limited value, as variations cannot be explained. On the other hand, using data from only one analysis will ensure consistent data collection, reducing other sources of uncertainty. Differences between waste composition analyses can include analysis methods, such as sample size and location, or types and number of waste components (Dahlén & Lagerkvist 2008). In conclusion, the consistent data collection was considered to compensate for the small number of observations, and the study deemed a good contribution to advance the understanding of how external factors can affect MSW source-sorting results. To strengthen the findings and increase generalisability, similar studies should be carried out.

3. Results and discussion

The results from the linear regression model for FW are shown in Table 4. The model explains (R^2) 85 % of the variation in the fraction, but only separate FW collection has a significant impact (p<0.01) on the amount of FW in the MW

Table 4 Results for food waste fraction regression model.

Dependent variable: **Food waste**					
R^2	85 %			95 % confidence intervals	
Obs.	20	*Coeff.*	*P-value*		
(Intercept)		35 %	< 0.01	32 %	39 %
sepFW		-9.7 %	< 0.01	-12.5 %	-7.0 %
kerbG&M		-3.3 %	0.13	-7.6 %	1.1 %
Settlement		1.7 %	0.45	-3.0 %	6.5 %

bin. With each increment in the separate FW collection variable, i.e. from no collection to optibag, or from optibag to dedicated FW bin, the share of FW in the MW bin is reduced by almost 10 percentage points. Due to how the sepFW variable is coded, this means the FW share goes from 45 % with no collection, to 35 % with optibag, to 25 % with dedicated bin, which amounts to an average 25 % reduction per step. The confidence intervals indicate a 95 % likelihood that the real reduction is between 7 and 12.5 percentage points. Kerbside collection of glass and metal packaging waste and the settlement type do not show any significant effects on the amount of FW in the MW bin.

The results from the linear regression model for correctly sorted MW are shown in Table 5. The model explains 82 % of the variation in the fraction, and both separate FW collection and kerbside G&M collection have significant effects (p<0.01) on the

Table 5 Results for mixed waste fraction regression model.

Dependent variable: **Mixed waste**					
R²	82 %			95 % confidence intervals	
Obs.	20	*Coeff.*	*P-value*		
(Intercept)		31 %	< 0.01	27 %	34 %
sepFW		6.7 %	< 0.01	3.9 %	9.4 %
kerbG&M		7.5 %	< 0.01	3.1 %	11.8 %
Settlement		-4.4 %	0.07	-9.1 %	0.3 %

dependent variable. With each increment in the separate FW collection variable, the share of correctly sorted MW increases by almost 7 percentage points, going from 24 % with no collection, to 31 % with optibag, to 37 % with dedicated FW bin. This amounts to an increase of about 25 % per step on average. The confidence intervals indicate an increase between 4 and 9.5 percentage points. For kerbside collection of G&M, the share of correctly sorted MW increases by 7.5 percentage points, from 31 % to 38.5 %, which amounts to a 24 % increase. There is a 95 % probability that the increase is between 3 and 12 percentage points.

The increase in the share of correctly sorted MW from kerbside collection of G&M is larger than can be explained by a reduction in G&M alone, as G&M are small weight fractions. This suggests that other factors may be at play, and indeed we found a significant effect on the share of plastic packaging waste in the MW bin: When there is kerbside G&M collection there is less plastic packaging waste in the MW bin. This interrelation cannot be explained by our dataset, but Mikkelborg (2017) also found a connection between sorting of different waste fractions, where better sorting in one fraction coincides with better sorting in other fractions. He points at communication and information as important factors behind this, further discussed below for other variables.

Despite size of settlement not having a statistically significant effect on the share of correctly sorted MW, it is not far off, with a p-value of 0.07. Mikkelborg (2017) showed that typical city traits, such as limited indoor and outdoor space, more households likely to have small children (citizens in 20s and 30s age groups), more citizens with foreign background and more lower income households, are related to a lower degree of source sorting. The results from Mikkelborg (2017) were further analysed in Fagernæs (2018), and better communication differentiated by target group is presented as a key measure to improve poor source sorting. One could hypothesize that targeted communication is easier and social control stronger in a smaller community. The insignificant effect of settlement type on correctly sorted MW indicates better sorting in smaller communities.

Based on the results we see that two factors are important for improved MSW sorting at source: (1) The possibility to sort at source. For FW, we see that having separate FW collection reduces the amount of FW in the MW bin with both systems. This is in many ways obvious, as the only option if there is no separate FW sorting, is to dispose of FW in the MW bin (or home composting). (2) The convenience of the sorting system. We see that kerbside pickup of G&M reduces the amount of G&M in the MW bin, with significant reductions of the glass and metal packaging waste fractions, as well as a significant increase in correctly sorted MW. The easier it is for the consumer to sort at source, and the less transport required to use the correct waste bin, the more likely it is that they sort better at source.

However, while a more convenient sorting system means more of the fraction is collected separately, studies have shown that it also leads to more missorting in the case of glass and metal packaging. Syversen et al (2019) show that while the amount of sorted G&M waste collected with a kerbside system is more than 25 % higher, the share of waste that is missorted more than doubles, from 4 % to almost 10 %. This means that the quality of the sorted fraction is worse, and it will need more processing before recycling.

A third factor, which was outside the scope of this study, is the transparency of the waste system and the perceived fate of the waste. For FW, we see that the optibag system gives inferior results compared to a dedicated bin collected separately. Based on discussions with industry partners in the CircWtE project, we believe the main reason for this is that the inhabitants do not know enough about how the waste is further treated downstream. When several waste fractions are collected in one bin, they believe all the waste goes to the same place, most likely to incineration, hence perceiving it as less important to sort correctly. Industry partners reported that they see the same effect when several bins are collected with the same multi-compartment truck (CircWtE workshop, 31 May 2022, Trondheim, Norway). It is however worth noting that the optibag system is a space saving solution, requiring a single bin for four separate fractions. In densely populated areas where space is limited, this can enable source sorting of more fractions. This does not apply to indoor space limitations though, where multiple bins are still needed. The optibag system also enables collection of four fractions with one single-compartment truck, which can be beneficial in sparsely populated areas with long transport distances.

The type of settlement was not found to be a significant factor for the MSW sorting at source in this study. This may be rooted in how the type of settlement was determined. As we did not have detailed knowledge about where the waste was collected, sample areas were defined as city or town/village based on the name of the settlement. Most Norwegian cities are small and not densely populated, implying that the differences between a city and a town may be minor. Demographic traits also vary between neighbourhoods and different degrees of urbanization, meaning the sample areas from cities may not represent typical city traits. Without detailed knowledge about the locations, we were not able to analyse this further.

Despite interesting findings, a limited number of variables were available for analysis. Demographic variables are found to have significant effects on degrees of sorting at source (Mikkelborg 2017), but the dataset did not allow for such variables. With only two sample areas per MSW company, quantitative analysis of differences between companies was impossible, despite anecdotal indications that e.g. communication strategies can affect results significantly. In the presentation of the waste composition analysis, one MSW company was highlighted for outstanding FW sorting results, despite launching separate sorting only recently. This was explained by an active communication strategy, combined with a strict policy of not collecting wrongly sorted waste. Bad sorting is photodocumented, and the documentation including an explanation of why the waste is not collected is sent to the customer immediately. According to the MSW company, this active communication led to good sorting results as well as high customer satisfaction, which was validated in that they have the two highest shares of correctly sorted MW of all 20 sample areas (CIVAC: Presentation of results from waste composition analysis, 7 December 2021, Stjørdal, Norway). In summary, many variables other than sorting systems affect the results of MSW sorting, and these should be considered when designing systems to enable effective sorting at source. As for the effect of the sorting systems, comparable results have been found in a similar study by the authors (Grytli & Birgen 2023), strengthening the findings of this study.

4. Conclusions

The results from our regression analyses show that the possibility to sort at source, as well as the user-friendliness and transparency of the waste system, are significant factors impacting MSW sorting behaviour. We recommend further in-depth studies of waste composition analyses to confirm and expand on these findings.

Despite no significant effects from type of settlement in this study, demographic variables have been found to have an effect in other studies. Targeted communication and sanction systems also appear to impact sorting results. These types of variables should be studied further, to aid MSW companies in developing strategies for increased sorting.

This study intends to better utilise the potential of waste composition analyses and the results obtained contribute to a deeper understanding of sorting behaviour and its drivers. The study can help MSW companies better design and select effective collection systems and policymakers in implementing measures to improve sorting and increase circularity.

Acknowledgements

This work is part of the CircWtE project co-funded by industry and public partners and the Research Council of Norway under the SIRKULÆRØKONOMI program (CircWtE - 319795). The authors thank the personnel at CIVAC for providing data and valuable information on the waste management system.

References

L. Dahlén, A. Lagerkvist, 2008, Methods for household waste composition studies, Waste Management, Vol. 28, 7, 1100-1112, https://doi.org/10.1016/j.wasman.2007.08.014.

Directive (EU) 2018/851, 2018, Directive (EU) 2018/851 of the European Parliament and of the Council of 30 May 2018 amending Directive 2008/98/EC on waste (Text with EEA relevance)

A.B. Fagerheim, E.L. Mikkelborg, S. Bjørnerud, 2021, Avfallsanalyser restavfall Midt-Norge 2021, https://civac.no/dokumentasjon/hvilke-ressurser-finnes-i-restavfallet/, acc. 2022-11-29

C.C. Fagernæs, 2018, Recycling and material recovery of plastic packaging and food waste in Oslo municipality - How to change behaviour and attitude through facilitated measures?, Master thesis, Norwegian University of Life Sciences.

T. Grytli, C. Birgen, 2023, A Critical Look at Waste Composition Analyses: Challenges and Opportunities, Proceedings of the 33rd European Symposium on Computer Aided Process Engineering (ESCAPE33), June 18-21, 2023, Athens, Greece. (Accepted)

E.L. Mikkelborg, 2017, Increased material recycling from household waste in Oslo Municipality – the role of socio-demographics and socio-psychological factors, Master thesis, Norwegian University of Life Sciences.

Statistics Norway, 2022, Table 12263: Selected key figures for household waste (M) 2015 - 2021

F. Syversen, K. Kirkevaag, S. Bjørnerud, 2019, Renere råvarer. Case: Optimalisering av verdikjeden for glass- og metallemballasje, https://avfallnorge.no/fagomraader-og-faggrupper/rapporter/renere-r%C3%A5varer, accessed 2022-11-29

C. Thomas, 2004, Effective use of data in waste strategy planning in the UK. In: ISWA World Environment Congress, 17-21 Oct 2004, Rome, Italy.

J. Wooldridge, 2009, Introductory Econometrics: A Modern Approach, 4th Edition, Mason, Ohio: South Western Cengage Learning.

Antonis Kokossis, Michael C. Georgiadis, Efstratios N. Pistikopoulos (Eds.)
PROCEEDINGS OF THE 33rd European Symposium on Computer Aided Process Engineering
(ESCAPE33), June 18-21, 2023, Athens, Greece

A Critical Look at Waste Composition Analyses: Challenges and Opportunities

Tuva Grytli,[a] Cansu Birgen,[b]

[a] *SINTEF Industry, P.O. Box 4760 Torgarden, Trondheim, 7465, Norway*
[b] *SINTEF Energy Research, P.O. Box 4760 Torgarden, 7465 Trondheim, Norway*

Abstract

This paper presents regression analysis results on time series data of waste composition analyses from a municipal waste company in Norway, with a discussion on methodological challenges. The aim is to investigate what affects municipal solid waste (MSW) source-sorting results, to facilitate achieving circular economy targets. The share of four MSW fractions in the mixed waste bin are studied considering five independent variables: Collection/sorting systems for food waste, glass and metal packaging waste, and mixed waste; year of analysis, and pre/post covid-19. We find that only the source sorting systems have significant impacts on the waste composition, and that user-friendliness matters. We recommend more in-depth studies to increase data quality to better utilise the potential of waste composition analyses for a more circular waste system.

Keywords: Municipal solid waste, sorting behaviour, waste management, regression analysis, circular economy.

1. Introduction

To improve circularity, recycling, and reuse, we need to improve municipal solid waste (MSW) management systems, where large quantities of materials have the potential to be recovered. The 2021 material recovery rate in Norway was 43 %, well below the 55 % target for 2025 (Statistics Norway 2020, Directive (EU) 2018/851 2018). More than 60 % of the waste in the mixed waste (MW) bin is recyclable, so understanding how households sort their waste is vital, and good data is key for effective waste strategy planning (Grytli & Birgen 2023). The aim of this study is to increase the insight into what affects sorting behaviour, to enable increased circularity in the waste treatment system.

This study is based on waste composition analysis data from a municipal waste company in Norway that serves more than 100 000 inhabitants. Waste composition analyses of the MW bin were carried out in 2016, 2017, 2019 and 2021, comprising a time series where in total almost 4 000 kg waste has been physically analysed.

2. Methodology

To better understand what affects waste sorting at source, the results from the waste composition analyses were investigated using multiple linear regression analysis. The number of sample areas analysed per year were two (2016-2017), three (2019) and four (2021), giving a total of 11 observations.

The waste composition analyses examine the contents of the MW bin, estimating the share, in weight percentage, of ten different waste fractions. Table 1 shows the sample area average share per fraction. A regression analysis was run for each fraction, with the share of the fraction in the bin as the dependent variable. Significant effects were found for five fractions, and this study focuses on four of them: Food waste (FW), glass packaging waste (GW), metal packaging waste (MeW), and MW which is the correctly sorted MW.

To identify independent variables, the differences between the 11 observations were assessed. Two sample areas have separate FW collection. Kerbside glass and metal packaging waste (G&M) collection was rolled out between 2017 and 2019, and five observations took place after this roll-out.

Table 1 Overview and average share of waste fractions in the mixed waste bin.

Waste fractions with average shares	
Paper and cardboard	7.65 %
Food waste incl. paper towel	43.6 %
Garden waste	4.83 %
Plastic packaging	9.43 %
Glass packaging	2.94 %
Metal packaging	1.43 %
Other metal	0.98 %
Recyclable textiles	3.03 %
Hazardous waste and WEEE	1.35 %
Mixed waste	24.8 %

Three sample areas have underground MW collection, where the waste is discarded in a large underground container via a waste inlet aboveground. The other eight have regular rolling bins for MW. Four observations were from late fall of 2021, i.e. after Covid-19. These differences gave rise to five potential independent variables, described in Table 2. As all independent variables (except year) are categorical, they were coded as dummy variables (Wooldridge 2009).

Table 2 Overview and coding of independent variables.

Independent variables	*Short name*	Values			
		0	1		
Separate FW collection	*sepFW*	No separate collection	Separate FW bin		
Kerbside G&M collection	*kerbG&M*	G&M to central recycling point or civic amenity site	Kerbside G&M collection		
Underground waste collection	*Undergr.*	Aboveground MW bin	Underground MW system		
Post Covid-19	*Covid19*	Before Covid-19	After Covid-19		
		1	2	4	6
Year	*Year*	2016	2017	2019	2021

There were some challenges with the identified independent variables. Most had some correlation with the time variable, as they either happened at a certain time (kerbside G&M collection and Covid-19) or were included only in later years (underground waste collection). This can cause a multicollinearity problem, as discussed later. The next challenge was a very small number of observations for a regression analysis. Each observation is however the aggregated waste from 30-40 bins. This means that the 11 sample areas represent more than 330 bins, and almost 400 households. For practical reasons this study was done on a sample area level. It is unlikely that additional resolution would have added much value, as we had no information about each household, thus could not have explained any additional variation. As each bin is a snapshot in time and may not be representative of the household's average waste composition, an average of a sample area may be considered more representative of an average waste composition. In general, a small number of observations increases uncertainty and means that the results should be interpreted carefully and may not be generalisable.

During analysis, unexpected results emerged. Close inspection of the data revealed large increases in one fraction, that could not be explained even after communication with the

MSW company. Thorough debugging revealed a summation error for a subfraction of the fraction in a spreadsheet, causing a hard to spot error that had significant impacts on the results. Data from waste composition analyses is registered on paper and punched manually into a spreadsheet template. As a copy of the template was used, the mistake spread to five sample areas over two years, causing a systematic error. Left undiscovered, this systematic error would have impacted the accuracy of the model significantly. Several random errors were also discovered, where numbers were punched incorrectly. Although most random errors were too small to affect overall results, they would have reduced the model's precision.

Due to Norway's sparse and spread population and complex geography, MSW companies are often small, which induces several of the methodological complications discussed. Small companies have limited resources for carrying out waste composition analyses, resulting in fewer observations, and for quality assuring results. The uncovered systematic error gave a faulty picture of the company's performance in collection, sorting and recycling, ultimately giving an incorrect base for their strategic development. This shows the benefit provided by research, as in addition to giving insight into effects of measures implemented, the MSW company's data is improved and quality assured.

3. Results and discussion

The results from the linear regression model for the FW fraction are shown in Table 3. The model explains (R^2) 91 % of the variation in the fraction, but the only variable with a significant effect (p<0.01) is separate FW collection. The model finds that the share of FW is reduced by 25 percentage points (pp) when there is separate FW collection, from 53 % to 28 % FW in the MW bin, amounting to a 48 % reduction. The confidence intervals indicate that there is a 95 % likelihood that the reduction is between 10 and 40 pp. None of the other variables were found to have a significant effect on the share of FW in the MW bin.

Table 3 Results for food waste fraction regression model.

Dependent variable: **Food waste**					
R^2	91 %	Adj. R^2	82 %	*95 % confidence*	
Obs.	11	*Coeff.*	*P-value*	*intervals*	
(Intercept)		53 %	< 0.01	43 %	63 %
sepFW		-25 %	< 0.01	-40 %	-10 %
kerbG&M		7.4 %	0.37	-12 %	27 %
Undergr.		0.3 %	0.95	-10 %	11 %
Covid19		-6.7 %	0.40	-25 %	12 %
Year		-1.5 %	0.40	-6 %	3 %

Multicollinearity (a correlation between several of the independent variables) may give unreliable results in a regression analysis. To assess whether multicollinearity was a problem in the model, the variance inflation factors (VIF) for the independent variables were analysed (Wooldridge 2009). There is no agreed-upon threshold for when multicollinearity becomes problematic, but threshold values of 5 or 10 are commonly used. The VIFs for the independent variables are never over 10, but for kerbside G&M collection, year and Covid-19, the VIF is over 5. This became a problem for the GW and MeW fractions, where results were significantly affected. The collection system for G&M has been shown to impact the sorting behaviour in other studies (Grytli & Birgen 2023, Syversen et al 2019), and it is therefore assumed to be more important than development over time, or the effect of a pandemic. To avoid the multicollinearity problem, the variables Post Covid-19 and Year were excluded from the models for the GW and MeW fractions, and a simplified model is presented.

The results for the simplified regression models are shown in Table 4 for GW and Table 5 for MeW. For GW we found that the adjusted R^2 increased when running the model with only kerbside G&M collection as independent variable, indicating that the simplified model may be overfitted. Results from both runs are thus included in Table 4, and the discussion will focus on the *Only G&M* model for GW. The underlying reasons will be discussed later.

Table 4 Results for glass packaging waste fraction regression models.

Dep. variable: **Glass pack. waste**	Simplified model				Only G&M			
	R^2 54 %		Adj. R^2 34 %		R^2 50 %		Adj. R^2 45 %	
Obs. 11	*Coeff.*	*P*	95 % CI		*Coeff.*	*P*	95 % CI	
(Intercept)	3.7 %	< 0.01	2.6 %	4.8 %	3.7 %	< 0.01	2.8 %	4.7 %
sepFW	0.8 %	0.51	-1.9 %	3.5 %				
kerbG&M	-2.2 %	0.04	-4.3 %	-0.1 %	-1.8 %	0.01	-3.1 %	-0.4 %
Undergr.	0.4 %	0.65	-1.6 %	2.5 %				

For GW the model explains (R^2) 50 % of the variation in the fraction, while the MeW model explains more of the variation, with an R^2 of 66 %. Only the kerbside G&M collection affects the share of G&M waste in the MW bin significantly in any of the models. For MeW the effect is more clearly significant (p<0.01) than for GW (p=0.01). The estimated reduction in MeW is 0.9 pp, from 1.6 % to 0.7 % MeW in the MW bin, amounting to a 55 % reduction. The confidence intervals indicate a true reduction between 0.3 and 1.5 pp. For GW the reduction is 1.8 pp, from 3.7 % to 1.9 % GW in the MW bin, or a 48 % reduction, with a true reduction estimated between 0.4 and 3.1 pp. The findings are in line with Syversen et al (2019), where the increase in collected metal is larger than the increase in collected glass when introducing a kerbside G&M collection. They found an increase in the amounts sorted at source of 11 % for glass and 157 % for metal. As metal packaging is lighter than glass packaging, we can assume that a smaller reduction in the share of metal in the MW can account for a larger increase in the amount collected from sorting at source.

Table 5 Results for simplified metal packaging waste fraction regression model.

Dependent variable: **Metal packaging waste**				
R^2	66 %	Adj. R^2 51 %	95 % conf. intervals	
Obs.	11	*Coeff.*	*P-value*	
(Intercept)	1.6 %	< 0.01	1.3 %	1.9 %
sepFW	0.7 %	0.07	-0.1 %	1.4 %
kerbG&M	-0.9 %	< 0.01	-1.5 %	-0.3 %
Undergr.	0.4 %	0.18	-0.2 %	0.9 %

For the MW, the results from the full regression model are shown in Table 6. The model explains (R^2) 90 % of the variation in correctly sorted MW in the MW bin. The only significant effect is

Table 6 Results for mixed waste fraction regression model.

Dependent variable: **Mixed waste**				
R^2	90 %	Adj. R^2 80 %	95 % conf. intervals	
Obs.	11	*Coeff.*	*P-value*	
(Intercept)	21 %	< 0.01	13 %	29 %
sepFW	18 %	0.01	6 %	29 %
kerbG&M	-2.7 %	0.66	-18 %	12 %
Undergr.	-0.02 %	1	-8 %	8 %
Covid19	9.1 %	0.16	-5 %	23 %
Year	-0.5 %	0.71	-4 %	3 %

separate FW collection (p=0.01), which leads to an increase in the share of correctly sorted waste in the MW bin of 18 pp, from 21 % to 39 %, or an 83 % increase. The confidence intervals indicate a real increase in correctly sorted MW with separate FW collection of 6 - 29 pp.

We see high explanatory powers (R^2) and large effects for both the FW and MW fractions, when separate FW collection is the significant variable. Two factors help explain this: (1) FW is heavy, constituting by far the largest fraction in the MW bin for households without separate FW collection, as shown in Figure 1. With separate FW collection, the correctly sorted MW is the largest fraction. (2) If there is no separate FW collection, the only option is to dispose of FW in the MW bin (or home composting). This implies that

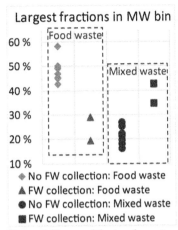

Figure 1 Shares of the two largest fractions in the mixed waste bin.

when we say *correctly sorted MW*, it does not mean that households without separate FW collection are doing anything wrong by discarding their FW in the MW bin, but rather that there should be separate FW collection, so this fraction can be recycled.

On the other hand, we saw how kerbside G&M collection did significantly affect the shares of GW and MeW in the MW bin, but not the share of correctly sorted MW. This is likely due to GW and MeW being small fractions in the MW bin. Effects may thus be obscured by variations in the larger fractions, such as food, plastic, and paper/cardboard. This shows the importance of looking at each fraction separately, to catch individual variations also in the smaller fractions.

Due to the weight of FW, we expected separate FW collection to impact the results of the shares of GW and MeW in the MW bin. As the heavy fraction is removed, the relative shares of all other fractions should increase. For MeW we saw this effect be close to significant, but we did not observe this for GW in the simplified model. The effect of the kerbside G&M collection was also less significant for GW in the simplified model. Through in-depth analysis of the background data, the reason for these results was found to be one outlier data point in a sample area after the rollout of kerbside G&M collection. The outlier lifts the sample area average above the lowest value before the rollout, and above the average of the areas with separate FW collection. This shows how the small number of sample areas can inflate the impact of noise from outliers. It is also a challenge when working with waste composition rather than absolute amounts. We found that by subtracting the FW from the waste composition, the problematic sample area's GW share falls below the lowest value before the rollout. This is because the earlier sample area had a very large share of FW, and the later (problematic) sample area a share below average, thus affecting the relative share of GW in the opposite direction. Both of these problems caused noise that made the simplified GW model seem overfitted, and as a consequence the choice to further simplify the model was made.

As discussed, the small number of observations caused some difficulties. However, using data from only one MSW company yields consistent data collection, which reduces other sources of uncertainty. Different methods for the waste composition analysis, in terms of for example sample size and location, or types and number of waste components, are good examples of such sources (Dahlén & Lagerkvist 2008). We assessed that the benefits from the data collection consistency outweigh the disadvantages of few observations and consider this study a good contribution to understanding the effects of external factors on the results of MSW sorting at source. To strengthen findings and make them more generalisable, similar analyses should be carried out for other MSW companies.

4. Conclusions

The regressions show that it is mainly separate sorting and user-friendly systems that influence the sorting results in the MW bin. The more fractions it is possible and easy to sort, the better the sorting results. We found no significant impacts over time, post covid-19 or from a different type of MW bin.

There are many factors that can influence sorting at source that we were not able to analyse in this study, such as demographic variables, communication, and information, which is further discussed in Grytli & Birgen (2023). To optimize the design of MSW systems for a circular future, it is crucial to identify and understand all relevant factors, and more research should be carried out to this effect.

This study has revealed challenges related to waste composition analyses due to manual data registration. As most MSW companies in Norway are small, the capacity for quality assurance is limited, increasing the risk of errors. Through this work we were able to provide debugging and data quality improvements for the MSW company. Improving the waste composition statistics will help the company better understand how their implemented measures are working and which new ones to consider.

Besides helping the specific MSW company, this study can help similar companies better understand and design their collection systems to increase circularity. We also aim to contribute to the necessary high-quality information and data needed for effective waste strategy planning by policymakers.

Acknowledgements

This work is part of the CircWtE project co-funded by industry and public partners and the Research Council of Norway under the SIRKULÆRØKONOMI program (CircWtE - 319795). The authors thank the personnel at the municipal solid waste company for providing data and valuable information on the waste management system.

References

L. Dahlén, A. Lagerkvist, 2008, Methods for household waste composition studies, Waste Management, Vol. 28, 7, 1100-1112, https://doi.org/10.1016/j.wasman.2007.08.014.
Directive (EU) 2018/851, 2018, Directive (EU) 2018/851 of the European Parliament and of the Council of 30 May 2018 amending Directive 2008/98/EC on waste (Text with EEA relevance)
T. Grytli, C. Birgen, 2023, Enabling circularity: Analysis of factors influencing MSW sorting behaviour in Central Norway, Proceedings of the 33rd European Symposium on Computer Aided Process Engineering (ESCAPE33), June 18-21, 2023, Athens, Greece. (Accepted)
Statistics Norway, 2022, Table 12263: Selected key figures for household waste (M) 2015 - 2021
F. Syversen, K. Kirkevaag, S. Bjørnerud, 2019, Renere råvarer. Case: Optimalisering av verdikjeden for glass- og metallemballasje, https://avfallnorge.no/fagomraader-og-faggrupper/rapporter/renere-r%C3%A5varer (accessed 2022-11-29)
J. Wooldridge, 2009, Introductory Econometrics: A Modern Approach, 4th Edition, Mason, Ohio: South Western Cengage Learning.

Antonis Kokossis, Michael C. Georgiadis, Efstratios N. Pistikopoulos (Eds.)
PROCEEDINGS OF THE 33rd European Symposium on Computer Aided Process Engineering
(ESCAPE33), June 18-21, 2023, Athens, Greece

Waste valorisation within the Energy-Water-Food Nexus: A hybrid techno-geospatial optimisation approach

Mohammad Alherbawi, Sarah Namany, Maryam Haji, Gordon McKay, Tareq Al-Ansari *

College of Science and Engineering, Hamad Bin Khalifa University, Qatar Foundation, Doha, Qatar
talansari@hbku.edu.qa

Abstract

Waste of multiple categories such as agricultural, municipal, and industrial waste are the most abundant biomass sources worldwide. The mismanagement of such waste results in severe impacts on the environment, human health, and resources availability for future generations. However, waste can be utilised within energy, water and food sectors. As such, evaluating the optimal processing pathway, in addition to the optimal supply chain is crucial to fulfilling an adequate balance in the Energy-Water-Food (EWF) Nexus considering national priorities. In this study, 10 waste biomass categories were located over 35 locations across the Qatar using the geographic information systems (GIS), while their quantities, calorific values, and possible associated product yields were defined. Meanwhile, three biomass processing plants representing the EWF industries have been considered, including: biorefinery, desalination plant, and livestock feed developer. An optimisation model was then developed to select optimal biomass and minimise the overall transportation costs for all three industries. Three scenarios were introduced to ensure an EWF Nexus balance throughout the biomass allocation process. The scenarios included: a) equal biomass-based production amongst EWF industries, b) equal biomass distribution, and c) equal biomass-based sales. The model was able to allocate biomass resources from almost all sites at different utilisation percentages. Biomass transportation cost was minimised to 1.98 \$/t in scenario (b), where 0.48 M t/y of biomass were utilised, 49.8 M t/y of products were generated and sold at 82.5 M \$/y. However, scenario (c) achieved higher utilisation rate of biomass at 1.49 M t/y, transported at a minimised cost of 4.33 \$/t. The established model provides an insight on advanced decision-making approaches for optimal waste biomass valorisation to meet the growing demand on EWF resources.

Keywords: EWF nexus; Optimisation; GIS; Decision-making; Supply chain; Qatar.

1. Introduction

The Energy-Water-Food (EWF) Nexus is an emerging approach that supports the security of these three vital sectors and the adequate management of their resources (Namany et al. 2019). The importance of this approach lies in its capability to trace the interlinkages between the EWF systems and subsystems and alert any possible nexus imbalances (Al-Ansari et al. 2015). Biomass presents itself as a crucial aspect of EWF Nexus systems, where the resource is mainly utilised for clean energy production, while it may significantly occupy lands and consume water at the expense of food and water sectors. Several EWF models have been presented earlier for biomass and biofuel industry.

López-Díaz et al. (2018) developed a framework to enhance the EWF resources' management efficiency for biodiesel-bioethanol industry in Mexico. The model targeted the generation of biofuel to fulfill the market demand at a high profit, while minimising water consumption. Whereas Moioli et al. (2018) developed a "Nexus Index" to examine different EWF resources needed for the production of 1 unit of energy. Moreover, Alherbawi et al. (2021) evaluated 11 processing pathways of biomass to select an optimal route that maximises profit and energy, while minimising emissions, lands and water consumption via genetic algorithm.

In this study, a novel framework was developed to ensure an EWF Nexus balance. A hybrid techno-geospatial approach was developed to allocate local biomass resources for different EWF industries, maximise the production and minimise the gate-cost of biomass. ArcGIS was utilised to map the biomass supplying sites and EWF representative industries and to define the distances amongst them.

A mathematical optimisation model was then developed to optimally allocate biomass supplying sites for each industry, given their location, quantity, calorific values, and possible processing pathways. Three EWF balancing scenarios were studied: a) equal EWF production, b) equal biomass distribution, and c) equal biomass-driven sales.

2. Methodology

2.1 Data Collection and Mapping

The required data was adapted from authors' earlier study (Alherbawi et al. 2022), where geospatial, quantitative, and qualitative data (i.e., heating value) for 10 different waste categories in Qatar were re-mapped at 35 sites in ArcGIS (10.7.1) using Universal Transverse Mercator (UTM) coordinates system.

In addition, three industries representing energy, water and food sectors were selected and located on the map, including an earlier proposed biorefinery site (Alherbawi et al. 2022), a desalination plant, and a livestock feed developing industry.

The biorefinery was assumed to produce syngas via gasification of biomass at a yield of 65% (Alherbawi et al. 2020). Whereas the desalination plant was assumed to be equipped with a biomass-based combined heat and power plant (CHP) at 34% net efficiency (IEA 2007) and consumes 11.4 MJ/m^3 (Rayburn 2012). Nevertheless, the feed developer was assumed to utilise food waste and agro-waste only for the manufacturing of livestock feed with a 75% nutrients recovery. All three plants operate for 8000 hours a year with a biomass feed capacity of 30 tonnes per hour. Moreover, the distances between the 35 biomass supply sites and the 3 industries were calculated in ArcGIS along with the corresponding transportation costs.

2.2 Optimisation Model Development

The optimisation model aims at minimising the overall biomass transportation costs to the 3 industries, while achieving an EWF Nexus balance via optimal biomass allocation. A summary of the mathematical model is presented in Table 1. Three constraints scenarios were introduced to restrict the model for the achievement of EWF Nexus balance. Scenario (a) leads to the production of equal quantities (i.e., syngas, water, and livestock feed) in (tonnes/y). While scenario (b) distributes biomass equally between the EWF industries, within the given capacities. Finally, scenario (c) achieves equal gross product sales, given that the maximum possible biomass quantity is valorised. The problem was solved using Microsoft Excel solver.

Table 1: Mathematical optimisation model for optimal biomass allocation.

Function	Description
Objective function:	
$$MIN\ TransCost = C * \sum_{i=1}^{3} \sum_{j=1}^{35} D_{ij} * \frac{Q_j}{L} * A_{ij}$$ C: Cost of transportation (\$/km). D_{ij}: Distance (km) between biomass site (j) to industry site (i). Q_i: Total quantity (t) of biomass at site (j). L: Freight load (t/y).	Minimising the total biomass transportation cost.
Decision variable: A_{ij}: Allocated biomass (t/y) from supply site (j) to industry (i).	
$$Distance\ (D_{ij}) = \sum_{i=1}^{3} \sum_{j=1}^{35} \sqrt{(X_i - X_j)^2 + (Y_i - Y_j)^2}$$ (X_i,Y_i): Coordinates of industry (i). (X_j,Y_j): Coordinates of biomass site (j).	Distance between biomass supplying site (j) to industry (i) based on UTM coordinates.
$$Gross\ Biomass\ (B_{ij}) = \sum_{i=1}^{3} \sum_{j=1}^{35} Q_j * A_{ij}$$	Total biomass utilised (t/y)
$$Gross\ Generation\ (G_{ij}) = \sum_{i=1}^{3} \sum_{j=1}^{35} B_{ij} * R_{ij}$$ (R_{ij}): Rate of production (t/t) at industry (j) using biomass (i).	Total products generated (t/y)
$$Gross\ Sales\ (S_{ij}) = \sum_{i=1}^{3} \sum_{j=1}^{35} G_{ij} * P_{ij}$$ (P_{ij}): Price of product at industry (j) using biomass (i).	Total products sales (\$/y)
Constraints:	
$\sum_{j=1}^{35} A_j \leq 1 \qquad \forall i$	Biomass consumed at site (j) should not be more than its capacity.
$A_{2,13}, A_{2,14}, \dots, A_{2,35} = 0$	Biomass from sites (13-35) is not suitable for industry (2).
$A_{i,1}, A_{i,2} \leq 0.578$	Only non-recyclable biomass can be utilised from sites (1,2).
$A_{2,1}, A_{2,2} \leq 0.427$	Only food waste can be utilised from these sites for industry (2).
$G_{ij} \leq CP_{ij}$	Gross production at industry (i) should not exceed its capacity (CP_{ij})
EWF Balancing Constraints:	
$\sum_{j=1}^{35} G_{1,j} = \sum_{j=1}^{35} G_{2,j} = \sum_{j=1}^{35} G_{3,j}$	Scenario (a): Equal products generation.
$\sum_{j=1}^{35} B_{1,j} = \sum_{j=1}^{35} B_{2,j} = \sum_{j=1}^{35} B_{3,j}$	Scenario (b): Equal biomass utilisation.
$\sum_{j=1}^{35} S_{1,j} = \sum_{j=1}^{35} S_{2,j} = \sum_{j=1}^{35} S_{3,j}$	Scenario (c): Equal products sales.

3. Results and Discussion

3.1 Data Mapping

The mapped sites of biomass resources and the three targeted EWF industries are presented in Figure 1. Most of the sites exist along the eastern coast towards the middle of the map, while the industries are distributed in the southern and western sides.

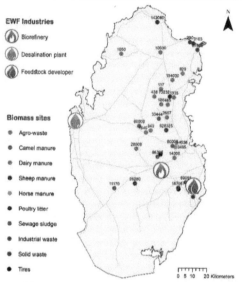

Figure 1: Quantities (t/y) and geospatial distribution of waste in Qatar.

3.2 Biomass Allocation

The Excel solver yielded the optimal biomass allocation shown in Table 2 and Figure 2. In scenario (a), given equal production for the three industries (at 0.24 M tonne/y), 1.15 M tonne of biomass were utilised at an average transportation cost of 2.45 $/t. While in scenario (b), equal biomass quantities were allocated for all industries (total: 0.48 M tonne/y), with the lowest biomass transportation cost attained at 1.98 $/t and highest gross production and sales at 49.8 M t/y and 82.5 M $/y, respectively. Whereas in scenario (c), equal products sales were achieved at 24 M $/y, where 1.49 M tonne of biomass were utilised at an average transportation cost of 4.33 $/t.

Table 2: Optimal allocation of biomass over EWF industries.

Biomass		\multicolumn{9}{c}{Allocated biomass (A_{ij})}							
Category	(j)	\multicolumn{3}{c}{Scenario (a)}			\multicolumn{3}{c}{Scenario (b)}			\multicolumn{3}{c}{Scenario (c)}		
		E	W	F	E	W	F	E	W	F
		i=1	i=2	i=3	i=1	i=2	i=3	i=1	i=2	i=3
Solid	1	12%	0.00%	88%	12%	18%	6%	0%	0%	100%
waste	2	4%	0.00%	96%	10%	23%	67%	0%	0%	100%
Agro-	3	0%	0.00%	25%	0%	88%	0%	31%	30%	31%
waste	4	0%	0.00%	100%	50%	50%	0%	31%	27%	32%
	5	0%	0.00%	100%	50%	50%	0%	31%	25%	32%
	6	0%	0.00%	100%	50%	50%	0%	35%	18%	37%
	7	0%	0.00%	47%	0%	88%	11%	31%	29%	31%

Biomass		Allocated biomass (A_{ij})								
		Scenario (a)			Scenario (b)			Scenario (c)		
Category	(j)	E	W	F	E	W	F	E	W	F
		i=1	i=2	i=3	i=1	i=2	i=3	i=1	i=2	i=3
	8	0%	0.00%	42%	0%	88%	8%	36%	22%	37%
	9	0%	0.00%	100%	50%	50%	0%	31%	25%	32%
	10	0%	0.00%	100%	50%	50%	0%	36%	25%	37%
	11	0%	0.00%	100%	50%	50%	0%	35%	14%	37%
	12	0%	0.00%	100%	50%	50%	0%	35%	0%	38%
Tyers	13	1%	0.00%	0%	1%	0%	0%	68%	0%	0%
Dairy	14	5%	0.00%	0%	4%	0%	0%	90%	0%	0%
manure	15	54%	0.12%	0%	0%	0%	0%	64%	9%	0%
	16	10%	0.01%	0%	6%	0%	0%	80%	0%	0%
	17	4%	0.00%	0%	0%	0%	0%	93%	0%	0%
Sheep	18	9%	0.01%	0%	6%	0%	0%	82%	0%	0%
manure	19	2%	0.00%	0%	0%	0%	0%	97%	0%	0%
	20	18%	0.01%	0%	0%	0%	0%	63%	0%	0%
Camel	21	21%	0.01%	0%	7%	0%	0%	58%	1%	0%
manure	22	21%	0.01%	0%	14%	0%	0%	58%	1%	0%
Horse	23	2%	0.00%	0%	1%	0%	0%	87%	0%	0%
manure	24	1%	0.00%	0%	1%	0%	0%	67%	0%	0%
Poultry	25	19%	0.01%	0%	16%	0%	0%	61%	0%	0%
litter	26	26%	0.02%	0%	0%	0%	0%	49%	1%	0%
	27	50%	0.03%	0%	0%	0%	0%	60%	1%	0%
Sewage	28	12%	0.00%	0%	4%	0%	0%	76%	0%	0%
sludge	29	23%	0.01%	0%	19%	0%	0%	53%	0%	0%
	30	24%	0.01%	0%	16%	0%	0%	51%	0%	0%
	31	4%	0.00%	0%	1%	0%	0%	92%	0%	0%
Industrial	32	22%	0.01%	0%	0%	0%	0%	56%	0%	0%
waste	33	31%	0.01%	0%	0%	0%	0%	48%	0%	0%
	34	26%	0.01%	0%	12%	0%	0%	48%	0%	0%
	35	0%	0.00%	0%	0%	100%	0%	36%	25%	0%

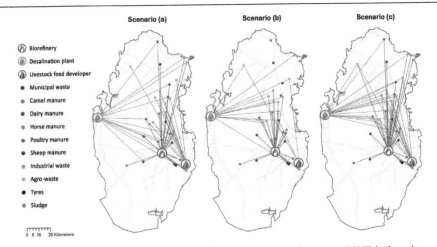

Figure 2: Representation of optimal biomass allocation over EWF industries.

4. Conclusion

An EWF Nexus balancing framework was developed in this study to optimally allocate local waste biomass over EWF industries using ArcGIS and Excel tools. The framework has considered geospatial, quantitative, and qualitative aspects of biomass supplying sites and the receiving EWF industries. The model aimed at selecting optimal biomass resources and minimising their overall transportation costs. Moreover, three scenarios to achieve a Nexus balance were evaluated: a) equal biomass-based production, b) equal biomass distribution, and c) equal biomass-based sales. Scenario (b) achieved the lowest biomass transportation costs at 1.98 \$/t, however, lower biomass quantity has been utilised (0.48 M t/y). Whereas the maximum biomass utilisation was attained in scenario (c), with 1.49 M t/y of biomass transported at a minimised cost of 4.33 \$/t. This model provides insights on possible means to optimally utilise biomass resources at lower costs, while ensuring an EWF Nexus balance.

Acknowledgement

The authors would like to thank Qatar National Research Fund (QNRF) for their support of this research through NPRP-11S-0107-180216, Hamad Bin Khalifa University (HBKU) and Qatar Foundation (QF).

References

C. Rayburn, 2012, *Water Reuse: Potential for Expanding the Nation's Water Supply through Reuse of Municipal Wastewater.*

E. Moioli, F. Salvati, M. Chiesa, R.T. Siecha, F. Manenti, F. Laio, and M.C. Rulli, 2018, "Analysis of the Current World Biofuel Production under a Water–Food–Energy Nexus Perspective." *Advances in Water Resources* 121:22–31.

IEA, 2007, *Biomass for Power Generation and CHP.*

López-Díaz, D. Celeste, L.F. Lira-Barragán, E. Rubio-Castro, M. Serna-González, M.M. El-Halwagi, and J.M. Ponce-Ortega, 2018, "Optimization of Biofuels Production via a Water–Energy–Food Nexus Framework." *Clean Technologies and Environmental Policy* 20(7):1443–66.

M. Alherbawi, A. AlNouss, G. Mckay, and T. Al-Ansari, 2020, "Optimum Utilization of Jatropha Seedcake Considering the Energy, Water and Food Nexus." Pp. 229–34 in *30 European Symposium on Computer Aided Process Engineering.* Vol. 48, edited by S. Pierucci, F. Manenti, G. L. Bozzano, and D. B. T.-C. A. C. E. Manca. Elsevier.

M. Alherbawi, A. AlNouss, G. McKay, and T. Al-Ansari, 2021, "Optimum Sustainable Utilisation of the Whole Fruit of Jatropha Curcas: An Energy, Water and Food Nexus Approach." *Renewable and Sustainable Energy Reviews* 137:110605.

M. Alherbawi, G. Mckay, H.R. Mackey, and T. Al-Ansari, 2022, "Multi-Biomass Refinery Siting: A Gis Geospatial Optimization Approach." *Chemical Engineering Transactions* 92:73–78.

S. Namany, T. Al-Ansari, and R. Govindan, 2019, "Sustainable Energy, Water and Food Nexus Systems: A Focused Review of Decision-Making Tools for Efficient Resource Management and Governance." *Journal of Cleaner Production* 225:610–26.

T. Al-Ansari, A. Korre, Z. Nie, and N. Shah, 2015, "Development of a Life Cycle Assessment Tool for the Assessment of Food Production Systems within the Energy, Water and Food Nexus." *Sustainable Production and Consumption* 2:52–66.

Antonis Kokossis, Michael C. Georgiadis, Efstratios N. Pistikopoulos (Eds.)
PROCEEDINGS OF THE 33rd European Symposium on Computer Aided Process Engineering
(ESCAPE33), June 18-21, 2023, Athens, Greece

Levelised Energy-Water-Food (LEWF) Indicator for Biomass Assessment and Optimisation

Mohammad Alherbawi, Ahmed AlNouss, Gordon McKay, Tareq Al-Ansari [*]

College of Science and Engineering, Hamad Bin Khalifa University, Qatar Foundation, Doha, Qatar
talansari@hbku.edu.qa

Abstract

The Energy-Water-Food (EWF) Nexus approach has recently become an important tool to ensure efficient and optimum management of the three vital resources by considering the inter-dependencies amongst them. However, EWF optimisation complexity increases at times when resources are prioritised during trade off-driven decision-making, coupled with differences between the three resources in terms of measurement units. Therefore, this study presents a novel 'Levelised-EWF indicator' (LEWF) which merges the net value of energy, water and food requirement for a system or a product into a single value in (Megajoule equivalent "MJ-e") unit. This tool enables a convenient comparison between biomass sources and processing technologies in the light of the EWF Nexus by converting multi-objective problems into single objective. Moreover, the study proposed a 10-level EWF resilience chart, to evaluate the system resilience to the changes in EWF resources supplies based on the obtained LEWF value. As a case study, the proposed LEWF tool is used to evaluate the various processing routes for *Jatropha curcas'* seedcake to produce various energy products. Required data are obtained via earlier Aspen Plus® simulations of potential processes including gasification, pyrolysis, hydrothermal liquefaction, incineration, fermentation, and anaerobic digestion. The results demonstrated that pyrolysis is the optimal processing route for *Jatropha* seedcake with the lowest LEWF value of 0.23 MJ-e/MJ (product), with an EWF Resilience Level of 9 out of 10. The importance of this tool lies in its ability to ensure an adequate management of EWF resources within biomass systems and ensure system resilience against possible changes in supply chains.

Keywords: LEWF; Resilience; Biomass; Optimisation; Simulation; *Jatropha*.

1. Introduction

Energy, water, and food are the three essential resources for human existence and nations development. The increasing demand on these resources has raised the need to efficiently manage their consumption in a balanced manner. However, efficient management becomes more pushing at times when complicated interlinkages between these resources exist. Such a relationship was officially introduced in the Rio+20 Summit as "EWF Nexus" (Galmés et al. 2014).

The EWF Nexus concept has emerged as a key tool to evaluate the EWF resources' synergies and tradeoff to ensure an adequate consumption balance. Several EWF Nexus frameworks have then been developed based on existing quantitative and qualitative approaches.

Al-Ansari et al. (2015) presented a Nexus framework based on lifecycle assessment (LCA) concept to study the possible tradeoffs amongst EWF systems and sub-systems. Meanwhile, Daher and Mohtar (2015) developed a scenario-based framework (WEF Tool 2.0) to trace the flow of EWF resources for different case studies towards reaching a self-sufficiency scenario. In addition, Alherbawi et al. (2021) developed a technology-based optimisation framework to evaluate the key biomass processing technologies and their interventions with EWF resources. Whereas Flammini et al. (2014) introduced a "Nexus Rapid Appraisal" method as a decision-making tool, through which the systems' associated EWF resources are plotted in spider charts against standard values. The chart's polygon area is considered a rapid visual representation of the possible intervention between the three resources. Moreover, Moioli et al. (Moioli et al. 2018) developed a "Nexus Index" as reference to evaluate the system's EWF requirement to produce one energy unit.

In this study, a new concept is introduced, named "The Levelised Energy, Water, Food Nexus (LEWF Indicator)" for biomass-biofuel systems. The indicator merges the system's associated EWF resources input into a single value, with energy equivalent unit (E-eq) based on the concept of resource payback that was introduced earlier by the authors (Alherbawi et al. 2020, Alherbawi et al. 2021). Merging the EWF into a single value facilitates a convenient comparison between the various biomass resources and the different processing technologies.

2. Methodology

2.1. Data Collection

Required data for the establishment of the LEWF concept and the development of its formulation have been collected from literature as displayed in Table 1. The table includes information on the interlinked requirements amongst the EWF resources and sectors.

Table 1: Data collected to assist the formulation of the LEWF Indicator.

Scope	Detailed item	Value[*]	Unit
Energy to water	Water requirement for coal-power generation	0.0790	(m^3/GJ)
	Water requirement for NG-power generation	0.0760	(m^3/GJ)
	Water requirement for nuclear-power generation	0.0180	(m^3/GJ)
	Water requirement for solar-power generation	0.0064	(m^3/GJ)
	Water requirement for wind-power generation	0.0002	(m^3/GJ)
	Water requirement for hydro-power generation	0.3000	(m^3/GJ)
	Water requirement for diesel fuel production	5.3000	(m^3/GJ)
Water to Energy	Energy requirement for rainfed irrigation	0.0000	(GJ/m^3)
	Energy requirement for groundwater provision	0.0013	(GJ/m^3)
	Energy requirement for FO desalination	0.0029	(GJ/m^3)
	Energy requirement for RO desalination	0.0108	(GJ/m^3)
	Energy requirement for MED desalination	0.0130	(GJ/m^3)
	Energy requirement for MSF desalination	0.0114	(GJ/m^3)
	Energy requirement for wastewater treatment	0.0022	(GJ/m^3)
Food to Energy	Energy requirement for cereal production	3.5600	(GJ/t)
	Energy requirement for beef production	44.7500	(GJ/t)

Land to Energy	Land requirement for coal-power generation	37.6040	(m²/GJ)
	Land requirement for NG-power generation	38.2200	(m²/GJ)
	Land requirement for nuclear-power generation	39.1440	(m²/GJ)
	Land requirement for solar-power generation	133.971	(m²/GJ)
	Land requirement for wind-power generation	217.557	(m²/GJ)
	Land requirement for hydro-power generation	970.815	(m²/GJ)
	Land requirement for diesel fuel production	0.2400	(m²/GJ)
Land to Food	Land requirement for cereal production	0.00023	(m²/t)
Fertilisers to Food	N-fertiliser requirement for cereal production	0.014	(t/t)
LS Feed to Food	Protein-feed requirement for beef production	0.13	(t/t)

* Resources:(Alherbawi et al. 2021; Dickinson 2018; FAO 2000; Jungbluth, Meili, and Wenzel 2018; Ladha et al. 2016; Stevens et al. 2017)

2.2. LEWF Indicator Development

The LEWF indicator is formulated as given in Equation 1. The mathematical formula translates the water consumed or produced in a system into an equivalent energy amount, that is the amount of energy required to generate an equivalent quantity of water. Similarly, the formula converts the amount of food consumed into energy equivalent value based on the amount of energy required to produce similar amount of food. Nevertheless, the formula considers the embodied water and land of 7 energy systems to comprehensively account for all EWF resources' consumption throughout the lifecycle of a biomass-based process. Meanwhile, the different conversion factors (CF) required for the conversion are calculated and reported in the results section based on the data collected in Table 1.

$$LEWF = 1 - \left[\frac{\left(\sum_{e=1}^{e'} E_e.CF_e + \sum_{w=1}^{w'} W_w.CF_w + \sum_{f=1}^{f'} F_f.CF_f \right)}{\left(\sum_{p=1}^{p'} Q_p.HV_p \right)} \right] \qquad (1)$$

Where:
LEWF: Levelised Energy-Water-Food indicator (GJ-e in/GJ out).
E: Total energy consumed throughout the process' lifecycle via different energy supplies (e).
W: Net water consumed throughout the process' lifecycle via different water supplies (w).
F: Net food consumed throughout the process' lifecycle via different food supplies (f).
CF: Conversion factor into energy equivalent value.
Q: Gross mass quantity of different energy products generated (p).
HV: Heating value of product (p).

The LEWF obtained value may offer an indication of how resilient the system is. Where LEWF values of less than 1 (MJ-e/MJ product) indicate that the biomass-based system can generate sufficient energy to payback for the water and food consumed, while the resource payback is performed using an energy currency rather than money. Thus, the system hedges towards the worst-case scenarios, where in case of EWF supply chain disturbance, the system must be able to generate the required resources based on the extent of energy generated. Therefore, a 10-step EWF Resilience Chart is formulated based on the LEWF values, with a step size of 0.2, as illustrated in Figure 1.

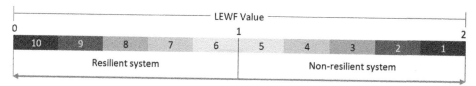

Figure 1: Proposed EWF Resilience Chart.

2.3. Case Study

To evaluate the proposed LEWF tool and Resilience chart, a case study from author's previous work has been considered (Alherbawi et al. 2021), in which *Jatropha curcas* seedcake biomass was processed into multiple energy-producing technologies as illustrated in Figure 2. Energy, water and food-related data (Table 2) were adapted and plugged into the LEWF formula (Equation 1).

Figure 2: *Jatropha* seedcake's processing technologies within EWF boundaries.

Table 2: Case study's data on EWF systems consumption and generation.

Technology	Energy Generated (GJ/y)	Energy consumed (GJ/y)	Net water consumed (10^3 m³/y)	land used (10^3 m²/y)	By-products credit (Fertilisers) (tonne/y)
Gasification	1,039,967	291,924	38,557	25,000	0
Pyrolysis	1,645,898	291,735	38,431	25,000	0
Liquefaction	1,072,398	292,112	38,552	25,000	0
Fermentation	180,661	292,559	38,533	25,000	21,712 (N: 5%)
Digestion	664,308	291,782	38,398	25,000	2,778 (N: 5%)
Incineration	475,820	293,154	38,375	25,000	0

3. Results and Discussion

3.1. LEWF Indicator's Database

Utilising the data collected in Section 2, a set of conversion factors from different EWF resources into an energy equivalent value is provided in Table 3. The database comprises 7 energy supply resources and 7 water supply resources, in addition to food-related resources as energy crops or lands that are diverted from food cultivation towards energy production.

Nevertheless, food-related by-products (i.e., fertilisers, livestock feed) were taken into consideration, and were associated with a negative conversion factor as they represent a credit to the system, which cuts from its net resources consumption.

Table 3: LEWF Indicator's conversion factor.

Source	Description		Energy equivalent conversion factor (GJ-e)
Energy (GJ)	Coal-Power		1.0313
	Natural Gas-Power		1.0317
	Nuclear- Power		1.0322
	Solar-Power		1.1097
	Wind- Power		1.1781
	Hydro- Power		1.7967
	Diesel fuel		1.0317
	Average		1.1731
Water (m³)	Rainfed		0.0000
	Groundwater		1.3E-03
	Forward osmosis		2.9E-03
	Reverse osmosis		1.1E-02
	Multi-effect distillation		1.3E-02
	Multi-stage flash		1.1E-02
	Wastewater treatment		2.2E-03
	Average		5.9E-03
Food-related (tonne)	Energy Crops		3.5600
	Land (m²)		8.2E-04
	By-products credit	Fertilisers (N content)	-0.0498
		Animal feed (Protein content)	-5.8175

3.2. Model Application

The proposed LEWF Indicator has been applied to a *Jatropha* seedcake case study. The breakdown of total equivalent energy requirement for the different processing pathways is plotted in Figure 3. Whereas the final LEWF values and the corresponding EWF Resilience levels are illustrated in Figure 4.

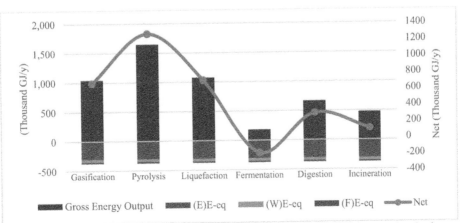

Figure 3: Energy equivalent analysis of *Jatropha* seedcake processing technologies.

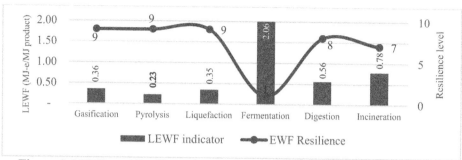

Figure 4: LEWFs and Resilience levels of *Jatropha* seedcake processing systems.

Pyrolysis achieved the lowest LEWF value (0.23 MJ-e/MJ) and highest EWF Resilience level (9/10), therefore, it is selected as an optimal processing technology. Similar conclusion has been reached through an EWF multi-objective optimisation conducted in MATLAB earlier (Alherbawi et al. 2021), which further validate the proposed LEWF Indicator.

4. Conclusion

A novel LEWF Indicator is introduced for the evaluation of EWF resources requirement for energy systems. The indicator merges the three EWF input values into a single figure in energy equivalent unit (E-eq), which enables a convenient evaluation of EWF interlinkages in a system and facilitates the comparison with other systems. The indicator was implemented in a case study to select optimal processing pathway of *Jatropha* seedcake, where pyrolysis achieved the lowest LEWF at 23 (MJ-e/MJ product).

References

A. Flammini, M. Puri, L. Pluschke, and O. Dubois, 2014, *Walking the Nexus Talk: Assessing the Water-Energy-Food Nexus in the Context of the Sustainable Energy for All Initiative*. Rome: Food and Agriculture Organization of the United Nations (FAO).

B.T. Daher, and R.H. Mohtar, 2015, "Water–Energy–Food (WEF) Nexus Tool 2.0: Guiding Integrative Resource Planning and Decision-Making." *Water International* 40(5–6):748–71.

E. Moioli, F. Salvati, M. Chiesa, R.T. Siecha, F. Manenti, F. Laio, and M.C. Rulli, 2018, "Analysis of the Current World Biofuel Production under a Water–Food–Energy Nexus Perspective." *Advances in Water Resources* 121:22–31.

E.W. Dickinson, 2018, *Solar Energy Technology Handbook*. CRC Press.

FAO, 2000, *The Energy and Agriculture Nexus*. Rome.

J.K. Ladha, A. Tirol-Padre, C. K. Reddy, K. G. Cassman, Sudhir Verma, D. S. Powlson, C. van Kessel, Daniel de B Richter, Debashis Chakraborty, and Himanshu Pathak, 2016, "Global Nitrogen Budgets in Cereals: A 50-Year Assessment for Maize, Rice, and Wheat Production Systems." *Scientific Reports* 6:19355.

L. Stevens, B. Anderson, C. Cowan, K. Colton, and D. Johnson, 2017, "The Footprint of Energy: Land Use of US Electricity Production." *STRATA: Logan, UT, USA*.

M. Alherbawi, A. AlNouss, G. McKay, and T. Al-Ansari, 2021, "Optimum Sustainable Utilisation of the Whole Fruit of Jatropha Curcas: An Energy, Water and Food Nexus Approach." *Renewable and Sustainable Energy Reviews* 137:110605.

M. Alherbawi, A. Ahmed AlNouss, G. Mckay, and T. Al-Ansari, 2020, "Optimum Utilization of Jatropha Seedcake Considering the Energy, Water and Food Nexus." Pp. 229–34 in 30 European Symposium on Computer Aided Process Engineering. Vol. 48, Manca. Elsevier.

N. Jungbluth, C. Meili, and P. Wenzel, 2018, "Life Cycle Inventories of Oil Refinery Processing and Products." *ESU-Services Ltd. Commissioned by BFE, BAFU, Erdöl-Vereinigung, Schaffhausen, Switzerland, Retrieved from: Www. Lc-Inventories. Ch.*

T. Al-Ansari, A. Korre, Z. Nie, and N. Shah, 2015, "Development of a Life Cycle Assessment Tool for the Assessment of Food Production Systems within the Energy, Water and Food Nexus." *Sustainable Production and Consumption* 2:52–66.

U.V. Galmés, R. Smith, I. Bokova, G.M. Candido de Castro, R. Toneto Jr., P. Brabeck-Letmathe, S. Cleasby, J. Korff, K. Beckett, S. Haye, and V. Somavilla, 2014, *The Road from RIO+20*.

Antonis Kokossis, Michael C. Georgiadis, Efstratios N. Pistikopoulos (Eds.)
PROCEEDINGS OF THE 33rd European Symposium on Computer Aided Process Engineering
(ESCAPE33), June 18-21, 2023, Athens, Greece

The controlled environment agriculture: a sustainable agrifood production paradigm empowered by systems engineering

Liang Wang[a,d], Guoping Lian[a,b], Zoe Harris[a], Mark Horler[c], Yang Wang[d], Tao Chen[a]

[a]Faculty of Engineering & Physical Sciences, University of Surrey, Guildford, GU2 7XH, UK
[b]Unilever R&D Colworth, Sharnbrook, Bedfordshire, MK40 1LQ, UK
[c]UK Urban AgriTech, 46 Jamaica Street, Liverpool, L1 0AF, UK
[d]National Innovation Center for Digital Fishery, China Agricultural University, Beijing, 100083, China

Abstract

Controlled environment agriculture (CEA) has some clear advantages over traditional farming, such as: reliable and consistent production capability; efficiency in water and space use; reducing the use and runoff of fertiliser and pesticides; etc. As such CEA can greatly benefit from the CAPE (computer-aided process engineering) approach – cross-fertilization of these two apparently distinct areas may result in new methods and applications to improve CEA and process engineering, with potentially significant contribution to circular economy. In this paper, we discuss several important aspects of CEA drawing from our own experiences in aquaculture and aeroponics, including product development, process design and process operation, and the potential contribution of CAPE. Finally, we postulate a systems platform for CEA, aiming to foster a long-lasting partnership between the two scientific communities.

Keywords: Controlled environment agriculture; Computer-aided process engineering Digital twin; Optimisation; Process control; Process modelling.

1. Introduction

A recent breaking news is that the global population has exceeded 8 billion people, (United Nations, 2022). This means that more and more agricultural products and raw materials are required in future. The traditional production methods face a tremendous challenge to meet the ever-increasing demand whilst aiming to achieve net zero and minimal environmental impact (Hemathilake et al., 2022). Thus, advanced agricultural systems with high efficiency, high production capacity and low emission need to be developed.

Controlled environment agriculture (CEA) refers to the production of food (e.g., crops, aquatic and livestock) and non-food (e.g., biomass, fur) agricultural products under conditions that are more controllable than traditional farming in the open field/ponds. It uses adjustable production environment and facility conditions to optimal and correct the growth system over real-time (Benke et al., 2017), provides a high-efficiency, high-yield, potentially low carbon and sustainable production. At scale, CEA can spare the use of land and water bodies with significant ecological benefits. However, CEA faces the challenges of high capital investment and operating costs, the need for crop/animal to adapt to the indoor environment, and complex cultural and political resistance. Some of these issues can be addressed by technological advancement including the use of systems engineering thinking and methods.

CEA encompasses a wide range of primary production methods, ranging from loosely controllable environment in polytunnel greenhouses, aquaculture pens, to totally controlled production systems in plant factories and recirculating aquaculture systems. Like any emerging and rapidly developing subject, this field has been given many names which often overlap in their connotation but have different focuses, such as plant factory, vertical farming, urban faming, facility agriculture, to name a few. We choose CEA to emphasise its key difference from traditional, uncontrolled farming.

In this paper, we discuss from the perspective of systems engineering the three tasks in CEA: product development, process design and process operation, with a focus on the latter. The narrative will mainly draw from our own experiences in aquaculture, aquaponics and aeroponics (Li et al., 2021; 2022; Guo et al., 2022; Zhou et al., 2021), supplemented by work reported in the open literature. Figure 1 illustrates two examples of CEA production systems. Other crop and livestock production systems, such as indoor poultry and insect farming, have their own special requirements in terms of product development, process design and operation. However, they share the same conceptual framework of using systems engineering approach to enable more sustainable production.

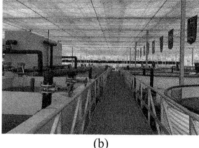

(a)   (b)

Figure 1. CEA examples: (a) Aeroponic cultivation of bioenergy willows in a greenhouse; (b) Industrialised aquaponics to produce fish and vegetables.

2. Systems engineering techniques for product and process design

2.1. Computer aided product design

Products play an important role on the production system. In CEA, product design / redesign is often needed to adapt to the spatiotemporal characteristics and the constraints of resources. For example, the willows growing in an indoor, aeroponic environment (Figure 1(a)) should ideally be dwarf, fast-growing, comfortable with aeroponic irrigation and nutrition, and resistant to disease. The same idea is also used in aquaponics factories, where the high-density production environment requires fish species with fast growth and disease resistance. Such controlled environment and the adapted crops or animals may also require new formulations of the feed.

Computer aided breeding is a well-established concept. Statistical design of experiments has been widely used in decades. Advanced machine learning / artificial intelligence methods, incorporating multi-faceted data such as imaging, genomics and phenomics, have shown promises (Harfouche et al., 2019; Marsh et al., 2021). More fundamental, systems biology models, such as gene regulatory networks, have been applied to obtain the favourable traits with mechanistic understanding of why and how (Lavarenne et al., 2018). Notably, these computational methods have the maximum potential when they are integrated with high-throughput experiments – CEA can provide such high-throughput facilities with close control of the environmental parameters.

2.2. Computer aided process design

CEA at industrial scale is analogous to biomanufacturing processes. The centrepiece is the "bioreactor" to produce the biomass (animal or plant). A wide range of facilities, equipment and devices are needed to support the microclimate environment, water quality, nutrient supply, disease control, downstream processing of products and/or wastes, etc. These decisions mirror process design in CAPE. For most facilities, the key design considerations are to do with HVAC (heating, ventilation and air conditioning), water supply, and power supply, to provide required temperature, humidity, ventilation, supplementary light, irrigation, etc.

Specific requirements are needed beyond the above general principle. For example, recirculating aquaculture systems (RAS) require design of not just the fish tanks, but also aeration, sterilisation, water quality adjustment, multi-stage filtration (physical and biological), etc. For soilless plant production, the design of nutrient supply and recycle units (Bittsánszky et al., 2016), supplementary lighting / shading facilities, etc., will be required in addition to HVAC.

It is necessary to fully consider the integration between product design (breeding) and process design, as opposed to treating them as two separate tasks. Computer aided process design in CEA is emerging (Guo et al., 2022; Holzworth et al., 2015), thought its full potential is yet to materialise, especially in whole-process simulation and design.

Energy consumption has always been a major factor in agricultural production. The requirement for heating during winter has restricted the use of many agricultural technologies in high latitude regions. This is particularly an issue for CEA. A potential solution is the use of a multitude of energy sources, including solar, wind energy, biomass, geothermal and tidal energy. In high latitude regions, distributed energy systems for combined heat and power hold promise to reduce reliance on the grid, reduce cost and reduce emission. We conducted large-scale aquaponics production experiments by integrating solar, geothermal and biomass energy in Northern China (Le et al., 2021), supported by computational fluid dynamic simulations (Le et al., 2020).

3. Systems engineering methods for process operation

Similar to chemical manufacturing, the operation of CEA processes can be organized in a hierarchy from regulatory control, optimisation to scheduling, planning and supply chain management. Agricultural production is inherently slow and in batch mode. Slow dynamics in biomass production suggests some level of tolerance to operation disturbances. For example, although the desired temperature for Salmon growth is around 12 °C, a short exposure to temperature moderately outside this range is usually tolerable, creating degrees of freedom for optimisation. On the other hand, the nutrient dynamics can be fast, such as nitrification in a fish tank, so is the dynamics of microclimate parameters. This disparity in dynamics may pose challenges to process monitoring, control, optimisation and scheduling.

3.1. Soft sensor for process monitoring

Effective process operation requires reliable measurement. In traditional agriculture, a key consideration is the cost and maintainability because of the often harsh environment in the open field. These challenges are alleviated to some extent in CEA. Nevertheless, since agricultural products are relatively low value, cost and maintainability are still big drivers. For example, many commercial ammonia sensors need to be calibrated once a month – this might be an affordable, standard practice in a wastewater processing plant in the UK (private communication) but impractical for agriculture.

Therefore, software based sensor self-calibration has been proposed as a solution (Yuri et al., 2015), for temperature, dissolved oxygen (DO) and ammonia (Yu et al., 2020). Liu et al. (2013) developed a multi-source heterogeneous data preprocessing technologies to improve the measurement accuracy. Li et al. (2021) applied a moving horizon estimation approach, widely used in CAPE, with a mechanistic nitrogen cycle model as a state estimator to improve the resolution and reliability of total nitrogen monitoring from infrequent sensor readings.

3.2. Process control & optimisation

Process control in CEA normally refers to the control of the microclimate parameters, i.e., temperature, humidity, water quality, nutrient supply, light intensity, CO_2 concentration (if CO_2 enrichment is possible in a totally controlled environment), etc. In the context of aquaculture DO control has been the primary focus, because fish health and viability is very sensitive to DO. Aeration is an effective manipulated variable to control DO, but it associates with significant energy cost. DO's importance has warranted a large number of reports of controller design, such as adaptive PID control enhanced by evolutionary algorithms (Zhou et al., 2021). In principle, model predictive control (MPC) should give improved results as shown in controlling DO in wastewater processing (O'Brien et al., 2011), though the application of MPC in agriculture is still under-explored. In addition, due to the difficulties in real-time measurement of other water quality indices, e.g. ammonia nitrogen, their feedback control is less common in practice.

Direct control and/or optimisation of the biomass growth has been less reported in the literature. In principle these are similar to batch process control/optimisation problems. The challenges are two-fold: the limitation on measurement and the incredibly long batch duration. Direct measurement of biomass growth could be achieved by manual sampling or using imaging based techniques, but the latter itself is an open research area. Long batch duration, from weeks to months, renders run-to-run control impractical. There have been reports of reference tracking using MPC (Chahid et al., 2021) and direct use of a mechanistic model in growth optimisation (Li et al., 2022). If computer models are developed to incorporate the prediction of both production and environmental parameters, they could be used to formulate economic indices and thus cast into an economic MPC framework for dynamic, direct optimisation of the economic measures. Currently such methods are limited to simulations and have a long way to be taken up in practice.

3.3. Scheduling, planning & supply chain

Due to the long lead time of agriculture produces, scheduling, planning and supply chain integration is key to commercial viability and success. At the process level, scheduling models could be developed at steady state to mixed integer linear programs (MILP) or even nonlinear ones, similar to those in the vast CAPE literature, and solved accordingly. With distributed energy systems, the objective is typically to maximise capacity utilisation with minimal energy costs, considering variable energy inputs/costs due to the use of renewables. Alternatively economic MPC, as mentioned above, might be useful to make dynamic decisions enclosing both the scheduling and optimisation levels. The key enabler is digitalisation throughout the production, and integrated with pre-production to post-production for optimising the fresh/cold supply chain. An especially interesting case is aquaponics, where the production of fish and crops should be scheduled in a staggered manner to meet variable market demand.

3.4. Big data, IoT and AI

Agriculture is traditionally data poor but has seen rapid transition to big data, enabled by massive connectedness, industrial internet-of-things (IoT), and data processing platforms such as the Apache Spark and Storm. Big data has allowed the application of advanced AI for the intelligent perception, prediction, and decision making in agrifood production (Wang et al., 2021). A particularly important application of big data AI in CEA is disease control. AI is well suited to integrating data from low-cost camera with environment data, and by using machine learning algorithms to providing early warning of pathogens (Wen et al., 2012; Zhang et al., 2020). We recently demonstrated a proof-of-concept of using acoustic signal to determine fish feeding intensity (Cui et al., 2022).

4. Concluding remarks

In this paper we provided an overview of CEA as an emerging method for efficient and sustainable agrifood production. Whether CEA can fulfill its perceived benefits is likely to depend on how the products are processes are designed, and how the processes are operated, all in a holistic approach encompassing commercial viability, energy consumption, environmental impact, and circularity. There is a strong case for adapting CAPE methods to the application in CEA, as shown in increasing research reported in the literature. However, challenges abound. Agriculture produces are of relatively low value per kg produced compared with many chemical products familiar to the CAPE community. Most processes are slow with a long time delay / time constant between input and output. Measurement of key process parameters, especially those to do with the biomass properties (e.g., growth and health of the animal/crop), is difficult. On the flip side, the live biomass being produced are often tolerant to minor environment changes, providing more space for optimisation. We foresee a fruitful and long-lasting partnership between CAPE and CEA to address these and emerging challenges in the near future.

References

Kurt Benke , Bruce Tomkins , 2017, Future food-production systems: vertical farming and controlled-environment agriculture, Sustainability: Science, Practice and Policy, 13:1, 13-26.

A. Chahid , I. N'Doye, J.E. Majoris, M.L. Berumen, T.M. Laleg-Kirati, 2021, Model predictive control paradigms for fish growth reference tracking in precision aquaculture, Journal of Process Control, 105, 160-168.

M. Cui, X Liu, J. Zhao, J. Sun, G. Lian, T. Chen, M.D. Plumbley, D. Li, W. Wang, 2022, Fish Feeding Intensity Assessment in Aquaculture: A New Audio Dataset AFFIA3K and a Deep Learning Algorithm," 2022 IEEE 32nd International Workshop on Machine Learning for Signal Processing (MLSP), 2022, pp. 1-6.

Y. Guo, S. Zhang, S. Li, M. Zhang, H. Benli, Y. Wang, 2022, Numerical investigation for effects of natural light and ventilation on 3D tomato body heat distribution in a Venlo greenhouse, Information Processing in Agriculture, in press.

A.L. Harfouche, D.A. Jacobson, D. Kainer, J.C. Romero, A.H. Harfouche, G.S. Mugnozza, M. Moshelion, G.A. Tuskan, J.J.B. Keurentjes, A. Altman, 2019, Accelerating Climate Resilient Plant Breeding by Applying Next-Generation Artificial Intelligence, Trends in Biotechnology, 37: 1217-1235.

D.M.K.S. Hemathilake, D.M.C.C. Gunathilake, 2022, Chapter 32 - High-productive agricultural technologies to fulfill future food demands: Hydroponics, aquaponics, and precision/smart agriculture,Future Foods,Academic Press,Pages 555-567.

D.P. Holzworth, V. Snow, S. Janssen, I.N. Athanasiadis, M. Donatelli, G. Hoogenboom, J.W. White, P. Thorburn, 2015, Agricultural production systems modelling and software: Current status and future prospects,Environmental Modelling & Software, 72: 276-286.

J. Lavarenne, S. Guyomarc'h, C. Sallaud, P. Gantet, M. Lucas, 2018, The Spring of Systems Biology-Driven Breeding, Trends in Plant Science, 23, 706-720.

A.T. Le, Wang, L.; Wang, Y.; Vu, N.T.; Li, D, 2020, Experimental Validation of a Low-Energy-Consumption Heating Model for Recirculating Aquaponic Systems, Energies, 13, 1958.

A.T. Le, L. Wang, Y. Wang, D. Li, 2021, Measurement investigation on the feasibility of shallow geothermal energy for heating and cooling applied in agricultural greenhouses of Shouguang City: Ground temperature profiles and geothermal potential, Information Processing in Agriculture, 8: 251-269.

H. Li, S. Chatzifotis, G. Lian, Y. Duan, D. Li, T. Chen, 2022, Mechanistic model based optimization of feeding practices in aquaculture, Aquacultural Engineering, 97: 102245.

H. Li, W. Li, M. McEwan, D. Li, G. Lian, T. Chen, 2021, Adaptive filtering-based soft sensor method for estimating total nitrogen in aquaponic systems, Computers and Electronics in Agriculture, 186: 106175.

S. Liu, L. Xu, D. Li, Q. Li, Y. Jiang, H. Tai, L. Zeng, 2013, Prediction of dissolved oxygen content in river crab culture based on least squares support vector regression optimized by improved particle swarm optimization,Computers and Electronics in Agriculture, 95: 82-91.

J.I. Marsh, Hu, H., Gill, M. et al., 2021, Crop breeding for a changing climate: integrating phenomics and genomics with bioinformatics. Theor Appl Genet 134, 1677–1690.

M. O'Brien, J. Mack, B. Lennox, D. Lovett, A. Wall, 2011, Model predictive control of an activated sludge process: A case study, Contorl Engineering Practice, 19, 54-61.

United Nations, 2022, Data Portal Population Division, https://population.un.org/dataportal/home.

Wang T, Xu X.,Wang C,Li Z., Li D, 2021,From Smart Farming towards Unmanned Farms: A New Mode of Agricultural Production,Agriculture, 11, 145.

H. Wen, L. Zhang, Z. Fu, X. Li, Y. Su, 2012, Agricultural disease-control scene determination based on audio-visual fusion, Journal of Food, Agriculture & Environment, 10: 867-870.

H. Yu , Y. Chen, S. Gul Hassan, D. Li, 2016, Prediction of the temperature in a Chinese solar greenhouse based on LSSVM optimized by improved PSO,Computers and Electronics in Agriculture,122: 94-102.

H.H. Yu, L. Yang, D. Li, Y. Chen, 2020, A hybrid intelligent soft computing method forammonia nitrogen prediction in aquaculture, Information Processing in Agriculture, (20)30034-2.

P. Zhang, L. Yang, D. Li, 2020, EfficientNet-B4-Ranger: A novel method for greenhouse cucumber disease recognition under natural complex environment,Computers and Electronics in Agriculture,176: 105652.

X. Zhou, D. Li, L. Zhang, Q. Duan, 2021, Application of an adaptive PID controller enhanced by a differential evolution algorithm for precise control of dissolved oxygen in recirculating aquaculture systems, Biosystems Engineering, 208, 186-198.

Acknowledgement

We thank funding support from the UK BBSRC (BB/S020896/1), the UK Biomass Feedstocks Innovation Programme, and the National Science Foundation of China (62339001).

Antonis Kokossis, Michael C. Georgiadis, Efstratios N. Pistikopoulos (Eds.)
PROCEEDINGS OF THE 33rd European Symposium on Computer Aided Process Engineering
(ESCAPE33), June 18-21, 2023, Athens, Greece

CO_2 Capture by Amine-Functionalized Magnesium Oxide: Experimental and Modelling Studies

Sagheer A. Onaizi

Department of Chemical Engineering and Center of Excellence in Nanotechnology, King Fahd University of Petroleum and Minerals, Dhahran 31216, Saudi Arabia

Abstract

The concentration of carbon dioxide in the atmosphere has reached an alarming level. CO_2 can be captured using the traditional amine absorption process. However, this process is an energy-intensive and environmentally unfriendly. Thus, we have developed in this work a novel adsorbent using magnesium oxide (MgO) that has been functionalized with 3-aminopropyl-triethoxysilane (APTES). This CO_2 adsorbent (APTES-MgO) is reported for the first time in this work to the best of our knowledge. APTES-MgO can significantly adsorb CO_2 at ambient conditions. It is fully regenerable and the regeneration process can be carried out at mild conditions (120°C and atmospheric pressure) under a nitrogen environment. This low regeneration temperature indicates that the energy consumption of CO_2 capture process using APTES-MgO could be low. The adsorption capacity of APTES-MgO is more than 1.5 mmol/g at 30 °C and 1 atm. The regeneration process of APTES-MgO can fully restore its CO_2 adsorption capacity. The adsorption process has been modelled using a first order adsorption model. The mathematical model can track the experimental data quite well, allowing the estimation of the key adsorption parameters. The effectiveness of APTES-MgO in adsorbing CO_2 at ambient conditions and its regenerability and stability suggest its potential commercial application in CO_2 capture industry.

Keywords: Amine-functionalized magnesium oxide, Adsorption, Regeneration, Ambient conditions, Carbon dioxide (CO_2) capture, Mathematical modelling

1. Introduction

Next paragraphs should start with heading as given in below example. The emission of CO_2 to the atmosphere is steadily increasing. One of the serious impacts of increasing CO_2, and other greenhouse gases, level in the atmosphere is global warming, which has several negative consequences. One of such negative consequence is the significant increase in the water level of oceans and seas and, thus, the submerging of some islands and coastal cities. Thus, CO_2 has to be captured and managed in a proper way. The current main process for CO_2 capture is the absorption using aqueous solutions of some amines such as methyldiethanolamine (MDEA) and monoethanolamine (MEA). In addition to environmental concerns on the utilization of these organic solvents (i.e., amines), amine absorption process is corrosive and the regeneration of the spent amine is energy intensive (i.e., expensive) [1-3]. Thus, alternative processes have been (and are still) actively sought by academia and relevant industries.

One of the attractive alternatives is adsorption. Nonetheless, several adsorbents reported in the literature are either not regenerable or/and require relatively high regeneration energy. Furthermore, some adsorbents are only effective at high pressures or sub-zero temperatures. Accordingly, there is a pressing need for developing CO_2 adsorbents that are effective at (or near) ambient conditions. Besides their effectiveness at ambient conditions, theses adsorbents should be stable and fully regenerable under mild conditions. Thus, MgO has been synthesized in this work and used for CO_2 capture. To enhance the CO_2 adsorption capacity, MgO was functionalized with APTES. The kinetics of CO_2 adsorption on both MgO and the APTES-modified MgO was fitted using three different adsorption kinetics models (pseudo-first order, pseudo-second order and Avrami), enabling the extraction of the values of the adsorption parameters.

2. Methodology

2.1. Preparation of adsorbents

MgO was synthesized using the following protocol. First, an aqueous solution (270 mM) of magnesium nitrate hexahydrate was prepared. After obtaining a homogenous solution of this magnesium salt via stirring at room temperature for 10 min, a certain volume of 30% ammonium hydroxide (NH_4OH) was added to make the molar ratio of Mg: NH_4OH as 1:5. The new mixture was stirred for few minutes to thoroughly mix its ingredients. After that, the glass bottle containing the mixture was closed and incubated in a hot water bath (60 °C) for 6 h. the mixture was stirred while incubated in the hot water bath. Then, the mixture was removed from the hot water bath, the bottle cap was removed (to allow some liquid to evaporate) and the bottle content was stirred for additional 24 h at room temperature, leading to the formation of a thick paste (consists of $Mg(OH)_2$ and NH_4NO_3) [4]. To remove the formed NH_4NO_3, which is water-soluble, the thick paste was washed extensively with distilled water. Such an extensive wash is expected to also remove any unreacted magnesium precursor. After drying at 60 °C, the material was ground and calcinated (under an air environment) at 400 °C for 10 h to yield MgO. The synthesized MgO was characterized using BET and energy-dispersive X-ray spectroscopy (EDX).

After the synthesis of MgO, it was functionalized with APTES according to the following procedure. An aqueous dispersion of MgO containing 0.83 g/L was prepared by vigorously stirring the MgO solution for 5 min. After that a certain amount of APTES (MgO/APTES mass ratio = 1/2) was added to the MgO suspension. Then, the resultant solution was heated at 80 °C and refluxed for one day. Following this step, the obtained amine-functionalized MgO was rinsed thoroughly with distilled water. The produced APTES-MgO was collected via centrifugation and then it was dried at 60 °C overnight. The synthesized APTES-MgO was also characterized using BET and EDX.

2.2. Adsorption kinetics of CO_2

The CO_2 adsorption kinetics on the bare and the APTES-modified MgO was monitored using a thermogravimetric analysis (TGA) technique. In each experiment, the TGA pan (made from alumina) was thoroughly cleaned and dried and then a certain amount (roughly 20 mg) of either MgO or APTES was placed in the TGA pan. To remove moisture, the adsorbent was degassed at 100 °C using a stream of nitrogen. The mass of the adsorbent was continuously monitored while degassing until it reached a steady value and remained constant for at least 30 min. After that, the temperature was dropped to the desired adsorption temperature (i.e., 30 °C). The temperature was continuously monitored until it stabilized at 30 °C and remained stable. Then, nitrogen gas valve was closed and the carbon dioxide (> 99.9% pure) valve was opened. CO_2 flowrate was kept constant at 100 mL/min. To ensure the attainment of adsorption equilibrium, the adsorption of CO_2 was monitored in time until no significant change in the sample mass (adsorbent + adsorbate) was observed.

2.3. Regeneration of spent amine-functionalized adsorbent

The regeneration of APTES-MgO (given the effectiveness of this adsorbent relative to the unmodified MgO as will be presented and discussed in the subsequent section) was studied. The regeneration was carried out using nitrogen gas flowing at 100 mL/min. The protocol of APTES-MgO regeneration is as follows. APTES-MgO was contacted with CO_2 until adsorption equilibrium was approached. Then, CO_2 flow (100 mL/min) was halted and N_2 valve was opened. Once, the adsorbent was fully regenerated (i.e., the mass of APTES-MgO returned to its original value before CO_2 injection), the flow of N_2 was stopped and CO_2 valve was opened. Four cycles of adsorption-regeneration were conducted. In each of these cycles, CO_2 adsorption took place at 30 C while the regeneration temperature was set to 120 °C. The pressure of adsorption and regeneration processes was fixed at 1 atm.

3. Results and discussion

3.1. Characterization of the synthesized MgO and APTES-MgO

After the synthesis of MgO and its modification with APTES, elemental analysis of the composition of these two adsorbents was carried out using EDX. The results obtained from the elemental analysis are summarized in Table 1. The MgO sample contains about 53.61 atomic% magnesium and 46.39 atomic% oxygen, indicating the formation of magnesium oxide. The hypothetical chemical formulas of the formed magnesium oxide is $MgO_{0.87}$, which is reasonably close to the chemical formula of magnesium oxide (MgO). Upon the modification of MgO with APTES, other elements (N, C, and Si) appeared in the elemental analysis results. These elements originate from the APTES molecules (molecular formula: $C_9H_{23}NO_3Si$). The atomic ratios of N, C and Si in the APTES-MgO adsorbent are close to their atomic ratios in the APTES molecules. The increase in the oxygen content of the amine-modified MgO relative to MgO stems from the presence of oxygen in the APTES molecules.

Table 1: Elemental compositions of the synthesized MgO-based adsorbents.

Sample	Mg atomic%	O atomic%	C atomic%	N atomic%	Si atomic%
MgO	53.61	46.39	–	–	–
APTES-MgO-A	15.85	44.75	31.83	3.01	4.56

Besides the elemental analysis, textural properties (i.e., BET surface area, pore volume and pore diameter) of MgO and APTES-MgO were studied. The textural properties of the synthesized MgO and the amine-modified MgO are shown in Table 2. The BET surface area and pore volume of MgO were about 246 m^2/g and 0.086 cm^3/g, respectively. Upon the functionalization of MgO with APTES, the BET surface area and the pore volume of the modified MgO dropped by a factor of ~5. Such a reduction in the BET surface area of MgO and its pore volume upon the functionalization with APTES is expected since the amine molecules are likely to fill some of MgO pores [5, 6]. Unlike the drop in the BET surface area and pore volume of APTES-MgO relative to the unmodified MgO, the pore size has increased by about 20% upon the attachment of APTES to MgO. The attachment of APTES to MgO-A was carried out at 80 °C under continuous stirring and refluxing for 24 h. Such a thermal treatment, despite being mild, coupled with APTES attachment to MgO-A might have assisted in expanding the size of the pores of the APTES-MgO-A sample [7]. The expansion in the pore size upon the functionalization of MgO with APTES does not contradict the observed reduction in pore volume and surface area. However, despite having higher pore size, the number of pores in the APTES-MgO sample is likely much lower than that in the unmodified MgO due to the blockage of some pores by APTES molecules.

Table 2: BET surface area, average pore size and average pore volume of the synthesized MgO-based adsorbents.

Samples	BET Surface area (m²/g)	Pore volume (cm³/g)	Pore size (nm)
MgO	246.421	0.086	1.389
APTES-MgO	48.651	0.020	1.644

3.2. Adsorption kinetics of CO_2

CO_2 adsorption on MgO and APTES-MgO was studied at ambient conditions (i.e., 30 °C and 1 atm). Before contacting CO_2 with these two adsorbents, they were degassed using nitrogen gas as described in the previous section. Figure 1 shows the kinetics of CO_2 adsorption on the unmodified MgO. The adsorption capacity approached about 0.68 mmol/g when the adsorption process approached equilibrium. The adsorption data were fitted to pseudo-first order, pseudo-second order and Avrami kinetics model as shown in Figure 1. Among these three models, the pseudo-second order kinetics model provides the best fit to the experimental data ($R^2 = 0.9996$). The estimated parameter values of the CO_2 adsorption kinetics using these models are presented in Table 3.

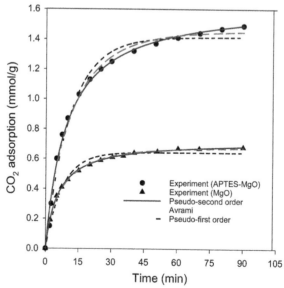

Figure 1: Adsorption kinetics of CO_2 on MgO and APTES-MgO. The solid curves represent the calculated values using the adsorption kinetics models.

The adsorption of CO_2 on APTES-MgO was also studied and modeled using the above adsorption kinetics models (see Figure 1). The estimated values of the adsorption parameters are shown in Table 3. The results displayed in Figure 1 and Table 3 (particularly the R^2 values) clearly demonstrate the superiority of the pseudo-second order in tracking the experimental data relative to the other two models. However, unlike the case of MgO where Avrami kinetics model provided a good data fitting, the fitting of CO_2 adsorption on APTES-MgO using this model is relatively poor. Nonetheless, Avrami model still provides a better data fitting for CO_2 adsorption on both MgO and APTES-MgO compare to the pseud-first order model.

Table 3: Values of the fitting parameters of CO_2 adsorption kinetics.

| | | MgO | | | |
Model	Model Equation	Adsorption rate constant (k)	Adsorption capacity at equilibrium (qe)	Avrami exponent (n)	R^2
Pseudo-first order	$q_t = q_e\left(1 - e^{-kt}\right)$	0.131 min⁻¹	0.643 mmol.g⁻¹	-----	0.9849
Pseudo-second order	$q_t = kq_e^2 t / \left(1 + kq_e t\right)$	0.241 g.mmol⁻¹.min⁻¹	0.725 mmol.g⁻¹	-----	0.9996
Avrami	$q_t = q_e\left(1 - e^{-(kt)^n}\right)$	0.117 min⁻⁰·⁷⁰⁶	0.675 mmol.g⁻¹	0.706	0.9980
		APTES-MgO			
Model	Model Equation	Adsorption rate constant (k)	Adsorption capacity at equilibrium (qe)	Avrami exponent (n)	R^2
Pseudo-first order	$q_t = q_e\left(1 - e^{-kt}\right)$	0.089 min⁻¹	1.414 mmol.g⁻¹	-----	0.9879
Pseudo-second order	$q_t = kq_e^2 t / \left(1 + kq_e t\right)$	0.064 g.mmol⁻¹.min⁻¹	1.647 mmol.g⁻¹	-----	0.9951
Avrami	$q_t = q_e\left(1 - e^{-(kt)^n}\right)$	0.082 min⁻⁰·⁸⁴⁶	1.454 mmol.g⁻¹	0.846	0.9921

3.3. Regeneration of spent amine-functionalized adsorbent

In addition to the good CO_2 adsorption capacity on the APTES-functionalized MgO, it has also an excellent regenerability as shown in Figure 2. The CO_2 adsorption capability of APTES-MgO is preserved with the repetitive use. This means that the adsorbent is stable and the regeneration process of the spent adsorbent does not compromise its adsorption capacity. This is an attractive characteristic that is highly required in industrial applications. Another attractive characteristics of APTES-MgO is the mild regeneration conditions (120 °C and 1 atm). Exposing the spent APTES-MgO to N_2 at 120 °C and 1 atm can fully restore the original CO_2 adsorption capacity of this adsorbent within a very short time (< 20 min). This regeneration time can shortened with increasing the flow rate of the regeneration gas.

4. Conclusions

The modification of MgO with APTES greatly enhances CO_2 adsorption by a factor of more than 2 despite the significant drop (more than 5 times) in the adsorbent surface area and pore volume. Kinetically, the adsorption of CO_2 on both adsorbents follows a pseudo-second order adsorption kinetics. The pseudo-first order model failed to reproduce the experimental data while the data fitting using the Avrami model was not as good as using the pseudo-second order. The full restoration of the adsorption capability of APTES-MgO upon the contact with N2 at mild conditions in addition to the stability of the adsorbent make it a good candidate for commercial CO_2 capture.

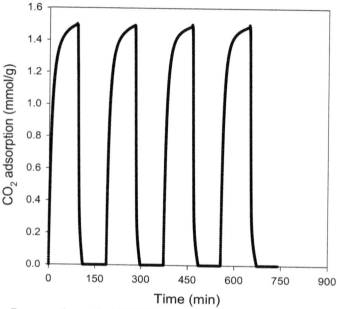

Figure 2: Regeneration of APTES-MgO using nitrogen gas. The regeneration temperature and pressure are 120 °C and 1 atm, respectively.

References

[1] P. Luis, 2016, Use of monoethanolamine (MEA) for CO2 capture in a global scenario: Consequences and alternatives, Desalination, 380:93-99.
[2] J. Liu, S. Wang, B. Zhao, H. Tong, C. Chen, 2009, Absorption of carbon dioxide in aqueous ammonia, Energy Procedia,1:933-940.
[3] R. Strube, G. Pellegrini, G. Manfrida G, 2011, The environmental impact of post-combustion CO_2 capture with MEA, with aqueous ammonia, and with an aqueous ammonia-ethanol mixture for a coal-fired power plant, Energy, 36:3763-3770.
[4] A. Hanif, S. Dasgupta, A. Nanoti A, 2016, Facile Synthesis of High-Surface-Area Mesoporous MgO with Excellent High-Temperature CO2 Adsorption Potential, Industrial & Engineering Chemistry Research, 55:8070-8078.
[5] X. Xu, C. Song, J. M. Andresen, B. G. Miller, 2002, Scaroni AW. Novel Polyethylenimine-Modified Mesoporous Molecular Sieve of MCM-41 Type as High-Capacity Adsorbent for CO2 Capture, Energy & Fuels. 16:1463-9.
[6] S.-H. Liu, W.C. Hsiao, W.-H. Sie, 2007, Tetraethylenepentamine-modified mesoporous adsorbents for CO2 capture: effects of preparation methods, Adsorption, 18:431-7.
[7] Y. Matsuo, Y. Nishino, T. Fukutsuka, Y. Sugie, 2007, Introduction of amino groups into the interlayer space of graphite oxide using 3-aminopropylethoxysilanes, Carbon, 45:1384-1390.

Antonis Kokossis, Michael C. Georgiadis, Efstratios N. Pistikopoulos (Eds.)
PROCEEDINGS OF THE 33rd European Symposium on Computer Aided Process Engineering
(ESCAPE33), June 18-21, 2023, Athens, Greece

Development of 'Green' LNG through a CO_2 allocation procedure within a CO_2 utilisation network in Qatar

Razan Sawaly, Mohammad Alherbawi, Tareq Alansari*

College of Science and Engineering, Hamad Bin Khalifa University, Doha, Qatar
Corresponding email: talansari@hbku.edu.qa

Abstract

Natural gas has a crucial role in minimising carbon emissions and is considered a transition fuel. Liquified natural gas (LNG), which is derived from natural gas is one of the most promising alternative fuels, and LNG producers are actively developing their methods to reduce or offset their carbon footprints. The allocation of carbon dioxide (CO_2) emissions is crucial in identifying who is responsible for reducing emissions at the economy or enterprise level. Different strategies for allocating CO_2 emissions have been developed and used during the last few decades. As such, this study introduces options for low carbon LNG within a CO_2 utilisation network in the state Qatar; where a linear programming (LP) model is developed to address the optimal CO_2 allocation based on environmental and economic objectives. The model considers a single CO_2 source (Qatar Gas), where a post combustion carbon capture technology (PCC) is utilised to capture the CO_2 emissions, and then allocated to 7 sinks that use CO_2 as input to their production (including Enhanced Oil Recovery (EOR)). Three scenarios are considered to allocate CO_2 to different sinks to obtain the optimal solution for each scenario. Scenario 1 had the highest total profit of 55 billion USD. The lowest CO_2 emissions were produced by scenario 2, which had zero emissions. Scenario 3 had a total profit of 20 billion USD and a total of 2 million CO_2 emissions. For all scenarios, EOR sinks had the highest CO_2 allocation. The total global warming potential through the lifecycle of LNG is reduced from 0.46 to 0.14 tonne CO_2e/tonne of LNG produced.

Keywords: LNG, CO_2 allocation, Sustainability, Carbon Dioxide, Carbon capture and utilisation (CCU)

1. Introduction

Due to the growing environmental concerns, there are indicators that expanding symbiotic linkages between industries within the same vicinity may significantly contribute to the development of sustainable industrial operations. Several energy corporations have recently revealed far-reaching carbon-neutrality goals, accompanied by more or less remote time horizons. Regardless of the fact that natural gas is the cleanest-burning hydrocarbon, regulators are paying closer attention to GHG emissions linked to the natural gas supply chain, including that of liquified natural gas (LNG) (Roman-White et al., 2021). To guarantee natural gas's place in the energy transition in a decarbonizing world with a rising proportion of renewable energy sources, natural gas emissions must be reduced, and this can be made possible through a systems approach on an industrial level. For resource allocation in industrial park, a variety of tools has been utilised previously, including optimisation, game theory, agent-based modeling, network analysis, etc. Due to the established technologies in many industries, which are mostly focused on the use of fossil fuels, the direct reduction of the emissions is currently complex in the short term. Therefore, the concept of Carbon Capture and Storage (CCS) arises as a compelling proposal for reducing emissions, where absorption has been demonstrated as an effective approach to produce a high capture yield with various solvents, such as piperazine (PZ) or monoethanolamine (MEA) (Li and Zhang, 2018). The requirement of large storage capacity of carbon dioxide gas is a challenge that affects the adoption of CCS technologies (IPCC, 2005). Furthermore, Carbon Capture and Utilisation (CCU) aims to gain an advantage by using CO_2 in various industrial processes in place of conventional raw materials (Aresta, 2010). Utilising CO_2 is a possibility for supplying a renewable energy source for the creation of numerous valued items. The optimum method of utilising CO_2 is thought to be its conversion into fuel. Some of the chemicals that can be generated by using captured CO_2 as an input or feedstock include methane, methanol, alkanes, and syngas (Marchese et al., 2021).

Research on energy conservation in an eco-industrial park (EIP) is mostly unexplored in the field of mathematical optimisation. Boix et al. (2015) stated that many objective functions, such as EIP energy efficacy, CO_2 emission reductions, waste heat recovery, and network payback period, have been proposed in the literature. However, the challenge of achieving multiple conflicting objectives at once is not efficiently examined yet as well as carbon focused eco-park resource allocation with an aim of producing low carbon and zero-carbon LNG. As such, this study is an extension of previous study conducted by Sawaly et. al, (2022), with the integration of a multi-objective model and Alshaheen field as an new EOR sink. A linear programming (LP) model is developed to address the optimal CO_2 allocation with multi-objective to maximise the overall profit of the system while minimising CO_2 emissions. The model considers a single CO_2 source (Qatar Gas), where a post-combustion carbon capture technology (PCC) is utilised to capture the CO_2 emissions, and then allocated to sinks that use CO_2 as input in their production. Thus, this study aims at introducing additional options for low-carbon LNG within a CO_2 capture and utilisation (CCU) network in the state Qatar to optimise and enhance eco-parks exchanges.

2. Model Design and Optimisation

2.1 Available data and assumptions

The objective is to facilitate a resource trade scheme for CO_2 emissions to produce low carbon LNG in Qatar Gas. With a total of roughly 21 LNG trains, Qatar Gas (QG) is the largest LNG producer in the world, operating 14 LNG trains with a 77 MT annual production capacity (Al-Yaeeshi, 2020). According to Mohammed, (2016), the corresponding CO_2 emissions are 35.8 MT/year (Scope 1 – Direct total GHG emissions). Therefore, a post combustion carbon capture technology (PCC) is utilised to capture 70% of the total CO_2 emissions emitted by QG (25 MT year), and then allocate to sinks that use CO_2 as input to their production (QAFCO, QAFAC, Pearl GTL, Oryx GTL, Dukhan Field Well (EOR), Alshaheen Field (EOR)).

$$\text{Enhanced Lifecycle GWP of Qatar Gas} = 30\% \times \frac{tonne\ of\ CO_2\ emitted/year}{tonne\ of\ LNG\ produced/year} \qquad (1)$$

With an OPEX of \$40 per tonne of CO_2, chemical absorption using amine solvents was expected to capture CO_2 with 100% efficiency. In order to prevent pressure reductions throughout the pipeline and to preserve the operational parameters of the sink, CO_2 must be compressed. For every 100 km along the pipeline, the unit cost for compression and subsequent transportation is roughly \$1.5/tonne CO_2 (Al-Yaeeshi, 2020; Sawaly et. al, 2022). It is assumed that the injection cost and emissions for EOR is neglected. Table 1 lists the sinks' products, production rate, CO_2 capacity, and emission rate. The suggested methodology uses a Linear Programming (LP) optimisation framework to address CO_2 allocation and find solutions for reducing CO_2 emissions at a realistic cost while also maximising the profits of value-added products. The proposed methodology is used to implement the model's outcomes using the Excel solver and MATLAB. Carbon capture and compression, along with transportation are the only expenditures taken into account in this study that have an influence on profit, whilst the only emissions included are process related emissions. Three scenarios are implemented in this study: (1) A scenario for allocating CO_2 from the source to the sinks with a single objective of maximising the profit of the overall network (2) A scenario for allocating CO_2 from the source to the sinks with a single objective to minimise the CO_2 emission of the overall network, and; (3) A scenario for allocating CO_2 from the source to the sinks with a multi-objective for profit maximisation and emission minimisation of the overall network. The commodity prices of the products are collected over the period 2014-2018 (Intratec Solutions, 2019).

Table 1: Characteristics of the sinks (Sawaly et al., 2022; Ghiat et al., 2021; IEA, 2018; Mohammed, 2016).

Sink	Process/ product	Emission Rate	Production Rate	CO_2 Capacity	Distance from source
Unit	-	Tonne CO_2 out/tonne CO_2 feed	Tonne of product/tonne of CO_2	Million tonne/year	Kilometre
QAFAC H_2	Methanol from H_2	0.02	5.9	1.83	100
QAFAC NG	Methanol from NG	2.09	3	0.60	100
QAFCO	Urea	0.23	1.35	1.01	100
Pearl GTL	ATR-GTL-Wax	2.33	0.884	2.19	5
Oryx GTL	ATR-GTL-Wax	9.75	0.884	0.36	3
Dukhan Field	EOR	0	0.2	15.6	111
AlShaheen Field	EOR	0	0.32	13.9	80

2.2 Model Formulation

$$\text{Maximise } \sum_{i=1}^{7}\sum_{j=1}^{5} x_{ij}[P_{ij}\, PR_{ij}\, Q - C_{operating,ij}] \tag{2}$$

$$\text{Minimise } \sum_{i=1}^{7}\sum_{j=1}^{5} x_{ij}\, Q\, ER_i \tag{3}$$

Subject to

$$\sum_{i=1}^{7}\sum_{j=1}^{5} x_{ij} Q\, PR_i \leq C_{Pij} \tag{4}$$

$$\sum_{i=1}^{7} x_{ij} = 1 \tag{5}$$

Where;

$$C_{operating,ij} = \sum_{i=1}^{7}\sum_{j=1}^{5}\left(C_{transport} + C_{pcc} + C_{OPEX}\right) \tag{6}$$

$$C_{transport} = Q x_{ij} \times \left(\frac{D_i}{100}\right) \times 1.5 \tag{7}$$

$C_{operating,ij} = Cost\ for\ pcc, transport, OPEX\ of\ sink\ i\ for\ year\ j$
$C_{P,ij} = Capacity\ of\ sink\ i\ for\ year\ j$
$Q = Carbon\ Captured\ from\ source\ (tonne/year)$
$x_{ij} = Fractional\ amount\ allocated\ to\ sink\ i\ for\ year\ j$
$PR_{ij} = Production\ Rate\ of\ sink\ i\ for\ year\ j\ as\ a\ function\ of\ carbon\ feed$
$ER_{ij} = Emission\ Rate\ of\ sink\ i\ for\ year\ j\ as\ a\ function\ of\ carbon\ feed$
$P_{ij} = Product\ price\ of\ industry/sink\ i\ for\ year\ j$
$D_i = Distance\ from\ the\ source\ to\ sink\ i$

3   Results and Discussion

The proposed methodology is used to implement the model's outcomes using the Excel solver for scenario 1 and scenario 2, and MATLAB for scenario 3. *Scenario 1* aims to determine the optimal solution for allocating 25×10^6 tonne/year of CO_2 to all 7 sinks with an objective to maximise the profit of the overall system. According to the findings, the market prices of the products and production rate are the key drivers of the techno-economic optimisation. The total annual profits for all sinks ranged from 8.1 to 14.6 billion USD and the total profit of the overall system throughout the specified time period is 55 billion USD. The annual amount of CO_2 emitted by the sinks is 10.1 million tonnes CO_2/year and the total emissions of the overall system throughout the specified time period is 50.7 million tonnes CO_2. Figure 1 represents the allocation across the sinks, with QAFCO, QAFAC H_2, QAFAC NG, Pearl GTL, and Oryx GTL being fully satisfied due to their higher commodity prices and smaller CO_2 capacity in comparison to Dukhan field well and Alshaheen field, which use CO_2 for enhanced oil recovery. Alshaheen field has a larger share of CO_2 allocation than Dukhan field, which is due to Alshaheen field's higher production rate and consequently generates more profit. The total CO_2 utilisation of the system is 25×10^6 tonne/year, which implies that the captured amount of CO_2 from Qatar Gas (25MT/year) is fully utilised by the sinks. The allocation ratio throughout the years is relatively in the same range for all years between 2014 and 2018.

This is due to the price linkage between products, which causes them to shift together.

In *Scenario 2*, the goal is to find the best way to allocate 25×10^6 tonnes of CO_2 per year among the seven sinks with the least amount of emissions possible for the overall network. The allocation's primary factor, as indicated by the results, was each sink's emission rate. During enhanced oil recovery, the majority of the CO_2 that is injected stays there trapped in the geological formation, causing little to no emissions of CO_2. Therefore, the allocation will shift toward the sink with the lowest emissions rate in the case of emission reduction. The captured CO_2 was allocated equally to Dukhan Field and Alshaheen field (figure 2) since their process generates zero emissions. The total annual profits for all sinks ranged from 1.43 to 5.09 billion USD and the total profit of the overall system throughout the specified time period is 14.6 billion USD.

Scenario 3 was generated using MATLAB for finding a set of solutions that define the best trade-off between the competing objectives. The model generated 70 different Pareto optimal solutions (figure 4), however, most of the solutions favored EOR sinks (both, Dukhan Field and AlShaheen Field), which is because they generate zero emissions, and got higher CO_2 capacity allowing for more profit as well regardless of the fact that they both have the lowest production rate. Furthermore, following the EOR sinks, QAFAC H_2 obtained the highest allocation percentage due to its high production rate and minimal CO_2 emissions. Out of the 70 solutions that were developed, the one with the highest profit had a GWP of 4.5 MT CO_2e and a profit of 22.6 billion USD. The solution that achieved the lowest GWP had a profit of 15 billion USD and zero GWP (zero emissions). The mean solution had a profit of 20 billion USD and a GWP of 2 MT CO_2-eq. Figure 3 illustrates the CO_2 allocation for the mean solution.

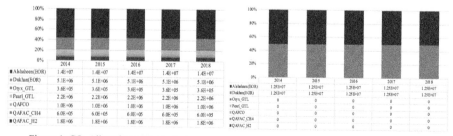

Figure 1: CO_2 Allocation of Scenario 1.

Figure 2: CO_2 Allocation of Scenario 2.

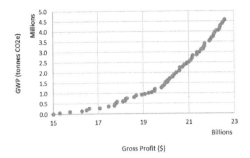

Figure 4: Pareto Front.

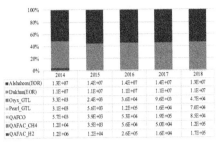

Figure 3: CO₂ Allocation of Scenario 3.

Figure 5 compares the total CO_2 emissions and total profit of scenario 1 and 2 with the mean solution of scenario 3. Scenario 2 witnessed a huge decrease in the profit and CO_2 emission compared to scenario 1; where CO_2 emissions were reduced by 100% and the profit was decreased by around 73%. Scenario 3 on the other hand aims to satisfy both objectives simultaneously. Therefore, CO_2 emissions are reduced significantly from 50.7 million tonnes CO_2 (scenario 1) to 2 million tonnes CO_2 (scenario 3), while the profit fell sharply from 55 billion USD to 20 billion USD for scenarios 1 and 3 respectively. When comparing scenario 2 and 3, the total profit increased slightly from 14.6 billion USD (scenario 2) to 20 billion USD (scenario 3). The total GWP through the lifecycle of LNG in Qatar Gas was 0.46 tonne CO_2e/tonne of LNG produced. This ratio has been enhanced through this model and reduced to 0.14 (70% cut of emissions) through CO_2 capture and allocation by substitution approach.

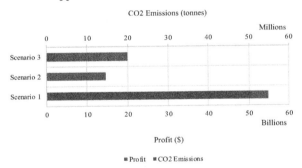

Figure 5: Comparison between scenario 1,2 and 3.

4   Conclusion

The findings of this study aim to provide the most efficient strategies for producing low carbon LNG within a CO_2 utilisation network in the state Qatar by converting carbon dioxide emissions into value added products. The model considers a single CO_2 source (Qatar Gas) and 7 sinks. Furthermore, the optimal CO_2 allocation based on environmental and economic objectives is addressed using a linear programming (LP) model where 3 scenarios where generated. Scenario 1 aimed to maximise the overall profit of the system and had a value of 55 billion USD; whereas scenario 2 aimed to minimise the CO_2 emissions which reached to zero emissions. Scenario 3 generated a profit of 20 billion USD and 2 million CO_2 emissions overall. According to the results of using allocation by substitution, only 0.14 tonne CO_2e/tonne LNG is emitted whereas 0.32 tonne CO_2e/tonne LNG is captured and utilised by the sinks. The lack of actual selling market price and plant data for each of the products studied between 2014 and 2018 are some of the limitations of this study. Also, the production rate is calculated based on assumptions derived from literature that may or may not represent real operating data for the relevant time period. Future work would include: 1) implementation of carbon tax to the model, which can have an effect on both the environmental and economic objective; 2) capture a larger amount of CO_2 from Qatar Gas (85% and 100%).

5   Acknowledgement

The authors acknowledge the support of Qatar National Research Fund (a member of Qatar Foundation) provided by GSRA grants no. GSRA9-L-1-0531-22069.

6   References

Al-Yaeeshi, A. A., Govindan, R., & Al-Ansari, T. (2020). Techno-economic-based dynamic network design for optimum large-scale carbon dioxide utilisation in process industries. Journal of Cleaner Production, 275, 122974. https://doi.org/10.1016/j.jclepro.2020.122974

Aresta, M. (2010). Carbon dioxide: Utilization options to reduce its accumulation in the atmosphere. Carbon Dioxide as Chemical Feedstock, 1–13. https://doi.org/10.1002/9783527629916.ch1

Boix, M., Montastruc, L., Azzaro-Pantel, C., & Domenech, S. (2015). Optimization methods applied to the design of Eco-industrial parks: A literature review. Journal of Cleaner Production, 87, 303–317. https://doi.org/10.1016/j.jclepro.2014.09.032

Ghiat, I., Alnouss, A., & Al-Ansari, T. (2021). Superstructure Optimisation in Various Carbon Capture and Utilisation Supply Chains, PSE 2021+ Conference.

Iea. (2018, November 1). Can CO2-Eor really provide carbon-negative oil? – analysis. IEA. Retrieved November 29, 2022, from https://www.iea.org/commentaries/can-co2-eor-really-provide-carbon-negative-oil

Intratec Solutions. (2019). Commodity prices - full sample - Historical & Forecast Data in several countries. Intratec.us. Retrieved February 1, 2023, from https://www.intratec.us/products/primary-commodity-prices

IPCC 2005. IPCC special report on carbon dioxide capture and storage. prepared by working group iii of the intergovernmental panel on climate change. In IPCC Special Report on Carbon Dioxide Capture and Storage 2:442. doi: 10.1002/anie.201000431.

Li, H., & Zhang, Z. (2018). Mining the intrinsic trends of CO2 solubility in blended solutions. Journal of CO2 Utilization, 26, 496-502.

Marchese, M., Buffo, G., Santarelli, M., & Lanzini, A. (2021). CO2 from direct air capture as carbon feedstock for Fischer-Tropsch chemicals and fuels: Energy and Economic Analysis. Journal of CO2 Utilization, 46, 101487. https://doi.org/10.1016/j.jcou.2021.101487

Middleton, R.; Bielicki, J. A Comprehensive Carbon Capture and Storage Infrastructure Model. Energy Procedia. 2009, 1, 1611-16.

Mohammed, S.S., 2016. Qatar's National Emission Inventory Repor. Qatar Environment and Energy Research Institute , Qatar Foundation/Hamad Bin Khalifa University, Figshare open access book license CC BY-4.0. http:// creativecommons.org/licenses/by/4.0.

Roman-White, S. A., Littlefield, J. A., Fleury, K. G., Allen, D. T., Balcombe, P., Konschnik, K. E., Ewing, J., Ross, G. B., & George, F. (2022). Response to comment on "LNG Supply Chains: A supplier-specific life-cycle assessment for improved emission accounting." ACS Sustainable Chemistry & Engineering, 10(41), 13552–13554. https://doi.org/10.1021/acssuschemeng.2c05197

Sawaly, R., Ghiat, I., Mohamed, A., Abushaikha, A., & Al-Ansari, T. (2022). Optimal CO2 allocation for enhanced oil recovery operations within Carbon Utilisation Networks in Qatar. Computer Aided Chemical Engineering, 493–498. https://doi.org/10.1016/b978-0-323-95879-0.50083-7

Antonis Kokossis, Michael C. Georgiadis, Efstratios N. Pistikopoulos (Eds.)
PROCEEDINGS OF THE 33rd European Symposium on Computer Aided Process Engineering
(ESCAPE33), June 18-21, 2023, Athens, Greece

Process Design and Bayesian Optimization of 5-Hydroxymethylfurfural Hydrodeoxygenation

Yuqing Luo[a], Zhaoxing Wang[a,b], Prahalad Srinivasan[a], Dionisios G. Vlachos[a,b], Marianthi Ierapetritou[a,b*]

[a]*Department of Chemical and Biomolecular Engineering, University of Delaware, 150 Academy Street, Newark, Delaware, 19716, USA*

[b]*Catalysis Center for Energy Innovation, RAPID Manufacturing Institute, and Delaware Energy Institute (DEI), University of Delaware, Newark, Delaware 19716, United States*

mgi@udel.edu

Abstract

The hydrodeoxygenation (HDO) of 5-hydroxymethylfurfural (HMF) is an essential step in lignocellulosic biomass conversion. In this work, we developed process flowsheet simulation models using the hydrodeoxygenation experimental conditions, yields, and separation requirements. Economic and environmental evaluations are then performed and incorporated with Bayesian optimization to search for reaction conditions that minimize production costs and greenhouse gas emissions. The Bayesian optimization demonstrates that the HDO reaction condition with the lowest production cost has a relatively high reaction temperature, no water content, high H_2 pressure, medium HMF loading, and short reaction time. However, the HDO condition with the lowest greenhouse gas emission has a lower temperature but a longer reaction time.

Keywords: Bayesian optimization, 5-Hydroxymethylfurfural, Techno-economic analysis, Life cycle assessment.

1. Introduction

5-Hydroxymethylfurfural has been established as a promising platform chemical from lignocellulosic biorefinery after hexose dehydration [Bozell and Petersen, 2010]. HMF typically further undergoes hydrodeoxygenation to produce the 2,5-dimethylfuran (DMF) intermediate for the downstream production of other renewable chemicals [Luo and Ierapetritou, 2020]. Over the Ru/C catalyst, HMF is first hydrogenated to 2,5-bishydroxymethylfuran (BHMF), whose alcohol group undergoes hydrogenolysis to produce DMF with the hydrogen produced from the dehydrogenation of a secondary alcohol as the solvent. The furans also undergo etherification, which slowly convert to DMF [Jae *et al.*, 2014]. Water inevitably exists in the feed and process streams as a component of biomass or a byproduct of dehydration and HDO reactions [Esteban *et al.*, 2020]. For instance, in the upstream HMF production (fructose dehydration), many water-organic biphasic systems are developed to prevent rehydration or condensation side reactions [Wang *et al.*, 2020]. As water is partially soluble in the organic phase, the HMF product stream could contain a significant amount of water. The presence of water is observed to impact the ether-alcohol-furan equilibrium and affects the reaction rate. However, removing water from intermediate streams may be energy-intensive and expensive due to the high heat of vaporization. Therefore, it is important to study the trade-off of the water effects based on process simulation.

Bayesian optimization is an effective active-learning method that combines building the surrogate model and proposing the next experiment/simulation points [Wang and Dowling, 2022]. It thus provides an efficient solution for various practical optimization tasks involving time-consuming computation or experiments, which expedites the discovery of optimal reaction conditions and new materials [Iwama and Kaneko, 2021]. For instance, Diwale *et al.* applied Bayesian optimization under noise to reduce the time spent on computationally expensive non-equilibrium molecular dynamics simulations for the search of additives in polyethylene nucleation. Bayesian optimization also finds its application in the model predictive control parameter tuning [Lu *et al.*, 2021].

In this work, we first built HMF conversion and DMF yield surrogate models as a function of reaction temperature, time, H_2 pressure, HMF loading, and water content using the adaptive design of HDO experiments based on Bayesian optimization. Next, Aspen Plus process flowsheets are simulated using the reaction yields/conversion and appropriate separation process design, which provided information for the techno-economic analysis (TEA) and life cycle assessment (LCA). Next, Bayesian optimization is applied again to search optimal reaction conditions that perform the HDO reaction with the lowest DMF minimum selling price (MSP) and global warming potential (GWP). The results also revealed the effects of water content and reaction conditions on the overall HDO process economics and environmental impacts.

2. HMF Hydrodeoxygenation Flowsheet Design and Analysis

First, the NEXTorch (Next EXperiment toolkit in PyTorch) python package was utilized to build accurate HMF conversion and DMF yield surrogate models from the preliminary bench-top HDO reaction experiments *via* Gaussian processes, which ensures the exploration of the design space while exploiting the current best conditions [Wang *et al.*, 2021]. In this case, a multi-variable (reaction temperature, time, H_2 pressure, HMF loading, and water content) ordinary least squares (OLS) regression demonstrated satisfactory fitting performance for DMF yield and HMF conversion. This OLS model was used in section 3 for the process design in each step of the Bayesian optimization.

Process flowsheets are then simulated in Aspen Plus V11 for 29 initial HDO reaction conditions and their corresponding yields/conversions. These conditions affect not only the reaction outcome but also the separation process. The universal quasi-chemical activity coefficient (UNIQUAC) model was chosen for the liquid-vapor and liquid-liquid phase behavior [Abrams and Prausnitz, 1975]. Four different flowsheet configurations are developed for various pentanol/water azeotrope and feedstock compositions of the HDO processes. Figure 1 illustrates the process flowsheet when no water exists in the reaction inlet. Hydrogen gas and HMF solution in pentanol were pressurized and heated to the reaction condition before sending to R-1. After the HDO reaction, flash drum F-1 cooled down the system to 65 °C and recycled hydrogen gas. Then, humins were removed by filtration. It was assumed that most of the liquid phase was retained with a 1% loss in the waste humins stream. The liquid stream was sent to two distillation columns in series (C-1 and C-2) to recycle pentanol and obtain a DMF/water stream. Two decanters cooled down both the bottom (D-1) and distillate streams (D-2) to 40 °C, so that excess water was removed by phase separations. The process capacity was chosen as 10,000 kg HMF/h, which was in line with the scale considered in a recent paper that produced HMF as an intermediate [Chang *et al.*, 2020]. It is noteworthy that separating the targeted DMF product from water/pentanol azeotrope is not always easy in the other flowsheets, as they have very close boiling points.

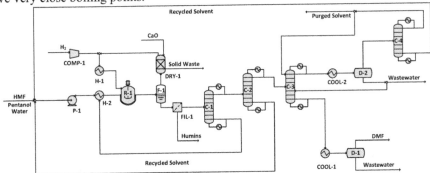

Figure 1: Hydrodeoxygenation process flowsheet without water in the feed

3. Bayesian Optimization

Bayesian optimization on the reaction conditions (i.e., water content, HMF loading, temperature, pressure, and reaction time) is conducted through iterative optimal search via NEXTorch and flowsheet simulation/evaluation. The NEXTorch package utilizes the Gaussian process surrogate model with the Matern covariance function kernel. In each Bayesian optimization iteration, new reaction condition samples are suggested based on the Monte Carlo q-expected improvement (qEI) acquisition function of the Gaussian process. The expected improvement acquisition function has the ability to effectively leverage exploration and exploitation aspects of searching [De Ath *et al.*, 2021]. The Monte Carlo sampling approach shown in Eq.(1) further reduces the computational complexity by approximating the integral over posterior distribution [Snoek *et al.*, 2012].

$$qEI(X) = \frac{1}{N} \sum_{i=1}^{N} \max_{j=1,\dots,q} \{max(\xi_{ij} - f^*, 0)\}, \; \xi_{ij} \sim \mathbb{P}(f(X)|D) \tag{1}$$

where $X = (x_1, \dots, x_q)$, representing the reaction conditions (e.g., temperature, pressure, reaction time, water, and HMF loading), q is the number of points considered jointly, f^* is the best value observed so far, and $\mathbb{P}(f(X)|D)$ is the posterior distribution of the function f (e.g., MSP or GWP) at X given the observed data $D = (X, f)$. Individual flowsheets are developed based on these trial points, whose economic and environmental performance is evaluated and added to observed data for the next iteration.

4. Results and Discussions

4.1. Bayesian Optimization of Minimum Selling Prices (MSP)

Techno-economic analysis (TEA) was performed to calculate each case's MSP, the selling price of the product when the net present value is zero at the end of the recovery period, was determined through discounted cash flow analysis for each condition [Athaley *et al.*, 2019]. A 15% internal rate of return (ROR) on investment and a 35% corporate tax was assumed. The straight-line method for depreciation with a 10% salvage value after 20 years was applied [Luo *et al.*, 2022]. The raw material, solvent, and catalyst prices are based on Athaley *et al.* (2019). For instance, the HMF price is taken as $0.88/kg, and the pentanol cost is assumed to be $1.5/kg. As illustrated in Figure 2, the best DMF MSP gradually decreases as the iterations of Bayesian optimization continue, reducing from $5,420/t to $4,022/t. In all cases, the HMF feedstock is the leading cost contributor. Hence, the MSP of DMF decreases significantly when the DMF yield increases and less humins byproduct is generated. This is typically the case with little or no water loading, which has high DMF selectivity. However, when the water content increased from 0% to 4% in the feedstock, the MSP only increased by $ 312/t.

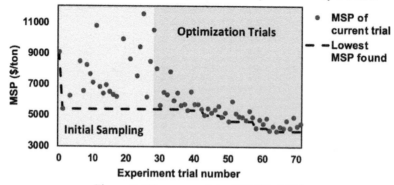

Figure 2: Discovery plot of DMF's MSP

On the one hand, high solvent usage leads to much larger equipment sizes and more solvent loss during filtration. On the other hand, reaction conversion and selectivity generally drop as a result of lower solvent usage. Hence, medium HMF concentration at 3 wt. % was favored by Bayesian optimization as a trade-off between these two effects.

4.2. Bayesian Optimization of GWP

Next, a "cradle-to-gate" LCA was performed to evaluate the environmental impacts by including the upstream activities of raw material extraction (e.g., HMF), utility generation, and the production stages. The functional unit was chosen as 1 kg of DMF produced, and the GWP was selected as the environmental impact of interest as it bears great importance in renewable technology development and policy making. The Ecoinvent v3.7 database was used to provide the background emission data, and the Tool for the Reduction and Assessment of Chemical and other environmental Impacts (TRACI) method was used for the impact assessment [Bare, 2011]. The trajectory of GWP under each explored reaction condition is demonstrated in Figure 3. During these Bayesian optimization runs, the lowest GWP was found to be 22.5 kg CO_2 eq/kg DMF, while the best GWP from the initial 29 experiments was 26.5 kg CO_2 eq/kg DMF.

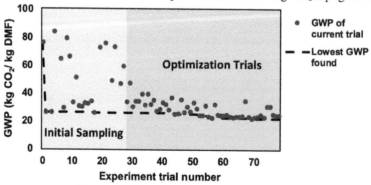

Figure 3: Discovery plot of DMF's GWP

Similar to the improvement trend in the MSP case, the GWP of each process generally matches the increase of HMF conversion and DMF selectivity. Medium HMF loading, high hydrogen gas pressure, and zero water content are still favored. However, as with other biomass-based chemical production processes, the utility cost only takes up a small fraction of the total MSP while it plays a vital role in the GWP. Based on the conversion and selectivity model, higher temperatures correspond with better yields at fixed water content, HMF loading, and reaction time. Nevertheless, a higher reaction temperature also means more hot utilities to heat up the feedstock and solvent before the reaction. Thus, the optimal temperature for the lowest GWP case (199 °C) is lower than that of the lowest MSP case (210 °C). Other contributors to utility usage include the steam required for distillation and electricity used in the compressor.

Although longer reaction time causes larger reactor sizes and higher capital investment, it has little impact on environmental performance. Thus, the lowest GWP case has increased the reaction time from 3 h to 4.3 h at a lower temperature to improve the HMF conversion without worrying about more expensive equipment. It should be noted that non-zero water content in the feedstock inlet causes a drastic increase in the GWP. The process with 4% inlet water content now has a GWP of 31.1 kg CO_2 eq/kg DMF, which is 38.4% higher than the lowest GWP without any water loading. This is the result of high utility usage during the product separation, as the azeotrope compositions significantly affect the distillation operating conditions. A summary of optimal reaction conditions for the MSP and GWP Bayesian optimization is listed in Table 1.

Table 1: Optimal reaction conditions suggested by MSP and GWP Bayesian optimization

Reaction Conditions	minimum MSP	minimum MSP (with H_2O)	minimum GWP	maximum DMF yield*
Temperature (°C)	210	208	199	195
H_2 pressure (bar)	10	10	10	10
Water loading (wt. %)	0	4	0	0
HMF loading (wt. %)	3	3	3	1
Time (h)	3	3	4.3	4.6

* "max DMF yield" is based on the best condition in the initial design of experiment

As illustrated in Table 1, the reaction conditions with the lowest cost and emission are close to the maximum DMF yield condition observed in bench-top experiments. Nevertheless, Bayesian optimization incorporates process design insights to select higher HMF loadings and shorter reaction times that would be otherwise not explored.

5. Conclusions

The current work incorporated process simulation, TEA, and LCA into the Bayesian optimization to determine economically and environmentally viable HDO reaction conditions in the early stage. The Bayesian optimization results illustrate that the reaction conditions that minimize the production cost have low water contents, medium HMF loadings, high H_2 pressure, and relatively high reaction temperatures because of low HMF feedstock usage and low solvent loss. The HDO reaction with the minimal GWP chooses a lower reaction temperature but longer reaction time to reduce the emissions associated with utility usage. Moreover, the existence of water has more significant impacts on the process's GWP than on its production cost. The integrated TEA, LCA, and Bayesian optimization provide candidate HDO conditions and process parameters to explore. However, detailed analysis and validation should be carried out after future scale-up experiments and process improvement. Furthermore, as feedstock and solvent usage are the main contributors to the cost and emissions, more efficient upstream HMF production and effective filtration technologies have great potential to further improve the HMF HDO process.

Acknowledgement

This work was supported by the Department of Energy's Office of Energy Efficiency and Renewable Energy's Advanced Manufacturing Office under Award Number DE-EE0007888-7.6. The Delaware Energy Institute gratefully acknowledges the support and partnership of the State of Delaware toward the RAPID projects.

References

Abrams, D. S. and Prausnitz, J. M., 1975 Statistical thermodynamics of liquid mixtures: A new expression for the excess Gibbs energy of partly or completely miscible systems, AIChE Journal, 21, 1, 116-128

Athaley, A., Saha, B. and Ierapetritou, M., 2019 Biomass-based chemical production using techno-economic and life cycle analysis, AIChE Journal, 65, 9, e16660

Bare, J., 2011 TRACI 2.0: the tool for the reduction and assessment of chemical and other environmental impacts 2.0, Clean Technologies and Environmental Policy, 13, 5, 687-696

Bozell, J. J. and Petersen, G. R., 2010 Technology development for the production of biobased products from biorefinery carbohydrates—the US Department of Energy's "Top 10" revisited, Green Chemistry, 12, 4, 539-554

Chang, H., Bajaj, I., Huber, G. W., Maravelias, C. T. and Dumesic, J. A., 2020 Catalytic strategy for conversion of fructose to organic dyes, polymers, and liquid fuels, Green Chemistry, 22, 16, 5285-5295

De Ath, G., Everson, R. M., Rahat, A. A. M. and Fieldsend, J. E., 2021 Greed is good: Exploration and exploitation trade-offs in Bayesian optimisation, ACM Transactions on Evolutionary Learning and Optimization, 1, 1, 1-22

Diwale, S., Eisner, M. K., Carpenter, C., Sun, W., Rutledge, G. C. and Braatz, R. D., 2022 Bayesian optimization for material discovery processes with noise, Molecular Systems Design & Engineering, 7, 6, 622-636

Esteban, J., Vorholt, A. J. and Leitner, W., 2020 An overview of the biphasic dehydration of sugars to 5-hydroxymethylfurfural and furfural: a rational selection of solvents using COSMO-RS and selection guides, Green Chemistry, 22, 7, 2097-2128

Iwama, R. and Kaneko, H., 2021 Design of ethylene oxide production process based on adaptive design of experiments and Bayesian optimization, Journal of Advanced Manufacturing and Processing, 3, 3, e10085

Jae, J., Zheng, W., Karim, A. M., Guo, W., Lobo, R. F. and Vlachos, D. G., 2014 The Role of Ru and RuO2 in the Catalytic Transfer Hydrogenation of 5-Hydroxymethylfurfural for the Production of 2,5-Dimethylfuran, ChemCatChem, 6, 3, 848-856

Lu, Q., González, L. D., Kumar, R. and Zavala, V. M., 2021 Bayesian optimization with reference models: A case study in MPC for HVAC central plants, Computers & Chemical Engineering, 154, 107491

Luo, Y. and Ierapetritou, M., 2020 Comparison between different hybrid life cycle assessment methodologies: a review and case study of biomass-based p-xylene production, Industrial & Engineering Chemistry Research, 59, 52, 22313-22329

Luo, Y., Kuo, M. J., Ye, M., Lobo, R. and Ierapetritou, M., 2022 Comparison of 4,4′ - Dimethylbiphenyl from Biomass-Derived Furfural and Oil-Based Resource: Technoeconomic Analysis and Life-Cycle Assessment, Industrial & Engineering Chemistry Research, 61, 25, 8963-8972

Snoek, J., Larochelle, H. and Adams, R. P., 2012 Practical bayesian optimization of machine learning algorithms, Advances in neural information processing systems, 25,

Wang, K. and Dowling, A. W., 2022 Bayesian optimization for chemical products and functional materials, Current Opinion in Chemical Engineering, 36, 100728

Wang, Y., Chen, T.-Y. and Vlachos, D. G., 2021 NEXTorch: A Design and Bayesian Optimization Toolkit for Chemical Sciences and Engineering, Journal of Chemical Information and Modeling, 61, 11, 5312-5319

Wang, Z., Bhattacharyya, S. and Vlachos, D. G., 2020 Solvent selection for biphasic extraction of 5-hydroxymethylfurfural via multiscale modeling and experiments, Green Chemistry, 22, 24, 8699-8712

Antonis Kokossis, Michael C. Georgiadis, Efstratios N. Pistikopoulos (Eds.)
PROCEEDINGS OF THE 33rd European Symposium on Computer Aided Process Engineering
(ESCAPE33), June 18-21, 2023, Athens, Greece

Investigation of biomass blending ratios for optimal biochar's soil application

Farah Obar, Mohammad Alherbawi, Gordon Mckay, Tareq Al-Ansari[*]

College of Science and Engineering, Hamad Bin Khalifa University, Qatar Foundation, Doha, Qatar
*talansari@hbku.edu.qa

Abstract

Due to the rapid growth of population, urbanization and economic development, global municipal waste generation is expected to increase by nearly 70% to 3.4 billion tonnes by 2050. The mismanagement of these wastes results in the deterioration of soil, air, and water quality, causing serious health problems. As such, investigating waste valorization routes becomes more essential to alleviate the concerns associated with the burden of wastes. In this study, 10 different types of biomass used to produce biochar through pyrolysis were investigated; namely camel manure (CM), date pits (DP), sewage sludge (SS), coffee waste (CW), cattle manure (CM), poultry litter (PL), cabbage waste (CB), cucumber waste (CU), tomato waste (TM), and carrot waste (CR). Techno-economic and environmental analyses were performed from "cradle to grave", covering the key biochar production stages from acquisition of biomass, transportation, plant's construction, pre-processing, pyrolysis, and lastly the end-use of biochar as soil enhancer. The cost and emissions savings using biochar as an alternative to commercial fertilizers were evaluated for the different biomass scenarios. Finally, an optimization model was developed to select the optimal biomass blending ratio for two scenarios to maximize the savings, which was then solved using MATLAB. The findings of this study indicated that the optimal biomass blend for maximizing cost, energy and emissions savings shall consist of the following proportions: 50% TM, 14-15% CW and PL, 5 – 10% CM and CU, 2.5% CT, 1.8-1.9% CB and DP, 1.2% CR and 0 – 0.6% SS. Future studies may involve more types of biomasses, additional optimization objectives, and may target multiple biochar's end-use applications.

Keywords: Pyrolysis; Biochar; Soil enhancement; Biomass.

1. Introduction

Global municipal solid waste generation is expected to increase from 6.94 million metric tonnes in 2021 to 3.4 billion metric tonnes in 2050 due to the rapid population growth and urbanization (NEA, 2016; Tiseo, 2021). This has accelerated the need to establish new waste-to-energy technologies including pyrolysis. Pyrolysis is a thermochemical process that works on converting biomass, which is a carbon based renewable organic material generated from animals and plants into value added products in the absence of oxygen to produce; biogas, bio-oil and biochar (BC) (Elkhalifa et al., 2019). BC is an incredibly rich source of carbon that can be employed in various disciplines to address the most significant environmental challenges as follows; (1) can be employed as a renewable energy source, (2) can be employed as a soil enhancer reducing the chemical fertilizers requirements, and (3) can be utilized as an innovative adsorbent for water and wastewater decontamination (Haider et al., 2022).

This study focused on biochar's end-use as a soil enhancer as it increases the crop yield, enhances the water retention of soil due to its porous structure, and sequesters carbon in the soil by absorbing it from the atmosphere to mitigate climate change (Mukherjee & Lal, 2013). Moreover, this study investigates the optimal biomass blends based on 10 different types of biomass feedstocks including; camel manure (CM), date pits (DP), sewage sludge (SS), coffee waste (CW), cattle manure (CT), poultry litter (PL), cabbage waste (CB), cucumber waste (CU), tomato waste (TM), and carrot waste (CR). The aim of this investigation is to target three objectives; maximize cost, energy and carbon emissions savings.

2. Methodology

The methodology of this study follows two main steps, which are the data collection from literature, and formulating the equations that measure the savings that can be used for optimizing the biomass blends using the genetic algorithm tool in MATLAB. The process flow diagram is shown in figure 1.

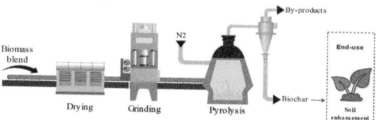

Figure 1: Process flow diagram

2.1 Data Collection

The production of biochar from the different biomass feedstocks has been simulated using Aspen. This model has been used to evaluate the cost, carbon and energy footprint when producing biochar from each single biomass. The operating parameters of the process have included an operating temperature of 300°C, 5L/min.tonne nitrogen flowrate, 90% drying efficiency and 30 minutes' reaction time. The elemental analysis for the biomass feedstocks was collected from literature as presented in Table 1, whereas this model requires the analysis to be on a dry ash and sulfur-free basis. The price of date pits was quoted locally at 100$/t. While the prices of the other biomass feedstocks were obtained from literature and was found to be 15$/t sewage sludge, 88$/t coffee waste, 25$/t for all types of animal manure and 10$/t for vegetable waste (Alherbawi et al., 2021; Kamil et al., 2019; Marufuzzaman et al., 2015; Yun et al., 2018).

Table 1: Elemental analysis of biomass feedstock.

Biomass	%C	%H	%O	%N	%S	Ash (%)	Moisture (%)	Ref.
CM	36.7	5.6	34.7	2.6	0.4	19.9	4.3	(Al-Ansari et al., 2020)
DP	46.5	6.5	44.7	0.9	0	1.4	0.3	
SS	46.9	6.8	23.2	7.4	0.5	2.6	82.2	(Poudel et al., 2015)
CW	79.3	5.0	12.5	2.1	1.1	11.1	4.2	(Uwaoma et al., 2021)
CT	41.1	5.9	49.9	2.7	0.4	15.4	10.1	(Yuan et al., 2017)
PL	42.9	5.6	33.4	5.5	0.7	12	18.1	(Tańczuk et al., 2019)
CB	44.5	4.9	35.0	2.3	0.5	11.9	8.6	(Zhang et al., 2022)
CU	56.7	6.5	32.9	3.9	0	1.9	8.1	
TM	49.7	7.4	39.3	3.7	0	4.4	10.0	(Elkhalifa et al., 2022)
CR	41.5	5.4	51.7	1.4	0	5.2	6.6	

A comprehensive comparison has been made between the application of NPK fertilizer and biochar as a soil amendment to evaluate the cost, energy and emissions for each. The model has only estimated the cost, energy and carbon footprint of BC production from each single biomass, whilst the rest of the data relative to BC and NPK application to soil and NPK fertilizer production was obtained from literature. It was found that 0.05 kg/m² of NPK fertilizer is required to grow tomato, which was selected as the reference crop for this comparison (Ziane et al., 2021). The production of NPK fertilizer costs 0.8$/kg, requires energy of 12.45 MJ/kg and emits 0.20 kgCO₂-eq/kg (Majumder et al., 2019; Samoraj et al., 2022; Wu et al., 2021). Energy and carbon emissions for BC application were found to be equal to 0.434 MJ/m² and 0.024 kgCO₂-eq/m², which are lower than the values of NPK fertilizer application which are 0.615 MJ/m² and 0.069 kgCO₂-eq/m² (Brentrup & Palliere, 2008; Jiang et al., 2021).

2.2 Problem Formulation

The data collected was inserted in equations 1-3 to find the cost, energy and carbon emissions when utilizing NPK fertilizer and BC, where X refers to the percentage of NPK fertilizer being used, k refers to the conversion factor between NPK fertilizer and BC, P and A represent production and application.

$$Cost = \left(X \cdot NPK_{required} \cdot cost_{NPK-P}\right) + \left(k \cdot (1-X) \cdot NPK_{required} \cdot cost_{BC-P}\right) \quad (1)$$

$$Energy = X \cdot \left(NPK_{required} \cdot energy_{NPK-P} + energy_{NPK-A}\right) + k \cdot (1-X) \cdot \left(NPK_{required} \cdot energy_{BC-P} + energy_{BC-A}\right) \quad (2)$$

$$GWP = X \cdot \left(NPK_{required} \cdot GWP_{NPK-P} + GWP_{NPK-A}\right) + k \cdot (1-X) \cdot \left(NPK_{required} \cdot GWP_{BC-P} + GWP_{BC-A}\right) \quad (3)$$

The values were later used to find the savings using equations 4-6:

$$Cost_{savings} = Cost_{NPK} - Cost_{BC} \quad (4)$$
$$Energy_{savings} = Energy_{NPK} - Energy_{BC} \quad (5)$$
$$GWP_{savings} = GWP_{NPK} - GWP_{BC} \quad (6)$$

Based on the savings' results from the previous section, a multi-objective mathematical optimization is performed using genetic algorithm tool within MATLAB. The objectives targeted by this optimization are; maximizing cost, energy and carbon emissions savings, where the mathematical formulation of the multi-objective optimization problem is represented by equations 7-9, and the constraint adopted is shown in equation 10.

Maximize cost savings: $\sum_{i=1}^{10} B_i C_i$ $\quad (7)$
Maximize energy savings: $\sum_{i=1}^{10} B_i E_i$ $\quad (8)$
Maximize emission savings: $\sum_{i=1}^{10} B_i G_i$ $\quad (9)$
Constraint: $\sum_{i=1}^{10} B_i = 1$ $\quad (10)$

The multi-objective optimization was conducted based on two scenarios as shown in Table 2 to provide broader insight for decision-makers.

Table 2: Objectives to be obtained in each scenario.

Scenario	Objectives obtained
1	Maximize cost and emissions savings

3. Results and Discussion

70 solutions were obtained fulfilling the objectives for each scenario by performing a trade-off between the desired objectives using the genetic algorithm tool in MATLAB. The results of the scenarios are as follows:

3.1 Scenario 1

This scenario aimed to achieve the maximum cost and emissions savings. The findings as illustrated in figure 2 show that all solutions lean more towards achieving higher emissions savings and focused less on the cost savings. Moreover, the solutions show that the most optimal biomass blends consisted of very high proportions of poultry litter, and lower proportions of the other biomass feedstocks except the camel manure that had zero contribution in achieving the objectives. Some optimal blends consisted of around 80% PL, 1.7-4% of CU, TM, SS, CW and CT, 0.2-1% DP, CB and CR, and 0% CM.

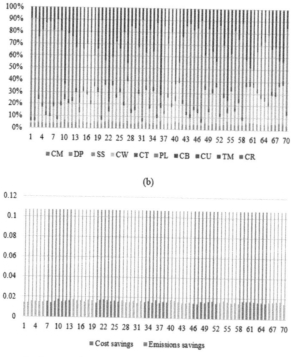

Figure 2: Scenario 1 – maximum cost and emission savings: (a) optimal biomass blends, and (b) optimal savings.

3.2 Scenario 2

This scenario was concerned with achieving three objectives; maximize energy, cost and emissions savings and to find the optimal biomass blends that fulfill these objectives. The results as illustrated in figure 3 show that all solutions have contributed mostly to achieving maximum energy savings, then maximizing emissions and cost, respectively.

For optimizing the biomass blends, most of the results show that the optimal blend consisted of around 50% TM, 14-15% CW and PL, 5 – 10% CM and CU, 2.5% CT, 1.8-1.9% CB and DP, 1.2% CR and 0 – 0.6% SS.

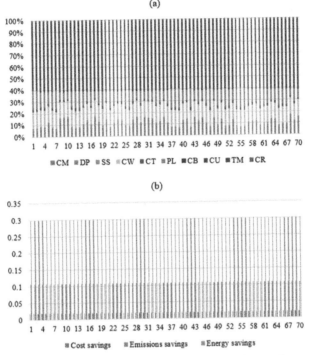

Figure 3: Scenario 2 – maximum cost, energy and emission savings: (a) optimal biomass blends, and (b) optimal savings.

4. Conclusion

Pyrolysis is a thermochemical process that converts biomass into value-added products, mainly biochar. This carbon-rich product has widely contributed into a diversity of applications including soil enhancement applications.

This study has provided insight into the sustainable solutions for researchers and decision-makers on selecting the most optimal biomass blends for biochar production from an economic and environment perspective. Two scenarios have been studied where each aimed to achieve specific objectives. Scenario 1 focused on maximizing cost and emissions only, where most solutions showed that the most optimal biomass blends consisted of very high proportions of poultry litter, lower proportions of the other biomass feedstocks, and no contribution of camel manure. On the other hand, the results of scenario 2 that aimed to achieve all three objectives, have shown that the most optimal biomass blends have consisted of very high proportions of tomato waste, less proportions of the others, and around 0% of sewage sludge. Future studies may involve more types of biomasses, additional optimization objectives, and may target multiple biochar's end-use applications.

Acknowledgement

The authors would like to thank Qatar National Research Fund (QNRF) for their support of this research through NPRP-11S-0107-180216, Hamad Bin Khalifa University (HBKU) and Qatar Foundation (QF).

References

Al-Ansari, T., AlNouss, A., Al-Thani, N., Parthasarathy, P., ElKhalifa, S., Mckay, G., & Alherbawi, M. (2020). Optimising Multi Biomass Feedstock Utilisation Considering a Multi Technology Approach. In *Computer Aided Chemical Engineering* (Vol. 48).

Alherbawi, M., Parthasarathy, P., Al-Ansari, T., Mackey, H. R., & McKay, G. (2021). Potential of drop-in biofuel production from camel manure by hydrothermal liquefaction and biocrude upgrading: A Qatar case study. *Energy, 232*, 121027.

Brentrup, F., & Palliere, C. (2008). GHG emissions and energy efficiency in European nitrogen fertiliser production and use. *Proceedings - International Fertiliser Society, 2*(December).

Elkhalifa, S., Al-Ansari, T., Mackey, H. R., & McKay, G. (2019). Food waste to biochars through pyrolysis: A review. *Resources, Conservation and Recycling, 144*, 310–320.

Elkhalifa, S., Mariyam, S., Mackey, H. R., Al-Ansari, T., McKay, G., & Parthasarathy, P. (2022). Pyrolysis Valorization of Vegetable Wastes: Thermal, Kinetic, Thermodynamics, and Pyrogas Analyses. *Energies, 15*(17).

Haider, F. U., Wang, X., Farooq, M., Hussain, S., Cheema, S. A., Ain, N. ul, Virk, A. L., Ejaz, M., Janyshova, U., & Liqun, C. (2022). Biochar application for the remediation of trace metals in contaminated soils: Implications for stress tolerance and crop production. *Ecotoxicology and Environmental Safety, 230*, 113165.

Jiang, Z., Zheng, H., & Xing, B. (2021). Environmental life cycle assessment of wheat production using chemical fertilizer, manure compost, and biochar-amended manure compost strategies. *Science of The Total Environment, 760*, 143342.

Kamil, M., Ramadan, K. M., Olabi, A. G., Shanableh, A., Ghenai, C., Al Naqbi, A. K., Awad, O. I., & Ma, X. (2019). Comprehensive evaluation of the life cycle of liquid and solid fuels derived from recycled coffee waste. *Resources, Conservation and Recycling, 150*, 104446.

Majumder, S., Neogi, S., Dutta, T., Powel, M. A., & Banik, P. (2019). The impact of biochar on soil carbon sequestration: Meta-analytical approach to evaluating environmental and economic advantages. In *Journal of Environmental Management* (Vol. 250).

Marufuzzaman, M., Ekşioğlu, S. D., & Hernandez, R. (2015). Truck versus pipeline transportation cost analysis of wastewater sludge. *Transportation Research Part A: Policy and Practice, 74*.

Mukherjee, A., & Lal, R. (2013). Biochar impacts on soil physical properties and greenhouse gas emissions. In *Agronomy* (Vol. 3, Issue 2).

NEA. (2016). Waste Statistics and Overall Recycling. *National Environmental Agency*.

Poudel, J., Ohm, T. I., Lee, S. H., & Oh, S. C. (2015). A study on torrefaction of sewage sludge to enhance solid fuel qualities. *Waste Management, 40*, 112–118.

Samoraj, M., Mironiuk, M., Witek-Krowiak, A., Izydorczyk, G., Skrzypczak, D., Mikula, K., Baśladyńska, S., Moustakas, K., & Chojnacka, K. (2022). Biochar in environmental friendly fertilizers - Prospects of development products and technologies. *Chemosphere, 296*.

Tańczuk, M., Junga, R., Kolasa-Więcek, A., & Niemiec, P. (2019). Assessment of the energy potential of chicken manure in Poland. *Energies, 12*(7).

Tiseo, I. (2021). *Global Waste Generation - Statistics & Facts*. Environment, Development and Sustainability.

Uwaoma, R. C., Strydom, C. A., Bunt, J. R., Matjie, R. H., Okolo, G. N., & Marx, S. (2021). Co-gasification reactivity and kinetics of municipality and coffee waste residue hydrochar and South African density separated coal blends. *Bioresource Technology Reports, 16*, 100877.

Wu, H., MacDonald, G. K., Galloway, J. N., Zhang, L., Gao, L., Yang, L., Yang, J., Li, X., Li, H., & Yang, T. (2021). The influence of crop and chemical fertilizer combinations on greenhouse gas emissions: A partial life-cycle assessment of fertilizer production and use in China. *Resources, Conservation and Recycling, 168*, 105303.

Yuan, X., He, T., Cao, H., & Yuan, Q. (2017). Cattle manure pyrolysis process: Kinetic and thermodynamic analysis with isoconversional methods. *Renewable Energy, 107*, 489–496.

Yun, Y. M., Lee, M. K., Im, S. W., Marone, A., Trably, E., Shin, S. R., Kim, M. G., Cho, S. K., & Kim, D. H. (2018). Biohydrogen production from food waste: Current status, limitations, and future perspectives. *Bioresource Technology, 248*, 79–87.

Zhang, Y., Cheng, X., Wang, Z., Tahir, M. H., Wang, Z., Wang, X., & Wang, C. (2022). Full recycling of high-value resources from cabbage waste by multi-stage utilization. *Science of The Total Environment, 804*, 149951.

Ziane, H., Hamza, N., & Meddad-Hamza, A. (2021). Arbuscular mycorrhizal fungi and fertilization rates optimize tomato (Solanum lycopersicum L.) growth and yield in a Mediterranean agroecosystem. *Journal of the Saudi Society of Agricultural Sciences, 20*(7).

Antonis Kokossis, Michael C. Georgiadis, Efstratios N. Pistikopoulos (Eds.)
PROCEEDINGS OF THE 33rd European Symposium on Computer Aided Process Engineering
(ESCAPE33), June 18-21, 2023, Athens, Greece

Methodology for estimating a country's production of biomethane from agricultural residues: a multiscale and holistic approach

Manuel Taifouris, Mariano Martin

[a]Salamanca, Department of Chemical Engineering, University of Salamanca, Plza. Caídos 1-5, 37008 Salamanca, Spain

Abstract

Dependence on fossil fuels and recent issues on the supply chain from major natural gas providers have highlighted the importance of energy security in the countries of the European Union. This work presents a multi-scale analysis to evaluate the implementation of the circular economy to reduce the region's dependence on fossil natural gas. For this purpose, a techno-economic analysis of technologies (gasification and anaerobic digestion), size, and location of the waste treatment facilities is performed, followed by a facility location problem. This analysis is applied to a case study in Spain, concluding that an investment of 9458M€ and an operating cost of 5000M€ per year are necessary to cover 35% of the natural gas demanded. Up to 19 provinces are totally independent with this investment, favoring the creation of decentralized networks of natural gas, which is a strong incentive to invest in waste management because of the social and economic importance of natural gas as an energy source.

Keywords: Agricultural residues, Sustainable process design, Energy, Circular economy

1. Introduction

Despite the implementation of decarbonization policies in European countries, the current dependence on fossil fuels together with the centralization of their production in a reduced number of countries put the energy security of most European countries at risk (Mišík, 2022). In addition, the increase in the world's population has led to the intensification of crop, meat, and milk production processes, creating areas of high waste generation. These areas can suffer from nutrient pollution, which can lead to the eutrophication of water bodies and soil deterioration. In addition, municipal solid waste (MSW) and sewage sludge also represent an issue when cities are densely populated and waste is treated inefficiently so in most cases, ending up in landfills or incinerated (European Commission, 2021). The implantation of a circular economy for waste can solve both problems, by using waste management technologies, such as anaerobic digestion and gasification. Studies that analyze the biomethane production potential of a country are especially useful for determining where biomethane production plants should be placed, as well as determining which regions can be energy-independent (Wang et al., 2018). However, these studies directly relate the biomethane production potential to the number of animals, crops, or people, through the empirical yields. This approach decouples the estimation of biomethane (composition and amount) from the processing technology, as well as from the type and composition of waste. On the one hand, the waste composition can be highly variable, even for the same species of animal (Council, 2000). On the other hand, the composition of the waste affects both the design of the treatment equipment and the amount and composition of the biomethane produced, and therefore, the economic and environmental evaluation of its valorization. A multi-scale approach allows for addressing the different scales of the waste management system (Floudas et al., 2016), from waste characterization and process design to treatment capacity and facility location. By analyzing all scales simultaneously, a better estimate of the biomethane production potential of a location can be determined, as well as the economic and environmental cost of the development and implementation of the process. This allows optimizing the selection of the location, the treatment process, and the type of waste. It also facilitates energy integration between the different stages of the process. This work presents a multiscale study, which uses an optimization framework, to design waste treatment plants (size and type of waste treated), select the best location, and estimate the investment and operating costs. The total amount of biomethane and its cost are also calculated. This framework is applied to the case study of Spain, where the management budget is the main optimization variable. The optimization framework aims at maximizing biomethane production and minimizing production costs.

2. Framework development

This section presents the optimization framework, which addresses waste production characterization (amount and composition), the modeling of processing technologies, the scale-up/down of the facilities, and the selection of the location.

2.1-Estimation of the production and composition of waste

Two types of waste are considered in this work, dry waste (lignocellulosic residue) and wet waste (manure, MSW, and sludge). Lignocellulosic residues of a region are estimated from the annual production of crops and the amount of residue grown by the type of crop. Manure is calculated from the amount of waste generated based on the age and the number of animals. The production of MSW and sludge are obtained through the annual production per capita (388 kg of MSW and 105 kg of sludge) and the population. The composition of the residues is obtained from the literature. Nevertheless, it is possible to perform studies for specific residues to increase the accuracy of the estimates.

2.2-Process analysis and design

A techno-economic analysis is proposed to analyze and compare gasification and anaerobic digestion as possible technologies to produce biomethane from agro-industrial wastes. In addition, a scale-up study is performed for each waste and for 50 different treatment capacities. Non-linear programming (NLP) is used to optimize each design and reduce the cost of methane production.

An indirect gasification process, which consists of a combination of a gasifier and a furnace, is used for the treatment of dry waste. Both gasification and biomethane purification are modeled following first principles, such as mass and energy balances, empirical yields and correlations, and thermodynamic equilibria. The heat necessary for the gasification process is supplied by the combustion of the char formed in that same gasification when the steady-state regime is reached. Olivine is used as heat transfer media (HTM) to transmit the thermal energy generated in the combustion process to the gasifier. Therefore, the system is energetically self-sustainable. The total combustion of all compounds in the furnace is considered to perform the mass and energy balances. The composition of the syngas is estimated from the gasification temperature (Phillips et al., 2007). Regarding the syngas purification stage, empirical yields are used to estimate the separation efficiency of char, olivine, and ash in the cyclones (99%) and electrostatic precipitators (99%), as well as in the adsorbed amount of H_2S in the ZnO beds (100%). The hydrocarbon reforming process, the water gas shift reactor, and the methanation

Methodoly for estimating a country's production of biomthane
from agricultural residues: a multiscale and holistic approach

2201

reactor are modeled by using equilibrium constants (Sánchez et al., 2019). Finally, the PSA system is used to reduce the amount of CO_2 down to 2% and remove NH_3 and H_2O. This system is modeled by empirical yields (León & Martín, 2016). An anaerobic digestion system is designed for the treatment of wet waste. The composition and amount of biogas are estimated from the carbohydrate, lipid, and protein composition of the waste, stoichiometric ratios, and biodegradability (Taifouris & Martín, 2018). The distribution of the gases between the liquid and gas phases is modeled from thermodynamic equilibria. The biogas purification phase consists of an iron bed to remove H_2S (100%) and a PSA system to reduce the content of CO_2 down to 2% and remove NH_3 and H_2O. Both systems are modeled from adsorption yields.

The gasification and anaerobic digestion systems are scaled up/down from a minimum to a maximum size of treatment capacity. The minimum corresponds to the minimum amount of waste generated per region, considering all available regions simultaneously, while the maximum can be established by the literature (anaerobic digestion) or by the maximum availability of waste per region (gasification). Up to 50 designs with different capacities for each waste and for each technology are considered. For each design, a techno-economic evaluation is carried out to determine the operating expenses (OPEX) and the capital expenses (CAPEX) following the procedures described in the literature (Sinnott, 2005) and the results of the modeling of the processes.

2.3-Facility location problem
A facility location problem is formulated including surrogate models for the facilities (yields, CAPEX, OPEX as a function of their size). A mixed integer linear programming (MILP) problem is solved to select the best location and size of the waste treatment facilities. The objective function aims at maximizing methane production based on the allocated budget. With the information on the amount of natural gas demand, it is possible to determine the fraction of self-sufficiency of each spatial unit.

3.Result
3.1-Case of study
The optimization framework is applied to the case study in Spain. The country is divided into agricultural districts (345 possible locations). Spain is selected by its high agricultural production (Gobierno de España, 2022), high dependency on foreign suppliers of natural gas (Enagas, 2022), and low production of biofuels. The optimization framework consists of two types of models, NLP and MILP. Both are solved in a computer with an Intel Core i7-7700, 32 Gb of Ram, using GAMS.

3.2-Properties of the different types of factories
The designs presented in section 2 are modeled and evaluated economically. The main results are shown in Table 1.

Table 1.- Main results of the techno-economic analysis of the residues (q: capacity of factory(t/ye), DW: dried waste)

Residue	Yield (kg$_{Biomethane}$/kg$_{DW}$)	Production Cost (€/kg$_{DW}$)	Min.Capacity (t/ye)	Max.Capacity(t/ye)
SLUDGE	0.003	Pcost=$1.814 \cdot 10^9 \cdot q^{-0.97}$	700624	49754
MANURE	0.012	Pcost=$1.603 \cdot 10^8 \cdot q^{-0.885}$	63072	367920
MSW	0.070	Pcost=$7.203 \cdot 10^7 \cdot q^{-0.97}$	19657	52560
LIGNO	0.285	Pcost=$6.060 \cdot 10^5 \cdot q^{-0.626}$	10000	820000

As it can be seen in Table 1, the relationship between the cost of biomethane production and the treatment capacity of the plant follows a power law, due to the strong economies of scale present in both OPEX and CAPEX. Besides, gasification has the best yield to produce biomethane from lignocellulosic waste while anaerobic digestion of sludge is the worst option.

3.3-Total potential of biomethane production in Spain

The application of gasification and anaerobic digestion to the total amount of waste available per year in Spain shows that the maximum percentage of natural gas that can be covered is 43%. To achieve this, a CAPEX of 21391M€ and an OPEX of 25852M€/ye are required. This allows up to 21 provinces of the 52 available in Spain, to be self-sufficient. It is observed that the areas with the highest biomethane production are those with the highest amount of lignocellulosic residues. This is because gasification has a higher yield than the anaerobic digestion of any of the wet waste. In addition, a significant mismatch is observed between the areas that consume natural gas (industrial and densely populated areas) and the areas that can produce biomethane (rural areas).

3.4-Determination of the optimal budget for the reduction of Spain's dependence on fossil natural gas

The facility location problem is used to establish the optimal selection of location, size, and type of facilities for different budgets, drawing the Pareto (see Figure 1). This curve shows that there are two sections divided by the point of 5000 M€ per year. In the first segment, each 100M€ increase in the budget represents a 2% increment in the self-supply rate. However, in the second section, the self-supply rate remains constant even if the budget is increased. Therefore, the point of 5000M€/ye is selected as the best budget to invest in waste treatment in Spain. This OPEX corresponds to a CAPEX of 9505M€. Besides, the biomethane cost is similar to the current market price of natural gas (10.19 vs 6.24 USD/MMBTU). With this budget, it is possible to cover up to 35% of the country's total natural gas demand. Therefore, through an optimal selection of the budget dedicated to waste treatment, it is possible to reduce the CAPEX and OPEX of the plants by 55.56% and 80.65%, respectively, with respect to the previous section, while the self-sufficiency ratio only decreases by 18.60%. Besides, 19 provinces are totally independent of natural gas from foreign suppliers. As discussed in Section 3.3, these provinces correspond mainly to rural areas where natural gas consumption is low and potential biomethane production is high. However, the areas with the greatest potential for biomethane production are those with a large concentration of lignocellulosic residues (see Figure 2). The results of the techno-economic analysis of the treatment processes show that economies of scale are strongly favored, prioritizing the selection of large treatment plants over small ones.

Methodoly for estimating a country's production of biomthane
from agricultural residues: a multiscale and holistic approach

2203

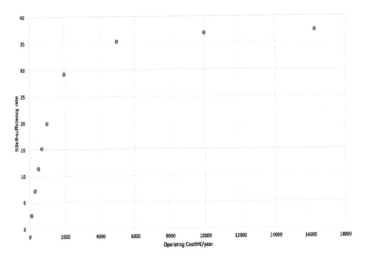

Figure 1.-Relation between OPEX and the self-sufficiency rate

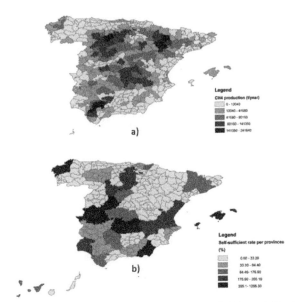

Figure 2.- (a) Production of Biomethane (ton/ye) with an OPEX of 5000 MM/ye and (b) self-sufficient rate of each province

4.Conclusion

The dependence on fossil natural gas and the growing problem of waste generation have pushed European countries to analyze the circular economy of waste as a possible solution to both problems. This work presents a multiscale and holistic analysis to evaluate the biomethane production potential from different types of waste. An optimization framework is formulated to determine the best process, size, and location for the facilities, as well as the system cost. This framework is applied to the case study in Spain, where with an OPEX of 5000M€, it is possible to cover 35% of natural gas consumption and up to 19 provinces can be independent of fossil natural gas. This opens the way to create decentralized structures to reduce the environmental and economic impacts of transporting natural gas from main pipelines to these rural regions. Therefore, this potential energy independence is a strong incentive to invest in infrastructure for waste treatment in these regions. Most of the biomethane comes from the treatment of dry waste through gasification since it has a performance of 4 times greater than anaerobic digestion. Therefore, the treatment of lignocellulosic waste is prioritized over the others. The optimization framework developed in this work can be applied to other countries, changing the crop, population, and animal databases. Some parameters, such as the composition of the residues, can be modified to improve the estimates, without modifying the models of the optimization framework.

Acknowledgement

The authors would like to acknowledge Salamanca Research for the optimization licenses, the funding received from the European Union's Horizon 2020 research, innovation program under the Marie Sklodowska-Curie grant agreement no 778168. M.T. ,and appreciates the FPI Ph.D. fellowship from the Junta de Castilla y León.

References

Council, N. R. (2000). Nutrient Requirements of Beef Cattle. In Nutrient Requirements of Beef Cattle. National Academies Pres.

Enagas. (2022). Boletín Estadístico . https://www.enagas.es/content/dam/enagas/es/ficheros/gestion-tecnica-sistema/energy-data/publicaciones/boletin-estadistico-del-gas/Boletin-Estadistico-Gas-agosto-2022.pdf

European Comision. (2021). Municipal waste statistics - Statistics Explained. Eurostat Statistics Explained. https://ec.europa.eu/eurostat/statistics-explained/index.php?title=Municipal_waste_statistics

Floudas, C. A., Niziolek, A. M., Onel, O., & Matthews, L. R. (2016). Multi-scale systems engineering for energy and the environment: Challenges and opportunities. AIChE Journal, 62(3), 602–623. https://doi.org/10.1002/AIC.15151

Gobierno de España. (2022). PERTE Agroalimentario. https://www.lamoncloa.gob.es/consejodeministros/resumenes/Documents/2022/080222-ResumenEjecutivoPERTEAgro.pdf

León, E., & Martín, M. (2016). Optimal production of power in a combined cycle from manure based biogas. Energy Conversion and Management, 114, 89–99. https://doi.org/10.1016/j.enconman.2016.02.002

Mišík, M. (2022). The EU needs to improve its external energy security. Energy Policy, 165, 112930. https://doi.org/10.1016/J.ENPOL.2022.112930

Phillips, S., Aden, A., Jechura, J., Dayton, D., & Eggeman, T. (2007). Thermochemical ethanol via indirect gasification and mixed alcohol synthesis of lignocellulosic biomass. https://doi.org/10.2172/1216397

Sánchez, A., Martín, M., & Vega, P. (2019). Biomass Based Sustainable Ammonia Production: Digestion vs Gasification. ACS Sustainable Chemistry and Engineering, 7(11), 9995–10007. https://doi.org/10.1021/acssuschemeng.9b01158

Sinnott, R. K. (2005). Chemical Engineering Design (6th Edition, Vol. 6, pp. 339–359). Elsevier.

Taifouris, M. R., & Martín, M. (2018). Multiscale scheme for the optimal use of residues for the production of biogas across Castile and Leon. Journal of Cleaner Production, 185, 239–251. https://doi.org/10.1016/J.JCLEPRO.2018.03.018

Wang, S., Jena, U., & Das, K. C. (2018). Biomethane production potential of slaughterhouse waste in the United States. Energy Conversion and Management, 173, 143–157. https://doi.org/10.1016/J.ENCONMAN.2018.07.059

Antonis Kokossis, Michael C. Georgiadis, Efstratios N. Pistikopoulos (Eds.)
PROCEEDINGS OF THE 33rd European Symposium on Computer Aided Process Engineering
(ESCAPE33), June 18-21, 2023, Athens, Greece

Energy management for controlled environment agriculture based on physics informed neural networks and adaptive linearization based data-driven robust model predictive control with AI

Guoqing Hu[a], Fengqi You[a]

[a]Cornell University, Ithaca, New York, 14853, USA

Abstract

Reducing stress on energy demand is crucial for promoting controlled environment agriculture, the most energy-intensive sector. Refining the modeling approach and control strategies can help reduce energy demand from crop cultivation facilities. This study presents a novel physics-informed neural network that considers both greenhouse energy demand and crop models to minimize weather forecast errors in controlled environment facilities. The adaptive linearization feedback scheme is used to handle nonlinear relationships between environment, crops, and control elements such as heating, ventilation, CO2 enrichment, and more. The proposed control inputs are determined by the data-driven robust model predictive control, which uses machine learning to address weather forecast uncertainties. The proposed control framework has been demonstrated to effectively reduce energy consumption while maintaining suitable growing conditions for crops, even in challenging conditions.

Keywords: controlled environment agriculture, model predictive control, neural network.

1. Introduction

Controlled environment agriculture provides a promising solution for the food demand arising from ascending global population (Costantino et al., 2021). Energy-intensive agricultural systems can contribute to the energy crisis and high levels of greenhouse gas emissions during the cultivation process (FAO, 2018). Therefore, effective control is needed to enhance energy usage efficiency and maintain a favorable cultivation environment for crop productivity and environmental sustainability. The conventional control strategies may not be able to adequately be used for energy management since they didn't comprehensively consider the nature of highly coupled nonlinear climate and crops dynamics and intensive environmental disturbances (Chen et al., 2022). The lack of consideration for these could result in excessive/limited energy applied and failure to maintain a favorable cultivation environment. Alternatively, Model predictive control (MPC) has been acknowledged as a suitable control strategy to solve the problem of regulating the climate and growth processes and guarantee the fulfillment of the constraints specification. There are two fundamental notions in tuning an MPC application: (1) high-fidelity model and (2) robustness of control in application (Garriga and Soroush, 2010). An accurate model could prevent the violation of the constraints caused by modeling error during the application and further control efforts to adjust the states.

We propose a novel control framework for regulating energy consumption and boosting crop yield. To construct the comprehensive high-fidelity model for control strategies, we leverage the physics-informed neural network (PINN) in estimating the nonlinear model, which considers heating/cooling, humidification/dehumidification, ventilation, CO_2 enrichment, lighting, irrigation, and fertilization. In this work, instead of using a single state-space model (SSM) to generalize the dynamics of the indoor production facility, adaptive linearization feedback is adopted to calculate each SSM based on the data predicted from PINN at each control interval. Afterward, each control interval's robust control strategy is determined from the RMPC framework based on the SSM and historical forecast uncertainties. Historical weather forecast data and historical weather measurement data are gathered to form the forecast uncertainties. The data-driven uncertainty set is constructed by machine learning methods, including principal component analysis (PCA) (Wold et al., 1987) coupled with kernel density estimation (KDE) (Zhang et al., 2006). Based on these methods, we further develop a data-driven polyhedron uncertainty set, in which quantile functions are used to describe confidence intervals. Using this modeling and control approach, the nonlinear correlations between environmental disturbances and control effort can be linearized with acceptable accuracy, and control energy can be developed more effectively based on a more accurate model.

2. Physics informed neural network and control framework formulation

The physics-informed layer contains all the differential equations to numerically approximate the values of the temperature states described in the section above. However, the nonlinear dynamics of the greenhouse contain stiff ordinary differential equations, which are numerically unstable in the traditional Euler method integration process (Gear et al., 1985). Therefore, to relieve this issue during the integration process, the Radau method (Guglielmi and Hairer, 2001) which is under fully implicit Runge-Kutta class and is essential for the numerical integration over the stiff problems (Hairer and Wanner, 1999), is included in the integration process. Then the values calculated from the physics-informed layer will be fed into the linear weighted-bias layer for fitting to the experimental values. We recognize that the PINN can retain a higher accuracy when the sample data is limited (only 30 samples were used for training). Meanwhile, PINN with stiff ordinary differential equation integration methods can achieve better simulation results than with the Euler methods and save 194% of the simulation time needed. Disjunctive uncertainty sets are built for learning the trend of the uncertainty data. The uncertainty set W can be formulated under the introduction of forward and backward deviation variable z^+ and z^- (Ning and You, 2018; 2019):

$$\mathbb{W} = \left\{ \mathbf{w} \in R^H \left| \begin{array}{c} \mathbf{w} = \hat{\mu} + Q\xi, \; \xi = \underline{\xi}z^- + \overline{\xi}z^+ \\ 0 \le z^+, z^- \le 1, \; z^+ + z^- \le 1, \; \mathbf{1}^T\left(z^+ + z^-\right) \le \Gamma \\ \underline{\xi} = \left[\hat{F}_1^{-1}(\alpha), ..., \hat{F}_1^{-1}(\alpha)\right]^T \\ \overline{\xi} = \left[\hat{F}_1^{-1}(1-\alpha), ..., \hat{F}_1^{-1}(1-\alpha)\right]^T \end{array} \right. \right\} \qquad (2)$$

With the data-driven uncertainty set, the MPC problem with ADF can be formulated as :

$$\min \sum_{i \in B_u} c_i u_i + \lambda^T L\lambda$$

s.t. $F_u[Mw + h] \le f_u,$ \qquad\qquad\qquad $\forall w \in \mathbb{W}^k$

$\qquad F_x[Ax_0 + B_u h + B_v v + (B_w + B_u M)w] \le f_x + \lambda, \quad \forall w \in \mathbb{W}^k$

$$\qquad\qquad\qquad\qquad\qquad\qquad\qquad\qquad\qquad\qquad\qquad\qquad\qquad (3)$$

where F_x, F_u, f_x, f_u represent the state variable constraints matrix, control input constraints matrix, constraints for state variables, and constraints for the input. L is the weighted cost matrix that penalizes the violation to the constraints. Λ is the slack variable that allows some extent of violation to the hard constraints. The data-driven disjunctive uncertainty

Energy managemetn for controlled environment agriculture based on physcis
informed neural networks and adaptive linearization based data-driven robust
model predictive control with AI

2207

sets are adopted in this work and incorporated into the RMPC framework, because one of the major challenges for control lies in the inherent uncertainty of weather forecasting, due to the stochastic nature of atmospheric processes and the imperfection of the weather model. However, a box uncertainty set can potentially lead to over-conservative solutions to the RMPC problems. Hence, the introduction of data-driven uncertainty sets can alleviate the over-conservatism issue confronted by the most commonly used single uncertainty set-based RMPC frameworks.

3. Performance profile of control framework

Six control strategies are applied to the semi-closed greenhouse year-round control. We can observe that CEMPC could have the violation to the constraints with the disturbance of forecast uncertainties. On the other hand, the RMPC and DRMPC without the PINN framework could still have large violations of the constraints due to the significant model errors in the state prediction, as those model errors could have a non-negligible impact on the system control even though the control scheme left some space for the violation. This result implies the importance of having an accurate prediction model. Alternatively, with the help of PINN in accurately predicting the states, all robust MPC frameworks with the PINN leave some space to prevent constraint violation when weather forecast errors are significant. Thanks to these margins, most constraints are not violated even under extreme scenarios or model errors strike. The results are demonstrated in Table 1. The control frameworks with PINN are observed to have comprehensive advantages over the control schemes which are simply based on the single linear model. Therefore, in comparison to the forecast uncertainties, the model errors are also influential during the control decision-making process. The over-conservatism issue of RMPC could make control frameworks leverage more energy instead of relying on the ambient environment because uncertain forecast errors and purposes of avoiding the violation penalty (Zhao and You, 2022). Therefore, in order to fulfill the state constraints, the control actuators are more often open to maintain the humidity, leading to excessive energy usage. Hence, even though RMPC can achieve a higher yield in the result, the higher energy and resources expenses prevent it from being the most profitable control framework. From this perspective, our control framework can be safely acknowledged as the optimal control strategy compared to other control frameworks by having the highest yield-energy ratio. Despite that the simulation is conducted within Ithaca, in which the climate is relatively hospitable for crop growth, harsh conditions such as Tucson should also be tested. The recent increment in magnitude and frequency of extreme climate in Tucson requires prompt and efficient control strategies to mitigate the effect of climate change (Ghasemi Tousi et al., 2021). To ensure the feasibility and wide applicability of our proposed control framework, the control of the crop cultivation facility is simulated with the summer weather data of Tucson. This location is chosen because of its harsh conditions for crop cultivation. The temperature in Tucson in 2021 was measured as high as 42 °C, and the humidity was less than 50% in summer. As these growing conditions pose tremendous challenges to controlled environment agriculture to minimize the total control cost in maintaining the ideal growing conditions (Alkhalidi et al., 2019), especially for growing tomatoes and leafy green. Therefore, the simulation of growing tomatoes in Tucson was also conducted and served as a case study to demonstrate the performance of our proposed control framework. Our proposed control framework can handle the model error effectively. Similar to the case study above, the control framework with the PINN can noticeably lower the violation rate of 58.07% for temperature control and 2.82% for humidity control. Another observation is that with more severe conditions, the model errors and forecast uncertainties become more influential in the control decision-making process. Therefore, in order to hedge against the sources of the errors, a more accurate predictive model and more robust control methods are needed. By comparing the results to those simulated with Ithaca weather conditions, we can observe that because of the high ambient temperature and sparse precipitation in Tucson, the climate within the cultivation facility tends to be hotter and drier (Hu and You, 2022). Despite of the harsh cultivation environment in Tucson (Chen et al., 2021), our controller still succeeds in maintaining the hospitable growing environment for the crops. Overall, under both scenarios, our control framework can achieve the highest yield-to-cost ratio, i.e., can achieve the same amount of the yield with the lowest cost, comparing to other control frameworks.

Table 1. The statistical summary of the control results of six controllers in the year 2020 in the indoor cultivation facility in Ithaca, New York. DRMPC is data-driven robust MPC with single linear models based on the training dataset; CEMPC is certainty equivalent MPC with single linear models based on a training dataset; RMPC is robust MPC with single linear models based on the training dataset.; DRMPC_PINN is data-driven robust MPC with the multiple linear models calculated from PINN; CEMPC_PINN is certainty equivalent MPC with the multiple linear models calculated from PINN; RMPC_PINN is robust MPC with the multiple linear models calculated from PINN.

	DRMPC	CEMPC	RMPC	DRMPC_PINN	CEMPC_PINN	RMPC_PINN
Temperature	47.67%	80.21%	45.78%	5.78%	65.21%	1.56%
Relative Humidity	34.56%	70.25%	30.89%	4.21%	60.45%	1.28%
Energy-yield ratio	$1.14/kg	$3.23/kg	$1.29/kg	$0.83/kg	$1.82/kg	$0.99/kg
CPU Time	1.77 s	0.07 s	0.99 s	1.88 s	0.12 s	1.09 s

Table 2. The statistical summary of the control results of six controllers in the year 2020 in the indoor cultivation facility in Tucson, Arizona.

	DRMPC	CEMPC	RMPC	DRMPC_PINN	CEMPC_PINN	RMPC_PINN
Temperature	30.67%	85.41%	49.18%	6.17%	80.21%	5.78%
Relative Humidity	3.56%	80.15%	5.14%	4.56%	70.25%	3.89%
Energy-yield ratio	$1.28/kg	$9.09/kg	$1.78/kg	$0.93/kg	$3.57/kg	$1.09/kg
CPU Time	1.51 s	0.08 s	0.97 s	2.01 s	0.11 s	1.01 s

4. Discussion

In this study, we proposed a novel control framework for minimizing the energy input and sustaining the yield under both forecast uncertainties and model errors. This model takes into not only the greenhouse energy demand model, which includes cover temperature, tray temperature, vegetation temperature, mat temperature, floor temperature, soil temperature, water vapor density, and CO_2 concentration, but also the crop model, which contains both mass of carbohydrate and relative growth rate in leaves, fruit, and stem. This model was incorporated into the physics-informed neural network and integrated with the Radau method to ensure stability throughout the simulation process. The validation results demonstrate that our proposed physics-informed neural network can have high-fidelity prediction over the simulated data which is disturbed under random model error. After the model construction, adaptive linearization feedback is adopted to relieve the computational effort from the nonlinearity within the greenhouse dynamics. Within each linear state-space model, we adopted the data-driven robust MPC to calculate the optimal control solution in each control interval. The case studies are conducted to test the controller's performance under the weather conditions in Ithaca, New York, USA, and Tucson, Arizona, USA. On average, our modeling approach can also help contribute to 85.76% of benefits boost under the same energy cost by balancing the accuracy from the nonlinear models and computational efficiency from the linear models. On the other hand, both results demonstrate that our proposed data-driven robust MPC framework uses minimal control efforts to achieve the highest yield-to-cost ratio.

References

A. Alkhalidi, M. K. Khawaja, D. Abusubaih, 2019. Energy efficient cooling and heating of aquaponics facilities based on regional climate. International Journal of Low-Carbon Technologies, 15, 287-298.

W.-H. Chen, N.S. Mattson, F. You, 2022, Intelligent control and energy optimization in controlled environment agriculture via nonlinear model predictive control of semi-closed greenhouse. Applied Energy, 320, 119334.

W.-H. Chen, F. You, 2021, Smart greenhouse control under harsh climate conditions based on data-driven robust model predictive control with principal component analysis and kernel density estimation. Journal of Process Control, 107, 103-113.

W.-H. Chen, F. You, 2022, Semiclosed Greenhouse Climate Control Under Uncertainty via Machine Learning and Data-Driven Robust Model Predictive Control. IEEE Transactions on Control Systems Technology, 30, 1186-1197.

A. Costantino, L. Comba, G. Sicardi, et al,, 2021. Energy performance and climate control in mechanically ventilated greenhouses: A dynamic modelling-based assessment and investigation. Applied Energy, 288, 116583.

FAO, 2018. The future of food and agriculture: alternative pathways to 2050.

J. L. Garriga, M. Soroush, 2010. Model Predictive Control Tuning Methods: A Review. Industrial & Engineering Chemistry Research, 49, 3505-3515.

C. W. Gear, B. Leimkuhler, G. K. Gupta, 1985. Automatic integration of Euler-Lagrange equations with constraints. Journal of Computational and Applied Mathematics, 12, 77-90.

T. E. Ghasemi, W. O'Brien, S. A. Doulabian, T. Shadmehri, 2021. Climate changes impact on stormwater infrastructure design in Tucson Arizona. Sustainable Cities and Society, 72, 103014.

N. Guglielmi, E. Hairer, 2001. Implementing Radau IIA Methods for Stiff Delay Differential Equations. Computing, 67, 1-12.

E. Hairer, G. Wanner, 1999. Stiff differential equations solved by Radau methods. Journal of Computational and Applied Mathematics, 111, 93-111.

G. Hu, F. You, 2022, Renewable energy-powered semi-closed greenhouse for sustainable crop production using model predictive control and machine learning for energy management. Renewable and Sustainable Energy Reviews, 168, 112790.

C. Ning, F. You, 2018. Data-driven decision making under uncertainty integrating robust optimization with principal component analysis and kernel smoothing methods. Computers & Chemical Engineering, 112, 190-210.

C. Ning, F. You, 2019, Optimization under uncertainty in the era of big data and deep learning: When machine learning meets mathematical programming. Computers & Chemical Engineering, 125, 434-448.

L. Sun, F. You, 2021, Machine Learning and Data-Driven Techniques for the Control of Smart Power Generation Systems: An Uncertainty Handling Perspective. Engineering, 7, 1239-1247.

S. Wold, K. Esbensen, P. Geladi, 1987. Principal component analysis. Chemometrics and Intelligent Laboratory Systems, 2, 37-52.

X. Zhang, M. L. King, R. J. Hyndman, 2006. A Bayesian approach to bandwidth selection for multivariate kernel density estimation. Computational Statistics & Data Analysis, 50, 3009-3031.

N. Zhao, F. You, 2022. Sustainable power systems operations under renewable energy induced disjunctive uncertainties via machine learning-based robust optimization. Renewable and Sustainable Energy Reviews, 161, 112428.

Antonis Kokossis, Michael C. Georgiadis, Efstratios N. Pistikopoulos (Eds.)
PROCEEDINGS OF THE 33rd European Symposium on Computer Aided Process Engineering
(ESCAPE33), June 18-21, 2023, Athens, Greece

Life Cycle Assessment on Chemical Recycling-aided Cascaded Use of Polypropylene Plastics

Xiang Zhao,[a] Fengqi You[a,b]

aSystem Engineering, Cornell University, Ithaca, New York 14853, USA

Abstract

Plastic pollution posed by material losses and their subsequent chemical emissions is pervasive in nature and varies with age. Cascaded polypropylene plastic use enables the re-manufacturing of virgin polymers from waste and minimizes its generation and environmental exposure. Here we showed the advantages of this cascaded use over other waste end-of-life management pathways by investigating the environmental consequences of plastic losses across the entire life cycle. Plastic losses could form volatile organic chemicals via photo-degradation and pose non-negligible global warming, ecotoxicity, and air pollution effects that worsen by at least 189% in the long run. These environmental burdens could be increased by above 9.96% under high ultraviolet radiation levels and participation rates, which facilitated plastic particulate compartment transport and degradation. Plastic cascaded use aided by fast pyrolysis technologies could effectively cut environmental losses and outperform landfills and incineration in reducing 23.35% ozone formation and 19.91% air pollution by replacing the external monomer and fuel production with onsite manufacturing.

Keywords: plastic losses, chemical recycling, cascaded use, life cycle assessment.

1. Introduction

Plastics commonly used by humans always come with an environmental cost (Thomas, 2022). Limiting the reuse of plastics can release tons of waste into natural ecosystems (Chin et al., 2022), if not properly addressed through effective waste recovery technologies (Wen et al., 2021), including incineration and recycling, which aim to recover virgin material from waste (Zhao and You, 2021a). These technologies are still not dominantly used in end-of-life (EoL) waste treatment (Fan et al., 2022), and most waste will end up in landfills, posing an environmental burden due to the loss of plastic materials (Min et al., 2020). These chronic plastic losses in the form of micro- and nanoparticles are can easily ingested by organisms and produce volatile organic chemicals through photodegradation (Wu et al., 2022). Explicit assessment remains unmet research needs due to the lack of knowledge about the compartment transport of plastic particles. The cascade use of plastics enables material and waste used in tandem, including re-manufacturing virgin materials from monomers aided by chemical recycling (Zhao et al., 2022), and effectively minimizing material loss release and the derived environmental impacts. However, given the existing knowledge gaps regarding the environmental burden of plastics loss throughout their life cycle, these advantages over other waste management pathways have not been evaluated. This life cycle assessment (LCA) work investigated the advantages and disadvantages of cascade plastic use assisted by chemical recycling processes, which enable material re-production from monomers yielded from effectively cracking plastic waste. Polypropylene (PP) plastic was studied because it embodies similar chemical properties to other plastics (Bora et al., 2020) and is a major source of plastic contamination (Rosenboom et al., 2022). The environmental burden of plastic losses was assessed based on a novel post-disposal fate model (Wang et al., 2019), and a time-dynamic assessment was considered, reflecting the short- and long-term environmental burden of plastic pollution. Then, a holistic life-cycle assessment approach was used to evaluate the environmental performance of polypropylene cascade use, involving the EoL fate at the life-cycle stage from raw material and resource extraction to plastic loss. Full-spectrum environmental impact results of cascade plastic use over 20, 100, and 500 years were evaluated to reflect pros and cons, point out the future technological innovations, and inform policy implications.

2. LCA Methodology

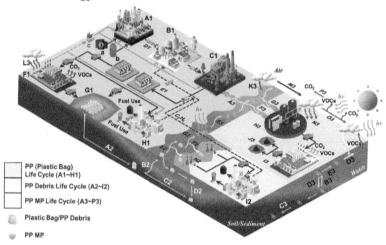

Figure 1. This LCA study's system boundary of PP plastics cascading system.

This LCA study aims to evaluate the life cycle environmental impacts of PP plastics, which are widely used in packaging materials and are also well-known major (micro)plastic (MP) pollution sources (Zhao and You, 2022). The upstream life cycle stages corresponding to the PP manufacturing, use, and EoL waste management, including incineration, landfill, and polypropylene chemical recycling, are included in the system boundary shown in Figure 1, while the downstream processes are related to plastic losses formation, degradation, aquatic waste collection and reclamation by fast pyrolysis-based chemical recycling, MP removals via drinking water treatment (DWTP) (Tian et al., 2020), and incineration. The functional unit was chosen as one-ton waste polypropylene plastic bag treated to align with the mass and energy balance across the entire life cycle (Lal and You, 2022). The life cycle inventories (LCIs) of all polypropylene pre-disposal life cycle stages are extracted from the well-archived Ecoinvent V3.8 Database based on their mass- and energy balance data. The LCIs of onsite energy and chemical production, including the mid-voltage electricity and process heat co-generation from polypropylene waste incineration and virgin chemicals produced from polypropylene chemical recycling, were modelled as their market processes avoided by their internal production, referring to the avoided burden approach (Curran, 2013). The PP waste chemical recycling chemical composition is extracted from relevant PP fast pyrolysis LCA studies (Zhao and You, 2021b). The mass flow rates of the plastic losses and their derived chemical emissions via degradation are evaluated based on the mass balance-based EoL fate model, which is developed based on the random sampling methodology. This methodology has more than 95% accuracy represented by less than

5% relative errors (Wang et al., 2003) in the multimedia transport and exposure dose model. The LCIs of plastic debris removal processes concerning coastal waste collection and drinking water treatment processes are modelled based on the Ecoinvent V3.8 Database (Niaz and You, 2022). The collected LCIs are then interpreted into environmental impact assessment results based on the global warming potential (GWP), ReCiPe 2016, EF3.0, and USEtox indicators in 20, 100, 500 years (Thomassen et al., 2019). The elucidation of the environmental assessment results should help reinforce the effective plastic EoL waste management choices towards pollution reduction by achieving the following goals: 1. Understanding the extent of environmental pollution caused by polymeric material losses and its effect factors; 2. Identification of the environmental pros and cons of current EoL waste management pathways. 3. Comparing the full-spectrum environmental performances of various EoL pathways.

3. Result and Discussion

We evaluate the environmental consequences of plastic losses over 20, 100, and 500 years to assess the time-dependent environmental effects across the entire life cycle and inform the disadvantages of plastic pollution. Figure 2 shows that global warming, ecotoxicity, air pollution corresponding to particulate matter (PM) formation, and ozone formation corresponding to photochemical ozone formation and its specific effects on human health and terrestrial ecosystems are four major environmental consequences of the chemical and MP releases. In the long run, the plastic particulates break up and yield smaller MPs and worsen the air pollution effects associated with PM formation by 15 times, while the ozone formation impacts that harm human health and the terrestrial ecosystem shift by 2,105%. The ecotoxicity effects also increase in the long run by at least four times.

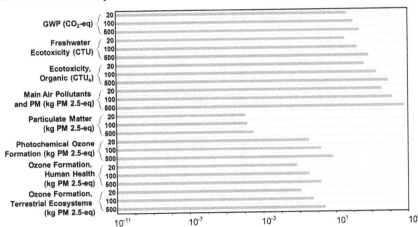

Figure 2. The absolute environmental impacts of plastic losses in 20, 100, and 500 years.

Figure 3 indicates that temperature and UV radiation dose can determine plastic losses and their subsequent chemical releases. High UV radiation levels in summer can enhance the degradation rate of the airborne polymer particulates by two times, resulting in at least 41.48% increments in full-spectrum environmental impacts. During the winter season, these environmental impacts will decrease by over 10.90% at 0 °C. Therefore, as the temperature varies, the plastic degradation rates and their derived chemical emissions can be influenced by the season change and the subsequent long-term environmental burdens.Based on the investigated environmental burdens of the plastic losses, we then evaluate the environmental effects of the PP plastic wastes treated by diverse EoL waste management pathway, which includes landfill, incineration, and recycling. We consider three typical plastic EoL waste management pathways: 1. U.S. EoL waste management pathway; 2. 50% incineration with 50% recycling; 3. All chemical recycling aided by fast pyrolysis processes. The U.S. pathways are denoted by 2.73% recycling, 19.96% incineration, and 78.31% treated by landfills.

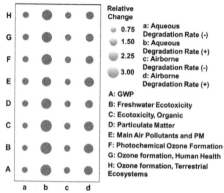

Figure 3. The effect factors of the absolute environmental impacts posed by plastic losses and the relative changes of the environmental impacts with parameter variations.

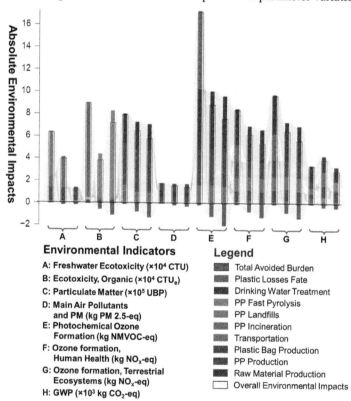

Figure 4. Environmental breakdowns of the life cycle processes of the polypropylene waste treated by U.S. EoL waste management pathway, 50% incineration with 50% recycling, and 100% chemical recycling.

Figure 4 shows that the plastic losses and their environmental impacts can be significantly reduced if other EoL waste management technology options, including incineration or recycling, replace landfilling. Moreover, plastic losses can be cut by above 99.99% if 50% of solid wastes go to incineration and another half by chemical reclamation. However, the onsite incineration processes can contribute 32.15% to the overall GHG emissions due to the intensive CO_2 emissions. On the other hand, fast pyrolysis processes can allocate the carbons to fuels involving gasoline, diesel, and ethylene and propylene to reduce direct GHG emissions from their onsite production. This process can also reduce the environmentally hazardous plastic losses and their derived environmental burdens, as illustrated by 92.58%, 40.87%, and 25% decrements in total ecotoxicity posed by life cycle organic chemical emissions, photochemical ozone formation, and ozone formation that harms terrestrial ecosystems, respectively. Although high organic solvent use in the aromatic extraction process within polypropylene chemically recovery processes can increase ecotoxicity from life cycle organic emissions, this EoL waste management process still outperforms other EoL options, including incineration and landfills, in terms of the full-spectrum environmental impacts. Onsite monomer and fuels and energy production aided by polypropylene chemical recycling can decrease environmental burdens from offsite manufacturing by reducing 19.91% PM formation, 26.49% photochemical ozone formation, and 23.35% ozone formation. The PP can then be used multiple times because the propylene produced from chemical recycling can be re-polymerized into PP plastics. The environmental benefits of this PP cascaded use, as illustrated by Figure 4, are shown as a decrement of 6.74% photochemical ozone formation, 5.70% ozone formation that harms human health, 5.86% ozone formation harming the terrestrial ecosystem, and 4.57% GWP. Moreover, this chemical recycling-aided PP cascaded use can save the raw material and fossil fuels by 11.25% and 25.75%, respectively. Further improving the monomer recovery rate, limiting energy use, and specialized air pollution controls on the exhaust gas can better hinder environmental consequences from plastic losses and reduce fossil resource consumption.

4. Conclusion

This work investigated the environmental burdens posed by the plastic losses from PP waste mismanagement in current practices to inform the future EoL disposal towards environmental sustainability. The environmental advantages of PP cascaded use, which is aided by chemical recycling, over other waste end-of-life management pathways can be illustrated by an above 99.99% reduction in plastic losses and their derived environmental burdens. Specifically, the volatile organic chemicals formed by plastic losses via photo-degradation can pose non-negligible global warming, ecotoxicity, and air pollution effects that worsen by at least 189% in the long run. These environmental burdens could be increased by above 9.96% under high ultraviolet radiation levels and participation rates. Based on the investigated environmental burden results, the environmental advantages of plastic cascaded use aided by fast pyrolysis technologies are specified as a reduction of 23.35% ozone formation and 19.91% air pollution compared to landfills and incineration. By replacing the external monomer and fuel production with their onsite manufacturing in PP cascade use, 25.75% of fossil fuel is saved.

References

R.R. Bora, R. Wang, et al., 2020, Waste Polypropylene Plastic Recycling toward Climate Change Mitigation and Circular Economy: Energy, Environmental, and Technoeconomic Perspectives. ACS Sustainable Chemistry & Engineering, 8, 16350-16363.

H.H. Chin, P.S. Varbanov, et al., 2022, Plastic Circular Economy Framework using Hybrid Machine Learning and Pinch Analysis. Resources, Conservation and Recycling, 184, 106387.

M. A. Curran, 2013. Life cycle assessment: a review of the methodology and its application to sustainability. Current Opinion in Chemical Engineering 2(3), 273-277.

Y.V. Fan, P. Jiang, R.R. Tan, et al., 2022, Forecasting plastic waste generation and interventions for environmental hazard mitigation. Journal of Hazardous Materials, 424, 127330.

A. Lal, F. You, 2022. Targeting Climate - Neutral Hydrogen Production: Integrating Brown and Blue Pathways with Green Hydrogen Infrastructure via A Novel Superstructure and Simulation - based Life Cycle Optimization. AIChE Journal, e17956.

K. Min, JD. Cuiffi, RT. Mathers, 2020. Ranking environmental degradation trends of plastic marine debris based on physical properties and molecular structure. Nature Communications 11(1), 1-11.

H. Niaz, MH. Shams, JJ. Liu, et al., 2022. Mining bitcoins with carbon capture and renewable energy for carbon neutrality across states in the USA. Energy & Environmental Science 15(9), 3551-3570.

J-G. Rosenboom, R. Langer, G. Traverso, 2022. Bioplastics for a circular economy. Nature Reviews Materials 7(2), 117-137.

K. V. Thomas, 2022. Understanding the plastics cycle to minimize exposure. Nature Sustainability 5(4), 282-284.

G. Thomassen, M. Van Dael, S. Van Passel, et al., 2019, How to assess the potential of emerging green technologies? Towards a prospective environmental and techno-economic assessment framework. Green Chemistry, 21, 4868-4886.

X. Tian, R.E. Richardson, J.W. Tester, et al., 2020, Retrofitting Municipal Wastewater Treatment Facilities toward a Greener and Circular Economy by Virtue of Resource Recovery: Techno-Economic Analysis and Life Cycle Assessment. ACS Sustainable Chemistry & Engineering, 8, 13823-13837.

Z. Wen, Y. Xie, M. Chen, CD. Dinga, 2021. China's plastic import ban increases prospects of environmental impact mitigation of plastic waste trade flow worldwide. Nature Communications 12, 1-9.

X. Zhao, JJ. Klemeš, M. Saxon, F. You, 2022. How sustainable are the biodegradable medical gowns via environmental and social life cycle assessment? Journal of Cleaner Production, 380, 135153.

X. Zhao, F. You, 2021, Consequential Life Cycle Assessment and Optimization of High-Density Polyethylene Plastic Waste Chemical Recycling. ACS Sustainable Chemistry & Engineering, 9, 12167-12184.

X. Zhao, F. You, 2021. Waste high-density polyethylene recycling process systems for mitigating plastic pollution through a sustainable design and synthesis paradigm. AIChE Journal, 67, e17127.

X. Zhao, F. You, 2021. Waste respirator processing system for public health protection and climate change mitigation under COVID-19 pandemic: Novel process design and energy, environmental, and techno-economic perspectives. Applied Energy 283, 116129.

X. Zhao, F. You, 2022, Life Cycle Assessment of Microplastics Reveals Their Greater Environmental Hazards than Mismanaged Polymer Waste Losses. Environmental Science & Technology, 56, 11780-11797.

S-W. Wang, PG. Georgopoulos, G. Li, H. Rabitz, 2003. Random sampling− high dimensional model representation (RS− HDMR) with nonuniformly distributed variables: application to an integrated multimedia/multipathway exposure and dose model for trichloroethylene. The Journal of Physical Chemistry A 107(23), 4707-4716.

Y. Wang, JW. Levis, MA. Barlaz, 2019. An assessment of the dynamic global warming impact associated with long-term emissions from landfills. Environmental Science & Technology 54, 1304-1313.

X. Wu, X. Chen, R. Jiang, J. You, G. Ouyang, 2022. New insights into the photo-degraded polystyrene microplastic: Effect on the release of volatile organic compounds. Journal of Hazardous Materials 431, 128523.

Antonis Kokossis, Michael C. Georgiadis, Efstratios N. Pistikopoulos (Eds.)
PROCEEDINGS OF THE 33rd European Symposium on Computer Aided Process Engineering (ESCAPE33), June 18-21, 2023, Athens, Greece

Enviro-economic analysis of tandem and direct processes for ethylene electrosynthesis

Grazia Leonzio, Benoit Chachuat, Nilay Shah[*]
Sargent Centre for Process Systems Engineering, Department of Chemical Engineering, Imperial College London, London SW7 2AZ, UK

[*]*n.shah@imperial.ac.uk*

Abstract
Ethylene is the most important organic chemical in terms of global demand and production capacity. Of the sustainable alternatives to conventional ethylene production based on steam cracking of natural gas and naphtha, both direct electrochemical reduction of CO_2 as well as a tandem process consisting of CO_2 electro-reduction to CO followed by CO electro-reduction to ethylene have attracted attention. This conference paper presents a comparison between the tandem and direct CO_2 electro-reduction processes both from an economic and environmental point of view, including a global sensitivity analysis of key process parameters on production cost and climate change impact. The results depict a clear trade-off between the economic and environmental performance of both electrochemical routes, although the tandem process remains more favorable at the current carbon price of the EU emission trading system (ETS).

Keywords: ethylene, tandem process, CO_2 electro-reduction.

1. Introduction

There is an urgency to reduce carbon dioxide (CO_2) emissions to limit the impact of global warming and climate change. This motivates the development of low-carbon technologies based on CO_2 utilization for the production of major chemicals, with a view to feeding a circular chemical economy.

With global demand and production capacity of 172 and 200 Mt per year in 2021, and a forecast that these values will keep increasing in the near future, ethylene remains the most important organic chemical (Statista, 2022). Most ethylene production is currently via steam cracking of natural gas or naphtha, which is energy intensive (up to 40 GJ heat per ton of ethylene) and emits between 1.8–2 ton of CO_2 per ton of ethylene. About 30% of the total energy consumed by the chemical industry is used to meet the global ethylene demand, leading to the emission of 0.26 $GtCO_2eq$ (Haribal et al., 2018).

Of the alternatives to produce ethylene in a more sustainable way, direct CO_2 electro-reduction (CO_2ER) has attracted significant attention (Berkelaar et al., 2022). Nevertheless, the performance and sustainability of this technology at scale is still hindered by low current densities, significant CO_2 loses by crossover in neutral membrane electrode assemblies (MEA), and carbonate formation in alkaline flow cells (AFC). In response to these challenges, Sisler et al. (2021) proposed a tandem process, whereby CO_2 is first reduced to carbon monoxide (CO) in solid-oxide electrolytic cells (SOEC) followed by CO reduction to ethylene using MEA or AFC. This paper presents a comparison of the tandem and direct CO_2ER processes. The techno-economic and environmental performance of the two routes is evaluated through scaling up ethylene production based on current performance of electrochemical cells, following Sisler et al. (2021) for the tandem process and Jouny et al. (2018) for the direct process. Multiple CO_2 feed and electricity sources are furthermore considered and uncertainty quantification is conducted for key process parameters using global sensitivity analysis (GSA).

2. Methodology

2.1. Process modelling

The tandem process is based on two electrolysis stages: CO_2 is first reduced to CO in an SOEC, followed by the reduction of CO to ethylene in an AFC, thereby avoiding CO_2 losses through carbonate formation. Modeling of these electrolytic cells is based on simple material balances and considers the optimistic conditions for the SOEC and AFC, respectively (Sisler et al., 2021): cell voltage of 1.3 V and 1.8 V, Faradaic efficiency of 100% and 90%, current densities of 1000 mA/cm^2, single-pass conversion of 60% and 53%. Pressure swing adsorption (PSA) units are used after each stage for gas separation. Modeling of the electrolytic cell for direct CO_2ER to ethylene is also based on simple material balances, considering alkaline conditions and the following assumptions (Jouny et al., 2018): current density of 300 mA/cm^2, cell voltage of 2 V, Faradaic efficiency of 90%, CO_2 conversion of 50%. As with the tandem process, a PSA unit is used for gas purification.

2.2. Techno-economic analysis

The analysis considers a plant producing 100 ton of ethylene per year, with the mainland UK as the geographical location. The CO_2 sources include: natural gas processing/coal to chemicals, ammonia/bioethanol/ethylene oxide, methane steam reforming, iron and steel, cement, power generation, and direct air capture (DAC) with respective costs of $23.7, $35.6, $77.1, $83, $106, $88.9 and $325 per ton of CO_2 (IEA, 2022); and the electricity sources include: solar, wind, conventional nuclear, and small modular nuclear reactor (SMR), with respective costs of $0.14, $0.09 (IRENA, 2019), $0.059 (NAMRC, 2022) and $0.049 per kWh (BEIS, 2016).

For the tandem process, capital (CAPEX) and operating (OPEX) expenditures are estimated using the same methodology as Sisler et al. (2021), assuming values of $1,067 and $300 per kW for the SOEC and AFC, respectively. For the direct process, the methodology reported in Jouny et al. (2018) is used for the overall CAPEX costs ($920 per m^2 of electrolyser), while the OPEX costs follow Peter & Timmerhaus (1991). These costs are then used to estimate the production cost of ethylene.

2.3. Environmental assessment

The life-cycle assessment (LCA) follows the four phases defined in the ISO 14040 standards: (i) goal and scope, (ii) inventory analysis, (iii) impact assessment, and (iv) interpretation. The aim is to compare the environmental impacts of the two ethylene production routes, considering 1 kg of ethylene production as the functional unit. The system boundaries are defined for a cradle-to-gate analysis. The foreground inventory data are provided by the process models, for a plant located in the UK. The carbon footprint of the different CO2 sources is taken from Muller et al. (2020). The impact assessment relies on the Environmental Footprint 2.0 method, as recommended by the European Commission (2018), through the software SimaPro interfaced with ecoinvent.

2.4. Global sensitivity analysis

A GSA is conducted to quantify the overall uncertainty in the KPIs of interest (ethylene production cost and global warming impact) by simultaneously varying all of the uncertain factors through their entire range, then apportioning a KPI's uncertainty to each factor. The uncertain factors for each route assume triangular probability distributions, shown collectively in Table 1. The GSA is conducted using the software SobolGSA, which constructs high dimensional model representation (HDMR) surrogates by drawing 1000 Sobol samples (Sobol, 2001) from the uncertainty distribution and reevaluating the process model incentories, then estimates the first- and total-order Sobol sensitivity indices (Rabitz & Salis, 1999) for the ethylene production cost and global warming impact.

Table 1: Uncertain factors in the GSA of the CO$_2$ER direct and tandem processes

Factor	Distribution	Lower	Mode	Upper
Electricity price ($/kWh), wind	Triangular	0.0675	0.09	0.1125
CO$_2$ price ($/ton), NG processing	Triangular	14.22	23.7	33.18
Direct CO$_2$ER to ethylene:				
Cell voltage (V)	Triangular	2	2	2.3
Current density (mA/cm^2)	Triangular	200	300	300
Tandem CO$_2$ER to ethylene:				
CO$_2$ conversion in SOEC (%)	Triangular	50	60	60
Current density in SOEC (mA/cm^2)	Triangular	475	1000	1000
CO conversion in AFC (%)	Triangular	43	53	53
Current density in AFC (mA/cm^2)	Triangular	144	1000	1000
Cell voltage in AFC (V)	Triangular	1.8	1.8	2.32
Faradaic efficiency in AFC (%)	Triangular	38	90	90

3. Results and discussions

3.1. Economic and environmental analyses

The ethylene production costs are compared in Figure 1 for the various electricity and CO2 sources. They range between $1.3–3.7 per kg of ethylene with the tandem process and between $2.1–8.2 with the direct process. The lowest cost corresponds to nuclear energy from SMR followed by conventional nuclear, wind and solar given the assumed electricity prices. The contribution of CO2 price to the production cost is small in comparison to electricity, except in the case of CO2 from DAC that leads to significantly higher production costs. These production costs are also to be compared with the business-as-usual (BAU) technology, naphtha cracking, at a cost of $1.3 per kg of ethylene (ICIS, 2022). Under nominal operating conditions, only the tandem process is economically competitive with the BAU, which concurs with the findings by Sisler et al. (2021).

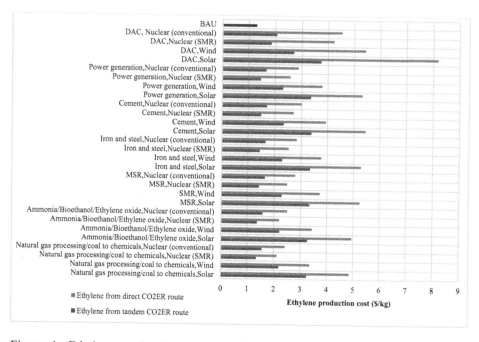

Figure 1: Ethylene production costs for the tandem and direct CO_2ER processes considering different CO_2 and electricity sources, and comparison with the BAU process

The global warming impacts compared in Figure 2 for the same electricity and CO_2 sources show a different reality, whereby the direct and tandem routes have predicted impacts between -2.5–1.8 and 0.8–2.3 kgCO_2-eq per kg of ethylene, respectively. Note that the direct route has negative emissions (from a cradle-to-gate accounting) in most electricity and CO_2 source scenarios, which concurs with the findings by Pappijn et al. (2020). Emission from the tandem route are consistently below 1 kgCO_2-eq per kg of ethylene (for electricity sources other than solar), still making it a positive emitter compared to the direct route, but a significant improvement in comparison with the average global warming impact of naphtha cracking (BAU) at around 1.6 kgCO_2-eq per kg of ethylene. Interestingly, even with emissions allowances (EUA) traded on ETS at a historically high carbon price of $100 per ton of CO_2 the direct process fails to be competitive with the tandem process or naphtha cracking. By contrast, the tandem process could be competitive with naphtha cracking in the nuclear scenario, although still 50% more expensive than BAU with wind energy.

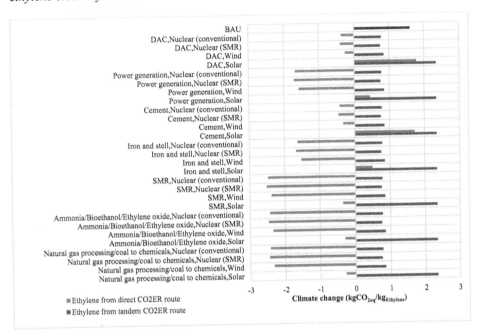

Figure 2: Global warming impact for the tandem and direct CO_2ER processes considering different CO_2 and electricity sources, and comparison with the BAU process

3.2. Global sensitivity analysis

The GSA considers the scenario of wind energy and CO_2 captured from natural gas processing/coal to chemicals (Table 1). For the direct CO_2ER route, the interquartile (IQ) range for the ethylene production cost is $3.78–4.19 per kg of ethylene, with electricity price as the dominant factor (total-order Sobol index (TSI): 90%) followed by cell voltage (TSI: 9%); while the global warming impact is seldom affected by the uncertainty with an IQ range of -2.34--2.32 $kgCO_2eq$ per kg of ethylene. For the tandem CO_2ER route, the interquartile (IQ) range for the ethylene production cost is $2.54–3.28 per kg of ethylene, now with Faradaic efficiency in AFC (TSI: 73%) as dominant factor followed by electricity price (TSI: 20%) and cell voltage in AFC (6%); while the uncertainty in global warming impact is again small with an IQ range of 0.862–0.886 $kgCO_2eq$ per kg of ethylene. Some of these factors provide relevant targets for future technology improvements.

4. Conclusions

Economic and environmental analyses have been conducted for two CO_2 electroreduction processes producing ethylene from CO_2. A comparison under different CO_2 and electricity sources has shown that the tandem process is economically superior to the direct route, and could even be competitive with naphtha cracking with nuclear energy. Nevertheless, the direct process presents a lower global warming impact than the tandem process, itself lower than naphtha cracking. These results depict a clear trade-off between the economic and environmental performance of both CO_2ER routes, although the tandem process remains more economically favorable even at a high carbon price of the EU's emission trading system.

Acknowledgments: Funding by the Engineering and Physical Sciences Research Council (EPSRC) under grants EP/V011863/1 and EP/V042432/1 is gratefully acknowledged.

References

L Berkelaar, J van der Linde, J Peper, A Rajhans, D Tiemessen, L van der Ham, H van den Berg, 2022, Electrochemical conversion of carbon dioxide to ethylene: Plant design, evaluation and prospects for the future, *Chemical Engineering Research & Design* 182, 194–206

BEIS, 2016, *Electricity Generation Costs*

European Commission, 2018, *Product Environmental Footprint Category Rules Guidance* (version 6.3, May 2018)

P Haribal, VY Chena, L Neal, F Li, 2018, Intensification of ethylene production from naphtha via a redox oxy-cracking scheme: Process simulations and analysis, *Engineering* 4, 5, 714-721

ICIS, 2022, *Ethylene*

IEA, 2022, *Is carbon capture too expensive?*

IRENA, 2019, *Solar Costs*

M Jouny, W Luc, F Jiao, 2018, General Techno-Economic Analysis of CO_2 Electrolysis Systems, *Industrial & Engineering Chemistry Research*, **57**, 2165–2177

NAMRC, 2022, *Small Modular Reactors*

LJ Muller, A Katelhon, S Bringezu, S McCoy, S Suh, R Edwards, V Sick, S Kaiser, R Cuellar-Franca, AE Khamlichi, JH Lee, N von der Assen, A Bardow, 2020, The carbon footprint of the carbon feedstock CO_2, *Energy & Environmental Science* 13, 2979

CAR Pappijn, M Ruitenbeek, MF Reyniers, KM Van Geem, 2020. Challenges and opportunities of carbon capture and utilization: Electrochemical conversion of CO2 to ethylene, *Frontiers in Energy Research* 8, 557466

MS Peter, KD Timmerhaus, RE West, 1991, *Plant design and economics for chemical engineering*, McGraw-Hill

H Rabitz, OF Alis, 1999, General foundations of high-dimensional model representations, *Journal of Mathematical Chemistry* **25**, 197–233

J Sisler, S Khan, AH Ip, MW Schreiber, SA Jaffer, ER Bobicki, CT Dinh, EH Sargent, 2021, Ethylene electrosynthesis: A comparative techno-economic analysis of alkaline vs membrane electrode assembly vs CO_2–CO–C_2H_4 tandems, *ACS Energy Letters* **6**, 997-1002

IM Sobol, 2001, Global sensitivity indices for nonlinear mathematical models and their Monte Carlo estimates, *Mathematics & Computers in Simulation* **55**, 271 - 280

Statista, 2022, Ethylene demand and production capacity worldwide from 2015 to 2022

Antonis Kokossis, Michael C. Georgiadis, Efstratios N. Pistikopoulos (Eds.)
PROCEEDINGS OF THE 33rd European Symposium on Computer Aided Process Engineering
(ESCAPE33), June 18-21, 2023, Athens, Greece

Modeling and Experimental Validation of Poly(3-Hydroxybutyrate-co-3-Hydroxyvalerate) Chain Length Distribution

Stefanie Duvigneau[a,*], Annette Wilisch-Neumann[a], Robert Dürr[b], Achim Kienle[a,c]

[a] *Otto von Guericke University, Germany*

[b] *Magdeburg-Stendal University of Applied Sciences, Germany*

[c] *Max Planck Institute for Dynamics of Complex Technical Systems, Germany*

*stefanie.duvigneau@ovgu.de

Abstract

Biodegradable and bio-based polyhydroxyalkanoates (PHAs) are ecological alternatives to conventional crude oil-based polymers. PHAs can be produced by microorganisms using waste streams from the food industry and domestic households, CO_2, or other biogenic residues. To increase the current market share of PHAs, producing biopolymers with tailor-made properties by adjusting process parameters is a desirable goal.

In the biotechnology community, much is known about different cultivation conditions, e.g., bioreactor concepts or carbon sources, affecting the thermal and mechanical properties of PHAs (McAdam *et al.*, 2020). Those properties depend mainly on chain length distribution and polymer composition. Mathematical modeling can help investigate the effect of different process conditions on chain length distribution quantitatively.

This contribution outlines a multiscale model that describes the dynamics of chain length distribution and composition of the copolymer Poly(3-Hydroxybutyrate-co-3-Hydroxyvalerate) produced in a fed-batch process with *Cupriavidus necator* using fructose and propionic acid as carbon sources. Theoretical predictions are compared with experimental findings.

Keywords: Biopolymer, Multiscale Modelling, Polymerization, *Cupriavidus necator*, Poly(3-Hydroxybutyrate-co-3-Hydroxyvalerate)

1. Introduction

Compared to conventional crude oil-based polymers, biopolymers like PHAs are a promising alternative raw material for producing plastic compounds. These polymers can be produced by various bacteria and archaea using different, mostly inexpensive substrates. They are biodegradable under environmental conditions and are open to a wide range of possible applications. In comparison to conventional plastics, PHAs are still expensive. One way to reduce production costs is using organic waste from agriculture and food industries as carbon sources. In addition, polymer properties are essential. For this, monitoring the distribution and composition of PHAs is desirable to assess polymer quality and thus wanted (or unwanted) properties directly within the production process. However, for some important quantities, e.g., the chain length distribution that cannot be measured online, model-based sensors must be applied for the abovementioned goals

(Dürr *et al.*, 2021b). Furthermore, mathematical modeling is a powerful tool for investigating the dynamics of this distribution. Previous works introduced multiscale modeling approaches to simulate chain length distribution of the homopolymer Poly(3-Hydrocybutyrate) (PHB), the best-known representative of the PHA group (Penloglou *et al.*, 2012; Duvigneau *et al.*, 2022).

The current article focuses on the production process of the copolymer Poly(3-Hydroxybutyrate-co-3-Hydroxyvalerate) (PHBV). PHBV outperforms PHB because of better processability and a broader range of possible applications caused, for instance, by decreased crystallinity compared to PHB. The considered production process is a fed-batch process with *Cupriavidus necator* using fructose and propionic acid as carbon sources to produce the copolymer. Here, we present an updated metabolic model based on the model published in Duvigneau *et al.* (2021a) that leads to improved dynamic concentration profiles. By coupling this metabolic model with polymerization kinetics (Dürr *et al.*, 2021a) and calculating an average monomer concentration, we can qualitatively describe changes in PHBV composition and chain length distribution. To demonstrate this, characteristic values of chain length distribution are calculated and compared with measurements to evaluate the multiscale approach.

2. Experimental Methods

2.1. Microorganism and cultivation

Cupriavidus necator (H16, DSM 428) obtained from DSMZ GmbH Braunschweig was used to produce PHBV. The preculture of the bacteria before transfer in a bioreactor is described in detail in Duvigneau *et al.* (2021a). The bioreactor experiment was performed in a DASGIP parallel bioreactor system with an initial optical density of 0.4 at 600 nm. As carbon sources, 21.75 g/L fructose and 0.48 g/L propionic acid are supplemented to an M81 media. During the experiment, the dissolved oxygen (DO) level was kept at 70 % and the temperature was 30 °C. A 20 g/L propionic acid solution and 2 M sodium chloride were used to stabilize the pH at 6.8.

2.2. Substrate concentrations

Ammonium chloride and fructose were determined from the supernatant of the samples using an enzymatic test kit (Kit No. 5390 and No. 10139106035, R-Biopharm AG, Germany). Propionic acid concentrations were measured with an Agilent 1260 high-performance liquid chromatography (HPLC). After preparation of the supernatant, 10 µL was loaded on a reverse phase column (Inertsil 100A ODS-3, 5 µm pore size, 250 × 4.6 mm, MZ-Analysentechnik GmbH, Germany) and eluted isocratically with 1 mL/min and 0.1 M $NH_4H_2PO_4$ at pH 2.6 and 40 °C. The propionic acid was detected with a diode array detector (Agilent, Germany) at 210 nm.

2.3. Determination of PHBV and biomass concentrations

To validate our modeling approach, it is necessary to determine the distribution of the PHBV polymers. For this, 5 mL of trichloromethane was added to a fixed value of lyophilized biomass (> 4 mg). The mixture was incubated for 48 h at 37 °C and 150 rpm. After extraction, the polymer was filtered and precipitated using 5 mL ice-cold ethanol. The glass tubes were left open under the laboratory hood to ensure solvent evaporation.

The extracted polymer was dissolved again in trichloromethane to achieve a concentration of 0.4 g/L. 100 µL of the solution was separated by gel permeation chromatography (GPC, Agilent + PSS, Germany) using four poly(styrol-co-divinylbenzene) columns (PSS, Germany) at 40 °C and trichloromethane as eluent. Detection of the polymer chains was done by conventional calibration with a refractive index detector (Agilent, Germany). As reference material, 12 polystyrene standards were measured with a weight average molecular weight between 576 Da and $1.42 \cdot 10^7$ Da.

The differential distributions $W(\log M)$ as well as the weight average molecular weight M_w, the number average molecular weight M_n, and polydispersity PDI of the samples were calculated with WinGPC UniChrom Software (PSS, Germany).

The biomass concentrations were determined gravimetrically, as described in Duvigneau *et al.* (2021a). A detailed description of the procedure for determining the total 3-Hydroxybuyrate and 3-Hydroxyvalerate monomer concentrations in the PHBV polymer can be found in Duvigneau *et al.* (2021b).

3. Mathematical Model

The metabolic model used in the current work is based on Duvigneau *et al.* (2021a) and adapted as described below. Within the metabolic model, state equations for the dynamics of the substrates and products are given.

One key element in the advanced metabolic model approach is using online data for exhaust CO_2 to follow the dynamics of substrate and product concentrations during the process. This is achieved in Duvigneau *et al.* (2021a) by introducing a factor called metabolic activity coefficient:

$$b_{CO_2} = \frac{CO_{2,out}(t)}{CO_{2,in}}.$$

(1)

In this contribution, we use a slightly adapted version of the metabolic activity coefficient that can include the relative amount of residual biomass:

$$b_{CO_2} = \frac{CO_{2,out}(t)}{CO_{2,in} \, P_{res}(t)}.$$

(2)

The relative amount of residual biomass P_{res} can be defined as follows:

$$P_{res}(t) = \frac{c_{res}(t)}{c_{res}(t) + c_{hb}(t) + c_{hv}(t)}.$$

(3)

The relative amount of residual biomass P_{res} is the ratio between the concentration of residual biomass and the total biomass consisting of residual biomass and the total HB and HV monomer concentrations.

Besides the update of the metabolic activity coefficient, we decided to include the dependency not only for biomass growth but also for the uptake of fructose to produce HB (terms with k_4 as rate constant). The updated state equations for fructose and HB are:

$$\frac{dc_{fru}}{dt} = -b_{CO_2} \, c_{res} \, c_{fru} [\, k_1 \, c_n \, inh_1 + k_4 \, inh_2 \, inh_3 + k_7 \,] - D \, c_{fru} \, ,$$

(4)

$$\frac{dc_{hb}}{dt} = c_{res} \, inh_2 \, inh_3 [\, b_{CO_2} k_4 \, c_{fru} + k_5 \, c_P \,] - k_3 \, c_{res} \, c_n \, c_{hv} - D \, c_{hb} \, .$$

(5)

The remaining equations for propionic acid, HV, ammonium chloride, and biomass can be found in our previous publication (Duvigneau *et al.*, 2021a).

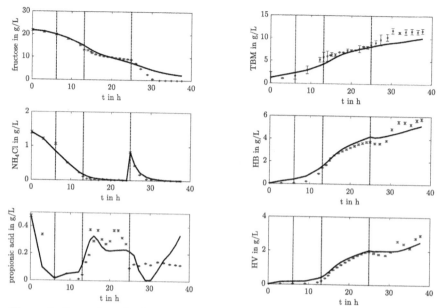

Figure 1: Dynamics of substrate (left) and product concentrations (right) for a PHBV production process. Propionic acid is added after 10 hours by pH-depended feeding strategy. 3-Hydroxyvalerate (HV) can be included in the chain using this additional odd carbon source. The total biomass (TBM) consists of 3-Hydroxybutyrate (HB), HV, and residual biomass.

An algebraic condition can be applied as described in Duvigneau *et al.* (2022) to couple the metabolic model and polymerization kinetics:

$$\frac{1}{MW_{HA}^{P}} m_{HA} = \sum_{i=1}^{\infty} i\left([LP]_i + [DP]_i\right). \tag{6}$$

This implies that the number of all HA monomers in all polymers is the same in both model parts. For the polymerization kinetics, all active $[LP]$ and inactive chains $[DP]$ are multiplied by the corresponding chain length i and added up to calculate the total number of monomers. To calculate the total monomer mass m_{HA}, the total masses of HB (m_{HB}) and HV (m_{HV}) monomers from the metabolic model were joined:

$$m_{HA} = m_{HB} + m_{HV} \tag{7}$$

The determination of the average molecular weight of a monomer unit MW_{HA}^{P} in equation (6) is much more complex than in the case of the homopolymer PHB. For example, copolymer chains with the same number of monomer units can have different molecular weights caused by varying HB:HV ratios from chain to chain. Conversely, this means that with known molecular weights of the polymer chains, the composition and the number of monomers in the individual chains cannot be determined. For simplification, we assume that the HV fraction in all chains is the same and corresponds to the measured total HV fraction. With these assumptions, an average molecular weight for the monomer can be calculated:

$$MW_{HA}^{P} = x_{HB} MW_{HB}^{P} + x_{HV} MW_{HV}^{P}. \tag{8}$$

Therein, x_{HB} and x_{HV} are the fractions of HB and HV in the polymer measured by HPLC. The molecular weight for an HB or HV unit in a polymer (MW_{HB}^{P}, MW_{HV}^{P}) can be found in literature. All other equations to describe the dynamics of all chain and monomer concentrations can be found in Duvigneau *et al.* (2022). Important characteristic values of the chain length distribution to estimate the potential processing properties of the

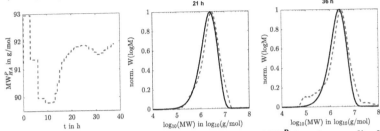

Figure 2: Average dynamical molecular monomer weight (MW_{HA}^P) and normalized chain length distribution $W(logM)$ for chosen time points. Simulated distributions are given as solid black lines and measured distributions are shown as dashed red lines.

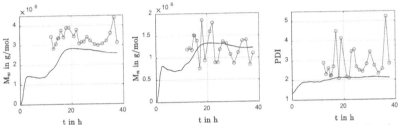

Figure 3: Characteristic values of the chain length distribution. The characteristic values are weight average molecular weight (M_w), number average molecular weight (M_n), and polydispersity (PDI). Simulated values are given as solid black lines, and measured values as red circles.

polymer are the weight average molecular weight (M_w), the number average molecular weight (M_n), and the polydispersity index (PDI). The calculation of the values is shown in Duvigneau *et al.* (2022).

4. Results

The updated metabolic model based on Duvigneau *et al.* (2021a) is adapted to datasets from experiments with fructose and propionic acid as carbon sources and another experiment with fructose as single carbon source (not shown here). With that model, it is possible to depict the dynamics of substrate and product concentrations. This is shown in Figure 1 for the setup with fructose and propionic acid as carbon sources. Propionic acid is essential to produce the copolymer PHBV but has an inhibitory effect on bacterial growth. Hence, the acid is added after 10 hours with a variable feed rate coupled to pH control to ensure that inhibitory concentrations cannot be reached (Kim *et al.*, 1992). To observe not only accumulation but also the degradation process, ammonium chloride is added after 24 h by the operator. After that, the polymer concentration decreases while consumption rates of the external carbon sources fructose and propionic acid are accelerated. The metabolic model can generally capture the concentration profiles with good agreement. Only in the experiment's last phase (>24 h) mismatch between simulation and experiments can be seen, especially for the carbon sources. One possible explanation for the gap could be the unsecure propionic acid measurement by HPLC that is covered by more apoptotic fragments in the late phase of the experiment. In addition to mapping concentration curves, the chain length distribution is essential for evaluating polymer quality. The dynamics of the chain length distribution can be simulated after coupling the metabolic model to a polymerization kinetics. Using this simple and less detailed approach in terms of the polymerization process, the monomers' average molecular weight must be assumed to simulate the chain length distribution of a PHBV copolymer. Figure 2 shows the average molecular weight of the monomers, as well as two normalized differential distributions during the PHBV accumulation at 21 h and the end of the process at 36 h. It can be seen that the simulated differential distributions have a similar order of magnitude as the experimental distributions. However, the model simulation cannot reproduce partially occurring chains in the range of 10^5 g/mol. The same applies to polymers with long chain lengths. This discrepancy is also reflected in the comparison of the characteristic values shown in Figure 3.

5. Conclusion

In this paper, a multiscale model for the PHBV production process using fructose and propionic acid was presented. Using the model, it becomes possible to examine product yields as well as the composition and quality of the copolymer. Future work should deal with adjusting the parameters for the polymerization kinetics to achieve higher accuracy for the prediction of chain length distribution. Furthermore, the preliminary results presented here for the experimental chain length distribution should be repeated and evaluated with universal calibration to examine the influence of measurement noise on the distribution mean values (Figure 3) and simultaneously determine the exact concentrations of the chain lengths. Finally, the influence of certain cultivation conditions, such as pH, temperature, or feed composition, should be experimentally validated and integrated into the multiscale model (Pederson *et al.*, 2006; McAdam *et al.*, 2020). Such knowledge is essential to achieve online control of chain length distribution during microbial PHBV production.

Acknowledgments The funding of the EU-program ERDF (European Regional Development Fund) for the projects DIGIPOL (ZS/2018/11/95489) and the Research Center Dynamic Systems (CDS, ZS/2016/04/78155) is greatly acknowledged. Further, we acknowledge the financial support of the state of Saxony-Anhalt within the SmartProSys initiative.

References

R. Dürr, S. Duvigneau and A. Kienle. 2021a. "Microbial production of polyhydroxyalkanoates – modeling of chain length distribution." 1975–81. doi: 10.1016/B978-0-323-88506-5.50306-5.

R. Dürr, S. Duvigneau, C. Seidel, A. Kienle, A. Bück. 2021b. "Multi-rate data fusion for state and parameter estimation in (bio-)chemical process engineering." *Processes* 9(11):1–16. doi: 10.3390/pr9111990.

S. Duvigneau, R. Dürr, J. Behrens and A. Kienle. 2021a. "Advanced kinetic modeling of bio-co-polymer poly(3-hydroxybutyrate-co-3-hydroxyvalerate) production using fructose and propionate as carbon sources." *Processes* 9. doi: https://doi.org/10.3390/pr9081260.

S. Duvigneau, R. Dürr, M. Wulkow and A. Kienle. 2022. "Multiscale modeling of the microbial production of polyhydroxyalkanoates using two carbon sources." *Comput. Chem. Eng.* 160:107740. doi: 10.1016/j.compchemeng.2022.107740.

S. Duvigneau, A. Kettner, L. Carius, C. Griehl, R. Findeisen and A. Kienle. 2021b. "Fast, inexpensive, and reliable HPLC method to determine monomer fractions in poly(3-hydroxybutyrate-co-3-hydroxyvalerate)." *Appl. Microbiol. Biotechnol.* doi: 10.1007/s00253-021-11265-3.

J. H. Kim, B. G. Kim and C. Y. Choi. 1992. "Effect of propionic acid on poly(β-hydroxybutyric-co-β-hydroxyvaleric)acid production by *Alcaligenes eutrophus*." *Biotechnol. Lett.* 14(10):903–6. doi: 10.1007/BF01020626.

B. McAdam, M. B. Fournet, P. McDonald and M. Mojicevic. 2020. "Production of polyhydroxybutyrate (PHB) and factors impacting its chemical and mechanical characteristics." *Polymers (Basel).* 12(12):1–20. doi: 10.3390/polym12122908.

E. N. Pederson, C. W. J. McChalicher and F. Srienc. 2006. "Bacterial synthesis of PHA block copolymers." *Biomacromolecules* 7(6):1904–11. doi: 10.1021/bm510101.

G. Penloglou, C. Chatzidoukas and C. Kiparissides. 2012. "Microbial production of polyhydroxybutyrate with tailor-made properties: an integrated modelling approach and experimental validation." *Biotechnol. Adv.* 30(1):329–37. doi: 10.1016/j.biotechadv.2011.06.021.

Antonis Kokossis, Michael C. Georgiadis, Efstratios N. Pistikopoulos (Eds.)
PROCEEDINGS OF THE 33rd European Symposium on Computer Aided Process Engineering
(ESCAPE33), June 18-21, 2023, Athens, Greece

Predictive Modeling and scale-up of Wet Oxidation for Hydrothermal Liquefaction Process Water treatment

Carolin Eva Schuck,[a,b] Thomas Schäfer,[b] Konstantinos Anastasakis[a,*]

a Department of Biological and Chemical Engineering, Aarhus University, Hangøvej 2, Aarhus 8200, Denmark
b Department of Chemical Engineering and Biotechnology, Hochschule Darmstadt – University of applied sciences, Stephanstraße 7, Darmstadt 64295, Germany
*kanastasakis@bce.au.dk

Abstract

The present work proposes a general kinetic model for wet oxidation (WO) of hydrothermal liquefaction (HTL) process water (PW) that can aid towards upscaling, integration with HTL and overall optimization of the system. Aspen Plus is used to build a model by incorporating WO kinetics of 13 components selected to represent the composition of HTL-PW derived from sewage sludge. The model showed very good fit with experimental data for WO temperature of 350°C and residence times greater than 2min, with absolute error in COD and TOC removals prediction below 8%. The validated kinetic model was then used for the design of an upscaled WO process to examine the heat performance and energy requirements of the system. The results showed that WO can be operated autothermally with the only external energy input being related to oxygen compression and further cooling of the WO products.

Keywords: hydrothermal liquefaction; process water; wet oxidation; kinetics; Aspen plus

1. Introduction

Hydrothermal liquefaction (HTL) is not only one of the technologies with the greatest potential for the production of advanced biofuels but is also considered as an advanced technology for efficient waste valorization. However, being a wet thermochemical process, it involves processing of slurries with high water content (typically 80wt.% water), hence resulting in a large fraction of process water (PW) as a by-product stream. The formed PW has high COD (chemical oxygen demand), TOC (total organic carbon) and NH_3 content, hindering direct treatment by conventional biological technologies for wastewater treatment (WWT). Nevertheless, it contains between 20 and 40% of the organics present in the biomass feed (Si et al., 2019) offering the potential for an untapped energy source, that if utilized can potentially improve the overall efficiency of HTL technology.

Several different technologies, including anaerobic digestion, microbial electrolysis, electrochemical oxidation, hydrothermal gasification, etc. have been considered for the efficient valorization/treatment of HTL-PW (SundarRajan et al., 2021). All pose certain advantages and disadvantages related to energy efficiency, cost, catalyst stability, toxicity, inhibition, etc. Recently, we proposed wet oxidation (WO) as a pre-treatment technology to significantly reduce the organic loading of HTL-PW prior directing it in a conventional WWT plant for the final polishing (Silva Thomsen et al. 2022).

During WO a gaseous form of oxygen (oxygen, air) is used to oxidize the organics present in PW. The process takes place at elevated temperatures and pressures, maintaining PW in its liquid state and enhancing oxygen's solubility in water under these conditions. Upon total oxidation the carbon and nitrogen present in PW are oxidized to CO_2 and NH_3, respectively. However, partial instead of total oxidation usually takes place giving rise to acetic acid as an intermediate product via a free radical mechanism (Debellefontaine and Foussard, 2000). Both partial and total oxidation however result in significant reduction of the COD, TOC and toxicity contents of the PW enabling the seamless return to the WWT plant for final treatment (**Figure 1**). At the same time, the exothermicity of the oxidation reactions releases heat at an estimated flux of 435 kJ/mol O_2 reacted (Debellefontaine and Foussard, 2000).

The heat release during WO opens opportunities for heat integration either 'in-situ' the WO process by utilizing excess heat for minimizing the overall energy needed to operate the process, or for process-to-process heat integration between HTL and WO. Until now WO has not been subject of flowsheeting and process simulations, both necessary steps for further examining the integration and upscale possibilities.

Hence, the aim of this work is twofold; first to develop a kinetic model in Aspen Plus that can give satisfactory prediction of the WO process of HTL-PW derived from sewage sludge and second, to upscale the model to industrial relevant size and examine the thermal behavior and heat integration options. **Figure 1** depicts the general concept of application of an integrated HTL-WO system in wastewater treatment plants and the specific focus of the present study.

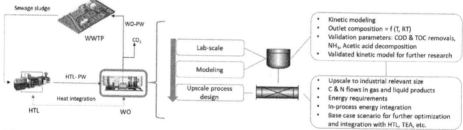

Figure 1. *General concept of application of wet oxidation (WO) for hydrothermal liquefaction (HTL) process water (PW) treatment and specific focus of present study (in green).*

2. Methodology

A rigorous kinetic model is built in Aspen Plus (v12) to give realistic prediction for WO of HTL-PW. The PW considered in this study comes from HTL of sewage sludge. Typical organics present in PW from sewage sludge include alcohols, carboxylic acids, aldehydes, ketones, aromatics, N-components and N-containing aromatics (Silva Thomsen et al. 2022). Published kinetic data for WO of individual compounds present in HTL-PW were implemented in individual models in Aspen Plus and validated according to the original experimental kinetic data. Subsequently the individual kinetic reactions were combined to build a multi-reaction kinetic model by incorporating the relevant reaction mechanism. In total 13 compounds (namely: acetic acid, acetaldehyde, acetamide, acetone, butyric acid, cyclopentanone, ethanol, formic acid, formaldehyde, methanol, phenol, propanoic acid and pyridine) were selected to represent the PW stream. Typical products of WO included CO_2, H_2O, NH_3, $C_2H_4O_2$, CO and H_2. Own experimental data from batch-scale WO of sewage sludge HTL-PW at different temperatures and residence times (Silva Thomsen et al. 2022) were used for the validation of the multi-reaction kinetic model. The composition of the inlet PW stream was adjusted to match the Chemical Oxygen Demand (COD), ammonia and total N (TN) contents of the PW in the study used for validation. **Table 1** shows the characterization of PW by Silva Thomsen et al. 2022 vs the one used in the present study according to the adjusted model compounds composition. For the simulation in Aspen Plus, RBATCH was used for the representation of the batch reactor used in the experiments, while the Predictive Soave–Redlich–Kwong (PSRK), which is adequate for water and hydrocarbon mixtures in the presence of light gases, was selected as the property method. An excess amount of O_2, equivalent to 1.5 times the theoretical O_2 needed for the complete oxidation of the organics present in PW, was provided for the oxidation.

Table 1. *Characteristics of PW from HTL of sewage sludge as determined experimentally and as represented in the model*

	Experimental (Silva Thomsen et al. 2022)	Model
COD (mg/L)	28,300	28,598
NH3 (mg/L)	208	208
TN (excluding NH_3) (mg/L)	477	485
TOC (mg/L)	11,900	9,205

After validation, the model is scaled up to represent mass flows of PW resulting from HTL processing of sewage sludge slurry of 20 wt.% dry matter (DM) content at 1 t/h. The process flowsheet is shown in **Figure 3** where an RPlug reactor and a heat exchanger are implemented to model the scaled-up WO process. The property methods, reaction kinetics, composition of model-feed and reaction conditions remain mostly the same compared to the batch-model. Mass and energy balances have been calculated at 350°C and 15 min residence time (RT). These process conditions were selected since the WO process will likely take place at similar temperature and residence time as the HTL process and represent the base case scenario that can subsequently be further optimized.

3. Results & Discussion

3.1. Kinetic multi-reaction model validation

The kinetic multi-reaction model, represented by RBatch reactor, was run under the exact conditions the experimental data were produced to check its validity. **Figure 2** presents predicted (solid lines) versus experimental (data points) COD and TOC removals, as well as final NH_3 concentration and acetic acid decomposition in post-WO PW at two different temperatures (300°C and 350°C) and residence times (0-180min). As it is clear from Figure 2 the model gave satisfactory predictions at 300°C and very good predictions at 350°C for all indicators and RT examined. The absolute error for both COD and TOC removal ranged between 6%-17% at WO temperature of 300°C and between 1%-15% at WO temperature of 350°C and different residence times. For WO temperature of 350°C maximum absolute errors for both COD and TOC removals were observed at 2min RT (9% and 15%, respectively) while the errors were significantly lower (less than 8%) for longer than 2min RT. In general, both COD and TOC removal rates predicted by the Aspen model are lower than the experimental ones for most of the RTs examined, however the general trends are consistent with the experimental data. A similar behavior is observed for NH_3 concentration and acetic acid decomposition in post-WO PW (figure 2-c and 2-d, respectively) with the absolute errors decreasing with increased WO temperature. Overall, the highest agreement with experimental data is reached for WO temperature of 350°C and residence times longer than 2min for all indicators examined. The two prevailing reasons for the deviation between experimental and modeling results (especially at lower temperature) are the assumed composition of the modeled PW and the implemented kinetic parameters of some of the considered compounds. A significant fraction of the components present in PW remains unidentified by conventional analysis

2232

C. E. Schuck et al.

methods employed such as GC-MS. However, even if the complete composition was known, a model feedstock consisting of 13 compounds was assumed for the present study that might not represent the actual PW composition completely. Even though the concentration of the 13 compounds in the model feedstock was adjusted to match the experimental COD, NH₃ and TN contents of PW, the respective TOC content of the model feedstock was just 78% of the one experimentally measured (**Table 1**). Furthermore, the implemented kinetics for the selected 13 compounds were derived during WO of pure components. These can slightly defer when the compounds are present in a multi-component mixture as in the case of the present study. Nonetheless, despite these discrepancies, the developed model showed satisfactory (at 300°C) to very good (at 350°C) predictions for the key performance indicators (KPIs) examined. It was therefore used for the design of an up-scaled WO process that can be used as the base case scenario for further optimization of the overall system.

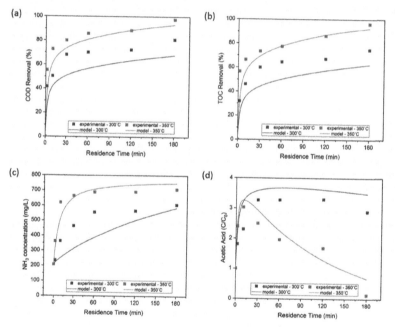

Figure 2. *Predicted (solid lines) vs experimental (data points) post-WO PW (a) COD removal, (b) TOC removal, (c) NH₃ concentration and (d) acetic acid decomposition at 300°C (black lines and symbols) and 350°C (red lines and symbols) and 2min<RT<180min.*

3.2. Process design of the scaled-up WO reactor

Figure 3 presents the flowsheet of the scaled-up WO system. 800 kg/h of HTL-PW (assuming an upstream HTL plant with capacity of 1t/h of sewage slurry at 20wt.% dry matter content) at 60°C is pressurized at 220 bar and mixed with a compressed oxygen stream (at 220 bar) before entering the cold side of the heat exchanger (HEATX). Oxygen is fed at approximately 1.5 times the stoichiometric oxygen needed for the complete oxidation of the organics present in the modeled HTL-PW stream and is compressed in a 5-stage compressor at 220bar. The combined inlet stream (COLD-IN) enters the cold side of the heat exchanger at 63°C and is being heated to reaction temperature (350°C) by exchanging heat with the post-WO stream (HOT-IN). The overall heat transfer coefficient

(U) is assumed as 500 W/m² °C. Subsequently the heated stream (COLD-OUT) enters an adiabatic PFR reactor (WO-PFR) where the previously validated kinetics have been implemented. The PFR reactor consists of 15 tubes with 10m Length and 0.07m Diameter which ensures residence time of approximately 15min. The hot outlet stream from WO (HOT-IN) is cooled in two stages; first by exchanging heat with the inlet cold stream in HEATX to 137°C and then by a simple cooler to 40°C, before separating the stream into vapor (F-VAP) and liquid (F-LIQ) products in a flash drum (FLASH). The HEAT block between the heat exchanger and the WO reactor is used to ensure that the desired reaction temperature is reached and is not being used for the present case.

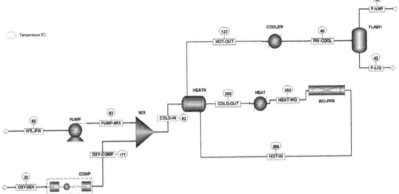

Figure 3. *Process flowsheet of scaled-up WO process, with stream temperatures indicated on the flowsheet, in Aspen Plus (v12).*

The results from the upscaled system indicate that WO can be operated autothermally with the only external energy input being related to oxygen compression and further cooling of the WO products. Autothermal operation is achieved due to the exothermicity of the oxidation reactions, increasing the temperature of the WO outlet stream to 366°C in the present case. Heat capacity of liquid water increases substantially at elevated temperatures, especially above 340°C. As an indication c_p of liquid water increases from 135.9 J/mol*K at 350°C to 210.7 J/mol*K at 365°C (Lemmon et al., 2022). Additional external utilities are estimated as 5.6 kW/m³ of HTL-PW for oxygen compression (electricity), -4.6 kW/m³ of HTL-PW for compressor cooling and -109.9 kW/m³ of HTL-PW for final cooling of WO products, for the present case with the assumed organic loading of HTL-PW. The results suggest a unique potential for heat integration with HTL process which needs to be further examined. The developed model sets strong fundamentals for the subsequent investigation of the heat integration options.

Carbon and nitrogen flows in HTL-PW assumed in this study and in WO products (gas and WO-PW) are depicted in **Figure 4** (4-a and 4-b, respectively) in the form of a Sankey diagram. The majority of carbon (~70%) present in HTL-PW is oxidized to gaseous products upon WO (figure 4-a). The gaseous stream is composed almost entirely of CO_2 (~98.5%) offering a relative pure-CO_2 stream that can be potentially further utilized to produce synthetic fuels in a power-to-X approach. The liquid stream from WO still contains approximately 30% of the inlet carbon with most of the carbon (~24% of initial TOC) found in the form of acetic acid. Acetic acid is an exceptionally refractory intermediate and is also formed during oxidation reactions of several other organic components (Shende and Levec, 1999). As also shown previously (Figure 2-d) higher temperatures and residence times are required for its decomposition. COD and TOC removals for upscaled WO at 350°C and 15min were 73.5% and 70.8%, respectively, in line with both modeling and experimental results obtained from batch scale WO shown in **Figure 2**. Unlike carbon, that is mostly distributed in the gas phase as CO_2, nitrogen, present in the N-containing components assumed in HTL-PW (pyridine, acetamide), end up as ammonia in WO-PW.

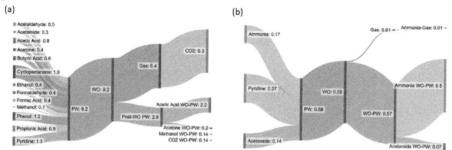

Figure 4. (a) Carbon and (b) Nitrogen flows (kg/m^3 PW) before and after WO at 350°C and 15min RT as predicted by the developed model.

4. Conclusions

A kinetic multi-reaction model was developed in Aspen Plus to predict the wet oxidation of organics present in process water from hydrothermal liquefaction of sewage sludge. The model showed satisfactory to very good predictions with batch-scale experimental data on COD and TOC removals, NH_3 and acetic acid concentrations in post-WO PW. Especially at higher oxidation temperatures (350°C) and longer than 2 min RT, the absolute error was less than 8% in all key performance indicators examined. Process up-scale revealed that WO can be operated autothermally, leveraging on the excess heat generated by the oxidation reactions. At the same time, the results revealed further potential for heat-integration with HTL technology. The developed kinetic model sets the base case scenario to assist in a subsequent process upscale optimization design, in terms of tuning WO process conditions to achieve both the desired organic loading reduction and the desired heat release for the different integration options and HTL-PW composition.

References

H. Debellefontaine, J.N. Foussard, Wet air oxidation for the treatment of industrial wastes. Chemical aspects, reactor design and industrial applications in Europe, Waste Manage. 2000, 20 (1), 15–25.

E.W. Lemmon, I.H. Bell, M.L. Huber, M.O. McLinden, 2022, Thermophysical Properties of Fluid Systems, in NIST Chemistry WebBook, NIST Standard Reference Database Number 69, Eds. P.J. Linstrom and W.G. Mallard, National Institute of Standards and Technology, Gaithersburg MD, 20899

R.V. Shende, J. Levec, 1999, Wet Oxidation Kinetics of Refractory Low Molecular Mass Carboxylic Acids, Ind. Eng. Chem. Res., 38 (10), 3830–3837

B. Si, L. Yang, X. Zhou, J. Watson, G. Tommaso, W.T. Chen, Q. Liao, N. Duan, Z. Liu, Y. Zhang, 2019, Anaerobic conversion of the hydrothermal liquefaction aqueous phase: fate of organics and intensification with granule activated carbon/ ozone pretreatment, Green Chem. 21 (6), 1305–1318.

L.B. Silva Thomsen, K. Anastasakis, P. Biller, 2022, Wet oxidation of aqueous phase from hydrothermal liquefaction of sewage sludge, Water Research, 209, 117863.

P. SundarRajan, K. Gopinath, J. Arun, K. GracePavithra, A.A. Joseph, S. Manasa, 2021, Insights into valuing the aqueous phase derived from hydrothermal liquefaction, Renew. Sustain. Energy Rev. 144, 111019.

Antonis Kokossis, Michael C. Georgiadis, Efstratios N. Pistikopoulos (Eds.)
PROCEEDINGS OF THE 33rd European Symposium on Computer Aided Process Engineering
(ESCAPE33), June 18-21, 2023, Athens, Greece

Process Feasibility for Converting the Waste Plastics to Hydrogen and Methanol

Ali A. Al-Qadri [a], and Usama Ahmed [a,b]*

aDepartment of Chemical Engineering, King Fahd University of Petroleum and Minerals, Dhahran, 31261. Saudi Arabia
b Interdisciplinary Research Center for Hydrogen and Energy Storage, King Fahd University of Petroleum & Minerals, Dhahran 31261, Saudi Arabia
usama.ahmed@kfupm.edu.sa

Abstract

Hydrogen and methanol are among the most vital fuels and chemicals nowadays. Hydrogen is considered as clean fuel because it produces zero carbon emissions. Comparably, methanol is regarded as a low-carbon alternative fuel. The resources of the two fuels are mainly crude oil, coal, and natural gas. Alternative sources of the two fuels are highly needed to meet the increasing energy demand. On the other hand, plastic waste is also a pollutant and drastically affects the environment. Thermochemical recycling and processing of plastic waste could lead to hydrogen and methanol production. Two comparative models have been developed in Aspen Plus to produce hydrogen and methanol from the waste plastics. In base model (case 1), the waste plastic is first gasified using steam as a gasification agent and the produced syngas containing H_2, CO, and CO_2 is introduced to the water gas shift unit to maximize hydrogen production. Finally, the CO_2 was captured to recover the pure hydrogen. In the alternative model (case 2), the integration between plastic gasification and steam methane reforming (SMR) is proposed to increase the syngas yield while minimizing the overall process energy requirements. The two cases have been technically and economically evaluated to determine the best case. The results showed that the process efficiency of case 2 is higher than base case by 3%. The case 2 also showed a potential to reduce the fuel production cost by 51.8% as compared to case 1. The CO_2 specific emissions in case 2 are reduced by 34% as compared to case 1. Overall, it has been seen that integrating the SMR process with plastic gasification for the dual hydrogen and methanol production is less energy intensive and represents better project feasibility.

Keywords: Gasification; SMR; Waste Plastic; Methanol production; GHG Emissions.

Introduction

The global production of the plastics over the past 70 years is more than 6.3 billion tons (Okunola A et al., 2019). The huge amount of plastic waste has been contributing to huge environmental problems and distributing the earth ecosystem. The thermochemical conversion of waste plastics is 12%, whereas, 9 % is usually recycled (Okunola A et al., 2019). Therefore, proper recycling of the plastics should be considered to minimize its massive environmental impacts and to utilize them in beneficial applications. The thermochemical recycling of plastic waste is a good option because it eliminates the need for plastic classification and simultaneously produce syngas that could be used to produce several fuels and chemicals with high purity. Plastics in general have a high heating value with high hydrogen and carbon content, thus, syngas with high heating value can be produced by thermochemical conversion. This route offers alternatives to produce clean fuels from carbon-based raw materials. Hydrogen and methanol are some of the essential fuels that could be produced from waste plastics with a minimum carbon emission. Therefore, the recycling of the plastics should be focused in this approach synthesizing hydrogen and methanol from plastic waste via thermochemical route. AlNouss et al. (2020) performed a techno economic analysis on the conversion of carbonaceous fuels to hydrogen testing several feedstocks and several gasification agents. It was concluded that steam as a gasification agent performed better than others in terms of maximizing the hydrogen production with considerable cost. Another related study conducted by Al-Qadri & Ahmed (2022) on the conversion of waste plastics (polyethylene and polypropylene) to hydrogen via steam gasification. They performed the whole technical and economic analysis investigating an alternative model that integrates the steam methane reforming with plastic gasification to maximize the hydrogen production, enhance the syngas quality, and reduce the cost. They found a net increase in the process efficiency by 4% and a reduction in the levelized hydrogen cost by 29%. Ahmed (2021) also performed a technoeconomic analysis on the dual production of hydrogen and methanol from coal, integrating the steam methane reforming with coal gasification. The study was succeeded in enhancing the fuel production by 22% and reduce the fuel cost by 13% when compared to the conventional methodology. Since there is an interest to produce hydrogen and methanol from alternative resource (plastic waste) to meet the energy demand, the integrated design between plastic gasification and steam methane reforming is an option. Therefore, the objective of this work is to develop two case studies that produce hydrogen and methanol from waste plastics (polyethylene blended with polypropylene) and then performing the technoeconomic analysis for different process alternatives.

Modelling and Simulation

Aspen Plus (V12) was used to build the simulation models for the dual production of hydrogen and methanol. The thermodynamic package used in the study is Peng Robinson because it is applicable to high temperature and pressure, and several studies had used it in the gasification field. The composition of polyethylene and polypropylene were defined on the basis of their approximate and ultimate analysis. Table 1 represents the assumptions and the modules used for developing the simulation model. The plastic gasification results were validated with the literature (Saebea et al., 2020). Additionally, the methanol synthesis from syngas was also confirmed in terms of methanol yield (Mccaul, 2019). The steam methane results were also validated based on hydrogen to carbon monoxide ratio (Ghoneim et al., 2016).

Table 1: Design Assumptions for case 1 and case 2

Equipment	Aspen Model	Assumption
Flowrate of plastics	RYield	Plastics = 100 kg/h; Steam/Plastic = 1.25
Gasifier	RGibbs	Entrained flow gasifier 1500 °C and P = 25 bars
Pre-reformer	RStoic	Methane/Plastics (mass basis) = 1.5
Reformer	RGibbs	1200 °C and 25 bar, steam: methane (mass basis) = 1.6; with nickel-based catalyst
H$_2$S Removal	RadFrac	Rectisol process; 0 °C, and 25 bar; H$_2$S remaining = 1 ppm using 20 stages.
Methanol Synthesis	Plug flow reactor	25 bars and 270 °C; Using Graaf.et al kinetics

Water Gas Shift	REquil	Two reactors, Molar steam: CO = 2 using γ-Al$_2$O$_3$-α-Ga$_2$O$_3$ catalysts
CO$_2$ Removal	RadFrac	Rectisol process; 0 °C, and 1 bar CO$_2$ Removal = 99% using 10 stages.

3. Process Description

Two case studies have been developed in this study for dual production of H$_2$ and methanol. Figure 1 represents the base case (case 1) which consists of several units containing steam plastic gasification, H$_2$S removal, methanol synthesis unit, water gas shift (WGS), CO$_2$ removal unit, and solvent regeneration. The syngas produced from gasification unit is treated in the acid gas removal unit to remove the H$_2$S followed by its conversion into methanol synthesis unit, whereas, the remaining gases were fed to the WGS to maximize the hydrogen content followed by the CO$_2$ removal.

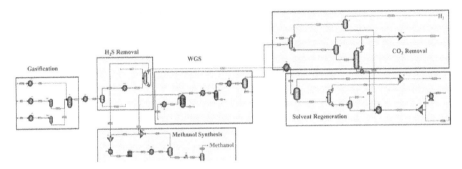

Figure 1: Aspen Plus Model for the Production of H$_2$ and Methanol from Waste Plastics

The case 2 shown in Figure 2 also represents the dual production of H$_2$ and methanol. It is similar to case 1, however, this case integrates the steam methane reforming (SMR) technology with plastic gasification, where, the produced syngas from gasification and reforming unit is mixed to enhance the overall syngas production. Then H$_2$S is removed from the syngas feed to make it suitable for the methanol synthesis unit and the unconverted syngas is recycled back to WGS unit to get the maximum conversion of unconverted CO to H$_2$.

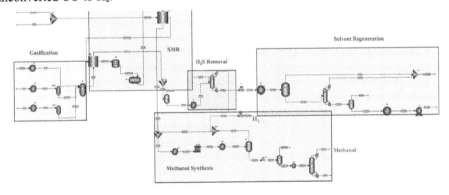

Figure 2: Aspen Plus Model of Waste Plastic Gasification and Methane Reforming models for the Production of H$_2$ and Methanol

4. Results and Discussion

4.1. Hydrogen and Methanol Production

The results for the two cases were analyzed in terms of syngas compositions, hydrogen production, methanol production, CO_2 emissions, and overall process efficiency. The case 2 produces syngas with higher hydrogen content of 9% and higher heating value by 11% compared to case 1. The hydrogen production rate on mass per unit feedstock basis (hydrogen yield) has been calculated as 0.06 and 0.10 for case 1 and case 2, respectively. On the other hand, the methanol production rate on mass per unit feedstock basis for case 1 and case 2 has been calculated as 2.12 and 2.18, respectively. From the results, it has been seen that the second case is more efficient than case 1 in term of producing H_2 and CH_3OH.

4.2 Process performance analysis

The following equation (Eq 1) is used to calculate the overall process efficiency.

$$\text{Process Efficiency} = \frac{H_2 \text{ thermal energy [MWth]} + CH_3OH \text{ thermal energy [MWth]}}{\text{Feed stock thermal energy [MWth]} + \text{Energy consumed [MWth]}} \times 100\% \quad (1)$$

Case 2 utilizes an additional feedstock namely, methane, intaking more feedstock thermal energy of 1995.92 MW. In addition, the second case required 1070.13 MW more utilities for the process needs. While comparing the total energy content, it has been seen that the thermal energy content in terms of produced hydrogen and methanol was higher in the case 2 than case 1 by 2392.96 MW. Developing the heat exchanger network and calculating all the energy requirements using Equation 1, the overall process efficiencies for Case 1 and case 2 is estimated as 73% and 76%, respectively, as represented in the Figure 3. The results also showed that the yield of H_2 and methanol in case 2 is higher than case 1. Additionally, the second case reduces the specific carbon dioxide emissions by 34%. Therefore, case 2 found to be more effective and feasible for dual hydrogen and methanol production from waste plastics. Furthermore, the economic analysis is performed to analyse the overall process feasibility.

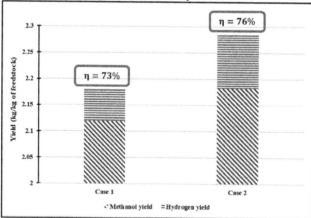

Figure 3: The fuel production and process efficiency for the two cases

5. Project Feasibility

The project feasibility is determined via economic analysis which basically included the capital expenditures and operating expenditures for the two cases. The economic analysis was performed based on several assumptions specified by the previous paper (Al-Qadri et al., 2022).The capital expenditure (CAPEX) mainly encompasses the equipment cost, installation and piping costs, instrumentations, and civil work costs. On the other hand, the operational expenditure (OPEX) consists of utilities, catalyst replacement maintenance, labor costs, and administration costs. The power law (order of magnitude) analysis has been used to estimate the CAPEX, where, the CEPCI of 620 has been used for the year 2022 in the equation (Eq 2).

$$Cost_{New} = Cost_{Old} \times \left(\frac{Capacity_{New}}{Capacity_{Old}}\right)^X \times \frac{CEPCI_{New}}{CEPCI_{Old}} \tag{2}$$

The total investment cost of case 2 is higher than case 1 by 40%. Nevertheless, the total investment cost (TIC) per ton of fuel produced is lower in case 2 than case 1 by 46% which shows the feasibility of the second case in terms of higher fuel yield. The fuel levelized cost over 30 years was reduced in the case 2 by 52% when compared to the base case. The net present value for case1 and case 2 were calculated to be 2.734 M€ and 26.201 M€ respectively. Additionally, the present value ratio for the two cases were 1.614 M€ and 5.301 M€. Table 2 represents the main parameters for the technoeconomic analysis. The economic analysis has been presented in terms of cash flow diagram in Figure 4.

Table 2: Technoeconomic Analysis

Parameter	Units	Case 1	Case 2
Produced hydrogen per feedstock	mass basis	0.06	0.10
Produced methanol per feedstock	mass basis	2.12	2.18
Process Efficiency	%	73	76
Total Investment Cost	€ (10^3)	4070	5690
TIC per ton of H_2 and MeOH	M€/ton	18530	9970
Total OPEX/Year	€ (10^3) / Year	839.1	1032.2
Total OPEX/ton H_2 and MeOH	€ (10^3) / ton	16.4	4.9
Revenue	M€/Year	1.7	5.2
Hydrogen and MeOH Cost	€/kg	0.534	0.257

Case 2 is more attractive compared to the case 1 in terms of cash flow and return on investment. This significant enhancement in the return on investment is attributed to the higher revenue generation capacity of case 2 since it represented higher annual income of 3.526 M€ compared to base case. Case 2 can pass the breakeven point in the fifth year, whereas, case 1 requires 12 years to return the initial investment. Therefore, case 2 has been found to be more economically feasible for dual production of hydrogen and methanol.

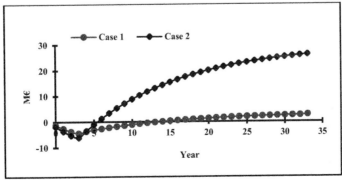

Figure 4: The cash flow diagram for base case and alternative case

Conclusions

The dual production of hydrogen and methanol from waste plastics has been represented in two different cases. The first case represents on plastic gasification design for the production of hydrogen and methanol. On the other hand, case 2 integrates the steam methane reforming with plastic gasification to enhance the syngas heating value with higher hydrogen content. The results revealed that, case 2 produces 65% and 3% more hydrogen and methanol, respectively, compared to the case 1. The overall process efficiency of the case 2 showed an increment of 3% and reduction of CO_2 specific emissions by 34% compared to case 1. Moreover, Case 2 offers an annual revenue of 5.2 M€ compared to case 1, which is only 1.7 M€. The fuel (hydrogen and methanol) production cost has been found to be lower in the case 2 by 51% as compared to case 1. Keeping in view all the preliminary analysis, case 2 has been found to be more techno-economically attractive compared to the case 1.

Acknowledgements

The authors would like to acknowledge support provided by the Deanship of Research Oversight and Coordination (DROC) at the King Fahd University of Petroleum & Minerals (KFUPM) for funding this work through project no. DF201017.

References

Ahmed, U. (2021). Techno-economic analysis of dual methanol and hydrogen production using energy mix systems with CO2 capture. *Energy Conversion and Management, 228*, 113663. https://doi.org/10.1016/J.ENCONMAN.2020.113663

AlNouss, A., McKay, G., & Al-Ansari, T. (2020). Enhancing waste to hydrogen production through biomass feedstock blending: A techno-economic-environmental evaluation. *Applied Energy, 266*. https://doi.org/10.1016/j.apenergy.2020.114885

Al-Qadri, A. A., & Ahmed, U. (2022). Techno-Economic Analysis of the Conversion of Waste Plastics to Hydrogen Fuel. *Computer Aided Chemical Engineering, 51*, 421–426. https://doi.org/10.1016/B978-0-323-95879-0.50071-0

Al-Qadri, A. A., Ahmed, U., Abdul Jameel, A. G., Zahid, U., Usman, M., & Ahmad, N. (2022). Simulation and Modelling of Hydrogen Production from Waste Plastics: Technoeconomic Analysis. *Polymers, 14*(10). https://doi.org/10.3390/POLYM14102056

Ghoneim, S. A., El-Salamony, R. A., & El-Temtamy, S. A. (2016). Review on Innovative Catalytic Reforming of Natural Gas to Syngas. *World Journal of Engineering and Technology, 04*(01), 116–139. https://doi.org/10.4236/wjet.2016.41011

Mccaul, S. C. (2019). *Simulation and Techno-Economic Analysis of the AChT Green Methanol Process.*

Okunola A, A., Kehinde I, O., Oluwaseun, A., & Olufiropo E, A. (2019). Public and Environmental Health Effects of Plastic Wastes Disposal: A Review. *Journal of Toxicology and Risk Assessment, 5*(2). https://doi.org/10.23937/2572-4061.1510021

Saebea, D., Ruengrit, P., Arpornwichanop, A., & Patcharavorachot, Y. (2020). Gasification of plastic waste for synthesis gas production. *Energy Reports, 6*, 202–207. https://doi.org/10.1016/j.egyr.2019.08.043

Antonis Kokossis, Michael C. Georgiadis, Efstratios N. Pistikopoulos (Eds.)
PROCEEDINGS OF THE 33rd European Symposium on Computer Aided Process Engineering
(ESCAPE33), June 18-21, 2023, Athens, Greece

Process development for efficient hydrogen transportation and analyses: technical, economic, and environmental perspectives

Byeongchan Ahn,[a] J. Jay Liu,[b] Wangyun Won[a],*

[a] Department of Chemical Engineering (Integrated engineering), Kyung Hee University, 1732 Deogyeong-daero, Giheung-gu, yongin-si, Gyeonggi-do 17104, Republic of Korea
[b] Department of Chemical Engineering, Pukyong National University, 45, Yongso-ro, Nam-gu, Busan 48513, Republic of Korea
*corresponding author (e-mail: wwon@khu.ac.kr)

Abstract

Here, we have designed an integrated hydrogen transport process including hydrogenation and dehydrogenation processes using a toluene-methylcyclohexane (MCH) system. The toluene-MCH system is one of the promising liquid organic hydrogen carrier (LOHC) candidates because it is relatively less toxic and an economical option. For efficient energy utilization, we have carried out heat integration using pinch analysis. To determine the minimum hydrogen transport cost, techno-economic analysis was performed using discounted cash flow analysis. It was estimated that the minimum hydrogen (H_2) transport cost is \$2,083/kg when 400,000 tons of hydrogen is transported per annum. The sustainability of the proposed process was determined through life-cycle assessment quantifying the direct and indirect carbon emissions. This study can give insight into technological development for the realization of the hydrogen economy.

Keywords: Eco-friendly, Hydrogen storage-transportation, Clean energy, Separation process, comprehensive analysis

1. Introduction

Recent climate crisis has prompted researchers to take an interest in the transition from a petrochemical-based energy society to a renewables-based energy society. However, regional inconsistency and intermittent energy generation make the commercialization of renewable energy complicated. To resolve this drawback, hydrogen is attracting attention as a carbon-free energy carrier. Although the high gravimetric energy density of hydrogen (120 MJ/kg) is suitable for an energy carrier, its low volumetric energy density (10.9 kJ/L) is a major contributor to increase transport costs (Markiewicz et al., 2019).

There are two major methods for efficient hydrogen storage: physical storage and chemical storage. For physical storage, compression and liquefaction are mainly used. Although they have the advantage of relatively simple processes, their strict storage conditions (400~700 bar and -253°C, respectively) entail safety accident risk and require lots of energy causing indirect carbon emissions. Chemical storage means storing hydrogen in other chemicals by forming chemical bonds. Chemical storage requires sophisticated separation processes to recover high-purity hydrogen, but it consumes relatively low energy (Singh et al., 2021; Teichmann et al., 2011).

Liquid organic hydrogen carriers (LOHCs) are promising candidates for chemical storage systems due to their high hydrogen storage capacity. Furthermore, LOHCs can store hydrogen in the liquid state at ambient conditions with relatively low energy loss and has similar properties with petroleum derivates, which makes it available to use existing petrochemical infrastructures. LOHCs are well known in industrial set-ups and so they are easy to produce in large quantities.

Among LOHCs, the toluene-methylcyclohexane (MCH) system was selected for the designed process due to its relatively high hydrogen storage capacity (6.2 wt.%) (Niermann et al., 2019). It also has a relatively affordable price and reversible hydrogenation reaction which makes toluene recycle after dehydrogenation to reduce the hydrogen transport cost (Preuster et al., 2018). However, toluene and MCH have high volatilities that make it hard to separate hydrogen from unwanted organics.

In this study, we have designed the integrated hydrogen transport process composed of two sections (hydrogenation and dehydrogenation sections) based on the experiment data from the literature. Especially, the sophisticated separation process including the extraction column and pressure swing adsorption (PSA) column was designed considering the volatile property of toluene and MCH. We have also made heat networks to use waste energy efficiently reducing indirect carbon emissions. Then, techno-economic analysis (TEA) and life-cycle assessment (LCA) were performed to evaluate economic feasibility and sustainability.

2. Technology Overview

Toluene in the proposed design is converted to MCH by a hydrogenation reaction and converted back to toluene by a dehydrogenation reaction. For the hydrogenation reaction, we have used a commercially available Ni-SiO$_2$ catalyst at 120°C with excess hydrogen condition. Although the experimental data from the literature was originally based on ambient pressure, we designed it to react at higher pressure (370 psi) (Atsumi et al., 2020). Generally, hydrogenation reaction at higher pressure results in higher conversion, but we assumed that it has the same yield at ambient pressure due to no experimental data at higher pressure. Instead, the high-pressure reaction can reduce the cost required for pressurization when separating unreacted hydrogen. Unlike the hydrogenation reaction, the dehydrogenation reaction has a higher conversion at lower pressure. MCH dehydrogenates under 1atm at 310°C (Okada et al., 2006). More details including stoichiometric equations and yields are specified in Table 1.

Table 1. Summary of reaction condition.

Reaction	Reaction	Yield
Hydrogenation	Toluene + 3 H$_2$ → MCH	0.99
	Toluene + H$_2$ → CH$_4$ + Benzene	0.01
Dehydrogenation	MCH → Toluene + 3 H$_2$	0.9491
	MCH → CH$_4$ + Benzene + 2 H$_2$	0.0002
	MCH + H$_2$ → Others	0.0008

3. Method

3.1. Process synthesis

The process is developed to target the white birch handling of 2,000 dry tons per day which is the commercial scale of biorefinery. The process is simulated using the Aspen Plus process simulator V11 and the Soave-Redlich-Kwong (SRK) equation as the thermodynamic method. The heat networks are created based on the pinch analysis method using the Aspen Energy Analyzer V11.

3.2. Techno-economic analysis

The cost of equipment is calculated using the Aspen Process Economic Analyzer V11, and the equipment and material costs are scaled to unify the base year of 2018 using the Chemical Engineering Plant Cost Index. Using the scaled costs, the minimum transport cost is estimated as an indicator of economic feasibility by discounted cash flow method when the internal rate of return is assumed to be 10%. The economic parameters are derived from the National Renewable Energy Lab report, and major economic parameters are summarized in Table 2 (Davis et al., 2018).

Table 2. Major assumed economic parameters.

Economic parameter	Value	Economic parameter	Value
Plant financing by equity	40.0%	Plant life	30.0 years
Income tax rate	21.0%	General plant depreciation	7.0 years
Interest rate for dept financing	8.0%	Steam/electricity plant depreciation	20.0 years
Term for dept financing	10.0 years	On-stream percentage	90.0%

3.3. Life cycle assessment

LCA is performed by following the international guideline presented by the International Organization for Standardization, which is constituted of four steps: 1) goal and scope definition, 2) inventory analysis, 3) impact analysis, and 4) interpretation. The system boundary is defined from the production of input materials and energies (cradle) to the production of hydrogen from the dehydrogenation section (gate), and the functional unit is defined as 1 kg of transported hydrogen. The environmental impacts were estimated using SimaPro V9.1 with the EcoInvent 3.6 database with ReCiPe 1.13V of mid-point from the hierarchical perspective. The environmental impacts were calculated based on the functional unit and allocated based on the mass production rate of the products.

4. Process design

4.1. Process synthesis

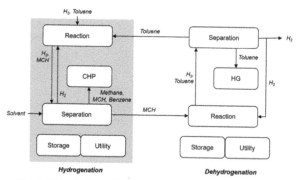

Fig 1. Block flow diagram of the proposed process.

In this study, we designed the integrated process using the Aspen Plus process simulator assuming that 400,000 tons of hydrogen are transported per annum. The integrated process consists of two sections such as hydrogenation and dehydrogenation sections which are 10,000 km apart. including five subsystems respectively. The hydrogenation section is constituted of 1) Reaction, 2) Separation, 3) Combined heat and power (CHP), 4) Storage, and 5) Utility. The dehydrogenation section has the same subsystems except for CHP. Instead of CHP, the dehydrogenation section has a heat generation (HG) subsystem because electricity is not produced in the dehydrogenation section.

4.1.1. Hydrogenation section

Toluene supplemented to compensate for the toluene loss and transported from the dehydrogenation section is hydrogenated to MCH in the reaction subsystem. Since the hydrogenation reaction occurs in the excess hydrogen condition, hydrogen and MCH need to be separated to reduce hydrogen loss. Hydrogen can be separated by depressurization and cooling using the boiling point difference between hydrogen and MCH. Benzene produced as a byproduct from the hydrogenation reaction is easily accumulated in the proposed process. However, eliminating benzene is challenging because its physical property is similar to MCH. Therefore, the extraction column was introduced using N-formylmorpholine (NFM) as a solvent (Chen et al., 2007). The benzene-removed MCH is transported to the dehydrogenation section, and the used NFM is recycled to the extraction column after separating from the benzene using a distillation column.

The separated byproducts are utilized as energy source in the CHP subsystem. After using the heat required for the hydrogenation section, the remaining heat is used to produce the electricity which can be sold as a byproduct or used to produce hydrogen through water electrolysis (Sen et al., 2012). To reduce CO_2 emissions, carbon capture units are also installed in the CHP subsystem (Lee et al., 2021)

4.1.2. Dehydrogenation section

The transported MCH is dehydrogenated in the reaction subsystem with some hydrogen. Since hydrogen (5-20 mol% of MCH) can improve the life period of the catalyst and the yield, the purified hydrogen from the separation subsystem is recycled to the dehydrogenation reactor. Further, the produced hydrogen is purified in the separation subsystem by pressurization and PSA. Firstly, volatile toluene is liquefied at high pressure and low temperature to recover toluene, then the remaining organics are removed by PSA.

Byproducts are utilized as a heat source like the hydrogenation section to supply the heat required for the dehydrogenation section. However, since the dehydrogenation reaction is endothermic, it is difficult to meet the heat required by byproducts alone. Therefore, some of the produced hydrogen is also burned in the HG subsystem. The HG subsystem also has carbon capture units like the hydrogenation section.

4.2. Heat integration

To lower the operating costs and to efficiently use energy, we have carried out heat integration using pinch analysis. As we assumed that the hydrogenation and dehydrogenation sections are far away, heat integration is conducted separately for each section, and the results are summed up and shown in Figure 2.

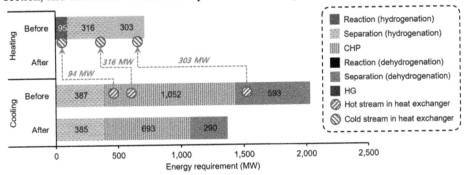

Fig. 2. The result of heat integration and major heat networks

The total heating requirement for the proposed process was 715 MW before heat integration. Most of the heating requirement (44%) was used for separation in the hydrogenation section. However, the heating requirement became zero after heat integration. The hydrogenation section has sufficient heat sources such as the heat from CHP and the hydrogenation reaction which is exothermic. The heat required for the dehydrogenation section is satisfied by burning produced hydrogen.

For the cooling requirement, most of the energy (52%) was required for the power cycle of the CHP subsystem. To liquify the steam used to generate electricity, a huge amount of cooling water was required. Furthermore, the large flow rate of the power cycle made it difficult to form heat networks completely. As a result, only 33% of the cooling requirement was decreased.

5. Result and Discussion

5.1. Techno-economic analysis

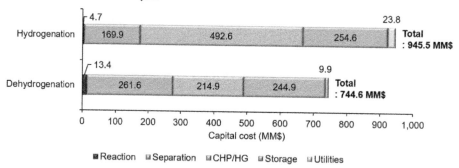

Fig. 3. The result of capital cost breakdown.

As shown in Fig. 3, the hydrogenation section requires a larger capital cost than the dehydrogenation section mainly because of the CHP subsystem. The generator in the power cycle is the most cost-intensive equipment due to the large flow rate. For the dehydrogenation section, the separation subsystem is the most cost-intensive subsystem. Multi-compressor was used to recover toluene, but the high flow rate of a mixture stream required a large-size compressor accounting for a major fraction of the capital cost in the separation subsystem.

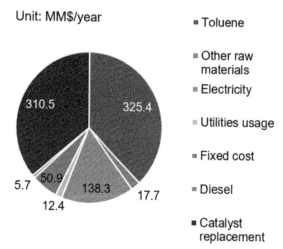

Fig. 4. Operating cost breakdown

As shown in Fig. 4, most of the operating costs come from toluene supplementation and catalyst replacement. Although toluene was recycled in the proposed process, some of the toluene was lost in the separation process of each section. Most of the catalyst replacement cost (89%) comes from the dehydrogenation catalyst because of the high price of platinum. The cost of electricity also accounted for a large portion. While electricity can be self-sufficient in the hydrogenation section because it generates electricity on its own in the CHP subsystem, all electricity required for the dehydrogenation section have to be purchased externally.

5.2. Minimum transport cost

We have calculated the minimum hydrogen transport cost which is the value when the total cost of the process and revenues including byproduct (electricity and captured CO_2) incomes are identical. Fig. 5 shows the total cost and revenue, and the portion of each inventory.

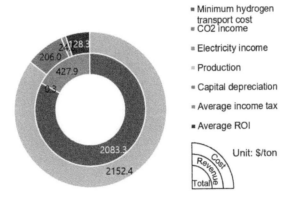

Fig. 5. Minimum transport cost of the process

The total cost of the proposed process is \$2,511/ton-H_2, and the byproduct incomes are \$428/ton-H_2. Therefore, the minimum hydrogen transport is found to be \$2083/ton-H_2.

5.3. Life cycle assessment

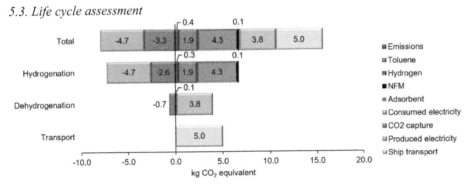

Fig. 6. The result of life-cycle assessment.

The life-cycle assessment was performed to determine the environmental impact of each section. Fig. 6 shows each section's contribution to climate change based on CO_2 emission. The hydrogenation section produced the most carbon emissions mainly because of the solar energy needed for hydrogen production. Though solar energy is sustainable, noticeable amounts of carbon emissions occur when constructing solar power plants. However, since produced electricity and captured from the CHP subsystem are environmentally beneficial, most of the carbon emissions in the hydrogenation section can be offset. The major cause of carbon emission is electricity used in the separation subsystem. The multi-compressor for pressurization consumes 98% of total electricity consumption.

Therefore, carbon emission would be further reduced if pressurized in a less energy-intensive manner.

6. Conclusion

In this study, we have proposed an integrated process for long-distance hydrogen transport using a toluene-MCH system and designed heat networks reducing 100% of heating requirement and 33% of cooling requirement. Based on the designed process and heat networks, TEA and LCA were conducted to study economic feasibility and sustainability of the proposed process. It was found that the hydrogenation section incurs higher capital costs than the dehydrogenation section mainly due to the power cycle of the CHP subsystem. It also found that toluene supplementation and catalyst replacement are the major cost contributors. From a sustainable point of view, although more carbon emissions occur in the hydrogenation section compared to the dehydrogenation section, most of the carbon emissions are offset because of the environment friendliness of the produced electricity and captured CO_2 from the CHP subsystem. The carbon emissions of the dehydrogenation section can be decreased by reducing electricity consumption.

References

R. Atsumi, K. Kobayashi, C. Xieli, T. Nanba, H. Matsumoto, K. Matsuda, and T. Tsujimura, 2020, Effects of steam on toluene hydrogenation over a Ni catalyst, Applied Catalysis A: General, 590, 117374.

D. Chen, H. Ye, and H. Wu, 2007, Liquid–liquid equilibria of methylcyclohexane–benzene–N-formylmorpholine at several temperatures, Fluid phase equilibria, 255, 2, 115-120.

R. E. Davis, N. J. Grundl, L. Tao, M. J. Biddy, E. C. Tan, G. T. Beckham, D. Humbird, D. N. Thompson, and M. S. Roni, 2018, Process Design and economics for the conversion of lignocellulosic biomass to hydrocarbon fuels and coproducts: 2018 biochemical design case update; biochemical deconstruction and conversion of biomass to fuels and products via integrated biorefinery pathways. NREL/TP-5100-71949.

S. Lee, H. S. Kim, J. Park, B. M. Kang, C.-H. Cho, H. Lim, and W. Won, 2021, Scenario-Based Techno-Economic Analysis of Steam Methane Reforming Process for Hydrogen Production, Applied Sciences, 11, 13, 6021.

M. Markiewicz, Y.-Q. Zhang, M. T. Empl, M. Lykaki, J. Thöming, P. Steinberg and S. Stolte. 2019, Hazard assessment of quinaldine-, alkylcarbazole-, benzene- and toluene-based liquid organic hydrogen carrier (LOHCs) systems, Energy & Environmental Science, 12, 1, 366-383.

M. Niermann, A. Beckendorff, M. Kaltschmitt, and K. Bonhoff, 2019, Liquid Organic Hydrogen Carrier (LOHC)–Assessment based on chemical and economic properties, International Journal of Hydrogen Energy, 44, 13, 6631-6654.

Y. Okada, E. Sasaki, E. Watanabe, S. Hyodo, and H. Nishijima, 2006, Development of dehydrogenation catalyst for hydrogen generation in organic chemical hydride method, International Journal of Hydrogen Energy, 31, 10, 1348-1356.

P. Preuster, Q. Fang, R. Peters, R. Deja, L. Blum, D. Stolten, and P. Wasserscheid, 2018, Solid oxide fuel cell operating on liquid organic hydrogen carrier-based hydrogen–making full use of heat integration potentials, International journal of hydrogen energy, 43, 3, 1758-1768.

S. M. Sen, D. M. Alonso, S. G. Wettstein, E. I. Gürbüz, C. A. Henao, J. A. Dumesic, and C. T. Maravelias, 2012, A sulfuric acid management strategy for the production of liquid hydrocarbon fuels via catalytic conversion of biomass-derived levulinic acid, Energy & Environmental Science, 5, 12, 9690-9697.

R. Singh, M. Singh, and S. Gautam, 2021, Hydrogen economy, energy, and liquid organic carriers for its mobility, Materials Today: Proceedings, 46, 5420-5427.

D. Teichmann, W. Arlt, P. Wasserscheid, and R. Freymann, 2011, A future energy supply based on Liquid Organic Hydrogen Carriers (LOHC), Energy & Environmental Science, 4, 8, 2767-2773.

Antonis Kokossis, Michael C. Georgiadis, Efstratios N. Pistikopoulos (Eds.)
PROCEEDINGS OF THE 33rd European Symposium on Computer Aided Process Engineering (ESCAPE33), June 18-21, 2023, Athens, Greece

Integration of metabolic models in biorefinery designs using superstructure optimisation

Lucas Van der Hauwaert[a]; Alberte Regueira[a,b], Miguel Mauricio-Iglesias[a]

[a] CRETUS, Department of Chemical Engineering. Universidade de Santiago de Compostela, 15782, Santiago de Compostela, Spain.
[b] Center for Microbial Ecology and Technology (CMET), Ghent University, B-9000, Ghent, Belgium.

Abstract

In this work the integration of metabolic (structured) models, which describe the entire metabolic network of microorganisms, into a superstructure optimisation framework is presented. These models are of particular interest because they can predict the outcome of a fermentation with different substrates, and without the need of experimental data. In this contribution, a workflow for 2 types of structured models is described: i) genome-scale metabolic models (GEMs) and ii) community models where product yields can be influenced by environmental factors (e.g., pH). To showcase this methodology a simple case study is presented of a superstructure optimisation problem. With this case study it is demonstrated that the described workflow can aid the design of novel biorefineries, by screening potential cultures using metabolic models without the need of experimental data.

Keywords: Structured models, biorefinery design, superstructure optimisation, surrogate models, Genome-scale metabolic models

1. Introduction

A current hurdle towards the implementation of biorefineries are the various process alternatives which lead to a multitude of possible refinery designs. To aid the design of biorefinery networks, superstructure optimisation tools, like O2V have been created (Gargalo et al., 2022). A superstructure represents all the possible links between process units (and their possible substitutes) formed by flowchart and logic rules, where alternative topologies of processing networks can be visualized. Optimisation techniques can be applied to this superstructure to generate an optimal processing network according to a desired objective (e.g., maximizing profit) (Bertran et al., 2017).

In a superstructure the unit processes are represented by mathematical models, describing the flow and/or transformation of mass. For biorefineries, key unit processes are bioreactors, responsible for the chemical transformation of feedstocks. Metabolic models, simulating intra- and extra-cellular spaces, are commonly used for simulating biotransformations. These models can predict the microbial products from different substrates and can be linked to several inputs, or outputs of other reactors, thereby increasing the amount of new possible network topologies. Additionally, the creation of these models is not dependant on data from time consuming fermentation experiments.

A common type of metabolic model are genome-scale metabolic models (GEMs). A GEM comprehensively reconstructs the organism's entire metabolic network using its genome (Lieven et al., 2020). Another type of metabolic model are community models, such as the bioenergetic model of Regueira et al. (2020). These models are of interest because of their ability to describe open and co-culture fermentations with different substrates and at different operational conditions (e.g., different pH values). Additionally open culture fermentations are interesting to include because they have economic and operational advantages, such as avoiding sterilisation and being able to handle complex substrates (Kleerebezem et al., 2015).

Unfortunately, structured bioreactor models are often too complex to feasibly incorporate into a superstructure and thus are often not considered. In other words, the integration of GEMs or community models in a superstructure problem is not strait forward. For this reason, the aim of this contribution is to create a workflow to incorporate these models in a superstructure optimization problem. To showcase this workflow a simple superstructure optimization problem is presented to produce propionate and acetate where GEMs and community models are incorporated as surrogate models.

2. Methodology

The workflow for the integration of two types of models into a superstructure optimisation problem, by creating surrogate models, will be discussed. The first type of surrogate model is derived from GEMs, where the yield of a product is only dependant on the selected substrate. The second are derived from community models where the yield is dependent on external influences (e.g., pH) other than the type of substrate.

2.1. Workflow

A flowchart to incorporate the 2 types of structured models into a superstructure optimization problem is shown in Figure 1. The goal of this flowchart is to find a mathematical expression (i.e., a reaction equation) that can be used in a superstructure optimisation problem, starting from a community model or a GEM. These reaction equations represent the chemical transformation of mass, hence linking the inputs to the outputs (i.e., a yield) of the bioreactors.

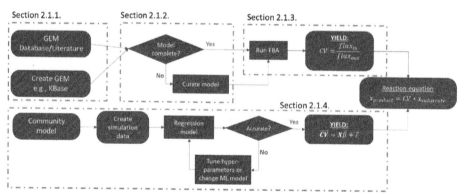

Figure 1. Flowsheet to make reaction equations (i.e., surrogate models) from structured models for superstructure optimisation problems

2.1.1. Making/finding GEMs

To generate a GEM of a particular microorganism(s), software tools and genetic data available on platforms such as KBase and MetaNetX can be used, which enable the creation of metabolic network reconstructions (Arkin et al. 2018). Alternatively, already existing GEMs can be downloaded from databases such as BiGG Models, BioModels, or published work in literature.

2.1.2. Validating GEMs

Many GEMs often lack the required quality to be used reliably and need to undergo a quality assurance and further manual curation. Here the software tool MEMOTE which is an open-source software, representing a unified approach to ensure the formally correct definition of GEMs, was used to check the quality (Lieven et al., 2020). The criteria used to assure the quality of the GEMs are the following: 1) the percentage of metabolic reactions where the chemical elements are correctly balanced; 2) the percentage of metabolic reactions that have a correct charge balance; 3) the percentage of dead-end metabolites (i.e. internal metabolites which do not participate in the internal metabolism); and 4) a leak test, which tests if biomass is created without any substrate. If any of the criteria are not met, the model must be manually curated. For this curation the reader is referred to the work of Thiele & Palsson (2010).

The percentage of correct mass and charge balanced reactions should obviously be as high as possible. In this work a threshold of 95% of correctly mass and charge balanced reactions is chosen as a selection threshold for GEMs. This threshold was taken to minimise the time needed to curate models, as manually correcting the reactions is a very time-consuming procedure. The percentage of dead-end metabolites should be kept as low as possible but are not *per se* indicators of low-quality models. However, a large proportion of dead-end metabolites (e.g., >50%) may indicate problems in reconstruction that need solving.

All non-balanced reactions and the dead-end metabolites need to be checked, and manually corrected if they take part in essential reactions (i.e., essential for the formation of a product). Finally, models producing biomass without substrates indicate an incorrect model formulation and need to be checked thoroughly.

2.1.3. Incorporating GEMs into the superstructure problem

For GEMs, yields are found by solving a flux balance analysis (FBA). An FBA returns the flux of all the metabolites flowing through the metabolic network at steady state (Orth et al., 2010). Using the flux data from the model solution, yields are determined as the ratio of the produced extracellular metabolites and the substrate.

2.1.4. Incorporating community models into the superstructure problem

For community models where the yield is dependent on environmental influences, machine learning (ML) techniques can be used to find simplified algebraic relationships between the yield and the operational variables (e.g., varying pH) (Vollmer et al., 2022).

To create the data needed to calibrate such models, simulations need be to run under varying operational conditions, hence covering the input space. With this data, polynomial equations describing the yield are fitted using lasso regression. This regression technique is used because it effectively avoids overfitting and has a simple algebraic formulation which can easily be handled by superstructure optimization algorithms (such as BARON). To validate the equations, parity plots are made to compare the observed and predicted values of the yields. The normalised mean standard error (NMSE) is also used as a validation metric to make sure the surrogate models reproduce the original model sufficiently. If an insufficient quality of the surrogate model is attained, calibration parameters should be tuned (e.g., the hyperparameter α determining the degree of regularisation) or the number of polynomials can be changed. If the obtained model still does not sufficiently reproduce the original model, a different ML technique should be used (Vollmer et al., 2022).

2.2. Superstructure model

The objective, the flow of mass and logic rules (i.e., the selection of a path in the process network) of the superstructure are represented as mathematical equations (Eq. 1 to Eq. 6):

$$\max f(x, y) \quad (1) \qquad g(x, y) \geq 0 \quad (2) \qquad h(x, y) = 0 \quad (3)$$

$$x \in X \quad (4) \qquad x^{LO} \leq x \leq x^{UP} \quad (5) \qquad y \in \{0; 1\}^n \quad (6)$$

where f is the objective function (e.g., maximising profit), x is the vector of continuous variables representing operating variables (e.g., the flow of mass) defined by their lower and upper bounds x^{LO} and x^{UP} in a continuous feasible region X, y is the vector of binary variables representing the selection of a unit-process, g and h are the vectors of inequality and equality constraints, representing the process models and process specifications of a unit process. The equations for the process models are represented by generic process intervals. Every interval in the superstructure is modelled with the same set of generic equations, representing a sequence of processing tasks, namely mixing, reaction, waste removal and product separation, as well as utility consumption (Bertran et al. 2017). The reaction equations are mass conversions between inputs and outputs to a reaction step and are represented by the surrogate models of the structured models.

2.3. Case study

The goal of this simplified case study is to find out which type of microbial culture(s), producing propionate and acetate, can generate the largest profit from 2 alternative feedstocks being fructose or glucose. Here 5 different *Propionibacterium* species *(P. acidipropionici, P. acnes, P. avidum, P. freudenreichii,* and *P. propionicum)* and an open fermentation are considered (Figure 2). The surrogate models for the pure cultures are obtained from GEMs from the works of McCubbin et al. (2020), while the open fermentation surrogate model was derived from the bioenergetic model of Regueira et al. (2020). All the cultures use either glucose or fructose as substrate, except for the bioenergetic model that can only consider glucose. In this example, the profit is defined as the revenue generated through sales of the products (i.e., acetate and propionate), minus the operating cost being the cost of raw material (i.e., glucose and fructose). The wholesale prices of the raw materials were found on SELINA WAMUCII (https://www.selinawamucii.com) while the wholesale prices for the products were found on ECHEMI (https://www.echemi.com). Furthermore, logic constraints were defined to ensure the selection of only one substrate and one type of culture. To further simplify the case study, no utility consumption, waste removal or mixing was considered, and product separation is assumed to be perfect after all reactors.

The formulation of the superstructure was created using PYOMO (version 6.4.2), in the Python environment (version 3.10.1). The formulation of this problem results in a multiple integer non-linear program (MINLP) and was solved using outer-approximation and generalized Benders decomposition (DICOPT) which is accessible through GAMS (GAMS Development Corporation, 2021).

3. Results and Discussion

3.1. Validating the GEMs

The first step after finding the GEMs form literature, is to assess their quality. The models have been previously curated by McCubbin et al. (2020) and appear to have a good quality where at least 95 % of all reaction are mass and charge balanced. The few metabolic reactions which were not mass or charged balanced were found to not affect essential reactions producing the products. Furthermore, the percentage of dead-end metabolites are all low (< 5%) and no biomass was created from any of the models without substrate.

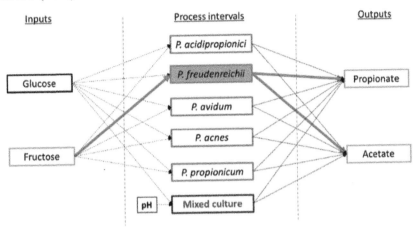

Figure 2. Superstructure representation of the case study. □: process intervals derived from GEMs; □: process intervals derived from community models; → and ■: optimal processing route.

3.2. Creating the surrogate reaction models
With the quality of the GEMs assured, the yields can be found (Table 1) by taking ratio of the substrate and product fluxes obtained by solving the FBA for each GEM.

Table 1. Yields (g-carbon$_{product}$/g-carbon$_{substrate}$) of the different substrates (G: Glucose, F: Fructose) and products (P: Propionate, A: Acetate) resulting from the flux balance analysis

Species\yield	P/G	A/G	P/F	A/F
P. acidipropionici	0.32	0.11	0.32	0.11
P. acnes	0.46	0.07	0.48	0.03
P. avidum	0.36	0.05	0.33	0.11
P. freudenreichii	0.34	0.13	0.42	0.16
P. propionicum	0.29	0.15	0.27	0.17

To find the reaction equations for the open fermentation, the bioenergetic model was run to create datapoints of the fermentation outcome at different pH, which were sampled uniformly between 4 and 8.5. Using lasso regression, 5th order polynomial equations were obtained to express the yields, with pH being the sole variable of the equations. The NMSE values of the fitted equations were low (0.04 g-carbon$_{acetate}$/g-carbon$_{glucose}$ for the yield of acetate and 0.05 g-carbon$_{propionate}$/g-carbon$_{glucose}$ for propionate) and the parity plots of the predicted and observed yields closely resemble each other. With these observations it can be concluded that the used regression technique sufficiently reproduces the bioenergetic model and can thus be used in the superstructure model.

3.3. Solution of the superstructure
The solution of the superstructure (Figure 2) resulted in fructose being the preferred feedstock with *P. freudenreichii* being the optimal culture to produce acetate and propionate. In this process network 100 kg-C of fructose is transformed into 42.45 kg-C of propionate and 15.56 kg-C of acetate resulting in a profit of 32.39 €.

4. Conclusion

A workflow to incorporate structured models into superstructure optimization problems was developed and successfully showcased using a simplified/conceptual case study. With this workflow more intricate superstructure problems can be defined (starting from different available feedstocks) which can result in the design of novel biorefineries. More specifically this workflow allows for a screening of different microbial cultures in biorefinery designs. This includes open fermentations which are key to handling complex substrates in a biorefinery design.

References

Arkin, A. P., Cottingham, R. W., Henry, C. S., Harris, N. L., Stevens, R. L., Maslov, S., Dehal, P., Ware, D., … Yu, D. (2018). KBase: The United States department of energy systems biology knowledgebase. In *Nature Biotechnology* (Vol. 36, Issue 7, pp. 566–569). Nature Publishing Group. https://doi.org/10.1038/nbt.4163

Bertran, M. O., Frauzem, R., Sanchez-Arcilla, A. S., Zhang, L., Woodley, J. M., & Gani, R. (2017). A generic methodology for processing route synthesis and design based on superstructure optimization. *Computers and Chemical Engineering*, 106, 892–910. https://doi.org/10.1016/j.compchemeng.2017.01.030

GAMS Development Corporation. General Algebraic Modeling System (GAMS) Release 36.1.0, Fairfax, VA, USA, 2021

Gargalo, C. L., Rapazzo, J., Carvalho, A., & Gernaey , K. v. (2022). Optimal Conversion of Organic Wastes to Value-Added Products: Toward a Sustainable Integrated Biorefinery in Denmark. *Frontiers in Chemical Engineering*, 4, 54. https://doi.org/10.3389/fceng.2022.837105

Kleerebezem, R., Joosse, B., Rozendal, R., & van Loosdrecht, M. C. M. (2015). Anaerobic digestion without biogas? *Reviews in Environmental Science and Bio/Technology*, 14(4), 787–801. https://doi.org/10.1007/s11157-015-9374-6

Lieven, C., Beber, M. E., Olivier, B. G., Bergmann, F. T., Ataman, M., Babaei, P., Bartell, J. A., Blank, L. M.., … Zhang, C. (2020). MEMOTE for standardized genome-scale metabolic model testing. *Nature Biotechnology*, 38(3), 272–276. https://doi.org/10.1038/s41587-020-0446-y

McCubbin, T., Gonzalez-Garcia, R. A., Palfreyman, R. W., Stowers, C., Nielsen, L. K., & Marcellin, E. (2020). A pan-genome guided metabolic network reconstruction of five propionibacterium species reveals extensive metabolic diversity. *Genes*, 11(10), 1–26. https://doi.org/10.3390/genes11101115

Orth, J. D., Thiele, I., & Palsson, B. O. (2010). What is flux balance analysis? In *Nature Biotechnology* (Vol. 28, Issue 3, pp. 245–248). https://doi.org/10.1038/nbt.1614

Regueira, A., Lema, J. M., Carballa, M., & Mauricio-Iglesias, M. (2020). Metabolic modeling for predicting VFA production from protein-rich substrates by mixed-culture fermentation. *Biotechnology and Bioengineering*, 117(1), 73–84. https://doi.org/10.1002/bit.27177

Thiele, I., & Palsson, B. (2010). A protocol for generating a high-quality genome-scale metabolic reconstruction. *Nature Protocols*, 5(1), 93–121. https://doi.org/10.1038/nprot.2009.203

Vollmer, N. I., Al, R., Gernaey, K. v., & Sin, G. (2022). Synergistic optimization framework for the process synthesis and design of biorefineries. *Frontiers of Chemical Science and Engineering*, 16(2), 251–273. https://doi.org/10.1007/s11705-021-2071-9

Antonis Kokossis, Michael C. Georgiadis, Efstratios N. Pistikopoulos (Eds.)
PROCEEDINGS OF THE 33rd European Symposium on Computer Aided Process Engineering
(ESCAPE33), June 18-21, 2023, Athens, Greece

Environmental Impacts of Bio-derived Silicon: Uncertainty in the Benefit of Industrial Transition

Ethan Errington,[a] Miao Guo,[b] Jerry Heng,[b]

[a] Department of Chemical Engineering, Imperial College London, London, SW7 2AZ, UK
[b] Department of Engineering, King's College London, London, WC2R 2LS, UK
miao.guo@kcl.ac.uk (MG); jerry.heng@imperial.ac.uk(JH)

Abstract

Millions of tonnes of silicon are produced each year as part of a global supply chain supporting the manufacture of steel, automobiles, electronics and silicones. However, the supply chain does have its challenges – particularly the high energy intensity of silicon purification methods, and the lack of global supply chain diversity. Consequently, this study investigates the benefit of producing high purity silicon metals (>98 wt% Si) from agricultural biomass wastes from an environmental perspective using Life Cycle Assessment methodology. Taking rice husk as a case study, findings show a reduction in carbon emissions is achievable for co-recovery of silicon with bio-energy. The robustness of these findings is also addressed through uncertainty modelling, which provides further confidence in the major findings. Future agricultural sources of biomass-derived silicon have also been projected for staple crops including corn, sugarcane and wheat.

Keywords: silicon, rice husk, life cycle assessment, global warming potential, uncertainty.

1. Introduction

8.5 M-tonnes of silicon is produced annually to meet demand for iron alloys in the steel industry, aluminum alloys in the automotive industry, semiconducting materials for electronics products, and a range of wider industrial products such as silicones, silicon carbide and silanes (USGS, 2021). Consequently, a large silicon supply chain (SSC) exists to meet global demand by recovering silicon from mineral deposits of silica (SiO_2). However, the SSC has its challenges. First, mineral excavation introduces a range of environmental impacts. Second, commercial methods for silicon refinement are energy intensive (Maurits, 2014), making sectoral decarbonization challenging. Third, the SSC lacks diversity as ~70 wt% of global production comes from China alone (USGS, 2021). For these reasons, gaps exist for the development of the economic and environmental sustainability of the SSC. One way this could be achieved is by using biomass-derived silica ($bSiO_2$) as a substitute for mineral-derived silica ($mSiO_2$). This appeals as it: 1) avoids mineral excavation, 2) provides a pathway for decarbonizing the energy used in silicon purification (via bio-energy co-recovery), and 3) promotes SSC diversity.

Yet, the benefit of using $bSiO_2$ to produce bio-derived silicon (bSi) is still unknown. Consequently, this work provides the first evaluation of that benefit from the perspective of carbon emissions reductions. Reflecting the SSC, three purity grades of silicon are considered – metallurgical grade (MG-Si), solar grade (SG-Si) and electronic grade (EG-Si). Rice husk-derived bSiO2 (RH-$bSiO_2$) is used as a case study due to its production volume (100s of Mtonnes per year), the availability of relevant impact models

(Errington et al., 2022 and 2023), and because it may be considered an agricultural waste product.

2. Overview of the Silicon Supply Chain

Global demand for silicon comes from the functionality that it provides to wider products. Consequently, the SSC involves intermediate products concerning mainly industrial actors. Thus, silicon materials are often categorised based on their purity by weight (wt%). This includes: ferrosilicon (>50 wt%), MG-Si (>98 wt%), SG-Si (>99.9999 wt%), and EG-Si (>99.9999999 wt%) (USGS, 2022; Chigonodo, 2018) - see Figure 1.

Of the 8.5 Mtonnes produced, approximately 70% (5.95 Mtonne) is used in the production of ferrosilicons, with the remainder (2.55 Mtonne) being used for silicon metals (USGS, 2022). Recent statistics suggest as much as 21% of silicon metal use (0.54 Mtonne) may be attributed to the combined production of polycrystalline silicon (p-Si), EG-Si and SG-Si – with the remainder (2.01 Mtonne) finding use as MG-Si in aluminium alloys, silicones and other products (USGS, 2019).

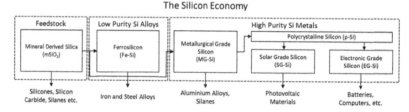

Figure 1: A summary of the major categories of silicon as considered within this work.

3. Agricultural Biomass as a Source of Silicon

Many investigations have reported over the past 40 years on the recovery of silicon dioxide from agricultural biomass. This has included a strong focus on rice husk (RH) – a natural waste from Rice Grain Agriculture (RGA) – due to its high silica content. Accordingly, the present authors have recently reported a model for assessing the environmental impact of RH-bSiO$_2$ in a process including RH combustion for co-recovery and use of RH bio-energy (Errington et al., 2022 and 2023). Having found a potential environmental benefit for such a method, and given the high purity (>90%) of RH-bSiO$_2$ typically produced, it seems a logical extension that bSi may be produced by using RH-bSiO$_2$ as a feedstock by extending the method as is summarised in Figure 2.

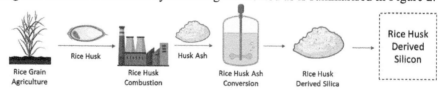

Figure 2: The process for recovery of silicon using rice husk biomass as a resource.

4. Method

This work aims to evaluate the benefit of producing MG-Si, EG-Si and SG-Si from bSiO$_2$. The "benefit" has been defined based on greenhouse gas emissions reductions associated with producing each grade of Si from bSiO$_2$ instead of mSiO$_2$. This has been done using a Life Cycle Assessment (LCA) methodology to: 1) assess the environmental impact of individual mSi and bSi silicon metal products, 2) calculate the benefit of bSi by comparison of mSi and bSi material impact predictions, and 3) quantify the uncertainty associated with each benefit prediction. Further information on methods is given below.

4.1. Impact Assessment: Life Cycle Inventories and Impact Characterisation Factors

A summary of the scope and relationship of Life Cycle Inventories (LCIs) for silica and silicon products considered is shown in Figure 3. LCIs for each silicon grade produced by existing industrial methods were taken from the Ecoinvent database (v3.6, Allocation at Point-of-Substitution basis). Inventories for RH-bSiO$_2$ were adapted from models provided by Errington et al. (2023) - in which RH combustion is followed by the recovery of RH-bSiO$_2$ via the "wet method". All LCIs were modelled on a global level.

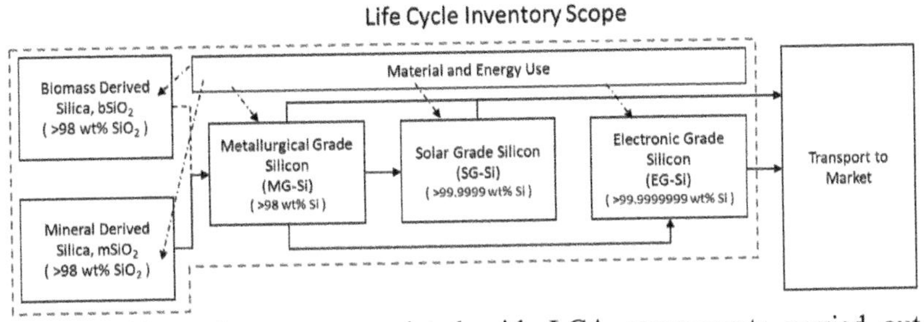

Figure 3: Life cycle scope associated with LCA assessments carried out for the comparison of mineral and biomass derived silicon in this work. Values for energy and material demands are taken from relevant inventories for each metal grade within the EcoInvent Life Cycle Inventory Database.

The environmental impact, EI (impact per kg$_{Si}$), of each product was calculated as the sum of impacts from all activities over the entire product life cycle as described by Errington *et al* (2022). Assessment was based on the Global Warming Potential (GWP) of the ReCiPe (H) method (Huijbregts *et al.*, 2017) using calculation cut-off of 0.1%.

4.1.1. Calculation of the Benefit of bio-derived silicon (bSi)

The benefit of bSi production, B^{bSi} (GWP/kg) was calculated by region based on the assumption that bio-energy co-recovered during bSiO$_2$ production, $\hat{Q}^b$ (MJ/kg$_{Si}$), was used to substitute for local electricity in a region with an associated environmental impact, EI^e (GWP/MJ) – see Equations 1 and 2. Benefits were also converted to a decarbonization extent, D^{bSi} (%), using Equation 3.

$$B^{bSi} = EI^{bSi} - \hat{Q}^b EI^e \tag{1}$$

$$\hat{Q}^b = (\eta^{BC}/\eta^{SR}) \times \left([\hat{Q}^{LHV}(1 - x^{H_2O}) - \lambda x^{H_2O}]/x^{SiO_2}(1 - x^{H_2O}) \right) \tag{2}$$

$$D^{Si} = B^{bSi}/EI^{bSi} \tag{3}$$

Where η^{RHC} is the electrical efficiency of bioenergy recovered during combustion (%), η^{SR} is the efficiency associated with recovering usable bSi from a dry biomass feedstock (%), $\hat{Q}^{LHV}$ is the lower heating value of the biomass used (MJ/kg$_{feedstock}$), x^{H_2O} is the moisture content of the biomass feedstock prior to combustion (wt%, wet basis), and x^{SiO_2} is the silica content of the biomass used (wt%, dry basis). LCIs for the impact of medium-voltage regional electricity were based on the ecoinvent LCI database (Wernet *et al*, 2016). Regions Europe, USA, Japan and China were considered as case studies for EI^e due to the extent of contribution that they currently supply to the SSC.

4.1.2. Quantification of Uncertainty in Impact Predictions
The sensitivity of impact evaluations to uncertainty in LCIs (arising from data quality) was quantified by the pedigree matrix approach - in which, the quality of inventory data was rated against five criteria: reliability, completeness, and temporal, geographic and technological representativeness. Ratings were quantified as log-normal probability distributions based on the method of Frischknecht et al. (2004). Monte-Carlo simulations were used to estimate prediction uncertainty (sample size = 10,000).

5. Results and Discussion

5.1. Impact Predictions and Implications for Industry Decarbonisation by Region
Predicted environmental impacts of different purity grades of bSi and mSi are shown alongside associated decarbonization extents for Europe, USA, Japan and China in Figure 4. Associated estimates for the regional availability of rice husk and regional demands for silicon metal are shown in Table 1.

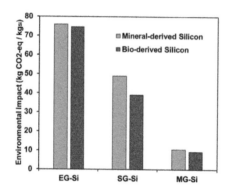 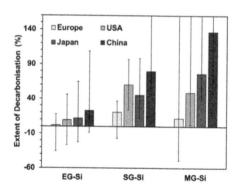

Figure 4: Impact predictions for producing 1 kg of mSi and bSi (left) and the decarbonization expected from a transition to production of bSi (right). Error bars represent a 99% prediction interval.

Table 1: Feasible annual supply of bSi and associated decarbonisation extent

Statistic	Europe	Japan	USA	China	World
RH Production[a] (Mtonne)	0.81	1.94	2.06	42.72	151.35
Available RH-bSiO$_2$[b] (Mtonne)	0.10	0.23	0.25	5.09	18.02
Benefit of bMG-Si (kg$_{CO2}$/kg$_{Si}$)	-1.2	-8.1	-14.4	-5.2	-1.2
Benefit of bEG-Si (kg$_{CO2}$/kg$_{Si}$)	-10.0	-17.7	-24.9	-14.4	-10.0
Benefit of bSG-Si (kg$_{CO2}$/kg$_{Si}$)	-1.3	-8.6	-15.4	-5.5	-1.3
Silicon Metal Production[c] (Mtonne)	0.37	N.G.	0.09	1.69	2.40
Possible bMG-Si Supply[d] (Mtonne)	0.04	0.09	0.09	1.88	6.67
Possible bSG-Si Supply[d] (Mtonne)	0.03	0.08	0.09	1.77	6.26
Possible bEG-Si Supply[d] (Mtonne)	0.03	0.08	0.08	1.67	5.91

[a] based on paddy production reported by the United Nations Food and Agriculture Organization (UNFAO, 2022) assuming 20 wt% of paddy is RH, [b] based on a mass ratio of 8.4 kg$_{RH}$/kg$_{SiO2}$ reported by Errington *et al.* (2023), [c] based on values from U.S. Geological Survey (USGS, 2022), [d] based on silica-to-silicon ratios present for each silicon purity grade in the ecoInvent LCI database (Wernet *et al.* 2016), N.G.: not given

Results in Figure 4 indicate that significant amounts of decarbonization may be expected to occur in all regions where bSi is produced as an alternative to mSi. However, the extent of decarbonization as well as uncertainty in the extent of decarbonization appear to be sensitive to both region and the silicon grade being produced. These correlations are understandable for two main reasons.

Firstly, correlations between the extent of decarbonization and the region considered are driven by differences in the GWP of electrical energy available in each region. This is due to differences in energy generation fuel blends – as shown by Errington *et al.* (2023).

Secondly, correlations between the magnitude of extent of decarbonization and the silicon grade considered (as well as associated uncertainty in the extent of decarbonization) are attributable to differences in the energy demand associated with each silicon grade considered. Particularly, the demand for energy rises significantly more quickly than the demand for $bSiO_2$ when increasing the purity of a silicon metal. Consequently, the fraction of bio-energy available to offset regional energy use is significantly diminished for EG-Si and SG-Si when compared to MG-Si – causing the magnitude and certainty in decarbonization to also reduce.

Finally, results in Table 1 imply a reduction in carbon emissions of 25 $Mtonne_{CO2-eq}$ for silicon metal production in the USA and China (both of which show a reasonable extent of certainty in decarbonization in Figure 4). This is equivalent to approximately 0.05% of global greenhouse gases emitted in 2019 (World Resources Institute, 2022). However, the assumption of complete utilization of global RH waste is an unrealistic scenario. Thus, alternate sources of agricultural $bSiO_2$ are detailed below.

5.2. Alternatives to RH for the recovery of $bSiO_2$ and bSi from biomass at scale

A summary of estimates for the availability of $bSiO_2$ associated with five possible alternatives to RH for $bSiO_2$ and bSi is shown in Table 2. These are: rice straw, wheat straw, corn stalk, corn cob and sugar cane straw. Results in Table 2 indicate 88 Mtonne of $bSiO_2$ may be practically recoverable from a combination of the agricultural wastes of rice, wheat, corn and sugar globally. This equates to more than 10 times the annual demand for silicon metal in the SSC (assuming 3 kg_{SiO2}/kg_{Si} based on the LCI of EG-Si), which may act as a more reasonable scenario for recovery of $bSiO_2$ from waste biomasses.

Table 2: Amounts of $bSiO_2$ available annually in major agricultural biomass wastes.

Statistics	Rice Husk	Rice Straw	Wheat Straw	Corn Stalk	Corn Cob	Sugar Cane Straw
Global Production[a] (MT/yr)	757	757	761	1162	1162	1870
Product-to-Biomass Ratio[b] (wt%)	20	20	50	50	38	20
Silica Content[b·] (wt% dry basis)	20	10	6	3	2	2.4
High Heat Value[c] (MJ/kg)	14	19	19	18	18	14
Moisture Content[b] (wt%)	12	12	7.5	16	16	12
Theoretical $bSiO_2$[d] (MTonne)	27	13	21	15	121	4
Biomass Use (kg_{RH}/kg_{SiO2}, wet basis)	8	23	37	81	121	170
Bioenergy Recovery[d] ($MJkg_{SiO2}$)	22	86	144	269	403	298

[a] based on 2020 crop production reported by the United National Food and Agriculture Organisation (UNFAO, 2022), [b] assumed values, [c] assumed values, [d] based on a 69% mass recovery efficiency as reported by Errington *et al.* (2023) for RH and assuming a conservative recovery efficiency of 49% (i.e. $\eta^{SR} = 0.49$) for other crops.

5.3. Financial Incentives for using $bSiO_2$ as an alternative to $mSiO_2$

Assuming a cost of 100 and 250 $ tonne^{-1} for RH and $mSiO_2$ respectively, the effective cost of RH-$bSiO_2$ (800 $ tonne^{-1}, scaled based on *"Biomass Use"* in Table 2) is larger than that of $mSiO_2$. However, the value of co-recovered RH-$bSiO_2$ bioenergy (22 MJ $kg_{bSiO2}$$^{-1}$, Table 2), equates to 5148 $ tonne$_{RH-bSiO2}$$^{-1}$ - assuming a 0.065 $ MJ^{-1} grid energy cost, EU Commission (2023). Thus, estimates suggest that the cost of RH-$bSiO_2$ is compensated for in a grid energy substitution scenario. Future work should better account for bioenergy co-recovery costs and different energy substitution scenarios.

6. Conclusion

In conclusion, this work aimed to assess the benefit of using biomass-derived silica as an alternative feedstock in production of silicon metals. This has been achieved using rice husk as a biomass case study and led to three main findings. Firstly, results show that a benefit is expected for all silicon grades considered. However, the extent of decarbonization is sensitive to the grade of silicon produced and the region in which it is produced. This is due to differences in the energy required for each grade of silicon, fuel sources used to produce regional energy, and the fraction of energy replaced with co-recovered bioenergy. Secondly, results show a transition to biomass-derived silicon could reduce global emissions by ~25 Mtonne per year (i.e. ~0.05% of global annual carbon emissions), which is driven by the yearly volume of metallurgical grade silicon (2 Mtonne). Finally, preliminary estimates show that as much as 10 times the current demand for silicon metals may be accounted for from agricultural products of rice, wheat, corn and sugarcane. Findings of this study generate new insights into the resource recovery potential from agriculture biomass and sustainable development of the SSC.

Acknowledgements

The authors acknowledge funding from Scottish Water and the UK EPSRC Science and Solutions for a Changing Planet Doctoral Training Programme. Thanks also to Kimberley Pavier and Alex Bowles for insight on the silicon supply chain and resource recovery.

References

Chigondo, F. 2018. From metallurgical-grade to solar-grade silicon: an overview. *Silicon*. 10 (3), 789-798.

Errington, E., *et al.* 2022. Environmental Impacts of Rice Husk-Derived Silica under Uncertainty: Is "Bio" better? in Proceedings of the 32nd European Symposium on Computer Aided Process Engineering (eds. Montastruc, L. & Negny, S.) 1615–1620.

Errington, E. *et al.* 2023. Synthetic amorphous silica: environmental impacts of current industry and the benefit of biomass-derived silica. *Green Chemistry*. Advance Article.

EU Commission. [Online]. Electricity industry retail prices 2008-2019. (2023). Available at: www.energy.ec.europa.eu/

Food and Agriculture Organization of the United Nations (UNFAO). 2022. Food and Agriculture Data: Crops and Livestock Products– "Rice, Paddy", "Maize (corn)", "Wheat", "Sugar Cane"

Frischknecht, N. *et al.* 2004. The ecoinvent Database: Overview and Methodological Framework . The International Journal of Life Cycle Assessment 2005 10:1 10 (1), 3–9

Huijbregts, M. *et al.* 2017. "ReCiPe2016: a harmonised life cycle impact assessment method at midpoint and endpoint level." The International Journal of Life Cycle Assessment 22.2 138-147.

Jungbluth, N. *et al.*. Photovoltaik. Sachbilanzen von Energiesystemen: Grundlagen für den ökologischen Vergleich von Energiesystemen und den Einbezug von Energiesystemen in Ökobilanzen für die Schweiz. ecoinvent report No. 6-XII (2009).

Maurits, J. E. A. Silicon Production. Treatise Process Metall. 3, 919–948 (2014)

Siddique, R. & Cachim, P. 2018. Waste and Supplementary Cementitious Materials in Concrete: Characterisation, Properties and Applications. Elsevier, Woodhead Publishing.

U.S. Geological Survey (USGS). Mineral Commodities Summary 2022. (2022).

U.S. Geological Survey (USGS). Mineral Commodities Summary 2019. (2019).

Wernet, G., Bauer, C., Steubing, B., Reinhard, J., Moreno-Ruiz, E., and Weidema, B., 2016. The ecoinvent database version 3 (part I): overview and methodology. The International Journal of Life Cycle Assessment, 21(9), pp.1218–1230.

World Resources Institute. 2022. [Online]. Climate Analysis Indicators Tool (CAIT), Climate Watch: "Historical GHG Emissions". Available at: www.climatewatchdata.org

Antonis Kokossis, Michael C. Georgiadis, Efstratios N. Pistikopoulos (Eds.)
PROCEEDINGS OF THE 33rd European Symposium on Computer Aided Process Engineering
(ESCAPE33), June 18-21, 2023, Athens, Greece

Technoeconomic and Life Cycle Assessment of an mRNA Vaccine Integrated Manufacturing Plant

Shang Gao[a] , Brahim Benyahia[a*]

[a]*Loughborough University, Department of Chemical Engineering, Epinal Way, Loughborough, Leicstershire, LE11 TU, UK*
B.Benyahia@lboro.ac.uk

Abstract

In this study, a technoeconomic and Life Cycle Assessment (LCA) methodology was developed to identify the best compromises between costs and environmental performance of an integrated mRNA vaccine production plant. The plant-wide simulator was obtained and adopted from the literature which helped generate the technoeconomic data along with the life cycle inventories. The LCA was developed using SimaPro software based on the Ecoinvent database and ReCiPe 2016 method. A prediction tool based on artificial intelligence was used to estimate the environmental footprint of several raw materials. A base case scenario helped identify the environmental hotspots which laid the ground for a new set of scenarios to explore ways to lower the overall environmental footprint. Consequently, several recycles were implemented which helped reduce the endpoint categories by up to 10% while reducing the operating expenditures by 13%. The results gave key insights into the optimization of mRNA vaccine production and waste minimization.

Keywords: Plant-Wide Simulator, mRNA vaccine, Life Cycle Assessment (LCA), Technoeconomic Assessment, Recycles.

1. Introduction

Over the last few years mRNA technologies have gained a resurgent interest, due to recent technical advances and growing industrial investment putting mRNA vaccines in the position to substitute protein based conventional vaccines. Compared to the conventional vaccines, such as live attenuated and killed virus vaccines, subunit, and DNA-based vaccines, mRNA vaccines are safe, more efficient, and can be produce at large scales (Pardi et al.,2018). The upstream processing of mRNA vaccines is relatively straight forward and exhibits less sources of variability compared to cell-based vaccines. Furthermore, mRNA vaccines production process requires smaller bioreactors (from 30 L to 50 L) compared to cell-based vaccines where typically 2000 L bioreactors are required. With a relatively smaller production scale reduced facility footprints, the capital costs of the mRNA plant may be much lower. In addition, mRNA plants are flexible and versatile and can be rapidly deployed to produce new vaccines to responded for new variants, because different RNA sequences that used into different vaccines can be produced in the same process (Pardi et al., 2018). However, the production of mRNA requires complex recipes and a large set of raw materials, excipients and solvents which may exhibit significant environmental impacts.

The new mRNA vaccines played a prominent role during the covid-19 pandemic. The global mRNA production capacity is skyrocketing and is expected to exceed 8 billion doses per year (Kis et al., 2021). Consequently, it is a crucial time to evaluate the available performance compromises, which include the economic, quality and environmental criteria, to make well informed decision at this early stage of a new era of mRNA vaccines. Flowsheet and plant wide models are critical in all design stages to help identify the most economic and environmentally friendly options optimize process operation and develop plant wide control strategies (Benyahia et al, 2012; Benyahia 2018; Ramin et al., 2018).

This work is focused on the technoeconomic and Life Cycle Assessment (LCA) of an integrated end-to-end mRNA production plant. The plant-wide simulator was obtained and adopted from the literature (Kis et al., 2021). It is worth emphasizing that process data are still very scarce and so are the environmental data associated with the chemicals, and raw materials. Here, the environmental footprint of the chemicals not available in the data bases were estimated using artificial intelligence-based methods (Wernet et al., 2009). Firstly, a base case LCA was conducted to identify the environmental hot spots which informed on the best opportunities for the implementation of recycles and resulted in 2 new design alternatives. Additional LCA were conducted for the new scenarios along with inherent economics to help evaluate and compare the performance of different design alternatives.

2. Material and Methods

2.1. Process description

The process was modelled and simulated by using SuperPro Designer V10. The initial SuperPro simulation version was obtained from the literature (Kis et al., 2021) then adopted to the current needs. The mRNA vaccine production process starts with a biochemical reaction called *in vitro* transcription where the mRNA is synthesized, with a series of additions of raw materials including different nucleotides, DNA templates and enzymes to the bioreactor as shown in the upstream processing in Figure 1. This is followed by a series of separation processes to purify the solution and sperate solid and liquids. Firstly, a tangential flow filtration (TFF) along with diafiltration unit is applied to remove RNA fragments, with the ultrafiltration membrane size of 500 kDa and KCl buffer to wash the filter to increase the efficiency. This stage is followed by a chromatography unit which separates the content based on size, total charge, surface hydrophobic groups. After the chromatography unit, another TFF unit is used with different buffer sodium citrate followed by sterile filter to prepare the formulation step. Finally, the mRNA is encapsulated within lipid nanoparticles (LNPs) to deliver the final injectable suspension.

2.2. Life cycle assessment

LCA is a holistic approach that can be used to capture the environmental footprint of a process, a product or a service across its life cycle. It provides a comprehensive tool to evaluate the environmental burdens based a large set of impact indicators referred to as midpoint and endpoint categories which are far beyond the capabilities of the green engineering metrics. Here, a gate-to-gate approach is used to evaluate different design and recycle options. The LCA requires the inventory data which can be obtained from the simulators under different scenarios. The implementation of the LCA requires 4 key steps namely:

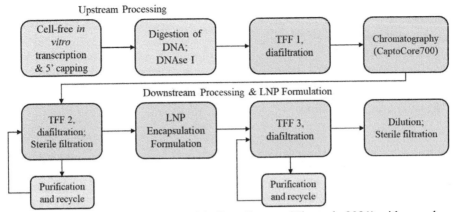

Figure 1. Simplified mRNA plant wide flow diagram (Kis et al., 2021) with recycles.

- *Goal and scope*
  Here, the first objective is to perform a LCA for the base case mRNA vaccine production process using a functional unit of 1 kg of mRNA product. The second objective is to evaluate alternative designs by implementing different recycles.

- *Life cycle inventory*
  LCI data were obtained from the plant wide simulation of the process flow diagram above using SuperPro Designer.

- *Impact assessment*

  In this study, the impact assessment is based on the method ReCiPe 2016 in SimaPro with Ecoinvent 3, environmental footprint database and European and Danish Input/Output data base. ReCiPe 2016 method has three endpoint categories (human, ecosystems, and resources) and seventeen midpoint categories. These different categories will help compare the different design options.

- *Interpretation of results*
  Finally, the results obtained from the previous steps are interpreted and the environmental hotspots are identified.

To capture the environmental footprint of the different processing steps, the plant was divided into six main stages: Upstream processing, first filtration TFF1, Chromatography, second filtration TFF2, Encapsulation, and final filtration TFF3. It should be emphasized that several chemicals used in the process are not available in the Ecoinvent database or any other environmental databases. To address this limitation, a prediction tool, based on machine learning Finechem was used to predict the global warming potential (GWP), cumulative energy demand (CED) and Eco-Indicator 99 score (EI99) based on the chemical structure of the molecules (Wernet et al., 2009). These data were then fed to Simapro to obtain the overall impact results.

The proposed LCA was implemented for the following scenarios:

Scenario 1: Base case - the initial design presented in Figure 1 without recycles.

Scenario 2: TFF2 Recycle. In this design option, the waste from the second filtration is purified using an ultrafiltration and a distillation unit to retrieve pure solvent.

Scenario 3: TFF2 and TFF3 Recycles. Here, the wastes from both hotspots are purified and recycled.

To evaluate the economic performance of each of the scenarios, the operating and capital expenditures/costs (OpEX and CaPex) were estimated by using the built-in tool within SuperPro Designer. The economic data were obtained based on the current plant/equipment capacity and material and energy usage for each scenario.

3. Results and Discussions

3.1. LCA results

Figure 2 shows the normalized impacts categories for each of the 6 processing stages in the base case scenario. The second tangential flow filtration stage (TFF 2) exhibits the largest contribution in all impact categories. This can be explained by the large amount of solvent used and wastes generated at this stage along with large quantities of sodium citrate which is used as a buffer to adjust the pH of the product solution. The second largest contribution is associated with the diafiltration (TFF3) followed by the encapsulation. These results clearly confirm the environmental hotspots and justify the alternative scenarios based on different recycle options.

3.2. Assessment of scenarios with recycles

As described above, TFF2 is the most important environmental hotspot among all stages. To purify then recycle the wate stream TFF2, a filtration and a distillation unit are proposed and integrated to the flowsheet simulator in SuperPro designer. The reason to choose these units is that in this stream the objective is to recycle pure sodium citrate, so the first ultrafiltration is used to desalt the solution and remove waste RNA transcripts by adding water as solvent to wash the waste. This step is followed by a distillation unit used to sperate water and sodium citrate. TFF3 is the second hot spot based on the base case LCA results. In this case, a new recycle scenario is proposed to remove lipids left from encapsulation process and desalt the waste solution to retrieve the buffer from the waste.

With the new recycle scenarios associated with TFF2 and TFF3 process, the LCA were conducted. The corresponding midpoint and endpoint indicators along with the cost

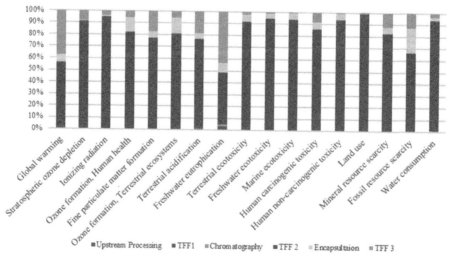

Figure 2. Normalized midpoint categories for the base case scenario.

performance are summarised in Figure 3. Compared to the base case, both recycle scenarios exhibit dramatic reduction in all midpoint categories particularly, the global warming, stratospheric ozone depletion, and land use (Figure 3a). When comparing both recycle seniors, their performances are close with a visible advantage for the senior with combined recycles particularly on the freshwater eutrophication impact category.

The economic performance indicators along with endpoint indicators are shown in figure 3b. Again, the environmental advantages of both recycle scenarios are clear. Although both recycle scenarios require increased capital expenditures associated with the waste purification and recycle equipment, their operating expenditures are reduced by nearly 8% and 13% respectively for the simple and combined recycles respectively. This is a very important outcome which clearly highlights the advantages of the recycle far beyond the environmental consideration and confirms the practicality and feasibility of the recycles in pharma and biopharma which are both notorious for their costly chemicals and additives.

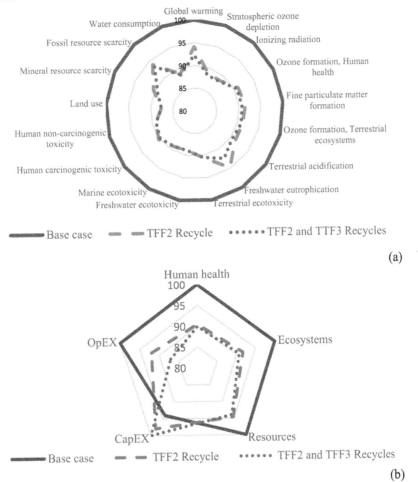

(a)

(b)

Figure 3. Summary of the environmental and economic key performance indictors for all scenarios. (a) Midpoint indicators. (b) End-point indicators and economic criteria.

4. Conclusions

A technoeconomic analysis and Life Cycle Assessment of a COVID-19 mRNA vaccine production plant was developed. The integrated end-to-end mRNA simulator was developed in SuperPro designer. A first LCA was conducted using the base case scenario to identify the hotspots, based on which new designs with recycles were proposed and simulated. Consequently, new inventory and economic data were obtained for the proposed new designs which helped build new LCA and economic evaluations. As expected, the recycles reduced dramatically the environmental footprint of the process which translates into a decrease in the endpoint indicators, namely human health, ecosystem and resources, from 6% to 10%. Most importantly the results show that the implementation of recycles can also reduce the operating expenditures by up to 13% making the whole process more viable and more environmentally friendly. These results highlight the benefits of the recycles in pharma and biopharma which allows recycling precious or very expensive pharma grade chemicals and reduce inherent wastes. These benefits can be even more important when the production scale or the number of batches is large which is true for the current mRNA vaccine production plants.

References

Benyahia, B., Lakerveld, R., Barton, P.I., 2012 A plant-wide dynamic model of a continuous pharmaceutical process, Industrial and Engineering Chemistry Research, 51(47), pp. 15393–15412.

Benyahia, B., 2018. Applications of a plant-wide dynamic model of an integrated continuous pharmaceutical plant: Design of the recycle in the case of multiple impurities. Computer Aided Chemical Engineering, Elsevier B.V. 41, 141-157.

Kis, Z., Kontoravdi, C., Shattock, R., Shah, N,. 2021. "Resources, Production Scales and Time Required for Producing RNA Vaccines for the Global Pandemic Demand" *Vaccines* 9, no. 1: 3. https://doi.org/10.3390/vaccines9010003

Kis, Z., Zain, R,. 2021. 'How to Make Enough Vaccine for the World in One Year', *Public Citizen.*

Pardi, N,. Hogan, M,. Porter, F,. *et al*, 2018. mRNA vaccines — a new era in vaccinology. *Nat Rev Drug Discov* **17**, 261–279.

Ramin, P., Mansouri, S.S., Udugama., I.A., Benyahia, B., Gernaey, K.V., 2018. Modelling continuous pharmaceutical and bio-based processes at plant-wide level: A roadmap towards efficient decision-making. Chemistry Today 36 (2), 26-30.

Wen, E.P., Ellis, R., Pujar, N.S., 2015, *Vaccine Development and Manufacturing*, Wiley Series in Biotechnology and Bioengineering (Hoboken, New Jersey, United States: John Wiley & Sons, Inc.)

Wernet, G,. Stavros P., Stefanie, H., Konrad, H,. 2009. "Bridging data gaps in environmental assessments: Modeling impacts of fine and basic chemical production." *Green Chemistry* 11 (2009): 1826-1831.

Antonis Kokossis, Michael C. Georgiadis, Efstratios N. Pistikopoulos (Eds.)
PROCEEDINGS OF THE 33rd European Symposium on Computer Aided Process Engineering
(ESCAPE33), June 18-21, 2023, Athens, Greece

Obtaining biofuels from agroindustrial waste of *Fragaria spp* and *Mangifera indica spp*

Ernesto Barrera-Reyes[a], Luis Alberto Bretado-Aragón[a], Guillermo Adolfo Anaya-Ruiz[a], Pascual Eduardo Murillo-Alvarado[a]*, José Luis Baltazar-Álvarez[a], José Joel Román-Godínez[a] and Gabriela Guadalupe Esquivel-Barajas[a].

aUniversidad de La Ciénega del Estado de Michoacán de Ocampo. Av. Universidad Sur 3000, Lomas de Universidad, Michoacán 59103, México.

Abstract

Population growth has advanced by leaps and bounds, which has caused the agricultural sector to see the need to meet this demand. Conveniently, the practices of the agro-industrial sectors have directly affected natural resources. Within the discharges from agro-industries, they are mainly conditioned by pollutants with organic and inorganic compounds, being the objects of study of the present investigation organic waste because they are a potential source that compromises human health. In such a way that, the purpose of the present investigation consisted of collecting residues from strawberry and mango agro-industries in Zamora-Jacona, Michoacán, Mexico, such residues were synthesized as biodiesel and were characterized by Fourier transform infrared spectroscopy (FT-IR) and UV-VIS, so it was possible to identify absorption bands, corresponding to the functional groups, which are similar to those of hydrocarbons. Thus, it is concluded that the conversion of agro-industrial remnants can be used in the field of biofuels.

Keywords: biodiesel, agroindustry, remnants.

1. Introduction

Mexico is one of the main producing countries of both strawberries (*Fragaria spp*) and mango (*Mangifera indica spp*), with the central region of the country being the area with the highest production, this is due to the fact that they were produced in 2016, 468.25 and 1.89 million tons respectively of these fruits, and it is estimated that by 2030 production will increase between 26 and 42% (SAGARPA 2016).

Berries, strawberries or forest fruits, refer to the various fruits of different botanical groups, which are characterized by their short shelf life. They tend to be small, tart-sweet, juicy, and brightly colored, making them very attractive to eat. The production of berries at the national level is concentrated in four states, highlighting the state of Michoacán as the most important producer, followed by Baja California, Jalisco and Guanajuato. The country exported close to 1,000 million dollars of fresh berries to the world in the 2015, which projects this group of fruits as one of those with the greatest growth potential in the agricultural sector (González et al 2019).

In the case of the present investigation, the site of interest focuses on 2 municipalities in the state of Michoacán, which are Zamora and Jacona, since they are the main producers of berry and mango crops nationwide. In such a way that the objective was to elaborate the synthesis of biofuels from the organic waste material generated by agro-industries of Zamora and Jacona of the State of Michoacán de Ocampo.

2. Case study

2.1 *Geographical characteristics of Zamora-Jacona*

Zamora R., & Salazar in 2019, describe in their manuscript that the municipality of Zamora is among the 6 most important in the state of Michoacán, because it produces 4.6% of the total gross production. It has a population of 170,748 inhabitants (4.3 percent of the total population), of which 26,997 are employed persons, with 6,778 economic units. The city represents one of the three conurbations of the state; Zamora-Jacona-Tangancícuaro. One of the great advantages that Zamora has in strawberry production is its proximity to several of the main trade distribution points nationwide, such as the markets of the Federal District. In addition to its proximity to several of the freezing and packing companies that are involved with the strawberry agro-industrial activities, located mainly in the cities of Irapuato and Celaya, they are also dedicated to freezing and storing strawberries, located in several trading and industrializing companies' leaders in the export of strawberries.

2.2 *Contaminating sources of strawberry and mango crops*

At present there is an indiscriminate demand for the consumption of different fruits, among them red fruits stand out, unfortunately high amounts of waste materials produced by these fruits are produced annually.

Fraire-Cordero M., et al (2003), carried out an investigation whose purpose was based on identifying pathogenic fungi in strawberry fruit and their relationship with varieties and the cultivation system, five postharvest samplings were made: three in the region of Zamora, Michoacán, México, in December 2001, and in February and May 2002, in the *Camarosa* and *Aromas* varieties, cultivated with the traditional system, and with plastic cover. In their research they describe the identification of nine pathogenic fungi which were identified from national and imported fruit: *Aspergillus, Botrytis, Colletotrichum, Geotrichum, Mucor, Penicillium, Pestalotiopsis, Phytophthora,* and *Rhizopus stolonifer*; and more than eight saprophytic fungi.

Estrada-Loera et al (2018), carried out various samples in mango crops, subsequently carried out microbiological tests, thus in this study they allowed to identify the presence of *Klebsiella spp*, and non-enterohemorrhagic *E. coli*.

Due to the aforementioned, waste from agro-industries is produced, which have different places where they are stored and if they do not have an appropriate treatment, they are a potential source of microorganism reservoirs and these, in turn, can compromise human health. In this way, the purpose of the investigation lies in identifying freezers in the Zamora-Jacona region and sampling the generated waste so that it can be converted into a biofuel.

3. Methodology

For the development of the present investigation, it was divided into 4 stages, which consist of: 1) identification of agro-industries, 2) sampling of strawberry and mango agro-industrial residues, 3) extraction of essential oils and obtaining biodiesel, and 4) characterization of oils and biodiesel

Stage 1. Identification of agro-industries. Agro-industries that process and distribute strawberry and mango were identified in the municipalities of Zamora-Jacona, Michoacan, Mexico.

Stage 2. Strawberry and mango remnants from agro-industries were collected, where they were considered: strawberry pulp and stem, on the other hand mango seeds were collected, the raw material was stored in sterile plastic bags, later they were transferred to a desiccator (Novatech), with the aim of dehydrating the biomass.

Stage 3. Extractions of essential oils. For the development of this stage, an adaptation was made to the methodology proposed by Sánchez I. & Huertas K. (2012), which consisted of: 1) dehydrating the biomass of both fruits, 2) grinding the mango seeds and the dehydrated strawberry pulp, and 3) the soxhlet method (KIMAX brand) was used, which consisted of pouring 200 ml of isopropyl alcohol (MEYER), subjecting it to a constant temperature and stirring of 300°C and 1,400 rpm, for 15 cycles. Afterwards, a rotoevaporator was used to separate the solvent from the essential oil and finally the transesterification method was carried out, which was carried out by centrifuging the essential oil for 30 minutes at 6,000rpm and eliminating the degumming or residues to be able to be mixed with the concentration. to 1% of the potassium iodide solution (Reasol brand) with an addition of 3% (p/p) of distilled water and place on the grill to keep stirring for 30 minutes at 60°C to finish it was brought to the centrifuge at 6,000rpm for one hour.

Stage 4. Characterization of biodiesel. Biodiesel was characterized by Fourier transform infrared spectroscopy (FT-IR), Perkin Elmer Frontier model in the 4,500-400cm^{-1} region and by Perkin Elmer Lambda 25 model UV-VIS spectrophotometry, in the absorbance region of 300-800nm.

4. Results and discussions

Sampling points for agro-industry waste were identified. Once the samples were prepared, the fruits were taken to the drying oven, in which they were dehydrated with the aim that a biomass screening could be carried out (Fig. 1).

Fig.1. Fruits considered as waste in agro-industries in Zamora-Jacona. a) Decaying fruits of Mangifera indica spp, b) Decaying fruits of *Fragaria spp.* and c) Screening for *Mangifera indica spp*.

Subsequently, the extraction of essential oils was carried out (Fig. 2), carrying out an adaptation to the methodology proposed by Sánchez I. & Huertas K. (2012). It started by obtaining essential oils using the Soxhlet method (Fig. 2a and 2b), having the essential oils biodiesel was synthesized (Fig. 2c) which was subjected to refining with the purpose of reducing impurities (Fig. 2d).

Once the biodiesel was obtained, it was characterized by both UV-VIS and FT-IR (Fig. 3 and Fig. 4). Regarding the characterization by UV-VIS, the band corresponding to the reading of mango biodiesel (MD2), presents a greater intensity and absorbance, compared to the band of strawberry biodiesel (FD2), this is attributed to the fact that in the upper part mango seed stores a greater amount of oil.

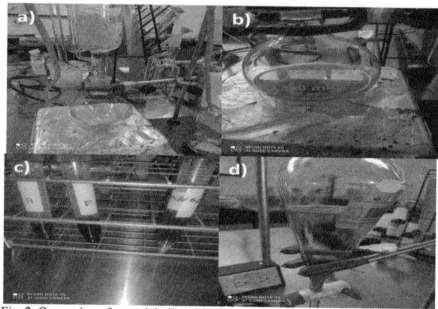

Fig. 2. Conversion of essential oils to biodiesel. a) and b) Essential oil of *Mangifera indica spp* and *Fragaria spp*, c) and d) Refined biodiesel.

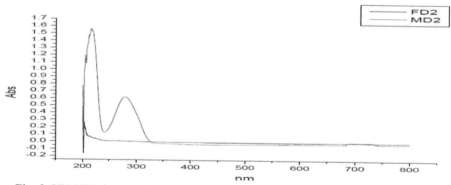

Fig. 3. UV-VIS characterization of strawberry (FD2) and mango (MD2) biodiesel.

Regarding the characterization by FT-IR (Fig.4), it was possible to identify the functional groups, characteristic of the absorption bands of strawberry biodiesel, within the identified groups are: OH, CH2, CH and PH.

In the same way, a proposal was developed where this technology can be migrated on an industrial scale, for which the design at the industry level is proposed (Fig. 5).

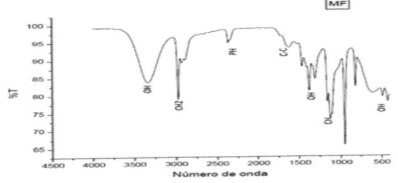

Fig. 4. FT-IR characterization identify the absorption bands of strawberry biodiesel

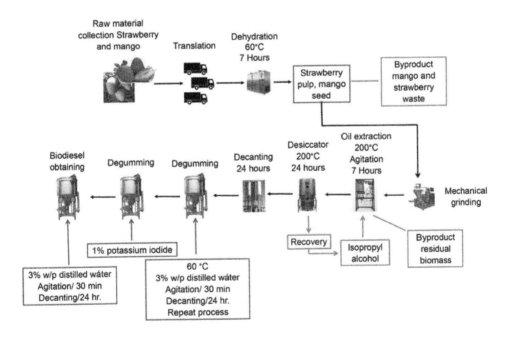

Fig. 5. Biodiesel production scheme from agro-industry remnants.

In the year 2021 the state of Michoacan had a production of 180,871.84 tons of mango, for the proposed methodology only the residual part due to the processing of the fruit is considered, in the case of mango only the interior seed that corresponds to 17.5% is considered of the fruit. In this sense, considering mango production, there are 31,652.57 tons of residual biomass. Based on the data obtained at the experimental level and the process proposed in Fig. 5, it is possible to obtain 7.0339×10^5 Lt of biodiesel.

In the case of strawberries, in 2021 a production of 326,191.10 tons was obtained, in this case the largest amount of the fruit is pulp, so the residual biomass obtained during processing corresponds to 10% of the total available. Therefore, a total of 32,619 tons of residual biomass is available. From this flow, it is possible to obtain a total of 1.63×10^5 Lt of biodiesel. Through the residual biomass it is possible to obtain a total of 8.664×10^5 Lt of biodiesel, considering an approximate price of $1 dollars, a profit of 8.664×10^5 USD.

5. Conclusions

The remnants generated in the agro-industries can be possible sources of infection for society, in such a way it is necessary to identify ways of using such residues, in this way the alternative of using the obtaining of biofuels and these biofuels can be used by society. The highest yield and production of biodiesel occurs in the mango; this is due to the fact that a greater amount of residual biomass is generated during its processing. The strawberry presents a considerable production, since it is one of the fruits with the highest production and demand in Mexico.

References

Estrada-Loera Rosa María, Gallegos-Robles Miguel Ángel, Orona-Castillo Ignacio, García-Hernández José Luis, Osuna-García Jorge Alberto, Sánchez-Lucio Roberto, Ríos-Plaza Juan Luis, Vázquez-Vázquez Cirilo (2018) Prevalence Of Pathogenic Bacteria In Mango (Mangifera Indica L) Cv. Tommy Atkins. Revista de Ciencias Biológicas y de la Salud. P.p. 6-9.

Fraire-Cordero María de Lourdes, Yáñez-Morales María de Jesús y Nieto-Ángel Daniel (2003) Hongos Patógenos en Fruto de Fresa (Fragaria x ananassa Duch.) en Postcosecha. Revista Mexicana de FITOPATOLOGIA. P.p. 285-290.

González Razo Felipe de Jesús, Rebollar Rebollar Samuel, Hernández Martínez Juvencio, Morales Hernández José Luis y Ramírez Abarca Orsohe (2019) Situación Actual Y Perspectivas De La Producción De Berries En México. Revista Mexicana de Agronegocios, vol. 44. P.p. 261.

SAGARPA (2016) Planeación agrícola nacional 2017-2030. Mango Mexicano. P.p. 2.

SAGARPA (2016) Planeación agrícola nacional 2017-2030. Fresa Mexicana. P.p. 2.

Sánchez Medina y Huertas Greco Karina (2013) Obtención Y Caracterización De Biodiesel A Partir De Aceite De Semillas De Ricinus Communis. (Higuerilla) Modificadas Genéticamente Y Cultivadas En El Eje Cafetero. Universidad Tecnológica De Pereira.

Zamora Jacobo Ricardo y Salazar Mosqueda (2019) Importancia De La Producción De Fresa En El Sector Agrícola En Zamora, Michoacán. Rev. Realidad Económica. P.p. 108.

Antonis Kokossis, Michael C. Georgiadis, Efstratios N. Pistikopoulos (Eds.)
PROCEEDINGS OF THE 33rd European Symposium on Computer Aided Process Engineering
(ESCAPE33), June 18-21, 2023, Athens, Greece

Life Cycle Assessment of Cellulose Nanocrystals Production in Sugarcane Biorefineries

Gustavo Batista[a], Ana Carolina Borges Silva[a], Cristiane Sanchez Farinas[a,b], Antonio José Gonçalves da Cruz[a]

[a] Graduate Program of Chemical Engineering, Federal University of São Carlos, 13565-905, São Carlos, SP, Brazil.
[b] National Nanotechnology Laboratory for Agribusiness (LNNA), Embrapa Instrumentation, Rua XV de Novembro 1452, 13560-970, São Carlos, SP, Brazil.

Abstract

The purpose of using Life Cycle Assessment (LCA) in this work was to evaluate the cellulose nanocrystals (CNCs) production in sugarcane biorefineries, identifying hotspots that can direct to process design changes to reduce environmental footprint. Modeling and simulation of an incremental unit for CNCs production from sugarcane bagasse (the main residue from the sugarcane industry) was carried out. In order to proceed the LCA, an inventory of raw-materials, products and emissions was made. The analysis employed the cradle-to-gate approach. The Global Warming Potential (GWP) metric was estimated using the SimaPro® software and using the EcoInvent v.3.3 database. LCA results showed that daily production of 7.73 ton of CNCs occurred with accumulated GWP of 23.6 tons of CO_2 equivalents. Ethanol from organosolv pretreatment, sulfuric acid from hydrolysis, and sugarcane bagasse burning for steam and energy generation were the main contribution inputs for GWP. An estimated GWP of 0.23 kg of CO_2 equiv. per kg of dry equiv. CNCs (or 13.43 g of CO_2 equiv. / MJ of process outputs) were accounted after energetic allocation. In other impact categories aside from GWP, the environmental impacts of H_2SO_4 were generally the most relevant, especially in the abiotic depletion.

Keywords: Life Cycle Analysis, Cellulose Nanocrystals, CNCs, Sugarcane Biorefineries.

1. Introduction

Life Cycle Assessment (LCA) is a standardized tool to quantify the environmental sustainability performance of emerging technology products along all phases of their life cycles, allowing the comparison of different process designs (ÖGMUNDARSON et al, 2020a; ÖGMUNDARSON et al, 2020b; FOROUGHI et al., 2021). If the methodology is used during a new product development phase, especially during the process methodology planning, it can indicate the process stages or technologies with the highest environmental impacts, and thus provide a guide for improvements in the implementation of the technology (IBICT, 2014). ISO 14040:2006 and ISO 14044:2006 established a methodological framework for conducting LCA studies.

With the implementation of the carbon credits market, the quantification of the Global Warming Potential (GWP) and of other metrics of environmental impacts have now also a strong impact on investment decision-making, once high environmental impact production methodologies will be neglected due to environmental and economic aspects as well (ÖGMUNDARSON et al., 2020a). Nanocellulose is an example of an emerging, still under development material for which a reduced environmental impact is expected when comparing to other existing materials (ARVIDSSON et al., 2015; PICCINNO et al., 2018). Cellulose nanocrystals (CNCs) are crystalline and highly ordered materials of small diameter, elongated length, and high surface area whose main applications are as reinforcement in polymeric materials and the applications in biomedicine. Consisting in a high added-value renewable biocomposite, but still in limited availability and presenting low yields of obtainment and isolation, the nanocellulose market is in continuous and strong growth (MARKETS AND MARKETS, 2021).

Due to some challenges such as data unavailability and limitations related to the end-of-life applicability and the solvent recycle treatments in lab-scale processes, there are only a few LCA studies related to nanocellulose products. A cradle-to-gate environmental impacts study of three production routes for cellulose nanofibrils (CNFs) was carried out by ARVIDSSON et al. (2015). The authors used wood pulp as feedstock, and three production routes investigated were: enzymatic pretreatment, carboxymethylation pretreatment and no previous pretreatment. The results obtained showed that processes without pre-treatment and using enzymatic treatment had relatively low environmental impact in terms of GWP (0.79 and 1.2 kg of CO_2 equiv. / kg of CNFs produced, respectively). NASCIMENTO et al. (2016) evaluated the LCA of the CNCs production from coconut fibers by extraction with high powered ultrasound, while using the residual lignin as power source for energy cogeneration to the process. The authors quantified the GWP of the process as 0.207 g of CO_2 equiv. / g of CNC produced. Still, to the best of our knowledge, the technical-environmental analysis of the production of CNCs from sugarcane biorefinery residues is a topic not yet well investigated in literature. Thereby, the purpose of using LCA in this work was to simulate CNCs production in sugarcane biorefineries attached to a second-generation ethanol (E2G) biorefinery, and to quantify the GWP of this nanocellulose unit in a 100-year horizon through LCA, identifying hotspots that can direct to process design changes to reduce environmental footprint.

2. Methodology

The simulated methodology of production proposed in this work employed sugarcane bagasse as feedstock. The process started with organosolv treatment (190 °C, 2h, 1:1 $V_{ethanol}/V_{water}$, 1:10 dry mass / volume of solution) with ethanol recycle by flashing and distillation, followed by purification with H_2O_2 (50 °C, 1 h, H_2O_2 7% vol, NaOH 5% vol, 10:1 m/m relative to input) and subsequent acid hydrolysis with H_2SO_4 (45 °C, 45 min, 10:1 m/m relative to dry nanocellulose output, 70% yield). After washing and filtering of the produced CNCs, sulfuric acid was recycled by a system of evaporators operating in three stages, two of which operate at vacuum pressure (1atm, 300 mbar and 100 mbar). The nanocrystals downstream process then proceeded with residual H_2SO_4 neutralization, dialysis, sonication, and drying of the material to purity of 95% in a spray-dryer. The annual sugarcane bagasse utilization as feedstock was set at 72000 tons per year. The mass and energy balances for discontinuous equipment were discretized, i.e., the input and output streams had mass and energy contents distributed over the time of the equipment usage cycles. Organosolv pretreatment with ethanol was selected over other treatment options because it provides high rates of removal of lignins and residual hemicelluloses, whilst being environmentally less aggressive than the chemical processes (NASCIMENTO et al., 2016). In its turn, complementary alkaline delignification was selected for its high effectiveness in removing residual lignin content, generating biomasses with high-purity cellulose. Acid hydrolysis by concentrated H_2SO_4 was chosen because it is the state-of art hydrolysis process in the CNCs production (VANDERFLEET & CRANSTON, 2021).

Several sources of literature data, external sources and laboratory data were used as input parameters of the Case Studies analyzed in this work. Mass and energy balances were performed on Microsoft Excel® electronic spreadsheets. In the energy balances, the thermodynamic calculations for the equipment were performed using the Aspen Plus® software and the results were transposed to the electronic spreadsheets. The NRTL thermodynamic model was used to represent the non-idealities of liquid phase mixtures in each analyzed system. The vapor phases were considered ideal due to the low pressure in the processes. Thermodynamic data related to sugarcane biomasses that was absent in the Aspen Plus® were inserted into this simulator using values obtained in DOMALSKI et al. (1986). Figure 1 shows a box diagram that represents the simulated process.

In order to proceed the LCA, an inventory of raw-materials, products and emissions were made based on the process modeling stage. The analysis employed the cradle-to-gate approach, and the functional unit was considered as 1 kg of cellulose nanomaterials produced. The GWP coefficient quantified the Greenhouse Gas Emissions (GHG) as grams of equivalent CO_2 emitted per each produced kg of CNCs for 100 years of operation. The carbon intensity of residues from any process stage was equaled to zero in the cradle-to-gate approach.

In addition to the GWP metric, other environmental assessment categories were also evaluated, as the *CML-IA Baseline v3.04 2000* method was selected in *SimaPro® 9.0.0.35 PhD* software and using the *EcoInvent® v.3.3* database. For the allocation of environmental impacts to the product and the by-product from the process (lignin), energetic allocation was carried out as the impact's distribution factor.

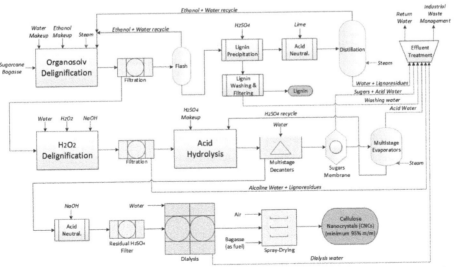

Figure 1 - Block diagram for the CNCs production methodology proposed in this work.

3. Results and Discussion

Modeling and simulation results showed that the CNCs production from sugarcane bagasse was water-use intensive, as 209 m^3 water / ton CNCs were consumed. LCA results showed that daily production of 7.73 ton of CNCs occurred with accumulated GWP of 23.6 tons of CO_2 equivalents. Ethanol from organosolv pretreatment, sulfuric acid from hydrolysis, and sugarcane bagasse burning for steam and energy generation were the main contribution inputs for GWP.

An estimated GWP of 0.23 kg of CO_2 equiv. per kg of dry equiv. CNCs were accounted after energetic allocation. This result corresponds to 13.43 g of CO_2 equiv. / MJ of process outputs. In other impact categories aside from GWP, the environmental impacts of H_2SO_4 were generally the most relevant, especially in the abiotic depletion (96% of the total 1.37 kg of Sb equiv. per day of production), ozone layer depletion, soil acidification (52.8% of total 683 kg of SO_2 equiv. per day of operation) and water ecotoxicity (Figure 2), indicating that research studies may be directed to optimize the use of this acid in the nanocellulose production chain. Other compounds for acid hydrolysis (as organic acids) may represent a hindrance in the economics of the process but are also alternatives to be environmentally analyzed.

Life Cycle Assessment to produce nanocellulose varieties using recycled hydrolysis solvents are scarce studies in literature. Previous works that evaluated LCA of nanocellulose production, such as those by HOHENTHAL et al. (2012) (750 - 3100 g of CO2 equiv. / kg of CNCs by TEMPO oxidation / high-pressure homogenization of sulfite pulp) presented higher GWP impacts per kg of nanocellulose produced. The relatively low GWP values when compared to processes that use derivatives of fossil fuels (and other biochemicals) indicate that the CNCs production from sugarcane bagasse is an option of interest for industries in the sector that aim to expand their portfolio with a high added-value product while reducing the emission of environmental impacts.

Biochemical production faces economic and environmental challenges that need to be overcome to enable a viable and sustainable bioeconomy. There are combined analysis frameworks indicated in literature (ÖGMUNDARSON et al., 2020b) that consistently combine environmental and economic indicators to support optimized biochemical production at early development stages. These frameworks propose the monetization of environmental impacts that, added to the economic indicators of the process, can generate process indexes that be used in the investment analysis. Given the global appeal for government policies to reduce environmental damage, it is expected that the adoption of this type of combined framework will be important for the expansion of bioeconomy.

4. Conclusions

Producing biochemicals from renewable resources is a key driver for moving towards a sustainable society where energetic efficiency and the use of agricultural residues are performance indicators (ÖGMUNDARSON et al., 2020b). A LCA of a nanocellulose production unit annexed to a sugarcane biorefinery was performed. The results showed that the GWP potential of the proposed process is concentrated in specific steps as the organosolv pretreatment and the sulfuric acid from hydrolysis. The identification of these steps can orientate future researchers for the demand of new methods of lignocellulosic biomass purification which can be environmentally friendly withal economically competitive for the nanocellulose production.

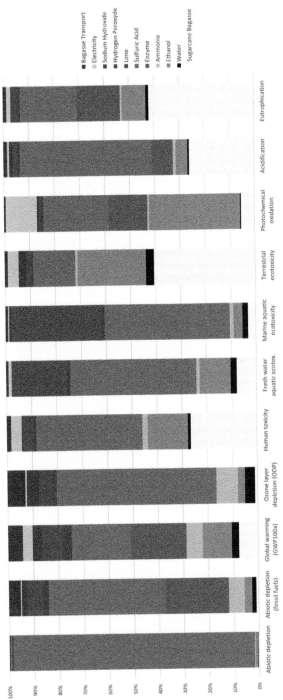

Figure 2 - Relative environmental impacts of each process input for obtained thorough Life Cycle Assessment (*CML-IA Baseline v3.04 2000* method, *SimaPro®* software).

References

C. Hohenthal; M. Ovaskainen; D. Bussini; P. Sadocco; T. Pajula; H. Lehtinen; J. Kautto; K. Salmenkiv. Final assessment of nano enhanced new products. In SUNPAP (Scale-up Nanoparticles in Modern Papermaking), Oct.31, 2012. CTT Technical Research Center of Finland, InnovHub-SSCCP, Poyry Management Consulting Oy. Funded by European Community's 7th Frame Work Programme under Grant Agreement n° 228802; VTT Technical Research Centre of Finland: Espoo, Findland, 2012.

D. M. Nascimento, A. F. Dias, C. P. Araújo Junior, M. F. Rosa, J. P. S. Morais, M. C. B. Figueirêdo, 2016. A comprehensive approach for obtaining cellulose nanocrystal from coconut fiber - part II: environmental assessment of technological pathways, Industrial Crops and Products, volume 93, pages 58-65.

E. S. Domalski, T. L. Jobe Jr., T. A. Milne, 1986. Thermodynamic data for biomass conversion and waste incineration, 326 pages, National Bureau of Standards, US Department of Energy.

F. Foroughi, E. R. Ghomi, F. M. Dehaghi, R. Borayek, S. Ramakrishna, 2021. A review on the life cycle assessment of cellulose: from properties to the potential of making it a low carbon material, Materials, volume 14, pages 714-736.

F. Piccinno, R. Hischier, S. Seeger, C. Som, 2018. Predicting the environmental impact of a future nanocellulose production at industrial scale: application of the life cycle assessment scale-up framework, Journal of Cleaner Production, volume 174, pages 283-295.

IBICT - Brazilian Institute of Information in Science and Technology, 2014. Desenvolvimento Sustentável e Avaliação do Ciclo de Vida, 38 pages, CNI, Brasília, Brazil.

Markets and Markets, 2021. Nanocellulose Market worth $661.7 million by 2023, available in: <https://www.marketsandmarkets.com/PressReleases/nanocellulose.asp>, acessed on August 04, 2021.

O. M. Vanderfleet, E. D. Cranston, 2021. Production routes to tailor the performance of cellulose nanocrystals, Nature Reviews Materials, volume 6, pages 124-144.

O. Ögmundarson, S. Sukumara, M. J. Herrgård, P. Fantke, 2020a. Combining environmental and economic performance for bioprocess optimization, Trends in Biotechnology, volume 38, issue 11, pages 1203-1214.

O. Ögmundarson, M. J. Herrgård, J. Forster, M. Z. Hauschild, P. Fantke, 2020b. Addressing environmental sustainability of biochemicals, Nature Sustainability, volume 3, pages 167–174.

R. Arvidsson, D. Nguyenand, M. Svanström, 2015. Life cycle assessment of cellulose nanofibrils production by mechanical treatment and two different pretreatment processes, Environmental Science and technology, volume 49, pages 6881-6890.

Acknowledgements

The authors would like to thank the Brazilian research funding agencies Fundação de Amparo à Pesquisa do Estado de São Paulo - FAPESP (Grants 2016/10636-8 and 2019/25261-8) and Conselho Nacional de Desenvolvimento Científico e Tecnológico - CNPq (Grant 140761/2017-9). This study was financed in part by Coordenação de Aperfeiçoamento de Pessoal de Nível Superior – CAPES, Brazil (Finance Code 001) with Grant 88887.364443/2019-00.

Antonis Kokossis, Michael C. Georgiadis, Efstratios N. Pistikopoulos (Eds.)
PROCEEDINGS OF THE 33rd European Symposium on Computer Aided Process Engineering
(ESCAPE33), June 18-21, 2023, Athens, Greece

Decarbonization of energy-intensive industries: sustainable implementation of CO₂ recycling within the industrial symbiosis

Marta Rumayor*, Javier Fernández-González, Antonio Domínguez-Ramos,
Angel Irabien
*Universidad de Cantabria, Departmento de Ingenierías Química y Biomolecular,
Av. Los Castros s/n, Santander, Spain*
*rumayorm@unican.es

Abstract

Several decarbonization strategies have been proposed for the cement sector including fuel switching (FS), carbon storage (CS), and carbon recycling (CR). FS and CS are expected in the short-term whereas CR is expected by 2050. Between the CR options, electrochemical reduction (ER) provides the opportunity to close the anthropogenic CO_2 cycle producing chemicals while promoting industrial symbiosis (IS). However, it requires high penetration of renewables. This study evaluates three decarbonization scenarios emphasizing the IS benefits provided by ER. Since ER is at low TRL, we elucidate the prospects to ensure its environmental feasibility within the European cement sector. We analyze the prospects of ER performance to compete with FS, which is viable even under low TRL. However, ER cell energy efficiencies are higher than 50% and methanol concentrations higher than 40%wt. are needed to compete with CS. From the defossilization perspective CR exhibits a wider margin of opportunity to compete even with CS.

Keywords: carbon dioxide, electrochemical reduction, decarbonization, defossilization, cement

1. Introduction

Since the consequences of climate change become more apparent, industries, scientists, and governments are committed to meet stringent greenhouse gas (GHG) emissions targets. These climate goals together with the recent geopolitical and energy market reality require us to drastically accelerate the energy transition increasing Europe's energy independence from unreliable suppliers and volatile fossil fuels.

Cement, which is an important building material, will pave the way to achieve a climate-resilient infrastructure. However, the success of the energy transition hinges on the effectiveness of its decarbonization. During cement production, carbon dioxide (CO_2) is emitted directly from the chemical conversion process of clinker production (calcination) but also because of fossil fuel combustion (mainly petcoke, coal, and oil). Given the nexus of cement production with the world's population, cement demand has been projected to increase by 2 or 3 times during the next few decades (Favier et al., 2018). This trend will lead to drastic consequences in terms of CO_2 emissions and fossil resource depletion.

Today, the decarbonization of energy-intensive industries as the cement sector is not straightforward. It employs extremely high temperatures, which can be achieved cost-effectively only by burning fossil fuels. The challenge of finding new, carbon-free approaches to produce low-carbon cement is just as complicated as well as expensive.

The recent cement decarbonization roadmap has predicted an amount of 7.7 Gt CO_2 of cumulative direct CO_2 savings by 2050 under the proper application of reduction strategies (International Energy Agency, 2018). This could be only achieved through improving energy efficiency, fuel switching (FS) to those that are less carbon-intensive, reducing clinker content in cement, and implementing innovative technologies such as Carbon Sequestration (CS) and/or Carbon Recycling (CR). In this sense, innovative CR technologies have arisen in the last decade as promising options to boost the so-called industrial symbiosis (IS) concept. The IS concept was introduced in the fifth Assessment Report from the Intergovernmental Panel on Climate Change (IPCC) as a way to reduce raw material consumption while mitigating GHG emissions on a process level, industry park level, and national level. CR implementation in the cement sector as an IS strategy can contribute not only to reaching the projected cumulative CO_2 emissions reduction but also to reducing the consumption of fossil fuels elsewhere.

The present study evaluates the benefits of the CR strategy in the cement sector in comparison with FS and CS. We have chosen methanol as the CO_2 conversion product given its market capacity and potential uses as fuel and chemical (Olah, 2005). Several CO_2 conversion pathways to methanol have been proposed in the latest decades, including catalytic conversion (CC), thermochemical (TC), and electrochemical reduction (ER) (Roh et al., 2020). We have selected ER route given its potential scalability, mild temperature, and pressure conditions, as well as the final possibility of using sunlight directly in the long-term. In a previous environmental assessment carried out by the authors (Rumayor et al., 2019), it was analyzed the influence of the selected key performance parameters (KPPs) in the methanol carbon footprint (CF), concluding that the feasibility of the ER technology is still hampered by the low methanol production rates, low current densities, and Faraday efficiencies. This fact results in high-energy requirements and therefore, CO_2 savings are restricted to the utilization of a low-carbon electricity source. This study aims to provide a comprehensive perspective on the technical prospects of ER technology to maximize the decarbonization/defossilization synergy in the European cement sector including the renewable requirement. The findings can promote the development of CR within the climate and defossilization transition.

2. Methodology

The present study aims to elucidate the environmental trade-offs for a feasible implementation of a CR strategy based on the electrochemical reduction (ER) of CO_2 into methanol within the European cement sector. The assessment is conducted in terms of carbon footprint (CF), fossil abiotic depletion (f-ADP) as well as renewable electricity requirements. The proposed CR scenario is compared with two scenarios, which are expected in a shorter timeframe. These are based on: i) FS; and ii) CS. As the objective is to emphasize the IS benefits involved in the co-production of cement and methanol in the CR scenario, the combined conventional production of both cement and methanol is used as a benchmark for comparison purposes. A Life Cycle Assessment (LCA) tool is applied according to international standards ("ISO 14040:2006," 2006b; "ISO 14044:2006," 2006). The LCA is conducted using the software GaBi v9.5 (Sphera) and the Ecoinvent database v3.8. The functional unit (FU) is set as 1 ton of cement plus 429 kg of methanol, which is the stoichiometric amount of methanol that can be converted from the direct emissions of cement (672 kg CO_2/ton cement) by ER. Figure 1 shows the system boundary of the scenarios analyzed in the present study. The baseline scenario considers the production of 1 ton of cement and the conventional synthesis of 429 kg of methanol according to their Ecoinvent datasets. Briefly, the production of 1 ton of cement involves the consumption of resources such as limestone, gypsum, metals, and fossil fuels including natural gas (0.158 m³), heavy fuel oil (23 kg), petcoke (3.53 kg) and hard coal (32 kg). The CO_2 direct emissions associated with the process account for 672 kg of CO_2. The production of 429 kg of methanol involves mainly natural gas (280 m³), which is the main raw material, together with metals and water, being 2.8 GJ the heating needs that come from natural gas. The CR scenario captures CO_2 assuming membrane separation technology, which has a 90% of efficiency and an energy requirement of 300 kWh/ton CO_{2in} (Giordano et al., 2018). The ER of CO_2 yields methanol by direct CO_2 reduction with water and electricity. ER process comprises 4 steps: 1) capture of the CO_2; 2) electroreduction of CO_2 into methanol and subproducts (O_2 and H_2); 3) recovery of unreacted CO_2; and 4) distillation of methanol up to commercial grade. The inventory for the electroreduction is obtained by a mathematical process model described in previous studies (Rumayor et al., 2019, 2020). Given the low technology readiness level (TRL) of the ER route into methanol at its current stage, we have carried out a series of sensitivity analyses. The energy efficiency of the ER cell is ranged between 10% (current) and 90% (optimistic) while the methanol concentration at the outlet stream of the ER cell ranges between 10%wt. and 60%wt. The bivariate analyses show the influence in the categories of CF and f-ADP for the selected FU. It is expected a rapid installation of wind/PV solar in projected national plans as well as the economic incentives for carbon reduction in industries. Therefore, the analyses are conducted using two electricity impacts: i) EU-PV solar mix (2021); and ii) EU-wind-mix (2021). Their impacts may be representative of the environmental performance of renewable electricity in the medium and long-term timeframes, respectively.

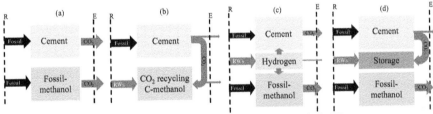

Figure 1. System boundary of a) Baseline; b) CR; c) FS; and d) CS

In the FS scenario, electricity is assumed renewable and the heat to produce 1 ton of cement (3.8 GJ) and 426 kg of methanol is considered to be supplied by 100% green H_2. Green H_2 production is proposed to be synthesized by the polymer electrolyte membrane (PEM) technology, which is currently at the commercial stage. The electricity consumption of PEM is fixed at 50 kWh/kg H_2, which is an optimistic value for the medium term. It must be considered that other resources such as limestone, gypsum, metals, and natural gas, which are used as raw materials, are the same that those considered in the Baseline. The CS scenario includes the capture of CO_2, the compression and its sequestration. Energy consumption for the compression and injection is 117 kWh/t of CO_2 according to the literature (Koornneef et al., 2008). Note that the production of 429 kg of methanol will be kept as in the Baseline. Authors will display FS and CS in the result section only as benchmarking, being the prospective exploration of a combination of CCUS out of the scope of the present study because of the differences in TRL of the proposed technologies. Finally, it is calculated the resources and carbon emissions of an hypothetical decarbonized scenario. This scenario assumes the decarbonization of 10% of EU cement production by the proposed CR pathway. For a fair comparison, the impacts are also calculated for the Baseline scenario (conventional). Therefore, the resources and impacts are calculated for 10% of the overall EU cement capacity (20 Mton/yr from the overall 200 Mton/yr). The decarbonized scenario is performed according to the figures of merits found in the sensitivity assessment as they will ensure the feasibility of the proposed CR pathway.

3. Results and discussion

Figure 2 displays the CF and f-ADP results obtained from the sensitivity analysis for the FU (1 ton of cement+429 kg of methanol) in the CR scenario. Since ER conversion in the CR scenario consumes a significant amount of electricity, the carbon intensity of the electricity to power the conversion process is critical. The CF and f-ADP results include the influence of the environmental impacts of renewable electricity to power the processes using the EU-PV solar mix (Figure 2 (a) and (c)) and the EU-Wind mix (Figure 2 (b) and (d)). There are also shown the CF and f-ADP of the CS (orange line) and FS (yellow line) scenarios as a function of the environmental impacts of renewable electricity. Furthermore, the baseline scenario is presented as a white dashed line and it is calculated keeping fixed the current EU grid mix. It should be mentioned that the contribution of the electricity to the overall CF value per FU (1120 kg CO_2e) as well as to the overall f-ADP per FU (16500 MJ) in the Baseline are negligible. On one hand, direct CO_2 emissions are the main contributor to the overall CF value representing 85% according to the datasets. On the other hand, natural gas, which is the raw material in methanol production, is the main contributor to the overall f-ADP. From the CF perspective (Figure 2 (a) and (b)), the CR pathway exhibits a wide margin of opportunity to compete with the baseline scenario. CR may be competitive with the FS scenario using renewable electricity with a carbon intensity lower than 0.07 kgCO_2e/kWh for methanol concentrations higher than 40%wt and energy efficiencies higher than 40%. However, the window of opportunity to compete with CS is narrower. The results indicate that CR may be competitive with CS only using renewable electricity with a CF lower than 0.01 kg CO_2e/kWh and efficiencies higher than 50%. Note that the current energy efficiency of CO_2 ER is near 10-15%. As can be seen in Figures 2 (c) and (d), the benefits in the f-ADP are clearer. The results indicate a wider window of opportunity even under low-TRL performance of the ER technology compared with Baseline and CS scenarios. Note that 280 m^3 of natural gas is saved in the CR scenario as the amount of methanol (429 kg) is synthesized from the CO_2 direct emissions from the production of 1 ton of cement. According to the results, energy efficiencies higher than 40% and 50% and methanol concentrations higher than 40% could be a positive scenario to compete with CS and FS, respectively.

Figure 3 shows an overview of the resources, in terms of f-ADP, and the carbon emissions, in terms of CF, of the current cement production. It has also analyzed a hypothetical decarbonized scenario based on CR. The values shown in the figure correspond to the production of 10% of the overall EU cement capacity (20 Mton/yr from the overall 200 Mton/yr). The decarbonized scenario is performed according to the figures of merits found in the sensitivity analysis. It is assumed a methanol concentration of 40%wt. and a cell energy efficiency of 50%. EU-Wind-mix electricity impacts have been used to calculate the environmental impacts of the CR plant. According to the results, f-ADP could be reduced by a percentage of 70% with the CR strategy because of the resource savings (natural gas) within the conventional methanol production. On the other hand, CF is reduced by around 80% because CO_2 recycling substitutes natural gas in methanol production. Since ER is an electricity-intensive process, the decarbonization of

10% of EU cement production would require $15 \cdot 10^{13}$ kWh of low-carbon renewable electricity.

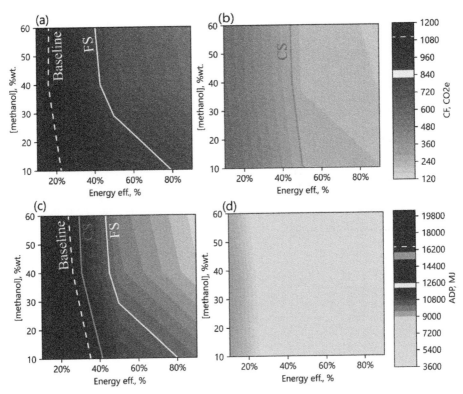

Figure 2. Influence in the CF and f-ADP (per FU) during the sensitivity analysis of CR scenario using electricity based on EU-PV-Mix (a, c) and EU-Wind-Mix (b,d)

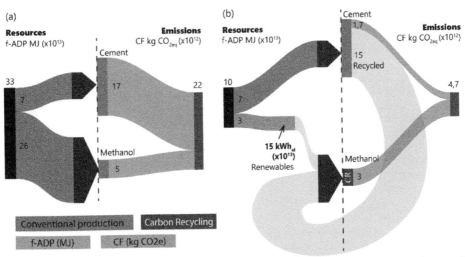

Figure 3. Environmental impacts (f-ADP and CF) of: (a) conventional production; and (b) decarbonized scenario considering 10% of European cement capacity

4. Conclusions

This work has provided the environmental prospects that CR strategy through ER could bring to the EU cement sector within the climate and defossilization transition. We have applied a prospective life cycle assessment tool to compare the CF and f-ADP of CR with FS and CS strategies. Since ER conversion is an electricity-intensive process, the environmental impacts of renewable electricity to power the conversion process is critical. In the nearer term, implementing CS seems more beneficial to reach the decarbonization goal of the cement sector. However, the CR strategy, which is an industrial symbiosis approach may boost the circular economy. The pathway involves a 70% of fossil resource savings and a reduction of 80% of carbon emissions.

Acknowledgments

Authors thank the Spanish Ministry of Science and Innovation for the financial support through the project PID2020-112845RB-I00 funded by MCIN/ AEI /10.13039/501100011033.

References

Chauvy, R., & de Weireld, G. 2020. CO_2 Utilization Technologies in Europe: A Short Review. In Energy Technology, 8, 12

Giordano, L., Roizard, D., & Favre, E. 2018. Life cycle assessment of post-combustion CO_2 capture: A comparison between membrane separation and chemical absorption processes. International Journal of Greenhouse Gas Control, 68, 146–163.

International Energy Agency. 2018. Technology Roadmap - Low-Carbon Transition in the Cement Industry. www.wbcsdcement.org.

Koornneef, J., van Keulen, T., Faaij, A., & Turkenburg, W. 2008. Life cycle assessment of a pulverized coal power plant with post-combustion capture, transport and storage of CO_2. International Journal of Greenhouse Gas Control, 2(4), 448–467.

Olah, G. A. (2005). Beyond oil and gas: The methanol economy. Angewandte Chemie - International Edition, 44(18), 2636–2639.

Roh, K., Bardow, A., Bardow, A., Bardow, A., Bardow, A., Bongartz, D., Burre, J., Chung, W., Deutz, S., Han, D., Heßelmann, M., Kohlhaas, Y., König, A., Lee, J. S., Meys, R., Völker, S., Wessling, M., Lee, J. H., Mitsos, A., Mitsos, A. 2020. Early-stage evaluation of emerging CO_2 utilization technologies at low technology readiness levels. Green Chemistry, 22(12), 3842–3859.

Rumayor, M., Dominguez-Ramos, A., & Irabien, A. 2019. Innovative alternatives to methanol manufacture: Carbon footprint assessment. Journal of Cleaner Production, 225, 426–434.

Rumayor, M., Dominguez-Ramos, A., & Irabien, A. 2020. Toward the Decarbonization of Hard-To-Abate Sectors: A Case Study of the Soda Ash Production. ACS Sustainable Chemistry and Engineering, 8(32), 11956–11966.

Rumayor, M., Fernández-González, J., Domínguez-Ramos, A., & Irabien, A. 2022. Deep Decarbonization of the Cement Sector: A Prospective Environmental Assessment of CO_2 Recycling to Methanol. ACS Sustainable Chemistry and Engineering, 10(1), 267–278.

Antonis Kokossis, Michael C. Georgiadis, Efstratios N. Pistikopoulos (Eds.)
PROCEEDINGS OF THE 33rd European Symposium on Computer Aided Process Engineering
(ESCAPE33), June 18-21, 2023, Athens, Greece

Gray and hybrid green ammonia price sensitivity to market fluctuations: the Russia-Ukraine war case

Andrea Isella[a], Alberto Lista[a], Gabriele Colombo[b], Raffaele Ostuni[b],
Davide Manca[a]*

[a] *PSE-Lab, Process Systems Engineering Laboratory, Dipartimento di Chimica, Materiali e Ingegneria Chimica "Giulio Natta", Politecnico di Milano, Piazza Leonardo da Vinci 32, 20133 Milano, Italy*
[b] *Casale SA, Via Giulio Pocobelli 6, 6900 Lugano, Switzerland*
davide.manca@polimi.it

Abstract

Ammonia synthesis is the biggest emitter of carbon dioxide among chemical processes. Emission regulations will become even stricter over the coming decades and call for innovative sustainable pathways to meet the decarbonization targets. An answer to such a transition may be retrofitting (*i.e.* hybridization) the existing plants by implementing low- or zero-emission technologies. One of the most promising retrofits is introducing green hydrogen from electrolyzers into the synthesis loop. This paper compares the operational expenditures (OPEX) of a conventional gray ammonia plant and a 10%-decarbonized electrolysis-based hybrid-green ammonia plant. The OPEX terms depend primarily on natural gas costs, renewable electric energy prices, and carbon tax values. Specifically, hybrid-green ammonia proves to be competitive at current EU natural gas quotations, which have increased more than five-fold since the first semester of 2021 due to the ongoing Russia-Ukraine war.

Keywords: decarbonization; hybridization; electrolysis; renewables; hydrogen.

1. Introduction

Ammonia synthesis is the most carbon-intensive process in the chemical industry, with emissions above 440 Mt_{CO2eq}/y in 2020 (Isella and Manca, 2022). Indeed, current ammonia synthesis heavily relies on fossil fuels to produce the hydrogen needed in the ammonia converters and generate the heat and compression required for the process. This explains the blooming of decarbonization strategies to mitigate the carbon footprint of conventional "gray" ammonia. Among them, we have: (i) blue ammonia, which implements various solutions of carbon capture and utilization or storage (CCUS); (ii) green ammonia, which fully decouples the ammonia production from fossil feedstocks (*e.g.*, hydrogen production through an electrolyzer powered by renewable electricity); (iii) hybrid-green ammonia, which consists in revamping the existing conventional plants by installing an electrolyzer to replace part of the gray, fossil-based hydrogen feedstock with a green, water-based one (MPP, 2022). Concerning the latter technology, it is worth clarifying that within the perspective of a fully decarbonized ammonia scenario, such a production method should be considered a transitional technology due to its reduced CO_2 abatement potential, which is partial by nature. Nonetheless, until near-zero-emissions paths become commercially ready in the following decades, an available-to-date alternative such as hybrid-green ammonia can be the best way to meet emissions reduction in ammonia synthesis. The main purpose of the paper is to identify the price volatility of hybrid-green ammonia. Specifically, its market trends are reproduced and compared with the far more widely-known and better-characterized conventional ammonia prices. Indeed, the cost of hybrid-green ammonia is shown to be, just as its gray counterpart, a function of several market variables (*e.g.*, the price of fossil fuels and electrical energy). To our knowledge, this assessment is unprecedented in the scientific literature and represents a first attempt to formally describe the market evolution of hybrid-green ammonia. Moreover, the global geopolitical conditions given by current events (*in primis*, the Russia-Ukraine war), add even more novelty to our study.

2. Simulation of the hybrid-green ammonia plant

A medium-high capacity 2000 t/d stand-alone natural gas-based ammonia plant was simulated in UniSim® Design R491 to assess its economic hybridization potential. The decarbonization target was set to 10% (a distinctive value of such plants resulting mainly from design considerations, as discussed in the following). The injection of green hydrogen from an alkaline electrolyzer (AEL) powered by renewable electric energy can achieve that target. Indeed, this strategy aims to reduce the "front-end" (*i.e.* the whole of the reforming, shift, and syngas purification sections) load and, consequently, to lower both the natural gas fuel and feed demands of the process (see Figure 1).

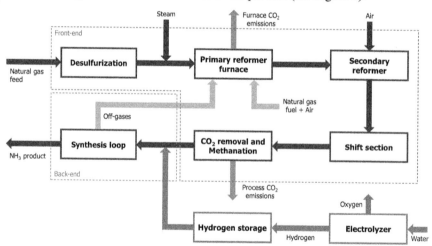

Figure 1: Block flow diagram of the simulated hybrid-green ammonia plant.

An equivalent amount of hydrogen fed to the ammonia synthesis loop must be generated by electrolysis, assuming a 10% average reduction in carbon dioxide emissions ($DeCO_2 = 10\%$). However, the hydrogen production capacity of the electrolyzer is strongly affected by the fluctuating trends of renewable electric energy profiles. Indeed, renewable energy production is firmly susceptible to the varying availability of energy sources (such as sunlight and wind in the case of solar and wind farms). Since the electrolyzer converts such fluctuating electricity profiles into hydrogen flows that are thus intermittent, it follows that the target green hydrogen productivity should be considered as an average of the oscillations that span from a minimum of zero (when no green electricity is available) to a maximum, which is strictly related to the capacity factor of the renewable power station (that determines the peak power available to the user) and also to some design considerations, precisely the reformers' load. Indeed, considering to keep constant the ammonia productivity of the plant, the injection of an excessive amount of green hydrogen into the synthesis loop results in substantial drops in the front-end flow rates (*i.e.* the ones linked to the conventional gray route), which may affect a few critical aspects, such as the operability of the fired tubes in the primary reformer and the rise in the temperature of the secondary reformer. To tackle this, hydrogen storage is needed downstream of the electrolyzer to set the maximum amount of green hydrogen fed to the synthesis loop. Specifically, it was assumed that the upper limit of green hydrogen input should be such to require a decrease in the front-end "throughput" (*i.e.* its total processed flow rate) to 80% maximum turndown to ensure reliable operative conditions of the primary reformer. Respecting such a directive leads to an instantaneous decarbonization degree of about 20% ($DeCO_2 = 20\%$) at the highest. This is a somewhat limiting constraint since, by ignoring all the above design recommendations and basing the calculations exclusively on the peak availability of green electricity (which depends only on the capacity factor of the renewable power source), the maximum achievable decarbonization degree could be, in principle, higher.

Table 1 reports the leading process variables of the simulated hybrid-green ammonia plant in its three main operating conditions: S0, which corresponds to the minimum (*i.e.* null) green hydrogen availability; S1, which corresponds to the average working conditions (which should guarantee a 10% decarbonization target, as requested); and S2, which corresponds to the upper limit of green hydrogen input.

Table 1: Process variables of the simulated hybrid-green plant in its three main operative conditions. The electrolyzer has a specific electricity consumption of 4.5 kWh/Nm^3_{H2}, a specific water consumption of 0.9 kg_{H2O}/Nm^3_{H2}, and a hydrogen output temperature and pressure of 35 °C and 30 bar, respectively (Nel Hydrogen, 2021). Note that a natural gas-to-electricity conversion factor of 3 MWh_{LHV}/MWh_{el} was used to evaluate the net plant energy consumption.

Process variable		*Status "S0"* (0% $DeCO_2$)	*Status "S1"* (10% $DeCO_2$)	*Status "S2"* (20% $DeCO_2$)
Primary reformer throughput	[t/h]	163	146	129
Natural gas feed to the front-end	[t/h]	38.0	33.9	29.9
Natural gas fuel to the furnace	[t/h]	12.9	11.9	10.7
Green hydrogen production	[t/h]	0	1.74	3.55
Green electricity consumption	[MW]	0	86.8	177
Net plant energy consumption	[MWh_{LHV}/t_{NH3}]	8.14	10.4	12.9
Furnace CO_2 emissions	[t/h]	38.9	34.8	30.7
	[t_{CO2}/t_{NH3}]	0.467	0.418	0.368
Process CO_2 emissions	[t/h]	97.4	87.9	78.1
	[t_{CO2}/t_{NH3}]	1.17	1.06	0.937
Total CO_2 emissions	[t/h]	136.3	122.7	108.8
	[t_{CO2}/t_{NH3}]	1.64	1.48	1.31

As expected, an increase in the green hydrogen input to the ammonia synthesis loop leads to a significant decrease in both the furnace and process carbon dioxide emissions at the expense of the net energy consumption of the plant. Indeed, as the heating demand associated with natural gas streams declines, electricity demand rises significantly. However, also other implications occur, namely: (i) the increase in the operating temperature of the secondary reformer (from 996 °C in S0 to 1043 °C in S1 and 1104 °C in S2) and (ii) the drop in the methane slip (from 0.244 mol% in S0 to 0.098 mol% in S1 and 0.032 mol% in S2) and the inert gases content (from 11.7 vol% in S0 to 8.92 vol% in S1 and 7.28 vol% in S2) in the ammonia synthesis loop. Consequence (i) is mainly due to the choice of keeping the process air flowrate constant (to meet the nitrogen demand of the plant and guarantee the stoichiometric H_2/N_2 ratio at the inlet of the first ammonia converter). This leads to an increase in the oxygen-to-carbon molar ratio at the secondary reformer inlet (from 0.757 in S0 to 0.848 in S1 and 0.963 in S2) and, therefore, in its operative temperature. Concerning consequence (ii), instead, it is strictly related to the first one, (i). Indeed, such a result is due to the higher temperatures in the secondary reformer, which shift the steam reforming reaction towards the products. Finally, it is worth noting that while consequence (i) is very detrimental, as excessive overheating of the secondary reformer and the downstream process boiler must be avoided, consequence (ii) is beneficial to the process as it reduces compression costs and improves the reaction thermodynamics and kinetics in the ammonia converters.

3. Gray and hybrid-green ammonia production costs

Through the previous simulations, hybrid-green ammonia proves to be an available-to-date, feasible (although conspicuously electricity-demanding) technique to decarbonize ammonia synthesis. However, the actual testing benchmark to assess the competitiveness of an alternative production route to the conventional one is the cost of the final product. Indeed, since the cheapest process is expected to be the most appealing, the present section focuses on evaluating both gray and hybrid-green ammonia production costs. Precisely, the following assumptions hold:

- both facilities operate 8000 h/y;
- the hybrid-green plant has an average annual decarbonization target of 10%;
- the costs for cooling and demineralized water are 0.04 and 0.75 €/m³, respectively (Intratec, 2022);
- the revenues from the medium-pressure steam production within both plants are estimated by considering the natural gas duty that an auxiliary boiler would need for the same task (Noelker and Ruether, 2011). Specifically, using pure water at 35 °C and 40 bar leads to a steam energy export equal to 0.13 MWh_{LHV}/t_{NH3} from both facilities;
- the decarbonization degree of the hybrid-green plant is assumed to fluctuate in the 0-20% interval, depending on the instantaneous availability of renewable electricity;
- the installed electrolyzer capacity must be 177 MW to meet the hydrogen demand at the peak of green electricity availability (see Table 1, S2). On top of this, such power capacity value is overdesigned by 10%;
- the electrolyzer capital expenditures (CAPEX) must be accounted for. Indeed, since the electrolyzer is the most expensive unit that the hybridization retrofit calls for, its cost affects the price of the final product. Specifically, the electrolyzer exhibits a 10-year depreciation time and a fixed fee of 700 €/kW_{el} (including engineering, housing, and balance-of-plant components such as the compression and the water and hydrogen purification units). This is a rather low (yet conservative) value compared to the average ones reported in the literature (which typically span from 800 to 1000 €/kW_{el}).

Indeed, it is recognized that the economy of scale notably affects AEL modules with
an installed power capacity higher than 100 MW (Fraunhofer ISE, 2021).

It is worth noting that the main contributions to the final prices of both gray and hybrid-
green ammonia are given by natural gas, carbon tax, and (only for the hybrid plant)
renewable electricity, whose values are continuously changing at rather high rates. This
aspect is very significant since specific fluctuations of such variables affect the costs of
both production pathways, eventually overturning their mutual relationship. For example,
decreasing renewable power purchase agreement (PPA) costs and increasing values of
carbon taxes are expected to promote the shift to environmentally mitigating technologies
despite the conventional, carbon-intensive (and otherwise far cheaper) ones. Anyway, the
aspect that has predominantly affected the recent evolution of the price trends in ammonia
production is the runaway behavior of natural gas quotations. Indeed, with the outbreak
of the Russia-Ukraine war, its already rising trend, which started in the summer of 2020,
reached extraordinary peaks in a few weeks. Specifically, the remarkable surge in natural
gas prices made the hybrid-green ammonia production pathway cheaper than the
conventional gray one. Indeed, the reduced fossil fuel consumption in the hybridized
facility makes it less susceptible to changes in natural gas quotations.

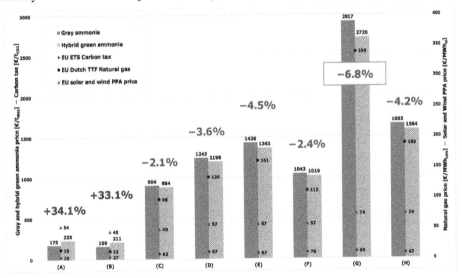

Figure 2: Gray and hybrid-green ammonia production costs and carbon tax values (left y-axis);
natural gas and renewable PPA prices (right y-axis) for the A-H scenarios. The percentages at the
top of the bars quantify the difference between the hybrid-green vs. gray ammonia total
production costs. Input data from Trading Economics (2022), for natural gas and carbon tax, and
LevelTen Energy (2022), for PPAs.

To follow the ever-changing panorama of incidents that significantly affected the main
prices/costs of both process typologies, we identified eight scenarios that occurred in the
last three years (see also Figure 2):

    A. Pre Covid-19 era (30-Sep-19);
    B. After 6 months of Covid-19 (30-Sep-20);
    C. After 18 months of Covid-19 (30-Sep-21);
    D. Start of Russia-Ukraine war (24-Feb-22);
    E. One week into the conflict (03-Mar-22);
    F. One month into the conflict (24-Mar-22);

G. Highest peak of natural gas price (26-Aug-22);
H. First EU protective measures (30-Sep-22).

Figure 2 shows the economic assessment, where the prices of gray and hybrid-green ammonia were considered by utilizing as input data the values of natural gas price, renewable PPA price, and carbon tax for the European market at the A-H scenarios.

As previously mentioned, the sensitivity of ammonia production prices to market fluctuations is so elevated that the remarkable increase in natural gas quotations has made hybrid-green ammonia cheaper than gray ammonia in the recent C-H scenarios (*i.e.* since 30-Sep-2021). Indeed, hybrid-green ammonia has gradually become cheaper than gray ammonia, despite starting from a 34% higher price than the conventional route in the pre Covid-19 era. This is still true after the highest recorded peak in natural gas quotations (on 26-Aug-2022), with a discounted price of 4.2% to gray ammonia.

4. Conclusions

This paper introduced a novel method to evaluate ammonia market price, aimed at discussing the economic attractiveness of hybrid-green vs. gray ammonia production. The simulation of a 2000 t/d hybridized ammonia plant highlighted the main problems related to the operating limits of that facility. These are the significant demand for renewable power and operability issues in the reformers as far as excessively high decarbonization degrees are concerned. Regarding the economic assessment, the production costs of gray and hybrid-green ammonia proved particularly sensitive to market fluctuations, primarily to natural gas quotations. The steep increase in the natural gas price due to the outbreak of the Russia-Ukraine war made the hybrid-green production route cheaper than the gray one, although being far more expensive just a few months before. This outcome matches the increased adoption of electrified technologies to decarbonize ammonia synthesis.

References

Fraunhofer ISE (2021). Cost Forecast for Low-Temperature Electrolysis. Fraunhofer Institute for Solar Energy Systems ISE.

Intratec (2022). Cooling Water Current Costs, Historical Series & Forecasts. Available at: https://www.intratec.us/products/water-utility-cost-database/commodity/cooling-water-cost (accessed on November 17th, 2022).

Intratec (2022). Demineralized Water Current Costs, Historical Series & Forecasts. Available at: https://www.intratec.us/products/water-utility-cost-database/commodity/demineralized-water-cost (accessed on November 17th, 2022).

Isella, A., & Manca, D. (2022). GHG Emissions by (Petro)Chemical Processes and Decarbonization Priorities—A Review. Energies, 15(20), 7560.

LevelTen Energy (2022). Europe's Renewable Energy Shortage Causes PPA Prices to Rise More Than 11% in Q3 and 51% Year Over Year, According to LevelTen Energy. Available at: https://www.leveltenenergy.com/post/2022q3-ppa-europe (accessed on November 11th, 2022).

MPP (2022). Making Net-Zero 1.5°C-Aligned Ammonia Possible. Mission Possible Partnership.

Nel Hydrogen (2021). Nel Hydrogen Electrolyzers The World's Most Efficient and Reliable Electrolysers. Available at: https://nelhydrogen.com/wp-content/uploads/2020/03/Electrolysers-Brochure-Rev-D.pdf (accessed on October 30th, 2022).

Noelker, K., & Ruether, J. (2011). Low Energy Consumption Ammonia Production: Baseline Energy Consumption, Options for Energy Optimization. Presented at the Nitrogen & Syngas 2011 International Conference & Exhibition, Duesseldorf, Germany.

Trading Economics (2022). EU Carbon Permits. Available at: https://tradingeconomics.com/commodity/carbon (accessed on November 11th, 2022).

Trading Economics (2022). EU Natural Gas. Available at: https://tradingeconomics.com/commodity/natural-gas (accessed on November 11th, 2022).

Antonis Kokossis, Michael C. Georgiadis, Efstratios N. Pistikopoulos (Eds.)
PROCEEDINGS OF THE 33rd European Symposium on Computer Aided Process Engineering
(ESCAPE33), June 18-21, 2023, Athens, Greece

Paving the way for the integration of synthesis, assessment, and design tools within an ontological framework

Adrián Pacheco-López[a,b], Kristiano Prifti[a,b], Flavio Manenti[b], Ana Somoza-Tornos[c], Moisès Graells[a], Antonio Espuña[a*]

[a] *Department of Chemical Engineering, Universitat Politècnica de Catalunya, Escola d'Enginyeria de Barcelona Est, C/ Eduard Maristany 16, Barcelona 08019, Spain*
[b] *CMIC Department "Giulio Natta", Politecnico di Milano, Piazza Leonardo da Vinci 32, Milan 20133, Italy*
[c] *Department of Chemical Engineering, Delft University of Technology, Van der Maasweg 9, 2629 HZ Delft, Netherlands*
*antonio.espuna@upc.edu

Abstract

The constant development of new alternatives to treat waste aids in closing material loops towards the circular economy and improving sustainability through the use of new renewable materials and energy. This fact leads to the increasing need for decision-making tools for process synthesis and assessment, which can be addressed with an integrated framework that employs ontologies for knowledge management and optimization tools to perform a hierarchical assessment of alternatives. The systematization of these procedures raises the need for tools to automate techno-economic and life cycle analyses. In this work, such a challenge is addressed through the additional integration of add-on modules such as the CapEx-Opex estimation tools and surrogate modeling within this framework. A case study on plastic waste is proposed with the inclusion of several pyrolysis and gasification alternatives. Results show pyrolysis, followed by the subsequent purification of its products, as the best alternative and helped identify main drivers for technologies feasibility such as feedstock purity and energy consumption.

Keywords: circular economy, integrated modeling, sustainable development, machine learning, economic optimization.

1. Introduction

The Circular Economy appears as a subject of paramount importance toward economically, environmentally, and socially sustainable development. Thus, many entities across the world are working hard to find alternatives that not only bring economic profit but also are environmentally benign and socially responsible. One of the most urgent matters is waste management and, in this specific line, many technical efforts are currently devoted to finding new waste-to-resource processes, recovering valuable resources, and closing material loops. Consequently, decision-makers face the challenge to determine which alternatives are most suitable for each kind of waste. To address this task, Pacheco-López et al., (2022) developed a framework of integrated tools to connect waste sources with valuable products and find the best routes to close material loops. The application of this framework requires performing economic and environmental analyses and developing models suitable for each one of the alternatives included. To undertake these tasks more systematically, this contribution aims to integrate new tools to make the framework more versatile and minimize the required human intervention for synthesizing and assessing new sustainable approaches. Within the proposed framework, the ontology is used to manage the information and data that are needed as inputs on each module, as well as storing the outputs, therefore centralizing all the knowledge required and produced in the system (orange arrows represent information exchange in Figure 1).

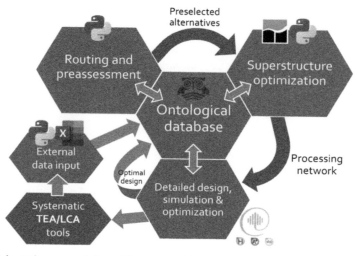

Figure 1. Schematic representation of the proposed framework with an additional module (green).

2. Methodology

The methodology used in this contribution is an extended version of a previously proposed one, in which several synthesis, assessment, and optimization tools were integrated to build a decision-making framework (Pacheco-López et al., 2022). There, a sequential approach is used to build alternative paths between waste materials and marketable products and to assess them from several points of view as well as to find the optimal configuration. First stage uses a new proposed algorithm and metric based on different objectives to pre-select alternatives, then a second stage consisting on a Multi-objective network optimization to find best routes based on each objective (for more details of this procedure see (Somoza-Tornos et al., 2021)). One of the identified challenges in this work was the need to perform previous techno-economic (TEA) and life cycle (LCA) assessments manually for every single process in order to introduce their results in the framework to be used by the other modules. To address this challenge, the need for systematization tools was identified to reduce human intervention and standardize the results. In this direction, some tools were already developed and are being implemented in the framework. Prifti et al. (2022) developed a CapEx-OpEx Robust Optimizer (CORO) framework that was intended to assess the cost evaluation of a plant including some sustainability targets, thus, it provides the framework with more dynamism and systematization in the TEA tasks. CORO automatically takes a simulation file (Hysys or ProII) as an input and generates the capital and operational expenditures for each process. Using this tool, the cost estimations for all processes in the ontology have been performed in order to obtain standardized results. Figure 1 shows the original methodology and the proposed additional module for techno-economic/life-cycle systematic assessment tools.

As of now, only systematic TEA is implemented for illustrative purposes, although future developments are intended to include LCA systematization in the same way.

On the other hand, process simulation is also a critical step for the design and optimization of new processes, which, in turn, may require considerable computational effort. Surrogate modeling offers the possibility to speed up simulation convergence times by substituting rigorous models with machine learning methods (Granacher et al., 2021), as the one proposed by Galeazzi et al., (2022) to predict the behavior of several process variables of existing steady-state digital twins of industrial plants. This tool can be accessorily used in the third stage of the methodology to ease the simulation step or to study the processes' sensitivity to variations in feed composition or operating conditions.

Table 1. Deployed process paths for MPW treatment, outputs, and GPI for 24 alternatives. Sor.: Sorting; FBR: Fluidized bed reactor; HGHTR: horizontal gas heated tube reactor; ZSM5: ZSM-5 zeolite; FCC: fluid catalytic cracking; R1: FCC-R1 commercial FCC equilibrium catalysts; HUSY: Ultrastabilized Y zeolite; SAHA: amorphous silica-alumina; HZSM: acid zeolite; SA9Z1: hybrid catalyst (SA: Zeolite, 9:1); LPG: C1-C3 aliphatics.

	Processes	Outputs	GPI
1-3	Pyrolysis 500°C R1/Red Mud/ZSM5/	C1-C5 alkanes, ethylene, H2, hexene, toluene, ethylbenzene, styrene, naphthalene, xylene & C9-C14	723-642
4	Pyrolysis 600°C /Y zeolite/	Methane, ethylene, propene, ethane, butane, butane, hydrogen, benzene, toluene, ethylbenzene & styrene	558
5-8	Acid FCC 390°C /HUSY//SAHA/R1/ZSM5/	Gasoline, LPG, light aromatics mixture, C9-C14 mixture, char & HCl	435-425
9	Gasiforming	Methanol & CO2 credits	288
10	Sor. + PE pyrolysis 520°C HGHTR /ZSM5/	Gasoline, ethane, propane, butane & char	216
11	Sor. + PE pyrolysis 740°C	LPG, ethylene, benzene, toluene, indane & pyrene	183
12-13	Sor. + PE pyrolysis FBR 375°C /HZSM/SA9Z1/	Gasoline, ethane, propane, butane & char	151-116
14	Sor. + PP pyrolysis 350°C	WPPO (diesel substitute) & char	96
15	Sor. + PE pyrolysis 1000°C	Methane, ethylene, propene, butadiene & benzene	87
16	Sor. + PP pyrolysis 760°C	LPG, ethylene, benzene, toluene & naphthalene	79
17	Idem + Ethylene hydration	LPG, ethanol, benzene, toluene & naphthalene	70
18	Co-gasification 850°C	Methane, ethane, syngas & char	63
19	Gasification 850°C	Methane, ethane, syngas & char	59
20	Electrified Gasiforming	Methanol & CO2 credits	54
21	Sor. + PS pyrolysis 425°C	Toluene, styrene, cumene & 1,3,5-triphenylbenzene	42
22	Sor. + PE pyrolysis 550°C	WPPO (waste plastic pyrolysis oil – diesel substitute)	35
23	Incineration	Energy Recovery	0
24	Landfill	None	0

3. Case Study

The framework is tested on a case study dealing with the recovery of plastic waste materials and the conversion/purification of the resulting products into other valuable carbon-based products. New alternatives are added to the set of alternatives available in the ontology from previous implementations using the same assumptions as presented in Pacheco-López et al., (2022). These new processes consist of gasification followed by methane reforming and conversion to obtain methanol as a final product, either using traditional energy sources or electrified ones (Prifti et al., 2021). A total of 66 processes (see Figure 2) are implemented in this case, where 200 paths were created and assessed, from which 24 were selected in the first stage (see Table 1).

The preselected alternatives were sent to the multi-objective network optimization module obtaining a set of configurations along with the representation with the corresponding Pareto fronts. The newly added alternatives are compared against those already available in the ontology. The feedstock considered corresponds with a mixture of plastic waste with the following mass composition: 45% polyethylene, 32% polypropylene, 20% polystyrene, and 3% of PVC. A plant to treat an annual amount of 20,000 tonnes of MPW (2.5 tonnes/hour for a plant operating 8000h a year) was considered and all processes were scaled accordingly.

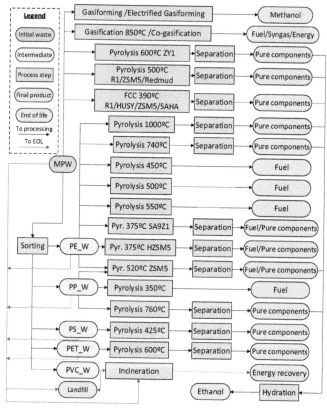

Figure 2. Implicitly generated graph in the pre-assessment stage with all tentative connections. PE: polyethylene, PP: polypropylene, PS: polystyrene, PET: polyethylene terephthalate, PVC: polyvinyl chloride; Pyr.: Pyrolysis; EOL: End of life.

4. Results

After updating the ontology with all the current and new process parameters, the algorithm is run and all the possible connections are made with an input-output matching procedure, resulting in a graph as seen in Figure 2. With the assessment procedure presented in Pacheco-López et al., (2021), all the paths are built and a list of prioritized alternatives is obtained and preselected according to their global performance indicator (see Table 1, for brevity similar routes have been combined). This prioritized list of alternatives is passed to the network optimization stage where a multi-objective optimization is performed. A set of different process configurations is obtained according to different objectives and intermediate alternatives as well as a set of Pareto points using the ε-constraint method. As an example, the configuration with the best economic performance is shown in Figure 3 along with the Pareto fronts obtained for the economic profit against each one of the environmental endpoint indicators as shown in Figure 4.

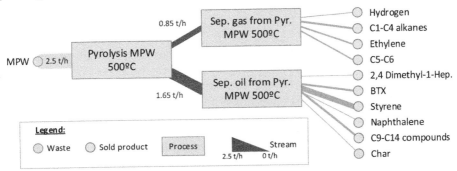

Figure 3. Configuration obtained for maximized profit. Point number 10 in Figure 4.

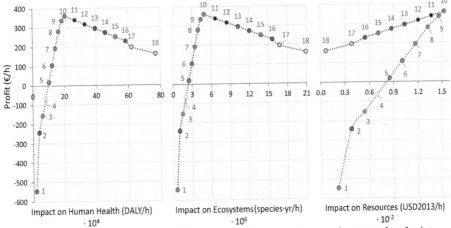

Figure 4. Pareto fronts for the trade-off between profit and the three environmental endpoint indicators. The points are numbered and color-coded to represent different configurations. Each point (as the one seen in Figure 3) consists of a different combination of processes from Figure 2.

Results in this contribution were consistent with those obtained in previous runs of the algorithm (Pacheco-López et al., 2022, 2021). Newly introduced gasification alternatives were selected in the first stage in the positions 9th, for the traditionally powered route, or Gasiforming™, and 20th, for the Electrified Gasiforming™. In addition, they were not among the best-performing alternatives in the second stage for any of the four obtained anchor points, which may suggest they need further development and more efficient and economically competitive energy sources. It was identified that capital expenditures play an important role in the implementation of these technologies. Electrical energy consumption has been also identified as a critical part since it entails most of the operating expenditures. On the other hand, results show that pyrolysis of mixed plastic waste with or without catalysts at a temperature of 500°C is still among the best-performing alternatives for most objectives since they are the most economically and environmentally competitive or beneficial options for this kind of waste.

5. Conclusions

Results show the framework's ability to synthesize and assess routes for the upcycling of waste accounting for economic and environmental objectives. The capability of the framework to be easily expanded and complemented with the integration of other tools to satisfy processes modeling, design, and assessment needs has also been proven. The use of the CapEx-OpEx Robust Optimizer (CORO) has aided in systematically generating the economic performance indicators for all the processes asserted in the ontology, providing a standardized procedure to avoid user-introduced errors and keeping the assessment parameters consistent among all alternatives. The third stage of the methodology, detailed simulation, and design of a chosen alternative is one of the most work-intensive tasks of the framework and stays out of the scope of this study. The use of surrogate modeling has been envisaged as a useful alternative to ease this stage and might be implemented in future developments to study the effect of feedstock composition changes on the economic and environmental performance of the upcycling technologies. Some identified challenges have been approached with the inclusion of new ideas and tools that help advance the automation and systematization of the framework, mainly in terms of TEA/LCA and simulation efforts. The possible inclusion of carbon emission credits has also been included as a potential source of income, therefore improving the overall economic performance of the considered alternatives.

6. Acknowledgments

Financial support received from the Spanish "Ministerio de Economía, Industria y Competitividad" and the European Regional Development Fund, both funding the research Project CEPI (PID2020-116051RB-I00) is fully acknowledged. A.P-L thanks the financial support received from the "Ministerio de Ciencia e Innovación" (grant ref. PRE2018-087135) and the Erasmus+ programme of the EU 2021-27 (KA131 call).

References

Galeazzi, A., Prifti, K., Gallo, F., Manenti, F., 2022. A Methodology for The Optimal Surrogate Modelling of Digital Twins Using Machine Learning. Comput. Aided Chem. Eng. 51, 1543–1548. https://doi.org/10.1016/B978-0-323-95879-0.50258-7

Granacher, J., Kantor, I.D., Lopez, M., Maréchal, F., 2021. Self-learning surrogate models in superstructure optimization, in: Computer Aided Chemical Engineering. Elsevier, pp. 439–444. https://doi.org/10.1016/B978-0-323-88506-5.50069-3

Pacheco-López, A., Somoza-Tornos, A., Graells, M., Espuña, A., 2022. Integrated synthesis, modeling, and assessment of waste-to-resource alternatives, in: Computer Aided Chemical Engineering. Elsevier, pp. 787–792. https://doi.org/10.1016/B978-0-323-95879-0.50132-6

Pacheco-López, A., Somoza-Tornos, A., Graells, M., Espuña, A., 2021. Synthesis and assessment of waste-to-resource routes for circular economy. Comput. Chem. Eng. 153. https://doi.org/10.1016/j.compchemeng.2021.107439

Prifti, K., Galeazzi, A., Barbieri, M., Manenti, F., 2022. A Capex Opex Simultaneous Robust Optimizer: Process Simulation-based Generalized Framework for Reliable Economic Estimations, in: Computer Aided Chemical Engineering. Elsevier, pp. 1321–1326. https://doi.org/10.1016/B978-0-323-95879-0.50221-6

Prifti, K., Galeazzi, A., Margarita, I., Papale, A., Miele, S., Bargiacchi, E., Barbieri, M., Petea, M., Manenti, F., 2021. Converting end-of-life plastic waste into methanol: the gasiformingtm process as new, efficient and circular pathway. Environ. Eng. Manag. J. 9, 1629–1636.

Somoza-Tornos, A., Pozo, C., Graells, M., Espuña, A., Puigjaner, L., 2021. Process screening framework for the synthesis of process networks from a circular economy perspective. Resour. Conserv. Recycl. 164, 105147. https://doi.org/10.1016/j.resconrec.2020.105147

Antonis Kokossis, Michael C. Georgiadis, Efstratios N. Pistikopoulos (Eds.)
PROCEEDINGS OF THE 33rd European Symposium on Computer Aided Process Engineering
(ESCAPE33), June 18-21, 2023, Athens, Greece
© 2023 Elsevier B.V. All rights reserved. http://dx.doi.org/10.1016/B978-0-443-15274-0.50366-8

Pyrolysis of livestock manures: Optimal operating conditions and feedstock blending ratios

Prakash Parthasarathy, Mohammad Alherbawi, Snigdhendubala Pradhan,
Gordon McKay, Tareq Al-Ansari *

*Division of Sustainable Development, College of Science and Engineering, Hamad Bin
Khalifa University, Qatar Foundation, Doha, Qatar.*
*talansari@hbku.edu.qa

Abstract

Livestock manure contributes to around 18% of global greenhouse gas (GHG) emissions, while the mishandling of manure further pollutes water and soil. As such, the valorisation of these wastes becomes vital for the mitigation of global warming impact. In this context, anaerobic digestion (AD) is commonly used to generate biogas and digestate from manure; however, pyrolysis has recently emerged as a viable alternative to AD for producing cleaner products in much shorter processing times. Therefore, this study investigates the pyrolysis of four types of livestock manures including poultry (PM), dairy (DM), camel (CM) and sheep manure (SM). The air-dried manure samples are characterised with reference to literature reported data, while the process is simulated using existing empirical prediction models. In addition, the process performance in terms of economic feasibility and environmental impact is evaluated from cradle to gate for different manure feedstocks. A mathematical optimisation model is then developed to evaluate optimal blending of manure for the commercial production of bio-oil, considering different operating conditions, manure compositions and availability in the State of Qatar. The model is solved to fulfill seven different objectives independently. The results revealed that PM is linked to the highest amount of bio-oil (41%) at 600°C, while the lowest yield is associated to CM. Finally, the optimal manure blending ratio for highest bio-oil yield and lowest cost is achieved at 52% PM: 2% DM: 3% CM: 43% SM. Whereas the optimal blending ratio for lowest carbon footprint is realised at 42% PM: 21% DM: 30% CM: 7% SM. The designed model provides a thorough insight on optimal technical and environmental valorisation of livestock manures by pyrolysis.

Keywords: Pyrolysis; Livestock manures; Optimisation; GHG; Bio-oil; Biochar.

1. Introduction

Several billion tonnes of livestock waste/manure are produced worldwide every year. The mismanagement and misuse of manure is severely polluting the environment and contaminating the water supply. As a matter of fact, the livestock sector is estimated to contribute 18% of total Greenhouse gas (GHG) emissions globally (Moran and Wall, 2011). Biomass, including livestock manure, has piqued the interest of many because its carbon neutrality offers a strategic measure to reduce GHG emissions. Biological treatment of livestock manures, such as anerobic digestion (AD), has traditionally been used. The use of pyrolysis method to convert livestock manure into energy has been emerging as a viable alternative to AD because they can be used to produce value-added products in a much shorter timeframe than AD. Pyrolysis also significantly reduces waste volume and eliminates pathogens.

This study considers four types of livestock manures: poultry (PM), dairy (DM), camel (CM), and sheep (SM). Due to the large numbers of aforementioned livestock raised in the State of Qatar, these manures were considered. The study also attempts to find an optimal blending ratio of the aforementioned manures for commercial production of bio-oil, considering different operating conditions, manure compositions, and availability in the State of Qatar

In order to predict the pyrolysis product distribution, researchers have created a number of models. In the current study, pyrolytic kinetics model proposed by Song is applied (Song, 2016). The model represents the general trends in product distribution using empirical equations that cover mass and energy balances as well as empirical relationships (based on temperature). Only a few empirical models are based on both feedstock attributes and pyrolysis operation parameters, with the majority of models being based solely on the properties of feedstock materials (Abhijeet et al., 2020; Al-Rumaihi et al., 2022). One of the few studies that considers both the pyrolysis conditions and feedstock qualities is the current one (Parthasarathy et al., 2022). Prediction model-based economic analysis is uncommon, as such, this study is distinctive as it concentrates on both technical and economic assessment. The commercialisation of bio-oil will benefit from the techno-economic analysis of the pyrolysis of the aforementioned manures.

2. Methodology

2.1. Materials and methods

The proximate and ultimate analyses results of the manures under study are provided in **Table 1**.

Table 1: Proximate and ultimate analyses findings of the feedstock manures.

Analysis	PM	DM	CM	SM
Moisture content (%) (as-received basis)				
Moisture	40.00	85.00	58.00	50.00
Proximate analysis (% wt.) (Dry basis)				
Volatile matter	63.58	57.18	60.51	57.30
Fixed carbon*	13.31	10.11	22.74	15.56
Ash	23.09	32.69	16.73	27.13
Ultimate analysis (% wt.) (Dry and ash free basis)				
Carbon	36.67	27.61	37.11	33.14
Hydrogen	5.33	3.48	4.07	4.48
Nitrogen	3.54	1.88	2.27	2.65
Oxygen*	30.61	33.91	39.54	32.21
Sulphur	0.73	0.41	0.24	0.37
Chlorine	0.00	0.00	0.00	0.00

*Calculated by difference

2.2. Model Development

Figure 1 depicts a simplified process flow diagram of the pyrolysis process. The following assumptions are made for the development of the prediction model. Feedstocks are pyrolysed in an isothermal reactor with a nitrogen environment. As a pre-treatment, the feedstocks are dried, and then they are pyrolysed and transformed into water, bio-oil, tar, pyrogas (CO_2, CO, CH_4, H_2, hydrocarbons, etc.), biochar, and ash, as denoted by **equations 1-3** (Swagathnath et al., 2019). Further, as the temperature proceeds, the bio-oil is further decomposed into gas components.

$$Bio - char\ yield = 0.106 + 2.43 * \exp(-0.66 * T * 10^{-2}) \tag{1}$$

$$Bio - oil\ yield = Y_{bio-oil,F} + Y_{H_2O,F} + samples'moisture\ composition \tag{2}$$

$$Pyrogas\ yield = Y_{CO_2,F} + Y_{CO,F} + Y_{H_2,F} + Y_{CH_4,F} \tag{3}$$

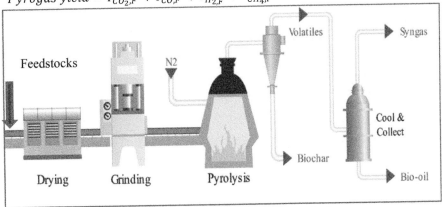

Figure 1. Process flow diagram of the pyrolysis process.

2.3. Economic Analysis

The economic analysis of pyrolysis is investigated based on the assumptions listed in **Table 2**. The primary pyrolysis equipment costs and labor requirements have been adapted from previous technical reports (Humbird et al., 2011). Using the Chemical Engineering Pant Cost Index, all prices are scaled up and inflated to the year 2019 (Chemical Engineering Plant Cost Index (CEPCI)).

Table 2: Economic analysis assumptions.

Parameters	Values
Pyrolysis plant location	Qatar
Plant capacity	20 t/h
Analysis base year	2019
Plant lifetime (years)	25
Discount rate (%)	20
Feedstock price (USD $/t)	50

2.4. Environmental Analysis

Environmental analysis covers the concept of carbon footprint, which is the total amount of CO_2 emissions that have been generated over the lifecycle of bio-oil, from cradle to gate.

2.5. Optimisation

Using the objective functions and the constraints presented below, an optimisation model is created using Excel Solver. For the selection of optimal blends of the four different manures for the pyrolysis process, 7 objectives are adopted independently. The objectives seek to maximise the yields of bio-oil, syngas, and biochar, maximise the heating value of bio-oil and syngas, and minimise cost and carbon footprint. Additionally, the blending must use no less than 10% no more than 50% of each available manure, considering the plant's capacity.

Objective functions:

Bio-oil yield (Z_1) = Max $\sum_{i=1}^{4} BO_i X_i$
Syngas yield (Z_2) = Max $\sum_{i=1}^{4} SG_i X_i$
Biochar yield (Z_3) = Max $\sum_{i=1}^{4} BC_i X_i$
Bio-oil HV (Z_4) = Max $\sum_{i=1}^{4} BHV_i X_i$
Syngas HV (Z_5) = Max $\sum_{i=1}^{4} SHV_i X_i$
CO_2 emissions (Z_6) = Min $\sum_{i=1}^{4} E_i X_i$
Bio-oil Cost (Z_7) = Min $\sum_{i=1}^{4} C_i X_i$

Constraints:

$\sum_{i=1}^{4} X_i = 1$
$\sum_{i=1}^{4} Q_i X_i \leq feed\ capacity$

$0.5 \geq X_i \geq 0.1$

Where: X_i: blending ratio % of manure (i); Q_i: available quantity (t/y) of manure (i); BO_i: bio-oil yield (t/y) for manure (i), SG_i: syngas yield (t/y) for manure (i); BC_i: biochar yield (t/y) for manure (i); BHV_i: Bio-oil heating value (GJ/t) via manure (i); SHV_i: syngas heating value (GJ/t) via manure (i); E_i: process emissions in kg CO_2-e /kg manure (i); C_i: Bio-oil cost ($/kg) using manure (i); $i=1$-4 represent manure types, whereby: *1*: Poultry manure, *2*: Dairy manure, *3*: Camel manure, *4*: Sheep manure.

3. Results and Discussion

3.1. Products Distribution

Figure 2. depicts the effect of temperature on the distribution of pyrolysis products in feedstock samples. In terms of bio-oil and syngas yields, increasing the temperature from 300 to 600°C increased the bio-oil and syngas yields in all feedstocks. However, the increased temperature resulted in a decrease in the biochar yield of all feedstocks. The highest bio-oil and syngas yields were achieved at 600°C, while the highest biochar yield was achieved at 300°C. Among the feedstocks, CM produced the most syngas (55%), DM produced the most biochar (33%), and PM offered the highest bio-oil yield (41%).

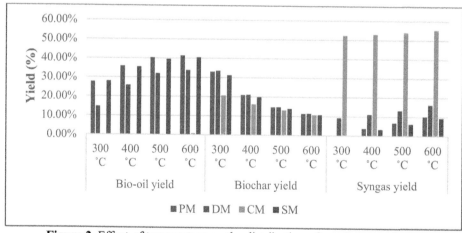

Figure 2. Effect of temperature on the distribution of pyrolysis products.

3.2. Bio-oil Levelised Cost

The bio-oil levelised cost of all manures with respect to temperature is presented in **Figure 3a**. The levelised cost for the bio-oil production from CM is not considered since the bio-oil production from the manure is little. For all manures, the levelised cost for the production of bio-oil is minimum at 500°C, whilst the cost is maximum at 300°C. Of all the manures, PM contributed for the lowest bio-oil levelised cost (0.34 USD/kg of bio-oil). This suggests that pyrolysing PM would result in greater financial gain than pyrolysing other manures.

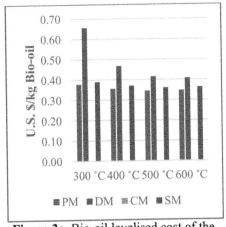

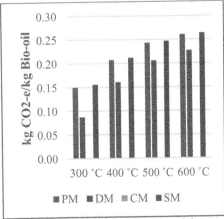

Figure 3a. Bio-oil levelised cost of the manures.

Figure 3b. Carbon footprint of the manures.

3.3. Carbon Footprint

Figure 3b. illustrates the carbon footprint (kg CO_{2-e}/kg Bio-oil) of all manures in relation to different temperatures. The carbon footprint value for all manures is lowest at 300°C and highest at 600°C. Because the CM produces a meagre amount of bio-oil, the carbon footprint of the manure is negligible. DM had the lowest carbon footprint of the feedstocks, while SM had the highest carbon footprint. This suggests that DM pyrolysis is less polluting than other manures.

3.4. Optimal Blend

The effect of manure blending ratio with respect to products' yields, heating value, carbon footprint, and production cost is illustrated in **Figure 4**. The best manure blending ratio for the maximum bio-oil yield and the lowest cost is achieved at 52% PM: 2% DM: 3% CM: 43% SM. The optimal blending ratio for the lowest carbon footprint is accomplished at 42% PM: 21% DM: 30% CM: 7% SM.

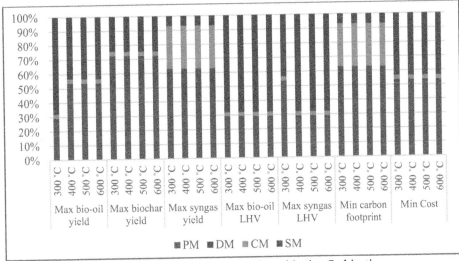

Figure 4. Optimal blending ratios for achieving 7 objectives.

4. Conclusion

The pyrolysis of four different types of livestock manures—poultry (PM), dairy (DM), camel (CM), and sheep manure is examined in this study (SM). The process is simulated using empirical prediction models that are already well established. Additionally, the process performance for the manure feedstocks is assessed from cradle to gate in terms of economic viability and environmental impact. The best manure blend for commercial bio-oil production is then evaluated using a mathematical optimisation model that considers various operating conditions, manure compositions, and availability in the State of Qatar. According to the findings, PM is connected to the highest yield of bio-oil (41%) at 600°C, while CM is responsible for the lowest yield. Finally, 52% PM: 2% DM: 3% CM: 43% SM is found to be the ideal manure blending ratio for the highest bio-oil yield and lowest cost. While 42% PM: 21% DM: 30% CM: 7% SM is the blend that results in the lowest carbon footprint. The created model offers a thorough understanding of the pyrolysis's ideal technical and environmental valorisation of livestock manures.

Acknowledgement

The authors would like to thank Qatar National Research Fund (QNRF) for their support of this research through NPRP-11S-0117-180328, the Supreme Committee for Delivery and Legacy (SCDL) and Hamad Bin Khalifa University (HBKU) and Qatar Foundation (QF).

References

Abhijeet, P., Swagathnath, G., Rangabhashiyam, S., Asok Rajkumar, M., Balasubramanian, P., 2020. Prediction of pyrolytic product composition and yield for various grass biomass feedstocks. Biomass Convers. Biorefinery 10, 663–674.

Al-Rumaihi, A., Alherbawi, M., Parthasarathy, P., R.Mackey, H., McKay, G., Al-Ansari, T., 2022. A techno-economic assessment of biochar production from date pits in the MENA region. Comput. Aided Chem. Eng. 51, 805–810.

Alherbawi, M., Parthasarathy, P., Mckay, G., Mackey, H., Al-Ansari, T., 2022. Investigation of optimal blending of livestock manures to produce biocrude via hydrothermal liquefaction. Comput. Aided Chem. Eng. 51, 1243–1248.

Humbird, D., Davis, R., Tao, L., Kinchin, C., Hsu, D., Aden, A., Schoen, P., Lukas, J., Olthof, B., Worley, M., Sexton, D., Dudgeon, D., 2011. Process design and economics for conversion of lignocellulosic biomass to ethanol: : Dilute-acid pretreatment and enzymatic hydrolysis of corn stover. United States.

Moran, D., Wall, E., 2011. Livestock production and greenhouse gas emissions: Defining the problem and specifying solutions. Anim. Front. 1, 19–25.

Parthasarathy, P., Alherbawi, M., Shahbaz, M., Mackey, H., Mckay, G., Al-Ansari, T., 2022. Conversion of oil palm waste into value-added products through pyrolysis: a sensitivity and techno-economic investigation. Biomass Convers. Biorefinery 2022 1, 1–21.

Song, B., 2016. Biomass pyrolysis for biochar production: kinetics, energetics and economics. Biochar 227–238.

Swagathnath, G., Rangabhashiyam, S., Parthsarathi, K., Murugan, S., Balasubramanian, P., 2019. Modeling Biochar Yield and Syngas Production During the Pyrolysis of Agro-Residues, in: H, D., Pillai RG, Tharian MG, M.A. (Eds.), Green Buildings and Sustainable Engineering: Proceedings of GBSE 2018. Singapore, pp. 325–336.

Antonis Kokossis, Michael C. Georgiadis, Efstratios N. Pistikopoulos (Eds.)
PROCEEDINGS OF THE 33rd European Symposium on Computer Aided Process Engineering
(ESCAPE33), June 18-21, 2023, Athens, Greece
© 2023 Elsevier B.V. All rights reserved. http://dx.doi.org/10.1016/B978-0-443-15274-0.50367-X

Investigating the feasibility of biomass integrated gasification combined cycle (BIGCC) as a power alternative in Gaza Strip

Rawan Shihada,[a] Rania Salem,[a] Amanda Munia,[a] Nashat Naim,[a] Mohammad Alherbawi [b*]

[a] *Faculty of Engineering and Information Technology, Al-Azhar University, Gaza, Palestine.*
[b] *College of Science and Engineering, Hamad Bin Khalifa University, Qatar Foundation, Doha, Qatar*
malherbawi@hbku.edu.qa

Abstract

Gaza strip suffers from severe electricity deficit of 62%, due to the lack of fuel needed to operate the local power plant. As a result, local authorities adopted a rationing plan, through which the power is often cutoff during half of the day. Meanwhile, more than 2,000 tonnes of wastes are generated every day in Gaza, while its municipal capabilities to recycle wastes are limited, leading to the landfilling of most of the generated wastes in a highly populated region. Therefore, this study explores the feasibility of the biomass integrated gasification combined cycle (BIGCC) in Gaza scenario in utilising local wastes and independently generating power at reasonable costs. The BIGCC plant is simulated using the advanced system for process engineering software (Aspen Plus®). Furthermore, an economic feasibility analysis is conducted considering the actual costs of similar existing units, as well as local operating costs. The obtained results are encouraging, where the designed BIGCC plant was able to generate ~207 MW, which can fulfill 46% of Gaza's electricity demand, and 86.25% of its current power deficit. The levelised cost of electricity (LCOE) is estimated at 0.045 $/kWh, which represents 26% only of the current market price. The proposed scenario may contribute to diversifying the power supplies in Gaza and compensate for the power deficit. In addition, a diversion from landfilling can be achieved through wastes valorisation, which reduces the associated land and carbon footprints.

Keywords: IGCC; Techno-economics; Clean power; LCOE; Aspen Plus; Gaza.

1. Introduction

Gaza Strip has been suffering from a severe electricity shortage since 2007 due to the lack of diesel fuel. The present demand is approximately 450 MW, while the current supplies are no more than 210 MW, ending up with more than 60% shortage of power (PENRA 2022). However, there are 8,900 TJ/year of biomass potential in the Gaza Strip, where considerable amount of biomass is produced every day, including municipal waste (MSW), agricultural residues, and sewage sludge (Al-Najjar et al. 2020). More than 2,000 tonnes of MSW are generated every day, which is expected to reach 2,660 tonnes during the year 2025 (MDLF 2017).

Therefore, this study investigates the feasibility of BIGCC establishment in Gaza to utilise local biomass for the generation of electricity to fill up the current power shortage. In comparison to existing coal power plants, BIGCC plants are known to have a higher thermal efficiency as they convert biomass into syngas and then use its flue gas and heat to generate electricity through a heat-integrated system (Zang et al. 2020).

Considering local biomass, a BIGCC plant is modelled and evaluated in Aspen Plus (V.9), while an economic assessment is conducted based on local prices of materials and international equipment costs.

The model in hand may provide insights on power generation alternatives in developing countries through the utilisation of local waste at reasonable costs. Nevertheless, the deployment of BIGCC technology may also contribute towards land filling diversion and emissions mitigation.

2. Methodology

2.1. Model Development

An integrated gasification combined cycle using municipal solid waste (MSW) is modelled in Aspen Plus V.9. A simplified process flow diagram is shown in Figure 1. Throughout the simulations of BIGCC, mass and energy balance and chemical equilibrium were observed, considering an isothermal and steady state systems.

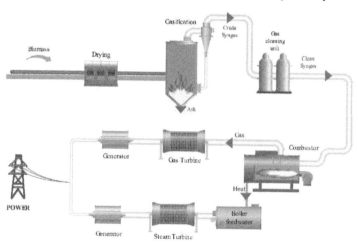

Figure 1: Process flow diagram of BIGCC.

MSW is introduced into the model based on its proximate and ultimate analysis reported in Table 1 (Phyllis 2022). The gasification part of the model was simulated using four Aspen Plus blocks based on Alherbawi et al. (Alherbawi, AlNouss, et al. 2020) model. The process starts by processing the feedstock in a drying unit (RStoic) to reduce its moisture content. The dry MSW which is defined with zero moisture was then passed through an RYield reactor for pyrolysis. where non-conventional components are converted into their equivalent conventional elements and compounds (i.e., H_2, N_2, O_2, C, S, ash). To simplify the model, ash is separated using a solid separator, while useful components are processed along with a gasifying agent (i.e., steam) into two consecutive blocks representing the oxidation and reduction processes following the reactions in Table 2. The reactor calculates the syngas composition by minimising the Gibbs free energy to achieve a chemical and state equilibrium (Alherbawi, Al-Ansari, and Mckay 2020).

Investigating the feasibility of biomass integrated gasification combined cycle (BIGCC) as a power alternative in Gaza Strip

2305

Table 1: Proximate and ultimate analyses of MSW.

Proximate analysis	(wt. %)	Ultimate analysis	(wt. %)
Moisture	7.56	Carbon	48.23
Ash	17.8	Hydrogen	5.16
Volatile matter	57.99	Nitrogen	1.21
Fixed Carbon	24.21	Oxygen	27.31
		Sulfur	0.29

Table 2: Key gasification reactions.

Reaction	Equation	No.
Carbon oxidation	$C + O_2 \leftrightarrow CO_2$	(1)
Partial oxidation	$C + \frac{1}{2} O_2 \leftrightarrow CO$	(2)
Steam generation	$H_2 + \frac{1}{2} O_2 \leftrightarrow H_2O$	(3)
CO oxidation	$CO + \frac{1}{2} O_2 \leftrightarrow CO_2$	(4)
Boudouard reaction	$C + CO_2 \rightarrow 2CO$	(5)
Methanation	$C + 2H_2 \rightarrow CH_4$	(6)
Steam gasification	$C + H_2O \rightarrow CO + H_2$	(7)
Methane reforming	$CH_4 + H_2O \leftrightarrow CO + 3H_2$	(8)
Water gas shift reaction	$CO + H_2O \leftrightarrow CO_2 + H_2$	(9)
Ammonia formation	$N_2 + 3H_2 \rightarrow 2\,NH_3$	(10)
Hydrogen sulfide formation	$S + H_2 \rightarrow H_2S$	(11)

Following the gasification reaction, traces of unconverted char are separated from the syngas using solid separator. While a flash unit is used to remove tar and water contents before processing the raw syngas into a methanol-based absorption unit (Alherbawi et al. 2021). To remove CO_2, NH_3 and H_2S, cool methanol is introduced into a RadFrac absorber against the flow of syngas, where gas impurities are dissolved in methanol and then stripped in a regenerator column at a higher temperature.

The power generation stage is simulated as a combined cycle system to further utilise the generated heat into electricity production. Where syngas is pumped into a combustion chamber, in which the relevant combustion reactions are defined with adequate flow of air to observe a complete combustion.

The hot and high-pressure gas resulting from the combustion is introduced into a gas turbine and a generator to generate power. In parallel, the combustion's resulting heat is utilised to create steam using a boiler feedwater, where the high-pressure steam is passed through a steam turbine to generate additional power.

2.2. Economic Assessment

An economic analysis is conducted based on pre-COVID19 values. An BIGCC plant located in Gaza (Palestine) is considered with a feed capacity of 2000 tonnes/day. The analysis is conducted based on the given parameters and assumptions listed in Table 3. The biomass collection charges, wages, raw materials and utility prices are quoted locally. As for capital expenses (capex), the key equipment prices are adapted from high-level technical reports (National Energy Technology Laboratory 2007; National Renewable Energy Laboratory 2011), scaled up and inflated to the base year of analysis using the Chemical Engineering Plant Cost Index (CEPCI) as expressed in Equation 12 and 13. While the minimum selling price of electricity is calculated using Equation 14.

Table 3: Economics parameters and assumptions used in this study.

Parameter	Value/ assumption
Location	Gaza strip (Palestine)
Analysis year	2019
Currency	U.S. $
Plant feed capacity	2000 t/d
Feedstock gate price	30 $/t
Water- price	0.26 $/m^3
Waste disposal charges	30 $/t
Electricity price	0.14 $/kWh
Operating	25% of the labour
Maintenance	2% of equipment
Overhead	60% of the labour
Discount rate	10%
Taxes	25%
Insurance	1.5% of total installed cost (TIC)

$$Cost_{desgin} = Cost_{base} \cdot \left(\frac{Capcity_{desgin}}{Capacity_{base}}\right)^{Scalling\ Factor} \tag{12}$$

$$Cost_{desgin\ \$\ 2019} = Cost_{desgin\ \$i} \cdot \left(\frac{CEPCI_{2019}}{CEPCI\ i}\right) \tag{13}$$

$$LCOE\left(\frac{\$}{kWh}\right) = \frac{Capex + \sum_{1}^{Lifespan}(Opex(1+Discount\ Rate^{-Lifespan}))}{\sum_{1}^{Lifespan} generated\ power\ (1+Discount\ Rate\ )^{-Lifespan}} \tag{14}$$

3. Results and Discussion

3.1. Technical Analysis

At gasification stage, an enhanced syngas is generated with a molar ratio of hydrogen to carbon monoxide at 1.3. A small amount of methane is produced (<0.1%). While the gas impurities (CO_2, NH_3, H_2S) are perfectly removed using methanol absorption. The final molar and mass composition of cleaned syngas is presented in Table 4.

In addition, at combined-cycle power generation stages, 207 MW is generated a year, where 80% of the generated power is obtained from the gas turbine, while the steam turbine contributed with the remaining 20%. The attained power can fulfill 46% of Gaza's electricity demand, and 86.25% of its current power deficit.

Table 4: Cleaned syngas composition.

Component	Mass fraction	Mole fraction
H_2	8.3%	55.7%
CO	88.5%	43%
CO_2	2.2%	0.6%
Others	1%	0.7%
Sum	100%	100%

3.2. Economic Analysis

The key findings of the BIGCC economic analysis are presented in Table 5. The capital expenses (capex) are estimated at 608 M U.S.$, while the plant requires annual operating expenses (opex) of 16.5 M U.S.$. Whereas 190 M U.S.$ of profit (after tax) can be attained. The return on investment is at 31%, while the investment cost can be paid in 3.2 years. Furthermore, the levelised cost of electricity is achieved at 0.045 $/kWh, which represents only 26% of the current market price.

Figures 2 and 3 reveal the breakdown of capex and opex, where purchased equipment occupied 40% of capex, while opex is dominated by raw materials cost at 42%.

Table 5: Key economics parameters of Gaza's BIGCC project.

Item	Value	Unit
Levelised cost of electricity	0.045	$/kWh
Project capex	608,229,938	$
Project opex	16,489,484	$/y
Capacity	207.13	MW
Sales	254,018,100	$/y
Annual profit	237,528,616	$/y
Profit after tax	190,022,893	$/y
Payback time	3.2	years
ROI	31	%

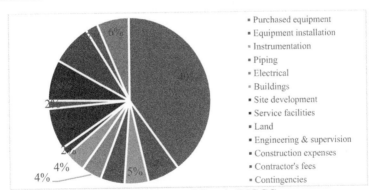

Figure 2: Breakdown of Gaza's BIGCC capex.

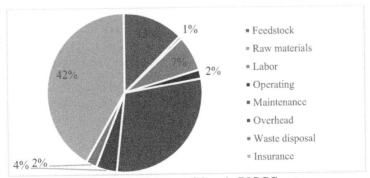

Figure 3: Breakdown of Gaza's BIGCC opex.

4. Conclusion

This study has investigated the feasibility of BIGCC establishment in Gaza Strip which is suffering from severe power shortage. The plant was assumed to operate based on local solid waste at 2000 tonnes per day and was simulated and evaluated in Aspen Plus software. The proposed plant generated 207 MW a year, which may fulfill nearly 50% of net Gaza's power demand, and 86% of the present power shortage.

The economic analysis indicated that the plant requires an investment cost of around 600 M U.S.$, and that it can generate power at a minimum cost of 0.045 $/kWh, which represents quarter of the current market price.

The model provides insights on local resources-based power generation as a feasible and clean alternative for countries suffering from severe fossil fuel shortage.

References

G. Zang, J. Zhang, J. Jia, E.S. Lora, and A. Ratner, 2020, "Life Cycle Assessment of Power-Generation Systems Based on Biomass Integrated Gasification Combined Cycles." Renewable Energy 149:336–46.

H. Al-Najjar, P. Christoph, R. Al Afif, and H.J. El-Khozondar, 2020, "Estimated View of Renewable Resources as a Sustainable Electrical Energy Source, Case Study." Designs 4(3):32.

M. Alherbawi, T. Al-Ansari, and G. Mckay, 2020, "Potential Integrated Pathways for Jet Biofuel Production from Whole Fruit of Jatropha." Pp. 235–40 in 30 European Symposium on Computer Aided Process Engineering. Vol. 48, edited by S. Pierucci, F. Manenti, G. L. Bozzano, and D. B. T.-C. A. C. E. Manca. Elsevier.

M. Alherbawi, A. AlNouss, G. Mckay, and T. Al-Ansari, 2020, "Optimum Utilization of Jatropha Seedcake Considering the Energy, Water and Food Nexus." Pp. 229–34 in 30 European Symposium on Computer Aided Process Engineering. Vol. 48, edited by S. Pierucci, F. Manenti, G. L. Bozzano, and D. B. T.-C. A. C. E. Manca. Elsevier.

M. Alherbawi, G. McKay, H.R. Mackey, and Tareq Al-Ansari, 2021, "A Novel Integrated Pathway for Jet Biofuel Production from Whole Energy Crops: A Jatropha Curcas Case Study." Energy Conversion and Management 229:113662.

MDLF. 2017. Gaza Solid Waste Management Project (GSWMP).

National Energy Technology Laboratory, 2007, Baseline Technical and Economic Assessment of a Commercial Scale Fischer-Tropsch Liquids Facility.

National Renewable Energy Laboratory. 2011. Process Design and Economics for Conversion of Lignocellulosic Biomass to Ethanol.

PENRA, 2022, "Gaza Power Needs." Retrieved August 20, 2022 (http://www.penra.ps/index.php?option=com_content&view=article&id=759:2012-11-14-09-22-07&catid=27:2012-11-14-08-37-30&Itemid=61).

Phyllis, 2022, "ECN Phyllis Classification." Retrieved August 15, 2022 (https://phyllis.nl/Browse/Standard/ECN-Phyllis).

Antonis Kokossis, Michael C. Georgiadis, Efstratios N. Pistikopoulos (Eds.)
PROCEEDINGS OF THE 33rd European Symposium on Computer Aided Process Engineering
(ESCAPE33), June 18-21, 2023, Athens, Greece
© 2023 Elsevier B.V. All rights reserved. http://dx.doi.org/10.1016/B978-0-443-15274-0.50368-1

The Infinity Reactor: A new conceptual design for a more cost-efficient CO₂ to methanol route

Hilbert Keestra, Edwin Zondervan, Wim Brilman

University of Twente, Drienerlolaan 5, 7522 NB Enschede, The Netherlands

Abstract

With conventional methanol reactor designs for CO_2-based methanol synthesis, there is the risk of rapid catalyst deactivation with commercial $Cu/ZnO/Al_2O_3$ catalysts at high water partial pressures, related to the use of CO_2 as carbon feed. As catalyst reactivity and exothermicity are influenced by the switch to a CO_2-rich feed, also heat integration options are affected. A new reactor/process design, the Infinity Reactor, is proposed to both improve catalyst lifetime and reduce catalyst usage. The Infinity Reactor is a shell and tube type of reactor, with catalyst on both sides, utilising the difference in temperature between the inlet and outlet of the reactor to operate as a quasi-gas-cooled reactor. The gas loops over both sides of the reactor, with intermediate cooling and condensation. This modelling study shows around 35% savings in catalyst volume in comparison with adiabatic operation.

Keywords: methanol, intensification, infinity, reactor, CO_2

1. Introduction

Methanol serves as an important feedstock for about 30% of industrial chemicals [1]. Methanol is commercially synthesized from fossil-derived syngas (a mixture of CO, CO_2 and H_2) via the hydrogenation of CO_2 (Equation 1), which is an exothermic reaction. The role of CO at high conversions is to remove excess water via the water-gas shift reaction (WGS) (Equation 2), thereby enhancing CO_2 conversions to methanol [2,3]. However, this mechanism also increases the exothermicity of the process.

$$CO_2 + 3H_2 \rightleftharpoons CH_3OH + H_2O \qquad (\Delta H_R^0 = -49.16\ kJ\ mol^{-1})\ ^4 \qquad (1)$$

$$CO + H_2O \rightleftharpoons CO_2 + H_2 \qquad (\Delta H_R^0 = -41.21\ kJ\ mol^{-1})\ ^4 \qquad (2)$$

Many companies have understood that developing CO_2-utilizing technologies are an essential step towards a sustainable industrial world [4-6]. However, using pure CO_2 as a feedstock (e.g. obtained from flue gases to reduce CO_2 emissions) reduces the contribution of the WGS reaction. This reduces the exothermicity of the system but also increases water production during methanol synthesis.

Catalyst deactivation is one of the major hurdles in green methanol production due to a higher water concentration in the process that induces ZnO agglomeration and ZnAl-spinel formation [7,8]. Whereas commercial catalysts have a lifetime of 4-6 years [3,9,10] with CO-rich syngas-based feeds, methanol production with CO_2-rich feeds suffers from more rapid deactivation in terms of a few months or even weeks [7,11,12]. It is anticipated that this is especially related to higher water partial pressures at high temperatures. Model-based tools like process simulations are necessary for investigating and comparing performances and costs. Therefore, this study has investigated new plant configurations and reactor concepts to improve catalyst lifetime, reduce catalyst usage, and make, through this, the CO_2 to methanol process more cost-efficient.

2. Methods

Boiling water reactors (BWR) and quench reactors are the most utilised reactors in commercial methanol synthesis [13,14] and were therefore used as a starting point for this study. Marlin et al.[15] from Carbon Recycling International (CRI) utilise a tube-cooled reactor as a way to save on capital investment because the reaction is less exothermic and progresses slower compared to syngas-based methanol production. The effective wall area for heat exchange between the cooling medium and reaction gas can be decreased by using larger tubes and, thus, heat transfer rates are reduced while still sufficient to remove excess reaction heat.

The models of all the above plant configurations were simulated in Aspen HYSYS and first validated with available plant data from Lurgi [14,16], which utilises a BWR, and Johnson Matthey [14,17] (formerly ICI), which uses the quench design. All simulations include a recycle stream, heat exchanger, cooler, and flash unit for separating the liquid product mixture.

In this work, an improvement to the above existing reactor concepts and plant configurations is proposed, using in-situ direct heat integration between two different catalyst zones in one (shell and tube) reactor unit to intensify the process, enabled by a lower exothermicity when using CO_2. Two gas-gas heat exchangers, a combined air cooler, two flash vessels, and a blower complete the (Infinity) reactor loop. Detailed reactor calculations for this reactor were done using Python.

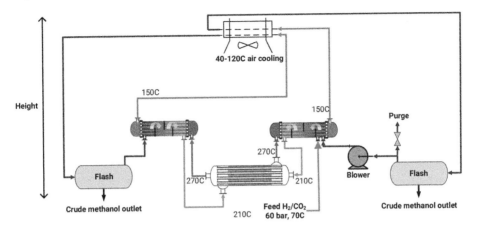

Figure 1. Proposed flowsheet of the methanol synthesis loop for the Infinity Reactor.

Figure 2. Sketch of an Infinity Reactor. The reactor has catalyst (not drawn) on both sides of the tubes and consists of two stationary heads and two rear heads. The overall flow contacting pattern is counter-current.

In the proposed flowsheet, the fresh feed is mixed with recycle gas and fed to a heat exchanger where is it heated up to 210 °C and then send to the reactor. The reactor inlet is heated up by the hot reactor outlet (270 °C) of the gas stream on the other side of the tubes, thereby accelerating the reaction rate at the cold feed side and preventing too high temperatures at the hot side. At the other end of the reactor, the reverse happens. Therefore, the reactor is a type of gas-cooled reactor, effectively utilising the temperature gradients inside the reactor. The hot reactor outlet streams are used in a separate heat exchanger to preheat the cold recycle stream from the condenser. The cooled reactor effluent with an outgoing temperature of ~150 °C is subsequently cooled down in a combined air cooler. Any condensed product will be separated in a flash vessel from the unreacted gas. The process can be designed for a low pressure drop to enable the use of a blower instead of a recycle compressor for the recirculation of unconverted syngas.

2.1. Model development

The kinetic model of Vanden Bussche and Froment [18] (VBF) was chosen because it is one of the most utilised models and serves as a benchmark in the industry [14,19]. Due to input restrictions in Aspen HYSYS, the equilibrium constants as proposed by Graaf et al.[20] were refitted as Arrhenius equations. The kinetic model was first validated with data from VBF [18] and industrial plant data [16,17,21,22]. For heat transfer calculations the lumped value UA was used, which is the combination of the overall heat transfer coefficient U_{ov} (W/m^2.K), multiplied by the tubular wall area A_{wall} (m^2).

Table 1. Main process parameters used in the simulations.

Operating pressure (bar)	60
Cooling temperature (°C)	70
Molar flow rate (kmol/hr)	957
H_2/CO_2 ratio	3
Catalyst particle diameter (mm) [14,21]	5.7
Catalyst particle density (kg/m^3) [18]	1775
Bed voidage [14,21]	0.39

3. Results & discussion

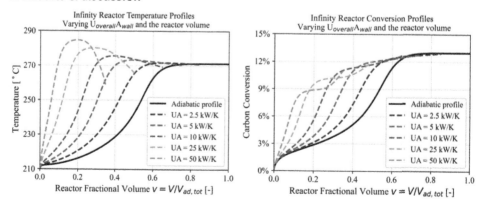

Figure 3. Temperature and conversion profiles for the Infinity Reactor with varied UA and reactor volume but constant inlet temperature. For the adiabatic profile $V_{tot}=V_{ad,tot}$, UA=2.5 kW/K has $V_{tot}=0.85V_{ad,tot}$, UA=5 kW/K has $V_{tot}=0.75V_{ad,tot}$, UA=10 kW/K has $V_{tot}=0.65V_{ad,tot}$, UA=25 kW/K has $V_{tot}=0.5V_{ad,tot}$, UA=50 kW/K has $V_{tot}=0.35V_{ad,tot}$.

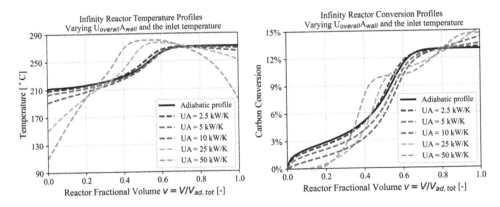

Figure 4. Temperature and conversion profiles of the Infinity Reactor with varied UA and inlet temperature but constant reactor volume. For the adiabatic profile T_{in}=210 °C, UA=2.5 kW/K has T_{in}=207 °C, UA=5 kW/K has T_{in}=202 °C, UA=10 kW/K has T_{in}=191 °C, UA=25 kW/K has T_{in}=150 °C, UA=50 kW/K has T_{in}=110 °C.

Figure 3 shows that the required catalyst amount can be drastically lowered if there is sufficient tube wall area for heat transfer. The average temperature in the reactor is increased and, therefore, the reaction rate is higher on average, resulting in less catalyst required to reach equilibrium. However, too much area for heat transfer will result in a higher peak temperature due to the added heat and results in faster catalyst degradation. Figure 4 shows that the inlet temperature can be lowered while slightly increasing conversions. However, if the temperature is lowered too much, the reaction rate is nearly zero and then it acts only as a cooling medium to lower the reactor outlet temperature of the other reactor. Conversions are ~13% per pass in the Infinity Reactor and adiabatic reactor at the simulated conditions while the BWR and quench reactor are ~20%. Therefore, the water partial pressure is ~35% lower in the proposed concept.

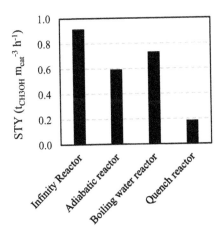

Case-specific conditions and parameters	
Inlet temperature (°C) of the BWR	240
Boiler feed water temperature (°C)	250
UA (kW/K) of the BWR	263
Inlet temperature (°C) of the other designs	210
UA (kW/K) of the Infinity Reactor	10
4-bed volume ratio of the quench reactor	0.2-0.23- 0.23-0.34

Figure 5. Space-time yield per simulated reactor concept and the case-specific assumptions.

Figure 5 shows that the Infinity Reactor has a higher catalyst utilisation compared to the other simulated reactor concepts. For the chosen conditions and parameters, catalyst savings are 35% compared to adiabatic operation, 20% compared to BWR, and ~80% compared to a 4-bed quench reactor. In addition, the catalyst lifetime is improved due to the lower water partial pressure compared to a BWR and quench reactor as the single-pass conversion is lower (13% compared to 20%). Furthermore, a multi-stage design with water removal reduces the recycle size and, thus, reduces operational costs.

4. Conclusions

The Infinity Reactor requires significantly less catalyst than adiabatic, boiling water, and quench reactors. It is expected that this will translate to a lower pressure drop, lower investment and lower operational costs. The operational costs are lower due to a smaller gas recycle ratio, caused by a higher conversion per full cycle of the gas loop, related to the two reactor passes per cycle. The somewhat lower single pass conversion leads to a lower water partial pressure inside the reactor compared to the boiling water and quench reactor concepts, by which the catalyst lifetime is improved.

References

1. Tountas, A. A. *et al.* Towards Solar Methanol: Past, Present, and Future. *Advanced Science* **6**, (2019).

2. Liaw, B. J. & Chen, Y. Z. Liquid-phase synthesis of methanol from CO2/H2 over ultrafine CuB catalysts. *Applied Catalysis A: General* **206**, 245–256 (2001).

3. Dieterich, V., Buttler, A., Hanel, A., Spliethoff, H. & Fendt, S. Power-to-liquid via synthesis of methanol, DME or Fischer–Tropsch-fuels: a review. *Energy and Environmental Science* **13**, 3207–3252 (2020).

4. Ott, J. *et al.* Methanol. *Ullmann's Encyclopedia of Industrial Chemistry* (2012) doi:10.1002/14356007.a16_465.pub3.

5. Mitsui Chemicals. CSR Report 2010. https://jp.mitsuichemicals.com/en/sustainability/report/pdf/csr2010_e.pdf.

6. Bayer. 2010 Sustainable Development report. 2 (2010).

7. Lunkenbein, T. *et al.* Bridging the Time Gap: A Copper/Zinc Oxide/Aluminum Oxide Catalyst for Methanol Synthesis Studied under Industrially Relevant Conditions and Time Scales. **55**, 12708–12712 (2016).

8. Prašnikar, A. & Likozar, B. Sulphur poisoning, water vapour and nitrogen dilution effects on copper-based catalyst dynamics, stability and deactivation during CO 2 reduction reactions to methanol. *Reaction Chemistry & Engineering* **7**, 1073–1082 (2022).

9. Cul, Y. *et al.* Latest catalyst provides more methanol for longer. (2022).

10. Hirotani, K., Nakamura, H. & Shoji, K. Optimum catalytic reactor design for methanol synthesis with TEC MRF-Z® reactor. *Catalysis Surveys from Asia 1998 2:1* **2**, 99–106 (1998).

11. Schlögl, R. Chemical Batteries with CO2. *Angewandte Chemie International Edition* **61**, e202007397 (2022).

12. Liang, B. *et al.* Investigation on Deactivation of Cu/ZnO/Al2O3 Catalyst for CO2 Hydrogenation to Methanol. *Industrial & Engineering Chemistry Research* **58**, 9030–9037 (2019).

13. Dieterich, V., Buttler, A., Hanel, A., Spliethoff, H. & Fendt, S. Power-to-liquid via synthesis of methanol, DME or Fischer–Tropsch-fuels: a review. *Energy & Environmental Science* **13**, 3207–3252 (2020).

14. Bisotti, F. *et al.* Impact of Kinetic Models on Methanol Synthesis Reactor Predictions: In Silico Assessment and Comparison with Industrial Data. *Industrial and Engineering Chemistry Research* **61**, 2206–2226 (2022).

15. Marlin, D. S., Sarron, E. & Sigurbjörnsson, Ó. Process Advantages of Direct CO2 to Methanol Synthesis. *Frontiers in Chemistry* **6**, 446 (2018).

16. Chen, L., Jiang, Q., Song, Z. & Posarac, D. Optimization of Methanol Yield from a Lurgi Reactor. *Chemical Engineering & Technology* **34**, 817–822 (2011).

17. Al-Fadli, A. M., Soliman, M. A. & Froment, G. F. Steady State Simulation of a Multi-Bed Adiabatic Reactor for Methanol Production. *Journal of King Saud University - Engineering Sciences* **7**, 101–132 (1995).

18. Vanden Bussche, K. M. & Froment, G. F. A steady-state kinetic model for methanol synthesis and the water gas shift reaction on a commercial Cu/ZnO/Al2O3 catalyst. *Journal of Catalysis* **161**, 1–10 (1996).

19. Mbatha, S. *et al.* Power-to-methanol process: a review of electrolysis, methanol catalysts, kinetics, reactor designs and modelling, process integration, optimisation, and techno-economics. *Sustainable Energy & Fuels* **5**, 3490–3569 (2021).

20. Graaf, G. H., Sijtsema, P. J. J. M., Stamhuis, E. J. & Joosten, G. E. H. Chemical Equilibria in Methanol Synthesis. *Chemical Engineering Science* **41**, 2883–2890 (1986).

21. Rahmatmand, B., Rahimpour, M. R. & Keshavarz, P. Introducing a novel process to enhance the syngas conversion to methanol over Cu/ZnO/Al2O3 catalyst. *Fuel Processing Technology* **193**, 159–179 (2019).

22. Keramat, F., Mirvakili, A., Shariati, A. & Rahimpour, M. R. Investigation of anti-condensation strategies in the methanol synthesis reactor using computational fluid dynamics. *Korean J. Chem. Eng.* **38**, 2020–2033 (2021).

Antonis Kokossis, Michael C. Georgiadis, Efstratios N. Pistikopoulos (Eds.)
PROCEEDINGS OF THE 33rd European Symposium on Computer Aided Process Engineering
(ESCAPE33), June 18-21, 2023, Athens, Greece

Optimal design of a microbial electrolysis cell for biohydrogen production

Uriel Roberto Pedroza-Medina[a], Luis Felipe Cházaro-Ruiz[b], Ricardo Femat[b], Alicia Román-Martínez[a*]

[a]*Facultad de Ciencias Químicas, UASLP, M. Nava 6, San Luis Potosí, 78210, MÉXICO*
[b] *Division de Ciencias Ambientales, Instituto Potosino de Investigación Científica y Tecnológica A. C., Camino a la Presa San José 2055, San Luis Potosí, 78216, MÉXICO*
*alicia.romanm@uaslp.mx

Abstract

Biohydrogen is considered as an alternative to replace fossil fuels and is being researched as a possible alternative to reduce conventional non-renewable hydrogen technologies. Towards a circular economy, microbial electrolysis cells have been suggested as a tool to both reduce organic pollutants from wastewater as well as to produce desirable hydrogen gas. Research efforts have focused on experimental design and computational simulation. The work presented here aims to develop an optimization model capable of estimating an optimal design and operational criteria to maximize cell's hydrogen output.
Keywords: microbial electrolysis cell, hydrogen, modeling, optimization

1. Introduction

The Microbial Electrolysis Cell (MEC) is a fairly new technology that has emerged and got scientific attention. The cell exploits the conductor-like properties of electrogenic microorganisms that catalyze the electrochemical oxidation of organic substrate, converting electrical energy into chemical energy stored in the H_2 molecule that is produced through the electrochemical reduction of the protons of the electrolytic medium. The main advantage of this technology is the ability to both treat wastewater effluents (source of organic matter) and synthetize biohydrogen. Numerous wastewater effluents have been demonstrated as potential substrate sources for MEC operation with varying degrees of efficiency in both organic-waste removal and hydrogen generation.

MEC-reactor design is known to be influenced by effluent's substrate concentration, reactor geometry and sizing, electrode material and configuration, anode-cathode distance, presence of a separator between them, composition of the electrolytic medium, as well as the microbial communities in the anodic chamber (Rousseau, 2021). Experimental studies have been performed to comprehend and optimize the processes occurring in these systems, however, so far there is no consensus to determine a feasible reactor configuration that would lead to competitive hydrogen production rates.

One approach that could be used to oversee these hindrances is mathematical modelling, to which fewer studies have been carried out. The first model proposed for a MEC was reported by Pinto *et al.* in 2011. This work studied the effect of several microbial communities: electrogenic, methanogenic, acetoclastic and hydrogenotrophic distributed in bioanode layers to understand their competition for the substrate in a continuous MEC. Later, Karimi, *et al.* in 2015 reported a generalized model that assumes coexisting microbial populations and the effects of a dynamic biofilm growth on substrate-potential distribution within it considering a conduction-based biofilm model in a batch reactor. A similar model was proposed by Flores-Estrella, *et al.*, in 2019 considering a continuous operation MEC. Hernández-García, *et al.*, in 2020 reported a model that considers the influence of geometry with primary and secondary current-substate distributions in 2D, and was re-studied, in 3D considering the effects of hydrodynamic behavior in a fed-batch reactor. Other works have developed models, which include stability and sensibility analysis and process control. As for optimization, some studies have been performed by trial-error and online methods, with little focus in a mathematical problem statement.

Design optimization is an engineering tool that utilizes a mathematical formulation of a design problem to support the selection of an optimal design among several alternatives. This work proposes a process model for a MEC based in a conduction-based bioanode considering operational and topological constraints, stated as discrete decision variables in order to find a design that maximizes hydrogen production rate.

2. Methodology

1.1. Modeling framework

A systematic modeling approach has been adapted from Aguiñaga-Morales *et al.* (2017) which allows the construction of a process model that will be set as equality constraint in an optimization model, described in Figure 1.

1.2. Problem statement

A Microbial electrolysis cell (MEC) consisting of two chambers, separated by a cation exchange membrane, contains acetate as substrate in the anodic chamber, a carbonaceous material anode colonized by electroactive bacteria and an abiotic platinum cathode. The electrodes are linked by an external circuit. The Ag/AgCl/KCl$_{(sat)}$ system is used as the reference electrode. The substrate oxidation releases CO_2, electrons, and protons, which will be electrochemically reduced at the cathode. The system is illustrated in Figure 2.

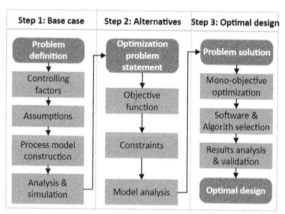

Figure 1. Modeling approach adopted to solve the optimal MEC design problem.

1.2.1. Identification of limiting factors and assumptions

(1) Substrate mass transport is mediated by diffusion, here described in 1D from and to the bioanode. (2) The biofilm acts as a porous conductor with conductivity κ_{bio}. (3) Electroactive bacteria exhibit extracellular electron transfer mechanisms though conductive pili and cytochrome C in their cell wall. (4) Cell operates in batch mode with uniform substrate distribution in bulk liquid. (5) Biofilm is mediated by advection. (6) Cathodic chamber is free of microorganisms. (7) An external power supply provides

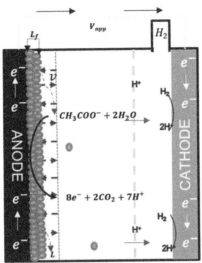

Figure 2. Schematic representation of phenomena in the modeled MEC. On the bioanode the model substrate (acetate) is diffused and oxidized by electroactive bacteria for cell growth. The generated protons go through the membrane to be reduced at the cathode.

the cell with a voltage that drives the electrochemical production of H_2. (8) The substrate oxidation generates the electrons needed to reduce the protons at the cathode, in terms of produced current or consumed electric charge. (9) This model is limited to hydrogen generation only, and does not consider methane production by other bacteria.

1.3. Construction of process model

1.3.1. Substrate mass balance in biofilm domain

Substrate accumulation in biofilm is mediated by diffusion and the consumption term, expressed as the following equation:

$$\frac{\partial S_d}{\partial t} = D_{S,f}\frac{\partial^2 S_d}{\partial z^2} - X_{f,a}q \quad (1)$$

With boundary conditions in anode surface (z=0) and biofilm-liquid interface ($z=L_f$):

$$D_{ED,f}\frac{\partial S_d}{\partial z}\Big|_{z=0} = 0 \qquad\qquad D_{ED,f}\frac{\partial S_d}{\partial z}\Big|_{z=Lf} = \left(\frac{D_{S,l}}{L}\right)\left(S_{d,l} - S_{d,n}\right)$$

Where S_d is the substrate concentration, $D_{S,f}$ is the substrate diffusion constant at the biofilm, $X_{f,a}$ is the electroactive biomass density and q is the substrate consumption rate. q is expressed as Monod-like kinetics derived from the Nernst equation, named Nernst-Monod kinetic equation:

$$q = q_{max}\phi_a\frac{S_d}{S_d+K_S}\frac{1}{exp[-F\eta/RT]+1} \quad (2)$$

q_{max} is the maximum rate of substrate consumption, ϕ_a is the volumetric fraction of electroactive biomass, F is the Faraday constant, R is the ideal gas constant, T is the anodic chamber temperature and η is the local potential, and K_S is the Monod's half-saturation constant.

1.3.2. Charge balance in biofilm

Electron transfer through the biofilm matrix is allowed by the local potential difference across the biofilm; the biofilm potential's change over time is expressed by this conduction (derived from the Ohm's law) and the addition of electrons generated by substrate oxidation and endogenous respiration:

$$\frac{\partial \eta}{\partial t} = \kappa_{bio}\frac{\partial^2 \eta}{\partial z^2} + \frac{F\gamma_1}{\tau}L_f fe^0 X_{f,a}q + \frac{F\gamma_2}{\tau}L_f X_{f,a}r_{res} \quad (3)$$

With boundary conditions in anode surface (z=0) and biofilm-liquid interface (z=L_f, no conduction):

$$\eta|_{z=0} = V_{anode} \qquad\qquad \frac{\partial \eta}{\partial z}\Big|_{z=L_f} = \frac{j|_{z=L_f}}{\kappa_{bio}} = 0$$

Here η is the local potential, κ_{bio} is the biofilm's conductivity, γ_1 is the yield of electrons from substrate oxidation, L_f is the biofilm thickness, $X_{f,a}$ is the biomass density, fe^0 is the fraction of electrons used for electrosynthesis, γ_2 is the electron yield from endogenous respiration, F is the Faraday's constant, τ is the time conversion and r_{res} is the endogenous respiration rate:

$$r_{res} = b_{res}\frac{1}{exp[-F\eta/RT]+1} \quad (4)$$

1.3.3. Biomass balance in biofilm

The microbial population exists as a consortium of electroactive bacteria, capable of metabolic extracellular electron transfer, and non-electroactive bacteria, which are capable of substrate consumption. In volume terms, the sum of their fractions is given by:

$$\phi_a + \phi_i = 1 \quad (5)$$

ϕ_a and ϕ_i are the electroactive fraction and the non-electroactive fraction, respectively. Their mass balances are stated as:

$$\frac{\partial \phi_a}{\partial t} + \frac{\partial(v\phi_a)}{\partial z} = Yq - r_{res} - r_{rina} = \mu_a \quad (6)$$

Which represents the transient electroactive fraction with respect to its position (a function of the advective velocity), v, by metabolic growth, loss by endogenous respiration and loss by inactivation.

$$\frac{\partial \phi_i}{\partial t} + \frac{\partial(v\phi_i)}{\partial z} = \frac{X_{f,a}}{X_{f,i}}r_{rina} = \mu_i \quad (7)$$

This equation gives the transient volumetric fraction of non-electroactive biomass, and its growth is given by the rate of inactivation, r_{rina}, which is stated as:

$$r_{rina} = b_{ina}\phi_a \quad (8)$$

Where b_{ina} is the inactivation constant. The advective velocity may also vary on biomass growth, so we can state:

$$\frac{\partial v}{\partial z} = \mu_a + \mu_i \quad (9)$$

This can be integrated along the biomass specific growth rate to get the advective velocity at any point z', considering that the net growth enhances the advective velocity.

The biofilm growth is thus a function of the advective velocity and the detachment phenomenon (which reduces biofilm's thickness) and can be formulated as:

$$\frac{dL_f}{dt} = v(t, L_f) - b_{det}L_f \quad (10)$$

Where b_{det} is the detachment constant.

1.3.4. Mass balances in liquid bulk

The substrate fed to the cell is accumulated on its volume, and fluxes to the biofilm domain as it's also consumed. The macroscopic mass balance can be stated as:

$$\frac{d(S_l V_b)}{dt} = A_s\frac{D_{S,l}}{L}(S_{d,n} - S_l) - V_b q_b \quad (11)$$

Here S_l represents the substrate concentration in the bulk liquid, V_b is the liquid volume in the anodic chamber, A_s is the anode surface area, $D_{S,l}$ is the diffusion coefficient in the liquid, L is the thickness of the diffusion layer, $S_{d,n}$ is the substrate concentration in the biofilm's surface and q_b is the substrate consumption rate by suspended biomass. For the suspended biomass, we express:

$$\frac{dx_b}{dt} = Y_b q_b - b_b x_b \quad (12)$$

x_b is the biomass concentration in the anodic chamber, b_b is the death rate and Y_b is the substrate/biomass yield. The consumption rate law for suspended biomass, which doesn't contribute to electron transport, is:

$$q_b = q_{max,b} x_b \frac{S_l}{K_{s,l} + S_l} \quad (13)$$

1.3.5. Current generation and hydrogen productivity

The anode potential is function of the applied cell potential by the external power supply.

$$V_{an} = V_1 V_{app} + V_2 \quad (14)$$

Where V_{an} is the anode potential, V_{app} is the applied potential, *and V_1* and V_2 are constant relationships of linear applied potential.

By Ohm's Law, the current density (j) might be stated as the product of the biofilm's conductivity times the quotient of the potential/biofilm thickness gradients, and the cell current (I) is known by the product of the anode's surface area times j:

$$j = -\kappa_{bio} \frac{\Delta \eta}{\Delta z} \quad and \quad I_{MEC} = A_s j \quad (15)$$

So, we can state the hydrogen productivity as a function of the cell current that conducts to the cathode:

$$Q_{H_2} = \frac{RT}{V_b P} \left(\frac{\tau I_{MEC}}{nF} Y_c \right) \quad (16)$$

Q_{H_2} is the hydrogen volumetric productivity (m³ H₂ m⁻³ anode day⁻¹) and Y_c is the cathode's dimensionless efficiency.

1.4. Analysis and simulation

Kinetic and electrochemical parameters are obtained from previous MFC and MEC modeling works (Pinto, 2011; Karimi, 2015). To implement, discretize (1) and (3) and their B.C., in space through finite differences, solve non-linear system through Newton's method. Calculate advective velocity (9) by numerical integration (trapezoid rule) then solve for biofilm thickness (10). Biomass balances (5, 6, 7) at the biofilms are solved by Backward-Euler. Subtract detachment rate from biofilm thickness. Interpolate the remaining biomass onto a new grid and iterate. Use a similar approach for (11) and (12).

2.5. Optimization problem statement

2.5.1. Objective function

The present work aims to optimize the energy efficiency of an MEC considering spatial and operational factors. The cell efficiency is stated as the energy load achieved by hydrogen gathered, a function of its enthalpy of combustion *versus* the applied energy (applied electric potential) and any other applied energy sources as mixing and pumping. We can express this mathematically as:

$$\eta_{H_2} = \frac{W_{H_2}}{W_E} = \frac{Q_{H_2} \rho_{H_2} \Delta H_{H_2}}{I_{MEC} V_{app}}$$

Thus, we can state our objective function as:

$$\max \eta_{H_2} = \frac{Q_{H_2} \rho_{H_2} \Delta H_{H_2}}{I_{MEC} V_{app}}$$

$$s.t. \ \mathbf{h(x,y) = 0} \ (equality \ constraints); \ \mathbf{g(x,y) \le 0} \ (inequality \ constraints)$$

$$\mathbf{x \in X \subseteq \mathbb{R}^n; \ y \in Y} \ integer$$

3. Process model simulation results

Thus far, the process model returns spatial distributions of substrate, voltage, and current density in the z direction across the biofilm. Figure 3 shows that the substrate concentration is highest in biofilm's vicinity with liquid, opposite to the potential which decreases with biofilm length, while the current density behaves similarly, achieving a current density of 2.9 A/m² at tested substrate load of 1.5×10⁻³ mol/L. The max. current reached in the 10 cm²-bioanode is 3.1×10⁻⁴ A, so the max. hydrogen productivity is 0.102

m^3H_2/m^3cell day at 32 days, when microbes reach stationary phase and substrate starts depleting. For validation, potential/substrate distribution and H_2 productivity profiles in Figure 3 are similar to those reported by Karimi *et al.* (2015).

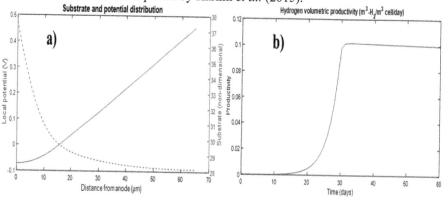

Figure 3. (a) Distribution of potential (dashed) and substrate (solid) across biofilm at max length 70 μm. Simulated time, 60 days, initial substrate concentration: 1.5×10^{-3} mol Acetate/L. (b) Hydrogen productivity, reached max. at 32 days.

4. Conclusions

The developed process model is suitable to be formulated as an equality constraint for the objective function and may be tested in different parameter scenarios. It demonstrates microbial processes are limiting to MEC performance. Further alternative analysis will be carried out to complete the optimization problem formulation.

5. Acknowledgments

U.R. Pedroza-Medina gratefully acknowledges CONACyT in Mexico for granting the scholarship 1143427 to pursue his M. Sc. Degree.

References

1. R. Rousseau, 2021, Microbial electrolysis cell (MEC): Strengths, weaknesses and research needs from electrochemical engineering standpoint, Applied Energy, V. 257, 113938.
2. F. Aguiñaga-Morales, 2017, Model identification, calibration and validation of the aerobic stage in water remediation of a paper mill effluent, Computer Aided Chem. Eng., V40, 547-552.
3. R. Pinto, 2011, Multi-Population Model of a Microbial Electrolysis Cell, Environ. Sci. Technol., 45, 5039–5046.
4. M. Karimi-Alavijeh, 2015, A Generalized Model for Complex Wastewater Treatment with Simultaneous Bioenergy Production Using the Microbial Electrochemical Cell, Electrochim. Acta, V. 167, 84-96.
5. R.A. Flores-Estrella, 2019, A Dynamic Biofilm Model for a Microbial Electrolysis Cell, Processes, 183, 7, 1-20.
6. K.M. Hernández-García, 2020, Modeling 3D current and potential distribution in a microbial electrolysis cell with augmented anode surface and non-ideal flow pattern, Biochem. Eng. Journal, V. 162, 107714.

Antonis Kokossis, Michael C. Georgiadis, Efstratios N. Pistikopoulos (Eds.)
PROCEEDINGS OF THE 33rd European Symposium on Computer Aided Process Engineering
(ESCAPE33), June 18-21, 2023, Athens, Greece

Simulation and Exergoeconomic Analysis of Brewer's Spent Grains convective drying process

Mero-Benavides M.,[a] Enriquez-Posligua J.,[a] Barcia-Quimi A.F.,[a] Tinoco-Caicedo Diana L.[a,b*]

[a] *Facultad de Ciencias Naturales y Matemáticas (FCNM), Escuela Superior Politécnica del Litoral Ecuador, 090903 Guayaquil, Ecuador*
[b] *Centro de Energías Renovables y Alternativas (CERA), Escuela Superior Politécnica del Litoral Ecuador, 090903 Guayaquil, Ecuador*
[*]*dtinoco@espol.edu.ec*

Abstract

A low-cost and highly exergetic efficient drying process assures a sustainable transformation of Brewer's Spent Grains (BSGs) into value-added product. This work aims to perform an exergoeconomic analysis at a level component of the convective simulated drying process of BSGs. To validate the simulated process, characterization, and a drying kinetic study of BSGs obtained from a factory in Guayaquil, Ecuador, were carried out. The simulated drying process was validated, reaching a maximum relative error of 4.64%. The exergy destruction rate varied from 12.6 MJ/kg to 3.42 MJ/kg when the air inlet temperature varied from 60 °C to 120 °C. Results suggest that the increment of the air inlet temperature could reduce the specific cost of the dry BSGs up to a 68.3% and increase the exergetic efficiency by 24.6%.

Keywords: kinetic study, exergoeconomic analysis, process simulation, modeling, Brewer's Spent Grains (BSGs)

1. Introduction

Currently, reducing waste is considered a critical need in the brewing industry. The main waste in this production process is the brewer's spent grains (BSGs) (Rachwał et al., 2020). Usually, it is necessary to dry the BSGs due to the high moisture levels before converting them into value-added products (Mallen & Najdanovic-Visak, 2018). Hence, the drying process needs to be sustainable and have low operating costs. The exergoeconomic analysis is considered a vital decision-making tool to reduce production costs and increase the sustainability of a process by reducing the exergy destruction rate (Bejan & Tsatsaronis, 2021). Although there are some drying experimental studies of this waste at temperatures from 40 °C to 95 °C (Capossio et al., 2022) (Castro et al., 2022), these studies do not analyze the best-operating conditions to diminish the exergy destruction rate and the operating cost of the drying process. Thus, there has been no exergoeconomic analysis of the drying process of BSGs. In this context, this work presents an exergoeconomic analysis of a convective drying process of BSGs. The characterization and drying kinetic study were performed using BSGs from a factory located in Guayaquil, Ecuador. The drying process was simulated using Aspen Plus V12.1, and the obtained experimental data was used to validate the process simulation.

The exergetic, economic, and exergoeconomic balances were performed for the dryer using Engineering Equation Solver (ESS) software following the methodology from (Bejan et al., 1996). The exergy destruction rate, investment and operational costs were determined for drying temperatures from 60 °C to 120 °C.

2. Materials and Methods

2.1. Experimental Studies

2.1.1. Characterization of BSGs
The BSGs were obtained from a brewing company in Guayaquil, Ecuador. The samples were dried at 105°C based on ISO 18123:2015. The proximate analysis (ash and volatile material (VM)) was done according to ISO 18122:2015, respectively. The difference between ash and volatile material determined the fixed carbon (FC) in dry basis. The ultimate analysis was performed by a CHNS Elemental Analyzer (Elementar). The particle size distribution was analyzed according to the ASTM D1511-10 procedure.

2.1.2. Drying Kinetics
The drying kinetic study was carried out at temperatures 60, 70, 80, 90, 105, and 120°C. Higher temperatures were not selected, because a previous study shows that the degradation rate increases (Mishra et al., 2017). Dry BSGs were submerged in distilled water at room temperature during 2 h to reach their initial moisture (Cuevas et al., 2019). Samples were placed on a plate up with thickness of 1.30 cm inside a convective dryer (UNB 500, Memmert, Germany). Periodically the samples were removed from the dryer, weighing them on digital balance in order to measure their average water content.

The critical moisture was determined as transition moisture, where the first constant speed dry deviation was observed. The equilibrium moisture was estimated as the constant behavior for an extended period (Srikiatden & Roberts, 2007). The effective diffusivity was determined from the slope of the natural logarithm of the moisture ratio (M_R) versus drying time according to the second diffusion Fick Law for the infinite dry plate (Cuevas et al., 2019). The heat and mass coefficients were determined from the dimensionless numbers based on Incropera et al. (Incropera et al., 2007).

2.2. Process Simulation
Figure 1 shows the BSGs simulated drying system. It consists of an air preheater (HX-101) and a batch- type convection drying chamber (D-101). Flue gases at 300 °C (stream 6) with a mass composition of 18.5% CO_2, 6.8% H_2O , 7.09% N_2, and 3.8% O_2 enter the HX-101. Then, ambient air enters the HX-101 with a relative humidity of 73% at 25 °C (stream 1) and exits at the drying temperature (stream 2), which varies between 60 °C to 120 °C. Then, the flue gases exit the HX-101 (stream 7) and they are discharged to the environment. The hot air enters the D-101 to remove the water from the BSGs (stream 3), and the moist air exits the dryer (stream 4). After the process, the dry BSGs (stream 5) is discharged.

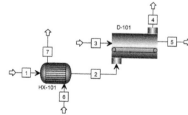

Figure 1. Process flow diagram of the drying system

The process simulation was performed in Aspen Plus V12.1. BSGs were simulated as an unconventional component. The ultimate and proximate analysis results obtained in section 2.1 were entered into the HCOALGEN and DCOALIGT models. First, the ideal thermodynamic model was used since it worked at atmospheric pressure, and the MIXCNCPSD for handling unconventional solids configured the flow class. Next, the dryer was simulated using a convective dryer. Finally, the simulation was configured in batch mode, where the total drying time was set as the maximum iteration and operation time. Drying data such as heat and mass transfer coefficients, drying curve, equilibrium, and critical moisture were defined in the component.

2.3. Model Validation

The absolute moisture of the BSGs obtained in the simulation was compared with the results obtained experimentally for each drying air temperature. The same operating conditions were maintained to ensure the validity of the modeled process.

2.4. Exergoeconomic analysis

The exergoeconomic analysis was developed according to the methodology offered by (Bejan et al., 1996) using equations Eq.(1)-(4). Engineering Equations Solver (EES) software was used to determine the thermophysical properties for the calculation of enthalpy and entropy of each state.

$$\dot{E}_{D,k} = \dot{E}_{F,k} - \dot{E}_{P,k} \quad (1) \qquad \dot{C}_{F,\,k} + \dot{Z}_k = \dot{C}_{P,\,k} \quad (3) \qquad PEC_{HX-101} = 1900 + 2500 \cdot S \quad (5)$$

$$\dot{Z}_k = \dot{Z}_k^{O\&M} + \dot{Z}_k^{CI} \quad (2) \qquad \dot{C}_{D,k} = c_{F,k}\dot{E}_{D,k} \quad (4) \qquad PEC_{D-101} = 10000 + 7900 \cdot S^{0.5} \quad (6)$$

The economic analysis was carried out following the revenue requirement method. The economic parameters were taken from a previous study by authors (Tinoco-Caicedo et al., 2021). The purchase equipment costs (PEC) of the HX-101 and the D-101 were determined using Eq. (5)-(6) (Towler & Sinnott, 2012), respectively, for 2010 (CEPCI of 532.9). Where S is the area of each tray from the dryer (20 m^2) and the heat exchange area for the HX-101 (obtained in the simulator). The chemical engineering plant cost index (CEPCI) of 607.5 was used to adjust the PEC for 2019.

Table 1 shows the fuel exergy ($\dot{E}_F$), product exergy ($\dot{E}_P$), fuel cost ($\dot{C}_F$), product cost ($\dot{C}_P$) and auxiliary equations at the component level.

Table 1. Equations used for the exergoeconomic analysis

Component	$\dot{E}_{F,k}$	$\dot{E}_{p,k}$	$\dot{C}_{F,k}$	$\dot{C}_{p,k}$	Auxiliary equation
					$c_6 = 0$
HX-101	$\dot{E}_6 - \dot{E}_7$	$\dot{E}_2 - \dot{E}_1$	$\dot{C}_6 - \dot{C}_7$	$\dot{C}_2 - \dot{C}_1$	$c_1 = 0$
D-101	$\dot{E}_2 - \dot{E}_4 + \dot{E}_3$	$\dot{E}_5$	$\dot{C}_2 - \dot{C}_4 + \dot{C}_3$	$\dot{C}_5$	$c_3 = 0$
Overall System	$\dot{E}_F = (\dot{E}_6 - \dot{E}_7) + \dot{E}_3 + (\dot{E}_2 - \dot{E}_4)$				
	$\dot{E}_P = \dot{E}_5$				

3. Results and Discussion

3.1. Experimental studies

The characterization and drying kinetic study was done for the BSGs. Table 2 shows the results obtained for the ultimate and proximate analysis and the particle size distribution.

Table 2. BSGs characterization required for convective dryer simulation

Ultimate analysis (% in dry basis)		Proximate analysis (% in dry basis)		Particle size distribution (mm, %)	
C	47.55	Ash	4.03	PS>2	68.48

H	8.09	VM	88.31	1.4<PS<2	16.24
N	5.56	FC	7.66	1<PS<1.4	6.03
O	38.13	Moisture (wb)	81.77	0.85<PS<1	1.28
S	0.67			PS< 0.85	7.97

Figure 2 a) shows the drying kinetic of BSGs in a batch type convective dryer at different drying temperatures between 60 °C and 120 °C. The drying time reduction from 4 to 1.25 hours was observed when the temperature increased from 60 °C to 120 °C. This change occurs because the increase in temperature causes a faster water evaporation rate from 0.43 to 2.45 g/kg-s, which can be graphically seen in the slope of the curve from the first period of constant rate. The critical and equilibrium moistures vary from 53.9 % to 72.4 % and 0.63 % to 3.86 % in wet basis, respectively.

Figure 2 b) shows the mass and heat transfer coefficients estimated from the results of the drying kinetic study. It is observed that the heat and mass transfer coefficients increased at high temperatures because the increase in the inlet drying temperature intensifies the water evaporation rate (Beigi et al., 2017a). Similar results were obtained using a tray dryer for pumpkin (Guiné et al., 2012) in which the mass transfer coefficient increased from 0.1×10^{-5} m/s to 4.1×10^{-5} m/s when the inlet air drying temperature increased from 30 °C to 70 °C.

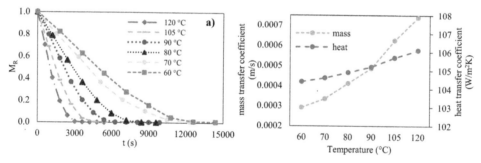

Figure 2. a) Drying kinetic curves for BSGs, b) mass and heat transfer coefficients at different temperatures.

3.2. Model validation

The drying process was simulated in Aspen Plus using the data obtained from the characterization of the BSGs and the kinetic drying study. The final moisture obtained in the simulation and in the drying experiments were compared to validate the model developed for the dryer. Table 3 shows that the simulation results have relative errors between 3.0 and 4.6%. Therefore, the developed drying model can describe the behavior of the drying process between 60 °C and 120 °C.

Table 3. Model validation of the batch type convective dryer.

Temperature (°C)	Moisture (wet basis)		Relative Error (%)
	Experiment	Simulation	
60	2.610	2.729	4.56
70	3.861	3.979	3.05
80	0.635	0.656	3.28

90	0.819	0.856	4.55
105	0.715	0.749	4.64
120	1.231	1.279	3.90

3.3 Exergoeconomic Analysis

The exergoeconomic indicators of the simulated BSGs convection drying process at different drying air temperatures were determined. Figure 3 a) shows that the exergetic efficiency increases significantly between 70 °C and 80 °C, while exergetic efficiency varies insignificantly at higher temperatures. This behavior is similar to the overall exergy destruction rate. The greatest reduction of destroyed exergy (52.3%) occurs between 70 °C and 80 °C because the required drying air flow decreases by 66%, according to Figure 3

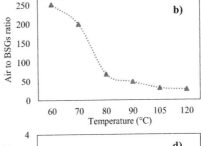

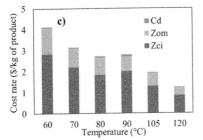

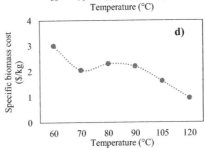

b). Moreover, at higher temperatures, less drying air is required. Therefore, the drying airflow is directly proportional to the destroyed exergy rate. (Beigi et al., 2017b) obtained similar results, where rice is dried in a convective dryer and the exergetic efficiency increased from 33% to 68% when the drying temperature increased from 40 °C to 60 °C.

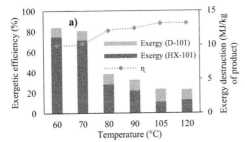

Figure 3. The effect of drying temperature in a) global exergy destruction and exergetic efficiency per component, b) air to BSGs ratio, c) exergy destruction, investment, operational and maintenance cost rates of the drying system, and d) specific biomass cost.

Figure 3 c) shows that at 120 °C the operating costs are the lowest and the cost per destroyed exergy. This change occurs because, at 120 °C, the lowest air/flue gas ratio was used. Thus, there is a need to burn less fuel for flue gas production. Also, previous studies have demonstrated that implementating other technologies, such as microwave-assisted dryers, would reduce operating costs (Ranjbaran & Zare, 2013). Furthermore, (Agbede et al., 2020) recommended implementating renewable energy sources such as solar energy in the air heater of drying processes due to their lower energy requirement. (Atalay, 2019) compared conventional heat pump drying with solar-assisted heat pump and found that this reduces investment and operating costs by 20.3% and 34.5%, respectively. Figure 3 d) shows that the specific cost of biomass decreases significantly with the temperature increment. At 120 °C, it is possible to reduce the cost of dry biomass by 69.3%. This cost reduction responds to the low operating costs achieved at 120 °C. The minimum cost of BSGs was $0.95/kg, which is half that obtained in a previous study for dry spent coffee grounds (Tinoco-Caicedo et al., 2021).

4. Conclusions

This study aimed to simulate and perform an exergoeconomic analysis of a batch type convective dryer of BSGs based on experimental studies, carried out for air drying temperatures between 60 °C to 120 °C. This research shows that the exergetic efficiency varies between 66.4% and 88.0% when increasing the temperature from 60 °C to 120 °C. Production costs can be reduced by 69.1% by increasing the temperature from 60 °C to 120 °C. The specific cost of biomass varies between 3.00 $/kg and 0.95 $/kg and is at 120 °C when the lowest cost is reached. Finally, the results suggest that convective drying should operate at high temperature because it benefits exergetic efficiency and operating costs. Future studies should focus on experimental studies using other drying technologies that increase the drying rate and reduce both costs and the rate of destroyed exergy.

References

Agbede, O. O., Oke, E. O., Akinfenwa, S. I., Wahab, K. T., Ogundipe, S., Aworanti, O. A., Olatunde Arinkoola, A., Agarry, S. E., Ogunleye, O., Nihinlola Osuolale, F., & Babatunde, A. (2020). Thin layer drying of green microalgae (Chlorella sp.) paste biomass: Drying characteristics, energy requirement and mathematical modeling. *Bioresource Technology Reports Journal, 11.*

Atalay, H. (2019). Comparative assessment of solar and heat pump dryers with regards to exergy and exergoeconomic performance. *Energy, 189.*

Beigi, M., Tohidi, M., & Torki-Harchegani, M. (2017a). *Exergetic analysis of deep-bed drying of rough rice in a convective dryer.*

Beigi, M., Tohidi, M., & Torki-Harchegani, M. (2017b). Exergetic analysis of deep-bed drying of rough rice in a convective dryer. *Energy, 140,* 374–382.

Bejan, A., & Tsatsaronis, G. (2021). Purpose in thermodynamics. *Energies, 14*(2).

Bejan, A., Tsatsaronis, G., & Moran, M. (1996). *Thermal Design & Optimization.* John Wiley & Sons, Inc.

Capossio, J. P., Fabani, M. P., Reyes-Urrutia, A., Torres-Sciancalepore, R., Deng, Y., Baeyens, J., Rodriguez, R., & Mazza, G. (2022). *Sustainable Solar Drying of Brewer's Spent Grains: A Comparison with Conventional Electric Convective Drying.*

Castro, L. E. N., Matheus, L. R., & Colpini, L. M. S. (2022). Optimization of brewers' spent grain drying process. *Brazilian Journal of Development, 8*(2), 14481–14488.

Cuevas, M., Martínez-Cartas, M. L., Pérez-Villarejo, L., Hernández, L., García-Martín, J. F., & Sánchez, S. (2019). Drying kinetics and effective water diffusivities in olive stone and olive-tree pruning. *Renewable Energy, 132,* 911–920.

Guiné, R. P. F., Henrriques, F., & Barroca, M. J. (2012). Mass Transfer Coefficients for the Drying of Pumpkin (Cucurbita moschata) and Dried Product Quality. *Food and Bioprocess Technology*, *5*(1), 176–183.

Incropera, F. P., Dewitt, D. P., Bergman, T. L., & Lavine, A. S. (2007). *Fundamentals of Heat and Mass Transfer* (J. Hayton (ed.); 6th ed.). John Wiley & Sons, Inc.

Mallen, E., & Najdanovic-Visak, V. (2018). Brewers' spent grains: Drying kinetics and biodiesel production. *Bioresource Technology Reports*, *1*, 16–23.

Mishra, P. K., Gregor, T., & Wimmer, R. (2017). Utilising brewer's spent grain as a source of cellulose nanofibres following separation of protein-based biomass. *BioResources*, *12*(1), 107–116.

Rachwał, K., Waśko, A., Gustaw, K., & Polak-Berecka, M. (2020). Utilization of brewery wastes in food industry. *PeerJ*, *8*, e9427.

Ranjbaran, M., & Zare, D. (2013). Simulation of energetic- and exergetic performance of microwave-assisted fluidized bed drying of soybeans. *Energy*, *59*, 484–493.

Srikiatden, J., & Roberts, J. S. (2007). Moisture transfer in solid food materials: A review of mechanisms, models, and measurements. *International Journal of Food Properties*, *10*(4), 739–777.

Tinoco-Caicedo, D. L., Mero-Benavides, M., Santos-Torres, M., Lozano-Medina, A., & Blanco-Marigorta, A. M. (2021). Simulation and exergoeconomic analysis of the syngas and biodiesel production process from spent coffee grounds. *Case Studies in Thermal Engineering*, *28*(October), 101556.

Towler, G., & Sinnott, R. (2012). *Chemical Engineering Design - Principles, Practice and Economics of Plant and Process Design* (2nd ed.). Elsevier Inc.

Antonis Kokossis, Michael C. Georgiadis, Efstratios N. Pistikopoulos (Eds.)
PROCEEDINGS OF THE 33rd European Symposium on Computer Aided Process Engineering
(ESCAPE33), June 18-21, 2023, Athens, Greece

Incorporating negative emissions technologies with policy instruments for net-zero emissions

Elizabeth J. Abraham, Patrick Linke, Dhabia M. Al-Mohannadi[*]

Department of Chemical Engineering, Texas A&M University at Qatar, Education City, PO Box, 23874, Doha, Qatar

Abstract

The ongoing climate crisis requires the reduction and removal of carbon to avoid global catastrophic consequences to the environment. As such, efforts presently focus on understanding the impact of negative emissions technologies in actively removing carbon dioxide from the atmosphere. While these emerging technologies bring a distinctive set of feasibility challenges, their integration into existing systems is essential due to the inability of decarbonization strategies to solely mitigate greenhouse gas emissions. The objective of this work is to develop strategies for integrating negative emissions technologies with carbon capture utilization and storage options, spurred by the increasing relevance of circular economies. The multi-period resource integration approach will determine the optimal technology portfolios best suited to a specific industrial landscape to achieve net-zero emissions in the presence of environmental policy instruments across time. The analysis will provide decision makers with a preliminary understanding of the viability of these technologies and the progress needed in regulatory frameworks to enable their effective implementation. The applicability of the proposed approach is demonstrated through the deployment of negative emissions technologies, such as bioenergy carbon capture and storage, direct air carbon capture and storage, enhanced weathering of minerals, and utilization of biochar, in an oil and gas based industrial setting. The optimal strategies determined indicate that carbon caps and taxes are fundamental in reducing the creation of emissions, while carbon credits encourage the active removal of carbon dioxide from the atmosphere.

Keywords: Negative emissions technologies, Process integration, Optimization, Carbon management

1. Introduction

With the climate crisis at the forefront of the sustainable development agenda, oil and gas economies that account for 15-20% of global greenhouse gas emissions (Gatto et al., 2021) will play a critical role in achieving net-zero emissions by 2050. To meet these targets, negative emissions technologies or NETs will need to be integrated into these economies along with more rampant decarbonization efforts that focus on the improvement of process efficiencies, transition to renewable energy, and deployment of carbon capture and storage (CCS) options. NETs fundamentally remove greenhouse gases from the atmosphere through various mechanisms, with their primary target being carbon dioxide (CO_2) (McLaren, 2012). Some examples of NETs include technological solutions such as bioenergy with carbon capture and storage (BECCS) and direct air carbon capture and storage (DACCS), natural solutions such as afforestation and reforestation, and enhanced natural solutions such as the use of biochar and enhanced weathering (IEA, 2020). Recently, these technologies have gained significant traction in industrial and academic research and development domains due to the major roles they play in the scenarios analyzed by integrated assessment models towards attaining carbon neutrality, where both carbon removal and reduction are of paramount importance (Tavoni & Socolow, 2013).

A significant portion of the extensive literature in this direction belongs to BECCS and DACCS, detailing the critical feasibility challenges that prevent their large-scale deployment and the measures to overcome them. Of the two, DACCS shows promising potential for deployment in oil and gas economies due to their large energy reserves, which can power the capture process and provide a means through the depleted reserves to store the captured carbon (Babonneau et al., 2022). While storage concerns for BECCS can also be managed similarly, there are concerns surrounding the sustainable procurement of the biomass required that additionally needs to be addressed. However, if obtained in a viable manner, the bioenergy it produces is a valuable commodity that can provide power for various operations such as those of other NETs, as suggested by Abraham et al. (2022c), who investigated the feasibility of such synergistic exchanges between NETs. Alternatively, instead of storing the carbon removed in geological storage sites, principles of circular economy suggest that it can be redirected for utilization in CCUS options (Abraham et al., 2022a). Therefore, to create an integrated system that embodies circular economy, all material and energy resources in a system must be analyzed. Towards enabling such circularity in systems, Abraham et al. (2022b) developed a multi-period optimization approach based on the resource integration method proposed by Ahmed et al. (2020) for designing processing clusters. This work adopts the multi-period resource integration approach to analyze the viability of integrating NETs with CCUS options and policy instruments in energy-rich economies to further devise deployment strategies over time where the carbon removed is specifically stored in environmental compartments in their customary sense to achieve net-zero emissions.

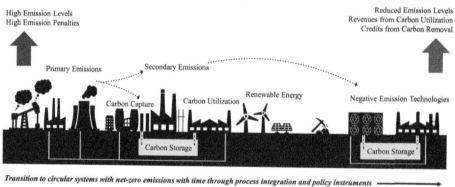

Figure 1: Overview of the proposed strategy for net-zero emissions

2. Approach

The multi-period resource integration model adopted in this work modifies the approach proposed by Abraham et al. (2022b) to develop planning strategies set to achieve carbon neutrality using networks of NETs and CCUS options. The model formulation takes the form of a mixed-integer linear program (MILP) to develop an integrated deployment strategy where further synergistic opportunities that are unapparent when solely focusing on carbon can be explored. For a given carbon removal and reduction target, it identifies the optimal network that achieves the maximum net present value (NPV) over a planning horizon as given by Eq(1):

$$NPV_{Deployment\ strategy} = \sum_{Period} \frac{\sum_{Process} Revenue - Operating\ Cost - Capital\ Cost}{(1 + Discount\ Rate)^{Number\ of\ periods\ considered}} \quad (1)$$

The optimization determines which processes become active in a period and the capacities they operate at to meet the specified target. The carbon removal and reduction options are each represented as processes that consume certain resources to produce certain others using explicit resource parameters. Each of the resources is then represented through a resource line defined by a unique set of specifications to facilitate integration between processes that share common resources. The total costs incurred depend on the capital and operating costs of the processes, along with any stipulated inflation rates, resource penalties, and credits considered. Furthermore, equality and inequality constraints can be placed on the operational capacities and resource flows throughout the network according to the specific details of the scenario analyzed, along with resource balances that monitor the flow of resources to, from, and between processes.

3. Illustrative Example

To illustrate the application of this approach, a hypothetical scenario is investigated over a planning horizon that consists of three periods spanning 10 years each. The processes considered for capturing, storing, utilizing, and removing CO_2 are listed in Table 1, along with their reference products and CAPEX parameters. The process parameters for the carbon removal options, namely, the BECCS, biomass pelletization unit, biochar production unit, DACCS, and EW mineral processing unit, were retrieved from Abraham et al. (2022c), while those of the sequestration unit and the production and capture units of methanol and urea, were determined from Ahmed et al. (2020). Each process operates at a maximum capacity of 25 million tons of their respective reference products in the first period, after which the capacities of the NETs and biomass pelletization unit increase by 20% from the previous period.

Table 1: Reference products and CAPEX parameters

Processes (Process ID)	Reference Product	CAPEX Parameter (Million $)
Bioenergy Carbon Capture and Storage (BECCS)	CO_2 removed	2,478
Biomass Pelletization Unit (BMP)	Biomass pellets	130
Biochar Production Unit (BCP)	CO_2 removed	1,601
Direct Air Carbon Capture and Storage (DACCS)	CO_2 removed	2,175
EW Mineral Processing Unit (EW)	CO_2 removed	71
Fischer-tropsch Unit (FTU)	Fischer-tropsch fuels	248
Methanol Production Unit (MPU)	Methanol	262
Methanol Carbon Capture Unit (MCC)	CO_2 captured	7
Synthetic Natural Gas Production Unit (SNG)	SNG	39
Sequestration (SQU)	CO_2 captured	12
Urea Production Unit (UPU)	Urea	135
Urea Carbon Capture Unit (UCC)	CO_2 captured	7

The data for the SNG production unit was obtained from Chauvy et al. (2021), while those for the FT unit were obtained from Panahi et al. (2018) and Al-Yaeeshi et al. (2020). The resource parameters for these processes are summarized in Table 2, along with the prices of their associated resources. The prices per ton of the resources not listed here were set at $294 for ammonia, $1,383 for biochar, $112 for biomass pellets, $509 for calcium carbonate, $14 for low pressure steam, $305 for methanol, $484 for propane, $24 for residues, $26 for rock in, and $288 for urea. All prices were set to vary at an inflation rate of 5%.

Table 2: Resource parameters of the FT and SNG units along with their prices (2020 US$)

Processes (Resource unit)	FT Unit	SNG Production Unit	Prices ($/ Resource unit)
CO_2 emitted (tons)	3.85	-	-
CO_2 utilized (tons)	-2.07	-2.41	-
Cooling water (tons)	-	-23.69	0.03
Electricity (kWh)	-	-17.14	0.03
FT Fuels (tons)	1.00	-	1,029
High pressure steam (tons)	3.88	-	16
Hydrogen (tons)	-	-0.45	1,178
Medium pressure steam (tons)	4.70	3.71	15
Natural gas (tons)	-2.01	-	100
Saturated steam (tons)	-3.56	-	15
Synthetic natural gas (tons)	-	1.00	100
Wastewater (tons)	-	23.69	-
Water (tons)	-	-3.71	5
Water out (tons)	4.22	1.86	-

Certain resources in the processes considered contain CO_2 that must be accounted for to ensure climate targets are met in time. As such, the compositions of these resources, specified in Table 3, will also be incorporated into the overall carbon flow analysis.

Table 3: CO_2 compositions of the carbon carrying resources

Resource	CO_2 Composition
CO_2 emitted, captured, utilized, removed	1.00
Methanol process emissions	0.30
Methanol capture unit emissions	0.06
Urea process emissions	0.26
Urea capture unit emissions	0.03

Two cases were investigated with this setup to maximize the NPV of the deployment strategy analyzed at a discount rate of 10%. Every period receives a feed of CO_2 captured from various industrial processes associated with the oil and gas industry. With each period, this feed decreases at a rate of 25% as changes to conventional processes are expected to lower the emissions they release. To achieve net-zero emissions at the end of the planning horizon, constraints are placed on the overall flow of CO_2 from the system. While the first period has no specific constraints, the CO_2 that leaves the network in the second and third periods, defined as the difference between the carbon created and removed by it, are constrained with carbon caps, as specified in Table 4.

Table 4: Carbon feed and cap limits over the three periods

Periods	Period 1	Period 2	Period 3
CO_2 fed (tons)	175,000,000	131,250,000	98,437,500
CO_2 cap (tons)	No constraint	65,625,000	0

The processes activated for the two cases and their operational capacities in each period are given in Table 5. In the first case, net-zero emissions are achieved by transitioning from focusing solely on carbon utilization to additionally incorporating carbon removal. The optimal strategy determined has an NPV of $32 billion and removes nearly 128 million tons of CO_2 using the BECCS, DACCS, and EW processes. These NETs only become active when constraints are placed on the emissions leaving the system. In contrast, periods such as the first, without any constraints, focus on generating revenues through carbon capture and utilization through the production of FT fuels, methanol, and urea. In later periods, there is a decrease in carbon utilization processes with high operating costs, such as SNG production, and those which lead to further emissions, like methanol production. This decrease, when simultaneously implemented with an increase in carbon removal through NETs, permits the system, over time, to lower sequestration capacities and leads to the cumulative release of about 295 million tons of CO_2.

Table 5: Operational capacities in million tons of the reference product determined for the three cases in each period and their set carbon tax and credit rates

Case	Case 1			Case 2		
Time periods	tp1	tp2	tp 3	tp1	tp2	tp 3
Carbon tax	0	0	0	150	300	450
Carbon credit	0	0	0	200	400	600
Processes						
BMP	-	22.21	26.65	18.51	22.21	26.65
BECCS	-	30.00	36.00	25.00	30.00	36.00
DACCS	-	-	27.10	25.00	27.12	27.10
EW	-	1.26	33.66	25.00	30.00	36.00
FTU	25.00	25.00	25.00	25.00	0.10	0.10
MPU	25.00	25.00	17.95	25.00	25.00	25.00
MCC	-	2.56	1.84	2.56	2.56	2.56
SNG	14.98	0.10	0.10	16.69	19.92	6.31
SQU	25.00	21.23	0.10	25.00	25.00	25.00
UPU	25.00	25.00	25.00	25.00	25.00	25.00
UCC	0.10	1.66	1.66	1.66	1.66	1.66

In the second case, carbon taxes and credits that doubled and tripled from their rates in the first period are implemented as described in Table 5, along with the emission caps detailed previously. The increasing carbon tax rates further limit the release of emissions, while the increasing carbon removal credits incentivize the deployment of NETs with the overall network emitting and removing about 99 and 261 million tons of CO_2, respectively. Over time, emissions were drastically reduced from 97 million tons in the first period to nearly only a million tons of CO_2 in the second and third periods. The reduction was realized by decreasing carbon utilization in the FT unit, increasing carbon capture from the two capture units, and increasing sequestration to operate at its maximum capacity throughout the periods considered. The operations of the FT unit were specifically reduced while methanol and urea production remained constant as these processes have capture units to manage the emissions they create. On the other hand, the NETs operating at or near their maximum capacities enable the network to achieve the increased carbon removal required. The only exception here is the biochar process, which fails to be deployed in any period as it releases some emissions. An overview of the different CO_2 flows in the two cases explored through the proposed approach is illustrated in Figure 2.

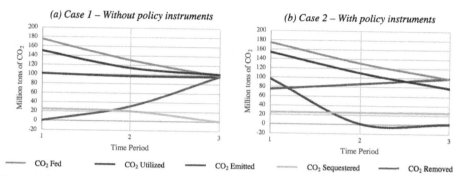

Figure 2: Overview of CO₂ flows (a) without and (b) with policy instruments

With time, the carbon caps ensure that the amount of carbon removed from the atmosphere increases while carbon utilization decreases as it still leads to the creation of emissions. Additionally, the carbon tax and credit implemented in the second case further reduce emissions and promotes the deployment of NETs. Of the policy instruments in place, the carbon credits were particularly pivotal in achieving net-zero emissions and generating profits in the second case with an NPV of $45 billion. On further analysis, the network can break even with carbon credits set as low as $53.37 per ton CO_2 removed in the first period. Furthermore, with the great uncertainty encompassing the capital required for the NETs and their associated processes, adjusting these carbon credit rates appropriately can encourage their deployment. In fact, with their CAPEX parameters doubled, the second case can break even at rates as low as $68.73 per ton CO_2 removed in the first period.

4. Conclusion

The approach described in this work enables the assessment of policy instruments such as carbon caps, taxes, and credits with CCUS and NETs options while identifying potential for material and energy integration between them. Future work should integrate renewable energy with these technologies and additionally consider their land and supply chain feasibility. Furthermore, the non-linear expressions used in pricing the carbon taxes and credits can be simplified to allow for the analysis of more network configurations to achieve climate targets. Thus, the multi-period multi-resource approach described here allows policymakers to propose guidelines to achieve net-zero emissions.

References

E.J. Abraham, D.M. Al-Mohannadi, P. Linke, 2022a, Integrating Carbon Negative Technologies in Industrial Clusters, Computer Aided Chemical Engineering. 49, 763–768.

E.J. Abraham, D.M. Al-Mohannadi, P. Linke, 2022b, Resource integration of industrial parks over time, Computers & Chemical Engineering. 107886.

E.J. Abraham, P. Linke, D.M. Al-Mohannadi, 2022c, Optimization of low-cost negative emissions strategies through multi-resource integration, Journal of Cleaner Production, 372, 133806.

R. Ahmed, S. Shehab, D.M. Al-Mohannadi, P. Linke, 2020, Synthesis of integrated processing clusters, Chemical Engineering Science, 227, 115922.

A.A. Al-Yaeeshi, A. AlNouss, G. McKay, T. Al-Ansari, 2020, A simulation study on the effect of CO_2 injection on the performance of the GTL process, Computers & Chemical Engineering. 136, 106768.

F. Babonneau, M. Benlahrech, A. Haurie, 2022, Transition to zero-net emissions for Qatar: A policy based on Hydrogen and CO_2 capture & storage development, Energy Policy, 170, 113256.

R. Chauvy, D. Verdonck, L. Dubois, D. Thomas, G. De Weireld, 2021, Techno-economic feasibility and sustainability of an integrated carbon capture and conversion process to synthetic natural gas, Journal of CO_2 Utilization, 47, 101488.

A. Gatto, W. Loewenstein, E.R. Sadik-Zada, 2021, An extensive data set on energy, economy, environmental pollution and institutional quality in the petroleum-reliant developing and transition economies, Data in Brief, 35, 106766.

IEA, 2020. Going carbon negative: What are the technology options?

D. McLaren, D, 2012, A comparative global assessment of potential negative emissions technologies, Process Safety and Environmental Protection, 90, 489–500.

M. Panahi, E. Yasari, A. Rafiee, 2018, Multi-objective optimization of a gas-to-liquids (GTL) process with staged Fischer-Tropsch reactor, Energy Conversion and Management, 163, 239–249.

M. Tavoni and R. Socolow, 2013, Modeling meets science and technology: an introduction to a special issue on negative emissions, Climatic Change, 118, 1–14.

Antonis Kokossis, Michael C. Georgiadis, Efstratios N. Pistikopoulos (Eds.)
PROCEEDINGS OF THE 33rd European Symposium on Computer Aided Process Engineering
(ESCAPE33), June 18-21, 2023, Athens, Greece

Waste valorization of non-commercialized edible mushrooms

Brenda I. Lazaro-Molina,[a] Teresa Lopez-Arenas[a]

[a] *Departamento de Procesos y Tecnología, Universidad Autónoma Metropolitana – Cuajimalpa, Vasco de Quiroga 4871, Mexico City 05348, Mexico*
mtlopez@cua.uam.mx

Abstract

The production of edible mushrooms generates two types of waste: the spent substrate from growing mushrooms and a loss of non-marketed mushrooms. Currently, great interest has been placed in the recovery of the spent substrate through the development of biorefineries for bioethanol production, mainly. However, the use of non-marketable mushrooms as raw material has not been extensively investigated. Therefore, the main objective of this work is the proposal for the synthesis, design and evaluation of a biorefinery of non-marketable edible mushrooms to obtain chitin, citric acid and animal feed, as high value-added products. The methodology proposes a technical-economic-environmental evaluation for decision-making. As a result, a viable conceptual design is obtained, with positive profitability and reduction of environmental impact factors.

Keywords: biorefinery, edible mushrooms, waste valorization, process simulation.

1. Introduction

Edible mushrooms constitute a functional food with nutritional and medicinal properties that promote health, which can strengthen food security in times of crisis. Their production has increased more than 30 times since 1978, reaching a world production of 34 million tons (Royse et al., 2017). The most widely cultivated edible mushroom worldwide is *Agaricus bisporus* (common white mushroom), followed by *Lentinus edodes, Pleurotus spp.* and *Flammulina velutipes* (Kumar et al., 2021), which are regulated according to different standards to guarantee organoleptic qualities and consumer satisfaction. Depending on its quality, there is an amount of non-marketable waste that ranges between 5 - 20% of the production.

It is important to highlight that there are proposals for the use of waste lignocellulosic substrates (such as cereal straws) employed in mushroom cultivation (Awasthi et al., 2022), but there are no biorefineries that take advantage of mushroom waste, since they are currently used directly as animal feed or fertilizer. Thus, the objective of this work is to design a biorefinery to produce high value-added products from these (non-marketable) edible mushroom residues, under a circular economy perspective.

Agro-industrial residues are unavoidable and should be considered as a new opportunity for their reuse within a biorefinery, using environmentally friendly and profitable bioprocess technologies. In such a way that these wastes from biorenewable resources can be used for the production of food, animal feed, fertilizers, fuels, energy, amino acids, biopolymers, among other industrial chemical products. From a review of the literature, it was found that the product with the highest value that can be obtained from edible mushrooms is chitin, which, together with its derivatives, show low immunogenicity and therefore have a great use in biomedical applications. As a second product of the biorefinery, citric acid is considered due to its wide use as an additive in the food, pharmaceutical, cosmetic, and cleaning industries. And since citric acid can be processed by microbial culture, then the biomass resulting from cell growth with high nutritional value can be processed for animal feed.

In this work, first, a biorefinery conceptual design is proposed for the waste valorization of non-commercialized edible mushrooms to produce chitin, citric acid and animal feed. Then the biorefinery flowsheet diagram is implemented and evaluated in a modular process simulator. Afterwards, a parametric sensitivity analysis is performed, considering equipment dimensions, amount of waste fed, design configurations, etc., as a result: the technical evaluation demonstrates the feasibility of using the waste, while the economic evaluation presents a positive profitability of the proposal using as indicators the unit production cost, the return on investment and its payback time; and the environmental impact is evaluated considering the consumption of water, utilities and energy.

2. Methodology

The methodology consists first of determining the composition of the edible mushroom residues, then identifying the best reaction routes to obtain value-added products, then determining a process flow diagram, and finally evaluating the biorefinery in a comprehensive manner considering technical, economic and environmental impact aspects. Each of these stages is briefly described below.

2.1. Mushroom composition

The bodies of the fruits of *A. bisporus* are rich in proteins (29.14%), carbohydrates (51.05%) and fats (1.56%); and also include proteins, lipids, vitamins and minerals (Ahlawat et al., 2016). For simulation purposes, the composition is established according to the compounds of interest (chitin and glucans present), while the rest of the compounds were grouped under the name of other solids. Therefore, the composition used was: 90% water, 0.47% chitin, 1.01% glucans, and the rest of other solids (Nitschiske et al., 2011; Khan et al., 2017).

2.2. Conceptual design of the biorefinery

A biorefinery is proposed to obtain three high-value products, through the following processing stages:

(a) Conditioning of the mushroom waste, which consists of washing, drying, grinding and sieving the mushrooms to obtain dry flour.

(b) Chitin extraction, where the deproteinization process is first carried out by enzymatic hydrolysis using proteases. Immediately a thermal hydrolysis allows extracting compounds such as lipids, minerals, vitamins and remaining proteins. As a result, supernatant rich in soluble glucans and a precipitate rich in chitin are obtained (Urbina-Salazar et al., 2020). The latter is sent to dry to obtain the chitin.

(c) Saccharification and fermentation, where first the soluble glucans are converted to glucose by enzymatic hydrolysis. Then, glucose is converted to citric acid by fermentation with Aspergillus niger. Subsequently, the solid phase (residual biomass) is separated, and the liquid phase is subjected to a citric acid recovery process.

(d) Animal feed production, by using the residual biomass from fermentation, which is dried and packaged for its sale.

(e) Production of citric acid, where this product is first precipitated as calcium citrate, then treated with sulfuric acid and the gypsum precipitate formed (calcium sulfate) is removed. Finally, the recovery of the product is carried out through a crystallization process to obtain dry crystals of citric acid.

Figure 1 shows the process flow diagram for the biorefinery of non-marketable edible mushrooms, according to this proposed conceptual design. Here it is important to mention that all liquid waste is sent to a wastewater treatment plant (WWT), which is not included but is considered in the evaluation.

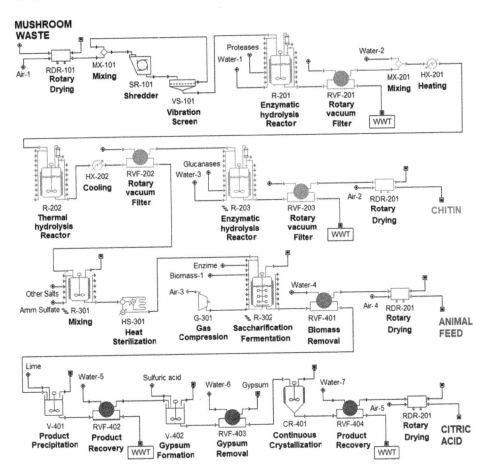

Figure 1. Process flow diagram of the proposed biorefinery.

2.3. Biorefinery assessment

The comprehensive evaluation of the biorefinery includes technical, economic and environmental aspects, as detailed below.

2.3.1. Technical evaluation

The implementation of the conceptual diagram of the biorefinery considers technical and design aspects for an industrial scale, with which the mass and energy balances are evaluated taking into account the reactions in the processes and their conversions; the sizing of equipment and its efficiency; the working times of the equipment and processing cycles; as well as the installed capacity of the plant in terms of the amount of raw material fed and products generated.

2.3.2. Economic evaluation

Undiscounted cash flow profitability criteria provide an immediate assessment of process economics and are often used for the early stages of process synthesis and design (Seider et al., 2009). So, in this work, the economic criteria used to determine if a project is profitable are the unit production cost (UPC), the return on investment (ROI) and the payback time (PT) of the investment.

2.3.3. Environmental evaluation

A successful economic development also depends on the rational use of resources and the reduction of the adverse environmental impact of the development project (Dixon and Pagiola, 1998). Therefore, this work evaluates (a) the consumption of water in the process, due to problems of water scarcity and damage to ecosystems; (b) energy consumption, as it is a key element in climate change; and (c) the demand for heat transfer agents, such as steam and cooling water.

3. Results

The biorefinery flow diagram was implemented in *SuperPro Designer*, a modular process simulator to evaluate the mass and energy balances that allow to calculate the technical-economic-environmental criteria. It is important to mention that the evaluation of the economic criteria depends on the unit production revenues, which were estimated at 200, 2.0, 0.2 USD/kg for chitin, citric acid and animal feed, respectively (Pighinelli et al, 2019; Heinzle et al. al., 2017; Anaya-Reza and Lopez-Arenas, 2018). With the purpose of carrying out a profitability analysis of the biorefinery, first, the feeding of edible mushroom waste was set at 50 T/batch, and later, a sensitivity analysis was carried out based on the number of reactors installed.

3.1. Base Case

Considering a 50 T/batch mushroom feed and an annual operating time of 336 days, the recipe batch time was calculated at 14.7 days, processing 50 batches per year and producing 301.6 kg of chitin, 21.3 kg of animal feed and 178.6 kg of citric acid per batch. The direct fixed capital, working capital, and startup cost were estimated as 13 176 000, 212 000, and 659 000, USD, respectively, giving a total investment of 14 047 000 USD for the biorefinery installation. The calculated value for UPC was 390.13 USD/kg, which was too high compared to the unit production revenue, so the ROI was -2.88% and it was not possible to assess the PT. The latter are summarized in the first line of Table 1.

Table 1. Profitability analysis based on the number of reactors installed in R-203, 301 and 302

Number of reactors	UPC (USD/kg)	ROI (%)	PT (years)	Amount of mushroom waste (T/year)
1	390	-2.9	NA	2 500
2	305	2.6	37.8	4 950
3	278	6.7	15.0	7 450
4	261	9.6	10.4	9 900
5	252	11.9	8.4	12 400
6	256	10.7	9.4	12 450

3.2. Profitability analysis

Since the initial base case was not profitable, the equipment occupancy was analyzed and it was determined that the reactors R-203, R-301 and R-302 generate a bottleneck as shown in Fig. 3(a). So, a sensitivity analysis was carried out increasing the number of these reactors installed, managing to increase the profitability of the biorefinery as shown in Table 1. As can be seen by increasing the number of reactors it is possible to increase the profitability of the process, achieving a maximum with 5 reactors. The increase in profitability is explained by the fact that dead times decrease, reaching an optimal equipment occupancy with 5 reactors. The comparison of processing times for 1 and 5 reactors can be seen in Fig. 3. So, increasing to 6 reactors, a bottleneck starts again and the annual amount of mushroom waste increases little (as observed in the last line of Table 1), so that cost effectiveness tends to decrease.

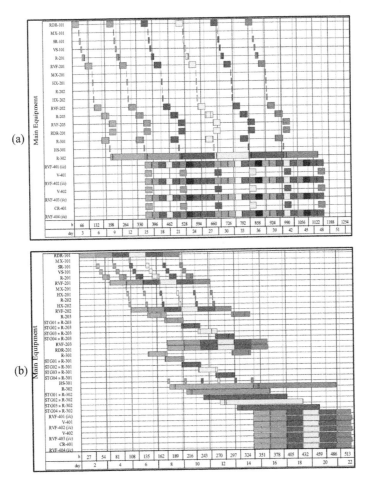

Figure 3. Equipment occupancy (R-203, R-301, R-302) considering (a) one and (b) 5 reactors..

3.3. Environmental results

Regarding the environmental impact indicators, Table 2 shows the values obtained for the base case (with 1 reactor) and the improved case (with 5 reactors). According to these results, it can be seen in general that a large amount of utilities is required, however, by increasing the number of reactors it was possible to reduce the use of heat transfer agent resources and energy consumption. Regarding the process water, it is important to highlight that although this water is required for all processing, the residual water is sent to a WWT plant and later recirculated. Therefore, the consumption of water after start-up is minimal, only to recover losses due to venting and evaporation.

Table 2. Environmental criteria for the base and improved cases.

Case study	No. reactors	Process water (T/kg chitin)	Utilities		
			Steam (T/kg chitin)	Cooling water (T/kg chitin)	Std. power (kW-h / kg chitin)
Base	1	349	0.45	22.8	51.5
Improved	5	271	0.35	17.3	46.4

4. Conclusions

The need to create alternatives for the synthesis, design, and operation of biorefineries for the processing of high value-added products from non-commercialized edible mushrooms is a promising, profitable, and sustainable alternative. In this work, a conceptual design of a biorefinery has been proposed, and a comprehensive evaluation methodology has been presented that allows decision-making at an early design stage. Although the results show a positive profitability of the project, there are still various aspects to be studied, such as the modification of the installed capacity, the synthesis of other bioproducts (for example, lactic acid instead of citric acid), mixing various types of edible mushroom waste, alternatives to reduce the consumption of utilities (such as heat integration), etc.

References

O. P. Ahlawat, K. Manikandan, M. Singh, 2016, Proximate composition of different mushroom varieties and effect of UV light exposure on vitamin D content in *Agaricus bisporus* and *Volvariella volvacea*. Mushroom Res, 25(1), 1-8.

O. Anaya-Reza, T. Lopez-Arenas, 2018, Design of a sustainable biorefinery for the production of lactic acid from sugarcane molasses, Revista Mexicana en Ingeniería Química, 17(1), 243-259.

M.K. Awasthi, S. Harirchi, T. Sar, et. al, 2022, Myco-biorefinery approaches for food waste valorization: Present status and future prospects, Bioresource Technology, 360, 127592.

J. Dixon, S. Pagiola, 1998, Economic analysis and environmental assessment. Environmental assessment sourcebook update, 23, 1-21.

E. Heinzle, A. Biwer, C. Cooney, 2007, Development of Sustainable Bioprocesses: Modeling and Assessment. Chichester: Wiley

A.A. Khan, A. Gani, F.A. Masoodi, U. Mushtaq, A.S. Naik, 2017, Structural, rheological, antioxidant, and functional properties of β–glucan extracted from edible mushrooms *Agaricus bisporus*, *Pleurotus ostreatus* and *Coprinus attrimentarius*. Bioactive Carbohydrates and Dietary Fibre, 11, 67-74.

H. Kumar, K. Bhardwaj, R. Sharma, E. Nepovimova, N. Cruz-Martins, D.S. Dhanjal, R. Singh, et al., 2021, Potential Usage of Edible Mushrooms and Their Residues to Retrieve Valuable Supplies for Industrial Applications. Journal of Fungi, 7(6), 427. MDPI AG.

J. Nitschke, H.J., Altenbach, T. Malolepszy, H. Mölleken, H., 2011, A new method for the quantification of chitin and chitosan in edible mushrooms. Carbohydrate Research, 346(11), 1307-1310.

L. Pighinelli, J. Broquá, J., B.G. Zanin, A.M. Flach, C. Mallmann, F.G.D. Taborda, L.E.L. Machado, S.M.L. Alves, M.M. Silvia, R.J.S.P. Dias, 2019, Methods of chitin production a short review. Am. J. Biomed. Sci. Res, 3(4), 307-314.

D.J. Royse, J. Baars, Q. Tan, 2017, Current Overview of Mushroom Production in the World, in Edible and Medicinal Mushrooms, 5-13.

D.W. Seider, J. D. Seader, D. R. Lewin, S. Widagd, 2009. Product and Process Design Principles: Synthesis, Analysis and Design. John Wiley & Sons, New York.

A.R. Urbina-Salazar, A.R. Inca-Torres, M.P. Carbonero-Aguilar, J. Bautista-Palomas, J., 2020, Preparación de quitina fúngica a partir de subproductos de hongos comestibles (*Agaricus bisporus*), Polo del Conocimiento: Revista científico - profesional, 5, 115-140.

Antonis Kokossis, Michael C. Georgiadis, Efstratios N. Pistikopoulos (Eds.)
PROCEEDINGS OF THE 33rd European Symposium on Computer Aided Process Engineering
(ESCAPE33), June 18-21, 2023, Athens, Greece

Enviro-economic assessment of sustainable aviation fuel production from direct CO_2 hydrogenation

Andrea Bernardi[a,b,*], Daniel Bagan Casan[b], Andrew Symes[c], Benoit Chachuat[a,b]

[a]*The Sargent Centre for Process Systems Engineering, Imperial College London, UK*
[b]*Department of Chemical Engineering, Imperial College London, UK*
[c]*OXCCU Tech Ltd, Oxford, UK*
*a.bernardi13@imperial.ac.uk

Abstract

The aviation industry is responsible for 2% of the total GHG emissions and 10% of the fuel consumption worldwide and sustainable aviation fuel (SAF) is considered a key step towards achieving net-zero aviation. In this work, we carry out an enviro-economic comparison of a one-step Fischer-Tropsch process (1sFT), based on a novel Mn-Fe-K catalyst, whereby CO_2 and H_2 are directly converted to liquid hydrocarbons, with a two-steps FT process (2sFT), in which a reverse water gas shift reactor is used to produce syngas, followed by a conventional FT process. Our analysis considers 1 MJ of liquid fuel as functional unit and the following key performance indicators: levelized cost of production, global warming potential, and monetized end-point environmental impacts. Our results suggest that the fuel blend from 1sFT has a minimum selling price 20% lower than the fuel blend from 2sFT, due to a lower capital cost and a higher selectivity towards liquid hydrocarbons. 1sFT is also found to be superior to 2sFT from an environmental point of view, with 30% lower GWP and 70% lower externalities cost.

Keywords: sustainable aviation fuel, technoeconomic analysis, LCA, Fischer-Tropsch

1. Introduction

Decarbonisation of the transport sector is a key challenge in the path towards net-zero and a sustainable future. The aviation industry is responsible for 2% and 10% of the global GHG emissions and fuel consumption, respectively. Sustainable aviation fuel (SAF) is called upon to play a pivotal role in a net-zero aviation, with up to 80% GHG emissions reduction (IEA 2022a). According to IEA (2020) low-carbon hydrogen-derived liquid fuels will gain significant market share in the sector, as batteries and hydrogen are not feasible alternatives, especially for long-rage flights, due to the low energy content per unit of mass and volume respectively (Karadotcheva *et al.* 2021). Moreover, hydrogen-fuelled aircrafts would require a complete redesign and advanced materials to avoid hydrogen embrittlement and leakage (Airbus 2020).

Fischer-Tropsch (FT) synthesis is a well-established chemical route to produce hydrocarbons of different chain length distributions from syngas, a mixture of CO and H_2, and can produce a drop-in alternative fuel for the aviation industry. However, syngas is nowadays produced from steam methane reforming or coal gasification, carries a high environmental burden. Significant research efforts have been devoted to identifying routes to use CO_2 as the carbon source in the FT process and deliver sustainable fuels. One option is to convert CO_2 and H_2 in a reverse water gas shift (RWGS) reactor and use the resulting syngas in a traditional FT process (Zang *et al.* 2021). The main limitations of this two-step approach are: (i) RWGS reaction requires high temperatures to achieve a good conversion (600-1000 °C); and (ii) the production of wax, a mixture of heavy hydrocarbons, requires additional upgrading units, such as a hydrocracking (HC) reactor to improve the yield of liquid fuels. The direct hydrogenation of CO_2 to liquid fuels in a single step is more appealing, but the selectivity towards liquid fuels is usually low, and the formation of short chain hydrocarbons is favored (Chen *et al.* 2018; Gao *et al.* 2017). Recently, Yao *et al.* (2020) synthesized a novel Mn-Fe-K based catalyst capable of converting CO_2 and H_2 with excellent yield and selectivity towards liquid hydrocarbons in the jet fuel range (C8–C16) and minimal wax production.

The main objective of this paper, therefore, is to conduct a comparative assessment of one-step (1sFT) and two-step (2sFT) FT synthesis from CO_2 and H_2 to produce liquid fuels, also with a fossil-based alternative, in terms of economic and environmental performance. The methodology is described in the next section, followed by results and discussions, before concluding the paper.

2. Methodology

Figure 1 presents block flow diagrams of the two processes considered in this paper, which are designed to utilize 100 t/h of CO_2 coming from a direct air capture (DAC) plant and H_2 from water electrolysis. We further assumed that the fuel blend is sent to an existing refinery.

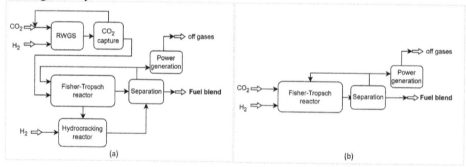

Figure 1: block flow diagrams of the two-step FT synthesis (a) and one-step FT synthesis (b). A wastewater stream is produced by both processes but is not represented in the diagram.

2.1. Process simulation and economic analysis

Flowsheet of both 1sFT and 2sFT routes were developed using Aspen HYSYS (version 11) and simulated in steady state, considering the Peng Robinson fluid package. Heat integration was performed using the Aspen Energy Analyzer. In the following the processes are described in more details.

Feed streams. We consider CO_2 from DAC as carbon source, available at 25 °C and 1 bar. According to IEA (2020), the cost of DAC CO_2 capture ranges between 135–345 $/t$CO_2$ (note that all costs throughout the paper are expressed in US$[2021]), in our base case the cost is 160 $/t$CO_2$ as reported in Keith *et al.* (2016). H_2 is obtained from water electrolysis using wind energy, and it is available at 20 bar and 25 °C (Al Qahtani *et al.* 2021). The H_2 price is 4.2 $/kg as reported in IRENA (2019) for wind-based water electrolysis. Both feed streams are assumed to have no impurities. An optimistic best-case scenario is also briefly discussed, considering 2050 predicted prices to compute the raw material costs (1.5 $/kg[H2], IRENA 2019; 50 $/t[CO2], IEA 2022).

One-step FT synthesis (1sFT). The FT reactor uses the new Mn-Fe-K based catalyst and operates at 260 °C and 10 bar (Yao *et al.* 2020). A two-stage compressor with intercooling is used to bring the CO$_2$ at 10 bar, while no compression is needed for H$_2$. The H$_2$ is fed to the process in a H$_2$:CO$_2$ molar ratio of 3:1. The outlet stream enters a sequence of two 3-phase separators operating at 70 °C and 40 °C under 10 bar. A small (2%) fraction of the gaseous stream coming out of the second separator is burned to produce heat and electricity in a gas and steam cycle, while the rest is recycled to the FT reactor. The other two streams from the 3-phase separators are wastewater and liquid hydrocarbons. The liquid hydrocarbons are mixed, expanded to 1 bar and cooled down to 35 °C. Lastly, the vapor phase from a flash separator (CO$_2$ and C1–C4 hydrocarbons) is burned to produce additional electricity, the liquid phase contains the fuel blend product to be sent to the refinery. This blend contains 29%$_{wt}$ gasoline, 48%$_{wt}$ SAF, 19%$_{wt}$ diesel, 2.5%$_{wt}$ light hydrocarbons, and traces (<1%$_{wt}$) amount of wax.

Two-step FT synthesis (2sFT). The RWGS reactor operates at 600 °C under 30 bar (Zang *et al.* 2021). A three-stage compressor with intercooling is used to bring the CO$_2$ to 30 bar, and a single stage compressor is used for the H$_2$ feed. An amine-based capture system is used to capture and recycle the unreacted CO$_2$ (Danaci *et al.* 2020). The hydrogen inlet flowrate is adjusted to have a CO:H$_2$ ratio of 2 in the syngas. The syngas is fed to the FT reactor operating at 220 °C and 25 bar. The wax produced by the FT reactor is sent to the HC reactor where it reacts with H$_2$ fed in stoichiometric amount at 350 °C and 25 bar. The downstream separation units are identical to the 1sFT, except for the pressure of the 3-phase separators set to 25 bar. The purge rate is set to 5% to avoid excessive accumulation of CO$_2$ in the FT section. The hydrocarbon mixture from the 3-phase separators is mixed with the HC reactor outlet stream before expanding to 1 atm and cooling to 35 °C, and feeding to a flash separator. The liquid fuel blend produced contains 25%$_{wt}$ gasoline, 40%$_{wt}$ SAF, 33%$_{wt}$ diesel, 1.4%$_{wt}$ light hydrocarbons, and traces (<1%$_{wt}$) amount of wax.

Economic analysis. The RWGS reactor is modeled as an equilibrium reactor, while the FT and HC reactors are modeled as simple conversion reactors in Aspen HYSYS. In 2sFT, yields and selectivities of the FT and HC reactors as well as the cost correlations for RWGS, FT and HC reactors are obtained from Zang *et al.* (2021). In 1sFT, the reactor yields and selectivities are obtained from Zao *et al.* (2020), with correlations from Zang *et al.* (2021) used to estimate the cost. The levelized cost of production (LCP) is calculated assuming a 15% interest rate and 20-year lifetime. LCP is compared with the fossil fuel cost, which is assumed to be 14 \$/GJ$_{fuel}$ in the base case, and 23 \$/GJ$_{fuel}$ in the optimistic best-case scenario considering a carbon tax of 150\$/tCO$_2$ (IEA 2020).

2.2. Life cycle analysis

The functional unit (FU) for the LCA analysis is chosen as "*1 MJ of liquid fuel blend*" and the assessment is done considering a Well-to-Tank (WtT) approach. GWP is also analyzed from a Well-to-Wake (WtW) perspective, assuming that all the carbon in the fuel blend is released as CO$_2$ during the use phase. Following the recommendation by von der Assen *et al.* (2013), system expansion via substitution is adopted to deal with multi-functionality herein.

Data collection and life-cycle inventories. The life-cycle inventories (LCIs) of 1sFT and 2sFT are computed from the simulated flowsheets described in Section 2.1 and are summarized in Table 1. The LCIs for DAC are taken from Keith *et al.* (2018) and those for H$_2$ are taken from Al-Qahtani, *et al.* (2021). The electricity source for H$_2$ production is assumed to be onshore wind from the UK, and the credits given for the electricity sold as a byproduct are attributed based on the electricity production mix of the UK.

Table 1: life-cycle inventories of 1sFT and 2sFT processes. FH = Fired Heater, CW = Cooling water, WW = Wastewater. The FU is 1 MJ of liquid fuel blend.

	1sFT	**2sFT**			**1sFT**	**2sFT**	
	Inputs	**Units**			**Outputs**	**Units**	
CO_2	$8.25 \cdot 10^{-2}$	$9.99 \cdot 10^{-2}$	kg	FU	$1.00 \cdot 10^{0}$	$1.00 \cdot 10^{0}$	MJ
H_2	$1.14 \cdot 10^{-2}$	$1.23 \cdot 10^{-2}$	kg	Electricity	$6.70 \cdot 10^{-3}$	$6.00 \cdot 10^{-3}$	kWh
FH	-	$3.04 \cdot 10^{-1}$	MJ	CO_2	$1.03 \cdot 10^{-2}$	$3.27 \cdot 10^{-2}$	kg
CW	$2.29 \cdot 10^{-4}$	$3.12 \cdot 10^{-4}$	m³	CO	$6.01 \cdot 10^{-4}$	$2.32 \cdot 10^{-4}$	kg
				WW	$6.41 \cdot 10^{-5}$	$6.95 \cdot 10^{-5}$	m³

Environmental impact assessment. The environmental assessment is conducted in SIMAPRO 9.3, using Ecoinvent 3.8 Cut-Off database for the background process inventories. The business-as-usual (BAU) fossil-based liquid fuel alternative is assumed to be "Petroleum {GLO} | market for", a liquid fuel with a LHV of 43.2 MJ/kg. The environmental KPIs are GWP100 alongside the monetized endpoint impacts to human health, ecosystem quality and resource scarcity, all computed with the ReCiPe2016 impact assessment methodology (Huijbregts *et al.* 2017) and using the monetization factors from Dong *et al.* (2019) converted to $\$_{2021}$.

3. Results and discussion

3.1. Midpoint environmental impacts – global warming potential
Figures 2a and 2b compare the GWP of 1sFt and 2sFT against BAU, from a WtT (2a) and WtW (2b) perspective. Both 1sFT and 2sFT have negative GWP because of the credits from DAC, with 1sFT outperforming 2sFT by c. 30%.

The molar-based carbon efficiency[1] is c. 68% in2sFT and 76% in 1sFT. Therefore, 2sFT has both a higher negative contribution from DAC ($-8.5 \cdot 10^{-2}$ kg_{CO2-eq}) and a higher positive contribution from direct emissions ($3.3 \cdot 10^{-2}$ kg_{CO2-eq}) compared to 1sFT. The positive contributions to the overall GWP are dominated by the direct emissions (65%), followed by H_2 production (19%) and heat generation (16%). In 1sFT credits from DAC are $-7.0 \cdot 10^{-2}$ kg_{CO2-eq}, while the total positive emissions ($1.9 \cdot 10^{-2}$ kg_{CO2-eq}) are determined by direct emissions (54 %) and H_2 production (46%)

Since the FU is a fuel blend, it is useful to assess the WtW GWP as well. We can observe from Figure 2b that c. 95% of the total GWP for BAU ($7.6 \cdot 10^{-2}$ kg_{CO2-eq}) is attributed to the use phase. Both 1sFt and 2sFt are significantly better then BAU, with 1sFT outperforming 2sFT, with a 75% and 57% GWP reduction, respectively.

3.2. Endpoint analysis and total cost
Figure 2c compares the monetized values of all three endpoint environmental impacts. The total externalities cost of 1sFT and 2sFT is lower than BAU, but the reduction achieved by 1sFT is significantly larger (80% vs. 30%). For all the alternatives impact on resource scarcity (RS) represents the largest share of externalities cost. Compared to

---

[1] ratio between moles of carbon in the liquid fuel blend and total moles of CO_2 fed to the process

BAU, 1sFT shows 40% reduction and 2sFT a 15% reduction. In both 1sFT and 2sFT the DAC energy requirements are an environmental hotspot, being responsible for 93% and 80% of the total impact on RS, respectively. In terms of damage to ecosystem quality (EQ) both processes have a negative impact, while damage to human health (HH) is almost zero for 1sFT and slightly larger than BAU for 2sFT.

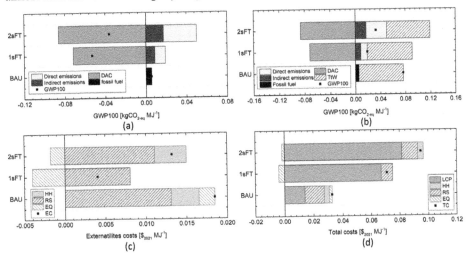

Figure 2: Results of the enviro-economic analysis for the 1sFT, 2sFT and BAU routes. (a) Well-to-Tank (WtT) global warming potential; (b) Well-to-Wake (WtW) global warming potential; (c) monetized externalities cost (EC); and (d) total cost (TC).

Turning to economic considerations, both options have a higher LCP compared to the fossil alternative, as expected. The LCP of 1sFT is 20% lower than that of 2sFT, but 5 times higher than the LCP of fossil fuel, nonetheless. In terms of total cost including externalities, 1sFT outperforms 2sFT by 25 % but has a total production cost 2 times higher than the BAU. It is worth noting that 1sFT has a 50% lower CAPEX than 2sFT, but the LCP is largely determined by the raw materials cost, where the difference between 1sFT and 2sFT per MJ$_{fuel}$ is only 10% (Table 1). In the base case, H₂ constitutes 75-78% of the raw material costs in 2sFT and 1sFT, respectively, with CO₂ accounting for the rest. However, in an optimistic scenario where prices drop to 1.5 $/kg$_{H2}$ and 50 $/t$_{CO2}$, the LCP of 1sFT would decrease to 25 $/GJ, only slightly higher than the cost of fossil fuels with a CO₂ tax of 150 $/t (23 $/GJ).

4. Conclusions

This paper presented an enviro-economic assessment of two FT-based processes to produce liquid hydrocarbon fuels from CO₂ and H₂, considering DAC and water electrolysis using wind energy as source of CO₂ and H₂ respectively. A one-step process using a novel catalyst was compared to a two-step process, assuming that the product blend is sent to an existing refinery to produce gasoline, SAF, and diesel. Our results show that the one-step process is superior both in economic and environmental terms to the two steps process, due to a lower capital cost, higher selectivity towards liquid hydrocarbons and lower energy requirements. Both processes have a negative GWP potential from a well-to-tank perspective, which compares favorably against fossil fuel production. The one-step process achieves a 75% reduction of well-to-wake GHG emissions, which is aligned with the reduction targets for SAF reported in the literature.

Our environmental assessment also considered an endpoint impacts analysis, confirming a better environmental performances of the synthetic fuels compared to fossil fuels, with an 80% reduction for the one-step process. The largest share of externalities cost in both the one-step and two-step processes is attributed to the energy requirements for DAC, and therefore alternative DAC processes should be considered to move further away from the fossil-dependency of the aviation sector. Nevertheless, the proposed low-carbon synthetic fuels are significantly more expensive than fossil fuels, and even if the prices of H_2 and CO_2 were to decrease, policy interventions such as carbon taxation, or reduced taxes on SAF would likely remain necessary for the low-carbon fuel to gain market shares. Future work will focus on adding more options for H_2 and CO_2, including blue hydrogen, alternative DAC processes, biomass gasification and other biogenic carbon sources. On the methodological side, an uncertainty analysis will be carried out to confirm the environmental and economic results.

Acknowledgements: AB and BC are grateful to the Engineering and Physical Sciences Research Council (EPSRC) for funding the research under the UKRI Interdisciplinary Centre for Circular Chemical Economy programme (EP/V011863/1).

References

Airbus, 2020, Airbus reveals new zero-emission concept aircraft. https://tinyurl.com/AirbusNet0 [Accessed 29 11 2022].

A. Al-Qahtani *et al.* 2021. Uncovering the true cost of hydrogen production routes using life cycle monetisation. Applied Energy, 281, 115958.

L. Chen *et al.* 2018. Entropy generation minimization for CO_2 hydrogenation to light olefins. Energy 147, 187–196.

D. Danaci *et al.* 2020. Exploring the limits of adsorption-based CO 2 capture using MOFs with PVSA–from molecular design to process economics. Molecular Systems Design & Engineering, 5(1), 212-231.

Y. Dong *et al.* 2019. Evaluating the monetary values of greenhouse gases emissions in life cycle impact assessment. Journal of Cleaner Production, 209, 538-549.

P. Gao *et al.* 2017. Direct conversion of CO_2 into liquid fuels with high selectivity over a bifunctional catalyst. Nature Chemistry, 9, 1019.

M.A. Huijbregts *et al.* 2017. ReCiPe2016: a harmonised life cycle impact assessment method at midpoint and endpoint level. The International Journal of Life Cycle Assessment, 22(2), 138-147.

IEA 2020, CCUS in Clean Energy Transitions, IEA, Paris https://www.iea.org/reports/ccus-in-clean-energy-transitions, License: CC BY 4.0

IEA 2022a, Aviation, IEA, Paris https://www.iea.org/reports/aviation, License: CC BY 4.0

IEA 2022b, Direct Air Capture 2022, IEA, Paris https://www.iea.org/reports/direct-air-capture-2022, License: CC BY 4.0

IRENA 2019, Hydrogen: A renewable energy perspective, International Renewable Energy Agency, Abu Dhabi

E. Karadotcheva *et al.* 2021. Structural power performance targets for future electric aircraft. Energies, 14(19), 6006.

D. W. Keith *et al.* 2018. A process for capturing CO_2 from the atmosphere. Joule, 2(8), 1573-1594.

B. Yao *et al.* 2020. Transforming carbon dioxide into jet fuel using an organic combustion-synthesized Fe-Mn-K catalyst. Nature Communications, 11(1).

G. Zang *et al.* 2021. Performance and cost analysis of liquid fuel production from H_2 and CO_2 based on the Fischer-Tropsch process. Journal of CO_2 Utilization, 46.

Antonis Kokossis, Michael C. Georgiadis, Efstratios N. Pistikopoulos (Eds.)
PROCEEDINGS OF THE 33rd European Symposium on Computer Aided Process Engineering
(ESCAPE33), June 18-21, 2023, Athens, Greece

Ammonia powered solid oxide fuel cells for general aviation propulsion systems: challenges and opportunities

Giampiero Di Legge,[a] Seyed Ali Nabavi,[a] Lorenzo Mazzei,[b] Riccardo Da Soghe,[b] Cosimo Bianchini,[b] Amirpiran Amiri,[c] Theoklis Nikolaidis,[d] Soheil Jafari,[d]

[a]School of Water, Energy and Environment (SWEE), Cranfield University, Cranfield, MK43 0AL, United Kingdom
[b]Ergon Research srl, 50127, Firenze, Italy
[c]Aston University, Birmingham, B4 7ET, United Kingdom
[d]School of Aerospace, Transport and Manufacturing (SATM), Cranfield University, Cranfield, MK43 0AL, United Kingdom

Abstract

The aviation sector contributes 3% of global CO2 emissions as it still relies heavily on fossil fuels. The use of low carbon hydrogen as a fuel, in both compressed or liquified forms, would drastically reduce its emissions. However, hydrogen has low volumetric energy density, and imposes a mass penalty due to storage tank requirements if used in compressed form. On the other hand, the liquefaction of hydrogen is energy-intensive and can consume ~40% of chemical energy stored in hydrogen. Ammonia, as a carbon-free hydrogen carrier, benefits from ease of storage and transport, and can be potentially used as an energy vector to enable advanced propulsion systems. Given the possibility of achieving zero-NOx in fuel cells, recent advances in fuel cell electric-powered aircraft, as well as fuel flexibility of Solid Oxide Fuel Cells (SOFCs), this work will explore the feasibility and challenges of direct ammonia-fed SOFCs as the power source for air aviation propulsion system. A model based on semi-empirical equations is developed for ammonia-powered fuel cells. A 130 kW general aviation case study is modelled and simulated in the paper to confirm the effectiveness and potential of the ammonia-powered fuel cell technology to be used as the next generation of propulsion systems in this size of aircraft. The comparative analyses demonstrated that the use of ammonia can decrease the mass of fuel consumed by 5.9% (in comparison with kerosene fueled aircraft) in a one-hour mission and allow achieving the emissions level targets set by Advisory Council for Aeronautics Research in Europe (ACARE) Flight Path 2050. Finally, the remaining practical challenges that need to be addressed to enable this important technology to be adopted by air vehicle manufacturers have been identified and discussed.

Keywords: Ammonia, Fuel cells, General aviation, Net zero targets, Flight Path 2050, Emission

1. Introduction

Anthropogenic greenhouse gas emissions from fossil fuel use are among the main contributors to climate change. It is estimated that the aviation sector, which is still highly dependent on petroleum-derived products, contributes to 3% of total carbon dioxide emissions [1], with an annual production of nearly 1000 million tons. Additional non-carbon pollutants are generated during the combustion process in gas turbines, such as nitrogen and sulfur oxides (NOx and SOx), which can contribute into contrail formation and global warming [2]. Not to mention the particulate matter (soot), known for its detrimental effects on human health, causing breathing issues, including asthma, bronchitis, coronary heart disease, and even cancer. Therefore, numerous efforts are being made to make the aviation sector more sustainable through its electrification. There are currently three main ways to obtain electrical energy on board. The first, represented by batteries, is currently limited by energy density. Bills et al. [3] estimated the distributions of specific energy required by different classes of aircraft. The mean requirements were determined as 600 (regional), 820 (narrow-body) and 1280 Wh/kg-pack (wide-body). However, despite the steady improvement over the last two decades, the specific energy of current generation Li-ion batteries is about 250 Wh/kg-cell [3]. Future developments are considering novel active materials and lithium metal anode, pushing the specific energy up to 400-500 Wh/kg-cell [4]. Post-lithium-ion technologies such as lithium/sulfur (Li/S) or lithium/oxygen (Li/O2) batteries might provide a further improvement up to 550 and 750 Wh/kg-cell [4]. Despite that, it is clear that batteries will not represent a viable solution for regional aircraft even in the short term and will require several decades to be deployed in long-range aircraft. Combustion engines (and in particular gas turbines), which are the second option, have low efficiencies in power production (around 45% [5]) and are currently relying on liquid fossil fuels. Despite the plans to develop hydrogen combustors, storage and safety are still concerning aspects. sustainable aviation fuel (SAF) is the most promising solution in the short term. Despite recent tests by Airbus CITE [6] and Boeing CITE [7] proved its feasibility, limited availability and cost (two to eight times as much as kerosene) are currently hindering its massive deployment. Onboard fuel cells represent the last choice, which benefits from higher energy efficiency and enables fuel flexibility. The use of fuel cells such as solid oxide fuel cells (SOFC) in aviation is receiving much attention but is still in the early stages of development. Given the issues associated to hydrogen storage, aviation industries are considering the exploitation of ammonia, which offers no carbon content, slow reactivity and the possibility to liquify it below -33 °C at atmospheric pressure. This paper explores the feasibility of using onboard direct ammonia-fed SOFCs as one of the potential solutions for the next generation of disruptive propulsion technologies.

Currently, most studies of SOFCs to power air vehicles are based on the integration with a gas turbine. These simulations differ in architecture (new components can be implemented into the process), type of fuel used, and the type of system powered (all-electric aircraft or electric APU). Many of these studies concern large aircraft and consequently long-range missions that require fuel cells with the power higher than 400 kW. One of the main obstacles to the use of SOFC onboard is represented by the weight of the stack that causes and excessive fuel consumption. Many studies, in fact, in addition to analysing the performance, have focused on the power density of the system [8-12]. Tornabene et al. [8] delivered a preliminary estimation of the impact on the system's mass and volume of all the components required for an APU powered by a SOFC-GT. Whyatt and Chick [9] provided a similar analysis for a Boeing 787 and concluded that the system weighed 1500 kilograms beyond the break-even point. Steffen, Freeh and Larosiliere [10] developed a model in which a SOFC-GT powered the APU of a 300-passenger aircraft. Their results, which showed an improved efficiency, but an insufficient energy density compared to combustion engines, are in line with other studies. Dollmayer et al.[11] analysed the performance of a two-engines all-electric aircraft powered by hydrogen for a 4000 nautical miles mission. They examined the mass penalties and fuel consumption, discussing strategies to lower fuel consumption such as cabin air for processing, a partial thrust recovery, or water re-utilisation. The influence of the exhaust temperature was also examined because it can enable considerable energy recovery. Their results also established that 4kg/kWh would be the starting mass-to-power ratio for achieving fuel consumption that would justify the presence of SOFC. Lim et al. [12] designed, built, and compared the performance of a SOFC as a stand-alone unit and a SOFC-GT. The study, which also included a comparison between hydrogen and methane, showed that high pressures have a positive effect on efficiency and power density. In a National Aeronautics and Space Administration (NASA) technical publication, Srinivasan et al. [13] replaced the APU of a Boeing 777-200ER with a SOFC-GT (450 kW) and explored different levels of integration with the electrical power supply (EPS), thermal management (TMS), and environmental control (ECS) systems. SOFC-GT integration reduced nitrogen oxides (NOx), carbon monoxide, and unburned hydrocarbons (UHC) emission by 15.45% (NOx), 3.25% (CO), and 4.72% (UHC) respectively for ground operations. In addition, the authors showed that higher integration increased fuel economy for long-range missions by up to 0.77%. However, the highest reduction in fuel consumption (3%) was obtained in another analysis that targeted short-range aircraft, supporting the hypothesis that SOFCs are currently best suited for small aircraft or Unmanned Air Vehicles (UAVs). In this regard, Ji et al. [14] modelled a UAV powered by a propane-fuelled SOFC hybrid engine, achieving a thermal efficiency of 57.6%.

As discussed above, most of previous studies used the SOFC as an APU. This paper is now exploring the feasibility of applying this technology to primary propulsion. In other words, to bridge the above-mentioned research results and the stringent targets and limitations set in the ACARE's flight path 2050 for emissions level reduction, this study will explore the feasibility of using ammonia-powered SOFC as the main power source for General Aviation (GA) applications. For this purpose, an experimentally validated ammonia-fed SOFC model will be developed in section two. The developed model will be used as the main source of the power for propulsion system in section three where a small aircraft (Cessna 172 size) will be simulated as the case study. The simulation results are presented and discussed in this section followed by the identified remaining research challenges in the way of using onboard ammonia powered SOFCs as the propulsion systems for the future air vehicles. Conclusion remarks are summarized in section four. It should also be mentioned that although the GA contributions to the overall climate impact of aviation is not huge, it does not question the methodological approach presented in this paper.

2. Ammonia Powered SOFC Modelling

This section describes the developed tool for performance simulation of the ammonia powered SOFC which could be used to generate the required power for future air vehicles. First, the mathematical modelling of the SOFC will be detailed through a semi-empirical physics-based approach. The developed process model in ASPEN Plus® software will then be explained, and the hierarchies of the calculation method are presented in a flowchart format. Finally, the developed model will be validated against experimental data to confirm the validity and effectiveness of the proposed modelling approach.

2.1. Mathematical Model

The common technique for modelling SOFCs' performance in steady conditions starts from the Open Circuit Voltage (OCV) obtained from the Nernst equation as follow:

$$H_2 + \frac{1}{2}O_2 = H_2O \tag{1}$$

$$E_N = \frac{RT}{2F} * \ln(\frac{p_{H_2} * p_{O_2}^{0.5}}{p_{H_2O}}) \tag{2}$$

Where R is the gas constant, T is the temperature, F is Faraday's constant. The polarisation losses (so-called overpotentials) occurring within the FC (i.e., activation, ohmic and concentration losses) are then subtracted from the OCV to determine the actual voltage. The activation overpotential (ηact) is due to the irreversibility of the charge transfer that take place at the electrodes. The ohmic overpotential (ηohm) is caused mainly by the electrolytic resistance to the ions flow. Finally, the concentration overpotential (ηconc) depends on the mass transfer limitation to the kinetics of the electrochemical process.

$$V_{final} = E_N - \left(\eta_{act} + \eta_{ohm} + \eta_{conc}\right) \tag{3}$$

In addition to being influenced by the working conditions of the fuel cell, these potential variations are closely dependent on the geometric and chemical-physical characteristics of the fuel cell. Therefore, this modelling approach requires parameters that regard the activation energy, interaction factors, anodic and cathodic composition, and cell structure. The model adopted in this work is based on semi-empirical equations derived from interpolations of experimental data. Besides providing greater flexibility, this method excludes the need to determine the microscopic characteristics of the FC and the mentioned parameters. This study is based on a generator produced by Siemens-Westinghouse and composed of multiple tubular SOFCs with internal reforming connected in bundles. The single cell has an active area of 834 cm2 and can produce about 130 kW at a temperature of 900-950 °C. YSZ was used as material for the electrolyte, while the anode is made of Ni-YSZ cermet [15].

For validation case, when natural gas is used, fuel is reformed with the steam coming from the anode exhaust gas recirculation. The products, mainly CO and H2, are sent to the anode and oxidised by oxygen coming from the cathode. The fraction of exhaust gas that is not recycled is sent to a combustion plenum (called afterburner), where the total oxidation of the remaining syngas or unburnt hydrocarbons takes place. Before being released into the atmosphere, the afterburner exhaust gases are used to pre-heat the incoming air. For ammonia, this compound, being carbon-less, is not subjected to SR but undergoes thermal cracking generating hydrogen and nitrogen ($2NH_3 \rightarrow N_2 + 3H_2$). Since the process is carried out with Ni-based catalysts, it can occur directly at the FC inlet. For this reason, the temperature used to model the reforming is 950 °C, even though previous studies have established that the decomposition of ammonia is already complete at 800 °C. The thermal cracking does not require steam; hence there is no need for recirculation. It should be noted that since the ammonia is thermochemically decomposed to hydrogen and nitrogen, and hydrogen undergoes electrochemical reaction, Eq. (2) is written based on hydrogen.

2.2. ASPEN Model

The described process has been reproduced in Aspen Plus using the model illustrated in figure 1. This section gives a detailed description of the technical specifications used to design the model. Aspen terminology is represented in italic

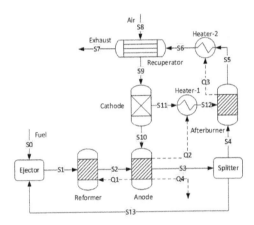

FIGURE 1 – ASPEN PLUS MODEL

The reforming and reactions occurring inside the anode were simulated using the *Rgibbs* block. This equilibrium reactor determines the final concentrations based on the chemical potential of the species involved. In other words, the Gibbs reactor predicts the product compositions by minimising the total Gibbs free energy *(Gtotal)* (Equation 4)

$$G_{total} = \sum_{i=1}^{N} n_i \Delta G_{fi}^0 = \sum_{i=1}^{N} n_i RT ln \frac{f_i}{f_i^0} \tag{4}$$

Where fi0 and fi represent the fugacity of the species i in the reactor and standard conditions, respectively. Two different algorithms have been developed for the model: the first (Figure 2a) allows the power produced by the fuel cell to be calculated from the amount of fuel consumed. The second (Figure 2b) determines the potential, current and fuel required starting from the expected power. These two algorithms are based on the same equations, which are rearranged to follow the structures described in Figure 1(In a) V, I and P are determined starting from the amount of fuel consumed. In b) V, I and required fuel are obtained from a pre-set power [16]).

It should be mentioned that in case b), after an initial estimation of the voltage from the process parameters, the current is easily calculated using the relationship:

$$Current = Power / Voltage \tag{5}$$

The next step is to calculate the amount of fuel required (mol/h) to generate that current. In the case of using hydrogen as the fuel, the required fuel could be calculated as follow:

$$H_{2,required} = \frac{Current}{2*U_f*F} * (3600) = \frac{0.018655*I}{U_f} \tag{6}$$

Where U_f is the utilisation factor determining the percentage of hydrogen consumed at the anode, and F is the Faraday constant. By calculating the amount of hydrogen required, the consumption of any suitable fuel can be determined just by knowing its composition. Finally, the efficiency can be obtained through the low calorific value (LHV_{fuel}) of the used fuel and its molar flow (n_{fuel}) (Equation 7)

$$Efficiency = \frac{Power}{(n_{fuel}*LHV_{fuel})} \tag{7}$$

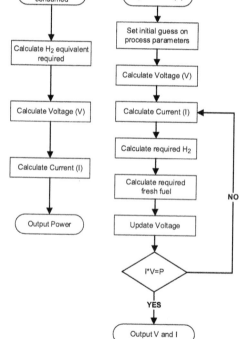

FIGURE 2 – HIERARCHIES OF CALCULATION USED IN THE MODEL

2.3. Validation of the Model

The developed model is validated with the experimental results of Zhu and Kee [17] SOFC fed with natural gas at a temperature of 910°C and pressure of 1 bar). The utilisation factor and steam to carbon ratio were set to 0.85 and 2.5, respectively. The natural gas used as fuel consists of the following components: methane 81.3%, ethane

2.9%, propane 0.4%, butane 0.2%, carbon monoxide 0.9% and nitrogen 14.3%. The proximity between experimental data and simulation results confirms the validity of the model (Table 1).

TABLE 1- VALIDATION OF THE DEVELOPED SOFC MODEL

	Literature data [17]	Simulation results
Power (kW)	120	120
Current density (mA/cm2)	180	175
Reforming temperature (°C)	550	537
Anode outlet composition (mol%)		
Water	48%	51.1%
Carbon dioxide	28%	25.1%
Hydrogen	14%	11.4%
Carbon monoxide	5%	7.3%
Nitrogen	5%	5.1%
Exhaust stream composition (mol%)		
Nitrogen	77%	77.3%
Oxygen	16%	16.1%
Water	5%	4.3%
Carbon dioxide	2%	2.2%

3. Results and Discussion

After having the validated SOFC model up-and-running, the next step is to assess the performance of the ammonia-powered SOFC to be used as the power source for general aviation. As a case study, a light aircraft with the take off power of 130 kW (Cessna 172 size aircraft) is used for simulation purpose. A short-range mission lasting a total of one hour is utilized in the simulations as shown in figure 3. The mission was simulated in all its phases, from take-off to landing, considering the different power requirements and duration of the phases. To calculate the required power for the propulsion system at each flight phase, the mission is simulated in Turbomatch (TM) - a Cranfield in-house developed software – for the Cessna 172 with Kerosene as the fuel. As a state-of-the-art gas turbine performance simulation and diagnostics code, TM is a key proprietary tool of the Propulsion Engineering Centre (PEC) at Cranfield University. This has been widely used in aircraft engine research projects and within the Cranfield postgraduate community [18]. The required power calculated by TM at Take-Off, Climb, Cruise, Descend, approach, and taxi phases are 130, 104, 83.2, 65, 52, and 19.5 kW respectively.

The mission with associated required power is then implemented in the developed SOFC model with ammonia as the fuel. Table 2 shows the results of the fuel consumption for different flight phases. The total fuel burn during the mission is 25.87 kg for ammonia powered SOFC. It is 5.9% less than the fuel consumption for the Kerosene powered aircraft with similar size (27.51 kg). The other advantages of using ammonia as the fuel is it does not generate any carbon dioxide. So, it is a high potential candidate for next generation of sustainable propulsion systems. However, the remained research challenges should be explored and addressed in order to enable this technology to be adopted by air vehicle manufacturers.

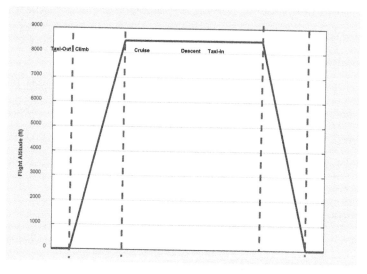

FIGURE 3 – DEFINED FLIGHT MISSION

TABLE 2 – SIMULATION RESULTS FOR 1-HOUR MISSION. FUEL: AMMONIA, T=950°C.

	Power required (kW)	Duration (s)	Ammonia		
			kg/h	L/min	Fuel consumed (kg)
Take off	130	30	46.37	1.06	0.39
Climb	104	600	36.49	0.83	6.08
Cruise	83.2	1800	28.96	0.66	14.48
Descend	65	480	22.46	0.51	2.99
Approach	52	210	17.88	0.41	1.04
Taxi	19.5	480	6.62	0.15	0.88
Total					25.87

In summary, the feasibility and potential of using ammonia-fed SOFC as the propulsion system of future light air vehicles is studied and the simulation results reported in table 2. This technology is pretty aligned with Net Zero 2050 and Flight Path 2050 targets as it reduces the fuel consumption and does not produce carbon dioxide. However, it should be mentioned that this is just the first shot in this topic, and the study just took the power required of a kerosene-powered Cessna 172 and limited the comparison to the fuel used by a SOFC providing said power profile. In other words, different aspects of mass and aerodynamics considerations should be studied in future work to fully explore the pros and cons of using SOFC as the main source of power in propulsion systems. It may result in either increasing power required for a given performance or limiting performance when maintaining power levels. Other details like drag penalties and tank-system design and integration should also be discussed accordingly. Moreover, the following challenges are still remained:

- The first challenge is the thermal management system (TMS) architecture design and development. As the temperature of the ammonia powered SOFC is 950°C, an advanced and effective TMS should be designed and implemented to manage the thermal loads generated in the propulsion system during the mission.
- NOx can be generated, besides from ammonia oxidation, also by thermal processes that take place at temperatures above 1300°C and convert diatomic nitrogen. It is worth noting that, in some simulations, the exhaust gases leaving the afterburner contain nitrogen and oxygen at a temperature close to 1300°C. However, introducing a heat exchanger to slightly cool the gases entering the combustion chamber can solve the problem.
- Most of the global production of ammonia is based on steam methane reforming (SMR) to produce (grey) hydrogen, followed by Haber–Bosch synthesis. Although the state-of-the-art SMR technology offers the most economical approach, the process is associated with significant CO2 emissions. So, potential solutions for green ammonia production are welcomed.
- The conventional SOFCs are developed mainly for stationary applications, in which the system weight may not be a key indicator. However, for air transportation, their energy-to-weight ratio is far below that of turboprops. Although the higher efficiency of SOFC may potentially compensate for this to some extent, for further deployment of SOFC-powered air vehicles, the energy-to-weight ratio should match or exceed that of turbines.
- Synchronization of the SOFC with other elements of the propulsion system should also be considered to address the implementation challenges like slow response time of SOFCs.

4. Conclusion

This research represents a first step in the feasibility study and conceptual design for introducing SOFCs with reforming onboard air vehicles. This technology makes it possible to overcome the obstacles posed by difficult hydrogen storage conditions. A physics-based model based on semi-empirical equations is developed for ammonia-powered SOFC and validated against experimental data. The performance of the developed model is assessed as the propulsion system for a light aircraft in Cessna 172 size with the Take-Off power of 130kW. Simulation results for a 1-hour mission confirm that using SOFC with ammonia fuel will reduce the mass of fuel consumed by the aircraft by 5.9% as well as omitting the carbon dioxide. So, the high potential of using ammonia powered SOFC as a disruptive propulsion technology for future aircraft is obvious. However, there are still remaining research challenges including thermal management system design, dealing with NOx production, and green ammonia production that need to be addressed in order to enable this important technology to be adopted by air vehicle manufactures in the future.

References

[1]. Grewe V., Gangoli Rao A., Grönstedt T., Xisto C., Linke F., Melkert J., et al. Evaluating the climate impact of aviation emission scenarios towards the Paris agreement including COVID-19 effects. Nature Communications 2021 12:1. Nature Publishing Group; 22 June 2021; 12(1): 1–10. Available at: DOI:10.1038/s41467-021-24091-y

[2]. Lee D.S., Fahey D.W., Skowron A., Allen M.R., Burkhardt U., Chen Q., Doherty S.J., Freeman S., Forster P.M., Fuglestvedt J., Gettelman A., De León R.R., Lim L.L., Lund M.T., Millar R.J., Owen B., Penner J.E., Pitari G., Prather M.J., Sausen R., Wilcox L.J., The contribution of global aviation to anthropogenic climate forcing for 2000 to 2018, Atmospheric Environment, Volume 244, 2021, 117834, ISSN 1352-2310, https://doi.org/10.1016/j.atmosenv.2020.117834.

[3]. Bills A., Sripad S., Fredericks W. L., Singh M., and Viswanathan V., "Performance Metrics Required of Next-Generation Batteries to Electrify Commercial Aircraft," ACS Energy Lett., vol. 5, no. 2, pp. 663–668, 2020, doi: 10.1021/acsenergylett.9b02574

[4]. Placke, T., Kloepsch, R., Dühnen, S. et al. Lithium ion, lithium metal, and alternative rechargeable battery technologies: the odyssey for high energy density. J Solid State Electrochem 21, 1939–1964 (2017). https://doi.org/10.1007/s10008-017-3610-7

[5]. Langston, L. S. "Aspects of Gas Turbine Thermal Efficiency." ASME. Mechanical Engineering. September 2020; 142(09): 54–55. https://doi.org/10.1115/1.2020-SEP4

[6]. This A319neo is the latest to test 100% SAF. An in-flight study is focusing on emissions performance of single-aisle Aircraft. 30 October 2021. https://www.airbus.com/en/newsroom/stories/2021-10-this-a319neo-is-the-latest-to-test-100-saf

[7]. 2 Aerospace Pioneers Partner For Sustainability. 2021. https://www.boeing.com/features/2021/11/aerospace-pioneers-partner-for-sustainability.page

[8]. Tornabene R., Wang X-Y., Steffen CJ., Freeh JE. Development of Parametric Mass and Volume Models for an Aerospace SOFC/Gas Turbine Hybrid System. Proceedings of the ASME Turbo Expo. American Society of Mechanical Engineers Digital Collection; 11 November 2008; 5: 135–144. Available at: DOI:10.1115/GT2005-68334

[9]. Whyatt GA., Chick LA. Electrical Generation for More-Electric Aircraft Using Solid Oxide Fuel Cells. Richland, WA (United States); 1 April 2012; Available at: DOI:10.2172/1056768

[10]. Steffen CJ., Freeh JE., Larosiliere LM. Solid Oxide Fuel Cell/Gas Turbine Hybrid Cycle Technology for Auxiliary Aerospace Power. Proceedings of the ASME Turbo Expo. American Society of Mechanical Engineers Digital Collection; 11 November 2008; 5: 253–260. Available at: DOI:10.1115/GT2005-68619

[11]. Dollmayer J., Bundschuh N., Carl UB. Fuel mass penalty due to generators and fuel cells as energy source of the all-electric aircraft. Aerospace Science and Technology. Elsevier Masson; 1 December 2006; 10(8): 686–694. Available at: DOI:10.1016/J.AST.2006.08.001

[12]. Lim TH., Song RH., Shin DR., Yang J il., Jung H., Vinke IC., et al. Operating characteristics of a 5 kW class anode-supported planar SOFC stack for a fuel cell/gas turbine hybrid system. International Journal of Hydrogen Energy. Pergamon; 1 February 2008; 33(3): 1076–1083. Available at: DOI:10.1016/J.IJHYDENE.2007.11.017

[13]. Srinivasan H., Yamanis J., Welch R., NASA ST-., CR-2006 undefined., 2006 undefined. Solid oxide fuel cell APU feasibility study for a long range commercial aircraft using UTC

ITAPS approach. ntrs.nasa.gov. 2006; Available at: https://ntrs.nasa.gov/api/citations/20070004892/downloads/20070004892.pdf?attachment=true

[14]. Ji Z., Qin J., Cheng K., Liu H., Zhang S., Dong P. Thermodynamic analysis of a solid oxide fuel cell jet hybrid engine for long-endurance unmanned air vehicles. Energy Conversion and Management. Pergamon; 1 March 2019; 183: 50–64. Available at: DOI:10.1016/J.ENCONMAN.2018.12.076

[15]. Fuerte A., Valenzuela RX., Escudero MJ., Daza L. Ammonia as efficient fuel for SOFC. Journal of Power Sources. Elsevier; 1 July 2009; 192(1): 170–174. Available at: DOI:10.1016/J.JPOWSOUR.2008.11.037

[16]. W. Zhang, E. Croiset, P.L. Douglas, M.W. Fowler, E. Entchev, Simulation of a tubular solid oxide fuel cell stack using AspenPlusTM unit operation models, Energy Conversion and Management, Volume 46, Issue 2, 2005, Pages 181-196, ISSN 0196-8904, https://doi.org/10.1016/j.enconman.2004.03.002.

[17]. Zhu H., Kee RJ. Thermodynamics of SOFC efficiency and fuel utilization as functions of fuel mixtures and operating conditions. Journal of Power Sources. Elsevier; 27 October 2006; 161(2): 957–964. Available at: DOI:10.1016/J.JPOWSOUR.2006.05.006

[18]. Apostolidis, A, Sampath, S, Laskaridis, P, & Singh, R. "WebEngine: A Web-Based Gas Turbine Performance Simulation Tool." Proceedings of the ASME Turbo Expo 2013: Turbine Technical Conference and Exposition. Volume 4: Ceramics; Concentrating Solar Power Plants; Controls, Diagnostics and Instrumentation; Education; Electric Power; Fans and Blowers. San Antonio, Texas, USA. June 3–7, 2013. V004T08A007. ASME. https://doi.org/10.1115/GT2013-95296

Antonis Kokossis, Michael C. Georgiadis, Efstratios N. Pistikopoulos (Eds.)
PROCEEDINGS OF THE 33rd European Symposium on Computer Aided Process Engineering
(ESCAPE33), June 18-21, 2023, Athens, Greece

Simulation-based design of regional biomass thermochemical conversion system for improved environmental and socio-economic performance

Leonardo L. Corradini,[a] Aya Heiho,[b] Yuichiro Kanematsu,[b] Ryoko Shimono,[b] Satoshi Ohara,[c] Yasunori Kikuchi[a,b,c]

[a]Department of Chemical Sytem Engineering, the University of Tokyo, 7-3-1 Hongo, Bunkyo-Ku, Tokyo 113-8656, Japan
[b]Presidential Endowed Chair "Platinum Society", the University of Tokyo, 7-3-1 Hongo, Bunkyo-Ku, Tokyo 113-8656, Japan
[c]Institute for Future Initiatives, the University of Tokyo, 7-3-1 Hongo, Bunkyo-Ku, Tokyo 113-8654, Japan
ykikuchi@ifi.u-tokyo.ac.jp

Abstract

Utilizing regional biomass instead of imported fossil fuels can support sustainability; but the presence of multiple options for products and conversion methods creates a complex scenario. In this study, we are tackling the simulation-based design of a regional system utilizing local biomass by thermochemical conversion technologies. By choosing a remote island, Tanegashima, in Japan, as a case study, the availability of local biomass and the demand for its products was determined. A model on Aspen Plus™ was developed to examine the possible yield of biomass products in different scenarios. Life cycle assessment (LCA) and input-output analysis (IOA) were conducted to calculate environmental and socio-economic indicators. Almost all scenarios considered saw greenhouse gas emission reductions and improved economic circularity, with byproduct reutilization proving beneficial for improving resource self-sufficiency.

Keywords: Biofuels, process simulation, sugarcane bagasse, woody biomass

1. Introduction

Japan's self-sufficiency ratio is only around 12.1% (IEA, 2020). In 2020, almost 90% of the country's primary energy supply was fulfilled by fossil resources and only about 3.9% of Japan's primary energy mix originated from biomass. Although local biomass use in Japan could make the country more self-sufficient and environmentally friendly (Goh et al., 2020), biomass must be converted into multiple different products of interest, like liquid fuels and chemicals. Hydrogen, ethanol, methanol, charcoal, and hydrocarbons usable as diesel are just a few examples of biomass-derived products (Kang et al., 2021). There are also various existing processes that can be utilized to perform this valorization; in some cases, the same end product can be created using different processes. In particular, thermochemical methods are the ones in which the biomass is decomposed by heat in addition to auxiliary chemicals such as gasification and pyrolysis (Wang et al., 2022). In addition to that, when designing a biomass valorization system, there is very likely to be a mismatch in the supply and demand quantities. For these reasons, if local governments and residents in a region make the decision to employ biomass as a fuel source due to the benefits outlined previously, a complex situation arises: a certain biomass-derived product may have to be prioritized over the others and it is necessary to determine what final product and technology choice would maximize the potential benefits to the population (Sun and Fan, 2020). Previous works in the literature have explored the matching of supply and demand of biomass within regions (Ayoub and Yuji, 2012; Sun and Fan, 2020). To quantify such "potential benefits", life cycle assessment (LCA) can be employed as a quantitative technology assessment tool (Kikuchi et al., 2010). Input-output analysis (IOA) (Leontief, 1936) could be applicable to illustrate how different sectors of a biomass-based economy interact with each other and to quantify the economic benefits of regional resource utilization (Kikuchi et al., 2020).

In this study, we are tackling the simulation-based design of a regional system utilizing local biomass by thermochemical conversion technologies. By choosing a remote island, Tanegashima, in Japan, as a case study, the availability of local biomass and the demand for its derived products was determined. A model on Aspen Plus™ was developed to examine the possible yield of biomass products in different scenarios. LCA and IOA were conducted to calculate environmental and socio-economic indicators.

2. Materials and methods

2.1. Tanegashima case study: biomass supply and fuel demand

In order to discuss the design of a biomass valorization system, the technologies and products considered in this study are applied to a case study in Tanegashima. The system boundary in this study is represented on Figure 1.

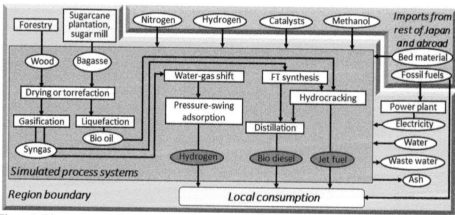

Figure 1: Biomass conversion system and its boundaries

Actual values were investigated and applied for estimating the supply of idle biomass from local forestry and sugarcane industries (Kikuchi et al. 2020; Kikuchi et al. 2016). As for the end product demand, hypotheses were made regarding the adoption of three biomass-derived fuels: hydrogen, jet fuel and biodiesel. Ten representative scenarios were generated to evaluate the effects of different technology and product choices, as listed in Table 1. In the first five, a single product is produced, and different technologies are used. In the other five scenarios, multiple products are manufactured. Scenarios Mix 1, 2 and 3 assume one third of the biomass is allotted to each product, altering the conversion technology between liquefaction, Fischer-Tropsch (FT) synthesis and liquefaction plus torrefaction pre-treatment, respectively. In Mix 4, 70% of the biomass is used for jet fuel production via liquefaction and 30% for hydrogen. In Mix 5, half of the raw materials are allocated for jet fuel production via liquefaction and the rest for diesel production via FT synthesis.

When possible, byproducts like char or flue gases are recycled into different unit operations, or combusted to generate steam and subsequently electricity. Thus, while electricity is always consumed to run some equipment, in some cases there is net generation of electricity.

Table 1: Scenarios of products and technologies applied for local biomass resources (Gas.: Gasification, FT: Fischer-Tropsch, Tor.: torrefaction, Liq.: liquefaction)

Allocation (wt%)	H_2	Jet (FT)	Jet (Liq)	DL (FT)	DL (Liq)	Mix 1	Mix 2	Mix 3	Mix 4	Mix 5
H_2	100					33	33	33	30	
Jet Fuel		100	100			33	33	33	70	50
Bio Diesel				100	100	33	33	33		50
Technology	Gas.	FT	Liq.	FT	Liq.	Liq.	FT	Liq. with Tor.	Liq.	Liq. And FT

2.2. Modeling of biomass conversion technologies in Aspen Plus^{TM}

The process inventory data was extracted from the process simulation in Aspen Plus (Kikuchi et al., 2010). Biomass is initially described as a "non-conventional" component.

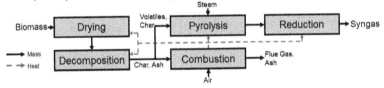

Figure 2: Gasification model scheme

As an example of biomass conversion modeling, the gasification model shown on Figure 2 is explained here. Dual fluidized bed gasification (Müller, 2013) was adopted in this study. The first stage, drying, was modeled with the RStoic reactor. (Patrocínio, 2017) The next block is the decomposition of biomass into conventional components modeled by an RYield reactor according to elemental balance (Abdelouahed et al., 2012). The pyrolysis occurs after decomposition process. The steam utilized as the gasifying agent is also an input in an RGibbs reactor at 400°C and atmospheric pressure. In this case, the final gasifier reactor is modeled as a CSTR in which reduction reactions occur in the gas phase. The combustion bed is also simulated on Aspen Plus by an RStoic reactor. Additionally, liquefaction is modeled with the RYield reactor taking data from previous studies (Wright et al., 2010). The torrefaction model developed by Peduzzi et al. was adopted (Peduzzi et al., 2014). Downstream processes were modeled by more conventional models (Saeidi et al., 2017).

2.3. Settings for LCA and IOA

The LCA boundary encompasses the conversion process itself, with biomass as the main input and the three potential products as the main outputs. Other auxiliary inputs necessary for operation such as electricity are also included. The inventory data was from IDEA v3.1 (AIST, 2022), a Japanese LCA database. The functional unit was defined as the utilization of local biomass for energy. The reference flow was set as unit mass of dry biomass input. IOA was performed based on the IO table for Tanegashima (Kikuchi et al., 2020). IOA in this study was executed by adding a new sector(s) to expand the IO table representing the new products being generated in the island. This requires calculating all the monetary inputs to the new sector (money spent by the manufacturing units) and the outputs; product sales. This is done by estimating the fixed and operational costs (Turton, 2009). The costs of plant construction are divided evenly throughout the 20-year lifetime of the project. Selling price was calculated assuming internal return rate of 11%. The circularity of the economy before and after the changes is then determined.

3. Results and Discussion

3.1. LCA results

The life cycle greenhouse gas (GHG) emissions are shown on Figure 3. The contribution of electricity represents a large share of the impact for the single product scenarios.

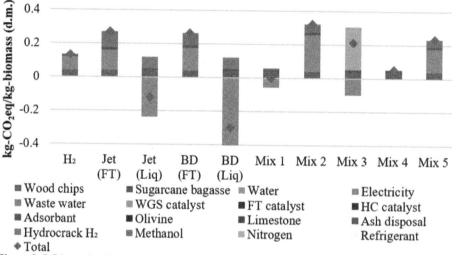

Figure 3: LCA results, life cycle

For FT processes, a relatively large amount of electricity is required in order to compress the syngas before it undergoes CO_2 removal and FT synthesis. Additionally, the operating temperature of gasification is high (>900°C). Therefore, all of the off gases released which could potentially produce electricity were re-directed into the gasifier to supply heat. On the other hand, for liquefaction, a smaller amount of the byproducts (char and flue gases) is required to heat up the pyrolyzer, since it operates at 480°C. Thus, net electricity generation is possible. Besides electricity, it can be seen that the liquefaction routes also stand out due to the large effects of hydrogen used in hydrocracking.

3.2. Input-Output Analysis

The first step of the IOA was the calculation of the monetary flows due to raw materials and utilities induced by the technology implementation. These values are ascribed to pre-existing sectors in the Tanegashima IO table, and utilized to describe the monetary flows in the renewable fuel production sector. After calculating the final effects on the economy, the change in internal and external flows before and after the production of fuels from regional biomass resources are presented on Figure 4. The overall value of the imports into Tanegashima actually only decreases on scenarios H_2 and Mix 2. In all other scenarios, after establishing these new sectors, the monetary value of imports increases.

It can be seen that, if limited to fossil fuels, the values of the imports decrease in all scenarios, however there are increases in other sectors such as petrochemicals.

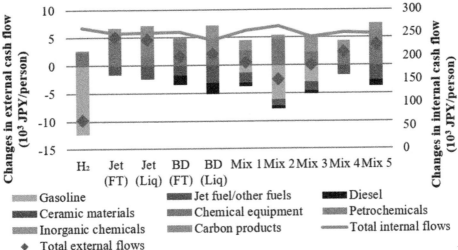

Figure 4: IOA results, where the bar and dots show the changes in external cash flows (left vertical axis) and the line shows those in internal cash flows (right axis).

One of the constants in all scenarios is the chemical equipment category. The capital costs for the plant, as would be expected, proved to be a significant factor in cash outflow. However, overall, the changes in internal flows were much larger, as the high added value of the fuels trickles down throughout many other sectors in the economy.

3.3. Interpretation of the multiple indexes for technology implementation

Using the results, three indexes were examined for interpretation: resource self-sufficiency (ratio of product demand fulfilled by the local biomass), GHG reduction potential and economic circularity improvement (increase in the ratio of internal monetary flows to total monetary flows).

Scenario H_2 was the optimal one in this study regarding resource self-sufficiency. This speaks to the efficiency of this conversion route itself and also to the lack of extra inputs required when compared to other scenarios. In fact, the Mix 4 scenario, in which mostly jet fuel is produced with the rest of the biomass generating hydrogen which fulfills the in-house demand for hydrogen in the hydrocracking process, performed better than the Jet (Liq) scenario, also due to the fact that demand for external auxiliary inputs was lower.

As for the GHG reduction potential, the Jet (Liq) scenario stands out as the best performing one. This illustrates that, in addition to low GHG emission results from LCA, the process yield is also important as the amount of imported fuel substituted also plays a large role in the GHG emissions reduction. For circularity, scenarios with a reduction in export flows performed better but the difference between scenarios is not as large.

4. Conclusion

Production of multiple fuels in mixed scenarios was shown to be more efficient due to the synergy between the thermochemical conversion processes. In general, all cases, especially the ones more reliant on liquefaction, showed potential to cause significant reduction in the GHG emissions in the energy sector. Particularly, electricity was shown to be a significant factor in GHG emissions, thus integrating this sort of biomass valorization with other projects for a greener energy grid could also be beneficial.

Acknowledgement

We thank Dr. Yuko Oshita for the discussion on the input-output analysis on Tanegashima and Dr. Shoma Fujii for discussions throughout the project. This work was supported by MEXT/JSPS KAKENHI Grant Number JP20K20016, 21K17919, and JST COI-NEXT JPMJPF2003. Activities of the Presidential Endowed Chair for "Platinum Society" at the University of Tokyo are supported by Mitsui Fudosan Corporation, Sekisui House, Ltd., the East Japan Railway Company, and Toyota Tsusho Corporation.

References

L. Abdelouahed, O. Authier, G. Mauviel, J. P. Corriou, G. Verdier, A. Dufour, Detailed Modeling of Biomass Gasification in Dual Fluidized Bed Reactors under Aspen Plus, Energy and Fuels, 26, 3840 (2012)

N. Ayoub, N. Yuji, Demand-driven optimization approach for biomass utilization networks, Computer and Chemical Engineerin, 36, 129 (2012)

C. S. Goh, T. Aikawa, A. Ahl, K. Ito, C. Kayo, Y. Kikuchi, Y. Takahashi, T. Furubayashi, T. Nakata, Y. Kanematsu, O. Saito, Y. Yamagata, Rethinking sustainable bioenergy development in Japan: Decentralised system supported by local forestry biomass, Sustainability Science, 15, 1461-1471 (2020)

IEA, World Energy Balances: Overview (2020)

K. Kang, N. B. Klinghoffer, I. Elghamrawy, F. Berruti, Thermochemical conversion of agroforestry biomass and solid waste using decentralized and mobile systems for renewable energy and products, Renewable and Sustainable Energy Reviews, 149 (2021)

Y, Kikuchi, K. Mayumi, M. Hirao, Integration of CAPE and LCA Tools in Environmentally-Conscious Process Design: A Case Study on Biomass-Derived Resin, Computer-Aided Chemical Engineering, 28, 1051-1056 (2010)

Y. Kikuchi, M. Nakai, Y. Kanematsu, K. Oosawa, T. Okubo, Y. Oshita, Y. Fukushima, Application of technology assessments into co-learning for regional transformation: A case study of biomass energy systems in Tanegashima, Sustainability Science, 15, 1473-1494 (2020)

Y. Kikuchi, Y. Kanematsu, M. Ugo, Y. Hamada, T. Okubo, Industrial Symbiosis Centered on a Regional Cogeneration Power Plant Utilizing Available Local Resources: A Case Study of Tanegashima, Journal of Industrial Ecology, 20(2) 276-288 (2016)

W. W. Leontief, Quantitative Input and Output Relations in the Economic Systems of the United States, Review of Economics and Statistics, 18, 105 (1936)

S. Müller, Hydrogen from Biomass for Industry, Vienna University of Technology (2013)

LCI database IDEA version 3.1; National Institute of Advanced Industrial Science and Technology (2022)

F. Patrocínio, Aspen simulation of biomass conversion processes, IST (2017)

E. Peduzzi, G. Boissonnet, G. Haarlemmer, C. Dupont, F. Maréchal, Torrefaction modelling for lignocellulosic biomass conversion processes, Energy, 70, 58 (2014)

S. Saeidi, F. Fazlollahi, S. Najari, D. Iranshahi, J. Klemeš, L. Baxter, Hydrogen production: Perspectives, separation with special emphasis on kinetics of WGS reaction: A state-of-the-art review, Journal of Industrial and Engineering Chemistry, 49, 1-25 (2017)

O. Sun, N. Fan, A Review on Optimization Methods for Biomass Supply Chain: Models and Algorithms, Sustainable Issues, and Challenges and Opportunities, Process Integration and Optimization for Sustainability, 4, 203 (2020)

R. Turton, R. C. Bailie, W. B. Whiting, J. A. Shaeiwitz, Analysis, Design and Synthesis of Chemical Processes (2009)

W. C. Wang, Y. C. Liu, R. A. A. Nugroho, Techno-economic analysis of renewable jet fuel production: The comparison between Fischer-Tropsch synthesis and pyrolysis, Energy, 239, 121970 (2022)

M. M. Wright, J. A. Satrio, R. C. Brown, D. E. Daugaard, D. D. Hsu, Techno-Economic Analysis of Biomass Fast Pyrolysis to Transportation Fuels (2010)

Antonis Kokossis, Michael C. Georgiadis, Efstratios N. Pistikopoulos (Eds.)
PROCEEDINGS OF THE 33rd European Symposium on Computer Aided Process Engineering
(ESCAPE33), June 18-21, 2023, Athens, Greece

Conceptual Design of a Biorefinery to Use Brown Seaweed Sargassum

Omar Flores-Mendoza,[a] Teresa Lopez-Arenas[a]

[a]*Departamento de Procesos y Tecnología, Universidad Autónoma Metropolitana –
Cuajimalpa, Vasco de Quiroga 4871, Mexico City 05348, Mexico*
mtlopez@cua.uam.mx

Abstract

The massive arrival of sargassum to the Caribbean beaches has generated a problem for
its collection, treatment, disposal, and use. Therefore, the recovery of sargassum has
become a topic of interest from the social, economic, and environmental point of view.
The objective of this work is to propose a conceptual design of a sargassum biorefinery
to produce three value-added products: alginate, a biopolymer and a biofertilizer. The
methodology consisted mainly of the synthesis, design, and techno-economic evaluation
of the biorefinery, using modeling and simulation tools. Comprehensive evaluation of the
biorefinery design shows its feasibility (with moderate product yields) and profitability
(with satisfactory return on investment and payback time) on an industrial scale.

Keywords: sargassum, biorefinery, circular economy, process simulation.

1. Introduction

Since 2011, the arrival of certain brown seaweed in the Caribbean, known as Sargassum,
has been a problem not only for the tourism and fishing sectors, but also for the
environment, given that the leachate produced by the seaweed saturates the habitat of
nutrients and its presence in the sea prevents the photosynthesis of seagrasses. Moreover,
the sargassum degradation by microorganisms in the open air generates greenhouse gases
(GHG). Currently, the collected sargassum is stored in landfills and is used mainly as
fertilizer. However, in some areas of the Caribbean, sargassum is also used to make
bricks, cosmetics, and paper products; but they must be viewed with caution due to their
ability to accumulate toxic metals. Thus, the implementation of the circular economy
through sargassum biorefineries (Azcorra-May et al., 2022) can help cushion the
secondary effects of its arrival, as well as obtain a benefit for society by transforming
seaweed into useful bioproducts such as precursors for food, biofuels, enzymes, or
biopolymers, generating a minimum of waste and emissions (Amador-Castro et al.,
2021).
The species of *Sargassum natans* and *Sargassum fluitans* reproduce in the subtropical
West Atlantic, which includes the Mexican Caribbean coasts, and in the North Equatorial
Recirculation Region, which is located in the tropical area between Brazil and Africa
(Wang, 2019). The main components of the cell wall of this brown seaweed are anionic
polysaccharides, of which are alginates and sulfated polysaccharides that contain fucose;
as well as cellulose, hemicellulose, lignin, arabinogalactan proteins and polyphenols
(Kloareg, 2021). On the one hand, alginate is a compound widely used in various
industries such as food (to maintain viscosity or retard crystallization), dental (as an
impression material), paper (to provide smooth, continuous films with great ink and print

retention quality), and medical (for the creation of biomedical and clinical materials such as tissue engineering). While on the other hand, hemicellulose and cellulose are a source for the generation of sugars (pentoses and hexoses) that can be fermentable to produce biofuels, such as bioethanol, biogas, or biodiesel, as well as other bioproducts with high added value such as biopolymers, organic acids, amino acids, among others.

This work presents the design and evaluation of a sargassum biorefinery to produce alginate, PHB (polyhydroxybutyrate) and a biofertilizer, using process modeling and simulation tools. The case study considers the brown seaweed sargassum that arrives at one of the main beaches of the Riviera Maya, Mexico. First, a conceptual design of the biorefinery is proposed, then the process implementation and the technical evaluation are carried out in a modular process simulator, and then simulations are performed to determine the best process configuration, the equipment dimensions, the operating conditions, as well as the productivity and performance of the products. At this early stage of the process design, a non-cash flow economic evaluation is performed to determine the profitability of the process.

2. Simulation methodology

The methodology consisted mainly of the synthesis, design, and techno-economic evaluation of the biorefinery. To carry out the synthesis of the biorefinery, first a bibliographic review was carried out to determine the composition of the sargassum and the best processing alternatives, identifying the physical, chemical, and biological phenomena. So that the unitary operations, the reaction routes, physicochemical properties of the components, etc. were established. Subsequently, the design of the biorefinery was implemented in a process simulator, considering an industrial scale, and taking into account equipment sizing and design, operating conditions, production programming, cost information (raw materials, equipment, operation, investment), etc. And finally, the simulation of the process was carried out through material and energy balances to assess technical-economic aspects, such as: product yields, installed production capacity, utilities consumption and profitability.

2.1. Availability and composition of sargassum

Currently the sargassum is collected on the beaches, followed by a drying process for storage and use. In particular, the amount of sargassum on the beach of Tulum, in the Mexican Caribbean, in recent years has ranged between 4 500 and 9 000 tons per year. The changes in its composition of cellulose, calcium alginate, hemicellulose and lignin vary depending on the season in which the sargassum is harvested. For simulation purposes, an average of each of the compounds was obtained (Bertagnolli et al., 2014; Lopez et al., 2020), corresponding to a composition on a dry basis of: 23% calcium alginate, 11.15% cellulose, 9.6% lignin, 8.2% hemicellulose, and 48.05% other solids.

2.2. Selection and selling cost of products

Alginate has been selected as the main product due to its diversity of applications and its high selling cost, which was set in the simulation at 20 USD/kg (Galindo et al., 2007). The PHB was selected as the second product, since from the point of view of the circular economy, the reduction of plastic consumption and its replacement with biopolymers should be encouraged. The PHB income was established at 4 USD/kg (Lopez-Arenas et al., 2017). While organic solids, a by-product of the PHB fermentation process, can be recovered as a biofertilizer due to its nutrient content. The unit selling cost was set at 0.35 USD/kg (Praveen and Singh, 2019).

2.3. Synthesis and design of the biorefinery

As part of the synthesis of the conceptual design, five processing sections of the biorefinery were identified: raw material conditioning, alginate extraction, alginate purification, PHB production, and PHB and biofertilizer purification. The corresponding flowsheet diagram of the sargassum biorefinery is shown in Figure 1, and each section is briefly described below.

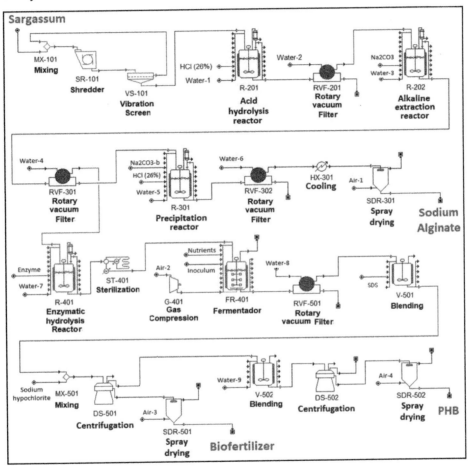

Figure 1. Process flowsheet diagram of the sargassum biorefinery.

2.3.1. Sargassum conditioning

First, the size of the dried seaweed is reduced to increase the exposed area and therefore the mass transfer. This is done by shredding (SR-101) and subsequent screening (VS-101) to ensure maximum particle size.

2.3.2. Alginate extraction

The crushed seaweed enters an acid hydrolysis reactor (R-201) to separate the cellulose and hemicellulose from the lignin, and to transform the calcium alginate into alginic acid. After removing the acidic liquid phase through a rotary filter (RVF-201), the alkaline extraction is carried out in the reactor R-202 with the addition of calcium carbonate, converting the alginic acid into sodium alginate. Subsequently, the liquid phase containing the soluble alginates is separated from the rest of the lignocellulosic biomass.

2.3.3. Alginate purification

The liquid phase rich in alginates is introduced into the reactor R-301, where hydrochloric acid and calcium carbonate are added to precipitate sodium alginate. Finally, the product is filtered (RVF-302) and dried (SDR-301).

2.3.4. PHB production

Lignocellulosic biomass is converted into glucose through an enzymatic hydrolysis reaction (R-401) using glucanases. Then, the hydrolyzed product is sterilized (ST-401) to avoid any contamination and production of unwanted compounds. The hydrolyzate passes to a fermenter (FR-401) for the intracellular production of PHB, using *Azohydromonas australica* as bacterial strain.

2.3.5. PHB and biofertilizer purification

Cell granules are recovered by filtration (RVF-501), and then PHB is extracted (V-501) by cell lysis by adding dodecyl sulfate surfactant. A sodium hypochlorite wash is then performed to achieve maximum cell disruption. Residual biomass and PHB are separated (DS-501). On the one hand, the residual biomass is dried for sale as a biofertilizer. And on the other hand, the stream with PHB is washed to remove the surfactant and sent for centrifugation (DS-502). Finally, the product is dried (SDR-502) to obtain the PHB in solid form.

2.4. Technoeconomic evaluation

Technoeconomic evaluation plays a crucial role in helping to select the best biorefinery conceptual design alternatives. From the technical point of view, product yields, processing times, installed capacity in terms of raw material and production rate are verified. In particular, the amount of required utilities is calculated, such as the consumption of heat transfer agents (cooling water and steam) and standard energy. While, to carry out the profitability analysis, an economic model without discounted cash flow was used to calculate the unit production cost, the rate of return on investment (ROI) and its payback time (PT). These cost effectiveness criteria provide a snapshot of the economics of the process and are often used to make preliminary estimates during the synthesis stage of process design. However, more rigorous profitability criteria could be used, in which cash flow is considered (such as the net present value method, NPV, or the internal rate of return, IRR) (Seider et al., 2009).

3. Simulation results

In this work, the biorefinery flowsheet diagram was implemented in *SuperPro Designer*, a commercial modular process simulator that allows the evaluation of material and energy balances, as well as the economic balance. This modular process simulator (like other commercial simulators such as *Aspen Plus*) includes databases of physicochemical properties of components, thermodynamic models, unit operations models, costs of raw materials and equipment, economic models, etc.

The biorefinery was simulated in batch mode and considering an annual operating time of 48 weeks. Each process equipment takes into account the times for material loading, process operation, material unloading and equipment cleaning, according to the case. The conceptual design proposal for the biorefinery considers a feed of 15 tons of sargassum per batch. Each process batch has a duration of 161.6 hours; however, each new production cycle can start after 37.3 hours of the previous cycle. Such that 212 batches are produced per year. A summary of the technical-economic evaluation is presented in Table 1.

Table 1. Technoeconomic report for the sargassum biorefinery.

Technical assessment			**Economical assessment**		
Concept	Value	Units	Concept	Value	Units
Sargassum consumption	3 180	t/yr	Direct fixed capital	32 822 000	USD
Sodium Alginate production	764.3	t/yr	Working capital	314 000	USD
PHB production	153.1	t/yr	Startup cost	1 641 000	USD
Biofertilizer production	0.472	t/yr	Total investment	34 777 000	USD
Sodium Alginate yield	0.2403	kg/kg S^*	Operating cost	8 941 000	USD/yr
PHB yield	0.0485	kg/kg S^*	ROI	19.98	%
Biofertilizer yield	0.0001	kg/kg S^*	PT	5.0	yr

Water and utilities consumption		
Concept	Value	Units
Process water	142.9	kg/kg SA^*
Steam	55.3	kg/kg SA^*
Cooling Water	5.5	t/kg SA^*
Standard Power	6.8	kW-h/kg SA

*S and SA to Sargassum and Sodium Alginate, respectively.

According to the technical results, it can be seen that the installed capacity of the plant in terms of annual sargassum consumption is within the availability of the Mexican beach of Tulum, as reported in the introduction section. The best product yield is obtained for sodium alginate, since it is the first product extracted; followed by the PHB yield, and finally, the yield of biofertilizer is quite low, which means that little residual biomass is recovered and its contribution to profitability is minimal.

Regarding the economic results, it can be observed that despite the estimated total investment being high, the biorefinery design shows positive and satisfactory profitability values for ROI and PT. According to Seider et al. (2009), an acceptable profit margin should be between 15-20% for ROI and 3 years for PT. It is worth mentioning that the operating cost considers various cost items, whose distribution is shown in Fig. 2(a). As can be seen, the utility cost is 15% of the annual operating cost, corresponding to the utility consumption reported in Table 1. The utility cost distribution is shown in Fig. 2(b), where the standard power (i.e., electricity) has the largest contribution. Therefore, alternatives to reduce energy consumption should be studied to reduce the cost and its environmental impact.

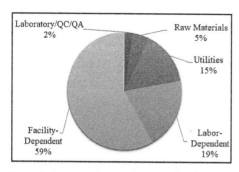
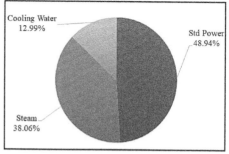

Figure 2. (a) Annual operating cost distribution and (b) utility cost distribution.

4. Conclusions

The results showed that the implementation of a biorefinery from the perspective of the circular economy allows addressing the problem of sargassum, not only to mitigate the secondary effects of its arrival, but also that a benefit could be obtained by transforming this seaweed into useful bioproducts such as precursors for biomaterials and food, biopolymers, fuels, among others. In particular, the profitability of the proposed biorefinery is supported by the production of sodium alginate, which is a product with high added value and great potential for use in biomedical areas. Even though the conceptual design presented shows satisfactory profitability, several aspects can be improved and studied further, such as analyzing other process configurations, optimizing operating conditions, evaluating different installed capacities, and more.

References

F. Amador-Castro, T. García-Cayuela, H.S. Alper, 2021, Valorization of sargassum biomass into sustainable applications: Current trends and challenges, J. Environ. Manage., 283, 112013.

K. J. Azcorra-May, E. Olguin-Maciel, J. Dominguez-Maldonado, 2022, Sargassum biorefineries: potential opportunities towards shifting from wastes to products, Biomass Convers. Biorefin.

C. Bertagnolli, A.P. Espindola, S.J. Kleinübing, L. Tasic, M.G. da Silva, 2014, Sargassum filipendula alginate from Brazil: Seasonal influence and characteristics. Elsevier.

E. Galindo, C. Peña, C. Núñez, D. Segura, G. Espín, 2007, Molecular and bioengineering strategies to improve alginate and polydydroxyalkanoate production by Azotobacter vinelandii. Microb Cell Fact 6, 7 (2007).

B. Kloareg, Y. Badis, J. M. Cock, G. Michel, 2021, Role and Evolution of the Extracellular Matrix in the Acquisition of Complex Multicellularity in Eukaryotes: A Macroalgal Perspective, Genes, 12 (7), 1059.

L.B. Lopez, J.J. Alvarado, J.C. Corral, A. Aguilera, R.E. Rodríguez, S.J. Guevara, J.V. Alcaraz, J.G. Rutiaga, J. Zárate, M.L. Ávalos, M. Morales, 2020, A prospective study of the exploitation of *pelagic Sargassum spp.* As a solid biofuel energy source. Applied Sciences, 10(23), 8706.

T. Lopez-Arenas, M. González-Contreras, O. Anaya-Reza, M. Sales-Cruz, 2017, Analysis of the fermentation strategy and its impact on the economics of the production process of PHB (polyhydroxybutyrate), Computers & Chemical Engineering, 107, 140-150.

K.V. Praveen, A. Singh, 2019, Realizing the potential of a low-cost technology to enhance crop yields: evidence from a meta-analysis of biofertilizers, Agricultural Economics Research Review, 32, 77-91.

D.W. Seider, J. D. Seader, D. R. Lewin, S. Widagd, 2009. Product and Process Design Principles: Synthesis, Analysis and Design. John Wiley & Sons, New York.

M. Wang, C. Hu, B. Barnes, G. Mitchum, B. Lapointe, J.P. Montoya, 2019, The great Atlantic Sargassum belt, Science, 365(6448), 83-87.

Antonis Kokossis, Michael C. Georgiadis, Efstratios N. Pistikopoulos (Eds.)
PROCEEDINGS OF THE 33rd European Symposium on Computer Aided Process Engineering
(ESCAPE33), June 18-21, 2023, Athens, Greece

Design and Production of Biodiesel and Porous Filaments from Endemic Species of the Cienega de Chapala Region

María Fernanda Torres-Magaña[a], José Luis Baltazar-Álvarez[a], Pascual Eduardo Murillo-Alvarado[a] and Gabriela Guadalupe Esquivel-Barajas[a]

aUniversidad de La Ciénega del Estado de Michoacán de Ocampo. Av. Universidad Sur 3000, Lomas de Universidad, 59103 Sahuayo de Morelos, Michoacán México.

Abstract

The production of biofuels has contributed to reducing the use of fossil fuels, however, in the means of obtaining biofuels, waste materials are generated, which must be given extra treatment for their storage. In such a way that the objective of the present investigation was to obtain biodiesel through the transesterification method, using the seeds of 3 plants (*Ricinus communis, Pithecellobium dulce* and *Prosopis laevigata*), which are endemic to the Cienega Region of the State of Michoacán, Mexico. The generated residual biomass served as raw material for the synthesis of bioplastics. Both biodiesel and bioplastics were characterized by Fourier Transform Infrared Spectroscopy (FT-IR) and UV-VIS, thus it was possible to identify absorption bands, corresponding to the functional groups, which are similar to those of hydrocarbons. Thus, it is concluded that the remnants obtained during biodiesel production processes can be used for a second use as bioplastics.

Keywords: biodiesel, bioplastic, seeds.

1. Introduction

Due to anthropogenic actions, the planet has manifested various changes, due to the effect of greenhouse gases produced by the consumption of fossil fuels, which emit CO_2 and sulfur, for which reason it has ventured into the use of alternative fuels, such is the case of obtaining and applying biofuels.

Biofuels are fuels produced from raw materials of vegetable or animal origin. Another way of obtaining them is through the processing of agro-industrial products or organic waste. Arellano C. (2015), describes biofuels as an alternative of interest to reduce the accumulation of greenhouse gases and thus reduce dependence on the use of fossil fuels, in the same way in his manuscript he describes that when using biomass from In the food sector, an alternative is the production of lignocellulosic materials, since these do not compete directly with food intended for both human and animal consumption, which makes it possible to take advantage of biomass generated from agro-industrial processes.

Mentioned the above, the objective of the research has been based on the use of 3 endemic plants of the Ciénega region, which are: castor plant (*Ricinus communis*), mezquite (*Pithecellobium dulce*) and guamúchil (*Prosopis laevigata*), of which two processes are derive, the first is to obtain biodiesel from plants and the second is to take advantage of the lignocellulosic remnants so that they have become biofilms, which have properties similar to bioplastics.

2. Biopolymers

Biopolymers are macromolecular materials, which can be synthesized naturally or synthetically. The classification of biopolymers is based on three families, which are: proteins, polysaccharides and nucleic acids. Among the most abundant biopolymers are cellulose and starch, which are of interest for the present investigation.

2.1 *Bioplastics*

Bioplastics are materials that have an organic composition, the main polymers that characterize them being both starch and cellulose (Riera A., & Palma R., 2018). It is necessary to have chemical structures that allow the degradation of materials from native biota where they have been discarded (mainly they must be degraded by soil microorganisms).

2.2 *Endemic plants and trees of the Ciénega de Michoacán Region*

The use of the material that is used for the generation of a biofuel must be careful not to break the fine line of balance between the use of organic waste and the use of organic resources for human consumption (Merino P., & Nonay M., 2009) that is why for this project in the generation of two bioproducts. In biodiesel, the decision was made to use seeds of fruits that have been considered as plants of low nutritional value for humans such as Guamúchil and Mezquite which are characterized by be abundant trees in the Ciénega de Chapala region.

3. Methodology

For the development of the present investigation, it was divided into 5 stages which consist of: 1) sampling of endemic plant remnants, 2) extraction of essential oils and obtaining biodiesel, 4) characterization of oils and biodiesel 5) synthesis and characterization of thin films from residual biomass.

Stage 1. Identification of endemic plants. Castor weed plants were identified, in addition to mezquite and guamúchil trees, in the municipalities of Sahuayo-Jiquilpan, Michoacan, Mexico.

Stage 2. Castor berries were collected, in addition to sampling the remnants of guamúchil and mezquite berries, approximately 2 kg of each berry, which were stored in sterile plastic bags, later stored in a desiccator (Novatech), with the aim of facilitating the extraction of seeds.

Stage 3. Extractions of essential oils. For the development of this stage, an adaptation was made to the methodology proposed by Demirbas 2008, which consisted of: grinding the seeds and using the soxhlet method (KIMAX brand) in which 200ml of isopropyl alcohol (MEYER) were poured, subjecting at constant temperature and stirring at 200°C and 200 rpm. In order to extract the essential oil from the seeds, it was necessary to recirculate the alcohol in the extractor 15 times with a duration of approximately 45 minutes each. Upon completion of the 15 cycles, the sample was removed from the grill and placed in a desiccator for a period of 24 hours to remove as much water and alcohol as possible to subsequently obtain the pure essential oil from the seed. To obtain biodiesel, the essential oil was taken to the centrifuge (Hetttich brand, EBA 20 model) for 30 minutes at 3,500 rpm and removing the degumming or residues to be able to be mixed with the 1% concentration of the potassium iodide solution (Reasol brand) with an addition of 3% (p/p) of distilled water and place on the grill to maintain agitation for 30 minutes at 60°C to finish it was placed in the centrifuge at 60 rpm for one hour.

Stage 4. Characterization of biodiesel. Biodiesel was characterized by Fourier transform infrared spectroscopy (FT-IR), Perkin Elmer Frontier model in the 4,500-400 cm^{-1} region and by Perkin Elmer Lambda 25 model UV-VIS spectrophotometry, in the absorbance region of 300-800 nm.

Stage 5. Synthesis and characterization of biofilms. For the development of this stage, an adaptation was made to the methodology proposed by Salazar, 2013. To make use of the waste generated from obtaining biodiesel, a bioproduct has been generated that is formed from the remains of seeds that passed by the transesterification method with an adaptation to the methodology carried out by Salazar M., & Valderrama A. (2013) in which the previous conditioning was carried out in an extraction hood (Tecnolab brand) with a grill that preheats a Porcelain evaporation dish (Duve brand) at 75°C in which 30 ml of distilled water (Golden Bell) and 3 ml of natural glycerin are added, shake for 5 minutes; add 0.9 ml of a 1 molar solution of glacial acetic acid (Golden Bell brand) keep stirring for 10 minutes and then pour it into a glass petri dish to take it to the preheated desiccator at 75 °C for 8 hrs. And finally they were characterized by UV-VIS and FT-IR.

4. Results and discussions

Seeds and berries were collected in the municipalities of Sahuayo and Jiquilpan Michoacan Mexico (Fig.1), later the essential oils were obtained from each of the samples, which were characterized by UV-VIS, to identify the band of greater absorption, which oscillated for the 3 samples in the interval of 700-800 nm (Fig. 2).

Fig.1. Seeds and berries of endemic plants of Sahuayo-Jiquilpan, Mich. a) *Prosopis laevigata*, b) *Ricinus communis* and c) *Pithecellobium dulce*.

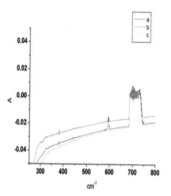

Fig.2. UV-VIS characterization of endemic plants from Sahuayo-Jiquilpan, Mich. a) *Prosopis laevigata*, b) *Ricinus communis* and c) *Pithecellobium dulce*.

Biodiesel was synthesized through the transesterification method, which was characterized by FT-IR, with the aim of identifying its functional groups (Fig. 3). Where it can be seen that both in *Ricinus communis* and in *Pithecellobium dulce* there is a greater predominance by the OH group, since their absorption bands are greater compared to that of *Prosopis laevigata*.

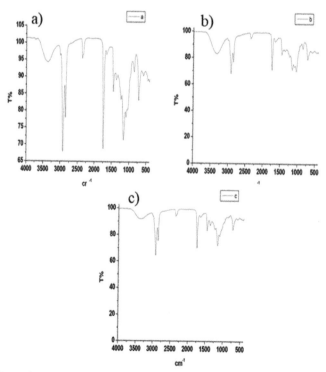

Fig.3. FT-IR characterization of endemic plants from Sahuayo-Jiquilpan, Mich. a) *Prosopis laevigata*, b) *Ricinus communis* and c) *Pithecellobium dulce*.

However, during the synthesis of biodiesel, remnants are generated during its elaboration, in such a way that it was decided to give this residual material a conversion, for which bioplastics were synthesized from it (Fig. 4).

Fig. 4. Bioplastics obtained from biodiesel remnants. a) *Prosopis laevigata*, b) *Ricinus communis* and c) *Pithecellobium dulce*.

The bioplastics were characterized by FT-IR (Fig. 5), where it can be observed that there is a decrease, compared to the characterization of biodiesel, in the absorption bands belonging to the esters and amides groups.

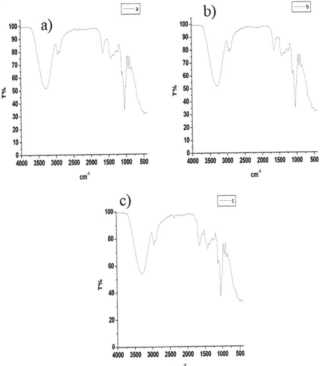

Fig. 5. FT-IR characterization of bioplastics, Mich. a) *Prosopis laevigata*, b) *Ricinus communis* and c) *Pithecellobium dulce*.

In the same way, a proposal was developed where this technology can be migrated on an industrial scale, for which the design at the industry level is proposed (Fig. 6).

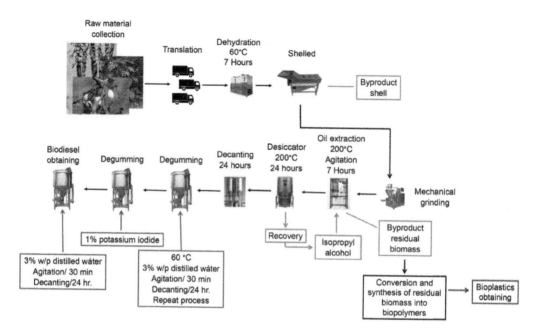

Fig. 6. Biodiesel and bioplastics industrial production scheme.

5. Conclusions

The transesterification method is suitable for obtaining biofuels, in particular it is the case of biodiesel. However, the synthesis process involves various reagents, which in turn produce materials of waste, for which it is considered viable to use them to obtain bioplastics, which can serve as environmentally friendly materials.

6. References

Arellano Perales Carla (2015) Obtención de bioetanol a partir de materiales lignocelulósicos sometidos a hidrolisis enzimática. Facultad de ciencias químicas. P. p. 3-5.

Demirbas Ayhan 2008, Biodiesel from Triglycerides via Transesterification, Biodiesel A Realistic Fuel Alternative for Diesel Engines, P.p. 121-139.

Merino Garcia Pedro Antonio y Nonay Domingo Maria Teresa (2009) Descripción, evolución y retos del sector de los biocombustibles. Boletin ICE Economico. No. 2971. P.p. 11-20

Riera Maria Antonieta and Palma Ricardo R., 2018, Obtención de bioplásticos a partir de desechos agrícolas. Una revisión de las potencialidades en Ecuador, Avances en Química, Vol. 13, No. 3, P.p. 69-78.

Salazar Max Carlos and Valderrama Negrón Ana, 2013, Preparation And Characterization Of Depolymerised Chitosan Films And Crosslinked With Sodium Tripolyphosphate. Rev. Soc. Quím. Perú Vol.79 No.3. P.p. 197.

Antonis Kokossis, Michael C. Georgiadis, Efstratios N. Pistikopoulos (Eds.)
PROCEEDINGS OF THE 33rd European Symposium on Computer Aided Process Engineering
(ESCAPE33), June 18-21, 2023, Athens, Greece

Real-time control and scheduling decision making in fermentation-based food process operations

Isuru A. Udugama[a,*], Keegan K. Hall[a], Christoph Bayer[b], Brent R. Young[c], Timothy G. Walmsley[a]

[a]Ahuora – Centre for Smart Energy Systems, School of Engineering, University of Waikato, Hamilton, 3240, New Zealand
[b]Department of Process Engineering, TH Nürnberg, Nürnberg, 90489, Germany
[c]Industrial Information and Control Centre, Department of Chemical & Materials Engineering, The University of Auckland, Auckland, 1010, New Zealand
isuru.udugama@waikato.ac.nz.

Abstract

With government policy encouraging decarbonisation, the New Zealand food processing industry is rethinking its process operations and their improvement. This work focuses on identifying opportunities to decarbonise industrial sites using simultaneous multi-process control and scheduling optimisation methods. A common issue for multi-process sites is the disconnection between the unit operation level process controls that operate under closed loop regulatory, or advanced regulatory, principles and the scheduling layer that ideally operates under real-time optimisation principles. As the first step towards solving this issue, an illustrative case of a cream cheese production process consisting of a pasteurisation unit, two fermentation units and a separation unit are studied by constructing a plant wide scheduling model in Python, integrated with a fermentation model, to explicitly model the interactions between two layers. Compared to typical plant operation, controlling the dosing of substrate in the fermenter allows for more uniform fermenter end times leading to improved scheduling of successive batches which results in less discarding of batches due to poor quality. Future work will embed the trade-off between overall energy usage, product quality and product throughput constraints (supply chain) within a holistic objective function centred on carbon footprint.

Keywords: process control, fermentation modelling, real-time decision making

1. Introduction

Food and dairy production processes account for over 40% of New Zealand export revenue but represent a significant proportion of New Zealand's natural gas and coal consumption [1]. Like many other countries [2], the New Zealand government has implemented the Emission Trading Scheme to encourage the process industries to decarbonise while supporting them to maintain their competitive edge through co-investment in low-emission technology. From an engineering perspective, tackling the decarbonisation challenge requires a mix of capital investments into new technologies, e.g., integrating heat pump technologies in supplying process heat for milk spray dryers [3], and real-time control optimisation of industrial site operations.

Food production processes are often operated batch-wise where a given batch is fed through multiple unit operations over time. As such, the control of these food production processes requires both process control (carried out at the unit operation level) and scheduling of unit operation utilisation throughout multiple batches. However, the simultaneous multi-process control and scheduling optimisation methods are not an established practice in the process systems engineering domain. At present, a hierarchical process control approach is taken where plant-wide scheduling and unit operation level control are executed at different layers, i.e., sequentially. In this approach, instructions are passed down the process control hierarchy from the plant-wide scheduling layer to advance process control layer and onwards to regulator controls. The idea being that the process control layer will focus on ensuring aspects such as product quality while the scheduling layer focuses on process optimisation [4].

However, food processing has large inherent production variations due to availability of the natural raw materials (e.g., milking seasons for dairy processes) and changes in consumer demands. As a result the schedules need to be constantly updated, for example, using short term and continuously updating schedules [5]. Another option of tackling these variations is to change set-points in the underlying supervisory and regulatory level control structure in "real-time". This entails speeding up or slowing down the unit operations such that the overall production schedule is not influenced. In practice, a combination of both "real-time" control actions and timely re-scheduling will be needed to ensure a production process maintain quality, production targets while actively ensuring decarbonisation goals are met.

The aim of this study is to investigate the decarbonisation potential of simultaneous optimisation of process control and multi-plant scheduling actions in the food industry. Figure 1 expresses the current industry practice in contrast to the proposed approach.

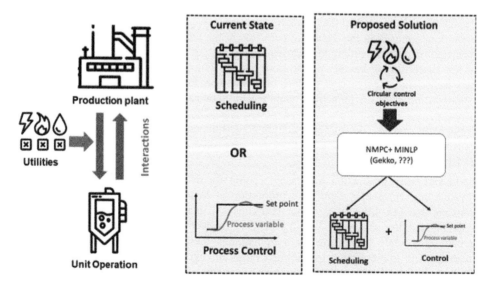

Figure 1: Visualisation of the challenge, current state and the proposed solution

Solvers such as active-set based APOPT have the potential to be a useful tool in bridging the gap between sequential and simultaneous methods for scheduling and control optimisation. For example, by translating advanced process control concepts such as model predictive control from differential algebraic equations to algebraic form allows the simultaneous solution of the scheduling and the control problem as a mixed-integer non-linear optimisation problem [6]. As a realistic test case, this paper presents a fermentation-based process to illustrate the potential of simultaneous decision-making.

2. Model Development

This section details the current state of the test case model of a simple cream cheese production process with a pasteurisation, fermentation and downstream processing unit operations. The test model is implemented in Python and solved using the GEKKO library with its APOPT solver.

2.1. Plant level model

A sequential scheduling model is used to represent the dynamic dependency of unit operations at the plant level. The scheduling model consists of three-unit operations; pasteurisation, fermentation and separation. Two fermenters are used in parallel while only a single pasteuriser and separation unit are present. This resembles industrial plant design since the fermentation time is longer than the processing times of the other two units. The pasteurisation and separation processing times are assumed to be static while the fermentation processing time varies between batches. A fermentation model is used to determine fermentation time where model parameters are randomly varied to simulate variation within a real process. More detail on the model is provided in Section 3.2. The model also considers cleaning in place (CIP) procedures as well as transfers between unit operations. These procedures are also assumed to have static times.

The model is initially setup to replicate reality where the fermentation end time is not known at the beginning, which means successive operations can only be scheduled when the fermentation ends. This can then be compared with a scenario where the fermentation is perfectly controlled to finish within a target time, and the next batch can begin preparation in advance allowing for maximum production. The time series data was stored in data frames from the Pandas package which is then visualised in a Gantt chart using the Plotly package in Python.

2.2. Fermentation model

The mechanistic fermentation model reported in [7] is used as the basis of modelling the fermentation process. The model tracks substrate (lactose), biomass (Lactococcus lactis) and product (lactic acid) evolution over time. The model utilises a Monod type biomass growth kinetics with substrate and product inhibition thus production rates are dependent on substrate concentration and availability. To represent inherent variabilities that are present in fermentation processes, the model parameters are randomly sampled from a uniform distribution within the reported upper and lower parameter bounds in [7]. The initial concentration of substrate, product and biomass were assumed constant throughout batches to represent feedstock standardisation. To this end, all variation in fermentation concentration profiles between batches is due to random variations in model parameters.

The fermentation model is implemented into Python as a series of ordinary differential equations (ODE) representing time dependent evolution of substrate (S), biomass (B) and product (P). In the case of no fermentation control the model is solved simultaneously with the Scipy numerical integration package across a series of discretised time points. Integration is performed until the product concentration reaches 5 g/L which represents experimental observations in [7] that ensures the target product quality is achieved.

To allow for fermentation end time control the model must be formulated as a dynamic non-linear programming (NLP) that is solved in Python GEKKO using the IPOPT solver [6]. The time dependent concentration growth rates of the species are governed by Equations 1-3 where the substrate dosing flowrate, m_s, has been included in Equation 2 as a manipulated variable. However, the flowrate must be divided by the fermenter volume to convert mass flow to change in concentration per unit time. A terminal constraint is applied in Equation 4 which enforces that the product concentration at the target final time, t_f, is equal to the target concentration, P_{target}. The model is solved across a consecutive series of time points ranging from 0 to t_f. Lastly the objective function is applied to maximize the final product concentration pursuant to the terminal constraint.

$$\frac{dX}{dt} = \mu_{max}\left(\frac{S}{K_{sx}+S}\right)\left(1-\frac{P-P_{ix}}{P_{mx}-P_{ix}}\right)\left(\frac{K_{ix}}{K_{ix}+S}\right)X \tag{1}$$

$$\frac{dS}{dt} = -q_{s,max}\left(\frac{S}{K_{ss}+S}\right)\left(1-\frac{P-P_{is}}{P_{ms}-P_{is}}\right)\left(\frac{K_{is}}{K_{is}+S}\right)+\frac{m_s}{V} \tag{2}$$

$$\frac{dP}{dt} = \alpha\frac{dX}{dt} + q_{p,max}\left(\frac{S}{K_{sp}+S}\right)\left(1-\frac{P-P_{ip}}{P_{mp}-P_{ip}}\right)\left(\frac{K_{ip}}{K_{ip}+S}\right) \tag{3}$$

$$P \times t_{final} - P_{target} = 0 \tag{4}$$

$$t = [0, t_{final}] \tag{5}$$

$$\max\{P \times t_{final}\} \tag{6}$$

2.3. Model integration and decarbonisation

The scheduling and fermentation model are integrated such that simultaneous decisions can be performed between changing manipulated variables in the fermentation process, and schedule changes at the plant wide level. More specifically, the fermentation end time applied to the scheduling model is dynamically updated from the fermentation model which is influenced by control actions. Regarding decarbonisation, the combined modelling of the schedule and fermentation enables the calculation of energy intensive actions such as CIP and material flows in unit operations. In future work, further investigation of explicit minimization of CO_2-e will be investigated. Explicitly considering decarbonisation allows identification of effective actions both in the scheduling layer and unit operation layer (e.g., mitigating scheduling actions that may require batch disposal or situations where CIP must be redone).

3. Model output

Figure 2 illustrates the process schedule for the two use cases and Figure 3 shows the concentrations of species within each fermenter batch. In both use cases 10 fermentation batches are processed. Figure 2a illustrates the progression of production batches through the cream cheese production process with no fermentation control and Figure 3a shows the corresponding fermentation concentrations.

Each batch is first heat treated in the pasteuriser prior to transfer to one of the two fermenters, then the batch is sent for separation. The inherent variations in the fermentation time clearly influence the schedule as the next batch cannot be pasteurised until a fermenter is available, nor separated until the separator is available. For example, in batch number 3 (second batch on fermenter 1), the fermentation process has finished early but cannot be transferred out for another two hours until the separator is available. This can result in product quality degradation leading to batch disposal. In contrast, Figure 2b shows the production schedule where the fermentation end time is perfectly controlled by substrate dosing to finish (i.e. reach the target lactic acid concentration) at 6 hours. The concentration profile in each batch is shown in Figure 3b where the substrate concentration changes considerably between batches due to control induced dosing.

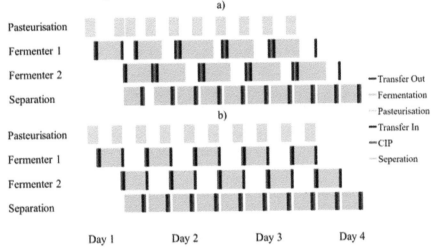

Figure 2: a) Current schedule of cream cheese process where fermentation end times are not controlled b) Improved schedule where the fermentation is controlled to finish at a specified end time.

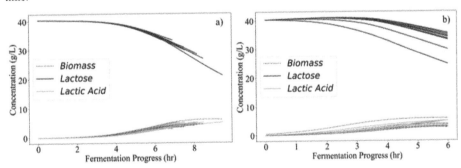

Figure 3: a) Concentration profiles in a cream cheese fermenter with random variation and no control b) Concentration profiles in a fermenter with random variation but controlled to achieve the target lactic acid concentration at 6 hours.

Fermentation control leads to improved scheduling since fermentation end times are known so successive batches can be scheduled so that fermentation ends as soon as a separator is available. As such, batches can immediately be transferred for separation resulting in no potential loss of product. It should be noted that in the improved case, the fermentation operation is changed from a batch to fed-batch.

Despite likely improvements in the form of reductions in batch disposal, the optimized schedule finishes the 10 batches the same time as the un-optimized schedule due to the separator unit availability. However, tightly controlling the fermentation end time means extra substrate dosing is needed. From a decarbonisation perspective, this optimisation does not consider the carbon trade-off from substrate dosing (which increases the CO_2-eof a given batch) that may be needed to save the batch, and the cost of losing the batch which has implications on successive batch scheduling. As an immediate future step, the simultaneous optimization of batch scheduling and fermentation substrate dosing must be considered. For example, re-looking at batch number 3 in Figure 2a (un-optimized case), a fed-batch fermentation could reduce the initial dosage of substrate thereby reducing the biomass growth rate and lactic acid production rate. As such, the end time of the fermentation will be prolonged which reduces the waiting time for separation without product degradation. In this example, the reduced substrate dosing will not only improve the overall CO_2-e of the specific batch, but also reduce the overall CO_2-e of a unit of finished product.

4. Conclusions and Future work

This work presented a cream cheese fermentation process where a plant wide scheduling model was integrated with a fermentation model to illustrate the propagation of variations in fermentation times at unit operation level and scheduling level. This work is the first step towards solving the key issue involving the disconnection between the operation level process controls to maintain product quality and the scheduling layer that attempts to maximise product throughput. It was seen that unit operation level decisions such as dynamic dosing of substrate can reduce the variation experienced at the scheduling layer thereby allowing for improved scheduling of successive operations. In future work, the test case will be used to develop a simultaneous multi-process control and scheduling optimisation methods that explicitly consider the CO_2-e trade-off for all decisions.

References

[1]   I. & E. MBIE NZ (Ministry of Business, "Dairy Manufacturing – Process Heat and Greenhouse Gas Emissions."

[2]   B. K. Sovacool, M. Bazilian, S. Griffiths, J. Kim, A. Foley, and D. Rooney, "Decarbonizing the food and beverages industry: A critical and systematic review of developments, sociotechnical systems and policy options," *Renew. Sustain. Energy Rev.*, vol. 143, p. 110856, Jun. 2021, doi: 10.1016/j.rser.2021.110856.

[3]   Q. Chen, D. J. Cleland, J. K. Carson, and T. G. Walmsley, "Integration of desiccant wheels and high-temperature heat pumps with milk spray dryers," *Appl. Therm. Eng.*, vol. 216, p. 119083, Nov. 2022, doi: 10.1016/j.applthermaleng.2022.119083.

[4]   I. A. Udugama *et al.*, "The Role of Big Data in Industrial (Bio)chemical Process Operations," *Ind. Eng. Chem. Res.*, vol. 59, no. 34, pp. 15283–15297, Aug. 2020, doi: 10.1021/acs.iecr.0c01872.

[5]   M. G. Ierapetritou and C. A. Floudas, "Effective Continuous-Time Formulation for Short-Term Scheduling. 1. Multipurpose Batch Processes," *Ind. Eng. Chem. Res.*, vol. 37, no. 11, pp. 4341–4359, Nov. 1998, doi: 10.1021/ie970927g.

[6]   L. Beal, D. Hill, R. Martin, and J. Hedengren, "GEKKO Optimization Suite," *Processes*, vol. 6, no. 8, p. 106, Jul. 2018, doi: 10.3390/pr6080106.

[7]   B. Li, Y. Lin, W. Yu, D. I. Wilson, and B. R. Young, "Application of mechanistic modelling and machine learning for cream cheese fermentation pH prediction," *J. Chem. Technol. Biotechnol.*, vol. 96, no. 1, pp. 125–133, 2021, doi: 10.1002/jctb.6517.

Antonis Kokossis, Michael C. Georgiadis, Efstratios N. Pistikopoulos (Eds.)
PROCEEDINGS OF THE 33rd European Symposium on Computer Aided Process Engineering
(ESCAPE33), June 18-21, 2023, Athens, Greece

Production of $Ca(HCO_2)_2$ from formate solution through CO_2 hydrogenation

Taeksang Yoon,[a] Hayoung Yoon,[b] Sungho Yoon,[b] Chul-Jin Lee[a,c]

[a]School of Chemical Engineering and Material Science, Chung-Ang University, 84 Heukseok-ro, Dongjak-gu, Seoul, Republic of Korea
[b]Department of Chemistry, Chung-Ang University, 84 Heukseok-ro, Dongjak-gu, Seoul, Republic of Korea
[c]Department of Intelligent Energy and Industry, Chung-Ang University, 84 Heukseok-ro, Dongjak-gu, Seoul, Republic of Korea

Abstract

With growing levels of CO_2 emission and severity of global warming, interest in carbon capture and utilization (CCU) has increased. However, the amount of CO_2 reduction that can be realized is lower than that of CO_2 emissions, and most of the existing CCU technologies cannot be commercialized owing to their limited economic feasibility. Therefore, it is necessary to identify new chemicals as CCU products. When CaO is added to a formate aqueous solution, it reacts with the formate to produce calcium formate $(Ca(HCO_2)_2)$. In this study, $Ca(HCO_2)_2$ was synthesized using dolomite and CaO ash, an industrial waste. Process design and evaluation were performed based on the experimental data. When CaO ash and dolomite were used as the raw material, the production costs were 16% lower and 8% higher than those of the conventional process, respectively. Moreover, for both materials, the CO_2 emissions were 20% lower than those of the conventional processes.

Keywords: CO_2 hydrogenation, Calcium formate, Carbon capture and utilization

1. Introduction

Carbon capture and utilization (CCU) technologies, which convert CO_2 to value-added chemicals, have emerged as promising solutions to address the growing severity of global warming related issues (Aresta and Dibenedetto 2007; Peters et al. 2011). However, most of the existing CCU processes have not been commercialized because of their low economic feasibility compared with the conventional process (Chauvy et al. 2019; Lindsey and Jeskey 1957). Furthermore, the amount of CO_2 that can be eliminated through CCU is lower than that of the CO_2 emissions. Therefore, novel CCU chemical products must be identified (Rafiee et al. 2018).

Recently, Park et al. (2020) developed a method for producing formic acid by hydrogenating CO_2 through a heterogeneous Ru catalyst and an aqueous amine solution. This method can overcome the disadvantage of heterogeneous catalysts, i.e., a low conversion rate. However, the process involves multiple distillation processes in separation (Figure 1(a)).

Calcium formate, $Ca(HCO_2)_2$, is used in various applications such as animal feed, cement production, and leather tanning as an additive (EFSA 2015). $Ca(HCO_2)_2$ is typically produced by the carbonylation of calcium hydroxide with carbon monoxide (Ma et al.

2016). Recently, Yoon et al. (2022) reported the production of $Ca(HCO_2)_2$ using CaO ash (waste from a desulfurization process) and CO_2 through hydrogenation. This method can instantaneously convert CaO ash, an industrial waste, to $Ca(HCO_2)_2$ and separate products through simple evaporation in mild conditions. Moreover, the main raw material, CaO, is readily produced through the calcination of dolomite, an abundant natural resource.

In this study, $Ca(HCO_2)_2$ was synthesized through CO_2 hydrogenation, with process design performed based on experimental data (Figure 1(b)). The feasibility of the process was confirmed through a techno-economic analysis (TEA) and lifecycle assessment (LCA), in comparison with the conventional process (Ma et al. 2016).

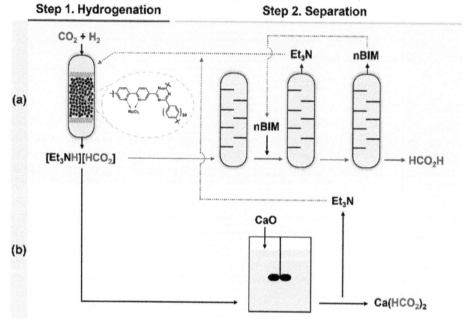

Figure 1 Schematic of (a) existing and (b) proposed (Yoon et al. 2022) processes for HCO_2H and $Ca(HCO_2)_2$ production

2. Experimental

2.1. Hydrogenation of CO₂ to [Et₃NH][HCO₂]

$$CO_2 + 0.74H_2 + Et_3N + 0.26H_2O \rightarrow 0.74[Et_3NH][HCO_2] + 0.26[Et_3NH][HCO_3] \quad (1)$$

CO_2 hydrogenation was performed in a stainless-steel tubular reactor packed with 1.5 g of Ru/bpyTN-30-CTF catalyst. The reactor was pressurized to 120 bar and heated to 120 °C before the reaction. H_2O and Et_3N were supplied to the reactor, and an Et_3N concentration of 3 M was maintained using a high-pressure liquid pump. The liquid product was analyzed to determine the formate concentration. The results indicated that the solution contained 2.03 M [Et₃NH][HCO₂] and 0.73 M [Et₃NH][HCO₃] (Equation (1)). The solution was heated at 90 °C and ambient pressure for 2 h to remove CO_2 that remained as the HCO_3 adduct and could react with CaO to produce $CaCO_3$.

2.2. Synthesis of Ca(HCO₂)₂ using CaO ash

$$2[Et_3NH][HCO_2] + CaO + CaSO_4 \rightarrow Ca(HCO_2)_2 + CaSO_4 + H_2O + 2Et_3N \quad (2)$$

Flue gas desulfurization ash (CaO/CaSO₄ ash) can be generated as a product of the desulfurization process using CaO. Because the solubility of CaSO₄ (0.2 g per 100 mL H₂O at 20 °C) is considerably lower than that of Ca(HCO₂)₂ (16.6 g per 100 mL H₂O at 20 °C), CaO and CaSO₄ are expected to be separated in a facile manner. When CaO/CaSO₄ ash was added to the [Et₃NH][HCO₂] solution at 90 °C for 5 min, a clear solution with a white precipitate, CaSO₄ (Equation (2)), was obtained. CaSO₄ and Ca(HCO₂)₂ were separated from the solution through filtration and evaporation, respectively.

2.3. Synthesis of Ca(HCO₂)₂ using dolomite

$$CaMgCO_3 \rightarrow CaMgO_2 + CO_2 \quad (3)$$

$$2[Et_3NH][HCO_2] + CaMgO_2 \rightarrow Ca(HCO_2)_2 + MgO + H_2O + 2Et_3N \quad (4)$$

First, dolomite was calcined in a furnace at 950 °C and ambient pressure for decomposition to CaMgO₂ (Equation (3)). Results of an X-ray diffraction analysis indicated that all of the dolomite was completely decomposed to CaO and MgO. Allen et al (2009) highlighted that reaction kinetics between Ca and Mg ions are considerably different. Therefore, when calcined dolomite was added to the [Et₃NH][HCO₂] solution at 90 °C, and a reaction was allowed to occur for 5 min, only CaO was expected to react with formate and produce Ca(HCO₂)₂ while MgO remained as the oxide (Equation (4)). Because MgO is sparingly soluble in water, it remained as a precipitate in the solution, which could be filtered to yield a high purity product. This filtrate could be separated into Ca(HCO₂)₂ and solvents by simple evaporation, as described in Section 2.2.

3. Process design and analysis

3.1. Process design

Based on the experimental results, two Ca(HCO₂)₂ processes were designed using CaO ash and dolomite. Figure 2 illustrates the process flow. Gaseous CO₂ and H₂ were compressed and heated to 120 °C at 120 bar before being supplied to the hydrogenation reactor. The unreacted gases were separated through degassing and recycled. The degassed solution was introduced to the reactor with a CaO mixture, thereby synthesizing Ca(HCO₂)₂. Because the reaction kinetics and solubility of CaO and MgO were considerably different, only CaO transformed to Ca(HCO₂)₂, and MgO remained as solid in the solution. The solid component, as a byproduct with high purity, was separated by filtering. The remaining solution was separated into the solvent and Ca(HCO₂)₂ through evaporation.

3.2. TEA

The economic feasibility of the process was evaluated in terms of the production cost. Table 1 lists the prices of the raw materials and utilities, and Table 2 summarizes the results of the economic analysis. When dolomite and CaO ash were used as the raw materials, the production cost was approximately 8% higher and 16% lower than that of the conventional process, respectively. Figure 3(a) shows the breakdown of the production cost for each process component: raw material, utility, pretreating and reaction, and separation. The cost components are defined in the following text.

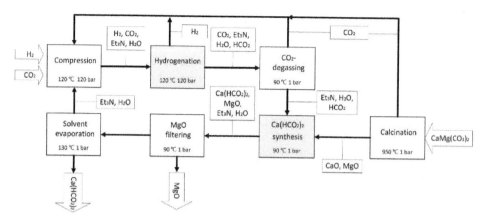

Figure 2 Block flow diagram of $Ca(HCO_2)_2$ production using dolomite.

- Raw material: Summation of all costs associated with the raw materials.
- Utility costs: Summation of all costs for utilities.
- Pretreating and reaction: Summation of capital expenditures (CAPEX) and fixed operational expenditure (fixed OPEX) for the pretreatment of material (e.g., compression and heating) and reaction.
- Separation: Summation of CAPEX and fixed OPEX for separation equipment/processes such as a column.

The ratio of utility costs was higher than that of the raw material costs, which indicated that $Ca(HCO_2)_2$ production is energy-intensive. The high energy consumption is attributable to the evaporation of water (Yoon et al. 2022).

Table 1 Material and utility costs for the economic analysis

Component	Price (unit)	Reference
CO_2	35 (USD/tonne)	(Kim et al. 2011)
H_2	1630 (USD/tonne)	(Lee et al. 2018)
LP steam	13.28 (USD/GJ)	(Turton and Bailie 2012)
Electricity	0.066 (USD/kWh)	(Turton and Bailie 2012)
Cooling water	0.354 (USD/GJ)	(Turton and Bailie 2012)
Natural gas	11.34 (USD/GJ)	(Turton and Bailie 2012)

3.3. LCA

LCA was performed to evaluate the CO_2 emissions resulting from the processes. The Intergovernmental Panel on Climate Change 2013 GWP 100a method was applied for the analysis. Data from Ecoinvent 3.8 database and the literature were used to calculate the influence of the product and CO_2 emission factors (Lee et al. 2018; Wernet et al. 2016). The functional unit of analysis was 1 kg of $Ca(HCO_2)_2$. Because the processes associated with CCU are multi-functional and yield more than one product, the CO_2 emission of the byproduct was excluded from the total CO_2 emission. Table 2 summarizes the LCA results. The CO_2 emissions associated with both the proposed processes were 20% lower than those of the conventional process. Figure 3(b) shows the breakdown of the cost and CO_2 emissions. Similar to the cost, the CO_2 emissions corresponded to the following four components:

- Raw material: Summation of CO_2 emissions from raw materials.
- Utility: Summation of CO_2 emissions for utilities consumed in the process.

- Purge: Summation of CO_2 emission from the purge gas.
- Byproduct: Summation of CO_2 emissions from the byproducts, excluded from the total CO_2 emission.

For all processes, the contribution of the utility to the total CO_2 emission was the most notable (~50% of the total value), and it accounted for more than 80% of the total CO_2 emissions in the CCU processes.

Table 2 Results of economic and lifecycle analysis of each process.

	Conventional	Proposed (dolomite)	Proposed (CaO ash)
Production cost (USD/tonne)	678	729	567
CO_2 emission ($kgCO_{2eq}/kgCa(HCO_2)_2$)	3.97	3.18	3.19

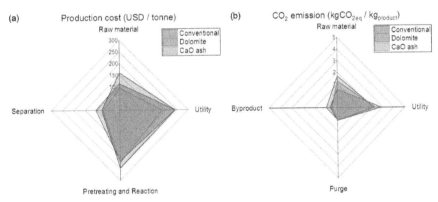

Figure 3 Breakdown of (a) production cost and (b) CO_2 emission for $Ca(HCO_2)_2$ production.

4. Conclusions

This paper proposes two processes for $Ca(HCO_2)_2$ production based on CCU, using CaO and formate adduct. The novel processes were compared with the conventional process. The proposed processes could selectively produce $Ca(HCO_2)_2$ and byproducts with high purity while utilizing CO_2 as a value-added chemical. Although the production cost of the CCU process using dolomite was 8% higher than that of the conventional process, it emitted 20% lower CO_2, demonstrating environmental feasibility. When CaO ash was used as the CaO source, the economic and environmental feasibility were comparable to those of the conventional approach. The results highlight the potential of $Ca(HCO_2)_2$ production based on CCU method as a replacement for conventional processes. The energy consumption can be reduced to enhance the economic feasibility of the CCU-based processes.

References

EFSA Panel on Additives and Products or Substances used in Animal Feed (2015). "Scientific opinion on the safety and efficacy of calcium formate when used as a technological additive for all animal species." EFSA Journal **13**(5): 4113.

Allen, J. P., S. C. Parker and D. W. Price (2009). "Atomistic simulation of the surface carbonation of calcium and magnesium oxide surfaces." The Journal of Physical Chemistry C **113**(19): 8320-8328.

Aresta, M. and A. Dibenedetto (2007). "Utilisation of CO2 as a chemical feedstock: opportunities and challenges." Dalton Transactions (28): 2975-2992.

Chauvy, R., N. Meunier, D. Thomas and G. De Weireld (2019). "Selecting emerging CO2 utilization products for short- to mid-term deployment." Applied Energy **236**: 662-680.

Kim, J., C. A. Henao, T. A. Johnson, D. E. Dedrick, J. E. Miller, E. B. Stechel and C. T. Maravelias (2011). "Methanol production from CO2 using solar-thermal energy: process development and techno-economic analysis." Energy & Environmental Science **4**(9): 3122-3132.

Lee, D.-Y., A. Elgowainy and Q. Dai (2018). "Life cycle greenhouse gas emissions of hydrogen fuel production from chlor-alkali processes in the United States." Applied Energy **217**: 467-479.

Lindsey, A. S. and H. Jeskey (1957). "The Kolbe-Schmitt reaction." Chemical Reviews **57**(4): 583-620.

Ma, H., X. Feng, Y. Yang, Z. Zhang and C. Deng (2016). "Preparation of feed grade calcium formate from calcium carbide residue." Clean Technologies and Environmental Policy **18**(6): 1905-1915.

Park, K., G. H. Gunasekar, S.-H. Kim, H. Park, S. Kim, K. Park, K.-D. Jung and S. Yoon (2020). "CO2 hydrogenation to formic acid over heterogenized ruthenium catalysts using a fixed bed reactor with separation units." Green Chemistry **22**(5): 1639-1649.

Peters, M., B. Köhler, W. Kuckshinrichs, W. Leitner, P. Markewitz and T. E. Müller (2011). "Chemical technologies for exploiting and recycling carbon dioxide into the value chain." **4**(9): 1216-1240.

Rafiee, A., K. Rajab Khalilpour, D. Milani and M. Panahi (2018). "Trends in CO2 conversion and utilization: A review from process systems perspective." Journal of Environmental Chemical Engineering **6**(5): 5771-5794.

Turton, R. and R. C. Bailie (2012). Analysis, synthesis, and design of chemical processes. Upper Saddle River, NJ, Prentice Hall.

Wernet, G., C. Bauer, B. Steubing, J. Reinhard, E. Moreno-Ruiz and B. Weidema (2016). "The ecoinvent database version 3 (part I): overview and methodology." The International Journal of Life Cycle Assessment **21**(9): 1218-1230.

Yoon, H., T. Yoon, H.-J. Yoon, C.-J. Lee and S. Yoon (2022). "Eco-friendly and techno-economic conversion of CO2 into calcium formate, a valuable resource." Green Chemistry **24**(4): 1738-1745.

Antonis Kokossis, Michael C. Georgiadis, Efstratios N. Pistikopoulos (Eds.)
PROCEEDINGS OF THE 33rd European Symposium on Computer Aided Process Engineering
(ESCAPE33), June 18-21, 2023, Athens, Greece

Techno-Economic Process Analysis of the Chemical Recycling of Nylon 6 Using Phosphoric Acid

Ann-Joelle Minor[a,*], Ruben Goldhahn[a], Liisa Rihko-Struckmann[a],
Kai Sundmacher[a,b]

[a]*Max Planck Institute for Dynamics of Complex Technical Systems, Process Systems Engineering, Sandstorstr. 1, 39106 Magdeburg, Germany*
[b]*Chair for Process Systems Engineering, Otto-von-Guericke University, Universitätsplatz 2, 39106 Magdeburg, Germany*
*E-mail address: aminor@mpi-magdeburg.mpg.de

Abstract

Chemical recycling has increasingly gained attention in order to counteract environmental pollution and support the transition to a circular economy. Today, only around two percent of polyamides are chemically recycled each year. This is mainly due to the economic performance of the process. The conventional depolymerisation process of polyamide 6 using phosphoric acid was simulated to analyse the techno-economic potential and identify its existing constraints. A conceptual process design was carried out using Aspen Plus and Python. The results show that the process becomes profitable at 4600 t/year of caprolactam. Assuming a reasonable plant capacity of 12.6 kt/year, the payback time would be between three and seven years in the best- and worst-case scenarios. The disposal of the waste stream containing unreacted polyamide 6, oligomers and phosphoric acid, as well as phosphoric acid raw material costs together contribute 47% of the total production cost. Accordingly, a sensitivity analysis showed that the total annualised costs depend significantly on the amount of phosphoric acid raw material used. Overall, the results indicate great industrial potential for the depolymerisation of polyamide 6. Still, the focus of further research on this conventional process should rely on phosphoric acid and its efficient recovery along with the treatment of the waste stream.

Keywords: chemical recycling, circular economy, Nylon 6, process simulation, techno-economic analysis.

1. Introduction

The massive production of plastics caused undeniable issues such as environmental pollution and depletion of petroleum-based resources. To meet the recycling target of 50% set by the European Commission for 2025, chemical recycling has increasingly gained attention (Pohjakallio et al., 2020). As opposed to the most popular mechanical recycling strategy, chemical recycling avoids quality loss of the polymer (also called downcycling), thereby acting in favour of a circular economy with multiple polymerisation–depolymerisation cycles. Chemical recycling involves breaking down the polymer chain into monomers through a high-temperature and solvolytic environment (Coates and Getzler, 2020).

According to a recent review, polyamide 6 (PA6), also known as Nylon 6, is an ideal candidate for chemical recycling as it is not thermodynamically and kinetically limited (Coates and Getzler, 2020). Despite having been investigated since the early

1960s, still today, only a minor fraction of the annual PA6 production is recycled each year (Coates and Getzler, 2020). Reportedly, the reasons for this are the low profitability and economic competitiveness of depolymerisation processes (Kamimura et al., 2014). Several companies have been employing the chemical recycling of PA6. The acid-catalysed hydrolysis of PA6 to its respective monomer caprolactam (CL) using phosphoric acid (H_3PO_4) and steam can be considered as a conventional process which is covered by several patents, e.g. of BASF (Corbin et al., 1999). Furthermore, acid-catalysed hydrolysis of polyamides has been investigated experimentally (Chen et al., 2010). However, the cost contributions, bottlenecks, most sensitive parameters, and the reason for lacking economic performance of the process have not been identified yet. In order to improve the understanding and find possible opportunities for optimisation of the process, the chemical recycling process of PA6 using H_3PO_4 was simulated, and a techno-economic analysis was performed, both of which have never been published in the literature. Subsequent to the next chapter giving the background information and methods, in the results section, the scale for profitability, the payback period, as well as the leading fixed and variable cost contributions are estimated. Additionally, the effect of the most influential parameters is studied within a sensitivity analysis and case study.

2. Background and Methodology

In the following, the continuous conventional process described by BASF will be referred to as the "representative process" (Corbin et al., 1999). This representative process was simulated and analysed with a model connecting Aspen Plus and Python. To study it in the most realistic manner, the changes to this process were kept at a minimum.

2.1. Polyamides and Depolymerisation Background Knowledge

PA6 is most commonly synthesised through hydrolytic ring-opening polymerization, involving reversible reactions, e.g. the hydrolysis of CL to aminocaproic acid (ACA) and polycondensation and addition reactions (Penczek et al., 1985). However, due to dipole-dipole and hydrogen bonding interactions of the amide groups, required depolymerisation temperatures are really high. The ceiling temperature of 277 °C can be lowered by adding a solvent such as H_3PO_4 in the conventional process (Coates and Getzler, 2020).

2.2. Thermodynamics and Kinetics

Vapour-liquid equilibrium (VLE) data of water and CL are retrieved from the NIST database by AspenPlus, and the NRTL model was used to describe the VLE behaviour. H_3PO_4 and PA6 were assumed to be non-volatile.

Owing to the excess of water and catalyst, the literature approximates the hydrolytic depolymerisation of PA6 to CL by a pseudo-first-order reaction in PA6 (Chen et al., 2010). This is displayed in Eq. 1, where [PA6], k' and X are the PA6 concentration, the rate constant and conversion, respectively.

$$-\frac{d[PA6]}{dt} = k' \cdot [PA6] \quad \Leftrightarrow \quad dt = \frac{dX}{k' \cdot (1-X)} \qquad \text{with } [PA6] = [PA6]_0 \cdot (1 - X) \qquad (1)$$

The reaction kinetics for chemical recycling using H_3PO_4 are not available. Therefore, the required kinetic parameter is derived from Eq. 1 using the batch data point with the final time and conversion given by the representative process (Eq. 2).

$$k' = \frac{1}{t_f} \cdot \ln\left(\frac{1}{1-X_f}\right) \qquad (2)$$

2.3. Data Base for Component Input Into Aspen Plus

A database needed to be established for the component input of PA6. Scalar properties, such as heat of formation and degree of polymerisation, as well as temperature-dependent properties, such as solid and liquid heat capacity values, were taken from the Polymer Handbook, and parameters were regressed when necessary (Brandrup et al., 2003). As reaction enthalpy, the negative of the standard polymerisation enthalpy (-12.5 kJ/mol) was assumed (Brandrup et al., 2003).

2.4. Process and Equipment Design

The process is designed to be as similar as possible to the representative process. The rate-based continuous stirred tank reactor (CSTR) model was chosen because of the liquid-phase reaction and slow reaction kinetics. The type of pumps, heat exchangers, vacuum equipment, columns or vessels was determined by heuristics. For the economic evaluation, equipment dimensions such as the vessel height, diameter, wall thickness or weight need to be known. The equipment design was done by a combination of empirical equations, e.g. the determination of heat exchanger areas or column diameters, and heuristics such as tray sizing or the choice of the heat exchanger tube length (Seider et al., 2009).

2.5. Utilities

Cooling water (CW), low-pressure steam (LP), medium-pressure steam (MP), high-pressure steam (HP) and No. 2 fuel oil (OIL) were chosen as utilities. The temperature (T), enthalpy of evaporation (h_e), or higher heating value (HHV) of those are displayed in Table 1.

Table 1: Details of the chosen utilities.

CW	LP	MP	HP	OIL
T_{in} : 30°C	T : 139.9°C	T : 186°C	T : 250°C	HHV :
T_{out}: 45°C	h_e : 2144 kJ/kg	h_e : 1991 kJ/kg	h_e : 1715.2 kJ/kg	43046.4 kJ/kg

Near-optimal approach temperatures were defined according to temperature levels, e.g. 5 °C for ambient temperature levels. The pipeline and valve pressure drop was neglected, and the heat exchanger pressure drop was set dependent on the phase of the stream.

2.6. Economics

The capital investment expenditures (CAPEX), total operational expenditures (OPEX) and cash flow were calculated in U.S. dollars. The CAPEX was computed using the Lang factor method with an accuracy of 35%. The Lang factor F_L is multiplied by the sum of the free onboard (f.o.b.) equipment purchase costs C_{pi} and accounts for all other direct and indirect investment costs (Eq. 3).

$$\text{CAPEX} = 1.05 \cdot F_L \cdot \sum_i \frac{I_i}{I_{bi}} \cdot C_{pi} \qquad (3)$$

The purchase costs were calculated using regressed correlations as a function of equipment size factors and updated to the current year by applying the chemical engineering plant cost index I (Seider et al., 2009).

The OPEX comprise variable costs, manufacturing costs, and general and administrative (G&A) costs. The variable costs, such as raw materials, utility and waste treatment costs, were determined using current industrial market prices and a plant operating factor of 0.9 was assumed. Direct wages and benefits were calculated by accounting for the typical labour requirements dependent on the type of equipment and assuming five shifts (J. Harmsen et al., 2018). The other manufacturing costs, such as maintenance, overall plant overhead and fixed costs, as well as the G&A costs, are typical fractions of the labour costs, capital investment expenses or sales (Seider et al., 2009).

For the cash flow analysis, net earnings have been calculated with the sales (S) and the income tax rate according to the U.S. federal income tax rate schedule for corporations. Subsequently, the annual cash flow was calculated by incorporating the depreciation (D) and assuming one year for the plant construction (Eq. 4 and 5) (Seider et al., 2009).

$$\text{Net earnings} = (S - OPEX) \cdot (1 - \text{income tax rate}) \tag{4}$$

$$\text{Annual Cash Flow} = (\text{Net earnings} + D) - CAPEX \tag{5}$$

The total annualised costs (TAC) are shown in Eq. 6 and include a reasonable return in investment i_{min} of 0.2 (Seider et al., 2009).

$$TAC = OPEX + i_{min} \cdot CAPEX \tag{6}$$

3. Results and Discussion

3.1. Choice of Separation System and Process Flowsheet

In contrast to the representative process, pure PA6 enters the depolymerisation reactor. Hence, the required separation steps after depolymerisation to purify CL from water need to be adapted to this simplification. It was investigated whether a flash evaporator would be sufficient to achieve the patents' CL recovery constraints (see Figure 1).

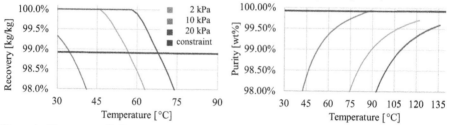

Figure 1: CL recovery (left) and purity (right) versus temperature at different flash unit pressures.

The graphs show that within the pressure range, the recovery and purity constraints cannot be met at the same time. Hence, a distillation column was chosen with an operating pressure of 16 kPa, which allows the use of a single liquid ring pump as well as CW as a cold utility. The process flowsheet for a reasonable CL production capacity of 14.0 kt/year, including the equipment and utility choices, is shown in Figure 2.

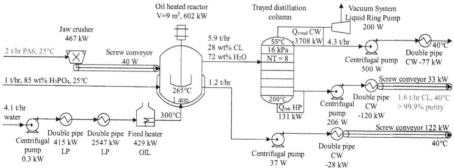

Figure 2: Schematic representation of process flowsheet and choice of equipment and utility.

As can be seen, crushed PA6 reacts to CL using H_3PO_4 and superheated steam in a jacketed and agitated reactor with a conversion of 80%. The formed CL is obtained at the top as a vapour stream along with steam. Superheated steam functions in several ways, as it shifts the reaction equilibrium by carrying CL, supports mixing and supplies heat. As described in chapter 2.1, H_3PO_4 acts as a catalyst by activating the carbonyl group and facilitating the attack of water. However, it is also consumed in the simultaneous protonation of ACA and salt formation (Češarek et al., 2020). Those products are discharged at the bottom of the reactor and have to be treated as hazardous waste.

The product stream is obtained at the bottom of the vacuum distillation column with a recovery of 99.9% and a CL purity of over 99.9 wt%. Subsequently, the product stream is cooled and solidified to 40 °C by a heat exchanger and screw conveyor.

3.2. Economic Analysis

It was found that the described process is profitable, as the break-even point is already reached at a CL production of 4.6 kt/year.

3.2.1. Cost Contributions

To identify the most cost-intensive process units, the OPEX and CAPEX structure was investigated (Figure 3).

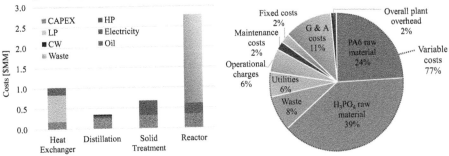

Figure 3: Costs of the process units in $MM (left bar chart) and OPEX (right pie chart).

Despite the fact that the CAPEX of $5.5M is rather small compared to the OPEX of $26.3M per year, none of these process units stands out regarding their capital expenditures. On the other hand, the graphs reveal that variable costs are dominant. Even more interesting is that H_3PO_4 raw material costs and its waste treatment as reactor discharge make up 47% of the overall OPEX. The cost contribution of H_3PO_4 is, therefore, further investigated through a sensitivity analysis.

3.2.2. Sensitivity Analysis and Case Studies

Next to the amount of H_3PO_4, possible effects of electricity price volatilities were analysed. Additionally, the kinetic pre-exponential factor was varied to study the consequences of potential inaccuracies on the results, which are displayed in Figure 4.

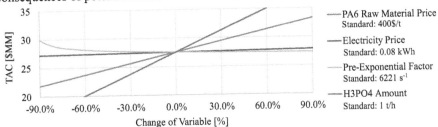

Figure 4: Sensitivity analysis where TAC is displayed over the percentual change of certain variables.

The change of the H_3PO_4 amount has notably the most significant impact on the TAC, which aligns with the results of the previous chapter. This is because increasing the amount of H_3PO_4 does not only result in high raw material costs but also in huge waste treatment costs. This further proves that H_3PO_4 is the main bottleneck of the process. On the other hand, the electricity and OIL price variations barely have any impact on the TAC. Similarly, in a range of 30%, any inaccuracy of the pre-exponential factor would have minor effects. However, the profitability of the process heavily depends on the PA6 waste raw material purchase cost, with an upper profitability boundary of $650 per tonne.

3.2.3. Cash Flow Analysis in Best and Worst Case

To account for inaccuracies and model errors, a best and worst-case scenario was established for the profitability and cash flow analysis (see Figure 5). Utility prices and the pre-exponential factor were all altered by 20%. Additionally, the Lang-Factor method for calculating CAPEX has an accuracy of ±35%, which was incorporated as well.

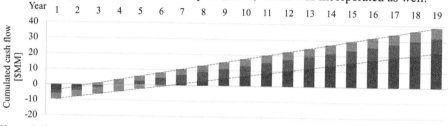

Figure 5: Cumulative cash flow in $MM for 19 years in the best, standard, and worst-case scenarios.

It can be predicted that the process becomes profitable within a minimum of three to a maximum of seven years, and cumulative cash flow stays within certain boundaries. This proves that the established model and process analysis are solid.

4. Conclusions

In this work, a techno-economic analysis of the conventional chemical recycling route of PA6 was performed. It showed that the process stays profitable even after the incorporation of realistic parameter inaccuracies and worst-case assumptions. The results indicate that neither the reaction nor the purification part of the process has a significant impact on the TAC. Instead, the amount of H_3PO_4 and the waste stream containing unreacted PA6, oligomers and H_3PO_4 are the critical variables. Eventually, it can be concluded that depolymerisation processes of PA6 can be very promising and have a great potential for optimisation, especially regarding the catalyst and solvent amount.

References

M. Pohjakallio, T. Vuorinen, A. Oasmaa, 2020, Chemical routes for recycling -dissolving, catalytic, and thermochemical technologies, Plastic Waste and Recycling, pp. 359– 384.

G. W. Coates, Y. D. Y. L. Getzler, 2020, Chemical recycling to monomer for an ideal, circular polymer economy, Nature Reviews Materials, Vol. 5, No. 7, pp. 501–516.

A. Kamimura, K. Ikeda, S. Suzuki, K. Kato, Y. Akinari, T. Sugimoto, K. Kashiwagi, K. Kaiso, H. Matsumoto, M. Yoshimoto, 2014, Efficient conversion of polyamides to ω-hydroxyalkanoic acids: a new method for chemical recycling of waste plastics, ChemSusChem, Vol. 7, No. 9, pp. 2473–2477.

T. F. Corbin, A. Handermann, R. Kotek, W. D. Porter, 1999, Reclaiming epsilon-caprolactam from nylon-6 carpet, BASF, US005977193A.

J. Chen, Z. Li, L. Jin, P. Ni, G. Liu, H. He, J. Zhang, J. Dong, R. Ruan, 2010, Catalytic hydrothermal depolymerisation of nylon 6, J Mater Cycles Waste Manag, Vol. 12, No. 4, pp. 321–325.

W. D. Seider, J. D. Seader, D. R. Lewin, 2009, Product and Process Design Principles: Synthesis, Analysis and Evaluation, New York:Wiley.

S. Penczek, P. Kubisa, K. Matyjaszewski, 1985, Chapter 11: Polyamides, Cationic Ring-Opening Polymerization, Advances in Polymer Science, Springer, Vol. 68/69, pp 201–208.

U. Češarek, D. Pahovnik, E. Žagar, 2020, Chemical Recycling of Aliphatic Polyamides by Microwave-Assisted Hydrolysis for Efficient Monomer Recovery, ACS Sustainable Chemistry & Engineering, Vol. 8, No. 43, pp. 16274–16282.

J. Harmsen, A. B. de Haan, P. L. J. Swinkels, 2018, Product and Process Design: Driving Innovation, De Gruyter.

J. Brandrup, E. H. Immergut, E. A. Grulke, 2003, Polymer Handbook, Wiley.

Antonis Kokossis, Michael C. Georgiadis, Efstratios N. Pistikopoulos (Eds.)
PROCEEDINGS OF THE 33rd European Symposium on Computer Aided Process Engineering
(ESCAPE33), June 18-21, 2023, Athens, Greece

Multiscale High-throughput Screening for Membrane-based Green Hydrogen Separation Process

Nahyeon An[a,b], Seongbin Ga[a,c], Hyungtae Cho[a], Boram Gu[d,†], Junghwan Kim[a,b,†]

[a]Green Materials and Processes RD Group, Korea Institute of Industrial Technology, 55, Jongga-ro, Jung-gu, Ulsan, 44413, Republic of Korea
[b]Department of Chemical and Biomolecular Engineering, Yonsei University, 50, Yensei-ro, Seoul, 03722, Republic of Korea
[c]Department of Chemical Engineering, College of Engineering, University of Ulsan, 93, Daehak-ro, Nam-gu, Ulsan, Republic of Korea
[d]School of Chemical Engineering, Chonnam National University, 77, Yongbong-ro, Gwangju, 61186, Republic of Korea
† Corresponding authors.

Abstract

Hydrogen purification is essential to use ammonia as a carrier of green hydrogen since the green hydrogen produced from the transported ammonia inevitably results in nitrogen as a byproduct. The membrane-based separation process has been considered an option for green hydrogen purification because of its various advantages. The selection of proper membrane materials is an important task to efficiently separate gas mixtures. Because it is difficult to compare numerous membranes through experiments, the membrane has been selected using the molecular simulation-based high-throughput screening (HTS) method. However, in the previous studies, the physical properties used to evaluate the membrane materials were not directly related to the separation process performances. To address this limitation, this study entailed the use of multiscale HTS to select the covalent organic frameworks (COFs) best performing in the membrane-based green hydrogen separation process. The proposed method combines the process and molecular simulations to evaluate the performances of COF membranes. The proposed multiscale HTS method was applied to a COF database, and the 648 COFs in the database were explored to select the COFs with the highest performance in hydrogen separation.

Keywords: green hydrogen, membrane separation process, covalent organic framework, high-throughput screening

1. Introduction

The recent enthusiasm for green technology has put green NH_3 as a hydrogen transportation medium in the spotlight. Separation of N_2 and H_2 is a critical step in converting NH_3 transported from another location into hydrogen for use (Morlanés et al., 2021). Among the various hydrogen purification methods, the membrane-based hydrogen separation process is an active topic of research owing to its ease of use and low energy cost (Ahmad et al., 2012). Various polymer membranes have been designed for efficient gas separation, but the trade-off between selectivity and permeability remains challenging. To address this issue, new nanoporous materials, such as metal organic frameworks and covalent organic frameworks (COFs), have been actively developed. Due to their chemical diversity and structural tunability, COFs have the highest potential for use as membranes (Aydin et al., 2022).

Selecting an appropriate material among the many porous materials developed to date is one of the most important tasks in the design of a membrane-based separation process because membranes significantly impact the membrane separation process performance. Previous studies on molecular simulation-based high-throughput screening (HTS) have evaluated various materials (Gulbalkan et al., 2022). However, the inability to evaluate the performance in the actual membrane process is a limitation in the previous HTS studies because the computational cost was too high and because only molecular simulation was used.

Therefore, to provide a practical evaluation of the membranes, this study proposes a multiscale evaluation framework. Using the multiscale evaluation framework, this study aimed to select high-performing COF membranes for the membrane-based green hydrogen separation process. In this multiscale evaluation framework, molecular and process simulations were performed to evaluate each COF with process performance. The 20 COFs with the highest H_2 recovery were selected based on the evaluation results. Finally, by examining the H_2 recovery and purity of the membrane process for various process conditions, the performances of the selected COFs were presented for different situations.

2. System description

The concept of ammonia as hydrogen transportation media is a future technology for transporting green hydrogen from renewable resource-rich countries to other countries. Green hydrogen is produced in areas with abundant renewable energy resources, and green hydrogen is transported to other areas in the form of ammonia. In the destination, the ammonia is decomposed into hydrogen. The hydrogen should be purified to high purity since its applications as an energy source or as a chemical stock require high purity of hydrogen.

To produce high purity of hydrogen from the transported green ammonia, NH_3 decomposition, NH_3 separator, and H_2 purification processes are required (Figure 1). First, the NH_3 decomposition reactor decomposes NH_3 into N_2 and H_2. The trace amount of unreacted NH_3 is assumed to be completely removed in the NH_3 separator. The mixture of N_2 and H_2 gases is used as the feed into the membrane separation process. The process conditions are set at a temperature of 313.15 K and a pressure of 8 bar. The ratio of hydrogen to nitrogen (H_2:N_2) in the composition was assumed to be 3:1, as in NH_3. In this study, a co-current membrane separator is used, assuming the same flow directions in the feed and permeate sides. Other required parameters are listed in Table 1.

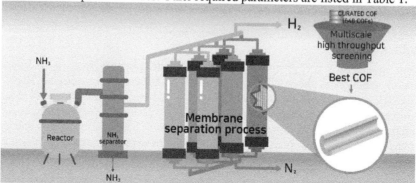

Figure 1. The schematic of integrated NH_3 decomposition and H_2 purification

Table 1. Model parameters and simulation conditions

Definition	Symbol	Value (unit)
Feed flowrate	$F_{f,0}$	3.02 mol/s
Feed composition, $H_2 : N_2$	$x_{H_2,0}, x_{N_2,0}$	0.75: 0.25
Membrane inner diameter	D_i	2×10^{-3} m
Membrane outer diameter	D_o	2.5×10^{-3} m
Membrane module diameter	D_m	0.1 m
Number of fibers	N	600
Fiber length	L	0.6 m
Mass transfer coefficient	k	1×10^{-4} m/s

3. Multiscale membrane screening method

Figure 2 illustrates the steps involved in the screening method used in this study to discover the membranes with the highest process performances. The proposed multiscale HTS method consists of structural analysis, molecular simulation, process simulation, and results analysis under a wide range of operating conditions (i.e., assess the H_2 recovery and purity of each component and compare the different membranes). To select the best-performing COF membranes, the proposed methodology is applied to the CURATED COF database. Each step of the methodology is explained in the following section.

3.1. Structural properties analysis

This step involves screening work based on the structural property. For gas molecules to permeate, the pore limiting diameter (PLD), which is the smallest pore in the nanoporous material, must be greater than the gas molecule diameter. The COFs with a PLD larger than 3.31 Å, the diameter of the N atom, were chosen because this study requires that H_2 and N_2 be permeable. The PLDs were calculated using the Zeo++ software (ver. 0.3).

3.2. Molecular simulation

In this step, the Grand Canonical Monte Carlo (GCMC) and molecular dynamic (MD) simulations were performed to acquire the parameters for membrane permeance calculations. Permeance, defined by the ratio of membrane permeability (P) to thickness (d), is an essential parameter to evaluate the process performance. Permeance is obtained using the following equation:

$$a_i = \frac{P}{d} = \frac{q_i \cdot \mathcal{D}_i}{p_i \cdot \rho \cdot d} \tag{1}$$

where q, $\mathcal{D}$, p, and ρ denote gas uptake, self-diffusion coefficient, partial pressure, and density of COF, respectively. The subscript i denotes each gas component. Parameters q and $\mathcal{D}$ are calculated from the GCMC and MD simulations, respectively. The simulations were conducted using the RASPA 2.0 package.

The N_2 and H_2 partial pressures were set to 2 and 6 bar, respectively, and the temperature at 313.15 K for the GCMC simulations. The GCMC simulation used 10,000 and 20,000 cycles for equilibration and ensemble averages, respectively. The van der Waals (vdW) interaction, which was modeled using the Lennard–Jones (LJ) 12-6 type equation, was used to approximate the interaction between atoms in the systems (Altintas and Keskin, 2019). The LJ parameters for N_2 and H_2 molecules were obtained from the transferable potential for the phase equilibria (TraPPE) force field. A vdW cutoff distance of 12.8 Å was used to truncate the vdW interactions between atoms.

MD simulations were performed to calculate $\mathcal{D}$ for both H_2 and N_2 (Azar et al., 2019). Thirty molecules for each gas component were simulated in infinite dilution. The number of initialization, production, and total cycles was set to 1000, 10,000, and 1,000,000,

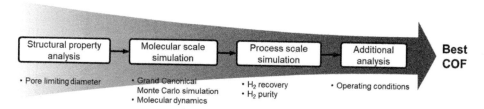

Figure 2. The multiscale high-throughput screening method

respectively. MD simulations were performed in the NVT ensemble with a time step of 0.5 fs. The mean square displacement of gas molecules was used to calculate $\mathcal{D}_i$.

3.3. Process simulation

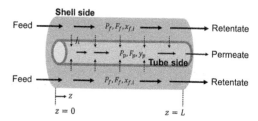

Figure 3. An illustration of a hollow fiber membrane separation module

Using the permeance obtained in the previous section, the H_2 recovery was calculated through process simulation in this step to evaluate each COF membrane. To simulate the membrane separation process, the system is configured as shown in Figure 3. P, F, and y represent pressure, flowrate, and mole fraction, respectively. The subscripts f, p, and i represent the feed, permeate side, and gas component, respectively. The partial pressure difference between the feed and permeate sides, which is calculated using Eq. (2), determines the flux of gas passing through the membrane.

$$J_i = a_i\left(P_{f,i} - P_{p,i}\right) \qquad (2) \qquad P_{j,i} = x_{j,i}P_j \quad \text{for } j = f \text{ or } p \qquad (3)$$

where J denotes the permeate flux in mol/(m²·s), a denotes the membrane permeance in mol/(Pa·m²·s) derived from the molecular simulations, and P denotes the partial pressure in Pa. Partial pressure is calculated using the total pressure and molar fraction of each component, as shown in Eq. (3), where $x_{j,i}$ denotes molar fraction of component i in j stream. The component balances for each feed and permeate side are expressed as follows:

$$\frac{dF_{f,i}}{dz} = -J_i W_{int} \qquad (4) \qquad \frac{dF_{p,i}}{dz} = J_i W_{int} \qquad (5)$$

where W_{int} denotes the interfacial membrane width in m, calculated using $W_{int} = \pi D_o N$. The detailed calculation method is described in the literature (Gu, 2022). The flowrate and pressure for each side and component are derived from solving the above ordinary differential equation along with z-axis. Then, the H_2 purity and recovery are defined as follows:

$$Purity = \frac{F_{p,H_2}}{\sum F_{p,i}} \times 100 \ (\%) \qquad (6) \qquad Recovery = \frac{F_{p,H_2}}{F_{f,H_2}} \times 100 \ (\%) \qquad (7)$$

The H_2 product must be at least 97% pure to be used as an energy source; the higher the recovery, the better. A parametric study with varied operating conditions was carried out to identify the optimal conditions for maximized H_2 recovery while satisfying the purity standard.

4. Results and discussion

4.1. Multiscale high-throughput COF screening results

In this section, the results of each screening step are reported. Figure 4(a) presents the PLD distribution of the 648 COFs in the CURATED COF database, where the PLD threshold of 3.31 Å is indicated by a red dashed line. Among the target COFs, 16 COFs were identified to have PLD values below the threshold. GCMC, MD, and process simulations were performed for the remaining 632 COFs.

The process simulation results are displayed in Figure 4(b). While most COF membranes can produce H_2 with a high purity greater than or equal to 80%, the H_2 recovery varies widely. The red dots in the graph represent the top 20 COFs that achieved the highest recovery with H_2 purity above the 97% threshold.

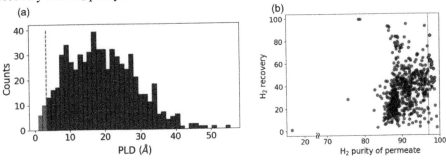

Figure 4. Pore limiting diameter distribution of CURATED COFs database (a), and H_2 purity and recovery results from process simulation (b).

4.2. Analysis of membrane module performance

This section analyzes the influence of feed flow rate on the membrane process performances, which reveals the correlation between H_2 recovery and H_2 purity of the 20 selected COFs. Figure 5(a) displays the variations in H_2 purity and H_2 recovery for varied feed flowrates into each fiber. It is shown that the recovery decreases with increasing feed flowrate, while the purity is higher at a higher flowrate. In other words, by increasing the feed flowrate, the purity constraint can be met at the expense of recovery. When each COF was set to have an H_2 purity of 97% by adjusting the feed flowrate individually, Tp-DBD turned out to achieve the highest H_2 recovery (synthesized by 2,5-diamino-1,4-

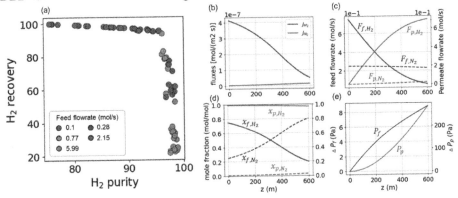

Figure 5. Process simulation results with different feed flowrate and (a) the spatial profiles of fluxes, (b) flowrates, (c) mole fractions (d) pressure drop, and (e) in the feed and permeate sides in the module

benzenediol dihydrochloride and 4,6-triformylphloroglucinol) (Cui et al., 2021). The spatial profiles of the state variables are provided in Figure 5(b)–5(e). Therefore, Tp-DBD COF, which demonstrated an increase in H_2 recovery (46.54% to 92.11%) with a change in feed flowrate (3.02mol/s to 1.00 mol/s), is the highest-performing COF.

5. Conclusion

This study demonstrated the use of the multiscale HTS method to select the best membrane material for the membrane separation process for green hydrogen purification. The proposed multiscale HTS method, consisting of structural analysis, GCMC simulation, MD simulation, and process scale simulation, is used to evaluate the process performance of each membrane. The method was applied to the CURATED COF database to identify the top 20 COFs. By analyzing the H_2 purity and H_2 recovery for different feed flowrates, the COF membrane with the highest recovery was determined. The parametric study with varied feed flowrate demonstrated the trade-off relation between H_2 recovery and purity. The feed flowrate was reduced from 3.02 mol/s to 1.00 mol/s, yielding the highest performance (H_2 recovery of 92.11%).

Acknowledgement

This work was supported by the National Research Foundation of Korea(NRF) grant funded by the Korea government(MSIT) (No. 2021R1C1C2011236) and by the Korean Institute of Industrial Technology within the framework of the following projects: "Development and application of carbon-neutral engineering platform based on carbon emission database and prediction model [grant number KM-22-0348]".

References

Ahmad, F., Lau, K.K., Shariff, A.M., Murshid, G., 2012. Process simulation and optimal design of membrane separation system for CO 2 capture from natural gas. Computers and Chemical Engineering 36, 119–128. https://doi.org/10.1016/j.compchemeng.2011.08.002

Altintas, C., Keskin, S., 2019. Molecular Simulations of MOF Membranes and Performance Predictions of MOF/Polymer Mixed Matrix Membranes for CO 2 /CH 4 Separations. ACS Sustainable Chemistry and Engineering 7, 2739–2750. https://doi.org/10.1021/acssuschemeng.8b05832

Aydin, S., Altintas, C., Keskin, S., 2022. High-Throughput Screening of COF Membranes and COF / Polymer MMMs for Helium Separation and Hydrogen Puri fi cation. https://doi.org/10.1021/acsami.2c04016

Azar, A.N.V., Velioglu, S., Keskin, S., 2019. Large-Scale Computational Screening of Metal Organic Framework (MOF) Membranes and MOF-Based Polymer Membranes for H2/N2 Separations. ACS Sustainable Chemistry and Engineering 7, 9525–9536. https://doi.org/10.1021/acssuschemeng.9b01020

Cui, W., Zhang, C.-R., Xu, R.-H., Chen, X.-R., Yan, R.-H., Jiang, W., Liang, R.-P., Qiu, J.-D., 2021. Low Band Gap Benzoxazole-Linked Covalent Organic Frameworks for Photo-Enhanced Targeted Uranium Recovery. Small 17, 2006882.

Gu, B., 2022. Mathematical Modelling and Simulation of CO2 Removal from Natural Gas Using Hollow Fibre Membrane Modules. Korean Chemical Engineering Research 60, 51–61. https://doi.org/10.9713/kcer.2022.60.1.51

Gulbalkan, H.C., Haslak, Z.P., Altintas, C., Uzun, A., Keskin, S., 2022. Assessing CH4/N2 separation potential of MOFs, COFs, IL/MOF, MOF/Polymer, and COF/Polymer composites. Chemical Engineering Journal 428, 131239. https://doi.org/10.1016/j.cej.2021.131239

Morlanés, N., Katikaneni, S.P., Paglieri, S.N., Harale, A., Solami, B., Sarathy, S.M., Gascon, J., 2021. A technological roadmap to the ammonia energy economy: Current state and missing technologies. Chemical Engineering Journal 408. https://doi.org/10.1016/j.cej.2020.127310

Antonis Kokossis, Michael C. Georgiadis, Efstratios N. Pistikopoulos (Eds.)
PROCEEDINGS OF THE 33rd European Symposium on Computer Aided Process Engineering
(ESCAPE33), June 18-21, 2023, Athens, Greece

Discrete Element Simulation and Economics of Mechanochemical Grinding of Plastic Waste at an Industrial Scale

Elisavet Anglou[a], Yuchen Chang[a], Arvind Ganesan[a], Sankar Nair[a,b], Carsten Sievers[a,b], Fani Boukouvala[a]

[a]Department of Chemical and Biomolecular Engineering, Georgia Institute of Technology, Atlanta, GA 30332 USA

[b]Renewable Bioproducts Institute, Georgia Institute of Technology, Atlanta, Georgia 30332, United States

Abstract

Efficient and sustainable chemical recycling pathways for plastics are vital for addressing the negative environmental impacts associated with their end-of-life management. Mechanochemical depolymerization in ball mill reactors is a new promising route to achieve solid-state conversion of polymers to monomers, without the need for additional solvents. Physics-based models that accurately describe the reactor system are necessary for process design, scaling up, and reducing energy consumption. Motivated by this, a Discrete Element Method (DEM) model is developed to investigate the ball milling process at laboratory and industrial scales. The lab-scale model is calibrated and validated with data extracted from videos using computer vision tools. Finally, scaled-up ball mill designs capable of depolymerizing varying feeds of PET waste were simulated, and their capital and operating costs are estimated to assess the economic potential of this route.

Keywords: ball milling, plastic recycling, discrete-element-method, technoeconomics

1. Motivation and background

Traditional economies rely on linear production processes in which raw materials are manufactured into useful products and later discarded as waste. However, this economic model places a significant strain on natural resources and the environment, particularly in the case of single-use plastics. In 2018, 360 million tons of plastics were produced globally, with 80% ending in landfills or in the sea, whereas only 10% were recycled (Tricker et al., 2022). To alleviate the environmental stress on natural resource degradation, there is increasing interest in transitioning to a circular economic model in which waste materials, such as plastics, will be recycled back into the economy.

Plastic recycling methods can be categorized into mechanical and chemical pathways. Most recycling infrastructures currently focus on mechanical processes in which waste is physically reshaped into new products from a melt. However, mechanical and thermal stresses acting on the polymer melt during processing degrade the integrity of the plastic product; therefore, each plastic product can only be recycled a limited number of times, and only if mixed with large quantities of virgin polymers (Tricker et al., 2022). A promising alternative is the chemical recycling of polymers, which bypasses the limitation of material degradation entirely, by converting polymers directly to monomeric molecules through depolymerization. Common chemical recycling pathways such as pyrolysis and solvolysis involve high energy consumption or the use of solvents, therefore hindering their economic viability. The development of alternative depolymerization processes for plastic waste is crucial for implementing sustainable practices and reducing negative environmental effects. In a previous study (Tricker et al., 2022), it was shown that mechanochemical depolymerization in ball mill reactors is a promising route for converting solid polymers into monomers without the need for additional solvents. Mechanochemical depolymerization pathways utilize the mechanical energy supplied by collisions between grinding bodies (balls and walls) of a ball mill to drive the reaction between solid reactant particles. Thus, accurately simulating the movement and interactions between these entities can provide crucial information for explaining mechanisms of mechanochemical depolymerization and for optimizing process design.

Two modeling frameworks have predominantly been used to simulate ball mill systems: semi-empirical population balance and mechanistic Discrete Element Method (DEM) models. DEM models for ball milling have received increasing attention owing to their ability to describe the kinematics of moving entities and the corresponding energy involved in their collisions during ball milling. Additionally, ball milling operations at the industrial scale can be energy-intensive; hence, even a marginal improvement in efficiency can lead to a significant reduction in expenses. For instance, approximately 110 kWh of electric energy is consumed during the production of 1 ton of cement, of which 70% is required for comminution, whereas only 1-5% is explicitly used for particle breakage; the remaining energy is wasted (Boemer & Ponthot, 2017; Tavares, 2017).

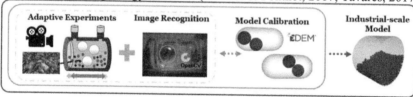

Figure 1: The main components of this study include: (a) experimental design and machine learning computer vision tools for the calibration of a lab-scale DEM model, and (b) DEM simulations of scaled-up ball mill configurations for various operating settings to estimate capital and operating costs

In this work, a DEM model was developed to investigate the depolymerization of poly(ethylene terephthalate) (PET) powder. First, high-speed video recordings of a lab-scale ball mill were obtained at operating conditions that allowed for quantitative PET depolymerization in corresponding experimental studies in the same mill (Tricker et al., 2022), and a computer vision algorithm was used to track the moving balls in the reactor. The extracted data were used to parameterize the DEM model and predict the kinematic interactions of moving entities in the mill. Once the mechanistic model parameters were optimized, the geometry of an industrial-scale ball mill reactor was created, and various operating settings were simulated. The resulting DEM simulation was used to investigate the tradeoffs between achieved PET yields and energy requirements at the industrial scale. Finally, capital, and operating costs of the process were calculated for various scenarios of waste feeds, using data obtained from US state levels of plastic waste collection.

2. Methods

2.1. High-fidelity model for the laboratory ball mill reactor

PET depolymerization experiments were performed in a Retch MM400 vibratory ball mill with stainless steel grinding bodies. PET in powder form was milled with a stoichiometric amount of sodium hydroxide (NaOH) to form ethylene glycol (EG) and disodium terephthalate salt (Na_2TPA). Details regarding experimental conditions and results have been reported in (Tricker et al., 2022). A transparent PMMA milling vessel identical to that used in the depolymerization experiments was manufactured and used in motion-tracking experiments, but without reactants (PET + NaOH) present. The mill's operation at various conditions (ball numbers and sizes, milling frequencies) was filmed using a high-speed camera (2134 fps) to study the collision kinematics between balls and the vessel wall. A Python script based on the OpenCV computer vision library was used to analyze the video frames and track the position of the ball(s). The coordinates of the ball(s) in space and time were identified and used to calculate their velocity, which is defined as the change in position between two consecutive frames.

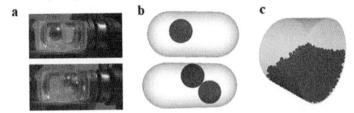

Figure 2: (a) Experimental setup including a vibratory 25mL reactor and stainless-steel balls (b) Replication of the experimental setup in DEM simulation with one and two balls (c) The geometry of the industrial-scale reactor as simulated in the DEM software

Next, the geometry of the laboratory setup was developed in Solidworks while the DEM model was developed in EDEM. The Hertz-Mindlin contact model was applied, as it is most appropriate for non-cohesive spherical shapes (Metta, Ierapetritou, & Ramachandran, 2018). Typical steel properties were used as inputs to the DEM model (density 7900 kg/m^3, Poisson's ratio 0.3, coefficient of restitution 0.5, shear modulus 77 GPa). The coefficient of static and rolling friction are taken as 0.5 and 0.01 respectively based on (Metta et al., 2018). For PMMA-steel interactions, a grid search of the coefficient of restitution and the friction parameters was performed, and optimal values that match the velocity and the number of collisions between the simulation and the motion-tracking experiments were identified. The laboratory ball milling simulation was validated based on the recorded experiments under different operating conditions. Once the lab-scale simulation is validated by experiments, the identified parameters that govern ball-to-ball and ball-to-wall interactions can be used to simulate a scaled-up geometry.

2.2. Scaling up and technoeconomic analysis of ball mill reactor

A typical industrial ball milling setup comprises a rotating cylindrical vessel filled with balls and PET particles. Electric energy is used to satisfy the energy requirements necessary to drive the electric motor that rotates the mill and grinding media. To establish the operating costs, a new geometry and scale of DEM simulations was developed. A rotating cylindrical vessel was designed, and the simulation was developed (Figure 2c) using the validated material and contact parameters. The resulting torque requirements loads were extracted from the DEM simulation, and the operating costs were evaluated based on Equation (1) for 7920 hours of operation annually.

$$\text{Electricity Cost} = \text{Price} \times \text{Power} = Price \times (\text{T}_M \times \omega_M) \qquad \text{Equation (1)}$$
$$\text{C}_{\text{Ball Mill}} = 64640 \times \text{W}^{0.64} \qquad \text{Equation (2)}$$
$$\text{Y}_{\text{PET}}(\%) = 6.38 \, \text{BPR} - 46.35 \text{ for } 10 \leq \text{BPR} \leq 23 \qquad \text{Equation (3)}$$

Here, T_M is the torque associated with the rotation of the reactor vessel while ω_M is the mill's angular velocity, which in this work is equals to 15 rpm. The angular velocity was chosen such that it lies between 60-80% of the critical speed depending on the reactor volume (Wang, Yang, & Yu, 2012). The fill level of the ball mill was kept constant at 30 %, and the radius of the balls was set to 80 mm for all cases. An electricity cost of $0.07/kW-h was used while the purchase cost of the ball mill equipment was estimated based on the feed flowrate (W [ton/hr]) according to Equation (2) (Seider, Seader, Lewin, & Widagdo, 2004), and then adjusted to the 2022 dollar value via the CEPCI Index. According to our previous kinetics study on the PET depolymerization reaction (Tricker et al., 2022) for the lab-scale ball milling system, monomer yield follows a linear relationship (Equation (3)) with the ball-to-powder mass ratio (BPR) parameter, that is, the ratio of the ball mass to the mass of the reactants. Assuming that the same BPR values can be maintained at the industrial scale, the corresponding conversion, and costs can be calculated. Higher BPR ratios results in higher PET monomer yields, and full conversion is achieved when BPR = 23 with a 20 min reaction time (Tricker et al., 2022). The BPR parameter was used to scale-up the reactor and to evaluate the corresponding conversion, number of balls, and reactor volume for a certain PET waste feed.

3. Results and Discussion

3.1. Laboratory-scale model validation

The average ball velocity is compared in Figure 3a, which reveals that the values are in near-perfect agreement for the different operating conditions (vibration frequencies, diameters) tested and can thus validate the lab-scale model. Stainless-steel balls with diameters equal to 17.6 mm and 20.6 mm were used for the model validation while the milling frequency ranged from 22.5 to 30 Hz. Once the DEM parameters were estimated, the simulation can be executed for different conditions and reactor sizes, and the results can be used to further analyze the ball milling system.

3.2. Simulations at the industrial scale

DEM simulations were used to define the scaling-up procedure for PET waste depolymerization in ball mills and to evaluate the associated energy requirements. To illustrate the resulting capital and operating costs of plastic waste depolymerization at the industrial scale, two separate cases were considered.

3.2.1. Case-study 1: Influence of BPR to the operating costs

The BPR ratio significantly influenced the extent of PET depolymerization. To this end, the first case study depicts how the operating costs change with respect to the BPR value for the same feed of 1000 kg/hr PET waste (+400 kg/hr NaOH). The reactor volume and number of balls required to depolymerize the PET waste were calculated based on the BPR parameter.

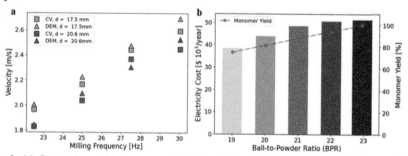

Figure 3: (a) Comparison of the average velocities of the stainless-steel ball at different vibration frequencies (22.5 – 30 Hz) extracted from the DEM simulations (triangles) and the recorded ball mill experiments (squares), (b) Comparison of annual electricity cost and achieved monomer yield for reactors processing 1000 kg/hr of PET (+400 kg/hr NaOH) at different BPR ratios.

The results of this analysis are shown in Figure 3b, which illustrates how the electricity cost changes with different BPR values. For higher BPRs, more balls are added inside the mill to grind the same amount of waste, resulting in increased power requirements for the mill rotation. More specifically, to achieve a 75% monomer (Na_2TPA) yield (BPR=19), $1.62/tonne_{PET} are spent on electricity, whereas for full depolymerization (BPR=22.9) the cost is increased to $2.16/tonne_{PET} for 20 min milling time (Tricker et al., 2022). The purchase cost of the ball mill vessel was estimated to be $119,719 and was constant for all BPR combinations (assuming a negligible cost for the additional grinding balls). However, as PET conversion is reduced, subsequent unit operations will be necessary to separate and recycle unreacted PET waste which will increase capital and operating expenditures.

3.2.2. Case study 2: Influence of PET waste feedstock flowrate

The second case study illustrates the dependence of the total cost of ball milling to the mass flow rate of PET waste. Different PET waste recycling feed flow rates were evaluated, and the corresponding costs were estimated for a BPR value of 20 resulting in 81.2 % of PET converted to monomers. The PET waste for different US states were used as input to the DEM simulation based on recycling quantities for 2018 as described in the report of the National Association for PET Container Resources (*NAPCOR 2018*). The study region encompasses the South Atlantic region of the US, which in turn comprises nine states (Florida, Georgia, North Carolina, Virginia, Maryland, South Carolina, West Virginia, and District of Columbia). It is assumed that the PET recycling quantities are equally distributed to each state in the region based on the corresponding population for the same year. The resulting PET waste flowrates for each state are used as inputs to the DEM simulation using stoichiometric NaOH feed.

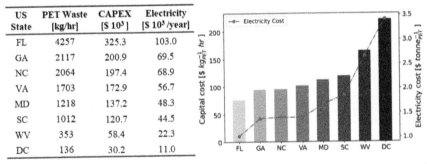

US State	PET Waste [kg/hr]	CAPEX [$ 10³]	Electricity [$ 10³ /year]
FL	4257	325.3	103.0
GA	2117	200.9	69.5
NC	2064	197.4	68.9
VA	1703	172.9	56.7
MD	1218	137.2	48.3
SC	1012	120.7	44.5
WV	353	58.4	22.3
DC	136	30.2	11.0

Figure 4: PET waste collected at each state in 2018 scaled based on population, and the resulting capital and operating costs. The graph illustrates the capital and operating costs per kg of PET waste fed in the ball mill reactor. The bars denote the capital cost per feed flow rate in $ kg$_{PET}^{-1}$hr, whereas the electricity cost per tonne is shown on the secondary axis for 20 min operation (red dotted line).

The results of this analysis are shown in Figure 4, which illustrates how the required expenditure per unit feed changes for different PET waste feeds. As expected, both capital and operating costs increased with feed flow rate, because maintaining the same BPR value requires larger numbers of balls and a larger reactor volume, which would require greater motor power to achieve depolymerization conditions for that entire mass of grinding bodies. In contrast, per unit costs decreases as the feed rate increases highlighting the benefits of economy of scale.

The economic potential of the mechanochemical route based solely on raw material (PET bale: $150/tonne, NaOH: $608/tonne, H_2SO_4: $96/tonne) and product (vTPA: $1143/tonne, EG: $961/tonne) prices (Singh et al., 2021) was evaluated at $978/tonne of

TPA, which is an indication of the upper bound of the potential profits. This value, in combination with the low electricity costs that were estimated in both case studies at $2-3/tonne$_{PET}$, reveals that there is still a significant profit margin to account for mechanical pretreatment, required to grind PET waste to micro sizes, as well as downstream processes to recover the products of mechanochemical hydrolysis of PET waste that can be capital- or energy-intensive, i.e., rTPA crystallization, EG distillation, salt recovery (Singh et al., 2021).

4. Conclusions

In this work, we built a DEM simulation required to design and cost PET depolymerization via ball milling for a variety of feedstock flow rates and operating settings. Two case studies demonstrate the dependence of the operating and capital costs on the achieved PET conversion and feed flow rates. The results indicate that this can be viable process economically, especially for states that recycle more PET where there would be a benefit from economy of scale. Also, there are economic trade-offs to be considered in the future with respect to more costly depolymerization at high conversions, versus partial depolymerization and design of subsequent separation/purification steps and recycling streams. In future work, a more detailed technoeconomic analysis of plastic waste depolymerization via mechanochemical reactions will be performed to account for pre-treatment costs to shred plastic waste into powder form, as well as downstream treatment to purify the mechanocatalysis products. This analysis will be critical for comparing this technology with alternative recycling routes and aid policymakers and industry stakeholders in making informed decisions regarding its overall potential.

5. Acknowledgments

This work is supported by Kolon Industries, Inc., through the Kolon Center for Lifestyle Innovation at Georgia Tech and the U.S. National Science Foundation Emerging Frontiers in Research and Innovation program under grant 2028998.

References

Boemer, D., & Ponthot, J.-P. (2017). DEM modeling of ball mills with experimental validation: influence of contact parameters on charge motion and power draw. *Computational Particle Mechanics, 4*(1), 53-67.

Metta, N., Ierapetritou, M., & Ramachandran, R. (2018). A multiscale DEM-PBM approach for a continuous comilling process using a mechanistically developed breakage kernel. *Chemical Engineering Science, 178*, 211-221.

Postconsumer PET Recycling Activity in 2018. Retrieved from https://napcor.com/wp-content/uploads/2021/07/Postconsumer-PET-Recycling-Activity-in-2018.pdf

Seider, W. D., Seader, J., Lewin, D. R., & Widagdo, S. (2004). Product and process design principles: synthesis. *Analysis and Evaluation, 4.*

Singh, A., Rorrer, N. A., Nicholson, S. R., Erickson, E., DesVeaux, J. S., Avelino, A. F., . . . Avery, G. (2021). Techno-economic, life-cycle, and socioeconomic impact analysis of enzymatic recycling of poly (ethylene terephthalate). *Joule, 5*(9), 2479-2503.

Tavares, L. M. (2017). A Review of Advanced Ball Mill Modelling. *KONA Powder and Particle Journal, 34*(0), 106-124. doi:10.14356/kona.2017015

Tricker, A. W., Osibo, A. A., Chang, Y., Kang, J. X., Ganesan, A., Anglou, E., . . . Sievers, C. (2022). Stages and Kinetics of Mechanochemical Depolymerization of Poly (ethylene terephthalate) with Sodium Hydroxide. *ACS Sustainable Chemistry & Engineering.*

Wang, M., Yang, R., & Yu, A. (2012). DEM investigation of energy distribution and particle breakage in tumbling ball mills. *Powder technology, 223*, 83-91.

Antonis Kokossis, Michael C. Georgiadis, Efstratios N. Pistikopoulos (Eds.)
PROCEEDINGS OF THE 33rd European Symposium on Computer Aided Process Engineering
(ESCAPE33), June 18-21, 2023, Athens, Greece

Multi-scale Modelling and Experimental Investigation of Hydrogen Sulphide Thermal Decomposition

Anna Nova,[a,b] Francesco Negri,[a,c] Flavio Manenti[a,b]

[a]Politecnico di Milano, CMIC Dept. "Giulio Natta", Piazza Leonardo da Vinci 32, Milan 20133, Italy
[b]Consorzio Interuniversitario Nazionale per la Scienza e Tecnologia dei Materiali, Via Giusti 9, 50121 Firenze, Italy
[c]Itelyum Regeneration spa, Via Tavernelle 19, Pieve Fissiraga 26854, Lodi, Italy
*flavio.manenti@polimi.it

Abstract

The splitting of hydrogen sulphide into hydrogen and elemental sulphur results to be a promising route for the treatment of this dangerous waste gas and, at the same time, for the sustainable production of hydrogen. To reach relevant production rates a comprehensive approach should be applied to the modelling of the process. This work investigates how the thermal decomposition of hydrogen sulphide can be described at different scales: the kinetic scale, the reactor scale, and the chemical process scale. In particular, a detailed kinetics is implemented into DSmoke, a reactor simulation suite for complex kinetics and non-ideal reactive systems. The obtained reactor model is validated through an experimental lab campaign. At the process scale, the splitting step is integrated into a complete plant design (considering a 10 t/h H_2S feed, the mass flow of H_2 produced from the simulated plant is equal to 543.8 kg/h, with a production cost of 2.39 $/kg).

Keywords: hydrogen sulphide, H_2S decomposition, multi-scale model, orange hydrogen.

1. Introduction

Hydrogen sulphide (H_2S) is a compound that is naturally present in many resources (both fossil and renewable) and is also artificially produced in some industrial processes. The main source are the processes for the treatment and desulphurization of oil and gas. Since hydrogen sulphide is pollutant, corrosive and flammable, it cannot be released in the environment, and it must be conveniently processed.

At present, H_2S is usually sent to sulphur recovery units (e.g., Claus process), which perform a partial oxidative conversion able to transform H_2S into sulphur and water via sulphur dioxide (SO_2) as intermediate and, at the same time, to recover thermal energy. The process leads to the production of elemental sulphur, a more environmentally friendly and less hazardous compound than H_2S. In this case hydrogen is not recovered and is completely wasted in the form of water. An interesting alternative could be its conversion into sulphur and hydrogen (Burra et al., 2018), according to the overall splitting reaction ($\Delta H_r^0 = 84.9$ kJ/mol):

$$H_2S \rightarrow H_2 + \frac{1}{2}S_2 \tag{1}$$

The production of hydrogen results in great interest since this compound is a highly valuable product with a wide range of applications, especially in the energy and chemicals industry. Hydrogen is usually identified through different colours, according to the production process. Black refers to coal gasification, grey to natural gas reforming, turquoise to natural gas pyrolysis, blue to reforming/gasification + CCUS, green to electrolysis of water with renewable electricity, purple to electrolysis of water with electricity from nuclear plants and yellow to electrolysis of water with mixed-origin grid energy). H_2S thermal splitting represents a promising path to produce relevant amounts of hydrogen. This H_2 has not been yet classified in the chromatic scale, therefore, it will be named "orange hydrogen". Obtaining this product from the H_2S generated in hydrotreating processes may enable the recovery and recycling of H_2, significantly reducing the need to acquire new H_2 sources and, in general, reducing costs and environmental impacts. In addition, H_2 has the potential to be integrated into the fuel cycle for energy storage carriers, leading to the complete suppression of air pollutants emissions.

A variety of splitting techniques aimed at producing hydrogen and sulphur from H_2S, have been studied. To date, the developed H_2S splitting methodologies include thermal (catalytic/membrane-assisted), photocatalytic, non-thermal plasma, thermochemical, electrolysis and biological processes (De Crisci et al., 2019). At 1200 °C, 1 atm and with a steady gas flow rate of 50 mL/min, H_2S conversion was 35.6 % (97.5 % of its thermodynamic limit). To decrease the high-temperature requirements, catalytic processes have been investigated. Low-temperature processes have been performed with the use of MoS_2-based catalysts, resulting in 95 % conversion to H_2 at 500 °C. Other materials have also been studied, including, for example, perovskites (Burra et al., 2018), alumina oxides (Cha et al., 2022) and vanadium-decorated titanium carbides (Zhou et al., 2021). Photocatalytic reactions have also been considered. TiO_2 is the most studied semiconductor because of its nontoxicity, low cost, and high stability (Uesugi et al., 2022).

Nevertheless, H_2 generation from H_2S has not yet been scaled up to bulk production due to economic and technical reasons (Gupta et al., 2016). Some of the open issues are related to the complete capture of pure H_2S, its energy-intensive regeneration step, the management of sulphur vapours, and the H_2S effective conversion into hydrogen and elemental sulphur. The problem requires a multi-scale approach, able to model the process from the kinetic scale, to the reactor scale, to the complete process scale. In this work, the hydrogen sulphide thermal splitting is analysed at different scale levels, proposing in this way a comprehensive approach to the study of a process.

2. Materials and methods

2.1. Experimental set up

The study of H_2S splitting is first approached by conducting an experimental campaign. An ex-novo laboratory apparatus has been set up. It consists of an open circuit system fed by a gaseous H_2S/N_2 mixture with a H_2S concentration up to 8 vol%. The reactor is made of quartz and is designed according to the ideal Plug Flow Reactor (PFR) model. To ensure the adequate thermal conditions required by the process, the reactor is placed inside a tubular oven. Upon exiting the reactor, the sulphur is condensed thanks to the bubbling of the reaction gases into a flask containing distilled water. The gaseous stream is then sent to a stilling chamber to remove all the entrained droplets. A filter ensures then the complete removal of any solid particle from the stream. The gases are finally analysed with a Micro GC (Micro Gas Chromatograph).

2.2. DSmoke simulation

DSmoke is a reactor network simulation suite for complex kinetics and non-ideal reactive systems, developed by the research group at Politecnico di Milano. The software allows the modelling of sequences of reactors (in series or in parallel), possibly coupled with mixers and/or splitters. It is composed by a chemical interpreter, that processes the kinetic mechanism and the thermodynamic properties of the species, and by a reactor model developer.

The lab reactor employed in the experimental campaign has been modelled in DSmoke with a network of three reactors in series. This sequence is composed by a first isothermal plug flow reactor, corresponding to the homogeneously heated length of the quartz tube inside the oven, a second plug flow reactor, displaying a linear temperature profile (going from the temperature value inside the oven to room temperature) and corresponding to the part of the tube at the exit of the oven, where the temperature drops rapidly, and finally a third isothermal plug flow reactor, simply corresponding to the remaining length of the tube. Table 1 shows the details of the three reactors composing the series.

Table 1 - Details of the three reactors composing the DSmoke model sequence.

	Length [mm]	Diameter [mm]	Volume [mm³]
Reactor 1	180	22	68389.2
Reactor 2	5	22	1899.7
Reactor 3	355	22	134878.7

The kinetic scheme introduced in the program for the thermal splitting of hydrogen sulphide has been developed by the group in the past years (Manenti et al., 2013). It is a revised model, introducing some modifications on H₂S pyrolysis literature schemes (Binoist et al., 2003). The complete mechanism, validated on literature and industrial data, is presented in Table 2.

Table 2 - Kinetic parameters for H₂S pyrolysis. Rate eq.: $k=A*T^{\beta}*\exp(-E_a/RT)$ [m-kmol-s-K] (Manenti et al., 2013).

n	Reactions	A	β	Ea	n	Reactions	A	β	Ea
R1	S+H+M=SH+M	.62E+17	-.6	.0	R10b	HSS+H=SH+SH	.11E+14	.353	210.0
R2	S+H₂=SH+H	.14E+15	.0	19300	R11	HSS+H=S₂+H₂	.12E+09	1.653	-1105.0
R3	S₂+M=S+S+M	.48E+14	.0	77000	R12	HSS+H=H₂S+S	.44E+14	.000	6326.0
R4	S₂+H+M=HSS+M	.40E+15	2.84	1665	R13	HSS+S=S₂+SH	.42E+07	2.200	-600.0
R5	SH+SH=S₂+H₂	.50E+12	0.0	.0	R14	HSS+SH=H₂S+S₂	.63E+04	3.050	-1105.0
R6	SH+S=S₂+H	.30E+14	0.0	.0	R15	HSS+HSS=HSSH+S₂	.96E+01	3.370	-1672.0
R7	H₂S+M=S+H₂+M	.16E+25	2.613	89100	R16	HSSH+M=SH+SH+M	.14E+16	1.000	57030.0
	N₂/1.5 SO₂/10 H₂O/10				R17	HSSH+H=HSS+H₂	.50E+08	1.933	-1408.0
R8	H₂S+H=SH+H₂	.35E+08	1.94	904	R18	HSSH+H=H2S+SH	.20E+15	.000	.0
R9	H₂S+S=SH+SH	.83E+14	0.0	7400	R19	HSSH+S=HSS+SH	.29E+07	2.310	1204.0
R10a	HSS+H=SH+SH	.97E+08	1.62	-1030	R20	HSSH+SH=HSS+H₂S	.64E+04	2.980	-1480.0

2.3. Process simulation

The detailed process simulation has been developed using Aspen HYSYS V11 suite. As a thermodynamic model, the SRK (Soave-Redlich-Kwong) equation of state has been chosen for the reactive section of the process, while the Aspen HYSYS amine package has been chosen for the separation section. The reactors present in the process are modelled as Gibbs Reactors.

The reaction is highly endothermic and takes place with an increase in the number of moles. So, it is favoured at high temperatures and low pressure. An H₂S mass flow of 10 t/h enters the splitting reactor at 1000 °C and 180 kPa. Sulphur is removed by condensation. Residual sulphur vapours are then hydrogenated to produce H₂S. The obtained gas stream (containing hydrogen in an amount equal to 83 mol% and unreacted H₂S and H₂S produced by the hydrogenation of sulphur vapours in an amount equal to 17 mol%) is treated in an amine scrubbing process employing 2×10^6 kg/h of a DEA (25 wt%) water solution (Jamekhorshid et al., 2021). After a water removal step, the sweet stream contains pure H_2 and can be send to the compression train to reach the market pressure of 200 bar. H_2S is instead recirculated to the splitting reactor. The detailed process flow diagram is reported in Figure 1.

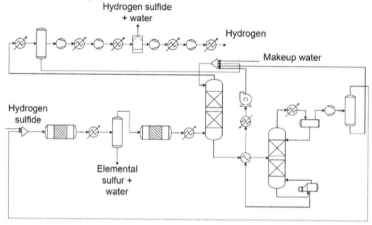

Figure 1 - Process flow diagram for H_2S conversion.

3. Results and discussion

3.1. Experimental results

The reaction has been performed at 1 bar and with a stream of 2 L/h H_2S (1.5 wt%) in N_2 as feed (to ensure a long residence time and therefore a thermodynamic regime of reaction). Tests at different temperatures has been carried out, obtaining the results summarized in Table 3. The conversions achieved are quite in line with the expected thermodynamic conversions (Kaloidas & Papayannakos, 1987). The error is due to the recombination that takes place at the exit of the reactor: above 800°C H_2 reacts with sulphur to give back H_2S.

Table 3 - Experimental H_2 concentrations and conversions.

Temperature [°C]	H_2 concentration [mol%]	Experimental conversion [%]	Conversion at thermodynamic equilibrium [%] (Kaloidas & Papayannakos, 1987)
640	0.06	5	6
740	0.18	14	18
831	0.35	28	31
929	0.54	43	47
1016	0.71	56	61
1107	0.81	64	73

3.2. DSmoke lab reactor simulation

The models realized are simulations of the H_2S splitting reaction performed in the lab reactor. Two temperatures have been considered: 1016 and 1107 °C. The results reported in the previous section, 3.1, are an experimental validation of the models obtained with the software.

In Figure 2 and 3 it is possible to notice that along the length of the reactor the concentration of H_2 increases and reaches a plateau, corresponding to the thermodynamical equilibrium. This condition is reached at a short length of the reactor (equal to 12 mm at 1016 °C and 4 mm at 1107 °C), suggesting that, in the considered operative regime, a more compact reactor could be a convenient design choice. At the point of temperature drop (after a residence time inside the reactor of 30 s), the concentration of H_2 starts to decrease, due to the recombination reaction. The temperature drop takes place in a 5 mm length; therefore, the quench is rapid, and the recombination reaction interrupts, fixing a final value for the H_2 concentration.

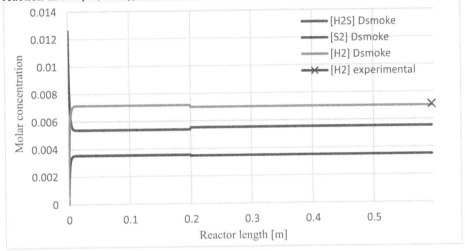

Figure 2 - Molar concentrations of H_2S, S_2, H_2 at 1016 °C.

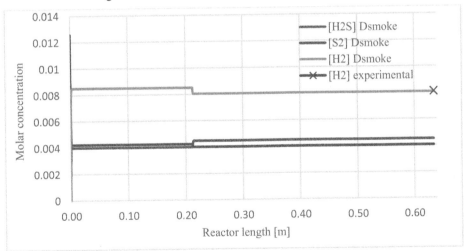

Figure 3 - Molar concentrations of H_2S, S_2, H_2 at 1107 °C.

3.3. Process design

Considering the process scale, a conceptual design for a plant integrating a H_2S splitting step results to be significantly different from that of classical H_2S sequestration and conversion processes. This is due to the high concentration of H_2S that characterises the process streams. In particular, H_2S is present as a pure component in the feed. This results in a difficult separation of the unreacted H_2S and in an extremely energy-intensive regeneration step.

In a simulation with the characteristics described in section 2.3, the production is equal to 8666 kg/h of elemental sulphur and 543.8 kg/h of hydrogen. The process is characterized by a considerable H_2S recycle, consisting of 48340 kg/h. This aspect has a huge impact on the operative costs of the plant, but at the same time makes the complete conversion of H_2S into H_2 and S_2 possible. Water as well can be recirculated inside the plant, exploiting the streams coming from the condensation steps. This allows to reduce the needed amount of makeup water to 110.5 kg/h. The steam consumption, particularly the one related to the regeneration column reboiler, is high, equal to 1.843×10^5 kg/h. A possible point of improvement could be represented by an optimisation able to reduce the energy consumption at the regeneration step.

An economic analysis has also been performed, by calculating the cost of the utilities associated with the treatment of H_2S in the simulated plant. The total OPEX are 1301.18 \$/h. Considering the obtained H_2 productivity of 543.8 kg/h, H_2 production cost is 2.39 \$/kg. In Europe, this price is lower than the average price of H_2 produced through electrolysis (3.6 \$/kg) and is almost equal to the price of H_2 produced from natural gas (EIA, 2019).

4. Conclusions

In this work a multi-scale approach has been applied to the development of a process layout for H_2S thermal splitting, including the kinetics, the reactor modeling and the plant design. As future development, DSmoke will be integrated into Aspen Hysys as CAPE-OPEN tool, in order to better combine the different scales and to optimize the process. H_2S splitting could really contribute to improving the performances of refineries, biorefineries and other industrial plants, since it changes the role of H_2S from waste to commodity.

References

M. Binoist, B. Labégorre, F. Monnet, P. D. Clark, N. I. Dowling, M. Huang, D. Archambault, E. Plasari, P. M. Marquaire, 2003, Kinetic study of the pyrolysis of H_2S, Industrial and Engineering Chemistry Research, 42(17), 3943–3951.

K. R. G. Burra, G. Bassioni, A. K. Gupta, 2018, Catalytic transformation of H_2S for H_2 production, International Journal of Hydrogen Energy, 43(51), 22852–22860.

B. J. Cha, J. Y. Choi, Y. Ji, S. Zhao, S. Y. Kim, S. H. Kim, Y. D. Kim, 2022, Fe-oxide/Al_2O_3 for the enhanced activity of H_2S decomposition under realistic conditions: Mechanistic studies by in-situ DRIFTS and XPS, Chemical Engineering Journal, 443(April), 136459.

A. G. De Crisci, A. Moniri, Y. Xu, 2019, Hydrogen from hydrogen sulfide: towards a more sustainable hydrogen economy, International Journal of Hydrogen Energy, 44(3), 1299–1327.

EIA, 2019, The Future of Hydrogen, Report prepared by the IEA for the G20 Japan, Issue June.

A. K. Gupta, S. Ibrahim, A. Al Shoaibi, 2016, Advances in sulfur chemistry for treatment of acid gases, Progress in Energy and Combustion Science, 54, 65–92.

A. Jamekhorshid, Z. Karimi Davani, A. Salehi, A. Khosravi, 2021, Gas sweetening simulation and its optimization by two typical amine solutions: An industrial case study in Persian Gulf region, Natural Gas Industry B, 8(3), 309–316.

V. E. Kaloidas, N. G. Papayannakos, 1987, Hydrogen production from the decomposition of hydrogen sulphide. Equilibrium studies on the system H_2S/ H_2/Si, (i = 1,...,8) in the gas phase, International Journal of Hydrogen Energy, 12(6), 403–409.

F. Manenti, D. Papasidero, E. Ranzi, 2013, Revised kinetic scheme for thermal furnace of sulfur recovery units, Chemical Engineering Transactions, 32(January), 1285–1290.

Y. Uesugi, H. Nagakawa, M. Nagata, 2022, Highly Efficient Photocatalytic Degradation of Hydrogen Sulfide in the Gas Phase Using Anatase/TiO_2(B) Nanotubes, ACS Omega, 7(14), 11946–11955.

J. Zhou, Z. Ao, T. An, 2021, DFT Study of the Decomposition Mechanism of H_2S on V-Decorated Ti_2CO_2 Single-Atom Catalyst, Wuli Huaxue Xuebao/ Acta Physico - Chimica Sinica, 37(8), 1–9.

Antonis Kokossis, Michael C. Georgiadis, Efstratios N. Pistikopoulos (Eds.)
PROCEEDINGS OF THE 33rd European Symposium on Computer Aided Process Engineering
(ESCAPE33), June 18-21, 2023, Athens, Greece

Reuse and recycling of waste material: design of a pyrolysis-based process

Gabriel Rodríguez Garrido[a], Pablo Marinangeli[b], Jorge A. Ressia[c,d],
Maria A. Volpe[c], Patricia M. Hoch[c]

[a]Instituto Petroquimico Argentino, Avda. Cordoba 629, CABA, Argentina
[b] CERZOS – Departamento de Agronomía – UNS, Bahia Blanca, Argentina.
[c]PLAPIQUI, Chem Eng, Dept., Universidad Nacional del Sur. Bahia Blanca, Argentina
[d] Comision de Investigaciones científicas de Buenos Aires, La Plata, Argentina.
p.hoch@plapiqui.edu.ar

Abstract

This paper addresses the use of pyrolysis to reduce carbon footprint of several waste material as agricultural waste plastic (AWP). The destination of such residues was usually landfills, but there is a high potential for the obtention of high added value products after the pyrolysis process is completed, with the added benefit of reducing the carbon footprint and then accomplishing objectives of circular economy. Argentina´s energetic matrix relies on the use of natural gas, which is abundant, so there is no compromise with the energy demand, however, the production of fuels from waste brings financial and environmental benefits, thus being highly attractive. Pyrolysis is a multifunctional process, and the indirect "avoided" impacts associated with the production of energy that substitute primary one can be considered (Giugliano et al., 2011).
Different paths for the treatment of the pyrolyzed material can be taken, as well as different pretreatment paths. Also, the possibility of mixing the feedstocks is considered, being the composition one of the key variables for the economics of the process. The objective function for the optimization considers environmental aspects as the reduction of CO_2 emissions. The potential carbon footprint emissions of pyrolysis process are calculated and compared to incineration and recycling processes. Results show that pyrolysis rises as the best alternative for reducing the carbon footprint, not only because CO_2 emissions are highly reduced, but also because the products are a fossil fuel replacement when the appropriate paths for pre and post treatment are used. Limitations are related to the composition of the residue to be treated.

Keywords: Pyrolysis, GHG emissions, circular economy

1. Introduction

The management and disposal of agricultural waste plastic (AWP) in the southwest region of Buenos Aires, Argentina, has become an environmental challenge. This waste is mainly composed of empty agrochemical containers and silo bags. As it is relatively easy to separate both sources of residues, silo bags can be physically recycled as they are made of polyethylene, and they do not contain any toxic component. Agrochemical containers, however, cannot be recycled with ease because they are contaminated with the pesticides or fertilizers used and no real solution has been found for the end of life of this waste.
Recently, industry was interested in chemical recycling of waste plastic, mainly to decrease the carbon footprint and to be in line with circular economy (Kruger, 2020).

According to ISO 15270 (2008), pyrolysis is one of the four methods of chemical recycling, which transforms plastics back into basic chemicals by thermal decomposition at elevated temperatures, where usually all the plastic additives and contaminants are also converted back into basic chemicals. The pyrolysis products, liquid and gas, could be used as feedstock in chemical industry substituting conventional fossil-based energy or product, but technologies are still in early stages for industrial scale use. The advantages are the potential to improve the recycling rate of plastics, as well as reducing fossil fuels demand for energy and GHG emissions. All of these is aligned with circular economy principles.

The goal of this study is to analyze pyrolysis of AWP as an end-of-life treatment, comparing it with other popular options, i.e. incineration. The carbon footprint of each option will also be calculated, depending mainly on the temperature of the thermal process if applicable (Gasification, incineration, pyrolysis) but the carbon footprint for disposing the AWP in landfills will also be calculated. For the pyrolysis process the optimal process temperature, the amount of utilities used for heating (e.g. natural gas for the pyrolysis reactor, if the pyrolysis gas results insufficient as fuel for this heating requirement) and for the condensing units will be found solving a nonlinear programming problem. Being the feedstock a residual material from other economic activities the price is very low or can be neglected in some cases. A comparison between two other different treatments for the residues (incineration and mechanical recycling) is presented. Between the main objectives, the calculation of CO_2 emissions and carbon footprint together with energy production and economic aspects was considered. Calculations of emissions were made using the 2006 Intergovernmental Panel on Climate Change (IPCC) guide, published data and data obtained in the context of the present study. Manfredi et al (2011) show that GHG emissions are lowered using pyrolysis opposite to other thermal treatments.

2. Materials and methods

Experimental data are required to adjust parameters of the models used for the simulation and optimization of the process. Pyrolysis of AWP is carried out in a bench scale reactor, at 500°C , under 5cc/min of N_2 flow (Figure 1).

The products are a liquid, a gas, and a carbonaceous co-product. The products are characterized in the *context* of their use for generating thermal and electrical energy. The solid co-product was studied in order to determine the possible presence of toxics (metals, Polyinsaturated Aromatic Hydrocarbons, Polycyclic Aromatic Hydrocarbons PHAs) The characterization tools are thermal analysis (DSC/TGA), elemental analysis, lower heating value (LHV) and viscosity measurements and gas chromatographic associated to mass spectroscopy detection

Figure 1: Pyrolysis bench scale reactor.

(GC/Mass) analysis. CO_2 equivalent emissions are calculated using these results. Pyrolysis technology is compared to both mechanical recycling and incineration of plastic, considering CO_2 emissions. In addition, the energetic efficiency of the pyrolysis process of AWP is calculated, considering the liquid and gas products as alternatives to fuel- oil and natural gas respectively.

The liquid and the char were analyzed to find their composition. The carbonaceous solid was practically free from transition metals, with only 120 ppm of iron. The presence of PAHs was not detected.

Liquid and gas are rich in saturated and non-saturated hydrocarbons, showing only trace amounts of oxygenated compounds. While gas is postulated to be a thermal source for the pyrolysis endothermic process, liquid is envisaged as a substitute of heavy fuel-oil, or at least for the formulation of blends with liquid fossil-based fuels.

From experimental values a simulation model is built for the pyrolysis reactor together with the mill used to chip the containers and the separation sector to obtain char, liquor and gas. Mass and energy balances are posed for all the equipment. This model is used for the optimization of the process. To model the process, yields of liquid, gas and char were correlated vs. temperature of reaction, so this temperature could be used as optimization variable.

The yields % as a function of the temperature are then used to find the flowrates of char, liquid and gas products. Those are found from the experimental values as functions of temperature T[°C] as:

Liquid yield %: $y_{LIQ} = -1.50E\text{-}03 \ T^2 + 1.47 \ T - 2.99E\text{+}02$ (1)

Char yield %: $y_{CHAR} = -1.00E\text{-}05 \ T^3 + 1.63E\text{-}02 \ T^2 - 8.85 \ T + 1.60E\text{+}03$ (2)

Gas yield %: $y_{GAS} = 1.65E\text{-}03 \ T^2 - 1.61 \ T + 4.29E\text{+}02$ (3)

Char yield has been approximated with a cubic function as it presents a plateau between 500-550 °C, to represent with fidelity the experimental behavior. A process schematic is shown in Figure 2. For illustrative purposes only one condenser is shown, but due to the high temperatures of the gas three units are used, with different cooling media (Dowtherm© for the first heat exchanger, where the gas is cooled with no phase change, Dowtherm© to condense up to 100°C, and cooling water for the subsequent unit. Final temperature of the process streams leaving the condenser is approximately 80°C).

Global mass balances for the pyrolysis products:

$M_j = M_{feed} \ y_j / 100; \ j = LIQ, \ GAS, \ CHAR$ (4)

being M_j the amount of liquid, gas or char obtained from the pyrolysis reactor after the condensation step takes place.

Energy balances for the process are also posed, together with the equations that allow calculating the energy requirements of the rest of the equipment.

3. Results

The analysis of the scenario corresponding to the accumulation and location of AWP in the region under study indicates that a plant for processing 4t per day is feasible, so this will be used as a calculation basis (M_{feed} = 4 t/day).

Equipment dimensions were calculated for the requirements of the process. The condensing unit is composed of several heat exchangers as the gas leaves the pyrolysis unit at the pyrolysis temperature (between 450-650 °C) as different refrigeration utilities will be used in each step. This total cost of the process can be calculated using this information.

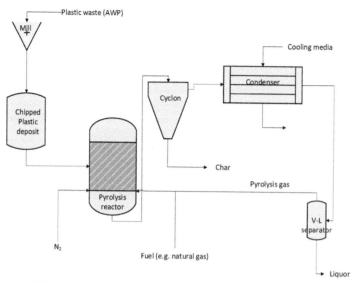

Figure 2: Process schematic for a standard pyrolysis process.

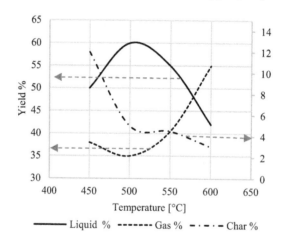

Figure 3: Yield of products from pyrolysis

Figure 3 shows the global liquid, gas and char yields for different pyrolysis temperatures as presented in Eqs. 1 to 3. The liquid product is maximized for a temperature around 500°C. It will be assumed for the purpose of obtaining the process model that the yields for a pilot scale reactor will be like the experimental ones. For approximately 500 °C, which resulted to be the optimum temperature for maximizing the liquid production, it was obtained experimentally that LHV was 49,6 MJ/kg for the liquid and 46,9 MJ/Kg for gas. It was found that those values do not change significantly between process temperatures ranging from 450 to 600°C as the amount of products change with temperature, but not the composition. Gas LHV is used to calculate the heat generated for the pyrolysis process. This could replace partially the equivalent of fossil-based fuel-oil traditionally used for energy production in thermoelectric plants in the region, with a consequent decrease in CO_2 eq. emissions (7.74 t of CO_2 avoided per day).

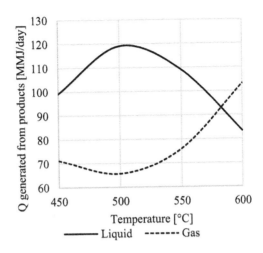

Figure 4: Energy obtained from the products

Table 1: Products distribution for the whole process

T (°C)	Liquid [kg/day]	Gas [kg/day]	Char [kg/day]
450	2000	1520	480
500	2400	1400	200
550	2200	1620	180
600	1680	2200	120

The incineration of AWP, without energy recovery produces 4,61 [t CO_2/t of plastic]. (O. Eriksson and G. Finnveden, 2009). Thus, the pyrolysis gas, used as the energy source of the recycling process originates 3.76 t CO_2 / t AWP per day, being the net CO_2 emission -4.0 t CO_2 per day.
For our process it is found that 4 t CO_2/day are avoided compared to incineration.
Regarding the use of the products as energy source, from Figure 4 it is shown the compromise between the production of liquid (which can be further used to produce higher added value products or can be injected in oil ducts as its characteristics are similar to light crude) and the production of gas that can be used as a fuel substitute to process the AWP. In any case, it must be noticed that producing plastics requires less energy than disposing them. Regarding the Sustainable Development Goals, it is tempting to conclude that the solution for the problem of pollution due to plastics should be banning them. But this will collapse with another objective of Zero hunger, as the petrochemical industry is tightly linked to the food related industry, apart from being counterproductive (Creadore and Castaldi, 2022)

4. Alternatives for processing

The study has been conducted using chipped agrochemical containers, made from polyethylene. Alternative sources for the feedstock can be considered:

1. Gas can be used to generate energy for the reactor instead of using fossil-based fuel, as shown in this paper
2. Liquid can be used as a fuel for the reactor, in which case there will be a remnant of gas.
3. As the amount of energy required for the pyrolysis process can be between 4 and 10 MJ/t, an analysis can be made of the external requirements as the energy required increases (for example when more gas is produced to decrease or even eliminate the need of an external source for fuel).
4. If there is biomass present in the residue (sometimes the containers are labeled and the paper is not separated beforehand, so it is chipped together with the plastic giving a mixed residue) LHV decreases. For example, for 10% biomass, LHV decreases 5%.

Several scenarios can be included within the model to compare the emissions and the cost of the process. If the objective is to decrease emissions, the objective function is to minimize t CO_2/t AWP as shown in this work.

5. Conclusions

Chemical recycling by pyrolysis represents a solution to the environmental problem associated with the end of life of AWP.

The impacts on environment are positive since liquid and gas products generate energy with a negative CO_2 emission balance.

Emissions due to incineration are avoided and at the same time, replacement of fossil-based fuel by pyrolysis liquid originates a significant decrease in CO_2 emissions.

Apart from the numerical results, from the experiments the carbonaceous co-product has no toxic components, thus it could be disposed in soils without producing negative impacts. These results can be compared to the other possible processes to dispose AWP,

- Landfills: 0,25 kg CO_2/kg plastic,

- Waste-to-energy processes 2,20 kg CO_2/kg plastic.

- Mechanical recycling emissions are alike to those of landfills, 0,2 kg CO_2/kg plastic.

If the pyrolysis gas is used to generate electricity, emissions are neglected as it is used to replace a fossil fuel.

Studies are being conducted to analyze the impact of this solid product from pyrolysis on the crops when used to improve soil which are beyond the scope of this work. Future work is also related to the development of processes for adding value to the liquor. Also, studies for scaling up the process are being conducted, because of the availability of data.

References

Creadore, L, Castaldi M, 2022 Quantitative Comparison of LCAs on the Current State of Advanced Recycling Technologies. Report from Earth Engineering Center Chemical Engineering Department, The City College of New York.

Eriksson O., G. Finnveden, 2009, "Plastic waste as a fuel - CO_2- neutral?", Energy Environ. Sci, 2, 907-914

Giugliano, M.,Cernuschi,S.,Grosso,M.,Rigamonti,L.,2011. Waste Management.31,2092–2101.

ISO 15270, 2008.

Kruger C., BASF (Sphera Solutions GmbH), 2020, Evaluation of pyrolysis with LCA: 3 case studies.

Manfredi, S.,Tonini,D.,Christensen,T.H.,2011.,Conserv.Recycl.55,995–1004.

Antonis Kokossis, Michael C. Georgiadis, Efstratios N. Pistikopoulos (Eds.)
PROCEEDINGS OF THE 33rd European Symposium on Computer Aided Process Engineering
(ESCAPE33), June 18-21, 2023, Athens, Greece

Life Cycle Assessment of bio-oil produced through lignocellulosic biomass liquefaction

Ana S. Augusto[a], Catarina G. Braz[a], Duarte M. Cecílio[a,b], Maria M. Mateus[a,b], Henrique A. Matos[a]

[a]*Centro de Recursos Naturais e Ambiente, Instituto Superior Técnico, Universidade de Lisboa, Av. Rovisco Pais 1, 1049-001 Lisboa, Portugal*
[b]*Fábrica Secil - Outão, Apartado 71, 2901-864 Setúbal, Portugal*
catarina.braz@tecnico.ulisboa.pt

Abstract

The objective of the present work was to assess the impact on the environment of a biofuel produced through lignocellulosic biomass liquefaction through a Life Cycle Assessment (LCA). Using *SimaPro* software, the Product Environmental Footprint (PEF) methodology was used to calculate the global environmental impact. The system was expanded to consider the credits of petroleum coke replacement. The effect of the type of solvent used in the liquefaction process was studied, as well as the possibility of operating the pilot plant in continuous mode.

Keywords: Biomass liquefaction, biofuels, environmental impact, Life Cycle Assessment, *SimaPro*.

1. Introduction

In response to the challenges of climate change, the European Green Deal (2019) set the goal of transforming the European Union (EU) into a resource-efficient and carbon-neutral economy, mobilising the industry for a clean and circular economy. In this context, the EnerGreen project, developed by Cimentos Maceira e Pataias, S.A., built a pilot plant to produce advanced biofuels from biomass-derived waste to replace the usage of petroleum coke in a clinker's kiln.

The main objective of this work was to carry out an LCA study that analysed the environmental impact of bio-oil production through the liquefaction of lignocellulosic biomass using the *SimaPro* software. The effect of the life cycle stages of the bio-oil was studied and evaluated to define which steps of the process are the main contributors to the environmental impact and which impact categories are the most affected. Two different biomass sources were compared, eucalyptus and sludges from the paper industry, at BATCH and later at continuous operating conditions.

The LCA method used for this purpose was the Product Environmental Footprint (PEF), available in *SimaPro* software. This methodology is the most suitable to use in this study since its results consider the European citizens' average impact as a reference.

2. Methods

2.1. Goal and Scope

The LCA study presented in this work has been conducted according to the ISO 14040/44 guidelines (2006a, 2006b), following an attributional approach. The study was conducted in a cradle-to-gate scope, considering the operation of the units (material and electric and thermal energy usage). On the other hand, energy and materials related to the construction of the processing units are not included in data inventories.

The inventory data was obtained from the biomass liquefaction pilot plant installed at Secil – Maceira e Pataias to produce 1GJ of energy. The Ecoinvent 3.6 database is used for the background data assuming the Portuguese context.

2.2. Life Cycle Inventory

In the liquefaction process (Figure 1), the biomass is fed to an endless screw, undertaking a pre-treatment process. Here, the biomass is pre-heated and sprayed with the solvent, 2-ethyl hexanol (2EH), undergoing a process of swelling, where the volume of the cells increases, breaking the cellulosic and lignocellulosic membranes, facilitating the access of the catalyst to all components of the residue. The liquefaction process takes place inside the stirring reactor at a temperature between 130 and 250 °C in the presence of an acid catalyst, p-toluene sulfonic acid.

The reactor is equipped with an external jacket, where the gases from liquefaction (mainly steam) circulate. The gases also go through a hollow conveyor outside the endless screw to pre-heat the feed and are condensed afterwards.

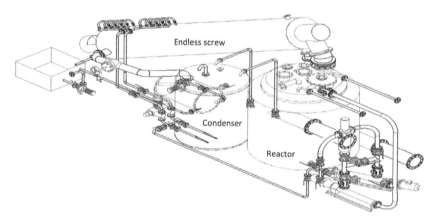

Figure 1 – Biomass liquefaction facility installed in Secil – Maceira e Pataias.

The lower calorific value (LCV) of the liquefied biomass was calculated as the pondered average of the components' LCV present in the reactive mixture and used to assess the amount of bio-oil equivalent to 1 GJ of energy. The results are presented in Table 1, together with the values for petroleum coke.

Table 1 - Mass of fuel to obtain 1 GJ of energy.

	LCV (GJ/ton)	Mass (kg) to 1 GJ
Bio-oil from Eucalyptus	21.1	47.5
Bio-oil from Paper Sludges	15.7	63.7
Petcoke	31.3	32.0

Table 2 presents the data inventory introduced in *SimaPro* to produce bio-oil from eucalyptus and paper sludges in BATCH and continuous operating conditions. Since biomass sources are residues, they were considered to have zero impact and weren't introduced in *SimaPro*.

Table 2 – Data introduced in *SimaPro* for the liquefied biomass from eucalyptus and paper sludges for BATCH and continuous mode.

Inputs and outputs	Eucalyptus		Paper Sludges	
	BATCH	Continuous	BATCH	Continuous
Inputs				
Solvent (kg)	29.8	3.7	208.7	3.9
Catalyst (kg)	1.1	0.5	4.0	0.4
Fuel (kg)	0.8	6.8	1.2	0.8
Transport (tkm)	5.3	43.3	15.9	10.1
Electricity (kWh)	2.0	16.3	3.0	1.9
Outputs				
Bio-oil (kg)	47.5	47.5	63.7	63.7
Condensed water (l)	19.9	210.3	139.1	132.8
Avoided products				
Petroleum coke		32.0		

2.2.1. Introduction of p-toluene sulfonic acid and 2-ethyl hexanol in SimaPro

As p-toluene sulfonic acid and 2EH are not available in the Ecoinvent database, these two components' environmental impact must be estimated. To create this molecule, the RREM method was used (Huber et al., 2022), which consists of four steps: general research on the chemical process, setting up the reaction equation, researching the required thermal energy, and modelling the dataset. Toluene sulfonic acid, the catalyst, is synthesised from a reaction between toluene and sulfuric acid with a molar ratio of 1:1.

$$C_6H_5CH_3 + H_2SO_4 \rightarrow C_7H_8O_3S + H_2O \tag{1}$$

The values introduced in *SimaPro* (Table 3) correspond to their respective molar masses, considering a yield of the reaction of 100 %. As the energy for this reaction was not found in the literature, it was assumed that 1 kg of product needed 1.8 MJ to be created (Huber et al., 2022).

Based on the synthesis of 2EH via butanol (BuOH), the same procedure applied to the catalyst was used. The synthesis of 2EH via butanol through the Guerbet reaction starts with a dehydrogenation of butanol followed by aldol condensation and hydrogenation of the unsaturated condensation product. A bifunctional catalyst, characterised by a dehydrogenating/hydrogenating metal species and a basic component, is required for this reaction to occur. Homogeneous catalysts at 120°C based on phosphine complexes of different transition metals (Rh, Ru, Ir, Pt, Pd, and Au) are used in the Guerbet self-condensation of BuOH.

$$C_4H_{10}O \leftrightharpoons C_4H_8O + H_2 \tag{2}$$
$$C_4H_8O \rightarrow C_8H_{14}O + H_2O \tag{3}$$
$$C_8H_{14}O \rightarrow C_8H_{18}O \tag{4}$$

Based on stoichiometric production reactions of p-toluene sulfonic acid and 2EH and molar masses, the following inputs in Table 3 were inserted in *SimaPro*.

Table 3 – Inputs and outputs of p-toluene sulfonic acid and 2-ethyl hexanol in *SimaPro*.

p-toluene sulfonic acid system		2-ethyl hexanol system	
Inputs		*Inputs*	
Toluene	92.14 g	Palladium	0.58 mg
Sulfuric acid	98.08 g	1-butanol	74.12 g
Heat from steam	0.05 MJ	Heat from steam	0.7 MJ
Outputs		*Outputs*	
p-toluene sulfonic acid	172.02 g	2-ethyl hexanol	130.23 g
Wastewater (water emission)	1.25 ml	Hydrogen (air emission)	2 g
		Wastewater (water emission)	18 g

3. Results

The environmental impact of the liquified biomass was evaluated using the PEF methodology, and the results are presented in Figure 2. The total impact score of both cases is positive. Since the petcoke replacement credits were considered, this means that the impact of both bio-oils is higher than that of petcoke, and thus, they are not suitable as a substitute.

Bio-oil derived from eucalyptus presents a much lower impact score than the one derived from paper sludges. This fact is justified mainly because the first presents a much higher LCV than the latter, meaning that more significant amounts of paper sludge-derived bio-oil are necessary to produce 1 GJ.

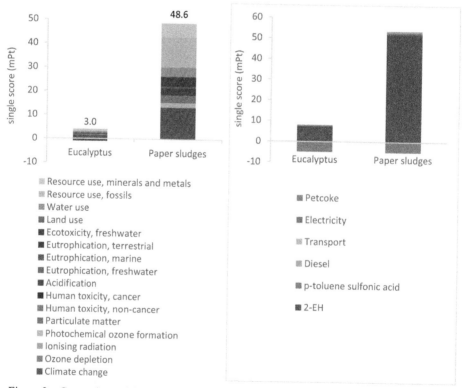

Figure 2 – Comparison of the weighting values of impact categories of liquefied biomass with PEF method detailed by impact categories and contribution of processes with the petcoke credits.

In both cases, the impact categories that stand out the most are the use of fossil resources, climate change and the use of minerals and metals resources. These results are due to the manufacturing process of 2EH, which uses minerals, metals and fossil resources, and consequently will release prejudicial compounds to the environment that contribute to climate change.

3.1. Influence of glycerine in the solvent

Since the solvent used in the liquefaction process is the main contributor to the environmental impact of the bio-oil produced, new alternatives were searched to substitute 2EH. A component that may be part of the solvent is glycerine, which influences the solvent's behaviour. Using glycerine with 2EH improves the homogenisation of the medium and makes it possible to increase the process conversion and obtain added-value products with nitrogen. Glycerine derived from waste cooking oil biodiesel plants was selected since it had the lowest impact of all the available options in *SimaPro*.

Figure 3 presents the results of the environmental impact score of bio-oil with different proportions of 2EH and glycerine mixtures.

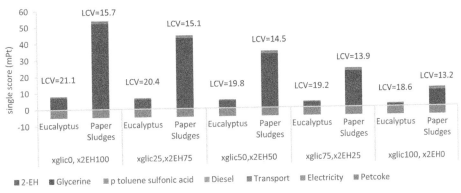

Figure 3 – Impact of glycerine from waste cooking oil on the bio-oil environmental impact. LCV is in GJ/ton.

As expected, the substitution of 2EH is beneficial to decrease the environmental impact of the liquified biomass. On the other hand, adding glycerine reduces the LCV of the resulting bio-oil, which might limit its utilisation in a clinker's kiln.

3.2. Continuous operation

A new environmental impact evaluation was performed, this time considering the liquefaction process operates in continuous mode (currently, the liquefaction pilot plant operates in BATCH), where less solvent is necessary per kg of bio-oil.

The results obtained in continuous operation (Figure 4) have a lower impact, associated with other benefits like less labour-intensive or less processing and holding time.

Considering the environmental impacts of bio-oil, the liquefied biomass from eucalyptus has a higher impact. This is a consequence of the smaller amount of solvent used per kg of bio-oil.

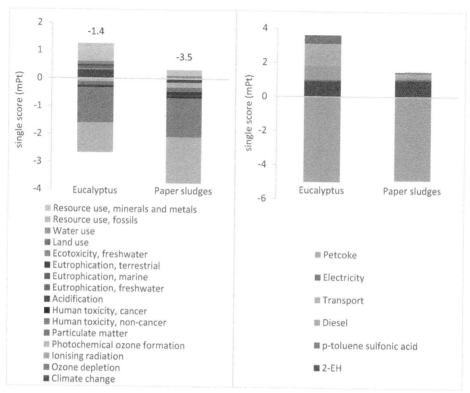

Figure 4 – Comparison of the weighting values of impact categories of liquefied biomass with PEF method detailed by impact categories and contribution of processes with the petcoke credits.

4. Conclusions and Future Work

The main objective of this work was to evaluate the impact of the liquefied biomass production process. Different scenarios, such as working conditions, biomass source and glycerine presence, were studied. Based on the analysis, it can be concluded that it is possible to obtain a bio-oil suitable to replace petroleum coke if the unit is operating continuously. Eucalyptus residues are preferred as biomass sources because they form a bio-oil with a higher LCV. Adding glycerine as solvent up to 75 % is beneficial for eucalyptus-derived liquefied biomass, reducing the environmental impact while maintaining a reasonable LCV of the bio-oil. In the future, more studies on the impact of p-toluene sulfonic acid and 2EH should be done to decrease the uncertainty of the results.

References

2006a, ISO 14040:2006 Environmental Management – Life Cycle Assessment – Principles and Framework, International Standardization Organisation, Geneva.

2006b, ISO 14044:2006 Environmental Management - Life Cycle Assessment – Requirements and Guidelines, International Standardization Organisation, Geneva.

2019, The European Green Deal

C. Carlini, A. Macinai, A. M. Raspolli Galletti, and G. Sbrana, 2004, Selective synthesis of 2-ethyl-1-hexanol from n-butanol through the Guerbet reaction by using bifunctional catalysts based on copper or palladium precursors and sodium butoxide, J Mol Catal A Chem, vol. 212, no. 1–2, pp. 65–70

E. Huber, V. Bach, P. Holzapfel, D. Blizniukova, M. Finkbeiner, 2022, An Approach to Determine Missing Life Cycle Inventory Data for Chemicals (RREM), Sustainability, 14, 3161

Antonis Kokossis, Michael C. Georgiadis, Efstratios N. Pistikopoulos (Eds.)
PROCEEDINGS OF THE 33rd European Symposium on Computer Aided Process Engineering
(ESCAPE33), June 18-21, 2023, Athens, Greece

Impacts of defossilising downstream derivatives in petrochemical clusters – MTBE case study

Inna Stepchuk[a], Mar Pérez-Fortes[a], Andrea Ramírez Ramírez[a]

[a]*Department of Engineering Systems and Services, Faculty of Technology, Policy and Management, Delft University of Technology, Jaffalaan 5, 2628 BX Delft, The Netherlands*
i.stepchuk@tudelft.nl

Abstract

Using alternative carbon sources (ACS) to produce downstream derivatives (DDs) is a promising option for defossilising the chemical industry. However, the potential consequences of using ACS in interconnected petrochemical clusters are generally overlooked. This paper aims to develop a methodological approach for systematically analysing defossilisation impacts at the value chain level. For this, a single value chain for producing methyl-tert-butyl-ether (MTBE) was used as a case study. The individual components of the value chain were modelled in Aspen Plus v12. Both ACS- and fossil-based value chains were compared in terms of (i) changes in the structure of the value chain and (ii) the magnitude of the impacts. The results show that the defossilisation of a single value chain causes additional impacts at the cluster level.

Keywords: Industrial defossilisation, downstream derivatives, interconnected cluster.

1. Introduction

The chemical sector is highly dependent on the usage of fossil carbon as feedstock for the production of chemicals (Nesbitt 2020). The launch of the EU chemicals strategy for sustainability has triggered the chemical industry to look for innovative solutions toward a green transition. One of the possible solutions is to defossilise the chemical industry by replacing fossil-based feedstock with alternative materials (EC 2020). Alternative carbon sources (CO_2, biomass, waste) are increasingly considered as alternative feedstocks for bulk chemicals production (i.e. ethanol, methanol, ammonia). However, there is less research on the potential of defossilising downstream derivatives (DDs) (Moncada, Posada, and Ramírez 2015). DDs are the chemicals present at the end of value chains before they are distributed to potential customers, for instance, styrene, acetone, or polyvinyl chloride (Kramer et al. 2017).

DD's value chains are usually integrated into large petrochemical clusters (Porter 1998), through symbiotic relationships inside and between value chains within the clusters (Chertow 2000). The defossilisation of one DD process will likely affect other processes inside the value chain and within the cluster. Thereby, making uncertain what the total impact on the performance of the value chain and cluster could be. The goal of the current work is to develop and test a methodology to analyse how using alternative carbon materials for defossilising a single DD value chain can affect its existing interactions and performance in a petrochemical cluster. For this paper, MTBE (methyl tert-butyl ether) was selected as a case study.

2. Methodology

This paper departs from an in-house model of a representative petrochemical cluster. The model mimics a real petrochemical cluster and it is based on the 6 industrial sites of the Port of Rotterdam (PoR) petrochemical cluster (Ramirez-Ramirez 2019). The model includes a total of 22 plants with the production of 52 chemicals classified in 9 subclusters. Each plant was modelled in Aspen Plus v12, and the interconnections are modelled in an interface. The subclusters are chlorine, methanol, ethylene, propylene, ammonia, aromatics, olefins, bio-based and other chemical-based subclusters.

The representative cluster contains a diverse range of chemicals. In this paper, chemicals produced in the cluster are presented as follows (Figure 1): feedstock (F), auxiliary chemicals (AC), chemical building blocks (CBBs), commodity chemicals (CCs), and downstream derivatives (DDs). All these chemicals are interconnected by mass, energy and or waste flows. The connectivity of the value chain is represented by the number of mass flows interconnections within value chain boundaries (i.e. horizontal connections) and the number of subclusters involved in the production of a single DD (i.e. vertical connections). The methodology contains four main steps which are discussed below.

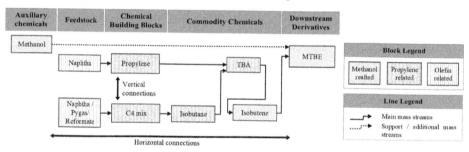

Figure 1: Schematic representation of MTBE value chain with existing mass flow interconnections.

2.1. Selection of key downstream derivatives

The representative cluster contains 13 DDs but not all of them have a large potential for defossilisation. Therefore, the selection of the most promising DDs was made based on three criteria: (i) embodied carbon, (ii) connectivity of the components in the value chains and (iii) availability of data of the ACS-based routes in the literature.

First, the amount of embodied carbon was assessed for each DD to identify the most promising derivatives for defossilisation. Embodied carbon (EC) is defined as the elemental carbon embodied in material flows of chemicals entering specific stages of production to meet process needs (Eq.(1)). For the assessment of embodied carbon, data was collected from each DDs' Aspen Plus model of the representative petrochemical cluster. The embodied carbon of each DD was calculated as follows:

$$EC_{DD} = \sum_{i=1}^{n} M_i * C_{wt\%_i} \qquad (1)$$

where $i = (1 \dots n)$ − number of material flows entering the production process of the DD; M_i − inlet capacity of particular material flow, ktonne/year; $C_{wt\%_i}$ − weight percent of fossil-based carbon embodied in the chemical, %.

The next step involved identifying the existing connectivity of mass flows within value chains and representative clusters. This was done by mapping the network, i.e., identifying the interconnections (i.e. mass flows) between chemicals, and the connectivity of each value chain was assessed in terms of horizontal and vertical interconnections.

Last, a literature screening was performed to identify ACS-based production routes (i.e. from CO_2, waste, biomass). Information was collected for each ACS-based route with a technology readiness level (TRL) of 4 or higher.

Based on three selection criteria, potential DDs were selected for further research. To evaluate the available defossilisation pathways (i.e. ACS-based routes) and develop the methodology to assess subsequent impacts at the value chain level, first, a single DD value chain (i.e. case study) was selected.

2.2. Selection of ACS-based route for the case study

This step involved the selection of the most promising ACS-based production routes for the case study. Routes were further screened in terms of data availability (i.e. the more data, the better), type of options (i.e. thermochemical, electrochemical, biochemical and the process conditions, e.g., mild conditions were preferred over stringent conditions), complexity (i.e. the number of process steps, the fewer steps the better), and scalability. The routes were then ranked and the most promising option was selected for further study.

2.3. Process modelling of ACS-based route

The selected ACS-based route was then modelled at the scale (i.e. capacity) and purity of the current fossil process. Here it is assumed that the alternative route will replace one-to-one the amount of product produced by the fossil-based route. For this, the development of the process flow diagram and Aspen Plus v12 model was done based on data collected for the selected ACS-based route. The model was assumed to be in continuous mode with a whole year of operation (i.e. 8000 hours per year). Properties for biomass components were introduced into Aspen Plus simulation by the usage of biomass property databases developed by the National Renewable Energy Laboratory (NREL).

2.4. Defossilisation impacts analysis

Value chains can be represented graphically as a network, with nodes and links. Nodes refer to chemical processes involved in the value chain, while links represent material, waste and or energy flows between the nodes. Networks were developed for fossil- and ACS-based value chains of the case study, which allow to study changes that would occur when replacing a fossil-based process, a node, in the network with ACS-based process. Structural changes in nodes and links were categorized, as (i) removed - no longer required, (ii) affected - required, but capacities are changed, (iii) unchanged - required, capacities remain the same, and (iv) added – required, to be added (i.e. new chemicals/flows).

It is expected that structural changes in the value chain will affect its total performance. To quantify the magnitude of these impacts the following techno-economic and environmental KPIs were selected: (i) production rates, (ii) fossil-based carbon offset (i.e. the amount of fossil embodied carbon replaced by alternative carbon), and (iii) additional investment needed. Based on these indicators, the ACS-based value chain of the case study was assessed and compared to the fossil-based counterpart.

3. Results and discussion - MTBE case study

3.1. Selection of case study – MTBE

Based on the methodology for the selection of DDs presented in Section 2.1, 6 DDs were selected (see Table 1). First, priority was given to DDs with embodied carbon of

> 40 ktonne/year. Then, chemicals were examined in terms of connectivity and availability of ACS-based routes. DDs with a higher number of vertical connections were preferred because changes in the value chains of those DDs will more likely result in a higher probability of impacts at the cluster level. MTBE was selected as a case study. As can be seen from Table 1, it has high EC, is highly interconnected within the value chain and cluster, and has a wide range of ACS-based routes present in the literature.

Table 1: Key downstream derivatives.

| Downstream Derivatives | EC* | Connectivity | | ACS-based routes | | | Research |
name	kt/y	vertical	horizontal	CO₂	biomass	waste	selected
Styrene	914	3	11	-	+	+	+
Methyl tert-butyl ether (MTBE)	281	2	8	+	+	+	+
Polyvinyl chloride (PVC)	181	2	6	+	+	+	+
Polyethylene terephthalate (PET)	155	2	8	+	+	+	+
Carbon Black	132	0	1	+	+	+	-
Phthalic anhydride	71	0	3	-	+	-	-
Propylene glycol ether (PGE)	58	4	14	+	+	+	+
Polyols	44	4	13	+	+	-	+
Dimethyl ether (DME)	35	0	2	+	+	+	-
Acetone	33	0	4	+	+	+	-
Biodiesel	4	0	4	n.a.	n.a.	n.a.	-
Hydrochloric acid (HCl)	0	0	8	n.a.	n.a.	n.a.	-
Bioethanol	0	0	2	n.a.	n.a.	n.a.	-

secondary data for calculations is taken from Aspen models

3.1.1. Fossil-and ACS-based MTBE

MTBE is used in the fuel industry as an additive to increase the octane number. Conventionally, it is produced from a chemical reaction of methanol and isobutene. In the representative cluster, MTBE is produced in the amount of 400 ktonne/year. For its production, methanol and C4 mix are purchased from outside the cluster. Propylene is obtained from the cluster's olefin plant. Isobutane and tert-butyl alcohol (TBA) are produced within the propylene subcluster of the representative cluster (Section 2).

From the literature, the biomass-to-isobutanol-to-MTBE synthesis process was selected based on economic and technical constraints. The process involves different stages: (i) conversion of biomass to sugars through ethanol-based organosolv, (ii) enzymatic hydrolysis and fermentation of isobutanol, (iii) dehydration of isobutanol to isobutene, and (iv) conversion of isobutene to MTBE (Moncada, Posada, and Ramírez 2017). Ethanol is required as a solvent in a quantity of 5 L per kg of dry biomass and is completely recycled in the process. Simplified block diagrams of both fossil- and ACS-based value chains of MTBE are presented in Figure 2.

3.2. Defossilisation impacts analysis

Figure 2 shows a qualitative comparison of major changes in the structure of mass flows of MTBE value chain after defossilisation. Nearly 60% of the structure of MTBE value chain is changed. TBA, isobutane, C4 mix, and propylene are no longer required inside value chain and therefore their use is reduced by 100%. However, an additional chemical, ethanol, is needed as a solvent for isobutanol production. Although the quantity of ethanol is high, it is fully recovered within the value chain. Here, it was assumed that it does not affect the value chain carbon offset rate. The fossil-based carbon offset was calculated as the amount of alternative carbon replacing fossil embodied carbon of MTBE. The latter is a sum of fossil-based carbon embodied in material flows of isobutene and imported methanol entering the MTBE process. The fossil-based carbon offset is 80 %, as fossil-based isobutene is completely replaced by an alternative one. Economic performance is also affected. Based on ongoing modelling work, additional investment is roughly 15% of the bare equipment costs of MTBE fossil-based value chain.

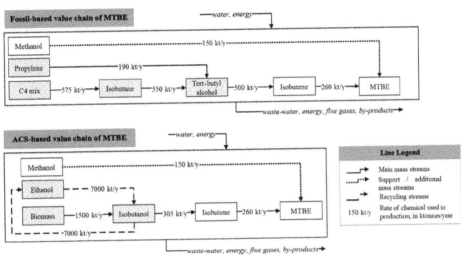

Figure 2: Simplified MTBE value chains with main mass flow interconnections for the production of 400 ktonne/year.

The majority of chemicals involved in fossil-based MTBE production are currently produced within the cluster and are involved in the value chains of other DDs (Table 2). Therefore, structural changes in the MTBE value chain will affect other value chains. For instance, tert-butyl alcohol is derived as a coproduct of propylene oxide production, requiring both C4 mix and propylene as feedstock materials. The defossilisation of the MTBE value chain will therefore affect the amount of C4 mix imported into the cluster as well as the isobutane and propylene capacities. Thereby impacting other DDs, such as styrene, and propylene glycol ether which also require propylene. The latter are highly interconnected within the cluster, resulting in further propagation of defossilisation impacts.

Table 2: Structural changes at the value chain and cluster level.

Chemical	Present in the cluster				Required for MTBE value chain		Defossilisation impacts	
					fossil	ACS	value chain	cluster
name	yes/no	*	**	kt/y	kt/y	kt/y	change	change
Methanol	yes	B	no	150	150	150	unchanged	unchanged
Propylene	yes	P	yes	497	190	-	removed	affected
C4 mix	yes	B	yes	575	575	-	removed	affected
Isobutane	yes	P	yes	550	550	-	removed	affected
Tert-butyl alcohol	yes	P	no	603	500	-	removed	affected
Isobutene	yes	P	no	260	260	260	unchanged	unchanged
Biomass	yes	B	yes	2000	-	1500	added	affected
Ethanol	no	B	no	-	-	7000	added	added
Isobutanol	no	-	-	-	-	305	added	added

*B –is bought from the market, P- is produced within the representative cluster;
** involved in value chains of other DDs, yes/no.

The impact assessment in this paper was done based on the changes in mass flows. However, to develop the full picture of potential impacts, it is crucial to also include impacts on energy and water flows due to defossilisation. This is part of the current research, which will be available at the time of the conference.

4. Conclusions and future work

The defossilisation of a single DD can significantly have impacts beyond the process itself, especially in highly interconnected petrochemical clusters. This paper proposed a novel methodology for systematically analysing defossilisation impacts at the value chain level. The findings highlight the need for methodologies to be able to assess changes at the value chain level and their consequences at the cluster level (e.g. impacts on fossil-based carbon offset of other DDs due to defossilisation of MTBE). In future work, the ACS-based MTBE value chain will be assessed in terms of changes in energy and water flows and the methodology will be further developed to identify where the largest impacts in terms of defossilisation can be obtained in different DD value chains.

Acknowledgements

This publication is part of the project "Unravelling the impacts of using alternative raw materials in industrial clusters", project number VI.C.183.010 of the research programme VICI which is financed by the Dutch Research Council (NWO).

References

Chertow, Marian R. 2000. "INDUSTRIAL SYMBIOSIS : Literature and Taxonomy." *Industrial symbiosis* 25(November): pp 313-337. https://www.annualreviews.org/doi/pdf/10.1146/annurev.energy.25.1.313.

EC. 2020. Brussels, 14.10.2020 COM(2020) 667 final *Chemicals Strategy for Sustainability Towards a Toxic-Free Environment*. Brussels.

Kramer, Jan-Philipp et al. 2017. no *Trilateral Strategy for the Chemical Industry*. Düsseldorf. www.mwide.nrw.de.

Moncada, Jonathan, John A. Posada, and Andrea Ramírez. 2015. "Early Sustainability Assessment for Potential Configurations of Integrated Biorefineries. Screening of Bio-Based Derivatives from Platform Chemicals." *Biofuels, Bioproducts and Biorefining* 9(3): 722–748. https://onlinelibrary.wiley.com/doi/epdf/10.1002/bbb.1580.

Moncada, Jonathan, John A. Posada, and Andrea Ramírez. 2017. "Comparative Early Stage Assessment of Multiproduct Biorefinery Systems: An Application to the Isobutanol Platform." *Bioresource technology* 241: 44–53. http://dx.doi.org/10.1016/j.biortech.2017.05.074.

Nesbitt, Elizabeth R. 2020. 1 Working Paper ID-065 *Using Waste Carbon Feedstocks to Produce Chemicals*.

Porter, Michael. 1998. "Clusters and the New Economics of Competition." *Harvard business review* Reprint 98(November-December): 3–8. http://backonline.apswiss.ch/6001/porter_clusters_and_the_new_economics_of_competition.pdf.

Ramirez-Ramirez, C. 2019. "Unravelling the Impacts of Using Alternative Raw Materials in Industrial Clusters." *NWO*. https://www.nwo.nl/en/projects/ttw-vic183010 (June 2, 2022).

Antonis Kokossis, Michael C. Georgiadis, Efstratios N. Pistikopoulos (Eds.)
PROCEEDINGS OF THE 33rd European Symposium on Computer Aided Process Engineering
(ESCAPE33), June 18-21, 2023, Athens, Greece

Methane-to-X: an economic assessment of methane valorisation options to improve carbon circularity

Ben Lyons,[a] Andrea Bernardi,[a] Nilay Shah,[a] Benoît Chachuat[a,*]

[a] *The Sargent Centre for Process Systems Engineering, Department of Chemical Engineering, Imperial College London, SW7 2AZ, UK*
*b.chachuat@imperial.ac.uk

Abstract

Methane side streams are produced in many different chemical processes and are normally combusted to provide process heat or to generate electricity. However, this practice is becoming less and less attractive as the industry strives towards net-zero targets and increasing the circularity of chemicals. Methane could instead be recovered and used as a valuable feedstock to produce other platform chemicals, such as H_2 or ethylene, which could be beneficial both for the economic performance and the carbon circularity of the system. In this work, seven different methane valorisation routes to produce additional chemicals are investigated. The considered routes include: i) five syngas-based routes combined with methanol synthesis and a methanol-to-olefins process; ii) plasma methane pyrolysis; and iii) oxidative coupling of methane. The results suggest that oxidative coupling of methane is the most profitable, with methane pyrolysis, tri-reforming and autothermal reforming also being more profitable in the base case. All routes have lower scope 1 and 2 emissions than the base case, however, dry-reforming and bi-reforming have the lowest emissions thanks to credited CO_2 feed streams.

Keywords: methane, olefins, hydrogen, techno-economic analysis, circular chemicals

1. Introduction

Methane is a by-product of many different chemical processes, including steam cracking for the production of olefins, pyrolysis and gasification processes, aromatic dealkylation and the anaerobic digestion of wastewater sludge. The most common use of these methane side streams is as a fuel, where it is combusted to either directly provide process heat or to produce additional electricity in a turbine. This use of methane helps chemical plants to close their energy balance and aids in their profitability. However, the chemical industry is moving towards a more sustainable future with both net-zero ambitions and carbon circularity at the forefront. The combustion of methane therefore becomes the antithesis of these goals due to the release of CO_2 and thus loss of carbon from the system (Lange, 2021). Additionally, methane itself is a potent greenhouse gas and a recent study discovered that flaring is less efficient than expected for the destruction of methane in oil and gas operations (Plant et al., 2022). The combination of these two factors reveals a critical need for alternative outlets for methane in the chemical sector.

A potential solution for this problem is using the methane as a feedstock to produce additional chemicals, such as H_2 or olefins. This would avoid the direct emissions from burning the side stream, but it would also maximise the value obtained from the feedstock in the original chemical process and improve the overall carbon circularity. In this paper, seven alternative processes are compared that convert methane into additional chemicals.

The routes considered are: i) five syngas-based routes which comprise of a methane reforming technology, methanol synthesis and finally a methanol-to-olefins conversion section; ii) a methane pyrolysis plant which decomposes methane into solid carbon and hydrogen; and iii) oxidative coupling of methane to produce ethylene. The main aim of this work is to conduct a comparative assessment of these seven methane valorisation routes by considering both their economic performance and their scope 1 and 2 emissions.

2. Methodology

Process flowsheets for the syngas-based routes, methane pyrolysis and oxidative coupling of methane were simulated using Aspen HYSYS® and are summarized in Figure 1. The results from these models are then used to perform a techno-economic analysis to compare the profitability of each route. A methane feed flowrate of 8 t/h was used in all routes, which is approximately the methane flowrate in an ethane steam cracker producing 700kt/y of ethylene (Belohlav et al., 2003). Note that a critical assumption made in this work is that separation units are already present and available in the base plant where the methane feed originates. Separation requirements were therefore assumed to be provided by the base plant and were not considered in this analysis.

2.1. Process Modelling

Syngas-Based Routes: These routes are comprised of three main sections: methane reforming; methanol (MeOH) synthesis; and MeOH-to-Olefins (MTO). In the reforming section, methane reacts with different oxidants to produce syngas, a mixture of carbon monoxide (CO) and hydrogen (H_2). In this work, five different methane reforming options were considered: steam reforming of methane (SRM); dry-reforming (DRM); bi-reforming (BRM); tri-reforming (TRM); and auto-thermal reforming (ATR) (Arora et al., 2016). These reforming routes produce syngas at different CO:H_2 ratios as shown below:

SRM	$CH_4 + H_2O \rightarrow CO + 3H_2$	(1)
DRM	$CH_4 + CO_2 \rightarrow 2CO + 2H_2$	(2)
BRM	$3CH_4 + 2H_2O + CO_2 \rightarrow 4CO + 8H_2$	(3)
TRM	$20CH_4 + H_2O + CO_2 + 9O_2 \rightarrow 21CO + 41H_2$	(4)
ATR	$4CH_4 + 2H_2O + O_2 \rightarrow 4CO + 10H_2$	(5)

All five of these reforming reactions were modelled in Gibbs reactors with the oxidants fed in their respective stoichiometric ratios. After the syngas has been produced, it is sent to the MeOH synthesis section, which is based on the work by Van-Dal and Bouallou (2013). Here, the syngas is cooled and then compressed to 78 bar via multistage compression with interstage cooling. The high-pressure syngas is sent to a packed bed isothermal reactor filled with a copper-based catalyst, where the syngas is converted into methanol via the following primary reaction:

$$CO + 2H_2 \rightarrow CH_3OH \tag{6}$$

Methanol and unconverted syngas are separated in a series of two flash vessels, with the syngas recycled to the reactor's inlet and the methanol sent to the MTO section. A purge is also present on the syngas recycle, which is sent to the base plant to recover any excess H_2. The MTO section is based on the UOP/Norsk Hydro process, which consists of a recirculating fluidized bed reactor with a SAPO-34 catalyst along with a regenerator (Kuechler et al., 2000). Ethylene and propylene are the main hydrocarbon products, along with higher olefins, coke, H_2, CO, CO_2, alkanes and water. The higher olefins are sent to an additional olefin cracking process (OCP), which consists of a catalytic packed bed reactor that converts the olefins into additional ethylene and propylene. The yields of the MTO and OCP reactors were taken from Hannula et al. (2015).

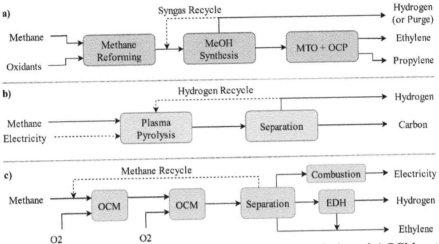

Figure 1: Schematic diagrams for the: a) syngas; b) plasma pyrolysis; and c) OCM routes

Methane Pyrolysis (MP) Route: In this route, methane is decomposed into its elemental components: solid carbon and H_2. Multiple methods exist for the pyrolysis of methane and include using plasma, molten metals, catalytic conversion and thermal decomposition. In this work, the plasma pyrolysis route was modelled as it is the technology closest to commercialisation (Schneider et al., 2020). Methane is injected into a reactor vessel between carbon electrodes, which are fed with recycled H_2 and electricity to generate a plasma gas. The plasma reacts with the injected methane and decomposes into solid carbon and H_2 and was modelled using a conversion reactor achieving a 94% per-pass conversion (Hardman et al., 2020). Temperatures within the plasma reactor reach approximately 2100°C so the products are first cooled and then separated using cyclones and filters. If it is of sufficient quality, the solid carbon produced can be sold as 'carbon black', which is commonly used as a pigment and a reinforcing filler for tyre production.

Oxidative Coupling of Methane (OCM): This route has long been the subject of high industrial and academic interest for being able to produce ethylene directly from methane and for overcoming the highly endothermic nature of steam cracking. OCM is an exothermic process that is conducted over metal oxide catalysts at temperatures between 500-1000°C. The two primary reactions in OCM are first the formation of ethane via the oxidative coupling of methane (Eq. 8) and then the formation of ethylene via the oxidative dehydrogenation of ethane (Eq. 9) (Stansch et al., 1997):

$$2CH_4 + 0.5O_2 \rightarrow C_2H_6 + H_2O \qquad (8)$$
$$C_2H_6 + 0.5O_2 \rightarrow C_2H_4 + H_2O \qquad (9)$$

OCM also forms CO, CO_2 and H_2 as unwanted side products and has typically low yields and selectivity towards both ethane and ethylene, which has restricted its commercialisation potential. In this work, the reaction flowsheet developed by Penteado et al. (2018) is used as a basis. Methane is fed into a series of two packed bed adiabatic reactors with oxygen in a $CH_4:O_2$ molar ratio of 10:1, which improves selectivity towards C_2 components at the expense of per-pass-conversion rate. The reactors were sized so that

O_2 was completely consumed at the end of each reactor. Additional O_2 was added before the second reactor to restore the 10:1 ratio. The products are cooled and separated, and unreacted methane is recycled back to the start of the process. Unreacted ethane is sent to an ethane dehydrogenation reactor (EDH), where it is catalytically dehydrogenated into additional ethylene and hydrogen. For this reactor, 60% was assumed for both the per-pass-conversion and selectivity towards ethylene (Salerno et al., 2016). The H_2 and CO side products are combusted and used to generate electricity using a steam cycle.

2.2. Assessment

Purchased equipment cost curves and the factorial method were used to obtain the total capital investment (TCI) for each route and a plant life of 20 years and interest rate of 10% were used to annualize the TCI (Towler et al., 2022). Due to the lack of reliable data on the cost of a plasma pyrolysis reactor, the MP route was costed using the six-tenths rule on the TCI of a similar plant (Keipi et al., 2016). The levelised profit, which is defined as the profit obtained per kg of methane feed, was used to determine the economic performance of each route. A business-as-usual (BAU) case was used as a benchmark, which is the price of methane if it was sold to the National Grid as a natural gas substitute (0.17 USD/kg, Feb 2021). Utility and raw material costs were taken as market prices in 2021, with heating provided by blue H_2 at 2.01 USD/kg (Al-Qahtani et al., 2021). The methane feed is assumed to be free as it originates from the base plant. Sensitivity analysis on the natural gas price will be considered in future work to account for its volatility. The scope 1 and 2 emissions were calculated considering direct CO_2 emissions from each route as well as the carbon intensity of the electricity supply (UK 2021 average at 0.265 kg_{CO2}/kWh) and the blue H_2 for heating (4.3 kg_{CO2}/kg_{H2}, Al-Qahtani et al., 2021). Credits were given for any CO_2 that was consumed as a feedstock.

3. Results and Discussion

The levelised profit and scope 1 and 2 emissions of each route are presented below in Figure 2 (a) and (b) respectively. For the economic assessment, initial observation reveals that four out of the seven routes are more profitable than the BAU case. OCM shows the highest profit at 0.38 USD/kg, followed by MP and TRM at 0.30 USD/kg and ATR at 0.23 USD/kg. The promising results for OCM are due to both its relatively low capital investment and the electricity generation from the side product combustion, which negates the utility cost. However, its total revenue is limited by its low ethylene selectivity as only 47% of the carbon from the methane feed ends up in ethylene. This carbon efficiency becomes 75% when ethane is included, however it presents a negligible contribution to the route's total revenue. Like OCM, MP benefits from a simple flowsheet with few units leading to a low capital investment, but it is hindered by the high electricity requirement in the plasma reactor. It is also dependent on the carbon black sale to remain more profitable than the BAU case. TRM produces syngas with a H_2:CO ratio slightly greater than 2:1, which meets the minimum stoichiometric requirement for methanol synthesis. This maximises both the production of methanol from syngas and thus olefin yield in the MTO and OCP section. TRM also has the lowest utility requirement out of the syngas-based routes due to its very low heat of reaction in the reforming section, which in turn enables it to be less vulnerable to the high H_2 heating costs. ATR has a higher total revenue than TRM thanks to the excess H_2 produced; however, this is not sufficient to offset the increase in capital and utility costs due to a higher heating requirement in the reforming section and a higher cost for the heat exchange network.

The other three syngas-based routes are less profitable than the BAU case: SRM at 0.11 USD/kg; BRM at 0.08 USD/kg; and DRM at -0.24 USD/kg. All three of these routes have highly endothermic reforming reactions and thus have significant heating requirements. Despite SRM and BRM having the highest revenues, this is not sufficient to offset the large utility cost. BRM performs slightly worse than SRM due to the additional cost of the CO_2 feed. Finally, DRM is the only route that is unprofitable. Since it generates syngas in a 1:1 ratio, most of the carbon in the methane feed is lost in the purge as excess CO. Further analysis revealed that only 42% of carbon in the DRM feed was converted into olefins. Another contributing factor to DRM's poor performance is the 1:1 CH_4:CO_2 molar feed requirement, which generates a substantial total raw material cost.

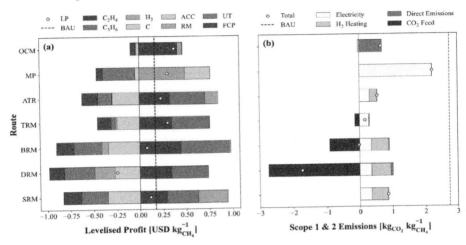

Figure 2: (a) Levelised profit and (b) Scope 1 & 2 emissions of each methane valorisation route. (LP = levelised profit; BAU = business-as-usual case; ACC = annual capital charge; RM = raw materials; UT = utilities; FCP = fixed costs of production)

Figure 2 (b) reveals that all routes have lower scope 1 and 2 emissions compared to the BAU case. However, the relative performance of the routes has somewhat reversed when compared to the economic results. MP has the highest emissions at 2.21 kg CO_2, followed by SRM and OCM at 0.86 and 0.66 kg CO_2 respectively. All the emissions for MP come from the electricity source, whilst OCM is the only route that has a substantial contribution from direct emissions due to the side-product combustion. ATR and TRM are the other routes with positive scope 1 and 2 emissions at 0.54 and 0.17 kg CO_2 respectively. As seen in the economic assessment, the minimal heating requirement for TRM results in better performance compared to ATR. BRM achieves net-zero operation at -0.01 kg CO_2 and DRM shows net-negative emissions at -1.74 kg CO_2 thanks to the substantial CO_2 feed requirement. However, the promising results for DRM will likely be negated when scope 3 emissions are considered, as the unconverted CO stream would likely be combusted. Note that accounting for the product separations may impact both the economic and emissions results and will be considered in future work.

4. Conclusions

This paper presented a model-based comparison between seven different methane valorisation routes to produce additional chemicals. The routes considered were i) five syngas-based routes which used different methane reforming technologies (steam, dry, bi, tri and autothermal reforming) followed by methanol synthesis and methanol-to-olefins; ii) methane plasma pyrolysis; and iii) oxidative coupling of methane. The results show that OCM has the highest profit due to its low capital and operating costs, despite its low ethylene selectivity. MP, TRM and ATR are also more profitable than the base case, whilst SRM, BRM and DRM are less profitable in part due to their highly endothermic reforming reactions. However, DRM and BRM have the lowest scope 1 and 2 emissions owing to their large CO_2 feed streams, whilst MP has the highest thanks to its high electricity requirement. Nevertheless, all seven routes have lower scope 1 and 2 emissions compared to the base case. Further investigation is required to assess the overall environmental impact of each route (including analysing alternative heating and electricity sources), the robustness of each route to fluctuations in market conditions, and the scale effect of the methane side stream. The low-capital OCM and MP routes are expected to remain feasible at lower methane flowrates whilst the syngas routes will only be applicable to large operations to leverage economies of scale.

Acknowledgements: The authors are grateful to the Engineering and Physical Sciences Research Council (EPSRC) for funding the research under the UKRI Interdisciplinary Centre for Circular Chemical Economy programme (EP/V011863/1). BL thanks the Dpt. of Chemical Engineering at Imperial College London for a PhD scholarship.

References

A. Al-Qahtani et al., 2021, Uncovering the true cost of hydrogen production routes using life cycle monetisation, Applied Energy, 281, 115958

S. Arora et al., 2016 An overview on dry reforming of methane: strategies to reduce carbonaceous deactivation of catalysts, RSC Advances, 6, 110, 108668-108688

Z. Belohlav et al., 2003, The kinetic model of thermal cracking for olefins production, Chemical Engineering and Processing: Process Intensification, 42, 6, 461-473

I. Hannula et al., 2015, Light olefins and transport fuels from biomass residues via synthetic methanol: performance and cost analysis, Biomass Conversion and Biorefinery, 5, 1, 63-74

N. Hardman et al., 2020, Carbon black from natural gas, US10808097B2 (US Patent)

T. Keipi et al., 2016, Techno-economic analysis of four concepts for thermal decomposition of methane: Reduction of CO_2 emissions in natural gas combustion, Energy Conversion and Management, 110, 1-12

K. Kuechler et al., 2000, Process for converting oxygenates to olefins with direct product quenching for heat recovery, US6121504A (US Patent)

J. P. Lange, 2021, Towards circular carbo-chemicals – the metamorphosis of petrochemicals, Energy & Environmental Science, 14, 8, 4358-4376

A. Penteado et al., 2018, Techno-economic evaluation of a biogas-based oxidative coupling of methane process for ethylene production, Frontiers of Chemical Science and Engineering, 12, 4, 598-618

G. Plant et al., 2022, Inefficient and unlit natural gas flares both emit large quantities of methane, Science, 377, 6614, 1566-1571

D. Salerno et al., 2016, Techno-Economic Evaluation of an Oxidative Coupling of Methane Process at Industrial Scale Production, Computer Aided Chemical Engineering, 38, 26, 1785-1790

S. Schneider et al., 2020, State of the Art of Hydrogen Production via Pyrolysis of Natural Gas, ChemBioEng Reviews, 7, 5, 150-158

Z. Stansch et al., 1997, Comprehensive Kinetics of Oxidative Coupling of Methane over the La_2O_3/CaO Catalyst, Industrial & Engineering Chemistry Research, 36, 7, 2568-2579

G. Towler, R. Sinnott, 2022, Chemical engineering design: principles, practice and economics of plant and process design, Elsevier

E. S. Van-Dal, C. Bouallou, 2013, Design and simulation of a methanol production plant from CO_2 hydrogenation, Journal of Cleaner Production, 57, 38-45

Antonis Kokossis, Michael C. Georgiadis, Efstratios N. Pistikopoulos (Eds.)
PROCEEDINGS OF THE 33rd European Symposium on Computer Aided Process Engineering
(ESCAPE33), June 18-21, 2023, Athens, Greece

Purification of Raw Bio-Methanol from Renewable Feedstocks: A Flexibility Assessment

Matteo Fedeli[a,b], Alessandro di Pretoro[b], Ludovic Montastruc[b], Flavio Manenti[a]

[a]*Politecnico di Milano, Dipartimento di Chimica, Materiali ed Ingegneria Chimica "Giulio Natta", Piazza Leonardo da Vinci 32, 20133 Milano, Italy*
[b]*Laboratoire de Génie Chimique, Université de Toulouse, CNRS/INP/UPS, Toulouse, France*
flavio.manenti@polimi.it

Abstract

This paper investigates the production of bio-methanol from different origins feedstocks such as renewable and biological ones. The focus is on the analysis of the raw methanol composition's perturbations, dependent on the chosen feedstock, which impacts the design, the economics, and the environmental emissions of the downstream section. A thermodynamic analysis is computed to determine the final raw methanol purification depending on the utilized feedstocks. The conventional distillation column configuration is chosen as the process design for methanol purification. The purified methanol is AA grade, according to the ASTM D1152-97 standard specification. The entire section is simulated with ProSim software with a chosen Non-Random-Two-Liquid (NRTL) thermodynamic package. A two-directional flexibility analysis is performed to assess the impact of different perturbations under uncertain conditions in raw methanol compositions. The results of the flexibility analysis highlight optimal plant flexibility with around 7 % of the total cost deviation.

Keywords: Flexibility analysis; Biomethanol; Crude Methanol Purification.

1. Introduction

Methanol is one of the most produced bulk chemicals worldwide, with a global production of 160 Mton/y (Fedeli et al., 2022). Its production is a well-established technology which involves basically (i) syngas production; (ii) CO_x hydrogenation (depicted in Eqs. 1-2); and (iii) raw methanol purification. The most utilized feedstocks for production are coal and natural gas. Recently biomethanol is gaining interest since is a promising advanced fuel for gasoline blends and especially for the disposal of marine fuels. Thus, could be helpful for the actual European policies and dictates. European RED II is currently set, to tackle the problem of increasing GreenHouse Gas (*GHG*) emissions, a target share of biofuel's final consumption of energy in the transport sector shall be at least 3,5 % in 2030 (EU SCIENCE HUB, 2022). Whatever feedstock wastes contain a good content of carbon and hydrogen is exploitable for the biomethanol production. Biogas, products waste gasification, and flue gases are considered the first candidates for this purpose. These new process technologies will increase worldwide biomethanol production despite several problems that are to figure out. The most highlightable obstacle of the feedstock wastes is the seasonability fluctuations and the produced raw biomethanol composition. Above mentioned feedstocks have a high CO_2 content compared to the natural gas one. Thus, will led to a higher production of methanol with the CO_2 hydrogenation pathway (Eq.2). Consequently, the produced raw methanol will have higher water content depending by the feedstock quality. Even a slight variation could impact the design of the plant purification section.

$$CO + 2H_2 \leftrightarrow CH_3OH \qquad\qquad \Delta H°298 \text{ (kJ/mol)} = -90.85 \qquad (1)$$

$$CO_2 + 3H_2 \leftrightarrow CH_3OH + H_2O \qquad \Delta H°298 \text{ (kJ/mol)} = -50.1 \qquad (2)$$

This work aims to analyze the effects of the perturbation due to different feedstock utilization in the conventional raw methanol purification. Flexibility analysis is introduced to supply an optimum design which includes the probability that a perturbation could change the process performance.

2. Materials and Methods

The flexibility assessment consists of three different steps: (i) a thermodynamic analysis to classify feedstock quality (based on CO_2 content) and the expected raw methanol composition; (ii) the design and the model of the biomethanol purification section in pre-defined nominal conditions; (iii) the evaluation of the process flexibility introducing the bi-directional system perturbations. In the first task, an Aspen HYSYS v11 process simulation is computed to understand how the CO_2 content in the feedstock affects the outlet composition. The feedstock composition is split into three main compounds: H_2, CO, and CO_2. The hydrogen molar fraction is kept constant while CO and CO_2 are changed respectively respecting the constraint of constant overall flowrate. The thermodynamic model selected is SRK-Twu. An isothermal Gibbs reactor with a temperature of 260°C is used as a model to predict the final methanol composition. The final target is to correlate the CO_2 to CO ratio with the final water content in the crude methanol. In this step, the by-products of the reaction such as higher alcohol, DME, and acetone are not considered. However, this final assumption is neglected in the second step where is important to model, with a good detail grade, the purification section.

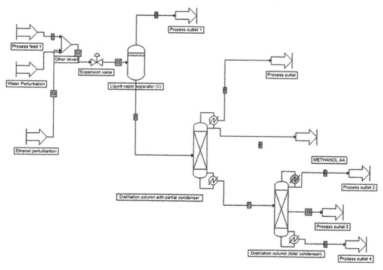

Figure 1: ProSim layout of raw methanol purification.

This is developed with ProSim software with a chosen Non-Random-Two-Liquid (NRTL) thermodynamic package. The typical impurities data is taken from the work o. The crude biomethanol is characterized by 19 different compounds (Xue et al., 2021).

The purification process is modelled with the most used industrial applications: distillation column trains. Figure 1 depicts the process flow simulation. The purified methanol target is AA grade, according to the ASTM D1152-97 standard specification (ASTM, 1998). As a nominal case study to design the unit is chosen the raw methanol from biogas processing. Finally, in the third step, two-directional perturbations are applied to system to assess the plant flexibility. Water and ethanol content in the crude are selected as perturbation variables. Resilience index (*RI*) and Swaney and Grossmann (F_{sg}) are selected to evaluate the impact concerning economical assessment. *RI* expresses the largest total disturbance load and is independent of the direction of the perturbation. F_{sg} represents the maximum fraction of the expected deviations δ that can be accommodated by the systems (Di Pretoro et al., 2022).

3. Results and Discussion

The thermodynamic analysis is useful to characterize and forecast the crude methanol composition among a wide range of syngas quality. This property is identified with the ratio between the inlet flowrates of CO_2 and CO. Figure 2 shows the expected water production (in molar fraction) varying the CO_x ratio. It is quite clear that the chemical pathway CO_2 hydrogenation is favoured increasing the carbon dioxide content. Considering syngas produced from biogas-based reforming processes, the CO_x ratio is close to the unity (Chein et al., 2021). In this case study, the estimated water molar fraction will be around 15% mol/mol$_{tot}$. When the molar ratio tends to infinite values, i.e., methanol produced from flue gas and renewable hydrogen, the water produced overcomes the value products. In this case, is useful to add in the process a pre-treatment with a CO-shift unit to increase the feedstock CO quantity.

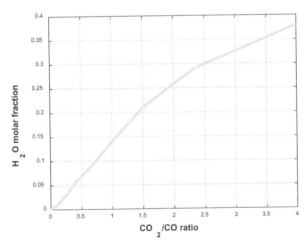

Figure 2: Thermodynamic analysis results, the x-axis represents the CO_2/CO ratio while the y-axis is the expected water molar fraction in the produced crude biomethanol.

The second step of this work is to develop a process simulation of the methanol purification section which includes all the crude impurities and the useful unit operation to reach the AA-grade methanol specification. The nominal case study is chosen as production of biomethanol from biogas-based reforming process. The syngas is laminated through a valve and passes through a separator which removes the uncondensable gases (CO, H_2). These are recycled back to enhance methanol production. After this operation, the crude mixture enters a stabilizer column to remove the CO_2. The last unit operation is a rectifier, with a total condenser, useful to refine until the desired specification. Both distillation columns work in atmospheric conditions. Table 1 describes the results of the process simulation.

Table 1: Main output variables of the ProSim process simulation.

Variables	Stabilizer	Rectifier
Temperature	64-67°C	64-67°C
Pressure	1 atm	1 atm
Reflux ratio	16.74	2.17
Reboiler duty	80.01 kW	575.25 kW
Condenser duty	30 kW	575.09 kW
Recovery ratio	100 %	70 %
Number of trays	5	51
Specifications	---	Methanol purity 0.99878 w/w Ethanol recovery 98%

The actual operating conditions are low-energy demanding, the utilities are Cooling Water (CW) for the condensers and Low Pressure Steam (LPS) for the reboilers. The first column (stabilizer) works with only five trays to remove the CO_2 from the crude biomethanol. The rectifier requires fifty-one equilibrium stages to achieve very high purity separations. This one presents a stage with an ethanol spill useful to reach the solution convergence. CAPital EXpenditure (CAPEX) and OPerational Expenditure (OPEX) related to the nominal case study have been estimated using Guthrie's method (Bailie and Whiting, 1998). 10 years of plant lifetime is chosen as parameter to normalize the fixed costs in a specified time range. In this way is possible to compute the Total Annual Cost (TAC) following Eq. 3:

$$TAC = \frac{CAPEX}{10 \; years} + OPEX \tag{3}$$

The main OPEX contributions are the CW and LPS utility costs. The final step of the work is the economical flexibility assessment of the purification section in different operating conditions. As mentioned in section 2, the perturbations variables are the water and the ethanol content in the crude mixtures. The aim is to study how the perturbations affect the nominal design and its flexibility. The output variables to analyze are the column duties which are directly correlated to the CAPEX/OPEX voices. If a distillation system requires more LPS to work, this is reflected in higher OPEX or a different sizing of the kettle reboiler (CAPEX). The perturbation variables are written as vectors as reported in Table 2, δ is the step deviation for each vector iteration.

Table 2 Property of the perturbation vectors

Perturbated variable	Value	δ	Size
Water feed	250 kg/h	± 50 kg/h	11
Ethanol feed	4 kg/h	± 0.8 kg/h	11

The combination of each value associated with the vectors represents a case study. The size of the perturbation matrix is 121 case studies. Each of them is evaluated in the process simulation with the tool "case studies", reporting convergence in the entire domain. The TAC is evaluated for each cell of the matrix, and the corresponding deviations are calculated concerning the nominal case study TAC. Figure 3 is the surface plot which correlates the TAC deviation (%) with the two flowrate perturbations.

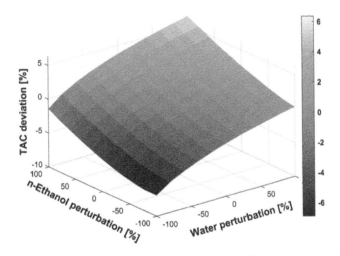

Figure 3: Matrix of TAC deviation with water and ethanol flowrates perturbation.

The maximum deviation value corresponds to 6.34% of the nominal TAC, the associated case study is related to the maximal perturbation of both vectors. The perturbation matrix and its relative TAC deviation are useful to assess the purification section flexibility via deterministic indexes computation. The plot represented in Figure 4 correlates the deterministic index, F_{sg}, and RI, in the function of the relative cost deviation. It is easier to observe the economic investment to effort different probability of perturbations are quite feasible. The proposed and simulated system has an optimistic value of flexibility following the deterministic indexes. Considering the F_{sg}, with a total investment of 6.3% the plant could afford the entire range of selected perturbations. The RI value is lower than F_{sg} because considers only the effort load with a specific perturbation direction. The negative TAC deviations mean less investments in case of feedstock with minor content of water and ethanol. That area of TAC saving includes the conventional process plants, where natural gas and coal are utilized for the production.

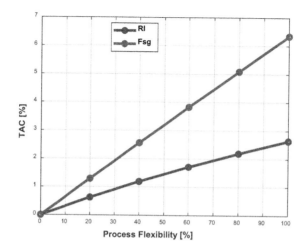

Figure 4: Deterministic index comparison to additional costs.

4. Conclusions

New processes to produce biomethanol consider the utilization of renewable and biological feedstock replacing conventional ones such as natural gas and coal. Different material input significantly affects the production introducing oscillating perturbations. Water and ethanol composition mostly impacts the crude methanol purification section. This work highlights the wide range of flexibility under uncertain conditions of conventional distillation columns. The results of flexibility analysis are quite good indicators of the potentiality to retrofit, revamp, and build process plants able to afford feedstock perturbations with relatively small investments.

References

ASTM, 1998. Standard Specification for Methanol (Methyl Alcohol).

R.C. Bailie, W.B.Whiting, 1998. Analysis, synthesis, and design of chemical processes, Choice Reviews Online, https://doi.org/10.5860/choice.36-0974

R.Y. Chein, W.H. Chen, H. Chyuan Ong, P. Loke Show, Y. Singh, 2021, Analysis of methanol synthesis using CO2 hydrogenation and syngas produced from biogas-based reforming processes, Chem. Eng. J. 426, 130835, https://doi.org/10.1016/j.cej.2021.130835

A. Di Pretoro, M. Fedeli, F. Ciranna, X.Joulia, L. Montastruc, F. Manenti, 2022, Flexibility and environmental assessment of process-intensified design solutions: A DWC case study, Comput. Chem. Eng. 159, https://doi.org/10.1016/j.compchemeng.2022.107663

EU SCIENCE HUB, 2022, URL https://ec.europa.eu/jrc/en/jec/renewable-energy-recast-2030-red-ii

M. Fedeli, F. Negri, F. Manenti, 2022, Biogas to advanced biofuels: Techno-economic analysis of one-step dimethyl ether synthesis, J. Clean. Prod. 376, 134076, https://doi.org/10.1016/j.jclepro.2022.134076

X. Xue, Q. Gu, H. Pascal, O.M. Darwesh, B. Zhang, Z. Li, 2021, Simulation and optimization of three-column triple-effect methanol distillation scheme, Chem. Eng. Process. - Process

Antonis Kokossis, Michael C. Georgiadis, Efstratios N. Pistikopoulos (Eds.)
PROCEEDINGS OF THE 33rd European Symposium on Computer Aided Process Engineering
(ESCAPE33), June 18-21, 2023, Athens, Greece

Life cycle assessment of Pine needle-based electricity generation in India

Pratham Khaitan,[a] Ankush Halba,[a] Pratham Arora,[a,*]

[a]*Hydro and Renewable Energy Department, Indian Institute of Technology Roorkee, Roorkee-247667, India*
*pratham.arora@hre.iitr.ac.in@Indian Institute of Technology Roorkee, Roorkee, India.
+91-1332284921*

Abstract

Electrification in the rural and remote areas and constant forest fires due to Pine needles are the major problems of the Uttarakhand state of India. Pine forests comprise 16.15% of the state's total forest area, producing 1.91×10^6 tonnes of Pine needles annually. Gasification of Pine needles could be employed to generate localized electricity, thereby diminishing forest fire frequency. Nevertheless, a life cycle assessment (LCA) must be conducted to comprehend the environmental advantages of this process. Therefore, using the GaBi software, the LCA of Pine-needles to the electricity gasification supply chain is performed for two biomass pretreatment cases (1-shredding and 2-briquetting) for the Almora district of Uttarakhand state. The predicted global warming potential (GWP)-100-year emissions for case 1 and case 2 were 0.679 and 0.578 kg CO_2-eq for 1 kWh of electricity production, respectively, compared to the national grid (0.86 kg CO_2-eq kWh).

Keywords: LCA, Pine needles, Gasification, Electricity, Supply chain.

1. Introduction

Almost 3.5 million people live close to significant tracts of Pine forests, where firewood and electricity are in short supply in Uttarakhand (Dhaundiyal and Tewari, 2016). According to the Forest Department of Uttarakhand, Pine forests are spread over around 3,430 km² in 17 forestry divisions across 13 districts, producing two million tonnes of Pine needles (PNs) annually (Bisht et al., 2014). Since PNs are extremely flammable by nature, they can contribute to or even start forest fires, especially in the summer when temperatures are high and relative humidity is low. In addition, PN creates a covering on the earth that prevents rainwater from entering the soil and restricts nearby plants' growth, reducing the food available for cattle and resources such as wood, medicinal herbs, and flowers. Therefore, collecting PNs from forest floors and using them for energy production can promote energy sustainability in rural and remote parts of Uttarakhand, which will also aid in minimizing forest fires (Dhaundiyal and Gupta, 2014). Pine needle gasification is a promising route for harnessing its energy potential (Kala and Subbarao, 2017). Additionally, decentralized PN gasification could be a sustainable source of electricity in areas far from the main power system. However, an ineffective biomass supply chain is a major hindrance to utilizing biomass gasification-based plants for energy generation (Halba et al., 2022). Generally, the biomass supply chain consists of three major components: collection, transportation, and storage, where all three stages, directly or indirectly, impact the environment (González and Sandoval, 2020). Therefore, the environmental assessment of the biomass supply chain is mandatory to make any biomass-based energy project successful in producing green energy and reducing greenhouse gas (GHG) emissions.

Life cycle assessment (LCA) is a commonly employed approach used to evaluate the environmental performance of any system or process. Whittaker et al. (2016) carried out the LCA using biomass (short rotation coppice) to predict the GHG and methane emissions from outside biomass storage. The study concluded that there were up to 20% dry matter losses during biomass storage and the biomass used provided up to 90% savings in GHG emissions. Another study by Beagle and Belmont (2019) used the LCA to evaluate the effects of transporting biomass (wood chips and wood pellets) on the overall amount of GHG emissions produced by bioenergy production. It was found that an integrated biomass facility could cut emissions by as much as 76% compared to coal. The research concluded using either of the two biomasses for electricity generation would result in lower GHG emissions in comparison to coal-powered stations. Costa et al. (2022) conducted the economic, environmental, and technical performance of the complete woodchips-to-energy supply chain using a commercially available biomass gasification-based system in Southern Italy. The study summarized that the impact of combustion and gasification accounted for almost 50% of the overall environmental impact. Gupta et al. (2022) employed LCA for bioenergy conversion via pyrolysis to produce bio-oil from the PNs, including the pyrolysis co-products (non-condensable gases and biochar). They integrated LCA with parameter sensitivity and economic analysis to evaluate processes more thoroughly. The study's main drawback was that the data came from a lab setting experimental investigation and not a commercial plant. The catalytic pyrolysis used led to a 42% decrease in the overall GWP of the process.

To the best of the author's knowledge, no researcher has examined the LCA of the PNs supply chain for electricity generation, including the biomass pretreatment scenario such as shredding and briquetting. Therefore, the current study has been undertaken to conduct the LCA of the PNs-to-electricity gasification supply chain, including the two biomass pretreatment cases (1-shredding and 2-briquetting) to check their environmental viability.

2. Methodology

A "cradle-to-gate" LCA analysis approach is adopted for the present study, which considers the environmental impacts from a product's life cycle, which start from the Pine tree plantation (cradle) to the factory gate (electricity). The PNs are assumed to be collected on foot in the forests of Uttarakhand's Almora district. Since the fallen PNs contain very little moisture throughout the summer, it is presumed that the PN was gathered during the 90-day interval (April to June). The biomass storage duration is assumed to be nine months, so the plant can function continuously throughout the year. The collected PNs are sent to primary storage (S1) via truck transportation (T1) in the next step. The biomass pretreatment routes (shredding and briquetting) for PNs are performed in a preprocessing plant (P1) located in the primary storage. The primary distinction between shredding and briquetting is that biomass briquettes often have larger shapes and sizes, calorific values, and densities than shredded biomass. The preprocessed PNs are then transported to a secondary storage facility (S2), where a gasification plant (P2) is installed. The gasification power generation system consists of a cooling and cleaning system, a gasifier, and a gas-to-electricity generator (Halba et al., 2022).

2.1. Life Cycle Analysis

The LCA process consists of four main steps: setting objectives and boundaries, conducting an inventory analysis, evaluating the effects of the process, and interpreting the results.

2.1.1. Goal and scope definition

The LCA process begins with establishing its goals and scope in the first phase. According to ISO 14040, this phase must specify environmental impacts, functional units, assumptions, and system boundaries (Kalinci et al., 2012). The study's goal was to compare and evaluate the potential environmental footprints of the two different preprocessed PNs (Shredded and briquettes) for electricity production based on gasification. The LCA model was developed using GaBi (version: 10.6.0.110). For the present study, 1 kWh of electricity production from pine needle biomass has been considered the functional unit (FU) for estimating the environmental impacts. The following steps are considered for conducting the LCA: biomass transportation, pretreatment of biomass, storage of biomass, and power generation from gasification.

2.1.2. Inventory analysis

This LCA combines and quantifies major inputs and outputs for the FUs. Energy inputs and raw materials for the LCA model were compiled from various sources like the GaBi-integrated Eco-invent database, research literature, and information acquired from local vendors and people. The India-specific database (available in GaBi) is utilized for creating the LCA model.

The PNs are assumed to be collected by foot in the forests surrounding Almora, Uttarakhand. It is assumed that raw PNs are delivered by diesel-powered Bharat Stage IV trucks (with 7.5, 11.4, and 25-ton payload capacities). The transportation distance from the collecting location to the preprocessing and storage site is assumed to be 11 km, whereas the transportation distance from the preprocessing site to the plant site is assumed to be 56 km for a decentralized plant.

The PNs shredding requires around 7.15 kWh of electricity to process per tonne of PNs (Gupta et al., 2022). In contrast, briquetting uses around 380.69 kWh of electricity to process per tonne of PNs (Singh et al., 2021). The factor used to calculate methane emissions from biomass storage is 0.0175 kg of CH_4 per kg of biomass (C. Whittaker et al., 2016). Raw, shredded, and briquette PNs are assumed to have a bulk density of 80 kg/m^3 (Houssami et al., 2018), 175 kg/m^3 (Dhaundiyal and Tewari, 2016), and 1200 kg/m^3 (Mandal et al., 2019), respectively. GaBi's Northern-India grid mix provided all the electricity input needed for chopping and briquetting (Gupta et al., 2022).

2.1.3. Impact assessment

In this stage, the data of the inventory analysis (the measured inputs and outputs) are evaluated for their ecological impact. The life cycle impact assessment (LCIA) is used to identify potential environmental consequences of life cycle inventory (LCI) and to determine their magnitude. The current investigation applies an LCIA technique based on the CML 2016-Aug 2016 version.

2.1.4. Interpretation

During the final phase of the LCA process, interpretation, it is crucial to confirm that the assessment's data align with its stated objective. In this regard, the two processes that yield useful information are the analysis of inventories and the evaluation of their environmental consequences (Kalinci et al., 2012). The qualitative interpretation of impact analysis data is made possible by addressing each effect category separately. The environmental effects of producing electricity from pine needles are evaluated and compared using impact categories such as Global Warming Potential (GWP), Acidification Potential (AP), and Eutrophication Potential (EP).

3. LCA Results and discussion

3.1. GWP

It is observed that the shredded and briquette PNs gasification results in an overall GHG emission of 0.679 and 0.578 kg CO_2 equivalent per kWh, as illustrated in Figure 1. It is revealed that the overall GWP for both the gasification of briquettes and the shredding of PNs is less than that of India's Northern grid electricity (0.86 kg CO_2 equivalent per kWh). The Pine tree absorbs CO_2 for its growth to produce products (such as PNs) by photosynthesis, resulting in -3.24 kg CO_2 equivalent per kWh and -2.61 kg CO_2 equivalent per kWh emissions for shredded and briquette biomass, respectively. The disparity between the values could be explained by the variation in energy density between the shredded and briquette PNs. In both cases, the emissions of exhaust gases have the largest impact on the surrounding environment, with 2.23 kg CO_2 equivalent per kWh for shredded PNs and 1.36 kg CO_2 equivalent per kWh for briquette PNs. The reason for the difference in emissions between shredded PNs and PN briquettes is due to the difference in their bulk density and biomass quantity. 1.5 kg of shredded PNs were required to produce 1 kWh of electricity (Avani, 2016) while only 1.2 kg of PN briquettes were required for the same (Kala and Subbarao, 2017). Based on the methane emissions induced by the decomposition of the biomass over time, the storage phase is accounted for as the second greatest contributor to overall GWP. It is found that the emissions from transportation are minimal due to shorter travel distances (11 and 56 km) and higher bulk density of shredded and briquette PNs.

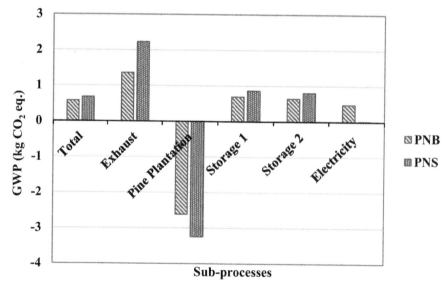

Figure 1: GWP results for shredded pine needles (PNS) and pine needle briquettes (PNB)

3.2. AP and EP

The potential for acidification across both LCA scenarios is shown in Figure 2. Shredded PNs have an overall AP of 0.35×10^{-3} kg SO_2 equivalent, with exhaust gases (0.25×10^{-3} kg SO_2 equivalent) as the primary contributor. Compared to shredded PNs, PNs briquettes have a higher AP of 4.59×10^{-3} kg SO_2 equivalent, with the most significant contribution resulting from the grid electricity required in the briquetting process. North Indian grid electricity has an AP of 7.76×10^{-3} kg SO_2 equivalent, which is higher in comparison to the PNs used. Briquetting requires a substantial amount of electricity, due to which the emissions of PN briquettes were higher than shredded PNs.

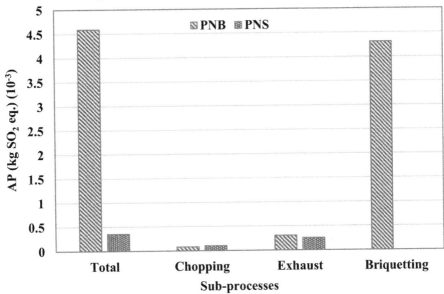

Figure 2: AP results for shredded pine needles (PNS) and pine needle briquettes (PNB)

In comparison to the Northern grid (4.8×10^{-4} kg PO_4 equivalent), the EP values of both PNs briquettes (3.23×10^{-4} kg PO_4 equivalent) and shredded pine needles (0.71×10^{-4} kg PO_4 equivalent) are lower. Figure 3 depicts the overall EP of both LCA scenarios. The difference in environmental impacts was due to the higher consumption of electricity in the case of briquetting.

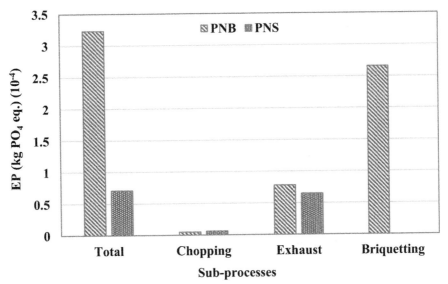

Figure 3: EP results for shredded pine needles (PNS) and pine needle briquettes (PNB)

4. Conclusions

The LCA approach was applied to carry out an analysis of the effects that either of the pre-treated biomasses has on the surrounding environment when gasified to produce electricity. In addition to this, the GWP, along with AP and EP, were investigated. For GWP, the exhaust gases from the gasification facility were responsible for the greatest amount of CO_2 emissions, while transportation had the least influence on the surrounding environment. Electricity used in the process was the leading contributor to acidification and eutrophication potentials. Both PN-based processes outperform the North-Indian grid in terms of GWP, AP, and EP for producing 1 kWh of electricity. Reduced emissions were a result of using plant biomass as an energy source instead of coal.

References

Y. Kalinci, A. Hepbasli, & I. Dincer, 2012. Life cycle assessment of hydrogen production from biomass gasification systems. International journal of hydrogen energy, 37(19), 14026-14039.

S. Gupta, P. Patel, & P. Mondal, 2022. Life cycle analysis (LCA) and economic evaluation of catalytic fast pyrolysis: implication of co-product's end-usage, catalyst type, and process parameters. Sustainable Energy & Fuels.

S. Mandal, G. V. Prasanna Kumar, T. K. Bhattacharya, H. R. Tanna, & P. C. Jena, 2019. Briquetting of pine needles (Pinus roxburgii) and their physical, handling, and combustion properties. Waste and Biomass Valorization, 10(8), 2415-2424.

P. Singh, T. P. Singh, R. K. Sharma, Y. K. Negi, & R. Pal, 2021. Potential of Pine Needle Biomass as an Alternative Fuel to Mitigate Forest Fire in Uttarakhand Himalayas-A Review. Journal of Agricultural Engineering, 58(2), 192-203.

M. El Houssami, A. Lamorlette, D. Morvan, R. M. Hadden, & A. Simeoni, 2018. Framework for submodel improvement in wildfire modeling. Combustion and flame, 190, 12-24.

M. Costa, D. Piazzullo, D. Di Battista, & A. De Vita, 2022. Sustainability assessment of the whole biomass-to-energy chain of a combined heat and power plant based on biomass gasification: biomass supply chain management and life cycle assessment. Journal of Environmental Management, 317, 115434.

L. D. Kala & P. M. V. Subbarao, 2017. Pine needles as potential energy feedstock: availability in the central Himalayan state of Uttarakhand, India. In E3S Web of Conferences (Vol. 23, p. 04001). EDP Sciences.

E. Beagle & E. Belmont, 2019. Comparative life cycle assessment of biomass utilization for electricity generation in the European Union and the United States. Energy Policy, 128, 267-275.

C. A. D. González & L. P. Sandoval, 2020. Sustainability aspects of biomass gasification systems for small power generation. Renewable and Sustainable Energy Reviews, 134, 110180.

C. Whittaker, W. Macalpine, N. E. Yates, & I. Shield, 2016. Dry matter losses and methane emissions during wood chip storage: the impact on full life cycle greenhouse gas savings of short rotation coppice willow for heat. BioEnergy Research, 9(3), 820-835.

A. Dhaundiyal & P. C. Tewari, 2016. Performance evaluation of throatless gasifier using pine needles as a feedstock for power generation. Acta Technologica Agriculturae, 19(1), 10-18.

A. Dhaundiyal & V. K. Gupta, 2014. The analysis of pine needles as a substrate for gasification. Hydro Nepal: Journal of Water, Energy, and Environment, 15, 73-81.

A. Halba, P. K. Vidyarthi, & P. Arora, 2022. Gasification as a potential solution for forest fires in the Indian Himalayan Region: A review. Bioresource Technology Reports, 101162.

A. S. Bisht, S. Singh, & M. Kumar, 2014. Pine needles a source of energy for Himalayan region. International Journal of Scientific & Technology Research, 3(12), 161-164.

Avani Bio Energy, 2016. Transforming Uttarakhand's Pine Needle Problem into an Innovative Energy Solution. <http://www.millenniumalliance.in/imgs/Avani_Case_study.pdf> (Accessed on 12/09/2022).

Antonis Kokossis, Michael C. Georgiadis, Efstratios N. Pistikopoulos (Eds.)
PROCEEDINGS OF THE 33rd European Symposium on Computer Aided Process Engineering
(ESCAPE33), June 18-21, 2023, Athens, Greece

Evaluation of alternative carbon based ethylene production in a petrochemical cluster: Technology screening & value chain impact assessment

James Tonny Manalal,[a] Mar Pérez-Fortes,[a] Paola Ibarra Gonzalez [a] and Andrea Ramirez Ramirez [a]

[A]Department of Engineering, Systems and Services, Faculty of Technology Policy and Management, Delft University of Technology, Jaffalaan 5, 2628BX, Delft, Netherlands.
E-mail: j.t.manalal@tudelft.nl

Abstract

Due to the heavy dependence on fossil-fuels as raw materials, the defossilization of feedstocks in the petrochemical industry represents a challenge. A large number of possible process routes that use alternative carbon sources (ACS) like CO_2, biomass, and waste are being developed for the feedstock replacement. For instance, to produce ethylene, more than 40 ACS process routes were identified. These multiple options make the selection of the promising process route a complex task. By replacing feedstocks, a process can change significantly and the impacts related to these changes in a highly interconnected industrial cluster can create cascading effects due to system interdependencies. This work aims to understand the cascading impacts in carbon flows and prices of implementing an ACS production process in an ethylene cluster. The results show that PVC will be the highest impacted and defossilizing one value-chain can have cascading effect on other value-chains as observed for PET.

Keywords: technology screening, feedstock defossilization, alternative raw materials, value chain impacts.

1. Introduction

Europe has set the ambitious goal to be carbon-neutral by 2050 and feedstock defossilization of carbon-based industrial processes is challenging (J.Rissman et al, 2020). For instance, the petrochemical industry depends on fossil fuels like naphtha, natural gas and ethane as carbon sources. Using alternative carbon sources (ACS) such as CO_2, biomass and waste is considered an option for the replacement of oil and gas feedstocks (D. Saygain et al, 2021). A large number of possible processes using ACS are being developed to produce chemicals through electrochemical, biochemical, and thermochemical routes (M. Stork, 2018). The existence of multiple processing routes makes the decision-making process of selecting the promising alternative process route a complex task that relies on multiple techno-economic and environmental criteria. It can be expected that replacing feedstocks might significantly change process requirements like equipment, energy, water, utilities and safety, and production outputs like products, by-products, and waste. These changes can affect the overall system , for instance in terms of product prices and import-export dependencies, due to system interdependencies. This work presents a systematic screening methodology for selecting promising ACS process technologies and aims to study cascading impacts (carbon and price), for the selected ACS ethylene production in an ethylene cluster as case study.

2. Methodology

For the production of ethylene using CO_2 and biomass as feedstocks, more than 40 different process routes were identified from the literature. In order to select the most promising alternative carbon-based ethylene production technologies a screening methodology was developed as explained in section 2.1. Then the value chain impacts of ACS technologies was studied for a base-case ethylene cluster, as explained in section 2.2 and section 2.3.

2.1. Technology screening

The screening methodology was developed based on a stage-gate concept and the selection was divided into five stages. The amount of information needed is reduced with this approach, as the processes are eliminated as stages progress. In stage-1, technology readiness level (TRL) is used for selection and technologies with TRL> 3 are selected to the next stage. In stage-2, the ideal stoichiometric reactions of each process route were used. Each stoichiometric reaction required to produce a chemical was assumed as individual unit operation or process step. Here, only ideal reactions are considered and side reactions are not taken into account. In stage-2, technologies with less than 4 process steps were selected. It was because as the number of steps increased, the number of routes as well as the level of complexity increased drastically. Then in stage-3, thermodynamic state functions; standard enthalpy change (ΔH^0), standard Gibbs energy change (ΔG^0) and standard entropy change (ΔS^0) for the overall reaction, were calculated at standard conditions using inputs from Aspen properties. Using these thermodynamic state functions, the theoretical overall heat need or generation and electricity need were calculated as:

$$Heat\ production/need = \sum_{Biochemical+Thermochemical+Catalytic\ steps} \Delta H^0 \qquad (1)$$

$$Electricity\ need = \sum_{Electrochemical\ step} \Delta G^0 \qquad (2)$$

At this stage, as the theoretical input and output components are known, the carbon utilization efficiency of the process route was calculated using Equation 3:

$$Carbon\ utilisation\ efficiency\ (CUE) = \frac{Moles\ of\ carbon\ atom\ in\ product}{Moles\ of\ carbon\ atom\ in\ feedstock} \qquad (3)$$

Based on the energy need and CUE, technologies were then ranked. Using a comparative assessment, technologies with electricity need <1500 kJ/mol ethylene and CUE >50% were selected to the next stage. In stage-4, a basic economic constraint was calculated (Equation 4), using mass flow, component price and energy requirements . An in-house compiled price database with prices adjusted to 2018 as base-year using the PPI (producer prices indices) and price data from ICIS chemicals outlook was used.

$$Economic\ constraint\ (EC)$$
$$= \frac{\sum_{reactants} mass\ flow * Component\ price + (\Delta H\ or\ \Delta G)_{endergonic} * Utility\ price}{\sum_{products} mass\ flow * Component\ price} \qquad (4)$$

Technologies with an economic constraint < 1 were selected. An EC ratio > 1 indicates that the input costs are higher than potential revenue and hence process is considered non-profitable for assumed product prices. In stage-5, the process technologies which passed the previous gates were ranked based on the number of process steps and economic ratio, and one process route from each feedstock category (CO_2 and biomass) were selected.

2.2. Base-case ethylene cluster model

This paper used an in-house developed ethylene cluster model part of the project "Unravelling the impacts of using alternative raw materials in industrial clusters", created in Aspen plus and based on existing processes in the Port of Rotterdam. The cluster model includes olefin, ethylene dichloride (EDC), vinyl chloride monomer (VCM), polyvinyl chloride (PVC), chlorine waste incineration (CKI), ethylene oxide (EO), ethylene glycol (EG), polyethylene terephthalate (PET), ethylbenzene (EB) and propylene oxide (PO)/styrene monomer (SM) synthesis units. The cluster has PVC, PET and SM value chains and these value-chains will be studied in this paper. The corresponding mass flows used as reference in this study are depicted in Figure 1.

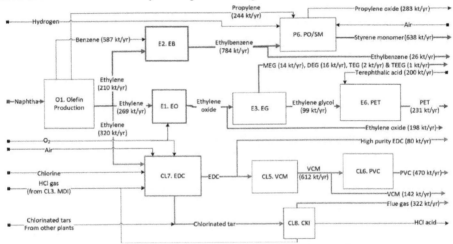

Figure 1: Simplified ethylene cluster with main mass flows

2.3. Value chain impact propagation

2.3.1. Feedstock defossilization impact on value chain

In this study, value chain defossilization is defined as the replacement of fossil-based raw materials with ACS carbon-based raw materials. To understand the extend of the defossilization impact on a value chain, Equation 5 and Equation 6 are used. The carbon contribution of a chemical building block (CBB) is defined as the amount of carbon in the value chain that originates from the CBB molecule (see Equation 5).

$$CBB\ carbon\ contribution = \frac{CBB\ carbon\ atoms}{Value\ chain\ product\ carbon\ atoms} \quad (5)$$

As in a chemical process, all the carbon used in a process does not end-up in the required product as by-products or waste are also formed. Hence, to understand the carbon flow along a value chain, the carbon utilization efficiency was also evaluated. The value chain carbon utilization efficiency (CUE) is defined as the ratio of the carbon mass flow in the value chain to the carbon mass flow of raw materials, as given in Equation 6. This equation helps to understand how efficiently the CBB carbon is used in the value chain to make the target product.

$$Value\ chain\ CUE = \frac{Value\ chain\ chemical\ carbon\ mass\ flow}{\sum_{CBB} Carbon\ mass\ flow} \tag{6}$$

2.3.2. Price change impact on value chain

It was assumed that the new ACS ethylene plant has the same capacity as the base-case ethylene production and the downstream units acquire the ethylene from the new ACS plant as it happens in the base-case scenario. For the downstream plants, it is considered their CAPEX remains the same (as the same product is produced and therefore there are no changes in equipment) but their OPEX changes due to changes in raw material price. Hence, in order to maintain the same gross margin, it is proposed that the downstream products will increase their corresponding prices (see Equation 7, and Equation 8).

$$Gross\ margin = Revenue_{base\ case} - OPEX_{base\ case} =$$
$$= Revenue_{ACS\ case} - OPEX_{ACS\ case} \tag{7}$$

$$\sum_{Products} \Delta Price * Mass\ flow = \sum_{Raw\ materials} \Delta Cost * Mass\ flow \tag{8}$$

For multi-product processes, the increase in a raw material cost needs to be allocated to different products. In this study, a constant revenue ratio between products was assumed for all of them and, based on Equation 9, product prices were allocated.

$$[\frac{Product\ Price_p * Product\ mass\ flow_p}{\sum_{All\ products} Product\ Price * Product\ Mass\ flow}]^{base\ case}$$
$$= [\frac{Product\ Price_p * Product\ mass\ flow_p}{\sum_{All\ products} Product\ Price * Product\ Mass\ flow}]^{ACS\ case} \tag{9}$$

3. Results and discussion

For the production of ethylene from CO_2 and biomass feedstocks, multiple technologies at different TRL were identified (see for example in Figure 2 the overview of CO_2-based routes).

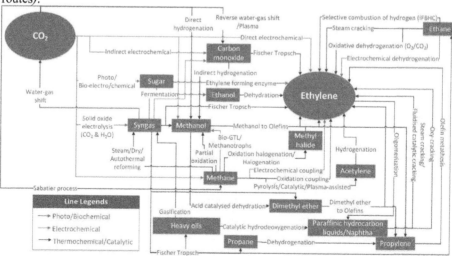

Figure 2: CO2-based process routes for ethylene production

The identified technologies were evaluated using the screening methodology and a comparative assessment for a total of 48 process routes was performed (see for example in Table 1 the comparison of four process routes). It was observed that for CO_2 –based routes, the needed theoretical electricity for the direct electrochemical route is lower than for the indirect water electrolysis route. This increase in electricity is reflected in the EC ratio, as the input cost is higher in the indirect route. For biomass routes, there is no electrochemical process step and hence the electricity need is zero. But these routes have lower CUE than electrochemical routes as biomass does not have sufficient inherent H_2 for complete conversion of carbon in biomass. Hence, carbon is not fully utilized although these routes theoretically seem economically better than electrochemical routes because of their lower EC ratio. Hence, based on the screening methodology as explained in section 2.1 using criteria: number of process steps, energy usage, CUE and EC; the selected process routes for the production of ethylene from biomass and CO_2 feedstocks respectively were: Biomass steam gasification (BSG) with Fischer-Tropsch (FT) process and direct electrochemical reduction (DER) of CO_2 to ethylene.

Table 1: Process route comparison using the described screening methodology for ethylene production (Abbreviations: DER-Direct electrochemical reduction, MTO-Methanol to olefins, BSG- Biomass steam gasification, FT- Fischer Tropsch)

Technology	No. of process steps	kJ/mol ethylene		CUE	EC
		Electricity need	Heat production		
DER of CO_2 to ethylene	1	1331	0	1.00	0.90
Water electrolysis + CO_2 to methanol + MTO	3	1423	-304	1.00	0.95
BSG + FT process	2	0	-82	0.67	0.14
BSG + syngas to methanol + MTO	3	0	-82	0.67	0.14

In this paper, only downstream plant impacts are studied using the preliminary results. The defossilization impact of ACS ethylene on an ethylene cluster, were studied as explained in section 2.3.1 and the results are tabulated in Table 2. The PVC value chain had the highest carbon impact due to the ACS ethylene production, as 100% of PVC carbon comes from ethylene. The least carbon impact was found for PET value chain, as significant part of the carbon in PET is provided by xylene, not ethylene. Hence, the impact of ethylene defossilization is non-identical for different value chains, despite sharing the same CBB. It means that different value chains based on the same carbon contribution molecule may require different defossilization strategies. The value chain carbon utilization efficiency shows that in the PVC value chain, only 66% of the ethylene entering ends-up in the PVC product. The CUE varies for different value chains based on the main reaction selectivity, conversion, product recovery rate and by-product demand. The significance of by-product demand can be observed in the PET value chain as some of the ethylene entering is used to make ethylene oxide (EO) which is an intermediate product used in other value chains. Therefore, for multi-product value chains, the defossilization impact is not just limited to the target product but can have wider impact on other dependent value chains.

Table 2: Ethylene cluster carbon impact

Value chain	Value chain theoretical carbon flow	CBB carbon contribution	Value chain CUE
PVC	$C_2H_4 + Cl_2$ $\rightarrow C_2H_4Cl_2 \xrightarrow{-HCl} C_2H_3Cl$	Ethylene= 100%	PVC= 66%, EDC=7%, VCM=20%, Waste=7%
PET	$C_2H_4 + 0.5\,O_2$ $\rightarrow C_2H_4O + H_2O$ $\rightarrow C_2H_6O_2 + C_8H_6O_4$ $\rightarrow C_{10}H_8O_4 + 2H_2O$	Ethylene= 20% P-xylene= 80%	PET=33%, PTA=19%, EG=3%, EO= 25%, Waste=20%

| SM | $C_2H_4 + C_6H_6 \rightarrow C_8H_{10}$ $\rightarrow C_8H_8 + H_2$ | Ethylene= 25% Benzene= 75% | **SM= 82%**, EB= 3%, Waste=15% |

The price impact of an ACS based ethylene plant on the value chains were studied as explained in section 2.3.2 and an ACS ethylene price increase of 67% was assumed as per literature (L.Berkelaar et al, 2022). As shown in Table 3, the PVC value chain will have the highest price impact due to the higher ACS ethylene price. This is because ethylene has the highest raw material purchase cost contribution in PVC due to its price and mass flow. However, in the PET and SM value chains, the respective product prices only increased by 17%. This is because ethylene is not the main raw material in these value chains, which can also be observed from the CBB carbon contribution values given in Table 2. This results also highlights how differently the price change impacts propagate in different value chains of the same CBB due to feedstock defossilization.

Table 3: Ethylene cluster price change impact

Value chain	Price change impact propagation (delta)
Ethylene→ EDC→ VCM→ **PVC**	67%→ 62%→ 62%→ **49%**
Ethylene→ EO→ EG→ **PET**	67%→ 47%→ 47%→ **17%**
Ethylene→ EB→ **SM**	67%→ 17%→ **17%**

4. Conclusion

A screening methodology based on stage-gate concept was developed to select promising ACS technologies for ethylene production. From 48 process routes, the selected technologies were: DER (for CO_2 feedstock) and BSG with Fischer Tropsch (for biomass feedstock). The methodology showed how the concept of stage-gate can be used to screen large number of process routes. Then the value chain impacts of feedstock defossilization for an ethylene cluster in terms of product carbon flow and price change was studied. It was observed that for ethylene value chain, PVC will have the highest carbon and price impacts due to the mass and price significance of ethylene in PVC production. It was also observed that as in the case of PET value chain, how defossilization of one value chain can effect multiple value chains due to system interdependencies.

Acknowledgements

This publication is part of the project Unravelling the impacts of using alternative raw materials in industrial clusters (with project number VI.C.183.010) of the research programme Vici DO which is (partly) financed by the Dutch Research Council (NWO).

References

J. Rissman et al., "Technologies and policies to decarbonize global industry: Review and assessment of mitigation drivers through 2070," Appl. Energy, vol. 266, no. March, p. 114848, May 2020, doi: 10.1016/j.apenergy.2020.114848.

D. Saygin and D. Gielen, "Zero-Emission Pathway for the Global Chemical and Petrochemical Sector," Energies, vol. 14, no. 13, p. 3772, Jun. 2021, doi: 10.3390/en14133772.

M. Stork, J. de Beer, N. Lintmeijer, and B. den Ouden, "Chemistry for Climate: Acting on the need for speed. Roadmap for the Dutch Chemical Industry towards 2050," Vnci, pp. 1–52, 2018.

L. Berkelaar et al., "Electrochemical conversion of carbon dioxide to ethylene: Plant design, evaluation and prospects for the future," Chemical Engineering Research and Design, vol. 182, pp. 194-206, 2022, doi: 10.1016/j.cherd.2022.03.034.

Antonis Kokossis, Michael C. Georgiadis, Efstratios N. Pistikopoulos (Eds.)
PROCEEDINGS OF THE 33rd European Symposium on Computer Aided Process Engineering
(ESCAPE33), June 18-21, 2023, Athens, Greece

Syngas fermentation to ethanol: the effects of gas recycling on economics

Haneef Shijaz,[a,b] Fausto Gallucci,[b] Adrie Straathof,[a] John Posada[a]

[a] *Department of Biotechnology, Faculty of Applied Sciences, Delft University of Technology, van der Maasweg 9, 2629 HZ, Delft, the Netherlands.*
[b] *Department of Chemical Engineering and Chemistry, Eindhoven University of Technology, 5600 MB, Eindhoven, the Netherlands*

Abstract

Syngas fermentation is a biochemical pathway to produce ethanol and has been commercialized successfully. The economic viability of this process could be further improved to become more competitive in the existing ethanol market. Improving gas utilization is the key, and can be done by recycling the unreacted syngas. This work is an early-stage techno-economic assessment of recycling in producing ethanol from Basic Oxygen Furnace (BOF) gas. Economic viability is measured in terms of Relative Competitive Percentage (RCP) and is a measure of closeness to the current market. Two scenarios, firstly a once-through process, and secondly a process with recycling (0.9 split ratio: recycle/purge) of gas is considered. None of them showed a positive RCP as compared to the current ethanol market. Comparing these scenarios, beyond the single pass conversion of 60%, the additional production costs due to recycling become dominating and lead to a lower RCP compared to once-through systems.

Keywords: Syngas fermentation, Ethanol, Basic oxygen furnace Gas, Recycling, Profitability

1. Introduction

Syngas fermentation is a ground-breaking industrial biotechnology platform to produce Ethanol from various feedstocks. Lanzatech, a carbon recycling company has Commercialized this technology at various locations around the world. At the same time, the produced ethanol is not competitive enough to the existing ethanol market based on thermo-catalytic processes (Benalcázar 2017). One of the promising ways to improve the profitability of ethanol is by increasing gas utilization. This is possible via improving the mass transfer, genetic engineering to enhance the microbial rate or recycling the unreacted reactants. The latter is a possible scenario at the commercial scale, but the same has not been studied in detail in the large-scale syngas fermentation processes. Therefore, this work is an attempt to give an early-stage techno-economic evaluation of the recycling of unreacted reactants in the gas fermentation process.

2. Methodology

The process concept for ethanol from BOF is shown in figure1. A bubble column fermenter and a distillation-based downstream process were considered as they are the most used types for ethanol production in large-scale syngas fermentation industries. Aspen Plus V8.8 was used to model this process concept in a steady state. The BOF gas with a composition (mole %) of 0.65% CO, 0.03% H_2, 0.16% CO_2, and 0.16% N_2 was the feedstock and a process reaction to ethanol was developed by the black box thermodynamic approach (Heijnen 1992). Gas-liquid mass transfer rates of the reactants were also integrated into the model using the empirical relations of the bubble column reactor for the mass transfer coefficient, and Henry's law for the compositions. The sizing of the units for economic evaluations was carried out either manually or by using Aspen Process Economic Analyzer. The economic viability of ethanol via syngas fermentation was checked in terms of relative competitive percentage (RCP), which is the measure of its marketability compared to the current ethanol market. The effect of recycling unreacted gas was studied via sensitivity analysis, and RCPs were measured in each case.

2.1. Process Description

As shown in figure 1, a mixture of fresh BOF gas and recycle gas is adjusted to the inlet pressure and temperature of the fermenter. In the bubble column fermenter, the gas (syngas + NH_3) and the medium (water+NH_4OH) are fed counter currently. The ethanol is fully stripped off by the gas leaving the fermenter, and it mainly contains product, water, and other unreacted components of syngas. Therefore, the gas is condensed first and sent to a phase separator. The liquefied stream mainly contains ethanol and water with other dissolved gases. Ethanol is further purified in a distillation-based downstream separation section. A part of the unreacted gas is purged (10%) to avoid the accumulation of H_2 in the reactor. The rest of the gas (90%) is sent back to the fermenter after separating N_2, and CO_2. The fermenter broth mainly contains water, biomass, & ammonium acetate, and the solids are separated before reusing the water in the fermenter. The fermenter operating conditions (Temperature at 37°C, pH at 6) are selected based on the optimal growth conditions for acetogens. The top pressure in the fermenter is 1 atm and the bottom pressure was calculated based on the hydrostatic pressure as follows;

$$p_b = p_t + \rho_{broth} * g * h_g \tag{1}$$

Where,

p_b	Bottom Pressure (atm)
p_t	Top Pressure (atm)
ρ_{broth}	The density of broth (kg/m^3)
g	acceleration due to gravity (m/s^2)
h_g	Height of the broth (m)

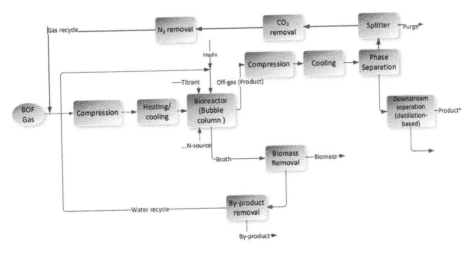

Figure 1 schematic block diagram of ethanol from BOFG via syngas fermentation

2.2. Economic evaluation

To calculate the economic viability of the process, the capital expenditure (CAPEX), and the annual operating cost (OPEX) are estimated using the factorial method proposed by Peters and Timmerhaus (1959). As given in Table 1, The CAPEX is estimated based on the bare equipment cost (BEC) of the units. These values are adjusted to the scale, and the base year of 2019 using six tenth rule and cost indices respectively. OPEX is evaluated based on the direct calculations from the process models and is called process-related cost (PRC). PRC consists of the cost of raw materials, utilities, and waste management. Raw material includes syngas, NH_3, NH_4OH, and deionized water. The cost of utilities is evaluated from the energy balances and average utility price in the Netherlands (2019). The cost of waste management is estimated using the avg. carbon releasing tax in Europe (5 Euro/ ton CO_2 release, 2019). For the depreciation cost, a straight-line depreciation method is used with 10% of the purchase cost as salvage value after a plant life of 15 years. The annual tax paid is calculated from the annual sales and the tax rate which is taken as 25%.

To measure the economic viability, the relative competitive percentage (RCP) is measured as follows;

$$RCP = 100 * \frac{(MP - MSP)}{MP} \tag{2}$$

Where MP is the market price of ethanol in 2019 and is 780 €/kg. And Minimum Selling Price (MSP) is evaluated for a payback time of 3 years.

Table 1 Calculation of CAPEX (Peters, 1959)

	Items	Factors on FCI (%)
Direct cost (DC)	Bare Equipment cost	21.8
	Installation	7.6
	Process piping	11.1
	instrumentation	4
	insulation	1.4
	electrical	4.7
	buildings	8.4
	yard improvements	3.1
	service facilities	7.2
	land	1.7
Indirect cost (IDC)	Engineering	7.5
	construction	9.7
	contractor's fee	4
	contingency	7.8
Fixed Capital Investment (FCI)	Sum up above	100
Start-up related cost	working capital	OPEX/12
	start-up cost	9
CAPEX	FCI + working capital + start-up cost	

Table 2 Calculation of OPEX (Peters, 1959)

	Items	Calculation
Facility-dependent costs	Depreciation	(DC-PC*0.1)/10
	maintenance	0.15 purchase cost
	insurance	0.01 TFC
	local taxes	0.01 TFC
	plant overhead	0.05 Revenue

Process related cost	Utility	
	Raw material cost	
	Waste management	
	(CO$_2$ tax)	
Labour-related cost	labour	0.07 (PRC+FDC)
	Laboratory charges	
		0.1 labour
OPEX	Sum up above	

2.3. Parametric studies

A sensitivity study was carried out for various conversions and the corresponding recycling ratios. The purge was fixed at 10%, and the rest 90% are recycled in all cases. In each case, the RCPs have been calculated.

3. Results

Figure 2 shows the relation between the single pass conversion of CO, and the recycle ratio (recycle flowrate/(fresh syngas+ recycle flowrate)). In all cases, the split ratio of recycle flow rate is fixed at 0.9. As we see, a lower conversion would lead to high recycle flow rates.

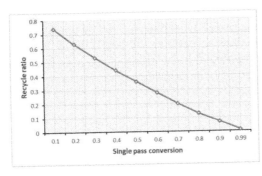

Figure 2 Single pass conversion vs Recycle ratio

The effect of recycling on the economic feasibility of the process is shown in figure 3. The economic feasibility is represented in RCP as indicated in the previous section. It shows the positive or negative economic viability of ethanol as compared to the present market. Two scenarios are compared in this figure. Firstly, the RCP of the once-through process (no recycling) and secondly RCP of the process with recycling of unreacted CO. None of the cases showed positive viability compared to the existing market, and therefore further optimization of the process, and support (subsidies) may be required to bring ethanol from syngas fermentation to the market.

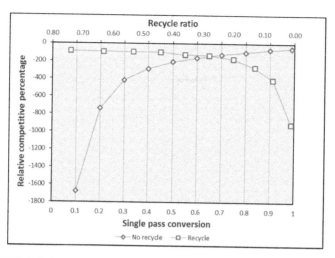

Figure 3 RCP (relative competitive percentage) for once through process and process with recycling in the production of ethanol from BOF gas via syngas fermentation.

Comparing the scenarios, recycling has improved the economic viability in multiple folds at lower conversions. It is due to the improved gas utilization, and production rates of ethanol. This is the trend up to around 60% of single pass conversion, and then the additional production cost of recycling is dominating. Therefore, at higher conversion rates, scenarios without recycling have better economic viability.

These results show a clear tradeoff between additional recycling costs and productivity improvement due to recycling. Therefore, to recycle the unreacted CO in improving the economic viability, the economic hot spots due to recycling must be identified, These units or variables must be optimized to minimize the recycling cost. Similarly, the limiting mechanism of improved production performance must be identified and must be tuned further to fully benefit from recycling CO back to the fermenter.

Acknowledgements

This publication is part of the project Microsync (with project number P16-10 project 7 of the research programme Perspectief Novel Approaches for Microbial Syngas Conversion to Chemical Building Blocks 2017 TTW. Project title: 'Process design and sustainability assessment' which is (partly) financed by the Dutch Research Council (NWO).

References

Benalcázar, Eduardo Almeida, Henk Noorman, Rubens Maciel Filho, and John Posada. 2017. Production of bulk chemicals from lignocellulosic biomass via thermochemical conversion and syngas fermentation: a comparative techno-economic and environmental assessment of different site-specific supply chain configurations. Biofuels, Bioprod. Bioref. 11:861–886.

Heijnen J.J et.al. 1992. A blackbox mathematical model to calculate auto- and hetrotrophic biomass yields based on Gibbs energy dissipation. Biotechnology and Boengineering. 40. 10:1139- 1154.

Peters, M. S. & Peters, J. I.1959. Plant design and economics for chemical engineers. *Eng. Economics.*

2019. DACE price booklet. 33rd edition.

2019. Carbon taxes in Europe. Tax foundation. *Taxfoundation.org*

Antonis Kokossis, Michael C. Georgiadis, Efstratios N. Pistikopoulos (Eds.)
PROCEEDINGS OF THE 33rd European Symposium on Computer Aided Process Engineering
(ESCAPE33), June 18-21, 2023, Athens, Greece

Modeling, simulation and techno-economic analysis of an integrated biorefinery based on halophytes

Tutku Taşçı Çilak,[a] Sanketkumar Raval,[a] Sylvia Fasse,[a] Mette H. Thomsen,[b] Axel Gottschalk,[a]

[a]*Institute of Process Engineering, Bremerhaven University of Applied Sciences, An der Karlstadt 8, 27568, Bremerhaven, Germany*
[b]*Department for Energy Technology, Aalborg University, Esbjerg Campus, Niels Bohrs Vej 8, 6700 Esbjerg, Denmark*

Abstract

Many existing biorefineries use limited feedstocks and technologies to produce biofuels. However, an integrated biorefinery provides a variety of feedstocks and highly marketable final products leading to better economic viability. For this reason, this paper aims to perform a techno-economic analysis of the whole integrated biorefinery with modeling of the process in a flowsheet simulation tool, here Aspen Plus®. After having mass and energy balances, the assessment of technical aspects, and the economics of the integrated biorefinery process is the step forward in the development of a sustainable and biobased industrial process. In that matter, capital expenditures (CAPEX), operational expenditures (OPEX) and production costs are estimated for the integrated biorefinery, processing lignified halophyte biomass.

Keywords: Biorefinery, second-generation feedstock, value-added compounds, flowsheet simulation, cost-estimation.

1. Introduction

To cover the rising demand for food and energy is a strengthening challenge in today's world. Additionally, there is a strong need for sustainable production processes. Therefore, the concept of circular economy is more and more applied in biorefineries. There are numerous options to design a biorefinery based on feedstocks used, processes applied and products yielded. In second-generation feedstock biorefineries, wastes from agriculture and lignocellulosic biomass are processed. Thus, the decline in agriculturally usable land can be circumvented (Kamm et al. 2000). But the lignocellulosic and recalcitrant structure of the biomass with lignin, cellulose, and hemicellulose requires more process steps than first-generation feedstocks. For the production of biofuels, the thus higher production cost can make the biorefinery not competitive with fossil sources. Integrated biorefineries are advanced biorefineries including multiple processes and products for different industries to maximize the biomass valorization (Patel and Shah 2021). Hence, the incorporation of the extraction of value-added compounds in integrated biorefineries can improve economic performance (Kamm et al. 2000).

In this paper, the biomass category in focus is halophytes. These are salt-tolerant plants, which, in addition to their capability of growing without freshwater supply, are also yielding a range of bioactive compounds. These secondary metabolites are synthesized as part of the halophyte's survival mechanism in saline environments. Those phenolics act as natural antioxidants and are of interest for their application in a variety of foods, cosmetics, and pharmaceuticals. This is additionally based on their antimicrobial activity against several pathogenic microorganisms (Jdey et al. 2017). Several halophyte species are traditional medical and nutritional plants, with the edible tips being part of gourmet foods (Antunes et al. 2021). But the dominant part of the plant is lignocellulosic biomass, which is to be valorized in circular economy biorefineries (Brown et al. 2014).

The halophyte species in the scope of this article is *Salicornia ramosissima*, which is already available on the market and can be harvested in saline areas across Europe. Further *S. ramosissima* biomass has elevated amounts of hydroxycinnamic acids and is therefore of interest for industrial valorization (Antunes et al. 2021). The lignified biomass contains extractable phenolic compounds such as hydroxycinnamic acids (HCAs) and polyphenolics, being of interest for the cosmetic and pharmaceutical industry.

Consequently, extraction methods are established also including the utilization of residual biomass for biogas and biochar production. Hence, it is a combination of different processing steps such as biomass fractionation and the production of biogas and biochar after the recovery of value-added compounds.

In the course of the process design of a whole integrated biorefinery, it is important to perform a techno-economic analysis to evaluate its technical feasibility and economic performance. Additionally, the choice of the technologies to be implemented, the operating conditions of the process units, and their sizing can be optimized through techno-economic assessment. (Bangalore Ashok et al. 2022).

2. Conceptual design of the integrated biorefinery

The conceptual process design of the integrated biorefinery with its individual process steps is shown in Figure 1.

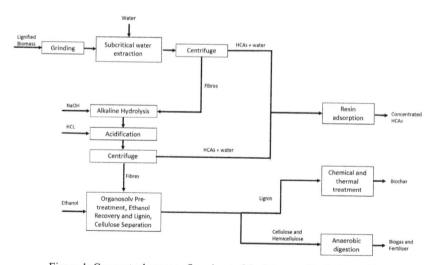

Figure 1: Conceptual process flowsheet of the integrated halophytes biorefinery

As a raw material for the process, lignified biomass was assumed to be sundried in the field, therefore grinding is the first step as pretreatment to reduce the particle size. It is followed by subcritical water extraction (SWE), a non-conventional extraction process (Taşçı Çilak et al. 2022). Downstream the SWE process, the remaining fibers are sent to alkaline hydrolysis to extract remaining HCAs. Alkaline hydrolysis at elevated temperatures is used as post-treatment of SWE. After alkaline hydrolysis, extracts are acidified by using hydrochloric acid (HCl aq.) to lower the pH value. Consequently, both extracts from acidification and SWE are combined and by ion exchange resin to concentrate HCAs and remove sugars, salts, and all water-soluble contaminants. To realize maximum benefits, residual lignified fibers are subjected to organosolv pretreatment to generate streams that are rich in cellulose and lignin. In the organosolv process, lignin is separated from cellulose and hemicellulose by using 50/50 v/v ethanol and water solution. After that, the cellulose rich stream is valorized in anaerobic digestion to generate biogas while the lignin stream is sent to a biochar production unit.

The process flowsheet modelling and simulation of the integrated biorefinery is carried out in the commercial computer-aided process engineering (CAPE) tool Aspen Plus® V11 to enable mass and energy balance calculations.

3. Methods

3.1. Process design and assumptions

In this study, the main focus is on the extraction of value-added compounds from the lignified biomass of the halophyte species, *S. ramosissima*. The respective approach offers the opportunity to expand the product portfolio by adding biogas and biochar as products along with high-value extracts. As a starting point to set the assumption regarding the capacity of the biorefinery, a production scenario of 1000 tons/year of lignified biomass is considered as the biorefinery capacity with 8000 h of operating hours. Furthermore, Germany is chosen for the biorefinery location. In the simulation tool (Aspen Plus®), the composition of the biomass is defined based on dry weight as 27 wt% ash, 5 wt% proteins, 3 wt% lipids, 9.5 wt% lignin, 11.5 wt% cellulose, 17 wt% hemicellulose, and 27 wt% extractives. Also, the moisture content in the feed is defined to be 15 wt%. In the process concept, recovery of lignin is assumed as 64 wt% whereas cellulose retention is 70 wt%. Also, the recovery of the targeted HCAs is estimated as 85 wt%. Assumptions, which are taken in this article, are inspired from the overall framework of the AQUACOMBINE project (AQUACOMBINE). In the flowsheet simulation model, the thermodynamic property packages 'Solid' and 'NRTL' have been applied. The physical properties such as boiling point, molecular weight and heat capacities of components being not available in the Aspen Plus® database (e.g. lignin, cellulose) have been entered by hand. Conventional compounds are selected from the AspenPlus® database. Moreover, the models that are used to define the unit operation of the process are given in Table 1.

Table 1: List of equipment and its model used in Aspen Plus®

Separation process for HCAs		Organosolv process and biochar generation	
Equipment	Model in Aspen Plus®	Equipment	Model in Aspen Plus®
Crusher	Gyratory Crusher	Mixer	Mixer
Reactor	Rstoic	Reactor	RStoic
Separator	Flash2	Separator	Flash2
Heat exchangers	Heater	Heat exchangers	Heater
Resin Adsorbtion	User Model	Washing	Swash
Centrifuge	Decanter	Separator	Sep
Pump	Pump	Distillation	RadFrac
		Precipitator	RStoic

3.2. Cost Estimation Method

In the simulation of the conceptual process, as shown in Figure 1, the various types of equipment such as heat exchangers, crushers, pumps, reactors, and vessels were used. The costs of each equipment have been calculated based on parameters like temperature, pressure, mass flow rate, and utilities which were compiled from the simulation. In addition to that, material of construction, efforts for installation, etc. were considered in the equipment costs. Furthermore, the whole economic assessment has been done with reference to the year 2019, using the chemical engineering plant costs index (CEPCI), a currency conversion rate of 0.87 USD/EUR and the estimated lifetime of the plant being 20 years. The cost estimation method based on the individual factor method of Guthrie (1969, 1974) is followed (Seider et al. 2009) to evaluate the total capital investment (TCI). Initially, the purchase costs of equipment are determined using empirical equations. Further, the respective cost estimation method is based on the bare module factors which are based on the type of equipment (e.g. pumps: 3.30, shell and tube heat exchanger: 3.17, crusher: 1.39). The sum of all bare module costs of equipment, process machinery, spares, costs of catalysts, etc. gives the total bare module investments (TBM). The direct permanent investment (DPI) includes the sum of the costs for site preparation, services facility and allocated cost for utility plant and related facilities as they are shown in Table 2.

Table 2: Assumptions for calculating total capital investment (Seider et al. 2009)

Item	Assumption
Total bare module investment TBM	Contains costs related to purchase and installation of equipment
Cost of site preparation	5 % of TBM
Cost of service facilities	1.5% of TBM
Allocated cost for utility plant and related facilities	14% of TBM
Total of direct permanent investment DPI	Contains costs of TBM, site preparation, service facility, utility plants and related facility
Cost of contingencies and contractor's fee	18% of DPI
Total depreciable capital TDC	Contains costs of DPI, contingencies and contractor's fees
Cost of land	2% of TDC
Cost of royalties	2% of TDC
Cost of plant start up	10% of TDC
Total permanent investment TPI	Contains costs of TDC, land, royalties and plant start up
Total permanent investment TPI corrected	120% of TPI
Working capital	18% of TPI corrected
Total capital investment TCI	Sum of costs of TPI and working capital

Along with the further costs such as costs of contingencies and contractor's fees, costs of land, royalties, plant start-up, and working capital are calculated based on the percentages of TDC Finally, the TCI is calculated as the sum of total permanent investment corrected and working capital (Seider et al. 2009).

4. Results

Based on the simulation results and selected operational parameters, the sizing of equipment has been performed. The cost estimation has been performed and the TCI has been calculated as 6869 k€ which covers the investment required for the considered biorefinery concept. The breakdown of the TCI is depicted in Table 3. The TBM is 2993 k€ whereas the total permanent investment corrected is 5821 k€. The addition of working capital (1048 k€) into the total permanent investment corrected provides the TCI.

Table 3: Break down of TCI

Costs	k€
Total bare module investment TBM	2993
Cost of site preparation	150
Cost of service facilities	45
Allocated cost for utility plant and related facilities	419
Total of direct permanent investment DPI	3606
Cost of contingencies and contractor's fee	649
Total depreciable capital TDC	4255
Cost of land	85
Cost of royalties	85
Cost of plant start up	426
Total permanent investment TPI	4851
Total permanent investment TPI corrected	5821
Working capital	1048
Total capital investment TCI	6869

Along with the estimation of total capital investment, the annual production costs are calculated and amount 4210 k€/year which is shown in Table 4.

Table 4: Break down of total production costs

Costs	k€/year
Raw material costs (excl. Biomass costs)	0.24
Utility costs	880
Direct wages and benefits (DW&B)	1300
Labour related operations (O)	1773
Maintenance cost (M)	343
Operation overhead	383
Property tax and insurance	85
Depreciation, D	167
Licensing fees	87
Cost of manufacturing	2869
General expenses	461
Total production cost	**4210**

The total production costs have been performed as the sum of variable cost items such as operational expenditures (OPEX) including raw material (e.g. water, ethanol, etc.) and utility costs (e.g. steam (2 bara, 120°C), cooling water (15°C), electricity, etc.) and other cost items as fixed operating expenses and general expenses. Fixed operating expenses include direct wages and benefits (DW&B) with consideration of 5 shifts, including illness and holidays, costs of labour related operations, maintenance costs, costs of operation overhead, cost of property tax and insurance (2% of total depreciable capital (TDC)), licensing fees (2% of sales), and cost of manufacturing. Moreover, the depreciation (D) is taken into consideration which is 4% of total depreciable capital (TDC). Also, selling expenses, research and development costs, management incentive compensation have been included in general expenses.

5. Conclusion

A techno-economic analysis on the integrated biorefinery processing lignified halophyte biomass was presented with the conceptual process design. The corresponding energy and mass balances have been calculated by the flowsheet simulation tool Aspen Plus®.

The estimated TCI and production costs amount to 6869 k€ and 4210 k€/year respectively. The selling prices of the high-value products are not included in this study because they are depending on the target sectors such as food, cosmetics and pharmaceutical industries. Hence, in this study total capital cost and production cost were aimed to be evaluated without focusing on revenue from the products. Nevertheless, in the context of the economics of the conceptual process, many factors such as the capacity of the biorefinery, production rate, type of feedstock, and technologies have an impact. In that regard, the sensitivity analysis and energy integration are required to reduce the cost in upcoming work. Therefore, the presented results in this study are preliminary, as they will be optimized and improved in the future.

6. Acknowledgements

This project has received funding from the European Union's Horizon 2020 research and innovation program under Grant Agreement No 862834. Any results of this project reflect only this consortium's view and the European Research Executive Agency is not responsible for any use that may be made of the information it contains.

References

M. Antunes; C. Gago; A. Guerreiro; A. Sousa et al., 2021, Nutritional Characterization and Storage Ability of *Salicornia ramosissima* and *Sarcocornia perennis* for Fresh Vegetable Salads, Horticulturae, 7, 1, p. 6. https://doi.org/10.3390/horticulturae7010006

R. P. Bangalore Ashok; P. Oinas; S. Forssell, 2022, Techno-economic evaluation of a biorefinery to produce γ-valerolactone (GVL), 2-methyltetrahydrofuran (2-MTHF) and 5-hydroxymethylfurfural (5-HMF) from spruce, Renewable Energy, 190, pp. 396–407. https://doi.org/10.1016/j.renene.2022.03.128

J. J. Brown; I. Cybulska; T. Chaturvedi; M. H. Thomsen et al., 2014, Halophytes for the Production of Liquid Biofuels, Sabkha Ecosystems, Springer, Dordrecht, pp. 67–72. DOI: 10.1007/978-94-007-7411-7_4

A. Jdey; H. Falleh; S. Ben Jannet; K. Mkadmini Hammi et al., 2017, Anti-aging activities of extracts from Tunisian medicinal halophytes and their aromatic constituents, EXCLI Journal, 16, pp. 755–769. DOI: 10.17179/excli2017-244

B. Kamm; P. R. Gruber; M. Kamm, 2000, Biorefineries-Industrial Processes and Products, Ullmann's Encyclopedia of Industrial Chemistry, Wiley-VCH Verlag GmbH & Co. KGaA, pp. 1–38. https://doi.org/10.1002/14356007.l04_l01

A. Patel; A. R. Shah, 2021, Integrated lignocellulosic biorefinery: Gateway for production of second generation ethanol and value added products, Journal of Bioresources and Bioproducts, 6, 2, pp. 108–128. https://doi.org/10.1016/j.jobab.2021.02.001

W. Seider; J.D. Seader; D. R. Lewin; S. Widagdo, 2009, Product and process design principles Synthesis, analysis and evaluation. 3. ed. Hoboken, NJ: Wiley.

T. Taşçı Çilak; S. Fasse; S. Raval et al., 2022, Extraction Methods of Value-Added Compounds from Biomasses, European Biomass Conference and Exhibition Proceedings, pp. 512–528 DOI: 10.5071/30thEUBCE2022-3CV.6.26

AQUACOMBINE, 2019-2023, EU H2020 IA project, Grant agreement ID: 862834, https://doi.org/10.3030/862834

Antonis Kokossis, Michael C. Georgiadis, Efstratios N. Pistikopoulos (Eds.)
PROCEEDINGS OF THE 33rd European Symposium on Computer Aided Process Engineering
(ESCAPE33), June 18-21, 2023, Athens, Greece

Optimal Control for Deriving Policies for Global Sustainability

Urmila Diwekar[1,2], Apoorva Nisal[2], Yogendra Shastri[3], and Heriberto Cabezas[4]

1. Vishwamitra Research Institute, Crystal Lake, IL, USA
2. The University of Illinois at Chicago, Chicago, IL, USA.
3. Indian Institute of Technology, Mumbai, India.
4. University of Miskolc, Miskolc, Hungary

Abstract

Inordinate consumption of natural resources by humans over the past century and unsustainable growth practices have necessitated a need for enforcing global policies to sustain the ecosystem and prevent irreversible changes. This study utilizes the Generalized Global Sustainability model (GGSM), which focuses on sustainability for the Food-Energy-Water (FEW) Nexus. GGSM is a 15-compartment model with components for the food web, microeconomic framework, energy, industry and water sectors, and humans. %It was validated based on historical data for global sectors and can predict population, global and regional water stress, GHG emissions, and the gross domestic product (GDP) for the next century. GGSM shows that an increased per capita consumption scenario is unsustainable. In this study, an optimal-control theory-based approach is devised to address the unsustainable scenario through policy interventions to evaluate sustainability by employing multiple global indicators and controlling them.

Keywords: GGSM, global sustainability, Fisher information, optimal control

1. Introduction

Global models depicting the state of the world help in the study of the complex interconnected dynamics of sustainable growth and provide a scientific basis for the decision-making process behind global policies. Some of the noteworthy global models are given below.

- WORLD 3 Model (Meadows et al., 1972[1,2])
- GUMBO Model (Boumans et al., 2002[3])
- Puliato et al. (2008)[4]: "Predator-Prey" model for population and gross domestic product (GDP)
- HANDY Model (Motesharrai et al., 2014[5])
- EARTH 3 Model (Randers et al., 2017[6])
- HARMONEY Model (King et al., 2020 [7])

However, these models have the following deficiencies.

- These models are mostly empirical models
- They are not focused on FEW nexus
- They are mostly descriptive and not Prescriptive.

To circumvent these deficiencies, we developed a new global model called Generalized Global Sustainability Model (GGSM). This model consists of differential-algebraic equations focused on FEW nexus, both descriptive and prescriptive.

2. The GGSM Model

In this section, the generalized global sustainability model and its various features will be described. The current model is conceptually similar to the previous model (Kotecha et al., 2013 [8]). This work's exception is that it is a global model and focuses on the food-energy-water (FEW) nexus. GGSM was created to represent a simplified global ecosystem with enough detail to allow the pursuit of further study. The 15-compartment model consists of a simplified ecological food web set in a macroeconomic framework and a rudimentary legal system. The different compartments in the model, shown in Fig 1, are three primary producers (P1, P2, and P3), three herbivores (H1, H2, and H3), two carnivores (C1 and C2), human households (HH), industry (IS), energy producer (EP), fuel source, water reservoir, resource pool (RP) and an inaccessible resource pool (IRP).

The food web is modeled by Lotka–Volterra type expressions, whereas the economy is represented by a price-setting model wherein firms and human households attempt to maximize their economic well-being. The model development is done using global-scale data for stocks and flows of food, energy, and water, which were used to parameterize this model. Appropriate proportions for some of the ecological compartments, like herbivores and carnivores, are used to model those compartments. The modeling of the human compartment was carried out using historical data for the global mortality rate. Historical data were used to parameterize the model.

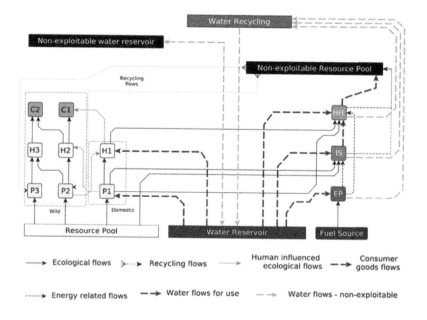

Fig 1. Generalized Global Sustainability Model (GGSM). Compartments: primary producers (P1, P2, P3), herbivores (H1, H2, H3), carnivores (C1, C2), human households (HH), industry sector (IS), energy producer (EP), fuel source, water reservoir, resource pool, inaccessible resource pool.

We used data for key variables like the human population, GDP growth, and greenhouse gases like CO2 and NOX emissions [9-12] to validate the model, as shown in Figure 2. The validated model becomes the foundation for the next part of the study and will be referred to in the following section as the "Base Case ."The model was then used to make long-term forecasts and study global sustainability over time.

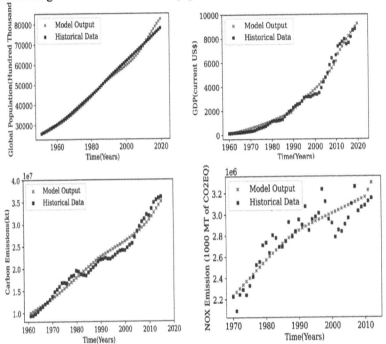

Figure 2: Results of Model Validation for Key Outputs

3. Scenario Analysis

The base case scenario represents the current state of the world. The other scenarios we considered are given below. We consider the forecast up to the year 2150.

- Scenario 1: Base Case: Business at the current level
- Scenario 2: Population Explosion: Human population grows at a rapid rate and then tapers off.
- Scenario 3: Consumption Increase: Consumption levels increase at a higher rate
- Scenario 4: Population Explosion with Consumption Increase: Both the population and consumption levels increase together

The ecological compartment results of scenarios 1 and 2 show that the planet can support the increase in population if human consumption does not increase significantly. However, for scenarios 3 and 4, the compartments start dying, as shown in Figure 3 for primary producer P1.

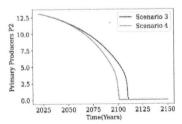

Figure 3: Scenarios 3 & 4, for P1

We know that these scenarios are the likely scenarios of our future. This means our planet will become unsustainable within this time frame. We are using optimal control theory to derive optimal techno-socioeconomic policies to make the planet sustainable in the above time frame.

4. Deriving Techno-Socio-Economic Policies

GGSM showed that an increased per capita consumption scenario is unsustainable. In this study, an optimal-control theory-based approach is devised to address the unsustainable scenario through policy interventions to evaluate sustainability by employing multiple global indicators and controlling them given below.

- Fisher Information (FI): Measurement of the stability of the system
  - Fisher Information (FI) has been used as a measure of sustainability (Doshi et al. 2015 [13])
  - Thus, the objective for the optimal control problem is to minimize the variance of FI between the unsustainable system and an ideal sustainable system or the base case
- Ecological Buffer
  - Equivalent land demand of the population or land required to support the annual average consumption and the waste production of an individual
- Green Net Product(GNP)
  - Estimation of sustainability from a social welfare standpoint
  - Sum of the real value of consumption and real value of net investments
- Gross Domestic Product (GDP)
  - Economic performance evaluation
- Green House Gases (GHG)
  - CO2 and NOX emissions to regulate total CO2EQ contribution by humans
- Global Water Stress
  - A measure of the stress on global water systems

The optimal control model assesses these multiple objectives by minimizing the variance in the Fisher Information. One significant result from this study is that optimizing for the Fisher Information-based objective is adequate to attain sustainability and manage the other objectives under consideration, as can be seen in Figure 4. Thus, forgoing a multi-objective problem framework. The results show that cross-dimensional policy interventions (Figure 5) such as increased vegetarianism (a), increased industrial efficiency (b), and increased penalty on industrial discharge (c) are shown to have a positive impact on scale. It should be noted that this increase in discharge fees for industries does not affect the GDP, as shown in Figure 4.

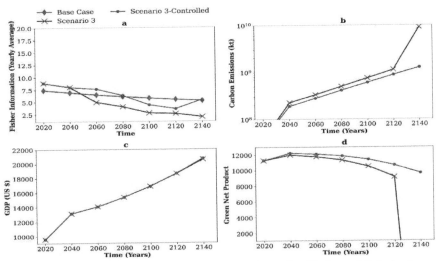

Figure 4: Behavior of Various Objectives when Fisher Information Variance Minimized

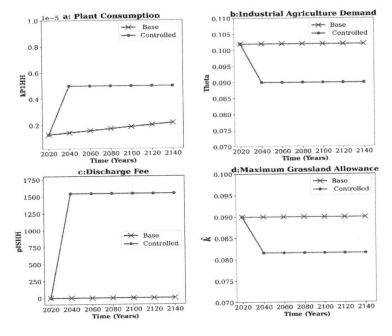

Figure 5: Optimal Profiles of Some of the Decision Variables

5. Summary

The Generalized Global Sustainability Model (GGSM), including ecological and economic dimensions to study the Food, Energy, and Water (FEW) Nexus is developed. Irrational human consumption would lead to an unsustainable future. Water availability is adequate, but regional water stress can be concerning. **Fisher Information**-based optimization is enough to manage other sustainability objectives. GGSM can provide **techno-socioeconomic policy solutions** to make sustainable changes globally

References

1. The World3 Model: Classic World Simulation | Insight Maker. The World3 Model: Classic World Simulation | Insight Maker; 2021.
2. Meadows D, Randers J, Meadows D. Limits to growth: The 30-year update. Chelsea Green Publishing; 2004.
3. Boumans R, Costanza R, Farley J, Wilson MA, Portela R, Rotmans J, et al. Modeling the dynamics of the integrated earth system and the value of global ecosystem services using the GUMBO model. Ecological economics. 2002; 41(3):529–560.
4. Puliafito SE, Puliafito JL, Grand MC. Modeling population dynamics and economic growth as competing species: An application to CO2 global emissions. Ecological Economics. 2008; 65(3):602–615.
5. Motesharrei S, Rivas J, Kalnay E. Human and nature dynamics (HANDY): Modeling inequality and use of resources in the collapse or sustainability of societies. Ecological Economics. 2014; 101:90–102.
6. Randers J, Rockstro"m J, Stoknes PE, Goluke U, Collste D, Cornell SE, et al. Achieving the 17 Sustainable Development Goals within 9 planetary boundaries. Global Sustainability. 2019; 2.
7. King CW. An integrated biophysical and economic modeling framework for long-term sustainability analysis: The HARMONEY model. Ecological Economics. 2020; 169:106464.
8. Kotecha P, Diwekar U, Cabezas H. Model-based approach to study the impact of biofuels on the sustainability of an ecological system. Clean Technologies and Environmental Policy. 2013; 15(1):21–33.
9. The World Bank. CO2 Emissions (kt) Data; 2020. Available from https://data.worldbank.org/indicator/EN.ATM.CO2E.KT.
10. The World Bank. GDP (current USD) Data; 2020. Available from: https://data.worldbank.org/indicator/ NY.GDP.MKTP.CD.
11. The World Bank. Nitrous Oxide Emissions (thousand metric tons of CO2 equivalent) Data; 2020. Available from:
 https://data.worldbank.org/indicator/EN.ATM.NOXE.KT.CE.
12. United Nations D. World Population Prospects Highlights, 2019 revision Highlights, 2019 revision. United Nations; 2019.
13. Doshi R, Diwekar U, Benavides PT, Yenkie KM, Cabezas H. Maximizing sustainability of ecosystem Technologies and Environmental Policy. 2015; 17(6):1573–1583.

Computer Code and All the data for the GGSM model is available from: https://zenodo.org/record/6331602.

Antonis Kokossis, Michael C. Georgiadis, Efstratios N. Pistikopoulos (Eds.)
PROCEEDINGS OF THE 33rd European Symposium on Computer Aided Process Engineering
(ESCAPE33), June 18-21, 2023, Athens, Greece

Application of CAPE Tools into Prospective Life Cycle Assessment: A Case Study in Feedstock Recycling of Waste Plastics

Yasunori Kikuchi,[a,b,c*] Yuki Nomura,[c] Takuma Nakamura,[c] Shoma Fujii,[a] Aya Heiho,[b] Yuichiro Kanematsu[b]

[a] *Institute for Future Initiatives, The University of Tokyo, 7-3-1 Hongo, Bunkyo-ku, Tokyo 113-8654, Japan*
[b] *Presidential Endowed Chair for "Platinum Society", The University of Tokyo, 7-3-1 Hongo, Bunkyo-ku, Tokyo 113-8656, Japan*
[c] *Department of Chemical System Engineering, The University of Tokyo, 7-3-1 Hongo, Bunkyo-ku, Tokyo 113-8656, Japan*
ykikuchi@ifi.u-tokyo.ac.jp

Abstract

In this study, we are tackling systems design with assessments for emerging technologies. Because of the data limitation on the systems and processes adopting emerging technologies, computer-aided process engineering (CAPE) tools such as process design heuristics, process simulation, optimization, parametric analysis for characterizing sensitivity and alternative generation. In this paper, we examine the feedstock recycling of waste plastics for raw materials in refineries. As for the pretreatment process of waste plastics, catalytic pyrolysis is applied. By modifying the condition of catalysts and reaction conditions, thermal decomposition of waste plastics into chemical feedstock was experimentally demonstrated to obtain low-molecular-weight hydrocarbons. This study converts the experimental results into process simulation models to conduct prospective life cycle assessment (LCA). For such purpose, the mathematical models on catalytic reactions and oil refinery were developed. The results showed that the performance of plastics pyrolysis could be linked with the improvements of environmental performances.

Keywords: catalytic pyrolysis, LCA, process modeling, process simulation.

1. Introduction

When carbon neutrality (CN) is achieved, assuming that no fossil carbon is used, carbon sources in the chemical industry will be largely limited to those from biomass, recycling, and CO_2 capture from the air (Meng et al., 2022). The current availabilities of these carbon sources are not sufficient to meet the demand for chemical raw materials due to their individual issues such as the limitation of the expansion of agriculture and forestry, the quality of waste-derived carbon, and the low technology readiness level (TRL). While the availability of biomass-derived carbon is limited (Energy Transition Commission, 2021), it should be utilized as carbon sources considering the effect of de-fossilization if waste biomass is included (Kikuchi et al., 2022) and the economic spill-over effects on agriculture and forestry (Kikuchi et al., 2020). Waste plastics as resources can be pyrolyzed to become hydrocarbons similar in composition to those used in conventional refineries (Koller et al., 2022). Although the qualities of waste plastics can be largely different from the collection routes and the purpose of use of plastic products, there is a possibility that they can be converted into hydrocarbons inputted to refineries through appropriate treatment.

In this study, we are tackling systems design with assessments for emerging technologies. Because of the data limitation on the systems and processes adopting emerging technologies, computer-aided process engineering (CAPE) tools such as process design heuristics, process simulation, optimization, parametric analysis for characterizing sensitivity and alternative generation. In this paper, we examine the feedstock recycling of waste plastics for raw materials in refineries. As for the pretreatment process of waste plastics, catalytic pyrolysis is applied. By modifying the condition of catalysts and reaction conditions, thermal decomposition of waste plastics into chemical feedstock was experimentally demonstrated to obtain low-molecular-weight hydrocarbons. This study converts the experimental results into process simulation models to conduct prospective life cycle assessment (LCA). For such purpose, the mathematical models on catalytic reactions and oil refinery were developed. For catalytic cracking, the yield of the reaction was set and the inventory per unit of plastics was estimated by process simulation using Aspen HYSYS™. The yields and complexity factors for each unit operation in the refinery were taken from previous inventory data, and the energy consumption of the entire refinery was distributed to each process by the throughput and complexity factors. Naphtha reforming was assumed to be equivalent to that in the refinery, and inventory data for naphtha cracking were taken from a LCA database.

2. Materials and methods

2.1. Application of CAPE tools into prospective LCA

Figure 1 shows the description of systems assessments applying CAPE tools for prospective LCA. In management activity and resource provider, data estimation and interpretations are assigned to CAPE tools considering the conditions in prospective LCA.

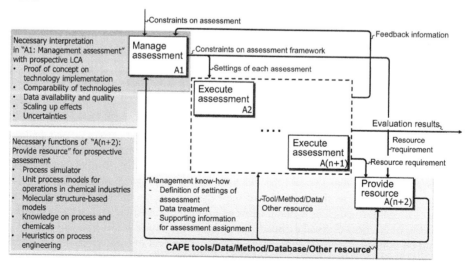

Figure 1 Description of multiple assessment activities with the necessary conditions for prospective LCA. n is the number of assessment methods. (Modified from previous studies (; Kikuchi, 2014; Kikuchi et al., 2010; Kikuchi and Hirao, 2009))

Life Cycle Thinking (LCT) is the process of considering the impact of a product throughout its entire life cycle, from the stage of raw material acquisition to its final disposal, when evaluating a product. LCA is a method for quantifying environmental impacts. There are two types of LCA: attributable LCA, which aims to visualize the environmental attributes of a product during its current life cycle, and consequential LCA, which aims to visualize the changes that occur when changes are made to the current life cycle (Ekvall, 2019). In the introduction of technological systems, where significant changes from the current system may occur, an analysis based on consequential LCT is necessary. The introduction of biomass-derived/waste-derived production technologies in chemical production and related life cycles, which are mainly based on the production of products derived from fossil resources, will bring about obvious changes and requires the design of the right type and scale of technical systems based on an assessment that takes into account the consequential impacts induced by these changes.

Conventional LCA can be carried out by collecting data on the target technology or system from actual production sites and combining other data extracted from LCA databases and other sources. On the other hand, in order to consider the development and introduction of CN and other sustainability-oriented technologies and systems, it is necessary to obtain information on non-existing life cycles and to carry out LCA. The applicability of LCA methods for the future of such technologies and systems is discussed (van der Giesen, 2020). LCA methods such as Consequential LCA, Dynamic LCA, Anticipatory LCA, Prospective LCA, and Ex-Ante LCA can be used for LCA of technologies under development, in other words technologies with low TRL, and systems after their introduction on a larger scale than at present, and for LCA of technologies that do not yet exist.

For the generation of life cycle inventory data, CAPE tools could be applicable. The previous review (Parvatker and Eckelman, 2019) demonstrated that the generation methods of inventory data should be selected in terms of the accuracy and the data/time requirements considering assessment objectives. For estimating inventory data. the applicable CAPE tools include process simulation, process input and output calculations, stoichiometry, and molecular structure-based models such as quantitative structure activity relationship (QSAR) models. Even if no stoichiometric/synthesis information is available, the inventory data of proxy chemical, which has already registered in LCA databases, can be selected with the understandings of chemistry of the technologies. While these CAPE tools have been developed mainly for chemical process systems, the inventory data of mechanical process systems such as metal processing and assembling might be estimated in the same manner.

2.2. Case study: Feedstock recycling of waste plastics

Figure 2 shows the life cycle boundary of plastics including catalytic pyrolysis as feedstock recycling of waste plastics for raw materials in refineries. Catalytic pyrolysis can produce hydrocarbons similar in compositions to those used in conventional oil refineries. The profiles of obtained hydrocarbons have various patterns based on the temperature of pyrolysis and compositions of resins in waste plastics. There is a distribution of carbon chains in the pyrolysis products, and it is not easy to extract only specific hydrocarbons. Therefore, if the pyrolysis products are fed directly into the refinery flow, they may be fed together with crude oil into an atmospheric distillation column. The catalytic cracking process can be followed by a separation process to separate the pyrolysis products into groups of certain carbon chains. This allows the products to be fed directly into certain unit operations of the refinery, thereby omitting some unit operations of the refinery.

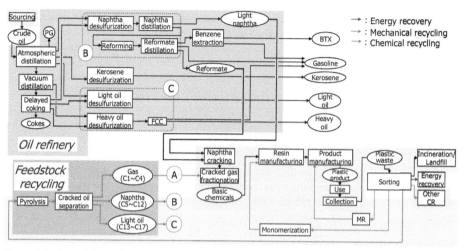

Figure 2 Life cycle boundary of plastics including catalytic pyrolysis as feedstock recycling of waste plastics for raw materials in refineries, where CR means chemical recycling, MR is material recycling, PG is purged gas, and FCC is fluidized-bed catalytic cracking.

The pyrolysis products were separated into three groups based on their carbon chains. Product A contains hydrocarbons (C1 to C4) which could be directly used as raw materials for basic chemical products through fractionation processes. This is similar to the stream after naphtha cracking in conventional petrochemical process systems. Product B is similar to naphtha, containing hydrocarbons (C5 to C12). This is fed to the unit operations around naphtha production in oil refinery. Product C contains heavy content and must be fed to FCC for further cracking. The unit operations and flows inside of oil refinery were extracted from the reports on existing refinery (Yoshitome et al., 2022).

Process simulation was applied as a CAPE tool to estimate inventory data. For catalytic cracking, the temperature dependence of pyrolysis products was collected from literature values, and the production yield from waste plastics was calculated (Keller et al., 2022). For the utility of catalytic cracking, inventory data for existing pyrolysis were used (Kikuchi et al., 2022). For inventory data in the separation system after catalytic cracking, the data were estimated using Aspen HYSYS™. For unit operations in the refinery, the inventory of the existing refinery was calculated from the complexity factors and the utility of the entire refinery. For unit operations where the flow rate increases due to the merging of waste plastic-derived components, the increase in energy input due to the increased flow rate was calculated using a linear approximation. For each unit operation, a facility capacity was set.

3. Results and discussion

Figure 3 shows the example of process simulation results; greenhouse gas (GHG) emission for oil refinery and petrochemical process for a conventional oil refinery in Japan, the throughput of which is 18,115 kL-crude oil per day. The functional unit is set as the operation of this oil refinery plant. The results of this prospective LCA

Application of CAPE Tools into Prospective Life Cycle Assessment: A
Case Study in Feedstock Recycling of Waste Plastics

2481

demonstrated that the catalytic pyrolysis of waste plastics could reduce the GHG emissions attributable to oil refinery by reducing refined crude oil. Based on the compositions of pyrolysis products, for example, the flowrates in oil refinery were changed, which resulted in the changes of the product ratios in oil refinery. If the pyrolysis products contain hydrocarbons similar to naphtha or ones from naphtha cracker, the carbon source from waste plastics can be easily circulated as raw materials of chemicals.

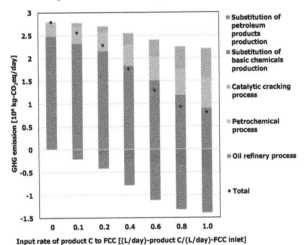

Figure 3 Example of process simulation results; GHG emission for oil refinery and petrochemical process for a conventional oil refinery in Japan, the throughput of which is 18,115 kL-crude oil per day. The inputted waste plastics were capped by the capacity of FCC. The background process inventories were extracted from Japanese LCA database (IDEA v3.2).

Converting waste plastics to oil through pyrolysis allows the reuse of existing refinery equipment, which may reduce the barriers to the introduction of new technologies. On the other hand, the composition of hydrocarbons obtained by catalytic cracking of waste plastics is different from that contained in crude oil. Therefore, when pyrolysis products derived from waste plastics are fed into refineries, the composition of the products obtained as petroleum derivative products will change. Although it is possible to adjust the composition of the product system by changing the temperature and catalyst in catalytic cracking, it is necessary to modify the refinery to meet future demand for petroleum products. These issues need to be analyzed through LCAs that take into account the consequences of other social scenarios, such as reduced gasoline consumption due to the electrification of automobiles and reduced demand for fossil fuels due to changes in the composition of power sources.

4. Conclusions

In this study, a mathematical model was constructed for catalytic pyrolysis of waste plastics and oil refinery receiving the pyrolysis products as raw materials as feedstock recycling of plastics. As a result, we confirmed the changes in GHG emissions from the production of various petroleum-based products and products due to the introduction of catalytic pyrolysis. The effect of conversion of waste plastics to chemical feedstock aimed at reducing fossil resource consumption cannot be fully assessed only by evaluating the GHG emissions resulting from the substitution of fuel oil and chemical feedstock. Changes in demands of energy and chemicals in society should be taken into account. As a sustainable or recyclable carbon sources, waste plastics should be able to become a stable source which can be gathered from industries and municipalities. However, the current waste plastics are contaminated with other materials such as papers, metals, sands, garbage, and so on. Additionally, the resin types are also mixed in waste plastics. The pretreatment process, i.e., sorters, is important for the quality of raw materials for catalytic pyrolysis. The prospective design of life cycle should be conducted with social systems and supported by LCA with CAPE tools.

Acknowledgement

This is based on results obtained from a project, JPNP20012, subsidized by the New Energy and Industrial Technology Development Organization (NEDO). This work was supported by MEXT/JSPS KAKENHI Grant Number JP20K20016, 21K17919, and JST COI-NEXT JPMJPF2003. Activities of the Presidential Endowed Chair for "Platinum Society" at the University of Tokyo are supported by Mitsui Fudosan Corporation, Sekisui House, Ltd., the East Japan Railway Company, and Toyota Tsusho Corporation.

References

T. Ekvall, 2019, Attributional and Consequential Life Cycle Assessment, Sustainability Assessment at the 21st century, https://www.intechopen.com/chapters/69212

Energy Transition Commission, 2021, Bioresources within a Net-Zero Emissions Economy: Making a Sustainable Approach Possible, https://www.energy-transitions.org/publications/bioresources-within-a-net-zero-economy/

van der Giesen, C., Cucurachi, S., Guinée, J., Kramer, G.J., Tukker, A., 2020, A critical view on the current application of LCA for new technologies and recommendations for improved practice, J. Clean. Prod., 259, 120904

F. Keller, R.L. Voss, R.P. Lee, B. Meyer, 2022, Life cycle assessment of global warming potential of feedstock recycling technologies: Case study of waste gasification and pyrolysis in an integrated inventory model for waste treatment and chemical production in Germany, Resour. Conserv. Recycl. 179, 106106

Y. Kikuchi, M. Hirao. 2009, Hierarchical Activity Model for Risk-Based Decision Making Integrating Life Cycle and Plant-Specific Risk Assessments, J Ind Ecol, 13(6) 945-964.

Y. Kikuchi, K. Mayumi, M. Hirao. 2010. Integration of CAPE and LCA Tools in Environmentally-Conscious Process Design: A Case Study on Biomass-Derived Resin, Computer-Aided Chemical Engineering, 27, 1051-1056.

Y. Kikuchi. 2014. Activity and Data Models for Process Assessment Considering Sustainability, Kagaku Kogaku Ronbunshu, 40(3) 211-223.

Y. Kikuchi, M. Nakai, Y. Kanematsu, K. Oosawa, T. Okubo, Y. Oshita, Y. Fukushima, 2020, Application of technology assessments into co-learning for regional transformation: A case study of biomass energy systems in Tanegashima, Sustai. Sci., 15, 1473-1494

Y. Kikuchi, N. Torizaki, L. Tähkämö, A. Enström, S. Kuusisto, 2022, Life cycle greenhouse gas emissions of biomass- and waste-derived hydrocarbons considering uncertainties in available feedstocks, Proc. Saf. Environ. Protec., 166, 693-703

F. Meng, A. Wagner, A.B. Kremer, D. Kanazawa, J.J. Leung, P. Goult, M. Guan, S. Herrmann, E. Speelman, P. Sauter, S. Lingeswaran, M.M. Stuchtey, K. Hansen, E. Masanet, A.C. Serrenho, N. Ishii, Y. Kikuchi, J.M. Cullen, 2022, Planet compatible pathways for transitioning the chemical industry, ChemRxiv, https://doi.org/10.26434/chemrxiv-2022-hx17h-v2

National Institute of Advanced Industrial Science and Technology, LCA-database IDEA v3.2. (2022)

A.G. Parvatker, M.J. Eckelman, Comparative evaluation of chemical life cycle inventory generation methods and implications for life cycle assessment results, ACS Sustain. Chem. Eng., 7 (2019), pp. 350-367,

T. Yoshitome, K. Saito, K. Tamura, 2022, Modeling of refinery equipment configuration and CO_2 emissions, PETROTECH, 45 (1), 21-28

Antonis Kokossis, Michael C. Georgiadis, Efstratios N. Pistikopoulos (Eds.)
PROCEEDINGS OF THE 33rd European Symposium on Computer Aided Process Engineering
(ESCAPE33), June 18-21, 2023, Athens, Greece
© 2023 Elsevier B.V. All rights reserved. http://dx.doi.org/10.1016/B978-0-443-15274-0.50395-4

Designing roadmaps for transitioning to value chains with net-zero emissions: Case of the chemical industry

Amrita Sen,[a] Vyom Thakker,[a] George Stephanopoulos,[b,c] Bhavik R Bakshi,[a]

[a] William G Lowrie Department of Chemical and Biomolecular Engineering, The Ohio State University, Columbus OH 43201, USA
[b] The Global KAITEKI Center, Arizona State University, Tempe, Arizona 85287, USA
[c] Department of Chemical Engineering, Massachusetts Institute of Technology, Cambridge, MA 02139, USA

Abstract

The global economy must restrict net greenhouse gas emissions to the environment to zero by 2050. However, in the absence of rigorous methods, the precise path to net-zero remains ambiguous, despite the overabundance of candidate solutions. Our formulation builds on the existing net-zero roadmapping literature by including technologies at various readiness levels within the life-cycle network superstructure. A case study for chemicals and plastics value chains demonstrates the framework. We model the current global plastics and chemicals industry and the promising low carbon solutions, based on electrification, biomass, carbon capture and circularization. Our framework allows the evolution of background emissions and maturities of technological solutions. We solve a multi-objective optimization problem at discrete time steps to roadmap for projected chemicals production till 2050. Our results show that net-zero value chains are only possible with a combination of energy efficiency and material recycling. Lower TRL technologies enable this transition, but only on upfront investments into their development. Our framework incorporates the realistic limitations of adopting emerging technologies with low TRL values into the decision-making process. The product specific insights gained may be used to guide future strategies.

Keywords: net-zero, circular economy, multi-objective optimization, life cycle

1. Introduction

The irreversible repercussions of incumbent climate change threaten the future of humankind. This has led policymakers across the world to seek to steer the global economy to operate with net-zero emissions. While most current research efforts in this direction are geared towards delivering cleaner energy with minimum emissions to industrial and residential systems, "net-zero" cannot be reached unless material value-chains are reinvented.[1] The direct emissions from the chemicals sector account for a humble 7% of the global emissions. However, as chemical energy carriers like hydrogen and methanol become increasingly important in other residential sectors, eliminating emissions from this hard-to-decarbonize sector becomes critical.

Many new technologies and innovations have been developed in the recent past to address this transition. However, in the absence of modelling techniques, choosing the most sustainable, economic and robust solutions is impossible. In addition to numerous life-cycle studies evaluating the emissions abatement potential of emerging technologies in isolation, research has compared the net-zero potential of technology combinations in product systems such as methanol.[2] Existing research has also evaluated the precise combination of technologies for the global chemicals industry to get to net-zero, while investigating possible cost drivers.[3] This steady state analysis, however, does not evaluate nascent innovations or the temporal evolution of technologies. Even though research, such as that by Zibunas et al. account for the capital investments required for technologies to become viable, the gap in developing roadmaps to reach net-zero chemical systems, remains unbridged.[4] The evolution of innovations through time is a crucial consideration in the design of roadmaps and remains unaddressed in current literature.

In this paper, we design a roadmap to guide the transition of the global chemical industry towards net-zero greenhouse gas emissions. Life-cycle assessment methods and the Sustainable Circular Economy (SCE) framework are used to model emerging technologies.[5] The temporal evolution of emerging technologies is utilized to design roadmaps i.e., value chains of chemicals and plastics with the least greenhouse gas emissions or generate the greatest possible value. The linear evolution model that we assume for innovations can be easily replaced by deterministic models and eventually surrogated with simpler functions in the optimization problem.

2. Methodology

2.1. Model of Chemical Industry

The global chemical industry is modelled as a network of individual processes. We use bottom-up linear steady state models to represent individual processes in the network. These processes are assumed to scale linearly to meet the global demand of chemicals and plastics. This assumption allows the scale-up of linear life-cycle models to meet demands at the economy scale. This version of the study does not model fractional yield and assumes stoichiometric conversion of reactants to products instead. Since the global demand of chemicals and plastics is used to solve for the optimal resource use, the calculated resource use via stoichiometric calculations essentially serves as a lower limit on resource use, which is the worst case scenario. Chemical value chains are modelled including refinery processes to convert fossil feedstock to olefins and aromatics, production of various chemical intermediates, solvents, monomers, fertilizers, and polymerization reactions to produce rubbers, plastics, and plasticizers. The thermal energy and electricity needs of these processes are modelled using global averages of industrial heating and electricity mixes. Our model covers all high-volume chemicals and plastics, and most minor flows in the industry. [9,10] Publicly available data are used to populate the network. These models are represented as a matrix of linear equations, similar to the LCA framework. Processes at the refinery stage are populated with globally averaged data over multiple refineries from industrial reports. Averaged composition of inputs to and outputs from refineries are used for the calculations. Further downstream, conversion and polymerization reactions are modelled using chemical reaction data. The model contains multiple production pathways and final demands are assumed to be fulfilled by appropriate contributions from these pathways. Energy demands of the processes are calculated from specific energy consumption data from IEA.[11] A detailed version of the approach can be found in the authors' past publication.[9]

2.2. Modelling emerging technologies

We supplement the baseline network of conventional technologies by models of emerging technologies to obtain a superstructure network with alternatives at all life-cycle stages.

At present, the sector of chemicals production is inherently dependent on carbonaceous fossil feedstock. These feedstocks can be bypassed by more circular pathways such as the utilization of carbon dioxide captured from other emission sources. In this work, we model the conversion of carbon dioxide by electrochemical and thermochemical means to ethylene and methanol, with and without the intermediate formation of syngas. We also model the electrolytic production of hydrogen via water splitting (green hydrogen).

Biomass feedstock may replace most hydrocarbons produced by refinery operations. Our model includes the production of syngas from woody biomass, such as pine and eucalyptus, as well as the catalytic biomass-to-aromatics processes discovered at the Catalysis Center of Energy Innovation, University of Delaware, referred to henceforth as the CCEI process. Since the emerging technologies modelled thus far only produce methanol as a platform chemical, we also include the acid catalyzed transformation of methanol to aromatics to allow production of aromatics derived products.

Mixed plastic waste generated from residential and industrial sectors may be landfilled or recycled to maintain some fraction of its original value. Our superstructure network includes the mechanical (primary) recycling of PET bottles, the chemical (secondary) recycling of PET viz enzymatic hydrolysis and methanolysis reactions, pyrolysis (tertiary recycling) of mixed plastic waste to produce pyrolysis gas as well as syngas, and the transformation of waste to energy (tertiary recycling) via incineration. We also model the recovery of monomers from polyethylene by pyrolysis to produce a mixture of hydrocarbons. Recycling is included in the superstructure network as circular flows which intrinsically determine the emissions associated with circular flows, instead of estimating them with stationary emission factors, using the SCE framework.[5]

Refinery operations, boilers and furnaces are responsible for the majority of the emissions from the chemical industry. We model steam methane reforming of natural gas, partial oxidation of coal and oil, steam cracking of natural gas and naphtha, and catalytic reforming of naphtha to be amenable to post combustion capture using Monoethanol amine (MEA), since flue gases from these processes have relatively high concentrations of carbon dioxide (5-15%).[6] The carbon dioxide thus captured is considered available for utilization i.e., conversion to value added products.

2.3 Cost Models

We model the costs of each module as the difference between product and resource costs. This formulation allows the estimation of emerging technological pathways with unknown costs, in terms of costs of input and output flows. We implicitly assume that emerging technologies replace conventionally made products at the same price point. This assumption requires investments into research and development, policy intervention and sharing of environmental burden across stakeholders.

2.4 Road-mapping Framework

$$\min_{s_t} Z(s_t) := -T_t = -C^{Module} s_t$$

$$\min_{s_t} Z(s_t) := h_{GWP} = \phi_{GWP} r_t$$

$$s.t. \ As_t \geq f_t$$

$$r_t \leq r_t^U$$

$$0 \leq s_t \leq s_t^U y_t$$

$$\mathcal{F}(s_t \geq 0)$$

$$where \ f_t = f_0 + f_0 \gamma t$$

$$r_t = B s_t$$

$$r_t^U = r_0^U + r_0^U \beta T$$

$$t \in T; \ T = \{0,1,2 \dots,10\}$$

$$y_t = \{y_{i,t}\};\ i \leq m\,;\ y_{i,t} \in \{0,1\}$$
$$y_{i,(t+1)} = 1\ if\ t \geq \left(11 - TRL_{i,t}\right) \forall\ i \in I$$
$$TRL_{i,t+1} = 1 + TRL_{i,t}$$

We perform a multi-objective optimization on our superstructure network, minimizing greenhouse gas emissions or maximizing the value generated by chemicals production. The epsilon constraint method is used to draw Pareto fronts through the optimal points. The underlying superstructure network is subjected to fulfillment of societal needs of chemicals. The global chemicals capacity f_t is used to impose a lower limit on production, which evolves linearly through subsequent time steps at current growth rates. γ is the vector of growth rates. An upper limit r_t^U is imposed on the availability of some resources like biomass, which also evolve linearly over time. The growth rate of biomass availability, β, is estimated using the current availability of 35 EJ of bioenergy and the assumption that, of the 150 EJ biomass required for sustainable bio-energy production by 2050, 90 EJ will be available for chemicals production by 2050.[7] Renewable electricity is also assumed to evolve linearly from generating at the most, 30% of the supply at the beginning, to 85% at the last time step.[8] Our formulation restricts the inclusion of technologies within the superstructure, depending upon their Technology Readiness Levels (TRL). The parameter $y_{i,t}$ determines whether a technology is available at time constraining the scale up of technologies until they are mature enough. All immature technologies (TRL<10) are assumed to reach the next TRL at each time step. $y_{i,t}$ is updated accordingly, A technology is considered to be mature and ready to be adopted once its TRL reaches 10. Finally, governing equations are imposed as inequality constraints, ensuring no non-negative product accumulation.

3. Results and Discussions

This optimization problem, formulated as a linear program with 136 variables and 259 constraints, yields globally optimal solutions within reasonable time ($\leq$1 min) on an Intel Core i7-1165G7 CPU 2.8 GHz with 12 GB RAM. We visualize the trade-offs between emissions and cost as Pareto curves at each time step in Fig 1. Individual Pareto fronts indicate the lowering of value addition as emissions are progressively decreased. Any point on a Pareto front represents a value chain, all of which are equivalent solutions and improving one of the objectives worsens the other. However, as more technologies mature with each time step, the Pareto curves shift, allowing improvement in both objectives. The technologies introduced at each time step are indicated on the figure. We observe greater value generation even as emissions decrease, as products of the same value can now be generated with consumption of less valuable resources and inputs. The introduction of mechanical recycling shows greater value creation with the retention of plastics within the economy, even though emissions do not decrease significantly. The first net-zero solution is obtained on the introduction of biomass-based processes, which allow the sequestration of carbon during growth phase of the biomass. The shift of Pareto fronts between time steps 3 and 4 is win-lose as thermochemical carbon dioxide conversion pathways can lower emissions but are unable to create greater value. A huge jump in value addition results when plastic waste pyrolysis is introduced in the superstructure, allowing value creation in the form of chemical products and diesel, from waste. Finally, as electrochemical processes mature enough to be adopted, the resource cost decreases significantly, along with emissions. We plot the adoption of individual technologies at each time step to better analyze how the value chains change with time.

Figure 2 shows the amounts of key hydrocarbons produced by technologies. Benzene is initially produced by a combination of steam cracking and toluene dehydrogenation and disproportionation. As soon as the recovery of monomers from polyethylene matures, it becomes economically viable to produce benzene primarily by the chemical recycling of polyethylene. The use of toluene as feedstock for benzene production is completely replaced. Methanol, initially produced by syngas from steam methane reforming, first switches to being produced thermochemically from carbon dioxide, and eventually to production from syngas. To track the generation of syngas, we look at hydrogen production. Initially, renewables support the production of green hydrogen. However, as more biomass becomes available, the system adopts syngas production from woody biomass. Finally, as the pyrolysis of mixed plastic wastes matures, the pyrolysis of plastics is adopted to supplement biomass. We also see the re-adoption of methane reforming when carbon capture becomes viable. Lastly, the production of xylene from catalytic reforming is phased out and replaced by production from methanol and biobased syngas when readiness levels permit.

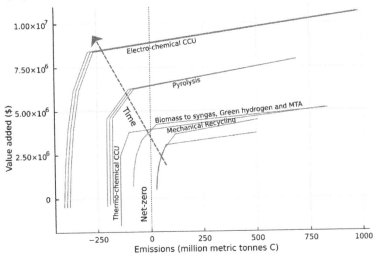

Figure 1: Improvement of both emissions and value addition objectives with adoption of new technologies over time

4. Conclusions and Future Work

Our results reconfirm the need to combine all low carbon solutions to cross the net-zero obstacle and get win-win solutions. Solutions like electrochemical conversion of carbon dioxide to hydrocarbons and bio-based chemicals emerge as robust solutions. Refinery operations such as catalytic reforming are replaced unless they can be retrofit with carbon capture technologies. Additionally, unless constraints are levied on landfilling for example, fossil based CCU solutions are preferred to the recycling of plastics. Ultimately, the ability of nascent technologies to ensure net-zero emissions is contingent on upfront investments into their research and development as well as policy interventions to ensure they are economically competitive with their fossil derived counterparts. The preliminary findings from our work indicate that emissions abatement, as well as cost, are highly sensitive to the constraints imposed on the network. Therefore, utilizing predictions of renewables and resource availabilities, assessing climate change scenarios to ensure resilience to background emissions, and utilizing historical data to predict technological advancements are crucial. This work will be extended in the near future, to include integrated assessment models, utilize research and development costs as well as patent literature to stochastically predict technology evolution.

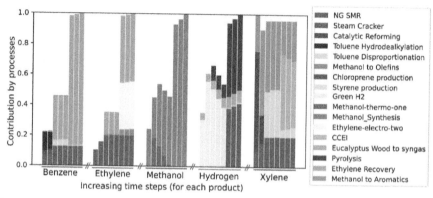

Figure 2: Contribution of processes to key hydrocarbon production. Bio-based and electrochemical processes are readily adopted as soon as they mature.

References

[1]Stephanopoulos, G., Bakshi, B.R. and Basile, G., 2022. Reinventing the Chemicals/Materials Company: Transitioning to a Sustainable Circular Enterprise. In Computer Aided Chemical Engineering (Vol. 49, pp. 67-72). Elsevier.

[2]Gabrielli, P., Gazzani, M. and Mazzotti, M., 2020. The role of carbon capture and utilization, carbon capture and storage, and biomass to enable a net-zero-CO2 emissions chemical industry. Industrial & Engineering Chemistry Research, 59(15), pp.7033-7045.

[3]Meys, R., Kätelhön, A., Bachmann, M., Winter, B., Zibunas, C., Suh, S. and Bardow, A., 2021. Achieving net-zero greenhouse gas emission plastics by a circular carbon economy. Science, 374(6563), pp.71-76.

[4]Zibunas, C., Meys, R., Kätelhön, A. and Bardow, A., 2022. Cost-optimal pathways towards net-zero chemicals and plastics based on a circular carbon economy. Computers & Chemical Engineering, 162, p.107798.

[5]Thakker, V. and Bakshi, B.R., 2021. Toward sustainable circular economies: A computational framework for assessment and design. Journal of Cleaner Production, 295, p.126353.

[6]Van Straelen, J., Geuzebroek, F., Goodchild, N., Protopapas, G. and Mahony, L., 2010. CO2 capture for refineries, a practical approach. International Journal of Greenhouse Gas Control, 4(2), pp.316-320.

[7]IEA (2021), What does net-zero emissions by 2050 mean for bioenergy and land use?, IEA, Paris https://www.iea.org/articles/what-does-net-zero-emissions-by-2050-mean-for-bioenergy-and-land-use, License: CC BY 4.0

[8]Global Energy Perspective, 2022, Mckinsey & Company

[9]Sen, A., Stephanopoulos, G. and Bakshi, B.R., 2022. Mapping Anthropogenic Carbon Mobilization Through Chemical Process and Manufacturing Industries. In Computer Aided Chemical Engineering (Vol. 49, pp. 553-558). Elsevier.

[10]Sen, A., Stephanopoulos, G. and Bakshi, B.R., 2022. Mapping the Chemicals and Materials industry. In preparation

[11] IEA, 2013. Energy and GHG Reductions in the Chemical Industry via Catalytic Processes.

Antonis Kokossis, Michael C. Georgiadis, Efstratios N. Pistikopoulos (Eds.)
PROCEEDINGS OF THE 33rd European Symposium on Computer Aided Process Engineering
(ESCAPE33), June 18-21, 2023, Athens, Greece

Data Envelopment Analysis of Ammonia and Methanol as Sustainable Energy Storage Vectors

Sebastiano C. D'Angelo,[a] Michael Bregy,[a] Philipp Steiner,[a] Raul Calvo-Serrano,[a] Gonzalo Guillén-Gosálbez[a]

[a]*Institute for Chemical and Bioengineering, Department of Chemistry and Applied Biosciences, ETH Zürich, Vladimir-Prelog-Weg 1, 8093 Zürich, Switzerland*
gonzalo.guillen.gosalbez@chem.ethz.ch

Abstract

The increasing concern about global warming presses for a transition to more sustainable energy sources such as solar and wind power. To tackle the associated intermittency, chemical energy storage is witnessing a surge in interest from investors and policymakers. The efficient development of such technologies requires comprehensive analyses encompassing: (1) power-to-chemicals (P2C), (2) storage, and (3) chemicals-to-power (C2P). This work aims to compare ammonia and methanol as chemical energy vectors with other energy storage technologies. Data envelopment analysis (DEA) was used to compare the assessed technologies in terms of economic and environmental life cycle assessment (LCA) indicators, ranking the different technologies and estimating their required improvement targets. Overall, this work sheds light on the potential role of chemicals in energy storage applications, analyzing their advantages and disadvantages, and proposing improvement strategies to facilitate their development and future implementation.

Keywords: ammonia, methanol, energy storage, life cycle assessment, data envelopment analysis.

1. Introduction

A sustainable transition of the power sector to a decarbonized economy calls for an increased penetration of renewables, primarily solar and wind power. However, the intermittency of such technologies constitutes a challenge: new energy storage solutions are urgently needed, featuring high scalability and long-duration support. Highly suitable vectors for such a purpose are platform chemicals able to effectively store hydrogen (H_2) in high energy density forms, such as ammonia (NH_3) and methanol (MeOH) (Davies et al., 2020). Accordingly, a large set of technologies to convert power to chemical storage vectors (P2C) and then back to power (C2P) has been recently investigated, giving rise to a plethora of routes with wide variations in economic and environmental performance and inherent trade-offs. Systematic approaches to quantify and assess such trade-offs could, thus, support informed decisions in developing more sustainable routes.

In this work, acknowledging the scarce literature using similar case studies (Rostami et al., 2022), we apply for the first time Data Envelopment Analysis (DEA) to a large set of routes for power-to-chemical-to-power (P2C2P) involving NH_3 and MeOH. The goal is to assess their efficiency by considering multiple criteria simultaneously and to identify margins of improvement for suboptimal scenarios. DEA originally stems from the field of economics (Charnes et al., 1979) and has been only recently applied to sustainability problems (Vázquez-Rowe et al., 2010). Unlike other multi-criteria decision-making tools

(Tzeng and Huang, 2011), DEA does not rely on user-defined weight preferences to quantitatively assess and classify sets of alternatives based on multiple decision criteria simultaneously. Here we shall use DEA to gauge the efficiency of a chemical storage route based on all the considered economic and environmental decision criteria, further ranking all the overall efficient alternatives through the use of super-efficiencies, and estimating the required improvement targets for the sub-efficient routes.

2. Methodology

2.1. Generation of scenarios

For the P2C section, the Haber-Bosch process was simulated in Aspen Plus® to produce NH_3, using H_2 from different sources and nitrogen from an air separation unit (D'Angelo et al., 2021). For MeOH, the single-step hydrogenation of CO_2 was assumed (González-Garay et al., 2019). Several H_2 routes were considered, including steam methane reforming (H2-SMR), H2-SMR using carbon capture and storage (CCS) (H2-SMR-CCS), biomass gasification with CCS (H2-G), and water electrolysis powered by a set of different sources: grid mix (H2-CE), nuclear (H2-N), wind (H2-W), solar (H2-S), bioenergy with CCS (H2-B), and hydropower (H2-H) (D'Angelo et al., 2021). CO_2 from carbon capture from combined cycle power plants (CC) or from direct air capture (DAC) powered by the same sources as for electrolytic H_2 was considered. For the C2P section, for NH_3, a power plant was simulated (Sánchez et al., 2021) considering three different downstream treatments for the flue gas: H_2 recovery with nitrogen-argon separation (HR-NAS), H_2 recovery with subsequent flue gas release (HR-FGR), and H_2 combustion with nitrogen-argon separation (HC-NAS). For MeOH, the C2P section considered direct combustion (MC), MeOH decomposition (MD), and steam reforming (MR) (Deng and Hynes, 2012), with CCS applied for the flue gas in all cases. The results of all the combinations of feedstocks and C2P alternatives were compared with an adiabatic compressed air energy storage with a modified Kalina cycle (CAES) (Liu and Wang, 2016) and pumped hydro energy storage (PHES) (Dones et al., 2007).

2.2. Environmental assessment

The life cycle assessment (LCA) was performed following the ISO 14040 (ISO, 2014). A cradle-to-gate study based on an attributional approach was adopted, considering the whole life cycle until the P2C step. The selected functional unit was 1 kWh of delivered electrical energy from the various storage routes.

In the second LCA phase, the life cycle inventories (LCIs) are modeled. The foreground system includes all the steps from P2C to C2P, according to the defined scenarios. At the same time, the underlying energy and raw materials suppliers belong to the background system, here modeled with Ecoinvent v3.5, accessed through SimaPro v9.2. All the inventories assumed global (GLO) regionalization where available and Rest-of-the-World (RoW) otherwise. For the NH_3 and MeOH P2C steps, the average global grid power mix was considered for the processes, excluding H_2 and CO_2 generation. Salt caverns were assumed for intermediate H_2 storage (D'Angelo et al., 2021), and the oxygen by-product from water splitting was considered vented. In the third LCA phase, ReCiPe 2016 with the hierarchical approach (Huijbregts et al., 2017) was selected, with a focus on the midpoint indicators, namely: global warming (GW), stratospheric ozone depletion (SOD), ionizing radiation (IR), fine particulate matter formation (FPMF), human health ozone formation (OFHH), terrestrial ecosystem ozone formation (OFTE), terrestrial acidification (TA), freshwater eutrophication (FWE), marine eutrophication (ME), water consumption (WC), land use (LU), mineral resource scarcity (MRS), fossil resource scarcity (FRS), marine ecotoxicity (MECOTOX), freshwater ecotoxicity

(FWECOTOX), terrestrial ecotoxicity (TECOTOX), human carcinogenic toxicity (HCT), human non-carcinogenic toxicity (HNCT). Finally, in the last step of LCA, the results are interpreted, and recommendations are drawn. Here, we applied DEA to provide additional insights into the environmental efficiency of the generated scenarios.

2.3. Mathematical models of DEA

DEA provides relative efficiencies of the different alternatives (n) as the maximum ratio between the weighted sum of outputs (j, criteria to maximize) to the weighted sum of inputs (i, criteria to minimize). To this end, it solves a linear programming (LP) model, which, when reformulated into its dual form, also provides improvement targets for the different sub-efficient alternatives. In this work, we consider the input-oriented variable return to scale dual model M1, presented below as described by Cooper et al. (2007):

$$\text{M1} \quad \min_{\theta_{n'}, S_i^-, S_j^+, \lambda_{n'n}} \quad \theta_{n'} - \varepsilon\left(\sum_i S_i^- + \sum_j S_j^+\right) \tag{1}$$

$$\text{s.t.} \quad \sum_n \lambda_{n'n} x_{ni} + S_i^- = \theta_{n'} x_{n'i}, \forall i$$

$$\sum_n \lambda_{n'n} y_{nj} - S_j^+ = y_{n'j}, \forall j$$

$$\sum_n \lambda_{n'n} = 1$$

$$\lambda_{n'n}, S_i^-, S_j^+ \geq 0, \forall n, i, j$$

where the model variables include $\theta_{n'}$, the efficiency value of alternative n', S_i^- and S_j^+, the input and output, respectively, slack variables, and $\lambda_{n'n}$, the calculated efficiency projection weights of alternative n' on the efficient alternatives n. The model parameters include x_{ni}, the input i value for alternative n, y_{nj}, the output j value for alternative n, and ε, a non-archimedean infinitesimal value.

Model M1 is solved for each of the considered alternatives n', obtaining for each one its respective efficiency and projection weights. In essence, the best technologies are those with an efficiency score equal to one, while the rest could match the best-performing units by attaining some improvement targets provided by DEA. More precisely, it is possible to estimate the relative input improvement targets $\tau_{n'i}$ for the different sub-efficient alternatives n' and input criteria i as the relative difference between the current input value $x_{n'i}$ and the value projected onto the efficient frontier $\sum_n \lambda_{n'n} x_{ni}$. The relative input improvement target can be calculated as follows:

$$\tau_{n'i} = 1 - \frac{\sum_n \lambda_{n'n} x_{ni}}{x_{n'i}} \tag{2}$$

In order to rank the alternatives defined as efficient in model M1, it is possible to use super-efficiency models to calculate the super-efficiency scores $\hat{\theta}_{n'}$ for the already efficient alternatives. These super-efficiency scores can be calculated by model M2, which only differs from model M1 in that the projection constraints consider only the other alternatives n and not the alternative being evaluated n', as seen in the equations below. As in model M1, model M2 is solved for each of the efficient alternatives n', obtaining the super-efficiency score $\hat{\theta}_{n'}$.

$$\sum_{n \neq n'} \lambda_{n'n} x_{ni} + S_i^- = \hat{\theta}_{n'} x_{n'i}, \forall i \tag{3}$$

$$\sum_{n \neq n'} \lambda_{n'n} y_{nj} - S_j^+ = y_{n'j}, \forall j$$

In essence, DEA is applied to identify the best technologies, which are further ranked according to their super-efficiency values, while providing valuable insight via improvement targets on how to make sub-optimal technologies efficient.

3. Case study: an application to chemical energy vectors route selection

The DEA method has been applied to 218 energy storage alternatives, encompassing 189 MeOH routes, 27 NH$_3$ routes, and 2 conventional technologies (compressed air energy storage, CAES, and pumped hydro energy storage, PHES). These alternatives have been analyzed considering their economic performance (i.e., relative cost of storage, RCS) and environmental performance. The latter has been evaluated through 18 LCA mid-point indicators in the ReCiPe2016 methodology (Huijbregts et al., 2017).

In total, these 19 economic and environmental indicators represent the input decision criteria (i.e., indicators to be minimized) for the DEA methodology. The electricity output of the system, 1 kWh for all alternatives, is the only output indicator considered. In order to ensure the numerical stability and robustness of the DEA results, all indicators have been linearly normalized between 0 and 1. The DEA model was implemented in GAMS 39.3.0 and solved with CPLEX 22.1.0.0 on an Intel® Core™ i7-4790 processor operating at 3.6 GHz. On average, each DEA model instance was solved with a zero optimality gap, taking between 0.03 and 0.14 CPU seconds. Models M1 and M2 consist of 240 variables and 22 constraints.

In addition to the considered economic and environmental indicators, the technical performance of each route was estimated through its total energy efficiency $\eta_E = \frac{1}{E_{In}+T_{in}}$, which considers the amount of electricity (E_{In}) and thermal energy (T_{In}) required through the entire route to obtain 1 kWh after releasing the stored energy. This indicator is not considered in the DEA, instead it is compared with the super-efficiency scores to determine whether super-efficient alternatives also present high energy efficiencies.

4. Results and discussion

Out of a total of 218 processes, only 52 (24% of them) were found to be efficient. Inspecting the efficient cases, no correlation was found between energy efficiency and super-efficiency scores (Figure 1). Thus, energy efficiency is not sufficient to establish the optimality level of an energy storage process from an economic and environmental perspective. The ten most super-efficient scenarios consist of eight NH$_3$-based scenarios, CAES, and PHES, with the NH$_3$ | H2-B+HC-NAS scenario performing best, followed by CAES. Although not shown in Figure 1, efficient MeOH routes represent the majority of the remaining 42 alternatives, albeit their super-efficiency scores are strikingly smaller than the NH$_3$ and conventional storage alternatives.

Moving to the inefficient scenarios, the distribution of the relative improvement targets along the different indicators (Figure 2A) highlights that the worst-performing indicators, especially environmental metrics such as FWE, TA, and FPMF, require sensible improvements (39% on average). Moreover, the average improvement targets of the 12 metrics with the worst results lie within a narrow range (28-39%), and the 13 worst-performing indicators, on average, require substantial improvements (75-86% improvement margin). Conversely, the three best-performing indicators (GW, WC, LU) require an improvement below 10%, on average, and lower than 25%, even when accounting for extreme cases. Notably, the RCS is the fourth metric requiring the lowest improvement on average (12%) and in the worst case (37%). Such a result highlights how most technologies present relatively similar operational costs and greenhouse gas emissions in their life cycle, as well as in other generally considered environmental indicators (e.g., water and land use). This finding reflects how particular emphasis should be placed on often disregarded environmental metrics (e.g., FEW, TA) rather than solely GHG emissions or economic indicators when aiming to design fully sustainable energy storage strategies.

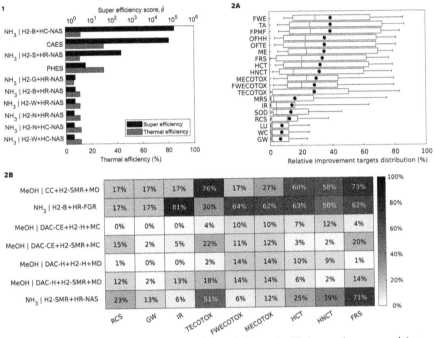

Figure 1. Super-efficiency (upper x-axis) and thermal efficiency (lower x-axis) scores for 10 alternatives with the highest super-efficiency scores.

Figure 2A. Relative improvement target distributions for all 19 input categories and their means (dots). Boxplots represent the 5th, 25th, 50th, 75th, and 95th percentiles.

Figure 2B. Relative improvement targets for sub-efficient alternatives with the best performance (not considering the efficient alternatives) in one or more input indicators.

Finally, Figure 2B presents the disaggregated relative improvement targets for the sub-efficient scenarios with the best performance in one or more of the considered 19 indicators. In contrast with previous results, MeOH-based technologies represent the majority of this set of sub-efficient technologies, with some of these alternatives presenting very high improvement targets in toxicity (e.g., TECOTOX) and resource scarcity (e.g., FRS) related indicators. This highlights how seemingly promising scenarios may still perform suboptimally, with only a few alternatives performing efficiently in these specific metrics. Among other scenarios, NH_3 | H2-B+HR-FGR would require significant improvements in toxicity-related categories due to the use of fertilizers in biomass cultivation. On the other hand, scenario MeOH | DAC-CE+H2-H+MC presents relatively low improvement targets (in some cases performing as other efficient alternatives), indicating how grid-powered DAC CO_2 capture and sequestration may be seen as a promising energy storage strategy based on methanol.

5. Conclusions

With the increasing number of developed alternative pathways involving energy storage, there is a spiking need for decision-support tools to systematically assess the current state of the art and optimally drive research efforts in this field. In this work, we apply a systematic methodology based on DEA to analyze a large representative set of P2C2P routes involving NH_3 and MeOH as energy storage vectors, considering both economic and environmental criteria. The approach here considered has been able to, given a set of economic and environmental metrics and without the use of user-defined weights, identify and rank the overall best-performing energy storage alternatives and provide necessary improvement targets for inefficient alternatives. Our results highlight how energy efficiency is not correlated with the overall economic and environmental performance of the different routes. Also, NH_3-based routes seem to be the most promising alternatives, along with CAES and PHES. Overall, this work aims to shed light on the potential of chemical energy storage by evaluating and comparing multiple alternative routes, highlighting the benefits that decision-support tools such as DEA can bring to the study and development of more sustainable energy systems.

Acknowledgments

This publication was created as part of NCCR Catalysis (grant 180544), a National Centre of Competence in Research funded by the Swiss National Science Foundation. SCD and GGG are authors associated with NCCR Catalysis.

References

A. Charnes, W.W. Cooper, E. Rhodes, 1979, Measuring the efficiency of decision-making units, Eur. J. Oper. Res., 3, 339.

W.W. Cooper, L.M. Seiford, K. Tone, 2007, Data Envelopment Analysis: A Comprehensive Text with Models, Applications, References and DEA-Solver Software, 2nd ed, Springer Science & Business Media, New York, NY.

S.C. D'Angelo, S. Cobo, V. Tulus, A. Nabera, A.J. Martín, J. Pérez-Ramírez, G. Guillén-Gosálbez, 2021, Planetary Boundaries Analysis of Low-Carbon Ammonia Production Routes, ACS Sustain. Chem. Eng., 9, 9740–9749.

J. Davies, F. Dolci, D. Klassek-Bajorek, R. Ortiz Cebolla, E. Weidner Ronnefeld, 2020, Current Status of Chemical Energy Storage Technologies, Joint Research Centre (European Commission).

S. Deng, R. Hynes, 2012, Advanced combined cycle systems based on methanol indirect combustion, J. Eng. Gas Turbines Power, 134.

R. Dones, C. Bauer, R. Bolliger, B. Burger, T. Heck, A. Röder, M. Faist, R. Frischknecht, N. Jungbluth, M. Tuchschmid, 2007, Life Cycle Inventories of Energy Systems: Results for Current Systems in Switzerland and other UCTE Countries, Dübendorf.

A. González-Garay, M.S. Frei, A. Al-Qahtani, C. Mondelli, G. Guillén-Gosálbez, J. Pérez-Ramírez, 2019, Plant-to-planet analysis of CO2-based methanol processes, Energy Environ. Sci., 12, 3425–3436.

M.A.J. Huijbregts, Z.J.N. Steinmann, P.M.F. Elshout, G. Stam, F. Verones, M. Vieira, M. Zijp, A. Hollander, R. van Zelm, 2017, ReCiPe2016: a harmonised life cycle impact assessment method at midpoint and endpoint level, Int. J. Life Cycle Assess., 22, 138–147.

International Organization for Standardization, 2014, ISO 14040:2006 - Environmental management - Life Cycle Assessment: Principles and framework, ISO.

J.L. Liu, J.H. Wang, 2016, A comparative research of two adiabatic compressed air energy storage systems, Energy Convers. Manag., 108, 566–578.

F. Rostami, Z. Kis, R. Koppelaar, L. Jiménez, C. Pozo, 2022, Comparative sustainability study of energy storage technologies using data envelopment analysis, Energy Storage Mater., 48, 412–438.

A. Sánchez, E. Castellano, M. Martín, P. Vega, 2021, Evaluating ammonia as green fuel for power generation: A thermo-chemical perspective, Appl. Energy, 293, 116956.

G.-H. Tzeng, J.-J. Huang, 2011, Multiple Attribute Decision Making : Methods and Applications, 1st ed, Chapman and Hall & CRC, New York, NY.

I. Vázquez-Rowe, D. Iribarren, M.T. Moreira, G. Feijoo, 2010, Combined application of life cycle assessment and data envelopment analysis as a methodological approach for the assessment of fisheries, Int. J. Life Cycle Assess., 15, 272–283.

Antonis Kokossis, Michael C. Georgiadis, Efstratios N. Pistikopoulos (Eds.)
PROCEEDINGS OF THE 33rd European Symposium on Computer Aided Process Engineering
(ESCAPE33), June 18-21, 2023, Athens, Greece
© 2023 Elsevier B.V. All rights reserved. http://dx.doi.org/10.1016/B978-0-443-15274-0.50397-8

Simulation of thermochemical recycling of waste plastic to produce dimethyl ether

Mohammaed Awwad [a] and Umer Zahid [a, b]

[a] Chemical Engineering Department, King Fahd University of Petroleum and Minerals, Dhahran 31261, Saudi Arabia.
[b] Interdisciplinary Research Center for Membranes & Water Security, King Fahd University of Petroleum and Minerals, Dhahran 31261, Saudi Arabia.

Abstract

In this study, the thermochemical plastic recycling process namely gasification has been investigated to process 10,000 kg/h of plastic waste. This study rigorously modeled and simulated the thermochemical processes that can recycle the plastic wastes efficiently to useful chemical such as di-methyl ether (DME). Currently, the cost of recycled plastic is not competitive compared to the fossil-based virgin plastics, which leads to the lack of industrial willingness for recycling. This study explored the technical feasibility to understand the processing of recycled plastics to useful chemical conversion. The results show that the gasification can provide a competitive solution with flexibility in terms of waste feed composition to produce DME at a large scale.

Keywords: Waste recycling; techno-economics; clean fuel; process design.

1. Introduction

Plastic production has readily invaded the global market by the middle of the previous century due to its wide range of applications in construction, textiles packaging, transportation, healthcare, and electronics sectors. Plastics have replaced the traditional materials such as wood, steel, and ceramics due to its versatile properties, i.e., low density, durability, low price, and corrosion resistance. Polyethylene (PE) and polypropylene (PP) represent nearly the half of production of plastics, on the other hand, polyvinyl chloride (PVC), polystyrene (PS), polyurethane (PUR), and polyethylene terephthalate (PET) are also produced significantly as shown in figure 1 (Lopez et al., 2018). Currently, the plastic production rate is increasing approximately at 4% per year, reaching an annual production of more than 353 million metric tons in 2019 (OECD, 2022). Owing to this mega-scale utilization of plastics in everyday domestic and industrial products poses a dramatic problem of plastic waste management that needs firm decisions be taken. According to Organization for Economic Co-operation and Development (OECD), only 9% of global plastic waste is recycled, while 49% of waste plastic was discarded, mainly in landfills, 19 % wastes were incinerated, and 22% of the waste remained uncollected and mismanaged (OECD, 2022). Due to plastics low degradability, landfilling of plastics represents a significant problem to the planet's ecosystem. Besides, plastic incineration has several environmental issues, such as formation of dioxins, fly ash, sulfur, nitrous oxides, and other hazardous chemicals. At the same time, the world is facing a real challenge from climate change, where plastics contribute around 3.4 % of greenhouse gas emissions globally (OECD, 2022).

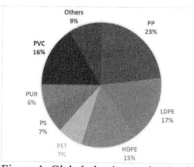

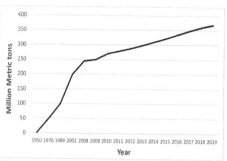

Figure 1: Global plastics production by type.   Figure 2: Global plastic production.

Some developed countries have started a waste plastic management plan; however, the required efforts are not enough, and mandate local, regional, and global actions be taken. Currently, different waste plastic recycling methods are already under investigation. Mechanical recycling is an efficient method that reconstruct the plastic into simpler shape while preserving the chemical structure. Although mechanical recycling of plastics is the most common plastic waste management method, it has some limitations. The feedstock needs first to be purified before proceeding further to achieve a high product quality; also, recycled product's market feasibility is still not proved (Ragaert et al., 2017). Recently, several countries are evaluating alternative fossil fuel options, and waste plastic is one such opportunity that is seen as lost resource that could be useful economically to produce valuable chemicals. Thermochemical plastic recycling is a technology where the heat is used to break a larger molecule to a smaller one that already made the plastic. Hence, chemical recycling is a potential route to enable a sustainable life cycle of plastic, and it is more flexible with feedstock than mechanical recycling [1,5]. Both pyrolysis and gasification processes fall under the category of thermochemical recycling. Gasification is a process where the waste plastic is heated at a high temperature with a limited amount of O_2. On the other hand, pyrolysis is a process where the plastic is being heated at a relatively lower temperature in the absence of O_2. The main advantage of gasification over pyrolysis is less sensitivity to different feed compositions or mixtures of plastics waste. Gasification of plastic mainly produces a gas stream that is rich in H_2, CO_2, CO, and CH_4. This syngas stream can then be employed either to produce energy or can serve as a raw material to produce a wide-range of important chemicals such as methanol, ammonia, di-methyl ether (DME) and others. DME has gained significant attention due to its clean combustion properties and easy handling as a fuel (Sansaniwal et al., 2017). DME can be synthesized from syngas in two ways, indirect and direct method. In the indirect method, methanol is first produced from syngas followed by the dehydration process. The direct method employs a bifunctional catalyst where methanol synthesis reaction, water gas shift reaction, and dehydration reaction co-occur as shown below in Eq. (1) – (4):

Eq.(1): Methanol formation $CO + 2H_2 \rightleftharpoons CH_3OH$
Eq.(2): Water gas shift $CO + H_2O \rightleftharpoons CO_2 + H_2$
Eq.(3): Methanol dehydration $2CH_3OH \rightleftharpoons CH_3OCH_3 + H_2O$
Eq.(4): Overall reaction $3CO + 3H_2 \rightleftharpoons CH_3OCH_3 + CO_2$

Direct synthesis of DME is a potential route as it can overcome the strong thermodynamic limitations of syngas conversion to methanol by shifting the equilibrium. This study is focused on the DME production from the gasification of waste plastic. Some previous works studied the plastic gasification. Shehzad et al., (2016) investigated the municipal solid waste (MSW) performance by varying gasification operational conditions. They reported that a higher gasification temperature has a significant impact on syngas composition. The results showed that gasification process has lower efficiency when utilizing MSW for electricity generation application, compared to the incineration; however, since chemicals could be produced from syngas, further economic analysis to make a suitable judgment was recommended. Weiland et al., (2021) studied the gasification of complex plastics and found that high temperature gasification processes that are usually assisted with high purity oxygen can generate syngas with heating value in the range of $10 - 13$ MJ/Nm^3. Similarly, many studies explored the DME synthesis through a one-step route. For example, Chen et al., (2012) studied the DME direct route synthesis from the syngas by developing a bifunctional Cu–ZnO–Al$_2$O$_3$/ZSM5 catalyst. They found that the DME yield decreases as the CO$_2$ concentration in the syngas increases. Guo et al., (2022) studied the DME one-step synthesis from the syngas produced by the biomass. This study was mainly focused on the catalyst development and reported that a hybrid catalyst comprised of CuO-ZnO-Al$_2$O$_3$ and HZSM-5 has a synergetic effect on the DME synthesis. Most of the previous works shed light on the waste plastic gasification experimentally and one-step direct DME synthesis independently. Besides, there are no remarkable studies that evaluated the DME production from waste plastic gasification. Therefore, the aim of this study is to design and simulate a commercial scale DME production unit considering waste plastic as a feedstock.

2. Process and Simulation details

The process contains three main sections: waste plastic gasification, acid gas removal (AGR) unit, and DME production and purification, as shown in figure 3.

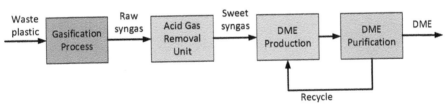

Figure 3: Simplified block diagram of plastic waste to DME production process.

2.1. Gasification Process

The waste plastic is first shredded before feeding it to the gasification process. The gasification reactor operates at 887°C and 1 bar in the presence of high purity oxygen. The gasifier effluent produces syngas rich in H$_2$, CO$_2$, CO, and CH$_4$ and slag which comprise of ash and unconverted carbon. One operational trouble with plastic gasification is the sticky nature of plastic material, however, this problem can be minimized by operating gasification at higher temperatures. A multi-stage cooler system is used to cool the reactor effluent before separating the syngas from the slag. Waste plastic is simulated as a non-conventional component in Aspen Plus by defining its ultimate and proximate analysis. Peng-Robinson thermodynamics property was utilized to model the gasification process. The base assumes the plastic waste feed rate of 10000 kg/h, while the oxidant is assumed to be 98 mol.% pure oxygen. The gasifier has been modeled using a RGIBBS subroutine that works on the principle of minimization of Gibbs free energy.

2.2. Acid gas removal (AGR) unit

A cleaning unit is an important process needed to reduce the CO_2 concentration before directly utilizing the raw syngas. In general, absorption with amines is the most convenient method to decrease the CO_2 level. A typical AGR unit consists of an absorber where the acid gases are absorbed by the amine, and a regenerator column where the rich amine is regenerated by stripping of the absorbed acid gases from the amine solvent. In this study, mono-ethanol-amine (MEA) has been utilized for the removal of acid gases. The AGR process has been modeled using AMINES thermodynamic package. The absorber is modeled using a RADFRAC block that removes 98% of the CO_2 in the syngas stream coming from the gasifier. The rich amine solvent is then sent to the regenerator (also modeled using RADFRAC) where the MEA solvent is regenerated, cooled, and recirculated to the absorber column. The calculator block in the Aspen plus is utilized to calculate the make-up solvent.

2.3. DME synthesis and purification unit

Sweet syngas from the AGR unit is first compressed in a multi-stage compressor stage with intercooling to reach the favorable reaction conditions (250°C and 60 bar). The compressed syngas is then fed to the DME synthesis reactor where a bifunctional catalyst yield DME via a one-step process explained earlier. The reactor effluent mainly contains unreacted syngas, DME, and methanol. The reactor effluent is sent to a series of separation units for the separation of product from the unreacted syngas and by-products. First, a flash drum is employed to separate most of the gas phase which mainly consists of unreacted syngas. The liquid phase leaving at the bottom of the flash drum is then sent to a series of distillation column where CO_2 is removed in the first column, product DME is recovered in the second column, and by-product methanol is separated in the third column which is recycled back to the DME synthesis reactor. DME synthesis and purification section has been modeled using Soave-Redlich-Kwong (SRK) equation of state. DME synthesis reactor is modeled using RPLUG by employing the Langmuir–Hinshelwood–Hougen–Watson (LHHW) kinetics. Figure 4 shows the process flow diagram of the overall plastic waste to DME process.

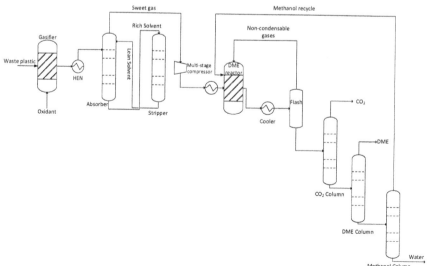

Figure 4: Process flow diagram of waste plastic to DME production process.

3. Results and discussions

3.1. Model validation

Model Validation is a crucial step in the design process that should not be overlooked before analyzing or extending the model to different operational conditions. Therefore, model validation is performed to ascertain how rigorous the designed model is. Figure 5 shows the validation of the DME synthesis reactor compared to the experimental data reported in the literature. The results show an amicable agreement between the simulation results and experimental data. In addition, composition of DME synthesis reactor effluent stream has been compared using RGIBBS and REQUIL reactors, and the results show a fair agreement with RPLUG reactor results as shown in Figure 6.

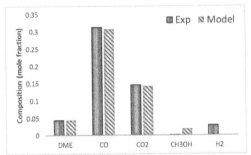

Figure 5: Validation of DME synthesis reactor.   Figure 6: Comparison of different reactor models for DME synthesis.

3.2. Energy analysis

A detailed energy analysis has been performed to explore the energy consumption in various sections of the process. The results for the main sections of the plastic waste to DME process are presented in Figure 7 and 8. The results show that the DME synthesis and purification section consumes the most energy in the process amounting to around 52% of the complete process. This is obvious because of intensive separation system in the DME purification section. AGR unit and gasification section contribute around 34.4% and 13.6% of the energy consumption respectively.

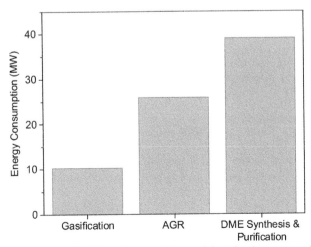

Figure 7: Energy consumption in various sections of the plastic waste to DME process.

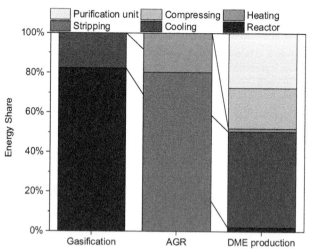

Figure 8: Energy share in various section of the plastic waste to DME process.

4. Conclusion

Plastic wastes are produced in significant amounts every year and can be utilized to produce useful clean chemicals such as DME. In this work, a conceptual design has been proposed for the production of high purity DME from the waste plastic gasification. This study showed the technical feasibility of DME production from waste plastic. The results showed that the proposed DME production process requires around 2.88 MJ per kg of DME production.

Acknowledgement

The author(s) would like to acknowledge the support provided by the Deanship of Scientific Research at King Fahd University of Petroleum & Minerals (KFUPM).

References

Chen, W.-H., Lin, B.-J., Lee, H.-M., Huang, M.-H., 2012. One-step synthesis of dimethyl ether from the gas mixture containing CO2 with high space velocity. Appl. Energy 98, 92–101.

Guo, X., Liu, F., Hua, Y., Xue, H., Yu, J., Mao, D., Rempel, G.L., Ng, F.T.T., 2022. One-step synthesis of dimethyl ether from biomass-derived syngas on CuO-ZnO-Al2O3/HZSM-5 hybrid catalyst: Combination method, synergistic effect, water-gas shift reaction and catalytic performance. Catal. Today.

Lopez, G., Artetxe, M., Amutio, M., Alvarez, J., Bilbao, J., Olazar, M., 2018. Recent advances in the gasification of waste plastics. A critical overview. Renew. Sustain. Energy Rev. 82, 576–596.

OECD, 2022. Global Plastics Outlook.

Ragaert, K., Delva, L., Van Geem, K., 2017. Mechanical and chemical recycling of solid plastic waste. Waste Manag. 69, 24–58.

Sansaniwal, S.K., Pal, K., Rosen, M.A., Tyagi, S.K., 2017. Recent advances in the development of biomass gasification technology: A comprehensive review. Renew. Sustain. energy Rev. 72, 363–384.

Shehzad, A., Bashir, M.J.K., Sethupathi, S., 2016. System analysis for synthesis gas (syngas) production in Pakistan from municipal solid waste gasification using a circulating fluidized bed gasifier. Renew. Sustain. Energy Rev. 60, 1302–1311.

Weiland, F., Lundin, L., Celebi, M., van der Vlist, K., Moradian, F., 2021. Aspects of chemical recycling of complex plastic waste via the gasification route. Waste Manag. 126, 65–77.

Antonis Kokossis, Michael C. Georgiadis, Efstratios N. Pistikopoulos (Eds.)
PROCEEDINGS OF THE 33rd European Symposium on Computer Aided Process Engineering
(ESCAPE33), June 18-21, 2023, Athens, Greece
© 2023 Elsevier B.V. All rights reserved. http://dx.doi.org/10.1016/B978-0-443-15274-0.50398-X

Sustainable recycling of refrigerants: Analysis of alternatives

J. Moreno[a], I. Karpov[a], A. Ahmed[a], J. Foglia[a], H. Wu[a], A. Gadekar[a], P.L.J. Swinkels[a], R. Gani[b,c]

[a]Chemical Engineering, Delft University of Technology, Van der Maasweg 9, 2629HZ, Delft, The Netherlands
[b]PSE for SPEED Company, Ordtup Jagtvej 42D, 2920 Charlottenlund, Denmark
[c]Sustainable Energy and Environment Thrust, The Hong Kong University of Science and Technology (Guangzhou), Guangzhou, China

Corresponding author: [a]P.L.J.Swinkels@tudelft.nl

Abstract

The use of refrigerants has been continuously increasing in thermal control systems employed in residential premises, offices, storage, process operations and many more. Widely used refrigerants such as R-410A have become a target for regulations to prohibit their use because of environmental issues. R-410A is a blend of R-32 (Difluoromethane) and R-125 (Pentafluoroethane). Because of high global warming potential (GWP) of R-125, it has become a target for removal from R-410A and reuse. An additional issue is the recovery of the refrigerant from current and decommissioned thermal control units. This paper reports the results of an investigation related to the separation of refrigeration blend compounds, recycle and reuse of the refrigerant mixture compounds and alternative blend compositions, with emphasis on environmental, health and physical hazards as well as optimal refrigeration cycle operation. Two alternatives for feasibility of operating a continuous separation-recycling process are studied.

Keywords: Refrigerant blends, R-32, R-125, R-410A, recycling, global warming potential, safety-health-environmental hazards.

1. Introduction

Hydrofluorocarbons (HFCs) are known for their excellent performance as refrigeration fluids. Their durability, inertness and thermal stability have made them the selection of choice in air-conditioning (AC) units. Nevertheless, their stability makes their presence persistent when released to the atmosphere, with a lifetime between 15 and 29 years (Xu, 2013). Furthermore, their capacity to absorb infrared light make them potent greenhouse gases (Castro et al. 2021).

R-410A is an HFC-based refrigerant, initially invented and patented by Allied Signal (now Honeywell) in 1991, that became popular as a substitute for R-22 (Chlorodifluoromethane) after the Montreal Protocol (UNEP, 2021) took place. The commercial R-410A consists of 50 wt% of R-32 and R-125 whose individual global warming potential (GWP) values are 677 and 3170, respectively, and giving the blend GWP value of 1923. Note that the mixture of R-32 and R-125 forms an azeotrope at 91 mol% R-32 (Castro et al. 2021). The objective of this paper is checking the feasibility of removal of R-125 from R-410A as well as recycling and reuse of the separated compounds. Monjur et al. (2022) have reported the separation of R-32 and R-125 by a continuous extractive distillation process with ionic liquids as the solvent. The feasibility of this process is compared with a pressure-swing distillation process, which also generates various R-410A blend alternatives. The performance of the generated alternatives is compared in terms of coefficient of performance (COP) of the refrigerant within the refrigeration cycle. Also, the effect of hazardous properties (health, physical and environmental) is considered from the points of view of safe and optimal refrigeration cycle operation.

Because of limited space, calculation details, hazardous properties of chemicals, etc., are not given in this manuscript. They can be obtained from the corresponding author.

2. Data-based Analysis of Opportunities for Recycling of F-gases

According to EU F-Gas legislation, all HFC refrigerants from refrigeration systems and air-conditioning systems must be recovered (European Parliament and Council of European Union, 2014) and replaced. If recovered material is too contaminated for further recycling, it should be decomposed. According to the European Environmental Agency, by 2020, the refrigerant HFC-23 account for nearly 90% of F-gases destroyed in the EU. Considering limitations of resources as well as undesirable impacts, the recovery and reuse of refrigerants is an option worth considering (Castro et al. 2021), provided they satisfy the safety, health and environmental concerns as well as operability of the refrigerant cycles. For R-410A and its components (R-32, R-125) requirements mentioned below are applied:

- Vapor phase contaminants: Air and other non-condensables must be < 1.5% by volume at 25.0 °C
- Liquid phase contaminants: Water must be <10 ppm by weight; all other volatile impurities must be < 0.5 wt%; high boiling residue must be < 0.01 wt%

Note that the above requirements are not an issue as R-410A is considered to be a blend of only R-32 and R-125.

2.1. R-410A recycling capacity limitations

The main use for R-410A as a refrigerant is in air-conditioning and heat pumps. In 2013 around 86% of all R-410A produced was used for HVAC systems only. Therefore, this study has selected R-410A as a representative example for the study of recycling of refrigeration fluids.

The global production of R-410A increased from 193 kMT in 2012 to 238 kMT in 2016, and the estimated production in 2022 will be 292 kMT. China is the leading producer of R-410A, with about 45% of the global production in 2016, followed by North America, with a production share of around 30% (360 Research Reports, 2020). The share of R-410A produced in EU in the years 2009 and 2013 was around 12% (Makhnatch & Khodabandeh, 2019).

Analysis of the technical data on HVAC systems indicate a life cycle of an AC unit is in the range of 10-15 years. Therefore, in China, the USA and the EU, the total maximum capacity of R-410A available for recycling in the years 2022-2027 is expected to be around 12, 8, and 3 MT/h, respectively (assuming 8000 h/yr). To understand the amount of bulk material (AC units) to be recycled, the assumption made is that R-410A can be extracted only from small-scale AC units (less than 150 kg). Based on the available information on commercial AC units, the average unit weights of 70 kg and as a mass fraction of R-410A of 4 wt%. Therefore, to obtain 1 ton of refrigerant, recovery from 26 tons or 370 units of decommissioned cooling devices should be done. Considering an average recovery time from one unit of around 20 min, and costing USD 16 per hour as a median manufacturing operator salary in USA, together with the average mass of refrigerant mentioned above, recovery costs for R-410A in USA would be 2200 $/MT.

2.2. R-410A separation

In the case of R-410A, separation of the pure compounds is complex due to an azeotrope for a R-32 purity of 91% mol. Monjur et al. (2022) proposed separation via extractive distillation with an ionic liquid in a continuous process.

An alternative to solvent-based extractive distillation is pressure-swing distillation if the azeotrope composition changes when the pressure is changed. Figure 1 shows the process flowsheet for separation of R-32 from R-125, along with the changes in the Txy diagrams. In the first column, operation at 17 bara yields a distillate with a near azeotropic composition of 90 mol% R-32 (80 wt%) at a temperature of 25 °C, allowing the use of cooling water in the condenser. Then, pressure is lowered to 1 bara, shifting the system to the right side of the azeotrope and resulting in a bottom product which is 99.5 mol% R-32 (99 wt%). Details regarding the process and the comparative study can be obtained from the authors. Operational expenses derived from consumption of utilities are 18 times higher for distillation at pressure. Since distillation at pressure is more energy intensive than extractive distillation, this result was expected. Nonetheless, when considering the cost of replenishing the ionic liquid BMIN PF_6, processing costs for extractive distillation increase from 0.004 $/hr to 2.4 $/hr, while remaining at 0.07 $/kg for the proposed process. An additional advantage of not using an ionic liquid is that, as seen in section 3.2, the coefficient of performance of the recycled R-32 could be affected considering the low vapour pressures which characterize ionic liquids, even at small concentrations.

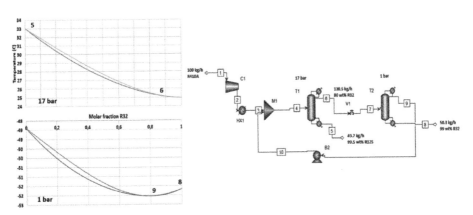

Figure 1: Flowsheet of pressure-swing distillation for separation of R-410A compounds, with Txy diagrams

Note, that the discussion in Section 2.1 indicates that reclaiming enough refrigerant to sustain a continuous process in a processing plant is not feasible, unless a periodic operational policy of collection followed by short periods of operation is applied. The design of the continuous process is sufficiently flexible to allow changes in production rates.

3. Evaluation of Alternatives: Recycle of F-gases

The high-pressure azeotropic distillation method provides opportunities for separation of R-32 and R-125 with individual purities of up to 95 and 99.5 wt%, respectively. In addition to the pure component product streams, the pressure swing distillation process allows intermediate product fractioning resulting in R-32 / R-125 mixtures of different compositions, which are investigated below

3.1. R-32 / R-125 mixture as a blend

Recently, several options were proposed as a substitute for phasing-out of R-410A. Among the options are:
- Non-flammable (A1 rating) blends: R-448A, R-449A,B,C, R-458A, and R-407C,F with high GWP rating (>1000)
- A2L safety class blends such as R-447A (GWP = 572), R-447B (GWP = 714), R-452B (GWP = 674), with a fraction of R-32 and R-125 in the range of 89.5 – 95.1 and 4.9 – 10.5 wt% respectively

A1-rated refrigerants are available ranging from light commercial systems (0.15 – 5kW) to industrial solutions (up to 10MW). In contrast, A2L refrigerants due to safety concerns have limited use in low-power capacity units (Danfoss A/S, 2022). The full list of patented and approved blends can be obtained from the authors. Tailor-made blends can also be generated (Udomwong et al. 2021).

3.2. Evaluation of performance of R-32 / R-125 blend systems

To investigate the effect of the R-125 composition in the refrigerant mix on the mixture's GWP and coefficient of performance COP a refrigerant cycle (shown in Fig 2a) has been simulated with Aspen HYSYS using the Peng-Robinson equation of state as the thermodynamic model for phase equilibrium. The refrigerant cycle has four operations: compressor, evaporator, condenser, and valve. The ProREFD software (Udomwong et al. 2021) and ChemSub (Syeda et al. 2022) software also allows a detailed cycle simulation, hazardous effects analysis and refrigerant substitution.

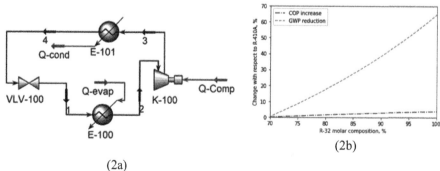

(2a)

(2b)

Figure 2: Refrigeration cycle and effect of R-32 composition on COP of refrigeration cycle

At first, the standard composition of R-410A, which is 70 mol% of R-32, was simulated resulting in a GWP of 1924 and a COP of 7.5. It is shown that increasing the composition of R-32 results in a lower GWP but does not decrease the COP (see Fig 2b). A total replacement of R-125 with R-32 towards a pure compound product, results in a GWP reduction of 65% and a COP increase of 4%. Nevertheless, there is an inherent safety risk associated to R-32 as it is classified with a safety level of A2L, meaning that it does not pose a toxicity risk at concentrations of less than or equal to 400 parts per million, but it is a mildly flammable gas. Its flammability limits are between 14.4% to 29.3%, which poses a risk of explosion if leakage occurred in the AC system (Osuagwu, 2022) Conversely, R-410A is classified as an A1 refrigerant (highest rank in safety classification), meaning that it is non-toxic, non-flammable and relatively safe to store, use and distribute (Osuagwu, 2022).

3.3. Evaluation of second life of recycled R-125

R-125 has recently found limited application in other sectors rather than thermal control systems, even though 99% of R-125 produced worldwide by the reporting companies is used as a blend component for commercial refrigerants. It could also be used as a fire suppression agent because it is odorless, colorless, electrically non-conductive, non-corrosive, and leaves no residue. It was also reported to act as a foam-blowing agent (Yelisetty & Visco, 2009) for making polyurethane, as the polyurethane foams are made by reacting a di-isocyanate molecule with a polyol in the presence of a blowing agent. R-125 is to substitute chlorofluorocarbons and hydrochlorofluorocarbons, because of the ozone-depleting potential of the latter ones. It only works as an interim solution because of R125's GWP concerns. According to Karecki (2000), R-125 can also be used as a substitute for 1,1-difluoroethane for a Dielectric Etch Application in an Inductively Coupled Plasma Etch Tool. A comparison of the emissions from R-125/R-152a processes and a reference C_2F_6-based process showed that emissions reductions of 68 to 76% were attainable, in large part due to much higher feed gas conversion efficiency (Karecki, 2000).

4. Conclusions

It is found that, while the separation by ionic liquid based extractive distillation is theoretically feasible, its implementation is not practical as the amount of R-410A available for recovery would not sustain a continuous process. For a processing capacity of 100 kg/h of R-410A, the number of average-size units to be recovered per day is 900. This is equivalent to 270,000 average-size AC units per year, or 4% of the R-410A units that are estimated to go out of service in China in 2022. Considering the vast area for which the availability of refrigerant is calculated, the process capacity is deemed infeasible, while pressure swing distillation is found to give 34 times lower operating costs. Increasing the pressure shifts the azeotrope to 91 mol% of R-32 at 17 bara (from 80 mol% R-32, at 1 bara), making separation of both components possible by shifting between both sides of the azeotrope in two distillation columns operating at 1 and 17 bara. However, both continuous operations need to be operated in a periodic (or batch) mode with periods of collection and short separation operation. Additionally, the presence of ionic liquid in recycled R-32, even at small concentrations, presents a health hazard for humans. Regarding the use of R-32 as a stand-alone refrigerant, coefficient of Performance (COP) value is found to increase by up to 4% with respect to R-410A, while displaying a 65% reduction in GWP. However, the flammability and health hazard concerns for R-32 also need to be considered and a stricter monitoring of the safety issues would need to be implemented. As the ionic liquid does not vaporize, the batch operation mode for ionic liquid based extractive distillation may need new batch operations designs, unlike the organic solvent-based batch extraction operations.

References

P. J. Castro, J. M. Aráujo, G. Martinho, A. B. Pereiro, (2021). Waste Management Strategies to Mitigate the Effects of Fluorinated Greenhouse Gases on Climate Change. *Applied Sciences*, 4367.

Danfoss A/S. (2022). *Refrigerant options now and in the future.* Danfoss A/S. Retrieved from https://assets.danfoss.com/documents/211728/AD224586434178en-001001.pdf

Difference Between R410A and R32. (2022). Retrieved from Acontech PLT: https://acontech.com.my/difference-between-r410a-and-r32/

European Environment Agency. (2020). *Fluorinated greenhouse gases 2020.* Luxembourg: European Environment Agency.

European Parliament and Council of European Union. (2014). Regulation (Eu) No 517/2014 of the European Parliament and of the Council of 16 April 2014 on fluorinated greenhouse gases and repealing Regulation (Ec) No 842/2006—European Environment Agency [Policy Document]. *Official Journal of the European Union, L 150*, 195–230.

360 Research Reports (2020). *Global R410A Market Research Report 2020.*

S. M. Karecki, (2000). *Development of novel alternative chemistry processes for dielectric etch applications.* PhD Thesis, Massachusetts Institute of Technology publications.

P. Makhnatch, R. Khodabandeh, (2019). Future Refrigerant Mix Estimates as a Result of the European Union Regulation on Fluorinated Gases. *International Institute of Refrigeration*, 2318–2325.

M. S. Monjur, A. Iftakher, M. M. F. Hasan, (2022). Separation Process Synthesis for High-GWP Refrigerant Mixtures: Extractive Distillation using Ionic Liquids. *Industrial and Engineering Chemistry Research*, 4390-4406.

S. Osuagwu, (2022). *R410a vs R32 – A Comparative Study of Features & Benefits.* Retrieved from Rx Mechanic: https://rxmechanic.com/r410a-vs-r32/

V. Panato, D. F. Pico, E. P. Filho, (2022). Experimental evaluation of R32, R452B and R454B as alternative refrigerants for R410A in a refrigeration system. *International Journal of Refrigeration, 135*, 221-230.

S. R. Syeda, E. A. Khan, N. Kuprasertwong, O. Padungwatanaroj, R. Gani, (2022), A Model-Data Driven Chemical Analysis System for Products and Associated Processes, Computer Aided Chemical Engineering, 49, 181-186

K Udomwong, A Robin, N Kuprasertwong, O Padungwatanaroj, AK Tula, ...R Gani, (2021), ProREFD: Tool for Automated Computer-Aided Refrigerant Design, Analysis, and Verification, Computer Aided Chemical Engineering 50, 457-462

R. Velgara, (2003, August). What You Need to Know When Recovering R-410A. *Yellow Jacket, 1*(3).

Y. Xu, D. Z. (2013). The role of HFCs in mitigating 21st century climate change. *Atmos. Chem. Phys.*

S. S. Yelisetty, D. P. Visco, (2009). Solubility of HFC32, HFC125, HFC152a, and HFC143a in Three Polyols. *Journal of Chemical & Engineering Data*, 781–785.

UNEP (2021). *About Montreal Protocol.* Retrieved from https://www.unep.org/

Antonis Kokossis, Michael C. Georgiadis, Efstratios N. Pistikopoulos (Eds.)
PROCEEDINGS OF THE 33rd European Symposium on Computer Aided Process Engineering
(ESCAPE33), June 18-21, 2023, Athens, Greece

Thermodynamic study and simulation of the process of separation of the IPA+water mixture by heterogeneous azeotropic distillation

Elvira Spatolisano[a], Camilla Barbieri[a], Laura A. Pellegrini[a], Stefania Moioli[a,*]

[a]GASP, Group on Advanced Separation Processes & GAS Processing, Dipartimento di Chimica, Materiali e Ingegneria Chimica "Giulio Natta", Politecnico di Milano, Piazza Leonardo da Vinci 32, I-20133 Milano, Italy
stefania.moioli@polimi.it

Abstract

The market of 2-Propanol is expected to reach a value of 4.8 billion dollars by 2032, considering that one possible use of this alcohol is as substitute or as additive to fossil fuels, with the aim of employing biofuels to increase the use of renewable energies. To dehydrate alcoholic mixtures containing azeotropes advanced processes are needed, as a heterogeneous azeotropic distillation process. To select the thermodynamic method best representing the binary mixture IPA+water for the simulation in ASPEN Plus®, new experimental data of Vapor-Liquid Equilibrium (VLE) have been collected at the Process Thermodynamics laboratory (PT lab) of Politecnico di Milano at conditions for which no information was found in the literature and used with other literature points for the thermodynamic description of the system IPA+water. The selected model has then been employed for the simulation of the novel azeotropic distillation process, that has been optimized.

Keywords: biofuels, isopropanol, azeotrope, distillation, PT lab.

1. Introduction

The industrial relevance of 2-Propanol (IsoPropyl Alcohol, IPA) is outstanding. It can be not only used as substitute or as additive to fossil fuels, but also as solvent and cleaning agent in a high number of industries as the medical (Guang et al., 2019) and the semi-conductor ones (Arifin and Chien, 2008). Its market, in 2021 of about 3.2 billion $, is estimated to grow reaching about 1% of the consumption of organic chemicals by 2032 (Fact.MR, 2022).

In several IPA production processes as the indirect hydrogenation or the fermentation of cellulosic materials, the raw product stream contains water and needs to be dehydrated to obtain pure IPA (De Guido et al., 2021). The process of water removal cannot be performed by employing one standard distillation column because the IPA+water mixture forms an azeotrope (at 101.3 kPa of about 87.5% (Moioli et al., 2021)), and the system, as other alcohol+water systems (CRC Handbook of Chemistry and Physics), is strongly non ideal. Advanced processes are therefore needed, as the one proposed by GASP and studied elsewhere (Spatolisano and Pellegrini, 2022).

To select the thermodynamic model for carrying out the simulation of the dehydration process, experimental data of Vapor-Liquid Equilibrium (VLE) are needed. The experimental activity carried out by GASP at the Process Thermodynamics laboratory (PT lab), part of the Process Design & Process Thermodynamics laboratory (PD&PT lab) of Politecnico di Milano, strives to fill the gaps of experimental data that the literature presents and it is a further development of the research started in 2021 (Moioli et al., 2021). In particular in this work VLE points of the 2-Propanol+water mixture have been collected at the pressure of 800 mbar. The new equilibrium data, together with other points at different pressures from the literature, have been used for the thermodynamic description of the system IPA+water. Then, the selected model has been employed for the simulation of a novel heterogeneous azeotropic distillation process of the mixture of interest. The process has been optimized taking into account several variables as the solvent flowrate, the number of theoretical stages of the column and the reflux flowrate.

2. Experimental set up and protocol

2.1. Chemicals

Demineralized water (ISO 3696 Q3, ASTM D 1193 TYPE 4) was used and isopropanol (CAS 67- 63 - 0) was supplied by Sigma-Aldrich, with a purity of at least 99.8% wt., molecular weight equal to 60.10 g/mol and specific gravity < 0.7840. Nitrogen (>99.99%, mole), coming from a tank storing N_2 in the liquid phase, was used to create the vacuum condition inside the VLLE unit (Fischer Labodest VLLE 602) up to the desired pressure and as a carrier for the gas chromatograph.

2.2. Apparatus

The Fischer® Labodest® VLLE 602 unit has been employed to collect the VLE data at PT lab. The condensed vapor and the liquid samples at the equilibrium conditions have been drawn from this unit and analyzed with the Agilent 7820A Gas Chromatograph equipped with a Thermal Conductivity Detector (TCD).

The details of the experimental protocol are reported in Moioli et al. (2021). In this work the mixture has been prepared under the hood in a container provided with a cap (400 mL) and, after mixing, part (100 mL) of it has been fed to the feed tank of the VLLE unit. After sampling, the unit is emptied and then refilled by another portion of the already prepared mixture. In order to ensure the repeatability of the test and to check the accuracy of the results obtained, each test (complete procedure) was repeated at least three times.

3. Results and Discussion

3.1. Experimental results and thermodynamic model

Figure 1 reports the experimental points collected at PT lab for a pressure of 800 mbar. The minimum temperature azeotrope occurs at about 68% mol of 2-Propanol. The bubble point curve becomes very close to the dew point curve on the right side of the azeotrope and this underlines the strong non-ideality of this mixture. However this behavior is expected also on the basis of the experimental data already collected at PT lab (Moioli et al, 2021) and from the literature at different pressures.

To describe the thermodynamic behavior of the system, the performance of the NRTL model (Renon et al., 1968) and the one of the UNIFAC-Dortmund modified (Gmehling et al., 1993) model have been tested, by considering 48 sets of isobaric or isothermal VLE data from NIST databank (detailed in Barbieri, 2022) and 8 sets of data collected at the PT lab at different pressures. As from Figure 1 and Table 1, reporting the percent Absolute Average Deviation (%AAD) of the two considered models, the UNIFAC-Dortmund modified (Gmehling et al., 1993) model best represents the system.

The UNIFAC (Dortmund modified) has therefore been selected as the thermodynamic package to be used for the simulation of the novel azeotropic distillation process in ASPEN Plus®.

Thermodynamic study and simulation of the process of sepration of the IPA+ water mixture by heterogeneous azeotropic distillation

2511

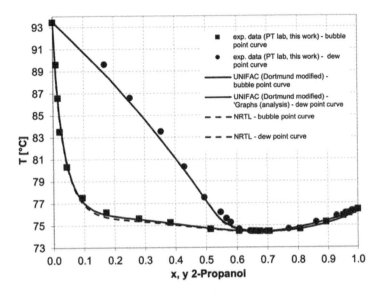

Figure 1: VLE experimental data for the IPA and water system collected at PT lab at 800 mbar and results obtained with the NTRL model with ASPEN Plus® V11.0 default parameters and the UNIFAC Dortmund modified model.

	Isobaric VLE data			Isothermal VLE data		
	AAD% T	AAD% $y_{2\text{-Propanol}}$	AAD% y_{water}	AAD% P	AAD% $y_{2\text{-Propanol}}$	AAD% y_{water}
NRTL	0.17395	7.77114	4.93413	18.722	50.383	11.214
UNIFAC	0.15138	6.63100	4.47493	10.113	19.061	7.256

Table 1: %AAD of the temperature for isobaric data or the pressure for isothermal data and the mole fraction of IPA and water in the vapor phase estimated with NRTL and UNIFAC (Dortmund modified) for all the considered sources of data.

3.2. Solvent selection

The literature offers two main alternatives in order to carry out the separation of the IPA+water mixture in its pure components: the extractive distillation (Arifin and Chien, 2008) and the heterogeneous azeotropic distillation (Arifin and Chien, 2007; Cho and Jeon, 2006).

In the extractive distillation the entrainer alters the separation factor of the mixture without introducing azeotropic behaviors with all the components of the system to be separated. On the other hand, in the case of the heterogeneous azeotropic distillation, the solvent is added in order to introduce a new lower boiling azeotrope which for its part can easily be separated. The solvent selection is crucial and, if properly done, can significantly affect the process performances. Among the high number of possibilities, disclosed in literature (Gmehling et al., 1998), the methyl butyl ether has been selected in this work. In fact it obeys to the criterion disclosed in a recent article by GASP (Spatolisano and

Pellegrini, 2022), where it is demonstrated that the best ideal entrainer, thus the one that minimizes the vapor flowrate and, ultimately, the reboiler duty, is the one that forms a binary azeotrope with water, as the methyl butyl ether does. In addition, its boiling point is lower than the boiling point of the IPA, which can be recovered as the bottom product. The other big advantage is that if some traces of methyl butyl ether are present in the IPA product, they are considered acceptable for the use of 2-Propanol as bio-fuel.

3.3. Process scheme and simulation

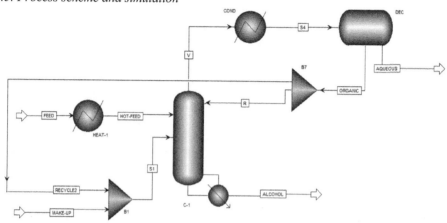

Figure 2: Scheme of the heterogeneous azeotropic distillation process with binary azeotrope for the 2-Propanol+water mixture in ASPEN Plus® V11.0.

The proposed flow sheet for the IPA-water separation exploiting methyl butyl ether as solvent is reported in Figure 2. The IPA-water mixture (FEED) and the solvent (S1) are fed into the distillation column at stage 25 and 20 respectively (the stages are counted from top to bottom with the first tray as the first stage and the reboiler as the last stage). The feed, an equimolar mixture of IPA and water, is available at atmospheric pressure and 30°C and has a flowrate of 100 kmol/hr. In the base-case configuration the column has 50 stages and the solvent flow rate is 250 kmol/h. The column separates IPA as bottom product, whose water content has to be lower than 3000 ppm in weight, and the binary water-solvent heterogeneous azeotrope, from the top, which can be separated into two liquid phases. The organic phase, exiting from the decanter, very rich in methyl butyl ether, is refluxed back to the column. A solvent make up stream is provided in order to counterbalance unavoidable solvent losses through the product streams.

The base-case configuration has been optimized in ASPEN Plus® V11.0 taking into account several variables as the solvent flow rate, the number of theoretical stages of the column and the reflux ratio (here defined as the ratio between the reflux flow rate and the organic flow rate exiting the decanter). The objective function, to be minimized, was the reboiler heat duty, as it is the main cost item of the process.

The initial solvent flow rate of 250 kmol/h has been decreased during each simulation (a delta of 10 kmol/h among each simulation has been considered) and within each run of the simulator the IPA bottom purity has been checked. If the bottom product purity was not enough, the reflux ratio would have been increased. This means that the solvent flow

Thermodynamic study and simulation of the process of sepration of the
IPA+ water mixture by heterogeneous azeotropic distillation
2513

rate and the reflux ratio have been optimized simultaneously. The results of the simulation are reported in Figure 3a).

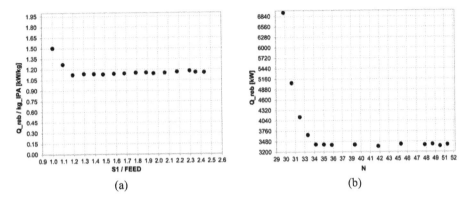

(a)   (b)

Figure 3: Duty of the reboiler as a function of a) the ratio between the solvent flowrate and the feed flow rate [100 kmol/h] and b) the number of theoretical stages, with a solvent flowrate equal to the optimum value of 120 [kmol/h].

The trend is almost constant (some little fluctuation are due to numerical issues) up to a solvent/feed ratio equal to 1.2. Then the reboiler heat duty tends to increase as the ratio further decreases. This means that the optimum solvent/feed ratio is 1.2, which corresponds to a solvent flowrate of 120 kmol/h. The optimum reflux ratio is equal to 0.56325.

Optimized the solvent flowrate and the reflux ratio, the theoretical number of stages was reduced with respect to the initial value of 50. Indeed the thermal profile and the composition profile of the vapor and liquid mole fractions of the IPA inside the column turned to be almost flat in a high number of stages inside the column, meaning that a lot of stages are not making any contribution to the separation process.

The results of the optimization are reported in the Figure 3b). Again the trend is almost flat up to a number of stages equal to 34 and then it tends to increase as the N decreases further. Therefore 34 is the optimum number of stages.

4. Conclusions

The isobaric vapor-liquid equilibrium for the system IPA+water at 800 mbar, conditions not already reported in the literature, has been experimentally investigated at the PT lab of Politecnico di Milano. The collected experimental points and the data from the literature have been used to select the thermodynamic model for the simulation and optimization in ASPEN Plus® of a novel azeotropic distillation process based on a binary water-solvent azeotrope. The future developments of this work could focus on the collection of more experimental data of phase equilibria in deeper vacuum conditions and on the economic evaluation of the simulated process of azeotropic distillation.

References

C. Guang, X. Shi, Z. Zhang, C. Wang, C. Wang, J. Gao, 2019, Comparison of heterogeneous azeotropic and pressure-swing distillations for separating the diisopropylether/isopropanol/water mixtures, ChERD, 143, 249-260.

S. Arifin, I. Chien, 2008, Design and control of an Isopropyl Alcohol dehydration process via extractive distillation using dimethyl sulfoxide as an entrainer, Ind. Eng. Chem. Res., 47, 790-803.

Fact.MR, 2022. https://www.factmr.com/report/836/isopropanol-market.

G. De Guido, C. Monticelli, E. Spatolisano, L. A. Pellegrini, 2021 Separation of the mixture 2-Propanol + Water by Heterogeneous Azeotropic Distillation with Isooctane as an Entrainer, *Energies 14*, 17 (2021): 5471.

S. Moioli, G. De Guido, M. Gilardi, L.A. Pellegrini, D. Bonalumi, G.G. Lozza, 2021, Isobaric Vapor–Liquid Equilibrium Data for the Isopropanol–Water System, J. Chem. Eng. Data, 66, 4148-4158.

CRC Handbookof Chemistry and Physics, 44th ed., 2143-2184.

E. Spatolisano, L.A. Pellegrini, Dehydration of IPA-H2O mixture: review of fundamentals and proposal of novel energy-efficient separation schemes, Chemical Engineering Science, (Article in press).

H. Renon, J. M. Prausnitz, 1968, Local compositions in Thermodynamic excess functions for liquid mixtures, AlChE Journal, 14, 135-144.

J. Gmehling, J. Li, M. Schiller, 1993, A modified UNIFAC model. 2. Present parameter matrix and results for different thermodynamic properties, Ind. Eng. Chem. Res., 32, 178-193.

C. Barbieri, Raccolta e analisi di dati sperimentali di equilibri di fase di sistemi non ideali e studio del processo di distillazione azeotropia, Master Degree Thesis, Politecnico di Milano, 2022.

S. Arifin, I-L. Chien, 2007, Combined preconcentrator/recovery column design for isopropyl alcohol dehydration process, Ind. Eng. Chem. Res., 46, 2535-2543.

J. Cho, J. Jeon, 2006, Optimization study on the azeotropic distillation process for isopropyl alcohol dehydration, Korean J. Chem Eng., 23, 1, 1-7.

J. Gmehling, C. Möllmann, 1998, Synthesis of distillation processes using thermodynamic models and Dortmund Data bank, Ind. Eng. Chem. Res., 37, 3112-3123.

Antonis Kokossis, Michael C. Georgiadis, Efstratios N. Pistikopoulos (Eds.)
PROCEEDINGS OF THE 33rd European Symposium on Computer Aided Process Engineering
(ESCAPE33), June 18-21, 2023, Athens, Greece

Sustainability assessment of poly(butylene succinate) production and End-of-Life options from wheat straw

Sofia-Maria Ioannidou[a], Dimitrios Ladakis[a], Ricardo Rebolledo-Leiva[b], Maria Teressa Moreira[b], Ioannis K. Kookos[c], Apostolis Koutinas[a],*

[a]Department of Food Science and Human Nutrition, Agricultural University of Athens, Iera Odos 75, 118 55 Athens, Greece
[b]Department of Chemical Engineering, School of Engineering, Universidade de Santiago de Compostela, D Rúa Lope Gómez de Marzoa, s/n, 15782, Santiago de Compostela, Spain
[c]Department of Chemical Engineering, University of Patras, 26504, Rio, Greece,
* akoutinas@aua.gr

Abstract

This study presents sustainability assessment of poly(butylene succinate) (PBS) production from wheat straw and post-consumer alternative End-of-Life (EoL) recirculation routes of the used biopolymer, under both techno-economic and environmental perspective. The processing steps for the whole life cycle of PBS production have been designed and simulated using the UniSim Design software (Honeywell). Techno-economic assessment employing discounted cash flow analysis has been implemented to compare the profitability of PBS production against the appropriate reference product, by estimating important techno-economic indicators. The EoL options for PBS recirculation have been evaluated, targeting the recycling of the material back into the production process, with emphasis on chemical recycling. The environmental impact of the production process and the EoL recirculation route have been evaluated via Life Cycle Assessment. A holistic evaluation of sustainability that takes into account the environmental and techno-economic pillars and combines them into a final life cycle cost has been presented.

Keywords: Biopolymer, Lignocellulosic residues, Life Cycle Costing, Sustainability assessment, Poly(butylene succinate)

1. Introduction

Global plastic production reached ca. 370 million t in 2019 with an annual increase of 2.5% [1]. The transition towards the bio-economy era necessitates the production of bio-based and biodegradable polymers to reduce the fossil resources used for the current production of the majority of plastics. Poly(butylene succinate) (PBS) is among the most widely used biopolymers, having an annual worldwide production capacity of 86,500 t [2], and constitute a biodegradable polyester produced via polycondensation of succinic acid (SA) and 1,4-butanediol (BDO). It has similar properties to polystyrene (PS), polypropylene (PP) and polyethylene terephthalate (PET) [3], fact that make it ideal candidate for the replacement of such polymers.

The development and implementation of the biorefinery concept are of great importance in order to reinforce the vision toward a sustainable economy based on bioresources. The goal within the integrated biorefinery concept is to start with a biomass feedstock and produce multiple end-products in a systematic and technologically feasible way so as to improve the overall financial performance. The combination of two principles: exploiting the residual biomass use and reducing the associated emissions are approached in the biorefinery concept [4]. Wheat is the most extensively cultivated crop worldwide, playing a vital crop in food supply chains [5]. The wheat residues are a valuable natural resource and their efficient management is essential for sustainable crop production. This study aims to assess potential valorisation of wheat residues, including major processing steps such as pre-extraction, pre-treatment, fermentation, downstream separation and purification (DSP) and polymerization, for the evaluation of the production process of PBS under the life cycle costing perspective.

2. Methodology

Process design has been carried out with the design software UniSim (Honeywell). The proposed biorefinery is based on the utilisation of wheat straw, as residue of wheat production. The chemical analysis of raw material used is based on the study of Ballesteros et al. (2006), while the pretreatment process is designed according to Al-Zuhair et al. (2013) [6-7] (Figure 1). Steam explosion and subsequent hydrolysis are utilized for the pretreatment of wheat straw and the production of the sugar rich hydrolysate. More specifically, the straw is mixed with saturated steam at 180°C for 10 min to break the fibers of the lignocellulosic material, and then the pressure is drastically reduced to atmospheric level. After cooling to 45°C, the enzymatic hydrolysis takes place where the lignocellulosic material are degraded into corresponding monomers with the aid of suitable enzymes.

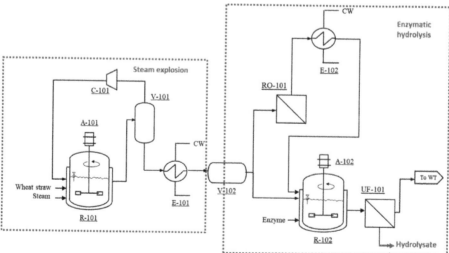

Figure 1: Process flow diagram for wheat straw pretreatment.

The sugar-rich hydrolysate produced from the pretreatment stage is then utilized for the fermentative production of PBS' monomers, e.g. SA and 1,4-BDO. Figure 2 presents the different stages of all production process. The bioconversion stage for both monomers includes a heat-sterilisation in a continuous process consisting of two heat exchangers and a holding tube. The fermentation medium is sterilized at 140°C and then cooled to the fermentation temperature before addition into the bioreactors. In case of SA, The selected parameters for the fermentation are selected from Ma et al. (2011) and are equal to 101.0 $g_{SA} \cdot L^{-1}$ concentration, 0.78 $g_{SA} \cdot g^{-1}_{sugars}$ yield and 1.18 $g_{SA} \cdot L^{-1}h^{-1}$ productivity [8]. CO_2 supply is used during fermentation due to metabolic requirements in the reductive TCA cycle to produce SA. Likewise, in case of 1,4-BDO the fermentation efficiency is reported by Burgard et al. (2016) using a genetically engineered *E. coli* strain. The final concentration of BDO is 125 $g_{BDO} \cdot L^{-1}$ with a yield of 0.40 $g_{BDO} \cdot g^{-1}_{sugars}$ and a productivity of 3.5 $g_{BDO} \cdot L^{-1}h^{-1}$, while microaerobic conditions are used during fermentation (0.02 vvm) [9].

The DSP of SA is designed according to Alexandri et al. (2019) [10]. The fermented broth is centrifuged (CF_{PBS}-201) to remove the bacterial biomass, fed into the activated carbon columns (V_{PBS}-202) for decolorisation and impurity removal and processed by cation-exchange resin columns (V_{PBS}-203) in order to transform organic acid salts into their corresponding organic acids. The acidified liquid is concentrated using the MVR-forced circulation evaporator system (EV_{PBS}-201) and the concentrated liquid is subsequently treated via crystallisation in continuous crystallizers (CR_{PBS}-201-202) at 4°C. The wet succinic acid crystals are dried in a spray dryer (DR_{PBS}-201), while the remaining liquid is recycled at the evaporation stage. The SA crystal purity achieved is higher than 99.5%, while the overall succinic acid recovery yield in the DSP is ca. 95% (w/w). In the DSP stage of 1,4-BDO, product is purified to 99.7% purity with recovery yield of 92%. The microbial biomass is initially removed via centrifugation (CF_{PBS}-202). The bacterial mass free liquid stream is processed through a series of cation- (V_{PBS}-206) and anion-exchange (V_{PBS}-207) resin columns to remove the minerals and organic acid salts that are present in the fermentation broth. The outlet liquid stream is subsequently concentrated using a MVR-forced circulation evaporator system (EV_{PBS}-202). BDO is purified via distillation (T_{PBS}-201) at atmospheric pressure and 180°C in order to separate the water and GBL.

The final stage of the production process is the polymerization stage to form the polymer PBS. The unit operations and the process conditions for the polymerization of PBS were taken from Kamikawa et al. (2013) [11]. This process is divided into three sections, the preparation of raw materials, esterification and polymerization. BDO and SA are initially mixed in a mixing tank (VPBS-301) at a molar ratio of 1.3:1 and 80°C using low pressure steam (LPS). The liquid outflow enters the esterification reactor (R_{PBS}-301), after it is heated to 180°C. The reaction takes place at 230°C and 1 bar in order for an appropriate rate of reaction to be achieved and the ester of succinic acid and BDO is formed after 3 h. The polymerization of ester is a polycondensation reaction in the presence of titanium tetrabutoxide as catalyst with a concentration of 2000 ppm (with respect to succinic acid) (R_{PBS}-302). The temperature of the reaction is 240°C and the vacuum applied is 2 torr. After 16.5h, the final product from the polycondensation reactor contains PBS with molecular weight of 70,000 Da.

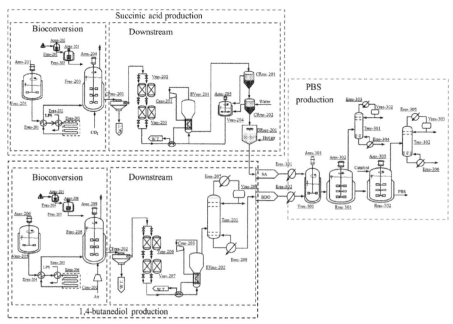

Figure 2: Process flow diagram for PBS production process.

The chemical recycling of PBS is finally performed to evaluate the potential recirculation of the monomer back to the process. The lipase-catalyzed degradation of PBS into cyclic 2-mer is designed according to Okajima et al. (2003) [18]. The reaction is carried out using a mixture of PBS and toluene having a solid to liquid ratio 1:1 (w/v) and 300% immobilized lipase CA (with respect to PBS) at 85 oC for 24 h. The repolymerization of the oligomer product obtained by the degradation reaction is also evaluated with employment of toluene under azeotropic dehydration conditions.

Techno-economic evaluation is based on preliminary economic analysis (accuracy up to ± 30%). Process design data were used for the estimation of fixed capital investment (FCI) and other important techno-economic indicators. Equipment sizing was based on established methodologies, while the estimation of free-on-board purchased equipment costs estimation was based on established textbooks and literature-cited reports [12-15]. The techno-economic indicators used in this study are the Cost of Manufacture (COM), Minimum Selling Price (MSP), Optimum Plant Capacity (OPC) leading to minimum COM, Discounted Payback Period (DPP) and Minimum Feedstock Requirements (MFR). MSP and DPP are estimated after a discounted cash flow (DCF) is performed. Life Cycle Assessment (LCA) is carried out using the GaBi software and the the ReCiPe 1.08 LCA methodology. The system boundaries for the analysis is "cradle to gate" and the functional unit is 1 kg of PBS. Finally, LCC is performed by estimating twelve different environmental categories followed by their conversion into monetized environmental externalities. The principal stages for the implementation of methodology are described by Bickel and Friedrich (2005) [16]. The monetization of the estimated impacts is based on the environmental prices reported by De Bruyn et al. (2018) for EU28 countries using a dollar to euro exchange rate of 0.856 [17].

3. Results and Discussion

Table 1 presents the results for the techno-economic evaluation of the production process of PBS, including monomers production and polymerization stages, without considering the cost of sugars production, as the wheat straw pretreatment is under development.

The FCI and COM per kg PBS are estimated at different plant capacities (10-120 kt/y) to identify the one where these indicators reach a constant value is reached (OPC 58.63 kt PBS per year). The MSP of PBS is lower than its current market price $4/kg$_{PBS}$ [19] with a decrease percentage of 45%. As regards the LCA, some representative results for the Global Warming Potential and the Abiotic Depletion Fossil potential, the most typical environmental indicators, are 2.31 kg CO_2-eq/kg$_{PBS}$ and 15.78 MJ/kg$_{PBS}$ when only the production process is taken into consideration. The results for both techno-economic and environmental assessment are expected to be varied when the pretreatment stage is also added in the estimations.

Finally, regarding chemical recycling, it was found that PBS having an Mw of 99000 was quantitatively degraded by lipase to produce cyclic oligomers mainly consisting of the cyclic diester of BS units.

Table 1. Techno-economic indicators for PBS and PLA production at the optimum plant capacity

Indicator	PBS
OPC (kt/year)	58,630
FCI ($/kg)	2.40
COM ($/kg)	1.85
NPV (M$)	649.8
MSP ($/kg)	2.18

Acknowledgements

This research has been supported by the project "Enhancing diversity in Mediterranean cereal farming systems (CerealMed)" funded by the General Secretariat for Research and Innovation of the Ministry of Development and Investments under the PRIMA Programme. PRIMA is an Art.185 initiative supported and co-funded under Horizon 2020, the European Union's Programme for Research and Innovation.

References

[1] Plastics Europe, 2019. The Circular Economy for Plastics - A European Overview. PlasticsEurope. https://www.plasticseurope.org/application/files/9715/7129/9584/FINAL_web_version_Plastics_the_facts2019_14102019.pdf, (accessed February 2021).

[2] European Bioplastics, 2019. Bioplastics facts and figures. Inst. Bioplastics Biocomposite, Nova-Institute. https://docs.european-bioplastics.org/conference/Report_Bioplastics_Market_Data_2020_short_version.pdf, (accessed February 2021).

[3] H.I. Moussa, S.B. Young, Y. Gerand (2012). Polybutylene succinate life cycle assessment variations and variables, in: AIChE Annual Meeting, Conference Proceedings.

[4] F. Cherubini (2010). The biorefinery concept: using biomass instead of oil for producing energy and chemicals. *Energy Convers. Manag.* 51 (7), 1412–1421.

[5] R. Rebolledo-Leiva, M.T. Moreira, S. González-García (2022). Environmental assessment of the production of itaconic acid from wheat straw under a biorefinery approach. *Bioresour. Technol.* 345, 126481.

[6] L. Ballesteros, M.J. Negro, J. M. Oliva, A. Cabañas, P. Manzanares, M. Ballesteros (2006). Ethanol production from steam-explosion pretreated wheat straw. In Twenty-seventh symposium on biotechnology for fuels and chemicals (pp. 496-508). Humana Press.

[7] S. Al-Zuhair, M. Al-Hosany, Y. Zooba, A. Al-Hammadi, S. Al-Kaabi (2013). Development of a membrane bioreactor for enzymatic hydrolysis of cellulose. *Renew. Energy*, 56, 85-89

[8] J.F. Ma, M. Jiang, K.Q. Chen, B. Xu, S.W. Liu, P. Wei, H.J. Ying, H.N. Chang, P.K. Ouyang (2011). Strategies for efficient repetitive production of succinate using metabolically engineered *Escherichia coli*. *Bioprocess Biosyst. Eng.* 34 (4), 411–418

[9] A. Burgard, M.J. Burk, R. Osterhout, S. Van Dien, H. Yim, (2016). Development of a commercial scale process for production of 1,4-butanediol from sugar. *Curr. Opin. Biotechnol.* 42, 118–125.

[10] M. Alexandri, R. Schneider, H. Papapostolou, D. Ladakis, A. Koutinas, J. Venus (2019). Restructuring the conventional sugar beet industry into a novel biorefinery: fractionation and bioconversion of sugar beet pulp into succinic acid and value-added coproducts. *ACS Sustain. Chem. Eng.* 7 (7), 6569–6579.

[11] M. Kamikawa, T. Matsuo, K. Oka, T. Kondo, Y. Sase, M. Tanto (2013). U.S. Patent No. 8,604,156. Washington

[12] M.S. Peters, K.D. Timmerhaus, R.E. West (2003). Plant Design and Economics for Chemical Engineers 5th edition, McGraw-Hill.

[13] R. Turton, R.C. Bailie, W.B. Whiting, J.A. Shaeiwitz, D. Bhattacharyya (2018). Analysis, Synthesis, and Design of Chemical Processes, Fifth Edition. Prentice Hall, New Jersey.

[14] E. Dheskali, K. Michailidi, A.M. de Castro, A.A. Koutinas, I.K. Kookos, (2017). Optimal design of upstream processes in biotransformation technologies. *Bioresour. Technol.* 224, 509-514.

[15] D. Humbird, R. Davis, L. Tao, C. Kinchin, D. Hsu, A. Aden, P. Schoen, J. Lukas, B. Olthof, M. Worley, D. Sexton, D. Dudgeon (2011). Process Design and Economics for Biochemical Conversion of Lignocellulosic Biomass to Ethanol: Dilute-Acid Pretreatment and Enzymatic Hydrolysis of Corn Stover, National Renewable Energy Laboratory.

[16] P. Bickel, R. Friedrich (2005). ExternE Externalities of Energy: methodology 2005 update, Reproduction. Office for Official Publications of the European Communities, Luxembourg.

[17] S. De Bruyn, M. Bijleveld, L. de Graaff, E. Schep, A. Schroten, R. Vergeer, S. Ahdour (2018). Environmental Prices Handbook EU28 Version - Methods and numbers for valuation of environmental impacts. CE Delft.

[18] S. Okajima, R. Kondo, K. Toshima, S. Matsumura (2003). Lipase-catalyzed transformation of poly (butylene adipate) and poly (butylene succinate) into repolymerizable cyclic oligomers. *Biomacromolecules.* 4(6), 1514-1519.

[19] E4tech, Re-Cord, Wur. (2015). From the Sugar Platform to biofuels and biochemicals. Final Rep. Eur. Comm. Dir. Energy.

Antonis Kokossis, Michael C. Georgiadis, Efstratios N. Pistikopoulos (Eds.)
PROCEEDINGS OF THE 33rd European Symposium on Computer Aided Process Engineering
(ESCAPE33), June 18-21, 2023, Athens, Greece

Computer simulation of a plastic waste gasification-solid oxide fuel cell power generation system using Aspen Plus

Mahmudul Hassan Riyad,[a] Sadman Fakid,[a] Nahid Sanzida[a,*]

[a]Department of Chemical Engineering, Bangladesh University of Engineering and
Technology, Dhaka-1000, Bangladesh
*nahidsanzida@che.buet.ac.bd

Abstract

This paper presents a simulation study of a plastic-waste-based combined heat and power (CHP) generation system. The process was modeled using the Aspen Plus V10 simulation software. The proposed power generation system is divided into two sections. The first is the plastic gasification system containing a pyrolyzer and a gasifier. The second part is the electrical power generation process with a pre-reformer, a solid oxide fuel cell (SOFC), and a gas turbine. Sensitivity analysis is done on the gasification process to find the optimum condition for gasification. It was observed that the combinations of relatively moderate temperature and steam to PE mass ratio are beneficial to the performance of the overall gasification system in terms of producing more syngas with greater energy/ exergy efficiency. This study still in the simulation stage aims to pave a way for establishing the plastic-waste-based CHP system as an alternative source of energy.

Keywords: Gasification, waste plastic, power generation, solid oxide fuel cell.

1. Introduction

Plastic manufacturing began commercially in the 1950s and has grown rapidly to the current worldwide annual production of 368 million tons in 2019 (Errera, 2022). According to the Geneva Environment Network, plastic is one of the major environmental threats (Geneva Environment Network, 2023). Plastic has had a severe impact on the natural ecology, causing problems for humans and wildlife, such as hurting plant life and putting animals in danger (Kumar *et al.*, 2021). Plastic contamination in the oceans is a growing problem all around the world.

Furthermore, harmful chemicals are released when plastic is burned or buried in soil (Verma *et al.*, 2016). Recycling is probably the best way to cut down on plastic waste. Plastic wastes are used in thermo-chemical processes such as combustion, hydrogenation, pyrolysis, and gasification due to their high energy content. These processes have been shown to be effective waste treatment alternatives. They not only eliminate waste, but also convert it into fuel. Separation of feedstock is not required here, unlike in mechanical recycling. Different feed materials, for example, can be combined and transformed to fuel in the same gasifier. Chemical recycling is a potential approach for the treatment of plastic waste for these reasons. This technology can be used to convert a variety of plastic waste materials, such as polyethylene terephthalate (PET), polystyrene (PS), polymethylmethacrylate (PMMA), polypropylene (PP), polyethylene (PE), low and high-density polyethylene (LDPE and HDPE), and polyvinyl chloride (PVC) (McKeown and Jones, 2020).

On the other hand, there is a growing demand for electricity in present world. Fossil fuels are still the primary source of energy for electricity generation around the world. Using fossil fuels has resulted in global warming and pollution due to the release of greenhouse gases, acid gases, and aerosol particles (Wang et al., 2019). It is crucial to use alternative fuel to generate power in sustainable approach. A recent study by Zhao et al., 2021, on biomass-based combined heat and power (CHP) system consisting of solid oxide fuel cell and gas turbine fed by syngas from a gasifier is reported to have achieved high efficiency for the generation of electrical power. Since plastic waste gasification gives similar products as biomass gasification, a plastic waste based CHP system can be envisioned, which may also be able to generate electrical power.

The aim of this work is to simulate a plastic waste-based small-scale power generation system where all the conversion steps are optimized within the limits of the simulation. This work uses Aspen Plus as the simulation software and waste PE (polyethylene) as the fuel for the simulation. It provides some preliminary qualitative and quantitative data on the gasification process's overall behavior, as well as the sensitivity of process parameters. Moreover, in the plastic waste-based CHP system, along with the optimization of the gasification process, consequent electrical power generation from SOFC is also simulated in Aspen Plus. The overall energy and exergy efficiency of the process is calculated and evaluated in comparison with already existing data.

2. Model Development

2.1. Modeling the gasification process

Plastic wastes considered in this work consist of polyethylene (PE). The equilibrium model utilized in this study to investigate the simulation of PE gasification is based on the previous work by Mitta et al., 2006. The simplified PE gasification model's process flow sheet is shown in Figure 1. At the very outset, the non-conventional fuel polyethylene was defined, and the ultimate and proximate analyses were provided as input to the model based on previous literature (He et al., 2009). The polymer NRTL/Redlich-Kwong equation of state with Henry's law "POLYNRTL" and "POLYSRK" were chosen as parameter models to calculate the thermo-physical properties of the components. The simulation began with unit processes of drying and separation which was used on the polyethylene sample to rid it of any moisture. The dried fuel is first pyrolyzed by adding external heat to facilitate breaking it down into simpler components suitable for normal gasification. Before being transferred to the gasifier, the gasifying mediums, air and steam, are preheated and combined.

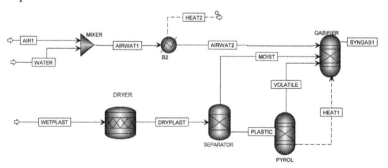

Figure 1: Gasification process flowsheet

2.2. SOFC Stack and Gas Turbine modelling

The SOFC stack model is developed based on the previous works by Doherty *et al.*, 2010. The process flow sheet is shown in Figure 2. Since ion cross-over across the electrolyte cannot be modeled in Aspen Plus, the overall oxidation of H_2 was studied rather than the cell half-reactions; and because only H_2 is electrochemically reacted, it is presumed that CO is shifted to CO_2 and CH_4 is reformed to H_2 (Panopoulos *et al.*, 2006). A fraction of processed air and unreacted CH_4 at SOFC output react in a burner. The high temperature flue gas is then introduced into a gas turbine to produce additional electrical power. In addition, coupling the syngas and air compressors to the gas turbine on the same shaft reduces energy consumption. The description of the simulation blocks in Figures 1 and 2 is given in Table 1.

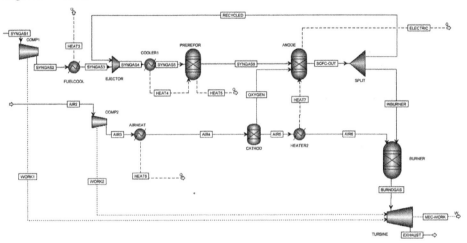

Figure 2: SOFC and Gas Turbine Process Flowsheet

Table 1: Description of Aspen Plus flowsheet unit operation blocks presented in Figure 1 and 2.

Simulation Model	Block ID	Unit	Operating Conditions
RStoic	DRYER	Dryer	Temperature: 110 °C, Pressure: 1atm
Sep	SEPARATOR	Separator	
	CATHODE	SOFC Cathode	
RYield	PYRO	Pyrolyzer	Temperature: 500 °C, Pressure: 1 atm
RGibbs	GASIFIER	Plastic Gasifier	Temperature: 850–1100 °C (Reactions were specified with equilibrium restricted), Pressure: 1 atm
	PREREFOR	Pre-reformer	Temperature: 571.4 °C (Doherty *et al.*, 2010), Pressure: 3 atm
	ANODE	SOFC Anode	Temperature: 850 °C (Doherty *et al.*, 2010), Pressure: 3 atm
	BURNER	Combustor	Temperature: 910 °C (Zhao *et al.*, 2021), Pressure: 3 atm
Compr	COMP1	Compressor	Isentropic, Pressure ratio: 3 (Doherty *et al.*, 2010) Mechanical Efficiency: 75%
	COMP2	Compressor	Isentropic, Pressure ratio: 3 (Doherty *et al.*, 2010) Mechanical Efficiency: 75%
	TURBINE	Gas Turbine	Isentropic, Pressure ratio: 0.33 Mechanical Efficiency: 75%

3. Results and Discussion

3.1. Sensitivity analysis of the gasification process

The simulation predicted equilibrium temperature for the gasification process to be 805°C. The effect of O_2 flow rate on the PE gasification process has been investigated in the range of 20 kg/h to 220 kg/h with a constant PE feed rate of 100 kg/h. As demonstrated in Figure 3, an O_2 flow rate of 99.78 kg/h maximizes CO and minimizes CO_2 production. It corresponds to the optimum O_2 to PE mass ratio of nearly 1. Similarly, the effect of steam flow rate is investigated in the range of 50 kg/h to 500 kg/h keeping the PE feed rate constant as before. Figure 4 shows that a steam flow rate of 50.04 kg/h maximizes CO and minimizes CO_2 production. It corresponds to the optimum steam to PE mass ratio of 0.5. Figure 5 shows that operating pressure does not affect the syngas composition much. For this reason, the operating pressure is selected to be 1 atmosphere.

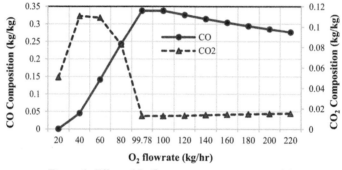

Figure 3: Effect of O_2 flow rate on syngas composition

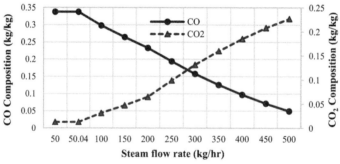

Figure 4: Effect of steam flow rate on syngas composition

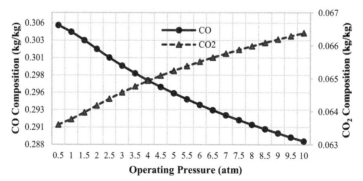

Figure 5: Effect of operating pressure on gasification products

3.2. Energy and Exergy analysis of the power generation system

In this study, to ensure maximum power generation, the power generation system is simulated at the optimum gasifier operating temperature of 805°C and the optimum steam to PE mass ratio of 0.5. The Sankey diagram of the process shown in Figure 6 demonstrates the energy flow for evaluating the plant's performance. In the figure, the widths of the different colored bars represent the quantity of energy in kilowatts. Using this system, plastic waste input equivalent to 2269 kW of higher heating value (HHV) results in an electrical power output of 721 kW. The electricity output from SOFC is 582 kW (for details please see Suther *et al.*, 2007) and the gas turbine can generate 139 kW. The energy efficiency is calculated to be 31.4% and the exergy efficiency of the process is 34.2%. These compare favorably to many gasification CHP systems studied in previous literature where the exergy ranged from 30-45% (Colpan *et al.*, 2010; Jia *et al.*, 2015; Mojaver *et al.*, 2019; Karimi *et al.*, 2020).

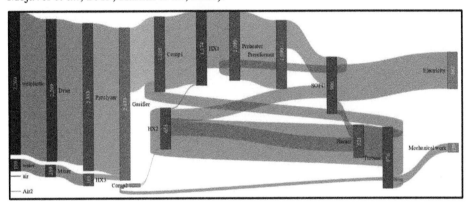

Figure 6: Sankey Diagram of the process

4. Conclusions

In this paper, a small-scale CHP system based on the concept of gasifying waste plastic to be used as fuel for power generation through solid oxide fuel cells (SOFC) has been developed using Aspen Plus. To determine the ideal operating conditions for the gasification system, effects of temperature, pressure, and O_2-steam flow ratio with the feed PE have been examined. The optimal operating conditions for gasification are determined to be 805 °C and 1 atm. The optimum O_2 to PE mass ratio is nearly 1, and the optimum steam to PE mass ratio is 0.5. In addition, the energy and exergy assessments of the complete power generation system are performed. This power generation system generates 721 kW of electrical power with an overall energy efficiency of 31.4 percent. The overall exergy efficiency is 34.2 percent. The results compare favorably to contemporary works and offer solutions to the prevailing plastic-waste recycling problem and alternative fuel crisis. Even though the results from this work are heavily dependent on the assumptions made, and it was within many limits of the simulation, it would help to establish plastic waste-based CHP system as a viable topic for research.

References

C. O. Colpan, F. Hamdullahpur, I. Dincer, and Y. Yoo, 2010, Effect of Gasification Agent on the Performance of Solid Oxide Fuel Cell and Biomass Gasification Systems, *International Journal of Hydrogen Energy*, 35(10), Pages 5001–5009. doi: https://doi.org/10.1016/j.ijhydene.2009.08.083.

W. Doherty, A. Reynolds, and D. Kennedy, 2010, Computer Simulation of a Biomass Gasification-Solid Oxide Fuel Cell Power System using Aspen Plus, *Energy*, 35(12), Pages 4545–4555. doi: 10.1016/j.energy.2010.04.051.

R. Errera, 2022, Surprising Facts about Plastic (Production, Use, and Recycling - Global Data 2023), Available at: https://www.tonerbuzz.com/blog/facts-about-plastic/#:~:text= Global % 20Plastic% 20Industry% 20Facts%20and%20Statistics&text=The% 20world% 20 produced %200.3%20percent,of%20118%20million%20metric%20tons. Accessed 30 January 2023.

Geneva Environment Network, 2023, Plastics and the Environment, Available at: https://www.genevaenvironmentnetwork.org/resources/updates/plastics-and-the-environment/. Accessed 30 January 2023.

M. He, B. Xiao, Z. Hu, S. Liu, X. Guo, and S. Luo, 2009, Syngas Production from Catalytic Gasification of Waste Polyethylene: Influence of Temperature on Gas Yield and Composition, *International Journal of Hydrogen Energy*, 34(3), Pages 1342–1348. doi: https://doi.org/10.1016/j.ijhydene.2008.12.023.

J. Jia, A. Abudula, L. Wei, B. Sun, and Y. Shi, 2015, Thermodynamic Modeling of an Integrated Biomass Gasification and Solid Oxide Fuel Cell System', *Renewable Energy*, 81, Pages 400–410. doi: https://doi.org/10.1016/j.renene.2015.03.030.

M. H. Karimi, N. Chitgar, M. A. Emadi, P. Ahmadi, and M. A. Rosen, 2020, Performance Assessment and Optimization of a Biomass-based Solid Oxide Fuel Cell and Micro Gas Turbine System Integrated with an Organic Rankine Cycle, *International Journal of Hydrogen Energy*, 45(11), Pages 6262–6277. doi: https://doi.org/10.1016/j.ijhydene.2019.12.143.

R. Kumar, A. Verma, A. Shome, R. Sinha, S. Sinha, P.K. Jha, R. Kumar, P. Kumar, Shubham, S. Das, P. Sharma, P.V. Vara Prasad, 2021, Impacts of Plastic Pollution on Ecosystem Services, Sustainable Development Goals, and Need to Focus on Circular Economy and Policy Interventions, *Sustainability*, 13, Page 9963. https://doi.org/10.3390/su13179963.

P. McKeown, and M. D. Jones, 2020, The Chemical Recycling of PLA: A Review, *Sustainable Chemistry*,1(1), Pages 1 –22. doi: 10.3390/suschem1010001.

N. R. Mitta, S. Ferrer-Nadal, A. M. Lazovic, J. F. Parales, E. Velo, and L. Puigjaner, 2006, Modelling and Simulation of a Tyre Gasification Plant for Synthesis Gas Production, in W. Marquardt, and C. Pantelides, (eds), *Computer Aided Chemical Engineering,* Elsevier 21, Pages 1771–1776. doi: https://doi.org/10.1016/S1570-7946(06)80304-4.

P. Mojaver, S. Khalilarya, and A. Chitsaz, 2019, Multi-Objective optimization using Response Surface Methodology and Exergy Analysis of a Novel Integrated Biomass Gasification, Solid Oxide Fuel Cell and High-Temperature Sodium Heat Pipe System, *Applied Thermal Engineering*, 156, Pages 627–639. doi: https://doi.org/10.1016/j.applthermaleng.2019.04.104.

K. D. Panopoulos, L.E. Fryda, J. Karl, S. Poulou, and E. Kakaras, 2006, High Temperature Solid Oxide Fuel Cell Integrated with Novel Allothermal Biomass Gasification: Part I: Modelling and Feasibility Study, *Journal of Power Sources* 159 (1), Pages 570-585. https://doi.org/10.1016/j.jpowsour.2005.12.024

T. Suther, A. Fung, and M. Koksal, 2007, Simulation of a Solid Oxide Fuel Cell-Gas Turbine System Using AspenPlus™, in *Proceedings of 3rd International Energy, Exergy and Environment Symposium*, Portugal 2007, Pages 1–18.

R. Verma, K.S. Vinoda, M. Papireddy, A.N.S. Gowda, 2016, Toxic Pollutants from Plastic Waste- A Review, *Procedia Environmental Sciences*, 35, Pages 701–708. doi: 10.1016/j.proenv.2016.07.069.

H. Wang, Z. Lei, X. Zhang, B. Zhou, and J. Peng, 2019, A Review of Deep Learning for Renewable Energy Forecasting, *Energy Conversion and Management*, 198, Page 111799. doi: https://doi.org/10.1016/j.enconman.2019.111799.

Z. Zhao, Y. A. Situmorang, P. An, J. Yang, X. Hao, J. Rizkiana, A. Abudula, and G. Guan, 2021, A Biomass-based Small-Scale Power Generation System with Energy/Exergy Recuperation, *Energy Conversion and Management*, 227, Page 113623. doi: https://doi.org/10.1016/j.enconman.2020.113623.

Antonis Kokossis, Michael C. Georgiadis, Efstratios N. Pistikopoulos (Eds.)
PROCEEDINGS OF THE 33rd European Symposium on Computer Aided Process Engineering
(ESCAPE33), June 18-21, 2023, Athens, Greece

Enhancement of hydrochar from coal discards and sewage sludge (co)-HTC for transition to a low-carbon economy

Gentil Kahilu Mwengula,[a] Orevaoghene Eterigho-Ikelegbe,[b] Jean Mulopo,[a]

Samson Bada[a]

DSI-NRF SARChI Clean Coal Technology Research Group, School of Chemical and Metallurgical Engineering, Faculty of Engineering and the Built Environment, University of the Witwatersrand, Wits 2050, Johannesburg, South Africa.
[b]Sustainable Energy and Environment Research Group, School of Chemical Engineering, University of Witwatersrand PO Box 3, Johannesburg, Wits 2050, South Africa.
Correspondence and requests for materials should be addressed to G.K (email: (1902723@students.wits.ac.za)

Abstract

Climate change scenarios highlight the significance of modifying production patterns in order to transition to a low-carbon economy, as well as the value of recycling innovations in this regard. This study examines the co-hydrothermal treatment (Co-HTC) of coal discard (CD) and sewage sludge (SS) in order to enhance the carbon content of the produced hydrochars for a sustainable circular and low-carbon economy. Using a coal-sewage sludge blend ratio of 5:1, the optimal hydrothermal carbonization (HTC) and Co-HTC parameters for producing hydrochars with the highest fixed carbon (FC) and lowest ash content (A) were 150°C, 27bar, 92.13 minutes and 208°C, 22.5 bar, 331 minutes, respectively. The physicochemical parameters of the optimized hydrochar were evaluated and compared to those of the raw materials. According to the results, HTC and Co-HTC raised the calorific value of CT and CS to 19.33 MJ/kg, 25.79 MJ/kg, and 24.31 MJ/kg, respectively.

Keywords: hydrothermal carbonization; coal tailings; coal slurry; sewage sludge, hydrochar, low-carbon economy.

1. Introduction

South Africa (SA) is one of the world's greatest coal producers and uses coal as its primary energy source (Bohlmann et al., 2018). According to the 2001 National Coal discard and slurry inventory compiled by the South African Department of Energy, about 65 million tons of coal wastes are produced annually. These coal by-products are disposed of in either tailings heaps or slurry dams (Van de Venter et al., 2021). Due to the solubilization of harmful pollutants from coal discard (CD) and the potential of spontaneous combustion, coal waste disposal is seen as a significant threat to the country's environmental waste management (Onifade and Genc 2020). Beneficiation procedures, such as physico-chemical processes and regeneration techniques, have evolved over time, however they are regarded as inefficient, not ecologically friendly, difficult, and costly (Petlovanyi et al., 2018). While the mining industry contributes to global CO_2 emissions, it may also be critical to lowering global CO_2 emissions, particularly in developing economies, not only by supporting responsible mineral extraction and processing, but also by ensuring the adoption of a circular economy approach based on mineral waste reuse and beneficiation as shown in Figure 1 for the cases of coal discards and sewage sludge. Conversely, South African wastewater treatment plants produce a substantial amount of sewage sludge (SS) (Grobelak et al., 2019). Current SS management procedures, including on-site land disposal and waste piling, are deemed unsustainable (Shaddel et al., 2019). Therefore, creative approaches to managing the CD and SS are thought essential. Hydrothermal carbonization technology (HTC) was used in this study to improve the physicochemical properties of CT, CS, and a blend of the two in order to develop suitable precursors for activated carbon, and other carbonaceous materials (value added products). The Co-HTC treatment of CD-SS mixture has the potential to enhance the carbon life cycle in coal waste while also offering an ecologically closed loop for WWTP effluent treatment, which is the cornerstone of this work (Figure 1).

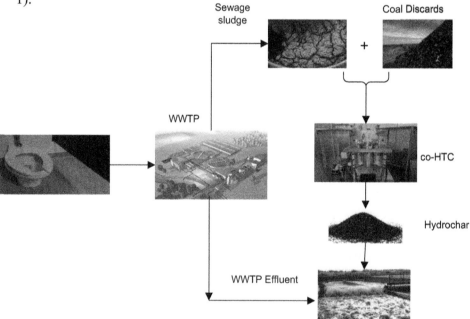

Figure 1: Co-treatment of coal waste and sewage sludge to improve the carbon life cycle in coal waste while providing an ecologically closed loop for WWTP effluent treatment.

2.   Materials and methods

2.1. Materials

The coal samples utilized in this study are coal tailing (CT) and coal slurry (CS) obtained from a coal beneficiation plant in Mpumalanga, South Africa. The sewage sludge (SS) obtained from a wastewater treatment plant (Ekurhuleni Water Care Company (ERWAT)).

2.2. Experimental Methods

The HTC and Co-HTC tests were conducted in the high-pressure tubular reactor using –1mm coal samples. HTC were performed, individually, on coal tailing (CT) and coal slurry (CS), while Co-HTC were performed on the coal-sewage sludge mixtures. 25g of the feedstock (coals, mixtures of coal and sewage sludge) was mixed with water as input solvent in a solid to liquid ratio of 1:4 in the reactor. The HTC and Co-HTC process parameters (temperature, pressure, residence time and CT+CS:SS ratio) were varied according to the design expert matrix. For experimental designs, the central composite design (CCD) and the custom design (CD) were utilized to know the number of runs that will be required to optimize the HTC and Co-HTC process respectively. Based on the CCD, 20 experimental runs were performed for the HTC of CT and CS, six center points, height factorial points and six axial points and was repeated five times to ensure reproducibility of the responses and to determine accurate experimental error. However, 40 experimental runs were performed according to the CD matrix of the Co-HTC process. The CD matrix predicted 20 edge* interior points and 20 vertex* interior points to accurately determine the process optimum conditions. The obtained solid products are identified as hydrochar coal tailing (HCT), hydrochar coal slurry (HCS) and hydrochar blended coal and sewage sludge (HCB).

3.   Results and discussions

The impact of operating conditions on the FC of the produced hydrochars (HC) is illustrated by the 3D plot of the response surface (Fig 2-3). The slopes observed on the surface plots of responses indicate the impact of reaction temperature on the FC of the produced hydrochars. The increase of temperature above 150°C decreases the FC of the produced HCT and HCS. However, the increase of temperature up to 208°C and (CT+CS): SS ratio up to 3:1 increases the FC of the produced HCB.

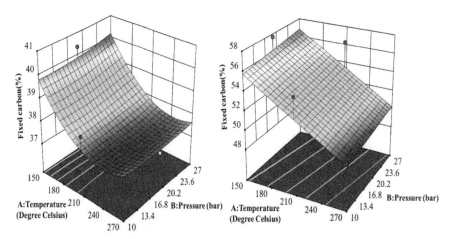

Figure 2: Surface plot showing the influence of temperature and pressure on the Fixed Carbon of Coal tailing's HC (left) and Coal slurry's HC (right).

The results indicate that 150°C, 27bar and 92.13 minutes were the optimum HTC conditions, while 208°C, 22.5 bar, 331 minutes and 20.02:4.98(CT+CS: SS) were optimum Co-HTC conditions. The process water (PW) analysis during the HTC and Co-HTC processes identified several chemical components that were dissolved in the liquid phase. For economic and environmental reasons, the PW generated by the HTC and Co-HTC processes was recycled back into the processes in this work.

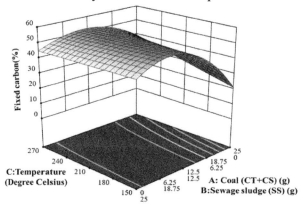

Figure 3: Surface plot showing the influence of temperature, pressure, time and the (CT+CS): SS ratio on the Fixed Carbon of produced hydrochar.

Additionally, the HTC and Co-HTC processes have shown the ability to upgrade the fixed carbon (FC) contents of coal discards and/or sewage sludge by releasing mineral matter (ash content), oxygen, and sulphur content from their molecular structures, which improved their porosity characteristics (Aliakbari et al., 2018; Saba *et al.*, 2020). The data obtained further show that the optimal HTC and Co-HTC conditions favored the hydrolysis of aliphatic components, the dissolution of mineral materials, and the prevention of carboxylate gas release. Table 1 shows the effect of the HTC and co-HTC processes under optimum settings on the physicochemical properties of the raw materials. Additionally, the HTC and Co-HTC processes raised the calorific value of raw CT, CS, and CB from 16.59MJ/kg, 24.40MJ/kg, and 17.68MJ/kg to 19.33MJ/kg, 25.79MJ/kg, and 24.99MJ/kg, respectively. Due to comparable interactions (acidic environment caused by the decomposition of SS) occurring between the feedstocks during the Co-HTC process, the CV increase of produced HCB (41.34%) was greater than that of HCT and HCS (16.51% and 5.69%). Table 2 compares the properties of the produced hydrochars to various previous study findings. The data confirmed that the optimal HTC and co-HTC conditions established in this study could be efficient for prospective carbonization of comparable feedstocks.

Table 1.Physicochemical properties of the raw materials and produced hydrochars

Analyses	CT	HCT	CS	HCS	SS	CB	HCB
Proximate Analysis (wt%, adb)							
Inherent Moisture	4.17	1.07	3.94	1.73	8.12	2.71	1.72
Ash content	38.64	37.96	23.22	22.78	36.07	37.75	19.56
Volatile Matter	21.44	21.04	21.28	18.38	47.07	21.92	20.31
Fixed Carbon	**35.75**	**39.94**	**50.98**	**57.11**	**8.75**	**37.62**	**58.41**
Ultimate Analysis (wt. %, adb)							
Carbon	**42.82**	**49.80**	**61.85**	**66.90**	**29.7**	**45.64**	**67.04**
Hydrogen	3.01	2.87	3.56	2.98	4.88	3.05	2.78
Nitrogen	1.14	1.72	1.39	1.99	4.15	1.82	2.43
Oxygen	8.79	5.83	4.78	2.75	15.22	8.2	6.08
Total Sulfur (wt. %, adb)	1.43	0.75	1.26	0.87	1.86	0.83	0.39
Calorific value (MJ/kg)	16.59	19.33	24.4	25.79	13.2	17.68	24.31
Surface area (m²/g)	6.06	11.88	6.37	14.35		6.17	20.35
CO_2 emission (mole)		0.16		0.1			0.6
Mass yield (%)	-	**97.66**	-	**94.74**	-	-	**88.38**

CT: coal Tailing. CS: Coal Slurry. HCT: Hydrochar coal tailing. HCS: Hydrochar coal slurry. Solid/liquid ratio: ¼. adb: air dried basis. Oxygen% = 100 − (Moiture + Ash + Total Carbon + Hydrogen + Nitrogen + Sulphur)

Table 2. Comparison of the produced hydrochars from this work with literature values

Material	HTC conditions		Proximate analysis (wt.%)			Ultimate analysis (wt.%)					Source
	Temperature (°C)	Time (Min)	Ash*	Fixed carbon*	Volatile matter*	C	H	N	O*	S	
Clarion coal (Cc)	-	-	11.7	54.0	34.3	63.8	4.1	1.5	26	4.6	Saba et al., 2017
HC	260	30	7.3	57.3	35.4	69.1	4.3	1.7	21.3	3.5	
Miscanthus (Ms)	-	-	2.7	12.8	84.5	48.7	5.7	0.2	45.4	0	
HC	260	30	0.7	47.8	51.5	69.6	4.4	0.3	25.6	0	
Blend (Cc + Ms)	200	90	4.2	33.1	62.7	58.2	5.3	0.8	34.4	1.2	
HC	260	90	3.5	53.4	43.1	68.7	4.1	1.1	24.4	1.6	
Bituminous Coal (CW1)	-	-	64.4	15.5	20.1	20.5	1.9	0.5	4.2	9.6	Reza et al., 2022
HC	230	30	62.8	18.7	18.5	22.5	1.5	0.5	5.4	8.4	
Bituminous Coal (CW)			67.4	14.2	18.4	18.6	1.8	0.6	4.1	8.5	
HC	230	30	65.4	17.6	17.0	20.2	1.6	0.7	5.2	7.9	
Food waste (FW)			9.6	18.8	71.6	39.3	6.0	1.5	44.0	0	
CW+FW	230	30	28.6	31.2	40.2	49.5	4.5	1.9	14.8	1.4	
Sub-bituminous coal (SbC)	-	-	25.5	40.5	34	75.1	5.1	1.1	18.4	0.33	Saqib et al., 2018
Coal			11.74	55.48	32.78	64.07	4.24	1.23	18.07	0.65	Zhao et al., 2021
HC (Coal+SbC)	340	60	7.16	65.12	27.72	74.81	3.5	1.88	12.15	0.56	
Coal discard (CD)	-	-	41.95	35.83	20.17	48.90	2.67	1.15	1.93	1.34	Setsepu et al., 2021
Searsia lancea Trees Grown (Ssl)	-	-	3.89	18.41	69.40	45.12	6.35	0.44	35.78	<0.10	
HC (CD + Ssl)	280	30	1.49	50.33	46.12	-	-	-	-	-	
Coal tailing (CT)	-	-	40.32	37.31	22.37	42.82	3.01	1.14	12.96	1.43	**This work**

HC	150	95	38.36	40.37	21.27	49.80	2.87	1.72	6.9	0.75	
Coal slurry (CS)			24.17	53.02	22.69	61.85	3.56	1.39	8.72	1.26	
HC	150	95	20.12	57.69	22.19	66.90	2.98	1.99	4.48	0.87	
Sewage sludge (SS)	-	-	39.25	9.52	51.23	29.7	4.88	4.15	23.34	1.86	
CT+CS+SS blend	-	-	38.8	38.67	22.53	45.64	3.05	1.82	10.91	0.83	
HC	208.10	360	19.71	58.82	21.46	67.04	2.78	2.43	7.8	0.39	

Ash: Ash content. FC: Fixed carbon. VM: Volatile matter. *: dried basis (moisture free). HC: Hydrochar. Oxygen%=100-(Moiture+Ash+Total Carbon+Hydrogen+Nitrogen+Sulphur)

4. Conclusion

The conversion of carbon-based wastes into energy and chemical products is a potential approach for waste treatment. In order to improve the hydrothermal carbonization (HTC) of coal discards from coal beneficiation operations and/or sewage sludge, this work optimized a range of process parameters to enable the synthesis of hydrochar with improved physicochemical properties. The best hydrothermal carbonization (HTC) and Co-HTC parameters for synthesizing hydrochars with the highest fixed carbon (FC) and lowest ash content (A) using a coal-sewage sludge blend ratio of 5:1 were 150°C, 27bar, 92.13 minutes and 208°C, 22.5 bar, 331 minutes, respectively. The enhanced hydrochar's physicochemical properties were assessed and compared to those of the raw materials. HTC and Co-HTC increased the calorific value of CT and CS to 19.33 MJ/kg, 25.79 MJ/kg, and 24.31 MJ/kg, respectively.

References

1. Aliakbari, Z., Younesi, H., Ghoreyshi, A. A., Bahramifar, N., & Heidari, A. (2018). Production and Characterization of Sewage-Sludge Based Activated Carbons Under Different Post-Activation Conditions. *Waste Biomass Valor*, *9*(3), 451–463. https://doi.org/10.1007/s12649-016-9823-7

2. Bohlmann, J. A., & Inglesi-Lotz, R. (2018). Analysing the South African residential sector's energy profile. *Renewable and Sustainable Energy Reviews*, *96*, 240–252. https://doi.org/10.1016/j.rser.2018.07.052

3. Grobelak, A., Grosser, A., Kacprzak, M., & Kamizela, T. (2019). Sewage sludge processing and management in small and medium-sized municipal wastewater treatment plant-new technical solution. *Journal of Environmental Management*, *234*, 90–96. https://doi.org/10.1016/j.jenvman.2018.12.111

4. Mazumder, S., Saha, P., & Reza, M. T. (2022). Co-hydrothermal carbonization of coal waste and food waste: fuel characteristics. *Biomass Conv. Bioref.*, *12*(1), 3–13. https://doi.org/10.1007/s13399-020-00771-5

5. Onifade, M., Genc, B., & Bada, S. (2020). Spontaneous combustion liability between coal seams: A thermogravimetric study. *International Journal of Mining Science and Technology*, *30*(5), 691–698. https://doi.org/10.1016/j.ijmst.2020.03.006

6. Petlovanyi, M. V., Lozynskyi, V. H., Saik, P. B., & Sai, K. S. (2018). Modern experience of low-coal seams underground mining in Ukraine. *International Journal of Mining Science and Technology*, *28*(6), 917–923. https://doi.org/10.1016/j.ijmst.2018.05.014

7. Saba, A., Saha, P., & Reza, M. T. (2017). Co-Hydrothermal Carbonization of coal-biomass blend: Influence of temperature on solid fuel properties. *Fuel Processing Technology*, *167*, 711–720. https://doi.org/10.1016/j.fuproc.2017.08.016

8. Saqib, N. U., Baroutian, S., & Sarmah, A. K. (2018). Physicochemical, structural and combustion characterization of food waste hydrochar obtained by hydrothermal carbonization. *Bioresource Technology*, *266*, 357–363. https://doi.org/10.1016/j.biortech.2018.06.112

9. Setsepu, R. L., Abdulsalam, J., Weiersbye, I. M., & Bada, S. O. (2021). Hydrothermal Carbonization of Searsia lancea Trees Grown on Mine Drainage: Processing Variables and Product Composition. *ACS Omega*, *6*(31), 20292–20302. https://doi.org/10.1021/acsomega.1c02173

10. Shaddel, S., Bakhtiary-Davijany, H., Kabbe, C., Dadgar, F., & Østerhus, S. (2019). Sustainable Sewage Sludge Management: From Current Practices to Emerging Nutrient Recovery Technologies. *Sustainability*, *11*(12), 3435. https://doi.org/10.3390/su11123435

11. Sultana, A. I., & Reza, M. T. (2022). Investigation of hydrothermal carbonization and chemical activation process conditions on hydrogen storage in loblolly pine-derived superactivated hydrochars. *International Journal of Hydrogen Energy*, *47*(62), 26422–26434. https://doi.org/10.1016/j.ijhydene.2022.04.128

12. *Van de Venter, F. F. Producing briquettes for domestic household fuel applications from coal tailings (Doctoral dissertation). (2021). (n.d.).*

13. Wang, L., Chang, Y., & Li, A. (2019). Hydrothermal carbonization for energy-efficient processing of sewage sludge: A review. *Renewable and Sustainable Energy Reviews*, *108*, 423–440. https://doi.org/10.1016/j.rser.2019.04.011

14. Zhao, P., Lin, C., Zhang, J., Huang, N., Cui, X., Tian, H., & Guo, Q. (2021). Moisture re-adsorption characteristics of hydrochar generated from the Co-hydrothermal carbonization of PVC and alkali coal. *Fuel Processing Technology*, *213*, 106636. https://doi.org/10.1016/j.fuproc.2020.106636

Antonis Kokossis, Michael C. Georgiadis, Efstratios N. Pistikopoulos (Eds.)
PROCEEDINGS OF THE 33rd European Symposium on Computer Aided Process Engineering
(ESCAPE33), June 18-21, 2023, Athens, Greece

Fully Electrified Conversion of Low-Quality Plastic Waste to Polymer Precursors

Kristiano Prifti [a,b], Andrea Galeazzi [a,b], Adrián Pacheco-López[a,c], Antonio Espuña[c], Flavio Manenti[a,b*]

[a]*Politecnico di Milano, CMIC Department "Giulio Natta", Piazza Leonardo da Vinci, 32, Milan 20133, Italy*
[b]*Consorzio Interuniversitario Nazionale per la Scienza e Tecnologia dei Materiali, Via Giusti 9, 50121 Firenze, Italy*
[c]*Department of Chemical Engineering, Universitat Politècnica de Catalunya, Escola d'Enginyeria de Barcelona Est, C/ Eduard Maristany 16, Barcelona 08019, Spain*
[*]*flavio.manenti@polimi.it*

Abstract

A new process design for plastic waste gasification and conversion to methanol is described in its main elements. An electrification approach is used to maximize the CO_2 credit generation of the plant. The overall process yields are described for three different feedstocks made of HDPE, PP, and PS. Electrical consumption is broken down by unit contributions to identify process bottlenecks. Preliminary profitability analysis is done with and without accounting for the possibility of selling credits CO_2 showing that the process is unprofitable unless CO_2 credits are accounted for. Depending on the feedstock, CO_2 credits between 89.5 and 103.2 €/t are shown estimated to reach a payback time of 3.5 years without accounting for labor and taxes.

Keywords: process simulation, circular economy, electrification, plastic waste, sustainable development, chemical recycling

1. Introduction

The issue of developing a fully circular supply chain for plastic materials is a crucial topic for waste management and the carbon economy (Pacheco-López et al., 2021). Chemical recycling of plastic waste has become a hot topic in recent years with pyrolysis being the main focus (Qureshi et al., 2020). Mechanical recycling has reached commercial maturity with the current bottleneck being feedstock purity and property degradation over multiple cycles (Schyns and Shaver, 2021). Gasification is a supplementary route for contaminated feedstocks that have become unsuitable for mechanical recycling and are available in large amounts. Continuous gasification processes are derived from biomass and asphalts treatment technologies to manufacture syngas which could be used to synthesize polymer monomeric precursors. The circular vision seeks to reach polymeric precursors that can be used to make polymers of virgin quality, which in turn can be recycled over multiple loops after usage, thus drastically extending the lifespan of the raw material in the market. This work develops and analyses a plant configuration that converts plastic waste through mild gasification (Kristiano Prifti et al., 2021) and fully electrifies the downstream of the gasification unit to make methanol with no CO_2 emissions assuming enough renewable electrical energy is available. The process was designed with the intent of maximizing CO_2 credit generation and carbon emission abatement. The boundaries are a mix of PE,

PS, PP waste, demi-water, and renewable electricity as an input to reach methanol of commercial purity as a final product. The process is divided into five conceptual blocks: gasification, light hydrocarbon bi-reforming, syngas conditioning, methanol synthesis, and methanol purification. The model was developed in Aspen HYSYS V11 using a thermodynamic equilibrium approach for the gasification and reforming reactors and kinetic models in methanol synthesis (Bisotti et al., 2021). Syngas conditioning is carried out with a commercial PEM electrolyzer (Carmo et al., 2013) which generates the hydrogen for CO_2 hydrogenation to methanol and the oxygen necessary for the gasification consuming 55 kWh/kg hydrogen. The environmental impact is evaluated with a simple CO_2 equivalent method while the economic performance was estimated using the CORO framework (Prifti et al., 2022) to compute both capital and operative expenditures. A sensitivity analysis on feed composition is carried out to create a surrogate model (Galeazzi et al., 2022). Which is intended to be introduced in the ontological framework developed by Pacheco-López et al (2021) to further expand such ontology in both applicability and flexibility, as well as evaluate the role this process might have in the value chain of plastic waste in the future.

2. Methods

The process was designed to start with a feedstock specification of 1000 kg/h of plastic waste mixtures of different compositions. Three feedstocks were analyzed:
- 48% HDPE, 32% PP, 20% PS: H/C ratio of 1.55, Mass Fraction of C: 88.59%
- 54% HDPE, 36% PP, 10% PS: H/C ratio of 1.73, Mass Fraction of C: 87.40%
- 60% HDPE, 40% PP: H/C ratio of 2, Mass Fraction of C: 85.71%

Changing the overall mass fraction of carbon greatly impacts the performance of the gasification and oxygen demand of the system. The PEM electrolyzer is assumed to have a consumption of 55 kWh/kg of hydrogen, and the specification of the module is given on hydrogen demand. The PEM produces all the hydrogen necessary for the plant to fully hydrogenate CO_2. Oxygen, if produced in excess is sold as a secondary product, while, if not produced in sufficient amount, is supplemented with an outside source (cylinders or Air Separation Unit). The PEM electrolyzer is a bottleneck for pressure in the upstream of the methanol loop. Pressure is set at 20 bar, which corresponds with the highest pressure supported by commercial solutions for a PEM electrolyzer considered in this study.

The steam methane reformer is electrified and follows the design commercialized by Topsoe A/S (Wismann et al., 2019). Thermal losses toward the environment are neglected and the outlet temperature of the syngas can reach 950°C due to the compact design and direct catalyst heating. Moreover, the unit uses $Ni\text{-}CeO_x\text{-}MgAl_2O_4$ catalyst instead of the conventional Nickel catalyst, which is active also for dry reforming reactions and to further help with the conversion of CO_2 (Ye et al., 2019).

The gasification unit is an entrained flow gasification reactor (Bader et al., 2018) working with pure oxygen coming from the electrolyzer and steam. A controller keeps the temperature at the outlet of the gasification at 850±1°C by regulating the oxygen demand and thus, adjusting the equivalence ratio. The dynamic variation of the equivalence ratio is made to keep the level of tars in the gasification chamber as constant as possible (the assumption is that tar degradation depends more on temperature than the composition of the feedstock, even if the higher aromatic content in PS affects tar concentration to some extent).

The methanol synthesis loop is done with two shell and tube reactors cooled with boiling feed water and with intermediate sequestration of the product. The kinetics model used for the reactor is based on Graaf kinetics (Bisotti et al., 2021) which was chosen due to its reliable performance at pressures close to 60 bar (Bisotti et al., 2022).

The distillation section works with a triple-column scheme. The scheme could be simplified since the kinetic scheme does not account for the formation of Dimethyl Ether, higher alcohols, and formaldehyde, but it was intentionally kept like the industrial designs to allow for a more accurate CapEx estimation.

The resulting process layout is a more detailed and improved design with regard to the conventional process layout as previously implemented in Prifti et al. (2021).

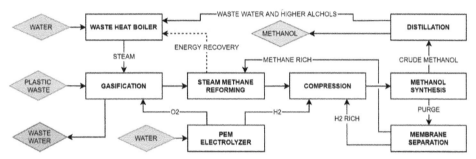

Figure 1: Block flow diagram of the process layout

3. Results

The overall input-output plant performance is reported in Table 1. As can be seen from the mass flow rates, the bottleneck of this process is the carbon content in the feedstock. This causes methanol production to rise as the H/C ratio of the feedstock decreases. This behavior can be explained by considering that, from the point of view of the process, there is an unlimited supply of hydrogen thanks to the electrolyzer's presence.

Table 1: Process input and output streams

		48PE32PP20PS	54PE36PP10PS	60PE40PP
	Hydrogen/Carbon ratio	1.55	1.73	2
INPUT	Electricity [kWh]	9650	9110	8349
	Plastic Waste [kg/h]	1000	1000	1000
	Demi water [kg/h]	1607	1468	1268
OUTPUT	Oxygen [kg/h]	249	139	-18
	Off gasses [kg/h]	26	26	28
	Wastewater [kg/h]	19	19	19
	Methanol [kg/h]	2316	2285	2238

Electricity shows the highest variability when changing the feedstock composition. Looking at the detailed breakdown of electricity consumption in Figure 2, the PEM electrolyzer is the main consumer of electricity as well as the most sensitive to feed changes. The PEM consumption from the most hydrogenated feedstock to the least hydrogenated one increases by 20%, while compression duty increases by 8% and the reformer one decreases by 8%. The resulting overall electricity consumption is 15.6% higher for the process with the most hydrogenated feedstock.

While the additional electricity consumption translates to higher operative costs, the plant has higher methanol and oxygen production, which could compensate for these higher operative costs. To understand the relative profitability of each feedstock an analysis of the operative margin was made and is reported in Table 2. First, the more hydrogenated the feedstock, the more profitable the process, even with lower methanol yield; second, assuming 8000 hours of operation a year, the plant can generate earnings between 940 k€/y in the best-case scenario, and 260 k€/y in the worst-case scenario, before taxes and without accounting for labor and maintenance. Using the framework described in Prifti et al. (2022) the plant CapEx was estimated at 15.7 M€. With these assumptions, the payback time would range from 16.7 to 60.4 years, being unfeasible in both cases.

Table 2: Process operative margin sensitivity made with EU October 2022 price assumptions.

	€/unit	[€/h] 48PE32PP20PS	[€/h] 54PE36PP10PS	[€/h] 60PE40PP
Methanol [kg/h]	0.58	1343.3	1325.3	1298.0
Electricity [kWh]	0.14	-1351.0	-1275.4	-1168.9
Oxygen [kg/h]	0.2	49.9	27.9	-3.6
Off gasses [kg/h]	0	0	0	0
Waste Water [kg/h]	0	0	0	0
Demi Water [kg/h]	0.006	-9.6	-8.8	-7.6
Net		32.5	68.9	117.9

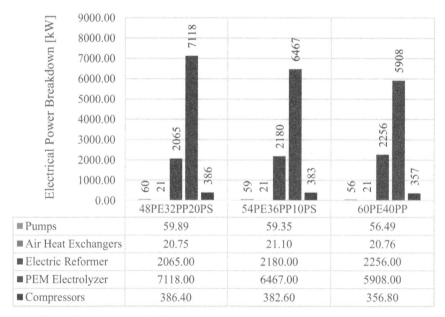

	48PE32PP20PS	54PE36PP10PS	60PE40PP
■ Pumps	59.89	59.35	56.49
■ Air Heat Exchangers	20.75	21.10	20.76
■ Electric Reformer	2065.00	2180.00	2256.00
■ PEM Electrolyzer	7118.00	6467.00	5908.00
■ Compressors	386.40	382.60	356.80

Figure 2: Electrical consumption breakdown by unit operation

Among the revenues, CO_2 credits have been neglected until now. Currently, methanol production has an average footprint of 110 gCO_2/MJ or converted in mass, 2.21 gCO_2/gCH_3OH (Hamelinck and Bunse, 2022). Since the plant does not emit CO_2 and is assumed to consume only renewable energy, for each kg of methanol, 2.21 kg of CO_2 credits are generated which translates into revenues.

As reported in Table 3, including CO_2 credits value at 67 €/t (Europe, October 2022), it considerably increases the process profitability but does not change the trend seen in the sensitivity analysis with the more hydrogenated feedstock still being the most profitable. In conclusion, the process feasibility relies heavily on CO_2 credits and thus the business model depends on subsidies, either local or regional. Therefore, the identification of an appropriate location with political stability and willingness to address the climate challenge is of key importance for the feasibility of the process.

Table 3: Process economics accounting for the CO_2 footprint

	48PE 32PP 20PS	54PE 36PP 10PS	60PE 40PP
CO_2 Credits from Methanol [kg/h]	5118.36	5049.85	4945.98
CO_2 credit revenues [€/h] October 2022 CO_2 credit value of 67 €/ton	342.93	338.34	331.38
Annual revenues from CO_2 credits sales [€/y]	2,743,440.96	2,706,719.60	2,651,045.28
Annual revenues by process sales [€/y]	259,984.00	551,616.00	943,744.00
Total revenues [€/y]	**3,003,424.96**	**3,258,335.60**	**3,594,789.28**
Payback time [y]	**5.23**	**4.82**	**4.37**
CO_2 credit value to have PBT of 3.5y [€/t]	103.2	97.4	89.5

Two other aspects that might come into play in the future are the revenues from plastic waste treatment that are intentionally underestimated to account for pre-treatment costs not discussed in the paper, and further subsidies on methanol produced from non-fossil sources. However, both terms require a legislative push towards chemical recycling against incineration and landfill, as well as economic commitments to the circular economy development model from governments.

4. Conclusions

The paper described a possible pathway to methanol synthesis from plastic waste through gasification, maximizing the generation of carbon credits by electrifying key unit operations and exploiting a PEM electrolyzer that generates hydrogen and oxygen to use on site. A sensitivity study on the process feed was carried out and used to obtain mass and energy balances using a detailed simulation. The results of the preliminary economic analysis show the process as unfeasible unless CO_2 credit sales are accounted for. The process becomes economically appealing for the highest price of CO_2 credits. To reach a payback time of 3.5 years, carbon credit values between 89.5 and 103.2 €/t are necessary, even though at current CO_2 credit values of 67 €/t, acceptable payback times lower than 6 years are achieved for every feedstock analyzed.

5. Acknowledgments

Adrian Pacheco-Lopez acknowledges the financial support received from the "Ministerio de Ciencia e Innovación" (grant ref. PRE2018-087135) within the research project CEPI (PID2020-116051RB-I00 granted by the Spanish "MINECO" and the European Regional Development Fund), and the Erasmus+ programme of the EU 2021-27 (KA131 call).

References

Bader, A., Hartwich, M., Richter, A., Meyer, B., 2018. Numerical and experimental study of heavy oil gasification in an entrained-flow reactor and the impact of the burner concept. Fuel Processing Technology 169, 58–70. https://doi.org/10.1016/j.fuproc.2017.09.003

Bisotti, F., Fedeli, M., Prifti, K., Galeazzi, A., Dell'Angelo, A., Barbieri, M., Pirola, C., Bozzano, G., Manenti, F., 2021. Century of Technology Trends in Methanol Synthesis: Any Need for Kinetics Refitting? Industrial and Engineering Chemistry Research 60, 16032–16053. https://doi.org/10.1021/acs.iecr.1c02877

Bisotti, F., Fedeli, M., Prifti, K., Galeazzi, A., Dell'Angelo, A., Manenti, F., 2022. Impact of Kinetic Models on Methanol Synthesis Reactor Predictions: In Silico Assessment and Comparison with Industrial Data. Industrial and Engineering Chemistry Research 61, 2206–2226. https://doi.org/10.1021/acs.iecr.1c04476

Carmo, M., Fritz, D.L., Mergel, J., Stolten, D., 2013. A comprehensive review on PEM water electrolysis. International Journal of Hydrogen Energy 38, 4901–4934. https://doi.org/10.1016/j.ijhydene.2013.01.151

Galeazzi, A., Prifti, K., Gallo, F., Manenti, F., 2022. A Methodology for The Optimal Surrogate Modelling of Digital Twins Using Machine Learning. Computer Aided Chemical Engineering 51, 1543–1548. https://doi.org/10.1016/B978-0-323-95879-0.50258-7

Hamelinck, C., Bunse, M., 2022. Carbon footprint of Methanol [WWW Document]. Methanol Institute. URL https://www.methanol.org/ (accessed 10.19.22).

Pacheco-López, A., Somoza-Tornos, A., Graells, M., Espuña, A., 2021. Synthesis and assessment of waste-to-resource routes for circular economy. Computers & Chemical Engineering 153, 107439. https://doi.org/10.1016/j.compchemeng.2021.107439

Pacheco-López, A., Somoza-Tornos, A., Muñoz, E., Capón-García, E., Graells, M., Espuña, A., 2020. Synthesis and Assessment of Waste-to-resource Routes for Circular Economy, in: Pierucci, S., Manenti, F., Bozzano, G.L., Manca, D. (Eds.), Computer Aided Chemical Engineering, 30 European Symposium on Computer Aided Process Engineering. Elsevier, pp. 1933–1938. https://doi.org/10.1016/B978-0-12-823377-1.50323-2

Prifti, K., Galeazzi, A., Barbieri, M., Manenti, F., 2022. A Capex Opex Simultaneous Robust Optimizer: Process Simulation-based Generalized Framework for Reliable Economic Estimations. Computer Aided Chemical Engineering 51, 1321–1326. https://doi.org/10.1016/B978-0-323-95879-0.50221-6

Prifti, Kristiano, Galeazzi, A., Margarita, I., Papale, A., Miele, S., Bargiacchi, E., Barbieri, M., Petea, M., Manenti, F., 2021. CONVERTING END-OF-LIFE PLASTIC WASTE INTO METHANOL: THE GASIFORMING PROCESS AS NEW, EFFICIENT AND CIRCULAR PATHWAY. Environmental Engineering and Management Journal 20, 1629–1636. https://doi.org/10.30638/eemj.2021.151

Prifti, K., Galeazzi, A., Margarita, I., Papale, A., Miele, S., Bargiacchi, E., Barbieri, M., Petea, M., Manenti, F., 2021. Converting end-of-life plastic waste into methanol: The gasiforming™ process as new, efficient and circular pathway. Environmental Engineering and Management Journal 20, 1629–1636. https://doi.org/10.30638/eemj.2021.151

Qureshi, M.S., Oasmaa, A., Pihkola, H., Deviatkin, I., Tenhunen, A., Mannila, J., Minkkinen, H., Pohjakallio, M., Laine-Ylijoki, J., 2020. Pyrolysis of plastic waste: Opportunities and challenges. Journal of Analytical and Applied Pyrolysis 152, 104804. https://doi.org/10.1016/j.jaap.2020.104804

Schyns, Z.O.G., Shaver, M.P., 2021. Mechanical Recycling of Packaging Plastics: A Review. Macromol Rapid Commun 42, e2000415. https://doi.org/10.1002/marc.202000415

Wismann, S.T., Engbæk, J.S., Vendelbo, S.B., Bendixen, F.B., Eriksen, W.L., Aasberg-Petersen, K., Frandsen, C., Chorkendorff, I., Mortensen, P.M., 2019. Electrified methane reforming: A compact approach to greener industrial hydrogen production. Science 364, 756–759. https://doi.org/10.1126/science.aaw8775

Ye, R.-P., Ding, J., Gong, W., Argyle, M.D., Zhong, Q., Wang, Y., Russell, C.K., Xu, Z., Russell, A.G., Li, Q., Fan, M., Yao, Y.-G., 2019. CO2 hydrogenation to high-value products via heterogeneous catalysis. NatCom 10, 5698. https://doi.org/10.1038/s41467-019-13638-9

Antonis Kokossis, Michael C. Georgiadis, Efstratios N. Pistikopoulos (Eds.)
PROCEEDINGS OF THE 33rd European Symposium on Computer Aided Process Engineering
(ESCAPE33), June 18-21, 2023, Athens, Greece

Deep neural network-based multi-objective optimization of NOx emission and profit by recovering lignocellulosic biomass

Kim Y.,[a,b§] Park J.,[a,b§] Lim J.,[a,b] Joo C.,[a,b] Cho H.,[a] Kim J.[a,b*]

[a]*Green Materials and Processes R&D Group, Korea Institute of Industrial Technology, 55, Jonga-ro, Ulsan, 44413, Korea*
[b]*Department of Chemical and Biomolecular Engineering, Yonsei University, 50, Yensei-ro, Seoul, 03722, Korea*
[§]*Kim Y. and Park J. contributed equally to this work as first authors.*
[*]*kjh31@kitech.re.kr*

Abstract

In the pulp and paper industry, the external energy and pulping chemical consumption have been reduced by recovering the lignocellulosic biomass (LB) produced during the pulping process. However, this involves the inevitable emission of thermal NOx owing to the high pyrolysis reaction temperature. Therefore, it is necessary to simultaneously optimize energy and pulping chemical recovery and minimize NOx emissions. Hence, this study focuses on the multi-objective optimization of maximizing the net profit from energy and pulping chemical recovery while minimizing NOx emissions in recovering LB. For multi-objective optimization, a deep neural network (DNN)-based optimization model for the net profit and NOx emissions was developed with the 1,071 simulation data points according to the operating conditions. Consequently, Pareto-optimal solutions with profits between 5,241,520 and 1,329,558 $/y and NOx emissions between 87.95 and 78.27 ppm were obtained. The proposed Pareto-optimal front can offer comprehensive solutions to decision-makers.

Keywords: Lignocellulosic biomass, Profit, NOx emission, Multi-objective optimization, Deep neural network

1. Introduction

The pulp and paper industry is energy-intensive, involving high operating costs because of the large amounts of chemicals used during the pulping process. Therefore, lignocellulosic biomass (LB) is recovered to reduce energy and pulping chemical consumption. LB is produced during the pulping process and is composed of lignin, inorganic compounds, and 80–85% water. The energy consumption in the pulp and paper industry is reduced by reusing lignin in LB as a primary energy source. Further, the consumption of pulping chemicals is reduced by recovering the pulping chemicals converted from the inorganic components in the LB via pyrolysis.

The recovery of energy and pulping chemicals is increased by decreasing the moisture content in LB via evaporation, followed by efficient combustion with multi-level injected air through the biomass boiler. Currently, to increase energy and pulping chemical recovery with LB reuse, research is actively conducted to reduce energy consumption by modifying the configuration of the evaporation process (Kim et al., 2021), or increase energy and chemical recovery by retrofitting the internal structure of the biomass boiler to improve combustion performance (Pérez et al., 2016).

However, increasing the efficiency of LB combustion leads to higher emission of thermal NOx owing to the high temperature during the pyrolysis reaction. Therefore, research has been conducted on reducing thermal NOx by combining an additional system with a biomass boiler (Tu et al., 2018).

Despite the considerable contributions of previous studies, some challenges remain to be solved. First, previous studies have focused on retrofitting the conventional evaporation process and biomass boiler (Kim et al., 2021; Pérez et al., 2016; Tu et al., 2018). The presented solutions have limitations in that on-site application to the existing pulp and paper industry is difficult due to structural changes or additional equipment installation. Second, previous studies have not simultaneously considered minimizing the energy consumption of the evaporation process, maximizing energy and pulping chemical recovery, and minimizing NOx emissions in the biomass boiler (Kim et al., 2021; Pérez et al., 2016; Tu et al., 2018). Removal of higher amounts of moisture in LB during evaporation leads to increased energy recovery through LB reuse; however, it also results in increased energy consumption during evaporation. In addition, more efficient LB combustion in a biomass boiler results in increased energy and pulping chemical recovery; however, it also leads to higher thermal NOx emissions. Therefore, it is necessary to simultaneously focus on minimizing the energy consumption during the evaporation process, maximizing energy and pulping chemical recovery, and minimizing NOx emissions in the biomass boiler.

To overcome these challenges, this study focused on the deep neural network (DNN)-based multi-objective optimization of NOx emission, and net profit from energy and pulping chemicals recovery by reusing LB. This study aimed to maximize net profit and minimize NOx emissions in an applicable on-site way by simultaneously considering the evaporation process and biomass boiler operating conditions. To this end, simulation models of the evaporation process and biomass boiler were developed to derive NOx emissions and net profit according to the operating conditions. Then, a DNN-based optimization model for the net profit and NOx emissions was developed with the 1,071 simulation data points to derive the optimal operating conditions.

2. Methodology

2.1. Model development

2.1.1. Process model

The simulation conducted in this study involved an evaporation process, followed by a biomass boiler. The evaporation process was simulated using Aspen Plus V11.0, ELECNRTL method, and industrial data. Lignocellulosic biomass (LB) with a concentration of 17% entered an intermediate evaporator and flowed partially forward and backward. Using this configuration, LB was efficiently concentrated to 69–85%. The higher the target LB concentration, the more energy sources are required for the evaporation process to remove more moisture. A schematic of the evaporation process is shown in Fig. 1.

The biomass boiler was simulated using Aspen Plus V11.0, Peng Robinson–Boston Mathias method, and industrial data. Tertiary-level air and concentrated LB were directed into the biomass boiler. Subsequently, dehydration, pyrolysis, gasification, reduction, reoxidation, and combustion reactions occurred. First, the concentrated LB was sprayed in the furnace in the biomass boiler, and the moisture in the LB was completely evaporated via dehydration. Second, the dried LB in the solid state dropped toward the bottom of the furnace and was combusted with secondary air. Lignin in the dried LB was decomposed into fixed carbon and volatiles via pyrolysis. Thermal NOx was also produced owing to the high pyrolysis reaction temperature. The volatiles (including thermal NOx) moved to the upper part of the furnace, and the fixed carbon and remaining inorganics in the LB were stacked in the form of a char bed at the furnace bottom. Third, the fixed carbon in the char bed was combusted with primary air and the energy required for the reduction reaction was produced. In the char bed, Na_2SO_4 in the inorganic components was converted to Na_2S via gasification with the remaining fixed carbon. Partial Na_2S was converted to Na_2SO_4 via reoxidation with the O_2 present in the primary air. Furthermore, a part of the fixed carbon was converted to CO_2 via gasification, and the converted CO_2 reacted with Na (the inorganic component in LB), forming Na_2CO_3. After the reaction, inert components such as Na_2S, Na_2SO_4, and Na_2CO_3 exited the furnace bottom in the form of smelt. Subsequently, Na_2CO_3 in the smelt was converted to NaOH by additional treatment, and Na_2S from the smelt and NaOH were recovered as pulping chemicals. Finally, all the volatiles moved to the upper part of the furnace and combusted with the tertiary air. In this combustion reaction, polluting gases such as H_2S, SO_2, CO, and NO_2 were reduced. The final flue gas exited the furnace, and steam was produced in the upper region (economizer, boiler bank, and superheaters) of the biomass boiler. A schematic of the biomass boiler is shown in Fig. 2.

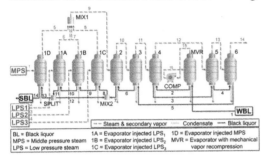
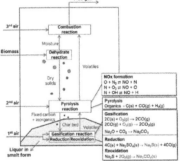

Fig. 1. Schematic of the evaporation process   Fig. 2. Schematic of the biomass boiler

The main operating parameters that affect the energy, pulping chemical recovery, and NOx emissions are the LB concentration and tertiary-level injected air distribution. The net profit by reusing the LB according to the operating conditions was calculated using Eqs. (1)–(4). The net profit was calculated by subtracting the energy consumption cost from the sum of the profit by recovering the energy (e.g., steam) and pulping chemicals, as shown in Eq. (1). The profit from steam recovery was calculated by multiplying the mass flow and specific enthalpy of the flue gas with the steam price, as shown in Eq. (2). The profit obtained by recovering pulping chemicals was calculated by multiplying the mass flow with the prices of Na_2S and NaOH, the main components of the pulping chemical, as shown in Eq. (3). Here, NaOH could be recovered only through additional treatment of Na_2CO_3; therefore, the NaOH profit was additionally considered in terms of Na_2CO_3 conversion and NaOH yield during the treatment. The steam consumption cost was calculated by multiplying the price of each type of steam consumed in the evaporation process with the consumed steam mass flow, as shown in Eq. (4).

$$Net\ profit = P^S + P^{PC} - C^S, \tag{1}$$

$$P^S = M^{flue\ gas} \times E^{flue\ gas} \times Price^S, \tag{2}$$

$$P^{PC} = Price^{Na_2S} \times M^{Na_2S} + Price^{NaOH} \times \alpha \times M^{Na2CO3}, \tag{3}$$

$$C^S = \sum_{i=1} C_i^S \times M_i^S. \tag{4}$$

2.1.2. DNN-based optimization model

Using the limited datasets derived from the process model makes it easy to fall into the local optimum problem. However, the DNN model enables global optimization by efficiently generating numerous datasets. Therefore, this study developed a DNN model for robust multi-objective optimization to maximize net profit while minimizing NOx emissions. First, the initial net profit and NOx emission datasets were generated through the process model. Subsequently, the DNN model was trained using these datasets. Among these datasets, 60% were used for training, 20% for validation, and 20% for testing. The input variables of the developed DNN model were the concentration of the LB exiting the evaporation process and tertiary-level injected air distribution ratio. The output variables of the model were net profit and NOx emissions, and the model consisted of 3 hidden layers and 20 hidden nodes. ReLu and R^2 were used as the activation function and evaluation index of the developed DNN model, respectively.

2.2. Multi-objective optimization

This section describes the multi-objective optimization methodology. There were several constraints in solving the multi-objective optimization problem of minimizing NOx while maximizing the net profit. First, the NOx emissions should not be higher than those of the existing operation. Second, the net profit should not be lower than the existing values. Third, the concentration of LB exiting the evaporation process was 69–85%. Finally, the sum of the distributed air was 100%. Multi-objective optimization was conducted using the datasets from the process and DNN models. Finally, this study proposed Pareto-optimal solutions that satisfy the specified constraints.

3. Results and discussions

3.1. Model validation

Conventionally, LB is concentrated to 70% during evaporation, and combusted with the distributed tertiary-level air at ratios of 24.89%, 49.06%, and 26.04% in the biomass boiler. The simulation results for the evaporation process and biomass boiler were validated using industrial data. The performance of the models showed little error compared to the industrial data ($\leq$0.24% and $\leq$4.55%, respectively). Consequently, the operating conditions of the evaporation process consumed 368,157 t/y of steam. 1,629,187 t/y of steam and 591,542 and 1,789,136 t/y of Na_2CO_3 and Na_2S, respectively, for converting to pulping chemicals were recovered by the existing biomass boiler. Therefore, the net profit was 1,329,641 $/y in the existing operation condition: 1,521,825 $/y of energy consumption cost, and 5,518,458 and 6,635,017 $/y by recovering energy and pulping chemicals, respectively. Furthermore, the NOx emissions from the existing operation were 87.95 ppm. The DNN model was trained using 1,091 data samples for profit and NOx emissions according to the operating conditions obtained by the process model. Consequently, the developed DNN model was confirmed to have high accuracy, with an R2 of 0.982 for profit and 0.981 for NOx emission.

3.2. Optimization of NOx emissions and net profit

This section first describes the changes in energy consumption, energy and pulping chemical recovery, and NOx emissions according to the operating conditions: the concentration of LB exiting the evaporator and tertiary-level injected air distribution ratio. Then, representative solutions are presented along with the results of multi-objective optimization for maximizing the net profit while minimizing NOx emissions.

The concentration of LB exiting the evaporator was changed to 69–85%, and the resulting steam consumption is shown in Fig. 3. The evaporation process consumed steam of 367,462 t/y at 69%, 368,157 t/y at 70%, and 387,810 t/y at 85%. The steam consumption of the evaporation process increased as the LB concentration increased, because the higher the target LB concentration, the more energy sources were required to remove the higher amount of moisture.

The effects of the LB concentration and tertiary-level injected air distribution ratio on steam, pulping chemical recovery, and NOx emissions are shown in Fig. 4 (a). The flue gas enthalpy, which is directly related to steam recovery, increased when the LB concentration and secondary air flow ratio increased. The higher the LB concentration, the less energy that is lost to remove moisture from the LB. In addition, as the secondary air flow ratio increases, more volatiles can be introduced into the flue gas by the activated pyrolysis reaction.

The Na_2S recovery for pulping chemicals increases as the secondary air flow ratio increases, and the primary air flow rate approaches 14.74%, as shown in Fig. 4 (b). As the secondary air flow ratio increases, the pyrolysis reaction actively occurs so that more substances for Na_2S conversion (C, CO) are decomposed from the LB. As the primary air flow ratio increases, more energy can be generated during gasification. However, it has an optimum point of 14.74% because it also promotes the reoxidation reaction of Na_2S.

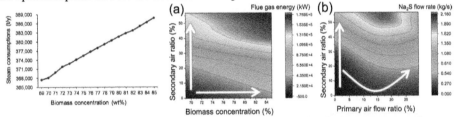

Fig. 3. Steam consumption according to LB concentration

Fig. 4. Steam and pulping chemical recovery according to operating conditions: (a) steam; (b) Na_2S

The Na_2CO_3 recovery for the pulping chemicals increases as the primary air flow ratio and biomass concentration increases, as shown in Fig. 5 (a). As the primary air flow ratio increases, CO_2 generation increases because of the activated gasification reaction. The Na_2CO_3 conversion is promoted by increased CO_2 generation. In addition, as the biomass concentration increases, the energy used for the Na_2CO_3 reaction increases by decreasing the energy lost to moisture removal from LB.

Finally, the NOx emissions decrease as the secondary air flow ratio decreases and the tertiary air flow ratio increases, as shown in Fig. 5 (b). As the secondary air flow ratio increases, the pyrolysis reaction becomes more active; accordingly, the amount of thermal NOx production increases with increasing temperature of the pyrolysis reaction. As the tertiary air flow ratio increases, NOx emissions decrease owing to the activated NOx-reducing combustion reactions.

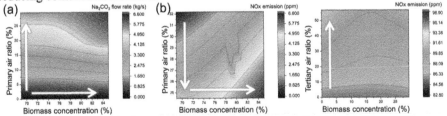

Fig. 5. Pulping chemical recovery and NOx emissions according to operating conditions: (a) Na_2CO_3; (b) NOx emissions

NOx emissions were derived. Fig. 6 shows the Pareto-optimal front with representative solutions. Representative solution 1 has NOx emissions equal to the existing operating conditions but can increase profits by up to 294.23%. Representative solution 3 has the same net profit as before but can reduce NOx emissions by up to 11.01%. Finally, Solution 2 is the average of Solutions 1 and 3, which can increase the net profit by 147.41% and reduce NOx emissions by 5.50%. In addition to these three solutions, decision-makers can select solutions that fit the situation from the proposed Pareto-optimal front.

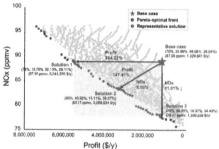

Fig. 6. Pareto-optimal front with representative solutions

4. Conclusions

This study focuses on multi-objective optimization of maximizing net profit and minimizing NOx emissions while recovering LB. This study makes two major contributions to the existing literature. First, the solutions proposed in this study do not include any retrofitting of the conventional evaporation process and the biomass boiler, and are thus applicable on-site. Second, robust multi-objective optimization was realized by simultaneously considering the net profit and NOx emissions in the overall processes in the pulp and paper industry using the DNN model. The derived representative solutions increased the net profit by up to 294.23% and decreased the NOx emissions by up to 11.01%. In addition to representative solutions, decision-makers can select solutions that fit the situation from the proposed Pareto-optimal front. Therefore, we believe this study can offer comprehensive solutions to on-site decision-makers.

Acknowledgments

This work was supported by the Korea Institute of Industrial Technology within the framework of the "Development and application of carbon-neutral engineering platform based on carbon emission database and prediction model" project [Grant Nos. KM-23-0098] and "Development of complex parameter smart analysis modules for color customering" project [Grant Nos. EH-23-0005].

References

Kim Y, Lim J, Cho H, Kim J. Novel mechanical vapor recompression-assisted evaporation process for improving energy efficiency in pulp and paper industry, Int J Energy Res, 46, 3, 3409-3427

Pérez MG, Vakkilainen E, Hyppänen T, Fouling growth modeling of kraft recovery boiler fume ash deposits with dynamic meshes and a mechanistic sticking approach, Fuel, 185, 872–85.

Tu Y, Zhou A, Xu M, Yang W, Siah KB, Subbaiah P, NOX reduction in a 40 t/h biomass fired grate boiler using internal flue gas recirculation technology, Appl Energy, 220, 962–973.

Lim J, Kim J, Optimizing ash deposit removal system to maximize biomass recycling as renewable energy for CO_2 reduction, Renew Energy, 190, 1006–1017.

Lim J, Cho H, Kwon H, Park H, Kim J, Reinforcement learning-based optimal operation of ash deposit removal system to improve recycling efficiency of biomass for CO_2 reduction, J. Clean. Prod., 370, 133605.

Park J, Kim Y, Lim J, Cho H, Kim J, Optimal operation of the evaporator and combustion air distribution system in a pulp mill to maximize biomass recycling and energy efficiency, J. Clean. Prod., 367, 133048.

Antonis Kokossis, Michael C. Georgiadis, Efstratios N. Pistikopoulos (Eds.)
PROCEEDINGS OF THE 33rd European Symposium on Computer Aided Process Engineering
(ESCAPE33), June 18-21, 2023, Athens, Greece

Decarbonization of gray ammonia production through the Allam power cycle: a techno-economic assessment and optimization

Andrea Isella, Mattia Forina, Mario Spagna, Davide Manca*

PSE-Lab, Process Systems Engineering Laboratory, Dipartimento di Chimica, Materiali e Ingegneria Chimica "Giulio Natta", Politecnico di Milano, Piazza Leonardo da Vinci 32, 20133 Milano, Italy
davide.manca@polimi.itw

Abstract

Ammonia is one of the leading products of the chemical industry worldwide. In addition, its manufacturing is significantly both energy- and emissions-intensive. As ammonia production is heavily based on fossil fuels, its synthesis integration with the Allam power cycle is a promising technology that can lead to substantial decarbonization of the whole process and an important abatement of the electrical energy costs (and indirect emissions). The Allam cycle is a system for high-efficiency power generation with supercritical CO_2 working fluid. Its integration into conventional ammonia plants allows for simultaneously producing electricity, water, and pipeline-ready carbon dioxide. An original, improved layout of a gray ammonia plant combined with the Allam power cycle is simulated to run a techno-economic assessment of the process and detect the optimal operating conditions that improve its economic and environmental sustainability.

Keywords: Haber-Bosch; carbon dioxide; sustainability; climate change mitigation.

1. Introduction

The rising energy transition era calls for the decarbonization of hard-to-abate sectors, with the chemical industry at the forefront. Specifically, in 2020 only, ammonia synthesis totaled more than 440 Mt_{CO2eq}/y, being the most carbon-intensive process of the whole sector (Isella and Manca, 2022). Considering a conventional ammonia plant, the core of its direct CO_2 emissions is the so-called "front-end". Such a term, which refers to the reforming, shift, and syngas purification sections as a whole, is opposed to the "back-end", which refers to the ammonia converters and the separation section that allows the final product recovery. This is because conventional front-ends heavily rely on fossil fuels (primarily natural gas) to meet both the hydrogen and energy demands of the process. In answer to this, the integration of a recently introduced power cycle known as the "Allam cycle" in the front-end of a conventional ("gray") ammonia plant is here proposed and performed through process simulations and techno-economic evaluations. Indeed, since the main peculiarity of the Allam cycle lies in exploiting supercritical carbon dioxide as a working fluid for electricity production (Allam *et al.*, 2015), the present work investigates how ammonia production may be decarbonized by diverting the whole CO_2 produced in the front-end of the production site to the Allam cycle. Such a strategy would lead to producing a sort of "blue" ammonia (to borrow the hydrogen color codes) and generating electric power (which would fulfill first the electricity demand of the whole plant and then export the surplus).

2. Process simulation

The simulations of this study were performed by UniSim® Design R491. Specifically, three distinct sections constitute the plant: the Front-end, the Air separation unit, and the Allam cycle. First, these sections were singularly modeled and finally joined into the whole process flow diagram reported in Figure 1.

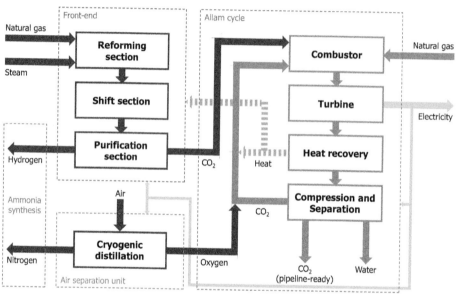

Figure 1: Block flow diagram of the simulated ammonia plant with the Allam cycle integration.

As evident from Figure 1, the main goal of the proposed integration is to fully abate the gaseous CO_2 emissions generated in the Front-end by routing all the tail gases (separated in the purification section from the hydrogen product) to the Allam cycle power plant. Such a CO_2-rich stream feeds the combustor of the Allam cycle. Specifically, this unit burns natural gas and oxygen to increase further the temperature above 1100 °C and produce even more CO_2 that feeds a turbine for electricity production. Next, a series of heat exchangers cool the exhaust stream and recover a significant amount of heat. One of them, the recuperator, has the crucial function of pre-heating the recycle stream before it returns to the combustor. After water removal, the CO_2 mass flow is compressed through a train of centrifugal compressors that make the expanded CO_2 supercritical again: a minor fraction is removed as pipeline-ready carbon dioxide, and the remaining part is recycled back to the combustor. It is worth noting that the Allam cycle contributes to an increase in the total amount of carbon dioxide produced by the whole process. However, unlike the gaseous emissions from the ammonia plant, the CO_2 product is collected in such a state that easily allows exporting it for industrial uses (*e.g.*, urea synthesis, enhanced oil recovery, food market). Indeed, the usage of carbon dioxide in the industrial sector records remarkable extents: around 230 Mt_{CO2}/y are currently used for the said purposes (IEA, 2022). Even water is among the products of the Allam cycle. Concerning the current integration layout, it can be used in the ammonia section either as a utility for the on-site steam generation required by the reformers or as cooling water. Finally, an air separation unit (ASU) that concurrently produces the nitrogen required for ammonia synthesis meets the oxygen demand of the Allam cycle combustor. Indeed, a nitrogen-to-hydrogen ratio equal to 3 is mandatory at the inlet of the ammonia converters.

The following subsections show each simulated section of the integrated plant.

2.1. Front-end

The Front-end model primarily comes from Soltani *et al.* (2014). It is a simplified model of a typical steam methane reforming (SMR) process, comprehensive of the furnace, the reformer and shift reactors, and the purification section.

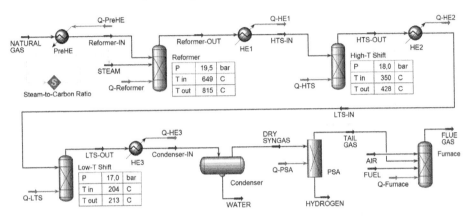

Figure 2: The simulated Front-end process flow diagram.

Precisely, all the reactors were simulated as "Equilibrium reactors" (*i.e.* reactors whose outlet streams achieve both chemical and physical equilibria according to the specified reaction set). In contrast, the furnace was modeled as a "Gibbs reactor" (*i.e.* a reactor that simply minimizes the Gibbs free energy of its products, suited to avoid complicated and inadequately accurate stoichiometric reactor models). Note that the furnace appears at the end of the flowsheet because it downstream evaluates the amount of fuel it needs. In this respect, the flue gases from the furnace virtually enter the reformer to exchange their enthalpy with the reacting flow.

2.2. Air separation unit

The Air separation model comes from Ebrahimi *et al.* (2015). Expressly, it consists of a conventional double-column cryogenic distillation process.

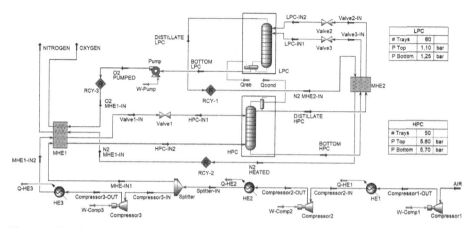

Figure 3: The simulated Air separation process flow diagram.

Concerning product purities, the nitrogen stream should preferably have purity above 99 mol% to avoid oxygen poisoning the catalyst of the ammonia synthesis loop. In contrast, the oxygen stream headed to the Allam cycle should be over 95 mol%, *i.e.* the near-optimal oxygen purity value for oxy-fuel combustion (Ye *et al.*, 2019).

2.3. Allam cycle

The Allam cycle model mainly relies on Allam *et al.* (2015) and Wang *et al.* (2021). The inlet tail gas and oxygen streams come from the Front-end and the Air separation unit sections, respectively.

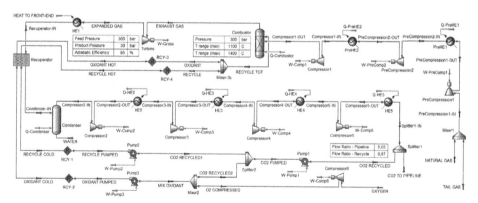

Figure 4: The simulated Allam cycle process flow diagram.

As happened for the furnace of the Front-end, even the combustor was modeled as a Gibbs reactor. Specifically, its temperature operability range spans from 1100 to 1400 °C. Also, note that an additional heat exchanger (missing in the original layout of the Allam cycle patent) was added to recover heat beneficial to the Front-end reforming section, as proposed by Wang *et al.* (2021). The CO_2 purge, ready for the pipeline, equals 3% of the processed flow, while the remaining 97% is recycled back to the combustor.

All the simulations implement the Peng-Robinson equation of state for thermodynamic computations. Finally, the three sections were combined in a single flowsheet to assess the performance of the retrofitted ammonia synthesis plant.

3. Results and discussion

Both CapEx (Capital Expenditure) and OpEx (Operational Expenditure) were evaluated for the economic assessment of the whole plant. The main hypotheses for such calculations consist in considering 8000 working hours in a year and a 10-year depreciation time for the equipment in each section of the plant. Regarding the Front-end, according to the rather simplified model, the CapEx and the OpEx terms rely on the indicative values of 390 USD/t_{H2} and 970 USD/t_{H2} (IEA, 2019), respectively. Conversely, concerning the Air separation unit, the CapEx estimation mainly derives from Guthrie's formulae (for columns and compressors) and the equipment costs reported by Young *et al.* (2021) for the two multi-pass heat exchangers provided in the process flow diagram. Finally, the costs of the Allam cycle originate from an indicative value of 685.7 USD/kWe_{output} (IEAGHG, 2015) for the CapEx, while market values of utilities (*i.e.* natural gas, 0.08 USD/kWh in the 1st semester of 2022) and products (*i.e.* electricity, 0.25 USD/kWh in the 1st semester of 2022, and ready-to-pipeline carbon dioxide, 10 USD/t_{CO2}) for the OpEx. Four ammonia plant capacities were considered: 750, 1000, 2000, and 3000 t_{NH3}/d. In this regard, a "0.8 power law" replaced the conventional "0.6 power law" for the economies of scale to run a conservatively safer economic assessment. Indeed, both the combustor and the turbine of the Allam cycle are rather peculiar units (especially as they have to deal with CO_2-rich streams), thus likely less liable to such a typical behavior of the conventional units of chemical plants.

Table 1: Main results of the techno-economic assessment. Key process variables, CapEx and OpEx, and the percentage cost increases (Δ%) of the performed Allam cycle integration as a function of the ammonia plant capacity. A combustor temperature equal to 1100 °C is considered.

Plant capacity [t_{NH3}/d]	CO_2 to pipeline [t_{CO2}/d]	Water output [t_{H2O}/d]	Net power [MWe]	Max pipe diameter [m]	CapEx [MUSD]	OpEx [MUSD/y]	Δ% CapEx [-]	Δ% OpEx [-]
750	1426	664	147	0.58	510	82.6	+82.4%	−34.5%
1000	1902	885	205	0.67	627	100.5	+85.9%	−28.4%
2000	3803	1770	438	0.95	1073	208.7	+88.8%	+5.4%
3000	5704	2655	670	1.16	1497	351.8	+87.4%	+37.7%

Table 1 shows the main process variables and the results of the CapEx and OpEx assessment performed on the integrated facilities. Specifically, increasing the operative temperature of the combustor does not show any beneficial effect on both the investment and operative costs of the plant. Thus, the table reports only the results corresponding to the simulated lowest feasible operative temperature at the combustor (*i.e.* 1100 °C). Nevertheless, as evident from Table 1, the integration of the Allam cycle to gray ammonia plants becomes increasingly unprofitable as the ammonia plant capacity increases. Indeed, although higher ammonia plant capacities allow producing considerable amounts of electric power, they also lead to extremely high flowrates in the Allam cycle loop (note that the CO_2-to-pipeline stream is just 3% of the total processed flow). In other words, this means the power cycle needs high piping diameters to process all the flowing mass within the loop. The maximum internal diameter of a standard stainless-steel tube (material needed to resist the corrosivity of the CO_2-rich stream: indeed, in presence of water, severe corrosion of the infrastructure may result due to the formation of carbonic acid) is around 0.8 m: thus, both 2000 and 3000 t_{NH3}/d plants would inevitably require at least a parallel configuration. As previously mentioned, the water produced by the Allam cycle could be exploited to generate steam as required by the reforming section: specifically, at each ammonia plant capacity scenario, it would reduce about 40 wt% of its demand. Finally, the oxygen demand of the Allam cycle is another crucial factor making the integration disadvantageous at high capacities. Indeed, an air separation unit producing about 1900 t_{O2}/d was simulated to deal with the nitrogen demand of the 750 and 1000 t_{NH3}/d ammonia plants but, in the case of 2000 and 3000 t_{NH3}/d facilities, such oxygen productivity just covers 25% and 50% of the final demand of the power cycle, respectively.

4. Conclusions

The integration of the Allam power plant into an ammonia synthesis process proved to be a feasible strategy to decarbonize ammonia production. The results of the presented techno-economic assessment showed that such a solution primarily suits low/medium-capacity ammonia plants. Indeed, concerning the 750 and 1000 t_{NH3}/d capacities, substantial reductions in OpEx complement the inevitable increases in CapEx. On the other hand, medium/high-capacities would suffer greatly from the extremely high flowrates circulating in the Allam cycle, although this would make it capable to export much more electricity. In addition, such higher capacities would call for further measures such as parallel configurations and additional raw materials and utilities entering the system. However, if one disregards any economic considerations, it is rather important to observe that every integrated layout was able to simultaneously prevent all the gaseous emissions originating from the front-end section of the ammonia plant and cover the electricity demand of the whole facility. Therefore, although occasionally proving less profitable than the original configurations, the retrofitting of conventional ammonia plants with the Allam cycle turned out to be an available-to-date strategy capable of fully abating both direct (*i.e.* CO_2 released on-site by the front-end) and indirect (*i.e.* CO_2 generated by the power stations for the production of electricity) emissions at all plant capacities. Finally, the significant amount of ready-to-pipeline CO_2 produced by the shown configurations may represent an attractive feed for several industrial applications. Indeed, although over 35 Gt_{CO2}/y are currently emitted into the atmosphere every year since the last decade (Our World in Data, 2022), shortages in industrial-grade carbon dioxide have been particularly recurrent in the past few months in several European countries, including Italy, Germany, and the United Kingdom (The Guardian, 2022). Moreover, by doing so, completely abating carbon dioxide emissions in the atmosphere allows to avoid paying any "carbon tax", which are expected to substantially raise in the next years.

References

Allam, R.J., Brown Jr., G.W., & Palmer, M.R. (2015). System and method for high efficiency power generation using a carbon dioxide circulating working fluid (U.S. Patent N° 8,959,887).

Ebrahimi, A., Meratizaman, M., Reyhani, H.A., Pourali, O., & Amidpour, M. (2015). Energetic, exergetic and economic assessment of oxygen production from two columns cryogenic air separation unit. Energy, 90, 1298-1316.

IEA (2019). The Future of Hydrogen. International Energy Agency.

IEA (2022). CO2 Capture and Utilisation. International Energy Agency.

IEAGHG (2015). Oxy-Combustion Turbine Power Plants. International Energy Agency GHG.

Isella, A., & Manca, D. (2022). GHG Emissions by (Petro)Chemical Processes and Decarbonization Priorities—A Review. Energies, 15(20), 7560.

Our World in Data (2022). Annual CO_2 emissions. Available at:
https://ourworldindata.org/explorers/co2 (accessed on November 30th, 2022)

Soltani, R., Rosen, M., & Dincer, I. (2014). Assessment of CO2 capture options from various points in steam methane reforming for hydrogen production. International Journal of Hydrogen Energy, 39, 20266-20275.

The Guardian (2022). UK food and drink sector sounds alarm over CO_2 shortage as plant halts output. Available at: https://www.theguardian.com/business/2022/aug/25/co2-producers-meet-food-needs-halt-production-energy-prices (accessed on November 30th, 2022).

Wang, S., Fernandes, D., Xu, Q., & Chen, D. (2021). New Conceptual Design of an Integrated Allam-Cycle Power Complex Coupling Air Separation Unit and Ammonia Plant. Industrial & Engineering Chemistry Research, 60(49), 18007-18017.

Ye, H., Zheng, J., & Li, Y. (2019). Feasibility analysis and simulation of argon recovery in low oxygen-purity cryogenic air separation process with low energy consumption. Cryogenics, 97, 109-121.

Young, A.F., Villardi, H.G.D., Araujo, L.S., Raptopoulos, L.S.C., & Dutra, M.S. (2021). Detailed Design and Economic Evaluation of a Cryogenic Air Separation Unit with Recent Literature Solutions. Industrial & Engineering Chemistry Research, 60(41), 14830-14844.

Antonis Kokossis, Michael C. Georgiadis, Efstratios N. Pistikopoulos (Eds.)
PROCEEDINGS OF THE 33rd European Symposium on Computer Aided Process Engineering
(ESCAPE33), June 18-21, 2023, Athens, Greece

Into the Valley of Death Rode the Green Transition

Robert Pujan[a,b] and Heinz A. Preisig[b]

aBNT Chemicals GmbH, PC-Straße 1, 06749 Bitterfeld-Wolfen, Germany
bNTNU Norwegian University of Science and Technology, Høgskoleringen 5,
7491 Trondheim, Norway
robert.pujan@bnt-chemicals.de

Abstract

Although the concept of Valley of Death (VoD) has been widely documented, also in the context of the slow uptake of the Green Transition (GT), the available literature rarely provides any contemplations from more than one point of view. By reviewing works from different viewpoints like economics, policies, research, and industries, this contribution discusses the phenomenon of VoD in the GT more comprehensively and outlines the extent of the issues at hand. The study argues that VoDs of individual biorefinery technologies are not solved independently but only by overcoming the systemic barriers that stranded the GT in its overarching VoD. In fact, the GT efforts may be the first venture that has to address such a multitude of technological, economic, social, political, and market aspects on its way to maturity and commercialization.

Keywords: biorefinery, assessment, sustainability, policy, industry

1. Motivation

Despite the interest and support for sustainable process industries, shared by many actors from the public, politics, industries and research, being higher than ever, the Green Transition (GT) is leaping behind considerably. With most biorefinery concepts still being stuck between the experimental and demonstration phases, unable to cross this so-called Valley of Death (VoD), the deployment of sustainable technologies on a full-industrial scale has been slow at best. The concept of VoD refers to a point in process or technology development where the sum of available research grants and existing commercialization funding is at its minimum. This is usually experienced after the technology readiness level (TRL) 4 has been achieved - the lab-scale validation, and before TRL 7 can be met, that is, the technology's prototype implementation in a relevant, operational environment. As depicted in Figure 1, research activities push new concepts until TRL 4, mainly aided by public funding. From TRL 6 on, those concepts have to be pulled by the industries since private investments become the predominant funding. TRL 5–7 are therefore the bridge between basic knowledge generation and industrial application. If, before this phase, the technology has not reached

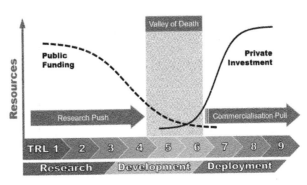

Figure 1: The Valley of Death in technology development

sufficient legitimacy and reliability for commercialization pulls, it strands in a calm belt of next to no push or pull; the VoD.

Since development processes are not linear and often even require the major scientific, technological and social disruptions of so-called paradigm shifts (Kuhn, 1970), the existence of VoDs is nothing new. Accordingly, the barriers involved are not only technical ones; those of economic, sustainable and socio-political nature continue to exist, although many technical issues may be long resolved. Identifying and understanding these barriers and how they can be overcome is thus essential for supporting the GT.

In an unfortunate plot twist concerning the GT, most of the VoDs observed with individual technologies can not be resolved independently - they are mere projections of the much larger, overarching VoD that the GT is currently stranded in. Analogously, the aforementioned barriers are rooted in significantly larger and systemic causalities, thereby necessitating the discussions on them to be lifted to a broader scale as well. The GT is probably the first genuinely global technical transformation humankind has ever aimed for; a new Industrial Revolution and an absolute at that. It is thus of no surprise, that the GT faces such a multitude of interrelated, global-scale barriers of scientific, technological, economic, social and political natures. However, the available literature on the concept of VoD and barriers of the GT rarely addresses more than one discipline and point of view. This contribution, therefore, bases its discussion on reviewing prior work from different sources like economics, policies, research and industries, thereby painting a comprehensive picture of the root causes and the extent of the issue at hand.

2. The Green Valley of Death

During the Industrial Revolution in the 18th and 19th centuries, hand-production workshops have largely been replaced with mechanized factories, thereby prompting such an unprecedented, radical technological shift that society was fundamentally altered. Considering the GT's scope and expected global impact, disruptive industrial shifts similar to those aforementioned are more likely than the comparably confined societal impacts of recent technological innovations. However, the GT is still often handled as a slow, iterative process of incremental changes and regime reconfigurations, which may hinder renewal and profound structural change if left unchallenged in policies and governance (Bauer, 2018). This notion is fortified by the current policies' emphasis on primarily supporting new technologies' developments, which seem to implicitly assume default commercialization of them as soon as sufficiently developed (Hansen & Coenen, 2017). Apparently, many actors still try to enforce traditional transformation measures on the GT instead of enabling much-needed paradigm shifts. Meanwhile, the private sector proves those notions wrong by being unwilling to finance biorefineries alone due to residual GT-associated risks (Philip, 2018). Even after over 20 years of research and development (R&D), the GT still struggles to move from a formative R&D phase towards one of commercial growth. Interrelated barriers that involve co-evolving changes in technologies, markets, policies and industrial strategies as well as capabilities still limit interests and abilities to mobilise resources around the emerging technologies, thereby embedding the GT in a typical case of systemic lock-in (Hansen & Coenen, 2017).

2.1. Industrial Constraints

Although innovation fosters economic growth, the GT depends on industries entering into uncertain environments, where they may be well advised to proceed and invest cautiously. Correspondingly, Nilsen (2017) describes that despite a few corporative strategy changes in recent years, many mature companies like Statoil ASA (since renamed Equinor ASA) remain highly dependent on the petroleum path as the main competitive advantage in the years to come. Industrial incumbents have lower incentives to prioritise radical innovations since efficiency and profit come from established, predictable and reliable technologies, and the organizational routines surrounding them. It also implies bureaucratic inertia and resistance to change. The Technology Innovation System (TIS) analyses by Hansen & Coenen (2017) and Frishammar et al. (2019) identified, among others, high capital intensity, a lack of strategic value-chain partnerships and market knowledge, as well as overall system weaknesses rooted in uncertainty, complexity, equivocality and ambiguity as the main limiters in the technological adoption of the GT. A limited industrial interest may also be inferred from present business models certainly having to drastically adapt in the GT. Contrary to traditional industrial characteristics like the economics of scale and competition based on market prices, businesses arising from the GT may be characterized by distributed, small-scale plants, economics dominated by local supply/demand, and competition in terms of sustainability. Growing new sectors like this, however, requires operational support via appropriate policies due to long formative periods in which investments can be substantial with little to no return and new businesses struggling to develop any legitimacy (Peck et al., 2009). Industrial initiatives require investment motivation and assistance by long-term, reliable policies.

2.2. Market Barriers

Instead of achieving the multi-product biorefinery configuration, already implemented biorefineries are mostly working as stand-alone facilities without any integral, well-designed process for the maximum use of an extensive feedstock portfolio or a range of different products (Cardona-Alzate et al., 2020). Consequentially, they are at the mercy of market fluctuations in the absence of operational support. Although renewable, the capitalism-inherent intention of unlimited growth cannot be met within the narrow constraints of biomass. Policies that aim on evolving the GT concurrent to the petroleum industries, increasing the industries' overall size instead of substituting fossil with sustainable resources, will thus fail due to many overlapping resource and market competitions. These competitions are, besides a lack of coordination and policy instruments in niche markets, already today a predominant cause for weak resource mobilization and market formation (Hansen & Coenen, 2017). There is a strong need to support the creation of bio-product markets through, for example, public procurement, price guarantees and biomass/bio-product production incentives (Fevolden et al., 2017). The implementation of carbon taxations and fossil fuel subsidy reforms would especially aid the creation of new markets, however, policies explicitly facilitating market adoption and commercialisation are weak or altogether lacking (Hellsmark et al., 2016). As a result, current market structures neither secure a sufficient supply of sustainable feedstocks nor generate ample demand for bio-products, thereby creating one of the conditions that deter investors (Philip, 2018). Supply chain analyses are therefore critical to the success of the entire GT. Their reliability may, however, be questioned: Seidenberger et al. (2010) and Schueler et al. (2016) demonstrate significant variability in published global biomass potentials due to several weaknesses and a lack of short-term estimates (i.e. 2030).

2.3. R&D Bottlenecks

Even though segueing to barriers within R&D may be appropriate at this point, most of the discussion can be found in Pujan & Preisig (2022), therefore keeping this paper's considerations on the topic to a minimum.

The conservative notion of forcing bio-products into the shapes of fossil products seems to be a predominant factor that is hindering innovations in R&D. However, despite early propositions by, for example, Marquardt et al. (2010) to use the GT as a unique opportunity for redesigning industrial value chains, current literature shows the majority of R&D trends still aiming at bio-feedstock conversions to products very similar to their fossil counterparts. In this regard, knowledge of biorefinery integrations into existing technical systems may remain critical for further developments, however, Bauer et al. (2017) reported a consensus in the literature that R&D may be necessary but investing more resources in it would certainly not enable biorefineries to cross the VoD.

2.4. Erratic Policies

Since the Industrial Revolution has been a century-long effort, one may conclude that the GT's slow uptake is par for the course. However, scientific innovation methods have tremendously improved since the 18th century and are accelerated by the rise of computing. This evolution is, however, barely visible since the GT faces barriers unknown to the Industrial Revolution. The latter was mainly driven by economic growth, disregarding environmental and social ramifications, and did not encounter fierce competition from deprecated industries. Additionally, the early years of the Industrial Revolution were confined to national scales, whereas the GT was always a global venture. Industrial boundaries often do not coincide with national borders and may vary between different industrial sectors. Accordingly, without supra-national policies that are stable in the long term, private investments in disruptive GT technologies remain too risky for consideration. Many EU and national policies still show insufficient levels of coherence and private sector incumbents question the coordination between EU, national and regional policy levels, especially in terms of biomass use (Peck et al., 2009). Due to the rush and hype regarding biofuels and bioenergy in recent years, many supra-/national policies are incentivizing the energetic utilisation of biomass, while almost neglecting material exploitation. Since the private sector must necessarily respond to market forces, a broad uptake of new businesses is generally dependent on significant political support with a consistent future vision. Regional developments, national R&D and global issues necessitate sufficiently flexible policies that are coherent across boundaries and minimum in duplications (Philip, 2018). Policies are of particular importance where significant positive externalities of the new system, like environmental and social benefits, are yet not priced by current market structures (Peck et al., 2009). However, although current political discussions centre on sustainability narratives, the resulting policies remain focused on efficiency, productivity and industrial competition instead of environmental benefits (Bauer, 2018). Standard policies like these do not yield the same effect in the GT as in other sectors, as demonstrated by Maes & Passel (2019). By investigating the policy frameworks for past technological transitions, Verbong et al. (2008) conclude that close adherence to industrial policies is problematic and that the innovation learning effects were not sufficiently guarded by past policy instruments. More than any other sector, the GT requires a combination of technology-push and demand-pull policies to stimulate sufficient levels of innovation and progress (Frishammar et al., 2019). Public-private partnerships, governmental loan guarantees and other innovative funding instruments like Green Bonds are necessary to de-risk private investments in the GT (Philip, 2018). This asks for large capital sums and long-term commitments since any biorefinery would need external support until the industry achieves critical mass and supply-demand stability. Relocating funding from fossil subsidies and introducing carbon taxations can be a huge source for these endeavours. However, their widespread realisations are still due.

2.5. Legitimacy Thresholds

Conducting the GT calls for eliminating more barriers than mere technological ones. It requires transitions in consumer behaviour and expectations, institutional norms, standards and regulations, as well as technological, economic and organizational innovations throughout supply and value chains – it is such a politically contested transformation where public perceptions are key (Bauer, 2018). Understanding public perceptions has been shown by Marciano et al. (2014) to be critical for managing the costs of emerging biorefineries. Peck et al. (2009) outlines how an understanding of the processes of building cognitive and socio-political legitimacy is relevant to the GT and how early legitimacy enhancements can reduce the duration and difficulties of formative phases in the evolution of innovative businesses. The socio-economic aspect influences most of the barriers to the critical adoption of the GT, which necessitates policy intervention (Ubando et al., 2020). Unfortunately, the public is largely oblivious to the environmental and social costs of fossil fuel subsidies, as well as the state-of-the-art in the GT. Accordingly, public and political opinions are erratic, and actors may even be confused by the aims and activities of the GT. An informant in the study of Frishammar et al. (2019) reported public backlash to the company when buying palm oil, even though it was certified – strangely, the public is less discerning with regard to crude oil. The impact and pace of climate change are just not obvious enough and too complex to grasp for most of the public. Similarly, the GT's complexity and its multitude of technological pathways are confusing even to scientists working in the field (Peck et al., 2009). Researchers still debate whether the bioeconomy is itself inherently sustainable or rather some weak ecological modernization and thus a threat to sustainability (Pfau et al., 2014). Truly sustainable solutions require much wider perspectives than presently considered. There are some mainstream claims that the GT's early adopters will mostly be interested in the green image, thus, technologies would not have to be refined or reasonably priced. Although those early adopters may provide initial finances, relying on them would certainly not enable an absolute GT. Furthermore, since the sustainability concept is often not objectively defined in current assessment methods (Cardona-Alzate et al., 2020), reaching public and political legitimacy in terms of sustainability may be a lost game.

Although R&D is proposing several technologies for the GT, hardly any evidence is provided to ensure widespread economic legitimacy. Evidence has to prove the feasibility and reliability of the GT technologies in development that is systematic scale-up studies, extensive simulation runs and reliable assessment analyses. After failed R&D hypes in recent decades, like, for example, microalgae, public funding measurements and investors understandably became very hesitant.

3. Concluding Remarks

As reasoned before by others as well, the GT can be successful in terms of economics, sustainability and society only if developed under certain conditions (Devaney & Henchion, 2018). However, as discussed here, the lack of these conditions as well as a multitude of interrelated, global-scale barriers of scientific, technological, economic, social and political natures keep the GT stranded in a VoD. Conflicting narratives dominate public opinion, policy-makers are acting too reluctant, R&D fails to establish legitimacy for GT technologies, and industries are enclosed in established routines and market constraints. Many barriers are rooted in notions that still try to undertake the GT without drastic shifts. Often, outdated measures as well as the profile of traditional fossil industries are enforced on the GT, thereby ignoring the largely different nature of this industrial revolution. More than 20 years of mainstream attention to the GT, yet, the nurture of "monkey see, monkey do" remains a dominating approach, trying to reproduce traditional chemical industries and old societal systems in the GT. It is about time to encourage that old monkey to take more risks and teach it some innovations.

Acknowledgements

This paper originated from ongoing work in the research projects Bio4Fuels (Norwegian Centre for Environment-friendly Energy Research (FME), project 257622) and MarketPlace (Horizon 2020, project 760173).

References

F. Bauer, 2018, Narratives of biorefinery innovation for the bioeconomy: Conflict, consensus or confusion?. Environ. Innov. Soc. Trans., 28, 96-107

F. Bauer, L. Coenen, T. Hansen, K. McCormick & Y.V. Palgan, 2017, Technological innovation systems for biorefineries: a review of the literature, Biofpr., 11(3), 534-548

C.A. Cardona-Alzate, S. Serna-Loaiza & M. Ortiz-Sanchez, 2020, Sustainable Biorefineries: What was learned from the design, analysis and implementation, J. Sustain. Dev. Energy Water Environ. Syst., 8(1), 88-117

L. Devaney & M. Henchion, 2018, Consensus, caveats and conditions: International learnings for bioeconomy development, J. Clean. Prod., 174, 1400-1411

A.M. Fevolden, L. Coenen, T. Hansen & A. Klitkou, 2017, The role of trials and demonstration projects in the development of a sustainable bioeconomy. Sustainability, 9(3), 419

J. Frishammar, P. Söderholm, H. Hellsmark & J. Mossberg, 2019, A knowledge-based perspective on system weaknesses in technological innovation systems, Sci. Public Policy, 46(1), 55-70

T. Hansen & L. Coenen, 2017, Unpacking resource mobilisation by incumbents for biorefineries: The role of micro-level factors for technological innovation system weaknesses, Technol. Anal. Strateg. Manag., 29(5), 500-513.

H. Hellsmark, J. Mossberg, P. Söderholm & J. Frishammar, 2016, Innovation system strengths and weaknesses in progressing sustainable technology: The case of Swedish biorefinery development, J. Clean. Prod., 131, 702-715

T.S. Kuhn, 1970, The structure of scientific revolutions (Vol. 111), University of Chicago Press

D. Maes & S. Van Passel, 2019, Effective bioeconomy policies for the uptake of innovative technologies under resource constraints, Biomass and Bioenergy, 120, 91-106

J.A. Marciano, R.J. Lilieholm, M.F. Teisl, J.E. Leahy & B. Neupane, 2014, Factors affecting public support for forest-based biorefineries: A comparison of mill towns and the general public in Maine, USA. Energy Policy, 75, 301-311

W. Marquardt, A. Harwardt, M. Hechinger, K. Kraemer, J. Viell & A. Voll, 2010, The Biorenewables Opportunity – Toward Next Generation Process and Product Systems, AIChE Journal, 56(9), 2228-2235

T. Nilsen, 2017, Innovation from the inside out: Contrasting fossil and renewable energy pathways at Statoil, Energy Res. Soc. Sci., 28, 50-57

P. Peck, S.J. Bennett, R. Bissett-Amess, J. Lenhart & H. Mozaffarian, 2009, Examining understanding, acceptance, and support for the biorefinery concept among EU policy-makers, Biofpr., 3(3), 361-383

J. Philp, 2018, The bioeconomy, the challenge of the century for policy makers, New Biotechnol., 40, 11-19

R. Pujan & H.A. Preisig, 2022, Biorefinery modelling is in tatters, and here is why. Comp. Aid. Chem. Eng., Vol. 51, 295-300

V. Schueler, S. Fuss, J.C. Steckel, U. Weddige & T. Beringer, 2016, Productivity ranges of sustainable biomass potentials from non-agricultural land, Env. Res. Letters, 11(7), 074026

T. Seidenberger, D. Thrän, R. Offermann, U. Seyfert, M. Buchhorn & J. Zeddies, 2010, Global Biomass Potentials. Investigation and assessment of data, remote sensing in biomass potential research, and country-specific energy crop potentials, energy [r] evolution - A sustainable world energy outlook, 3, 166-168

A.T. Ubando, C.B. Felix & W.-H. Chen, 2020, Biorefineries in circular bioeconomy: A comprehensive review, Bioresour. Technol., 299, 122585

Antonis Kokossis, Michael C. Georgiadis, Efstratios N. Pistikopoulos (Eds.)
PROCEEDINGS OF THE 33rd European Symposium on Computer Aided Process Engineering
(ESCAPE33), June 18-21, 2023, Athens, Greece

A novel industrial biotechnology approach to valorize fatty acids to bioplastics: scope for scale-up and process efficiency using an integrated approach

Pantelis Vasilakis[a], Kostas Pyrgakis[a], Melina Psycha[a], Antonino Biundo[b], Antonis Kokossis[a],*

[a]*Chemical Engineering National Technical University of Athens, Athens, Greece*
[b]*REWOW srl, Via Ciasca 9, 70124, Bari, Italy*
akokossis@mail.ntua.gr

Abstract

Waste Cooking Oils (WCO) are considered a harmful waste, though also constitute a valuable carbon source if treated correctly. This paper illustrates a novel WCO valorization chemistry to produce added-valued bioplastics. The core process is based on bio-based catalysis, in which *E.Coli* is fermented to produce oleate hydratase enzyme, which catalyzes FFAs conversion into the key building block of 10-Hydroxystearic-Acid (10-HA). The process is further integrated with upstream saponification-acidification stages for the pre-treatment of mixed WCOs, while downstream chemical processes (hydrogenation, esterification, polymerization) are upgrading the 10-HA building block into bio-based poly-10-HAME polymer. As upstream and downstream stages involve different scales, the paper addresses the scope to scale-up using a systems approach that builds efficiency and optimizes interactions. Overall, 3 tn of WCO yield in 1.5 tn of poly-10-HAME, CAPEX and OPEX are estimated to 141 M€ and 65 M€/yr, while emissions are estimated to 2.4 kg CO_{2_eq}/kg biopolymer.

Keywords: WCO, FFA, Bio-catalysis, Fermentation, *E.Coli*.

1. Scope of work

Waste Cooking Oils (WCO) constitute a valuable carbon source composed of unsaturated triglycerides (TGs) collected as wastes after frying with vegetable oils. They are usually disposed in landfills and/or along with wastewater resulting in huge environmental impacts; nevertheless, there are still options for recycling. The traditional solutions include the well-known transesterification chemistry that returns 1G biodiesel. Biofuel production is questioned in terms of social (edible sources), economic and environmental aspects (Lopes et al, 2019). Other emerging technologies use alternative renewable feedstocks (Pateromichelakis et al, 2022), or target to other bio-based products, like alcohols for vehicle and/or shipping fuels (Pyrgakis et all, 2016). Instead, a novel green solution based on a biocatalytic chemistry is proposed enabling the valorization of WCOs towards high-value bioplastics. The chemistry has been validated at laboratory scale (TRL4) and, in this work, it has been studied at modelling level to build knowledge and efficiencies for scaling up purposes.

The chemistry is based on the catalysis of the transformation of FFAs into 10-HA by means of oleate hydratase enzyme, which is produced by *Escherichia Coli*. The unsaturated fats of WCO are processed to prepare the required FFAs feedstock for the parallel enzymatic conversion to 10-HA, while, finally, 10-HA is upgraded to prepare the

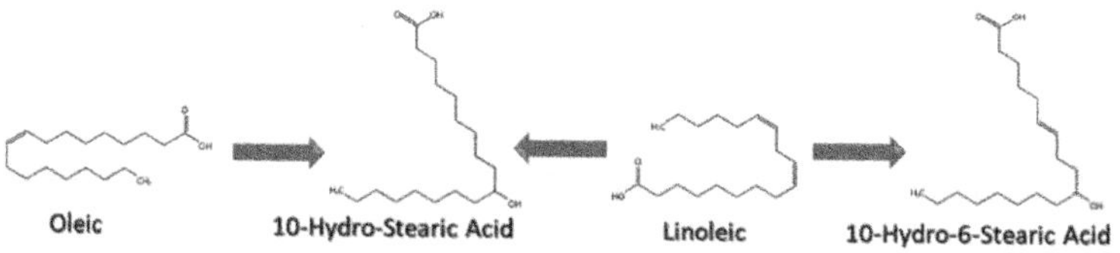

Figure 1 Reactions of FFAs transformation to biopolymer building blocks

10-HAME monomer for biopolymer production. Specifically, saponification is decomposing triglycerides (WCOs) into soaps, while acidification of soaps returns the exploitable FFAs co-substrate. *E. Coli* is fermented using glucose substrate and nutrients as cultivation media for growth and the production of oleate hydratase enzyme. The latter catalyzes the transformation of oleic and linoleic acids (FFAs) into 10-Hydroxyl-stearic acid and 10-Hydroxy-linoleic acid (10-HAs) following the reaction paths of Figure 1. Next, both 10-HAs and unconverted FFAs are hydrogenated generating their saturated fractions, which are next esterified forming the 10-HAME precursor and biodiesel (FAME). The latter closes the gap of carbon loses of the organic feedstock, which are not catalyzed to 10-HAs. The two fractions (10-HAME and FAME) are split by distillation returning drop in biodiesel and pure 10-HAME for polymerization purposes.

Based on experimental data for the core and surrounding chemistries, an integrated biorefinery flowsheet was developed from scratch in ASPEN Plus for analysis and simulation purposes.

2. Challenges in WCO valorisation

The studied chemistry was never addressed before at a modeling level, or as a complete biorefinery flow diagram. Experimental data concerning the biocatalysis cultivation media and its substrates and conversion efficiencies of FFAs are available for analysis. In this scope, challenges rise in terms of transferring the new-to-nature chemistry into a complete biorefinery flowsheet from raw (WCOs) and auxiliary (i.e., glucose) substrates to end-biopolymer products. Moreover, this plant utilizes 5 different chemical paths (saponification, acidification, fermentation, hydrogenation, polymerization) and challenging separation unit operations (e.g., spray drying). As far as concerns biocatalysis, the chemistry brings new challenges in modelling hybrid systems utilizing a fermentation-driven pathway and enzymatic conversion across the same system. Biocatalysis combines growth with glucose and production of primary metabolites (enzyme), which are exploited as catalyzing means of other reactions converting the FFAs co-substrate into desired products. These conditions increase the size and complexity of coupling different kinetic models that essentially require Monod-based kinetics for the *E. Coli* fermentation, and Michaelis-Menden-based kinetics for the enzymatic reactions. In addition, fermentation is affected by O_2 inhibition and temperature bounds that define fermentation and enzymatic efficiencies. Finally, the modeling procedure was challenged by lack of properties for special nutrients/auxiliaries, like the biochemical inducer IPTG.

3. Methodological steps

This work followed a 6-step approach to transfer chemistries at laboratory scale into a complete process biorefinery flowsheet for simulation, scaling up and analysis purposes of new-to-nature biotechnology concepts. The approach includes (1) estimation of biocatalysis kinetics; (2) process flowsheeting from scratch; (3) simulation and scaling up; (4) energy integration; (5) LCA for estimation of environmental impacts; and (6) techno-economic analysis of bio-based polymers.

3.1. Development of biocatalysis kinetics

An engineered strain of *E.coli* is incubated to achieve desirable growth rates for target scales. In parallel, the raw feedstock is prepared by means of saponification (KOH) and acidification (H_2SO_4) resulting treated FFAs; the conventional reaction systems are modeled by RSTOIC in ASPEN. Centrifuges are employed to remove water from organic phases. Next, experimental data including biocatalysis yields were used for kinetics modeling. Conditions such as growth, temperature and O_2 inhibition are considered resulting in an extended version of Monod kinetic model, which integrates the following terms for each limiting condition (Wang et al, 2014; Cruz et al, 2018):

Growth:
$$r_C = k^{cell} \cdot \frac{C_c \cdot C_S}{C_S + C_M^{Mon}} \Leftrightarrow \frac{dC_R}{dt} = k^{cell} \cdot \frac{C_c \cdot C_S}{C_S + C_M^{Mon}} \cdot \frac{C_R - C_{R0}}{C_c - C_{c0}} \quad (1)$$

Temperature:
$$r_C = r_C^{max} \cdot \frac{T_{max} - T}{T_{max} - T_{opt}} \cdot \left(\frac{T}{T_{opt}}\right)^{\frac{T_{opt}}{T_{max} - T_{opt}}} \quad (2)$$

O_2 inhibition:
$$k_{obs} = k^{cell} \cdot \left(1 - \frac{C_{O_2}}{C_{O_2}^*}\right)^n \quad (3)$$

Enzymatic conversion:
$$r_{10HA} = k^{enz} \cdot \frac{C_{E0} \cdot C_{FFA}}{C_{FFA} + C_M^{enz}} \quad (4)$$

where C_i and C_{i0} are the concentrations of component i at the end of each batch experiment and the beginning, respectively; $C_{O_2}^*$ the limiting concentration of inhibition; r_C the reaction rate of cells growth; k the maximum growth rate; C_M the Monod constant; k_{obs} the observed growth rate; n the exponential factor of inhibition; and T_{max}, T_{opt} and T the maximum, the optimum and the actual temperatures in cultivation medium. The indexes C, S, R and O_2 account for cells, Substrate (glucose), product (enzymes) and oxygen components that participate and affect the biocatalysis system.

3.2. Process flowsheeting of biocatalysis chemistry

The overall biorefinery diagram is represented in Figure 2. At preliminary steps of the plant, the WCOs feedstock is chemically pre-treated. The cell factories and the processed feedstock are mixed in conventional reactor systems generating 10-HAs (with and without hydroxyl groups) that are saturated by hydrogenation. The saturated FFAs are esterified with methanol yielding in the 10-HAME biopolymer precursor and FAME. Finally, 10-HAME is polymerized resulting in the end-product of poly-10-HAME.

The above process diagram of Figure 2 was designed in detail from scratch including all process operations. The flowsheeting procedure was followed to deliver all pre-treatment, bioconversion and formulation chemistries; including intermediate separation stages for cost-effective removal of by-products, co-products and water-organic streams resulting high purity poly-10-HAME. In Aspen, the "General" type and the "LHHW" class of reactions have been used to integrate the complex kinetics. Based on the calculated Monod parameters, the optimal cells and glucose concentrations that result in maximum production rate and conversion have been defined. Accordingly, based on the tuned fermentation and enzyme production rates, the optimal feed rate of FFAs and the retention time for maximum production of 10-HA are estimated. Accordingly, the reactor volume is specified and the RCSTR model (Aspen plus) was used for the biocatalysis system.

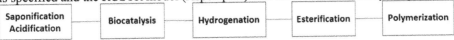

Figure 2 Biorefining stages

Based on the finalized process flowsheet, techno-economic and LCA analyses have been conducted to address the economic and environmental sustainability of biopolymers production in front of competitive synthetic and fossil-based polymers.

3.2.1. Scaling up and analysis

The developed process flowsheet was modeled in Aspen for the simulation of mass and energy balances at higher scales (3 tn/hr). Thereafter, energy integration has been performed to identify the maximum feasible energy savings at higher scales. Moreover, the biorefinery input-outputs (energy, materials and wastes) fed the LCA analysis for the estimation of mid- and end-point indicators; including the global warming potential (CO_2 equivalent) for biopolymers production. Finally, a techno-economic analysis focusing on capital and operating expenses was performed to identify profitability margins and potentials for market entry of alternative biopolymers.

4. Results

The process flowsheet that is presented in Figure 3 consists of 4 sections for (i) WCO pre-treatment, (ii) biocatalysis, (iii) FFAs upgrade and (iv) biopolymers production.

4.1. Saponification-Acidification stage (WCO pre-treatment)

Saponification (RSTOIC model) takes place at 80 °C using KOH (5 M) to break the bonds of triglycerides of WCO resulting in glycerol and potassium oleic, linoleic, palmitic and stearic acid at conversion rates of 78.3%. In the basis of 3 tn of WCO, 2349.7 kg of potassium-FFAs are produced, which are then acidified at 80 °C in presence of H_2SO_4 returning the treated FFAs for biocatalysis. A vapor-liquid separation stage is employed to recover volatile compounds in acidification effluents, while a downstream decanter is implemented to remove the water-glycerol content finalizing the biocatalysis feedstock.

4.2. Biocatalysis stage

The biocatalyst was prepared by inoculation to reach 3 tn of biomass ready for use in fermentation (37 °C). The biocatalysis step results in 1827.1 kg (77.7% conversion) of 10-Hydroxystearic-Acid. Based on experimental data, the kinetic parameters are accordingly estimated as follows: $k^{cell} = 0.022388 \ h^{-1}$, $C_M^{Mon} = 74.6269 \ mmol/L$, $T_{max} = 315.2 \ K$, $T_{opt} = 311.9 \ K$, $C_{O_2}^* = 0.12 \ mg/L$, $n = 1$, $k^{enz} = 1.0288 \ mol/L$, $C_M^{enz} = 0.025 \ mol/L$. The fermentation exit stream is treated with ethyl-acetate solvent (5% v/v) to facilitate mixing and extraction of the biocatalysis phase (desired products) and removal of the excess water content by centrifugation. Next, spray drying was implemented to recover an ethyl-acetate rich phase (43%), the unsaturated FFAs (47%), and the remaining water (10%).

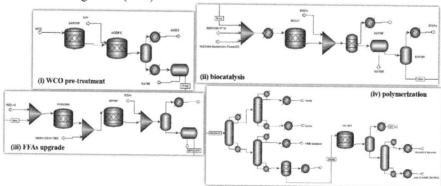

Figure 3 Biorefinery process flowsheet for biocatalysis implementation

4.3. Hydrogeneration and Esterification stage (FFAs upgrade)

The recovered organic phase was hydrogenated (fixed bed reactor) with the use of a nickel-based catalyst to saturate double carbon bonds with hydrogen reaching nearly 99.9% saturation. The esterification stage is next implemented to treat the saturated acids with methanol. The key component, 10-Hydroxystearic-Acid, is esterified into 10-Hydroxystearic-Acid-Methyl-Ester (10-HAME) by 99.4% (1816.72 kg), while the methyl-esters (FAME) of all remaining FFAs are produced. The effluent methyl-esters stream is mixed with ethyl-acetate solvent to entrain the organic phase from water, which are recovered through a 2 liquid phase separation.

4.4. Polymerization stage

Downstream esterification, a 3-distillation system is employed to recover the 4 key fractions from the ethyl-acetate entrainment: (i) unreacted methanol, (ii) FAME, (iii) 10-HAME and (iv) ethyl-acetate. The distillation separations have been designed to perform sharp splits at high purities ($\geq$99%). Methanol and ethyl-acetate can be recovered back to esterification and liquid-liquid separation stages, while FAME (1440.5 kg) constitutes the biodiesel co-product. The 10-HAME (1816.54 kg) is finally driven to the polymerization stage producing the poly-10-Hydroxystearic-Acid-Methyl-Ester (1453.19 kg), that is equivalent to an overall yield of 48.44%.

4.5. Evaluation: Energy, techno-economic and environmental analyses

Energy integration is applied to minimize operating and capital costs; the biorefining concept will fall short to compete with conventional processes or address sustainability targets [Kokossis et al, 2015; Pyrgakis et al, 2019]. Prior to energy integration, hot and cold utilities consumption have been at 7.4 MW and 14.1 MW, respectively. After integration the total utilities ware estimated to 2.24 MW (hot) and 8.9 MW (cold), which is equivalent to 70% and 37% of energy savings in heating and cooling demands. The optimal distribution over the utilities levels was accordingly estimated to 2.05 MW (LP steam), 0.19 (VHP steam) and 8.9 MW cooling water, as presented in the Grand Composite Curve of Figure 4.a.

Moreover, an LCA approach was followed using SimaPro to estimate the mid- and end-point indicators to produce biopolymers based on the input-output flows estimated by the biorefinery model. The Ecoinvent databases were used for the environmental indicators, while the ReCiPe methods was followed for the calculations. Accordingly, the contribution of the difference flows to the respective indicators are summarized in Figure 4.b. The emissions of CO_2 equivalent for the production of biopolymer were estimated

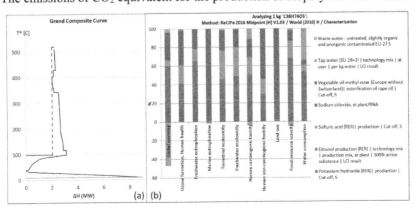

Figure 4 Energy Integration (a) and LCA (b) results

to 2.4 kg CO_{2_eq}/kg biopolymer, while the emissions for conventional fossil-based polymers range between 2-3.5 kg CO_{2_eq}/kg polymer.

Finally, the Aspen Process Economic Analyzer (APEA) was used to approximate the capital and operating expenses for production of 17,600 tn/yr FAME and 11,200 tn/yr biopolymers; accordingly, a mass allocation of 61-39% can be considered. The CAPEX was estimated to 141 M€, while OPEX was 65 M€/yr. As a result, for a depreciation period of 20 years (and the mass allocation), a production cost of 1.5 €/kg of biodiesel and ≈1 €/kg of biopolymer are estimated, while respective market costs are identified to 0.91 €/kg (biodiesel) and 1.8 €/kg (PVC equivalent cost). This is enough to consider that if another cost allocation strategy was accounted (i.e. by 34-64%), then an average price of 1.66 €/kg of biopolymer and 0.84€/kg of biopolymer could be achieved, resulting in improved prices (by 8%) compared to existing ones.

5. Conclusions

The work presented a new chemistry driven at industrial scales for production of biopolymers and biodiesel (co-product). The chemistry is attributed by complex kinetics in terms of cellular growth, enzymes production and enzymatic conversion of FFAs into valuable building blocks for biopolymers. Biodiesel was also considered to maximize utilization of residual carbon content of WCOs. The chemistry was presented at biorefining scales and achieved reductions at both costs and environmental impacts compared to existing solutions and products. Future work could include the investigation of alternatives auxiliaries (KOH, H_2SO_4) with lower cost and environmental impacts, as well as alternatives for downstream separations (e.g., membranes).

Acknowledgments

Financial support from the project Industrial Biotechnology Innovation and Synthetic Biology Accelerator (PREP-IBISBA) is gratefully acknowledged.

References

Cruz, M., Costa, E., Fonseca Almeida, M., da Conceição Alvim-Ferraz, M., Maia Dias, J., 2018. Recovery Of By-Products From The Olive Oil Production And The Vegetable Oil Refining For Biodiesel Production. Detritus In Press, 1.

Kokossis, A.C., Tsakalova, M., Pyrgakis, K., 2015, Design of integrated biorefineries, Computers & Chemical Engineering, 81, 40-56

Lopes, M., Miranda, S.M., Belo I., 2019. Microbial valorization of waste cooking oils for valuable compounds production – a review. Critical Reviews in Environmental Science and Technology, 50 (24), 2583-2616

Pyrgakis, K.A., Kokossis, A.C., 2019, A Total Site Synthesis approach for the selection, integration and planning of multiple-feedstock biorefineries, Computers & Chemical Engineering, 122, 326-355

Pateromichelakis, A., Psycha, M., Pyrgakis, K., Maréchal, F., Kokossis, A., 2022, The Use of GVL for Holistic Valorization of Biomass, Computers & Chemical Engineering, 164, 107849

Pyrgakis, K.A., de Vrije, T., Budde, M.A.W., Kyriakou, K., López-Contreras, A.M., Kokossis, A.C., 2016, A process integration approach for the production of biological iso-propanol, butanol and ethanol using gas stripping and adsorption as recovery methods, Biochemical Engineering Journal, 116, 176-194

Wang, H., Wang, F., Wang, W., Yao, X., Wei, D., Cheng, H., Deng, Z., 2014. Improving the expression of recombinant proteins in E. coli BL21 (DE3) under acetate stress: an alkaline pH shift approach. PLoS One. 9(11):e112777

Antonis Kokossis, Michael C. Georgiadis, Efstratios N. Pistikopoulos (Eds.)
PROCEEDINGS OF THE 33rd European Symposium on Computer Aided Process Engineering
(ESCAPE33), June 18-21, 2023, Athens, Greece
© 2023 Elsevier B.V. All rights reserved. http://dx.doi.org/10.1016/B978-0-443-15274-0.50408-X

On the re-deployment of lignocellulosic biorefineries with solvent-free designs and process-to-process integration

Iraklis Sarris, Kostas Pyrgakis, Antonis Kokossis*

School of Chemical Engineering, National Technical University of Athens, Iroon Polytechneiou 9, GR-15780, Greece
Corresponding author: akokossis@mail.ntua.gr

Abstract

This work follows a systems engineering approach to re-deploy a real-life lignocellulosic biorefinery for lower operating and capital costs, featuring benefits as solvent demands are reduced, and a better scope to integrate with first generation biorefineries. The redeployment focused on hot spots with high materials and energy use that affect cost-effective production of pulp, lignin, and bio-based products. The collaborative work investigates alternative designs for process, feedstocks and products portfolio and combines process-to-process integration to ensure optimal valorization of materials and energy. The reference real-life biorefinery case is modeled in ASPEN and involves high volumes of solvents for delignification, so re-deployment addressed a delignification-free case with parallel integration of downstream ethanol fermentation resulting in 22.5% lignin and 26% ethanol yields. The revised mass and energy balances of alternative scenarios were estimated and new insights for CAPEX (€ 39.6M), OPEX (€ 212M/yr) and energy costs (74% lower hot utilities) are obtained.

Keywords: biorefinery, delignification, bioethanol, techno-economic, energy integration.

1. Introduction

To date, there are 93 European lignocellulosic biorefineries (BBIC, 2022). Given the extended experience in 2G biorefining and the current social, political, and economic environments, the design contexts need to be revisited to face critical challenges related to cost and production efficiencies and serve replicable solutions for broader commercialization of the biorefining concept. State of the Art 2G biorefineries undertake the fractionation of lignocellulosic biomass into its main ingredients: C5, C6 sugars and lignin. All fractions can be integrated in bioeconomy markets and downstream chemistries for fuels, power, chemicals, and food additives production (Pyrgakis et al., 2016; Kokossis at al., 2015). This work is focusing on a real-life 2G biorefinery (CIMV, 2022), whose operation is based on processing wheat straw and poplar residues with organic acids (organosolv technology) returning C5, C6 sugars fraction and a functional biolignin based on the CIMV technologies (Delmas et al., 2011; Delmas at al., 2013; Nadja et al., 2014). Still, alternative organosolv technologies have been also recently appeared using biobased solvents and new fractions recovery concepts (Pateromichelakis et al., 2022).
Biolignin™ is a phenolic polymer with linear structure and low molecular weight. Moreover, the cellulosic polysaccharide is composed of a linear chain of β-1,4 linked d-glucose units, while the pentose sugar syrup consists of xylose monomers and oligomers (Mesa et al., 2016). The CIMV process takes advantage of the organosolv fractionation

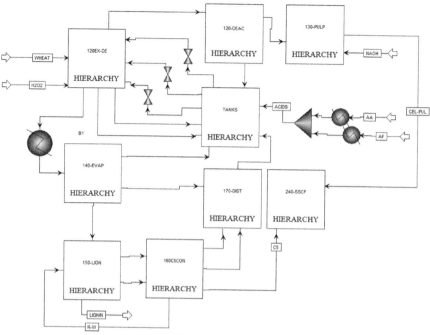

Figure 1 CIMV biorefinery flow diagram

concept utilizing a mix of organic acids as solvents of the three lignocellulosic components. The technology is applicable to both forestry and woody feedstocks and the overall biorefinery flow diagram is presented in Figure 1.

The biorefinery of CIMV consists of the main Biomass fractionation section (120EX-DE, in Figure 1), where C6 sugars and C5-soluble lignin are recovered from the raw lignocellulosic biomass. The C6 sugars are next delignified with a H_2O_2-organic acids mixture (Khama et al., 2005). After deacidification of the cellulosic fraction (120-DEAC), that is finally pulped (130-PULP) resulting the C6 sugars of the reference biorefinery. The C5-lignin mixture (from 120EX-DE) is dried through a thin film evaporation system (140-EVAP), while the lignin content is precipitated with water (150-LIGN), driving the residual C5 sugars towards the concentration section (160C5CON) where C5 sugars are collected. Any residual organic acids are recovered by distillation (170-DIST) and stored in the biorefinery acids tanks (TANKS) to be recycled across biorefinery.

Across the reference biorefinery, there are 3 main challenges identified that affect cost-effective valorization of biomass:

  i.   the high auxiliaries' costs for delignification (water and H_2O_2 solvents)
  ii. the co-production of added-value chemicals to face technoeconomic bottlenecks
  iii. the use of alternative feedstocks to face pricing seasonality and build flexibility

2. Problem description

As shown in Figure 2, the delignification process uses a 1:1 H_2O_2-water mixture to extract 2.6% of lignin trapped within pulp fraction returning 550 kg/hr of soluble Lignin from 21.000 kg/hr of the pulp feedstock. The high water and H_2O_2 demands (≈ 4.500 kg.hr) in combination with the high energy needs for downstream recovery of auxiliary organic acids and water crucially affect the sustainability of this section and profitability

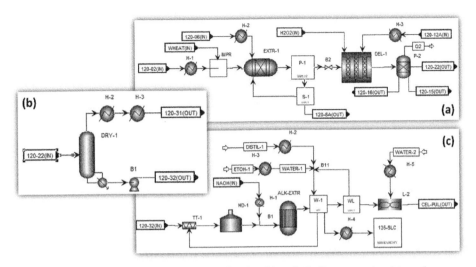

Figure 2 The delignification (a), deacidification (b) and alkaline extraction (c) sections

of the whole biorefinery, raising concerns of re-engineering, or even removing, this section. Such decision will incur significant effects on the quality of delivered C6 sugars as end-product and might require its valorisation through alternative, still mature, technologies to other products. Modelling of the design alternatives will reveal vital answers on the trade-offs and redevelopment of the biorefining concept towards the exploitation of the C6 biomass fraction. Under these conditions, the replacement of the delignification section with direct hydrolysis and co-fermentation of C5 and C6 sugars for bioethanol production was addressed; otherwise, pulp could attribute commercial value as lower-grade C6 sugars.

The case of co-producing biofuels is a radical swift on the reference biorefining concept in technological and commercial contexts by changing the products portfolio. Bioethanol production can be provided by regular glucose (C6) fermentation, while simultaneous valorization of the xylan and glucan substrates are also considered to boost ethanol production (Hassan et al., 2019; CIMV, 2022; Pyrgakis and Kokossis, 2019).

The implementation of alternative biomass feedstocks featuring alternative C5-C6 sugars fractions are capable of significantly affecting bioethanol and functional lignin productivities. Changes in feedstocks will infer a series of changes in mass and energy balances of the high complexity biorefinery raising questions about sustainability margins on the different lignocellulosic feedstocks. Biorefining is naturally affected by seasonality of the different woody and forestry feedstocks – with subsequent effects on integration potential and energy savings – thus, flexibility in operation is indispensable to face pricing volatilities and seasonality. As a result, besides wheat straw, corn stover is also examined to investigate new production potentials and sustainability margins. All above challenges map a set of alternative design options for re-designing the reference biorefinery and address trade-offs in terms of capital, operating and integrated energy costs.

3. Methodology

Dealing with above challenges, this work employs process systems engineering know-how and tools to identify better solutions than the reference biorefinery scenario. The whole and interconnected sections of the reference biorefinery has been modeled in Aspen in contexts of a previous work (Mountraki et al., 2016). This work elaborates the existing process flowsheet by reengineering the existing sections of Figure 2 providing a

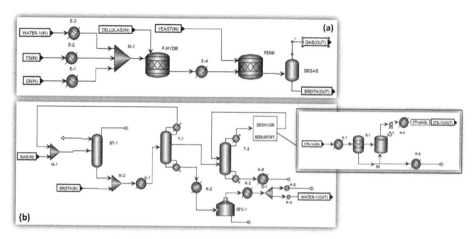

Figure 3 Ethanol production chemistry: Fermentation (a), products recovery (b)

set of alternative biorefining scenarios for analysis.

The case of removing the challenging delignification section, at cost of low-grade C6 sugars production has been examined. Such decision inherently affected mass and energy balances; let alone given the high complexity of the interconnected sections. The reference biorefinery has been appropriately amended and converged to the new balances. As an alternative, bioethanol production has been integrated along with hydrolysis, fermentation and downstream separations systems targeting to the production of high purity ethanol. The biorefinery section simultaneously valorizes the C5 (xylose) and C6 (pulp) fractions and the appropriate chemistry has been integrated (Figure 3). The analysis focused on all alternative combinations of using the delignification and ethanol chemistries resulting in 4 main scenarios with/without delignification, and with/without ethanol production. In addition, the 4 alternative biorefinery designs have been also tested in presence of two alternative feedstocks (corn stover and the reference-wheat straw) to adapt flows, unit operations and productivities to each carbon source. In this scope, 8 alternative biorefinery scenarios, and the corresponding simulations, have been developed and performed establishing a valuable portfolio of test cases for analysis.

The analysis involves energy integration and techno-economics to estimate the new sustainability margins of alternative scenarios. The energy analysis of multiple-process and multiple-product biorefineries is essentially challenged by the complex and variable integration routes (Pyrgakis and Kokossis, 2019). Furthermore, the investigation of alternative chemistries (e.g., bioethanol production) rises questions on the new integration patterns that need to be systematically addressed for all potential biorefinery scenarios. The guidelines of Pyrgakis and Kokossis (2019) have been followed to handle the energy integration alternatives, while the Aspen Process Economic Analyzer was used to estimate the new capital and operating expenses for each case.

4. Results

The process systems engineering analysis of the lignocellulosic biorefinery results a set of 8 alternatives scenarios to compare with the reference biorefinery in terms of capital and operating expenses. To this end, the design options account for (i) the use, or not, of the pulp delignification with sequent costs in the quality and quantity of end-pulp product;

Table 1: Production yields of the 8 scenarios

Scenario	Delignif. section	Ethanol production	Feedctosk	Mass flows [tn/hr]			
				C5	C6	lignin	ethanol
Reference	YES	NO	Wheat straw	5.82	15.79	8.96	
2	YES	YES	Wheat straw	0.57	1.48	8.96	9.44
3	NO	NO	Wheat straw	5.67	14.57	8.43	
4	NO	YES	Wheat straw	0.58	1.55	8.43	9.74
5	YES	NO	Corn stover	6.00	8.63	4.13	
6	YES	YES	Corn stover	0.56	0.75	4.13	11.63
7	NO	NO	Corn stover	5.31	7.88	3.75	
8	NO	YES	Corn stover	0.56	0.94	3.75	12.00

Table 2: Energy and economic results of the addressed scenarios

Design strategy	Hot [MW]	Cold [MW]	CAPEX [k€]	OPEX [k€/yr]
Del. (YES) - Eth. (NO)	133 (62%)	148 (54%)	31.8	215.6
Del. (YES) - Eth. (YES)	256 (60%)	64 (55%)	42.4	221.0
Del. (NO) - Eth. (NO)	126 (58%)	141 (50%)	28.9	207.0
Del. (NO) - Eth. (YES)	148 (74%)	57 (53%)	39.6	212.4

(ii) the integration, or not, of bioethanol production replacing profits from biomass C5 and C6 fractions with biofuels, at cost of additional equipment and (iii) the changes in biorefinery economic performance at use of different feedstocks. Accordingly, the production yields of end-products (Table 1) considered in each scenario have been calculated preparing the economic analysis for comparison. The 8 scenarios have been also addressed in terms of alternative energy integration patterns of the revised energy balances across biorefinery resulting in significant savings and cost reductions, as shown in Table 2; the value in parenthesis reflects the savings compared to the non-energy integrated biorefineries justifying advanced economics for all scenarios. In addition, Table 2 also summarizes the capital and operating expenses of each biorefinery scenario.

5. Conclusions

As indicated by the presented data, there has been a significant hot utilities reduction (up to 74%), after the successful bypassing of the delignification section and its consequent blocks. In addition, by the design of the downstream ethanol production unit, the insignificant amount of delignified lignin (0.53 tn/hr) and the sugars fractions are replaced by produced ethanol (9.74 tn/hr). With CIMV's starting point referring to a case scenario containing Delignification which was afterward deleted, the "energy needs" reduction was an expected change. Though the ethanol production integration has an apparent increase in CAPEX by 33%, there is a slight effect (3%) on the operating expenses (including the energy integrated costs) favoring payback of the new investment. In this context, the delignification section was observed to have a strong effect of operating costs by € 8.7M/yr, while the also reduces the efficiency of the ethanol production (ethanol production drops by 300kg/hr). As a conclusion, the most promising results are observed in the scenario of replacing the delignification section with ethanol production resulting in 22.5% lignin and 26% ethanol yields, which are the highest among all cases; CAPEX and OPEX indexes are accordingly estimated to € 39.6M and € 212M/yr. Feedstock seasonality is capable of affecting the biorefinery operations and production capacities by 2-55%, while corn stover appears better productivities and, thus, process economics than wheat straw.

Acknowledgements

Financial support from the Marie Curie European Research Program RENESENG-II is gratefully acknowledged.

References

Hassan, S.S., Williams, G.A., Jaiswal, A.K., 2019. Lignocellulosic biorefineries in Europe: Current State and prospects. Trends in Biotechnology, 37, 3, 231–234.

Pyrgakis, K.A. and Kokossis, A.C., 2019. Systematic synthesis and integration of multiple-effect distillation into overall processes: The case of biorefineries. AIChE Journal, 65, 7, e16631).

Mesa, L., Albernas, Y., Morales, M., Corsano, G., González, E., 2016, Integration of organosolv process for biomass pretreatment in a Biorefinery. Biomass Fractionation Technologies for a Lignocellulosic Feedstock Based Biorefinery, 229-254.

CIMV, France: http://www.cimv.fr/ (Accessed: December 2, 2022).

Delmas, G.-H., Benjelloun-Mlayah, B., Le Bigot, Y.s, Delmas, M., 2011. Functionality of wheat straw lignin extracted in organic acid media. Journal of Applied Polymer Science, 121, 1, 491-501.

Nadja, C., Séverinem C., Benjelloun-Mlayah, B., Jean-Stephane, C., Delmas, M., 2014. Esterification of organosolv lignin under supercritical conditions. Industrial Crops and Products, 58, 287-297.

Delmas, G.-H., Benjelloun-Mlayah, B., Le Bigot, Y.s, Delmas, M., 2013. Biolignin™ based epoxy resins. Journal of Applied Polymer Science, 127, 3), 1863-1872.

BBIC-Bio Based Industries Consoritum, Mapping European Biorefineries: https://biconsortium.eu/news/mapping-european-biorefineries (Accessed: December 2, 2022)

Khama, L., Le Bigot, Y., Delmas, M., Avignon, G., 2005. Industrial Crops and Products, 21, 9-15.

Mountraki, A.D., Mlayah, B.B., Kokossis, A.C., 2016. A Study on the Endogenous Symbiosis of First and Second Generation Biorefineries: Towards a Systematic Methodology. Computer Aided Chemical Engineering, 38, 2181-2186.

Pateromichelakis, A., Psycha, M., Pyrgakis, K., Maréchal, F., Kokossis, A., 2022, The Use of GVL for Holistic Valorization of Biomass, Computers & Chemical Engineering, 164, 107849.

Pyrgakis, K.A., Kokossis, A.C., 2019, A Total Site Synthesis approach for the selection, integration and planning of multiple-feedstock biorefineries, Computers & Chemical Engineering, 122, 326-355.

Pyrgakis, K.A., de Vrije, T., Budde, M.A.W., Kyriakou, K., López-Contreras, A.M., Kokossis, A.C., 2016, A process integration approach for the production of biological iso-propanol, butanol and ethanol using gas stripping and adsorption as recovery methods, Biochemical Engineering Journal, 116, 176-194.

Kokossis, A.C., Tsakalova, M., Pyrgakis, K., 2015, Design of integrated biorefineries, Computers & Chemical Engineering, 81, 40-56.

Antonis Kokossis, Michael C. Georgiadis, Efstratios N. Pistikopoulos (Eds.)
PROCEEDINGS OF THE 33rd European Symposium on Computer Aided Process Engineering
(ESCAPE33), June 18-21, 2023, Athens, Greece

Digital-twin development for a novel vibrating membrane aiming at fractionating fermentation broths

Oscar A. Prado-Rubio[a,b], Wai Fung Hui[b], Mads Stevnsborg[b], Manuel Pinelo[b], Jakob Kjøbsted Huusom[b]

[a]*Departamento de Ingeniería Química, Universidad Nacional de Colombia – 170003 Manizales, Colombia*
[b]*Department of Chemical and Biochemical Engineering, Technical University of Denmark (DTU), DK-2800 Lyngby, Denmark*
oapr@kt.dtu.dk

Abstract

In the near future, tools brought on by digitalization will play a more significant role in technology development and operation for debottlenecking challenges in the biochemical industry. In that regard, a novel vibrating membrane filtration system at pilot scale is currently under investigation to fractionate bio-succinic acid fermentation broth. Remarkably sustained flux has been obtained that is 3.6 times higher than the best results found in literature. Experimental data was used to develop a digital object that combine a robust communication system and hybrid model to predict hydrodynamics and fouling rate. The digital object is used to forecast membrane performance 25 min ahead, predicting the flux with high accuracy (4% error) and the onset of the irreversible fouling formation during step-up experiments. These tools will become the core of a digital twin for the adaptive operation of the ultrafiltration system under uncertainty.

Keywords: Digital-twin, Vibrating membrane ultrafiltration, Fractionation fermentation

1. Introduction

Extensive research in debottlenecking bioprocesses gains traction as manufacturers commit to more sustainable development practices (Ruales-Salcedo et al. 2022). Digitalization initiatives including advances in process monitoring, data storage, computing power, modelling approaches, and simulation tools join to improve bioprocess development. The concept of integrating digital and physical objects is not revolutionary and has existed for many years in areas such as aeronautics. However, in chemical- and biochemical processes lack of advanced process analytical tools to measure critical compounds and computationally expensive models hindered harnessing digitalization in these fields. Digital shadow and twin have been defined as a system of an interconnected physical process and in-silico/digital replica which have continuous information exchange. In a digital shadow, the virtual plant representation can assess process performance (past and present) to evaluate future scenarios, while the digital twin is capable to determine and implement open or closed loop actions given by an optimal objective function (Gargalo et al. 2021). The heart of a digital-twin is data-driven and hybrid modelling that has a great potential to exploit a priori system understanding complemented by nonparametric models. This approach is particularly relevant in emerging technologies where there is a lack of understanding of the underlying phenomena, thus, a data-driven or hybrid approaches could provide a more accurate system description (Azevedo et al. 2019).

This research is focused on optimizing downstream process for the fractionation of succinic acid fermentation broth. Succinic acid bioproduction is interesting since it is one of the most important building blocks in the biorefinery concept (Ruales-Salcedo et al. 2022). Despite that succinic acid bioproduction is already operated at a commercial scale, there is still has room for improvements to reach an even larger scale. In this context a novel vibrating membrane ultrafiltration system at pilot scale is currently under investigation as the first stage of a possible downstream processing. This step aims to separate cells, cell debris and proteins from the main dissolved organic acids (i.e., succinic, formic, lactic, and acetic). This separation is particularly challenging and conventional technologies like centrifugation reports around 90% recovery of cells and only 50% of proteins (Wang et al. 2013). Membrane ultrafiltration has been proposed as an alternative to obtain better separation performance with the possibility of integrating a membrane bioreactor. Despite the promising components recovery (i.e., 100% cells and 83% proteins), there is up to 75% throughput loss due to membrane fouling. A novel vibrating membrane technology is a possible alternative to conventional membrane ultrafiltration where fouling is mitigated by inducing a relatively high shear rate at the membrane surface (Prochaska et al. 2018). In that way, operation at higher throughput is attempted while reducing energy consumption. However, determining and understanding the maximum membrane throughput (so called critical flux), requires extensive experiments due to time-variant fouling nature plus potential inlet disturbances. Therefore, the design and optimal operation of the system will benefit by developing a digital twin aimed to predict the hydrodynamics and irreversible fouling onset. This research aims to present the challenges on building a digital twin for the vibrating membrane system including communication architecture, experimental campaign, digital object development and evaluation. The main outcome of this research is to build a tool which can estimate the performance of the novel membrane system operation and employ the model to optimize the energy demanding operation improving the sustainability of the process.

2. Methodology

2.1. Digitalization infrastructure

Last year, the Department of Chemical- and Biochemical Engineering at the Technical University of Denmark (DTU) heavily retrofitted the existing digital infrastructure aiming to centralize data collection, process monitoring and model validation activities across the portfolio of pilot scale laboratory equipment (Jones et al. 2022). A representation of the communication infrastructure is depicted in Figure 1. Process data and logged at-line tags from each unit are now forwarded to an IoT gateway software, vNode that facilitates online bidirectional communication to the process equipment through a secure connection to the DTU network. A Kubernetes cluster has been implemented in parallel to store historical data and contain PSE and modelling tools. Kubernetes is a container management architecture that manages a collection of independent internally connected server entities. One of the entities are designated as the "master" node, while the remaining entities are categorized as "worker" nodes (Gupta et al. 2021). The master-node orchestrates the available hardware across the collection of entities and designates resources depending on the demand of active applications. Local virtualization makes it possible to simulate several server entities and orchestrations on a local machine to test application deployment dataflow and database schemas before large scale implementation. vNode ensures that the OPC-UA signal received from the units are routed to the Kubernetes cluster during regular operation. Additionally, a relational SQL database standardized data collection for the process units and contextualizes data across different experimental runs. The cluster also acts as a deployment domain for containerized model application that can be used for prediction, control, and monitoring in real-time. Easy access to the process real-time data and historian improves the capabilities of utilizing data-driven methods and real-time optimization for these systems. The vibrating membrane module is being integrated with this new infrastructure.

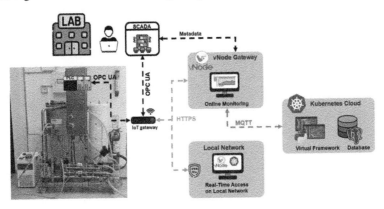

Figure 1. Schematic of the connectivity to the digital infrastructure at DTU Chemical and Biochemical Engineering. The OPC-UA signal from the PLC cabinet is sent to the IoT gateway software to be joint with metadata from offline measurements thus send to Kubernetes cluster where the tag values are stored and online accessible for monitoring or modelling tasks.

2.2. Experimental campaign

The equipment is a 7.5 m^2 Vibro Pilot membrane separation unit from SANI Membranes. For the separation, a polyethersulfone membrane (PES) is employed with a Molecular Weight Cut-Off of 30 KDa, which is enough to retain cells and most of the expected protein. Due to the experiment's scale, a considerable amount of fermentation broth is required. To circumvent this limitation, a simulated fermentation broth is used for tuning purposes. The simulated fermentation broth composition is extracted from literature for continuous operation including biomass (8 g/l yeast cells), proteins (0.25 g/L BSA) and organic acids (25 g/l succinic, 0.5 g/l acetic, 0.5 g/l formic acids). The experimental strategy to find the operating point is the well-known step-up method at constant transmembrane pressure. The feed channel pressure is increased stepwise, while the system pressures, flowrates and temperatures are monitored every 15 s. Because of the fouling, the membrane flux is expected to decline with time. According to the resistance in series model, the flux decline with time corresponds to the fouling rate which determine when the membrane should be cleaned. The influence of the crossflow velocity and fermentation broth composition are currently under investigation.

2.3. Developing the digital object

A multipurpose digital twin for design and control, consists of multiple digital objects that can combine different modelling approaches (i.e., first principles, data-driven and hybrid)(Grisales Díaz and Prado-Rubio 2022; Prado-Rubio, Grisales Diaz, and Kjøbsted Huusom 2022). Herein, a hybrid model is proposed with the potential to incorporate forthcoming process understanding provided by a more dedicated first principles models (e.g., CFD) and experiments. As initial step, the digital object approach uses a parallel implementation of a hydrodynamic and fouling rate models. The system hydrodynamics is modelled using the unsteady laminar flow model near an oscillating plate. This model is used to predict velocity profiles and shear rate in the zone adjacent to the membrane surface. This information is relevant to compare model prediction with literature. A MIMO ARX model with forgetting factor is then tuned using the online recursive least squared method, where the model structure and forgetting factors are defined by a sensitivity analysis using MAPE, FIT and NRMSE as performance indexes. This model is calibrated to predict the membrane flux and transmembrane pressure based on the pressure and recovery setpoints plus the crossflow rate. Thus, the model is used to forecast 25 min in the future the flux decline and its trend which represent the dynamic fouling rate.

3. Results

To date, five campaigns have been conducted including 10-15 experiments each. The results of the experimental campaigns showed remarkable performance where the vibrating membrane module achieved a maximum sustainable operation of around 53 LMH at TMP of 2 bars, and up to 100%, 89.4% and ~25% rejection of yeast, BSA and organic acids, respectively. Previous research has shown that ultrafiltration membrane with analogous properties and operating conditions achieved 14.74 LMH at laboratory scale (Wang et al. 2013), which represent a 3.6 fold improvement at a more larger scale with the potential to be further optimized. The best result from the campaigns is herein used for the modelling tuning and forecasting analysis. The hydrodynamic modelling results for the conventional operation show that the velocity profile decreases, as expected, linearly towards the membrane surface where the shear rate is zero.

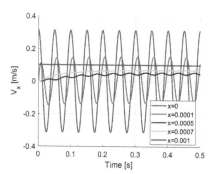

Figure 2. Velocity profiles for the conventional crossflow ultrafiltration and vibrating membrane operation at 20 Hz and an oscillation amplitude of 2.5 mm.

Figure 2 shows the velocity profiles at the boundary layer. The oscillating membrane at 20 Hz and with 2.5 mm amplitude generates a relative velocity at the membrane surface up to 0.31 m/s, which is 1.5 to 3 times higher than the crossflow. This corresponds to an average surface shear rate of 2239.5 1/s to mitigate fouling. Previous research with vibrating microfiltration of yeast and BSA at 20 Hz and 1.375 mm amplitude, generated a shear rate of 1230 1/s (Prip Beier & Jonsson, 2009). Under those conditions, a sustained flux of 30 LMH was obtained. The current results show how almost doubling shear rate has substantially enhanced flux.

The increased shear rate appeared to be more efficient in removing the pore blocking and cake-controlled fouling expected from fermentation broths.

The adaptive online system identification model was able to capture the time variant flux nature with high accuracy (Model quality: NRMSE = 0.0092, MAPE = 0.57% and FIT = 97.7%). Experimental data and model predictions are shown in Figure 3, where 97% of data have relative error lower than 1%.

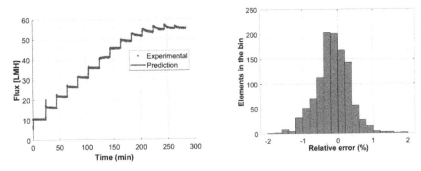

Figure 3. Online system identification predictive power during 14 experiments. Flux prediction and histogram of the relative model error

The model is also used in real time to forecast 100 samples into the future (25 min) and estimate the flux decline and fouling rate. A zoom-in is depicted in Figure 4. Note that there is a small deviation of ~4% in the flux forecast. Despite that the step-up disturbances during the experiment are mild (i.e., 0.2 bar at the time), there is a tremendous effect on the fouling rate which tends to stabilize after 10 min. This result is important since it defines that the experimental tuning should last at least 20 min to stabilize the fouling when vibration is on. Now, defining an affordable fouling rate of -0.05 LMH/min (corresponding to a 50% of max flux decline in 8 h operation), it can be predicted that experiments beyond 190 min (exp 10) the system consistently overcomes the threshold. Every step further leads to a higher fouling rate where the last represent 2.4 times the desired target. All these points then represent super-critical operation and should be avoided. This tool can then predict the onset of the irreversible fouling formation during experiments which will be tested in the near future when the full digitalization infrastructure is implemented.

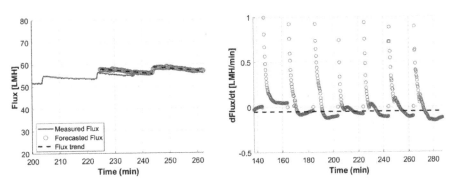

Figure 4. Moving horizon flux and fouling rate forecasting 25 min ahead using the last estimated ARX model

Conclusions

The department of Chemical and Biochemical Engineering at DTU is designing a robust pilot plant infrastructure capable of coping with the current and forthcoming challenges of biomanufacturing using a digitalization strategy. Herein, the development of a digital twin for a novel vibrating membrane is elaborated upon, discussing communication, experimental campaigns, and modelling approaches. This technology has shown remarkable performance during the fractionation of succinic acid fermentation broth by increasing the membrane flux 3.6 times compared to the best reported conventional ultrafiltration. Employing an online system identification, it is possible to predict the flux behaviour with high accuracy (MAPE = 0.57%). This was used to forecast the onset of irreversible fouling formation and detecting 4 super-critical experiments. The developed digital object plus the digitalization infrastructure will becomes the core of the implementation of digital twin for process optimization and adaptive system operation under uncertainty.

Acknowledgements
The authors of this work would like to thank the Novo Nordisk Foundation for their support of this project through grant NNF19SA0035474.

References

Azevedo, Cristiana Rodrigues et al. 2019. "Hybrid Semiparametric Modeling: A Modular Process Systems Engineering Approach for the Integration of Available Knowledge Sources." In *Systems Engineering in the Fourth Industrial Revolution*, eds. Ron S. Kenett, Robert S. Swarz, and Avigdor Zonnenshain. Wiley, 345–73.

Gargalo, Carina L. et al. 2021. "Towards the Development of Digital Twins for the Bio-Manufacturing Industry." *Advances in biochemical engineering/biotechnology* 176.

Grisales Díaz, Víctor Hugo, and Oscar Andrés Prado-Rubio. 2022. "Flux Estimation of Dynamic Ultrafiltration for Wastewater Treatment Using Combinations of Machine Learning Methods." In *Euromembrane 2022*, Sorrento, Italy.

Gupta, Manu, Konte Sanjana, Kontham Akhilesh, and Mandepudi Nobel Chowdary. 2021. "Deployment of Multi-Tier Application on Cloud and Continuous Monitoring Using Kubernetes." In *2021 5th International Conference on Electrical, Electronics, Communication, Computer Technologies and Optimization Techniques, ICEECCOT 2021 - Proceedings,*.

Jones, Mark Nicholas et al. 2022. "Pilot Plant 4.0: A Review of Digitalization Efforts of the Chemical and Biochemical Engineering Department at the Technical University of Denmark (DTU)." *Computer Aided Chemical Engineering* 49: 1525–30.

Prado-Rubio, Oscar Andrés;, Victor H.; Grisales Diaz, and Jakob Kjøbsted Huusom. 2022. "A Platform for Online System Identification of Dynamic Ultrafiltration Systems." In *Computer Aided Chemical Engineering*, , 1075–80.

Prochaska, K., J. Antczak, M. Regel-Rosocka, and M. Szczygiełda. 2018. "Removal of Succinic Acid from Fermentation Broth by Multistage Process (Membrane Separation and Reactive Extraction)." *Separation and Purification Technology* 192.

Ruales-Salcedo, Angela V. et al. 2022. "Production of High-Added Value Compounds from Biomass." *Biofuels and Biorefining: Volume 1: Current Technologies for Biomass Conversion*: 381–445.

Wang, Caixia et al. 2013. "Clarification of Succinic Acid Fermentation Broth by Ultrafiltration in Succinic Acid Bio–Refinery." *Journal of Chemical Technology & Biotechnology* 88(3): 444–48. https://onlinelibrary.wiley.com/doi/full/10.1002/jctb.3834 (November 30, 2022).

Antonis Kokossis, Michael C. Georgiadis, Efstratios N. Pistikopoulos (Eds.)
PROCEEDINGS OF THE 33rd European Symposium on Computer Aided Process Engineering
(ESCAPE33), June 18-21, 2023, Athens, Greece

Post-Combustion Carbon Capture and Utilization by Sodium Hydroxide Aqueous Solution for Bicarbonate Microalgae Cultivation

Jialin Liu,[a*] Yoong Kit Leong,[a] Jo-Shu Chang[a]

[a]Research Center for Smart Sustainable Circular Economy, Tunghai University,
No. 1727, Sec.4, Taiwan Boulevard, Taichung, Taiwan
jialin@thu.edu.tw

Abstract

The increasing amounts of CO_2 in the atmosphere, which is caused by the burning of fossil fuels, contribute greatly to global warming. In a standard post-combustion carbon capture (PCC) process, the energy penalty attributes to regenerate the CO_2 lean solvents. Microalgae have been established as the most promising candidate to the CO_2 capture and utilization (CCU) technology due to a high growth rate, oil productivity, and photosynthetic efficiency. However, the gaseous form of CO_2 is utilized by microalgae only to a very limited degree due to its low water solubility and high outgassing rate. In addition, the flue gases from a coal-fired power plant, which contains 10% to 16% of CO_2, cannot be supplied as the carbon source for the microalgae, directly. The bicarbonate-based microalgae cultivation may offer a promising alternative that relieves the burden of regeneration energy and improves the low carbon utilization efficiency of the gaseous form of CO_2 cultivation. In the present work, the rate-based model of Aspen Plus is adopted to simulate a flue gas in the flow rate of 1800 m^3/h with 14% CO_2 that is absorbed by aqueous NaOH solution. The absorbent is recycled from the mother liquor where most of bicarbonate is precipitated, and the fresh makeup with 30 wt% NaOH is applied. The column height to achieve 90% of CO_2 removal rate is investigated by varying the recycle rate of absorbent and the solid fraction in the bicarbonate product. In addition, the carbon utilization efficiency (CUE) by gaseous CO_2 and bicarbonate solution for the microalgae cultivation is compared. The results show that the CUE of the bicarbonate-based cultivation is far superior to that of the counterpart.

Keywords: carbon capture and utilization, chemical absorption, rate-based model, bicarbonate-based microalgae cultivation.

1. Introduction

In 2015, the United Nations General Assembly (UNGA) set out 17 sustainable development goals (SDGs) to be achieved by 2030. Microalgae have an increasing interest in various sectors in the SDGs, such as SDG-6 "clean water and sanitation", SDG-7 "affordable and clean energy", and SDG-13 "climate action". The microalgae can be cultivated in wastewater or seawater that provides a way for water reclamation (Wang et al., 2015). Moreover, microalgae could complete the entire growth cycle via photosynthesis reactions that convert light energy into renewable energy. The CO_2 emitted from the power generation and heavy industries is the largest contributor to the global warming problem. Vuppaladadiyam et al. (2018) reviewed the technical feasibility of combined carbon fixation and microalgae cultivation for carbon reuse. They concluded that the rate of carbon fixation by microalgae tends to be too slow to capture CO_2 from flue gases. Unless the rate of CO_2 uptake can be drastically improved, culturing algae may not be suited for the post-combustion carbon capture (PCC) from the power generation and heavy industries.

Amine scrubbing has been used to separate acid gas, such as H_2S, SO_2, and CO_2, from the flue or tail gases since 1930 (Bottoms, 1930). It is a robust technology and is ready to be tested and used on a larger scale for CO_2 capture from coal-fired power plants. However, the high regeneration energy of the amine-based solvents is the most critical challenge for extensively applying the PCC technology to thermal power plants. It is well known that the improvements of process configuration and solvent development may reduce the regeneration energy that is around 4.0 GJ/t-CO_2 for a standard PCC process configuration using 30 wt% monoethanolamine (MEA) aqueous solution (Rochelle, 2009). Solvent development has historically focused on reducing the enthalpy of absorption, increasing the absorption capacity, and enhancing the reaction kinetics by blending and synthesizing new amines. Recently, Benquet et al. (2021) reported that the second generation of amine-based solvent, named CO_2 Enhanced Separation and Recovery (CESAR1), was tested at large scale with a real flue gas at the pilot plants in Technology Centre Mongstad (TCM). The CESAR1 solvent, which is a blend of 27 wt% amino-methyl-propanol (AMP) and 13 wt% piperazine (PZ), that poses the lower regeneration energy (3.0 GJ/t-CO_2), higher stability, and lower corrosivity compared to those of traditional 30 wt% MEA solvent. The energy reduction of 25% can be achieved by blending amine-based solvents in the standard PCC process configuration. On the other hand, Le Moullec et al. (2014) reviewed available literature as well as the patent database and summarized 20 elementary modifications to save regeneration energy for the PCC. These methods can be categorized into three groups: absorption enhancement, heat integration, and heat pump. According to the survey by Le Moullec et al. (2014), up to 39.0% of energy savings could be achieved by modifying the PCC process configuration.

Recently, Zhu et al. (2022) suggested the integration of the bicarbonate-based microalgae cultivation with the carbon capture that significantly reduces the production costs of cultivation and the regeneration energy in the PCC. The authors estimated that the cost of bubble CO_2 gas for microalgae production ranges from \$1.47 to \$7.33 kg^{-1} and the carbon utilization efficiency (CUE) is in the range of 1.0% to 5.0%. In contrast, the carbon cost of bicarbonate-based Spirulina production is around \$0.359 kg^{-1} due to the high CUE around 90%, particularly. Kim et al. (2017) reported that only 3.59% of the gaseous CO_2 was used for biomass synthesis, bicarbonate was effectively incorporated into the biomass with 91.40% of CUE. The CUE of bicarbonate-based cultivation is significantly superior to that of the gaseous CO_2 because CO_2 supplied in a gaseous form easily escapes from the culture medium, rather than being captured and utilized by the cell.

Wang et al. (2022) developed a PCC process by chemical absorption using aqueous sodium glycinates solutions (SGS) to produce high-value $NaHCO_3$ nanomaterials. CO_2 was captured from a split flue gas stream emitted from the 600 MWe coal-fired power plant in the Wolverine Clean Energy Venture project. The operating expenditure (OPEX) and capital expenditure (CAPEX) were calculated as \$27.53 and \$7.96 per metric ton of CO_2, respectively. In this study, aqueous NaOH solvents are applied to capture CO_2 to form the precipitation of $NaHCO_3$, which can be diluted for bicarbonate-based microalgae cultivation. Subsequently, the PCC process design is briefed in Section 2. Section 3 compares the experimental results of cultivating a variant strain of Chlorella vulgaris by gaseous CO_2 and bicarbonate, respectively. Conclusions are presented in Section 4.

2. Process Design for Bicarbonate Precipitation

The rate-based model of Aspen Plus is applied to design an absorber to achieve 90% of CO_2 removal rate from flue gases by sodium hydroxide solutions. The Redlich-Kwong equation of state and the electrolyte-NRTL method were used to compute the properties of the vapor and liquid phases, respectively. In addition, all the ionic reactions and solid dissociation can be assumed in equilibrium reactions, as listed in (R1) − (R5), where the equilibrium constants are calculated from the Gibbs free energy changes. In the rate-based model, the mass transfer rate of CO_2 diffused into the liquid phase needs to be considered; therefore, the rate-controlled reactions of CO_2 with OH^- are concerned as following (R6) and (R7). The kinetic reaction rate is described by the reduced power law:

$$r = k \exp(-E/RT) \prod_{j=1}^{N} (C_j)^{a_j}$$ where the concentration basis is Molarity, N is the number

of components and a_j is the stoichiometric coefficient of component j; the rate constants of k and E are listed in Table 1 (Aspen Technology, 2014). The parameter settings for the rate-based model, such as Film resistance and Flow model options can be found in the technical document (Aspen Technology, 2014).

(R1)	$HCO_3^- + H_2O \leftrightarrow H_3O^+ + CO_3^{-2}$	(R5)	$NaHCO_3 \leftrightarrow Na^+ + HCO_3^{-2}$
(R2)	$2H_2O \leftrightarrow H_3O^+ + OH^-$	(R6)	$CO_2 + OH^- \rightarrow HCO_3^-$
(R3)	$NaOH \leftrightarrow Na^+ + OH^-$	(R7)	$HCO_3^- \rightarrow CO_2 + OH^-$
(R4)	$Na_2CO_3 \leftrightarrow 2Na^+ + CO_3^{-2}$		

Table 1. Parameters k and E in (R6) and (R7)

r (kmol·m^{-3}·s^{-1})	k	E (J·kmol^{-1})
(R6)	4.32×10^{13}	55.4
(R7)	2.83×10^{17}	123.2

The flow rate of the flue gas from a coal-fired boiler was 1800 m³/h, containing 14% CO_2 and the saturated water vapor at 40 °C. Assuming that SO_x in the flue gas was removed before entering the CO_2 absorber, the reaction of forming sodium sulfite can be neglected. In this study, the fresh makeup of sodium hydroxide solution was set to 1500 kg/h, which contains 30 wt% NaOH and the balanced water. The column diameter was determined by calculating 80% of the maximum flood capacity, so that the diameter of 0.85 m was applied. The column height was adjusted to fulfill the CO_2 removal rate at 90%. Figure 1(a) shows the process flow diagram for the PCC process where the flue gas and lean solvent enter the absorber from the bottom and top, respectively. The outlet temperatures of the absorber are higher than those of inlets, because the CO_2 absorption is an exothermic reaction. The temperature of the settling tank was maintained at 20 °C for the $NaHCO_3$ precipitates, in which the recycled flow rate and the weight fraction of $NaHCO_3$ drawn in the outlet flow are two adjustable variables. Figure 1(b) shows the packed height that needs to maintain the CO_2 removal rate at 90% under varying the recycle-to-outlet ratio (Recycle/Outlet) and the solid fraction (SF) of precipitates in the outlet flow. The results show that the packed height needed is under 2 m once the SF is larger than 0.7, whereas the operating condition of the Recycle/Outlet ranges from 0.5 to 3.0 that is a robust design for operations. In contrast, the feasible operating range of Recycle/Outlet limits from 0.8 to 2, once the SF is operated at 0.5, as shown in Figure 1(b), where the packed height is between 2 to 2.5 m. Figure 1(a) show the simulation results of the ratio of Recycle/Output at 1.8 and the SF operated at 0.7. The outlet flow contains 28 wt% of $NaHCO_3$, whereas 6.6 wt% of bicarbonate in the recycle flow can be dissolved after mixing with the makeup flow, by which the lean temperature after mixing was risen due to the exothermic reactions of mixing the recycle flow and the fresh makeup.

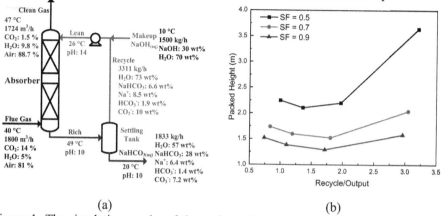

(a)	(b)

Figure 1. The simulation results of the makeup flowrate being 0.7 times flue gas, (a) process flow diagram and (b) the packed heights under varying the designed parameters. In this study, the makeup-to-feed flow ratio was set as 0.7 (kg/kg) for the proof of concept, which microalgae could be cultivated by captured CO_2 through chemical absorption. The CAPEX, OPEX, and levelized cost of carbon (LCOC) will be investigated under varying the makeup-to-feed ratio, whereas the robust operations are also taken into consideration for designing the PCC process in the future work.

3. Carbon fixation by Bicarbonate-Based Microalgae Cultivation

The microalga used in this study (Chlorella vulgaris SU-1) is a variant strain of Chlorella vulgaris, which is similar to that of the study by Wang et al. (2015). BG-11 medium was used as preculture medium with continuously supplied carbon dioxide (2%) at a flow rate of 0.2 L/min for the baseline cases. On the other hand, 5 g/L $NaHCO_3$ were supplemented as the carbon source at the beginning of cultivation and the 4th, 8th, and 12th day after sampling. The microalga was cultivated in a 1-L glass photobioreactor illuminated by external light sources (white fluorescent lamps) mounted on both sides of the photobioreactor at a light intensity of 200 W/m^2, and the room temperature was controlled at 25 ± 2 °C. A volume of 10 mL microalgae suspension was collected every day from each photobioreactor for nutrient analysis. Meanwhile, the pH value for each photobioreactor was adjusted to 7, 8, and 9, respectively, by adding hydrochloric acid on a daily basis for the bicarbonate-based cultivation. On the other hand, the pH value for cultivating microalgae by gaseous CO_2 was not controlled. The daily samples were centrifuged at 7000 rpm for 2 min. Dry cell concentrations were obtained based on the optical density (OD) values measured at 685 nm with a spectrophotometer (model U-2001, Hitachi, Tokyo, Japan). Figure 2(a) show the calibration curve of OD_{685}, in which the R^2 is 0.98, that was used to calculate the biomass concentration by the measured OD values. Figure 2(b) show the measured pH values during the experimental period where the pH values in the culture medium by the gaseous CO_2 were not controlled and the averaged pH value is around 8.5. The pH values of the other counterparts were measured and controlled to the corresponding values on the daily basis.

The growth curves are shown in Figure 3(a) where the biomass concentration (X, g/L) was calculated by the measured OD_{685} values and the calibration curve in Figure 2(a). Figure 3(a) shows that the growth curves of the bicarbonate-based cultivation were inferior to the gaseous CO_2 cases; the final biomass concentration under the gaseous CO_2 condition was over four times as high as those by the bicarbonate-based cultivation. In addition, the figure shows that the pH values maintained at 8 may favor the bicarbonate-based cultivation for this algae strain by contrasting to pH values controlled at 7 and 9. The biofixation of CO_2 can be calculated through the biomass concentration (Adamczyk et al., 2016): $CO_2 \ fixation = C \times X \times \left(M_{CO_2} / M_C \right)$ where C, M_{CO_2} and M_C are the carbon content in the biomass and the molecular weights of CO_2 and C, respectively. In this study, the carbon content of 47.1 wt% was applied for the strain of Chlorella vulgaris (Silva et al., 2016). Figure 3(b) shows the carbon utilization efficiency for each photobioreactor, in which the efficiency for the gaseous CO_2 cases is not over 6%, because the CO_2 gas was supplied at the steady flow of 0.2 L/min. Over 90% of CO_2 were slipped by directly exposing the gaseous form of carbon source that is impractical on the PCC perspective. On the other hand, the average CUE of the bicarbonate-based cultivation was around 40%, as shown in Figure 3(b); in some cases of the pH controlled at 8, the efficiency could be over 60%. In this study, the CUE of the bicarbonate-based cultivation is not as promising as the work of Kim et al. (2017) reported, which is over 90%, because the final biomass concentration was too low for the cases by bicarbonate cultivation. Kim et al. (2017) reported that the final biomass yields by gaseous CO_2 and bicarbonate were comparable.

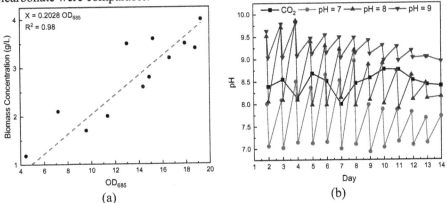

Figure 2. Cultivation of Chlorella vulgaris SU-1 with pH controlled, (a) calibration curve of OD_{685} to biomass concentration, (b) measured pH values.

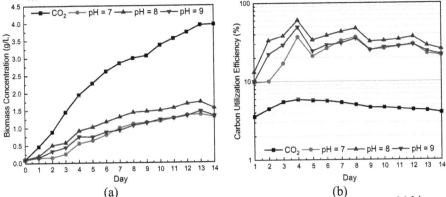

Figure 3. Growth curves of Chlorella vulgaris under different carbon sources, (a) biomass concentration (g/L) and (b) carbon utilization efficiency (%).

In this preliminary study, the biomass yield by bicarbonate is inferior to that by the gaseous form of carbon source. The alkaline culture medium may suppress the Chlorella vulgaris strain growth, although the pH values of the bicarbonate-based photobioreactors were maintained on a daily basis. However, the CUE of the bicarbonate-based cultivation is still superior to the counterpart. In the future work, the amount and frequency of the supplemented $NaHCO_3$ will be adjusted to enhance the productivity. More importantly, adopting a strain of microalgae that is resilient to the alkaline culture medium may improve the final biomass concentration, whereas increases the CUE.

4. Conclusions

Microalgae cultivation to recycle material and energy represents a significant circular economic approach. However, the flue gases from a power plant cannot be directly supplied to cultivate microalgae due to the high concentration of CO_2 in the flue gases. On the other hand, the energy penalty for regenerating absorbents, which is around 15%–20% reduction in overall electricity, is a major obstacle to implement the PCC for a coal-fired power plant. This study demonstrated that the penalty-free PCC through bicarbonate-based microalgae cultivation is a feasible solution for the circular economy. However, the results show that the biomass yield of bicarbonate-based cultivation is inferior to that of using gaseous CO_2 as the carbon source. The limitation will be investigated by cultivating the microalgae that can be found in hypersaline environments. In addition, the simulation results show that the robust design for the absorber can be achieved, for which the designed column height is insensitive to the recycle flow rate, by removing sodium bicarbonate from the settling tank as possible.

References

M. Adamczyk, J. Lasek, A. Skawińska, 2016, CO_2 biofixation and growth kinetics of Chlorella vulgaris and Nannochloropsis gaditana, Appl. Biochem. Biotechnol., 179, 1248-1261.

Aspen Technology, 2014, Rate-based model of the CO_2 capture process by NaOH using Aspen Plus.

C. Benquet, A. B. N. Knarvik, E. Gjernes, O. A. Hvidsten, E. Romslo Kleppe, S. Akhter, 2021, First process results and operational experience with CESAR1 solvent at TCM with high capture rates (ALIGN-CCUS Project), In Proceedings of the 15th Greenhouse Gas Control Technologies Conference, 15-18.

R. R. Bottoms, 1930, Separating acid gases, U.S. Patent 1783901.

G. Y. Kim, J. Heo, H. S. Kim, J. I. Han, 2017, Bicarbonate-based cultivation of Dunaliella salina for enhancing carbon utilization efficiency, Bioresour. Technol., 237, 72-77.

Y. Le Moullec, T. Neveux, A. A. Azki, A. Chikukwa, K. A. Hoff, 2014, Process modifications for solvent-based post-combustion CO_2 capture, Int. J. GreenH. Gas Control, 31, 96-112.

G. T. Rochelle, 2009, Amine scrubbing for CO_2 capture, Science, 325, 1652–1654.

C. M. Silva, A. F. Ferreira, A. P. Dias, M. Costa, 2016, A comparison between microalgae virtual biorefinery arrangements for bio-oil production based on lab-scale results, J. Clean. Prod., 130, 58-67.

A. K. Vuppaladadiyam, J. G. Yao, N. Florin, A. George, X. Wang, L. Labeeuw, Y. Jiang, R. W. Davis, A. Abbas, P. Ralph, P. S. Fennell, M. Zhao, 2018, Impact of flue gas compounds on microalgae and mechanisms for carbon assimilation and utilization, ChemSusChem, 11, 334-355.

R. Wang, H. E. Ashkanani, B. Li, B. I. Morsi, 2022, Development of an innovative process for post-combustion CO2 capture to produce high-value NaHCO3 nanomaterials, Int. J. Greenh. Gas Control., 120, 103761.

Y. Wang, W. Guo, H. W. Yen, S. H. Ho, Y. C. Lo, C. L. Cheng, N. Ren, J. S. Chang, 2015, Cultivation of Chlorella vulgaris JSC-6 with swine wastewater for simultaneous nutrient/COD removal and carbohydrate production, Bioresour. Technol., 198, 619-625.

C. Zhu, S. Chen, Y. Ji, U. Schwaneberg, Z. Chi, 2022, Progress toward a bicarbonate-based microalgae production system, Trends Biotechnol., 40, 180-193.

Antonis Kokossis, Michael C. Georgiadis, Efstratios N. Pistikopoulos (Eds.)
PROCEEDINGS OF THE 33rd European Symposium on Computer Aided Process Engineering
(ESCAPE33), June 18-21, 2023, Athens, Greece

Comparing different modelling approaches for metabolic network dynamic simulation under uncertainty

Oliver Pennington,[a] Dongda Zhang,[a]

[a] *Department of Chemical Engineering, University of Manchester, Oxford Road, Manchester, M1 3AL, UK.*
Dongda.zhang@manchester.ac.uk

Abstract

Modelling strategies in metabolic engineering cover a broad spectrum to simulate microbial cell behaviour and product formation. Understanding differences and limitations of various approaches is key for accurately simulating metabolic network activities and designing high-efficiency strains. One of the main challenges lies within the simulation of dynamic environments, where steady-state assumptions are invalid. However, modelling a dynamic operation is essential for developing batch processes, which are immensely common for fermentation and pharmaceutical industries. This study is to propose and compare different dynamic modelling approaches for metabolic network simulation. The overall aim is to identify the most cost-effective and robust approach to model a metabolic network under non-steady-state conditions.

Keywords: Dynamic metabolic flux analysis, kinetic modelling, metabolic network simulation, parameter estimation, dynamic modelling.

1. Introduction

Microbial fermentation modelling lies at the epicenter of biochemical engineering, as such systems can produce renewable biofuels, bioplastics, and many other high value bioproducts, and mammalian cell lines account for a large proportion of therapeutic recombinant proteins. The UK bioeconomy alone is worth around £220 billion (as of 2018), and is expected to double by 2030, making this a lucrative research field (Harrington, 2018). Such growth requires overcoming many challenges, including low yields in reactor scale-up, deficient metabolic and secretory phenotype for protein production, accumulation of by-products (increasing separation costs and product loss), and finally significant variation between batches, leading to quality-control challenges.

Understanding the metabolic pathways is essential to conquering such challenges for two reasons: 1) to develop industrially desired microbial strains for large scale fermentation; and 2) modify mammalian cells to overexpress recombinant proteins and improve process performance. Several modelling techniques can be employed to infer the behaviour of the metabolic network. These include two main approaches: 1) flux-based methods, which can be applied to both large or small-scale networks to either steady-state or dynamic systems; and 2) mechanistic kinetic models, which try to establish a mechanistic understanding of the behaviour of each reaction pathway. However, there are challenges associated with the various modelling approaches. For example, flux-based methodologies often utilize a fundamental steady-state assumption, as reviewed by Ahn et al. (2012), that is invalid for batch and fed-batch operation and cannot account for intracellular metabolite accumulation. Although existing dynamic MFA approaches can encapsulate the changing extracellular environment, most methodologies do not account for any intracellular accumulation, constant or dynamic (Antoniewicz, 2013). Regarding kinetic modelling, finding an appropriate model structure can be time-consuming and challenging – many structures also exhibit stiffness, causing trouble with parameter estimation. This study aims to propose a new dynamic metabolic flux analysis framework and then compare with other existing methods to determine the most reliable and robust (least susceptible to simulation deviations for a given input deviation) modelling approach, while considering minimal computational expense.

2. Methodology

2.1. Kinetic modelling

The first approach to be considered is the kinetic modelling, in which each reaction flux obeys the Michaelis-Menten model structure as shown in Eq. 1, as described in the study by (Goffaux et al., 2017):

$$v = v_{f,max} \prod_f \alpha_f \frac{R_f}{K_{R,f} + R_f} - v_{r,max} \prod_r \beta_r \frac{P_r}{K_{P,r} + P_r} \tag{1}$$

where α_f and β_r represent stoichiometric coefficients associated with the forward and reverse reactions, respectively, which have their corresponding maximum rates, $v_{f,max}$ and $v_{r,max}$, respectively. Equilibrium constants, $K_{R,f}$ and $K_{P,r}$, are associated with every reactant, R_f, and product, P_r, respectively. Although this structure follows basic assumptions, it is utilized to avoid high computational expense to remain competitive with other modelling approaches. Using a mass balance across each metabolite, the reaction fluxes determine the rate of change of each metabolite – integrating this generates concentration profiles that can be compared to recorded data. A sum of least squares objective function can be used to minimise the difference between the actual and simulated concentrations, $C_{j,meas_t}$ and $C(t)_j$, respectively, at the times measurements were taken, as shown in Eq. 2:

$$\min \sum_{t,meas} \sum_{j,meas} \left(\frac{C_{j,meas_t} - C_j(t)}{C_{j,meas_t}} \right)^2 \tag{2}$$

Parameters of the Michaelis-Menten model structure, $v_{f,max}$, $v_{r,max}$, $K_{R,f}$, and $K_{P,r}$, are constant with time, but vary for each reaction.

2.2. Dynamic metabolic flux analysis (DMFA)

For the first DMFA methodology, a traditional approach is undertaken in which it is assumed no changes in intracellular metabolite concentrations occur, unless proven otherwise by measurements. 2nd degree polynomial expressions can be used to fit measured intracellular and extracellular metabolite concentrations: $C_j(t) = a \cdot t^2 + b \cdot t + c$, where a, b, and c are parameters to be fitted. No higher order polynomials were utilized to avoid overfitting. Such polynomials construct mass balances, as shown in Eq. 3; an equality constraint. For extracellular metabolites, a similar concept is used to minimise the difference between the simulated and predicted rates of concentration at each timestep t, shown as Eq. 4. In this case, the reaction fluxes are the problem variables.

$$\frac{dC_{j,int}(t)}{dt} = \sum_i \alpha_i v_i \tag{3}$$

$$\min \left(\frac{X(t)\cdot\mu_t - \dfrac{dX(t)}{dt}}{\dfrac{dX}{dt}}\right)^2 + \sum_{j,meas,ex}\left(\frac{10^3\cdot X(t)\cdot\sum_i\alpha_{i,j}\cdot v_{i,t} - \dfrac{dC_{j,ex}(t)}{dt}}{\dfrac{dC_{j,ex}}{dt}}\right)^2 \qquad (4)$$

For the second DMFA methodology, the cell is assumed to behave as efficiently as possible. This is done by minimising the magnitude of each reaction flux at each timestep t, shown by Eq. 5, where ω_i is the weight associated with flux v_i. As with the previous MFA approach, the same polynomials are utilized to determine the intracellular concentration profiles. For extracellular metabolites, the mass balance forms equality constraints to avoid over-complication of the objective function, presented via Eq. 6.

$$\min \sum_i \omega_i v_{i,t}^2 \qquad (5)$$

$$\frac{dC_{j,ex}(t)}{dt} = 10^3\cdot X(t)\cdot\sum_i \alpha_{i,j}\cdot v_{i,t} \qquad (6)$$

Finally, we proposed a novel approach to DMFA where accumulation is considered. The accumulation is captured by an accumulation variable A. This coincides with the mass balance of each intracellular metabolite, $\dfrac{dC_{j,int,t}}{dt} = \sum_i \alpha_i v_{i,t} = A_{j,t}$. The aim is to recognize that the dynamic system is unlikely to be at complete steady-state but that the cell will ideally not excessively produce or consume any metabolites. Hence, it is appropriate to penalise accumulation in the objective function Eq. 7.

$$\min \left(\frac{X(t)\cdot\mu_t - \dfrac{dX(t)}{dt}}{\dfrac{dX}{dt}}\right)^2 + \sum_{j,ex}\left(\frac{10^3\cdot X(t)\cdot\sum_i\alpha_{i,j}\cdot v_{i_t} - \dfrac{dC_{j,ex}(t)}{dt}}{\dfrac{dC_{j,ex}}{dt}}\right)^2$$

$$+ w_{acc}\sum_{j,in}\left(\frac{A_{j_t}}{\max\left(\left|A_{j_{t-1}}\right|, 10^{-20}\right)}\right)^2 \qquad (7)$$

Metabolite concentrations are updated linearly, which is valid given a small time increment, Δt, between MFA calculations, $C_{j,in_t} = C_{j,in_{t-1}} + A_{j_{t-1}}\cdot\Delta t$.

2.3. Regularizing flux inconsistency

In the DMFA methodologies discussed, there is potential for consecutive fluxes to display severe inconsistency, which is unlikely to be a true representation of the system. To avoid such behaviour, a penalty function is implemented to penalise deviations from the flux at the previous timestep. This penalty has a corresponding weight, which is determined iteratively, to ensure it does not overrule the main objective function.

3. Case Study

A metabolic network of Myeloid-Derived Suppressor Cells (MDSCs) studied in a previous paper (Goffaux et al., 2017) was used as a case study given the availability of essential metabolites data. The metabolic network consists of 46 reactions, including biomass generation, and 44 metabolites. The data available contains measurements, taken every 8 hours for 96 hours, of 7 extracellular metabolites, biomass concentration, and 12 intracellular metabolites. The remaining 25 metabolites have an initial concentration measurement only. The metabolic reaction network is shown as Fig. 1. More details about the case study can be found in (Goffaux et al., 2017). Four modelling methodologies are to be considered and compared: a class kinetic modelling approach and three variations of dynamic metabolic flux analysis (DMFA).

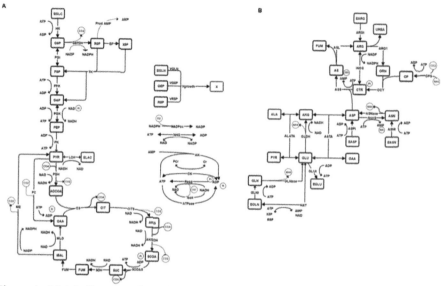

Figure 1: Metabolic network to be simulated (Goffaux et al., 2017), consisting of: (A) Central carbon metabolism and bioenergetics; (B) Urea cycle and amino acid catabolism.

4. Results and Discussion

When comparing the fitting results of the four approaches, the kinetic model is found to performed far worse (mean error 37.8%) than the other methodologies and the three DMFA approaches were almost inseparable (mean error 16.6%), due to the added flexibility provided by the data-driven elements.

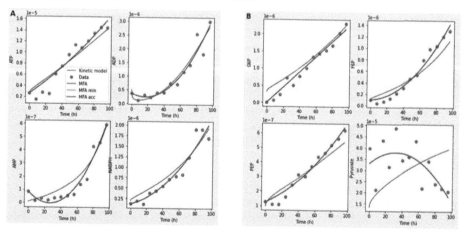

Figure 2: Compairson of experimental data (red) against the simulations of various modelling methodologies, DMFA (blue), DMFA with minimisaiton (cyan), and DMFA with accumulation (magenta), for both extracellular (A) and intracellular (B) metabolites.

For much of the extracellular metabolites, good fits were obtained – especially in instaces of low-level noise in the data. For intracellular metabolites, much of the same was generally observed. However, some datasets were extremely noisy, such as pyruvate, which led to poor fitting – especially for the kinetic model, as shown in Fig. 2B. The excessive noise causes great uncertainty in the initial pyruvate concentration. This is a potential cause of the stiff region simulated by the kinetic model across the initial timesteps (up to a time of 0.3 hours). The stiffness was alleviated by allowing the initial concentration of pyruvate to vary within a sensible region. In addition, the penalty function was successfully implemented, greatly reducing the inconsistency in several reaction fluxes without hindering the performance of the main objective function. When comparing the different methodologies, the DMFA and DMFA with flux minimisation showed the most agreement between any two methodologies. A few cases in which the kinetic modelling and DMFA with accumulation showed better agreement than the traditional DMFA approaches could highlight a metabolite that is in fact violating steady-state or a reaction that requires some form of accumulation to stabilise it. When comparing to DMFA studies and flux observations of similar networks (Tang et al., 2009 and Chan et al., 2003), it is apparent that DMFA and DMFA with flux minimisation simulate more realistic fluxes for both central carbon metabolism, and the urea cycle with amino acid catabolism. DMFA with accumulation and kinetic modelling have a tendency to allow large accumulation of some metabolites as opposed to a larger flux consuming said compound. Since the DMFA approaches used the maximum flux parameters found in the kinetic modelling approach for the flux constraints, it can be concluded that is likely that these parameters were not well estimated since the DMFA and DMFA with minimisation were unable to converge to a feasible solution with such tight constraints. Several constraints, especially in the cyclical structures, had to be relaxed, and literature agrees that the kinetic model and DMFA with accumulation did in fact underestimate said fluxes (Tang et al., 2009) (Chan et al., 2003). DMFA with accumulation was able to overcome these tight constraints by allowing excessive accumulation, despite being unrealistic. Literature containing intracellular metabolite concentrations were immensely scarce, but studies on mammalian cells (Yishitoshi et al., 2017) indicate that the kinetic model was generally predicting sensible concentrations, where data was available.

Finally, the correlation matrix of the kinetic parameters in the kinetic model is also estimated and presented in Fig. 3. About 5% of the correlation values exceed 0.7 ad the parameter pairs involved include 25% of the model parameters, meaning the kinetic model has an appropriate structure in most cases, but could benefit from fine-tuning.A high concentration of excessive correlation is observed in parameters associated with biomass growth, indicating it either does not obey the mechanism suggested, or the mechanism involves less relevant metabolites.

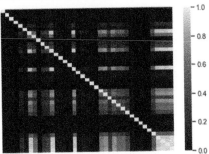

Figure 3: Correlation matrix of the kinetic model parameters as a heat-map.

5. Conclusion

Kinetic modelling can account for intracellular metabolite accumulation and is able to generate reasonable fits, although less accurate than DMFA. If such fits are improved, it could show a good understanding of the mechanisms within the metabolic pathway. However, it is time-consuming to construct a model and computationally expensive to solve. High stiffness meant that gradient-based optimisation failed, and stochastic search optimisation (Particle-Swarm) was utilized, thus not guaranteeing a local optimum being found. Literature also indicates that the fluxes are being underestimated, and this, along with the high stiffness, cases of high parameter correlation and highest average error, indicates that the correct structure was not found, which is a great challenge in kinetic modelling. The DMFA approaches yielded excellent fits to the experimental data and had a much lower computational cost due to the convex nature of each optimisation approach; allowing a global optimum to be found with gradient-based optimisaiton. However, there is always a high risk of overfitting when using purely data-oriented models for the measured metabolites and so it is essential to not over-complicate the model structure.

Penalizing inconsistent fluxes was effective for DMFA and DMFA with flux minimization. For DMFA with accumulation, the weights of the penalty functions were more sensitive and challenging to determine due to the presence of three terms in the objective function: 1) the main objective; 2) the accumulation penalty; and 3) the flux inconsistency penalty. All approaches can yield good results to simulate experimental data, but it is essential to study the metabolic network prior to experimental work and identify potential bottleneck metabolites (those leading into a cyclical structure) and try to measure these metabolites to help conclude the appropriate flux within these cycles. Traditional DMFA approaches appear to give more realistic flux simulations, which is intuitive as cells aspire to maintain a balance and not over-consume/produce any metabolites. DMFA with accumulation is the most robust approach due to its ability to overcome changes to the input via its allowance for accumulation.

As a result, based on the current research, an overall approach to modelling metabolic networks is suggested; begin with DMFA or DMFA with flux minimisation. If good results are unattainable, add accumulation terms as required. Apply kinetic modelling last; use the DMFA results to inform the kinetic model to reduce computational expense.

References

W. S. Ahn et al, 2012, Towards dynamic metabolic flux analysis in CHO cell cultures, Biotechnology Journal, Volume 7, Issue 1, Pages 61-74

M. R. Antoniewicz, 2013, Dynamic metabolic flux analysis—tools for probing transient states of metabolic networks, Current Opinion in Biotechnology, Volume 24, Issue 6, Pages 973-978

C. Chan et al, 2003, Metabolic Flux Analysis of Cultured Hepatocytes Exposed to Plasma, Biotechnology and Bioengineering, Volume 81, Issue 1, Pages 33-49

G. Goffaux et al, 2017, A Dynamic Metabolic Flux Analysis of Myeloid-Derived Suppressor Cells Confirms Immunosupproession-Related Metabolic Plasticity, Scientific Reports, Volume 7, Isuue 9850

R. Harrington, HM Government, 2018, Growing the bioeconomy: a national strategy to 2030.

Y. J. Tang et al, 2009, Metabolic Flux Analysis of Shewanellaspp. Reveals Evolutionary Robustness in Central Carbon Metabolism, Biotechnology and Bioengineering, Volume 102, Issue 4, Pages 1161-1169

E. Y. Yoshitoshi-Uebayashi et al, 2017, Modelling urea-cycle disorder citrinnemia type 1 with disease-specific iPSCs, Biochemical and Biophysical Research Communications, Volume 3, Issue 468, Pages 613-619

Antonis Kokossis, Michael C. Georgiadis, Efstratios N. Pistikopoulos (Eds.)
PROCEEDINGS OF THE 33rd European Symposium on Computer Aided Process Engineering
(ESCAPE33), June 18-21, 2023, Athens, Greece

Integrating hybrid modelling and transfer learning for new bioprocess predictive modelling

Sam Kay,[a] Harry Kay,[a] Alexander W. Rogers[a], Dongda Zhang,[a]*

[a]*Department of Chemical Engineering, University of Manchester, Oxford Road, Manchester, M1 3AL, UK.*
dongda.zhang@manchester.ac.uk.

Abstract

Hybrid modelling combines data-driven and mechanistic modelling, providing a cost effective solution to modelling complex (bio)chemical processes when the underlying mechanisms are not fully understood. While more data-efficient than pure data-driven methods, sufficient experimental data must still be collected through time-consuming experiments for hybrid model construction when applied to simulate a newly investigated process. To address this challenge and accelerate design of new systems, transfer learning has been recently implemented to transfer knowledge from a well-understood source domain to a less-studied target domain. Therefore, in this study, hybrid modelling and transfer learning are integrated for the first time to predict a novel dynamic process for microalgal lutein synthesis through photo-production. Specifically, a hybrid model was built to extract primary knowledge about a well-known lutein production strain, *Desmodesmus* sp., for which considerable process data and a high-fidelity kinetic model are available from the literature. Then transfer learning was used to update the hybrid model using limited process data for a recently isolated microalgal strain, *Chlorella sorokiniana*. Once built, the hybrid transfer model was tested by predicting biomass growth and lutein production of the new strain under various operating conditions. Furthermore, bootstrapping was adopted to estimate and propagate the uncertainty of the hybrid transfer models when predicting each batch process. The performance and efficiency of the hybrid transfer learning model were thoroughly demonstrated through experimental verification. It was found that hybrid transfer learning more efficiently modularised primary knowledge than previous pure data-driven transfer learning techniques, enabling more selective updates of source-specific knowledge to preserve generalisable knowledge. This study, therefore, provides a novel contribution to the predictive modelling and knowledge transfer of complex bioprocesses.

Keywords: Machine learning, hybrid modelling, transfer learning, microalgae photo production, small data problems.

1. Introduction

Small data problems are prevalent throughout bioprocess industries, where collecting samples of experimental data is both time consuming and resource intensive. Datasets classified as small data lack sufficient data points, so the system domain is sparsely explored hence, it is not possible to parameterise a model which fully represents the process. Recently, the limitations of small data have been discussed in the literature, an several techniques have been proposed to tackle this such as artificial data generation (Abdul Lateh et al., 2017). Another such approach is transfer learning, where modelling knowledge of another system can be applied to the new system in a controlled manner Literature reports significant advantages in data efficiency using transfer learning to impart knowledge from a previously studied (larger) dataset to another (similar) system characterised by a lack of data (Barman et al., 2019).

Traditionally, modelling is categorised into data driven and mechanistic approaches (i.e., models derived from statistics and those derived from first principles), however, in recent years there has been a rise in the use of hybrid modelling throughout the literature (Hodgson, 2005). Hybrid modelling combines the two traditional approaches, supplementing the current mechanistic knowledge with a surrogate statistics-based model able to capture additional complex non-linearities within the system. The inclusion of physical knowledge reduces the data required to construct an accurate model which, for systems where data is scarce, can present large benefits. In the bioprocess industries, it is common for first-principle models to be restricted in complexity due to limited mechanistic knowledge of an organisms metabolic reactions and the difficulties in deriving models that represent the immense number of possible metabolic pathways present within the biological system (Hodgson, 2005). Hybrid models, therefore, offer a unique solution to overcome the challenge of limited mechanistic knowledge and small data by integrating data driven techniques to represent the extremely non-linear system complexities that may be left unaccounted for in current first-principle models.

2. Problem statement

The focus of this study is directed at determining the possible effectiveness of combining hybrid modelling of a well understood biological system and transfer learning of the learned knowledge to a less studied target domain. The aim is to develop a framework for accelerating the design of new systems from scarce data, typical of that experienced in laboratory conditions for such newly researched systems. The source model is built for the microalgal strain, *Desmodesmus* sp., and the target model is constructed for a recently isolated strain, *Chlorella sorokiniana*. Sufficient data is available for the prior whereas little data is available for the latter. The biomass concentration in each case is recorded alongside the substrate concentration (nitrate) and the concentration of product (lutein) is monitored through the full experiment timespan. In the source domain (*Desmodesmus* sp.), a nitrate feed stream is introduced at the same hour for every experiment, the concentration of which is altered between experiments. This substrate feed stream is not present in the target domain. Furthermore, for both strains, light intensity is a controlled variable modified between experiments. More detailed introduction to the experimental setup and data collection of the case study can be found in (Rogers et al., 2022).

For a successful transfer of knowledge, the two systems under study must share similar underlying mechanisms such that information extracted from one system can be relevant to the other (Weiss et al., 2016). In this study, it is believed that the two microalgal strains share similar kinetics and metabolic pathways such that some level of mechanistic and statistical knowledge can be used to inform the target strain, reducing the data requirements of the hybrid model. The greatest advantages of transfer learning are observed when the target domain is severely lacking in data, which, in this study manifests itself as a lack of experiments available for model construction. The prerequisite to designing a successful transfer model is first to develop a robust, accurate source model from which information can be shared. Without a solid foundation, any information transferred to the target domain will be non-generalisable and will likely limit the performance of the target model.

Although the use of transfer learning and hybrid modelling has been well explored in recent studies, the combination of the two has not been explored. In this work, the data driven component of the hybrid models will be managed by an artificial neural network (ANN). Bootstrapping techniques are used to estimate the uncertainty associated with the batch predictions. Once a robust source model is designed, the optimised kinetic parameters and ANN parameters (weights and bias') are implemented in their respective objective functions to provide guidance to the target model during its construction.

3. Methodology

First, the kinetic model structure will be introduced. For the production of lutein from the *Desmodesmus* sp. strain, (del Rio-Chanona et al., 2017) proposes an informed kinetic model using modified Monod kinetics. The equations proposed are used in this study to represent the source domain and are as follows:

$$\frac{dX}{dt} = u_0(X, N, I) \cdot \frac{N}{N + K_N} \cdot X - \mu_d \cdot X \tag{1a}$$

$$\frac{dN}{dt} = -Y_N \cdot u_0(X, N, I) \cdot \frac{N}{N + K_N} \cdot X + F_{N0} \tag{1b}$$

$$\frac{dP}{dt} = k_0(X, N, I) \cdot \frac{N}{N + K_P} \cdot X - k_d \cdot P \cdot X \tag{1c}$$

Where $\frac{dX}{dt}$, $\frac{dN}{dt}$ and $\frac{dP}{dt}$ represent the rate of change of biomass, substrate (nitrate) and product (lutein) concentrations respectively. In equation 1a, u_0 is the specific growth rate, I is incident light intensity, K_N is the nitrate half velocity constant and μ_d is the specific decay rate. In equation 1b, Y_N is the nitrate yield coefficient and F_{N0} is the mass flow rate of nitrate influent. In equation 1c, k_0 is the product synthesis rate parameter, K_P is the nitrate half velocity constant for synthesis of lutein and k_d is the product consumption rate constant. Derivation and justification of the model composition is explained thoroughly in (del Rio-Chanona et al., 2017).

For this methodology, K_N, μ_d, Y_N, K_P and k_d are estimated as time constant parameters where as u_0 and k_0 are assumed time varying. Using experimental data for the *Desmodesmus* sp. system, the model parameters are estimated with finite element discretisation and optimised using interior point algorithms in python. In the parameter estimation problem, a least squares error function is implemented to optimise the constant and time-varying terms; for an input of n experiments and N timesteps,

$$obj = argmin\left(\frac{1}{n}\sum_{j=1}^{n}\sum_{i=1}^{N}\left(\theta_{j,i} - \theta_{j,i}'\right)^2\right); \; \theta_{j,i} = \{X, N, P\} \tag{2}$$

where, θ and θ' are the set containing the measured and estimated states respectively. Similarly, a least squares approach is employed in the ANN loss function. u_0 and k_0 are complex functions of X, N and I so, to reduce the non-linearity of the parameter optimisation problem, u_0 and k_0 are solved as independent parameters at each timestep by minimising the fitting error, then an ANN is implemented to alter the terms achieving a better model fit by better approximating the inherent complexities in the system. Let us address the structure and intent behind the ANN component of the hybrid model. The objective of the ANN is to perform alterations to the time varying parameters (u_0 and k_0) to the aim of providing accurate estimations of biomass, nitrate, and lutein concentrations. In addition to this, the uncertainty of the predictions must be determined to ensure confidence in the models' performance hence, bootstrapping is employed to this effect. Two-fold validation was employed to assess the predictive abilities of the model and optimisation of the model hyperparameters was carried out. The structure of the ANN is shown in Figure 1.

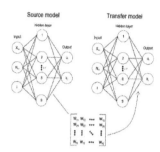

Figure 1: ANN component of the hybrid source and transfer models, detailing the structures, inputs and outputs.

To provide a meaningful comparison of results, the following metrics are defined for the mean relative percentage error (MRPE) and mean relative percentage uncertainty (MRPU) respectively:

$$MRPE\% = \frac{x' - x}{\frac{1}{2}(x' + x)} \cdot 100 \tag{3a}$$

$$MRPU\% = \frac{\sigma'}{\frac{1}{2}(x' + x)} \cdot 100 \tag{3b}$$

Where x' and x represent the prediction and experimental datapoint respectively, and σ' represents the uncertainty estimate obtained from bootstrapping. It is important to consider the relative error and uncertainty due to the proximity of some datapoints to zero values. If simply the mean average percentage error was taken, when the substrate concentration decreased close to zero, the denominator would then tend to zero yielding unrealistic percentage errors and uncertainties.

Herein, we will shift focus to the transfer of knowledge from the source domain to the target domain and the methodology that entails. It is known that the advantages offered by transfer learning are at their greatest when data is severely limited so, in this study only two experiments are used for estimation of model parameters and training of the target ANN. The target domain mechanistic structure is held almost identical to equations 1a, 1b and 1c, differing only by the F_{NO} term in equation 1b which is absent for the *Chlorella sorokiniana.* strain due to no addition of nitrate solution in the experimental datasets. The transfer of information between mechanistic structures is achieved via implementation of penalty terms for specific constant and time varying parameters estimated by the transfer model which penalises deviation from the source model values. The exchange of statistical knowledge was executed via the direct implementation of the ANN architecture and optimal hyperparameters determined for the source model. Furthermore, an additional penalty was enforced in the ANN loss function to discourage divergence between the weights and bias' calculated through backpropagation in the transfer ANN and the parameters of the finalised source ANN.

4. Results and Discussion

The source kinetic model was fitted using all available datasets. Single-fold cross validation on all 8 experiments of the *Desmodesmus* sp. dataset was used to determine the optimal hyperparameters of the source model ANN. This was achieved using the average MRPE per experiment alongside the determined training loss values to inform this optimisation procedure. Once the parameters were obtained, the source model was tested on the entire dataset to evaluate its performance with respect to both predictive accuracy and uncertainty estimation. The results in Figure 2 detail the model predictions of biomass concentrations (g/L), nitrate concentration (g/L) and lutein concentration (mg/L) for experiments 6 and 8.

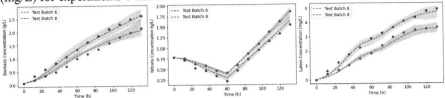

Figure 2 - Hybrid model predictions and corresponding uncertainty limits for test batches 6 and 8.

From Figure 2, it can be seen that the hybrid model captures the underlying mechanisms to a considerable extent for the evolution of biomass, substrate and product concentration through time, providing reasonable uncertainty estimations which cover the residuals between the predictions and experimental datapoints. It was therefore determined that the source model was robust in its construction and, due to its success in capturing the fundamental physics driving the system, it is deemed appropriate for use as the foundation to support knowledge transfer to the novel *Chlorella sorokiniana.* system. The transfer model was constructed using only two experiments, severely limiting the information available to the model for informed predictions. The kinetic model for the transfer domain was parameterised using experiments 3 and 6 however a modified objective function was implemented such that the constant parameters K_N and K_P were penalised the further they deviated from the source domain values. A similar approach was taken for the time varying parameter, u_0 however, the source model values were averaged at each timestep to obtain a single vector of values which the transfer kinetic model parameters could be compared against. Similarly, the ANN component of the hybrid model was then trained on experiments 3 and 6 and optimised using a MSE loss function adjusted to discourage significant variation of the weights and bias from the source model parameters. The predictions of the transfer model on testing experiments 5 and 7 are shown in Figure 3.

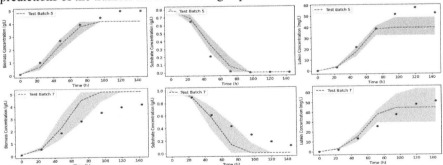

Figure 3: Transfer model predictions and corresponding uncertainty limits for test batches 5 and 7.

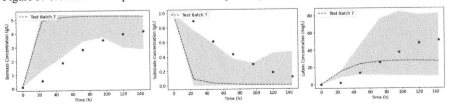

Figure 4: Benchmark predictions and corresponding uncertainty limits for test batch 7.

From Figure 3, it is observed that, with very small data, the transfer model successfully reproduced the trends within the process, while offering adequate estimations of state parameters as well as representative uncertainty bounds. To assess the potential of the transfer model, it is necessary to construct a benchmark model built under the same conditions of extreme data scarcity. The mechanistic structure of the benchmark and its corresponding ANN were once again parameterised however, no external influence from the source model was introduced. The results from this analysis are shown in Figure 4.

From Figure 4, it can be observed that the trends captured by the benchmark are generic and lacking in knowledge specific to the provided dataset. When compared to the transfer model, it is clear that the introduction of prior knowledge from the source domain provides notable advantages. It is also evident that the information held within the small training set is insubstantial and is not representative of the *Chlorella sorokiniana*. domain and, as such, knowledge transfer provides a surrogate for this missing information.

5. Conclusion

In conclusion, transfer learning offers a unique solution when used in conjunction with hybrid modelling to overcome the necessity of large datasets to achieve satisfactory model performances. The controlled introduction of prior mechanistic and statistical knowledge into a novel, less studied domain establishes a robust framework from which accelerated design can be accomplished. The combination of both data-driven and kinetic modelling simultaneously reduces the data intensity of the transfer learning process while extracting primary knowledge from which some degree of transferable information is contained. Furthermore, the use of hybrid modeling allows for the extension of the framework to systems in which mechanistic understanding is bounded, as increased reliance on the data-driven component can be used to account for this. Overall, this work demonstrates the novel combination of hybrid modelling and transfer learning and exhibits the potential entailed within the use of this framework for situations in which data availability is limited.

References

Abdul Lateh, M., Muda, A. K., Izzah Mohd Yusof, Z., Azilah Muda, N., & Sanusi Azmi, M. (2017). Handling a Small Dataset Problem in Prediction Model by employ Artificial Data Generation Approach: A Review. *Journal of Physics: Conference Series*, *892*(1), 012016.

Barman, R., Deshpande, S., Agarwal, S., & Inamdar, U. (2019). *Transfer Learning for Small Dataset*.

del Rio-Chanona, E. A., Ahmed, N. rashid, Zhang, D., Lu, Y., & Jing, K. (2017). Kinetic modeling and process analysis for Desmodesmus sp. lutein photo-production. *AIChE Journal*, *63*(7), 2546–2554.

Hodgson, B. J. (2005). *Hybrid modelling of bioprocesses*.

Rogers, A. W., Vega-Ramon, F., Yan, J., del Río-Chanona, E. A., Jing, K., & Zhang, D. (2022). A transfer learning approach for predictive modeling of bioprocesses using small data. *Biotechnology and Bioengineering*, *119*(2), 411–422.

Weiss, K., Khoshgoftaar, T. M., & Wang, D. D. (2016). A survey of transfer learning. *Journal of Big Data*, *3*(1), 1–40.

Antonis Kokossis, Michael C. Georgiadis, Efstratios N. Pistikopoulos (Eds.)
PROCEEDINGS OF THE 33rd European Symposium on Computer Aided Process Engineering
(ESCAPE33), June 18-21, 2023, Athens, Greece

Machine learning-supported cybergenetic modeling, optimization and control for synthetic microbial communities

Sebastián Espinel-Ríos,[a] Katja Bettenbrock,[a] Steffen Klamt,[a] José L. Avalos,[b] Rolf Findeisen[c]

[a]*Analysis and Redesign of Biological Networks, Max Planck Institute for Dynamics of Complex Technical Systems, Sandtorstraße 1, 39106 Magdeburg, Germany*
[b]*Department of Chemical and Biological Engineering, Princeton University, 08544 Princeton, United States*
[c]*Control and Cyber-Physical Systems Laboratory, Technical University of Darmstadt, Landgraf-Georg-Straße 4, 64283 Darmstadt, Germany*
*Correspondence: rolf.findeisen@iat.tu-darmstadt.de. †This work was supported by the International Max Planck Research School for Advanced Methods in Process and Systems Engineering.

Abstract

Synthetic microbial communities are promising production strategies that can circumvent, via division of labor, many challenges associated with monocultures in biotechnology. Here, we consider microbial communities as lumped metabolic pathways where their members catalyze different metabolic submodules. We outline a machine learning-supported cybergenetic strategy for manipulating the reaction rates of microbial consortia via dynamic regulation of key biomass population levels. To do so, we show a quasi-unstructured modeling approach for synthetic microbial communities with external regulation of intracellular growth regulatory components. Then, we formulate an optimal control problem to find the optimal initial conditions and dynamic input trajectories. We use model predictive control to address system uncertainty, which can be coupled to an observer based on moving horizon estimation. Using a two-member community with optogenetic control as a simulation example, we found the optimal initial biomass concentrations and light intensity trajectories to maximize naringenin production.

Keywords: synthetic microbial communities, machine learning-supported optimization, model predictive control, estimation, optogenetics.

1. Introduction

Monocultures dominate biotechnological production, i.e., processes catalyzed by single-cell types. While metabolic engineering may increase the production efficiency of cells, it often involves intense genetic modifications. As a result, the engineered cells may suffer from resource burden and disturbances in the energy and redox balances, thus portraying poor growth and low volumetric productivity rates. Synthetic microbial communities (SMCs) are possible solutions to these challenges as metabolic tasks are distributed among a designed consortium of microorganisms, hence division of labor is achieved (Roell et al., 2019). In this work, we outline a (machine-learning supported) modeling, optimization and control framework for maximizing the production efficiency of SMCs.

We visualize SMCs as lumped metabolic pathways consisting of several metabolic submodules, each catalyzed by a different member population. We consider that the reaction rates of the submodules in the SMCs can be dynamically adjusted by online modulation of the biomass population levels. Being able to do this in a structured and flexible manner can unlock a powerful optimization degree of freedom in biotechnology. To enable online modulation of the population levels, we propose to introduce tunable genetic systems into (some of) the community members for regulating the gene expression of key intracellular growth regulatory elements via external inputs.

The remainder of this paper is structured as follows. In Section 2, we describe a general modeling approach for SMCs in batch processes. We lump up most of the intracellular bioprocesses and focus on the dynamics of the external species and the externally manipulated intracellular growth regulatory components. With this, we aim to decrease the model complexity as much as possible, while still capturing the most relevant process dynamics. We outline the use of machine learning for improving the prediction capacity of the derived models by learning uncertain or unknown parts of the dynamic equations. In Section 3, we formulate a model-based optimal control problem to find the optimal initial conditions and dynamic input trajectories. In Section 4, system uncertainty is addressed via model predictive control (MPC), optionally coupled to a state observer based on moving horizon estimation (MHE) (Findeisen and Allgöwer, 2002; Rawlings et al., 2020). Finally, in Section 5, a two-member consortium for naringenin production (Lalwani et al., 2021) is used as a simulation example.

2. Modeling of synthetic microbial communities with cybergenetic inputs

Let $B \in \mathbb{R}^{n_B}$ refer to the biomass concentrations of the community members. We comprise in $p \in \mathbb{R}^{n_p}$ all the intracellular concentrations of growth regulatory components present in the SMC and in $z \in \mathbb{R}^{n_z}$ the concentrations of extracellular metabolites. The reaction rates of the model $V \in \mathbb{R}^{n_V}$ are divided into rates of production $V_p \in \mathbb{R}^{n_p}$ and degradation/dilution $V_d \in \mathbb{R}^{n_p}$ of intracellular growth regulatory components, rates of biomass growth $V_B \in \mathbb{R}^{n_B}$ and rates of metabolic conversion of extracellular metabolites $V_z \in \mathbb{R}^{n_{V_z}}$. Hence, $V := \left[V_p^{\mathrm{T}}, V_d^{\mathrm{T}}, V_B^{\mathrm{T}}, V_z^{\mathrm{T}}\right]^{\mathrm{T}}$. Without loss of generality, we focus on batch processes, thus the dynamics of the SMC follows

$$\dot{p}(t) = S_p V(t), \quad p(t_0) = p_0, \qquad (1)$$
$$\dot{B}(t) = S_B V(t), \quad B(t_0) = B_0, \qquad (2)$$
$$\dot{z}(t) = S_z V(t), \quad z(t_0) = z_0, \qquad (3)$$

where $S_p \in \mathbb{R}^{n_p \times n_V}$, $S_B \in \mathbb{R}^{n_B \times n_V}$ and $S_z \in \mathbb{R}^{n_z \times n_V}$ are matrices that map the reaction rates to the corresponding differential equations.

The control inputs are collected in $u \in \mathbb{R}^{n_u}$ and the model parameters in $\theta \in \mathbb{R}^{n_\theta}$. For simplicity, we assume that a given control input can affect the expression of at most one growth regulatory component. In addition, we consider that there is no more than one tunable growth regulatory element per community member. Hence,

$$V_{B_i} = \begin{cases} B_i \mu_{\mathrm{reg}_i}(p_i, z, \theta) & \text{for regulation via } p_i, \\ B_i \mu_{\mathrm{unr}_i}(z, \theta) & \text{otherwise,} \end{cases} \forall i \in \mathbb{B} \quad (4)$$

$$V_{p_i} = \eta_i\left(u_i, \mu_{\mathrm{reg}_i}, \theta\right), \forall i \in \mathbb{P}, \qquad (5)$$
$$V_{d_i} = d_i\left(\mu_{\mathrm{reg}_i}, p_i, \theta\right), \forall i \in \mathbb{P}. \qquad (6)$$
$$V_{z_i} = B_j q_i\left(\mu_j, \theta\right), \forall i \in \mathbb{Q}(j). \qquad (7)$$

where $\mathbb{B}$ is the set of community members, $\mathbb{P}$ be the set of growth regulatory elements and $\mathbb{Q}(j)$ is the set of conversion rates catalyzed by a member B_j. Furthermore, $\mu_{\mathrm{reg}_i} : \mathbb{R} \times \mathbb{R}^{n_z} \times \mathbb{R}^{n_\theta} \to \mathbb{R}$, $\mu_{\mathrm{unr}_i} : \mathbb{R}^{n_z} \times \mathbb{R}^{n_\theta} \to \mathbb{R}$, $\eta_i : \mathbb{R} \times \mathbb{R} \times \mathbb{R}^{n_\theta} \to \mathbb{R}$, $d_i : \mathbb{R} \times \mathbb{R} \times \mathbb{R}^{n_\theta} \to \mathbb{R}$ and $q_i : \mathbb{R} \times \mathbb{R}^{n_\theta} \to \mathbb{R}$ are appropriate kinetic functions.

Given the quasi-unstructured nature of the model, the kinetic functions in Eqs. (4)-(7) are often phenomenological relations. Although the latter generally offer good fitting properties, they usually lack good predictability (Zhang et al., 2019). In some cases, one might not even have a mechanistic or phenomenological relation at hand due to the complexity of a given bioprocess. Therefore, *if necessary*, we can learn the uncertain parts of the model using machine learning. In Section 5 we show an example of a *hybrid* model.

3. Optimal cybergenetic control of synthetic microbial communities

Let us define $x := [p^T, B^T, z^T]^T$, $x \in \mathbb{R}^{n_x}$. We formulate a model-based optimal control problem to maximize the production efficiency of a SMC (e.g., productivity, profit, etc.)

$$\max_{u(\cdot), x_0} \int_{t_0}^{t_f} l(x(t), u(t)) \, dt + e\left(x(t_f)\right), \qquad (8)$$

$$\text{s.t.} \quad \text{Eqs. } (1) - (7),$$
$$x(t_0) = x_0, \qquad (9)$$
$$0 \le g(x(t), u(t)). \qquad (10)$$

In the previous formulation, $u(\cdot)$ is the input function, $l: \mathbb{R}^{n_x} \times \mathbb{R}^{n_u} \to \mathbb{R}$ is the stage cost and $e: \mathbb{R}^{n_x} \to \mathbb{R}$ is the terminal cost. The system state and input constraints are captured by $g: \mathbb{R}^{n_x} \times \mathbb{R}^{n_u} \to \mathbb{R}^{n_g}$. The process time spans from the initial time t_0 to the final time t_f. Note that we consider both $u(\cdot)$ and the initial state concentration x_0 as degrees of freedom (DOF). We also refer to this formulation as the *open-loop* optimization as no online corrective actions are taken if the system is out of "specs" or if disturbances occur.

4. Predictive cybergenetic control of microbial communities with estimation

We compensate for possible system uncertainty such as model-plant mismatch and disturbances using online feedback or closed-loop control in the form of MPC (Findeisen and Allgöwer, 2002; Rawlings et al., 2020). For the batch case we use a shrinking horizon, i.e., the control problem is solved iteratively over a time horizon $[t_k, t_f]$, where $t_{k+1} = t_k + \Delta t$ and Δt is the sampling time size. Considering that $\tilde{x}_k$ is the measured state at the sampling time t_k, the MPC problem reads

$$\max_{u(\cdot)} \int_{t_k}^{t_f} l(x(t), u(t)) \, dt + e\left(x(t_f)\right), \qquad (11)$$

$$\text{s.t.} \quad \text{Eqs. } (1) - (7), (10),$$
$$x(t_k) = \tilde{x}_k. \qquad (12)$$

When not all the states can be measured, we can *reconstruct* them via MHE. We collect the measurements in $y \in \mathbb{R}^{n_y}$ and Eqs. (1)-(3) in the vector function $f(x, u, \theta)$. Also, let $h: \mathbb{R}^{n_x} \times \mathbb{R}^{n_u} \times \mathbb{R}^{n_\theta} \to \mathbb{R}^{n_y}$ be the measurement equations and $w \in \mathbb{R}^{n_x}$ the state noise. We denote with $(\cdot)_i$ and $(\tilde{\cdot})_i$ the optimization and measured variables at time t_i, respectively. We indicate with $(\bar{\cdot})$ the prior information of a variable. We assume that the input signal is piecewise constant over the sampling times. Given a time-discrete model, the MHE problem at time t_k reads (Rawlings et al., 2020)

$$\min_{x_{k-N}, \theta_k, w_{k-N}, \dots, w_k} \left\| \begin{bmatrix} x_{k-N} - \bar{x}_{k-N} \\ \theta_k - \bar{\theta} \end{bmatrix} \right\|_P^2 + \Sigma_{i=k-N}^{k} \|y_i - \bar{y}_i\|_R^2 + \|w_i\|_W^2 \qquad (13)$$

$$\text{s.t.} \quad x_{i+1} = x_i + \int_{t_i}^{t_i + \Delta t} f(x, u, \theta_k) \, dt + w_i, \qquad (14)$$

$$\text{Eqs. } (4) - (7), (10)$$
$$y(t_i) = h(x_i, u_i, \theta_k), \qquad (15)$$
$$i \in [k - N, k], k, N \in \mathbb{N},$$

where $\|a\|_A^2 := a^T A a$, N is the length of the estimation window and $P \in \mathbb{R}^{n_x + n_\theta, n_x + n_\theta}$,

$R \in \mathbb{R}^{n_y, n_y}$ and $W \in \mathbb{R}^{n_x, n_x}$ are weighting matrices. In general, the decision variables of the MHE are the states at the beginning of the estimation window x_{k-N}, the parameters and the state noise; with these we can reconstruct the states via Eq. (14). The estimated state vector $\hat{x}$ can be used to update the MPC. Parameters are assumed to be constant in the MHE horizon, but they can be updated at each iteration to enable an *adaptive* MPC.

5. Microbial community with optogenetic control: naringenin biosynthesis

A summary of the application example is presented in Fig. 1. The SMC consists of an engineered *Escherichia coli* that produces tyrosine, and an engineered *Saccharomyces cerevisiae* that consumes tyrosine to synthesize naringenin. *E. coli* expresses a toxin constitutively, while it expresses an antitoxin, MazE, in an inducible manner using blue light intensity I_b as control input. MazE counteracts the effect and halts the expression of the toxin, thereby enabling different growth levels (Lalwani et al., 2021). This prevents *E. coli* from outgrowing *S. cerevisiae* and turning the SMC into a monoculture.

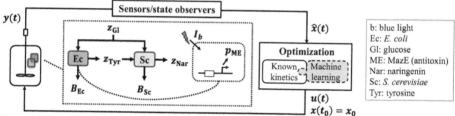

Figure 1. Cybergenetic strategy for naringenin production by a two-member SMC.

We consider that the dynamics of the SMC follows

$$\dot{p}_{ME} = V_{p_{ME}} - (k_d + \mu_{Ec})p_{ME}, \quad V_{p_{ME}} = \left(\eta_{ME_0} + \eta_{ME,m} \frac{I_b^{n_1}}{I_b^{n_1} + k_{ME}^{n_1}}\right)\mu_{Ec}, \quad (16)$$

$$\dot{B}_{Ec} = \mu_{Ec}B_{Ec}, \quad \mu_{Ec} = \mu_{m,Ec}\left(\frac{z_{Gl}}{z_{Gl} + k_{Gl,Ec}}\right)\left(\frac{p_{ME}^{n_2}}{p_{ME}^{n_2} + k_p^{n_2}}\right), \quad (17)$$

$$\dot{B}_{Sc} = \mu_{Sc}B_{Sc}, \quad \mu_{Sc} = \mu_{m,Sc}\left(\frac{z_{Gl}}{z_{Gl} + k_{Gl,Sc}}\right), \quad (18)$$

$$\dot{z}_{Gl} = -\left(Y_{Gl/Ec}\mu_{Ec}\right)B_{Ec} - \left(Y_{Gl/Sc}\mu_{Sc}\right)B_{Sc}, \quad (19)$$

$$\dot{z}_{Tyr} = \left(Y_{Tyr/Ec}\mu_{Ec}\right)B_{Ec} - \left(Y_{Tyr/Sc}\mu_{Sc}\frac{z_{Tyr}}{z_{Tyr} + k_{Tyr}}\right)B_{Sc}, \quad (20)$$

$$\dot{z}_{Nar} = \left(Y_{Nar/Tyr}Y_{Tyr/Sc}\mu_{Sc}\frac{z_{Tyr}}{z_{Tyr} + k_{Tyr}}\right)B_{Sc}, \quad (21)$$

with $x = \left[B_{Ec}, B_{Sc}, z_{Gl}, p_{ME}, z_{Tyr}, z_{Nar}\right]^T$, $x(t_0) = x_0$. We refer to this model as the *nominal* or *real plant* model. All the optimizations and the machine learning-related methods presented hereafter were performed with HILO-MPC (Pohlodek et al., 2022).

The open-loop optimization results are presented in Fig. 2. We maximized $z_{Nar}(t_f)$ using the optimal control problem in Section 3 based on the nominal model. We selected the parameter values of the nominal model such that biologically sound process dynamics under aerobic conditions could be simulated. Considering three DOF, $B_{Ec}(t_0)$, $B_{Sc}(t_0)$ and $I_b(\cdot)$, we obtained 12.0 mg/L naringenin. We also analyzed the case where *E. coli* does not possess the toxin/antitoxin system, i.e., $\dot{p}_{ME} = 0$ and $\mu_{Ec} = \mu_{m,Ec}(z_{Gl}/(z_{Gl} + k_{Gl,Ec}))$. In this second scenario, with $B_{Ec}(t_0)$ and $B_{Sc}(t_0)$ as DOF, we obtained 9.3 mg/L naringenin, 23 % less than with the previous optimization. Furthermore, the optimizer predicts a very low $B_{Ec}(t_0)/B_{Sc}(t_0)$ to reduce the risk of *E. coli* outgrowing *S. cerevisiae*.

In contrast, with dynamic growth regulation of *E. coli*, one can *safely* start the process at a higher *E. coli* concentration, which increases the tyrosine available for conversion to naringenin by *S. cerevisiae*, hence the better performance.

In the previous cases, the predictions of the optimization matched the plant simulations as we did not consider model-plant mismatch. However, often we do not know the exact plant model. In a third open-loop scenario, we assumed that $V_{p_{ME}}$ in Eq. (17) and the p_{ME}-dependent factor in the μ_{Ec}-kinetics (Eq. (18)) were unknown. A feedforward fully connected neural network was trained to learn these unknowns. We simulated eight batches with the nominal model subject to randomized input values and added 2.5 % Gaussian white noise to the *measurements*. We considered p_{ME} and I_b as features and the two unknown parts of the model as labels. From the eight batches, 20 % of the data was used for validation. Three additional independent batches were used to test the quality of the hybrid model (Fig. 3). Despite some model-plant mismatch, the hybrid model was able to predict overall trends. Nevertheless, using the hybrid model in the open-loop optimization with $B_{Ec}(t_0)$, $B_{Sc}(t_0)$ and $I_b(\cdot)$ as DOF rendered 8.5 mg/L naringenin (cf. Fig. 2), 29 % less than with the open-loop optimization based on the nominal model. The hybrid model-based optimization overestimated *E. coli* growth and tyrosine synthesis, thus negatively impacting on naringenin production.

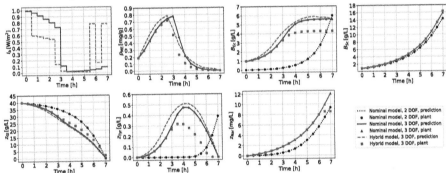

Figure 2. Open-loop optimization results of the SMC for naringenin production. Nominal model: $\mu_{m,Ec} = 0.825$ h^{-1}, $\mu_{m,Sc} = 0.418$ h^{-1}, $\eta_{ME_0} = 5.71 \times 10^{-5}$ g/g, $\eta_{ME,m} = 2.86 \times 10^{-3}$ g/g, $n_1 = 2.49$, $n_2 = 1.5$, $k_d = 1.39$ h^{-1}, $k_{Gl,Ec} = 6.8 \times 10^{-5}$ g/L, $k_{Gl,Sc} = 0.06$ g/L, $k_{ME} = 0.138$ W/m^2, $k_p = 5 \times 10^{-4}$ g/g, $k_{Tyr} = 1 \times 10^{-6}$ g/L, $Y_{Gl/Ec} = 2.2$ g/g, $Y_{Gl/Sc} = 2.0$ g/g, $Y_{Nar/Tyr} = 0.016$ g/g, $Y_{Tyr/Ec} = 0.164$ g/g, $Y_{Tyr/Sc} = 0.05$ g/g. The relevant product yields were inferred *at our best* from Lalwani et al. (2021). Similarly, *realistic* optogenetic-related parameters were inferred from Olson et al. (2014).

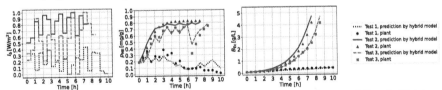

Figure 3. Testing of the hybrid model. We only show the dynamics of p_{ME} and B_{Ec}. Artificial neural network configuration: two hidden layers with ten neurons each, sigmoid activation function.

The hybrid model could be improved by increasing the amount of data used for training. However, experiments in biotechnology are generally expensive and time-consuming, which limits the available data. Therefore, in Fig. 4 we use MPC to address the model uncertainty of the hybrid model. First, we assumed that all the states can be measured online. This led to 11.5 mg/L naringenin, significantly compensating for the uncertainty of the hybrid model by adjusting the dynamic input trajectory. Despite the good results, online monitoring of intracellular components is challenging. Therefore, we also analyzed the case where p_{ME} cannot be measured online. A special case of MHE is when all the measurements from t_0 are considered; this is called full information estimation (FIE). We estimated p_{ME} via FIE, where we minimized the first two terms in Eq. (13) and did not consider the parameters as decision variables. Note that from the estimated state vector, we only passed $\hat{p}_{ME}$ to the MPC as we *trusted* the measurements. This resulted in 10.8 mg/L naringenin, representing a 21 % enhancement compared to the open-loop scenario with the hybrid model and only 6 % less efficient than the MPC with full state measurement. As expected, the FIE estimates improved as the estimation window grew due to its *memory* effect, i.e., more historical measurements available for the estimation.

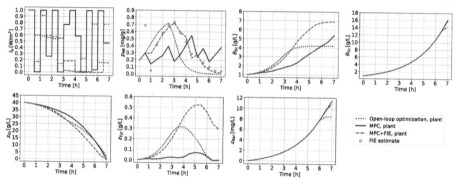

Figure 4. Machine learning-supported feedback control of the SMC for naringenin production. $B_{Ec}(t_0)$, $B_{Sc}(t_0)$ and the first step of I_b were found at the first MPC iteration, then the MPC only had I_b as a DOF. All the measurements had 2.5 % Gaussian white noise.

6. Conclusion

We outlined a (machine learning-supported) cybergenetic modeling, optimization and control strategy for exploiting biotechnological production by SMCs. We also presented a state observer for inferring unmeasured states. We demonstrated the potential of our framework with a two-member SMC for naringenin production with optogenetic control as a simulation example. Although we focused on batch setups, the framework can in principle be extrapolated to other modes of operation. Future work deals with the experimental implementation of the framework and additional hybrid modeling and control approaches, e.g., based on Gaussian processes, Bayesian neural networks, physics-informed machine learning and reinforcement learning.

References

Findeisen, R., Allgöwer, F., 2002. An introduction to nonlinear model predictive control, in: 21st Benelux Meeting on Systems and Control. pp. 119–141.

Lalwani, M.A., Kawabe, H., Mays, R.L., Hoffman, S.M., Avalos, J.L., 2021. Optogenetic control of microbial consortia populations for chemical production. ACS Synth. Biol. 10, 2015–2029.

Olson, E.J., Hartsough, L.A., Landry, B.P., Shroff, R., Tabor, J.J., 2014. Characterizing bacterial gene circuit dynamics with optically programmed gene expression signals. Nat. Methods 11, 449–455.

Pohlodek, J., Morabito, B., Schlauch, C., Zometa, P., Findeisen, R., 2022. Flexible development and evaluation of machine-learning-supported optimal control and estimation methods via HILO-MPC. arXiv:2203.13671.

Rawlings, J., Mayne, D., Diehl, M., 2020. Model predictive control: theory, computation and design, 2nd edition. ed. Nob Hill Publishing, LLC, Santa Barbara.

Roell, G.W., Zha, J., Carr, R.R., Koffas, M.A., Fong, S.S., Tang, Y.J., 2019. Engineering microbial consortia by division of labor. Microb. Cell Factories 18, 35.

Zhang, D., Del Rio-Chanona, E.A., Petsagkourakis, P., Wagner, J., 2019. Hybrid physics-based and data-driven modeling for bioprocess online simulation and optimization. Biotechnol. Bioeng. 116, 2919–2930.

Antonis Kokossis, Michael C. Georgiadis, Efstratios N. Pistikopoulos (Eds.)
PROCEEDINGS OF THE 33rd European Symposium on Computer Aided Process Engineering
(ESCAPE33), June 18-21, 2023, Athens, Greece

Model-based optimization of a recombinant *E. coli* fed-batch process for best time-space yield and conversion efficiency

Julian Kager[a], Nora Horst[b], Christoph Herwig[b] and Ioannis V.Skiadas[a]

aDepartment of Chemical and Biochemical Engineering, Technical University of Denmark, Building 228A, 2800 Kgs. Lyngby, Denmark
bInstitute of Chemical, Environmental and Bioscience Engineering, TU Wien, 1040 Vienna, Austria
jukager@kt.dtu.kt

Abstract

Within this contribution we use a mechanistic model to optimize the feed addition and harvest time point of the induction phase of a recombinantly produced protein in *E. coli*. The model includes the metabolic load of the forced product formation and its effects on the production rate as well as the biomass conversion yield. The model was fitted to three experiments with different constant substrate feed rates during the production phase. Under usage of the calibrated model, optimal substrate addition and harvest time points with the objective to improve the time-space yield were determined. In addition, a constant feed rate was compared to an increasing feed rate with the aim to maintain a constant biomass specific substrate uptake. Different optimal combinations were identified and compared considering time and resource efficiency. The work shows how recombinant protein production and the inevitable metabolic load can be incorporated into a simple mechanistic model. Once calibrated to a limited number of experiments the model is suited for the design of a resource and time optimal induction phase. Future experiments and uncertainty analysis will show how well the model predicted the real optimal operating points.

Keywords: bioprocessing, optimization, production phase modelling, simulation, recombinant protein production

1. Introduction

Fed-batch processes using genetically modified *E. coli* bacteria are widely used to produce high value recombinant proteins (Selas Castineiras et al., 2018). State of the art production processes are divided into three main phases, including a batch and fed-batch phase to produce a high amount of active biomass followed by the production phase (Yee and Blanch, 1992). The cell behavior during the first two phases is well understood and optimized to reach high biomass concentrations within short times (Lee, 1996) and to avoid by product formation (Abadli et al., 2021). Due to the metabolic load of the heterologous production of the target protein, cell metabolism changes in course of the induction phase (Neubauer et al., 2003). Therefore, modelling, optimization and control of the induction phase is a challenging task and harvest time points and feed addition are often static parameters derived from historic experiments, which often leads to suboptimal operation and low time space yields. Models can help to improve the exploration of the design space by simulating all different combinations and identifying optimal operation ranges. Möller et al. (2019) for examples used a model to explore the design space of a CHO cell process to identify optimal feed addition. Bano et al. (2019) extended this concept and recalculated the design space in real-time as soon novel information was available to the model. For recombinant protein production in *E. coli* strong influences on product formation from the feed rate was reported for different products (Mears et al., 2017; Huang et al., 2012; Chen et al., 2012), but due to the lack of proper, mechanistic models, successful examples for insilico exploration and optimization of the production phase are still missing. Although constant substrate feeds are often used during the production phase, recent works show possibilities and potentials to control the specific substrate uptake rate (Kager et al., 2022b) or even the product formation rate (Kager et al., 2022a) by manipulating the feed rate of the limiting substrate. Although the latter being very promising, the control of the specific substrate uptake is easier realizable and was already shown in different applications (Kager et al., 2020; Spadiut et al., 2013; Picon et al., 2005) and can be therefore regarded as a good alternative to constant substrate addition rates.

Within this contribution we show how a previously developed mechanistic model describing the production phase of a recombinantly produced protein can be used to determine optimal feed addition in combination with harvest time. Optimal constant feed addition will be compared to optimal set-points of the cell specific substrate uptake for their time and resource efficiency. After describing the three calibration experiments, the model and its fitting to the experiments are presented. The results show the design space of the analyzed process and the dynamics of the optimal process followed by the conclusion.

2. Methods

2.1. Experimental data

A modified K12 *E. coli* strain with a rhamnose-inducible expression system (rhaBAD promoter), producing a single chain antibody fragment with a transporter sequence to be transported into the periplasm was used. Fermentations were conducted in a DASGIP multibioreactor system (Eppendorf, Germany). Composition of the used minimal media was based on Wilms et al. (2001). Temperature was kept at 35°C, stirrer speed at 1400 rpm and aeration at 1.4 vvm for the whole process. The pH was controlled at 7.0 ±0.2 with addition of 12.5 % NH_4OH solution and the dissolved oxygen (DO_2) was kept over 25% by supplementing pure oxygen to the air. The reactors with an initial glycerol concentration were inoculated with 25 mL of an overnight shake flask culture (30°C and 170 rpm). The 20 g/L glycerol in the batch phase was consumed after approx. 12 h and a fed-batch phase with an exponential feed ramp with a μ of 0.14 according to the equation given in Lee (1996)) was applied to reach high biomass concentrations of 40 gL^{-1} after another 12 h runtime. Recombinant protein production was induced by a one-point addition of sterile filtrated rhamnose solution (1.5 g L-rhamnose in 10 mL H_2O) and the glycerol feed rate was set to different constant values (*high*= 14.5 mLh^{-1}; *medium* = 8.5 mLh^{-1}; *low* = 2.2 mLh^{-1}).

Periodical offline measurements were taken and analyzed. Biomass dry content was determined by centrifugation at 4500 g, 10 min, 4°C including one washing step before drying at 105°C for min. 72h. For product concentration, homogenized, centrifuged and gel filtrated (PD MiniTrap g-25) cell suspension was applied on a protein G affinity column (HiTrap ProtG (GE Healthcare; USA) with a flow rate of 2 $mlmin^{-1}$ (20 mM phosphate buffer) at 25°C and elution with a change of the pH from 7.4 to 2.5). Acetate and glycerol concentrations were quantified from the supernatant by enzymatic, photometric principle in a robotic system (BioHT, Roche, Germany) and were always under the detection limit during the induction phase.

2.2. Model equations

The system differential equations for the ideally stirred tank reactor in fed-batch mode are:

$$
\begin{aligned}
\frac{dc_X}{dt} &= \mu\, c_X & &- \frac{F_{in}}{V_R} c_X \\
\frac{dc_S}{dt} &= -q_S\, c_X & &+ \frac{F_{in}}{V_R} (c_{S,in} - c_S) \\
\frac{dc_P}{dt} &= q_P\, c_X & &- \frac{F_{in}}{V_R} c_P \\
\frac{dV_R}{dt} &= F_{in}; & \frac{dS_{met}}{dt} &= q_S
\end{aligned}
\tag{1}
$$

describing the concentration changes over time of the: biomass c_X, glycerol c_S, product c_P, metabolized substrate S_{met} and the reactor volume V_R. Glycerol is added to the system by the substrate inflow F_{in} with a concentration $c_{S,in}$. The biological conversion rates, namely the growth rate μ, the substrate consumption rate q_S and the product formation rate q_P are described by the following reaction kinetics:

$$
\begin{aligned}
q_S &= q_{S,max}\frac{c_S}{c_S + K_S} \\
\mu &= Y_{X/S}\,(q_S - m_S) \\
Y_{X/S} &= Y_{X/S,max}\exp(-S_{met}K_{Y_{X/S}}) \\
q_P &= q_{P,max}\frac{q_S}{q_S + K_{SqS}}\frac{S_{met}}{\frac{S_{met}^k}{K_{IqP}} + S_{met} + K_{SqP}}
\end{aligned}
\tag{2}
$$

The substrate uptake rate q_S is described by a Monod kinetic with $q_{S,max}$ being the maximum uptake rate in function of the glycerol concentration c_S and the half saturation constant K_S. Biomass growth μ is derived from consumed substrate q_S by substrate conversion yield $Y_{X/S}$ reduced by the substrate needed for cell maintenance m_S. Due to the metabolic load $Y_{X/S}$ decreases by an asymptotic decay ($K_{Y_{X/S}}$) in function of the metabolized substrate (S_{met}). The product formation q_P is composed by a Monod term describing the dependency of $q_{P,max}$ on the substrate uptake q_S with the half saturation constant K_{SqS} and a Haldane term to describe the underlying start up and decline phase with S_{met} as a trigger, K_{SqP} as the delay coefficient, K_{IqP} as the decay coefficient and k as the Haldane exponent determining the shape of the decay.

Table 1: Model Parameter

Parameter	Value	Unit	Description
$c_{S,in}$	850	gL^{-1}	feed concentration
$q_{S,max}$*	1.0	$gg^{-1}h^{-1}$	max. substrate uptake
K_S*	0.0050	gL^{-1}	half saturation constant
m_S	0.02	h^{-1}	maintenance constant
$Y_{X/S,max}$*	0.47	gg^{-1}	max. growth yield
$K_{Y_{X/S}}$	0.34	gg^{-1}	growth yield decay
$q_{P,max}$	0.0066	$gg^{-1}h^{-1}$	max. production rate
K_{SqS}	0.082	$gg^{-1}h^{-1}$	q_P affinity to q_S
k	4.6	–	Haldane exponent
K_{IqP}	3.8	$(gg^{-1})k^{-1}$	product decay
K_{SqP}	0.096	gg^{-1}	product formation delay

The model was fitted to three experimental data sets, with three different constant feed rates, displayed in figure 1. The model parameters are given in table 1. As fitting criterion the weighted sum of squared errors between measurements of biomass, product and sugar concentrations and the respective model simulations of all three experiments was minimized by a local optimizer (Fmincon:MATLAB). All calculations were done in MATLAB (R2022b, Mathworks, USA) and the differential equations were solved by ode15s. To keep the substrate uptake constant the following equation was used to derive the corresponding feed rate based on the current biomass concentration and aimed qs setpoint ($q_S << q_{S,max}$).

$$F_{in} = \frac{q_S \, c_X \, V_R}{c_{S,in} - c_S} \qquad (3)$$

3. Results and Discussion

3.1. Process & Model dynamics

In figure 1 the results of the model simulation of the three calibration experiments are shown in comparison to the measured product concentrations. Overall, with a normalized root mean square error (NRMSE) of 11,5 % the model has an acceptable accuracy considering the big feed rate range (2.2 – 14.5 mLh^{-1}) of the calibration experiments. From the model and the measurements, it can be seen that higher feed rates lead to a faster product formation but also to a faster stagnation. The lowest feed rate leads to highest product titers. Compared to the other processes it takes however longer to reach those concentrations, with a duration of up to 100 h.

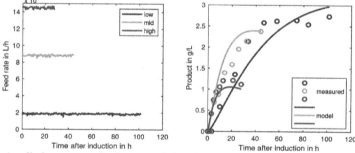

Figure 1: Applied constant feed rates of the three model calibration experiments and alignment of model output to measured product concentrations with NRMSE of 10,7% for low, 12.7 % for mid and 11.2 % for high experiment.

3.2. Design space exploration

Although the fitted model displayed in figure 1, is not perfect it can be used to explore the design space and to determine potentially optimal operating regions. In order to identify optimal combinations of feed rates, q_S setpoints and harvest time-points the time space yields (TSY) were calculated considering the duration of the induction phase in addition to batch (12 h), fed-batch (12 h) and setup times (7 h). The resulting design spaces for the constant feed rate (a) and the controlled q_S (b) are displayed in figure 2. According to the model the optimal TSY is reached by applying constant feed rates around 5 mL/h and harvesting after 23 h, which is between the *low* and the *mid* feed rate experiments. It seems that for this process, high constant feed rates are highly unfavorable

Model-based optimization of a recombinant E. coli fed-batch proces
for best time-space yield and conversion efficiency

2611

with very low TSY's. A similar pattern can be seen for the q_S controlled induction phases (figure 2 b). Also, here smaller setpoints (< 0.1 gg^{-1}h^{-1}) and harvest times between 18 - 40 hours, lead to best TSY's. In contrast to the constant feed rate two optimal regions are identified. A setpoint of 0.1 gg^{-1}h^{-1} in combination with a harvest time of 19.2 h and a smaller setpoint of 0.055 gg^{-1}h^{-1} with a harvest timepoint at 30.2 hours.

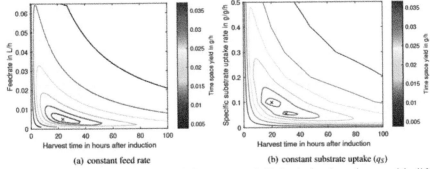

(a) constant feed rate   (b) constant substrate uptake (q_S)

Figure 2: Design spaces for optimal time space yield for induction phases with different constant feed rates (a) and specific substrate uptake (qs) setpoints (b) in dependence on the harvest time-point.

In figure 3 the trajectories of the three potential, optimal conditions are displayed with some performance criteria summarized in table 2. It can be seen that the different optimal conditions differ in their feed profiles, which are all in the lower range (medium feed = 8.5 mL/h). The constant feed rate results in a steadily decreasing q_S compared to the two optimal constant q_S setpoints. The product dynamics reveal a similar behavior as seen in the calibration experiments. Hereby, higher feed rates lead to a faster product increase but the lowest setpoint to the overall highest product titer (see also table 2). Although, the constant feed rate and the higher q_S setpoint have slighty higher TSY's, due to faster product formation dynamics, the low q_S setpoint reaches the highest product titer with overall lowest consumed glycerol and is therefore the most resource efficient.

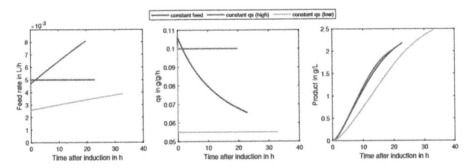

Figure 3: Model simulations of identified optimal constant feed and two q_S setpoints

Table 2: Output of the three potentially good performing processes

Feeding	Harvest (h)	TSY (g/h)	Product (g/L)	Biomass (g/L)	Glycerol (g)
$F_{in} = 0.005$ L/h	23.0	0.0389	2.19	58.3	97.8
$q_S = 0.10$ g/g/h	19.8	0.0386	2.03	61.0	108.3
$q_S = 0.055$ g/g/h	30.2	0.0372	2.47	54.2	91.4

4. Conclusion

In this contribution a mechanistic model was fitted to the production phase of three experiments with different feed addition. Based on the obtained model the design space could be explored and promising regions identified. Based on this methodology potentially optimal experiments can be planned and different control strategies can be bench-marked with only few available experiments. With a proper uncertainty analysis of the preliminary model, a more probability-based approach could be follow in future to predict the most probable design space rather than optima which are highly uncertain due to model-plant mismatches. To show the validity of the obtained results, verification experiments and a proper model uncertainty analysis are necessary. Future verification experiments are also the basis to further improve the underlying model.

Acknowledgements

This work obtained funding from the Novo Nordisk Foundation, start package grant tenure track Assistant Professor on bioprocess digitalization and control (Nr. 111279).

References

M. Abadli, L. Dewasme, S. Tebbani, D. Dumur, A. V. Wouwer, 2021. An experimental assessment of robust control and estimation of acetate concentration in escherichia coli bl21 (de3) fed-batch cultures. Biochemical Engineering Journal, 108103.

G. Bano, P. Facco, M. Ierapetritou, F. Bezzo, M. Barolo, Aug. 2019. Design space maintenance by online model adaptation in pharmaceutical manufacturing. Computers & Chemical Engineering 127, 254–271.

X. Chen, L. Liu, J. Li, J. Liu, G. Du, J. Chen, Feb. 2012. Optimization of glucose feeding approaches for enhanced glucosamine and N-acetylglucosamine production by an engineered Escherichia coli. Journal of Industrial Microbiology & Biotechnology 39 (2), 359–365.

C. Huang, Jr., H. Lin, X. Yang, 2012. Industrial production of recombinant therapeutics in Escherichia coli and its recent advancements. Journal of Industrial Microbiology & Biotechnology 39 (3), 383–399.

J. Kager, J. Bartlechner, C. Herwig, S. Jakubek, Mar. 2022a. Direct control of recombinant protein production rates in E. coli fed-batch processes by nonlinear feedback linearization. Chemical Engineering Research and Design, S0263876222001460.

J. Kager, N. Horst, J. Bartlechner, C. Herwig, S. Jakubek, Jan. 2022b. Generic model based control of different specific rates in recombinant E. coli Fed-batch processes. IFAC-PapersOnLine 55 (7), 756–761.

J. Kager, A. Tuveri, S. Ulonska, P. Kroll, C. Herwig, Mar. 2020. Experimental verification and comparison of model predictive, PID and model inversion control in a Penicillium chrysogenum fed-batch process. Process Biochemistry 90, 1–11.

S. Y. Lee, 1996. High cell-density culture of escherichia coli. Trends in biotechnology 14 (3), 98–105.

L. Mears, S. M. Stocks, G. Sin, K. V. Gernaey, Mar. 2017. A review of control strategies for manipulating the feed rate in fed-batch fermentation processes. Journal of Biotechnology 245, 34–46.

J. Moller, K. B. Kuchem¨uller, T. Steinmetz, K. S. Koopmann, R. P¨ortner, May 2019. Model-assisted Design of Experi-¨ments as a concept for knowledge-based bioprocess development. Bioprocess and Biosystems Engineering 42 (5), 867–882.

P. Neubauer, H. Lin, B. Mathiszik, 2003. Metabolic load of recombinant protein production: inhibition of cellular capacities for glucose uptake and respiration after induction of a heterologous gene in escherichia coli. Biotechnology and bioengineering 83 (1), 53–64.

A. Picon, M. J. Teixeira de Mattos, P. W. Postma, Apr. 2005. Reducing the glucose uptake rate in Escherichia coli affects growth rate but not protein production. Biotechnol Bioeng 90 (2), 191–200.

T. Selas Castineiras, S. G. Williams, A. G. Hitchcock, D. C. Smith, 2018. E. coli strain engineering for the production of advanced biopharmaceutical products. FEMS microbiology letters 365 (15), fny162.

O. Spadiut, S. Rittmann, C. Dietzsch, C. Herwig, Jan. 2013. Dynamic process conditions in bioprocess development. Engineering in Life Sciences 13 (1), 88–101.

B. Wilms, A. Hauck, M. Reuss, C. Syldatk, R. Mattes, M. Siemann, J. Altenbuchner, 2001. High-cell-density fermentation for production of l-n-carbamoylase using an expression system based on the escherichia coli rhabad promoter. Biotechnology and bioengineering 73 (2), 95–103.

L. Yee, H. Blanch, 1992. Recombinant protein expression in high cell density fed-batch cultures of escherichia coli. Biotechnology 10 (12), 1550–1556.

Antonis Kokossis, Michael C. Georgiadis, Efstratios N. Pistikopoulos (Eds.)
PROCEEDINGS OF THE 33rd European Symposium on Computer Aided Process Engineering
(ESCAPE33), June 18-21, 2023, Athens, Greece

Closed-loop optimization of high-throughput robotic platforms for reproducible bioprocess development

Federico M. Mione[a], Judit Aizpuru[b], Martin F. Luna[a], Pablo Rodriguez Bahamon[b], Jong Woo Kim[b], Ernesto C. Martinez[a,b], M. Nicolas Cruz B.[b]

[a] INGAR (CONICET-UTN), Avellaneda 3657, S3002GJC Santa Fe, Argentina
[b] KIWI-biolab, Bioprocess Engineering, TU Berlin, Ackerstrasse 76, 13355 Berlin, Germany
mariano.n.cruzbournazou@tu-berlin.de

Abstract

To confidently accelerate bioprocess development using high-throughput robotic platforms, parallel experiments should be reproducible whereas the generated data must be FAIR. To this aim, higher levels of automation for planning, redesigning, executing, and monitoring of related experimental-computational workflows are needed. In this work, an automation hierarchy for closed-loop optimization based on Bayesian update, belief propagation and Thompson sampling is proposed. For online redesign of parallel dynamic experiments, asynchronous *rolling out beliefs about beliefs* based on simulated data is used for counterfactual assessment of the effect of alternative (re)design decisions on the distributions of model parameters. A computational implementation of the Bayesian automation hierarchy using Apache Airflow is described. Simulation results obtained using an *E. coli* case study are presented to argue that parallel dynamic experiments made using the proposed Bayesian hierarchy are reproducible whereas experimental-computational workflows and data are FAIR by design.

Keywords: automation hierarchy, closed-loop experimentation, belief propagation, FAIR principles, reproducibility crisis.

1. Problem statement

Bioprocess development requires to explore efficiently huge design spaces for specific bioproducts and involves efficient strain selection, bioprocess optimization, scale-up, and optimal control strategies for reproducible industrial production. The shift towards high-throughput technologies and automation in bioprocess development (Cruz Bournazou et al., 2017; Bromig et al., 2022) demands a more autonomous approach for generating informative data sets to support model-based experimental design in the face of uncertainty. For reproducible bioprocess development and decision-making provenance, experimental data sets used for predicting the dynamic behaviour of genetically modified microorganism and the physiological effect of different combinations of process parameters on the productivity levels after scale-up must be obtained by following FAIR principles (Wilkinson et al., 2016; Mione et al., 2022). A significant impact is expected from using these principles in model-based autonomous experimentation for strain selection and the optimization of bioprocess parameters based on FAIR data, experimental protocols, and computational workflows. In this work, an automation hierarchy for integrating Bayesian inference with experimental (re)design and execution control using probabilistic models and belief propagation is proposed.

2. Bayesian automation hierarchy

First and foremost, automating the execution of an experiment requires that the specific objective for data gathering is clearly stated, the *a priori* available knowledge be explicit modeled, and the initial uncertainty quantified. Without any loss of generality, it is assumed here that a causal probabilistic model (see Martinez et al, 2021 for details) of the microorganism dynamic behavior is going to be used to guide experimental off-line design, online redesign, execution monitoring and control. The aim of the asynchronous parallel experiments would be to gather FAIR data to increase the model predictive power so that biomass production is maximized. Alternatively, the objective of the experiment would be strain screening for physiological robustness after scaling up. Gathering data for model selection would also be the aim of an experiment. In this work, it is assumed that the a priori uncertainty is summarized in the means and variances corresponding to the (normal) distributions of the parameters for the causal dynamic model.

A schema of the proposed automation hierarchy in the chosen Bayesian setting is given in Fig. 1. The top layer deals with off-line design where the *a priori* optimal sequence of control actions for a given objective is calculated alongside with the expectations regarding observations based on prior distributions for model parameters. In the middle layer, as new data is available, the *posterior* distributions are revised using variational Bayesian inference. Then, the optimal sequence of redesign decisions is re-estimated and future data expectations, which are needed to monitor the experiment execution, are updated accordingly. In the bottom layer, a predictive control strategy is used to safely implement (re)design decisions by properly handling both physiological and platform-related constraints using (stochastic) model predictive control.

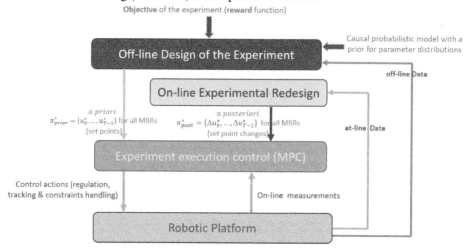

Figure 1. Bayesian hierarchy for closed-loop optimization of robotic platforms

At their three layers, the unifying approach is that by considering the counterfactual consequences of alternative (re)design decisions in generating alternative sampled data, belief propagation (Friston et al., 2021; Martinez et al., 2021) is used to update posterior distributions about the evolution of hidden states and the estimated sequence of optimal redesign decisions for each MiniBioReactor (MBR) in the platform based on simulated sampled data and Thompson sampling of the posterior distributions for model parameters. This recursive form of propagating beliefs over the shrinking horizon of a dynamic experiment effectively implements a deep tree search over alternative sequences of optimized (re)design decisions and their corresponding expectations regarding state evolution and sampled data at different times for dynamic parallel experiments.

For asynchronous *rolling out beliefs about beliefs* in each MBR, the same recursive procedure in Fig. 2 is used for both off-line experimental design and on-line redesign as new observations from all MBRs are available to update the distributions $f(\boldsymbol{\theta})$ for model parameters. The internal loop implements a forward pass where, for a given MBR, a *prior* density is increasingly converted into a *posterior* density for model parameters upon simulated data in a rollout where a stagewise greedy design procedure based on Thompson sampling (Russo et al., 2018) is recursively implemented. At each redesign iteration k, a decision is chosen to implement in the ith MBR by solving an optimization problem for the expected look-ahead reward based on the $r_t(\hat{y}_{t+1}^i)$ computed from simulated data $\hat{y}_{t+1}^i$ for alternative design decisions in a bounded set Ω and accounting for physiological and platform constraints $C(x_{i,t}, u_{i,t})$:

$$u_{i,t}^* = argmax_{u_i \in \Omega} \, r_t(y_{t+1}^i), \text{ subject to: } \quad C(x_{i,t}, u_{i,t}) \geq 0 \tag{1}$$

Inputs: $T, K, x_{i,0}$, prior $f(\boldsymbol{\theta})$, n_{MBR}, state evolution and observation functions f, g

  ▷ **For** $t = 1$ to $T - 1$

    Infer current state $\hat{x}_{i,t}$ using $u_{i,t-1}^*, \hat{x}_{i,t-1}$ and Thompson *Sampling* of prior $f(\boldsymbol{\theta})$

      ▷ **For** $k = 1$ to K

        $\hat{f}(\boldsymbol{\theta}_k) = f(\boldsymbol{\theta})$

        While $t < T$ (Forward Pass)

          ▷ **For** $i = 1, 2 \ldots n_{MBR}$ **do**

          Thompson *Sampling* of the prior $\hat{f}(\boldsymbol{\theta}_k)$: $\Theta_k = \boldsymbol{TS}[: \hat{f}(\boldsymbol{\theta}_k)]$

          $u_{i,t}^k = argmax_{u_i \in \Omega}(r_t(\hat{y}_{t+1}^i)|\Theta_k, \hat{x}_{i,t})$, S. to: f, g & $C(x_{i,t}, u_{i,t}) \geq 0$

          Simulate redesign using $u_{i,t}^k$ and predict $r_{t+1}(\widehat{y_{t+1}^k}|u_{i,t}^k, \hat{x}_{i,t})$

          Update prior: $\hat{f}(\boldsymbol{\theta}_k) \leftarrow \hat{f}(\boldsymbol{\theta}_k | u_{i,t}^k, \hat{x}_{i,t})$ using $(u_{i,t}^k, \hat{x}_{i,t})$

          Accumulate reward: $\mathcal{R}_k = \mathcal{R}_k + r_{t+1}(\widehat{y_{t+1}^k}|u_{i,t}^k)$

          ▷ **End for**

        End while

        Define the redesign policy: $\pi_t^k = \{u_t^k, \ldots, u_{T-1}^k\}$ with its corresponding $\mathcal{R}_k$

      ▷ **End for**

    Rank policies $\pi_{i,t}^k, k = 1, \ldots, K$, using $\mathcal{R}_k$

    Select the best policy $\pi_{i,t}^* = \{u_{i,t}^*, \ldots, u_{i,T-1}^*\}$ with the highest $\mathcal{R}_k$

    Redesign **the experiment** in the MBR i using only $u_{i,t}^*$ and measure $y_{i,t+1}$

    Update prior: $f(\boldsymbol{\theta}) \leftarrow f(\boldsymbol{\theta} | u_{i,t}^*, y_{i,t+1})$ using data $(u_{i,t}^*, y_{i,t+1})$ from all MBRs

  ▷ **End for**

Outputs: $\pi^* = \{u_{i,1}^*, \ldots, u_{i,T-1}^*\}, (u_{i,t}^*, y_{i,t+1}) \, i = 1, \ldots n_{MBR}, f(\boldsymbol{\theta})$

Figure 2. Asynchronous rolling out beliefs about beliefs for online redesign in an MBR.

where r is a *reward* function for observations and their underlying hidden states which accounts for the specific objective of the experiment. The predicted observation $\hat{y}_{t+1}$ and the estimated optimal action $u_{i,t}^*$ are then used to update the posterior distribution for model parameters $\hat{f}(\boldsymbol{\theta}_k)$ to be used at $t+1$. Thompson sampling is again applied and using

Eq. (1) the action $u_{i,t}^*$ is calculated and the corresponding simulated observation $\hat{y}_{t+2}$ is computed, and so on and so forth. This forward pass ends when $u_{i,T-1}^*$ is computed and the *optimal* redesign sequence $\pi_{i,t}^* = \{u_{i,t}^*, \ldots, u_{i,T-1}^*\}$ is completely defined for a given MBR. For each iteration of the outer loop, a different sequence π_k is computed. The generated sequences are then ranked based on their corresponding cumulative rewards $\mathcal{R}_k$ over the planning horizon $[t, ..T]$. From the top ranked redesign policy $\pi_{i,t}^*$, only the first action $u_{i,t}^*$ is used for redesigning the dynamic experiment in the *ith* MBR and, at the next sampling time t+1, the observation $y_{i,t+1}$ is obtained. Using sampled data $(u_{i,t}^*, y_{i,t+1})$ from all MBRs, the joint posterior distribution $f(\boldsymbol{\theta})$ for the parameters of the probabilistic model is updated using variational Bayesian inference (based on the Laplace approximation) and the master (external) loop begins re-estimating the optimal redesign sequence $\pi_{i,t+1}^*$ for each MBR from t+1 until the end of the experiment at time T using belief propagation. It is worth noting that online redesign in each MBR is only coupled with others' MBRs redesign by the global update of the posterior distribution $f(\boldsymbol{\theta})$.

3. Implementation in Apache Airflow

The FAIR principles (Wilkinson et al., 2016), describe a set of requirements for data management and stewardship to make research data **Findable, Accessible, Interoperable,** and **R**eusable. The FAIR principles are a set of guidelines that aim to maximize the value and usefulness of experimental data and highlight the importance of making experimental protocols and workflows digital objects findable and reusable by others. To this aim, the representation of the automation hierarchy in Fig. 1 and the (re)design algorithm in Fig. 2 using directed acyclic graphs (DAGs) foster their implementation in the Apache Airflow ecosystem to enforce FAIR principles by design in high-throughput experimentation facilities. For details on making experimental-computational workflows FAIR in Apache Airflow (Harenslak and de Ruiter, 2021), the interested reader is referred to (Mione, et al., 2022) for details.

In Fig. 3, an abstract DAG is used to represent the hierarchy in Fig. 1 as a pipeline for implementing computational tasks such as update of parameter distributions as soon as new sampled data are available from all MBRs, followed by asynchronous experimental redesign using the algorithm in Fig. 2 and then predictive control of redesign changes to the set-points for controlled variables to guarantee feasibility. It is worth noting the importance of execution control nodes since due Thompson sampling changes to the set-points of control loops may violate physiological constraints. For FAIRness and reproducibility, is key to save all seeds used for random number generation in posterior sampling at each iteration k in the belief propagation algorithm for each MBR. As a result, the parameter values used for solving optimization problems in Eq. (1) will be known.

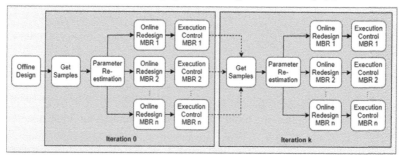

Figure 3. The Bayesian hierarchy with belief propagation as an Airflow pipeline.

4. Case study

The proposed method was tested using artificial data generated based on the model proposed in Nickel et al. (2017), consisting of six differential equations and eighteen parameters. Eight MBRs for Escherichia coli fed-batch cultivations were run over a six-hour period. Samples are taken every thirty minutes and biomass (X), glucose (Glc) and acetate (Ac) concentrations are measured at-line, whereas the dissolved oxygen tension (DOT) in each MBR is measured continuously using an online sensor. Based on new data, the sequence of glucose pulses for each MBR are computed to implement model-based optimized feeding strategies (see Fig. 2) so that the experiment is redesigned online to maximize the information content. The aim of the experimental design is to find *a posteriori* an optimal feeding profile that maximizes the amount of biomass obtained at the end of the run and prevents the DOT to drop below 20% throughout. For details regarding all the assumptions, models, software packages, parameters and hyper-parameters used to implement the automation hierarchy, the reader is referred to the repository:
https://git.tu-berlin.de/bvt-htbd/kiwi/tf2/experimental-computational-workflows/-/tree/escape
Once the parallel eight experiments have been completed, the probabilistic model is the updated using the complete data set. The Bayesian nature of the resulting probabilistic model allows for a chance-constrained formulation of the optimization problem

$$u^* = argmax_{u \in \mathcal{U}} E[m_{X,end}] \qquad (2.1)$$

Subject to $\Pr(DOT_{min} > 20) > 95\%$ $\qquad (2.2)$

where $m_{X,end}$ is the biomass amount obtained (in g/L) at the end of a testing experiment and DOT is in percentage of saturation. The decision variables are the duration of the batch phase and the feeding profile. Some results obtained using the *posterior* parameter distributions are shown in Fig. 4 to highlight the effectiveness of the proposed approach.

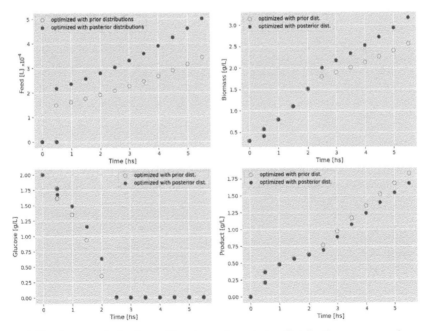

Figure 4. Feed rate optimization with prior and posterior distributions compared.

As it is shown, the *a priori* optimized feeding profile is more cautious than the one with the probabilistic model using the *a posteriori* distributions of model parameters. For the later, a higher amount of biomass is obtained which is aligned with the experiment objective which bias data gathering in the operating region where biomass is maximized whereas the dissolved oxygen tension is safely kept above the threshold value of 20% during the optimized run. The effect of data bias is clear in Fig. 5 which describes the evolution of the distributions for some sensitive model parameters as the iteration count k increases. The parameter distributions for Iter. #1 are the priors before the experiment.

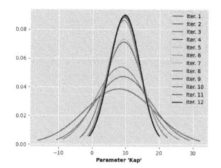 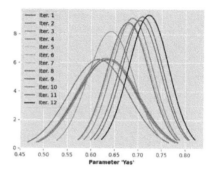

Figure 5. Evolution of two distributions of model parameters during the experiment.

5. Concluding remarks

A new automation hierarchy for closing the loop in high-throughput robotic platforms was proposed. The implementation of the hierarchy in Apache Airflow makes possible to reproduce the computational experiments and make sampled data FAIR. Finally, the asynchronous belief propagation algorithm in Fig. 2 is best suited for parallel processing.

References

L. Bromig, N. von den Eichen, D. Weuster-Botz, 2022, Control of parallelized bioreactors I: dynamic scheduling software for efficient bioprocess management in high-throughput systems, Bioprocess and Biosystems Engineering, 45,1927–1937.

M. N. Cruz Bournazou, T. Barz, D. B. Nickel, D. C. Lopez Cárdenas, F. Glauche, A. Knepper, P. Neubauer, 2017, Online optimal experimental re-design in robotic parallel fed-batch cultivation facilities. Biotechnol. Bioeng., 114, 610–619.

K. Friston, L. Da Costa, D. Hafner, C. Hesp, T. Parr, 2021, Sophisticated inference, Neural Computation, 33, 713–763.

B. Harenslak, J. de Ruiter, 2021, Data Pipelines with Apache Airflow. Manning Publications.

E. C. Martinez, J. W. Kim, T. Barz, M. Nicolas Cruz B., 2021, Probabilistic modeling for optimization of bioreactors using reinforcement learning with active inference, Proc. of ESCAPE-31 2021, 50 (Part A), 419-424.

F. M.Mione, A. N. Silva, M. F.Luna, M. N. Cruz B., E. C.Martinez, 2022, Managing Experimental-Computational Workflows in Robotic Platforms using Directed Acyclic Graphs, Computer Aided Chemical Engineering, 49, 1495-1500.

D. B. Nickel, M. N. Cruz-Bournazou, T. Wilms, P. Neubauer, & A. Knepper, 2017, Online bioprocess data generation, analysis, and optimization for parallel fed-batch fermentations in milliliter scale, Engineering in the Life Sciences, 17, 11, 1195-1201.

D. J. Russo, B. Van Roy, A. Kazerouni, Ia. Osband, Z. Wen, 2018, A Tutorial on Thompson Sampling, Foundations and Trends in Machine Learning, 11, 1, 1-96.

M. D. Wilkinson, et al., 2016, The FAIR guiding principles for scientific data management and stewardship. Nature 3:160018 DOI 10.1038/sdata.2016.18.

Antonis Kokossis, Michael C. Georgiadis, Efstratios N. Pistikopoulos (Eds.)
PROCEEDINGS OF THE 33rd European Symposium on Computer Aided Process Engineering
(ESCAPE33), June 18-21, 2023, Athens, Greece

Dynamic microbial-community metabolic modeling for yogurt fermentation based on the metagenome of starter culture

Sizhe Qiu[a,b], Zhijie Yang[b], Hong Zeng[b], Bei Wang[b*], Aidong Yang[a*]

[a] *Department of Engineering Science, University of Oxford, Oxford OX1 3PJ, United Kingdom*

[b] *Beijing Advanced Innovation Center for Food Nutrition and Human Health, Beijing Technology & Business University (BTBU), Beijing 100048, China*

Corresponding author

wangbei@th.btbu.edu.cn (B. Wang)

aidong.yang@eng.ox.ac.uk (A. Yang)

Abstract

Genome-scale metabolic models (GSMMs) and flux balance analysis (FBA) have been extensively used to model and design bacterial fermentation. However, models that can simulate co-culture fermentation dynamically with quantitative accuracy are relatively scarce. To investigate metabolic interactions in yogurt starter culture of *Streptococcus thermophilus* (ST) and *Lactobacillus delbrueckii subsp. bulgaricus* (LB), this study built a dynamic community-level metabolic model based on metagenomic data. We first assessed the quality of the model by comparing the predicted growth curve with reference experimental data, and then used it to simulate the impact of different initial ST:LB ratios and search for the optimal composition of starter culture for efficient acidification for yogurt fermentation. The modeling pipeline presented in this work provides a basis for an efficient tool for computer-aided process design of the production of fermented dairy products, contributing to the development of precision fermentation in the food industry.

Keywords: Metagenomics; Fermentation; Flux balance analysis; Genome scale metabolic model; Proteome allocation.

1.   Introduction

Yogurt is an important fermented dairy food, traditionally made by lactic acid bacteria (LAB). In industrial production of yogurt, the control of the fermentation process, in terms of acidification and flavor/probiotic compound production, largely depends on the composition of the starter culture. To investigate and engineer the microbial community of yogurt starter culture, a community-level metabolic model of LABs used in the production of yogurt is needed. Currently, several genome scale metabolic models (GSMMs) for dairy-origin LABs have already been reconstructed, and a dynamic co-culture metabolic model for cheese starter culture using those GSMMs has been built (Özcan et al., 2021). However, there is still a lack of metagenome-scale metabolic models (different from GSMMs of certain strains) that can simulate the growth and metabolism of the actual (as opposed to an assumed) LAB community in yogurt fermentation.

Furthermore, existing models could not simulate unique inter-species interactions in yogurt fermentation. In this work, we built a dynamic community-level metabolic model with reconstructed GSMMs of major species identified in the starter culture, i.e., *Streptococcus thermophilus* (ST) and *Lactobacillus delbrueckii subsp. bulgaricus* (LB), using the information obtained from metagenomic analysis. We then showed how the model can be used to design the control strategy of the yogurt fermentation process.

2. Methodology

Building the dynamic community-level metabolic model of the yogurt starter culture comprised two steps: 1. from metagenome to annotated protein coding genes; 2. from coding genes to GSMMs (**Figure 1**). With GSMMs, dynamic flux balance analysis (dynamic FBA, or dFBA) was implemented to simulate bacterial growth and metabolism.

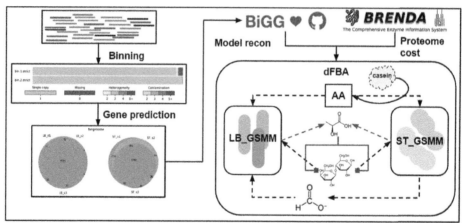

Figure 1. the workflow diagram of metabolic modeling from metagenomic analysis to GSMM reconstruction and dFBA.

2.1. *Metagenome assembly, binning, and protein coding gene identification*
Three parallel DNA samples of the commercial yogurt starter culture were sequenced by Illumina PE150. Reads were filtered for high quality ones, and then assembled by MEGAHIT.

MetaBAT2, MaxBin2 and CONCOCT were used to obtain individual genomes from the assembled metagenome. Bins were then processed using refinement and reassembly modules in metaWRAP. The final outputs are individual genomes of high quality, assessed by CheckM. GTDB-Tk performed taxonomic classification for refined individual genomes to identify major species.

Protein-coding gene prediction for each individual genome was done by Prodigal, generating the nucleotide sequence and translated protein sequence of each gene. Then, functional annotation provided information on cellular pathways, carbohydrate active enzymes and biosynthetic gene clusters.

2.2. *Reconstruction of genome-scale metabolic models*
Non-redundant protein sequences were filtered from concatenated protein sequences of different samples and used as the input for automatic GSMM reconstruction by CarveMe.

In addition to protein sequences, other inputs were chemically defined media (CDM) for gap filling and the universal bacterial template. The CDM was adopted from the co-culture metabolic model of cheese starter (Özcan et al., 2021). The universal bacterial template used for reconstruction was specialized for gram-positive bacteria.

The growth function (normalized to be the formation of 1gDW biomass) was adjusted based on biomass composition reported in previous studies and existing GSMMs of the same species. The non-growth associated ATP maintenance was set to be 0 to avoid infeasible solutions. Based on functional annotation of coding genes and existing GSMMs, reactions included erroneously were removed and missing reactions were added. In addition, to model the proteolysis activity in the co-culture system, a self-defined reaction decomposing casein protein into amino acids was added to the GSMM:

$$Casein \rightarrow \alpha \, alanine + \beta \, arginine + \ldots + \varepsilon \, valine$$

Stoichiometric coefficients in the proteolysis reaction were approximated from average amino acid distribution in the casein protein.

2.3.   *Dynamic flux balance analysis and proteome allocation constraint*

To simulate the growth of bacteria and the production of target metabolites in time, dFBA was adopted (Mahadevan et al., 2002). The intracellular metabolic fluxes, v_j, were computed by parsimonious FBA, maximizing the growth rate while minimizing the total sum of fluxes, based on the assumption that the cell minimizes the use of enzyme catalyzed reactions due to limited recourse (Eqs. 1, 2), subject to mass conservation (Eq. 3), in which S_j represented the stoichiometric matrix. Inter-species exchanges were not modeled directly, the concentration change of metabolites and biomass in the extracellular space was modeled by differential equations to account for biomass accumulation (Eq. 4) and exchange fluxes of metabolites M from major species (Eq. 5).

$$Maximize \; v_{j,Growth} \; \forall j \in major \; species \tag{1}$$

$$Minimize \; \sum_{i=0}^{\#reactions} v_{j,i} \tag{2}$$

$$S_j v_j = 0 \tag{3}$$

$$\frac{d[Biomass_j]}{dt} = v_{j,Growth}[Biomass_j] \tag{4}$$

$$\frac{d[M]}{dt} = \sum_{j=0}^{\#species} v_{j,EX\_M\_e}[Biomass_j] \tag{5}$$

Furthermore, proteome allocation was implemented to constrain reaction fluxes of central metabolism to model the flow of carbon source uptake (Zeng et al., 2020). Proteome was divided into sectors of inflexible housekeeping (Q), anabolism (A), transportation (T), catabolism (C), and the free sector. The upper bound of all flexible sectors combined was assumed to be 50% of the total proteome (Eqs. 6, 7). The proteome cost on each reaction was computed as the flux divided by the multiplicative product of enzyme activity, a_i, and saturation degree, σ_i (Eq. 8). The saturation degrees, σ_i, were assumed to be 0.5, except for the lactose transporter, which was considered as fully saturated by abundant lactose, $\sigma_{LT} = 1$. The activity of the ribosome, $a_{ribosome}$, for the anabolism sector was adopted from literature (Regueira et al., 2020). Other enzyme activities were obtained from BRENDA Enzyme Database. For the transportation of amino acids, the flux was constrained by michaelis-menten equation ($v_{max}\frac{[S]}{[S]+K_m}$), average v_{max} and michaelis-menten constant K_m were set based on parameters obtained from BRENDA and SABIO-RK.

$$\frac{P_{tot}}{1\ gDW} = p_Q(\sim 50\% \frac{P_{tot}}{1\ gDW}) + p_C + p_A + p_T + p_{Free} \tag{6}$$

$$0 < p_C + p_A + p_T \leq 50\% \frac{P_{tot}}{1\ gDW} \tag{7}$$

$$p_C = \sum \frac{v_i}{\sigma_i a_i}; \ p_A = \frac{v_{Growth}}{a_{ribosome}}; \ p_T = \frac{v_{LT}}{\sigma_{LT} a_{LT}} + \frac{v_{lacT}}{\sigma_{lacT} a_{lacT}} + \frac{v_{acT}}{\sigma_{acT} a_{acT}} \tag{8}$$

The activity of lactose uptake incorporated inhibition by undissociated lactate (LacH), the final product of glycolysis in anaerobic conditions. The exponential decay equation (Eq. 9) to model the inhibition of lactose transporter activity was adopted from Bouguettoucha et al. (2011). The concentration of undissociated lactate was computed with Henderson-Hasselbalch equation (Eq. 10), and pH was approximated as a linear function of lactic acid concentration, $pH = C_1[Lac] + C_2$.

$$a_{LT} = a_{LT}^0 e^{(-k_{LacH}[LacH])} \tag{9}$$

$$[LacH] = \frac{[Lac]}{10^{pH-pK_a}} \tag{10}$$

3.   Results

3.1.   Major species identified and status of reconstructed GSMMs

The taxonomic classification revealed that *Streptococcus thermophilus* (ST) and *Lactobacillus delbrueckii subsp. bulgaricus* (LB) were two major species in the starter culture. After binning and refinement, individual genomes of ST and LB showed good completeness and low level of contamination. There were 2499 and 1801 non-redundant protein coding genes predicted from ST and LB's genomes. In reconstructed GSMMs, reactions assigned with gene-protein-reaction relations were around 60% in both models. About 25% gap-filled reactions were added into the model for a complete metabolic network. The rest were boundary reactions for metabolite exchange.

3.2.   Simulation of growth and metabolism of ST/LB co-culture during yogurt fermentation

To assess the predictive power of the dynamic community-level metabolic model, dynamic simulation was performed using initial conditions of the fermentation system from reference data (Oliveira et al., 2012). For the biomass of ST and LB, the initiation of exponential phase and transition to stationary phase were adequately captured by the model (**Figure 2A**). Besides, our model also predicted the consumption of lactose and production of lactic acid with fairly good accuracy (**Figure 2B**).

In addition to lactic acid, acetic and formic acids are also important fermentative products of glycolysis. The predicted concentration curve of acetic acid showed a similar trend with that of lactic acid (**Figure 2C**). The accumulation of formic acid produced by ST initiated the exponential growth of LB, as it activated LB's synthesis of purine, whose concentration was low in milk. In return, the activation of growth and metabolism of LB with strong proteolytic activity enhanced the growth of ST by supplementing amino acids and small peptides, which was also limited in milk (**Figure 2AC**). Apart from end products of glycolysis, the model also predicted the secretion of glycolaldehyde, 4-hydroxy-benzyl alcohol, 3-methylbutanoic acid and succinic acid, among which 3-methylbutanoic acid appeared unassociated with cellular growth (**Figure 2DEF**).

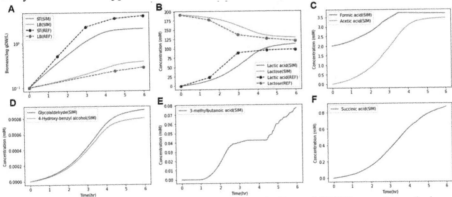

Figure 2. Model predictions of growth and metabolism of ST/LB co-culture during yogurt fermentation (solid line: simulation results, dashed line: reference experimental data; simulation results were obtained with a ST to LB ratio of 1, both 0.1g/L).

3.3.   Prediction of the impact of different ST:LB ratio on the fermentation behavior
Perturbations on the initial ratio of ST and LB in yogurt starter culture was conducted to investigate its impact on the fermentation behavior and search for the optimal starter culture composition. The simulated concentration profile of lactic acid indicated that the acidification rate (the rate of lactic acid accumulation) by starter cultures of different ST:LB ratios ranked as follows: ST:LB = 1 > ST:LB = 10 > ST:LB = 100 > ST:LB = 0.1 > ST:LB = 0.01 (**Figure 3A**). When the acidification rate is selected as the sole objective to maximize, an optimal ST:LB ratio is predicted to be between 1 and 10. The concentration profiles of formic acid and growth curves for different ST:LB ratios showed that LB's exponential growth would be activated when formic acid accumulated to a certain level (**Figure 3BC**). The supplement of nitrogen source (amino acids and small peptides) by LB was also necessary for ST. When ST:LB increased from 1 to 10, the proteolytic activity became weaker due to the lower biomass of LB, and thus the growth of ST was inhibited, leading to reduced acidification achieved in 6 hours.

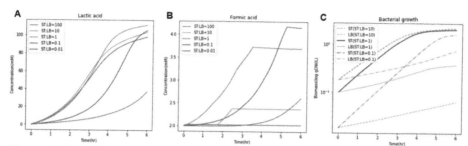

Figure 3. Simulation of yogurt fermentation by starter cultures of different ST:LB ratios (from 100 to 0.01, fixed total starting biomass of 0.2 g/L).

4.   Conclusion

Overall, this work presented the metabolic reconstruction of the classical yogurt starter culture, i.e., ST and LB, and provided a dynamic community-level GSMM modeling approach. The proposed model is able to capture the bacterial growth, the acidification of yogurt via lactic acid and predict active fluxes of flavor compounds like acetic acid, methyl butanoic acid and succinic acid during the yogurt fermentation process. Although several limitations remain to be overcome (e.g., the lack of calibration with experimentally determined biomass composition and ATP requirements), the model was shown to have the potential to offer an efficient tool to guide engineering decisions in the food industry, which could be used to address issues such as the optimal initial biomass ratio of ST and LB to maximize the rate of acidification and possibly other process targets.

References

E. Özcan, M. Seven, B. Şirin, T. Çakır, E. Nikerel, B. Teusink, E. T. Öner, 2021, Dynamic co-culture metabolic models reveal the fermentation dynamics, metabolic capacities and interplays of cheese starter cultures, Biotechnology and Bioengineering, Vol. 118, Issue 1, Pages 223–237

R. Mahadevan, J. S. Edwards, F. J. DoyleIII, 2002, Dynamic Flux Balance Analysis of Diauxic Growth in Escherichia coli, Biophysical Journal, Vol. 83, Issue 3, Pages 1331-1340

H. Zeng, A. Yang, 2020, Bridging substrate intake kinetics and bacterial growth phenotypes with flux balance analysis incorporating proteome allocation, Scientific Reports, Vol. 10, Issue 1, No. 4283

A. Regueira, J. L. Rombouts, S. A. Wahl, M. Mauricio-Iglesias, J. M. Lema, R. Kleerebezem, 2020, Resource allocation explains lactic acid production in mixed-culture anaerobic fermentations, Biotechnology and Bioengineering, Vol. 118, Issue 2, Pages 745-758

A. Bouguettoucha, B. Balannec, A. Amrane, 2011, Unstructured Models for Lactic Acid Fermentation – A Review, Food Technology and Biotechnology, Vol. 49, Issue 1, Pages 3–12

R. P. S. Oliveira, B. R. Torres, P. Perego, M. N. Oliveira, A. Converti, 2012, Co-metabolic models of Streptococcus thermophilus in co-culture with Lactobacillus bulgaricus or Lactobacillus acidophilus, Biochemical Engineering Journal, Vol. 62, Pages 62-69

Antonis Kokossis, Michael C. Georgiadis, Efstratios N. Pistikopoulos (Eds.)
PROCEEDINGS OF THE 33rd European Symposium on Computer Aided Process Engineering
(ESCAPE33), June 18-21, 2023, Athens, Greece

Multilevel Discrete Modelling of Microbial Cell Cultures using Population Balances

Menwer Attarakih[a], Hans-Jörg Bart[b], Khaled Rawajfeh[a]

aThe University of Jordan, Queen Rania Str., Amman, 11942, Jordan
bThe University of Kaiserslautern, Gottlieb-Daimler Str., 67663, Germany
m.attarakih@ju.edu.jo

Abstract

Microbial cell population balance models are used because of culture heterogeneity due to complex intercellular phenomena governed by cell growth, substrate utilization, and product formation. To solve these models, an efficient multilevel system approach based on conservation of cell distribution moments is introduced. This conservation is propagated along the characteristic curves of these hyperbolic conservation laws. This is assured by the careful design of numerical discrete schemes that pass the conserved discrete information from the initial state to its final extracellular state. The discrete model on each layer is derived with a guarantee to satisfy the conservation of cell total number and the mean cell biomass. Many reference solutions based on the Chebyshev-QMOM, the maximum entropy method, and the Lax-Wendroff finite difference scheme are used to validate the developed model. When compared to these methods, the model exhibits extreme accuracy and rapid convergence at a low CPU time cost.

Keywords: Microbial cells; Population Balances; Multilevel; Discrete modelling.

1. Introduction

The continuous population balance equation (PBE) with particle nucleation, growth, and breakage is a nonlinear integral-partial differential equation. For microbial cell cultures, the PBE presents a framework to describe the complex phenomena manifested by cell growth, division, substrate utilization, and product formation. With no general analytical solution, this calls for numerical modeling to understand the cell culture behavior and to predict, optimize, and control cell growth in biochemical reactors (Mantzaris et al., 2001; Ramkrishna et al., 2014; Jerono et al., 2021).

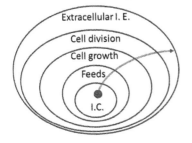
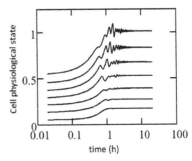

Figure (1): Left-Multilevel system discretization approach for microbial cell dynamics. Right-system characteristic curves as predicted by ChQMOM using 7 quadrature nodes.

Since the heterogeneity of microbial cultures results from cell size and its intercellular concentrations of DNA, we confine ourselves to the mono-variate cell probability density function f (x, t) with cell mass (x) as its internal coordinate. These cells are embedded in a continuous extracellular environment with a single substrate (S) (Durr & Buk, 2020; Jerono et al., 2021). Our numerical modeling method for this system starts by introducing a multilevel discrete approach that is evolved in five evolutionary layers, as shown in Fig. 1. This layer structure reflects the hyperbolic nature of the discrete conservation and population balance laws. The first layer in Fig. (1-Left) shows the initial conditions and the external feeds, which are considered an essential part of the internal inconsistency problem arising from the transformation of the continuous PBE into its discrete form. This is to prevent the propagation of internal inconsistency along the characteristic lines (Fig. 1-Right) associated with the growth, division, and extracellular information exchange layers, which if ignored would increase the computational cost. To explain this, given a discrete initial state vectors with cell boundaries x and centers χ which satisfy $x_i < \chi_i < x_{i+1}$ and initial cell population number concentration $N_i(0)$, i = 1, 2, …M. In this discrete framework, the integral properties such as total cell number and mass will never be the same as those calculated from its continuous probability density function except for only one integral property (Ramkrishna, 2000). For example, $N_i(0) = \int f(x,0)dx$ but $\Sigma x_i N_i(0) + \varepsilon = \int x f(x,0)dx$ where ε is called the error of moving the state from its continuous to its discrete states. Unfortunately, this error is of cumulative nature and is propagated from the initial state layer to the higher layers with increasing amplitude. Its elimination needs infinitely large number of discrete states of uniform grid: $x_i = i x_{min}$. However, this error can be eliminated for two selected integral properties starting from the initial state layers and moving forward to the growth and division layers. As the total number and mass are the two most important cell transport properties, we derive our discrete PBEs that satisfies these properties in each layer within this multilevel approach as shown in Fig.(1-Left).

2. Mathematical model

The heterogeneity of the cell population is adequately modelled using the population balance equation (Ramkrishna, 2014). In this equation, the cell density function f(x,t) accounts for variances in the cell population in a homogeneous physical space (CSTR):

$$\frac{\partial f(x,t)}{\partial t} + \frac{\partial [G(x,t)f(x,t)]}{\partial x} = -D(t)f(x,t) + R\{f(x,t),S\} \tag{1}$$

where x is the physiological state of single cell (biomass components) which depends on the intercellular growth rate G(x,S), where S is the single substrate concentration in the extracellular continuous medium, and D is the dilution rate. R is a volumetric source which represents the net number of cells per unit biomass volume due to birth of cells with the physiological state x because of binary division of all mother cells of size y:

$$R\{f(x,t),S\} = -\Gamma(x,S)f(x,t) + 2\int_{x}^{x_{max}} \Gamma(y,S)\beta(x,y)f(y,t)dy \tag{2}$$

In this equation $\Gamma(x,S)$ is the cell division rate and $\beta(x,y)$ is the cell partition probability function for a daughter cell of physiological sate x given a mother cell of state y. This function should satisfy the mass and number conservation conditions as well as the symmetrical properties (Ramkrishna, 2000). The cell PBEs (1) and (2) are coupled with the dynamic abiotic environment through the substrate mass balance as shown by Eq.(3). In this equation the source term represents the net flow of substrate through reactor boundary along with the total loss of substrate as utilized by the microbial cells.

$$\frac{dS}{dt} = D(S_f - S) - \int_{x_{min}}^{x_{max}} q(x,S)f(x,t)dx \qquad (3)$$

$$q(x,S) = \frac{1}{Y}\left(\frac{\mu_{max}S}{K_s + S}\right)x \qquad (4)$$

In these equations $q(x,S)$ is single cell consumption rate of a substrate (S), Y is a constant biomass product yield, μ_{max} is the maximum growth rate and K_s is the half saturation substrate concentration. The growth rate is given by $G(x,S) = Yq(x,S)$, and $\Gamma(x,S) = \gamma(x)G(x,S)$ where $\gamma(x)$ is expressed in terms of Gaussian probability density function (Mantzaris et al., 2001). The initial conditions are $f(x,0) = f_0(x)$ and $S(0) = S_0$ while S_f is the substrate feed concentration and the boundary condition is $f(x_{min},t) = f(x_{max},t) = 0$.

3. Multilevel model discretization

A multilevel model discretization approach that is consistent with that shown in Fig.(1-Left) is used here to ensure the consistency of the discrete model when compared to the continuous one as given by Eqs.(1-4).

3.1. Level 1: Initial conditions and feed discretization
The cell internal property is discretized into M contiguous numerical cells with geometric progression $x_i = x_{min}[r]^{i/M}$, with $r > 1$ and cell centers $\chi_i = (x_i + x_{i+1})/2$, $i = 0,1, \ldots M$.

$$N_i^0 = \alpha_0 N_i(0) + \beta_0 N_{i+1}(0), i = 0,1,.. \qquad (5)$$

where α_0 and β_0 are solutions of two linear equations to conserve the zero and first moments of the cell number concentration functions $f_0(x,0)$ and $f_1(x,0)$ in which 0 refers to the initial condition and 1 refers to the inlet feed distribution. This is a crucial step in this multilevel discrete framework to avoid the propagation of any discrete error along the characteristic lines of the system of Eqs. (1-4).

3.2. Level 2: Cell growth

$$\frac{dN_i(t)}{dt} = \frac{2}{x_i}\frac{1}{r^2 - 1}\left(G(\chi_i,S)N_i(t) - rG(\chi_{i-1},S)N_{i-1}(t)\right), i > 0 \qquad (6)$$

At this level, the discrete cell growth needs to guarantee the conservation of total number and mass concentrations using a new first order upwind scheme as given by Eq.(6). This conservation law can handle smooth and discontinuous numerical fluxes with suitable flux-vector splitting depending on the sign of $G(\chi,S)$ where $r = [x_{max}/x_{min}]^{(1/M)}$.

3.3. Level 3: Cell division

$$\frac{dN_i(t)}{dt} = -\eta_i\Gamma(\chi_i,S)N_i(t) + \sum_{j=0}^{M}\sigma_j\Gamma(\chi_j,S)\pi_{i,j}N_j(t), i = 0,1,\ldots M \qquad (7)$$

This equation is the discrete counterpart of Eqs.(1) and (2) without growth term where the loss and formation terms in Eq.(7) are modified by placing the two discrete functions (η and σ) to conserve exactly the zero and first moments of the cell population. This ensures passing the conserved information to higher levels without errors induced by the inconsistency problem. The two functions σ and η tends to unity as the grid is refined showing the consistency of these functions with the mean-value theorem of integrals over

$\beta(x,y)$. The discrete function $\pi_{i,j}$ calculates the total i^{th} fraction of daughter cells at any given mother of size y_j.

3.4. Level 4: Extracellular dynamics

$$\frac{dS(t)}{dt} = D(t)\big(S_f - S\big) - \sum_{j=0}^{M} q\big(\chi_j, S\big) N_j(t), i = 0,1,\dots M \tag{8}$$

This level provides a gate to pass the conserved discrete information from level 4 to the cell continuous environment and then returns any information from the continuous phase to the discrete one. The final discrete solution is then recovered at any time step as $n(\chi_i, t) = N_i(t)/\Delta x_i, \forall i, i = 0, M$.

4. Results and discussion

The multilevel discrete model that is given by Eqs.(5-8) provides a complete set of ODEs supplemented by consecutive equations for cell division rate, cell partition probability function, and cell intercellular growth rate.

Table (1): Parameters for test problem 2 as reported by Mantzaris et al. (2001).

μ_{max}	K_s	Y	D	S_f	S_0	Gaussian I.C.	
1.0/h	0.2 g/L	0.5 g/g	0.5/h	10 g/L	2 g/L	μ_0	σ_0
Gaussian division probability function				$\mu_1 = 0.575$	$\sigma_1 = 0.125$	0.2875	0.0675

The initial condition and the cell division probability functions are taken as Gaussian densities with mean and standard deviations as shown in Table (1). The cell partition probability function is taken as a beta distribution: $\beta(x,y) = c(q)y^{-1}(x/y)^{q-1}(1 - x/y)^{q-1}$ where c(q) is normalization factor and q = 40 (Mantzaris et al., 2001).

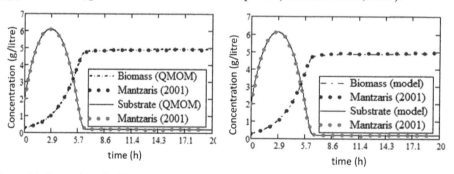

Figure (2): Dynamics of cell growth and division in a CSTR using Chebyshev-QMOM, the Lax–Wendroff scheme (Mantzaris et al., 2001, Toro, 1999) and the present multilevel discrete model.

4.1. Reference solutions

To validate the multilevel discrete model in terms of the conserved moments; namely, the zero (total cell number) and first moments (total cell mass) of the discrete cell distribution, two standard methods are used: The first one is the Chebyshev-QMOM (Upadhyay, 2012) and the conservative finite difference scheme FDS (Mantzaris et al., 2001). The Chebyshev-QMOM (ChQMOM) is superior to the standard QMOM which is based on the Product Difference Algorithm (PDA), and the Direct Quadrature Method Of Moments

(DQMOM) where ill-conditioned linear system of equations is encountered when the number of the quadrature nodes increases.

4.1.1. Chebyshev-QMOM reference solution

The system of Eqs.(1-5) can be written in compact form in terms of the moments of f(x,t) where the unclosed integrals are approximated using the Chebyshev-QMOM. The cell internal property integration domain is $\Omega = [x_{min}, x_{max}] = [0, 1]$ where the resulting system of ODEs was solved using the explicit Adams-Bashforth method with error tolerance = 10^{-3} and number of quadrature nodes Nq equals to seven.

4.1.2. Mantzaris et al. (2002) reference solution

The second reference solution for Eqs.(1-5) is regenerated from the work of Mantzaris et al. (2001) using conservative discretization finite difference schemes for a system of hyperbolic conservation laws. The results from the Lax–Wendroff scheme are generated with the same numerical parameters as shown in Table (1). The results are displayed in Fig.(2) where the number of numerical cells used by this scheme is 500.

4.1.3. Comparison of the present model predictions with reference solutions

Firstly, we compared the substrate and biomass dynamics as predicted by the ChQMOM and the reference solution of Mantzaris et al. (2001). The results are found indistinguishable using seven quadrature nodes which is obvious from the results shown in the left panel of Fig.(2). This result is not surprising since the ChQMOM is well-known for its high accuracy in predicting distribution moments (Attarakih et al., 2012, 2014). Secondly, and as gaining trust in the solution of the present model, the multilevel discrete model (Eqs.(5-8)) is solved in time using the explicit Adams-Bashforth method with error tolerance = 10^{-3} and number of grid points M=50. To ensure the correct propagation of information along the hyperbolic laws characteristic curves (Fig.(2-Right)), the Gaussian I.C. with parameters given in Table (1), is discretized according to the scheme shown in Eq.(5). This solution conserves exactly the total number and mass concentrations.

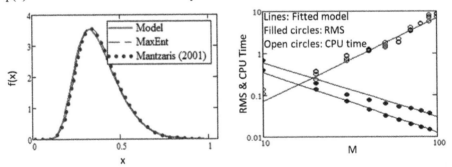

Figure (3): Left-Converged cell distribution. Right- CPU time (s) denoted by open circles and RMS (Root Mean Square) error in zero and 1st moment of cell distribution (Filled circles).

Starting from the first layer (see Fig.(1-Left)) and for M = 50, the factors α_0 and β_0 are found to be 0.99 and 0.01 respectively. This means 99.0 percent of the discrete population is assigned to the ith numerical cell center while only 1.0 percent is assigned to the next larger numerical cell center. This ensures the correct prediction of the population moments even if low number of numerical cells is used (e.g. 15). Based on this, the cell growth layer conserves exactly zero and first moments of the cell distribution which are then propagated correctly to the cell division layer in spite of the numerical diffusion associated with the used first-order upwind schemes (Toro, 1999). In the cell division layer, and as implied by Eq.(7), the computation of the two correction functions (η, σ)

are computed offline and are found to depend on the cell cell partition distribution function ($\beta(x,y)$) and the numerical grid structure. Once these correction vectors are available, the cell division is advanced in time and is then propagated to the extracellular dynamic layer without loss of any global information because of discretization. As a result of this system conservation approach, the biomass and substrate utilization are correctly predicted by our model as compared to the two reference solutions mentioned in sections (4.1.1 and 2). This is clear by referring to the right panel of Fig.(2). In a second level of validation, our model is compared with the FDS of Mantzaris et al (2001) and the Maximum Entropy (MaxEnt) method of Attarakih and Bart (2014) to predict the cell population distribution. Our present model converges rapidly to the FDS and the MaxEntM as the number of numerical cells increases. In Fig.(3)-Left, we used M = 50 as compared to the FDS solution using 500 numerical cells. On the other hand, five moments are used in the MaxEnt method to reconstruct the distribution at final time of 20 hrs. To quantify the convergence characteristics of the present model, the RMS error is calculated using the zero and first moments of the distribution as function of M. Fig.(3)-Right shows rapid decrease in the RMS as function of M with approximate order of $O(M^{-0.4})$. Concerning the computational cost of the present discrete model, and as can be seen from Fig.(3)-Right, the mean CPU time (four runs are used per M value) is increasing as $O(M^2)$.

5. Summary and conclusions

To conclude, the present multilevel discrete model is yet simple, with equations that are elegant in structure, efficient in implementation, and that handle the cell culture dynamics with high accuracy. These equations conserve the propagated information along the characteristic lines of the hyperbolic conservation laws through a hierarchical structure. When compared to the QMOM and the FDS, the present model converges rapidly as the number of numerical cells is increased. With more than four quadrature nodes, the present model computational cost is lower than that of the FDS or the ChQMOM.

Acknowledgments

This work was supported by the German Science Foundation (DFG).

References

M. Attarakih and H.-J. Bart, 2014, Solution of the population balance equation using the Differential Maximum Entropy Method (DMaxEntM): An application to liquid extraction columns. Chemical Engineering Science, 108, 123-133.

M. Attarakih, H. B. Jildeh, M. Mickler and H.-J. Bart, 2012, The OPOSPM as a nonlinear autocorrelation population balance model for dynamic simulation of liquid extraction columns, Computer Aided Chemical Engineering, 31, 1216-1220.

R. Durr and A. Buck, A., 2020, Approximate Moment Methods for Population Balance Equations in Particulate and Bioengineering Processes. Processes, 8, 1-18.

P. Jerono, A. Schaum & T. Meurer, 2021, Moment-based Kalman filter design for cell population balance models in batch fermentation processes. *IFAC-PapersOnLine, 54*, 19-24.

N. V. Mantzaris, P. Daoutidis, & F. Srienc, 2001, Numerical solution of multi-variable cell population balance models: I. Finite difference methods. Comp. Chem. Eng., 25, 1411–1440.

D. Ramkrishna and M. R. Singh, 2014, Population balance modeling: Current status and future Prospects. Annual Review of Chemical and Biomolecular Engineering, 5, 123-146.

D. Ramkrishna, 2000, Population balances:Theory and applications to particulate systems in engineering, Academic Press, San Diego.

E. F. Toro, 1999, Riemann solvers and numerical methods for fluid dynamics, Springer, Berlin.

R. R. Upadhyay, 2012, Evaluation of the use of the Chebyshev algorithm with the quadrature method of moments for simulating aerosol dynamics. Journal of Aerosol Science, 44, 11-23.

Antonis Kokossis, Michael C. Georgiadis, Efstratios N. Pistikopoulos (Eds.)
PROCEEDINGS OF THE 33rd European Symposium on Computer Aided Process Engineering
(ESCAPE33), June 18-21, 2023, Athens, Greece

Binary Classification of the Endocrine Disrupting Chemicals by Artificial Neural Networks

Zahir Aghayev,[a,b] George F. Walker,[a,b] Funda Iseri,[c,d] Moustafa Ali,[c,d] Adam T. Szafran,[e] Fabio Stossi,[e,f] Michael A. Mancini,[e,f] Efstratios N. Pistikopoulos,[c,d] Burcu Beykal[a,b]

[a]*Department of Chemical and Biomolecular Engineering, University of Connecticut, Storrs, CT 06269, USA*
[b]*Center for Clean Energy Engineering, University of Connecticut, Storrs, CT 06269, USA*
[c]*Artie McFerrin Department of Chemical Engineering, Texas A&M University, College Station, TX 77843, USA*
[d]*Texas A&M Energy Institute, Texas A&M University, College Station, TX 77843, USA*
[e]*Molecular and Cellular Biology, Baylor College of Medicine, Houston, TX 77030, USA*
[f]*GCC Center for Advanced Microscopy and Image Informatics, Baylor College of Medicine, Houston, TX 77030, USA*
burcu.beykal@uconn.edu

Abstract

We develop a machine learning framework that integrates high content/high throughput image analysis and artificial neural networks (ANNs) to model the separation between chemical compounds based on their estrogenic receptor activity. Natural and man-made chemicals have the potential to disrupt the endocrine system by interfering with hormone actions in people and wildlife. Although numerous studies have revealed new knowledge on the mechanism through which these compounds interfere with various hormone receptors, it is still a very challenging task to comprehensively evaluate the endocrine disrupting potential of all existing chemicals and their mixtures by pure *in vitro* or *in vivo* approaches. Machine learning offers a unique advantage in the rapid evaluation of chemical toxicity through learning the underlying patterns in the experimental biological activity data. Motivated by this, we train and test ANN classifiers for modeling the activity of estrogen receptor-α agonists and antagonists at the single-cell level by using high throughput/high content microscopy descriptors. Our framework preprocesses the experimental data by cleaning, scaling, and feature engineering where only the middle 50% of the values from each sample with detectable receptor-DNA binding is considered in the dataset. Principal component analysis is also used to minimize the effects of experimental noise in modeling where these projected features are used in classification model building. The results show that our ANN-based nonlinear data-driven framework classifies the benchmark agonist and antagonist chemicals with 98.41% accuracy.

Keywords: machine learning, artificial neural networks, estrogenic potential, high throughput analysis, nonlinear classification.

1. Introduction

There is a wide array of naturally occurring and synthetic chemical substances that interfere with the endocrine system and cause detrimental developmental, reproductive, neurological, and immunological consequences in humans which are referred to as endocrine disrupting chemicals (EDCs). The interaction of these substances with the estrogen receptor (ER) ligand binding domain induces the activation or blockage of the biological response, known as an agonist or antagonist chemical, respectively. Thus, rapid and accurate endocrine disrupting potential identification is a crucial task in environmental risk assessment.

By observing the mechanistic endpoints in the ER signaling pathway, it is possible to assess the estrogenic potential of various substances using cell-based or cell-free assays. While these approaches can be used to accurately assess a small number of substances, evaluating tens of thousands of chemicals and their mixtures becomes quickly infeasible. On the other hand, the availability of larger data sets from high throughput experimentation, increased computing capabilities, and cutting-edge machine learning models such as deep neural networks that can be trained to capture highly nonlinear behaviors in datasets make it possible to mathematically model and predict the expected biological activity of any chemical compound with prompt function evaluations.

Previous studies by Chierici et al. (2019) and Collins and Barton-Maclaren (2022) investigated a variety of machine learning models, including Support Vector Machines (SVMs), Random Forest (RF), ANNs, for predicting the ER potential of chemicals in binary and/or multi-class classification settings. However, the predictive performance of these studies has suffered from relatively low predictive balanced accuracies in certain blind testing sets with wide performance margins ranging from 59% to 93%.

In our earlier studies, we also presented a framework to categorize chemicals according to their ER activities using population-averaged high content imaging data. By analyzing different subsets of feature groups using correlation analysis, we formulated linear (Logistic Regression) and nonlinear classification models (RF) that achieved an accuracy of more than 90% on the benchmark chemicals (Mukherjee et al., 2020). Subsequently, we integrated RF and SVM algorithms with the single-cell ER data, where characteristic shape features of the non-central gamma distribution that were used to fit the cell populations (alpha, beta, and gamma) were used as descriptors in our predictive models (Ganesh et al., 2021). Most recently, we demonstrated a predictive modeling pipeline for the endocrine disrupting potential identification that directly uses the single-cell level information, along with principal component analysis, and SVM and RF classifiers (Aghayev et al., 2022). We were able to show that the SVM model classifies unseen testing chemicals with more than 96% accuracy and data preprocessing plays a key role in achieving favorable performance from the trained classifiers.

Different than our previous studies, here, we are proposing an ANN-based predictive modeling framework that will further improve the classification accuracy and sensitivity of the separation between agonist and antagonist chemicals using single-cell level data. By further incorporating a feature selection routine based on the Fowlkes-Mallows (FM) Index in the analysis pipeline, we analyze smaller feature subsets of the high-dimensional experimental data and quantify the predictive performance of the final models using standard evaluation metrics.

2. Methods and Materials

2.1. Experimental Methodology

Sixty reference EDCs with known ERα activities are analyzed with an engineered biosensor to collect the image-based high content data. The studied compound library contains 32 agonists with different activity levels (very weak, weak, moderate, strong), 15 antagonists, and 13 inactive chemicals. To record the estrogenic activity of these reference chemicals, biosensor cells were treated with varying concentrations of the chemicals for two hours, and high-throughput microscopy and image analysis pipelines detailed in Ashcroft et al. (2011) are followed. This analysis yielded 70 size, shape, and intensity attributes that relate to the various mechanistic endpoints on the ER signaling pathway with measurements recorded for every single treated cell (Szafran et al., 2017). Further details of the experimental data generation, processing, and quantification are provided in Aghayev et al., 2022.

2.2. Computational Methodology

2.2.1. Data Preprocessing

Once the multidimensional imaging data were generated and standardized using robust z-score, we conducted percentile analysis over cells that have visible nuclear spots (receptor-DNA binding, array positive cells). The goal of this step is to expand the dataset with engineering features that will potentially enable a better separation than the raw descriptors. We previously explored different percentile ranges and their effects on nonlinear classification algorithms (Aghayev et al., 2022). Based on our findings, only single-cell features that fall into the middle percentile (between 25th and 75th percentiles) were preserved and the dataset size was decreased from 165,525 to 22,286 observations.

The produced dataset was then put through missing data check and rigorous cleaning steps, where cell data representing the inactive compounds and control treatments were eliminated from the analysis. As a result, we end up with a preprocessed dataset size of 9420 observations × 36 features prior to the feature selection phase.

2.2.2. Feature Selection via Clustering Analysis

We used clustering analysis to identify promising subsets of the original features that will facilitate the separation of agonist and antagonist compounds. To this end, we created 11 different feature groups that capture different compartments and characteristics of the measured cells (e.g., cytoplasm, nucleus, nucleoplasm, shape, etc.), including the full set of features. We then performed hierarchical clustering on the mean cell response of each active compound and selected feature subsets using Euclidean distance and complete linkage. Each clustering dendrogram is then compared to a synthetic "ideal" cluster, where this cluster shows perfect separation between agonistic and antagonistic compounds. The cluster similarity of the subset features and the ideal dendrogram is quantified using the FM index (Onel et al., 2019). FM index evaluates the similarity between two dendrograms using the geometric mean of precision and recall (Eq. 1).

$$FM = \sqrt{\frac{TP}{TP + FP} \cdot \frac{TP}{TP + FN}} \tag{1}$$

A higher FM index value (1-perfect match, 0-no match) indicates greater similarity between the analyzed dendrograms which in this case shows that the studied set of features will either yield a similar clustering to the ideal case or not. This way we can make deductions about whether a feature subset can yield a perfect classifier and have superior performance compared to the full feature set. Additionally, we tested the correlation between the dissimilarity matrices of the clusters using the Mantel test across 10,000 permutations and computed Pearson's product-moment correlation coefficient to assess whether any observed relationships were due to random arrangement.

2.2.3. Principal Component Analysis (PCA)

After identifying promising feature subsets, these datasets were standardized to conduct PCA. As a result, the selected subsets of feature groups were projected to a new feature space - new orthogonal features that are the linear combinations of the original features – principal components (PCs) with descending explained variance. Fig 1 demonstrates the scree plot of % explained variance and the top 10 PCs for the dataset with all features. Based on the location of the elbow of the scree plot, the first 3 PCs are enough to retain around 91.07% of the variation in the original datasets.

Following the feature selection and dimensionality reduction steps, the transformed datasets are split for the training and testing of the ANNs. To train the model effectively, we considered a compound-wise split procedure by considering their ER activities. During this process, we kept the testing set unseen to assess the predicted performance of the model without any bias.

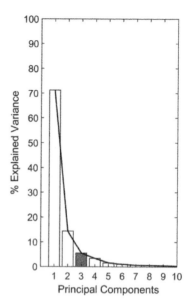

Fig 1. Scree plot of % explained variance and the top 10 PCs for the dataset with all features

2.2.4. The Deep Neural Network Model

The proposed neural network model is a feed-forward neural network with 3 nodes in the input layer and 2 nodes in the output layer. Every single partition of PCs is propagated through 3 fully connected hidden layers with 8, 16, and 20 nodes, respectively. For all the layers Rectified Linear Unit (ReLU) activation functions are used (Eq. 2).

$$g(z) = \begin{cases} 0 & if \ z < 0 \\ z & otherwise \end{cases} \quad (2)$$

To estimate the network parameters, we use the Adaptive Moment Estimation (Adam) optimizer which inherits characteristics of both the Adagrad (Adaptive Gradient Descent) and RMSprop (Root Mean Square Propagation) algorithms (Kingma and Ba, 2014). Regarding the loss function selection, as we are trying to categorize the compounds according to their agonistic and antagonistic behaviors (binary classification), we used binary cross-entropy. Finally, after training the model, its predictive capability is measured using accuracy, sensitivity, specificity, and balanced accuracy criteria by classifying the testing set that was reserved unseen to the model.

3. Results and Discussion

The results obtained from the feature selection phase are provided in Table 1. After thorough statistical analysis of the listed 11 feature groups, we found that the FM index value for 4 of them (highlighted with grey shade) is equal to 0.89, which means that they have high similarity to the ideal clustering of agonist and antagonist compounds. These 4 feature subsets, including all features, also have highly correlated distances matrices with respect to the ideal cluster and their p-values are below 0.05. This indicates that the observed correlations are not by random arrangement and the results are statistically significant. The other feature subsets are observed to have lower FM indices, and in one case p-value greater than 0.05, indicating that their dendrograms are more different than the ideal case with less distinct separation of agonist and antagonist compounds into two clusters. This shows that the underlying dataset might not yield a good classification model as the information carried on these features are not descriptive enough to have a generalized classifier. Hence, we only take the most promising feature subsets with the highest FM index to perform classification analysis. Although we have reduced dimensionality through this procedure, we still use PCA to decorrelate the experimental features and use the PCs to train the neural network. The training and testing results of the deep neural network are summarized in Table 2.

Table 1 The FM indices, Pearson's product-moment correlation coefficient (r), and the p-value resulting from the Mantel test for the studied feature subsets.

Feature Subset	FM Index	Mantel Test r	p-value
Array	0.80	0.93	0.0001
-Array	0.89	0.94	0.0001
Cell	0.89	0.92	0.0001
Cytoplasm	0.57	0.37	0.0001
Nucleus	0.89	0.94	0.0001
Nucleoplasm	0.58	0.93	0.0001
Intensity	0.74	1.00	0.0001
-Intensity	0.64	0.23	0.0104
Secondary	0.53	0.16	0.0534
Shape	0.59	0.31	0.0001
All	0.89	1.00	0.0001

According to the classification results, the dataset with all features outperforms the other 3 feature groups with 98.41% accuracy. This dataset has the highest accuracy, sensitivity, and balanced accuracy values, as well as a very high specificity value. However, datasets with only nucleus features and without array features have nearly perfect specificity values, which represent the fraction of correctly predicted antagonist compounds. Interestingly, although the dataset with only nucleus features mapped nearly all the antagonist compounds correctly (only 1 cell misclassified out of 1791 cells), its capability is very limited for agonist compounds (67.55%). We observe the opposite trend for the dataset with only cell features, where a deep neural network constructed with the projected set of these features yields a classifier with 97.54% sensitivity and 67.67% specificity in the testing set. Overall, the results show that the neural network classifier constructed from the dataset with all features outperforms other models and provides consistently very high performance scores in both training and testing steps.

Table 2 Performance metrics of deep neural network model for the 4 selected feature groups.

Feature Group	Set	Accuracy	Sensitivity	Specificity	Balanced Accuracy
W/o Array Features	Testing	92.42%	84.28%	**100.00%**	92.14%
	Training	98.64%	99.05%	98.27%	98.66%
Cell Features	Testing	82.07%	97.54%	67.67%	82.61%
	Training	94.35%	99.40%	89.77%	94.59%
Nucleus Features	Testing	84.33%	67.55%	99.94%	83.75%
	Training	98.09%	96.01%	99.97%	97.99%
All Features	Testing	**98.41%**	**97.66%**	99.11%	**98.39%**
	Training	98.96%	100.00%	98.02%	99.01%

4. Conclusion

We provided a classification framework based on multidimensional imaging data and deep neural networks to assess the endocrine disrupting potential of estrogen receptor-α (ERα) agonists and antagonists. To overcome the curse of dimensionality, both feature selection and dimensionality reduction techniques were incorporated. Feature selection is performed using hierarchical clustering, FM index, and Mantel test techniques, where the similarity between different feature subgroups was quantified with the ideal grouping of agonist and antagonist compounds. Four feature subsets, including the entire set of features, were identified as the most promising ones that will enable the agonist/antagonist separation. Principal component analysis is also performed to further reduce the dimensionality of the studied datasets and to remove any correlations between the features. The results show that the dataset with all features succeeded in predicting the estrogenic activity of unknown chemicals with 98.41% accuracy, which was followed by the dataset without array features (92.42%). Additionally, we observed an interesting trend in the sensitivity and specificity values of the other two datasets; while the dataset with only cell features could predict the ER activity of agonist compounds with very high accuracy (97.54%), the dataset with only nucleus features classified 1790 out of 1791 cells correctly (99.94%) for antagonist class. Overall, we demonstrated that highly accurate neural network models can be constructed to predict the biological activity of chemical compounds. This research was funded by the U.S. National Institutes of Health (NIH) grant P42 ES027704.

References

Z. Aghayev, A.T. Szafran, A. Tran, H.S. Ganesh, F. Stossi, L. Zhou, M.A. Mancini, E.N. Pistikopoulos, B. Beykal, 2022, Machine learning methods for endocrine disrupting potential identification based on single-cell data, Chemical Engineering Science (Under Review).

F.J. Ashcroft, J.Y. Newberg, E.D. Jones, I. Mikic, M.A. Mancini, 2011, High content imaging-based assay to classify estrogen receptor-α ligands based on defined mechanistic outcomes, Gene, 477(1-2), 42-52.

M. Chierici, M. Giulini, N. Bussola, G. Jurman, C. Furlanello, 2018, Machine learning models for predicting endocrine disruption potential of environmental chemicals, Journal of Environmental Science and Health, Part C, 36(4), 237- 251.

S.P. Collins, T.S. Barton-Maclaren, 2022, Novel machine learning models to predict endocrine disruption activity for high-throughput chemical screening. Frontiers in Toxicology, 4:981928.

H.S. Ganesh, B. Beykal, A.T. Szafran, F. Stossi, L. Zhou, M.A. Mancini, E.N. Pistikopoulos, 2021, Predicting the estrogen receptor activity of environmental chemicals by single-cell image analysis and data-driven modeling, Computer Aided Chemical Engineering, 50, 481-486.

D.P. Kingma, J. Ba, 2014, Adam: A method for stochastic optimization, arXiv preprint arXiv:1412.6980.

R. Mukherjee, B. Beykal, A.T. Szafran, M. Onel, F. Stossi, M.G. Mancini, D. Llyod, F.A. Wright, L. Zhou, M.A. Mancini, E.N. Pistikopoulos, 2020, Classification of estrogenic compounds by coupling high content analysis and machine learning algorithms, PLoS Computational Biology, 16(9), e1008191.

M. Onel, B. Beykal, K. Ferguson, W.A. Chiu, T.J. McDonald, L. Zhou, J.S. House, F.A. Wright, D.A. Sheen, I. Rusyn, E.N. Pistikopoulos, 2019, Grouping of complex substances using analytical chemistry data: A framework for quantitative evaluation and visualization, PloS one, 14(10), e0223517.

A.T. Szafran, F. Stossi, M.G. Mancini, C.L. Walker, M.A. Mancini, 2017, Characterizing properties of non-estrogenic substituted bisphenol analogs using high throughput microscopy and image analysis, PloS one, 12(7), e0180141.

Antonis Kokossis, Michael C. Georgiadis, Efstratios N. Pistikopoulos (Eds.)
PROCEEDINGS OF THE 33rd European Symposium on Computer Aided Process Engineering
(ESCAPE33), June 18-21, 2023, Athens, Greece

Digital Twins in Pilot Scale Fermentation: Non-Linear State Estimation for Improving Induction Timing

Mads Stevnsborg[a], Kurt Selle[b], Ryan Barton[b], Oscar A. Prado-Rubio[a,c], Carina Gargalo[a], Krist V. Gernaey[a], Gary Gilleskie[b], Jakob K. Huusom[a], *

[a]PROSYS, Dept. of Chemical and Biochemical Engineering, Technical University of Denmark, Søltofts Plads, Building 228A, 2800 Kgs. Lyngby, Denmark.
[b]Golden LEAF, BTEC, NC State University, 27695 Raleigh, NC, USA
[c]Departamento de Ingeniería Química, Universidad Nacional de Colombia – 170003 Manizales, Colombia
*jkh@kt.dtu.dk,

Abstract

In this work, a model is developed and implemented for GFP_{UV} production with aerobic fed-batch fermentation of *E coli* BL21 (DE3). The model parameters are estimated using historical process data and minimizing the prediction and measurement error. The model implements an extended Kalman filter for non-linear state estimation of biomass, glucose, and dissolved oxygen concentration. The filter includes an existing cascade feed-back loop for dissolved oxygen control which improves the predictive accuracy of the filter. The estimator is used during fermentation to predict the induction point based on a threshold glucose concentration which is otherwise determined exclusively with at-line measurements. The validation examples presented in this work show great agreement between the estimated and measured glucose concentrations, making it a useful tool for predicting the time until induction without requiring high-frequency at-line sampling.

Keywords: Digital Twin, Biobased Manufacturing, Operator Support, Scheduling

1. Introduction

Digitalisation in biomanufacturing aims to increase production efficiency and reduce time-to-market by tightly integrating physical and digital entities to enable smart manufacturing and faster characterization of fermentation kinetics through improved data-utilization [1,2]. Despite many biobased manufacturers are improving their digital infrastructures and data collection paradigms, few have addressed specific value generating activities associated with improved data access and -generation [2]. A challenge in biobased manufacturing is the generally low degree of automation compared to the petrochemical industry. The complex and non-linear nature of microbial and metabolic systems make it difficult and computationally expensive to predict the fermentation progression [1]. Understanding how to operate these processes and mitigate disturbances during operation often require operators with significant hands-on experience to ensure a steady operation thus making manufacturers heavily dependent on maintaining a stable of skilled operators (and clear procedures) to meet production targets [1]. Retaining the operator knowhow could help to reduce the time and cost of educating new employees and simultaneously guide the new operators during operation [3].

This work presents a protocol employing a standard host organism, *Escherichia coli* BL21 (DE3), commonly used to produce recombinant proteins in the biopharmaceutical industry [4]. The protocol is used in a course curriculum where students with little experience conduct a pilot-scale exercises. Here the operators are required to sample the fermentation broth and measure the glucose concentration and optical density at-line to determine when to induce. After induction, the operators initiate a fed-batch phase. Induction shifts the metabolic pathways to favour the formation of the recombinant protein, and mistimed induction has previously been shown to significantly affect the yield and productivity [5]. Tools that support the operators during the experimental run is therefore especially valuable to ensure a steady and predictable operation. Since the protocol is employed extensively as part of the course curriculum, there exists a large historian of process- and at-line lab data pertaining to the production of green fluorescent protein by *E. coli* aerobic fermentation. This data will be used to build a non-linear state-estimator to predict the induction timing based on concentration thresholds dictated by the protocol.

2. Materials & Methods

2.1 Fermentation Protocol

The presented protocol serves as a practical exercise in a short course curriculum on topics in advanced fermentation offered at the Golden LEAF Biomanufacturing Training and Education Center (BTEC), located at North Carolina State University. Recombinant green fluorescent protein (GFP_{UV}) exhibiting a high ultraviolet (UV) reflectivity is produced from a batch- & fed-batch aerobic fermentation of *E. coli* BL21 (DE3). Initially a seed culture is inoculated with a glucose- and yeast extract rich media and fermented in aerobic batch conditions at 50% dissolved oxygen (DO) saturation. The glucose is added in excess, resulting in the production and accumulation of Acetic acid during the lag and log phase. After running the batch condition for 5-6 hours, the glucose concentration has decreased to approximately 2 g/L. The operator then induces the broth with isopropyl β-D-1-thiogalactopyranoside (IPTG), shifting the metabolic pathways to favour the formation of the recombinant GFP_{UV}. The fermentation then enters a fed-batch phase, where the operator controls the feed flow rate of a 2 L glucose and yeast extract rich media until the supply runs out. As the organism enters the stationary phase and the feeding media is depleted, the accumulated Acetic acid is also consumed. Approximately 24 hours after inoculation the fermentation is terminated, and the product is harvested.

2.2 Equipment and Data Handling

BTEC employs 5 identical 30L Sartorius® bioreactors to run several instances of the protocol in parallel. Each bioreactor is augmented with high frequency sensors measuring pressure, temperature, DO, pH, and optical density (OD). The OD sensor measures near-infrared (NIR) wavelengths and correlates the results to OD_{600} absorbance. The absorbance at 600 nm corresponds to the dry cell weight of *E. coli* and functions as a soft sensor determination of the cell density. Additionally, the operator periodically takes samples from the broth and perform at-line determination of the glucose concentration and OD_{600} through spectrophotometric methods. The DO is controlled by a closed loop cascade that manipulates agitation and aeration. The cascade control strategy primarily manipulate agitation until it reaches a maximum value. The inner control loop will then start to manipulate air sparging flowrate to reach the DO setpoint.

Data collection, storage and querying is orchestrated by AVEVA PI hosted on a central database server located at BTEC. A human-machine interface (HMI) is developed using AVEVA's DeltaV software and hosted on the same central server. The operator can easily input at-line measurement results through the HMI ensuring that the low-frequency measurements are stored in the same concise database and historian as the high-frequency sensors. Online accessibility and tight integration between the digital- and physical entities allow for easy online monitoring and control of the bioreactors. This presents itself as a digitally mature system with a great potential for augmenting a digital twin as depicted in Figure 1. This can be used to support operators with control suggestions during run-time or as virtual laboratories to practice operation strategies before the exercise.

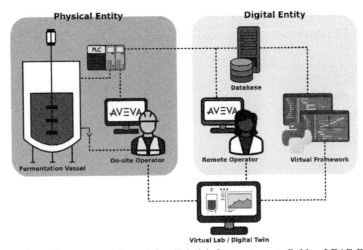

Figure 1: A schematic representation of the digital infrastructure at Golden LEAF BTEC. Data flows from the process unit to a central database and historian hosted with AVEVA PI and human machine interface (HMI) built in DeltaV. The HMI can be accessed remotely for control and monitoring when located on the BTEC network. The Virtual Laboratory / Digital Twin can be used to perform high-level model development and operator support during run-times.

A total of 53 fermentation runs have been extracted from the PI Vision database, conducted between 2020 and 2022. The measurements from these experimental runs are used in the objective function presented in Table 1 and include sensor- and at-line OD, DO sensor and at-line glucose.

2.3 Model & Nonlinear State-Estimator

This study employs an acetate overflow model to describe the metabolic behaviour for GFP$_{UV}$ formation in *E. coli* BL21 (DE3) this model is derived from a modification to the implementation originally presented by Christina et al. in 2003 [6]. The model by Christina et al has served as a foundation for recent studies in similar systems of *E. coli* BL21 (DE3) [5, 7]. The protocol only requires operators to collect endpoint titer data, which complicates parameter estimation for kinetic GFP production. Therefore, it is not currently possible to include product formation as a modelled state in the dynamic model. Since this study has focus on the pre-induction kinetics, the formation of GFP can be assumed to be reasonably small such that it should not skew the estimated yield coefficients significantly by excluding it from the mass balance constraints. Instead, the aim is to develop an accurate glucose concentration estimator to assist operators in determining when the glucose concentration is expected to reach the induction target given by the fermentation protocol. The number of parameters in the model explodes with increasing complexity. As a result, the final model is restricted to three rate expressions [4,5,7]. Parameter bounds are determined from literature review and heuristics to reduce the solution space [4,7]. Regardless, the problem is currently a concave optimization problem, and solutions will thus depend heavily on initial conditions. Good initial guesses are therefore necessary to extract dynamics that have realistic behaviour. The yield coefficients (Y_{ij}) are restricted to satisfy a mass balance for each rate expression. In the equations presented in Table 1. the species Glucose (S), Acetate (A), Oxygen (O), Carbon Dioxide (CO_2), Biomass (X) and Yeast Extract (YE) are modelled with Monod expressions, including the parameters, maximum rate ($q_{S,max}$, $q_{O,max}$, $q_{A,max}$), half-times and inhibitors (K and K_I).

Table 1: stoichiometric equations, rate expressions, objective function and parameters used in the presented model.

STOICHIOMETRY		
Glucose respiration	$S + Y_{OS1}O + Y_{YS1}YE \xrightarrow{\mu_1 X} Y_{XS1}X + Y_{CS1}CO_2$	$q_{S,max}$: 2.25
Glucose respiration	$S + Y_{OS2}O + Y_{YS2}YE \xrightarrow{\mu_2 X} Y_{XS2}X + Y_{CS2}CO_2 + Y_{AS}A$	$q_{S,max}$: 1.12
Glucose respiration	$A + Y_{OA}O + Y_{YA}YE \xrightarrow{\mu_3 X} Y_{XA}X + Y_{CA}CO_2$	$q_{O,max}$: 0.98

RATE EXPRESSIONS

Critical substrate uptake rate
$$q_{S,crit} = \frac{q_{O,max}}{Y_{OS}} \cdot \left(\frac{K_{IA}}{K_{IA}+A}\right)$$

Substrate uptake rate
$$q_S = q_{S,max} \cdot \left(\frac{S}{K_S+S}\right) \cdot \left(\frac{O}{K_O+O}\right)$$

Acetate uptake rate
$$q_A = q_{A,max} \cdot \left(\frac{A}{K_A+A}\right) \cdot \left(\frac{O}{K_{AO}+O}\right) \cdot \left(\frac{K_{IS}}{K_{IS}+S}\right)$$

REACTION RATES

$$\mu = \begin{bmatrix} \mu_1 \\ \mu_2 \\ \mu_3 \end{bmatrix} = \begin{bmatrix} min(q_S, q_{S,crit}) \\ max(0, q_S - q_{S,crit}) \\ max(0, q_A) \end{bmatrix}$$

PARAMETER FITTING

Objective function
$$obj = \min_\theta |y_{DO,OD,S} - \hat{y}_{O,X,S}(\theta)|_Q$$
s.t.

Mass balance
$$mb = \sum Y \cdot \mu$$

Redox balance
$$rb = \sum Y \cdot \mu \cdot \epsilon$$

Parameter	Value
$q_{S,max}$	2.25
$q_{S,max}$	1.12
$q_{O,max}$	0.98
K_S	1.00
K_{SO}	$3.97 \cdot 10^{-3}$
K_A	$5.33 \cdot 10^{-2}$
K_{AO}	$1.01 \cdot 10^{-4}$
K_{IS}	0.19
K_{IA}	$2.48 \cdot 10^{-3}$
Y_{XS1}	0.48
Y_{XS2}	0.67
Y_{XA}	0.41
Y_{AS}	0.29
Y_{OS1}	1.72
Y_{OS2}	0.95
Y_{OA}	1.36
Y_{CS1}	0.27
Y_{CS2}	0.42
Y_{CA}	0.35
Y_{YS1}	0.14
Y_{YS2}	0.16
Y_{YA}	0.15

The objective function is defined to minimize the error between sensor measurements and the simulated results. This includes online optical density and dissolved oxygen sensor values combined with at-line determination of OD_{600} and glucose concentrations. The weights are assigned to improve prediction accuracy for glucose and balance the contribution from at-line and online OD measurements. The optimization scheme employs a Nelder-Mead algorithm that penalises a scaled error between the measured- and simulated state, restricted by mass- and redox-balances. From the parameterized model, a non-linear state estimator is constructed. For this purpose, the Extended Kalman Filter (EKF) is formulated.

3. Results & Discussion

A preliminary assessment of the datasets shows high variance in total GFP titer (2.57 ± 1.05 g/L), product yield on the substrate (0.0743 ± 0.0011 g/g), glucose concentration at the induction point (2.57 ± 1.05 g/L), and induction time after inoculation (5.91 ± 0.27). These results support the hypothesis that inexperienced operators naturally infer high variance to a process despite following the same protocol. To validate the performance of the parameterized model, two batch runs are assessed that were not previously included in the parameter estimation scheme. Here an R^2-value for prediction of glucose and biomass were found to be 0.93 and 0.90 respectively, confirming a good accuracy between the model and the system.

Assuming that the most recent sensor measurements for the two cases are collected at ~5.25 hours and ~5.80 hours, it is possible to predict the evolution using the EKF. The measurements and mixture details at inoculation are passed to the EKF, which computes states and predict measurement trajectories. A qualitative assessment of the predictions presented in Figure 3 show a reasonable agreement between the predictions and sensor measurements, especially for determination of glucose concentration. The glucose predictions are then used to determine an estimate for timing the induction.

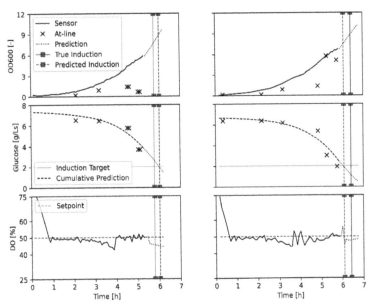

Figure 2: Measurements and predictions for two fermentation runs (left #1 & right #9) over time, beginning at inoculation of the fermentor. (left) Achieves induction close to the desired condition. (right) Overshoots the induction point by approximately 15 min reducing the concentration of glucose to 0.94 g/L.

As show that in Figure 2 the batch runs are expected to reach induction conditions approximately 6 hours after inoculation. In this way, operators would be able to receive the prediction in advance and adequately prepare for induction at the desired conditions without relying solely on at-line glucose measurements. Alternatively, it could assist operators in determining sections in the fermentation run where high-frequency at-line sampling has an added value. As evident by the case depicted on the right-hand side of Figure 2, operators miss the induction point by approximately 20 min. This could be caused by too infrequent at-line sampling during the phase with steep consumption of glucose leading to the operators being poorly prepared to perform the induction. From the results presented by Lopes et al. [5], this deviation should be identifiable in the yield and productivity by comparing the two batch runs at the end of the fermentation. However, assessing the final GFP$_{UV}$ titer at harvest after 24 hours shows that both fermentation runs achieve a similar titer and yield on substrate despite the run on the right-hand side initiating induction 20 min past the suggested glucose concentration listed in the protocol. Since the post-induction dynamics change and include feeding of the glucose and yeast extract solution, there are elements potentially explaining how the delayed induction run bounces back, which the current model cannot account for. Variations in later operating points could justify this observation, but another cause could be that the selected induction condition is not optimized to maximise the GFP$_{UV}$ yield or productivity. A more thorough study of final GFP$_{UV}$ maximisation and the inclusion of fed-batch fermentation behaviour post induction could address this caveat.

Conclusion

Existing historical data from pilot scale experiments were contextualised and used to estimate parameters in a suggested mechanistic model for *E. coli* BL21 based on state-of-the-art acetate overflow models. The system was validated against process data and used to develop an extended Kalman filter that predicts the evolution of biomass growth, glucose, and oxygen consumption with reasonable accuracy. In contrast to shake flask experiments investigating the same system, delayed induction did not appear to have a significant effect on the final fermentation performance. Since the agitation and aeration are controlled by the closed loop cascade with a DO setpoint at 50% for all fermentations, the root cause of the observed variation must be found elsewhere. The current implementation only account for the pre-induction dynamics, and variance in the glucose feeding strategy after induction might have greater influence on the performance than previously anticipated.

Acknowledgements

Nordisk Foundation in the frame of the 'Accelerated Innovation in Manufacturing Biologics' (AIMBio) project (Grant number NNF19SA0035474).

References

F. Bähner et. al., 2021, Challenges in Optimization and Control of Biobased Process Systems: An Industrial-Academic Perspective, Industrial and Engineering Chemistry Research, Vol. 60, 14985-15003

I. A. Udugama et. al., 2021, Towards Digitalization in Bio-Manufacturing Operations: A Survey on Application of Big Data and Digital Twin Concepts in Denmark: Frontiers in Chemical Engineering, Vol. 3, 727152

M. Muldbak et. al., 2022, Digital Twin of a pilot-scale bio-production setup, Computer Aided Chemical Engineering, Vol. 49, 1417-1422

M. Abadil et. al., 2021, An experimental assessment of robust control and estimation of acetate concentration in Escherichia coli BL21(DE3) fed-batch cultures, Biochemical Engineering Journal, 174, 108103

C. Lopes et. al., 2019, Improving the cost effectiveness of enhanced green fluorescent protein production using recombinant Escherichia coli BL21 (DE3): Decreasing the expression inducer concentration, Biotechnology and Applied Biochemistry, Vol. 66, 527-536

I. Cristina et. al., 2003, Model-based strategies process for computer-aided operation of a recombinant E. coli fermentation

C. Retamal et. al., 2018, Parameter estimation of a dynamic model of Escherichia coli fed-batch cultures, Biochemical Engineering Journal, 135, 22-35

Antonis Kokossis, Michael C. Georgiadis, Efstratios N. Pistikopoulos (Eds.)
PROCEEDINGS OF THE 33rd European Symposium on Computer Aided Process Engineering
(ESCAPE33), June 18-21, 2023, Athens, Greece

Optimal dosing of thyroid hormones in hypothyroid patients with an individualized compartmental model

Davide Manca*, Federico Appiani, Giovanni Colombo

PSE-Lab, Process Systems Engineering Laboratory, Dipartimento di Chimica, Materiali e Ingegneria Chimica "Giulio Natta", Politecnico di Milano, Piazza Leonardo da Vinci 32, 20133 Milano, Italy
*davide.manca@polimi.it

Abstract

Recent years have seen a growing interest in studying new methods to predict the correct dose of levothyroxine (LT_4) that is used as a synthetic drug for hypothyroidism treatment and supplements/substitutes the T_4 hormone secreted by the healthy thyroid. The paper proposes a quantitative procedure for the optimal administration of LT_4 through a model-based approach dependent on some individualized parameters of the patient to free the endocrinologist from the current rough guesses of LT_4 dosage. A dedicated algorithm receives as input data the patient features along with the major indicator of thyroid status, *i.e.* the serum concentration of TSH, the thyroid-stimulating hormone. Then the procedure calculates the Residual Thyroid Function (RTF), which quantifies the thyroid efficiency status, and eventually the optimal LT_4 dose. As a complement to the quantitative models available in the scientific literature, we also take into account the age of the patient to determine the optimal dose that can avoid any suboptimal prescriptions regarding hypothyroidism diagnosis for both young and elderly individuals.

Keywords: LT_4; TSH; individualized medicine; anthropometric parameters.

1. Introduction

The thyroid gland is the largest organ in our body capable of hormonal secretion, located in the lower part of the neck in front of the trachea, below Adam's apple. It is divided into two conical lobes linked by an isthmus. The physiological functional sites of the gland are the follicles. They comprise a central nucleus, a so-called colloid, which is surrounded by cuboidal cells. These cells exchange the necessary biological elements and products through the cardiovascular system that highly perfuses the thyroid gland. The thyroid is unique among all body organs thanks to its ability to store and concentrate iodine above forty times the normal plasma concentration. Iodine is the essential substance involved in the complex mechanism of hormone production. Elemental iodine is introduced into the body with nourishment, and it is later exploited by the thyroid to produce triiodothyronine (T_3) and tetraiodothyronine (T_4), the two main hormones secreted by the gland. The role of this organ is of vital importance since it can modulate processes such as thermoregulation, body growth, cellular metabolism, breathing, and heart rate. The two thyroid-secreted hormones are synthesized in the follicle in a complex series of metabolic interactions, mainly involving thyroglobulin protein, which is produced in the thyrocytes and then delivered into the colloid. Here, it interacts with iodine to form T_3 and T_4 precursors, containing three and four iodine atoms per molecule respectively, which are metabolized by specific enzymes to obtain the actual hormones eventually released into the bloodstream. A specific pituitary-secreted molecule called thyroid-stimulating hormone (TSH), which in turn is stimulated by the hypothalamic hormone thyrotropin-releasing hormone (TRH), promotes the cascade of events related to hormonal synthesis. From these intricate relations, it is evident the tight interconnections that link the hypothalamus, pituitary gland, and thyroid, in what is known as the H-P-T axis. These organs enable a peculiar control mechanism. TSH can stimulate the thyroid gland by increasing its T_3 and T_4 production, but the thyroid hormones, especially T_3, inhibit the TSH release by acting on the gene expression of TRH, which consequently hinders the TSH secretion (Werner *et al.*, 2005). Thyroid diseases can be traced back to the failure of critical links in the chain of events that have been described so far. These disorders can be divided into hypothyroidism and hyperthyroidism. Hypothyroidism is the most common thyroid disorder, where the patient experiences a low level of T_4 and a high level of TSH, caused either by pathologies directly linked with thyroid failure, referred to as primary hypothyroidism, or pathologies associated with hypothalamus or pituitary gland dysfunction, better known as central hypothyroidism. This disease has an estimated incidence of 4 to 5 individuals every 1000 per year in women and 0.6 to 0.9 individuals every 1000 per year in men. Prevalence exceeds 5% of the global population (Flynn *et al.*, 2004).

Researchers in the last six decades put constant efforts into establishing a mathematical model capable of mimicking the thyroid behavior. More recently, there has been an attempt to use the available models to predict the drug dosage for specific thyroid pathologies. The implementation of such models is motivated by two main controversies regarding the diagnosis of hypothyroidism and the prescription criterion of levothyroxine (LT_4), which is the active principle used to cure hypothyroidism and is the levorotatory enantiomer of the actual T_4 hormone in the form of sodium salt. The first issue concerns the evaluation methods of the hypothyroidism status, which are primarily based on the analysis of the TSH values, whose allowed range must fall between 0.45 and 4.5 mIU/l even though that range is mired in debates (Werner *et al.*, 2005, Chapter 46). Based on data belonging to patients under levothyroxine treatment, it is rather common to register miscalculations of the LT_4 dose, which is a major health risk. This issue should be considered of utmost importance. Health studies showed that 22% of analyzed subjects were improperly treated and that 18% of them were receiving higher LT_4 doses than necessary, with consequent negative side effects, namely dyslipidemia, cardiovascular disorder risks, and, particularly, iatrogenic thyrotoxicosis (Werner *et al.*, 2005, Chapter 46). In addition, LT_4 has been classified as a narrow therapeutic index drug, given the utter care that must be put into the titration of the dose, considering that even small changes can significantly affect the existing equilibrium of the H-P-T axis (Werner *et al.*, 2005, Chapter 13A). Current goal of the recently developed thyroid models is including individualized parameters in the attempt to set new standards for hypothyroidism diagnosis and LT_4 prediction. Among the available models, p-Thyrosim stands out for its capability of simulating the thyroid biological mechanism as a function of the patient's parameters such as gender, height, weight, residual thyroid function (*i.e.* quantification of the hypothyroidism status), and LT_4 administered dose (Cruz-Loya *et al.*, 2022).

2. Methods

Eisenberg and coauthors (2008) outlined the base structure of a multi-compartmental PBPK model to describe the physiological feedback control of the H-P-T axis. The reference model explains the H-P-T axis components by means of homogenous compartments. There are three main submodels: thyroid hormones (TH), Brain (B), and TH input (THI). The interconnection of each of them represents the interaction of the H-P-T axis that allows evaluating the trend of both TSH and the thyroid hormones in plasma. The TH submodel includes a set of ordinary differential equations that govern the thyroid hormones' fluxes from the plasma to the tissues and vice versa, occurring in the whole body. Then, the B submodel describes the endogenous circadian TSH secretion by lumping the interaction between the pituitary gland and the hypothalamus in a single equation, which is mainly a function of the T_3 level in the brain. Finally, the THI submodel contains the equations concerning the exogenous administration of LT_4 employing four alternative paths: intravenous, intramuscular, subcutaneous, or oral administration (Eisenberg *et al.*, 2008). This model saw a relevant development described in Cruz-Loya *et al.*, (2022) where the authors succeeded in implementing individual parameters such as body weight, height, and gender and proposed a suitable formulation to describe the circadian oscillation of TSH that affects the thyroid secretory pattern.

In case of patients with thyroid problems, it is common practice to consider the TSH concentration measured with a first blood analysis. If there is a positive match for hypothyroidism, the endocrinologist proceeds with an initial dose of LT_4, which can be determined in a variety of ways. The preferred LT_4 dose is 1.6 µg/kg per day but there is also an extremely simplified approach based on a one-size-fits-all of 100 µg/day. Afterwards, the patient has to take the remedy every morning before breakfast for at least six weeks after which a new blood sample provides the updated value of TSH that allows establishing if the thyroid levels belong to the recommended ranges. If they are not, the dose is empirically adjusted, and the procedure is iterated until normal hormone levels are reached (Werner *et al.*, 2005, Chapter 46). It is rather common to undergo a few iterations before an optimal prescription is achieved, which makes the whole procedure demanding for the patient and subject to the biased evaluation of the endocrinologist. Therefore, an optimized, individualized, and unbiased protocol is highly recommended.

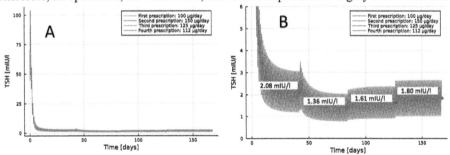

Figure 1: Panel (A) simulated endocrinologist trial and error approach in LT_4 dose administration for a 70-year-old hypothyroid female patient of 78 kg, 1.63 m, and TSH starting level of 71.44 mIU/l. Panel (B) zoomed trend. Red diamonds report the asymptotic TSH values after the iterative adjustments of LT_4 prescriptions.

Panel A of Figure 1 reproduces a hypothetical situation based on the traditional approach to the LT_4 dosing scheme where the endocrinologist wants the TSH to be far from the upper limit to prevent the persistence of hypothyroidal symptoms. Based on this assumption, the optimal value of TSH is set below the average of the euthyroid range (0.45-4.5 mIU/l), to a value of 1.9 mIU/l. A 70-year-old female, weighing 78 kg and 1.63 m tall undergoes a blood sample to measure her TSH concentration for the first time. TSH is 71.44 mIU/l, and the endocrinologist (neglecting any individualized features) prescribes a theoretical starting dose of 100 µg/day (Werner *et al.*, 2005, Chapter 46).

This produces a six-week asymptotic TSH value of 2.08 mIU/l, which is higher than the expected value and might make the symptoms persist. The dose is then increased to 150 µg/day and, after six weeks, TSH reaches 1.36 mIU/l, which is now too low to be acceptable and bearable for prolonged treatments. The endocrinologist then fine-tunes the dose to 125 µg/day after which TSH reaches 1.61 mIU/l. Since the LT₄ has a narrow therapeutic window, the endocrinologist chooses to further change the dose to 112 µg/day, which eventually corresponds to a TSH of 1.80 mIU/l which is a rather good value reasonably near to the expected one (1.9 mIU/l).

An alternative approach consists of the automated optimal LT₄ dosing by exploiting the power of quantitative deterministic models such as p-Thyrosim. With respect to the above iterative search for a suitable prescription by the human interaction between the endocrinologist and the patient, now the goal is to minimize the blood samples (and therefore the time to find the optimal dosage as each dose adjustment calls for 50-60 day more settling times) by getting the starting dose right the first time. To this purpose, we implemented an optimizing procedure that can return the optimal LT₄ tablet dose for hypothyroid patients, using p-Thyrosim and accounting for the patient's individual features such as weight, height, gender, age, and the TSH value from a recent blood sample. First, the proposed procedure evaluates the RTF (Residual Thyroid Function) value by zeroing the following function:

$$f(RTF) = TSH^{exp} - TSH(RTF) \tag{1}$$

Where the TSH^{exp} and TSH $[mIU/l]$ are the TSH blood sample and the p-Thyrosim asymptotic prediction of the in-silico individualized patient TSH plasma dynamics, respectively. Once RTF is computed, the procedure zeros Equation (2), by determining the optimal LT₄ dose that would yield to euthyroid values of both T_4 and TSH:

$$f(LT_4) = \left| \frac{TSH^{sp} - TSH(LT_4)}{4} \right| + \left| \frac{T4^{sp} - T4(LT_4)}{60} \right| \tag{2}$$

where TSH^{sp} $[mIU/l]$ and $T4^{sp}$ $[\mu g/l]$ are the optimal expected values (*i.e.* setpoints) for the corresponding euthyroid values. Initially, TSH^{sp} was set at 1.80 mIU/l for any hypothyroid patient as Cruz-Loya *et al.*, (2022) suggest a minor dependency from age as far as euthyroid conditions are concerned. However, given the high precision that must be kept in drug dosage, we also accounted for the dependency of TSH^{sp} from age. We derived an age-dependent trend for the upper limit of TSH euthyroid from Raverot *et al.*, (2020) that took to Equation (3).

$$TSH_{M,F}^{sp}(age) = \frac{TSH_{M,F}^{ul}(age) - TSH^{ll}}{2.7} \tag{3}$$

Where $TSH_{M,F}^{ul}$ and TSH^{ll} are the upper and lower limits of the TSH range respectively for either a male or female euthyroid patient. $TSH_{M,F}^{ul}(age)$ was evaluated by extrapolating the values from the TSH upper limit trends, both for males and females, from the work of Walsh (2022). Therefore, the data of Figure 2 were regressed with a nonlinear procedure. This methodology generates also the TSH upper limit formula from which it is possible to calculate, for every combination of age and gender, the associated upper limit and eventually derive the correct TSH^{sp}:

$$TSH_{Male}^{ul}(age) = 4.642520589559373 - 0.02039878317500956 \cdot age + 0.0002614981700062368 \cdot age^2 \tag{4}$$

$$TSH_{Female}^{ul}(age) = 4.1777986229928485 + 0.00463599130382207 \cdot age + 0.0001309278501772748 \cdot age^2 \tag{5}$$

Instead, the T_4^{sp} plasma values have been considered equal for both genders and fixed at 80 µg/l as the literature does not report any significant variability with age (Werner *et al.*, 2005, Chapter 11A).

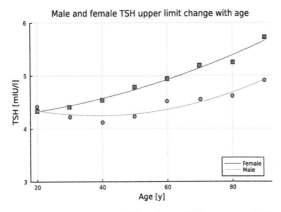

Figure 2: TSH upper limit trend for males and females at different ages derived by a nonlinear regression procedure.

3. Results and discussion

The proposed automated algorithm predicts the necessary LT_4 tablet dose (commercially available) to be administered to a hypothyroid patient by accounting for their individual features and endogenous TSH (before the treatment). The procedure, based on the p-Thyrosim model (Cruz-Loya *et al.*, 2022) can investigate how the optimal dose changes as a function of the patient's age. The dependency of TSH^{ul} with age, seen in Equation (3), affects TSH^{sp}. If the TSH upper limit is somehow increased, the LT_4 dose must decrease according to Equation (2). Thus, the implementation of the automated procedure of the TSH^{sp} age dependency ensures the achievement of different dosages for the same hypothyroid patient at different ages, allowing a better approach to individualized treatment.

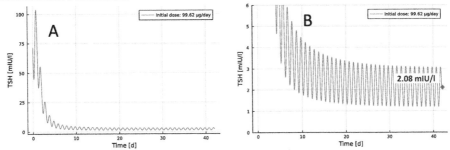

Figure 3: (A) Simulated TSH plasma value for a 70-year-old hypothyroid female patient with a weight of 78 kg, 1.63 m, and TSH starting level of 71.44 mIU/l under LT_4 treatment calculated through the double-step algorithm routine. (B) Simulated TSH plasma value for the same patient highlighting with a red marker the achievement of the correct TSH setpoint thanks to the theoretical optimal LT_4 dose (99.62 µg/day) evaluated by the optimizing procedure.

With reference to Figure 3, we tested our double-step algorithm for the same female hypothyroid patient in Figure 1. The main differences between the graphs must be searched in the diverse LT_4 dose treatment schemes and in the age-dependent approach related to the individualized TSH^{sp} value. Figure 1 shows how the iterative human-based procedure resulted in a total of 160 days and 4 adjustments before identifying the correct LT_4 dose centered on the endocrinologist's perspective. Conversely, Figure 3 shows the direct assessment of the optimal dose, which allows for reaching the expected TSH optimal value by a one-shot prescription. Furthermore, the TSH setpoint chosen in the dosing scheme of Figure 1 does not account for the TSH^{ul} dependency with age as the patient is assigned a constant setpoint at any stage of their life. Thus, the proposed optimizing procedure assigns the TSH setpoint as a function of both the patient's age and gender. For the considered 70-year-old female patient the correct TSH level should be 2.08 mIU/l, which is achieved by an LT$_4$ dose of 99.62 µg/day, found by the optimization algorithm. That dose is not available in commercial tablets and should be rounded to the nearest available formulation of 100 µg/day which would bring an asymptotic TSH value of 2.07 mIU/l, almost equal to the optimal one. It is worth observing that the scenario reported in Figure 1 calls for the involvement of an earnest physician. In real life, the endocrinologist might not look for those precise adjustments and might be satisfied with any reasonable dosages. Nonetheless, our automated and unbiased procedure can overcome the endocrinologist's sensitivity in the dose prescription and displays a lower time to reach the optimal dose based on a one-shot and more accurate treatment that takes into account the individual features of the patient.

4. Conclusions

The findings reported in the present paper were made possible by the decades-long studies published in the scientific literature. Our efforts concerned the application of compartmental models simulating thyroid secretion to replicate real-life scenarios, in the attempt to increment the precision of the first LT$_4$ tablet dose for hypothyroid patients. We developed an automated approach through a double-step procedure to support the endocrinologist's decisions in terms of optimal LT$_4$ dosage by exploiting the individualization features and unbiased calculations of p-Thyrosim. The attempt of implementing age as a further individualized parameter for hypothyroid patients allowed for outlining the dependency of TSH at different stages of life and improving the evaluation of the LT$_4$ optimal dose.

References

Cruz-Loya, M., Chu, B., Jonklaas, J., Schneider, D., & DiStefano, J. (2022). Optimized Replacement T4 and T4+T3 Dosing in Male and Female Hypothyroid Patients With Different BMIs Using a Personalized Mechanistic Model of Thyroid Hormone Regulation Dynamics. *Frontiers in Endocrinology, 13.*

Eisenberg, M., Samuels, M., & DiStefano J. (2008). Extensions, validation, and clinical applications of a feedback control system simulator of the hypothalamo-pituitary-thyroid axis. *Thyroid, 18*(10), 1071–1085.

Flynn, R. W. v, MacDonald, T. M., Morris, A. D., Jung, R. T., & Leese, G. P. (2004). The thyroid epidemiology, audit, and research study: thyroid dysfunction in the general population. *The Journal of Clinical Endocrinology & Metabolism, 89*(8), 3879–3884.

Hollowell, J. G., Staehling, N. W., Flanders, W. D., Hannon, W. H., Gunter, E. W., Spencer, C. A., & Braverman, L. E. (2002). Serum TSH, T4, and thyroid antibodies in the United States population (1988 to 1994): National Health and Nutrition Examination Survey (NHANES III). *The Journal of Clinical Endocrinology & Metabolism, 87*(2), 489–499.

Raverot, V., Bonjour, M., Abeillon du Payrat, J., Perrin, P., Roucher-Boulez, F., Lasolle, H., Subtil, F., & Borson-Chazot, F. (2020). Age-and sex-specific TSH upper-limit reference intervals in the general French population: there is a need to adjust our actual practices. *Journal of Clinical Medicine, 9*(3), 792.

Walsh, J. P. (2022). Thyroid Function across the Lifespan: Do Age-Related Changes Matter? *Endocrinology and Metabolism, 37*(2), 208.

Werner, S. C., Ingbar, S. H., Braverman, L. E., & Utiger, R. D. (2005). *Werner & Ingbar's the thyroid a fundamental and clinical text* (9th ed.). Lippincott Williams & Wilkins.

Antonis Kokossis, Michael C. Georgiadis, Efstratios N. Pistikopoulos (Eds.)
PROCEEDINGS OF THE 33rd European Symposium on Computer Aided Process Engineering
(ESCAPE33), June 18-21, 2023, Athens, Greece

Technoeconomic process modelling of waste remediation within a supply chain network

Robert Milton, Jagroop Pandhal and Solomon F. Brown

Department of Chemical and Biological Engineering, The University of Sheffield, Sheffield, S1 3JD, United Kingdom
r.a.milton@sheffield.ac.uk

Abstract

Technoeconomic modelling links engineering performance to economic outcomes. In this paper we present a method for embedding technoeconomic unit process models within a geographically distributed supply chain simulator. This enables the simultaneous optimization of how and where processing is performed. This is particularly suited to waste remediation, where the costs of transporting hazardous materials may outweigh those of performing some stages of treatment at the waste source. The approach is applied to a microalgal-bacterial remediation of hydrocarbon waste (sump oil mainly) produced by the MOD Sites (military bases) across the UK. Through this treatment, valuable bioproducts are recovered from algae which consumes hydrocarbon waste and CO2 as it grows.

Keywords: Technoeconomic analysis, supply chain modelling, waste remediation.

1. Introduction

Technoeconomic modelling has long been a basic step in process design, linking engineering performance to economic outcomes (Seider et al 2019). The operation and optimisation of individual processes within wider, distributed supply networks is typically considered with simplified models of the underlying processes, and may be independent of the scheduling of the operations contained therein (Garcia et al 2015). In this study, we address this by presenting software for embedding technoeconomic unit process models within a geographically distributed supply chain simulator. This enables the simultaneous optimization of how and where processing is performed. This is particularly suited to waste remediation, where the costs of transporting hazardous materials may outweigh those of performing some stages of treatment at the waste source.

As an example, the approach is applied to a microalgal-bacterial remediation of hydrocarbon waste. Microalgal-bacterial remediation is a multi-stage process to cultivate, extract and exploit the bioproducts produced by microalgae grown in hydrocarbon effluent. Through this treatment, valuable bioproducts are recovered from algae which consumes hydrocarbon waste and CO_2 as it grows (Zahra et al 2020). This offsets the costs, both economic and environmental, of our global abundance in hydrocarbon effluents, used oils and lubricants (Fregie et al, 2019).

Unit operations are represented by kinetic models for each stage that is linked to economic performance via cashflow modelling of revenue streams, resource costs (OpEx) and plant costs (CapEx). The unit processes (stages) are implemented as agents in an agent-based supply chain model. Incorporating transport costs, processing is then optimized across the entire network of waste producing and processing sites. This enables optimal choices between local and centralized processing.

2. Architecture

The basic problem architecture, embedding a hierarchy of agents (i.e. agents containing lower-level agents) in a geographically distributed supply chain network is depicted in Fig. 1.

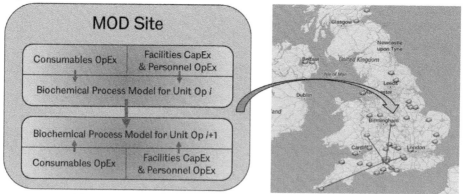

Fig.1. Embedding the layers of architecture – MOD Sites containing Unit Ops – in a supply chain network

2.1. Implementation

The Agent Based Model is implemented in AnyLogic (Borshchev 2014) extended with our own Java code). At every level, the representation in AnyLogic/Java is as generic as possible, the details of implementation being stored in a separate Excel spreadsheet for each layer of the agent hierarchy. Each spreadsheet ultimately lists every agent at that level of the hierarchy, providing for each a description specifying every parameter required to construct the agent. Upon model startup, the MOD Sites (military bases) are located by the postcode embedded in the (Excel) description of the site. This automatically exposes the road distance between any two sites, as a feature of AnyLogic.

2.2. MOD Site

The description of an MOD Site goes far beyond geographical location, populating all of the parameters in Fig. 2. These specify, amongst other matters, the frequency and amount of material to be processed at this site. Material is based on a single limiting consumable (e.g. hydrocarbon waste) by which to scale reagent amounts and energy requirements. The limiting consumable may be produced at another site, transported and received here, or be hydrocarbon waste produced onsite (the limiting consumable amount). Once processed, products 0-4 may be dispatched to destinations 0-4, which are MOD Sites specified in the description with product routing information. These matters are handled on the right of Fig. 2, by the receiver and dispatcher agents, which calculate scheduling (based on the trucks listed in this MOD Site's description), mileage, and thereby Haulage OpEx including fuel, depreciation per km and personnel.

The type of an MOD Site is used to populate the sequence of Unit Ops to the right of Fig. 2. For example, if the MOD Site is described as "Empty" it does nothing but produce and.

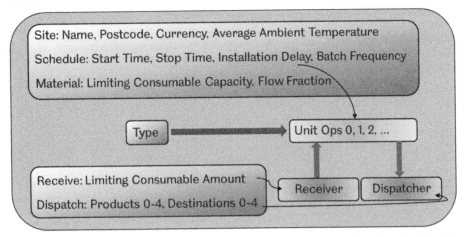

Fig. 2. A generic MOD Site, whose Unit Ops sequence is determined by the Site Type. Parameters in grey feed information according to black arrows. Material flows according to the red arrows.

dispatch hydrocarbon waste, whereas "Cultivation-Centrifuge" processes waste into biomass then dries it

2.3. Unit Op

In this implementation a Unit Op is an agent encapsulating the kinetics of a physio-chemical process together with its economics. Based on a generic template (base class or agent), there are dedicated sub-types of Unit Op for each stage of processing -- in this case study cultivation of biomass, dewatering and drying. Further Unit Ops could be added for downstream processing such as cell disruption, and bio-oil extraction and separation. Each Unit Op may operate in batch or flow mode, or a combination of the two. After each regular batch duration the reactor and the outlet are emptied into the product store, cleaned, and refreshed by recovering (from the product store), purchasing and/or receiving the consumables required for the next batch duration. The OpEx record is amended accordingly at this time.

The batch intervals are fixed by the Unit Op's start time, stop time, installation delay and batch frequency, which are inherited from its parent MOD Site. This information is simultaneously used to calculate CapEx cashflows, including any amortization or depreciation in the description. The product store is at any time available to supply the next Unit Op and/or the haulage dispatch agent at this MOD Site.

There remains the problem of sizing the facilities for any Unit Op, which dictates CapEx expenditure, and heat requirements via volume and surface area. For the initial Unit Op at any MOD Site, the size is determined by the limiting consumable annual capacity scaled by batch frequency. This is used to calculate a product capacity for this Unit Op. This product capacity is used to calculate the limiting consumable annual capacity for the the subsequent Unit Op, enabling us to size it, and calculate its product capacity. This reasoning is repeated along the sequence of Unit Ops at any MOD Site.

2.4. Economy

At the bottom of the agent hierarchy there is a simple economy layer, encapsulating cashflow schedules, amortization and depreciation, discounting, inflation and currency

LIMITING_CONSUMABLE	yieldCapacities		RATE_PRODUCT	RATE_CONSTANT
BIOMASS_2_PERCENT	BATCH	1.00	BIOMASS_15_PERCENT	17.8049
	FLOW	1.00		
	RECOVERY	0.00	BATCH_DURATION	BATCH_LAG
	LIMITING_CONSUMABLE_SCALE_FACTOR	0.00	0.1151	0.0575

RECOVERY	LIMITING_CONSUMABLE_SCALE_FACTOR	yieldFractions		kineticParameters		
					INERT	KS
0.00%	0%	OTHER_ELECTRICITY	0.00%	OTHER_ELECTRICITY	100%	1.0000E-12
0.00%	0%	LIGHT	0.00%	LIGHT	100%	1.0000E-12
0.00%	0%	HEAT	0.00%	HEAT	100%	1.0000E-12
0.00%	0%	STEAM	0.00%	STEAM	100%	1.0000E-12
98.00%	0%	WATER	86.67%	WATER	100%	1.0000E-12
98.00%	0%	BIOMASS_2_PERCENT	-100.00%	BIOMASS_2_PERCENT	100%	1.0000E-12
98.00%	0%	BIOMASS_6_PERCENT	0.00%	BIOMASS_6_PERCENT	100%	1.0000E-12
98.00%	0%	BIOMASS_15_PERCENT	13.33%	BIOMASS_15_PERCENT	100%	1.0000E-12
98.00%	0%	BIOMASS_20_PERCENT	0.00%	BIOMASS_20_PERCENT	100%	1.0000E-12
98.00%	0%	BIOMASS_25_PERCENT	0.00%	BIOMASS_25_PERCENT	100%	1.0000E-12
98.00%	0%	BIOMASS_92_PERCENT	0.00%	BIOMASS_92_PERCENT	100%	1.0000E-12
0.00%	0%	CHITOSAN	0.00%	CHITOSAN	100%	1.0000E-12
0.00%	0%	GENERAL_WASTE	0.00%	GENERAL_WASTE	100%	1.0000E-12
0.00%	0%	SEWERAGE	0.00%	SEWERAGE	100%	1.0000E-12

FACILITIES		facilities			
	27	11	FIXED_AMOUNT	BATCH_FACTOR	FLOW_FACTOR
VOLUME	0	VOLUME	0.0000	1.0000	1.0000
SURFACE_AREA	1	SURFACE_AREA	1.0000	0.0000	0.0000
LAND	100,000	LAND	1.0000	1.0000	0.6667
BUILDINGS	100,001	BUILDINGS	1.0000	1.0000	0.6667
GBP_SUPERVISOR	200,000	GBP_SUPERVISOR	0.0000	0.0000	0.0000
GBP_TECHNICIAN	200,001	GBP_TECHNICIAN	1.0000	0.0000	0.0000
CENTRIFUGE	5,000	CENTRIFUGE	0.0000	17.3810	0.0000
SPIRAL_PLATE	5,100	SPIRAL_PLATE	0.0000	0.0000	0.0000
PRESSURE_FILTER	5,200	PRESSURE_FILTER	0.0000	0.0000	0.0000
MEMBRANE_FILTER	5,300	MEMBRANE_FILTER	0.0000	0.0000	0.0000
VACUUM_FILTER	5,400	VACUUM_FILTER	0.0000	0.0000	0.0000
SPRAY_DRYER	5,500	SPRAY_DRYER	0.0000	0.0000	0.0000
DRUM_DRYER	5,600	DRUM_DRYER	0.0000	0.0000	0.0000

Fig 3. A fragment of the Excel description for a dewatering or drying Unit Op, showing some key inputs. Start and stop times are inherited from the parent MOD Site, not the Unit Op description.

Exchange rates. Consumables are stored in a price list that governs purchase, inflation, insurance and other consumables rates. Waste and sewerage are consumables that we sell for a negative price. Likewise, facilities and personnel are stored in a price list, which governs all economic matters including purchase, salvage, lifetime, inflation, contingencies, maintenance, Lang factors and scale factors. These default to industry benchmarks from (Fasaei et al 2018, Seider et al 2019) and market values circa 2021.

3. Results

The agent based supply chain model has been applied to preliminary studies of microalgal-bacterial remediation of hydrocarbon waste. Biomass cultivation is assumed to follow Monod kinetics (Monod, 1949), calibrated from laboratory work in a phycoflow reactor. The techno-economic inputs for dewatering and drying are based on literature review (Fasaei et al 2018). The cultivation and dewatering Unit Ops are flexibly distributed amongst 41 military bases across the UK. Scheduling consists of dispatching available products to the receivers of their destinations at the end of each batch duration. Transit times incorporating a loading/unloading delay are taken into account automatically, but must currently be scheduled manually in setting the start and stop times for each MOD Site. Optimal scheduling has not been automated.

A typical simulation result is shown in Fig. 5. This is summary information for the entire network, a complete breakdown of the time evolution of reactants and products, and all CapEx and OpEx cashflows is available down to the Unit Op Level (Fig. 6). This information has been used to show, for example, that centralized processing saves some £5.5m over six years in this network, as economies of scale outweigh transport costs. Surprisingly, drying locally cultivated biomass centrally saves £1.8m over local drying. So economies of scale dominate transport costs, even though wet biomass is very bulky.

Technoeconomic process modelling of waste remediation within a supply chain network

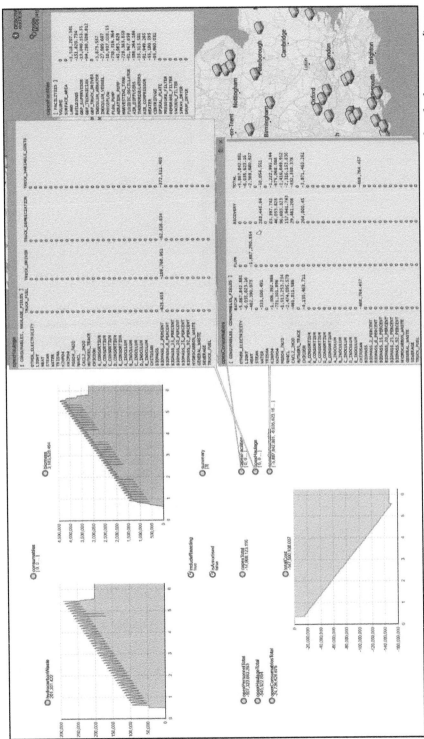

Fig. 5. Accumulated results for a six year nationwide project, showing hydrocarbon waste processed and biomass produced, above expenditure.

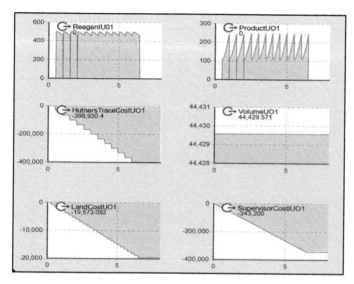

Fig 6. Drill-down results from a cultivation Unit Op.

4. Conclusion

In this paper we have presented a technoeconomic process model distributed within a supply chain network. The model is designed as a set of building blocks (Unit Ops) which may be distributed within the supply chain as desired. The many key parameters are neatly organized in a spreadsheet interface of descriptions, specifying location, timing, kinetics and economics. The software is highly extensible, and is almost agnostic to the nature of the Unit Ops. Adding new Unit Ops can be done in a few hours. The runtime for the 41 sites and 82 Unit ops considered is a few seconds. The software architecture is very satisfactory in these regards.

The major challenges only partially addressed concern optimal site capacity and transit scheduling. These could perhaps be optimized automatically in future work. Furthermore, the key outputs – namely material consumed, material produced and economic cost/benefit – should ideally be considered in some optimization scheme in future. Ultimately, it would be wise to analyse the sensitivity of any results to the techno-economic inputs, the future values of which may be quite uncertain.

References

W.D. Seider, J.D. Seader, D. R. Lewin and S. Widagdo. *Product and Process Design Principles: Synthesis, Analysis and Evaluation, 4th edition.* John Wiley & Sons Inc, 2019. ISBN 1119588006.

Garcia, D. J., & You, F. (2015), "Supply chain design and optimization: Challenges and opportunities", Computers & Chemical Engineering, vol. 81, pp. 153-170.

Z. Zahra, D. H. Choo, H. Lee, and A. Parveen, "Cyanobacteria: Review of Current Potentials and Applications", Environments, vol. 7, no. 2, p. 13, Feb. 2020.

S. B. Eregie and S. F. Jamal-Ally, "Comparison of biodegradation of lubricant wastes by Scenedesmus vacuolatus vs a microalgal consortium", Bioremediat. J., vol. 23, no. 4, pp. 277–301, Oct. 2019.

A. Borshchev, *The Big Book of Simulation Modeling: Multimethod Modeling with Anylogic 6*, Anylogic North America, 2014, ISBN 0989573176.

J. Monod, "The growth of bacterial cultures", Annual Review of Microbiology, vol 3, p.371, 1949. doi:10.1146/annurev.mi.03.100149.002103

F. Fasaei, J.H. Bitter, P.M. Slegers, A.J.B. van Boxtel, " Techno-economic evaluation of microalgae harvesting and dewatering systems", Algal Research, vol 31, pp. 347–362, 2018.

Antonis Kokossis, Michael C. Georgiadis, Efstratios N. Pistikopoulos (Eds.)
PROCEEDINGS OF THE 33rd European Symposium on Computer Aided Process Engineering
(ESCAPE33), June 18-21, 2023, Athens, Greece

Spatio-temporal analysis of milk safety under climate change

Lydia Katsini[1], Satyajeet S. Bhonsale[1], Styliani Roufou[2], Sholeem Griffin[2],
Vasilis Valdramidis[3], Simen Akkermans[1], Monika Polanska[1],
Jan F.M. Van Impe[1]

*[1]KU Leuven, Chemical Engineering Department, BioTeC+, Chemical & Biochemical
Process Technology & Control, Gebroeders De Smetstraat 1, Ghent 9000, Belgium*
[2]University of Malta, Department of Food Sciences and Nutrition, MSD 2080, Malta
*[3]National and Kapodistrian University of Athens, Department of Chemistry,
Panepistimioupolis Zografou, Athens, 157 84 Greece*
jan.vanimpe@kuleuven.be

Abstract

Bio-industrial processes can be optimized considering climate change using climate change impact assessments. These are based on models that describe the effect of climate on the affected sector. In this case, we focus on milk safety, thus we study Somatic Cell Counts (SCC), which is an indicator of mastitis, and Total Bacterial Counts (TBC), an indicator of the microbiological state of raw milk. The objective is to identify SSC and TBC patterns by looking into their spatial variability. For this purpose, traditional data mining methods are adopted to cover both space and time using data from 53 farms (2014-2019). According to the results, both SCC and TBC indicate spatial correlation. This means that apart from the already established temporal dependency, there is also a spatial dependency. In conclusion, collecting temporal data only from one farm may be misleading. Climate change-proof process design is essential for the dairy sector.

Keywords: milk, climate change, data mining, spatio-temporal analysis

1. Introduction

As climate change is expected to impact all aspects of life on the planet, it ranks high in both public and private decision making. The main tool needed in such decision making is climate change impact assessments, enabling a climate change-proof process design. It is based on impact models that describe the effect of multiple climate factors on affected sectors (Ahmad et al., 2001), here the dairy sector.

By utilizing these impact models, future food safety risks can be estimated under climate change (Katsini et al., 2021). For example, Gunn et al. (2019) estimated milk yield losses. By looking into raw milk quality, the already established processing standards can be revisited and re-optimized by taking into account the climate change impact on raw milk. A major consideration for process design is milk safety, i.e., the reduction of microflora in milk. Thus, the analysis involves the Somatic Cell Count (SCC), which is an indicator of the inflammation of the breast tissue of cattle, known as mastitis. In such a case, the milk is unsafe and not fit for human consumption. Moreover, the Total Bacterial Count (TBC) is studied, which is an indicator of the microbiological quality as it expresses the plate counts of the total amount of microflora in raw milk. TBC and SCC are recorded on a regular basis as they form regulatory thresholds for raw milk according to the EU legislation (EC 2073/2005).

The seasonal patterns of SCC and TBC have already been identified (Stürmer et al., 2018), nevertheless microbial dispersal takes place both in space and time. Thus, the objective of this work is to identify spatial patterns of milk contamination by analyzing the SCC and TBC temporal and spatial variability using data-driven modelling approaches.

2. Materials and Methods

2.1. Data set and data pre-treatment

Data from 682 dairy farms, located in Spain (A Coruña, Galicia), collected over the timespan 2014-2019 comprise the dairy data set. The first step of data pre-treatment was to screen the farms based on data completeness, e.g., cover all the years. After this screening, 53 farms were selected. Secondly, as the sampling frequencies of different farms were inconsistent, data was resampled to obtain the same sampling frequency for all farms, i.e., every 4 days. Existing outliers were removed and the gaps arising from the previous step were filled in using the median of the adjacent samples (10% interpolated data). Finally, the temporal trajectories for each farm were smoothed using a moving window weighted average. The address of each farm is represented in terms of longitude and latitude coordinates.

2.2. Methodology

The farms are clustered using non-linear unsupervised learning methods in terms of their temporal variability. The resulting clusters of farms are compared with their real topology in terms of coordinates. Farms that belong to the same cluster and are located close to each other indicate spatial correlation. SCC and TBC data are treated independently, which means that the above is done first for SCC and then repeated for TBC. On the one hand data-driven modelling techniques can be valuable to find patterns in data, however they remain black-box approaches, which means that the results are not always interpretable.

2.2.1. t-Distributed Stochastic Neighbor Embedding (t-SNE)

The first method applied in this paper is the non-linear t-distributed stochastic neighbor embedding (t-SNE), which was introduced by van der Maaten et al. (2008). t-SNE performs dimensionality reduction by resulting in a two-dimensional space, that enables the visualization of data in clusters. The method aims at keeping the probabilities of pair distances of the data in the original space as similar as possible to the corresponding probabilities in the reduced space. The clustering is done by visualizing data that have similarities in the original space as points close to each other in the reduced space. In this paper, this method is used, firstly, to identify the number of clusters of farms for SCC and TBC. Secondly, t-SNE results are exploited to assess any possible spatial correlations among pairs of farms. The results were obtained from the tsne function of MATLAB using perplexity equal to 4 and the distance metric as the Euclidean.

2.2.2. Self-Organizing Map (SOM)

The second data mining methodology exploited is self-organizing maps (SOMs), which is a type of artificial neural network (Kohonen 1982). It is an unsupervised learning method that performs dimensionality reduction by maintaining the topological and metric aspects of the input space. The high-dimensional input space is compressed in a two-dimensional space, which comprises neurons. The SOM algorithm assigns each data point to one of the neurons. Augustijn et al. (2013) explain in detail how SOMs can serve spatio-temporal analysis. In this paper, the SOM has the same number of neurons as the number of the resulting clusters from t-SNE for SCC and TBC. The selforgmap function of MATLAB is used with two girds. The grid is chosen to be hexagonal and the SOM batch algorithm is used to train the network for 1000 epochs.

3. Results and Discussion

3.1. Clusters based on t-SNE
The geographical coordinates act as the reference to assess the clustering of the farms as resulting from the t-SNE technique. t-SNE results are illustrated by the topology of the farms in the two-dimensional space. In the following figures, each dot represents one farm. The farms that are located close to each other for both the t-SNE reduced space and the geographical coordinates are connected by black lines. The dots representing those farms are accompanied by a label including the corresponding code of each farm to facilitate the comparison of their proximity in terms of geographical coordinates.

3.1.1. Results for SCC

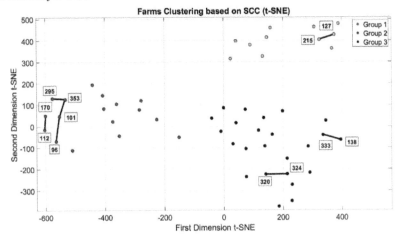

Figure 1. The t-SNE reduced space based on SCC.

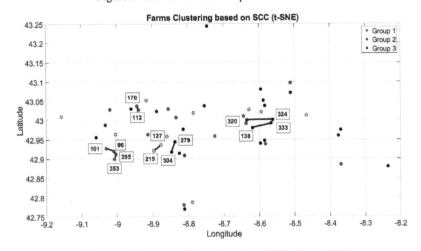

Figure 2. Clustering of farms based on SCC using t-SNE.

In the case of SCC, three clusters are identified in the reduced t-SNE space. By assessing the topology of each farm in the reduced space (Figure 1) with regard to their map coordinates (Figure 2), four pairs of farms come up, i.e., farm 127 & 215, 333 &138, 324 & 320, and 112 & 170, as well as a cluster of four farms, i.e., farm 96, 101, 353 & 295. This indicates spatial correlation among those farms in terms of SCC temporal variability.

3.1.2. Results for TBC

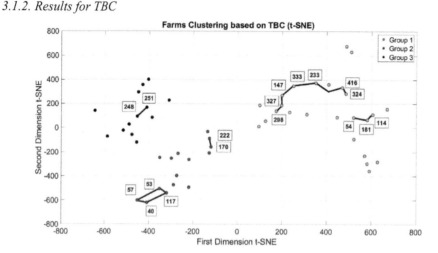

Figure 3. The t-SNE reduced space based on TBC.

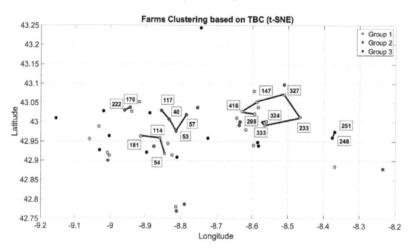

Figure 4. Clustering of farms based on TBC using t-SNE.

The t-SNE clustering based on TBC results in three clusters as well (Figure 3). Following the same rationale as in the case of SCC (Figure 3 & 4), two pairs of farms can be identified, i.e., farm 248 & 251 and 222 & 170. Furthermore, one group of three farms, i.e., farm 181, 114 & 54, along with one group of four farms, i.e., farm 40, 53, 57 & 117, are found. Lastly, the group formed by farm 298, 327, 147, 333, 233, 295, 416 & 324, are located close to each other (Figure 4), however in the t-SNE reduced space (Figure 3) they form a line instead of a group of points. This indicates spatial correlation among those farms in terms of TBC temporal variability as well.

3.2. Clusters based on SOM

As mentioned in the previous section, the SOM for each variable has the same number of neurons as the clusters resulting from the t-SNE analysis. In the case of SCC, three clusters of farms are formed in the t-SNE reduced space, thus the SOM is formed by three neurons which represent three groups of farms. However, for TBC, the clusters are less clear in the t-SNE reduced space. Thus, two SOM structures were assessed for TBC, one with two and one with three neurons or groups of farms. The results from the SOM with three neurons are presented here as it provides more insights. The clustering results are presented in terms of the geographical coordinates of the farms, i.e., each farm is depicted as a dot and colored based on the neuron it is assigned to after training the neural network.

3.2.1. Results for SCC

Results from the SOM trained based on SCC data with three neurons indicate clustering of farms with spatial characteristics (Figure 5). Groups 1 and 3 mostly consist farms that are located in the western part of the map, while Group 2 is in the eastern one. Furthermore, all the farms with an indication of spatial correlation following the t-SNE technique (Figures 1 & 2) are clustered accordingly based on the SOM methodology.

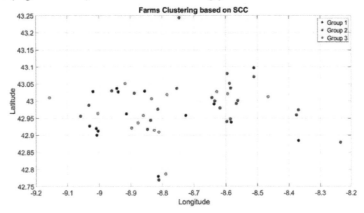

Figure 5. Clustering of farms based on SCC using SOM.

3.2.2. Results for TBC

In the case of the SOM trained based on the TBC data with three neurons, the geographical distinction is less clear compared with the SOM for SCC (Figure 6). Group 1 mostly consists of western while Group 3 of eastern farms and Group 2 shows no such geographical characteristic. Nevertheless, Group 3 appears to represent farms that are located around the longitude -8.6. Interestingly, some of these farms appear close to each other in the t-SNE reduced space (Figures 3 & 4) as well, i.e., the group of farms that forms a line in Figure 3.

The agreement between the t-SNE and SOM methodologies both based on SCC and TBC for those sets of farms makes a stronger argument regarding the hypothesis of this study, i.e., whether there is a spatial correlation among dairy farms in terms of SCC and TBC. Even though the results do not show a clear geographical clustering, the indication of its existence is presented, as resulting from these two methodologies.

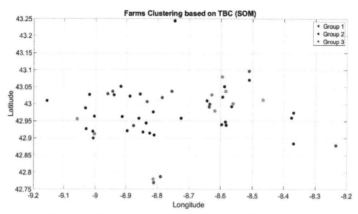

Figure 6. Clustering of farms based on TBC using SOM.

4. Conclusion

Processes such as milk pasteurization, can be optimized considering climate change using climate change impact assessments. As pasteurization aims to reduce the microbial load, such an impact assessment is based on a model that describes the effect of multiple climatic factors on milk contamination, which is expressed with SCC and TBC. This study aimed at identifying spatial correlations among dairy farms in terms of SCC and TBC by applying data mining techniques. According to the results, both SCC and TBC indicate spatial correlation among dairy farms. This means that apart from the temporal dependency between SCC and TBC and climate, there is also a spatial dependency among farms. Therefore, collecting data only from one farm may be misleading. In this way, predicting food safety aspects of milk under climate change is less uncertain. Assessing the application potential of such models to different locations may be realized by evaluating if predictive models developed from Spanish data can explain the spatio-temporal characteristics of raw milk produced in another region.

Acknowledgement

This work was supported by the PROTECT project funded by the European Union's Horizon 2020 Research and Innovation Programme [MSCA grant 813329]. SA was supported by the Research Foundation – Flanders (FWO), under grant 1224623N.

References

Q. K. Ahmad & R. A. Warrick, 2001. Methods and tools. In: J. J. McCarthy, O. F. Canziani, N. A. Leary, D. J. Dokken and K. S White. (eds) Climate change 2001: impacts, adaptation, and vulnerability. Cambridge University Press, 105

E.W. Augustijn & R. Zurita-Milla, 2013, Self-organizing maps as an approach to exploring spatiotemporal diffusion patterns, International Journal of Health Geographics, 12, 60

K. M. Gunn, et al., 2019, Projected heat stress challenges and abatement opportunities for US milk production. PloS one, 14,3, e0214665

L. Katsini, et al., 2022, Quantitative methods to predict the effect of climate change on microbial food safety: A needs analysis. Trends in Food Science & Technology, 126, 113

T. Kohonen, 1982. Self-organized formation of topologically correct feature maps. Biological cybernetics, 43,1, 59

M. Stürmer, et al., 2018. Relationship between climatic variables and the variation in bulk tank milk composition using canonical correlation analysis. International Journal of Biometeorology, 62,9, 1663

L. Van der Maarten & G. Hinton, 2008. Visualizing data using t-SNE. Journal of Machine learning research 9, 2579-2605.